Prosimian Biology

Prosimian Biology

edited by
R.D. MARTIN, G.A. DOYLE and
A.C. WALKER

Proceedings of a meeting of the
Research Seminar in Archaeology and Related Subjects
held at the Institute of Archaeology, London University

First published in 1974 by
Gerald Duckworth and Co. Ltd.
The Old Piano Factory
43 Gloucester Crescent, NW1

ISBN 0 7156 0672 7

Typeset by
Specialised Offset Services Limited, Liverpool,
Printed and bound in Great Britain by
Redwood Burn Limited, Trowbridge & Esher

CONTENTS

Section B: *Laboratory Studies of Behaviour*

Section C: *Olfactory Communication in Prosimians*

Section D: *Physiology of Behaviour in Captivity*

PART II: PROSIMIAN ANATOMY, BIOCHEMISTRY AND EVOLUTION

Section A: *General Studies of Prosimian Evolution*

Section B: *Anatomy and Function of Skull and Teeth*

Section C: *Morphology of the Brain*

Section D: *Prosimian Locomotion*

Section E: *Chromosomes, Proteins and Evolution*

Preface

On 14, 15, 16 and 17 April 1972, an international meeting was held at the Institute of Archaeology, University of London, to discuss the general field of prosimian biology. Following a suggestion made by Dr P.J. Ucko, coupled with much valuable advice and encouragement, the meeting was held within the framework of the series of Research Seminars in Archaeology and Related Subjects, and the organisers of the present meeting were accordingly able to benefit from the accumulated expertise of what has so far proved to be a highly successful pattern of seminars. One happy result was that the Research Seminar on Prosimian Biology profited greatly from the hospitality and ready assistance of the Institute of Archaeology, for which thanks are especially due to Professor W.F. Grimes. In particular, the success of the meeting depended heavily on the invaluable services of Miss S.E. Johnson, who undertook almost all of the secretarial work for the meeting, provided seasoned advice on the running of the Research Seminar, and assisted in the preparation of this book. Thanks are also due to Miss Lorna Mullings and Mrs Penny Wyatt for their willing assistance in the Research Seminar secretariat, which dealt with all the practical problems of running a meeting with over fifty active participants.

The meeting – like its three predecessors in the Research Seminar series – was designed to enable research workers concerned with different specialised aspects of the subject to meet in an attempt to bridge gaps and achieve a synthetic approach through personal contact and lively discussion. For this reason, emphasis was placed on informality of presentation and debate. Following the practice established in previous Research Seminars, all papers were circulated prior to the meeting in order to reduce the stifling influence of formal reading of papers during periods which could be more profitably used for informal discussion. Several sessions were arranged in which particular topics were thrown open to general discussion, and participants made full use of the opportunities provided. Judging from the various comments which were made after the meeting, the air of informality was generally greatly appreciated, and many participants felt that they had gained far more than would have been possible with more formal conference organisation. It must be admitted, however, that the general friendly tone of the meeting was set from the beginning by the warm official welcome provided by Professor Grimes in the form of a reception to open the Research Seminar.

The important task of chairing discussion sessions was undertaken by Professor R. Andrew, Professor N.A. Barnicot, Professor F. Bourlière, Dr W. Bishop, Professor M. Day, Dr K. Hiiemae, Dr J.R. Napier, Professor W.C. Osman Hill, Dr L. Rosenson, Dr E.L. Simons and Dr L. Wolin. All these people deserve our deep gratitude for encouraging constructive debate and ensuring that the participants did not wander too far from the subject of each session. In addition, the editors would like to express their thanks to all of the more than fifty participants who came from various parts of the world to contribute papers and to join in the discussion of prosimian biology. They have helped to produce a text which represents one of the first really broad-based attempts to consider the general biology of the prosimians. The success of this attempt is reflected in the fact that many of the participants extensively expanded and modified their contributions as a result of discussion held during the meeting, and that a number of new lines of research seem to have developed following the meeting.

The general organisation of the Research Seminar was dependent upon the assistance of Mr J.F. Dahl, who operated the tape-recorder, and of Mr D.G. Lewis, who was in charge of the projection facilities. Additional secretarial assistance was provided by Mrs P.M. Blair. The Institute of Archaeology provided many different services, which were invaluable to the smooth running of the meeting, and thanks are due to the many un-named people who provided assistance on behalf of the Institute.

Finally, a special note of thanks must be made to the initiating publishers, Duckworth & Co. Ltd., without whose help the Research Seminar could not have been held. Among the several unique features of this meeting, it was notable that our publisher attended the greater part of the proceedings himself, and in the preparation of the book even insisted on the inclusion of *more* material in order to make the text as comprehensive as it is.

R.D. Martin
G.A. Doyle
A.C. Walker

List of Contributors (present addresses)

B.C. Albright, Department of Anatomy, Medical College of Virginia, Health Science Division of Virginia Commonwealth University, Richmond, Virginia 23298, USA

R. Andrew, School of Biological Sciences, Sussex University, Falmer, Brighton, Sussex

N.A. Barnicot, Department of Anthropology, University College London, Gower Street, London WC1

K. Bauer, 69 Heidelberg 1, Postfach 100.560, West Germany

S.K. Bearder, Department of Psychology, University of the Witwatersrand, Jan Smuts Avanue, Johannesburg, South Africa

J.H. Bergeron, Department of Anatomy, Duke University Medical Center, Durham, North Carolina 27706, USA

A.C. Berry, Royal Free Medical School, 8 Hunter Street, London WC1

F. Bourlière, Faculté de Médecine, 45 Rue des St Pères, Paris 5e, France

F.M. Bush, Department of Anatomy, Medical College of Virginia, Health Science Division of Virginia Commonwealth University, Richmond, Virginia 23298, USA

M. Cartmill, Department of Anatomy, Duke University Medical Center, Durham, North Carolina 27706, USA

P. Charles-Dominique, Muséum National d'Histoire Naturelle, Ecologie Générale, 4 Avenue du Petit Château, 91 Brunoy, France

B. Chiarelli, Centro di Primatologia, Istituto di Antropologia, Università di Torino, Via Accademia Albertina 17, 10123 Torino, Italy

C.N. Cook, Department of Anthropology, University College, Gower Street, London WC1

H.M. Cooper, Department of Biological Sciences, Biology Unit I, Florida State University, Tallahassee, Florida 32306, USA

U.M. Cowgill, Department of Biology, Faculty of Arts and Sciences, University of Pittsburgh, Pittsburgh, Pennsylvania 15260, USA

M. Delombas, Muséum National d'Histoire Naturelle, Laboratoire

d'Anatomie Comparée, 55 Rue de Buffon, Paris 5e
R.F. Doolittle, Department of Chemistry, University of California, San Diego, Box 109, La Jolla, California 92037, USA
G.A. Doyle, Department of Psychology, University of the Witwatersrand, Jan Smuts Avenue, Johannesburg, South Africa
F. D'Souza, Wellcome Institute of Comparative Physiology, Regent's Park, London NW1
J. Egozcue, Instituto de Biologia Fundamental, Universidad Autónoma de Barcelona, Avda, San Antonio M.a Claret 167, Barcelona 13, Spain
Judy Epps, Department of Zoology, University of New England, Armidale, New South Wales, Australia
R.G. Every, Centre for the Study of Conflict, 25 Clifton Terrace, Sumner, Christchurch 8; and Dept. of Zoology, University of Canterbury, Christchurch 1, NZ
M.P.L. Fogden, Rancho Experimental La Campana, Apdo. 682, Chihuahua, Mexico
J. Gasc, Museum National d'Histoire Naturelle, Laboratoire d'Anatomie Comparée, 55 Rue de Buffon, Paris 5e
P.D. Gingerich, Division of Vertebrate Palaeontology, Peabody Museum, Yale University, New Haven, Connecticut 06520, USA
G.E. Goode, Medical College of Virginia, Health Science Division of Virginia Commonwealth University, Richmond, Virginia 23298, USA
M. Goodman, Department of Anatomy, School of Medecine, Wayne State University, Detroit, Michigan 48207, USA
A. Gorton, Department of Anthropology, New York University, Washington Square, New York, NY 10003, USA
C.P. Groves, Duckworth Laboratory, Faculty of Archaeology and Anthropology, Downing Street, Cambridge
Duane E. Haines, Department of Anatomy, West Virginia University School of Medicine, Morgantown, West Virginia 26506, USA
E.C.B. Hall-Craggs, Department of Anatomy and Embryology, University College, Gower Street, London WC1
J. Harrington, The Rockefeller University, New York, NY 10021
D. Hewett-Emmett, Department of Physiology, Wayne State University Medical School, Detroit, Michigan 48201, USA
M. Hladik, Muséum National d'Histoire Naturelle, Ecologie Générale, 4 Avenue du Petit Château, 91 Brunoy, France
K.M. Hiiemae, Unit of Anatomy with Relation to Dentistry, Anatomy Department, Guy's Hospital Medical School, London SE1
K. Holmes, Department of Physiology, Southern Illinois University, Edwardsville, Illinois 62025, USA
C.J. Jolly, Department of Anthropology, New York University, Washington, Square, New York, NY 10003, USA
F.K. Jouffroy, Muséum National d'Histoire Naturelle, Laboratoire

d'Anatomie Comparée, 55 Rue de Buffon, Paris 5e
W.P. Luckett, Department of Anatomy, Columbia University, College of Physicians and Surgeons, 630 West 168th Street, New York, NY 10032, USA
R.F. Kay, Museum of Comparative Zoology, The Agassiz Museum, Harvard University, Cambridge, Massachusetts 02138, USA
P.H. Klopfer, Department of Zoology, Duke University, Durham, North Carolina 27706, USA
R. Klopman, School of Biological Sciences, Sussex University, Falmer, Brighton, Sussex
G. Manley, Department of Zoology, Birkbeck College, Malet Street, London WC1
R.D. Martin, Department of Anthropology, University College, Gower Street, London WC1
H.M. Murray, Department of Anatomy, Medical College of Virginia, Health Science Division of Virginia Commonwealth University, Richmond, Virginia 23298, USA
S. Oblin, Département Audiovisuel, Université Paris 7, 2 Place Jussien, Paris 5e, France
W.C. Osman Hill, Oakhurst, Dixwell Road, Folkestone, Kent
M.G. Pariente, Muséum National d'Histoire Naturelle, Ecologie Générale, 4 Avenue du Petit Château, 91 Brunoy, France
M. Perret, Muséum National d'Histoire Naturelle, Ecologie Générale, 4 Avenue du Petit Château, 91 Brunoy, France
J.-J. Petter, Muséum National d'Histoire Naturelle, Ecologie Générale, 4 Avenue du Petit Chateau, 91 Brunoy, France
A. Petter-Rousseaux, Muséum National d'Histoire Naturelle, Ecologie Générale, 4 Avenue du Petit Château, 91 Brunoy, France.
L.B. Radinsky, Department of Anatomy, University of Chicago, 1025 East 57th Street, Chicago, Illinois 60637, USA
A. Richard, Department of Anthropology, Yale University, New Haven, Conn. 06520, USA
Y. Rumpler, Ecole Nationale de Médecine, Laboratoire d'Histologie, Embryologie et Cytogénétique, B.P.375 Tananarive, Madagascar
A. Schilling, Muséum National d'Histoire Naturelle, Ecologie Générale, 4 Avenue du Petit Château, 91 Brunoy, France
J. Schwartz, Department of Anthropology, University of Pittsburgh, Pennsylvania 15260, USA
D. Seligsohn, Department of Anthropology, Hunter College, 695 Park Avenue, New York, NY 10021, USA
E.L. Simons, Peabody Museum, Yale University, New Haven, Connecticut 06520, USA
R.W. Sussman, Department of Anthropology, Washington University, St Louis, Missouri 63130, USA
F. Szalay, Department of Anthropology, Hunter College, 695 Park Avenue, New York, NY 10021, USA
J. Tandy, c/o Mills Tandy, Department of Zoology, University of

Texas, Austin, Texas 78712, USA

I. Tattersall, Department of Anthropology, American Museum of Natural History, Central Park West at 79th Street, New York, NY 10024, USA

D. von Holst, Zoologisches Institut der Universität München, 8 München, Luisenstrasse 14, West Germany

A.C. Walker, Department of Anatomy, Harvard Medical School, 55 Shattuck Street, Boston, Mass. 02115, USA

L.R. Wolin, Laboratory of Neurophysiology, Research Division, Cleveland Psychiatric Institute, 1708 Aiken Avenue, Cleveland, Ohio 44109, USA

W.C. OSMAN HILL

Foreword

The foundation of our knowledge of the prosimians was laid during the latter half of the nineteenth century, and it is there that we first meet with monographers of the stature of Burmeister (1846) and Owen (1863) through their respective studies of the tarsier and of the aye-aye. There have also been numerous and outstanding contributions in the form of more comprehensive works, and here I should especially mention the labours of Murie and Mivart and the great explorers of Madagascar, such as Schlegel and Pollen (1867-77) and particularly Grandidier and Edwards (1876-1900), who produced a magnificent set of volumes on the natural history of the island,

While speaking of the giants of our discipline, it is opportune to include one of the more recent exponents – the late Sir Wilfred Le Gros Clark. Though he also made major contributions to many aspects of the anatomy, neuroanatomy and palaeontology of the Prosimii, he will perhaps best be remembered for his advocacy of the inclusion among them of the tree-shrews. These animals clearly still have a place in our deliberations, however controversial it may be, as is evidenced by several contributions to the present symposium.

It is now almost a quarter of a century since the first overall assessment of the status of the prosimian primates was made. At that time, I had the privilege of partaking in a discussion which was opened by my revered colleague F. Wood Jones, on the occasion of the meeting of the section on Anatomy and Anthropology at the annual meeting of the British Medical Association in Cambridge, 1948.

Since then, much has happened that is of vital importance in the field of prosimian biology, which is now not only wider, but has inevitably expanded to incorporate applications of disciplines almost unheard of at that time. It is, therefore, a matter of no little difficulty for any single author to write with authority on the whole gamut of prosimian biology. Thus the four-day research seminar on the subject held in April 1972 was particularly welcome, in that it enabled a wide range of prosimian biologists to take stock of their own advances and to profit from the researches of their colleagues from many different disciplines.

Needless to say, the problems posed by the position of *Tarsius* and its fossil allies proved to be as intriguing as ever, and they would have still delighted Wood Jones and engaged his powers of debate in no small measure, just as they did at the memorable meeting held at the London Zoo in 1918. But the problems set by the tarsiers are by no means the only ones that have engaged the interests of workers in recent decades. Our attention is automatically drawn towards 'la grande île' of Madagascar, where the activities of conservationists have emphasised time and again the pressing urgency for field-work on the behaviour of all of the lemurs. In spite of the laudable endeavours of the Malagasy Republic to protect their unique fauna, it may soon be too late to study at least some of the lemur species in the natural state. Fortunately, heed has been taken of the warnings, and there are now a number of studies that have been published or are in progress, as demonstrated by several of the contributions to the Research Seminar.

Many lemur species still demand special attention, however, and there is an even more pressing need for detailed study of their ecology so that regional schemes may be examined, which may permit the establishment of local breeding units. Madagascar is ideally situated as a natural workshop for the student of evolution, and it behoves us to adopt early measures for the protection of the existing fauna – an urgent necessity already recognised by the enlightened Malagasy government – and also to provide the means of enhancing the existing facilities for internal laboratories where local and visiting students may avail themselves of the unique opportunities to study the rich fauna.

That definite advances in prosimian biology have accrued in many different directions is amply shown by the contents of the present symposium. The wide range of problems which have now been considered makes this abundantly clear, and I have myself gained valuable knowledge both of a wealth of new facts and of reassessment of earlier problems. It would be invidious for me to single out any one paper for special praise, since all are remarkable in their various approaches and treatments; I will, therefore, deal with the different papers in an order that appears to link the subject matter in a logical fashion, as a guide to the mass of information brought together by the Research Seminar.

The Seminar was broadly divided into two sections: studies of prosimian behaviour, and studies of prosimian anatomy and evolution, including reference to cytogenetics and biochemical studies. Behaviour in the field justly received a major share of attention from the participants, and papers were given on representatives from all the major prosimian groups. Contributions by Frances D'Souza and

Michael Fogden brought in the two most controversial prosimians – namely, *Tupaia* and *Tarsius* – and thus provided welcome additions to the field of prosimian research.

In Madagascar, field studies on *Lemur catta* have been undertaken both by Alain Schilling and by Robert Sussman, the latter providing a very elegant study of the ecological distinctions between sympatric populations of *Lemur catta* and *Lemur fulvus.* Alison Richard elaborated on the patterns of mating in *Propithecus* and provided us with one of the first detailed accounts of the natural mating behaviour of any lemur species. Marcel Hladik and Pierre Charles-Dominique have significantly expanded our knowledge of lemur feeding ecology and social dynamics with their intensive study of *Lepilemur*, while Jean-Jaques Petter and A. Peyriéras have added to our understanding of *Indri*, specifically with respect to population density and home range. This latter paper is again one of particular interest, in that it deals with a little-known genus.

Turning to mainland Africa, the bushbabies and pottos were very comprehensively covered. All the prosimians outside Madagascar are nocturnal, and special techniques have been developed for studying the behaviour of these night-active forms over the last few years. These techniques are illustrated in the papers on *Tarsius* (Fogden) and *Lepilemur* (Hladik and Charles-Dominique); but in the papers on the African prosimians, the full possibilities of these techniques are displayed. Bearder and Doyle have given a comprehensive account of the South African species *Galago senegalensis* and *Galago crassicaudatus*, whilst Charles-Dominique has achieved the magnificent feat of considering the ecology and feeding behaviour of all 5 prosimian species in Gabon (*Galago demidovii, Galago alleni, Euoticus elegantulus, Arctocebus calabarensis* and *Perodicticus potto*). Finally, a link with laboratory studies is provided by Georges Pariente's able contribution dealing with two little-studied nocturnal Malagasy lemurs, *Phaner furcifer* and *Lepilemur mustelinus*, and illustrating the relationship of their activity patterns to ambient light intensities.

Laboratory studies of prosimian behaviour were equally well represented. The paper presented by Jan Bergeron on the Duke University Primate Facility, set up by John Buettner-Janusch, which I had the privilege of visiting a few years ago, summarises a very interesting large-scale project on laboratory observation of prosimians. We may expect a stream of far-reaching research projects on prosimian biology from this well-equipped facility. The lesser galago has been further studied in South Africa by Gerald Doyle in a very comprehensive programme aimed at obtaining quantifiable data over long periods, and his paper summarises our knowledge about the behavioural repertoire of this species. Judy Epps and Jocelyn Tandy, respectively, have given detailed attention to social behaviour of captive *Perodicticus potto* and captive *Galago crassicaudatus*, while

Peter Klopfer has looked specifically at aspects of lemur mother-infant relationships. Ursula Cowgill reported on some unusual observations indicating a possibility that pottos may exhibit a form of cooperative behaviour in the laboratory. On a more experimental slant, Howard Cooper investigated the possibility that *Lemur macaco* might exhibit learning sets in captivity, and reported on evidence that this capacity does in fact exist in at least some prosimians.

The functions of the glandular apparatus associated with the external genitalia of the two slow-moving African lorisoids (*Perodicticus* and *Arctocebus*) provided an interesting topic for Gilbert Manley, who exploited the subject to the full, drawing conclusions from behaviour, morphology and biochemistry. The perennial subject of urine-washing has been studied on a comparative basis by Richard Andrew and Robert Klopman, who have underlined the practical difficulties of providing a conclusive interpretation of a specific aspect of behaviour of this kind by presenting a wide-ranging examination of the various possibilities. The subject of olfactory communication, which had been considered in Alain Schilling's paper on *Lemur catta*, was also tackled in Jonathan Harrington's paper on *Lemur fulvus*.

A more physiological approach to the behaviour of captive prosimians was provided by two papers relating to a laboratory colony of Mouse Lemurs (*Microcebus murinus*). Arlette Petter-Rousseaux submitted a throughly competent study of the relationships between photoperiodicity, sexual activity and body-weight in this small nocturnal lemur, whose behaviour (like that of all other Malagasy lemurs) has a strict seasonal basis. The endocrine glands, especially the hypophysis, have been explored in detail by Martine Perret, who relates various changes to aspects of behaviour and the seasonal reproductive cycle. To round off the laboratory studies of behaviour, Dietrich von Holst provides a salutory warning for those who maintain animals of any kind in captivity. In his paper on social stress in tree-shrews (*Tupaia belangeri*), he provides ample evidence of the wide scale of pathological changes that may be brought about by interaction between conspecifics in captivity. He also raises the interesting possibility that such interaction between conspecifics – probably in a far milder form – may play a part in the natural situation.

In the second section, dealing with anatomy, biochemistry and evolutionary aspects, special praise is due to a number of papers which have a functional anatomical emphasis. This is evident in Tattersall's investigation of the mechanics of the temporo-mandibular joint and accessory structures of the subfossil *Archaeolemur*. Matt Cartmill casts a wide net in his behavioural and ecological consideration of anatomical features of such biological aberrants as *Daubentonia, Dactylopsila* and various woodpeckers, which are functionally linked by virtue of their adaptation to a wood-boring insectivorous regime, as a result of which they all exhibit some degree of klinorhynchy. P.D. Gingerich, in 'Dental function of the Palaeocene primate *Plesiadapis*', has produced cogent arguments for the retention of this genus within the Order Primates, basing his reasoning on imputed feeding behaviour, dental morphology and molar faceting. Dental occlusion mechanisms were likewise the subject of the joint contribution by Seligsohn and Szalay on Lemurinae, with its bearing on the systematics of *Lemur* and *Varecia*. They confirm the separation of *Varecia* from *Lemur* at the generic level and the relegation of the subfossil species *insignis* and *jullyi* to the former genus. Other significant odontological studies with a functional basis are the contribution on *Galago* presented by R.F. Kay and Karen Hiiemae, illustrating their cinefluorographic technique of analysis, and Ron Every's magnificently illustrated paper on thegosis. Jeffrey Schwartz, in a theoretical reinvestigation of premolar loss among primates, resorts to the 'field theory' to help resolve the problems posed by diminishing premolar representation in primate evolution.

Three contributions are devoted to cerebral anatomy in prosimians. One, by Leonard Radinsky, deals with functional and phylogenetic implications of prosimian brain morphology, and two others are concerned with the neuroanatomy of lorisids. These are Duane Haines' study of the cerebellum in various lorisids and the joint contribution by Haines and his colleagues from the Anatomy Department of the Medical College of Virginia, dealing with the external cerebral morphology of galagos, lorises and pottos.

There are also three papers dealing in various ways with the topic of prosimian locomotion. Clifford Jolly and Ann Gorton examined the proportions of the intrinsic foot muscles in some lorisid prosimians, while Mme F. Jouffroy used cineradiographic techniques to examine the biomechanics of jumping in *Galago alleni*. Combined in one paper, E.C.B. Hall-Craggs has collected both physiological and histochemical parameters of locomotor behaviour in *Galago* and the slow-moving lorisids, including data on the differing activities of red and white muscular tissue and the mechanics of saltation.

Passing now to evolutionary studies, there are a number of papers dealing with general palaeontological and comparative anatomical treatments of the prosimians. Elwyn Simons has provided notes on Early Tertiary prosimian fossils, thus bringing our knowledge of these forms up to date. Alan Walker tackles the other major category of fossil prosimians, presenting a review of the Miocene lorisids of East Africa, including reference to the postcranial skeleton. The new associated fossil lorisid material has confirmed the value of size analysis in allocating some of the remains. A particularly thought-provoking paper linking both behaviour and evolutionary mechanisms is that of Lee Wolin, with the intriguing title of 'What can the eye tell us about behaviour and evolution? or: The aye-ayes have it, but what is it?'. His only complaint is that *Tupaia* has, so far, not received enough attention. The aye-aye is also made the subject, among other notable matters, of Colin Groves' evocative paper on the taxonomy and phylogeny of prosimians. Finally, Patrick Luckett draws upon evidence from the morphogenesis of the foetal membranes and the placenta in order to establish the phylogenetic relationships of the prosimians, whilst Caroline Berry has utilised non-metrical features of the prosimian skull for evolutionary interpretation in the manner of Wood Jones' important review (1930-1).

A particularly valuable addition to the discussion of evolutionary relationships was provided by a series of papers on chromosomal and biochemical studies of prosimians. Chromosomal evolution in the prosimians is the subject of Josei Egozcue's paper, which provides further indication of the distinctiveness of *Tarsius*, in that it has the highest diploid number so far found in any primate. The results of Rumpler's study of lemur chromosomes with a view to revising their classification indicate that it would be justifiable to raise the cheirogaleines to family rank, with the family Cheirogaleidae containing two subfamilies – Cheirogaleinae and Phanerinae. Rumpler also confirms the distinctiveness of *Daubentonia*, suggesting that this genus should be separated from all other lemurs at least at the family level. A general survey of chromosomal features in prosimians, provided by Brunetto Chiarelli, rounds off this triplet of cytogenetic studies.

Nigel Barnicot and David Hewett-Emmet have contributed a paper on electrophoretic studies of red cell and serum proteins in prosimians, which covers a wide range of species. It was of great interest to learn that the haemoglobin, adenylate kinase and phosphatase patterns of *Tarsius* are so like those of man. The properties of the Cheirogaleinae studied are also of systematic value, emphasising the distinctiveness of *Microcebus* and *Cheirogaleus*. Electrophoretic properties of haemoglobins were also the subject of a paper by Francis Bush and colleagues. Other phylogenetically directed protein studies concern the amino acid sequences of individual proteins and the use of serological tests to determine affinity. Christopher Cook and David Hewett-Emmett provide a general paper examining the uses and limitations of tree-building techniques based on amino acid sequence data, while Russell Doolittle provided an account of recent work on the short-chain fibrinopeptides, which are of great significance because of their evident lability in evolution. Klausdieter Bauer reports on his recent work based on identification of the actual number of determinant sites on proteins which are involved in serological cross-reaction, and a paper by Morris Goodman and his colleagues rounds off the subject with a treatment of immunodiffusion systematics in *Tarsius*, Lorisidae and Tupaiidae, in which he introduces the concept of antigenic distance.

This collection of papers can thus be seen to cover an extremely broad spectrum, and the Research Seminar has accordingly provided a general basis for future discussion of the biology of prosimians and of their evolution. In conclusion, I can only commend to all the overall value of such a broad-based synthesis in dealing with a particular topic, such as that of prosimian biology.

W.C. Osman Hill
Folkestone, September 1973

PART I
PROSIMIAN BEHAVIOUR

G. A. DOYLE and R. D. MARTIN

The study of prosimian behaviour

While interest in the behaviour of the primates as a group can be traced back to Darwin's time, this interest was – until quite recently – manifested predominantly in studies of the 'higher' (simian) primates. It is now widely accepted that study of prosimian behaviour is important and of great interest in its own right, and that it is especially necessary for a reliable, overall view of the evolution of behaviour within the Order Primates. It is therefore doubly important that prosimian behaviour should now be intensively and adequately studied, both in the laboratory and in the field. One of the most significant goals of primate behavioural studies is the provision of a general framework for interpreting the evolution of primate behaviour – particularly that of man – and it is essential to recognise that extensive behavioural studies of the prosimians must play an integral part in any *comprehensive* investigation of primate evolution.

It was first pointed out by T.H. Huxley,[1] and subsequently re-emphasised by Le Gros Clark,[2] that the Order Primates is the only mammalian order in which selected extant members appear to represent a graded series approximating the sequence of evolution from early ancestral (prosimian) primates to man. The approximation to the evolutionary sequence permits us, to some extent, to use the living primates as a series of models for reconstructing our own behavioural past. This is a valuable asset, though it must be remembered that it is only an approximation. The living primates are, of course, the present end-products of an evolutionary radiation in which all species have, without exception, undergone some modification from the ancestral primate condition. This must be borne in mind when any attempt is made to reconstruct an actual evolutionary sequence in primate behaviour. The more that is known about primate behaviour in general, the more successful such attempts at reconstruction are likely to be, and due attention should therefore be directed to prosimian behaviour, as well as to that of simians.

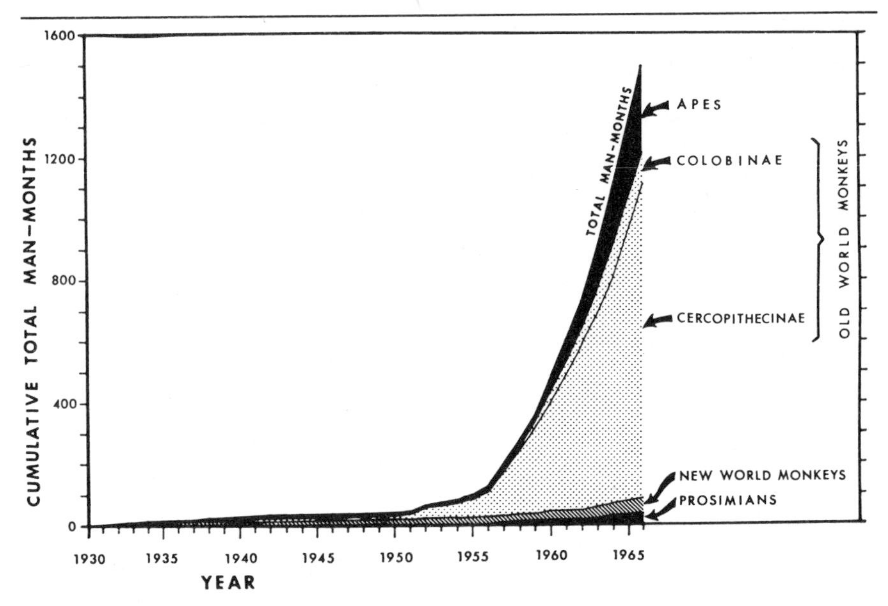

Figure 1. Cumulative man-months devoted to field studies of ecology and behaviour of non-human primates up to 1965 (after Altmann 1967).[3] The data include studies based on artificially established colonies maintained under natural or semi-natural conditions. Since 1965, the explosive growth of primate field studies has continued, and somewhat more attention has been devoted to the behaviour of promimians under field conditions. (Reprinted with the kind permission of the University of Chicago Press, © 1967 by the University of Chicago.)

Studies of primate behaviour have attracted growing interdisciplinary interest within the last few decades and, particularly from the early 1950s onwards, the growth rate in research output has increased enormously. In field studies alone the output has doubled every five years since 1955 (Fig. 1).[3] The prosimians, relatively speaking, have been generally neglected in this advancing front of research. There are many reasons for this. The simian primates are diurnal (except for *Aotus*) and live in a sensory world similar to our own, with vision predominating. Many species are terrestrial and thus more easily accessible to observation. Simians are not only easier to study; they are also closer to man in an evolutionary sense, and as a consequence more similar to man in morphology and behaviour. Altmann[3] notes that the focus of behavioural studies, within the 'higher' primates, has largely concerned the Old World monkeys and great apes, which satisfy these criteria more than do the New World monkeys. The prosimians, on the other hand, are almost exclusively arboreal, usually nocturnal, and live in a sensory world rather different from our own, with olfaction and audition still competing with vision as primary sensory modalities. They are, therefore, more difficult to study, and we cannot always be sure that our interpretations of observations are accurate because of the difference in sensory priorities. Since prosimians are farther removed from man in the evolutionary sense, studies of their behaviour are less likely to reveal interesting behavioural homologies with man than are studies of the simians. Yet an overall picture of primate evolution is just as incomplete without data on the prosimians as it would be without data on the simians.

The bias of primate behaviour studies towards the simian primates continued without modification until the early 1960s, after which time there was a notable increase of research interest in prosimians. Of the total number of papers in scientific journals that have been concerned wholly or partly with prosimian behaviour, approximately 80% have been published since 1960.

The 23 field and laboratory studies of prosimian behaviour represented in this present volume range over 19 different species (including 2 species of the Tupaiidae, regarded as 'honorary' prosimians for the purposes of the conference). Of a total of 18 generally recognised genera of Lemuriformes, Lorisiformes and Tarsiiformes, only 5 are not represented, although the lively discussions that followed all papers managed to bring in *all* genera. Many of the suspected difficulties involved in the study of small, nocturnal and arboreal primates in their natural habitat have been surprisingly easy to overcome, as is testified by a number of papers on field-studies in the present volume. Some of the difficulties have been overcome through indirect methods of studying behaviour (e.g. through various trapping techniques and analysis of stomach contents); but it has also proved possible to conduct extensive direct observation of behaviour by use of battery-operated headlamps to locate animals (through reflection of light from the tapetum of the eye) and to observe various aspects of ongoing behaviour. In captivity, because of their small size, the prosimians lend themselves more easily to accommodation under semi-natural conditions, and various devices (e.g. reversal of the light cycle and hence of the normal diurnal rhythm; use of weak red or blue observation lights) permit relatively easy observation of their behaviour in captivity.

In fact, the distinction usually made between field and laboratory studies of primates is often less valid for the prosimians than for the simians. In addition, although the distinction between studies conducted in the natural habitat and those conducted in the laboratory is nonetheless a vital one, consideration of the types of information sought indicates that a tripartite distinction is more relevant; namely, between studies in the natural habitat, investigations in semi-natural conditions, and studies in unnatural and restricted laboratory environments. This view is fully supported by Washburn and Hamburg[4] for 'higher' primates as well, the Rhesus monkeys of Cayo Santiago and the Japanese macaques in the large enclosed corral at the Oregon Primate Research Center providing good examples of semi-natural facilities for studying simian primates, intermediate between the typical field and laboratory situations.

Ideally, all three approaches to the study of prosimian behaviour should be utilised to the full in order to provide a sound basis for discussion. Each approach can be used to clarify particular aspects of behaviour for an eventual synthesis. However, it is essential to remember that all behavioural studies are subject to certain sources of error, and that great caution must be exercised in interpreting observations. Both field and laboratory studies are open to methodological errors, and in both areas the development of reliable techniques is a matter for priority. Some of these techniques are uniquely necessary for prosimians, but some have already been developed for field study of simians. For example, under field conditions it is becoming increasingly clear that long-term studies of prosimians and simians alike are necessary in order to take account of seasonal effects and in order to eliminate erroneous first impressions (e.g. underestimation of social group size; oversimplification of dominance relationships). It is also valuable to study the same species in different areas – as was the case with Alison Richard's study of *Propithecus verreauxi* – in order to determine natural variation in patterns of behaviour. Studies in captivity, on the other hand, are subject to a particular additional disadvantage. Whereas the field-worker is primarily concerned with the effect of his (or her) presence on the animals in their natural habitat, the laboratory investigator is faced with the dual danger of disturbing the captive animals directly through techniques of observation, and indirectly through provision of inadequate environmental conditions. There is also the combined drawback that animals confined in relatively small spaces may react far more violently to observer presence. A single example will place this in perspective: the lesser mouse lemur (*Microcebus murinus*) in its natural habitat in Madagascar[5] occupies a home-range which is probably comparable in volume to that of a cylinder of radius 25m. and height 15m., viz. approximately 30,000 cubic m. In the laboratory, it is customary to confine Lesser Mouse Lemurs in cages of 1-5 cubic m. capacity. Thus, the animal's effective living space is reduced by four orders of magnitude when it is confined in captivity. The laboratory observer must also choose what will be presented in the form of a social and physical environment within this restricted living-space, and the nature of this choice will necessarily influence the information which is collected.

This does not mean, however, that laboratory studies are less valuable than field studies; the two approaches provide different kinds of information which are equally valuable. Field observations are indispensable whenever *evolutionary processes* are considered, for the evolution of each prosimian species has taken place under a given set of environmental conditions, which may have changed gradually with time. Without knowledge of the present natural habitat, and some impression of the past history of that habitat, it is impossible to provide an accurate interpretation of the survival value of specific behaviour patterns or of a general evolutionary development. On the other hand, laboratory studies are necessary whenever any detailed analysis of *behavioural mechanisms* is required, and in such cases the research worker benefits greatly from the fact that the environmental conditions can be controlled. In this respect, studies conducted under semi-natural conditions perform a valuable intermediate role by indicating the behavioural effects achieved by the attempt to replicate the natural situation as far as is possible. Obviously, it is of great value to combine the results of field and laboratory studies in order to extract the special advantages of each approach and to eliminate sources of error. By bringing together both field and laboratory workers within the framework of the present conference, the authors hoped to further this kind of fruitful interchange of information about prosimian behaviour. The field worker can assist the laboratory worker by providing illustrations of the operation of behaviour patterns under natural conditions, while the laboratory worker can assist the field worker by indicating possible mechanisms and suggesting pertinent observations to be conducted under natural conditions.

Confinement *per se* away from the natural habitat may give rise to stress responses, as was suggested during the discussion (Martin). Stress can result both from the physical conditions of captivity and from the social environment of conspecifics caged in the same area. Of course, it is quite likely that stress responses – particularly when evoked by conspecifics – are present under natural conditions. However, confinement in captivity must almost always produce much higher stress levels, particularly since animals in caged groups are forced to interact far more frequently than would be the case under natural conditions. The laboratory observer should accordingly be on the lookout for symptoms of stress and consequent distortion of behaviour in caged animals. It was shown by von Holst that in *Tupaia belangeri* crowded conditions in captivity produce psychological stress which may result in 'ethological and physiological disturbances and diseases (e.g. renal disease)'. In particular, stress under caged conditions can, in this species, give rise to modification of mating and maternal behaviour. One of the best indications of captive stress in these tree-shrews is, in fact, a poor breeding record.[6,7]

It was also shown by Perret that in *Microcebus murinus* the adrenal glands are more developed under laboratory conditions, perhaps as a result of population density in captivity, and that this is associated with a very poor reproductive record. Again, as von Holst indicated for *Tupaia belangeri*, reproduction itself is probably a good index of the level of stress in the laboratory, and it can probably be assumed that prosimian species which breed successfully in captivity are not subject to high stress levels. Both of the present authors have attempted, with their laboratory colonies, to provide conditions as close as possible to those prevailing in the natural situation, despite the relatively small cages imposed by considerations of economy. With *Galago senegalensis* (Doyle), it has proved possible to attain a reproductive rate at least as high as that found in the wild. Behavioural considerations indicate that, for *G. senegalensis* at least, such semi-natural laboratory conditions are not notably stressful. With *Microcebus murinus* (Martin), maintenance of individuals in far smaller enclosures apparently gives rise to stress levels which are only partially offset by attempted replication of key features of the natural environment. Through constitution of social groups similar to those in the wild, replication of the natural diurnal and annual light-cycles, maintenance of natural temperature and humidity levels and provision of a varied and balanced diet, limited breeding successes have been achieved.[8] Recently (Martin, unpublished data) somewhat better breeding successes have been attained, though it is not certain whether this is a result of further adaptation of the animals to laboratory conditions, or to further improvements of those conditions (additional attention to diet; provision of a diurnal variation in temperature). Experience with *Microcebus murinus* generally indicates that this prosimian species is extremely sensitive to conditions in captivity, especially when housed in small cages (1 cubic m. capacity). Judging from the relative ease with which *Galago senegalensis* has been bred in captivity (Doyle), it seems likely that *M. murinus* would breed far more reliably if housed in cages of greater capacity (8 cubic m. and above).

Optimal conditions of hygiene must be at least as important as spatial, social and other factors, if animals are to be expected not only to survive but also to behave 'normally' in captivity. Doyle mentioned that the survival rate of his animals under semi-natural laboratory conditions is probably higher than in the wild, and the same probably applies to many laboratory colonies of prosimians. Bergeron and Buettner-Janusch's detailed description of the primate facility for prosimians at Duke University provides a model for the maintenance and care of prosimians under conditions which are eminently suitable for behavioural and other studies, and extensive veterinary precautions probably ensure that the animals are healthier in that laboratory than they would be in the natural habitat. Bergeron, in discussion, suggested that we might question the wisdom of this in the light of implicit and explicit attempts to duplicate natural conditions in the laboratory. Should we really be studying the behaviour of optimally healthy animals when their wild counterparts are not, on average, as healthy? However, close confinement of prosimians in captivity is once again a crucial factor, since the natural pattern of disease transference is completely altered by forced proximity. There is also the point that animals subject to social and environmental stress may be far more prone to suffer from diseases of various kinds. Thus, it would be difficult to ensure a natural balance of disease risk and hygiene under laboratory conditions. Therefore, with the usual aims of laboratory studies of behaviour in mind, there can be no doubt that extensive measures must be taken to ensure the health of laboratory animals: it will be the healthy, rather than the unhealthy, animal which will tend to display the full repertoire of both structural and functional characteristics that have been selected in the wild, and that account for successful adaptation to the ecological niche and hence for survival.

Turning to the various papers on prosimian behaviour, one can see how these various principles apply. Schilling's study of *Lemur catta* marking behaviour under natural conditions and Epps' study of *Perodicticus potto* social behaviour in captivity demonstrate the advantages and drawbacks of the two approaches in two cases where the central interest is naturalistic behaviour. The distinction between the two types of study lies mainly in the fact that in the first (field study) there is little manipulation of environmental variables,[9,10] whilst in the second (laboratory study) there is more manipulation of these variables provided primarily by the automatic constraints of a semi-natural environment, affecting (for example) living space, the arrangement of social groups, availability of food, and so on. While the intention of the field study is to leave the animal largely undisturbed so that its behaviour will be determined almost entirely by naturally occurring phenomena and can be observed over its entire range, the aim of the semi-naturalistic laboratory study is to observe the animal under conditions of minimal observer-interference consistent with the need to investigate particular aspects of natural behaviour at close range and in greater detail than is possible under natural conditions. Observations from both types of study are equally valid and useful; but care must be taken in interpretation of behaviour in terms of its survival value and adaptation to the natural environment. These two types of study, with their common aim of describing and interpreting naturalistic behaviour, are sharply distinct from laboratory investigations of specific aspects of behaviour where certain inherent mechanisms are the focus of interest. This applies, for example, to studies of intelligent behaviour and learning ability, where the nature of the investigation requires a situation almost totally unnatural to the animal, as was the case with Cooper's study of *Lemur macaco*. Petter-Rousseaux's study of the effect of photoperiod on sexual activity and body-weight of *Microcebus murinus* similarly involved the investigation of specific aspects of behaviour and physiology without particular reference to the natural situation, and this study dovetailed very neatly with Perret's investigation of endocrine gland variation in the same species. Laboratory investigations of this kind, devoted to specific aspects of behaviour and physiology, are extremely important in their own right, and they can also shed light on factors contributing to behaviour under natural conditions. However, special caution must be exercised in the general interpretation of such restricted laboratory studies in the absence of adequate information from the natural environment. The effect of confinement, directly or indirectly, on behaviour is invariably raised during discussions at conferences such as the one on which this present volume is based. Are we, in fact, studying the same animal in the laboratory – whether in small cages or in large semi-natural environments, whether indoors or outdoors – as that which exists in a natural state in the wild?

A good deal depends, of course, on the central interest of each study. For purely behavioural studies, Klopfer assures us that, with respect to his investigation of mother-infant relationships in *Lemur*, 'conditions of captivity were not unduly distorting the behaviour'. The same is probably true of many other studies where the research worker is interested in a fairly circumscribed aspect of behaviour, and where the conditions in captivity at least satisfy certain minimal conditions. For many far-reaching aspects of behaviour, however – notably such components as intra-group behaviour, dominance relationships and territoriality – confinement in the laboratory (even in a semi-natural environment) may seriously distort behaviour and create artificial pressures which might give rise to behaviour patterns not characteristic of relatively solitary animals in the wild. For this reason, consideration of social and territorial behaviour would normally be confined to field studies; yet there are some aspects of this very broad area of study which may legitimately be the subject of laboratory investigations. Tandy's study of social behaviour in an artificial group of *Galago crassicaudatus* in captivity, in which she demonstrates (among other things) a tendency towards solitariness and the importance of olfaction, provides a good example. Bearder (represented by Doyle) suggested that because even solitary animals must and do encounter one another in the wild, a good idea of the characteristic patterns of response to such encounters can be gained by arranging paired presentations under laboratory conditions, a procedure which he utilised with *Galago senegalensis*. Epps used essentially the same procedure with *Perodicticus potto* in her outdoor enclosure.

Nevertheless, it would not be true – except in a very limited sense – to think of laboratory studies as concerned exclusively with certain types of behaviour while field studies are concerned exclusively with other types. Although each type of study serves its own particular purposes, all studies are ideally directed at achieving a total descriptive picture of the species in question, and thus field and laboratory studies must be mutually complementary. This is particularly obvious from a comparison of Charles-Dominique's field studies relating to bush-baby locomotion[9] with the laboratory data reported below by Jolly and Gorton, Jouffroy et al, and Hall-Craggs. A synthesis of various kinds of data on locomotion[11] provides an extremely broad basis for considering the evolution of prosimian morphological adaptations for locomotor function. Very often, behaviour observed in one context cannot be fully appreciated without reference to observations from other contexts. Washburn and Hamburg[4] give many good examples of the inter-relationships between field and laboratory studies in simian primates, and of the extent to which each depends on the other for a full understanding of many complex aspects of behaviour. Several other examples among prosimians are provided in this present volume. For instance, the frequent carrying of very young infants in the mouth (characteristic of Galaginae) and the baby-parking behaviour found in many lorisids (characteristic of the Lorisinae, and apparently common among Galaginae) was mentioned by a number of participants in discussion. Such behaviour is, relatively speaking, much easier to observe under laboratory conditions, yet its significance is not readily apparent in the laboratory and only emerges when the natural situation is documented.

The field of olfactory communication provides a good illustration of the inter-dependence of studies conducted in the laboratory and in the wild, and this matter is specifically raised in Doyle's paper on *Galago senegalensis moholi* and in Harrington's paper on *Lemur fulvus*, which is exclusively devoted to olfactory communication. The need to appreciate the full range of marking behaviour in the wild is exemplified by Schilling's field study of *Lemur catta*, while the advantages of laboratory study for examining the fine details of specific aspects of marking behaviour are very neatly demonstrated by Manley's paper on two lorisine species.

Preliminary information on little-known species, often of great importance, is invariably provided by field-studies, which are generally devoted to aspects for which laboratory studies are not particularly suitable. There is also the point that the little-known prosimian species are almost always notoriously difficult to maintain in captivity. Such studies are typified by Petter and Peyriéras' paper on *Indri indri*, Hladik and Charles-Dominique's paper on *Lepilemur mustelinus* and Fogden's paper on *Tarsius bancanus*. All three of these species have a poor survival record in captivity, and the papers on their behaviour under natural conditions represent major contributions to our knowledge of these prosimians. Hopefully, these pioneer studies will lead to further field studies and eventually to appropriate laboratory studies. Detailed information on their behaviour in the wild may lead to solutions for the problems of maintaining these species in captivity.

Studies of factors accounting for adaptation to preferred habitats, as for instance in Bearder and Doyle's study of *Galago senegalensis* and *G. crassicaudatus* and D'Souza's field study of *Tupaia minor* and *T. glis*, and studies of ecological and ethological relationships between closely related species where they are sympatric in the wild, can (of course) only be undertaken in the natural situation, though some of the questions asked may have been generated out of laboratory studies. The broad ecological/ethological approach is exemplified by Charles-Dominique's study of live sympatric lorisid species, and by Sussman's detailed study of sympatric *Lemur catta* and *Lemur fulvus* in Madagascar. Sussman's study shows particularly clearly the kind of phenomena which may be observed where two closely-related species occur together in one given area, yet occur separately in other areas. Klopfer's study in the laboratory of *Lemur catta* and *L. fulvus* did, in fact, raise questions regarding differences in maternal-infant relationships, and Sussman's study provides partial answers to these questions from the natural situation, once again highlighting the inter-dependence of field and laboratory investigations.

Another area in which field and laboratory studies interact is that of seasonal variation in behaviour. Richard's study of mating activity in *Propithecus verreauxi* as a function of seasonal change, apart from providing surprising new information on the mating patterns of this species under natural conditions, provides a classical example of a field study of behaviour. Since reproductive behaviour in this lemur species depends upon seasonal variations in natural factors, such as availability of food, temperature, rainfall, foliage cover and day-length – which cannot be effectively varied in a large enough enclosure under laboratory conditions – its study calls for intensive field-work. The effects of certain diurnal and annual variations in environmental factors such as temperature, humidity and photoperiod, which can be individually controlled in the laboratory with respect to some elements of behaviour, do lend themselves to studies in captivity, however. Both Perret and Petter-Rousseaux reported on the effects of annual variation in daylength on endocrine activity and reproductive behaviour in *Microcebus murinus*, and provided us with a general basis for the interpretation of seasonal variation in the behaviour of the Malagasy lemurs. Along with Pariente's paper on the effect of diurnal and seasonal variation in light availability on the behaviour of *Phaner furcifer* and *Lepilemur mustelinus*, these different studies cover many aspects of seasonal variations in lemur behaviour.

As has already been stated, studies requiring closely controlled manipulation of experimental variables can, of course, only be undertaken under highly artificial laboratory conditions, as Cooper's study of learning sets in *Lemur macaco* demonstrates. Even in this case, however, the answers sought are related to questions raised by observations in the field as well as in the laboratory, such as those conducted by Jolly on object relationships.[12] Relevant questions for detailed research may also be raised through comparative considerations of brain function and behaviour and through comparative study of similar phenomena in simian primates and non-primate mammals.

The kinds of information that the student of prosimian behaviour seeks are, in general, much the same as those sought in the study of higher primates. Each of the behavioural studies presented in this volume has its counterpart in studies of simian primate behaviour. With the prosimians, perhaps, there is a greater implicit interest in more remote evolutionary developments. In the simians, particularly the Ponginae and the more terrestrial Cercopithecinae, there is an understandably greater interest in more recent evolutionary developments, particularly in such things as complex social group behaviour and communication, for example. All primate studies are, however, directed ultimately towards increasing our general knowledge of the entire mammalian Order to which man belongs. In the same way that field and laboratory studies should be regarded as mutually complementary, studies of prosimians are complementary to those of simians. One of the aims of this volume was to bring together a number of papers about prosimian behaviour as a first step in the long-outstanding drive towards a more comprehensive knowledge of prosimian biology.

NOTES

1 Huxley, T.H. (1863), *Evidence as to Man's Place in Nature*, London.

2 Le Gros Clark, W. (1962), *The Antecedents of Man*, Edinburgh.

3 Altmann, S.A. (1967), Preface, in S.A. Altmann (ed.), *Social Communication among Primates*, Chicago.

4 Washburn, S.L. and Hamburg, D.A. (1965), 'The Study of Primate Behaviour', in I. DeVore (ed.), *Primate Behaviour*, New York.

5 Martin, R.D. (1972), 'A preliminary field-study of the Lesser Mouse Lemur (*Microcebus murinus* J.F. Miller 1777)', *Z.f. Tierpsychol.*, Beiheft 9, 43-89.

6 Martin, R.D. (1968) 'Reproduction and ontogeny in tree-shrews (*Tupaia belangeri*), with reference to their general behaviour and taxonomic relationships', *Z.f. Tierpsychol.* 25, 409-532.

7 Von Holst, D. (1969), 'Sozialer Stress bei Tupajas (*Tupaia belangeri*)', *Z.vergl.Physiol.* 63, 1-58.

8 Martin, R.D. (1972), 'A laboratory breeding colony of the Lesser Mouse Lemur', in W.I.B. Beveridge (ed.), *Breeding Primates*, Basel, pp. 161-71.

9 Charles-Dominique, P. (1971), 'Eco-éthologie des prosimiens du Gabon', *Biol. Gabon.* 7, 121-228. This paper provides a very good example of a detailed prosimian field-study, though this particular investigation did involve some manipulation.

10 Even in field-studies, a degree of manipulation of the environment may occasionally be employed, as in Charles-Dominique's study[9] of various prosimian species in Gabon, where animals were forced to take certain aboreal routes or deliberately limited in their choice of routes in order to facilitate study of locomotor behaviour. Trapping techniques used under natural conditions, as in the separate investigations carried out by both Charles-Dominique and Fogden

(this volume) also represent a major intrusion into the natural situation.

11 Martin, R.D. (1972), 'Adaptive radiation and behaviour of the Malagasy lemurs', *Phil. Trans. Roy. Soc. (Lond.)*, B, 264, 295-352.

12 Jolly, A. (1964), 'Prosimians' manipulation of certain object problems', *Anim. Behav.* 12, 560-70. Jolly, A. (1966), *Lemur Behaviour: A Madagascar Field-Study*, Chicago.

PART I SECTION A
Field Studies of Behaviour and Ecology

F. BOURLIÈRE

How to remain a prosimian in a simian world

When compared with monkeys and apes, prosimians display definite behavioural handicaps. Jolly describes *Lemur* and *Propithecus* as 'hopelessly stupid toward unknown inanimate objects'[1] and she found that *Lemur catta* and *L.fulvus* were scoring 'far below both Old and New World monkeys on every type of problem, from the simplest of insight tests in object manipulation to object discrimination and delayed-response learning'.[2] Similar conclusions have been reached by Andrew,[3] Arnold and Rumbaugh,[4] Klüver[5] and Rumbaugh and Arnold.[6] Behaviourally speaking, as well as morphologically and physiologically, the Prosimii have therefore been considered as more 'primitive' than Anthropoidea and one has wondered how these two sub-orders could still coexist in nature.

The papers presented during the first two sessions of this Research Seminar, together with the ensuing discussions, provided a wealth of ecological and behavioural data which, when confronted with the findings of the morphologists, point towards a more balanced evaluation of the situation and lead to a better understanding of prosimian evolution. At last it has become possible to provide plausible answers to the following two questions: How did these primitive primates manage to survive, and in some cases to thrive, in a tropical arboreal world dominated by the 'more gifted' monkeys and apes? Why did some ancestral prosimians successfully radiate to give descendant forms parallel to their 'more advanced' relatives (i.e. in Madagascar)?

The answer to the first question now seems quite obvious. Wherever prosimians share the same habitat as monkeys and apes, they occupy entirely different ecological niches and therefore avoid any direct competition with higher primates. On the whole they have become nocturnal, arboreal fruit-and-insect-eaters, more or less narrowly specialised according to the presence or absence of sympatric competitors (e.g. arboreal marsupials, rodents or even carnivores). Furthermore, they did not content themselves with quietly living their obscure lives in obscure places – to paraphrase Harrison Matthews' evaluation of the Insectivoran way of life[7] – they have also taken advantage of some of their 'primitive' morphological and physiological characteristics to ensure a highly efficient occupation of their particular niches. For instance, the capacity of some lower forms to undergo prolonged seasonal torpor (Bourlière and Petter-Rousseaux[8]) plays the role of an 'energy-saving adaptation' enabling them to endure seasonal food shortages which would be incompatible with the mere survival of monkeys of similar size. The same comment also applies to the ability of some species to live on a low-calorie intake or a poorly-digestible diet, and also to many locomotor adaptations. Thus, the nocturnal life-styles of many nocturnal prosimians are not merely a way of 'making themselves forget'; they represent highly efficient adaptations to very particular niches – so successful indeed that they might well have prevented the emergence of competitive simian forms. This could explain why nocturnal monkeys only exist in the neotropics (*Aotus*), where prosimians disappeared long ago, and where the African galagine niche is only partly filled by some arboreal didelphids.

The major attributes of the typical African and Asiatic prosimian ecological niche emerge quite clearly from the recent field studies. Nocturnality ranks first; the circadian activity rhythms of sympatric species of prosimians and monkeys never overlap anywhere in continental Africa or tropical Asia. Although some species of the two sub-orders may frequent the same trees, they never actually meet; the temporal separation is complete. It is also worth mentioning that other mammalian competitors are few. Nocturnal arboreal didelphids are lacking in the Old World, and nocturnal arboreal rodents are rather scarce. The major potential competitors of prosimians in African and Asian forests are the nocturnal bats – mainly the frugivorous flying foxes – and some partly frugivorous scansorial carnivores. At this point, one may wonder whether temporal separation of activities is an absolute means of avoiding competition. This is certainly so in the case of insectivorous animals, as the kinds of insects (and other invertebrates) available by night to vertebrate predators are not the same as those available by day. The situation is slightly different for fruits, which remain available to frugivorous consumers after, as well as before, dusk. But in this case ecological separation between nocturnal and diurnal frugivorous species living in the same area is brought about by other factors, particularly differential location of fruits within the various forest layers, and on a variety of supports.

Preferences in the types of support used by nocturnal prosimians do indeed enable them largely to avoid direct competition with other vertebrate consumers, or even with sympatric species of their own sub-order. This is particularly obvious in the case of the five forest species studied by Charles-Dominique in north-east Gabon.[9,10] This is also apparently the case for animals living in more open environments (Sauer and Sauer, Bearder and Doyle).[11,12] The ability of *Galago senegalensis* to move with ease in extremely thorny bushes where very few other vertebrates dare to venture is also noteworthy (Doyle, verbal communications), as is the restriction of *Tarsier bancanus* to the forest understory (Fogden).[13] The use of small, and predominantly vertical, supports is made possible by specialised adaptations of the hands and, particularly, the feet. The occurrence of different species within different layers of the same forest environment is quite probably associated with their different abilities to use the various supports available and to negotiate obstacles (difference between 'leapers' and 'non-leapers').

The data on dietary specialisation of the different prosimian species presented at the Research Seminar are also particularly significant. Whereas overlap in dietary regimes can be the rule at times when food is plentiful (seasonality has been observed in most rain-forest types so far investigated), definite species-typical preferences for certain food items are observed during periods of food shortage. The exploitation of vegetable gums by both galagines and lorisines, which is apparently restricted to the African species according to available data (Rahm;[14] Charles-Dominique;[9,10] Bearder and Doyle[12]), is unusual among mammals. The preferences of African galagines and lorisines for different types of insect prey is also noteworthy – and clearly related to their different sensory abilities. The extremely active galagines hunt primarily by sight and sound and feed on a variety of active insects; passive echo-location is possibly used for detection of their animal prey by night. The sluggish African lorisines, on the other hand, have specialised upon prey generally considered as unpalatable by most predators: urticant caterpillars, centipedes, ants, etc., which are detected by olfactory means. In Ceylon, *Loris tardigradus* also feeds upon ants, spiders and bugs, in addition to more palatable insects (Petter and Hladik).[15] In so doing, the lorisines make use of food sources generally unexploited by reptiles, birds and mammals. They are, so to speak, earning their living upon a link of the food-chain more typically used by invertebrate predators and parasites.

The ecological niches of nocturnal prosimians are therefore both highly specialised and very efficiently filled. Though confined to rather marginal habitats and driven to nocturnality by the strong competition exerted by vegetarian monkeys and apes, prosimians have succeeded in maintaining their stand for millions of years by exploiting and developing a number of structural and functional adaptations.

One may be tempted at this stage to wonder which came first and what was the prime characteristic which enabled the nocturnal prosimians to survive so successfully to the present day; whether it was their special types of locomotion, their diverse sensory abilities or their nutritional plasticity which permitted them so efficiently to hold their own. Opinions apparently vary according to the particular interests of the various investigators, as exemplified during the seminar discussions. However, such a question might well be pointless. Primitive, 'generalised' prosimians were quite probably not so highly specialised as their modern relatives. Under various selective pressures they probably developed simultaneously their battery of present-day adaptations. After all, a living organism is an integrated whole, not merely a carcass, a set of teeth or a bunch of sense organs.

The answer to the second question is not so obvious as that to the first one. Recent field studies have definitely established that prosimians were able, on the island of Madagascar, to undergo an adaptative radiation which, in many ways, parallels the evolution of monkeys in Africa, Asia and America (see for instance Sussman).[16] But the environmental factors which initiated and made possible this impressive radiation have not yet been clearly established. However, one thing seems certain: the fauna of this mini-continent is largely 'unbalanced', or 'disharmonic'; a number of important taxonomic groups (e.g. Felidae; Ungulata, with the sole exception of the bushpig) being conspicuously absent. As a result of Madagascar's early separation from the African mainland and of the 'filtering effect' upon potential invaders exerted by the Mozambique Channel, the terrestrial and arboreal vertebrate groups have been able to evolve during millions of years without the competitive and predation pressures which were so strong elsewhere in the tropics. For instance, almost no other arboreal mammal was competing for food with the Madagascar prosimians, true arboreal rodents being virtually absent. Arboreal mammalian predators were also scarce. Even birds of prey were few (11 Aquilidae, 3 Falconidae, 6 Strigidae and Tytonidae) and the large lemurs of the island might have been expected to support some powerful 'monkey-eagle', such as *Stephanoaetus* of the African forests or *Pithecophaga* of the Philippines, as pointed out by Moreau.[17] Large arboreal lizards were lacking, as were large arboreal snakes. The predation pressure exerted upon prosimians in Madagascar ecosystems must therefore have been relatively slight. This situation is apparently quite general on this island: such anti-predator devices as cryptic coloration and mimicry are rare among insects here, and the local race of *Papilio dardanus* is the only one, within the entire Ethiopian region, which lacks pronounced sexual dimorphism and does not possess females which mimic aposematic butterflies (Paulian, Ford).[18,19] This apparent lack of strong predation pressure probably accounts in part for the amazing diversity of coat patterns and the conspicuous coloration of many diurnal species, as well as for the frequent occurrence of some unusual behaviour patterns, such as the sunning behaviour of *Lemur, Propithecus* and *Indri.* It seems highly likely that this sunning posture is also correlated with the imperfect homoeothermy of these animals and the necessity for them to warm up rapidly in the early morning; but such overt behaviour would have been terribly risky if large aerial predators hunting by sight had been present.

The emptiness of some ecological niches in Madagascar has, on the other hand, permitted their unexpected occupation by primates. Thus, the absence of any representative of the woodpecker family has made possible the extreme dietary specialisation of the aye-aye, which feeds at least partly upon wood-boring insect larvae (Petter and Peyriéras).[20] One may also wonder whether it is perhaps the absence of any endemic vertebrate grazer which has allowed *Hapalemur* to specialise upon certain grasses (reeds and bamboos; Petter and Peyriéras).[21]

The nocturnal lemurs of Madagascar have developed quite remarkable adaptations to the pronounced seasonality of their habitat. Some Cheirogaleinae are able to store fat in their tails before the dry season (Bourlière and Petter-Rousseaux).[22] In the *Didierea* bush of the South the folivorous *Lepilemur mustelinus* displays remarkable adaptations (economy of movement, enlarged caecum, caecotrophy) which permit an efficient utilisation of the scanty food resources (Charles-Dominique and Hladik).[23] In the Western coastal forest, at the end of the prolonged dry season, *Phaner furcifer* and *Microcebus coquereli* eat gums, resins and insect exudates (Petter, Schilling and Pariente).[24]

To sum up: present-day prosimians can no longer be considered as mere living fossils, more or less identical to their Eocene ancestors. Some of them have indeed kept up to modern times a 'primitive' morphology, but they have at the same time become so well-adapted physiologically and behaviourally to their marginal ways of life that they are very efficiently filling their role in tropical ecosystems. When given the chance to evolve in isolation from higher primates, under low predator pressure and in an unbalanced environment, as in Madagascar, other prosimians have been able to replicate extensively the ways of life of monkeys and apes. Sometimes they have even taken advantage of the emptiness of other vertebrate niches to give rise to such a bizarre and 'unlikely' primates as the aye-aye!

NOTES

1 Jolly, A. (1966), *Lemur Behaviour: A Madagascar Field Study*, Chicago.

2 Jolly, A. (1964), 'Choice of cue in prosimian learning', *Animal Behaviour* 12, 571-7.

3 Andrew, R.J. (1962), 'Evolution of intelligence and vocal mimicking', *Science* 137, 585-9.

4 Arnold, R.C. and Rumbaugh, D.M. (1971), 'Extinction: A comparative primate study of *Lemur* and *Cercopithecus*', *Folia primat.* 14, 161-70.

5 Klüver, H. (1933), *Behavioral Mechanisms in Monkeys*, Chicago.

6 Rumbaugh, D.M. and Arnold, R.C. (1971), 'Learning: A comparative study of *Lemur* and *Cercopithecus*', *Folia primat.* 14, 154-60.

7 Harrison Matthews, L. (1971), *The life of mammals*, London, vol. 2.

8 Bourlière, F. and Petter-Rousseaux, A. (1953), 'L'homéothermie imparfaite de certains prosimiens', *C.R.Soc. Biol.* 147, 1594-5.

9 Charles-Dominique, P. (1971), 'Eco-éthologie des prosimiens du Gabon', *Biol. Gabon.* 7, 121-228.

10 Charles-Dominique, P. (1972), 'Ecologie et vie sociale de *Galago demidovii* (Fischer 1808; Prosimii)', *Z.f. Tierpsychol.*, Beiheft 9, 7-41.

11 Sauer, E.G.F. and Sauer, E.M. (1963), 'The South West African bush-baby of the *Galago senegalensis* group', *J.S.W. Afri. Sci. Soc.* 16, 5-36.

12 Bearder, S.K. and Doyle, G.A. (this volume), 'Ecology of bushbabies, *Galago senegalensis* and *Galago crassicaudatus*, with some notes on their behaviour in the field'.

13 Fogden, M.P.L. (this volume), 'A preliminary field-study of the western tarsier *Tarsius bancanus* Horsefield'.

14 Rahm, U. (1960), 'Quelques notes sur le Potto de Bosman', *Bull. IFANA*, 22, 331-42.

15 Petter, J-J. and Hladik, C.M. (1970), 'Observations sur le domaine vital et la densité de la population de *Loris tardigradus* dans les forêts de Ceylan', *Mammalia* 34, 394-409.

16 Sussman, R.W. (this volume), 'Ecological distinctions in sympatric species of *Lemur*'.

17 Moreau, R.E. (1966), *The Bird Faunas of Africa and Its Islands*, New York.

18 Paulian, R. (1961), *'La zoogéographie de Madagascar et des iles voisines'*, Tananarive, Faune de Madagascar, 13, 1-485.

19 Ford, E.B. (1964), *Ecological Genetics*, London.

20 Petter, J.J. and Peyriéras, A. (1970), 'Nouvelle contribution à l'étude d'un lémurien malgache, le Aye-aye (*Daubentonia madagascariensis* E. Geoffroy)', *Mammalia* 34, 167-93.

21 Petter J.J. and Peyriéras, A. (1970), 'Observations éco-éthologiques sur les lémuriens malgaches du genre *Hapalemur*', *Terre et Vie* 24, 356-82.

22 Bourlière, F. and Petter-Rousseaux, A. (1966), 'Existence probable d'un rythme métabolique saisonnier chez les Cheirogaleinae (Lemuroidea)', *Folia primat.* 4, 249-356.

23 Charles-Dominique, P. and Hladik, C.M. (1971). 'Le *Lepilemur* du Sud de Madagascar: ecologie, alimentation et vie sociale', *Terre et Vie*, 25, 3-66.

24 Petter, J.J., Schilling, A. and Pariente, G. (1971), 'Observations éthologiques sur deux lémuriens malgaches nocturnes, *Phaner furcifer* et *Microcebus coquereli*', *Terre et Vie*, 25, 287-327.

C. M. HLADIK and
P. CHARLES-DOMINIQUE

*The behaviour and ecology of the sportive lemur (*Lepilemur mustelinus*) in relation to its dietary peculiarities*

Introduction

In the course of a field-visit to the south of Madagascar (September/ October 1970), *Lepilemur mustelinus leucopus* (F. Major 1894) was intensively observed in order to define the ecological characteristics of this nocturnal folivorous species (Fig. 1*a*), which is adapted to particularly arid conditions in this region. Study of one population provided us with an understanding of the social structure, permitting comparison with other prosimian species. Subsequently, a laboratory investigation was conducted on various samples from a digestive tract collected in Madagascar. This permitted more detailed examination of the phenomenon of *caecotrophy*, which had previously been observed in the field.

Field conditions

The sportive lemur,* which is still the subject of taxonomic debate, is found in almost all Malagasy forest areas. Most of the observations made during the present study were carried out in South Madagascar in a quite distinctive forest zone – Didiereaceae bush. This spiny bush forest is characterised by dense undergrowth composed of bushes, lianes and shrubs, dominated by species belonging to the Didiereaceae (a primitive endemic family intermediate between the Cactaceae and the Euphorbiaceae). *Alluaudia procera* and *A. ascendens*, which are the principal representatives of the family in

* The name 'sportive lemur' does not refer to the animal's motor activity, but to the defensive attitude which it adopts when threatened. The hands are used in the manner of a boxer to strike at the aggressor.

Figure 1 (*a*). The Sportive Lemur *Lepilemur mustelinus leucopus* (F. Major), photographed whilst active at night.

Figure 1 (*b*). The Didiereaceae bush occupied by the Sportive Lemur in the main study area (Berenty region; South Madagascar). The predominant plant representatives are the tall, taper-like *Alluaudia procera* and *A. ascendens.*

the Berenty region (S. Madagascar), constitute 65 per cent of the vegetation (Fig. 1*b*). These species resemble large, more-or-less ramifying tapers, about 12 m. in height, with a covering of fleshy leaves set at the bases of hard, sharp spines. Here and there, baobabs and some arborescent leguminous species project above the bush, which lacks a true canopy.

A different plant formation – *alluvial forest* or *gallery forest* – lines the rivers in the same area as a band 100-800 m. in width. This classical forest type is dominated by tamarind trees, which form the closed canopy at a height of 25-30 m., above a relatively scanty under-growth. Observations were conducted in this second type of forest purely for comparative reasons. The transition from gallery forest to Didiereaceae bush is very abrupt.

There is a long dry season, and rainfall (500 mm./year) occurs mainly during the austral summer (December/January/February). The monthly average of the maximum daily temperatures fluctuates between 37°C (austral summer) and 28°C (austral winter), and the monthly average minimum temperature fluctuates between 26°C (austral summer) and 16°C (austral winter).

In this xerophile forest formation, most of the plant species store reserves of water for the austral winter, which is particularly dry. In fact, our observations were conducted during part of the austral winter, and in that particular year (1970) there was an exceptional drought more pronounced than any experienced over the last ten years, at least.

Diet

In the Didiereaceae bush, the sportive lemur feeds primarily on the tough foliage of the two *Alluaudia* species, and on their inflorescences when the latter are available. *Alluaudia* flowers appear at the end of the dryest period, at a time when almost all of the leaves have fallen. Thus, the sportive lemur survives for some time on the basis of these flowers, which disappear when the first leaves begin to appear. The leaves of certain shrubs and lianes (*Salvadora augustifolia, Xerosicyos perrieri, Marsdenia cordifolia, Boscia longifolia*) are eaten in far smaller quantities. In addition, there may be a large number of other species whose leaves are utilised occasionally, along with certain fruits, since the present list is entirely derived from the end of the austral winter.

The sportive lemur is distinguished from other prosimians by its restricted motor activity during the night, which may obviously be correlated with the poor calorific value of the diet. As soon as the sun sets, the animal – which spends the entire day hidden in a tree-hollow, a fork of *Alluaudia ascendens*, or (more rarely) in a bundle of lianes or *Euphorbia* foliage – makes a number of rapid leaps to arrive at a leafy ramus, where it will 'browse' for short periods of 1-10 minutes (about 10 sessions per night). Apart from these feeding periods, the sportive lemur remains immobile almost continuously, rarely changing its place. When the animal does move, it makes a few rapid leaps and then remains immobile 10-20 m. away.

In the section on 'social life', it will be seen how we were able to study an entire population by capture, marking and release. The individual animals, which were exceptionally lacking in timidity, could be followed continuously with the light of a head-lamp. After a few days, we were able to observe the animals continuously from 3-6 m. without disturbing their activities. Such favourable conditions permitted us to observe the animals clearly and continuously. We were able to identify all of the food items taken, and vegetation samples taken back to the laboratory in France allowed us to conduct detailed analyses. In addition, continuous observation of individual animals and examination of digestive tracts collected at different times permitted us to determine the average dietary intake.*

* We are indebted to the Service des Eaux et Forêts de Madagascar for a special authorisation to capture a small number of specimens for subsequent laboratory analysis.

In collaboration with the Institut National de la Recherche Agronomique (Laboratorie d'Analyse et d'Essai des Aliments), we have analysed and described the average composition of the diet,[1] which is essentially composed of crassulescent leaves. This represents the 'poorest' diet hitherto observed in a primate: 13.6% vegetal proteins; 1.8% lipids; 4.9% reducing sugars; 15.1% cellulose and a 'complementary fraction' representing 64.6% (Fig. 2). The latter fraction is essentially composed of long-chain sugars (ligno-celluloses and hemicelluloses), which cannot be digested by most mammals, together with a large mineral component.

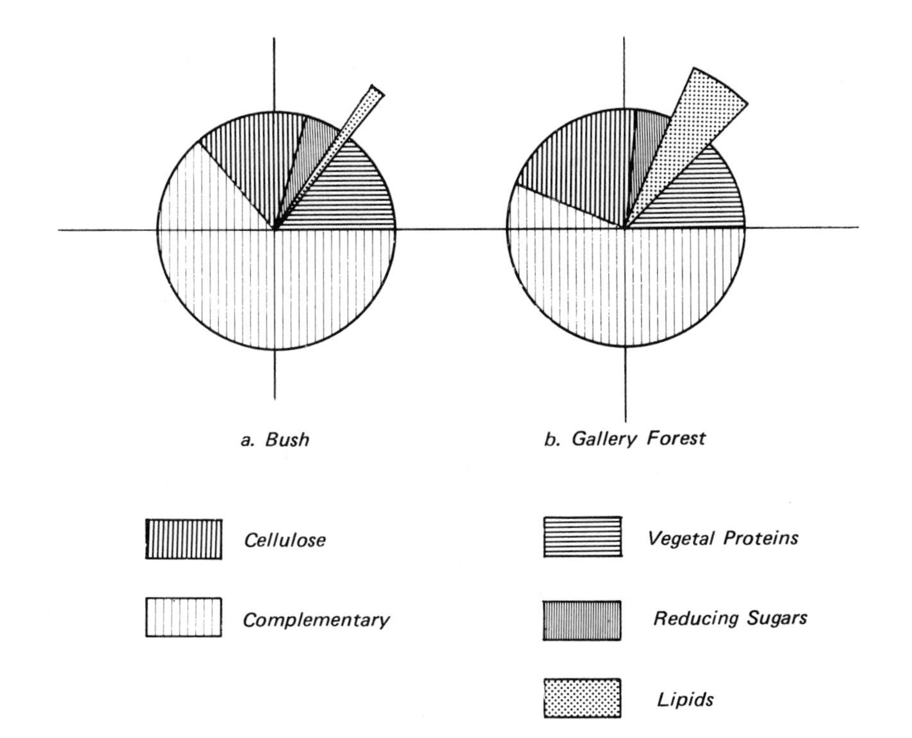

Figure 2. Diagrammatic representation of the dietary components of the Sportive Lemur in Didiereaceae bush and in gallery forest of South Madagascar.

In the gallery forest lining the river Mandrare, other individuals belonging to the same sportive lemur subspecies were, during the same period, feeding primarily on the leaves of *Tamarindus indica*, which contain 12.5% proteins, 6.0% lipids, 4.8% reducing sugars, 20.5% cellulose and a 'complementary fraction' of 56.2%. This composition is little different from that available to the animals feeding in the dryest areas of the bush (Fig. 2).

If this diet is compared with that of other known folivorous primate species, there is a considerable difference. In particular, the howler monkey (*Alouatta palliata*), which is the most folivorous of the South American monkeys, has a diet under natural conditions which includes a large proportion of fruits (60%), and thus has a relatively higher proportion of reducing sugars (21.7%), which are easily assimilable.[2] Similarly, among the Colobinae of the Old World the proportion of fruits included in the diet is quite high: 28% in *Presbytis senex* and 45% in *P. entellus.*[3] It is therefore certain that conditioning to soluble components of the diet, which plays an important part in the feeding behaviour of the 'higher' primates, is of negligible importance for *Lepilemur.* An immediate response to certain soluble components (principally short-chain sugars) can lead to rapid conditioning, which increases the efficiency of the animal in its natural environment, where it must detect its food with maximum yield. This constitutes motivation of the hedonic type. By contrast, conditioning through factors independent of the 'olfactory-gustatory' sense (physiological factors corresponding to an increased feeling of 'well-being' arising after the animal has taken its food) may play an important part in determining the sportive lemur's activity rhythm.

Dietary utilisation (caecotrophy) and the energy budget

Calculation of the energetic value of the dietary intake (about 60 gm. of fresh food per night), which was determined as part of the study already mentioned, shows that those nutritive elements which are readily assimilable (proteins, lipids and reducing sugars) represent only a weak energy source (13.5 Kcal. per day) for an animal weighing 600 gm. on average.

In fact, the cellulose fraction of the diet is degraded in the course of its sojourn in the caecum and colon of *Lepilemur.* The products of the components thus degraded are at least partially resorbed by the animal during a second passage of the food through the intestinal tract, which is achieved through *caecotrophy* (ingestion of certain faeces).

The behaviour pattern of re-ingesting certain faeces is reminiscent in several respects of similar behaviour in the rabbit, which is now well documented.[4] At about the mid-point of its diurnal resting period, the sportive lemur exhibits a phase of hyper-excitability. The animal begins by licking its fur and then concentrates on licking the anogenital region. When doing this, the thighs are spread apart and the tail is curled upwards, such that the pelvic area forms a kind of basin. The animal licks its anus and raises its head from time to time in order to swallow. This behaviour was observed directly on several occasions, always at the hottest time of the day (14.00-15.00 hrs.); but it would seem that it can also occur in the morning, as soon as the animal has returned to its daytime retreat. In fact, in the stomach contents of an animal captured at 7.00 hrs. we found fatty acids of bacterial origin, which could have originated from contamination by material re-ingested the day before, but which were more probably produced in the caecum and re-ingested as soon as the animal returned to its diurnal resting-place.

Analysis of the digestive tract contents of animals captured at different times of the day and night permitted us to follow the actual transformations which take place. In order to conduct this analysis, we profited from the period when the animals were feeding essentially on the leaves and flowers of the two *Alluaudia* species (which are very similar in composition), in order to have a 'basic diet' formed of a homogeneous mixture of these dietary samples. The differences in composition along the digestive tract are very marked, and there is no possibility of confusion. Although the basic diet only contained long-chain fatty acids, gas chromatography demonstrated that ramifying short-chain fatty acids ($C \leqslant 16$) with uneven numbers of carbon atoms were present in the digestive tract (analysis carried out at INRA).[5] The bacterial origin of these latter acids, and their abundance in the caecum, is not surprising. Electron micrographs of the caecal mucous membrane showed a dozen species of small-sized bacteria in contact with the microvilli. The structure of the caecum itself, which is lined with long villi supported by prolongations of the muscularis mucosae, shows that there must be mixing movements allowing rapid fermentation. The caecal contents become more and more fluid as the transformations progress.

The ingested food initially undergoes rapid absorption of the soluble fractions, which results in an apparent augmentation of the proportions of ligno-cellulose (rising from 21.4% to 43.0%). In the caecum, there is first of all hydrolysis of the cellulose fraction, which brings about a reduction in the ligno-cellulose level (17.5% in the caecal contents by the beginning of the night). The hemicelluloses are subjected to much slower degradation; their concentration amounts to 51.8% of the dry weight of the caecal contents, drops to 34.5% in the colon by the beginning of the night and reaches 20.6% in the faeces. Thus there is a progressive inversion of the hemicellulose/ligno-cellulose ratio, which changes from 3.9 in the caecum to 0.45 in the faeces.

The caecal contents are enriched with proteins (up to 36.0%, whereas the basic diet contents incorporate only 15.1%). The surplus protein is derived from desquamation of the mucous membrane and proliferation of bacteria. The importance of caecotrophy thus lies in limitation of nitrogen-loss. In the rabbit, the caecal matter which is re-ingested similarly contains a much higher concentration of proteins than the ingested food, and there is a marked difference in appearance between the caecotrophe (re-ingested caecal material) and the true faeces. In *Lepilemur*, however, we did not find any clear-cut morphological difference between the 'true' faeces and the caecotrophe.

In order to calculate an approximate energetic balance for *Lepilemur*, we selected an average example provided by an adult male which was continuously followed from the onset of activity until its return to rest. This individual ingested 61.5 gm. of leaves and flowers, corresponding to 8.0 Kcal. of proteins, 2.7 Kcal. of lipids and 2.8 Kcal. of carbohydrates, and thus to a total of 13.5 Kcal. of directly assimilable elements. The energetic contribution based on bacterial degradation of celluloses can be estimated at 10-15 Kcal. Thus, the overall energetic product is less than *30 Kcal/24 hr.*

In the example studied (Fig. 3), the animal which was followed covered 270 m. in the course of the night, making 180 leaps of about 1.5 m. Standard calculation of the energy expenditure indicates a *minimum muscular expenditure of 2.16 Kcal.*[6]

According to standard calculations of the basal metabolism, an animal of 600 gm. would require 40-60 Kcal. per day. In fact, it is known that many tropical and equatorial animals have imperfect thermoregulation and a basal metabolism inferior to the norm calculated for palaearctic animals adapted to cold climates. As has been demonstrated by Kayser (see also Bourlière and Petter-Rousseau)[7,8] the metabolic level of homoeothermic animals increases, with increasing perfection of thermoregulation, whilst some primitive homoeotherms in regions with a uniform climate have a basal metabolism at half the standard level. Research in progress in Gabon has confirmed this marked difference at the physiological level between primitive species derived from tropical and equatorial areas and species derived from temperate regions, on which the standard values have been based. For example, Hildwein[9] has shown that the basal metabolism of certain lorisids is inferior to that of 'true' homoeotherms, amounting to a deficit of 20-30% in *Arctocebus calabarensis*, and of 50% in *Perodicticus potto*.

Thus, one can expect that *Lepilemur* has a basal metabolism requiring in the order of *20-30 Kcal.*

This crude calculation is not intended as an exact indication of the energetic balance. It simply demonstrates, for an animal living under natural conditions, a system permitting it to survive with a minimal energy-input.

During the night, movement is fragmented into brief, regularly spaced periods (approximately twenty per night, see Fig. 3). Thus, chemical thermoregulation is augmented by the production of muscular heat energy at the time when it is most necessary. By contrast, during the diurnal resting phase, when the external temperature remains in the region of 30°C, energy expenditure must be minimal (bearing in mind the insulation provided by the dense fur).

The sportive lemur does not have to move far to find its food: a

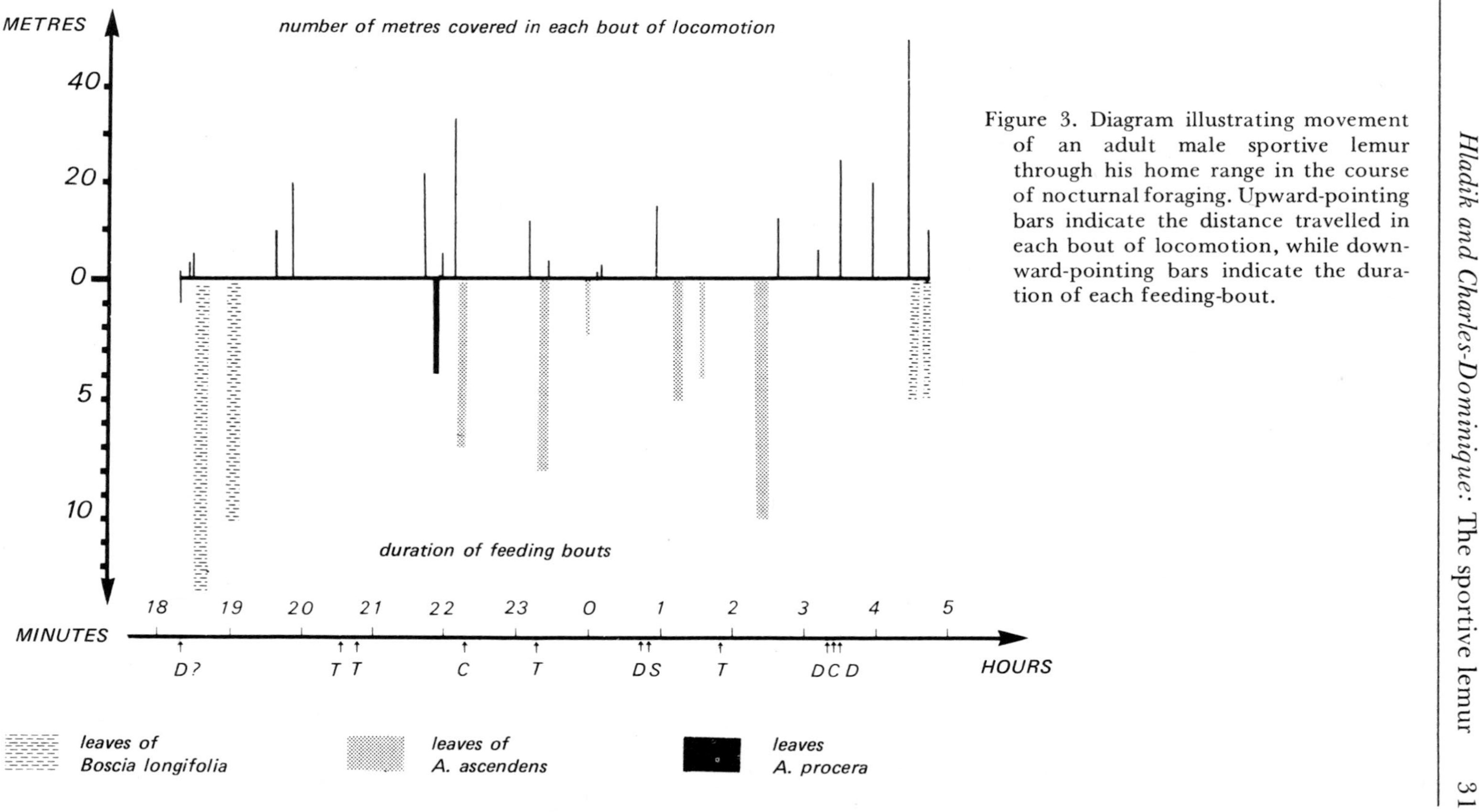

Figure 3. Diagram illustrating movement of an adult male sportive lemur through his home range in the course of nocturnal foraging. Upward-pointing bars indicate the distance travelled in each bout of locomotion, while downward-pointing bars indicate the duration of each feeding-bout.

single leaf-bearing ramus will suffice for one night. Thus, the animal performs a minimal muscular effort in moving around (scarcely more than 10% of the total energy expenditure). This economic way of life necessitates maximal utilisation of a naturally poor food-source, from which celluloses can be assimilated only after bacterial degradation. Accordingly, caecotrophy is a specialisation permitting utilisation of a food-source which normally yields little energy. This process, which is common among rodents and lagomorphs, seems to be exceptional among primates; nevertheless, it is less efficient than rumination, which permits even better utilisation of ligno-celluloses.[10]

Adaptation to the environment: population densities

The economic life-style of the sportive lemur is further expressed in the utilisation of extremely small home-ranges within which it can find its food. In the Didiereaceae bush, the average area of each home-range is 2,300 square m., and we were able to calculate under field conditions that the food available in the month of September could (theoretically) have provided for a maximum of 50 days. Bearing in mind the irregularity in distribution of plants and territorial demarcation (see next section), this figure can be regarded as a *minimum* which is explained by the drought present throughout the observation period. This exceptional drought, which was far more marked than any in the ten years preceding the study, represented a critical period, in the course of which there is doubtless a process of equilibriation between such populations and the environment.

The distribution of sportive lemurs in the Didiereaceae bush is not homogeneous, and the population counts that we carried out along the transects show that there are small population nuclei which are isolated to varying degrees. Various areas of the bush remain unoccupied, partly because of a lower density of edible plant species; but more frequently we observed that there was a scarcity of adequate diurnal shelters for the sportive lemurs to rest in.

In the gallery forest, the population density is greater than in the bush; it reaches 450 *Lepilemur* per square km., corresponding to a biomass of 2.7 kg. per hectare, as compared to 200-350 *Lepilemur* per square km. in the bush, which amounts to a biomass of 1.2-2.1 kg. per hectare. These population densities can be compared with those observed in a habitat entirely comparable to the gallery forest of southern Madagascar – that is, the alluvial forest of Ceylon. The biomass of a nocturnal prosimian (*Loris tardigradus*) found in the latter areas is extremely low, less than 0.25 kg. per hectare.[11] However, since *Loris tardigradus* is an insectivore, the 1:10 ratio of the biomasses is quite in agreement with standard ecological relationships found when passing from one trophic level to one immediately above. (*Microcebus murinus*, which is also partially insectivorous, has a biomass of 0.23 kg./hectare in the Didiereaceae bush). If we consider the folivores living in the alluvial forest of Ceylon (that is, the two colobines *Presbytis senex* and *P. entellus*), there are biomasses of 10 and 15 kg. per hectare.[3] These figures represent the maximum densities observed for animals whose 'dietary assimilation' is certainly better than that of *Lepilemur*, since their digestive system exhibits remarkable convergence with that seen in ruminants. In the gallery forest of South Madagascar, two large-bodied lemurs (the Sifaka, *Propithecus verreauxi*, and the Ringtail, *Lemur catta*) occupy ecological niches analogous to those of the colobines in Ceylon.

Therefore, the size of the *Lepilemur* home-range reflects, in the most arid area, more effective utilisation of the terrain. The biomass is very large in view of the low productivity of the Didiereaceae bush during adverse periods.

We carried out a complementary study of the distribution of the sportive lemur (*Lepilemur mustelinus mustelinus*) in the east coast rain-forest of Madagascar, which is dense and evergreen, and which doubtless has much greater primary production (study area: forestry station of Perinet). This other subspecies of *Lepilemur mustelinus*, which is 50-100% larger than the subspecies found in southern Madagascar, is sympatric with a nocturnal folivorous Indriid of comparable size – *Avahi laniger*. In the course of nocturnal counts, it was not always possible to distinguish between the two lemurs; but the overall density (of *Lepilemur* and *Avahi* together) is of the same order as that of *Lepilemur* in the dry forest of south Madagascar. Thus, one might ask whether the limiting factor governing such folivore populations is, in certain cases, something other than the level of food availability. However, the synecology of dense forests is much more complex – and hence much less well known – than that of dry forests. There are far more species in dense forests, the ecological niches are more specialised, and the conditions of observation are much more difficult.

Folivorous primate species share a large number of characters. They are the least active and the least mobile of all, though one must not confuse such slowness with that of the insectivorous lorisids (*Loris*, *Arctocebus* and *Perodicticus*), where it represents a cryptic mechanism.[12] It is obvious that the ease with which such folivorous species can find their leaf diet removes any necessity to move over large areas, and the home ranges are small. Among the gregarious 'higher' primate species the folivorous forms (*Alouatta*, *Presbytis*, etc.)[3,13] only defend small territories, compared with those of sympatric species which are frugivorous or insectivorous. Among the solitary nocturnal lemurs, individual territories are similarly smaller in folivorous forms than in frugivorous or insectivorous species. *Lepilemur* would appear to represent the lower limit of such home-range restriction.

Social life

The Didiereaceae bush constitutes a habitat which is particularly favourable for observation. During our study period, at the end of the dry season, the bush had the appearance of a forest of dry trunks and branches without leaves, in which visibility was very good. In addition to this, it was relatively easy to drive sportive lemurs from their retreats during the daytime, by beating the vegetation, and subsequently to capture them. When disturbed, sportive lemurs take refuge at a height of 5-8 m. on a ramus of *Alluaudia* and observe the intruder without moving. With a little patience it is possible to approach the animal very cautiously with a noose suspended at the end of a long pole and to pass the noose around the animal's neck. Some captures took only 10 minutes, whilst others required more than two hours; but we were nevertheless able to capture 12 of a population of 13 sportive lemurs. After a standard examination (genital organs, mammary glands, teeth, body-weight, etc.), the animals were marked by clipping the ears and shaving the tail according to various patterns, and then they were released and observed. In this study area, strings attached at a height of 1 m. in rows 10 m. apart, each marked with numbers, permitted us to localise any observation to within 2-3 m. on a map, and thus to delimit individual territories.

1. Territories

As has already been mentioned, the territories are small.[1] Adult females range over 0.18 hectares (0.15-0.32), adult males over 0.30 hectares (0.20-0.46), and juvenile females over 0.19 hectares (0.18-0.20). For comparison, one can consider the following values taken for 2 lorisid species (insectivorous and frugivorous prosimians): *Galago demidovii* – 0.8-2.7 hectares;[14] *Perodicticus potto* – 7.5-15 hectares.[15]

The female territories are distinct from one another. However, we noted a certain range overlap and mutual tolerance between an old female and two young females of one and two years of age. (All the females were gestating in October, and reproduction takes place only once a year). The territories of the males are similarly separate from one another, but we did observe one case of overlap between the two smallest males in the population. However, these two individuals were never observed together in this common area. Although there is quite clear separation of territories between males and between females, overlap of female territories with those of the males is the rule. The same applies, in fact, to the Lorisidae and to *Microcebus murinus.*[14,15,16]

The largest of the males was associated in this way with 5 females, whilst the other males were associated with two, one and one female, respectively.

2. *Territorial defence*

All adult males and females exhibit scars on their ears, snouts and tails as witness to intra-specific fights. We have never observed such traces of former wounds with young females at one year of age. Although olfactory marking constitutes the principal means of territorial defence with most nocturnal species, *Lepilemur* exhibits a process which is quite remarkable. In the sportive lemur, which has such small territories of approximately 50 m. diameter, the territory-owner can easily survey the limits, which are absolutely rigid. The occupant of a territory, usually whilst squatting on a high ramus of *Alluaudia*, surveys his immediate neighbours, which are also located on elevated branches. Such mutual surveillance is particularly frequent with the males, which spend hours observing one another, often with only a few metres between them. When one of them moves, it is common to see the neighbour move in his turn along the territorial limit and come to rest facing his opposite number once again.

From time to time, two – and sometimes three – males exchange vocalisations as a duet or a trio. A duet consists of a rapid series of 'Heh, heh . . . ' calls (aggression), followed by an abrupt, high-pitched 'Hiii' call (signalling). These two vocalisations are subsequently exchanged for a period of 10-20 seconds in a less precise order, and calm then returns. In fact, we counted many more duets of this kind when the moon was visible, and this indicates quite clearly that contact between neighbours is largely based on vision. In the course of reciprocal surveillance, we also saw a display consisting of powerful leaps against a support, or a simulation of jumping reminiscent (in a slow-motion form) of the rapid branch-shaking exhibited by monkeys. Side-to-side head shaking also occurs in territorial encounters.

The sportive lemurs deposited urine on the branches of the 'surveillance posts', but we never noticed any particular behaviour associated with such urination.

3. *Social relationships*

Even if there is nothing more than superposition of male and female territories, there must be relatively frequent encounters between individuals of opposite sex. When they do meet, one individual utters a 'Hiii' vocalisation, and then both continue on their way.

During our period of observation, it was seen that males and females sleep in separate retreats each night; but that the same retreat may be used by a male and a female on successive nights.

A few days before our observations were completed, we removed – by way of experiment – the largest male, whose territory overlapped with that of five females. The very next night, all of the other males penetrated into this zone, and each one began to associate with one of the females. It is important to note that the males were all in a sexually inactive state (regression of the testes) and that the females were either gestating or immature. Two days after the experiment, one female was already accepting the presence of one of the males, which was seen sniffing at her ano-genital region.

Courtship behaviour, as with surveillance of territories where females occur, is thus independent of direct sexual behaviour, since it can occur outside the reproductive period. Obviously, this represents some kind of social behaviour, as is found in other nocturnal prosimians.

Conclusions

Lepilemur, whose phylogenetic relationship to other lemurs is difficult to establish, is distinguished first and foremost by its extremely specialised diet. Alongside primitive anatomical and behavioural characters resembling those of the Cheirogaleinae and the Lorisidae (closure of the vulva during sexual inactivity; transport of the infant in the mouth; social organisation; overlapping of male and female territories), one can see a series of adaptations directly or indirectly associated with the dietary regime: specialised dentition, modified digestive tract, caecotrophy, small territories. Social organisation is of a primitive type; but the system of visual territorial surveillance, which is unusual for a nocturnal mammal, is doubtless related to the very small area of the defended territories.

Acknowledgments

This study was carried out in the private reserve of Mr H. de Heaulme in Berenty, where we benefited greatly from his generous hospitality. We would also like to thank our colleague, Dr R.D. Martin, who managed to find time to translate this article.

NOTES

1 Charles-Dominique, P., and Hladik, C.M. (1971), 'Le *Lepilemur* du sud de Madagascar: ecologie, alimentation et vie sociale', *Terre et Vie*, 25, 3-66.

2 Hladik, C.M., Hladik, A., Bousset, J., Valdebouze, P., Viroben, G. and Delort-Laval, J. (1971), 'Le régime alimentaire des Primates de l'île de Barro Colorado (Panama); résultats des analyses quantitatives', *Folia primat.* 16, 85-122.

3 Hladik, C.M. and Hladik, A. (1972), 'Disponibilités alimentaires et domaines vitaux des Primates à Ceylan', *Terre et Vie* 26, 149-215.

4 Morot, C. (1882), *Rec. Med. Vet.* 59, 635-46. [This was the first reference to caecotrophy in the rabbit; see reference 10 for more recent work.]

5 Hladik, C.M., Charles-Dominique, P., Valdebouze, P., Delort-Laval, J. and Flanzy, J. (1971), 'La caecotrophie chez un Primate phyllophage du genre *Lepilemur* et les correlations avec les particularités de son appareil digestif', *C.R. Acad. Sci. Paris* 272, 3191-4.

6 Kayser, C. (1963), 'Bioénergétique', in *Physiologie*, Paris, vol. 1.

7 Kayser, C. (1967), 'Evolution de l'homéothermie incomplète', in *Colloque pour le centenaire de Claude Bernard*, Fondation Singer-Polygnac, pp. 285-323.

8 Bourlière, F. and Petter-Rousseaux, A. (1953), 'L'homéothermie imparfaite de certains prosimiens', *C.R.S. Soc. Biol.* 147, 1594.

9 Hildwein, G. (1972), personal communication.

10 Thompson, H.V. and Worden, A.N. (1956), *The Rabbit*, London.

11 Petter, J.J. and Hladik, C.M. (1970), 'Observations sur le domaine vital et la densité de populations de *Loris tardigradus* dans les forêts de Ceylan', *Mammalia* 3, 394-409.

12 Charles-Dominique, P. (1971), 'Eco-éthologie des prosimiens du Gabon', *Biologia Gabonica* 7, 121-228.

13 Hladik, A. and Hladik, C.M. (1969), 'Rapports trophiques entre végétation et Primates, dans la forêt de Barro-Colorado (Panama)', *Terre et Vie* 23, 25-117.

14 Charles-Dominique, P. (1971), 'Eco-éthologie et vie sociale des prosimiens du Gabon', Doctoral thesis, no. A.O. 5816, Paris.

15 Charles-Dominique, P. (1972). 'Ecologie et vie sociale de *Galago demidovii* (Fischer 1808, Prosimii)', *Z.f. Tierpsychol. Suppl.* 9, 7-41.

16 Martin, R.D. (1972), 'A preliminary field-study of the lesser mouse lemur (*Microcebus murinus* J.F. Miller 1777)', *Z.f. Tierpsychol. Suppl.* 9, 43-89.

J.-J. PETTER and A. PEYRIÉRAS

A study of population density and home ranges of Indri indri *in Madagascar*

Introduction

The Indri (Fig. 1) is the largest of the extant Malagasy lemurs and the most specialised living representative of the family Indriidae, which also contains the Sifakas (*Propithecus verreauxi* and *P. diadema*) and the Avahis (*Avahi laniger*). All members of the family typically move around by leaping from trunk to trunk, with the body held vertical.

Although it is always difficult to observe Indri because of their unobtrusiveness, they are well known from their far-carrying, melodious vocalisations. They only live in mountainous areas in the northern half of the East Coast forest of Madagascar, which is characterised by an extremely humid climate. Because of the difficulties involved in observing these animals, largely as a result of their timid nature and the rugged terrain of their mountainous habitat, many features of their biology are still unknown. It is important that we should learn more about the Indri under natural conditions, while there is still time, particularly since such knowledge could assist us in protecting them more effectively. Data are particularly lacking for characteristics of diet, reproduction and home range.

In August 1968 and November 1969 we conducted expeditions into primary forest zones far from human influence, with the main aim of obtaining information about home range size. These two areas visited were: (1) a zone to the north-east of Maroantsetra and (2) the forest of Fierenana, to the north of Lake Alaotra. In 1971 and 1972, visits were made to zones close to forestry concessions in the forests of Perinet and Lakato.

Figure 1. The Indri (*Indri indri*), photographed in east coast rain-forest.

Working conditions

Indri occur in the forest of Analamazotra, at Perinet, which is a well-known forestry station on the road between Tananarive and Tamatave. In this area, they are more readily observed than anywhere else. In fact, the Service des Eaux et Forêts of Madagascar has recently created a special reserve in this zone for protection of the Indri. Numerous pathways, which are regularly serviced, permit easy penetration into the forest, and with a little luck it is possible to observe or film Indri under satisfactory conditions. Most of the photos, films and tape-recordings which have been obtained to date stem from the forest at Perinet.

Unfortunately, the forest at Perinet is already heavily exploited, and animals have been hunted in this area for some time. The fact that it is still possible to find Indri (whose Malagasy name 'Babakoto' can be colloquially translated as 'little father') is due to partial protection by local beliefs. However, collections of various kinds, exploitation and reforestation have modified their habitat, and the other diurnal lemurs have almost all been wiped out.

Thus, this area – despite the advantages which it offered for research on certain features of Indri biology – was not satisfactory for the basic requirements of this study. The same applied to various other forest zones in this region (around Mount Anketrambe on the road to Lakato, South of Perinet).

A number of surveys eventually indicated a favourable area, sufficiently remote from inhabited regions, roughly 20 km. to the east of Maroantsetra, in the forest of Fampanambo. The fringe of this forest could be approached by boat in August, following the River Antanambalana, and during the dry season (in November) it was accessible along a forest track from Maroantsetra. A three-hour march into the forest zone was necessary to reach the selected study area, where a camp-site was set up.

In July 1969, we worked there for eleven days in almost continuous rain. Subsequently, in November 1969, we profited from a period of more clement weather to work for seven days in alternating rainfall and sunshine.

On a map (Fig. 2) showing the topography of the region studied, one can distinguish a series of ridges which are mainly oriented in a north-south direction and have an average altitude of 400 m. They are separated by valleys not exceeding 200 m. in depth. Dense forest covers all the slopes, and vision is limited by foliage everywhere.

Another suitable area, though more difficult to investigate, was the forest of Fierenana, which is accessible along a pathway branching off the road from Moramanga to Lake Alaotra. Atmospheric conditions in this area are similarly extremely unsuitable for prolonged visits.

Methods of study

In the course of the first two expeditions, we were able to explore in some detail a vast forest area by spending the days, and parts of the nights, walking along pathways on the ridges in a rugged mountainous region. The animals were always very difficult to observe. In fact, over and above the inaccessibility of their habitat, Indri are almost invisible when they are resting in the fork of a tree, and they are almost impossible to detect if they remain motionless, as was recently observed by P. Charles-Dominique and M. Hladik (personal communication). The retracted dark limbs, surmounted by the white 'V' of the dorsal pelage, can easily be confused with the fork of a trunk seen against a cloudy sky, or with a patch of dark foliage beneath a fork covered with lichens. The white patch on the forehead, which gives the impression of a gap in the vegetation, adds to this effect.

We were able to locate the different Indri groups primarily from their vocalisations, by determining the direction with simultaneous estimation from a number of observation points. In this way, we were able to locate vocalising animals on a map by plotting and estimation of position (Fig. 2). In the course of this study, we were also able to collect information on the population density of *Varecia variegata*, whose powerful vocalisations are also easily recognised.

As a rule, after one Indri group had stopped calling, another neighbouring group took its turn, and successive groups took up the sequence into the distance. Thus, it is possible to make exact localisations, and there is no danger of counting a moving group twice.

Types of vocalisations, frequency and carriage

The vocalisations of the Indri are extremely characteristic; they are very resonant and are amplified by a dorsal laryngeal sac.[1] A number of observations on their structure and frequency have already been published.[2] Two types can be distinguished:

1. Repeated calls like blasts of a horn, uttered occasionally by disturbed or isolated individuals. These calls appear to be relatively powerful when one is close to these animals, but their carriage seems to be relatively weak.
2. Barks, followed by long, modulated howls (frequency range: 0.5-4 kilocycles), uttered by all individuals in a family. It is this latter type of vocalisation, which is extremely powerful, that primarily permitted localisation of Indri groups.

The number of calls is usually extremely variable on different days and at different times of the day. In general, more vocalisations are heard during good weather than during rainfall. The calls are most numerous in the morning, just after sunrise or towards midday, and they seem to be markedly more frequent in the period December/January.

The carriage of the calls is very difficult to define precisely, since the sound emitted is extremely variable in intensity and tonality. In addition, the animals can change position whilst vocalising, and the direction of the calling seems to vary.

We attempted to evaluate the carriage of vocalisations using two

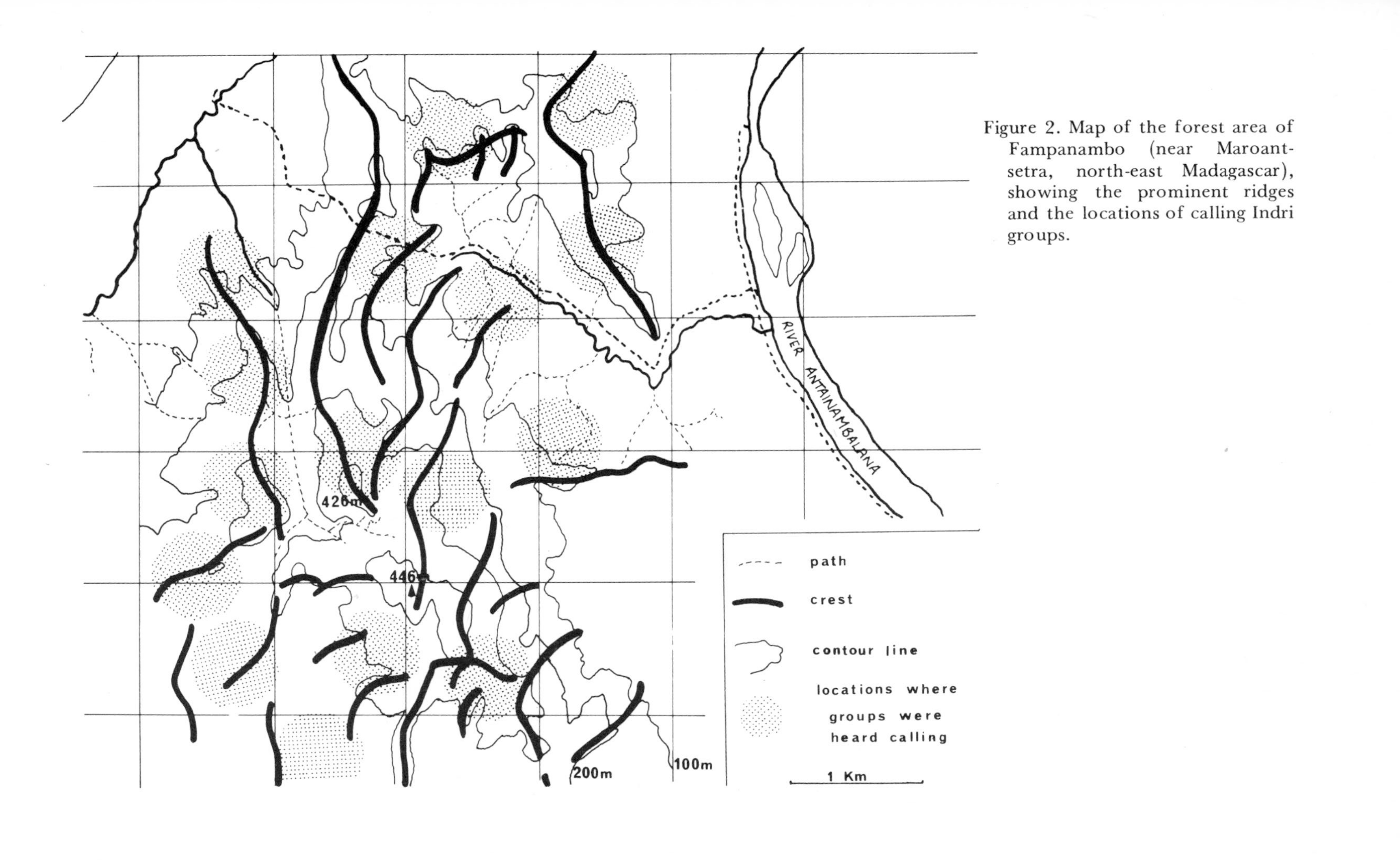

Figure 2. Map of the forest area of Fampanambo (near Maroantsetra, north-east Madagascar), showing the prominent ridges and the locations of calling Indri groups.

methods: the first method involved direct comparison of the impressions of several observers, who happened to be at differing distances from a well-localised group at the time of vocalisation. The second, indirect, method was based on recording of calls and subsequent comparison of the calls with others recorded under identical conditions, but with directly measured localisation.

Under normal conditions of population concentration, most of the Indri which we saw uttering the second type of call were near to the crests of the trees at the top of a ridge. It is possible that this position is optimal for vocalisation and reception of calls.

Under conditions where concentration is abnormally high, it seems that the Indri may emit their calls equally frequently from the depths of the valleys. Observation from one ridge under such conditions permitted localisation of three groups in three valleys surrounding the ridge.

When one is close to the ground in the middle of the forest, the vocalisations are generally audible over more than one kilometre in good weather, despite the density of the trees. When calls cross a valley, they may carry even farther, whilst calls emitted from a valley probably do not carry further than 500 metres.

The Indri possesses large ears with much better developed auricles than those found in other Indriids.

We have no idea of the auditory acuity of the Indri, but it is probably greatly superior to our own and sufficient for perception of the calls of other groups over great distances, despite the wind, the rain, and the screening effect of the rugged terrain.

The structure of the call, one part of which consists of a howl whose frequency varies in the course of emission, is certainly adapted for transmission over great distance in the forest environment. It is interesting to note that other primates living in dense forest, such as the gibbons (for example), have similar vocalisations, and that even certain birds in Madagascar (e.g. *Leptosomus discolor*) utter their calls from the tops of the trees and have a modulated, plaintive vocalisation. Incidentally, Betsimisaraka woodcutters communicate with one another in the forest – often over great distances – with modulated calls of this type.

Group density and composition

By plotting on a map (Fig. 2) the localisations obtained by listening to the calls, it is possible to calculate the approximate concentration of the Indri groups. Taking into account the imprecise nature of the localisation methods, one can trace circles covering a surface of about 100 hectares around each of the groups.

Given the present state of our knowledge, it is difficult to determine whether these 100 hectare areas correspond with a 'home-range' for each group, or whether they simply indicate reserved zones corresponding to the rather unusual relief of the eastern rain-forest. In this humid, rainy climate, it is possible that each group occupies an entire hill for thermo-regulation, seeking out optimal sun-exposure during the mornings and afternoons.

Estimations carried out in the forest zone to the east of Maroantsetra and in the forest of Fierenana are entirely comparable. They would therefore seem to be representative of the natural condition in primary forest which is still untouched by human interference.

Observations carried out at the forestry station of Perinet (P. Charles-Dominique and M. Hladik, personal communication) have shown that in this area there is a far greater density of Indri groups. The same would seem to apply to other zones, such as the region studied close to the Lakato road, where there has been – or still is – human interference (Fig. 3). This high density of groups would appear to be abnormal and associated with forest degradation. The Indri, fleeing from the presence of woodcutters, concentrate in quieter areas, and there may be a resulting temporary crowding of groups. In such areas, it is noticeable that the animals vocalise more often and reply to one another more frequently, as if continually excited by the presence of neighbours too close at hand.

In areas of high densities of Indri, the forest always exhibits a certain degree of degradation, at least in neighbouring zones. This is the case in Perinet, where – apart from the core area of the forestry reserve – there is not much left of the magnificent forest which once existed there. The same applies to the forest on the Lakato road. There is a vast zone streaked with pathways used for past or present exploitation, and virtually devoid of Indri. But on the fringe, where there is some forest which has scarcely been touched – if at all – one finds exceptionally high densities of these animals.

Near to the road to Anosibe, which once resounded with Indri calls, there are no Indri left at present. In the course of a 4-day stay in this forest, not a single vocalisation was heard. Probably, all of the Indri have been frightened away because of a heavily mechanised forestry company which is very active in this region.

Indri are indeed extremely sensitive to disturbance, and they flee rapidly when one tries to approach them; so this is probably the cause of the concentrations observed. Disturbances which occur too frequently (even in the absence of complete destruction of the biotope) induce the animals to flee. An emigrating group of this kind was observed and followed for 2 days near to the Lakato road. The Indri were fleeing from a disturbed zone where they had previously been resident. They moved along in gradual stages quite silently, and they covered about a kilometre in this way, eventually settling in an area with less disturbance, where they were subsequently observed on several occasions. In this context, it is interesting to note that

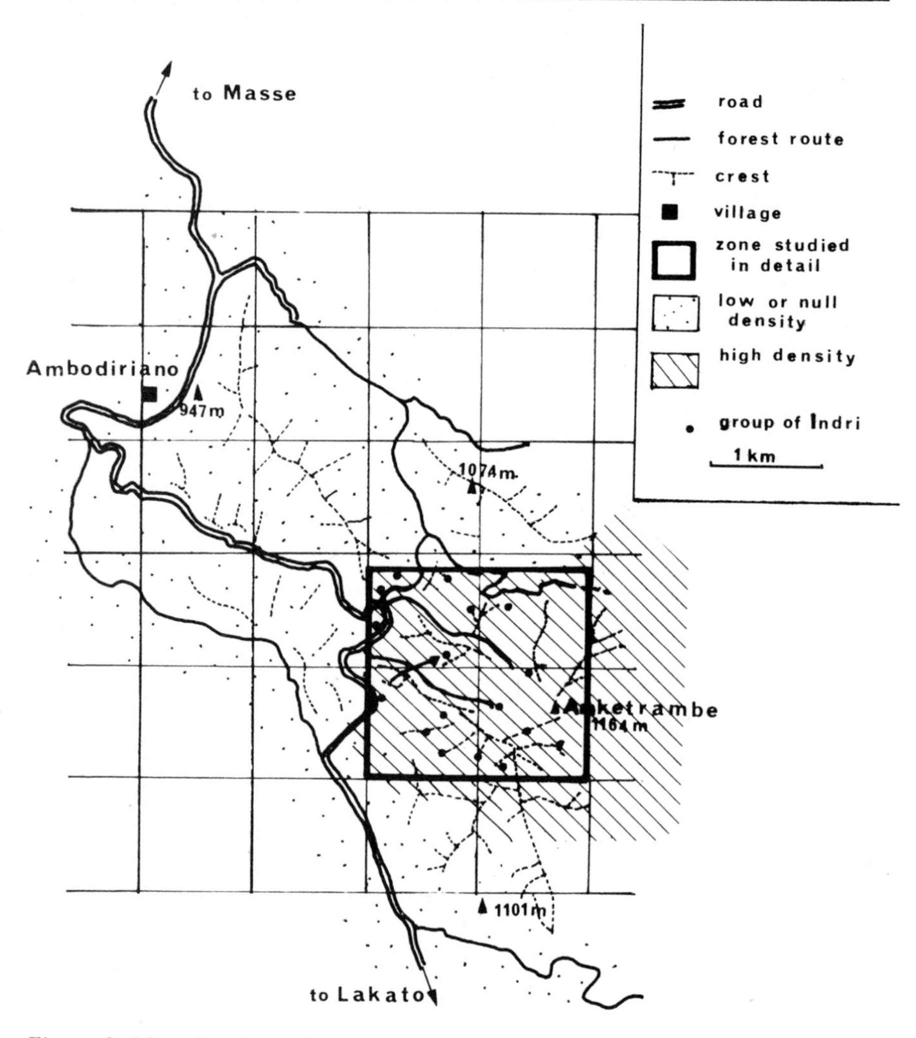

Figure 3. Map showing a forest area close to the Lakato road (near Perinet, east Madagascar), indicating the study area containing an unusually high concentration of Indri groups.

when one group of Indri crosses the 'zone of influence' of another, the invaders move along in silence, and it is the invaded group which can be heard calling frequently. Two groups observed in this region happened to come into confrontation. On ten successive occasions, they exchanged series of isolated barks, although this vocalisation is – under normal conditions – typically followed by long, modulated howls.

Another observation clearly demonstrates the sensitivity of the Indri. In the forest to the west of Mananara, Indri were once very common. The initiation of forest exploitation with heavy machinery belonging to the 'Moulins de Dakar' company drove the Indri completely from vast areas. Abandonment of exploitation two years ago has been followed by progressive re-establishment of these animals, and it seems likely that they were not wiped out by hunting forays made by forestry prospectors, but simply forced to concentrate further to the west in more elevated, calmer forest areas, after they had fled from noise and other disturbance.

Indri generally live in groups of two, three or (exceptionally) four in areas of natural forest, as we have observed. However, we have received reports of larger groups containing five to six animals. Such groups have apparently always been seen in areas disturbed by forestry exploitation, or in adjacent regions, and it would seem that these groups are abnormal products of hyperconcentration.

Conclusions

Our investigation of forest areas far from and near to zones of exploitation indicates that the Indri are normally extremely sensitive animals with respect to human presence. The 'home range' would appear to be extremely large for each group, covering about one square kilometre, and we do not yet know the survival value of such large ranges in view of the fact that the forest would seem to be rich enough to feed larger numbers. The limits of each range seem to be advertised by a powerful call which may carry further than one kilometre.

A greater concentration of such groups seems to be always associated with excessive frequency of human presence, and the animals flee from areas where they are disturbed. The number of animals in each group, normally strictly limited to three or four, seems to increase on occasions under abnormal conditions. It is interesting to compare these observations with similar reports on Sifakas in the west. With the Sifakas, as with the Avahi (i.e. in all Indriidae), the groups would normally have a strictly family basis, as we have observed in numerous areas in the west which are free of human presence. But in zones which are more-or-less degraded or transformed – notably in certain areas of the south – it is possible that the family groups are modified to form larger groups. This would explain the numbers of five to nine animals, instead of three to four, reported for groups by various observers. As with the Indri, it is degradation of the environment which favours the formation of larger groups. This modification probably results in reduced fecundity, since the family is probably the most efficient unit for favouring population growth, as J.H. Crook has shown.[3]

Acknowledgments

This study was carried out with the assistance of the Service des Eaux et Forêts of Madagascar and with the support of the World Wildlife Fund. Georges Pariente, Alain Schilling, Roland Albignac, N.S. Malcolm and Regis Prevôt all participated in this work at various times, and their assistance was extremely valuable. Translation of the article was carried out by Robert Martin.

NOTES

1 Milne-Edwards, A. and Grandidier, A. (1875), *Histoire Physique, Naturelle et Politique de Madagascar*, 6: *Mammifères*, Paris.

2 Petter, J.-J. (1962), 'Recherches sur l'écologie et l'éthologie des Lémuriens malgaches', *Mém. Mus. nat. Hist. nat.* (sér. A) 27, 1-146.

3 Crook, J.H. (1967), 'Evolutionary change in Primate societies', *Sci. J.* 2, 7.

A. RICHARD

Patterns of mating in Propithecus verreauxi verreauxi

Introduction

The earliest written account of *Propithecus verreauxi* was given by Sieur Etienne de Flacourt,[1] a French colonist who disembarked at the site of present-day Fort Dauphin in the seventeenth century. The species was not described in any detail until the 'Histoire naturelle des mammifères: Histoire physique, naturelle, et politique de Madagascar' was produced by Milne-Edwards and Grandidier in 1876, 1890 and 1896.[2] In the twentieth century, brief accounts of field observations have been published by various scientists, such as Kaudern and Rand.[3,4] However, the first systematic field study conducted was by Petter.[5] During that study, a general survey of lemur behaviour in the wild was undertaken, providing a valuable basis for further research on many species, including *P. verreauxi*. Petter-Rousseaux's laboratory work on prosimian reproductive behaviour[6] has also been of great assistance to the present study. Finally, the most recent major field study has concentrated on *P. verreauxi* and *Lemur catta* living in gallery forest alongside the River Mandrary at Berenty, in the south of Madagascar; this work, providing considerable insight into the social organisation and ecology of these two species, has been a most useful source of comparative material for the present study.[7] Nowhere, however, in the literature, is mating in *P. verreauxi* described, and indeed the only Malagasy prosimian species in which mating has been seen under natural conditions is *L. catta*.[7]

In this paper, a description of activities during the mating season of *Propithecus verreauxi verreauxi* is presented, indicating the extent to which they differ from activity patterns during the rest of the year. An attempt is made to interpret these results as part of a total mating pattern; but as the results from the two groups studied are very different, the final evaluation is highly tentative. I was not present during the mating season in the north of the island: in all

Malagasy prosimians, mating appears to be seasonal under natural conditions, and in *P. verreauxi* the timing of the birth period shows that mating must occur during January, February or March all over the island.

A field study was made in Madagascar between April 1970 and September 1971. The main aim of the study was to investigate the ecology and social organisation of *P. verreauxi* using quantitative techniques, with particular emphasis upon intraspecific variation and its ecological correlates. Two study areas were established, representing two extremes of environment in both of which *P. verreauxi* is abundant: thus divergent adaptations might be expected in each area comparable to the intraspecific variation found in many Old World primates living under different ecological conditions. One study area was located in the north west of the island, approximately one kilometre from the forestry station at Ampijoroa, in the ecologically rich region of the Ankarafantsika (National Reserve no. 7). *P. v. coquereli* is found in this region. The second study area, 1,500 km. from the first, lay about 80 km. from the south coast, in Reserve no. 11, 1½ km. south of the village of Hazafotsy. This extremely arid region is occupied by the sub-species *P. v. verreauxi* (Fig. 1).

In the north, with a mean annual rainfall of 1,600 mm., and maximum/minimum temperatures of 39.5°C and 14°C (recorded during the study), the forest is deciduous, growing on sandy soil in the hill-top study area. During the dry season, between April and September, which is also the cold season, most trees shed their leaves. There is no well-defined canopy, most trees reaching a height of 12-16 m., with emergents of 30-35 m. Although not as rich as some other parts of the Ankarafantsika, this particular area was selected because of the absence of hunting in the locality and consequent abundance of easily habituated animals, and because of its accessibility: the latter was vital for a study attempting to sample two areas in all seasons. The forest supports six other prosimian species: *Lemur fulvus, L. mongoz, Microcebus murinus, Cheirogaleus medius, Lepilemur mustelinus*, and *Avahi laniger*.

In the south, with a mean annual rainfall of 600 mm., and maximum/minimum temperatures of 44°C and 8°C (recorded during the study), the area is covered by xerophytic vegetation, rarely exceeding 13 m. in height. The forest is dominated by species of the Didiereacae family, particularly by *Alluaudia ascendens* and *A. procera*; in both, 1 cm. long spines stud the trunk, which itself divides to produce a candelabrum-like silhouette. Over 80% of the plant species in this forest are endemic, but species diversity overall is less in this region than in the northern study area, and only two other prosimian species were seen: *M. murinus* and *L. mustelinus*.

Two neighbouring groups of *P. verreauxi* were habituated in each area. Many of the activities discussed in this paper concern social communication between two of these groups. Ellefson[8] emphasised the importance of inter-group social communication in *Hylobates lar*, pointing out that this usage does not conform with Altmann's[9] definition of social communication as being intra- and not inter-group. It is thus stressed here that although extensive social communication takes place between *P. verreauxi* groups during the mating season, they are still considered to be discrete groups as defined by Struhsaker.[10]

Table 1 Composition of the groups studied

Northern Study Area		*Southern Study Area*	
Group I	*Group II*	*Group III*	*Group IV*
A♂ N (June-Aug. '70)	A♂ H	A♂ F	A♂ R (Jan-March '71)
A♂ GOP (June-Aug. '70)	A♀ NF	A♂ P	A♂ INT (April-June '71)
A♂ STR (Oct.-Dec. '70)	SA ♂D	SA ♂Y	SA♂
A♂ BE (Nov. '70-July '71)	Juvenile (♂)	A♀ FD	A♀ FNI
A♀ BT		A♀ NFD	A♀ FI
A♀ RT		Juvenile (♂)	
A♀ FT			
A♀ SL			

A=Adult, SA=Sub-adult. Initials after age/sex classes are for individual identification. Dates are given in parentheses for periods when animals did not move with groups throughout the study.

The composition of the four groups is given in Table 1. The method used to habituate the animals was similar to that used by Stoltz and Saayman with *Papio ursinus*.[11] When first contacted, an unhabituated group characteristically split up, its members fleeing in different directions. However, they rarely fled far, but rather hid themselves most effectively high up in tree forks. After constantly searching for and pursuing a group for a week, it generally stopped fleeing, and after three weeks animals would approach to feed within 1 m. of me. This was taken as the criterion of habituation, and the quantitative record was only begun after this point was reached. In Group IV the process took longer, two weeks passing before their immediate flight response disappeared. Animals were never provisioned, and although occasional glances at me demonstrated their awareness of my presence, they appeared largely to ignore and be unaffected by it.

In order to facilitate precise location of groups, the home-range of each (varying between groups from approximately 6.75 to 8.5 hectares) was divided up by a grid system of marked paths running north/south and east/west at 50 m. intervals. Quantitative observations were collected, over twelve months, covering three months in the dry season, and three months in the wet, in each area. Seventy-two hours of quantitative data were collected per group, per month, and observations were spread evenly between 06.00 hrs. and

Figure 1 (*a*). Verreaux's Sifaka, *Propithecus verreauxi verreauxi*. Female FD (Group III – Table 1) photographed together with her infant in September 1971 (Hazafotsy; South Madagascar).

18.00 hrs., and between the age/sex classes; in Groups II, III and IV, all animals were recognised individually. One animal was taken as the subject during each 12-hour observation period, in order to establish a continuous record of one individual's activities; a 72-hour sample taken in June 1970 and September 1971, made at half-hourly intervals on all group members, suggests that the timing of the various activities is closely synchronised between animals, so that the daily individual record is considered to be a fairly accurate indicator of group activities. At timed minute intervals during the observation period the subject was described in terms of a number of categories concerning, for example, its height above ground, activity, and the proximity of its nearest neighbour. A full description of recording techniques is given elsewhere.[12] As far as possible, this record was maintained during the mating season, and complemented by extensive descriptive notes.

Figure 1(*b*). Sub-adult male Y (Group III) standing bipedally on the ground whilst feeding. Note the very long hind limbs, associated with the typical vertical-clinging-and-leaping mode of locomotion.

Results

Nature of behavioural changes in the pre-copulatory period

There were two adult females in Group IV. Slight flushing of ♀ FNI(IV)'s* vulva coincided with a sudden, significant increase in certain activities in Group IV in late January 1971. Similarly increased frequencies were noted throughout February and during the first ten days in March, although there was no visible unusual coloration of either female's vulva in Group IV at this time. Copulation took place in March, and the six weeks preceding it are

* The age-class of the animal referred to is henceforth omitted, and the group to which it belongs has been added in brackets after the identifying initials. See Table 1.

henceforth referred to as the pre-copulatory period. The five activities described below occurred occasionally throughout the year, but only in the mating season were they common. The emphasis is thus upon a quantitative rather than a qualitative change in behaviour during these months. The five types of activity and their frequency changes are listed below. Where no figures are quoted, it is to be assumed that no change was found. It should be noted at this point that, although there are striking discrepancies between the data collected on each group throughout this section, infants were born to known females in both Groups III and IV in August 1971. (The implications of this are discussed below.)

1. 'Endorsing' by adult males

Although adult animals of both sexes scent-mark in a number of contexts, during this period there were increases only in scent-marking by males. While females mark either by rubbing the ano-genital region or by urinating on a trunk or branch, marking by adult males may include rubbing a branch or trunk with the scent gland on the ventral surface of the throat (Fig.2), then with the tip of the penis, usually urinating slightly as this is done, and finally with the perineal area. Although an adult male may perform any one part of this sequence when marking, the sequence was commonly performed in its entirety. Both sexes have a highly developed, almost tubular perineal area. The term 'endorsing' was applied when a male marked a spot within five minutes of a female having been there. Frequency changes in endorsing are shown in Table 2. The figures for Group IV are highly significant,* but those for Group III are not significant.

2 'Sniff-approach and mark' by adult males

This occurs when an adult male approaches a female and marks the tree trunk just below her tail: the male climbs the trunk under the female and touches her anus with his nose. He then throat-marks, and finally marks with his ano-genital areas. This sequence is frequently incomplete, for the female may lunge at the male as he thrusts his nose under her tail, forcing him to retreat without marking: in such cases endorsing usually follows when the female moves off. Summing the data, sniff-approach and mark sequences were recorded 6 times in 216 hours of quantitative observation of Group IV outside the pre-copulatory period. The sequence was observed (in complete or incomplete form) 65 times in 216 hours between 24 January and 15 March 1971. The difference between

* Unless otherwise stated, 'highly significant' means significant at the .001 level, and 'significant' means significant at the .05 level, using a Chi Square Test.

Figure 2 (*a*). Adult male R *Propithecus verreauxi verreauxi* (Group IV), showing the dark throat gland area (arrow).

Table 2 Changes in frequency of endorsing in Groups III and IV

Group	*Jan. 1971* *Vulval Change* *Pre-*	*Post-*	*Feb. 1971*	*March 1971*	*April 1971*	*Sept. 1971*
III	1.95	–	2.35	1.08	0.67	1.41
IV	1.90	6.75	6.26	7.00	1.04	–

Figures are expressed as a frequency per adult male hour. Only data recorded when an adult male was the day's subject were used in calculations.

Figure 2(*b*). Adult male R (Group IV – Table 1) marking a vertical trunk with his throat gland.

these figures is highly significant.

3. *'Roaming' behaviour*

Evidence from the study group males, and from the arrival of unknown males in the study area, indicates an increase in male 'roaming' behaviour at this time: males, singly or in pairs, detach themselves from their own groups and make long forays into the home-ranges of other groups. During the mating season, these excursions sometimes culminated in fierce fights between adult males and in copulation with adult females in the groups encountered. During the rest of the study, no solitary or paired males were encountered in the home-ranges of the study groups; on only four occasions was an adult male from Group III or IV recorded as probably being out of immediate audio-visual contact with all other members of his group for over two hours. However, since it was usually the absence of these animals that was noticed, it was rarely possible to confirm that they had detached themselves from the group rather than having moved to rest in a position out of my sight.

4. *Intra-group agonistic encounters*

Three features of intra-group aggression changed during the mating season: there was a highly significant increase in the frequency of aggressive encounters in both Groups III and IV during the mating season. Secondly, although most animals contributed to this increase (See Table 3), the Index of Increased Aggression shows that the proportionate increase was higher in some animals than in others. This index is an expression of the ratio between the number of aggressive incidents an animal initiates in the mating and non-mating

Table 3 Changes in frequency of aggression in the mating season

Group	*Initiator*	*Non-mating season*	*Mating season*	*Index of increased aggression*
III	A♂ F	7	38	5.4
	A♂ P	2	2	1.0
	A♀ FD	13	13	1.0
	A♀ NFD	5	24	4.8
	SA♂ Y	—	3	—
	Juvenile	—	2	—
IV	A♂ R)* A♂ INT)	6	49	8.1
	SA♂ Q	—	6	—
	A♀ FI	24	85	3.5
	A♀ FNI	9	12	1.3

* In order to provide data over the whole six months, results from these two adult males were combined.

Figure 3. Subordinate male P (Group III) being displaced by adult male F (Group III), from above. Male P (below) is displaying characteristic submissive gestures: a special facial expression ('grin'), rolled tail and hunched back.

seasons. Where the first value is zero, the value of the index is inevitably infinity and is consequently not shown.

In Group III the highest proportionate increase was seen in ♂ F. Comparison of the combined frequencies for ♂ R(IV) and ♂ INT(IV) between the mating and the non-mating season shows much higher frequencies during the mating season. It is of interest that there was no frequency increase in ♂ P(III), but that there was a substantial increase in ♂ Q, although an index cannot be calculated for this latter male.

Outside the mating season, a linear dominance hierarchy can be determined, incorporating all adults. The hierarchy was defined on the basis of agonistic encounters over access to feeding and resting stations. Initiation of aggression and displacements (Fig. 3) were found to occur consistently in one direction.[13] For convenience, this hierarchy is henceforth referred to as the non-mating season (NMS) dominance hierarchy. During the copulatory period, a breakdown in this structuring occurred in Group IV: although ♂ Q, an immature subordinate male, took little part in events during the period of copulation, it was the only time when he was seen initiating aggression against other members of the group. This period was also associated with persistent invasion of Group IV by 'roaming' males (see below). In Group III, ♂ F retained his position as NMS dominant male, unchallenged by intruders, and no reversal in the linearity of the NMS dominance hierarchy was seen, nor was there any increase in aggression by subordinate ♂ P within the group.

5. *Inter-group encounters*

During the pre-copulatory period and the mating season, Groups III and IV were both involved in many more encounters with neighbouring groups than during subsequent months. The difference was highly significant for both groups. Fig. 4 shows this change in the frequency of inter-group encounters for each group.

During the pre-copulatory period and mating season, 79% (11/14) of Group III's encounters were with Group IV, and 65% (11/17) of Group IV's encounters were with Group III. Since the home-ranges of at least two groups in addition to Group IV's were known to lie alongside that of Group IV, the frequency of Group III/IV interactions was significantly high, assuming that all neighbouring groups would otherwise interact equally often. There was no significant difference between the frequencies with which Groups III and IV interacted with their various neighbouring groups after the mating season; it is thus unlikely that my presence accounts for the discrepancy during the mating season, in that I was equally likely to inhibit the approach of unhabituated neighbouring groups throughout the study.

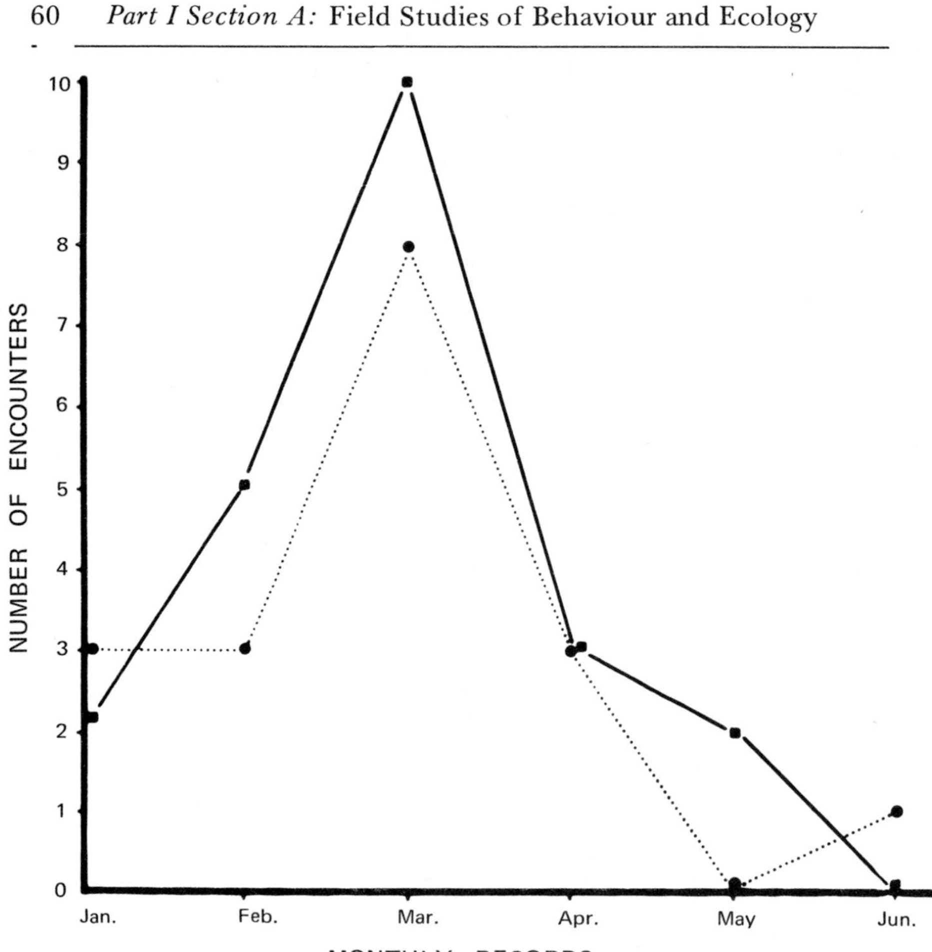

Figure 4. Changes in the frequency with which Groups III and IV encountered neighbouring groups.

Dotted line = Group III
Solid line = Group IV

This specific increase in the frequency of encounters between Groups III and IV is unlikely simply to have been a function of increased food availability in the overlap area of their home-ranges, resulting in both groups spending longer in this area. The Group III/IV overlap area was a strip about 100 m. wide, running about 150 m. along the edge of the home-ranges of the two groups; it constituted approximately 5% of the home-range of each group. Group III spent more total time in the overlap area during the pre-copulatory period and the mating season than in subsequent months, but their allocation of time to different activities did not increase uniformly. This is shown in Fig. 5. During the pre-copulatory period and mating season, 56% of the time Group III spent in the overlap area was devoted to activities other than feeding. This dropped to 38% after the mating season. Thus the results

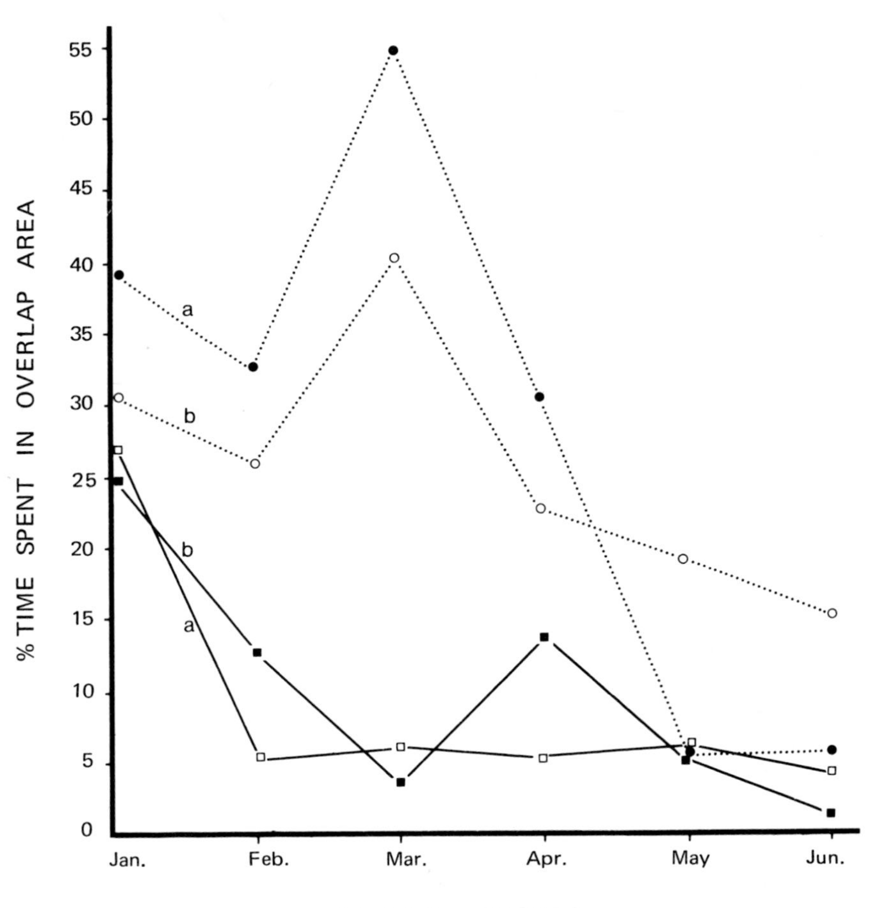

Figure 5. Number of minute records (expressed as a percentage of total number of such records) in which Group III and IV occupied the area of mutual overlap.
a = time spent in activities other than feeding
b = time spent feeding
Dotted line = Group III
Solid line = Group IV

suggest that Group III was spending longer periods in the overlap area not simply because there was abundant food there, but rather because they were involved in other activities there. Further, neither the vegetational analysis,[12] nor subjective impressions indicated that there was in fact an increase in food availability in that area at that time. Evidently, the results cannot be conclusive: it is possible that Group III entered the overlap area because of an abundance of food there, and that ensuing encounters with Group IV were incidental, although they did reduce the amount of time available for feeding.

Considering Group IV, only in January was significantly more time spent in the overlap area: when the January data are excluded, no significant difference remains between the other months (Kruskal Wallis One-Way Analysis of Variance, N=30). There was no significant difference in the proportion of time this group spent in feeding and activities other than feeding between the pre-copulatory period and mating season, and subsequent months. (Mann-Whitney U Test, N=36).

Throughout the study there was a highly significant difference between the amounts of time each group spent in the overlap area (Mann-Whitney U Test, $N_1=N_2=6$, U=0), Group IV spending much less time there than Group III.

No pattern of inter-group dominance emerged from inter-group encounters.[1 2] Group III did consistently spend longer than Group IV in the overlap area, particularly during the pre-copulatory period and mating season. The latter increase was associated with a disproportionate increase in activities other than feeding by Group III, and with an increased frequency of encounters between Groups III and IV. Thus, although a close association is apparent between the two groups during the mating season, the underlying mechanisms remain obscure.

Description of behavioural changes in the pre-copulatory period

The following account is of daily behavioural changes seen during the pre-copulatory period. Where no comment is made it can be assumed that activities were similar to those seen outside the mating season. Changes seen in Group IV are considered first.

Observation of Group IV was begun on *18 January*; in the following six days, ♂ Q(IV) was twice absent from the group for over two hours.

24 January: ♂ R(IV) endorsed after ♀ FNI(IV) nine times. On the four occasions he tried to sniff-approach and mark under ♀ FI(IV), she lunged at him and he retreated.

25 January: ♂ R(IV) endorsed 7 times after ♀ FI(IV) and 11 times after ♀ FNI(IV). There were 5 incomplete sniff-approach and mark sequences between ♂ R(IV) and ♀ FI(IV).

26 January: ♀ FNI(IV)'s vulva was slightly flushed. ♂ Q(IV) was found sleeping beside a strange male, ♂ INT, near the two females. This male, who had occasionally been seen with the group in September 1970, disappeared as soon as the group moved off. ♂ Q(IV) spent most of the day away from the group and ♂ R(IV) left it twice; the second time, he encountered Group III in the III/IV overlap area and headed straight for ♀ FD(III) until intercepted and chased away by SA ♂ Y(III). ♂ R(IV) left after three such attempts to reach ♀ FD(III). ♂ R(IV) endorsed 63 times after ♀ FI(IV) and

♀ FNI(IV). (Marking continued throughout the day, where in other months no marking was seen after 14.30 hrs.) Two incomplete sniff-approach and mark sequences were seen between ♂ R(IV) and ♀ FI(IV). Five complete sequences were seen between ♂ R(IV) and ♀ RNI(IV).

27 January: ♀ FNI(IV)'s vulva was still flushed. ♂ R(IV) left the group and returned to the locality of his encounter with Group III on the 26th, but did not make contact with Group III. ♂ INT approached the group and ♀ FNI(IV) marked beside him; he sniff-approached and marked under her once. ♂ R(IV), having lunged at ♂ INT when he first appeared, then moved away and only rejoined the group when ♂ INT moved off.

28 January: ♀ FNI(IV)'s vulva was no longer flushed. Endorsing and sniff-approach and mark frequencies were at a post-mating season level.

6 February: ♂ R(IV) had lost large chunks of fur. ♂ Q(IV) moved peripherally to the group and then approached with ♂ INT. The latter moved away again at once. Endorsing frequency by ♂ R(IV) was high; but no sniff-approach and mark sequences were seen.

7 February: ♂ R(IV) chased ♂ INT away from the group 3 times, each chase ending in a fight.

8 February: ♂ Q(IV) left the group at midday and was found in the evening grooming an unknown male, ♂ LCE, 10 m. from Group IV. ♂ LCE moved off as I approached and ♂ Q(IV) stayed with the group. Incomplete sniff-approach and mark sequences were seen three times between ♂ R(IV) and ♀ FI(IV).

13 February: Group IV were found sunning themselves with ♂ LCE and two unknown females. The three unhabituated animals fled when they saw me.

Cyclonic rain cut off access to the study area for the following 7 days.

20 February: ♂ R(IV) and ♂ Q(IV) were found in the middle of Group III's home-range, but no interaction with Group III was seen. ♂ R(IV) endorsed after ♀ FI(IV) and ♀ FNI(IV) 61 times. 11 complete sniff-approach and mark sequences between ♂ R(IV) and ♀ FNI(IV) were seen, and 4 incomplete ones between ♂ R(IV) and ♀ FI(IV): the latter cuffed and bit ♂ R(IV), as she did earlier in the month and in January; but for the first time, ♂ R(IV) did not retreat in the face of this aggression.

22 February: ♂ R(IV) endorsed 35 times in 3 hours' observations.

Thus, in summary, this period was notable in Group IV for the brief flushing of ♀ FNI's vulva, the heightened frequencies of endorsing and sniff-approach and mark sequences by ♂ R, and the frequency of roaming, both in non-group, and group males. Changes of this nature were absent in Group III, with two exceptions: twice during February, ♂ P(III) spent whole days away from Group III.

Description of behavioural changes and copulation in March 1971

Copulation was said to have occurred when ejaculation took place. (Mounting and intromission with no ejaculation are described separately.) Copulation by 3 males with the 2 females in Group IV was observed between 3 and 6 March. This was the only period during the field study when male/male and male/female mounting was observed, and on only two other occasions were males observed with erections: once during an inter-group interaction, and once during a play bout between a sub-adult and juvenile.

During the first observed mounting, ♂ F grasped ♀ FI with his hands around her legs, which were doubled up in a vertical squatting position (all copulation took place on vertical trunks), and he held the trunk below the female with his feet. She curled her tail up and held it slightly to one side during copulation. In subsequent mountings, there was some variation in posture, with the male grasping the female by her upper arms with his hands, and by her doubled-up legs with his feet.

The frequencies of the five activity patterns described above, namely, endorsing, sniff-approach and mark sequences, roaming behaviour, inter-male aggression, and inter-group encounters, reached a peak during the four days when copulation occurred with Group IV females FI and FNI.

The following account details events during this period.

3 March: after a brief early morning interaction with a group to the south, Group III fed, steadily moving west until 09.45 hrs. when they encountered Group IV. An extract from field notes is given below to describe the subsequent sequence of events:

09.54: A characteristic inter-group confrontation begins between Groups III and IV. ♀ FNI(IV) is not present.

09.55: ♂ F(III) mounts ♀ FI(IV); he dismounts almost at once and joins in the general chasing (characteristic of most group confrontations).

09.56: ♂ F(III) mounts ♀ FI(IV) again, but dismounts at once to chase ♂ F(IV).

10.01: ♀ FI(IV) is grasping a vertical trunk, with ♂ P(III), ♂ F(III), and R(IV), and SA ♂ Y(III) sitting on the ground in a circle round her. Each ♂ tries to approach her but she repels them by slapping, lunging and biting them. The rest of Group III are feeding 10 m. away.

10.05: ♂ R(IV) lunges at ♂ F(III) who moves off, followed at once by ♀ FI(IV). The other 3 males follow her.

10.09: ♂ F (III) mounts ♀ FI(IV) on a vertical trunk but SA ♂ Y(III) approaches and bites ♂ F(III)'s tail. He dismounts and chases

SA ♂ Y(III) down the trunk. This happens twice more.

10.14: ♂ F(III) mounts ♀ FI(IV) and ejaculates after 24 thrusts. No more than six consecutive thrusts had occurred in earlier mounting, and it was uncertain whether there was intromission. The other 3 males sit watching. ♂ F(III) then dismounts and sits beside ♀ FI(IV) licking his genitalia.

10.17: ♂ R(IV) sniff-approaches and marks under ♀ FI(IV), and then chases ♂ P(III) and SA ♂ Y(III).

10.29: ♀ FI(IV) leaps off followed by ♂ F(III). She cuffs him and he leaps off and is chased by ♂ R(IV). ♂ R(IV) then chases ♂ P(III) away. ♂ R sniff-approaches and marks under ♀ FI(IV) and she leaps off, closely followed by ♂ R(IV) and at a distance by ♂ F(III), ♂ P(III) and SA♂ Y(III).

10.35: ♀ FI(IV) sits self-grooming beside ♂ R(IV). ♂ P(III) and SA ♂ Y(III) are 5 m. away and ♂ F(III) much further still. ♂ R(IV) tries to mount ♀ FI and she cuffs him and moves off north, followed by him. ♂ P and SA ♂ Y(III) immediately endorse the sitting-place she has just left.

10.45: ♂ F(III) has moved off and is feeding with ♀ FD(III) and ♀ NFD(III).

This sequence was typical of the copulatory period, with one or other female acting as a focus for male attention, yet rejecting the large majority of male advances, and with persistent inter-male aggression. However, it was the only time a male was seen to mount and copulate with a female without first winning a prolonged and fierce battle with another male.

During the rest of the day, on 3 March, ♂ P(III) followed Group IV; he was chased away 47 times by ♂ R(IV) and several of these chases terminated in fights. SA ♂ Y(III) stayed with ♂ P(III) until midday, and then returned to Group III. ♂ P(III) mounted SA ♀ Y(III) 3 times, while they were trailing Group IV together, and each time SA ♂ Y(III) snapped and wriggled to free himself. This was the only occasion when male/male mounting was seen. Twice, ♂ R(IV) tried to mount ♀ FI(IV), and each time she snapped and wriggled and escaped.

Checks on Group III suggested that group activities were normal.

4 March: SA ♂ Y(III) and ♂ P(III), both in good condition, were found near Group IV. In contrast, ♂ R(IV) had lost large chunks of fur and had two long gashes on his thighs. Between 0600h. and 1030h., 9 incomplete sniff-approach and mark sequences were seen between ♂ R(IV) and ♀ FI(IV); ♂ R(IV) also chased and fought ♂ P(III) 45 times, and ♂ Q(IV) did so 8 times. However, ♂ R(IV) slowed up and tired visibly, and at 10.30 hrs. ♂ P(III) chased ♂ R(IV) away from the group: the latter retreated very slowly and quadrupedally across the ground; this was the only time an animal was ever seen moving quadrupedally on the ground. After a brief period of reciprocal chasing between ♂ Q(IV) and ♂ P(III), ♂ Q(IV) on his own did not chase ♂ P(III) again. At 11.36 hrs. ♀ FI(IV) leapt over to ♂ P(III) and they touched noses.

That afternoon, ♀ FI(IV) moved on to a vertical just above ♂ P(III)'s head and rolled her tail up. He mounted her and thrust slowly for about one minute, and then dismounted to chase ♂ Q(IV) away. Two further mountings were similarly interrupted, but finally he remained on the female for approximately 4 minutes, thrusting about 40 times. After this, when ♂ P(III) tried to approach, both ♀ FI(IV) and ♀ FNI(IV) started cuffing and biting. ♂ Q(IV) remained on the periphery of the group, alternately approaching and being chased away by ♂ P(III). At 17.42 hrs. ♀ P(III) chased ♂ Q(IV) west, where ♂ R(IV) was found feeding: ♂ R(IV) and ♂ Q(IV) fled together. ♂ P(III) mounted ♂ FI(IV) once more, but there was no intromission and he leapt off almost at once to chase ♂ R(IV) and ♂ Q(IV).

5 March: ♀ FI(IV), ♀ FNI(IV) and ♂ Q(IV) were found together, the latter with a swollen lower lip and a small cut on his arm. Neither ♂ P(III) nor ♂ R(IV) were present, and ♂ P(III) was not seen again until March 10th. In the late afternoon ♂ R(IV) appeared and ♀ FNI(IV) leapt over and touched noses with him briefly; ♂ R(IV) then approached ♂ Q(IV) and presented for grooming.* ♂ Q(IV) spat-called† and groomed ♂ R(IV) briefly, but suddenly appeared to start grappling with him; one of the two screamed and ♂ R(IV) leapt off at once. ♂ Q(IV) followed, but as he approached ♂ R(IV) tail-rolled, spat-called and leapt away. ♂ R(IV) did not approach the group again. That evening ♂ INT approached the group and ♀ FNI(IV) sexually presented to him (i.e. stationed herself on a vertical trunk just above his head and rolled her tail up). He mounted her and 53 thrusts were counted in 4 minutes before he dismounted. As after previous incidences of prolonged copulation, the two animals sat near each other licking their own genitalia.

6 March: ♂ INT mounted ♀ FNI(IV) 3 times in the early morning, twice dismounting to chase ♂ Q(IV). This was the last time mounting was seen. ♂ LCE appeared but was chased away by both ♂ INT and ♂ R(IV). ♀ FI(IV), ♀ FNI(IV), ♂ INT and ♂ Q(IV) moved together all day; ♂ R(IV) moved peripherally, retreating when any other animal came near him. Once ♂ Q(IV) groomed ♂ INT briefly: ♂ INT spat-called and tail-rolled as ♂ Q(IV) approached.

7 March: An inter-group encounter, typical of those seen outside

* Presentation for grooming is made by a non-sexually dominant animal to a subordinate, and involves holding out one arm to be groomed.

† The spat vocalization, described by Jolly,[7] is one of a repertoire of signals given by the submissive animal in an agonistic encounter. Other signals are the tense-lipped grin, hunching the back and tail-rolling. The latter is indistinguishable from female presentation tail-rolling prior to copulation, and involves rolling the tail up between the legs.

Figure 6. Adult male P of Group III photographed one month after the mating season. Note the healed – but still badly scarred – nose.

the mating season, took place between Groups III and IV (the latter group comprising ♂ INT, ♂ Q, ♀ FI and ♀ FNI).

10 March: ♂ P(III) joined Group III, with an injured nose: the nasal bones had been exposed, but the flesh still hung from his nose on a piece of skin; he persistently, but unsuccessfully, tried to lick his nose, and once licked his hand and then put it to his nose. After suppurating for two or three days, the wound began to heal and within a month it had healed totally, leaving him with flattened facial features and a wheeze as he breathed (Fig. 6). Throughout the 10th, and most of the 11th, ♂ P(III) moved away if approached by any member of Group III. On the evening of the 11th, however, ♂ P entered a tree where the juvenile was feeding; they touched noses, and ♂ P(III) began feeding in the same tree.

12 March: ♂ R(IV) approached J(III), who rapidly retreated towards other members of Group III, and SA ♂ Y(III) chased ♂ R(IV) away.

13, 14, 15 March: ♀ FI, ♀ FNI, ♂ INT and ♂ Q moved as a well-integrated group, and this composition was retained until observations ended in September 1971. ♂ R(IV) intermittently moved on the periphery of the group; he retreated from ♂ INT, but was not chased by him. ♂ LCE also moved peripherally to the group, attempting to approach the adult females; he was chased away by all group members and was not seen when observations were resumed in April.

The study area was abandoned on 16 March because of local political unrest. When observations were resumed in mid-April,

♂ R(IV) had disappeared and ♂ INT had largely taken over his role in Group IV. ♂ P(III) had resumed his former position as subordinate male in Group III.

A *summary* of events in this, and the pre-copulatory period, is given in Table 4.

Discussion and conclusions

Petter-Rousseau[5] has shown that, in the laboratory, the Cheirogaleinae are seasonally polyoestrous. *Lemur catta* is also seasonally polyoestrous in captivity, non-pregnant females cycling three times.[14] Jolly[7] produced evidence for *L.catta* in the wild having a 'pseudo-oestrous period' approximately one month earlier than the true breeding season: the vulval area of four or five out of nine females went through a pink phase 3-4 weeks before the week of mating, which faded and then flushed again just prior to mating. Two of these females were seen mating in the second period of flushing.

Data from Group IV in this field study indicate that *P. verreauxi* do not have more than one full oestrous period in the breeding season, but that there was a period analogous to Jolly's 'pseudo-oestrous period' in *L.catta.* It is suggested that the flushing of ♀ FNI(IV)'s vulva and associated activities, in late January, represent a partially suppressed oestrus 37 days before full oestrus. This did not occur in Group III, despite the presence of two adult females, and it seems likely that the suppression was total in this group: between 14 January and 13 February no more than three consecutive days passed without observations being made on Group III, so it is unlikely that (if it occurred) a 'pseudo-oestrous period' would not have been seen. Both females in Group III probably came into full oestrus once, at the same time, either between the 13 and 20 February, or after 15 March.

The timing of births provides further evidence for the extreme seasonality of breeding in *P. verreauxi*, in that it reflects the incidence of female receptivity in the mating season, and the degree of synchronisation between females. *P. verreauxi* has a gestation period of about 130 days.[6] The length of the oestrous cycle is not known, but it is reasonable to assume that it is of approximately the same duration as that of *L.catta*: in captivity, this lasts 39.3 days, with a range of 33-45 days. If females in the study groups had been polyoestrous, with a fully synchronised cycle similar in length to that of *L.catta*, two birth peaks might have been expected, the second about five weeks after the first, when females fertilised during the second oestrous period gave birth. However, extrapolating from the birth-period evidence in the wild, there was only one full oestrous period with considerable synchrony between females, during which

Characteristic events	*Pre-copulatory period*			*Copulatory period*	*Post-copulation*
	24-31 January	*1-13 February*	*20-28 February*	*3-6 March*	*7-15 March*
Vulval flush	Present in ♀ FNI(IV) on 26th and 27th	–	–	–	–
Endorsing	High frequency in ♂ R(IV)	⟶	⟶	High frequency in all adult males studied	–
Sniff-approach and mark sequences	High frequency in ♂ R(IV)	⟶	⟶	High frequency in all adult males studied	–
Roaming (movements by adult males)	1. ♂ INT approaches Group IV twice 2. ♂ Q(IV) leaves Group IV 3 times 3. ♂ R(IV) leaves Group IV 3 times (once to approach Group III)	1. ♂ INT approaches Group IV twice 2. ♂ LCE approaches Group IV twice 3. ♂ Q(IV) leaves Group IV once	1. ♂♂ R(IV) and Q(IV) leave Group IV and make a foray into home range of Group III 2. ♂ P(III) leaves Group III twice	1. ♂Y(III) approaches and follows Group IV 2. ♂ P(III) approaches and follows Group IV – mates 3. ♂ INT approaches & follows Group IV – mates 4. ♀ LCE approaches & follows Group IV – is chased away 5. ♂ R(IV) ousted from Group IV by ♂♂ P (III) and INT in turn.	1. ♂ P(III) rejoins Group III 2. ♂ R(IV) approaches Group III – and is chased away 3. ♂ LCE approaches Group IV and is chased away
Intra-group encounters	High frequency in both groups	–	–		
Inter-group encounters	High frequency in both groups	–	–	[♂ F(III) mates with ♀ FI(IV) during inter-group encounter on March 3rd.]	
Copulation	–	–	–	(In order of occurrence) ♂ F(III) with ♀ FI(IV) ♂ P(III) with ♀ FI(IV) ♂ INT with ♀ FNI(IV)	

all mating and fertilisation took place: at Ampijoroa in the north of Madagascar, births of *P.v.coquereli* were scattered over a maximum of 21 days in 1970 and 1971. Jolly[7] reports a ten-day birth season for *P.v.verreauxi* at Berenty, in the south of the island. In neither case was the distribution of births through time sufficient to indicate a polyoestrous breeding system.

The effect of the photo-period on *Microcebus* has been demonstrated by Petter-Rousseaux;[15] but similar work has yet to be done on *L.catta* and *P.verreauxi* to establish the influence of day-length and of temperature on sexual behaviour in these species. Since *L.catta* is polyoestrous in captivity and not in the wild, it is likely that, in this species at least, there are proximate ecological factors which repress and then induce oestrus in the wild. The absence of these factors in captivity would permit the appearance of a full polyoestrous system.

The effect of general physical condition on females' ability to come into oestrus is not well understood. Seasonal fluctuations in food availability in the two study areas are discussed in detail elsewhere;[12] in summary, food seems to be scarce at the end of the dry season, particularly in the south where the rain starts after a six-month drought. Food availability increases to a peak towards the end of the wet season in March, and then drops off again in the dry season. It is to be supposed that the physical condition of the animals mirrors this pattern to some extent and that full oestrus, if affected by the female's physical condition, would be most likely to occur towards the end of a period of optimal ecological conditions, in this case March. It is immediately apparent, however, that other phases, such as pregnancy, and suckling or weaning infants, may be more critical and that the system is timed to provide optimum conditions during one of these periods. In ultimate terms, the periodicity and control of the cycle are likely to be closely adapted to the environment in such a way as to ensure the maximum survival of populations. The timing observed in this study was as follows: mating occurred at the end of the wet season, gestation from the end of the wet season through the first half of the dry season, births in the middle of the dry season, suckling until the end of the dry season, and then the infants were almost completely weaned at the beginning of the following wet season.

The similarity between *L.catta* and *P.verreauxi*, postulated above from limited evidence, is substantiated in other aspects of their mating systems: the receptivity of Group IV females FNI and FI lasted a maximum of 12 and 36 hours respectively, as compared with a maximum of 36 hours for *L.catta* in the wild and 10-24 hours in captivity.[7,14] The incidence of marking behaviour by females in this study was not affected by the breeding season, and a similar stability is reported in *L.catta* females in captivity.[14] Finally, Evans and Goy[14] report that 'both long and short term fluctuations in gonadal activity were associated with changes in the frequency of expression of several non-sexual patterns', although these authors do not differentiate clearly between 'sexual' and 'non-sexual' patterns. In *P.verreauxi* it has been shown that changes in the frequency of five activity patterns occurred in the pre-copulatory period and that, as in *L.catta*, only two new patterns appeared: the act of copulation itself, and fierce fighting between adult males. It is probable that these frequency changes were associated with changes in gonadal activity in *P.verreauxi* in this study.

The close relationship between Groups III and IV has already been referred to. The basis of this relationship is not understood. Interactions were frequent, but never friendly, and it is unlikely that Groups III and IV had recently split up from one larger group since, in September 1970, Group IV seemed to be in the process of splitting away from another group to the south of its range.[12] The presence of three unknown animals sunning themselves in the same tree as Group IV, in the south of their range in mid-February, suggests that the split was not complete even by then. It is postulated that the frequent encounters between Group III and IV led to an across-group recognition of individuals, although not to a stable dominance hierarchy between the two groups considered as whole units. It is further suggested that ♂ R(IV) recognised ♂ F(III) as a dominant male. This contrasts with his reaction to ♂ P, a subordinate male in Group III: ♂ R(IV) fought ♂ P(III) for 24 hours to prevent him approaching either adult female in Group IV.

In the light of the circumstantial evidence, it seems improbable that events described in Group IV were atypical. There are many anecdotal accounts of fierce fighting between *P.verreauxi* in February and March, and a reliable first-hand account of one such fight was given to Jolly by S. de Guiteaud.[7] Further, the adult males in all four study groups had torn ears and old facial scars; since only one female had a slightly torn ear it is more likely that the males acquire these scars through fighting than through falling.

Given the various interpretations possible, the following tentative hypothesis is put forward as one explanation of the differential reaction observed of females towards males, and males towards each other: during the breeding season 'roaming' behaviour by adult males occurs at the same time as some degree of breakdown in group structuring; this breakdown is marked by the appearance of aggression directed at dominant males by subordinate males. This aggression may occur with respect to access to food, resting places, or females. As in *L.catta*,[7] frequency of mating – although not demonstrably of fertilisation – is not the prerogative of NMS dominant males. This does not mean that dominance becomes a meaningless concept during the mating season, but that the actual structure of the NMS dominance hierarchy changes at the onset of that period: *P. verreauxi* females allow only males dominant in the immediate situation, that is males dominant in the mating season (MS), to mount them. This MS dominance may be established through previous mutual knowledge and be a correlate of NMS dominance; however, MS dominance can also be achieved by previously NMS subordinate males, through fighting and ousting NMS dominant males. Unlike *L.catta*, where there is a total reversion to the previous NMS hierarchy at the end of the mating season,[7] the assertion of MS dominance in *P.verreauxi* appears to have enduring effects on the structure of NMS dominance within a group. A male dominant both in and out of the mating season mates with females in one or more other groups, but remains a member of the group in which he was dominant before the mating season. The NMS subordinate male who has fought his way to MS dominance in a group by ousting the resident NMS dominant male, may stay in that group, be it his own original group or one encountered while roaming, to become NMS dominant after the mating season. Fig. 7 further illustrates this idea.

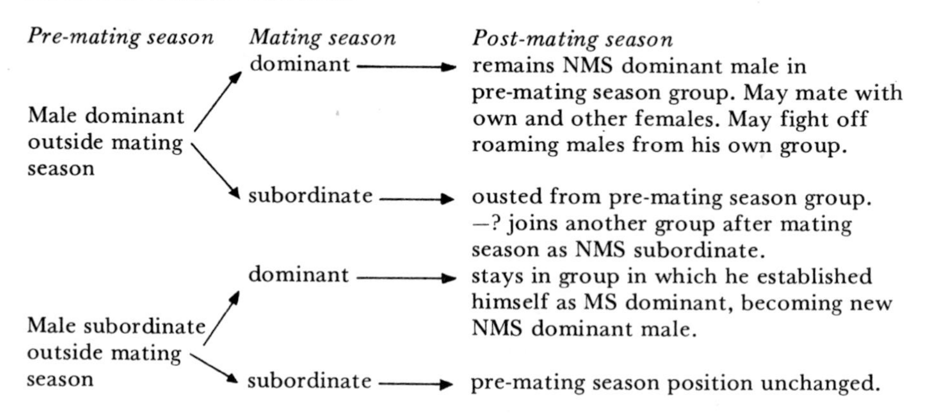

Figure 7. Possible changes in adult male status as a result of the mating season.

With reference to ♂ P(III), it is suggested that he returned to Group III only because he in turn was ousted from Group IV by ♂ INT. Under the interpretation above, ♂ R(IV) recognised ♂ F(III) as dominant both in and out of the mating season and hence as no threat to his own position of non-sexual dominance in Group IV; in contrast, ♂ P(III), a non-sexual subordinate, did constitute a challenge to ♂ R(IV)'s position, and fighting resulted.

It is not known whether ♂ F(III) mated with adult females FD(III) and NFD(III) or whether a roaming, MS and NMS dominant male did so. It is certain that ♀ FI(IV) and ♀ FNI(IV) refused to mate with ♂ R(IV), even when they were sexually receptive and prepared to mate with roaming males. According to the interpretation above, this was because the non-sexually dominant male, R, was being successfully challenged by intruders – ♂ F(III), ♂ P(III) and ♂ INT.

Finally, some consideration should be given to the selective advantage of the system. The evidence indicates that the system ensures some degree of outbreeding. Although probably this outbreeding is largely limited to neighbourhoods, it would be likely to produce some gene-flow through the population as a whole. This might be important in a species with a group size of less than 10, where chronic inbreeding might otherwise occur. New males did join Group I in the north outside the mating season, however, with no apparent fighting involved. It seems unlikely that the mating system, with its fierce inter-male fights, evolved simply as a mechanism to counter inbreeding, when more peaceful means of changing group were possible outside the mating season.

The system may also operate to produce intra-sexual selection between adult males. Access to females and, by inference, biological paternity, appears to be dependent upon the fighting ability, strength and endurance of the adult male. This may have been proved in the previous mating season or even earlier, or may only be manifested in the current mating season. Fighting ability, strength and endurance cannot be equated with overall fitness; however, it does seem unlikely that there should be intensive selection for fighting prowess specifically, during the mating season, when no fights were witnessed during the rest of the year. An alternative explanation is that the social upheaval and fights of the mating season do to some extent test the fitness of the males in terms of their ability to survive relatively prolonged periods of high energy output without a concomitantly increased energy input. This might be critical in a species subject to fluctuations in food availability.

Acknowledgments

I would like to thank J.R. Napier for his constant support throughout this research, and G. Manley, R.D. Martin, A. Jolly, and T. Clutton Brock for their comments on this paper.

The research was supported by a Royal Society Leverhulme Award, a NATO Overseas Studentship, and by grants from The Explorers Club of America, The Boise Fund, The John Spedan Lewis Trust Fund, the Society of the Sigma Xi, and Central Research Fund of London University.

NOTES

1 Flacourt, E. de (1661), *Histoire de la Grande Isle de Madagascar*, Paris.

2 Milne-Edwards, A. and Grandidier, A. (1875, 1890-1895), *Histoire naturelle des mammifères: histoire physique, naturelle, et politique de Madagascar*, Paris, vols. 6, 9, 10.

3 Kaudern, W. (1914), 'Einige Beobachtungen über die Zeit der Fortpflanzung der madagassischen Säugetiere'. *Ark. f. Zool.* 9, 1-22.

4 Rand, A.L. (1935), 'On the habits of some Madagascar mammals', *J. Mammal.* 16, 89-499.

5 Petter, J.-J. (1962), 'Recherches sur l'écologie et l'éthologie des lémuriens malgaches', *Mém. Mus. nat. Hist. nat.*, sér. A, 27, 1-146.

6 Petter-Rousseaux, A. (1962), 'Recherches sur la biologie de la reproduction des primates inférieurs', *Mammalia* 26, suppl. 1, 1-88.

7 Jolly, A. (1966), *Lemur Behaviour*, Chicago.

8 Ellefson, J.O. (1968), 'Territorial behaviour in the common white-handed gibbon, *Hylobates lar*', in Jay, P.C., *Primates: Studies in Adaption and Variability*, New York.

9 Altmann, S.A. (1962). 'A field study of the sociobiology of rhesus monkeys, *Macaca mulatta*', *Ann. N.Y. Acad. Sci.* 102, 460-84.

10 Struhsaker, T.T. (1969), 'Correlates of ecology and social organisation among African Cercopithecines', *Folia primat.* 11, 80-118.

11 Stoltz, L.P. and Saayman, G.S. (1970), 'Ecology and behaviour of baboons in the Northern Transvaal', *Ann. Transv. Mus.* 26, 99-143.

12 Richard, A.F. (1973), 'The social organization and ecology of *Propithecus verreauxi*', Ph.D. Thesis, London University.

13 Richard, A.F. and Heimbuch, R. (in press), 'An analysis of the social structure of three groups of *Propithecus verreauxi*', in Sussman, R.W. and Tattersall, I. (eds.), *Lemur Biology*.

14 Evans, C.S. and Goy, R.W. (1968), 'Social behaviour and reproductive cycles in captive ring-tailed lemurs (*Lemur catta*)', *J. Zool. Lond.* 156, 181-97.

15 Petter-Rousseaux, A. (1970), 'Observations sur l'influence de la photopériode sur l'activité sexuelle chez *Microcebus murinus* (Miller 1777) en captivité', *Ann. Biol. Anim.* 10, 203-8.

R. W. SUSSMAN

Ecological distinctions in sympatric species of Lemur

Introduction

In forests throughout Madagascar, there are many sympatric species of Lemuriformes. The dynamics of the interaction between these species are, however, as yet poorly understood. Generally, coexistence of related species is made possible by differential exploitation of the environment, which minimises competition. Differential exploitation of the environment, in most cases, is the result of habitat selection in which the species have particular preferences for different portions of the shared environment. Where species choose different habitats, the particular preferences may be the result of adaptations to different environmental conditions existing in places where the species are allopatric, or the result of divergence caused directly by interaction between the sympatric populations. With the latter phenomenon, referred to as *character displacement*,[1] sympatric populations of two species will tend to differ in one or more characteristics in which allopatric populations of the same two species may be similar.

A study of the ecological relations between two sympatric forms, therefore, may give indications as to the nature of the forces causing differentiation. Furthermore, once differences in ecological preferences are recognised, it is possible to formulate and test hypotheses concerning the relationship between ecology, morphophysiology, and social behaviour of the species in question.

In this paper, I will describe some of the results of a study on two species of *Lemur*: *Lemur fulvus rufus* and *Lemur catta* (Figs. 1-3). The study is discussed in more detail elsewhere.[2] Populations of *L. f. rufus* range from the north-west to the south-west of Madagascar. Populations of *L. catta* are found from the south-west to the more arid south. In many forests of the south-west, the two species are sympatric (Fig. 5).

Intensive studies were carried out in three forests: Antseran-

Figure 1. *Lemur fulvus rufus* male.

Figure 2. *Lemur fulvus rufus* female.

Figure 3. *Lemur catta.*

Figure 4. *Propithecus verreauxi verreauxi.*

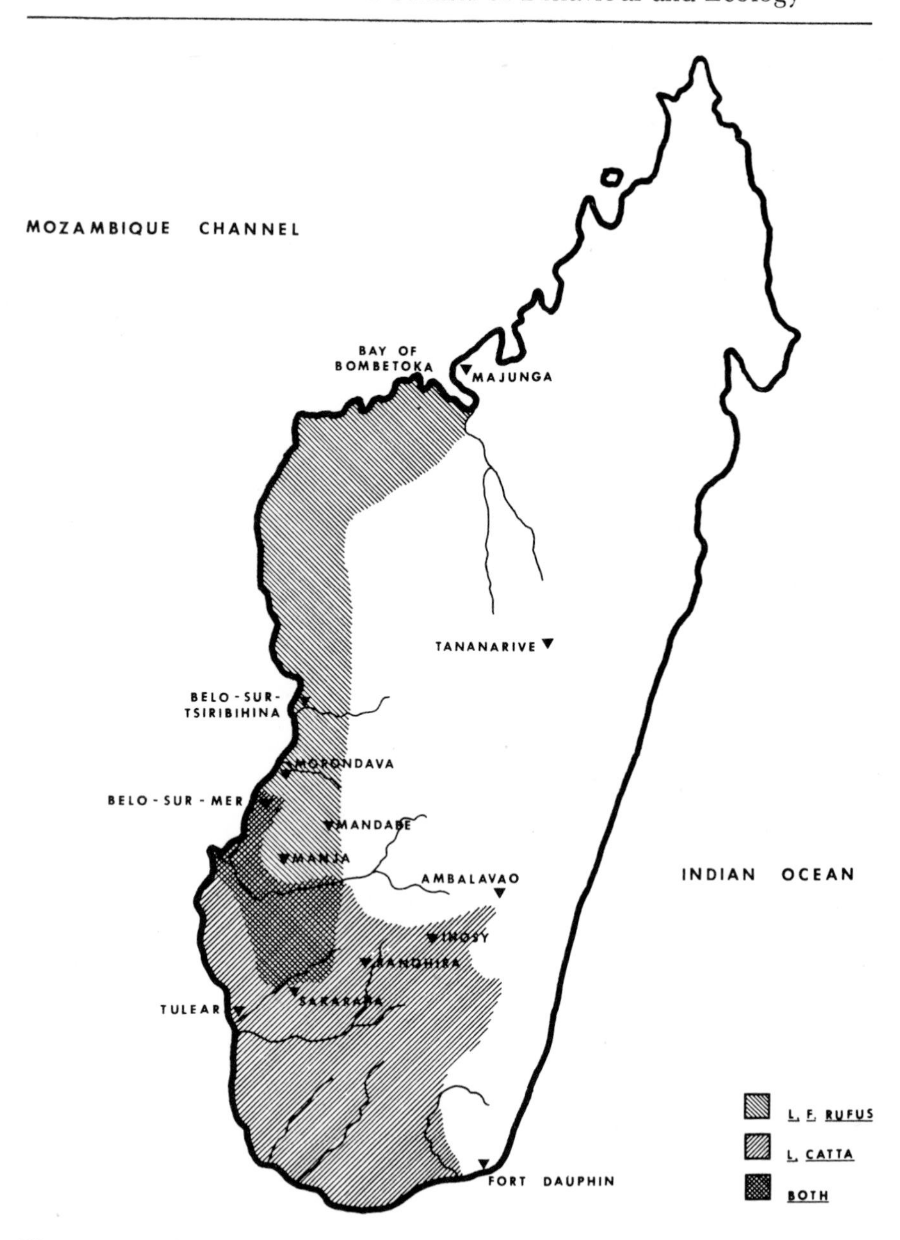

Figure 5. Geographical distribution of *Lemur fulvus rufus* and *Lemur catta.* Populations are not continuous within these areas, but are only found where suitable primary vegetation exists.

anomby, Tongobato, and Berenty (Fig. 6). At Antserananomby, *L. f. rufus* and *L. catta* are sympatric, and in the other two forests they are allopatric. Data on the utilisation of space and time, diet and social structure were collected on both the allopatric and sympatric populations of *L. f. rufus* and *L. catta.* The study was carried out

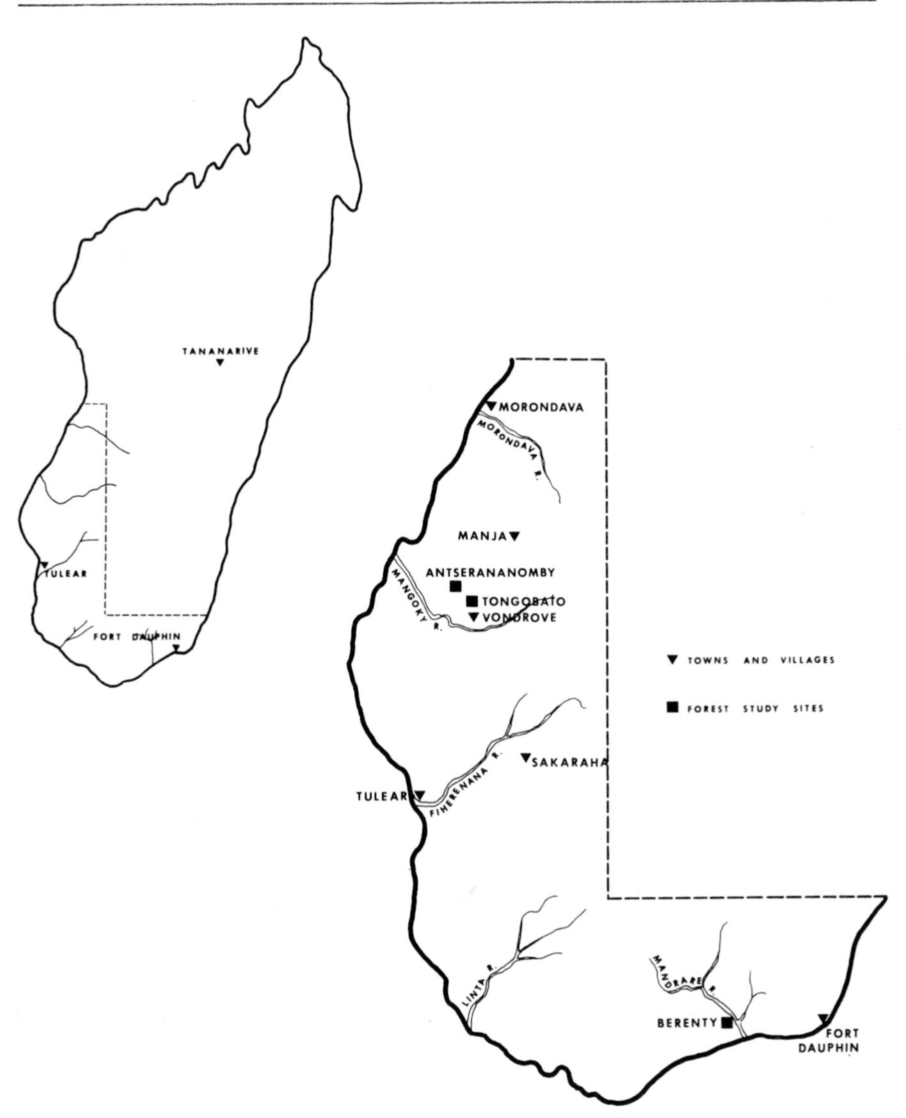

Figure 6. Study sites for intensive observations of behaviour.

between September 1969 and November 1970, during which period the animals were observed for a total of 830 hours.

General ecological distribution of the two species in the south-west of Madagascar

Populations of *Lemur fulvus rufus* and *Lemur catta* coexist in the region between 20° 44′ and 23° 12′ south latitude. These two species, along with *Propithecus verreauxi verreauxi* (Fig. 4), are the

Figure 7. The closed canopy portion of the forest at Antserananomby. Photograph taken from dry river bed.

only diurnal lemur species found in the south-west of Madagascar. The south-west is a region of transition between the moist, deciduous forests of the north-west and the desert-like vegetation of the south. In the south-west, the diurnal lemur species are found mainly in two types of primary forest: (1) closed canopy, deciduous forests and (2) brush and scrub forests. The deciduous forests are dominated by *Tamarindus indica* trees, which make up a continuous, closed canopy about 7 to 15 m. in height (Fig. 7). These forests have been classified by Perrier de la Bathie as 'bois des terrains silicieux'.[3] Forests of this type present essentially the same characters from the north-west to the south-west, and similar forests are also found along the large rivers in the south of Madagascar. The brush and scrub forests reach the height of only about 3 to 7 m. and the under-brush is very dense (Fig. 8). Perrier de la Bathie classified brush and scrub forests as 'bois des terrains calcaires'. The vegetation of Madagascar is described in detail by Perrier de la Bathie, Humbert, and Humbert and Darne.[3]

An initial survey was conducted in the south-west within the area in which populations of *L. f. rufus* and *L. catta* are sympatric. Data from the survey indicate that the ecological distribution of the two species differs within this area. *L. f. rufus* is found without *L. catta* only in small, circumscribed, continuous canopy forests. *L. catta* exists alone only in brush and scrub forests. *L. f. rufus* and *L. catta* coexist only in areas in which a continuous canopy forest merges directly into a primary brush and scrub forest. I have called these 'mixed' forests. Within the mixed forests, *L. f. rufus* was seen only in those portions with a continuous canopy, whereas *L. catta* could be found in all parts of these forests. *P. v. verreauxi* inhabits both

Figure 8. A brush and scrub forest near the Mangoky River.

continuous canopy and mixed forests, but was never seen in brush and scrub forests. This is probably due to the fact that the density and stratigraphy of these latter forests necessitate terrestrial locomotion.

The survey data, therefore, suggest that *L. f. rufus* and *L. catta* have preferences for different habitats. These preferences seem to be related to different locomotor adaptations in the two species. While *L. f. rufus* is limited to areas with a continuous, closed canopy, *L. catta*, because much of its travel is done on the ground, can exploit a number of regions which differ in ecological structure.

Intensive study

Three forests were chosen for intensive study: Antserananomby and Tongobato in the south-west, and Berenty in the south (Fig. 6). The basic physiognomy of the three forests is similar. Kily trees (*Tamarindus indica*) form a closed, continuous canopy in all cases. However, Antserananomby is a mixed forest in which only seven of the total ten hectares studied contain a closed canopy.

Lemur fulvus rufus and *Lemur catta* coexist at Antserananomby, whilst only *L. f. rufus* is found at Tongobato, and only *L. catta* is found at Berenty. *Propithecus verreauxi verreauxi* inhabits all three forests. The nocturnal lemur species found at Antserananomby are: *Lepilemur mustelinus ruficaudatus, Phaner furcifer, Cheirogaleus medius*, and *Microcebus murinus. L. m. ruficaudatus, P. furcifer*, and *M. murinus* were observed at Tongobato. *L. m. leucopus, C. medius*, and *M. murinus* are found at Berenty. Many of the same genera of mammals, birds, and reptiles are found in all three forests.

The study was conducted in December 1969 and in March and April 1970 at Tongobato, from July through September 1970 at Antserananomby, and in November 1970 at Berenty. The temperatures during the day were similar in all three forests. The evening temperatures, however, were higher at Tongobato and Berenty than at Antserananomby (Table 1). The days were also longer at Tongobato and Berenty (5.00 hrs. to 18.30 hrs.) than they were at Antserananomby (6.00 hrs. to 18.00 hrs.).

Table 1 Average maximum and minimum temperatures for the months of study at each forest*

Month	T_x	T_n
Tongobato		
December 1969	36.1	19.2
March 1970	36.2	19.9
April 1970	35.3	16.3
Antserananomby		
July 1970	32.1	11.8
August 1970	35.5	13.9
September 1970	35.5	15.5
Berenty		
November 1970	34.7	19.3

T_x = average maximum daily temperature in °C.
T_n = average minimum daily temperature in °C.

* The temperatures are taken from those recorded at the meteorological station closest to each forest.

1. *Utilisation of the forest strata*

Data were collected at five-minute intervals on the number of individuals engaged in each of six activities – feeding, grooming, resting, moving, travel, and other – and the levels of the forest at which the activities were performed.

Five forest levels were recognised: Level 1 is the ground layer of the forest, which includes herb and grass vegetation. Level 2 is the shrub layer, 1-3 m. above the ground. This layer is usually found in patches throughout closed canopy forests, but is much more dense and is the dominant layer in brush and scrub regions. Level 3 consists of small trees, the lower branches of larger trees, and saplings of the larger species of trees. This layer ranges from 3-7 m. above the ground. Level 4 is the continuous or closed canopy layer. It usually ranges from 5 to 15 m. in height. The dominant tree of the closed canopy, at all three forests, is the kily. Level 5 is the emergent layer and consists of the crowns of those trees which rise above the closed canopy and are higher than 15 m. All three forests in which I made intensive studies are primary forests; the tree layers are quite distinct. With observations in which the forest level could not be clearly distinguished, the level was not recorded.

Each observation of an animal constituted an individual activity record (IAR) collected in a given five-minute time sample.[4] The number of individual activity records for *L. f. rufus* was 7,084 at Antserananomby and 2,896 at Tongobato. The number of IARs for *L. catta* was 5,383 at Antserananomby and 6,861 at Berenty. The IARs were summed for each half-hour from 6.00 hrs. to 18.25 hrs. The day was thus divided into 25 half-hour periods. Comparisons were made between the quantitative data collected on the two species and between those collected on allopatric populations of the same species.

Fig. 9 shows the level at which the highest percentage of animals was observed during each of the 25 half-hour periods. Data on both species (from all three forests) are represented in this figure. Both at Tongobato and at Antserananomby, *L. f. rufus* was found in the continuous canopy for almost all of the half-hour periods. The vertical displacement of *L. catta* followed a pattern throughout the day which was related to the foraging and movement patterns of this species. The pattern for *L. catta* was similar in both forests.

Table 2 includes the mean percentage of animals observed at each level for each activity and at each level overall, regardless of activity, for the 25 half-hour periods. Again, the intraspecific comparisons show the populations of each species to be similar in their use of vertical habitat, regardless of the forest or the presence or absence of the other species. *L. f. rufus* was very specific in its choice of vertical habitat. A high percentage (over 90%) of the activities of *L. f. rufus* took place in the top layers of the forest (levels 3, 4, and 5). In all six activities, *L. f. rufus* spent the highest percentage of time in level 4, the continuous canopy of the forest. *L. catta*, on the other hand, was found in all forest levels. For the most part, *L. catta* moved and travelled on the ground, rested during the day in the low trees (levels 2 and 3), and rested at night in level 4. *L. catta* fed in all of the levels of the forest.

At both forests in which *L. catta* was studied, it spent more time on the ground than in any one of the other four levels (36% at Berenty and 30% at Antserananomby). Perhaps the most significant finding is that *L. catta* travelled on the ground 65% (Antserananomby) and 71% (Berenty) of the time. A high percentage of the individual movement of *L. catta* was also carried out on the ground. *L. f. rufus* travelled in level 4 88% of the time at Antserananomby and 83% of the time at Tongobato. *L. f. rufus* was seen on the ground in less than 2% of the observations overall.

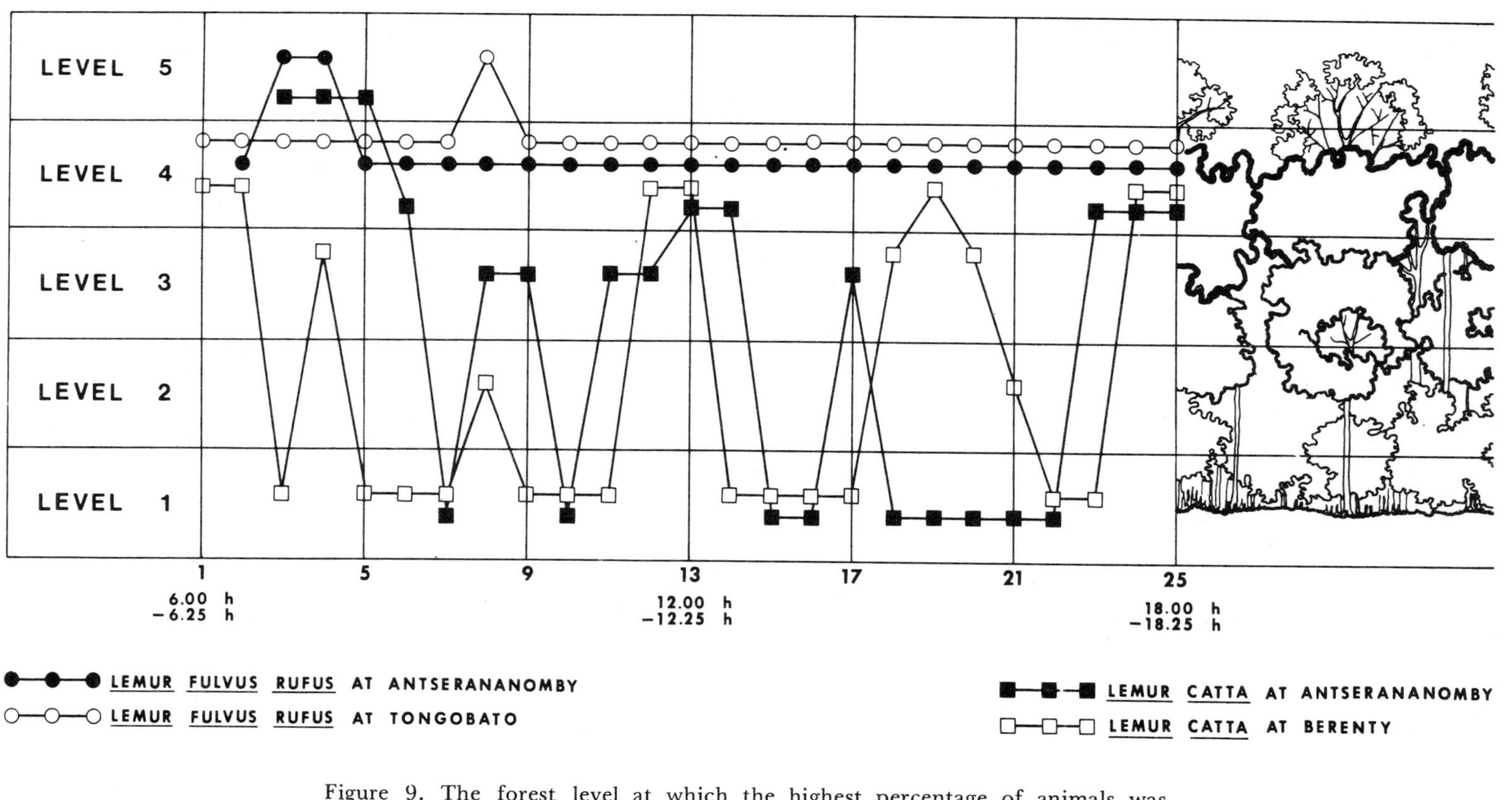

Figure 9. The forest level at which the highest percentage of animals was observed during each of the 25 half-hour periods for the species studied at each of the forests.

Table 2 Mean percentage of animals observed at each level for each activity, and regardless of activity

a. Lemur fulvus rufus at Antserananomby

Activity	Level 1	2	3	4	5
Feeding	0.00	2.58	15.60	59.66	22.16
Grooming	0.00	0.54	9.34	78.24	11.88
Resting	0.00	1.20	14.05	79.60	5.15
Moving	2.61	0.60	18.14	66.16	12.48
Travel	0.00	0.00	2.60	87.84	9.56
Other	0.00	8.33	5.95	54.70	31.02
All Activities	0.44	2.21	10.95	71.03	15.38

b. Lemur fulvus rufus at Tongobato

Activity	Level 1	2	3	4	5
Feeding	2.17	11.40	5.01	44.66	36.76
Grooming	0.00	0.60	9.93	79.99	9.49
Resting	0.34	0.83	8.03	79.91	10.89
Moving	6.02	1.27	4.14	78.64	9.93
Travel	0.95	3.69	5.92	82.94	6.50
Other	0.00	5.00	15.00	56.18	23.82
All Activities	1.58	3.80	8.00	70.39	16.23

c. Lemur catta at Antserananomby

Activity	Level 1	2	3	4	5
Feeding	27.58	12.78	25.23	19.01	15.40
Grooming	4.37	15.90	20.42	46.00	13.31
Resting	9.54	16.74	25.62	35.02	13.08
Moving	44.73	5.88	21.62	17.58	10.20
Travel	64.70	1.81	13.47	17.46	2.57
Other	28.70	5.02	27.31	22.50	16.47
All Activities	29.93	9.69	22.28	26.26	11.84

d. Lemur catta at Berenty

Activity	Level 1	2	3	4	5
Feeding	30.84	15.10	41.56	12.32	0.18
Grooming	10.16	23.35	27.97	38.38	0.14
Resting	14.01	26.14	28.18	31.68	0.00
Moving	55.32	13.51	15.82	15.35	0.00
Travel	70.93	6.44	5.48	17.15	0.00
Other	35.71	17.27	20.61	26.41	0.00
All Activities	36.16	16.97	23.27	23.55	0.05

When *L. catta* travelled in the trees, its locomotor behaviour differed from that of *L. f. rufus.* When groups of *L. f. rufus* moved horizontally through the trees, the animals ran along the fine terminal branches of the large trees. These branches were generally horizontal in relation to the ground and formed a continuous series of pathways throughout the closed canopy. *L. catta*, rather than moving along the fine, horizontal branches, would usually climb the large, oblique branches and leap from one of these branches to another. Thus, they moved through the trees by performing a series of runs and leaps, in each case running up an oblique branch and leaping down to another.

Morphological differences have been found between the foot of *L. f. rufus* and that of *L. catta.* In *L. f. rufus*, the heel is covered with hair, whereas in *L. catta* the heel is naked (Fig. 10). In fact, *L. catta* is the only species of prosimian in which the plantar pad extends proximally to the heel. In this feature, the foot of *L. catta* closely resembles that of monkeys. An examination of the skeletons of the feet of *L. f. rufus* and *L. catta* has shown that the mid-tarsal joints (calcaneo-cuboid and talo-navicular) are less mobile in *L. catta* and that the foot of *L. catta* may be adapted for what Morton has termed metatarsi-fulcrumation.[5] Further studies on the morphology and analyses of films of the locomotion of *L. f. rufus* and *L. catta* may allow more precise relationships to be found between the anatomy, locomotor behaviour, and vertical pattern of habitat preferences of the two species.

2. *Utilisation of the horizontal habitat*

Groups of *L. f. rufus* were identified and their locations were recorded throughout the study. Three groups of *L. f. rufus* were studied intensively at Tongobato and six groups at Antserananomby. At Antserananomby, there were twelve groups within the seven hectares of closed canopy forest. These groups were identified and their locations were noted on prepared maps whenever they were sighted. Groups of *L. f. rufus* usually could not be followed throughout the whole day because of the difficulty in remaining in continuous contact with this arboreal species. One group of *L. catta* at Antserananomby and two groups at Berenty were studied intensively. The groups of *L. catta* were usually followed throughout the day, their movements being recorded on prepared maps.

Figs. 11 and 12 illustrate the extreme differences in the size of the home-ranges and day-ranges between *L. f. rufus* and *L. catta* at Antserananomby. In this forest, the home-ranges of twelve groups of *L. f. rufus* (112 animals in all) were included within the home-range of one group of 19 *L. catta* (Fig. 11).

The day-ranges and home-ranges of *L. f. rufus* were very small, both at Antserananomby and at Tongobato. Although it was

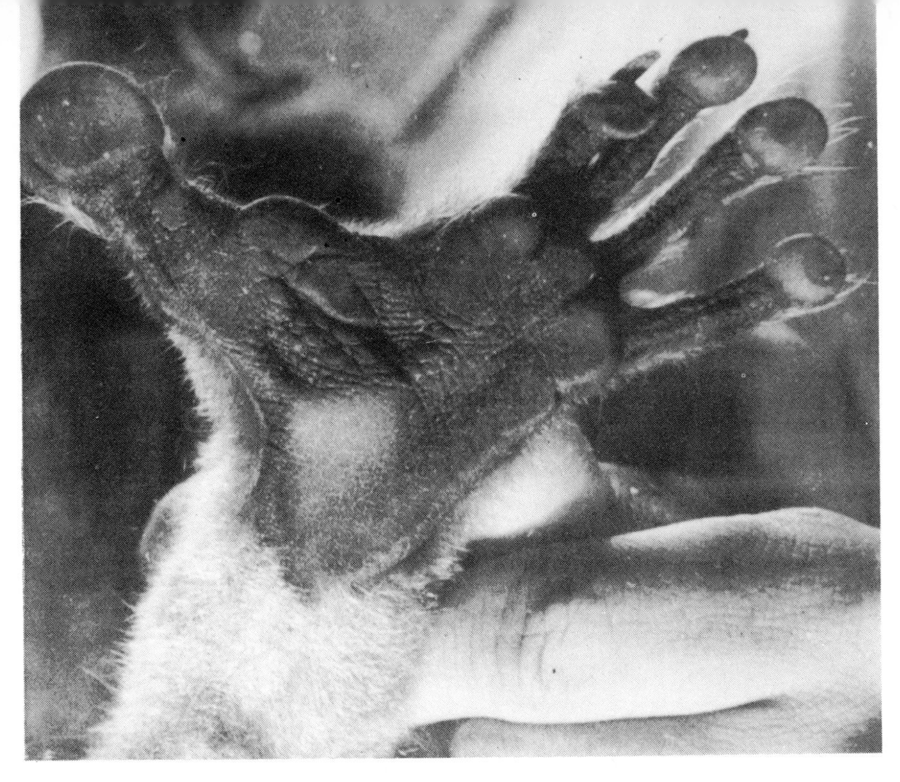

Figure 10. Above: foot of *Lemur fulrus rufus*. Below: foot of *Lemur catta*.

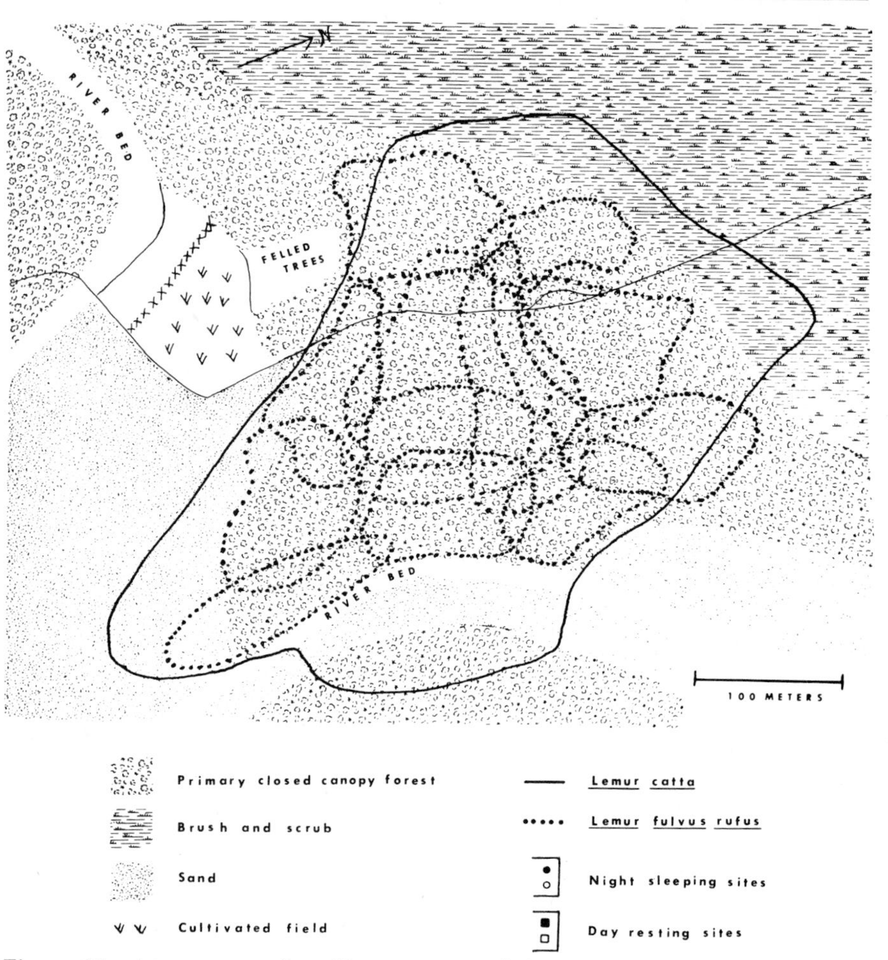

Figure 11. Antserananomby. Home-ranges of *Lemur catta* and *Lemur fulvus rufus.* This map includes the home-range of one group of *L. catta* (19 animals) and twelve groups of *L. f. rufus* (112 animals in all).

generally difficult to remain in contact with groups of *L. f. rufus*, at Antserananomby I was able to follow three groups for two days each (Fig. 12). The average day-range for the six days was between 125 and 150 m. The home-ranges of the groups studied at Antserananomby averaged 0.75 hectares (Fig. 11), while those of three groups at Tongobato averaged 1.0 hectares. The home-ranges of *L. f. rufus* were not rigidly defended, and the boundaries overlapped extensively.

L. catta had much larger day- and home-ranges than *L. f. rufus.* The average day-range for *L. catta* at Antserananomby was approximately 920 m. (Fig. 13). At Berenty, in a sample of four days, the average day-range was 965 m. The home-range of the group of *L. catta* at Antserananomby (group AC-1) was 8.8 hectares (Fig. 11). The home-range of group BC-1 at Berenty was 6.0 hectares. Jolly

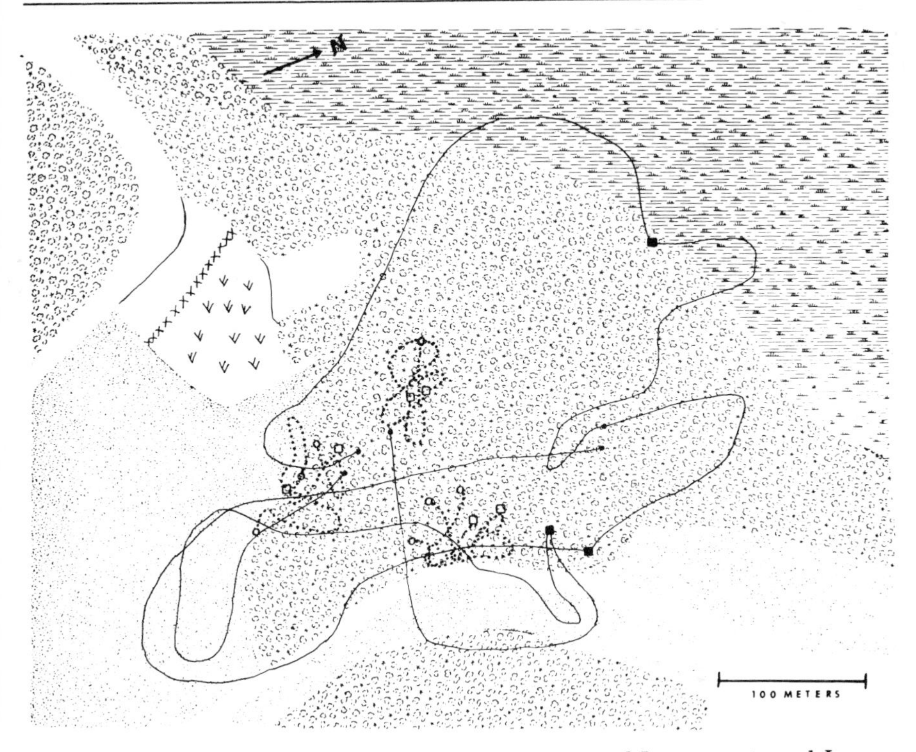

Figure 12. Antserananomby. Comparative day-ranges of *Lemur catta* and *Lemur fulvus rufus*. This map includes a sample of three day-ranges of one group of *L. catta* (group AC-1) and two day-ranges of three groups of *L. f. rufus*. (Key as for Fig. 11)

reported a home-range of 5.7 hectares for the group she studied at Berenty.[6] At Antserananomby, *L. catta* spent approximately 58% of its time in zones outside of the closed canopy area, even though these zones represented only 30% of its total home-range. By contrast, all of the home-ranges of the groups of *L. f. rufus* at this forest were located within the 7 hectares in which there is a closed canopy.

The population density of *L. f. rufus* was much higher than that of *L. catta*. The population density of each species was calculated from the average group size (see Tables 10 and 11) and the average area of the home-range. The population density of *L. f. rufus* at Tongobato was calculated only from data on groups TF2, TF3, and TF4.

Lemur catta	
Antserananomby	215 per square km.
Berenty*	250 per square km.
Average	233 per square km.

* Jolly estimated the density of *L. catta* at Berenty to be 320 per square km.[6] This would give an average density of 261 per square km. for populations of *L. catta* studied to date.

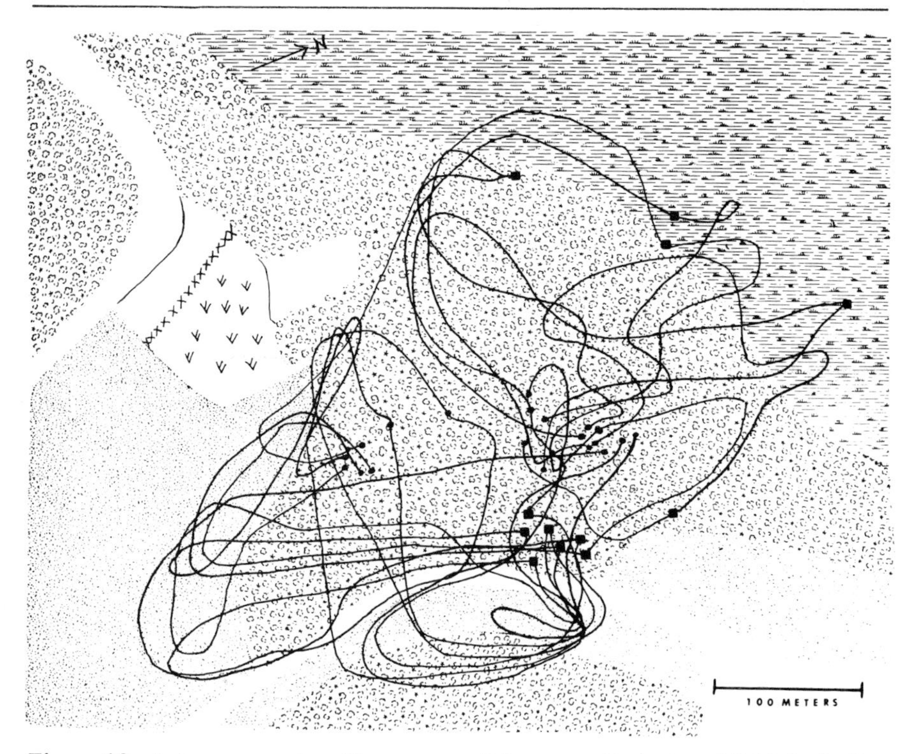

Figure 13. Antserananomby. Day-ranges of *Lemur catta* (group AC-1), including all of the ranges for which the group was followed throughout the day. (Key as for Fig. 11)

Lemur fulvus rufus

Antserananomby	1222 per square km.
Tongobato	900 per square km.
Average	1061 per square km.

3. *Diet*

The number of animals feeding and the level of the forest in which they were feeding was recorded during the collection of five-minute activity records. The plant and the part being eaten were recorded directly on the data sheet. Plant specimens were collected whenever possible and were later identified by Armand Rakotozafy of the Laboratoire Botanique of ORSTOM, Tananarive. No stomach content samples were obtained, because it was not feasible to shoot any of the animals during the field study. The dry weights of the plants and the nutritional values of the specimens have not yet been analysed. Therefore, the dietary schedules of the two species have been determined from direct observation only. Samples of faeces were collected, but have not yet been analysed.

Table 3 Plant species eaten by *Lemur fulvus rufus* at Tongobato

Acacia rovumae Olia*
Acacia sp.*
Alchornea sp.
*Flacourtia ramontchi**
Lawsonia alba Lamk
*Tamarindus indica**
*Terminalia mantaly**
*Vitex beravensis**

* Plant species eaten by *Lemur fulvus rufus* at both Tongobato and Antserananomby.

Table 4 Plant species eaten by *Lemur fulvus rufus* at Antserananomby

Acacia rovumae Olia*
Acacia sp.*
Ficus soroceoides Bak
*Flacourtia ramontchi**
Papilionaceae
Quisivianthe papinae
Rinorea greveana
*Tamarindus indica**
*Terminalia mantaly**
Tisomia sp.
*Vitex beravensis**

* Plant species eaten by *Lemur fulvus rufus* at both Tongobato and Antserananomby.

The differences between the diet of *L. catta* and *L. f. rufus* are related to the differences in habitat selection in the two species. *L. f. rufus* stayed in the continuous canopy of the forest and moved horizontally from tree to tree. It was rarely seen in areas that required movement on the ground. It fed over 90% of the time in the upper three forest levels (see Table 2). *L. catta*, on the other hand, was frequently observed feeding in all of the available forest layers. It spent about 58% of its time outside of the portion of the forest with a continuous canopy. *L. catta* fed over 65% of the time in levels 1, 2, and 3.

The restricted use of space by *L. f. rufus* is related to a less varied diet in this species than that found in *L. catta.* The plants which *L. f. rufus* was observed eating during the study are listed in Tables 3 and 4. I recorded *L. f. rufus* eating only 8 plant species at Tongobato and 11 at Antserananomby. There was a total of 13 different plant species eaten by *L. f. rufus* in the two forests. *L. catta*, on the other hand, had a much more varied diet in both of the forests in which it was studied. Both at Antserananomby and Berenty, *L. catta* was observed to feed on 24 different plant species (Tables 5 and 6). At Antserananomby, *L. catta* ate 12 plant species which were not eaten by *L. f. rufus.* Ten of these were ground plants or bushes, or plants found only outside the closed canopy portion of the forest.

L. f. rufus appears to be specialised in its choice of diet. A few species of plants made up a large proportion of its diet, and kily leaves were the main staple (Table 7). During the period of

Table 5 Plant species eaten by *Lemur catta* at Antserananomby

Acacia rovumae Olia	*Mimilopsis* sp.
Acacia sp.	*Paederia* sp.
Acalypha sp.*	*Paederia* sp.
Achyranthes aspera L.*	Papilionaceae
Adenia sp.	*Poupartia caffra* H. Perr
Alchornea sp.	*Quisivianthe papinae*
Commicarpus commersonii	*Tamarindus indica**
Ficus cocculifolia	*Terminalia mantaly*
Ficus soroceoides	*Vitex beravensis*
Ficus sp.	
Flacourtia ramontchi	
Grevia sp.	
Three species unidentified (small trees)	

* Plant species eaten by *Lemur catta* at both Antserananomby and Berenty.

observation at Tongobato, three species of plant constituted more than 80% of the diet of *L. f. rufus: Flacourtia ramontchi, Tamarindus indica*, and *Terminalia mantaly.* At Antserananomby, *Acacia rovumae, Ficus soroceoides*, and *Tamarindus indica* accounted for over 85% of the diet of *L. f. rufus.* Kily (*Tamarindus indica*) leaves made up 42% of the diet of *L. f. rufus* at Tongobato and about 75% of its diet at Antserananomby. This degree of specialisation is rare among those primate species that have been studied. It has been reported to occur, however, in *Lepilemur mustelinus leucopus.*[7]

The diet of *L. catta* was much less restricted than that of *L. f. rufus* (Table 7). At Antserananomby, the following species made up over 70% of the diet:

Achyranthes aspera	*Grevia* sp.
Alchornea sp.	*Mimilopsis* sp.
Ficus soroceoides	*Poupartia caffra*
Flacourtia ramontchi	*Tamarindus indica*

The kily tree provided 24% of the diet of *L. catta* in this forest. However, less than half of this (11%) consisted of leaves. Kily pods made up 12% of the diet, and ground plants (*Achyranthes aspera* and *Mimilopsis* sp.) 15%.

The diets of *L. catta* at Berenty and Antserananomby consisted of different plant species, but were quite similar in design. At Berenty, the following species accounted for over 80% of the observed plants eaten by *L. catta:*

Achyranthes aspera	*Phyllanthus* sp.
Boerhaavia diffusa	*Pithecelobium dulce*
Cassia sp.	*Rinorea greveana*
Melia azedarach	*Tamarindus indica*

The kily tree made up 23% of the diet. Kily leaves were eaten in 12% of the observations and kily pods in 10%. The fruit of two other

Table 6 Plant species eaten by *Lemur catta* at berenty

Acalypha sp.*	*Ehretia* sp.
Aizoaceae	*Hzima tetracantha* Lamk
Albizzia polyphylla Forven	*Mangifera indica* L.
Annona sp.	*Melia azedarach* L.
Achyranthes aspera L.*	*Opuntia vulgaris* Mill
Boerhaavia diffusa L.	*Phyllanthus* sp.
Cardiaspermum halicacabum	*Pithecelobium dulce* Benth
Cassia sp.	*Rinorea greveana* H. Bn.
Celtis philippensis Blanco	*Tamarindus indica**
Cissompelos sp.	*Zehneria* sp.
Combretum sp.	*Zizyphus jujuba* Lamk
Crateva excelsa Boj	One species unidentified (vine)

* Plant species eaten by *Lemur catta* at both Berenty and Antserananomby.

trees (*Rinorea greveana* and *Pithecelobium dulce*) together accounted for 40% of the observed feeding activity.

Both *L. catta* and *L. f. rufus* ate the fruit, leaves, flowers, bark, and sap of various species of plants (Table 8). The amount of time that they spent feeding on various parts of the plants depended upon the fruiting or blossoming seasons of the plants. As with differences in the number of plant species eaten by *L. catta* and *L. f. rufus*, the proportion of fruit or flowers fed upon by the two species is related to differences in the use of vertical and horizontal space. For example, a small tree (*Ficus cocculifolia*) produced fruits the size of large apples. I saw *L. f. rufus* feed on the fruit of this tree in a forest on the bank of the Mangoky River. There was only one of these trees in the forest at Antserananomby, and it was located on the side of a dry river bed opposite to the continuous canopy portion of the forest. *L. f. rufus* rarely crossed the river, and was never seen in this tree. *L. catta*, on the other hand, regularly crossed the river and foraged in this tree for a number of days while the fruit was ripe.

4. *Utilisation of time*

Once a group of animals was located, counts were made at five-minute intervals of the number of individuals engaged in each of six activities: feeding, grooming, resting, movement (i.e. movement of an individual), travel (i.e. movement of the group), and other (i.e. sunning, play, fighting, etc.). Each animal observed during each five-minute interval constituted an individual activity record (IAR). The IARs for each five minutes were combined for each half-hour period from 6.00 hrs to 18.25 hrs., and the percentage of IARs for each activity was calculated for each half-hour. The day was then divided into five phases as shown in Fig. 14, which includes the means of the percentages calculated for the half-hour periods within each of the phases.

Table 7 The number and percentage of individual activity records (IARs) for feeding on identified plant species

a. Lemur fulvus rufus at Antserananomby

Plant Species	*Number of IARs*	*Percentage of IARs*
Tamarindus indica	1802	75.68%
Leaves	1793	75.30
Fruit	9	0.38
Acacia sp.	156	6.55
Ficus soroceoides Bak	141	5.92
Acacia rovumae Olia	74	3.10
Terminalia mantaly	21	0.88
Quisivianthe papinae	18	0.75
Other	169	7.06
Total	2381	99.94%

b. Lemur fulvus rufus at Tongobato

Plant Species	*Number of IARs*	*Percentage of IARs*
Tamarindus indica	276	48.85
Leaves	237	41.95
Flowers	26	4.60
Fruit	10	1.77
Bark	3	0.53
Terminalia mantaly	127	22.47
Flacourtia ramontchi	69	12.21
Acacia rovumae Olia	38	6.72
Vitex beravensis	18	3.18
Other	37	6.54
Total	565	99.97%

The most striking differences between *L. f. rufus* and *L. catta* are seen in the data from Antserananomby (Figs. 14*a* and 14*b*). In this forest, *L. f. rufus* rested throughout the afternoon, whereas *L. catta* rested only during the midday phase (phase 3). Data from all three forests indicate that *L. f. rufus* rests more than *L. catta* during the day. The activity/rest ratios for the hours from 6.00 hrs. to 18.25 hrs. were calculated from the data in Table 9. The ratios are as follows:

Lemur fulvus rufus

Tongobato	50/50 = 1.00
Antserananomby	44/56 = 0.79

Lemur catta

Antserananomby	59/41 = 1.44
Berenty	61/39 = 1.56

c. *Lemur catta* at Antserananomby

Plant Species	*Number of IARs*	*Percentage of IARs*
Tamarindus indica	374	24.36%
[Fruit	[183	[11.92
Leaves	174	11.33
Flowers]	17]	1.11]
Small trees:	320	20.84
[*Alchornea* sp.		
Flacourtia ramontchi		
Grevia sp.		
Poupartia caffra H. Perr]		
Ground plants (all species)	225	14.65
Ficus soroceoides Bak	194	12.63
Vines (all species)	140	9.12
Quisivianthe papinae	94	6.12
Vitex beravensis	82	5.34
Ficus cocculifolia	40	2.60
Acacia rovumae Olia	18	1.17
Other	48	3.12
Total	1535	99.95%

d. *Lemur catta* at Berenty

Plant Species	*Number of IARs*	*Percentage of IARs*
Tamarindus indica	519	22.97%
[Leaves	[274	[12.13
Fruit	225	9.96
Bark]	20]	.88]
Rinorea greveana H. Bn.	474	20.98
Pithecelobium dulce Benth	433	19.16
Phyllanthus sp.	137	6.06
Melia azedarach L.	132	5.84
Ehritia sp.	130	5.74
Ground plants (all species)	124	5.48
Opuntia vulgaris Mill	84	3.71
Annona sp.	38	1.68
Vines (all species)	27	1.19
Other	161	7.10
Total	2259	99.91%

L. f. rufus fed very early in the morning and late in the afternoon and travelled little to obtain its food. *L. catta* began to feed later in the morning and stopped earlier in the evening than *L. f. rufus.* Groups of *L. catta* travelled greater distances to obtain food, foraging throughout the afternoon. At Antserananomby, *L. f. rufus* fed mainly during the first and fifth phases and *L. catta* during the second and fourth. In all phases but the third (12.00 to 14.55 hrs), when both species were resting, the patterns of activity were different.

Table 8 Number and percentage of individual activity records for feeding on identified parts of plants

a. Lemur fulvus rufus at Antserananomby

Part of plant eaten	*Number of IARs*	*Percentage of IARs*
Leaves	2123	89.16%
Fruit	161	6.76
Flowers	90	3.77
Bark	7	.29
Total	2381	99.98%

b. Lemur fulvus rufus at Tongobato

Part of plant eaten	*Number of IARs*	*Percentage of IARs*
Leaves	275	52.08%
Fruit	224	42.43
Flowers	26	4.92
Bark	3	.56
Total	528	99.99%

c. Lemur catta at Antserananomby

Part of plant eaten	*Number of IARs*	*Percentage of IARs*
Leaves	670	43.64%
Fruit	516	33.61
Herbs	225	14.65
Flowers	124	8.07
Total	1535	99.97%

d. Lemur catta at Berenty

Part of plant eaten	*Number of IARs*	*Percentage of IARs*
Fruit	1335	59.30%
Leaves	550	24.43
Flowers	137	6.08
Herbs	124	5.50
Bark, sap, cactus	105	4.66
Total	2251	99.97%

At Berenty, groups of *L. catta* began to feed earlier in the morning and the peak of their resting activity was earlier than at Antserananomby. They also began to feed earlier in the afternoon and stretched out their afternoon feeding activity over a longer period of time. These differences, which can be related to the early sunrise and high early morning temperatures at Berenty, make the behaviour pattern of *L. catta* at this forest superficially similar to that of *L. f.*

Table 9 Mean percentages of individual activity records for each activity each half hour

Species and forest	*Activity*					
	Feeding	*Grooming*	*Resting*	*Moving*	*Travel*	*Other*
Lemur fulvus rufus						
Tongobato	16.59	11.30	49.73	7.77	12.21	2.41
Antserananomby	26.22	5.25	56.58	3.09	2.47	6.40
Lemur catta						
Antserananomby	24.94	4.67	41.42	7.45	11.34	10.19
Berenty	31.12	10.97	38.63	6.30	6.91	6.07

rufus in the other study areas. Even though the temperature and time of sunrise at Tongobato were more similar to those at Berenty than to those at Antserananomby, the pattern of behaviour of *L. f. rufus* was similar in both of the forests in which it was studied. The high percentage of movement and travel of *L. f. rufus* at Tongobato resulted from the difficulty in habituating the animals to the observer, because local inhabitants frequently hunted the lemurs in this forest. Seasonal differences (in this case, time of sunrise and sunset and time of the maximum temperature during the day) seemed to affect the daily activity cycle of *L. catta* more than that of *L. f. rufus.*

L. f. rufus and *L. catta* both exhibited 'sunning' behaviour early in the morning at Antserananomby. This is reflected in the high percentage of 'other' activities during the first phase of the day. *L. catta* sunned twice as much as *L. f. rufus* during this phase, and sunning may be an important thermo-regulating mechanism in this species. Very little sunning was observed for either species in the other two forests. The mean minimum temperatures at night at Berenty and Tongobato were much higher than those at Antserananomby during the months of observation in these forests.

Significance tests were run on the resting and feeding activities of *L. f. rufus* and *L. catta* at Antserananomby. Since the samples for different five-minute periods are not independent, Z scores were computed for each five-minute interval (Figs. 15 and 16). The methods used for these calculations are described by Sussman[2] and Sussman, O'Fallon and Buettner-Janusch (in prep.).

Fig. 15 illustrates the fact that *L. f. rufus* fed significantly more in the early morning and late evening, and *L. catta* fed significantly more in the late morning and late afternoon. Feeding behaviour in the midday phase was not significantly different; at this time, both species were mainly resting. Throughout the day there are very few times (a total of 25 minutes) when *L. catta* rested significantly more than *L. f. rufus* (Fig. 16). There are only 26 five-minute periods in which a higher percentage of *L. catta* than *L. f. rufus* were observed resting.

Figure 14. Mean percentage of individual activity records for each of six activities during five phases of the day. F = Feeding; G = Grooming; R = Resting; M = Moving; T = Travel; O = Other.

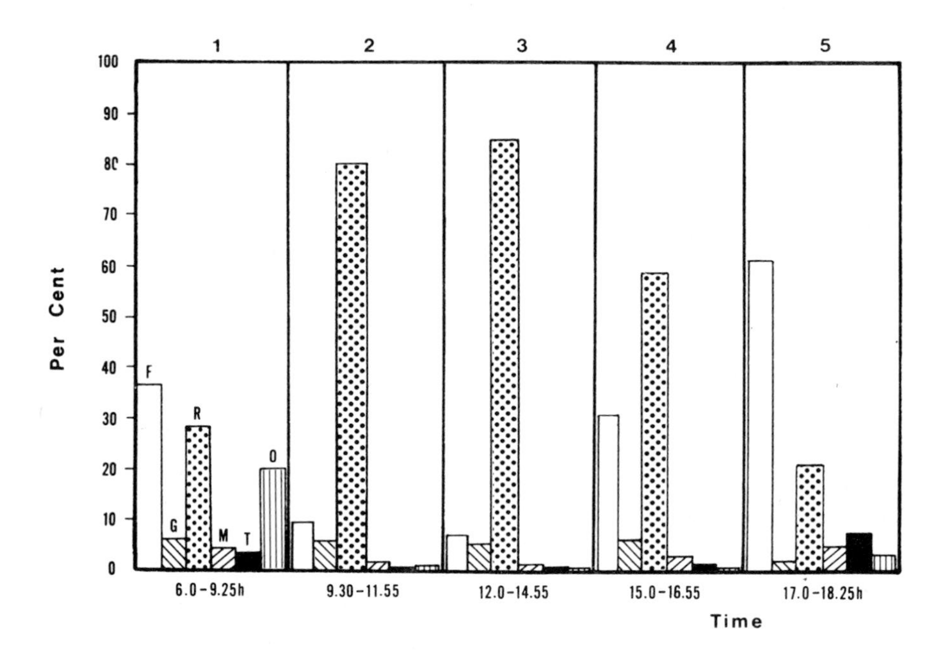

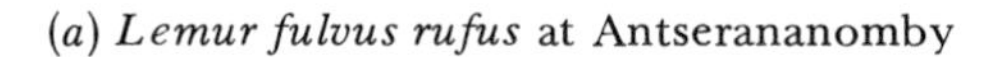

(*a*) *Lemur fulvus rufus* at Antserananomby

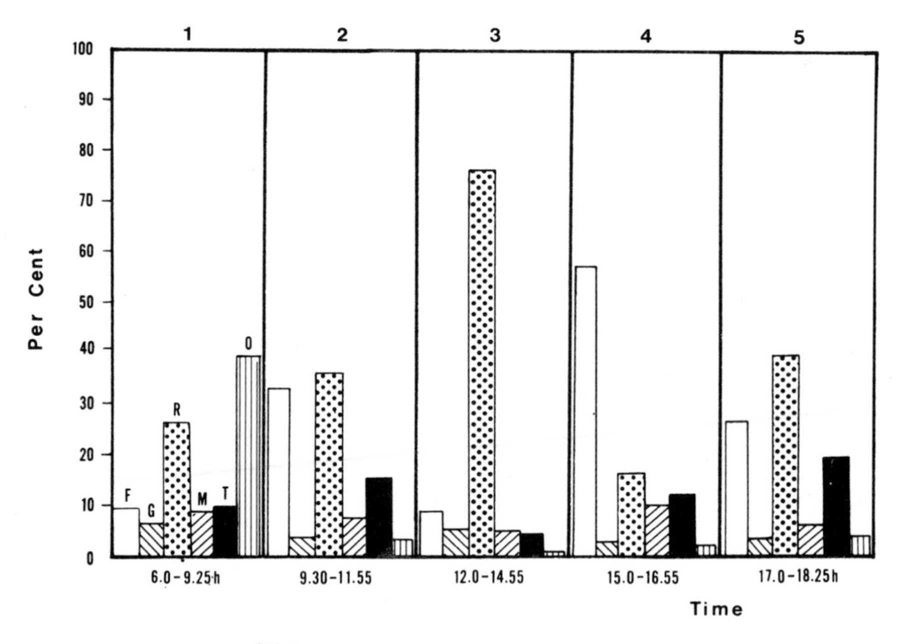

(*b*) *Lemur catta* at Antserananomby

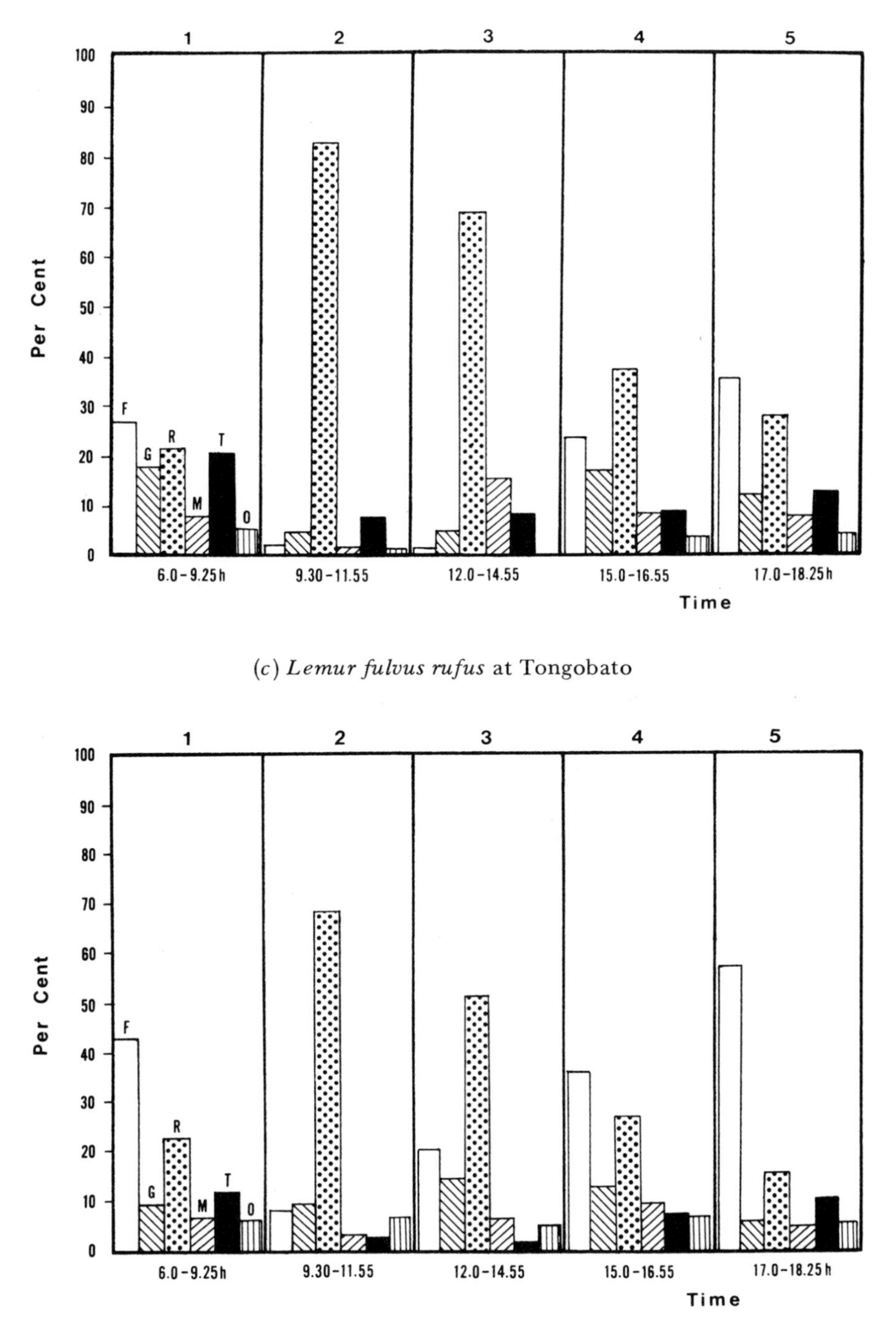

(*c*) *Lemur fulvus rufus* at Tongobato

(*d*) *Lemur catta* at Berenty

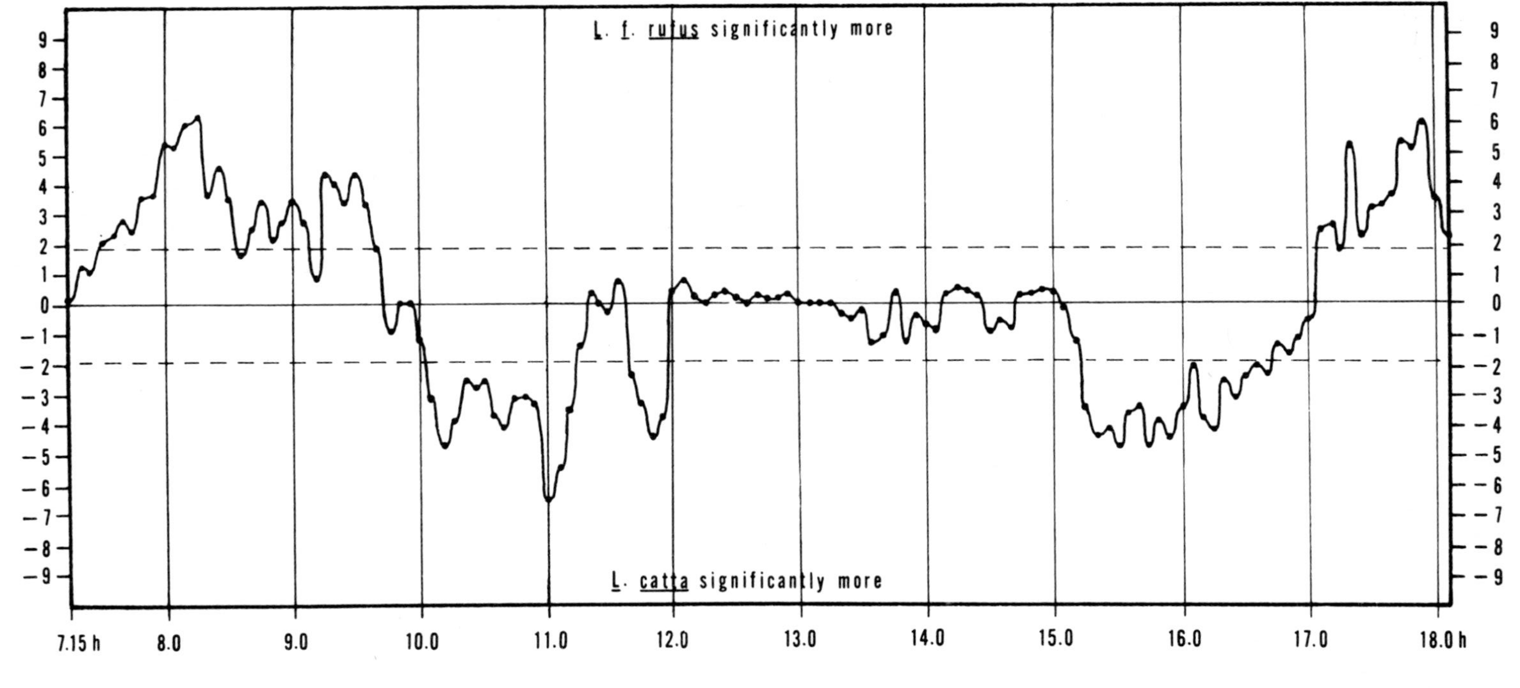

Figure 15. Z scores for the feeding activity of *Lemur fulvus rufus* and *Lemur catta* at Antserananomby computed for each five-minute observation period. A score equal to or more than 1.96 means that *L. f. rufus* feeds significantly more than *L. catta* ($p \leqslant .05$). A score equal to or less than -1.96 means that *L. catta* feeds significantly more than *L. f. rufus* ($p \leqslant .05$).

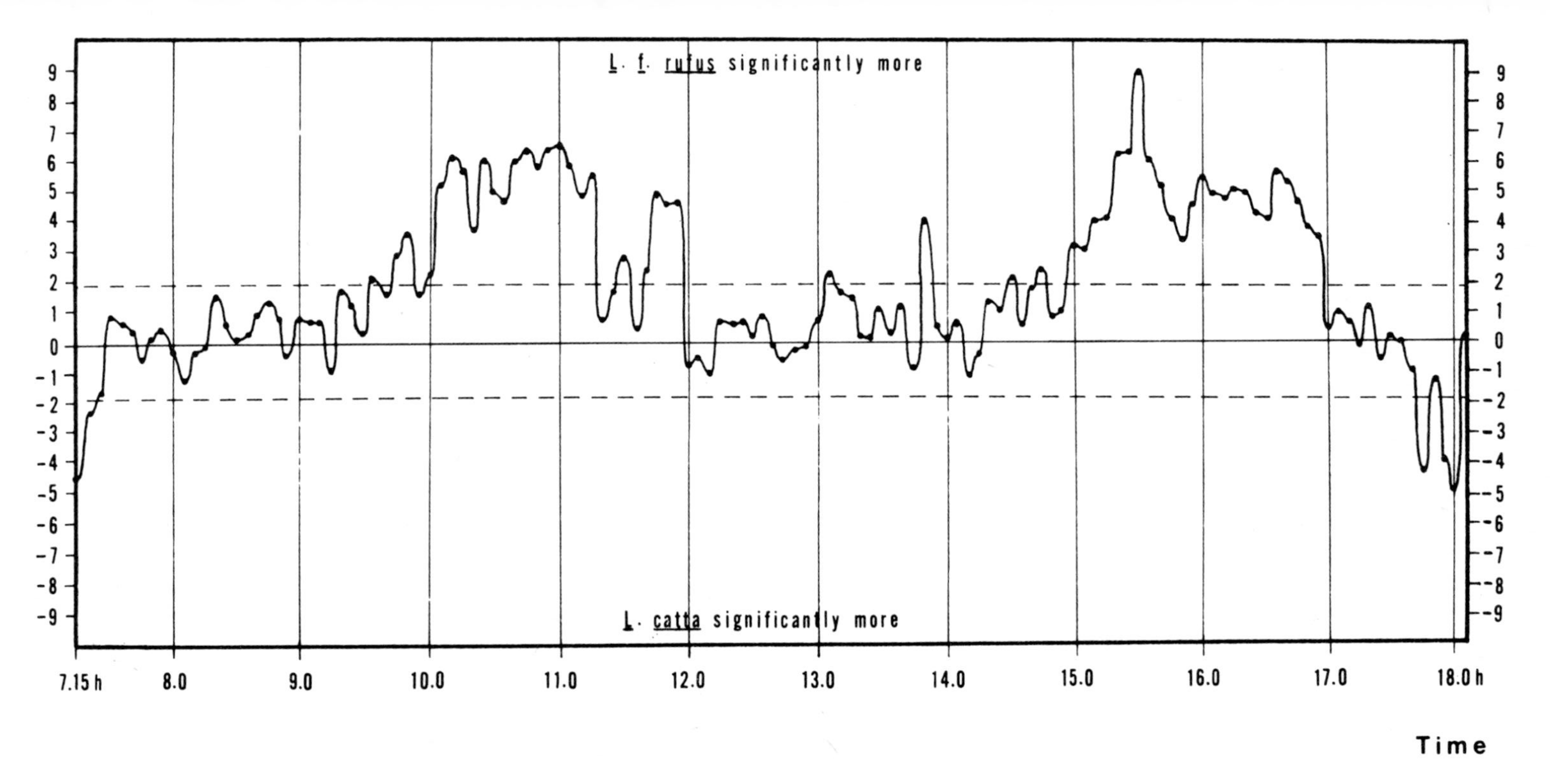

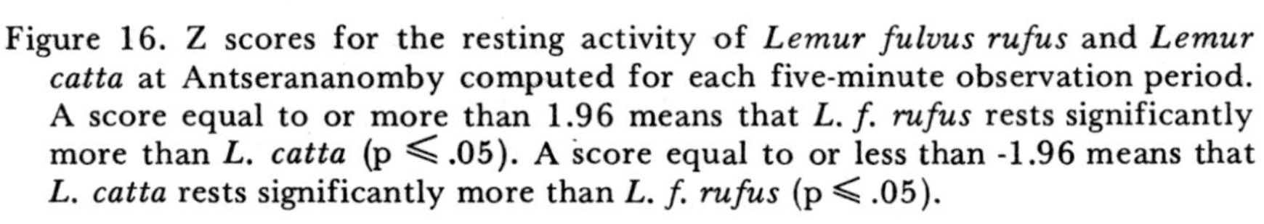
Figure 16. Z scores for the resting activity of *Lemur fulvus rufus* and *Lemur catta* at Antserananomby computed for each five-minute observation period. A score equal to or more than 1.96 means that *L. f. rufus* rests significantly more than *L. catta* ($p \leqslant .05$). A score equal to or less than -1.96 means that *L. catta* rests significantly more than *L. f. rufus* ($p \leqslant .05$).

Statistical analyses comparing data from different forests were difficult to interpret because of the differing forest conditions. Significance tests on the behaviour of populations in different locations only indicate whether or not the behaviour is significantly different. They do not give us any insight into whether differences are due to characteristics of the populations or to environmental conditions.

5. *Group size and composition*

Groups were counted throughout the study period at each forest, and census data on each group were checked and rechecked continually. In all three forests, census data were collected by more than one observer. In most cases, two observers worked together.*

The following criteria were used to distinguish between sex and age classes. There are conspicuous differences between the sexes in *Lemur fulvus rufus.* The male is grey and has a pronounced white face mask with a bright red-orange tuft of hair on the top of the head (Fig. 1). The female is red-orange, with white patches over her eyes and a black bar across the forehead and between the eyes (Fig. 2). These distinctions are obvious even in the juvenile animals. The coloration of infants, however, is often misleading, and sex determination of these young animals is not possible. In *Lemur catta* there are no differences in pelage between the sexes. In order to sex individuals of this species, it was necessary to get into strategic positions to observe the genitalia of the animals. It was not possible to determine the sex of juvenile or infant *L. catta.*

Each individual was placed into one of three age categories – infant, juvenile or adult. Infants were those animals that were still being carried by their mothers. This included animals up to the age of about four to five months. Juveniles were those animals that were no longer being carried, but had not yet reached full size. However, in the field it is often difficult to distinguish older juveniles from adults. Adult size is attained at about two years of age. In both *L. catta* and *L. f. rufus*, births occur once a year within a short period of time (usually a two-week period). The young all mature at approximately the same rate and the achievement of different stages of maturity is synchronous.

Tables 10 and 11 include the census data collected on *L. f. rufus* and *L. catta.* The 17 groups of *L. f. rufus* which I censused ranged in size from 4 to 17 animals. The average size of the groups was 9.5 animals. The average size of the three groups of *L. catta* which I censused was 18.0 animals. Including the counts of Jolly, and Klopfer

* Linda Sussman, Alain Schilling, and field guides Folo Emmanuel and Bernard Tsiefatao assisted me in collecting census data.

Table 10 Composition of groups of *Lemur fulvus rufus*

Name of group	*Adult* *Male*	*Female*	*Juvenile* *Male*	*Female*	*Infant*	*Total*	*Adult Sex ratio M:F*
Antserananomby*							
AF-1	4	6	1	1	0	12	1:1.50
AF-2	4	5	1	0	0	10	1:1.25
AF-3	4	3	0	1	0	8	1.33:1
AF-4	2	5	2	1	0	10	1:2.50
AF-5	4	6	0	2	0	12	1:1.50
AF-6	3	2	0	0	0	5	1.50:1
AF-7	3	4	1	1	0	9	1:1.33
AF-8	2	2	0	0	0	4	1:1.00
AF-9	2	2	1	0	0	5	1:1.00
AF-10	4	5	2	0	0	11	1:1.25
AF-11	5	7	1	2	0	15	1:1.40
AF-12	4	4	1	0	0	9	1:1.00
Totals	41	51	10	8	0	110*	41:51=1:1.24
Means	3.42	4.25	.83	.66	0	9.17	
Tongobato							
TF-1	5	8	0	0	4	17	1:1.60
TF-2	2	3	1	0	1	7	1:1.50
TF-3	4	5	0	1	2	12	1:1.25
TF-4	3	4	0	0	1	8	1:1.33
TF-5	3	2	0	1	1	7	1.50:1
Totals	17	22	1	2	9	51	17:22=1:1.29
Means	3.40	4.40	.20	.40	1.80	10.20	
Overall totals	58	73	11	10	9	161	58:73=1:1.26
Overall means	3.41	4.29	.64	.59	.53	9.47	

* There was also an extra male group consisting of two individuals. Therefore, the total number of animals censused at Antserananomby was 112.

Table 11 Composition of groups of *Lemur catta*

Name of Group	*Adult* *Male*	*Female*	*Juvenile*	*Infant*	*Total*	*Adult Sex ratio M:F*
Antserananomby						
AC-1	7	8	4	0	19	1:1.14
Berenty						
BC-1	4	5	2	4	15	1:1.25
BC-2	7	6	3	4	20	1.17:1
Totals	11	11	5	8	35	11:11=1:1.00
Means	5.50	5.50	2.50	4.00	17.50	
Overall totals	18	19	9	8	54	18:19=1:1.05
Overall means	6.00	6.30	3.00	2.70	18.00	

and Jolly,[6,8] all of the groups of *L. catta* which have been censused consist of 12 to 24 individuals and the average size of the groups is 18.8 animals. Thus, the sizes of groups of *L. catta*, on the average, are approximately twice those of groups of *L. f. rufus*.

Adult sex ratios were similar in groups of both *L. catta* and *L. f. rufus*. There were slightly more adult females than adult males in groups of both species. It is possible that groups of *L. catta* contain proportionately more infants per year than do those of *L. f. rufus*. The ratio of infants to non-infants for *L. f. rufus* at Tongobato was 1/4.66, while that for *L. catta* at Berenty was 1/3.38. Since there were no infants during the time of the census at Antserananomby, the ratio of juveniles (mainly one-year olds) to adults was calculated. This ratio was 1/5.11 for *L. f. rufus* and 1/3.75 for *L. catta*.

A characteristic which is not revealed by the census data is the tendency towards peripheralisation of subordinate males in groups of *L. catta*. This characteristic was associated with a well-defined dominance hierarchy. There was no noticeable hierarchy in groups of *L. f. rufus*. Jolly reported that peripheralisation of subordinate males occurred in the group of *L. catta* she studied intensively at Berenty.[6] The groups I studied also showed a tendency towards the peripheralisation of subordinate males. In one of the groups at Berenty, two adult males were actively kept away from the centre of the group. These two peripheral males were always together and sometimes intermingled with juveniles of the group, but were chased whenever they approached the centre of the group too closely.

Conclusions

Differences were found in the utilisation of space and time, the diet and the social structure of *Lemur fulvus rufus* and *Lemur catta*. These differences were related to the habitat preferences of the two species, and in each case were independent of the presence or absence of the other species. This indicates that the differences are not caused by the interaction between the populations of *L. f. rufus* and *L. catta*, but by adaptations to basically different environmental conditions which occur where the species are allopatric.

The quantitative data on the utilisation of the vertical habitat indicate that *L. f. rufus* and *L. catta* have distinctive preferences for different forest strata. *L. f. rufus* is very specific in its choice of vertical habitat; *L. catta*, on the other hand, is frequently found in all forest layers. In all three forests studied, vertical habitat preferences do not seem to be altered by the presence or absence of the other species. This indicates that the two species do not compete directly for the use of a particular vertical forest stratum. It also suggests that the differences in the choice of vertical habitat are not the result of interaction between the two species.

The day-ranges and home-ranges of *L. f. rufus* are much smaller than those of *L. catta*, and the population density is much higher. The home-ranges of groups of *L. f. rufus* at Antserananomby and Tongobato are restricted to portions of the forests with a continuous canopy. The home-ranges are not rigidly defended and have overlapping boundaries. The high density of the population and the small, overlapping ranges of the groups allow *L. f. rufus* to completely fill these restricted forest areas. This pattern of group distribution may be adaptive for the exploitation of small areas which are relatively rich in food supply and in which the food is evenly distributed.[9]

Populations of *L. catta* are found in the west and south of Madagascar in many areas in which there are no closed canopy forests but only brush and scrub forests. Many of these scrub forests are very dry and the vegetation is patchy and likely to be sparse for at least part of the year. The large home-ranges and day-ranges of groups of *L. catta* may be genetically fixed adaptive responses to these arid environments. Trends in population density and range size similar to those found for *L. f. rufus* and *L. catta* have been reported for forest monkeys and savannah or grassland monkeys, respectively.[9, 10, 11, 12, 13]

Differences in the diet of *L. f. rufus* and *L. catta* can also be related to differences in the habitat preferences of the two species. Although both seem to be opportunistic in their choice of food, the less restricted use of the vertical and horizontal habitat by *L. catta* allows this species a greater opportunity to obtain a more varied food supply. *L. f. rufus*, having a more limited vertical and horizontal range, has a more specialised relationship to the environment.

The foraging habits of *L. f. rufus* and *L. catta* seem to be adaptations to different econiches. *L. f. rufus* inhabits closed canopy forests, which in most cases are found in small, relatively uniform, circumscribed areas. In these forests, the population density of *L. f. rufus* is extremely high. The ability to exist almost exclusively on a diet of kily leaves, supplemented with the fruit of those trees which happen to be within the range of any one group, allows *L. f. rufus* to exploit efficiently this type of environment. A group of *L. catta* continuously forages within its range and exploits a number of plants in a large area. Within a period of seven to ten days, the group visits most of its total range. It is likely that this continuous foraging behaviour may be adaptive to arid environments in which the food supply is sparse. The constant surveillance of a large (or relatively large) range allows groups of *L. catta* to exploit a number of different plants which may have a patchy distribution over a wide area.

The utilisation of time by the two species further indicates that behavioural adaptations of *L. f. rufus* and *L. catta* have developed in response to different environmental conditions. *L. f. rufus* feeds early in the morning and late in the afternoon; it rests throughout the day and spends almost all of the warmest portion of the day in the closed canopy. The temperature of the micro-environments in which *L. catta* is found throughout the day is more likely to vary. At night, *L. catta* sleeps in the closed canopy portion of the forest. However, during the day it spends much of the afternoon in unshaded portions of the forest. In these areas, the ambient temperature is usually quite high. At Antserananomby, where the evening temperatures were low, *L. catta* sunned for long periods during the morning.

The distribution of *L. catta* within the forests in which it was studied, and the fact that it is found in the hot, arid regions of southern Madagascar, indicate that this species is well adapted to warm and dry climates. *L. f. rufus* is not found in the south of Madagascar, and it remains in the shaded portion of the forest throughout the day. Furthermore, *L. f. rufus* is active only during the coolest hours of the day. It seems likely that *L. catta* uses behavioural thermo-regulation (e.g. sunning) to help maintain its body temperature and in this way is able to utilise habitats with large ranges of ambient temperature.

The sizes of groups of *L. catta* are, on average, approximately twice those of groups of *L. f. rufus*. Terrestrially adapted species have generally been considered to form larger groups than arboreally adapted species.[9,10,11] There are, however, several exceptions, and phylogenetic relationships also play a role in the determination of group size.[12,14] Before relationships can be established between ecology and group size or social structure, further studies must be made on arboreally adapted species and their use of the environment.

There is a tendency in groups of *L. catta* towards peripheralisation of subordinate males. In a very dry environment, where food is sparse, the exclusion of excess males could be very adaptive for the survival of the group.[15,16] A study of *L. catta* in arid regions may reveal groups containing many more adult females than males.

In summary, *L. f. rufus* seems to be a very specialised, arboreally adapted species. *L. catta* is able to exploit a number of different habitats. However, the behavioural adaptations which have been found in *L. catta* seem to have developed in response to the more arid, warm environments in which this species is found (where resources are likely to be limited). *L. f. rufus* shows a configuration of behavioural adaptations which conform to those found in many arboreal species of New and Old World monkeys, whereas some behavioural adaptations of *L. catta* parallel those of many terrestrial primates living in savannah or grassland econiches.

Acknowledgments

I would like to thank Dr Brygoo of l'Institut Pasteur de Madagascar, Président M. Manambelona, Secrétaire Général du Comité National de la Recherche Scientifique, M. Ramanantsoavina, Directeur des Eaux et Forêts, and the staff of ORSTOM, Tananarive for their assistance and cooperation while I was in the field. I am also indebted to many people of the villages of Manja and Vondrove, whose kind hospitality is deeply appreciated. I am especially grateful for the continued assistance (both academic and otherwise) offered by Dr J. Buettner-Janusch throughout the period of this study. My wife, Linda, aided me in all aspects of the research and I am, of course, especially indebted to her. The study was supported in part by Research Fellowship MH46268-01 of the National Institute of Mental Health, United States Public Health Service and by a Duke University Graduate Fellowship.

NOTES

1 Brown, W.L. and Wilson, E.O. (1956), 'Character displacement', *Systematic Zool.* 5, 49-64.

2 Sussman, R.W. (1972), 'An ecological study of two Madagascan primates: *Lemur fulvus rufus* and *Lemur catta*', PhD thesis, Duke University.

3 Perrier de la Bathie, H. (1921), 'La végétation malgache', *Ann. Mus. Colon. marseille*, 3rd ser., 9, 1-268; Humbert, H. (1954), 'Les territoires phytogéographiques de Madagascar: leur cartographie', *LIX° Colloque International du Centre National de la Recherche Scientifique*, Paris, pp. 439-48; Humbert, H. and Darne, G.C. (1965), *Notice de la Carte de Madagascar*, Toulouse.

4 Crook, J.H. and Aldrich-Blake, P. (1968), 'Ecological and behavioural contrasts between sympatric ground-dwelling primates in Ethiopia', *Folia primat.* 8, 192-227.

5 Morton, D.J. (1924), 'Evolution of the human foot', *Amer. J. Phys. Anthrop.* 7, 1-52.

6 Jolly, A. (1966), *Lemur Behavior*, Chicago.

7 Charles-Dominique, P. and Hladik, C.M. (1971), 'Le *Lepilemur* du sud de Madagascar: écologie, alimentation et vie sociale', *Terre et Vie* 25, 3-66.

8 Klopfer, P.H. and Jolly, A. (1970), 'The stability of territorial boundaries in a lemur troop', *Folia primat.* 12, 199-208.

9 Crook, J.H. and Gartlan, J.S. (1966), 'Evolution of primate societies', *Nature* 210, 1200-3.

10 DeVore, I. (1953), 'A comparison of the ecology and behaviour of monkeys and apes, in Washburn, S.L. (ed.), *Classification and Human Evolution*, Chicago, 301-19.

11 Aldrich-Blake, F.P.G. (1970), 'Problems of social structure in forest monkeys', in Crook, J.H. (ed), *Social Behaviour in Birds and Mammals*, New York, 79-101.

12 Crook, J.H. (1970), 'The socio-ecology of primates', in Crook, J.H. (ed), *Social Behavior in Birds and Mammals*, New York, 103-66.

13 Denham, W.W. (1971), 'Energy relations and some basic properties of primate social organisation', *Amer. Anthrop.* 73, 77-95.

14 Struhsaker, T.T. (1969), 'Correlates of ecology and social organisation

among African cercopithecines', *Folia primat.*, 11, 80-118.

15 Kummer, H. (1967), 'Dimensions of a comparative biology of primate groups', *Amer. J. Phys. Anthrop.* 27, 357-66.

16 Kummer, H. (1971), *Primate Societies*, Chicago.

S.K. BEARDER and G.A. DOYLE

Ecology of bushbabies Galago senegalensis *and* Galago crassicaudatus, *with some notes on their behaviour in the field*

Introduction

The lorisoid primates of Africa and south-east Asia, unlike the lemurs of Madagascar, cohabit with monkeys, but – being exclusively nocturnal – do not compete ecologically with them. They have retained many characteristics of their primitive mammalian ancestors and remain similar to the most ancient primates. The two subfamilies are best characterised by essential differences in the mode of locomotion of their representatives; the Lorisinae being slow-moving climbers, while the Galaginae are described as active leapers. The pattern of locomotion of at least one species of the Galaginae, *Galago crassicaudatus*, is midway between the two types.

In Africa, the family Lorisidae comprises seven species, five of which may be found occupying the same habitat in tropical West Africa (Gabon).[1] Ecological localisation results in virtual absence of interspecific competition. The remaining two species, which are the subjects of the present study, are found over a large area of Africa south of the Sahara, and may also occasionally be found together, in regions where their characteristic habitats overlap.

Field research was begun in South Africa and Rhodesia in 1968, using red torchlight to make continuous direct observations of bushbabies at all hours of the night. Over 1,200 observation hours have been made to date, in a number of geographical regions and habitat types. The animals soon became habituated to the presence of the observer and allowed his close approach. Separate and detailed studies of each species during 12-month periods were made in two main study areas. Further information was collected in areas where both species occurred together. A method of trapping was devised which made it possible to mark and release a number of animals for long-term studies of population dynamics, individual movement, use of space and social behaviour.

Habitat and distribution

1. Galago senegalensis

The lesser bushbaby (maximum weight 300 gm.) is to be found in a wide range of habitats from sea level to 1,500 m. These include semi-arid, steep-sided valleys, open woodland, orchard bush or scrub and isolated thickets with grassland. This species also occurs in the primary forests of mountain regions, river valleys and coastal areas. Its distribution is restricted by open, or relatively treeless, areas and is often associated with *Acacia* thorn trees, particularly *A. karoo*.

Observations of a single sub-species were made in the northern Transvaal, including the Springbok Flats and the Waterberg mountains; the north-eastern Transvaal, comprising the lowveld in the east and the Drakensberg escarpment in the west; and the eastern Highlands of Rhodesia.

Mean temperature figures varied considerably over this entire range, with a minimum just below freezing and a maximum of around 37°C. The rainfall showed a marked fluctuation from year to year, with mean figures per annum of 599, 722 and 1463 mm. for the three areas, respectively. Rainfall was mainly confined to the summer months between October and April, although rain was often recorded in any month. The actual subspecies studied was *G.s. moholi* A. Smith (Moholi Galago): a large-eared race, mainly grey in colour, but with the lower back washed with otter-brown which distinguishes it from typical *senegalensis* and other grey-bodied northern races. It has a thin tail, showing no tendency to bushiness.[2] It is found throughout the wooded and bushveld areas of South West Africa, Angola, Zambia, Botswana, extending east through southern Transvaal, and north through Rhodesia as far as Tanzania.[3]

2. Galago crassicaudatus

The thick-tailed bushbaby (maximum weight 1,800 gm.) has a more restricted distribution than the former species. It occurs between sea level and 1,800 m., but is mainly confined to dense evergreen indigenous forest and riparian bush, where there is an abundance of trees bearing edible fruits. The species may also be found in stands of plantation timber such as blue gum, black wattle or pine, where these adjoin natural bush. Such vegetation occurs in well-watered coastal regions, or inland as montane forest in steep-sided valleys and on hillsides watered by run-off from the mountain catchment areas. Much of this habitat has been exploited by man in the past and replaced by comparatively sterile plantations, but a considerable number of Government-protected forest areas still remain. As with the former species, *G. crassicaudatus* has a wide range of temperature tolerance.

Comparative information was collected on three subspecies in three different habitat types: (*a*) *G.c. umbrosus*, (*b*) *G.c. garnetti* and (*c*) *G.c. lönnbergi*, as described by Osman Hill.[2]

(*a*) *G.c. umbrosus* (Thos.), Transvaal Dusky Galago.
This subspecies may be distinguished by its brown colouration, particularly dark on the dorsal surfaces of the hands and feet. A certain proportion of the study population (8%) was melanistic, having a slightly darker overall hue and dark tips to the tails. One dark female gave birth to light-coloured offspring. Limits of distribution of this race are undetermined. It inhabits the forest areas and riparian bush of the Drakensberg foothills in the north-east Transvaal, where long term observations were carried out.

(*b*) *G.c. garnetti* (Ogilby), Garnett's or Black Galago.
A brown race, only the digits being blackish and with 80% of the study population having dark tips to their tails to a varying extent. The subspecies is found in Natal and Zululand, but limits of the range remain undetermined. Observations were made in dense coastal Dune Forest with a heavy understorey and many climbers, and also in Temperate forest with a canopy approximately 20-30 m. high.

(*c*) *G.c. lönnbergi* (Schwarz), Lönnberg's Galago.
A light-coloured variety having no dark tip to the tail and being somewhat smaller in size. There is no dark colouration on the dorsal surfaces of the hands and feet. It occurs in south-east Africa, from the Zambesi in the north as far south as the Limpopo. The exact western limit of its range is undefined. This subspecies was observed in relatively open bush in and around Umtali, Rhodesia.

The three most southern forms of *G. crassicaudatus*, together with four other races, are larger in size than the northern group, which comprises four subspecies.[2]

Other primates which occur in the regions described are: the Vervet monkey, *Cercopithecus aethiops*; the Samango monkey, *Cercopithecus mitis labiatus*; and the baboon, *Papio ursinus*.

Population density

On the basis of direct counts made over several consecutive nights in each of the different habitats, calculations have been made on the population densities of the two species (Table 1). In open bush it was possible to count practically all the members of a population within a set area with little chance of error, due to the extremely reflective nature of the tapetum lucidum of the bushbaby eye. This could be seen at a distance of more than 100 m. Counts of the lesser bushbaby were also made during the day by finding their sleeping places. In forest habitats, counting was done along transect lines coinciding with forest paths. In these cases, allowances have been made for the effective scanning area of the torch beam in relation to the density and height of the vegetation.

Table 1 Population densities of the two bushbaby species

G. senegalensis

Area	*Habitat*	*Density / square km.*
N. Transvaal	*Acacia* thornveld	200
N. Transvaal	*A. karoo* thickets	500
N. Transvaal	Wooded valley	275
Rhodesia	Mixed woodland and plantations	87
N.E. Transvaal	Lowveld: riparian bush and savannah	103
N.E. Transvaal	Escarpment: riparian bush and scrub	95

G. crassicaudatus

Area	*Habitat*	*Density / square km.*
Zululand	Dune forest	125
Zululand	Temperate forest	112
Rhodesia	Mixed woodland and plantations	110
N.E. Transvaal	Lowveld: riparian bush and savannah	72
N.E. Transvaal	Escarpment: riparian bush and scrub	88

The density calculations are indications of the situation in areas where the vegetation was of a more or less uniform type, in which the animals were distributed fairly evenly. However, due to the uneven or interrupted nature of the bush, the bushbabies were frequently separated into small aggregations. Where the change of vegetation was even more marked, populations of bushbabies were virtually or completely isolated from other populations, and this situation may account for the great variety of subspecific characters which have been described.

There is a marked disparity between the population densities calculated here and those of the remaining five African lorisids calculated by Charles-Dominique in Gabon, using similar techniques.[4] This may be attributed to the fact that far fewer mammal species live sympatrically in South Africa and Rhodesia than in tropical West Africa.

In the north-east Transvaal, where the two species lived sympatrically their population densities were lower than for those regions where they occurred alone. This may be a result of interspecific competition or of the fact that the vegetation did not represent the optimum habitat for either species. In fact, competition was minimal since each species showed preference for different parts of the habitat. *G. senegalensis* spent more time in open 'orchard' bush, or at the periphery of the riparian bush and forest which was most utilised by *G. crassicaudatus.* No direct competition through antagonism or fear was shown by either species during interspecific encounters. Only in Rhodesia, where *G. crassicaudatus* had adapted to a much more open environment than elsewhere, were there indications of competition for food, and here the population density of *G. senegalensis* was surprisingly low.

The highest population densities of *G. crassicaudatus* were found in humid, forest regions, while *G. senegalensis* was most abundant in arid thornveld. Some of the factors which bring about ecological separation are outlined below.

Social organisation

The majority of nocturnal prosimians so far studied, in contrast to other primates, are predominantly solitary in their habits. Their social behaviour may therefore be considered as being less complex than in animals which live together in larger groups forming social units. Both *G. senegalensis* and *G. crassicaudatus*, however, formed relatively stable associations throughout the year and familiar individuals often slept together during the day. Social interactions included all grooming, play, sexual behaviour, and quite complex patterns of communication. Each species made approximately 20 calls of social significance.[5,6]

Both species of bushbaby slept either alone or in small family groups of 2-6 animals, which often included an adult male, an adult female, and the offspring of one or more generations. It was found that the males and maturing offspring would often sleep alone, while the females and youngsters would usually sleep together.

Occurrence of groups in: (*a*) *G. senegalensis* (N = 119), and (*b*) *G. crassicaudatus* (N = 148) was as follows:

Number of animals in group:		1	2	3	4	5	6
% frequency of occurrence:	(*a*)	40	30	23	5	1	1
	(*b*)	46	26	18	6	3	1

At night, the members of sleeping groups behaved differently in the two species:

1. G. senegalensis

Individuals which had spent the day together generally split up at night and sometimes did not come together again until just before dawn. Social contact, including that between a mother and her offspring, was typically brief, with exceptions during courtship and when juveniles had developed sufficient agility to be able to follow their mothers before their movements became independent. Records show that bushbabies of this species were alone on average for 70% of each night.[7]

2. *G. crassicaudatus*

This species was somewhat gregarious in that the female and her offspring moved together at night as a cohesive group. Even when the young were too small to follow the mother, she carried them from place to place, either transporting them in her mouth one at a time or with them clinging to the fur of her back. The adult male generally remained alone, but he occasionally joined the family group and moved with them. Even larger groupings could be formed by the addition of previous offspring or members of other groups, the latter usually being the result of attraction to a particular food source.

The social systems of *G. senegalensis* and *G. crassicaudatus* are comparable to those described for *Lepilemur mustelinus* by Charles-Dominique and Hladik,[8] and for *G. demidovii, G. alleni, Euoticus elegantulus, Perodicticus potto* and *Microcebus murinus* by Charles-Dominique and Martin.[9] The animals are loosely organised into 'family groups', which may be defined as comprising those bush-babies whose individual ranges coincided or overlapped to a large extent and who would share the same sleeping place, at least on occasions.

It was found that the size, shape and number of home-ranges changed gradually during the year. It appears that as the young males matured their ranges became increasingly separate from those of the parents and they began to show territorial behaviour in the form of olfactory marking, vocal advertisement and antagonism towards rival conspecifics. Pair bonds developed with females, which also showed territorial behaviour, and the larger and stronger males were usually able to establish their home-ranges over those of one or several females. In this way, social groups were formed, having communal home-ranges, while the smaller males were rejected to the periphery. Within an expanding population, such a social system would cause a species to become spread throughout the available habitat and prevent overcrowding in any one place. It is suggested by Charles-Dominique[10] that the social structures of monkeys and other social mammals may be derived from this primitive type.

Use of space

Measurements were made of the home-range sizes of 'family groups' in the two main study areas. *G. senegalensis* had an average range size of 2.8 hectares while that of *G. crassicaudatus* was 7 hectares. The

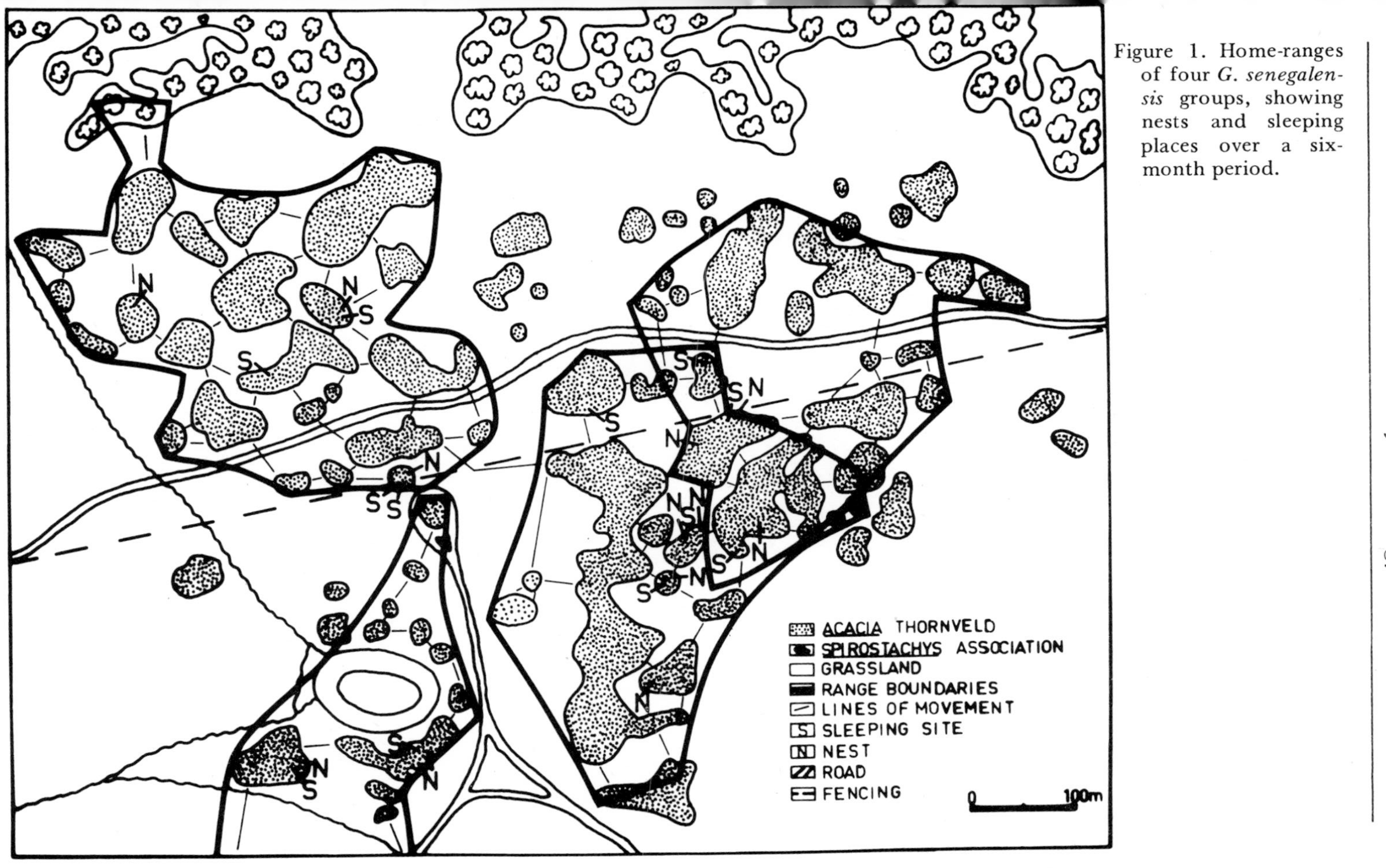

Figure 1. Home-ranges of four *G. senegalensis* groups, showing nests and sleeping places over a six-month period.

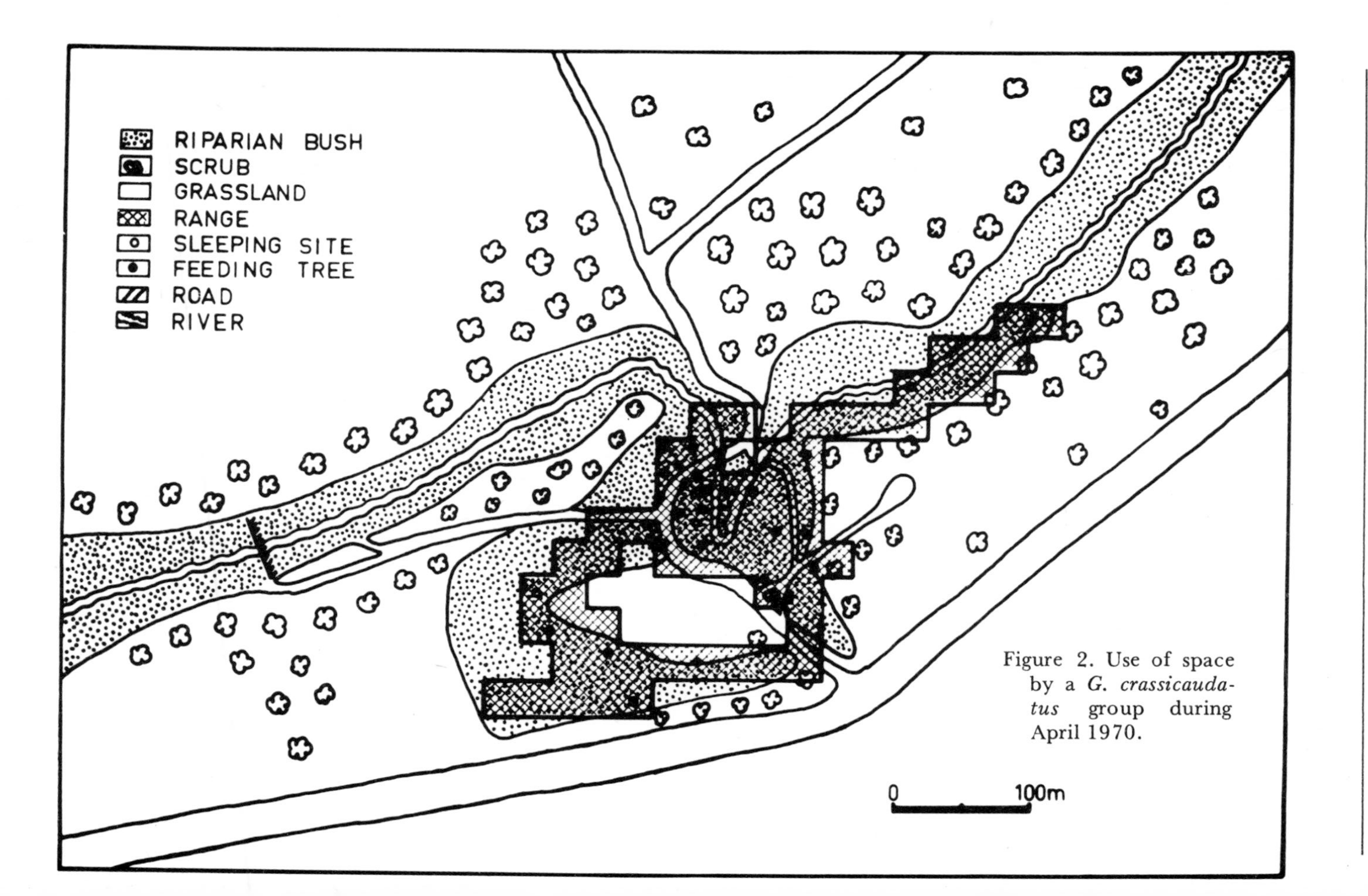

Figure 2. Use of space by a *G. crassicaudatus* group during April 1970.

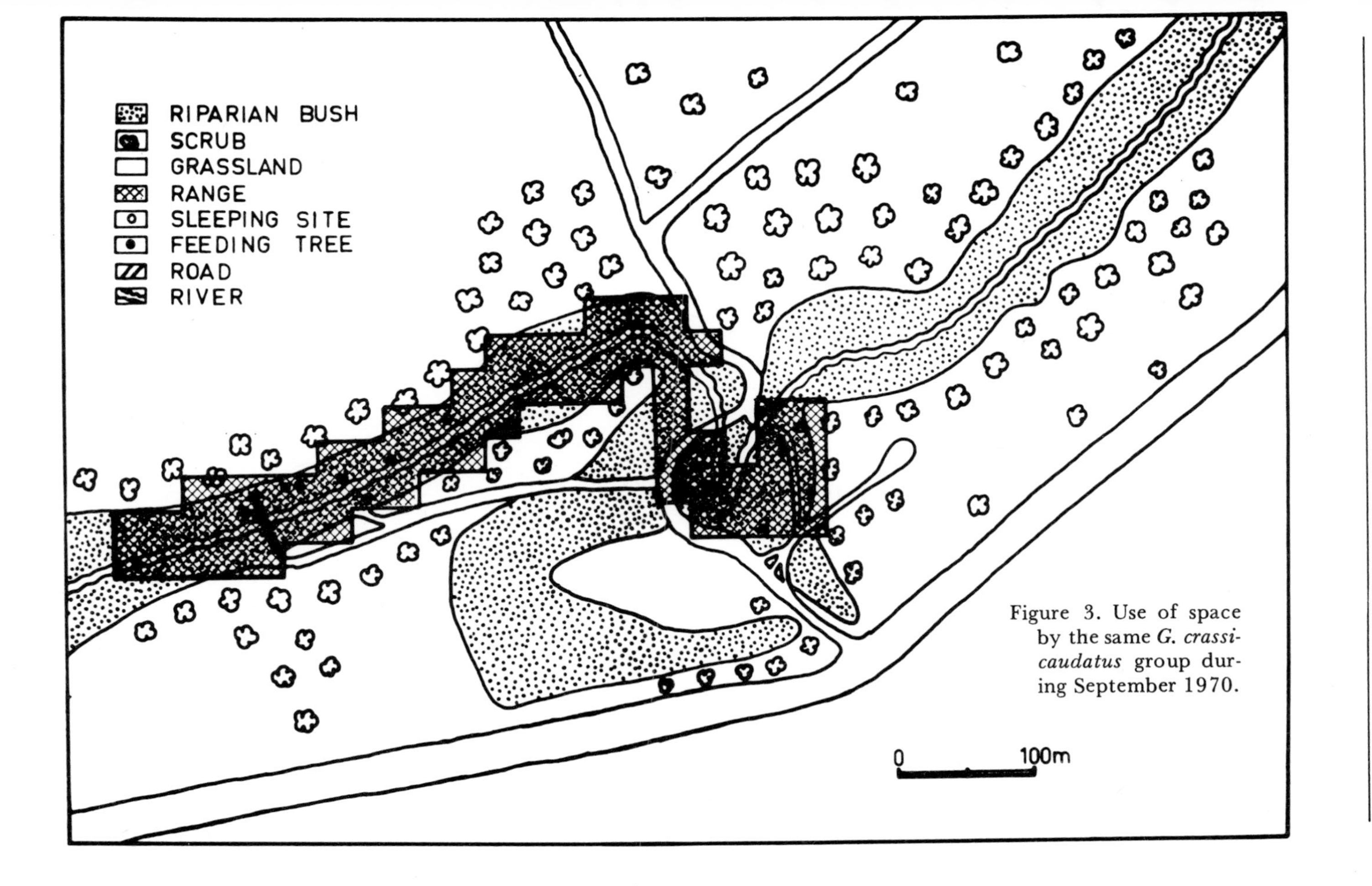

Figure 3. Use of space by the same *G. crassicaudatus* group during September 1970.

home-ranges of neighbouring groups were separate or overlapped to a small extent (Fig. 1).

The use of space by each animal may be analysed using a grid system (Figs. 2 and 3). The length of time which was spent in each square of the grid may be calculated for a single night and for consecutive nights at different times of the year. Activity is found to be centred around particular trees or groups of trees which provided sleeping sites, food sources or resting places, the use of which varied during the year depending upon their suitability. Such trees were not uniformly distributed throughout the range, and the centres of activity were therefore found at various unrelated points.

Bushbabies were active at all levels within their habitat, but the amount of activity at different levels varied between one habitat and another, depending on the spacing of trees and the availability of food. Those animals of both species which lived in open woodland descended to the ground readily in order to cross open spaces of up to 100 m. or to forage. In forest environments they only descended occasionally and with greater reluctance.

Each species used a variety of supports, but *G. senegalensis* – owing to its small size – could move along fine, closely-spaced and often very thorny branches without difficulty. It was thus able to exploit a large number of habitat types, and to a greater extent, than was *G. crassicaudatus.*

Movement and locomotion

Movements within the home-range were in accordance with the observations of Brown on other small mammals.[11] They were directional from one point to another, following habitual 'pathways'; but the routes taken and the exact movements made were often varied depending upon circumstances, indicating that the animals had a thorough knowledge of their entire range. One or more circular excursions were sometimes made during a night, starting at one point and returning to that point from the opposite direction some hours later. Alternatively, the same route was used to return as that used on the outward journey.

The movements of *G. crassicaudatus* differed from those of *G. senegalensis* in being generally more stereotyped. The pathways were followed more strictly, with similar movements being made on consecutive occasions. The more haphazard movements of *G. senegalensis* may be an adaptive feature related to its insectivorous diet. The food sources of *G. crassicaudatus*, which is predominantly frugivorous, were at fixed points within the home-range, and movements were made directly from one point to another. However, this did not prevent fairly continuous exploration of the home-range, so that food trees which were only productive at certain times of the

Table 2 A comparison of locomotion and diet in *Galago senegalensis*, *G. crassicaudatus* and *Perodicticus potto* [1,2]

Species	*Primary method of locomotion*	*Secondary method of locomotion*	*Movement on ground*	*Diet*
Galago senegalensis	Rapid leaping, using long hind legs as propulsive force	Slower 'stepping' movement within a tree	'Kangaroo-like', forelimbs not used	Insects and gum
Galago crassicaudatus	Quadrupedal running with occasional short jumps, or longer ones between trees, using hind legs as propulsive force	Slow, stealthy and quiet movement, sometimes beneath a branch. Weight transferred evenly forwards	1. 'Kangaroo-like', forelimbs not used 2. 'Galloping' action with both fore and hind feet together 3. Walk or run. Hindquarters and tail in the air	Fruit and gum Occasional slow-moving insects and birds
Perodicticus potto	Slow, stealthy and quiet movement, sometimes beneath a branch. Weight transferred evenly forward	Occasional running. No leaping	Walk. As along a branch	Fruit, gum, slow-moving insects and occasional small vertebrates

Figure 4. Locomotor postures in *G. senegalensis* (sketches 1 to 8) and *G. crassicaudatus* (sketches 9 to 14).

Figure 5. Potto-like movement in *G. crassicaudatus.*

year were not missed. Differences between the locomotor patterns of the two species may also be related to their diets.

If a comparison is made between the extreme form of active leaping locomotion shown by *G. senegalensis* and its associated methods of obtaining food with that of *G. crassicaudatus* and the slow moving *Perodicticus potto*.[12] it can be seen that the type of locomotion and feeding behaviour of *G. crassicaudatus* lies somewhere between that of the other two species, showing characteristics of both (Table 2; Figs. 4 and 5).

The most common method of locomotion in this species was not the rapid leaping gait characteristic of other galagos, but a 'monkey-like' running or walking along the tops of branches, with only short jumps when moving rapidly. Longer jumps were made when the occasion demanded, but these were generally in a downward direction with little forward propulsion and they differed from those of *G. senegalensis* in their execution.

It has been shown from the fossil remains of lorisoids from the Miocene of east Africa by Walker[12] that the leaping form of locomotion is a primitive state in the African prosimians, while stealthy locomotion is assumed to be a post-Miocene development, which makes the lorises dependent upon true forest conditions. It is of interest to note that while *G. crassicaudatus* may be found in relatively open areas it is more typically confined to dense bush and forest habitats, at least in the southern parts of its range.

Marking and rubbing behaviour

Six types of marking or rubbing actions have been observed which may carry olfactory clues of importance in territorial and social behaviour (Table 3). Some field notes are recorded here on the more conspicuous patterns, while the possible functions of marking actions are discussed by Doyle, and Andrew and Klopman.[13,14]

1. *G. senegalensis*

(*a*) *Urine-washing.*
This is a habit common to several nocturnal prosimian species and is described in detail elsewhere.[13,14] It was seen in all age classes and occurred several times in quick succession at the beginning of the active period, but thereafter only a few times during each hour of the night. Urine-washing at particular spots, apart from around the sleeping place, was not observed, and the animals showed a lack of obvious 'sniffing' when moving quickly through their home range. The idea of scent 'trails' has been discounted.[7,13] The deposit of urine on the branches may inform conspecifics of the presence, status and sexual condition of the marker, or it may be of significance to the animal itself. Adult males sometimes urine-washed in contact with conspecifics and in this way may have transferred their scent onto members of their 'family group'.

(*b*) *Chest rubbing.*
Chest rubbing was only performed by adult males, which had a triangular bald patch on their chests. It has yet to be established whether this patch incorporates glandular tissue. Rubbing, which was sometimes accompanied by biting of the branch, occurred in a similar fashion and context to that described for *G. crassicaudatus.* Rubbing of the head and mouth was also observed, but the significance of these movements is not known.

2. *G. crassicaudatus*

(*a*) *Urine-washing.*
Urine-washing was similar to that of *G. senegalensis.* It was performed by all age classes and presumably has the same significance. Both species were observed to urinate without urine-washing or rhythmically micturating.

(*b*) *Chest rubbing.*
Both adult males and females have a chest gland which may be seen as a longitudinal bald patch in the middle of the chest. The patch is larger and more conspicuous in the males. Chest rubbing was performed in a variety of postures, depending on the size and position of the branch or trunk which was being rubbed. The pattern was often executed beneath a branch in an upward-sliding movement. Particular places within the home-range were rubbed on repeated occasions, with a preference being shown for small upright projections and stumps of branches in large trees. Rubbing was frequently accompanied by sniffing of the branch, leg-rubbing, ano-genital rubbing and biting. It occurred during social encounters, when several places within a small area were sometimes marked one after another. Most chest rubbing observed was done by males, and it would appear to be one aspect of territorial behaviour in the species.

(*c*) *Ano-genital rubbing.*
Ano-genital rubbing was observed infrequently. The chest and whole underside of the body, including the genital region and sometimes the anus was rubbed along a branch in a forward sliding motion. It was sometimes performed beneath a branch with the legs and feet free on either side and the ano-genital region pressed against the branch by the pull of the arms. It was observed for both adult males and females and presumably has territorial or sexual significance.

Table 3 Marking and rubbing behaviour shown by age/sex classes of *G. senegalensis* and *G. crassicaudatus*

No.	*Behaviour*	*Species*	*Age class*	*Sex*
1	Urine-washing	*G. s.*	Ad. Juv. Inf.	Male and female
		G. c.	Ad. Juv. Inf.	Male and female
2	Rhythmic micturition	*G. s.**	Ad.	Male and female
		*G. c.**	Ad.	Male and female
3	Chest rubbing (median chest gland)	*G. s.*+	Ad.	Male
		G. c.	Ad.	Male and female
4	Head and mouth rubbing	*G. s.*	Ad.	Male and female
5	Ano-genital rubbing	*G. s.**	Ad.	Male and female
		G. c.	Ad.	Male and female
6	Leg rubbing	*G. c.*	Ad. Juv.	Male and female

* Not noticed under conditions of observation in the field, but observed in the laboratory.[5,6]

\+ Existence of glandular tissue not definitely established.

(*d*) *Leg rubbing.*
Leg rubbing was a common behaviour pattern for which it is difficult to find an explanation. It involved a rapid scraping of the basal part of the hind limb and sole of the foot against a branch or sloping trunk. This action was then often repeated with the other leg. The leg was moved noisily back and forward several times, sometimes with the foot clasped gently round the branch. Leg rubbing was done by juveniles as well as adults, but was seen most often with adult males.

Despite the similarity of this action with that of scent-marking, and its possible significance in transferring urine from the soles of the feet onto the support, it did not show any temporal relationship with scent-marking. It was common during social encounters, together with chest rubbing, and occurred when the subject was confronted by the observer or a potential predator, indicating that it may be a displacement activity. The sound of scraping may serve a communicatory function.

Sleeping habits

At the end of each night's activity, the members of a group returned to particular sleeping trees which were used at various times during the year, with some of them being favoured more than others. Some trees were used by several generations.

Use of sleeping places by (*a*) *G. senegalensis* (N = 168), and (*b*) *G. crassicaudatus* (N = 35), were as follows:

Type of sleeping site:		*Nest*	*Branch/Fork*	*Building*	*Hollow*
% frequency of use:	(*a*)	52	40	2	6
	(*b*)	8	91	1	0

1. *G. senegalensis*

Members of this species were found to sleep in nests or on branches and forks within a single habitat with equal regularity. Buildings and hollow trees were also used. Sleeping places were usually between 4 m. and 6 m. above the ground in dense, thorny trees (*Acacia tortilis; A. nilotica*), and the nests (which were of a size corresponding to the number of occupants) consisted of platforms constructed of flat leaves taken from neighbouring trees. The nests differed from those of *G. senegalensis* in South-West Africa, which were globular in shape and covered on top.[15]

Up to 13 sleeping places were used in any one home-range during a year. Nests were often renewed within the same tree or close by and the animals frequently slept in the nest tree but not on the nest, so that records of the positions of sleeping sites showed that they occurred in clusters.

During the year, with changes in temperature and leaf-cover, particular trees became more or less suitable for sleeping purposes, and this was reflected by a change in their use. It was not uncommon for more than one sleeping place to be used on consecutive days or by different individuals within one home-range.

The choice of sleeping place seems to depend on several factors, some of which may be sacrificed at the expense of others. These included: direct protection from predators and extremes of temperature; indirect protection from predators by concealment; and comfort.

Although activity at night was unaffected by weather conditions, voluntary movement during the day appeared to occur in response to changes in temperature. Cover was sought during the heat of the day, particularly in the summer, while the animals often slept in exposed conditions in winter.

Laboratory data indicates that nests are built by females during a period just before giving birth and for some time afterwards.[16] Nests were found in the field during the spring and summer at the beginning of the birth season when suitable nesting material became available. A few infants, which were born just before suitable nesting leaves could be obtained, were found in old birds' nests and hollow trees. The nests were probably used as platforms on which to give birth and to provide support and camouflage for the young, and thereafter they continued to be used even though they were no longer essential for support of the infants. Haddow and Ellice note that in east Africa nesting is rare and appears to be a vestigial habit.[17]

2. G. crassicaudatus

This species also used more than one type of sleeping place. The use of hollow trees was not observed, but has been recorded by Astley-Maberly.[18] In most cases, bushbabies concealed themselves in dense tangles of branches and creepers at a height of between 5 m. and 12 m. above the ground. A number of different trees were rendered suitable by the fact that they supported dense growths of climbers such as *Dalbergia armata* and *Pteralobium* sp. Only occasionally did the animals sleep on branches which left them exposed. Nests were only seen when a female had young infants, and consisted of inaccessibly leafy platforms, somewhat depressed in the centre and sheltered from above by foliage.

Up to 12 sleeping places were found within each home-range and, as with those of *G. senegalensis*, they occurred in clusters and their use followed a similar pattern. Movement away from the sleeping place during the day was not observed.

Field observations indicate that *G. senegalensis* relies on its ability to move rapidly away from danger should it be disturbed during the day. *G. crassicaudatus*, on the other hand, generally sleeps in confined places, relying on concealment to avoid detection by potential predators.

Birth periodicity

Climatic factors

Birth records of *G.s. moholi* were made in an area (Lat. 24°35′, Long. 28°47′) only 240 km. from the recording area for *G.c. umbrosus*, which lies a few degrees north of the Tropic of Capricorn. These areas have a seasonally arid climate (Table 4). The winters (June to September) are dry and sunny with a large daily temperature range (up to 17°C). The hot summer rainy season occurs between October and April, with thunderstorms contributing a large proportion of the total precipitation, although spells of continuous rain are also recorded. The distribution and annual total of rainfall is highly variable from year to year. The climate is affected by the varied local topography and aspect. All valleys are subject to winter temperature inversions and the daily temperature range may cause marked changes in the relative humidity, dew occurring in autumn following the summer rains. There is considerable dying-back of the vegetation during the winter and leaf growth begins again before the first good rains, with sometimes severe consequences if they are late. Long dry spells may occur between rains.

1. G. senegalensis

In the study area for this species, temperatures may range between -6°C and +38°C with an annual absolute mean of 15.5°C. Only 464 mm. of rain was recorded during the year of study (135 mm. below average), with 449 mm. falling between the months of October and April. Nesting material was not available until November, which may have been an important factor influencing the survival of infants.

The occurrence of births in the laboratory has been described by Lowther, Sauer and Sauer and Doyle et al.[15,16,19] The gestation period has been established as between 120 and 126 days, and twins are usually produced after the first birth, which is often a singleton. Post-partum oestrus is common. In the main study area, 13 sets of twins and 7 singletons were observed.

Restricted breeding has often been claimed or suggested for *G. senegalensis*, while Manley has shown the potentiality of this species for recurrent oestrus cycles.[20] Butler notes that *G. senegalensis* appears to have an oestrus cycle lasting four to six weeks recurring

Table 4 Mean temperatures (t, °C), precipitation (p,mm) and relative humidity (h,%) for the study areas of (a) *G. senegalensis* and (b) *G. crassicaudatus*

(a) *Month*	*t,°C* *Max.*	*Min.*	*p,mm.*	*h,%* *0800 hr.*
J	29.4	17.0	120	67
F	29.0	16.6	101	70
M	27.7	14.8	73	72
A	26.4	11.2	35	72
M	23.5	6.4	14	72
J	20.9	3.1	6	70
J	21.0	3.0	5	66
A	24.5	5.8	5	61
S	27.5	10.1	17	53
O	29.7	13.8	46	55
N	29.8	15.4	84	61
D	30.3	16.7	108	63
YEAR	26.7	11.2	614	65

(b) *Month*	*t,°C* *Max.*	*Min.*	*p,mm.*	*h,%* *0800 hr.*
J	29.1	17.7	352	82
F	28.6	17.7	307	85
M	27.7	16.4	280	88
A	26.8	13.4	95	87
M	25.1	8.4	44	86
J	23.2	5.1	20	85
J	23.1	5.0	29	83
A	24.8	6.8	28	79
S	27.0	10.0	49	73
O	28.8	13.4	109	73
N	29.0	15.4	189	78
D	29.8	16.8	266	80
YEAR	26.8	12.2	1768	80

throughout the year.[21] The intensity of sexual activity varies during the year, so as to give the appearance of one or more restricted breeding seasons.

The present data suggest that there are two separate birth and mating periods for this species in South Africa. New-born infants were found in October and early November and again at the end of January until March. The majority of infants were born during February. It can be deduced that matings occurred during mid-winter (June and July) and they were subsequently witnessed at the end of October and the beginning of November, which is in accordance with the timing of the second birth peak.

It has already been noted that the timing of births and nest-building activity coincided with the onset of the rainy season, when food supply, nest trees and nesting materials became abundant. It is possible that the onset of the first birth period varies from one geographical area to another. The existence of a double birth-peak may be partly explained by the fact that female bushbabies do not become sexually mature until approximately 200 days old, so that those which are born in February are not receptive until September, which precludes them from giving birth in October or November. On the other hand, those females born in November can come into oestrus by May, and with a post-partum oestrus are capable of giving birth twice during the following summer.

2. G. crassicaudatus

Within the study area, temperatures range between -2.8°C and 41.9°C with an annual absolute mean of 19.5°C. 81 mm. of rain was recorded between April and September 1970, while 854 mm. fell before the following April. During this time the evergreen bush became lush and almost impenetrable.

G. crassicaudatus had only a single birth season. All infants were born at the beginning of November and of 16 infants there were 6 sets of twins, one set of triplets and one singleton.

Buettner-Janusch records the gestation period as being approximately 18 weeks with no defined birth season in the laboratory.[22] Seasonal births in the wild may be largely the result of environmental factors. The period of birth and lactation again coincided with the summer months, when rainfall and temperature were high, while food and dense cover were readily available. It may be assumed that mating took place at the beginning of July.

Discussion

Those factors which seem to be important in determining the number and type of bushbabies which may be supported in a particular habitat, as with many other mammal species, include: the availability and suitability of cover, sleeping sites and food; the presence and abundance of predators; and competition with other species. Social behaviour, particularly territoriality, may affect the spacing of members of a population, while the number and frequency of births has a direct effect on population numbers. Local events, such as disease epidemics, veld fires and bush destruction for development schemes, may have a disastrous effect on the species, which lack the ability to move over very long distances.

G. senegalensis and *G. crassicaudatus* showed tolerance to a wide range of temperatures; they had a basically similar social structure, with family groups in both species showing territorial behaviour in the form of olfactory marking, vocal advertisement and aggression towards rival conspecifics. They both bred during the summer months when food and cover were readily available; each utilised all levels within the habitat, and their feeding habits overlapped to some extent. The adaptability of both species is illustrated by their occurrence in and around human habitations.

Comparison of the ecology and behaviour of allopatric populations of the two species indicate that they are primarily adapted to different habitat types for which they showed a distinct preference. The smaller bushbaby, *G. senegalensis*, was found in a greater variety of habitats and at higher population densities than *G. crassicaudatus*. It bred more rapidly, required less food of a type which was widespread in its occurrence, and had a smaller home range. In addition it had greater agility, less stereotyped movements, fewer requirements for dense cover and was able to make use of smaller branches than *G. crassicaudatus*.

The low population densities of one or both species in regions where they lived sympatrically may be accounted for by the fact that these areas did not represent the optimum habitat for either species. Yet *G. senegalensis* was absent from some apparently suitable forest areas where *G. crassicaudatus* was abundant, and direct competition cannot be discounted.

Acknowledgments

This research was supported by a grant from the National Geographic Society, Washington, D.C., as well as by grants from the University Development Foundation and the University Council, University of the Witwatersrand, the Human Sciences Research Council and the Department of Cultural Affairs of the Republic of South Africa.

NOTES

1 Charles-Dominique, P. (1971*a*), 'Eco-éthologie des prosimiens du Gabon', *Biol. Gabon.* 7, 121-228.

2 Osman Hill, W.C. (1953), *Primates: Comparative Anatomy and Taxonomy. Vol. I: Strepsirhini*, Edinburgh.

3 Shortridge, G.C. (1934), *The Mammals of South West Africa.* London.

4 Charles-Dominique, P., this volume.

5 Andersson, A. (1969), 'Communication in the Lesser Bushbaby (*Galago senegalensis moholi*)', unpublished MSc thesis, University of the Witwatersrand.

6 Bearder, S.K. (in prep.), 'Aspects of the ecology and behaviour of the Thick-tailed Bushbaby, *Galago crassicaudatus*', unpublished PhD thesis, University of the Witwatersrand.

7 Bearder, S.K. (1969), 'Territorial and intergroup behaviour of the Lesser Bushbaby (*Galago senegalensis moholi*, A. Smith), in semi-natural conditions and in the field', unpublished MSc thesis, University of the Witwatersrand.

8 Charles-Dominique, P. and Hladik, C.M. (1971), 'Le lepilemur du Sud de Madagascar: écologie, alimentation et vie sociale', *Terre et Vie* 25, 3-66.

9 Charles-Dominique, P. (1972), 'Ecologie et vie sociale de *Galago demidovii* (Fischer 1808; Prosimii)', *Z.f. Tierpsychol*, Beiheft 9, 7-41; Martin, R.D. (1972), 'A preliminary field study of the Lesser Mouse Lemur (*Microcebus murinus*, J.F. Miller 1777)', *Z.f. Tierpsychol.*, Beiheft 9, 43-89.

10 Charles-Dominique, P. (1971*b*), 'Sociologie chez les lémuriens', *La Recherche* 15, 780-1.

11 Brown, L.E. (1966), 'Home range and movement of small mammals', in P.A. Jewell and C. Loizos (eds.), *Play, Exploration and Territory in Mammals*, London.

12 Walker, A. (1969), 'The locomotion of lorises, with special reference to the potto', *E. Afr. Wild. J.*, 7, 1-5.

13 Doyle, G.A., this volume.

14 Andrew, R.J. and Klopman, R.B., this volume.

15 Sauer, E.G.F. and Sauer, E.N. (1963), 'The South West African Bushbaby of the *Galago senegalensis* group', *J. S. W. Africa scient. Soc.*, 16, 5-36.

16 Doyle, G.A., Pelletier, A. and Bekker, T. (1967), 'Courtship, mating and parturition in the Lesser Bushbaby (*Galago senegalensis moholi*) under semi-natural conditions', *Folia primat.* 7, 169-97.

17 Haddow, A.J. and Ellice, J.M. (1964), 'Studies on Bushbabies (*Galago* spp.) with special reference to the epidemiology of yellow fever', *Trans. Roy. Soc. trap. Med. Hyg.* 58, 521-38.

18 Astley-Maberly, C.T. (1967), *The Game Animals of South Africa*, 2nd ed, Cape Town.

19 Lowther, F. de L. (1940), 'A study of the activities of a pair of *Galago senegalensis moholi* in captivity, including the birth and post natal development of twins', *Zoologica*, N.Y., 25, 433-59.

20 Manley, G.H. (1965), 'Reproduction in lorisoid primates', *J. Reprod. Fert.* 9, 390-1.

21 Butler, H. (1960), 'Some notes on the breeding cycle of the Senegal galago, *Galago senegalensis senegalensis*, in the Sudan', *Proc. zool. Soc. Lond.* 135, 423-4.

22 Buettner-Janusch, J. (1964), 'The breeding of galagos in captivity and some notes on their behaviour', *Folia primat.* 2, 93-110.

P. CHARLES - DOMINIQUE

Ecology and feeding behaviour of five sympatric lorisids in Gabon

Introduction

This study was carried out in Makokou (Gabon) in the period 1965-1969, on the basis of four study visits of 7, 8, 14 and 3 months' duration, respectively. Field-work was carried out from the CNRS Laboratory in Gabon (originally known as the Mission Biologique au Gabon, under the direction of Professor P.P. Grassé), which is now referred to as the Laboratoire de Primatologie et d'Écologie Equatoriale (Director: A. Brosset).

The study region is located in the heart of the Congolese block of rain-forest (Ogoue-Ivindo basin). It lies 550 km. from the Atlantic coast and 0.4° latitude north of the equator. Along the access routes

Figure 1. A potto (*Perodicticus potto*) moving around on large-diameter lianes in the forest canopy. Note the strongly reflecting tapetum of each eye, and the reduction of the tail (characteristic of Lorisinae).

Figure 2. An angwantibo (*Arctocebus calabarensis*) moving around in dense undergrowth, using fine lianes attached to smaller trees. Note the reduction of the tail and the use of all four grasping extremities in locomotion.

Figure 3. Demidoff's Bushbaby (*Galago demidovii*) moving through dense vegetation composed of inter-twined branches and lianes.

(roads, certain water-courses, etc.), the forest has been degraded for cultivation which has been subsequently abandoned at various times, thus giving rise to areas of secondary forest at different stages of reconstitution. Primary forest is found a few km. away from these inhabited zones.

Five prosimian species live sympatrically in Gabon; two lorisines:

1. *Perodicticus potto edwardsi* (Bouvier, 1879). Body-weight 1100 gm. Head + body length 327 mm.; tail-length 52 mm. (Fig. 1).
2. *Arctocebus calabarensis aureus* (De Winton, 1902). Body-weight 200 gm. Head + body length 244 mm.; tail-length 15 mm. (Fig. 2).

and three galagines:

1. *Galago demidovii* (Fischer, 1808). Body-weight 61 gm. Head + body length 123 mm.; tail-length 172 mm. (Fig. 3).
2. *Galago alleni* (Waterhouse, 1837). Body-weight 260 gm. Head + body length 200 mm.; tail-length 255 mm. (Fig. 4).
3. *Euoticus elegantulus elegantulus* (Le Conte, 1857). Body-weight 300 gm. Head + body length 200 mm.; tail-length 290 mm. (Fig. 5).

(These figures represent averages based on 33, 30, 66, 17 and 39 specimens respectively.)

The Lorisinae and the Galaginae represent two quite distinct subfamilies: the former are exclusively slow climbers which never leap and always move around slowly and cautiously, whilst the latter are rapid and vigorous leapers.* These behavioural differences are paralleled by numerous anatomical adaptations, primarily affecting the limbs and the tail. In particular, the tarsus is extremely developed in the Galaginae, whilst the tail is very reduced in Lorisinae. By contrast, the skull, the dentition, the digestive tract and the reproductive organs exhibit only very slight differences which would not, in themselves, justify a separation into two subfamilies.

The leaping specialisation of the bushbabies is generally regarded as providing a means of rapid escape, at the same time permitting exploitation of an extensive home-range. On the other hand, the slow and deliberate gait of the lorises has been given different interpretations by different authors: '... high-grade specialisations for an arboreal existence' (Hill);[1] '... directly related to catching prey

* Many authors have exaggerated the locomotor performance of the bushbabies. Sanderson,[5] in particular, writes of a leap of over 10 m. made by *E. elegantulus*, with the animal purportedly gaining in height! We have measured leaps of 2 m. for *G. demidovii*, and of 2.5 m. for *G. alleni* and *E. elegantulus*, with take-off and landing at the same level. The present record is a leap of 5.5 m. (orthogonal projection of horizontal displacement) made by *E. elegantulus*, with a loss in height of 3.1 m.

Figure 4. Allen's Bushbaby (*Galago alleni*) clinging to a small-diameter, vertical support in the undergrowth. Note the long tail, which is used to effect minor corrections in leaps from one vertical trunk to another.

Figure 5. Needle-clawed Bushbaby (*Euoticus elegantulus*) in the process of eating gums on a branch in the forest canopy. Note the crouched body-posture and the application of the snout to the branch surface.

such as insects and roosting birds' (Walker).[2] Some authors have, in fact, expressed astonishment that these animals, which are apparently so vulnerable, have not been decimated by predators. From the author's observations in the forest,[3] it seems very likely that some kind of cryptic mechanism is involved, though this would of course operate only in the natural habitat of these species. (The eye, particularly that of raptors, is more sensitive to rapid movement than to slow progression, and many nocturnal arboreal predators are guided primarily by the auditory sense in localising their prey.) A cryptic mechanism would not, in any case, be the sole means of defence. As a last resort, when an encounter occurs, the lorises utilise active defence mechanisms which vary from species to species. The behaviour of the potto and the angwantibo towards predators has already been described, along with the morphological adaptations involved.[3]

In sum, the two lorisid groups have developed two radically different methods of escaping from predators. Whereas the bushbabies flee rapidly once detected by a predator, the lorises avoid detection by means of an elaborate pattern of slow locomotion. It will be shown that such adaptation has been associated with extensive modification of the feeding behaviour and the diet of the lorisines.

Ecology

The lorisids represent a tiny fraction of the forest mammalian fauna; there are 5 species among 120 mammal species living sympatrically at Makokou. Over and above this, these lorisid species live at relatively low population densities. Systematic counting along pathways and the results of trapping[3] indicate the following average densities per square km.:

Perodicticus potto	–	8 per square km.
Arctocebus calabarensis	–	2 per square km.
Galago demidovii	–	50 per square km.
Galago alleni	–	15 per square km.
Euoticus elegantulus	–	15 per square km.

Using the same techniques, we have calculated far higher densities for lemurs in Madagascar, where the prosimians represent the bulk of the mammalian fauna.[4] However, one must make allowance for heterogeneous distribution of populations, which occur as 'nuclei' ('noyaux') which can be separated to varying degrees. The extreme case seems to be that of the angwantibo, which is abundant in certain parts of the forest (7 per square km.) and virtually lacking over large areas. Nevertheless, we have observed the sympatric occurrence of three, four and even five of these prosimian species in some areas. In fact, numerous interspecific encounters were observed in the forest, without any sign of attack or escape behaviour. In general, any two lorisids of different species which encounter one another exhibit a brief bout of mutual observation and then continue on their way. Thus, any hypothesis of direct interspecific competition based on aggression must be discarded.

The lorisids are all nocturnal. Certain authors, on the basis of observations made in captivity, have suggested that they are partially diurnal or crepuscular.[5,6,7] Such observations must arise from artefacts of captivity, since all of the individuals followed in the forest showed themselves to be strictly nocturnal. In fact, although the lorises only move around during the night, the bushbabies exhibit some initial activity in twilight. But night falls so rapidly at the equator that the bushbabies are only moving around for ten minutes or so before the twilight has faded. The following values for the luminosity of the sky were measured at the times when activity began:

Euoticus elegantulus	–	300 to 100 lux
Galago demidovii	–	150 to 20 lux
Galago alleni	–	50 to 20 ux

Even then, it must be remembered that the light penetrating into the undergrowth of primary forest represents only one hundredth of the values measured above the canopy.

Quite erroneously – again as a result of observations in captivity – numerous authors have reached the conclusion that all lorisids use tree-holes as retreats. Of course, when placed in cages deprived of foliage, the animals do actually retreat into the only boxes placed at their disposition. However, of the five species studied, *Galago alleni* is the only one which sleeps in tree-holes under natural conditions. Usually, such tree cavities are in split hollow trunks, which can be entered from above. The bushbaby can spend the whole day clinging to the internal face of such a 'chimney'. Sometimes a rudimentary nest is built at the base with a few collected leaves. Interspecific competition for daytime retreats can be excluded; the observer has to examine a large number before finding one which is occupied. In addition, *Galago alleni* seems to be quite tolerant of other mammal species. We observed one individual sleeping a few metres away from an arboreal rodent (*Anomalurus erythronotus*) and a group of five bats in a hollow trunk of *Scyphocephalium ochocoa*. The two other bushbaby species sleep singly or in small groups on a small, leafy branch or in a tangle of lianes. *Galago demidovii* will also sleep in spherical nests constructed with green leaves.[3,8,9] The two lorisine species sleep singly on branches or lianes, protected by foliage.

The modern concept of the 'ecological niche' involves a large number of factors. Here, we will only examine the spatial localisation of the animals and their respective diets. These two factors cannot be dissociated, since food-seeking occupies the major part of the activity period of these lorisid species.

Spatial localisation

In order to define in an objective manner the localisation and nature of the supports utilised, we considered the following criteria:

height relative to the ground
orientation of the support
diameter of the support
nature of the support (ground, small trunk, liane base, large trunk, large branch, foliage, foliage mixed with lianes, lianes)

Every time a lorisid was encountered in the forest, we noted the various characteristics of the support on which the animal was *first* seen. (The presence of the observer could modify the selection of any subsequent pathway taken by the animal.) A large number of such observations (642 sightings of the five species) permitted statistical consideration of the data, which are discussed in detail in a previous publication.[3] All that will be given here is a brief description of the forest biotopes most frequently visited by each lorisid species (see Fig. 6):

1. *Perodicticus potto*

The potto inhabits the canopy (10-30 m. in primary forest; 5-15 m. in secondary forest). It is an exclusive climber, using supports with a wide range of sizes (1-30 cm. diameter). This permits the potto to pass from tree to tree by successively utilising large forks (20%) and small branches which interlock from one branch to another (21%). Lianes which permit short-cuts are also utilised (20%). The support orientations are: horizontals 39%, obliques 35%, verticals 26%. In exceptional cases, the potto descends to the ground (escape from a conspecific, crossing a deforested area, etc.). When this occurs, the potto is extremely cautious and heads directly for the nearest tree.

2. *Arctocebus calabarensis*

The angwantibo lives at 0-5 m., both in primary forest and in secondary forest, though it may rarely flee up to 10-12 m. when greatly alarmed. When progressing, this slow climber generally utilises small lianes passing between the small bushes in the undergrowth (43%). Quite frequently, the angwantibo is sighted in the foliage of

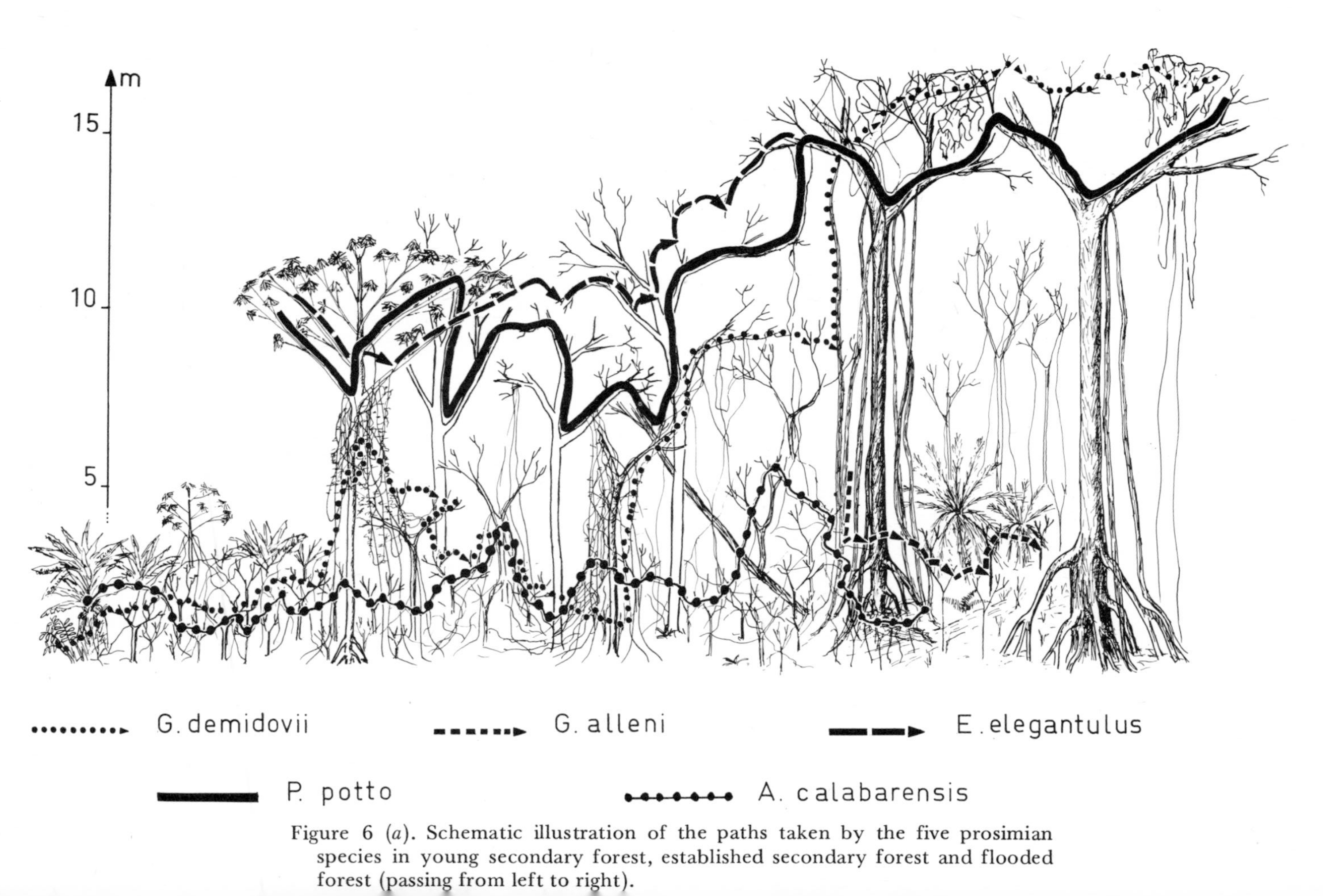

Figure 6 (*a*). Schematic illustration of the paths taken by the five prosimian species in young secondary forest, established secondary forest and flooded forest (passing from left to right).

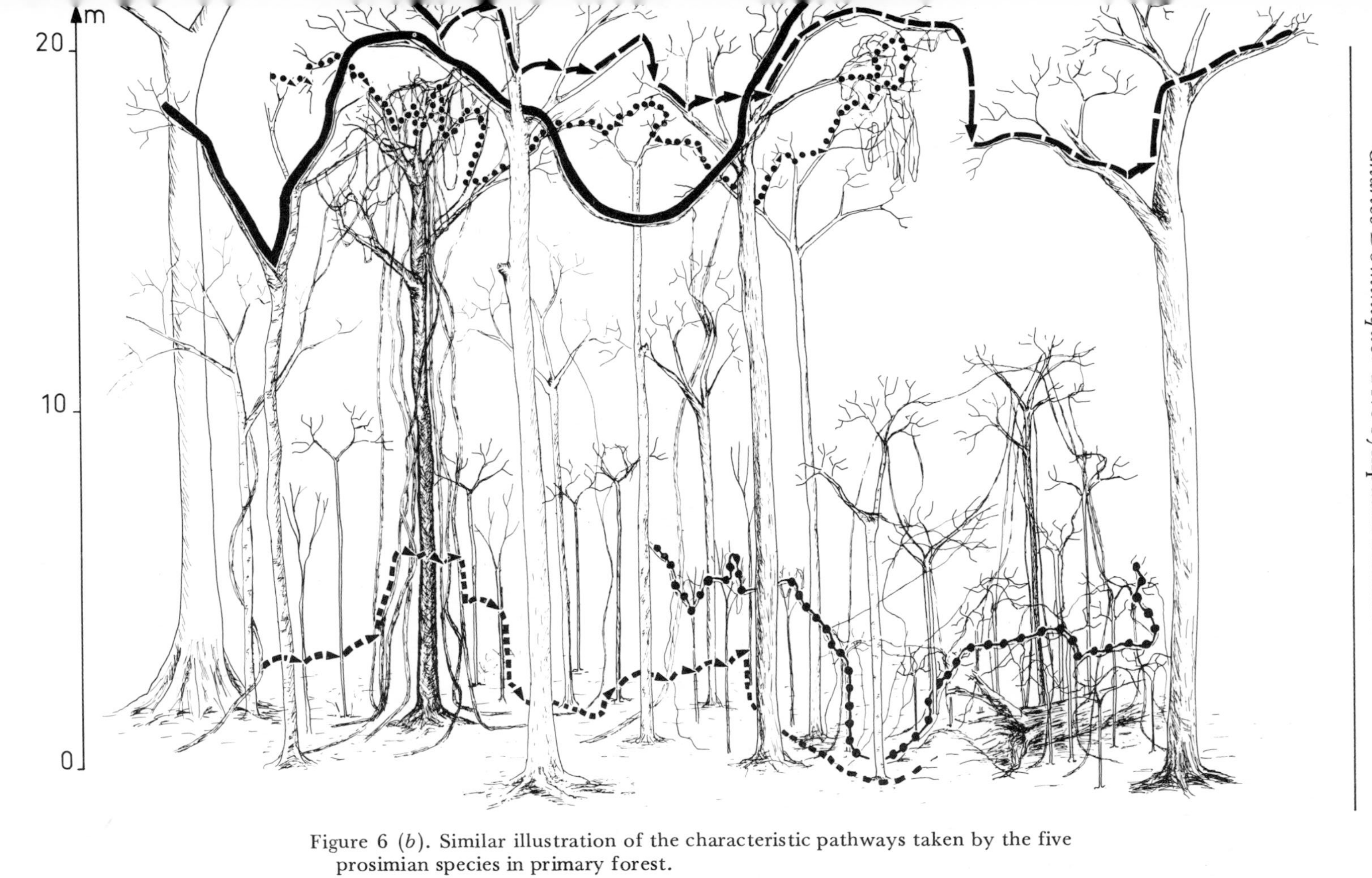

Figure 6 (*b*). Similar illustration of the characteristic pathways taken by the five prosimian species in primary forest.

these small bushes (43%), where it hunts for insects. This species will descend quite readily to the ground to eat fallen fruits or to follow the tracks made by terrestrial mammals (4%). 40% of the supports utilised have a diameter of less than 1 cm.; 52% are between 1 and 10 cm. in diameter. Orientations of supports are: horizontals 20%, obliques 30%, verticals 50%.

3. *Galago demidovii*

This bushbaby lives primarily in dense vegetation invaded by small lianes (35%) and in foliage (25%). Accordingly, it is found at a great height in primary forest (10-30 m.) and low down in secondary forest (0-10 m.); the vertical distribution follows the distribution of the biotopes utilised. 64% of the supports used are less than 1 cm. in diameter; 25% are between 1 and 5 cm. in diameter. The support orientations are: horizontals 22%, obliques 30%, verticals 48%.

4. *Galago alleni*

This bushbaby species is more or less restricted to primary forest, living at a height of 0-2 m. It hunts for animal prey at ground-level, at the same time collecting fallen fruits; but locomotion consists primarily of leaping from one vertical support to another (small tree-trunks; bases of large lianes). The supports utilised have the following diameters: 9% less than 1 cm., 60% between 1 and 5 cm., 19% between 5 and 10 cm. The support orientations are: horizontals 6%, obliques 13%, verticals 81%.

5. *Euoticus elegantulus*

This bushbaby, which scarcely ever descends to the ground, lives in the canopy up to 50 m. However, it may descend to a height of 3-4 m. on certain large lianes which provide droplets of gum. Much progression is along large branches (62%) or large lianes (30%), and along the larger tree-trunks. The special structure of the nails, which terminate in small 'claws', permits movement head-downwards or head-upwards along large, smooth trunks where no other lorisid species would be able to maintain a grasp. It will be seen that this adaptation is related to the diet. The support orientations are: horizontals 22%, obliques 51%, verticals 27%.

To summarise the above information, it can be said that within each sub-family the closest species are ecologically separated by the height of the forest stratum exploited. In the Lorisinae, the potto exploits the canopy and the angwantibo lives in the undergrowth. Among the Galaginae, *Euoticus elegantulus* principally exploits the canopy, whilst *Galago alleni* inhabits the undergrowth. *Galago demidovii* provides a special case: this species depends upon dense vegetation, in which it can move easily thanks to its small size, and it follows the distribution of such vegetation (high up in primary forest and low down in secondary forest).

This latter example provides a good illustration of the fact that the height of the stratum exploited is not due to a 'height preference' exerted by each animal. It is an indirect consequence of choice of particular vegetation zones in each case. We were able to observe that the same applies to *Galago alleni*. After hand-rearing two young animals captured at the age of 1 week (at which stage they can only move a few centimetres), we found that, from the age of 1 month onwards, they sought out all supports with a vertical orientation (especially chair and table legs). Such selection, which has an apparent hereditary basis, is accompanied by a specific leaping 'style' which is adapted to this kind of support.[3] In addition, *Galago alleni* captured as adults and placed in secondary forest (where they scarcely ever occur naturally) moved around at heights of 10-15 m. They utilised vertical or oblique branches at this height when progressing, whereas the average height utilised in primary forest is 1-2 m. With *Galago alleni*, it seems very likely that the vertical orientation of supports of a certain size contributes greatly to the localisation of this species in the undergrowth zone of primary forest.

It will be seen that spatial localisation of these lorisid species is no more than the bare framework of ecological separation permitting closely related species to avoid dietary competition. In parallel with such spatial localisation, there has been dietary specialisation, which has particularly affected feeding behaviour.

Diet

Direct observation of lorisids under natural conditions is difficult. When they are surprised by the light-beam of a headlamp they remain immobile for a few seconds and then disappear rapidly into the vegetation. Under these conditions, analysis of the diet could only be carried out through examination of stomach contents. In Gabon, the prosimians are virtually untouched by hunters, and it was therefore possible to collect a large number of specimens without endangering the local prosimian populations. The animals were collected with the aid of a rifle, samples being taken at different times of the night and in all months of the year. This permitted us to study alimentary rhythms during the night as well as annual rhythms (examination of 174 digestive tracts for the 5 species). The food is quite finely chewed before swallowing; but it is still quite easy to separate the different constituents, which are mixed only to a slight extent in the stomach. We were thus able to separate animal prey from fruit and gums and to obtain fresh weights for these constituent fractions for each stomach examined. The fruits can be identified when the kernels are ingested, but it is often impossible to obtain a precise identification for animal prey. The analysis was therefore restricted to the level of large taxonomic groups: Coleoptera, Lepidoptera (both caterpillars and moths), Orthoptera, Hymenoptera (ants), Isoptera (termites), Myriapoda (centipedes and millipedes), Arachnida, Gasteropoda (slugs), Batracia.

After calculating for each species the percentage of the three principal dietary categories, we obtained the following results (Table 1):

Table 1 Stomach contents

	Animal prey	*Fruits*	*Gums*
Perodicticus potto	10%	65% (+ some leaves and fungi)	21%
Arctocebus calabarensis	85%	14% (+ some wood fibre)	–
Galago demidovii	70%	19% (+ some leaves and buds)	10%
Galago alleni	25%	73% (+ some leaves, buds and wood fibre)	– (small amounts only)
Euoticus elegantulus	20%	5% (+ some buds)	75%

If this first result were taken at face value, one might conclude that the bushbabies include one insectivore, one frugivore, and one gummivore, whilst the lorises represent one frugivore and one insectivore. Since the most closely related animals are separated by stratification of the vegetational zones utilised, it would seem that there can be no dietary competition. However, the problem is in reality more complex than this. The percentage figures do not take into account differences in body size of the five prosimian species concerned – that is, they do not reflect the real weight of food ingested. By calculating the average weight of each of the three dietary categories per stomach, the following results are obtained (Table 2).

Table 2 Weights of dietary components

	Animal prey	*Fruits*	*Gums*
Perodicticus potto	3.4 gm.	21 gm.	7 gm.
Arctocebus calabarensis	2.0 gm.	0.3 gm.	0 gm.
Galago demidovii	1.16 gm.	0.3 gm.	0.15 gm.
Galago alleni	2.2 gm.	9.2 gm.	(negligible)
Euoticus elegantulus	1.18 gm.	0.25 gm.	4.8 gm.

Although the differences for fruits and gums are very clear, this is not the case with animal prey, which are consumed in virtually equal quantities by all five species. Insects (which represent the bulk of the animal prey) are dispersed in the vegetation, and a hunting animal must explore large areas in order to capture large numbers of prey. A small bushbaby and a large bushbaby cover approximately the same distance in the course of the night, and thus each has approximately the same number of opportunities to encounter insects. The same applies to the potto and the angwantibo.

This mechanism has already been described by Hladik and Hladik[10] for the platyrrhines of Panama, which consume roughly the same absolute quantities of insects regardless of their own body-size. The smaller platyrrhines obtain almost all of their food by hunting animal prey, whilst the larger species have to supplement their diet with plant food.

Among the bushbabies, *Galago demidovii* (60 gm.) relies almost entirely on hunting, whereas the two larger species supplement their diets with fruits (*G. alleni*, 260 gm.) or with gums (*E. elegantulus*, 300 gm.). Among the lorises, the angwantibo (200 gm.) derives most of its food by hunting, whilst the potto (1100 gm.) augments its diet of insects with fruits and gums.

In captivity, lorisids are often seen to 'ignore' fruits when they have a large quantity of insects available. The same applies in the forest, where pottos which begin the night with a 'good hunting session' do not go on to eat fruits afterwards. All the individual pottos dissected which had 8-15 gm. of insect matter in the stomach had failed to eat fruits or gums. On the other hand, all individuals with less than 1 gm. of insect matter in the stomach had eaten 20-60 gm. of fruits and/or gums. In *Galago demidovii* the diet is observed to be slightly less insectivorous in the early part of the night, when the animals are still hungry, than in the second half of the night (35% fruits and gums devoured before midnight, and 20% thereafter). Accordingly, in the dry season (when insects are rarer) this species consumes 50% fruits and gums, as against 30% during the rest of the year.

Thus it seems probable that the availability of animal food is the major factor influencing the diet of these lorisid species. Naturally, secondary specialisations affect the feeding behaviour of the 'large' species oriented towards gums or fruits.

1. *Animal food*

In this analysis, we have considered the lorisines and the galagines separately. However, it is conceivable that the members of the two subfamilies may compete with one another in hunting insects. An examination of the categories of prey taken by each species shows that this is not the case. Table 3 shows, in order of importance, the different prey categories found in the stomach contents. The most common categories are followed by percentage figures based on the relative weights of animal food types in the stomachs.

Among the galagines, 78% of the prey are beetles, nocturnal moths and grasshoppers, whereas in the lorisines caterpillars and ants represent 70% of the prey. Pottos particularly prey upon ants (*Crematogaster* sp.) which release large quantities of formic acid, on centipedes (*Spirostreptus* sp.) which release large quantities of iodine, and on 'criquets puants' (malodorous orthopterans), which also emit repellent substances. The angwantibo feeds primarily on caterpillars, most of which are covered with stinging hairs. (The caterpillars eaten by the bushbabies never bear such hairs.) Thus, it would seem that the two lorisine species are specialised to tolerate 'noxious' prey left untouched by the galagines.

This special tolerance of the lorisines for 'noxious' prey permits the capture of a sufficient quantity with a minimum of movement from place to place. In fact, these efficiently protected 'noxious' organisms almost always display certain forms, colours or (especially)

Table 3 Animal prey in order of preference

	1	*2*	*3*	*4*	*5*
Perodicticus potto * (N = 41)	ants (65%)	large beetles (10%)	slugs (10%)	cater-pillars (10%)	orthopterans (malodorous), centipedes, spiders, termites
Arctocebus calabarensis (N = 14)	caterpillars (65%)	beetles (25%)	ortho-pterans	–	dipterans, ants
Galago demidovii (N = 55)	small beetles (45%)	small nocturnal moths	cater-pillars (10%)	hemi-pterans	orthopterans, millipedes, homopterans, pupae
Galago alleni (N = 12)	medium-sized beetles (25%)	slugs	noc-turnal moths	frogs (8%) ants (8%)	grasshoppers, termites, millipedes, pupae caterpillars
Euoticus elegantulus † (N = 52)	grass-hoppers (40%)	medium-sized beetles (25%)	cater-pillars (20%)	nocturnal moths (12%)	ants, homopterans

* Examination of 41 stomachs from pottos did not reveal one case of vertebrate remains. However, whilst following a population of tame animals in the forest we were able to observe a female attempting to capture weaver-birds in a tree carrying a large number of nests. On another occasion, we surprised a female devouring a young frugivorous bat (*Epomops franquetti*). Capture of such prey must be relatively rare.

† No vertebrate remains were found in 52 *E. elegantulus* which were dissected; but we did once come across a tame individual eating a bird (*Camaroptera brevicauda*) in secondary forest.

odours, which normally signal their 'unpalatable quality' to potential predators. They are easily found, and their immobility permits easy capture. (In the case of ants of the genus *Crematogaster*, the potto follows moving columns, licking them up.) Over and above this, these prey organisms – which are ignored by many other predators – are abundant.

In this case, the lorisines are actually exhibiting a *tolerance* rather than a preference. If, in captivity, pottos or angwantibos are presented with their habitual prey alongside grasshoppers or moths normally eaten by galagos, they will eat the latter insects. In actual fact, angwantibos exhibit a particular form of behaviour prior to eating caterpillars. These animals are gripped by the head in the angwantibo's teeth, and the two hands 'massage' the prey for 10-20 seconds, with the result that many of the hairs are removed. Nevertheless, once it has eaten the caterpillar, the angwantibo spends some time wiping its snout and hands by rubbing them on a branch.

Despite their slow locomotion, which serves to protect them from predators, the lorisines succeed in capturing a sufficient quantity of animal prey to ensure a balanced diet. Proteins necessary for physiological equilibrium are found principally in animal prey or in the green parts of plants (leaves and buds). The primates are, in fact, either insectivorous/frugivorous (usually the more rapid species) or folivorous/frugivorous (usually the slower forms, with a few intermediate types).[10] Among the Lorisidae (insectivore-frugivores), the lorises exhibit an exceptional adaptation. These slow-moving animals have conserved the typical diet of the family Lorisidae by obtaining their proteins from a category of animal prey generally left untouched by insectivorous species.

One can therefore discount the idea of any competition between lorisines and galagines in predation on insects. A second contributory factor is important in this context: whereas the lorises capture resting, slow-moving prey, the bushbabies frequently capture rapid-moving prey, quite often when the latter are on the wing. In order to do this, the bushbaby projects its body forward with great rapidity, whilst maintaining a grasp on the branch with its hind-feet. The insect is trapped in flight with both hands, and the bushbaby returns to its starting position (immediately in the case of *Galago demidovii*, which returns like a spring, and only after a short time-lapse in the case of *Euoticus elegantulus*, which ends up suspended head-downwards after the capture).

On the one hand, the potto and the angwantibo – already separated by their height in the vegetational strata – hunt animal prey which are malodorous (potto) and irritating (angwantibo). On the other, the two large bushbaby species are separated by the heights at which they are active. *Galago demidovii* and *Euoticus elegantulus*, which both exploit the canopy in primary forest, are unlikely to compete with one another. The first species preys primarily upon small animals (beetles, moths), and the second feeds upon larger prey (grasshoppers, larger beetles).

2. Fruits

The dietary tables given above show that only the potto and Allen's bushbaby can be regarded as frugivorous. In fact, the other three species consume absolute quantities of fruits amounting to only 1/30th of that eaten by *Galago alleni* and 1/70th of that consumed by the potto. In general, all of the lorisids primarily select soft, sweet fruits (*Uapaca* sp., *Musanga ceroopioides, Ricinodendron africanus*). However, the potto can attack certain fruits with a hard exterior, and on two occasions we saw *Galago demidovii* profiting from the passage of a potto to eat the remains of a large fruit which the latter had opened. As far as *P. potto* and *G. alleni* are concerned, the former eats fruits in the canopy, whilst the second collects them on the ground. This excludes any competition between the two species. These field observations have been confirmed experimentally in captivity. In a cage containing a tree, the pottos preferred to eat fruits placed in the branches, whilst *G. alleni* preferentially ate those placed on the ground.

Trees which are in fruit not only attract animals which are seeking the fruit – they also attract predators. Frugivores are presented with an abundant source of food; but the more time they spend there, the more they expose themselves to predation. Thus, the optimal solution for them is to collect *rapidly* and *in large numbers* fruits which can be eaten in some protected place. These two necessities are met in different ways in different zoological groups: by the crop in birds, by the cheek-pouches in monkeys and rodents, by the rumen in ruminants, etc. The potto and Allen's bushbaby have particularly distensible stomachs which can contain up to 1/13th of the body-weight. With the other three species, by contrast, we have never found any individuals with more than 1/30th of their body-weights in the stomach. Whereas the species which are not specialised for a frugivorous diet primarily utilise their tooth-scraper for slow removal of small fruit morsels, the potto and Allen's bushbaby can swallow rapidly large pieces of fruit, often including the kernels. In 30-60 seconds, a potto can eat an entire banana, and Allen's bushbaby swallows cherry-sized pieces of fruit by pushing them into the mouth with both hands and keeping the head up. It is actually quite rare to find these two species in immediate proximity of trees in fruit. In general, they are found 30-50 m. away, digesting under cover.

In the forest, the sites of fruit production are diverse and continually fluctuating. Through systematic counts, we established that large trees with high productivity are roughly 50 times less numerous than small trees or lianes with medium or low productivity. This is of capital importance for the distribution of small territorial mammals with permanent, restricted home-ranges (in particular, murids and prosimians). When a large tree comes into fruit from time to time in such a home-range, it is exploited straight away; but for the rest of the year food is obtained from trees with low productivity. The latter suffice for mammals of small body-size, but they must be abundant enough to provide a permanent supply of ripe fruits in the existing home-ranges. At night, trees with high fruit productivity are rarely visited. Using traps, we never captured more than 2-3 individuals of any given mammal species in such a tree (i.e. just as many as around trees with low productivity). Conversely, trees with low fruit productivity are little exploited during the daytime – approximately ten times less, according to our counts. This is doubtless associated with the fact that diurnal animals detect fruits by sight (at long range, when large fruiting trees are involved), whereas most nocturnal mammals detect by smell fruit which is isolated and hidden in the vegetation.

In order to study the social life of the prosimians, we have conducted a great deal of trapping with banana as bait. The animals were marked and released, which permitted us – among other things – to investigate their natural feeding behaviour. A basket of lianes containing 10 bananas is discovered in 1-5 days by *Galago alleni* and in 1-10 days by *Perodicticus potto.* On subsequent nights, if the bananas are replenished as necessary, the animals will return regularly by direct routes, often immediately after waking. They spend a few hours close to the bait and then move on to other fruiting areas. If several baskets of bananas are placed in the home-range of one individual, they will all be discovered and 2-3 may be visited during one night. By dissecting numerous pottos and Allen's bushbabies which were collected at the end of the night, we were able to observe that they can eat 2-3 different types of fruit in one night, which must oblige them to visit several fruiting points every night.

From all of these observations, it would seem that frugivorous prosimians (and murids) exploit simultaneously several fruiting trees, and that they never cease to explore their home-ranges in search of new trees in fruit. This mechanism, which is based on memory and exploration, permits them to feed themselves even if habitually visited fruiting trees cease production, or if they have been visited and depleted by another animal. Through continuous exploration of the home-range, they can locate trees at the start of fructification which will replace those which are ceasing to bear fruit.

3. *Gums*

The tables show that the potto and the needle-clawed bushbaby (*E. elegantulus*) are the principal feeders on gums. (*G. demidovii* consumes only 1/50th of the quantity of gums eaten by *E. elegantulus*, whilst *G. alleni* and *Arctocebus calabarensis* scarcely eat gums at all.) Both the potto and *E. elegantulus* inhabit the canopy, so it might be expected that they compete for gums.

Gums form principally along trunks and large branches at the site of old wounds and holes made by the mouth-parts of homopterans. (Examination of stomach contents revealed the presence of certain Auchaenorhynch homopterans – Fulgorids, Membracids, Tibinicids, etc. – which were no doubt swallowed involuntarily along with the exudations of resins which they provoked.) In contrast to fruits, gums appear regularly at the same places throughout the year. However, their production – which is dependent upon the metabolism of the trees – may be diminished during the main dry season.

In equatorial West Africa, the main dry season (15 June-15 September) is characterised by almost complete absence of rain, continuous cloud cover during the daytime, and a reduction of 3-4°C in mean temperatures. During this period, we were able to identify a marked decrease in the biomass of insects and a reduction in fructification. In parallel, we have observed that the prosimians lose 1/10th to 1/14th of their body-weights during this critical period. The weight-loss is directly dependent upon food availability. For example, *G. demidovii* consumes an average of 0.65 gm. of insects per night during the main dry season, as against 1.28 gm. per night during the rest of the year (stomach contents taken between 20.00 hrs. and midnight). Under the same conditions, we found smaller quantities of insects and gums in the stomachs of *E. elegantulus* examined during the dry season than in those collected at other times of the year. Thus, any competition would be exaggerated during this critical period.

When the dietary habits of the lorisids are examined in greater detail, it emerges that the potto eats only fruits and insects during the dry season, whereas fruits, gums and insects are taken at other times of the year. Conversely, *E. elegantulus* examined during the dry season had not eaten any fruits, although they eat small quantities during the rest of the year. Throughout the critical period, apart from the animal prey consumed, the potto is thus strictly frugivorous and *E. elegantulus* is strictly gummivorous.

Observation of the feeding behaviour of these two species renders the mechanism of such competition easily comprehensible. Whereas the activity of the potto is primarily oriented towards searching for and visiting trees which are in fruit, *E. elegantulus* spend most of their time visiting trees which are gum-producers. Needle-clawed bushbabies have an excellent memory for gum-production sites, and they follow veritable 'rounds' which permit them to collect (with the aid of the tooth-scraper) tiny droplets of gum formed after their last visit. Visits are made almost every night, even though each animal must visit a very large number of production-sites (about 300) in order to collect sufficient quantities of gums. *E. elegantulus* rapidly covers its 'rounds' thanks to its powerful leaps, stopping only for a few minutes at each gum-exuding site. In addition, the 'claws' at the ends of the nails permit access to sites which are inaccessible to the other prosimian species, right along the largest trunks. During the dry season, when the gums accumulate more slowly, *Euoticus* must visit a larger number of production-sites, whereas smaller 'rounds' suffice at other times of the year. Under these latter conditions, large aggregations of gums collect on trees which are not often visited, and it is such gums which are eaten by the pottos. Indeed, the gums found in the stomachs of pottos are often harder and darker in colour than those found in stomachs of *E. elegantulus*, and they occur as large lumps.

Thus, one cannot really talk in terms of real competition for gums between these two species, since the gums eaten by the potto are those left untouched by *E. elegantulus.* Dietary specialisations are most evident during the critical period of the year, and it is doubtless during this time that natural selection for adaptive characters is most active.

Conclusions

In both the Lorisinae and the Galaginae, the different sympatric species are morphologically relatively similar, and their dietary requirements are quite comparable: insects, with a supplement of fruits and gums in the larger species. In captivity, the five lorisid species can be maintained easily without any provision of foods (e.g. gums, fruits or certain insects) of the kinds eaten under natural conditions. In our present captive colony, all five species have become perfectly adapted to the same diet: milk, banana, apple, pear and crickets. In the forest, it is primarily through exploitation of different vegetation strata that the various species avoid dietary competition. Their ecological delimitations (in particular, their stratification) follow from 'preferences' which orient each species towards a particular type of support. This orientation is complemented by certain behavioural and anatomical specialisations associated with utilisation of supports. These anatomical adaptations are relatively minor; it is primarily behavioural differences which permit the separation of ecological niches.

Acknowledgments

My thanks go to Professor P.P. Grassé and A. Brosset for their generous provision of facilities and assistance at the CNRS field laboratory in Makokou, Gabon, throughout this study. All the photographs were taken in primary forests by my colleague, G. Dubost.

I should also like to make a special note of thanks to my friend and colleague, R.D. Martin, who offered to translate the manuscript for this article, despite his heavy commitments with the organisation of the Research Seminar. Our discussions, which have been numerous and very rewarding, have permitted me to extract a number of valuable conclusions.

NOTES

1 Hill, W.C.O. (1953), *Primates: Comparative Anatomy and Taxonomy*, 1. *Strepsirhini*, Edinburgh.

2 Walker, A. (1969), 'The locomotion of the lorises with special reference to the potto', *E. Afr. Wild. J.* 7, 1-5.

3 Charles-Dominique, P. (1971), 'Eco-éthologie des prosimiens du Gabon', *Biol. Gabon.* 7 (2), 121-228.

4 Charles-Dominique, P. and Hladik, C.M. (1971), 'Le *Lepilemur* du Sud de Madagascar: écologie, alimentation et vie sociale', *Terre et Vie* 25, 3-66.

5 Sanderson, I.T. (1940), 'The mammals of the North Cameroons forest area', *Trans. Zool. Soc. Lond.* 24, 623-725.

6 Napier, J.R. and Napier, P.H. (1967), *A Handbook of Living Primates*, New York, 258.

7 Jones, C. (1969), 'Notes on ecological relationship of four species of lorisids in Rio Muni, West Africa', *Folia primat.* 11, 255-67.

8 Vincent, F. (1969), 'Contribution à l'étude des prosimiens africains: le Galago de Demidoff', Doctoral thesis, CNRS, AO 3575.

9 Charles-Dominique, P. (1971), 'Ecologie et vie sociale de *Galago demidovii*', *Z.f. Tierpsychol,*, Suppl. 9, 7-41.

10 Hladik, A. and Hladik, C.M. (1969), 'Rapports trophiques entre végétation et Primates dans la forêt de Barro Colorado (Panama)', *Terre et Vie* 23, 25-117.

M.P.L. FOGDEN

A preliminary field study of the western tarsier, Tarsius bancanus *Horsefield*

Introduction

Hill recognised three tarsier species, all of them closely related and referable to the single genus *Tarsius.*[1] *T. bancanus* occurs in eastern Sumatra and Borneo, *T. syrichta* in the Philippines, and *T. spectrum* in Celebes. Exceedingly little is known about the ecology and behaviour of any of the three, even though a knowledge of their biology, and that of other prosimians, is of great importance to a proper understanding of Primate evolution as a whole. There has been much recent research on the other prosimian groups – the Madagascar lemurs and loris/bushbaby group (see Charles-Dominique and Martin for references)[2,3] – and the tarsiers now remain as the most important of the unknowns in our knowledge of prosimian ecology and behaviour.

This paper is offered as a small contribution towards filling this gap and concerns the western or Horsefield's tarsier, *T. bancanus* (Fig. 1). The study came about by accident when a number of tarsiers were trapped in mist-nets during the course of an ornithological research programme. The opportunity was taken to mark them and many were subsequently retrapped or seen in the field. Other commitments prevented tarsiers becoming more than a subsidiary subject of research; nevertheless, in view of the general lack of information about tarsiers, enough data were accumulated on their ecology and behaviour to justify this present account.

Methods

The study was carried out between October 1964 and September 1966 in the Semengo Forest Reserve, 20 km. south of Kuching, the capital of Sarawak, Borneo. The study area has been described elsewhere (Fogden).[4,5] Briefly, it consisted of 25 hectares of

Figure 1. (*a*) Photograph taken under natural conditions at dusk, showing a western tarsier (*Tarsius bancanus*) clinging to a thin, vertical trunk in the forest understorey. Such vertical supports are abundant in relatively clear secondary forest zones.

Figure 1. (*b*) Close-up photograph showing the typical vertical clinging posture. Note the expanded, tactile discs at the finger-tips and the greatly elongated hind limbs (particularly extended in the tarsal region).

rainforest more or less equally divided between primary mixed dipterocarp forest, which is the climax vegetation type of lowland Sarawak, and secondary forest at a comparatively early stage in the succession, most of it being composed of trees 5-10 m. tall and perhaps 10-15 years old. Within this area, netting was carried out for about two weeks in every month of the study period except April 1965. Nets were set at ground level, apart from an occasional net set in the canopy, and cutting of vegetation was kept to a minimum by using natural sites wherever possible. Nets were 12 m. long and about 2½ m. high, and most sites were sufficiently long to take two nets in line. The trapping effort was about 500 net-days per month, and on any one day or night 30-40 nets were in use. The nets were rotated

Figure 1. (*c*) Close-up photograph showing an identification ring on the left hind limb (tarsus). Note the large eyes, with the pupil greatly reduced. There is no reflection from the rear of the eye, since the conspicuous tapetum of the lemurs and lorises is lacking in the tarsier.

around the 200 available sites so that the study area was more or less evenly covered every two months.

The 26 tarsiers that were trapped were marked with numbered metal bird rings placed on the tarsus, and were subsequently released at the spot at which they were trapped. Each was weighed and sexed, and particular note was taken of any physical peculiarities which might assist subsequent recognition in the field. For example, a female was noted to be missing the distal half of its tail, a male had lost the last joint of each of the three outer toes of its left foot, and several had characteristic notches in one or other of their ears. These natural recognition characters were supplemented by clipping away the tuft of hairs from the end of the tail of some individuals, and by noting whether the ring was placed on the left or right leg. By using various combinations of these characters it was possible to identify in the field under favourable conditions 19 of the 26 tarsiers which were trapped in the study area.

No special effort was made to observe tarsiers directly in the field, but in the course of the two years of the study they were seen on about 40 occasions (including some sightings by other observers), generally at dusk. Though this was an average of less than two sightings per month of the study, the information gained was considerable. It was often possible to approach to within 5-10 m. of a tarsier, and so establish its identity, and also to follow it for some time provided that the vegetation was not too dense.

The study of other prosimians has been greatly facilitated by the ease with which they can be located at night by the reflection of a light beam from the tapetum at the back of their eyes. It is a great disadvantage that the eyes of tarsiers lack a tapetum and that they cannot be located in this way. Many hours spent in tarsier habitat with torches at night failed to reveal any, even though civets and other animals were easily located. Tarsiers are comparatively silent so that they are not conveniently located by sound either. However, they do have a bird-like call (described by Hill, Porter and Southwick, and by Harrisson) which is probably associated with courtship.[6,7] On the few occasions that it was heard at Semengo it was usually possible to locate the animal that was calling. Tarsiers also have a strong and readily identifiable smell which occasionally led to the discovery of an individual close by.

The seasonality of insects and fruit in the study area

Tarsiers are largely insectivorous, so some knowledge of the seasonal fluctuation in the number of insects in the study area is essential. During the course of the study it became apparent that fluctuation in the amount of fruit in the study area also had an important effect on the behaviour of tarsiers (see below). The seasonality of both insects and fruit has been discussed elsewhere (Fogden).[4,5] Briefly, insect numbers are ultimately related to rainfall which, though heavy throughout the year, increases during the north-east monsoon lasting from December to February or March. The increased rainfall results in increased leaf production, and there is an associated flush of insects which is most marked from January to about June. This is the period during which insects were most abundant in both 1965 and 1966. In contrast to the insects, fruit is patchily distributed and erratic in its seasonality, particularly in secondary forest. There may be long periods when fruit is abundant, and others when there is little or no fruit over wide areas. There is no annually recurring period when fruit is always relatively abundant or scarce, as there is with insects. At Semengo, fruit was plentiful for the whole of the first year of the study, but generally scarce from October 1965 to August 1966.

Results

The basic trapping and observational data on which the following sections are based are recorded in Table 1 and Figs. 2, 3 and 4. In Table 1, a record refers to the trapping or sighting of an individually recognisable tarsier (identified by a number in the Table and Figures). The location of such records are plotted on a map in Figs. 3

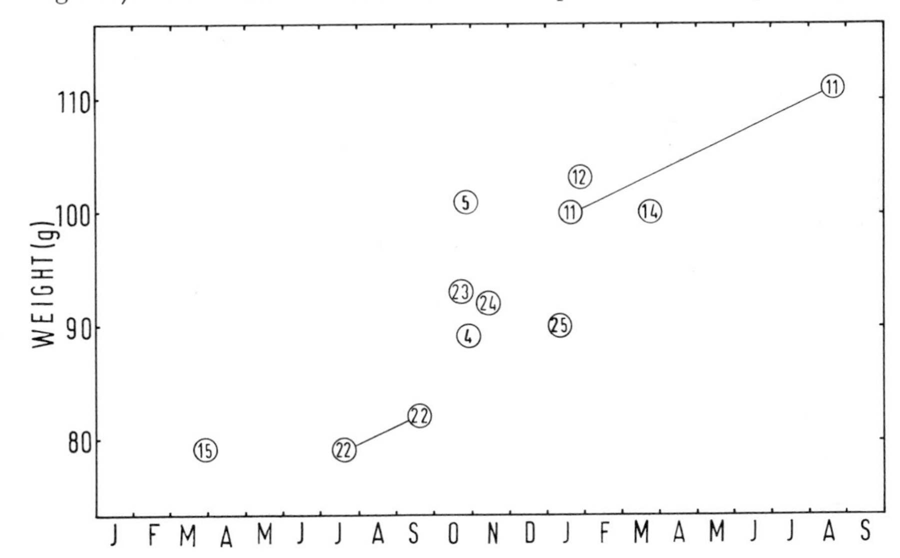

Figure 2. The pattern of weight increase of immature tarsiers during the 18-20 months after their birth in January or February. The records from two 20-month periods (January 1964-September 1965 and January 1965-September 1966) are plotted on the same scale. Each individual has an identity number which corresponds with those used in Table 1 and in the text.

and 4, except in the case of tarsiers recorded three or more times, in which case their range has been indicated by shading an area which just covers all points at which each was recorded. The 26 tarsiers involved were recorded 71 times (43 trapping records and 28 sight records), but the bulk of these records refer to only 12 individuals which were recorded 54 times. About a dozen sight records of

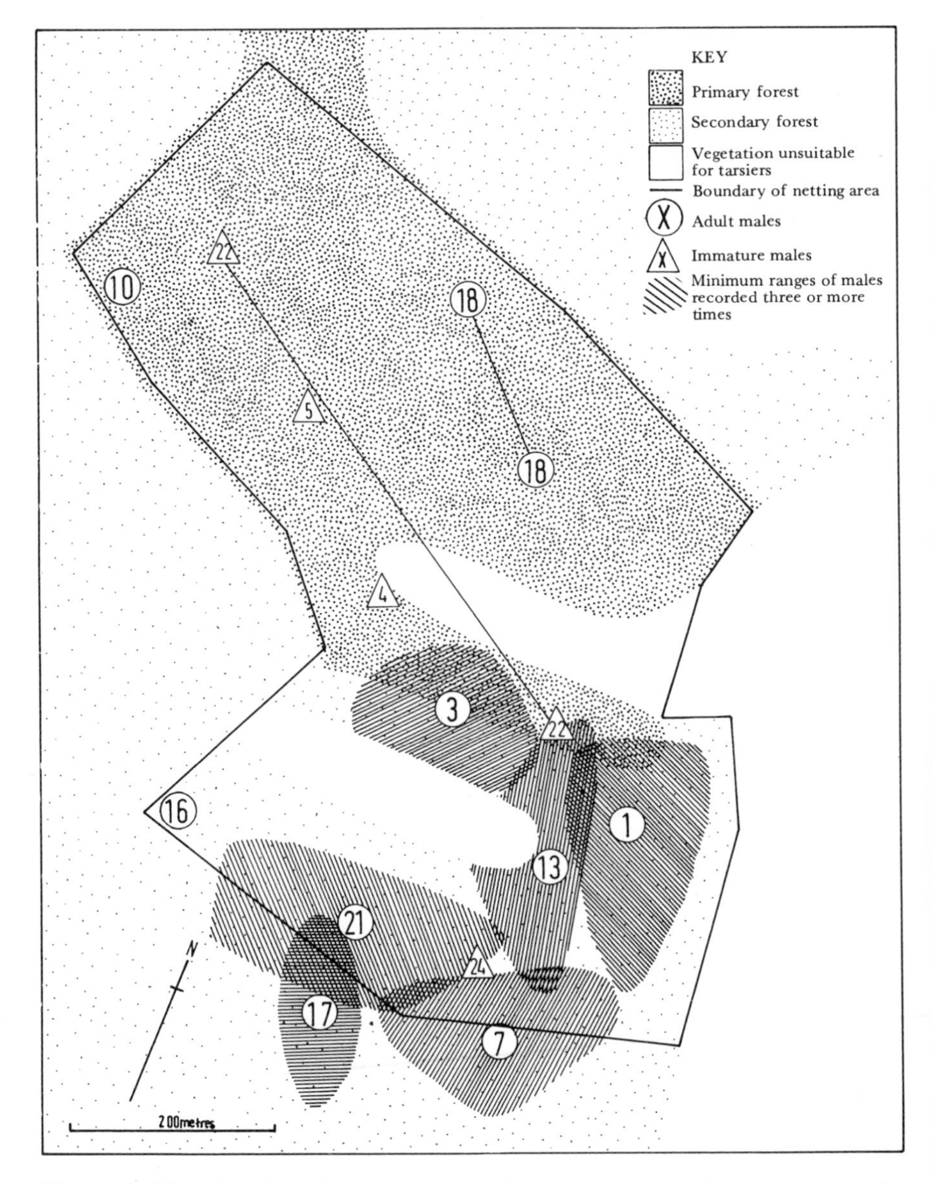

Figure 3. Map showing the distribution of male tarsiers in the Semengo study area. Each individual has an identity number which corresponds with those used in Table 1 and in the text.

unidentified tarsiers are included neither in Table 1 nor Figs. 3 and 4.

1. Sex ratio and body weights

The 26 trapped tarsiers consisted of 13 males and 13 females, so the sex ratio must be more or less equal.

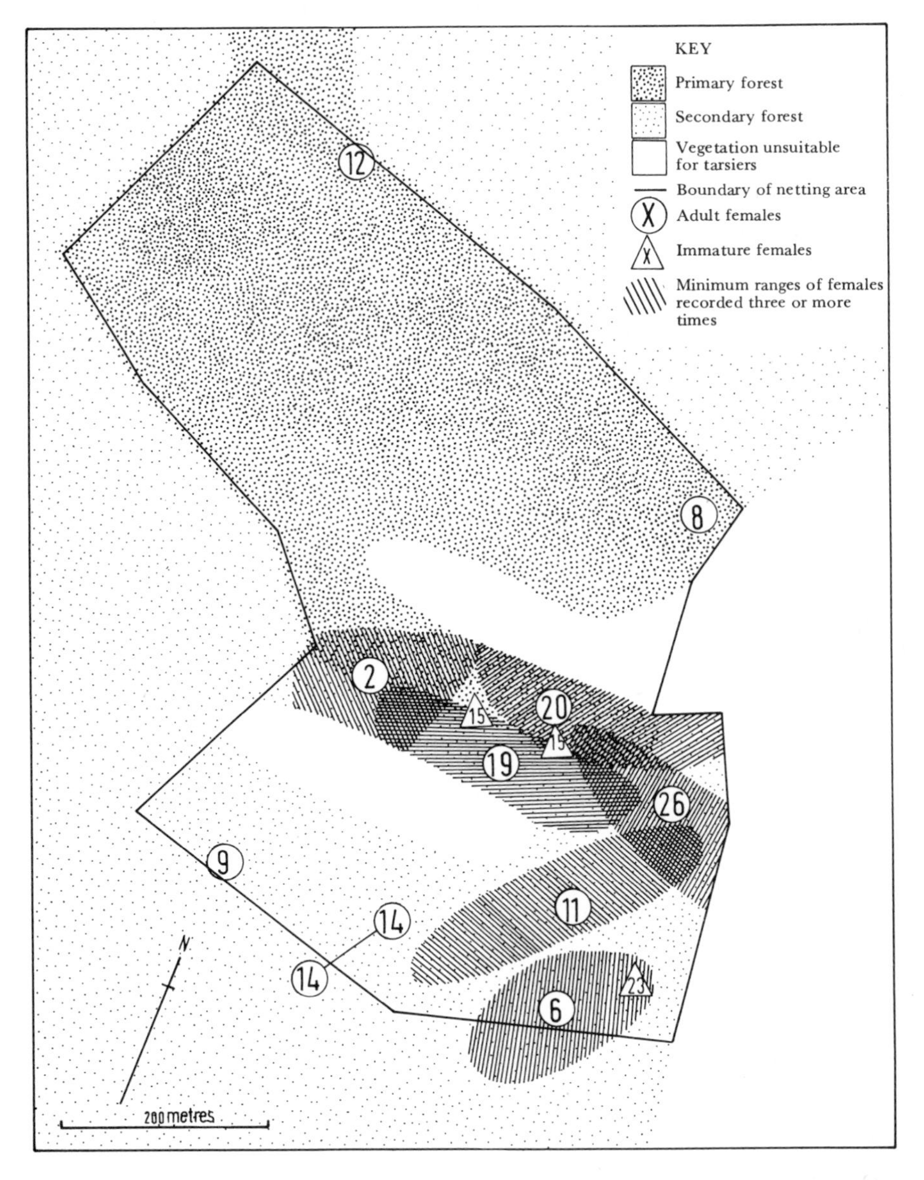

Figure 4. Map showing the distribution of female tarsiers in the Semengo study area. Each individual has an identity number which corresponds with those used in Table 1 and in the text.

Sexually mature males with large descended testes weighed 108-131 gm., with a mean of 120 gm. Such males had large testes throughout the year, though there may well have been some minor seasonal variation in their size. Individual body weights varied by up to 7 gm., but much of the variation could be attributed to animals being caught at different times of the night. There are insufficient data to show whether any of the variation in body weight was seasonal.

Mature females averaged lighter than males, weighing 100-119 gm. with a mean of 111 gm. However, the three lightest females (weighing 100 gm., 102 gm. and 103 gm.) might have been considered to be immature but for the fact that they included one

Table 1 Records of identified tarsiers

Identity No.	*Weight (gm.)*	*No. of records*	*Days between first and last record*
Males			
1	119	5	300
3	119-126	6	363
7	108-112	6	446
10	112	1	–
13	124-128	4	453
16	120	1	–
17	113	3	171
18	111-117	2	13
21	123-131	6	251
Immature males			
4	89	1	–
5	101	1	–
22	79-82	2	58
24	92	1	–
Females			
2	116	5	329
6	114	4	148
8	119	1	–
9	110	1	–
11	102-111	5	211
12	103	1	–
14	100	2	125
19	110	4	312
20	113-116	3	98
26	106	3	298
Immature females			
15	79	1	–
23	93	1	–
25	90	1	–

(no. 12) that gave birth while trapped in a net, and another (no. 14) that was lactating. It is unlikely that such females would be light because they were in poor condition, and the most likely explanation is that they were young females breeding when one year old and not yet fully grown. This view is compatible with the pattern of weight increase of young tarsiers (see below and Fig. 2), and is strengthened by the fact that the only one of the three to be re-weighed at a later date had increased in weight by 11 gm. in seven months (Fig. 2, no. 11). None of the females that were weighed was obviously pregnant, and it is probable that females attain much higher weights than the 119 gm. maximum recorded. New-born young are known to weigh 23-24 gm. (Le Gros Clark),[8] so female weights of 140 gm. or more are to be expected. In fact, Le Gros Clark recorded a weight of 142.5 gm. without giving the sex of the individual concerned.

Four males with very small testes weighing 79-101 gm. and three females weighing 79-93 gm. were considered to be definitely immature and not yet fully grown. The timing of their appearance in the trapped population is consistent with this view (see below and Fig. 1).

2. *Reproduction*

Tarsiers bear one young and have a sharply defined breeding season. In southwestern Sarawak, Le Gros Clark found that impregnation takes place from October to December and that the young are born in January and February, while Banks gave October to March as the breeding season.[8,9] Harrisson observed copulation between a pair of captive tarsiers in an outdoor enclosure in October,[7] while in this study a female was recorded to give birth in January and another to be lactating in March. All these observations are in good agreement. This sharply defined timing of breeding in south-western Sarawak results in coincidence of both the final stages of pregnancy and the period of maximum growth of the young with the time when insects are most abundant. There can be little doubt that the insect food supply is the ultimate factor determining the timing of the breeding season.

The timing of the appearance of light-weight and presumably immature tarsiers in the trapped population at Semengo is in good agreement with the timing of this restricted breeding season. The lightest individuals, weighing about 80 gm., were trapped in March and July. These probably represent early and late-born young. Thereafter immature tarsiers of gradually increasing weight were caught over a period of several months, adult weights being attained at an age of about 15 to 18 months. This progression of weights is illustrated in Fig. 1. It can be seen that the two females (nos. 12 and 14), weighing about 100 gm., which are presumed to have bred at an age of only one year, fit neatly into the pattern.

3. Habitat and general behaviour

Statements in the literature agree that *T. bancanus* is much more common in secondary than in primary forest (Le Gros Clark, Banks, Davis).[8,9,10] This is also borne out by this study, for 86% of records of tarsiers were either from secondary forest or from the narrow strip of primary forest cutting across the centre of the study area. Only seven tarsiers were caught in the block of primary forest in the northern half of the study area, and three of these were adults (male no. 10 and females nos. 8 and 12) trapped right on its edge, whose main ranges were probably in the adjoining secondary forest outside the study area. Of the four tarsiers that were trapped more than 50 m. inside primary forest, three were immature males, while the fourth (no. 18), which weighed 111-117 gm., was the only male tarsier weighing more than 110 gm. to have small testes. It seems probable that primary forest is a marginal habitat for tarsiers, and that it is utilised mainly by young males which are unable to compete with mature males for space in secondary forest.

Tarsiers have a greatly elongated tarsus and a low intermembral index (the ratio of forelimb length to hindlimb length) of about 0.58, and they are clearly specialised for vertical clinging and leaping. In accordance with this specialisation they were found to occur only in forest understorey with an abundance of vertical supports, particularly in the form of saplings 2-6 cm. in diameter and 1-2 m. apart (Fig. 1). They apparently avoided very dense tangled vegetation (many mist-nets were sited in such areas), the forest canopy in general, and areas of relatively open secondary vegetation with few saplings and a dense herbaceous cover.

Within the limits of the forest understorey, tarsiers have a wide vertical distribution, for they were seen from ground level up to heights of at least 8 m. There are indications that their vertical distribution varies seasonally (see below), and that at times they spend much of their time on or close to the ground. It is probable that most tarsiers were trapped while on the ground, for the majority were entangled in the very bottom of the net in positions which would have been impossible if they had been jumping between trees.

4. Food and feeding behaviour

T. bancanus is basically insectivorous, though given the opportunity it no doubt catches other small animals in the wild, particularly lizards, just as it does in captivity (Harrisson).[11] The stomach contents of seven specimens examined by Davis contained nothing recognisable apart from fragments of orthopteroid insects and a spider.[10]

At Semengo most of the locations at which tarsiers were seen and trapped were close to fruiting trees, or more specifically trees with ripe fruit lying on the ground beneath them. In fact, one of the best ways to see tarsiers was to wait quietly in the vicinity of such trees at dusk. Tarsiers were not attracted to the fruit itself, but by insects and other animals which were attracted to the fruit. These included large numbers of cockroaches, crickets, other orthopterans and large moths, as well as geckos and other lizards feeding on the host of smaller insects concentrated in the area. In these circumstances tarsiers captured most of their prey on the ground. They would scan the forest floor from a perch a metre or more above the ground, and having located prey (which was generally moving) leap directly on to it, killing it by biting with tight-shut eyes (see photograph in Harrisson).[7] Tarsiers were seen to catch prey in this way with leaps of up to 2 m., but most leaps were considerably shorter. The scanning phase of hunting sometimes lasted for up to ten minutes at a single perch, with the tarsier remaining more or less immobile; but more usually a failure to locate prey at one perch resulted in it moving on after only a minute or two. Some observations suggested that hearing as well as sight plays a part in locating prey.

On a number of occasions tarsiers were seen foraging on the ground by feeling in the leaf litter with their hands and apparently sniffing with their nose close to the ground. They appeared to be hunting by touch and/or smell, but the method also appeared effective in causing cryptic insects to move (this was observed in the case of stick insects and leaf-mimicking orthopterans), so that they could be seen and then pounced on in the normal way. Harrisson described captive tarsiers searching the floor of their large outdoor cage in an apparently similar way.[11]

5. *Seasonal variations in vertical distribution*

Tarsiers were neither trapped nor seen in the field at the same frequency throughout the two years of the study. They were trapped or seen 29 times (24 trapping records) in the first six months of the study, 34 times (13 trapping records) in the second six months, but only 16 times (seven trapping records) in the whole of the final year. This difference can probably be attributed to a change in the vertical distribution of tarsiers, resulting from the fact that fruit was abundant at Semengo for the whole of the first year, but very scarce for most of the second (see above). During the first year tarsiers were probably attracted to fruiting trees where they hunted low down and were easily trapped and observed, particularly as many mist-nets were sited, and much time was spent in visual observation, in such areas. During the second year there were no fruiting trees to act as focal points for either tarsiers or netting and observation. Consequently, tarsiers were neither easily trapped nor observed. Free-moving tarsiers were seen nine times during the second year and, though this sample is small, it is probably significant that they were at an average height of nearly 3 m., which is higher than the height of a mist-net.

The period during which tarsiers were only rarely recorded covered a time of insect abundance as well as scarcity, so it is unlikely that a proportion of the population left the area (in any case there was nowhere more suitable to go to), or went through a period of dormancy as do some lemurs when food is scarce (see Martin for references).[3]

It might also be argued that the decline in trapping records resulted from the tarsiers learning to avoid net sites. This may be true to a slight extent, but this explanation would not account for the fact that no tarsiers whatsoever were caught in areas of suitable habitat (to the south of the main study area) in which nets were sited for the first time during the second year of the study. Nor would this explanation account for the decline in sight records of tarsiers in the second year.

6. *Activity patterns*

Tarsiers are crepuscular and nocturnal. Most probably spend the day clinging to a suitable support in dense vegetation, though in the Philippines a small proportion of *T. syrichta* were found in hollow trees (Wharton).[12] Tarsiers were not systematically observed at night so nothing can be said about rhythms of nocturnal activity. However, it is clear that they become active at around dusk, for the vast majority of sight records were at that time, and a high proportion of tarsiers were trapped in the first two hours of the night. Several sight records were made a full hour before dusk on overcast evenings, and one as early as 17.15 hrs., nearly two hours before dusk (W. Corris, personal communication).

7. *Social organisation*

Table 1 and Figs. 3 and 4 show that adult tarsiers were sedentary, for the nine individuals for which most data are available (males nos. 1, 3, 7, 13 and 21, and females nos. 2, 11, 19 and 26) were recorded several times within relatively small areas over periods of 200 days or more. The ranges of individual tarsiers were about 2-3 hectares, and averaged larger in males than females. The ranges of individuals of the same sex appeared to be more or less exclusive, though there were small areas of overlap which were greater in females than in males, but the ranges of males and females overlapped extensively. Whether or not these ranges were territories in the sense of being defended is not clear. Marking with urine was frequent, but it is not known whether this had a territorial function.

Banks[9] stated that tarsiers usually occur in pairs, and statements by Harrisson[7,11] implied that established pairs are the rule in the wild. In this study two tarsiers were seen together on only eight occasions, two of which involved the same male with different females within the space of a few days. Pairings which were identified were male no. 3 with females nos. 2 and 19, male no. 13 with female no. 11, and male no. 21 with female no. 14. The known ranges of males generally overlapped with those of more than one female, though there is no evidence of any female actually associating with more than one. In the circumstances it is impossible to decide whether permanent associations between particular males and females were formed, but it does seem likely that males associate with more than one female. It is perhaps significant that the identified males that were seen with females were the three heaviest (Table 1). This might indicate that heavy males are dominant over lighter males in associating with females which overlap their ranges.

It has already been noted that immature male tarsiers differed from adults in being recorded mainly in primary forest. There is also a suggestion that they ranged more widely than adults, for one was re-trapped after moving nearly 600 m., twice the maximum recorded movement of an adult. It seems likely that immature males disperse into marginal habitat after becoming independent, and range widely until able to secure a permanent range in better habitat. It is possible that they only become sexually mature when they secure a permanent range, for one male (no. 18), which was caught twice in primary forest, had very small testes even though it weighed 111-117 gm. and must have been at least 15-18 months old.

Immature female tarsiers appear to secure a permanent range at an earlier age than males, for two females (nos. 12 and 14) bred when probably only one year old. Whether immature females disperse or remain in the area in which they were born is not clear, but it is noteworthy that no female tarsiers, adult or immature, were recorded far inside primary forest. Together with the fact that female ranges are smaller and overlap more than those of males, this might indicate that immature females establish ranges close to their parents.

Discussion

The restriction of tarsiers to secondary forest is a surprising feature of their ecology, for the understorey of primary forest has an abundance of suitable vertical supports and appears structurally more suitable than secondary forest for an animal that is specialised for vertical clinging and leaping. This apparent anomaly is emphasised by the fact that Allen's Bushbaby (*Galago alleni*) of West Africa, which resembles the tarsier in being specialised for vertical clinging and leaping, does occur in the understorey of primary forest and not in secondary forest (Charles-Dominique).[13] It is difficult to believe that the food supply in primary forest is unsuitable for tarsiers. The answer may lie in competition from other species, though there are no obvious competitors of the tarsier in primary forest, unless the various species of treeshrews *Tupaia* exploit the same food supply in spite of being diurnal. The only nocturnal treeshrew, *Ptilocercus lowii*, and the slow loris, *Nycticebus coucang*, occupy niches that are quite distinct from that of the tarsier, and neither can be regarded as a potential competitor.

The suggested social organisation of the tarsier has a number of possible points of similarity with the social organisation of Demidoff's bushbaby, *Galago demidovii* (Charles-Dominique) and the lesser mouse lemur, *Microcebus murinus* (Martin).[2,3] One can think in terms of heavy dominant males which pair with more than one female (the equivalent of the heavy 'A-central males' of *G. demidovii* and the 'central males' of *M. murinus*), lighter males which are in contact with females but subordinate to heavier males when it comes to pairing (the equivalent of 'B-central males' of *G. demidovii*), and immature males which inhabit marginal habitat and lack a permanent home range (the equivalent of the 'vagabond males' of *G. demidovii* and 'peripheral males' of *M. murinus*). On the other hand, there is no evidence that female tarsiers sleep socially in nests, as females do in the case of *G. demidovii* and *M. murinus*. Both the latter species are very vocal and apparently have calls which aid the reassembly of nesting groups each morning. The relative silence of tarsiers can probably be related to their not being highly sociable.

Acknowledgments

This study would not have been possible without the great help and encouragement of Tom Harrisson, Curator of the Sarawak Museum, and many members of the Museum staff, particularly Lian Labang, Gaun anak Sureng and Ambrose Achang. The mist-netting study was partly financed by an equipment grant which was kindly made available by Dr H. Elliot McClure in his capacity of Director of the United States Armed Forces Migratory Animal Pathological Survey. Thanks are also due to my wife who criticised the manuscript and drew the figures.

NOTES

1 Hill, W.C.O. (1955), *Primates: Comparative Anatomy and Taxonomy*, II: *Haplorhini: Tarsioidea*, Edinburgh.

2 Charles-Dominique, P. (1972), 'Ecologie et vie sociale de *Galago demidovii* (Fischer 1808)', *Z.f. Tierpsychol.*, Beiheft 9, 7-41.

3 Martin, R.D. (1972), 'A preliminary field-study of the Lesser Mouse Lemur (*Microcebus murinus* J.F. Miller 1777)', *Z.f. Tierpsychol.*, Beiheft 9, 43-89.

4 Fogden, M.P.L. (1970), 'Some aspects of the ecology of bird populations in Sarawak', D. Phil. thesis, University of Oxford.

5 Fogden, M.P.L. (1972), 'The seasonality and population dynamics of equatorial forest birds in Sarawak', *Ibis*, 114; 307-43.

6 Hill, W.C.O., Porter, A. and Southwick, M.D. (1952-3), 'The natural history, endoparasites and pseudoparasites of the tarsiers (*Tarsius carbonarius*) recently living in the Society's menagerie', *Proc. zool. Soc. Lond.* 122, 79.

7 Harrisson, B. (1963), 'Trying to breed *Tarsius*', *Malay. Nat. J.* 17, 218-31.

8 Le Gros Clark, W.E. (1924), 'Notes on the living tarsier *Tarsius spectrum*', *Proc. Zool. Soc. Lond.* 1924, 217-23.

9 Banks, E. (1949), *Bornean Mammals*, Kuching.

10 Davis, D.D. (1962), 'Mammals of the lowland rain-forest of North Borneo', *Bull. Singapore Nat. Mus* 31, 1-129.

11 Harrisson, B. (1962), 'Getting to know about *Tarsius*', *Malay. Nat. J.* 16, 197-204.

12 Wharton, C.H. (1950), 'The tarsier in captivity', *J. Mamm.* 31, 260-8.

13 Charles-Dominique, P. (1971), 'Eco-éthologie des prosimiens du Gabon', *Biol. Gabon.* 7, 121-228.

F. D'SOUZA

A preliminary field report on the lesser tree shrew Tupaia minor

Introduction

Tree shrews are small squirrel-like mammals occurring throughout South-East Asia. The family Tupaiidae is currently classified into two sub-families: the Tupaiinae, which includes five genera, and the Ptilocercinae having only one genus, the pen-tail tree shrew. Within the Tupaiinae, there is a range of species which are said to occupy different ecological niches. For example, *Tupaia minor* (Fig. 1) is reported as being almost wholly arboreal, *Lyonogale tana* (Fig. 2) as predominantly terrestrial, and *Tupaia glis* or *T. belangeri* as semi-arboreal. The morphological differences between these species, particularly in the skull region (Fig. 3), are assumed to be functionally related to the differing habits.[1-4]

The following table illustrates the broad 'extremes' within the sub-family Tupaiinae.

Species	*Assumed habit*	*Av. weight*	*Av. length of head and body*	*Tail length as % of head and body*[5]
T. minor	Arboreal	44 gms. Range 33-59	126 mm. Range 114-138	100-130%
T. glis	Semi-arboreal	145 gms. Range 88-173	177 mm. Range 156-197	75-95%
Lyonogale tana	Terrestrial	200 gms.	180-230 mm.	70-90%

As yet there have been no systematic field studies of tree shrew species; therefore the relationships between the morphological and behavioural differences of both Tupaiinae and Ptilocercinae and their precise ecological niches have yet to be established.

Figure 1. *Tupaia minor*; note the short snout.

Figure 2. *Lyonogale tana*; note the long snout.

Methods and study area

The following report is based on a short-term observational study carried out in a patch of primary forest, 17 km. north-west of Kuala Lumpur, Malaya. The forest contained, among other mammalian species, three tupaiid species: the diurnal *T. glis* and *T. minor*, and the only nocturnal tree shrew, *Ptilocercus lowii.* In addition to the field study data, records of the current Mark/Release Programme

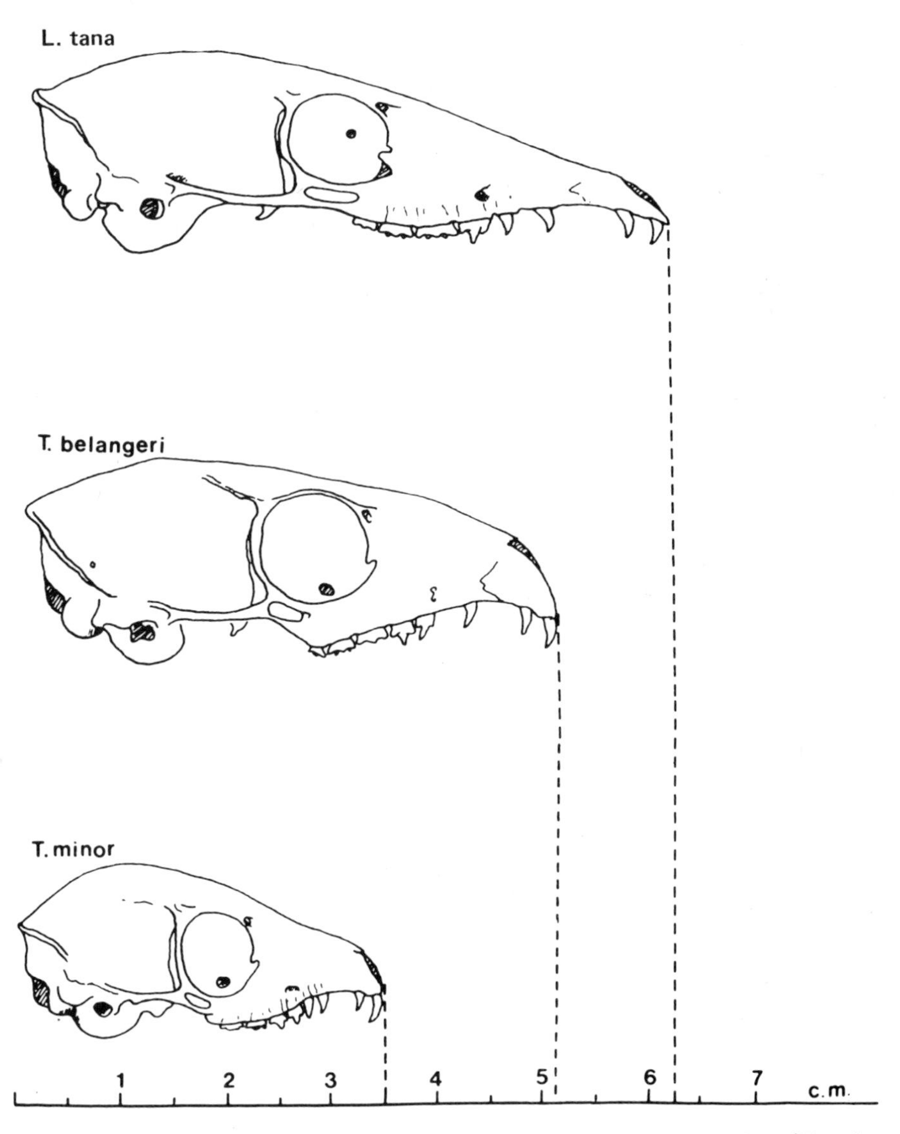

Figure 3. Scale drawings of the skulls of three tree-shrew species. (*Tupaia belangeri* is very similar to *Tupaia glis*).

were made available through the generosity of Dr Illar Muul of the United States Army Medical Research Unit, Institute for Medical Research, Kuala Lumpur.

Observations at canopy and sub-canopy level were made possible due to a system of ladders or transects built into the trees and interspersed with platforms. The transect system in the study area was 187 m. in length and had 7 viewing platforms varying in height above ground level from 9 to 26 m.[6] Binoculars were used to locate the animals initially and all observations were recorded on a cassette tape recorder and transferred the same day to data sheets supplied by the IMR. These data sheets incorporated maps of numbered trees surrounding the transect system. Thus it was possible to chart the progression of any one animal by tracing its movements from one numbered tree to another. Temperature and humidity were recorded on a thermohygrograph situated at a height of 8 m.

The broad habitat of tree shrews is that of tropical rain forest; however, within this category there are several relevant divisions. In Malaya, 5 vegetational zones can be distinguished: rice fields, lallang (long grasses), scrub, secondary and primary forest.[7,8] *T. minor* is said to inhabit primary forest predominantly, but it has been seen and trapped in both secondary forest and scrub. *T. glis* is a more wide-ranging species and has been recorded in primary and secondary forests as well as in scrub and cleared patches of land.

Bukit Lanjan, where this pilot study was carried out, is an area of approximately 25 square km. of predominantly primary forest bounded by villages, rubber plantations and roads. A good cross-section of both large and small mammals has been recorded in the forest, ranging from civets, leopards and bears to the minute bamboo mouse (*Chiropodomys gliroides*). Elevation ranges from sea level to approximately 1000′ (304 m.) and the area is defined as lowland rain forest. Temperatures in the lowlands are uniformly high, rarely falling below 21°C. The relative humidity ranges from 17%-96% during any one year. An analysis of variation in both temperature and humidity over a period of 18 months showed no significant fluctuations from one part of the year to another. The north-east monsoon and the south-west monsoon give rise to two slightly wetter periods during the year; however, this is not to imply that there are 'dry' periods or seasons. It is rare for rainfall to be less than 12 cm. in any one month.[9]

Primary forests are dominated by dipterocarp trees, which reach up to 30 m. in height, the crowns touching to form a continuous canopy. Above this, emergents rise to a height of 55-60 m. Below the canopy is a layer of saplings and smaller tree species which form a discontinuous under-canopy. The ground has a shrub and herbaceous layer which is generally sparse, although it can become dense where the canopy has been broken and the light penetrates to ground level. In the study area, clearings were not infrequent, as dead trees were uprooted by heavy rains and soil erosion. Thus, the pattern was of tall trees having a thick canopy and a lower interrupted layer of young trees and shrubs.

Intensive observation was concentrated within the area covered by the hill transect system. This observation station was situated well within the boundary of primary forest and at an elevation of 275 m. Occasional observations were carried out at the valley transect system, which again was just within the borders of primary forest, but located at sea level.

A third observation post, situated on the fringe of primary forest opening onto patches of secondary forest and lallang, was also selected. Only two *T. glis* were sighted in the secondary forest and only four in primary forest areas during the study period. Two *Ptilocercus lowii* were seen at night within the area of the hill transect, though the aborigines say it is mainly a primary forest creature. Thus the following report is mainly concerned with the lesser tree shrew *T. minor*, in a primary forest habitat. Apart from the lesser tree shrew, two other arboreal or semi-arboreal species were prevalent in the area; the black banded squirrel (*Callosciurus nigrovittatus*) and the slow loris (*Nycticebus coucang*), which is a nocturnal species. *T. minor* was distinguished from *C. nigrovittatus* and other small forest mammals (e.g. rodent species) by the following criteria:

(*a*) rapid 'jerking' type of locomotion, accompanied by tail-flicking
(*b*) light-coloured shoulder stripe
(*c*) pronounced length of tail in proportion to body size
(*d*) olivaceous colouring and ventral buff-coloured patches
(*e*) occasional distinctive vocalisations (recognised on the basis of recordings made in the laboratory).

C. nigrovittatus was recognised by its long lateral black band and *N. coucang* by the reflection from the tapetum.

Results

1. Arboreality

Since *T. minor* represents one of the 'extreme' species of the sub-family Tupaiinae, one of the main aims of the study was to determine the broad ecological niche and, in particular, the degree of arboreality. It is important to note that it is on the basis of structural similarities between the arboreal tree shrew species and some extant prosimian groups that the tree shrews have been included in the Order Primates. Thus, features such as the short rostrum, reduction of olfactory apparatus, relatively large eyes, forward rotation of the orbits, the development of a post-orbital bar and the structure of manus and pes, have been assumed to represent both arboreal adaptations and evidence of primate affinity.[1–2,10–14]

In order to assess the degree of arboreality, three observation points were selected from each of which there was a clear view of the ground, middle and canopy zones of the forest. Observation sessions of 15 minutes each were carried out three times daily for two weeks, between the hours of 09.00 and 10.30; 15.30 and 17.00; 20.00 and 21.30. Initial observations had confirmed that these times coincided with the main activity phases of the three most common species in the study area. Since the animals were not marked, it was not possible to assess the time spent at varying levels for *each individual.* Thus the animals were simply counted and the height above ground level recorded. During the 31.5 hours of observation, the sighting scores for the three species were as follows: *T. minor* = 35; *C. nigrovittatus* = 40; *N. coucang* = 21.

Of the 35 *T. minor* sightings, 11% (N = 4) were on the ground, 31% (N = 11) between 0.6 and 3 m., 34% (N = 12) between 3 and 7 m., and 23% (N = 8) between 7 and 12 m. Fig. 4 shows that both the squirrel and the loris appear to be restricted to heights above 3 m., and neither species was sighted on or near the ground.

These sighting scores may suggest that *T. minor* is less arboreal than usually supposed. If this is the case, it cannot be said that specialisations of the skull and hands and feet are all part of a complex adaptation to an exclusively arboreal way of life. This is

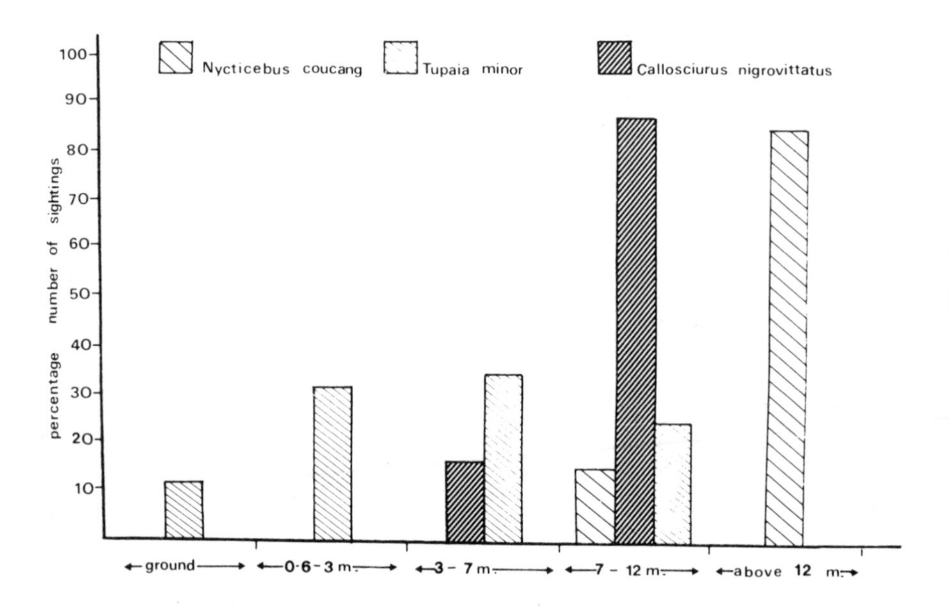

Figure 4. Sighting scores at varying heights above ground level for three primary forest mammal species.

particularly important if any comparison is to be made with the 'semi-arboreal' *T. glis*, which is presumed to show fewer morphological arboreal adaptations but which spends at least part of the time in the trees.

The structure of the primary forest in the study area was such that, in order to remain in the trees, an animal would be forced either to negotiate gaps of up to 6 m. or to exploit the 'fine branch' zone.

The slow loris, which has an average weight of 800 gm.,[12] moved through the canopy of the trees grasping extremely fine branches. The squirrel was able to negotiate gaps of up to 4 m., and one individual was seen to drop a vertical distance of 7 m. It would seem that *T. minor* shows neither of these types of locomotion and that it relies on scansorial adaptations in order to progress through the forest. The lack of extreme arboreal specialisation is correlated with the frequency of its occurrence on or near the ground (see Fig. 5).

Both the manus and pes of tupaiine species are narrow and elongated and have long, curved claws.[15,16] Palmar and plantar pads are well differentiated and the terminal digital pads have transverse ridges (except possibly in *T. javanica*).[14] *T. minor* has a rapid running and hopping type of locomotion; it balances on, rather than grasps, surfaces; though the curved claws and the frictional resistance of the palmar and plantar pads against the tree surfaces both contribute to the maintenance of balance. In climbing vertical branches, laboratory observations and tests show that use of the claws is more important than any grasping action of the digits.

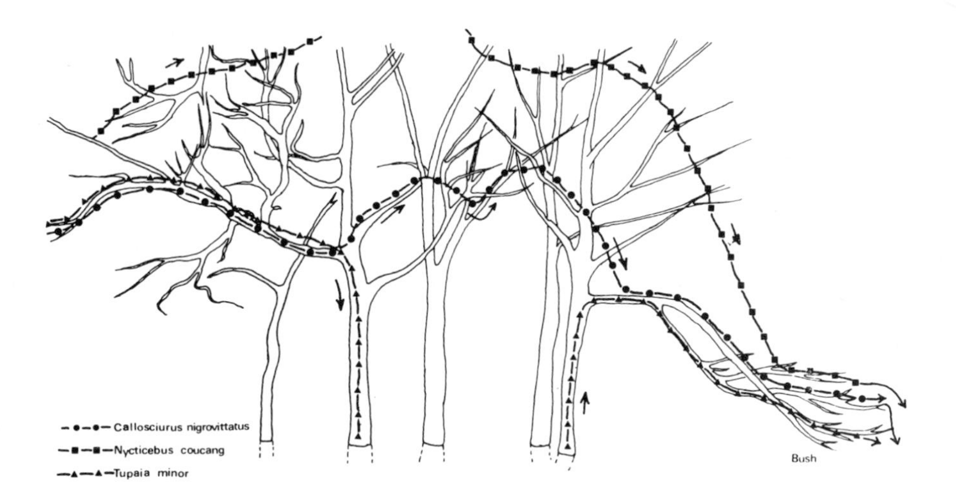

Figure 5. Characteristic routes through the forest taken by the three mammal species.

Tupaiine species do *not* show any marked divergence of either pollex or hallux, nor do they have truly grasping hands and feet.[4,12,17] Laboratory tests have shown that, when climbing, the limbs are well under the animal and not at an angle to the body; this is further evidence that the Tupaiinae, though excellent climbers, are not able to exploit the fine branch habitat, and thus their degree of arboreality is limited by the diameter of branches in the habitat.

In captivity, *T. minor* selects broad branches (i.e. 5 cm. or more in diameter) for activities such as eating, mating, grooming, defaecating and marking. Fine branches (i.e. approximately 1.5 cm. in diameter) are avoided, and this is markedly so for newly imported animals. If animals are given only fine branches, they will utilise the branches after some months as a means of getting from one cage area to another; however, all other activity in these cages takes place at ground level. There also seems to be a *preference* for height, both for general daily activity and for selection of nesting boxes. Given suitably broad branches at varying heights from ground level, individuals invariably select the highest nest boxes and topmost branches. Food pieces, for example, are taken in the mouth from the cage floor to preferred branches high up in the cage. This behaviour is markedly different from that observed in *Lyonogale*, which does not exhibit a *preference* for height, either for nesting or daily activities.

The ability to negotiate gaps between the trees depends to a large extent on the accuracy of visual perception of depth. Polyak (pp. 901-2) suggests that this ability is developed in Tupaiinae species (*Urogale everetti* and *T. glis*), but less highly than in some sciurid species.[18] Perhaps one of the limiting factors is the extremely small body size of *T. minor*. The type of locomotion exhibited by *T. glis* in the forest has not yet been studied, but its larger body size may indicate that it is able to leap across gaps more frequently and would thus be less scansorial than *T. minor*.

2. *Activity patterns*

Within the intensive study area, there was one main 'pathway system', consisting of a series of interconnecting broad branches. Another pathway, which was clearly visible from the ground, was located 150 m. from the transect system. These two pathways were used regularly by *T. minor* and *C. nigrovittatus*, and on two occasions by *T. glis*. The importance of these pathway systems lies in the fact that, once they had been located, peak activity periods could be established by counting the numbers of individuals crossing them. One such pathway system was within a few feet of the observer's platform, and it was possible to observe the animals from first light without causing any undue disturbance. Initial observations indicated that the earliest phase began at approximately 05.45 hrs. with the appearance of *C. nigrovittatus. T. minor* first appeared considerably later (approximately 07.45 hrs.), when the sun was well up.

Both pathway systems were observed for a total of 8 sessions. Each session began just before dawn at 05.20 hrs. and continued until just after sunset at 18.30 hrs. During two of the sessions, another observer, independently, scored the individuals sighted, in order to provide a check for accuracy. The species, exact time, and gross behavioural category (i.e. moving, feeding, grooming, mating, resting) were recorded. Laboratory observations on a captive colony have shown that maximum activity can be reliably estimated by both the total *amount* of movements made and the different *kinds* of activity engaged in at any one time. Movement from one part of the cage to another in addition to feeding would seem to characterise the height of the activity phase. Accordingly, in the field, movement or feeding were used as behavioural indicators of activity.

Only those days during which temperature, humidity and rainfall were uniform were selected. Rainstorms or even heavy clouds tended to curtail the activity of *Tupaia minor* and some other forest species. The requirements for an 'observation' session were as follows:

(*a*) The previous afternoon and early evening had been free of rainfall.
(*b*) The hill was not obscured by heavy mist at 05.30 hrs. on the morning of the observation day.

Fig. 6 shows that *T. minor* have a bimodal pattern of activity, with peaks at 08.30-11.30 hrs. and 16.00-17.30 hrs. These data correspond well to the activity phases recorded in the laboratory. The decline in activity during the middle part of the day is characteristic for *T. minor* and *T. belangeri*,[3,19] and in the field these hours are the hottest. Temperatures may reach 28°C, and the minimum temperature recorded during the observation sessions for the period 12.00-15.30 hrs. was 21°C (cf. Figs. 6 and 7).

A preliminary analysis of 4-day periods following a previous day of either (*a*) frequent and heavy, or (*b*) continuous rainfall, suggests that these activity phases are extended for both species by up to 1½ hours during the day. *C. nigrovittatus* was sighted as early as 05.15, and *T. minor* at 06.45. On these days, the termination of activity was characterised by frequent observation of the two species together on the pathway systems, which does not otherwise occur.

The predominant cone-type structure of the tupaiid retina (*Urogale everetti* and *T. glis*)[18,20] suggests that it is adapted to diurnal vision. Polyak (p. 823) writes: 'Tree-shrews must be considered as specialised almost exclusively for photopic conditions.'[18] The sciurid retina, however, is composed of both rods and cones and this structural difference may account for the differential activity phases. Martin reports that in the laboratory *T. belangeri*

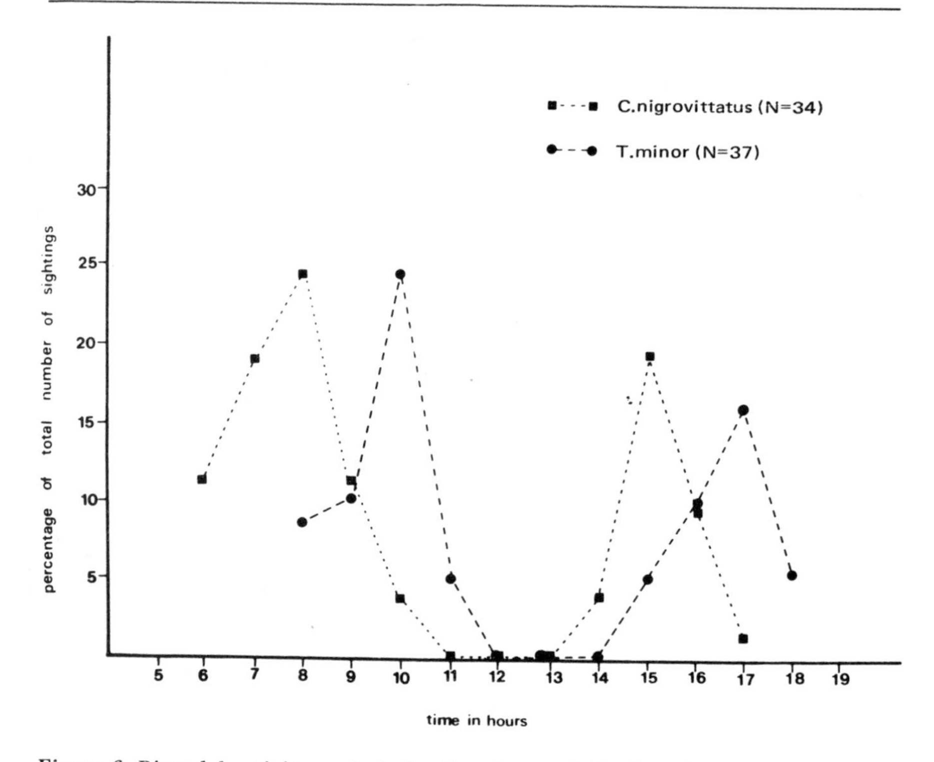

Figure 6. Bimodal activity periods for *T. minor* and *C. nigrovittatus.*

returned to the nest when heavy banks of cloud obscured the light.[3] On two occasions during the present field study, *T. minor* were seen disappearing into tree-hollows when the sun was only momentarily obscured by cloud, remaining there until the cloud dispersed. Captive *T. minor* appear to spend up to 5 minutes peering out of their nest boxes before emerging. This behaviour was also noted in the field with an individual *T. minor* whose nest entrance was within sight of the observer's platform. Observation of this nest entrance revealed that a period of simply looking out of the nest before emerging was a regular feature. The exact level of light intensity which promotes emergence and the onset of activity, or which determines return to the nest, has yet to be established, but it would seem that light intensity itself is an extremely relevant cue, as Pariente has shown for *Phaner* and *Lepilemur.*[21]

3. *Diet*

Laboratory observations show *T. minor* to be generally omnivorous, with a preference for live animal food (i.e. mealworms, insects and day-old mice). It is extremely difficult to observe the feeding habits

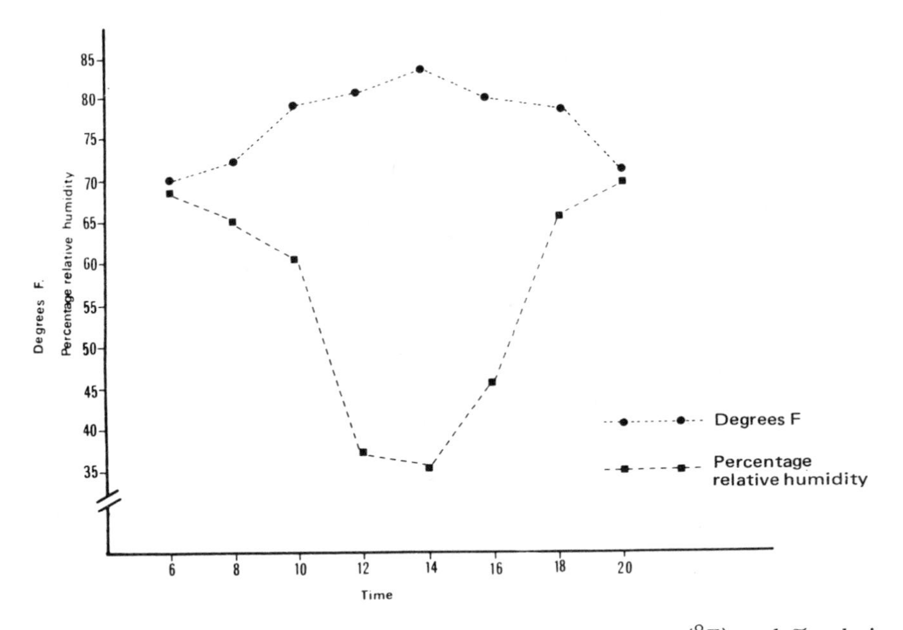

Figure 7. Average daily fluctuations of both temperature (°F) and % relative humidity during the study period.

of this species in the field. On two occasions, *T. minor* were seen carrying pieces of overripe bananas from the ground and feeding at a height of approximately 12 m. On another occasion, one individual was seen nosing and scratching the terminal branch of a dying tree, which was subsequently found to be infested with termites. Groups of *T. minor* have also been observed feeding on figs (*Ficus glabella*) (Dr I. Muul, personal communication). An analysis of the stomach contents of 10 shot *T. minor* showed fruit seeds, and the remains (e.g. wings and legs) of hymenopterans and other arthropods. The stomachs of two specimens contained several hairs, the structure of which indicated derivation from other mammal species. In the laboratory *T. minor* will tackle an adult mouse and kill it swiftly by grasping it in both hands and biting the head and neck. This method of killing appears to be typical for the genus *Tupaia*.[19] and it is quite possible that *T. minor* occasionally preys on small forest mammals.

4. *Nesting behaviour*

One *T. minor* nest was located and collected and two others inferred from the regular emergence and return of two individuals. The following table shows the location and height above ground level of the three nests:

Height	*Location*	*Zone*	*Remarks*
13 m.	In crevice of dried palm leaf	Middle of under canopy	Isolated tree surrounded by thick undergrowth. Nest constructed of tightly woven dried grass, leaves and strands from the nylon rope supporting the transect.[22]
19 m.	In hollow branch of 'medang' tree	Middle of under canopy	Entrance approximately 4 cm. in diameter. Isolated tree, not on pathway.
18-21 m.	'Medang' tree	Middle of under canopy	Isolated tree. Diameter of entrance approximately 7.5 cm., but situated on underside of branch and thus effectively protected from rain.

The first nest was only located, collected and examined on the last day of the study period, so it was not possible to observe the times of emergence and re-entry. This nest contained one female. The latter two nests were observed daily (when the entrances were not obscured by mists) at the beginning and end of both morning and afternoon activity phases. It appears that *T. minor* return to their nests either: (*a*) at the end of the afternoon activity phase at approximately 18.00 hrs., or (*b*) when the sun is obscured by cloud. They do not normally return to the nest during the midday decline in activity, which is also true of the captive colony. A fourth nest was inferred from the characteristic vocalisations of nestling *T. minor* and the regular disappearance of one adult into the tree hollow each morning at 09.00-09.15 hrs. (5 observations) and on one occasion at 18.00 hrs. If *T. minor* exhibit the 'absentee' system of maternal care reported for *T. belangeri*.[3,23,24] this latter behaviour pattern would suggest that the young had been deposited in a separate nest and were suckled at long intervals. Laboratory observations on the captive colony suggest that *T. minor* do in fact deposit their young in a separate nest box and that the inter-feeding interval is at least 24 hours and probably longer (D'Souza, in prep.).

Reproductive data

1. Seasonal weight changes

The Medical Ecology Division of the IMR initiated a field trapping programme in April, 1970, as part of the current research project into the transmission of diseases. Trapped individuals are taken back to the laboratory for examination, marked and then released the following day. In addition to examination for endoparasites and ectoparasites, any signs of pregnancy or lactation are noted for females, and testis descent and size are noted for males. Weights and body measurements are recorded, unless less than two weeks have elapsed since the last trapping date. During the course of this programme a total of 75 *Tupaia glis* (32 males and 43 females) and 36 *Tupaia minor* (17 males and 19 females) were trapped. Recapture dates for several *T. glis* are available, but none have been obtained for *T. minor.* Though the sample is as yet small, the data already collected permit some provisional statements as to the presence or absence of a breeding season in *T. glis.* Harrison demonstrated that male *T. minor* and *T. glis* show a significantly lower testicular weight during the period October-March than from April-September,[25] and this indicates a lower breeding potential during the former period. However, the number of pregnancies recorded for both species throughout the 4-year project was small (9), and did not show an absolute restriction to the April-September period. Martin suggests that the breeding potential of males as shown by a decline in testicular weight may vary according to environmental conditions and he concludes that pregnancy can generally occur throughout the year.[3]

In the IMR Mark/Release Programme there are only three recorded incidents of pregnancy, birth and suckling. One *T. glis* gave birth prematurely in the laboratory (26/3/71); a palpable foetus was noted for another individual (5/5/71); one *T. minor* was lactating (14/9/71). However, the recapture weights of several female *T. glis* show a greater incidence of fluctuation during the period March-August than from September to February. The average weight of wild-caught female *T. glis* is 140 gm. (43 individuals). Table 1 shows the weight changes for six females during the period May 1970-January 1972.

Rapid decreases in weight after capture may be a result of the stress involved in handling and anaesthetising the animal. However, there are four cases of substantial weight *increases* subsequent to capture and examination, one associated with birth and another with lactation (which occurs a few days prior to parturition in captive animals). It is thus reasonable to assume that major and rapid weight fluctuations in females are fair indications of pregnancy and birth. Furthermore, weight increases in males greater than 5 gm. are always associated with physical growth, as shown by gross changes in body measurements. Weight decreases in males appear to be rare; only one example is available, and the individual concerned died shortly after capture.

The greatest weight fluctuations, both in amount per individual and total recorded, occur in the period late March to the beginning of August, with only one exception (TG 33). The total number of animals trapped per month does not differ significantly. Weights of females captured during the period September-February do not generally show weight gains or losses greater than 7 gm. These data may support Harrison's evidence of a breeding peak, if not a well-defined breeding season.[25] It is unlikely that the weight changes

Table 1 Weight changes in female *Tupaia glis*

Animal number	*Trapping dates*	*Weight (grams)*	*Reproductive data*	*Weight change*
TG 11	19/5/70	129		
	12/4/71	179		+50
	21/4/71	147		-32
	14/5/71	144		-3
	6/6/71	153		+9
TG 23	11/12/70	158		-13
	30/12/70	145		
TG 27	17/12/70	149		
	11/1/71	137		-12
	26/3/71	179	Gave birth	+42
	13/4/71	162	3	-17
	5/5/71	153		-9
	19/7/71	180		+27
	6/8/71	103		-77
TG 33	29/12/70	155		
	20/1/71	113		-42
TG 62	3/6/71	170		
	7/6/71	130	Lactating	-40

are purely a result of variation in food sources, since males are not similarly affected.

2. *Testis retraction*

It has been stated that *Tupaia* exhibits non-seasonal permanent descent of the testes, whilst *Ptilocercus* exhibits seasonal descent of the testes;[2,11] however, the records from the Mark/Release Programme show that testis descent in *Tupaia* is extremely variable and is not restricted to any particular part of the year. Martin and von Holst have demonstrated that in captivity *Tupaia belangeri* can retract the testes into the abdominal cavity under conditions of social stress.[3,26] Martin (p. 421) writes: 'Dissection of the adult male *T. belangeri* demonstrated that the inguinal canal is open throughout life.'[3] The field records suggest that the incidence of testis-retraction is higher in animals confined in traps before examination than in those animals shot and examined immediately after death. Fig. 8 shows the percentage of testis descent and retraction in both shot and live-trapped individuals. It is interesting to note that 50% of the trapped *T. glis* exhibiting testis descent had extremely small testes, whereas the shot *T. minor* exhibited large testes. All individuals were adults, using body measurements as the criterion. Von Holst reports that under laboratory conditions, full testis-retraction occurs over a period of approximately 10 days.[26] It may be that in a stress situation as severe as confinement in a small trap for a period of some hours, testis-retraction is an extremely rapid stress response.

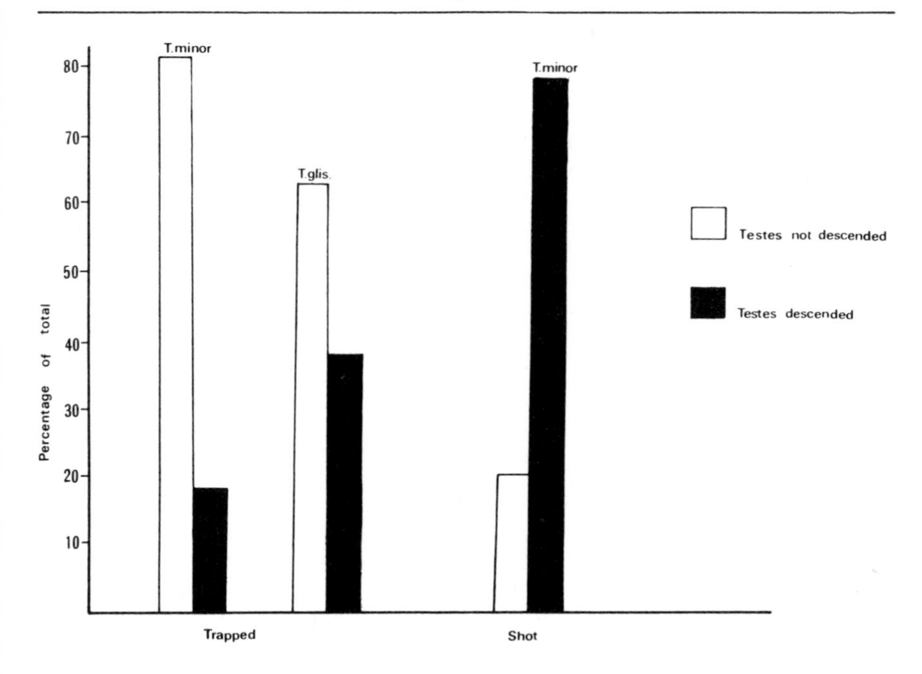

Figure 8. Histogram illustrating the percentages of testis-retraction in both shot and live-trapped individuals.

Acknowledgments

Many people have helped both in making the field study possible and in the preparation of the manuscript. In particular I should like to acknowledge the co-operation and generosity of the following persons during the field study period: Illar Muul, Lim Boo Liat, Michael Lee (IMR, Kuala Lumpur) and S.J. D'Souza, London. I am especially grateful to my supervisor Robert Martin, for his advice, criticism and encouragement, and to Dietrich von Holst and Alan Walker for their helpful discussions. Thanks are also due to Professor G.A. Doyle for his patience in awaiting the final draft of this paper. Finally, I should like to thank the following people for their help in preparation of the diagrams: Mrs Wendy Davies, Mrs Janet Maxwell, Ted Miller (University of Reading) and Miss Christine Roberts, London.

NOTES

1 Gregory, W.K. (1913), 'Relationship of the Tupaiidae and of Eocene lemurs, especially *Notharctus*', *Bull. geol. Soc. Amer.* 24, 247-52.

2 Le Gros Clark, W.E. (1962), *The Antecedents of Man*, 2nd ed., Edinburgh.

3 Martin, R.D. (1968*a*), 'Reproduction and ontogeny in tree-shrews (*T.*

belangeri) with reference to their general behaviour and taxonomic relationships', *Z.f. Tierpsychol.* 25, 409-532.

4 Martin, R.D. (1968*b*), 'Towards a new definition of primates', *Man*, (n.s.) 3, 377-401.

5 Data for *T. glis* and *T. minor* have been taken from the IMR field records. Data for *Lyonogale tana* are from Harrison (1964).[7]

6 Muul, I. and Lim, B.L. (1970), 'Vertical zonation in a tropical rain forest in Malaysia: method of study', *Science* 169, 788-9.

7 Harrison, J. (1964), *An Introduction to the Mammals of Sabah*, Jesselton.

8 Harrison, J. (1966), *An Introduction to the Mammals of Singapore and Malaya*, Singapore.

9 Banks, E. (1935), 'The change in climate and fauna with ascending altitude in Malaysia', *Sarawak Mus. J.* 4, 343-56.

10 Le Gros Clark, W.E. (1924), 'On the brain of the tree-shrew (*Tupaia minor*)', *Proc. zool. Soc. Lond.* 44, 1053-73.

11 Le Gros Clark, W.E. (1926), 'On the anatomy of the pen-tailed tree-shrew (*Ptilocercus lowii*)', *Proc. Zool. Soc. Lond.* 46, 1179-309.

12 Napier, J.R. and Napier, P.H. (1967), *A Handbook of Living Primates*, New York.

13 Wood Jones, F. (1917), *Arboreal Man*, London.

14 Wood Jones, F. (1929), *Man's Place Among the Mammals*, London.

15 Haines, R.W. (1955), 'The anatomy of the hand of certain insectivores', *Proc. zool. Soc. Lond.* 125, 761-77.

16 Haines, R.W. (1958), 'Arboreal or terrestrial ancestry of placental mammals', *Quart. Rev. Biol.* 33, 1-23.

17 Napier, J.R. (1961), 'Prehensility and opposability in the hands of primates', *Zool. Soc. Lond. Symp.* 5, 115-32.

18 Polyak, S. (1957), *The vertebrate visual system*, Chicago.

19 Sorenson, M.W. (1970), 'Behaviour of tree-shrews', in Rosenblum, L.A. (ed.), *Primate Behavior*, vol. 1, New York.

20 Wolin, L.R. and Massopust, L.C. (Jnr.) (1970), 'Morphology of the primate retina', in Noback, C.R. and Montagna, W. (eds.), *The Primate Brain*, vol. 1, New York.

21 Pariente, G., this volume.

22 Although the female inhabitant of this nest was seen taking nesting material to this nest, it could be that originally this 'woven' nest was inhabited by some other mammal or bird species. Laboratory data show that *T. minor* consistently builds a nest of loosely packed leaves and other fibrous material, but has never been noted 'weaving' the materials.

23 Martin, R.D. (1966), 'Tree-shrews: unique reproductive mechanism of systematic importance', *Science* 152, 1402-4.

24 Martin, R.D. (1969), 'The evolution of reproductive mechanisms in primates', *J. Reprod. Fert. Suppl.* 6, 49-66.

25 Harrison, J. (1955), 'Data on the reproduction of some Malayan mammals', *Proc. zool. Soc. Lond.* 125, 445-60.

26 von Holst, D. (1969), 'Sozialer Stress bei Tupajas (*Tupaia belangeri*). Die Aktivierung des sympathischen Nervensystems und ihre Beziehung zu hormonal ausgelösten ethologischen und physiologischen Veränderungen', *Z. vergl. Physiol.* 63, 1-58.

G. PARIENTE

Influence of light on the activity rhythms of two Malagasy lemurs: Phaner furcifer *and* Lepilemur mustelinus leucopus

Introduction

In the course of work on vision in Madagascar lemurs, it was considered necessary to obtain both ecological and ethological data for the species under investigation. The behaviour of the two species studied is known only from recent studies in which the author himself participated.[1,2] The author's study was concerned with the relationship between light and vision on the one hand and activity on the other.

The most striking feature of activity was its extreme regularity. Such a phenomenon requires either an 'internal clock' or a link with a regular physical parameter, like light or temperature. Since the meteorological conditions were particularly stable from one day to another at the time of the study, attempts were made to connect the activity of the animals with the cycles of light or temperature.

Description of species

1. Lepilemur mustelinus leucopus

The 'sportive lemur' (Fig. 1*a*), is a strictly nocturnal animal living in the south of Madagascar. By night its two big eyes strongly reflect the light of the observer's lamp, owing to the presence of a golden tapetum lucidum, which constitutes an efficient adaptation for nocturnal vision (Fig. 1*b*).[3] *Lepilemur* was observed in two different biotopes: the bush zone characterised by Didiereaceae, and the gallery forest along the Mandrare River. This study was carried out in

Figure 1 (*a*). *Lepilemur*: appearance of the animal in full light (note the vertical pupils in myosis).

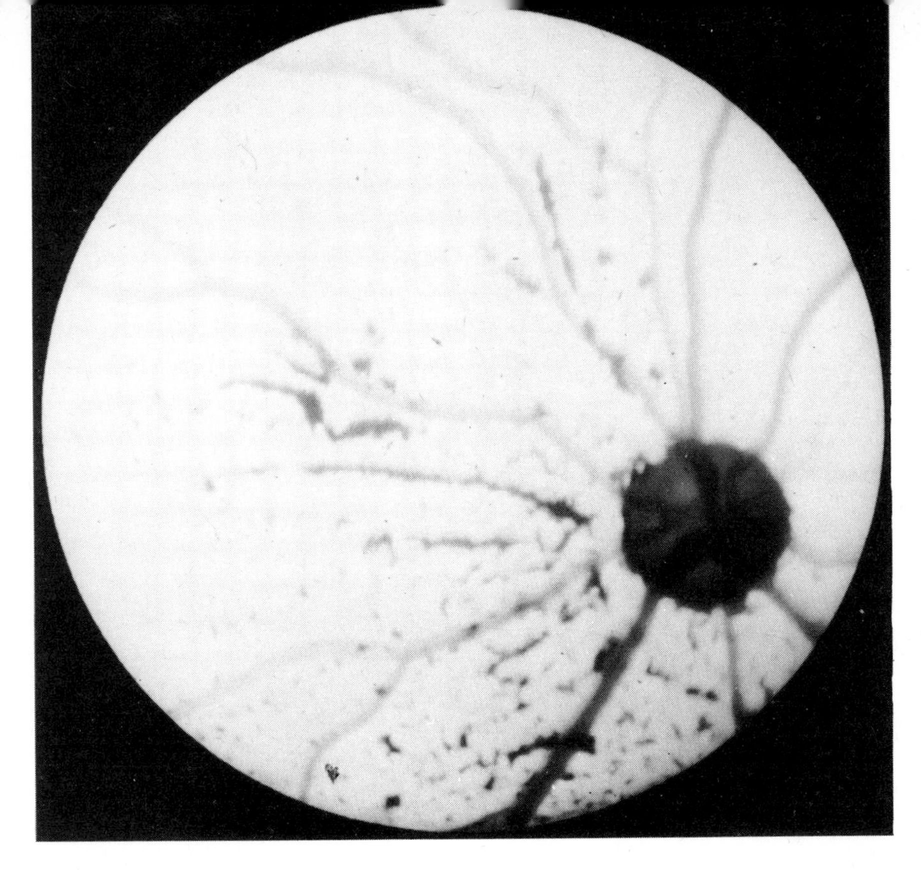

Figure 1 (*b*). Retinography of the papilla (blind spot) in *Lepilemur*. No foveal zone is seen on the golden tapetum.

Figure 1 (*c*). Pigmentation near the ora serrata (*Lepilemur*). No clear limit can be distinguished.

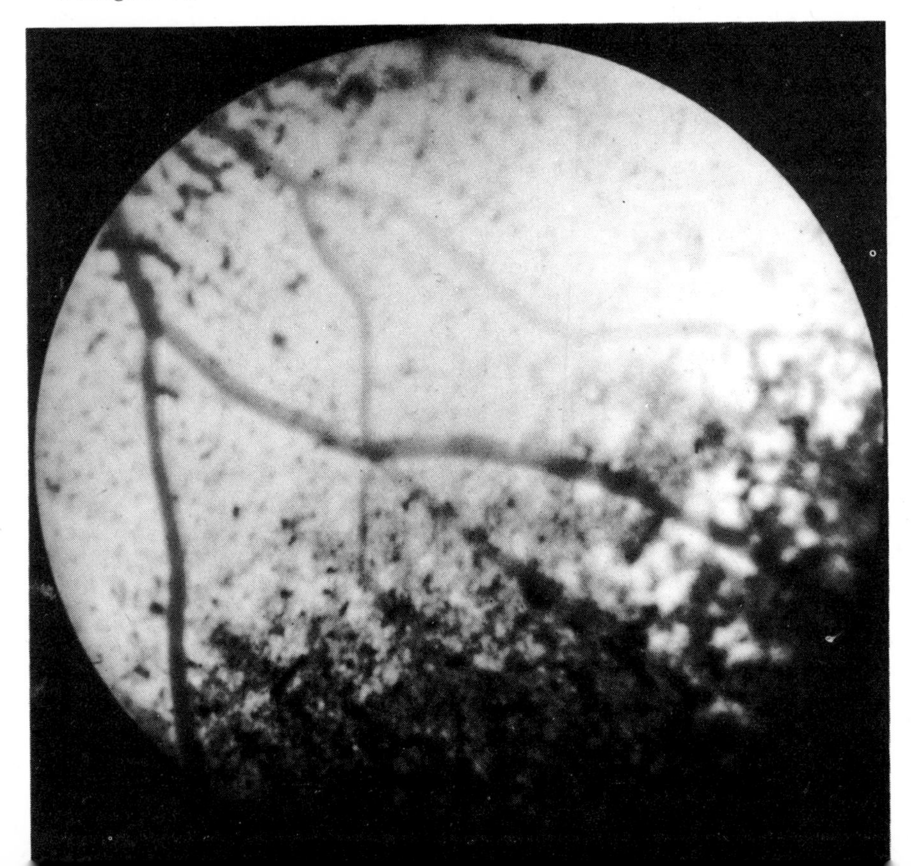

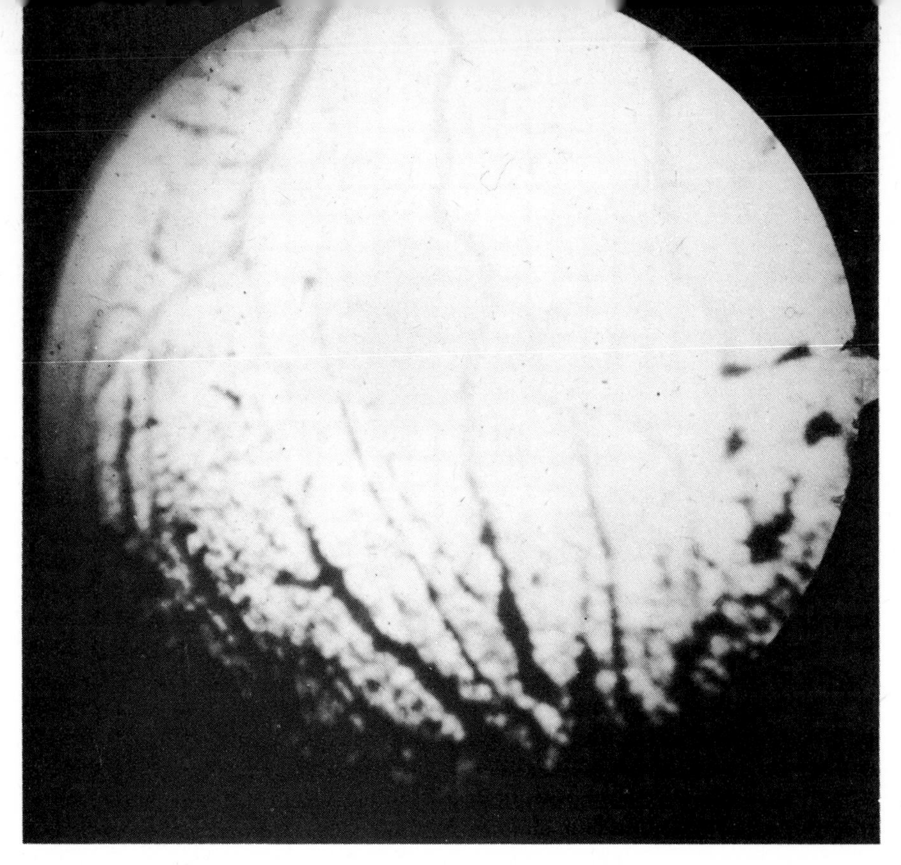

Figure 1 (*d*). Bright-gold appearance of the tapetum lucidum in *Lepilemur*.

Berenty (25°S, 46°24E), to the north-west of Amboasary, during the dry season when the temperature was 40°C at noon and about 14°C at 04.00 hrs. and the sky was cloudless most of the time.

2. *Phaner furcifer*

The 'fork-crowned lemur' (Fig. 2*a*) is also a strictly nocturnal animal, similarly possessing a tapetum lucidum.[3] It is as big as a squirrel with unbelievable agility in the trees. It was studied in Analabe, just North of Morondava (20°S, 44°33′E) in a typical forest zone where baobabs are very numerous. At the end of October and the beginning of November the weather was dry, very hot with a perfectly clear sky by night and by day with temperatures about 38°C at noon and 19°C at 05.00 hrs.

Apparatus and methodology

An autonomous and portable photometer IL 600 (International Light) was connected either to two different captors (PT 100 and PT 200 A) or to a photomultiplier system (Type PM 200 C).[4] Four filters were used with every captor (T_{max} at 350, 400, 520 and 640 nanometers). All measurements were taken at one metre above

Figure 2 (*a*). *Phaner furcifer*: the two eyeballs are very large in size.

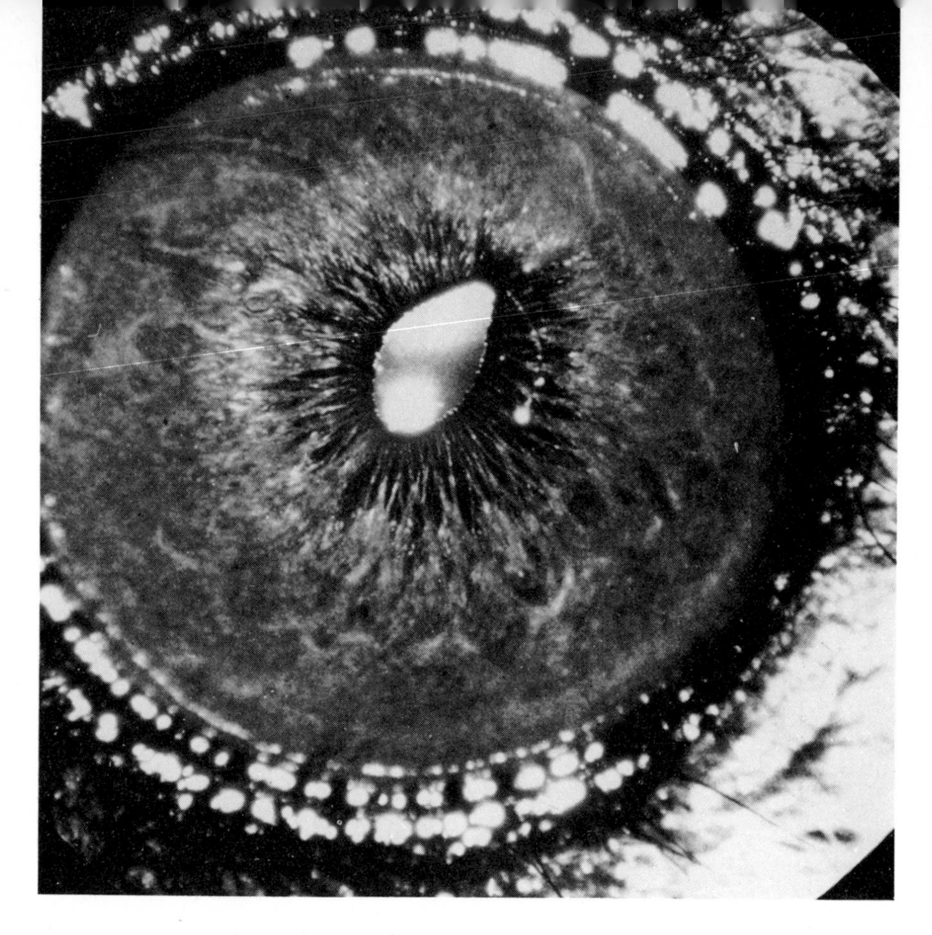

Figure 2 (*b*). Myosis showing the pear-shaped pupil of *Phaner*.

Figure 2 (*c*). *Phaner*: typical aspect of the ora serrata (it is quite different from *Lepilemur*, Fig. 1*a*).

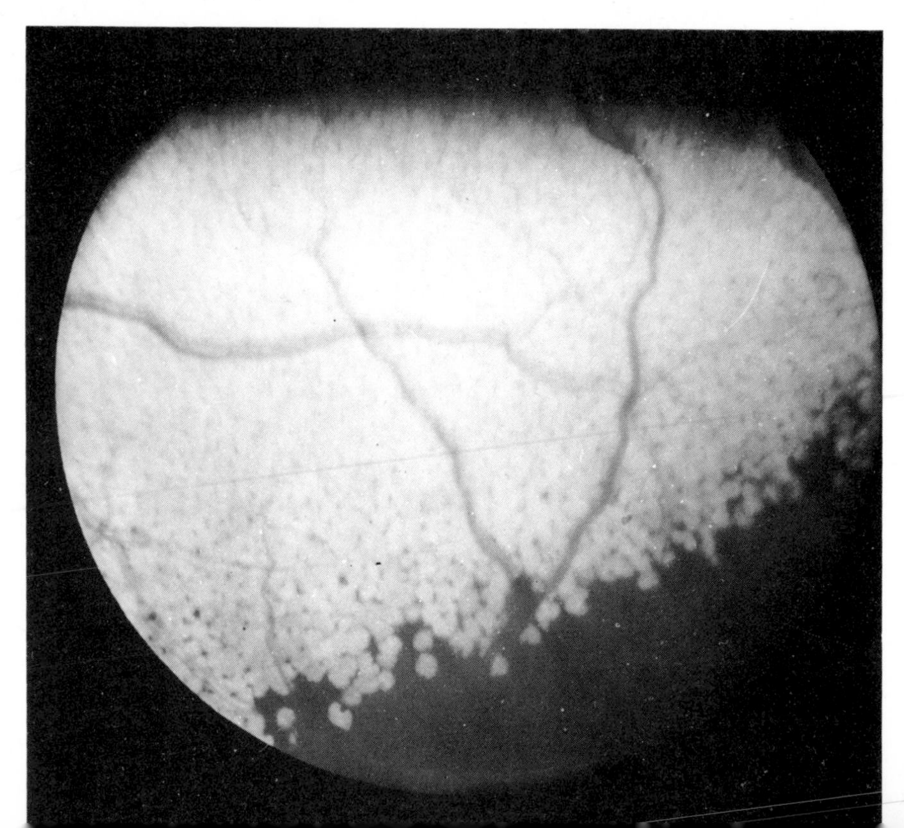

Figure 2 (*d*). Zenithal view showing the screening effect of vegetation on the sky in Analabe.

ground-level, every two hours and in five different spatial directions, north, south, east, west, azimuth (Fig. 2*d*). Orientation was always with reference to the geographic north (magnetic variation: W 18° 32 in Analabe; W 13° in Berenty). At dawn and twilight, continuous recordings of light were carried out.

As a reference, a series of measures was taken during the same period away from any vegetation (at airfields one or two miles from the forest). All the data obtained in this way are energy values expressed in μwatt/square cm., which have in some cases been converted into illumination units considering:[4]

(*a*) the spectral sensitivity curves of the three captors;
(*b*) the spectral distribution of energy of the sun for three colour temperatures;
(*c*) the transmission curve as a function of wavelength of the four filters;
(*d*) the spectral response of a standard observer (CIE) under two different conditions of adaptation (daylight adaptation for human eye = photopic; nocturnal vision = scotopic).

It must be stressed that the values in lux (Figs. 6-9) are not related to the animals (for their spectral sensitivity response is still unknown), but to the human observer.

Most of the measurements used for the graphs were south measurements (with the back towards the sun in the southern hemisphere). For nocturnal measurements it was also necessary to distinguish between the very low energies among the trees (illumination used for vision in real conditions) and the light coming from the sky through the trees.

Finally, the times of sunset and sunrise were calculated for the upper edge of the sun, taking into account the effect of atmospheric refraction:

(*a*) *Berenty* (24°59′ S; 46°24′ E)
27/9/1970 sunrise 05h. 40m. 36s. local time
meridian crossing 11h. 45m. 47s. local time
sunset 17h. 50m. 58s. local time

(*b*) *Analabe* (20°44′ E)
28/10/1970 sunrise 05h. 25m. 12s. local time
meridian crossing 11h. 47m. 48s. local time
sunset 18h. 10m. 24s. local time

Results

1. *Lepilemur mustelinus leucopus*

These animals have been observed in the Didiereaceae bush by Charles-Dominique and Hladik[1] and in the gallery forest by the author. Sleeping in tree holes or between branches, the animals in this area are relatively well protected from the environment and provided with shaded surroundings. At sunset, they habitually extend their heads out of their shelters and watch to see whether the light conditions are favourable. The regularity of emergence is remarkable, taking place between 18.02 hrs. and 18.23 hrs. (Fig. 3). Despite the relatively small number of observations, a distinction

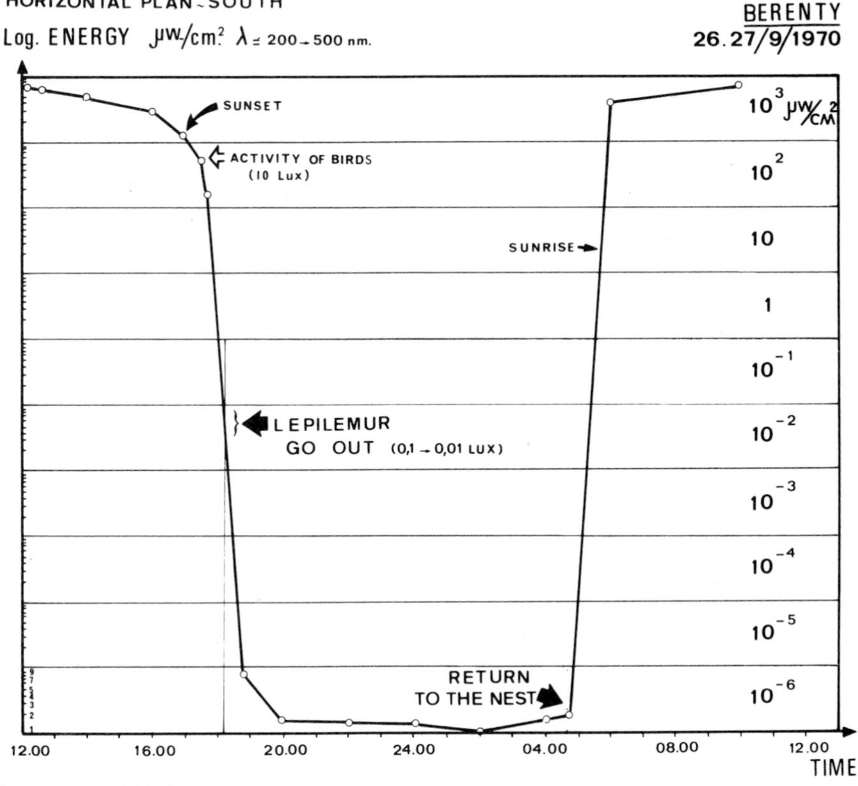

Figure 3. Diurnal variation in light energy on a south-horizontal direction during a 24-hour period at Berenty in September, showing the times *Lepilemur* emerges from and returns to the nest.

may be made between the animals living in the bush and those living in the gallery forest (Fig. 4). In terms of light penetration, the bush is much more open than the forest and is more similar to the open field. The average time of emergence is 18.08 hrs. in the forest and 18.16 hrs. in the bush.

This difference of 8 minutes permits all animals to leave their shelters when the light intensity is about 5×10^{-2} μw/square cm. When an intensity of 5×10^{-2} μw/square cm. is measured in the forest, the light is still about 1.5 μw/square cm. in the open field and 0.4 μw/square cm. in the bush. This illustrates the importance of light intensity for the activity rhythms of these animals. Temperatures and activity are plotted in Fig. 5; no obvious correlation between the two can be seen. Charles-Dominique and Hladik[1] consider that the rising of the moon during lunar periods induces the animals to vocalise; but no confirmation of this theory was sought in the present study.

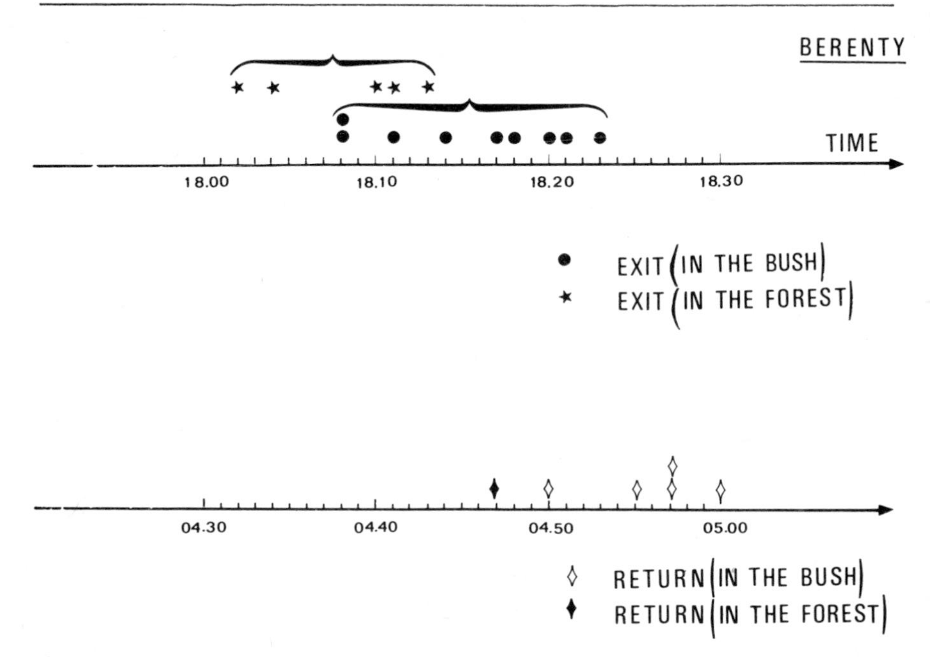

Figure 4. Comparison of nest-emergence time and nest-return time for *Lepilemur* in two different biotopes at Berenty, the bush and the forest.

During the night, the activity of *Lepilemur* is very variable. There is no doubt, however, that these lemurs have very good vision, since they are still able to leap several metres, even with the light level as low as 5 μlux (Fig. 6). The comparison of measurements in the open field and in the forest shows that the vegetation decreases the light almost 100 times.

The animals are less regular when returning to their retreat (cf. Fig. 4), but always do so when light is still at a very low level. The human observer is unable to detect a change in illumination at such times. In fact, the measurements show that the earlier returns correspond with an increase in light by at least 50% (100% in a number of observations). In all cases, the light conditions change extremely rapidly at about 04.45 hrs.

2. *Phaner furcifer*

In most cases, the first leap out of a spherical leaf-nest is accompanied by loud shrieks from the animal. In Analabe, this usually occurred with great regularity at 18.20 hrs. (only once did it happen at 18.10 hrs.). At 18.30 hrs. many animals started vocalising together. In fact, the timing of the first leap is different from that of true awakening; as with *Lepilemur* the animals wait for favourable environmental conditions before leaving the nest. It was thought at first that this mechanism was governed by temperature variations.

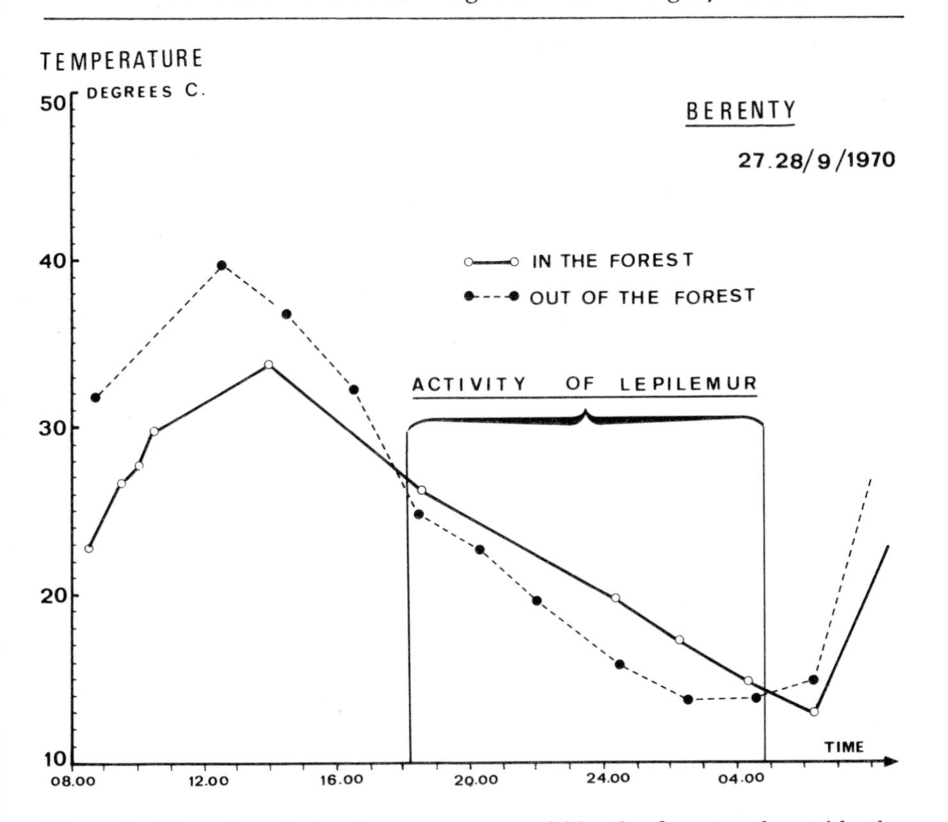

Figure 5. Diurnal variation in temperature within the forest and outside the forest at Berenty in September, showing the range of temperatures within which *Lepilemur* is active.

However, it can be seen from Fig. 7 that the temperature is 30°C at emergence and 18°C on return. Furthermore, temperature is decreasing progressively (1°C per hr.) when emergence takes place, whereas return to the nest precedes a rise in temperature by 30 to 40 minutes. Temperature variations do not therefore appear to be correlated with activity.

Fig. 8 shows that light variation is a very regular phenomenon. It can be seen from the same figure that activity starts as soon as the light falls below 2 lux. It is worth recalling here that in tropical zones (as in this study) the sun follows a half circle very high in the sky and sets along a line very similar to a perpendicular on the horizon. This is the reason why the change in luminosity is so pronounced at dawn and twilight.

For the observer's eyes, it is necessary to pass from photopic to scotopic conditions within a few minutes. Thus the time when the animals become highly active is a very characteristic one for human beings, as it corresponds to marked discomfort due to lack of visual adaptation.

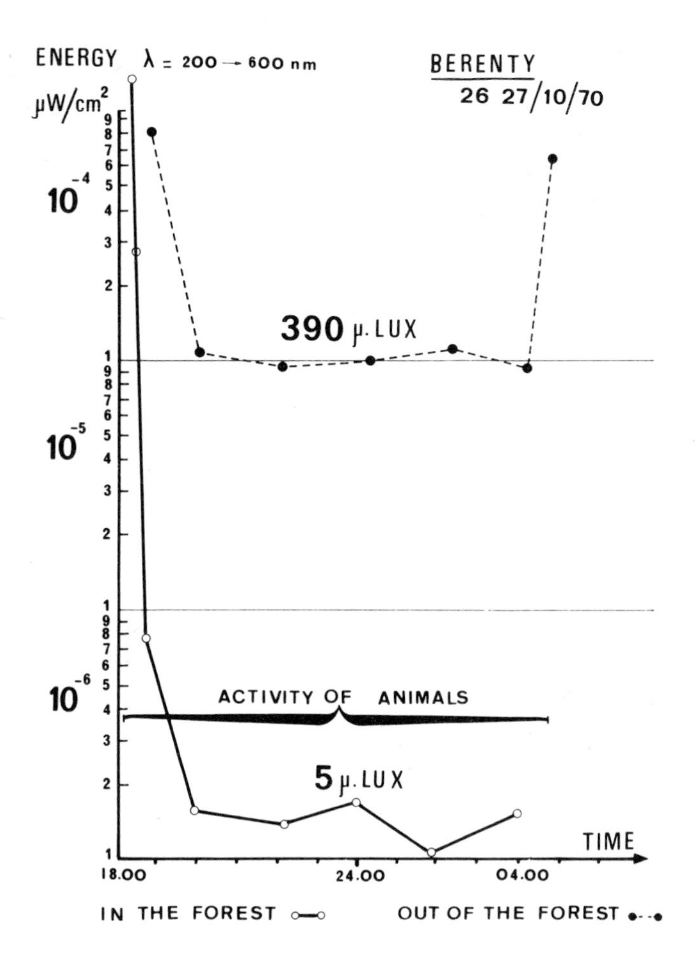

Figure 6. Nocturnal variation in light energy at Berenty in September, showing levels of light adequate for activity of *Lepilemur* both within the forest and outside the forest.

In the succeeding ten hours, *Phaner* shows evidence of intensive activity (running and leaping), while the light conditions remain at about 30 μlux, which represents very bad conditions for a human observer (Fig. 9). The time of returning to the nest is less regular. However, Petter et al.[2] never saw *Phaner* engaged in activity after 05.00 hrs. (a few lux on the curve). Most of the time, the returns were at 04.20 hrs. while luminosity was 0.01 lux, at which time the increase in light intensity is maximal (33 μlux at 04.00 hrs.; 30.000 μlux at 04.30 hrs.)

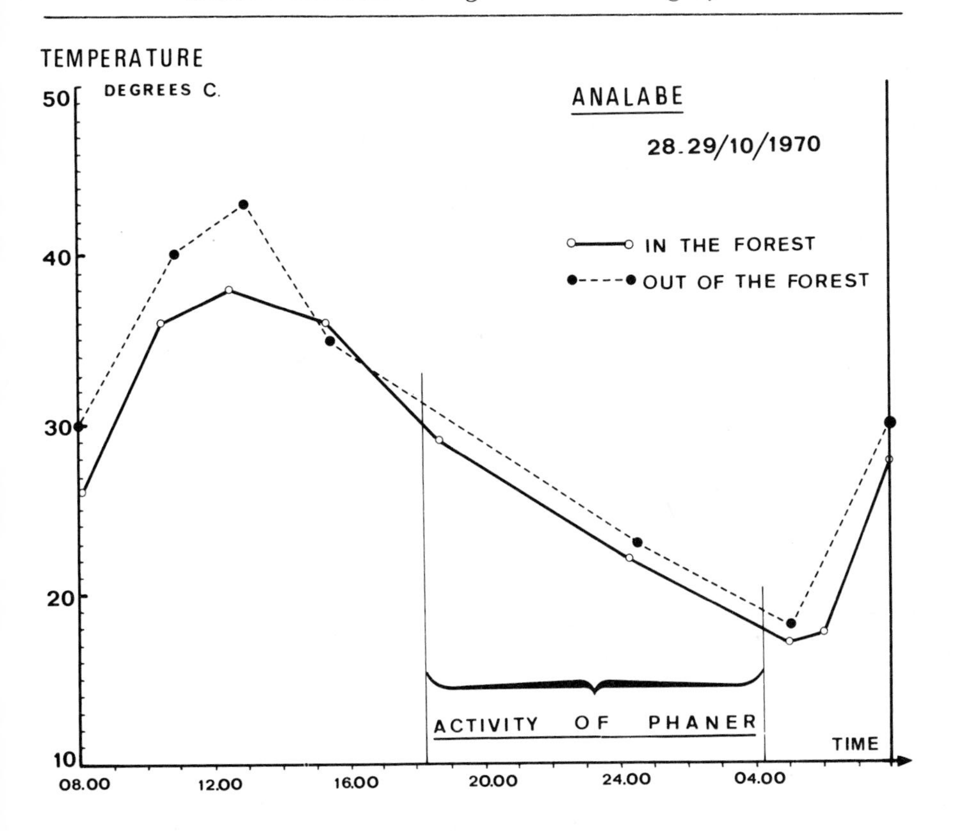

Figure 7. Diurnal variations in temperature within the forest and outside the forest at Analabe in October, showing the range of temperatures within which *Phaner* is active.

Conclusions

It can easily be seen that the activity rhythms of the two nocturnal Madagascan lemurs studied are perfectly adjusted to the daily cycle of the sun. Both species showed that they were able to leap several metres by night, which requires good judgment both of distance and of the strength of supports. Only vision can supply such information quickly and at a distance, since no echolocation seems to occur.

The regularity of emergence and return demonstrates that their eyes have the ability to discriminate, in decreasing and increasing light, a special level at a given moment. It is not known what mechanism is used, but it is conceivable that it would call for a duplex cone-rod system. Such a mechanism would be very surprising in nocturnal animals, which are reported to possess pure rod retinas.[5] The author is at present undertaking histological and behavioural

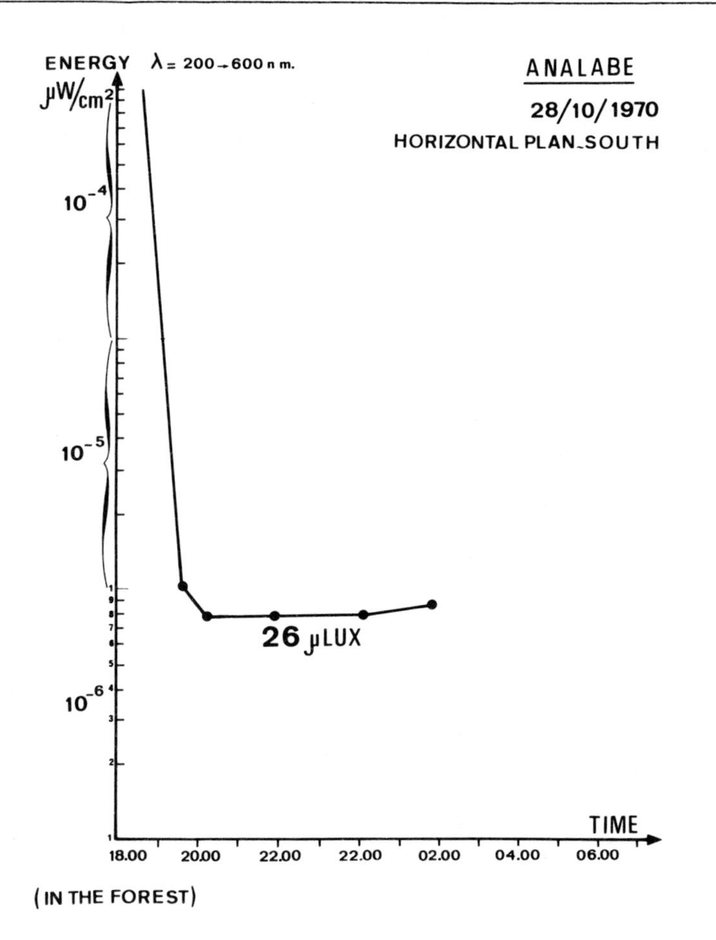

Figure 8. Nocturnal variations in light energy at Analabe in October. The level in μlux has been calculated as a reference.

investigations to determine whether colour vision is present in these two species.

If we consider the extremely low levels of radiation available by night in tropical forests, it is obvious that the vision of these animals presents very interesting and remarkable ecological/physiological adaptations. It should be added, as a final note, that these strictly nocturnal animals are still able to see quite well in full daylight. This fact demonstrates the perfect adaptation of their eyes, which can work right across the extraordinary scale of 10^{10} energetic units. Studies aimed at a better understanding of all these phenomena are at present underway in Brunoy, France (Laboratoire d'Ecologie Générale).

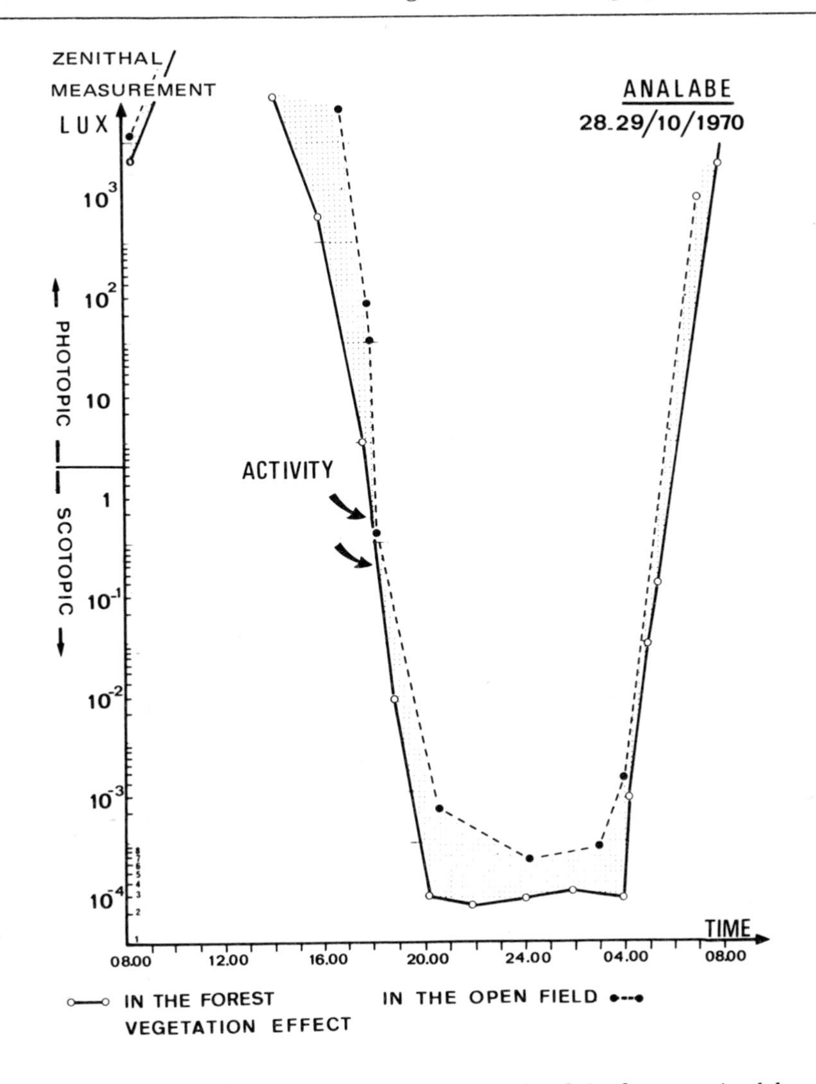

Figure 9. Visual conditions both inside and outside of the forest at Analabe, the difference between the two curves showing the screening effect of the vegetation.

Acknowledgments

This study was carried out with the help and generosity of Mr de Heaulme who made available the facilities both for living and working in the field at Analabe. The assistance of the Service des Eaux et Forêts of Madagascar is also gratefully acknowledged.

NOTES

1 Charles-Dominique, P. and Hladik, C.M. (1971), 'Le *Lepilemur* du sud de Madagascar: écologie, alimentation et vie sociale'. *Terre et Vie* 25, 3-66.

2 Petter, J.-J., Schilling, A. and Pariente, G. (1971), 'Observations éco-éthologiques sur deux lémuriens malgaches nocturnes: *Phaner furcifer* et *Microcebus coquereli*', *Terre et Vie* 25, 287-327.

3 Pariente, G. (1970), 'Retinographies comparées des lémuriens malgaches', *C.R. Acad. Sci.* 270, 1404-7.

4 Le Grand, Y. (1968), *Light, colour and vision*, London.

5 Wolin, L.R. and Massopust, L.C. (1970), 'Morphology of the primate retina', in Noback, C.R. and Montagna, W. (eds.), *Advances in primatology*. New York.

PART I SECTION B
Laboratory Studies of Behaviour

J. A. BERGERON

A prosimian research colony

Introduction

The Duke University Primate Facility contains about 180 prosimian primates used for research in anthropology, genetics, anatomy and animal behaviour. It is located in a secluded section of Duke Forest approximately two miles from the main campus of Duke University, adjacent to the Duke University Field Station for the Study of Animal Behaviour.

The Primate Facility staff consists of four faculty members and seven full-time non-academic employees. During the 1971-1972 academic year, seven graduate students and six undergraduate students conducted research at the Facility.

The Primate Facility was originally conceived by J. Buettner-Janusch while he was in the Department of Anthropology at Yale University. He founded the nucleus of the present prosimian colony in 1959, working with Professors William Montagna, then at Brown University, and Richard Andrew, then at Yale. On moving to Duke University in 1965, he brought with him the nucleus of the present colony. Peter Klopfer and others in the Zoology Department of Duke University helped to plan the present building. Research in genetic and evolutionary biology was the original impetus for the colony's existence; behavioural research developed as an adjunct to this primary interest.

Physical description

The Primate Facility building has 12,000 square feet of floor space. The basic floor plan of the building is three towers connected by intersecting corridors that form a T, as shown in Fig. 1. Two of the towers house animals exclusively; each contains five rooms.

The corridor that connects the two animal wings has an electronics

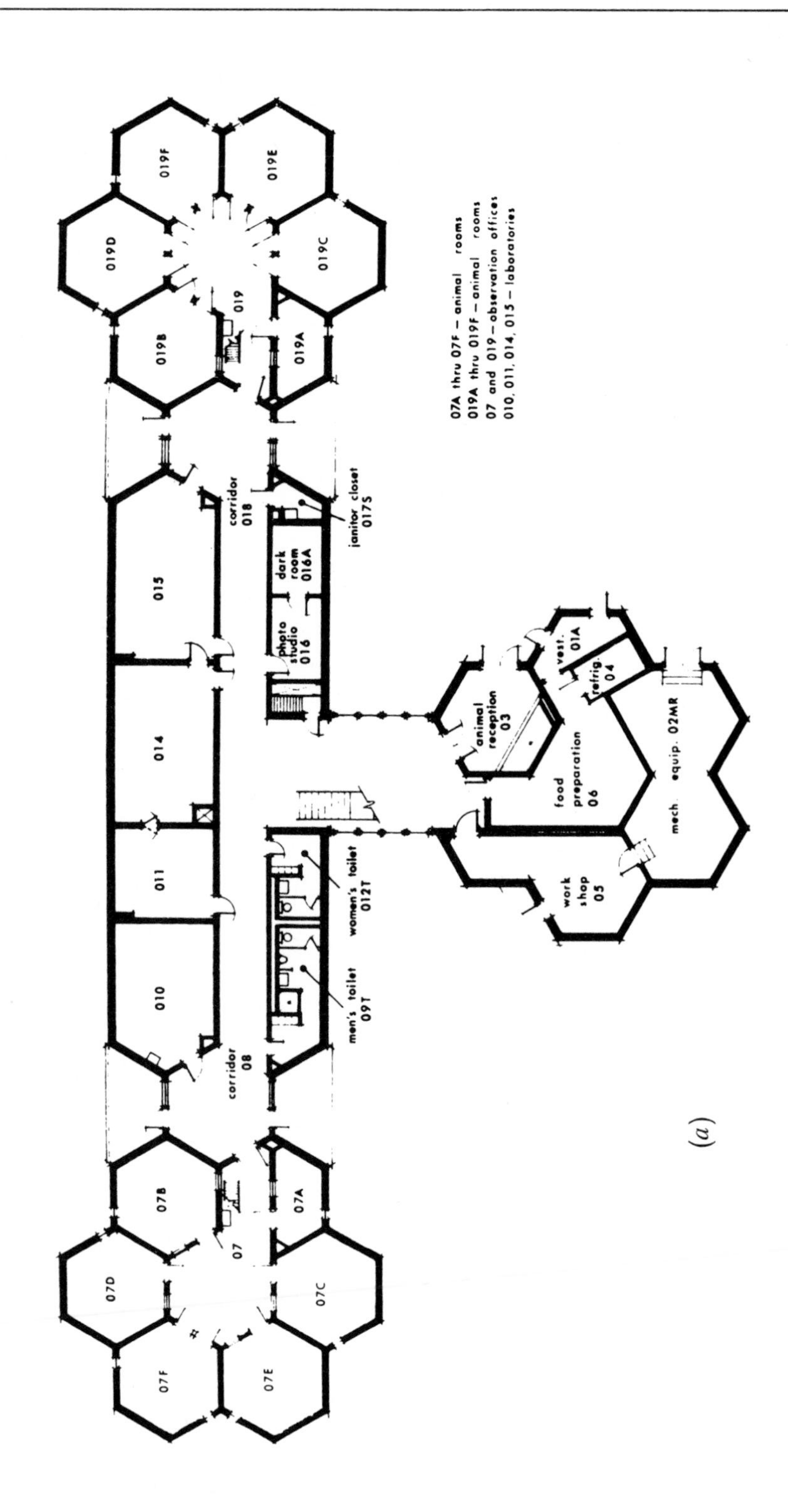
019F
019E
019D
019C
019B
019
019A
corridor 018
janitor closet 017S
015
dark room 016A
photo studio 016
014
011
010
animal reception 03
vest. 01A
refrig. 04
food preparation 06
mech. equip. 02MR
work shop 05
women's toilet 012T
men's toilet 09T
corridor 08
07B
07
07A
07D
07C
07F
07E
07A thru 07F – animal rooms
019A thru 019F – animal rooms
07 and 019 – observation offices
010, 011, 014, 015 – laboratories

(*a*)

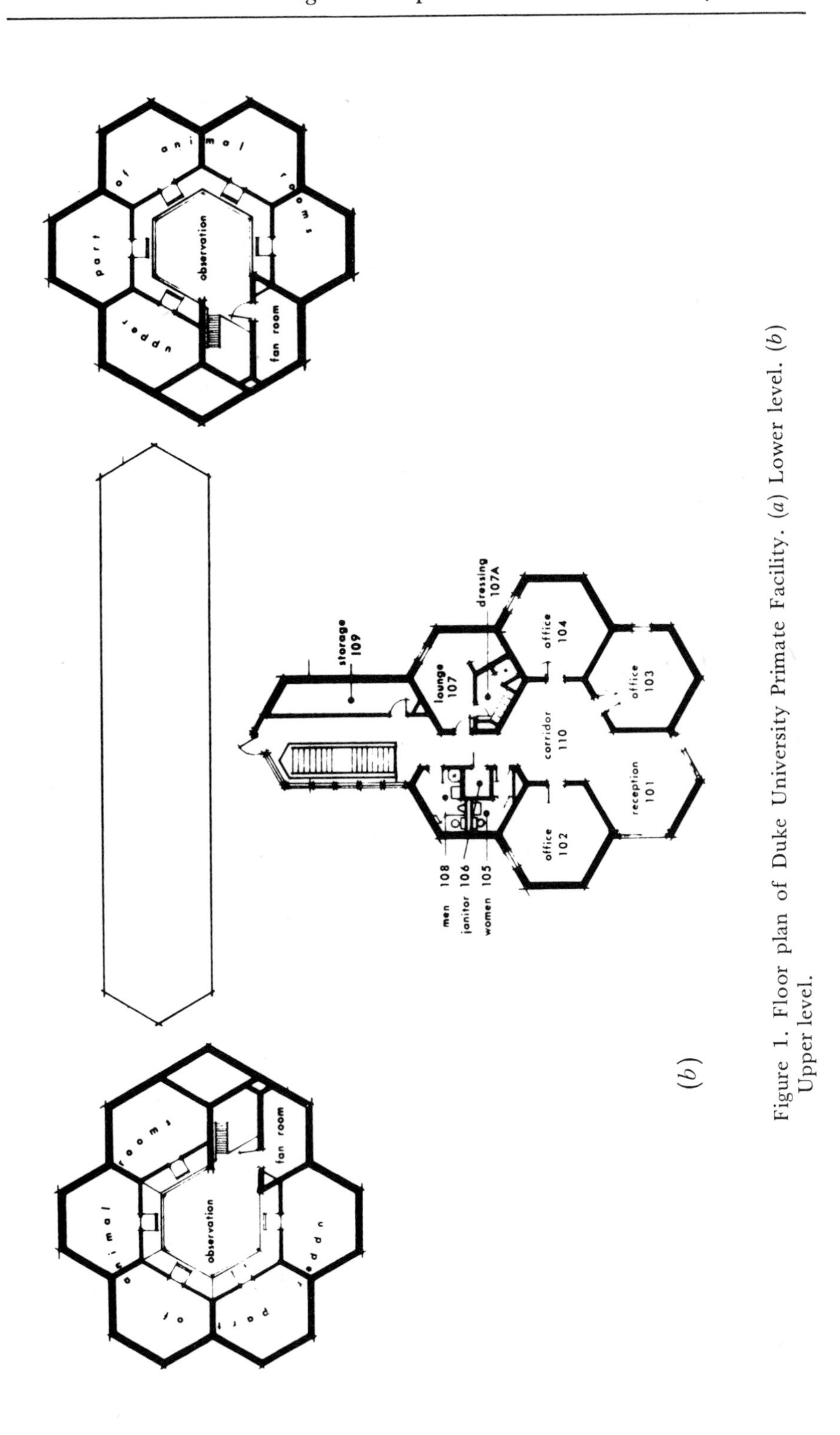

Figure 1. Floor plan of Duke University Primate Facility. (*a*) Lower level. (*b*) Upper level.

shop and individual laboratories for behavioural experiments, cytology, electron microscopy, and anatomy adjoining it. The remainder of the space off the main corridor is devoted to locker and shower rooms, a treatment room, and an isolation room for sick animals. The stem of the T connects the main corridor to the third tower, which contains cage-cleaning equipment, the kitchen, and a workshop on the ground floor, whilst there are administrative offices, a conference room and lounge on the upper floor.

In one of the hexagonal wings (019), two of the animal rooms have been divided into five individual compartments, and two of the rooms have been partitioned in half. The animal rooms are provided with climbing ropes, posts, dowels, and resting shelves (Fig. 2). The floors of the animal rooms are covered with pine shavings. The windows in the walls and ceilings are one-way glass to avoid distracting the animals when they are under observation. In addition to these larger rooms where the animals are housed in conditions resembling a free range, a number of animals are kept in cages of usual size.

There are also three outdoor holding areas. Two of these areas are large enclosures constructed of one inch mesh chain-link fencing. Each encloses about 1200 square feet and is between 8 and 10 feet high. Shelters are provided for the animals within these enclosures. Each shelter has plate glass windows on three sides and wood on the fourth. The animals come and go through a hole cut in the wooden side. The shelter is supported by posts and the floor (wood) is between 6 and 8 feet off the ground. Independent, thermostatically controlled heaters in each shelter maintain adequate temperatures even in coldest weather. Ambient temperature ranges from -18°C to +38°C. The third outdoor shelter, a commercial corn crib made of welded wire panels with a conical metal roof, is 16 feet in diameter and 24 feet high. There are resting shelves and climbing bars at several levels, and there are radiant heaters on the underside of the roof. The two larger outdoor enclosures have been in use since the autumn of 1968. One group of animals has been continuously housed in this enclosure since the summer of 1970. The animals which have lived in these enclosures include *Lemur fulvus* subspecies, *Lemur catta* and *Lemur macaco.*

The animals inside the building are kept under totally artifical light. Most of the animals are kept on a cycle of 12 hours of light and 12 hours of darkness. There are two separate sets of lights; fluorescent tubes for the light part of the cycle and red incandescent bulbs for the dark part of the cycle. In some rooms the light cycle is varied to correspond as closely as possible to the natural light cycle in North Carolina, with approximately nine hours of light in the winter and up to fifteen hours of light during the summer. There is, however, insufficient information to know whether this alteration in light cycle has any effect on breeding. The animals housed under a

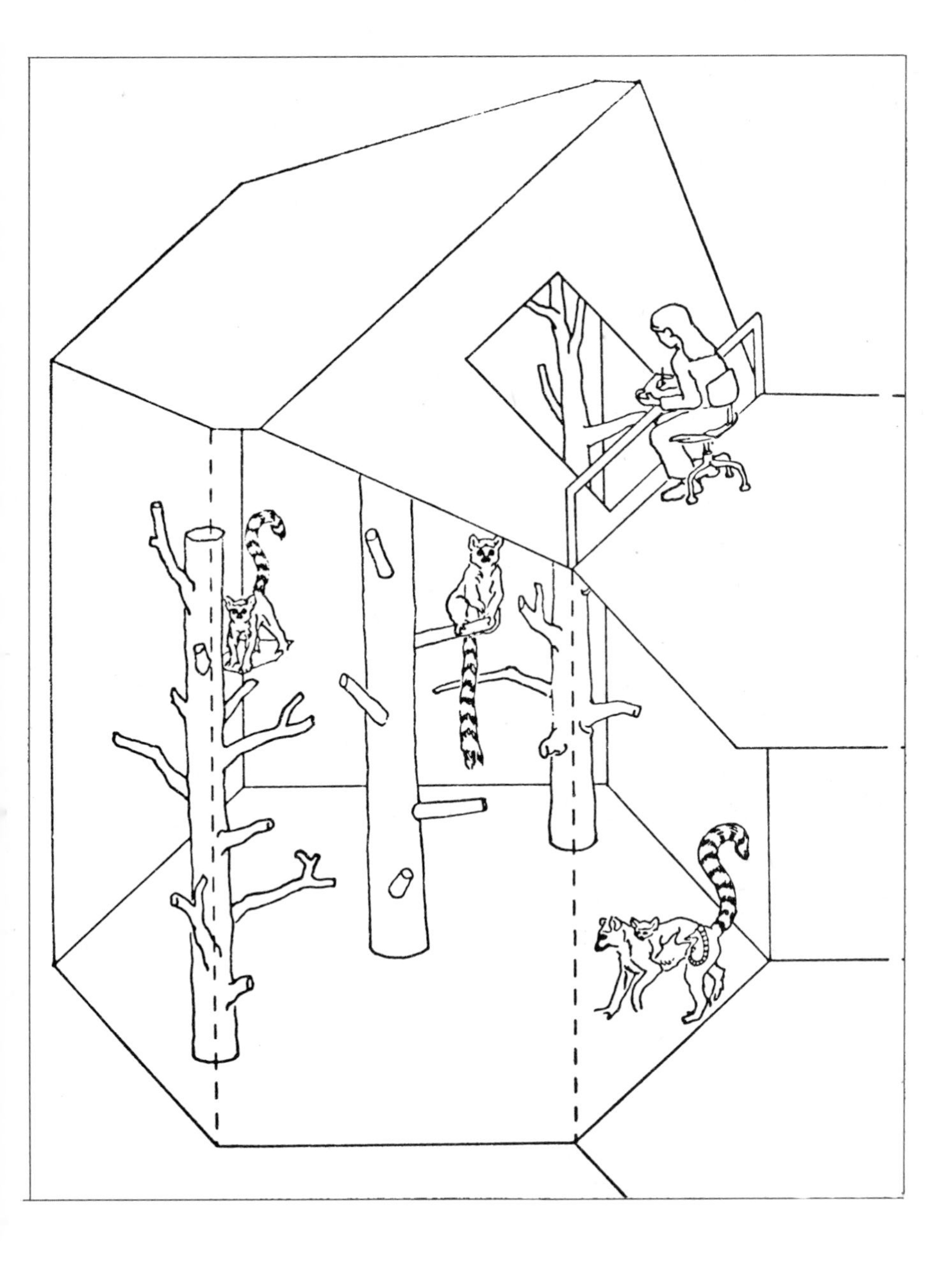

Figure 2. Typical behaviour observation room.

twelve hour light/twelve hour dark cycle breed at about the same season as those animals kept outdoors.

Diet

The basic diet fed to all animals at the Primate Facility is Purina Monkey Chow supplemented with a variety of fresh fruits and vegetables. Twice weekly the animals are fed a mixture of wheat cereal to which is added vitamins. In addition, mineral oil is added to the cereal mixture twice monthly; this helps prevent formation of hair balls in the digestive tract. Certain species require special dietary supplements. *Hapalemur* receives bamboo, which fortunately is obtainable quite readily from the grounds surrounding the building. *Propithecus verreauxi* are fed mango leaves and branches shipped weekly from Florida. These animals apparently need the large quantity of cellulose which comes from these leaves and branches. The Lorisidae and Cheirogaleinae are fed raw liver, canned dog food and hard-boiled eggs weekly as supplements. At times, the diet of these omnivorous animals has been supplemented with insects such as flour beetle larvae (mealworms) and crickets; however, it has not been found necessary to feed insects routinely to maintain their health.

The colony

The colony at present consists of about 180 prosimians: one genus of the infra-order Lorisiformes and six genera of the infraorder Lemuriformes. The galagos are *Galago crassicaudatus crassicaudatus* and *G.c. argentatus.* Sixty-six percent of the galagos in the colony have been born in captivity. Several of these animals are the fifth generation within the colony, and in several instances there are three living generations. The oldest captive-born galagos are now eleven years old. There are ten species of Lemuriformes, including four subspecies of *Lemur fulvus.* The animal population is shown in Table 1. Fifty-seven percent of the lemurs in the colony have been born in captivity, and at present there are four living generations. The oldest lemur born in the colony is eleven years old; another has lived in captivity for twelve years. The extensive pedigrees and the number of generations with known ontogenetic histories make this colony uniquely valuable for genetic and behavioural research (see notes).[1-12] Baseline physiological and haematological data on these animals have been compiled,[1-4,10,11] and information on problems of geriatric disease is beginning to emerge.[5,6]

Table 1 Prosimian population at the Duke University Primate Facility, 1972

Species	*Number of animals*
Galago crassicaudatus crassicaudatus	41
Galago crassicaudatus argentatus	26
Lemur catta	14
Lemur fulvus albifrons	7
Lemur fulvus collaris	3
Lemur fulvus fulvus	9
Lemur fulvus rufus	14
Lemur macaco	8
Lemur mongoz	1
Lemur variegatus	7
Lemur fulvus fulvus x *Lemur fulvus rufus*	19
Lemur fulvus rufus x *Lemur fulvus collaris*	2
Lemur macaco x *Lemur fulvus*	2
Hapalemur griseus	4
Lepilemur mustelinus	1
Cheirogaleus medius	3
Microcebus murinus	10
Propithecus verreauxi coquereli	6

Health and disease

The galagos have been of particular importance in studies of pathology. There is an exceptionally high incidence of chronic renal disease, in most instances taking the form of glomerulonephritis similar to that in man. Evidence of other age- and stress-related diseases similar to those which plague the human population has been found: atherosclerosis, hepatitis, bile stones, peptic ulcers, arthritis and cancer. One case of congestive heart failure and a case of auto-immune haemolytic anaemia in galagos have been seen. The most common health problem in galagos results from fights, in which severe bite wounds may be inflicted, generally on the head and tail. Often these bite wounds, particularly those on the tail, are extensive and cause severe blood loss resulting in shock and death. Bite wounds in galagos also tend to make the animals susceptible to bacterial infection.

Relatively few health problems occur with the lemurs. Several cases of acute haemorrhagic pancreatitis have been seen, but this is the only disease which has been seen on more than one occasion. One case of accidental death in a *Lemur catta* resulted from a fall in which two cervical vertebrae were fractured. Two lemurs died as a result of drug reactions. Complete post mortem examinations are performed, and the eviscerated corpses are frozen and saved for anatomical studies and skeletal preparations.

Special problems have occurred over the use of certain drugs. For antibiotic therapy, penicillin, penicillin-streptomycin combinations, tetracycline and chloramphenicol are used. Problems have occurred with procaine penicillin, particularly when used for galagos, and with a penicillin-streptomycin combination, and therefore this drug is not routinely used. Benzathine penicillin is used with no untoward effects. The antibiotic of choice for most infections is chloramphenicol (Chloromycetin). No adverse effects have been observed with this drug, and no evidence of bone marrow depression, as is reported in man, have been observed. The dose of chloramphenicol usually used is 25 mg. per kg./body weight. On some occasions oral antibiotic therapy has been used and it has been found that pediatric suspensions (generally fruit-flavoured) are quite useful for this purpose. The animals will usually take the drug readily when it is offered through the door of the cage.

For anaesthesia, intramuscular barbiturates have been highly successful. Hexabarbital sodium (Brevital) is used for short-acting anaesthesia and pentobarbital sodium (Nembutal) for longer anaesthesia. Doses of 30 to 35 mg. per kg. body weight are satisfactory. For most minor procedures, when sedation is sufficient, ketamine hydrochloride (Ketalar) is used. Several companies market ketamine in the United States. It is the drug of choice in sedation of cats, the recommended dosage ranging from 2.5 to 10 mg. per kg. For most routine use in prosimians, however, the smaller dose – 2.5 mg. per kg. – is adequate, and it has rarely been necessary to use higher dosages. Two lemurs have been lost following severe reactions to phencyclidine hydrochloride (Sernylan), and variable success has been obtained in using it as a sedative for galagos.

General management

Each of the animals in the colony has a three or four digit identification number. The first or second digit is constant for each genus and among galagos is constant within a breeding lineage. Each animal has a name, since naming the animals leads to better care and individual attention by members of the staff. Various physical characteristics are used to identify the animals visually, and when two animals look very much alike a system of ear notches is used. Ear notching as a means of identification poses a problem among galagos, since fighting often results in bite wounds on the ears and the obliteration of the original notches.

Each individual animal record contains a master identification sheet which contains ontogenetic information, a log sheet for all clinical laboratory data, and a chronological list of all manipulations made on that particular animal. A daily log is used to record manipulations and all animals in the colony are checked visually every day. Any that appear to be wounded, lethargic, or not eating well are examined thoroughly by the veterinarian. Progressive weight records are maintained on all animals, the animals being weighed once every three months; these records are particularly useful as indications of insidious disease.

Lemurs in small cages or enclosures are caught by hand, and those in large rooms or outdoor runs are caught in a nylon net. Heavy leather gloves are always worn when handling the animals. An animal is restrained by holding it around the neck at the base of the skull. Lemurs can also be restrained by a single individual who pulls the animal's arms behind its back and holds the upper arms together at the elbow.

Blood samples are collected from the animals without using sedation. They are physically restrained and blood is drawn from either the saphenous vein or the femoral vein. The saphenous vein is convenient for samples of 1 or 2 ml., and the femoral vein is used for samples of 5 or 10 ml.

Breeding

The management of breeding is centred around the genetic research. Galagos are routinely housed in pairs and occasionally in groups of one male and several females. The lemurs are housed in pairs, in groups of one male and several females, or in groups of several males and several females. The compositions of the groups are arranged so that the transferrin or acid phosphatase phenotype of any offspring can be used to identify unequivocally the male parent. All female lemurs are checked for pregnancy by abdominal palpation every two months. Pregnancy can be diagnosed as early as six weeks. When pregnancy is confirmed, the animal is checked once every four weeks until late in gestation, and then once weekly until parturition.

The large lemurs appear to mate in late autumn and early winter; births occur in late winter and early spring. The months in which most lemur births occur are March and April. The galagos do not appear to breed seasonally. The author does not believe that there is sufficient evidence to postulate seasonal breeding in the large lemurs; he has no data that indicate a breeding cycle correlated with environmental factors.

Cannibalism of newborns has occurred twice in lemurs. It seems likely that this has occurred either because the infant was the object of displaced aggression when adults were fighting or because the infant was not in good health when born. The latter explanation is the more probable. The ill or weak infant may trigger some response in the male or female parent that leads the parent to kill the infant. Cannibalism is much more frequent among galagos than among lemurs. Female galagos must be isolated prior to parturition, or the newborn infant will be killed by any other animal in the cage. Until the infant reaches a certain stage of social development, other adult galagos appear to react toward it as toward a bit of live food, according to experiments conducted by L.R. Rosenson. Rosenson suggested on the basis of these results that in the natural habitat female galagos isolate themselves at the time of parturition.[13]

The highlights of the present breeding programme include the births of several hybrid *Lemur* and three *Propithecus verreauxi coquereli.* Two of the hybrids are offspring of a male *Lemur fulvus collaris* (2N = 50)* and a female *Lemur fulvus rufus* (2N = 60). This pair has produced a male in March, 1970, and a female in July, 1971 (both 2N = 55). These hybrid offspring are thriving and appear to be normal in every way. There is no information about their fertility. Also, a male *Lemur macaco* (2N = 44) has been mated with a female *Lemur fulvus* (2N = 60), and they have had two female offspring (2N = 52), one in April, 1970, and one in May, 1971. The older of the two *Lemur macaco* x *Lemur fulvus* hybrids came into oestrus in the autumn of 1971, but there was no subsequent indication of pregnancy. There are a number of *Lemur fulvus fulvus* (2N = 60) x *Lemur fulvus rufus* (2N = 60) hybrids in the colony, and several of these have been back-crossed to *Lemur fulvus fulvus* or *Lemur fulvus rufus* and have had viable offspring.

The first *Propithecus verreauxi coquereli* was born in the colony in May, 1970. This infant was injured shortly after birth by an adult *Propithecus* in an adjacent cage, and it died within 24 hours of birth. A female offspring was born in December, 1970, from the same parents that produced the animal that died. A male offspring was born to the same parents in February, 1972. This family group, consisting of the parents, the year-old female offspring, and the male infant, are housed together.

The author feels that the success with the breeding programme and with the general management of the colony makes the Duke University Primate Facility a unique institution for breeding of prosimians for genetic, behavioural[7,8,9,12] and other biological research.

Acknowledgments

The author's attendance at the seminar was made possible by an international travel award from the National Science Foundation, No. GB-33212. The Primate Facility is supported by Duke University and U.S. Public Health Service grant RR-00388. Fig. 2 was prepared

* Diploid chromosome numbers.

by Miss Linda Hobbet, and Mrs Jane Hurlburt assisted in the preparation of this manuscript.

NOTES

1 Bergeron, J.A. and Buettner-Janusch, J. (1970). 'Hematology of prosimian primates: *Galago, Lemur* and *Propithecus.*' *Folia primat.*, 13, 155-165.

2 Bergeron, J.A. and Buettner-Janusch, J. (1971). 'Hematology of prosimian primates. II: A comparative study of Lemuriformes in captivity in Madagascar and North Carolina'. *Folia primat.*, 13, 306-313.

3 Buettner-Janusch, J. (1968). 'Man's place in nature'. *J. Invest. Dermatol.*, 51, 309-323.

4 Buettner-Janusch, J. and Wiggins, R.C. (1970). 'Haptoglobins and acid phosphatases of *Galago*'. *Folia primat.*, 13, 166-176.

5 Burkholder, P.M., Bergeron, J.A., Sherwood, B.F. and Hackel, D.B. (1971). 'A histopathologic survey of *Galago* in captivity'. *Virchows Arch. Abt. A Path. Anat.*, 354, 80-98.

6 Burkholder, P.M. and Bergeron, J.A. (1970). 'Spontaneous glomerulonephritis in the prosimian primate *Galago*'. *Am. J. Path.*, 61, 437-450.

7 Klopfer, P.H. (1970). 'Discrimination of young in Galagos'. *Folia primat.*, 13, 137-143.

8 Klopfer, P.H. and Jolly, A. (1970). 'The stability of territorial boundaries in a lemur troop'. *Folia primat.*, 12, 199-208.

9 Klopfer, P.H. and Klopfer, M.S. (1970). 'Patterns of maternal care in Lemurs: I. Normative description'. *Z. Tierpsychol.*, 27, 984-996.

10 Nute, P.E. and Buettner-Janusch, J. (1969). 'Genetics of polymorphic transferrins in the genus *Lemur*'. *Folia primat.*, 10, 181-94.

11 Nute, P.E., Buettner-Janusch, V. and Buettner-Janusch, J. (1969). 'Genetic and biochemical studies of transferrins and hemoglobins of *Galago*'. *Folia primat.*, 10, 276-287.

12 Roberts, P. (1971). 'Social interactions of *Galago crassicaudatus*'. *Folia primat.*, 14, 171-181.

13 Rosenson, L.R. pers. comm.

G. A. DOYLE

The behaviour of the lesser bushbaby

Since 1964 a colony of lesser bushbabies (*Galago senegalensis moholi* – Fig. 1) has been established under semi-natural conditions in our laboratory at the University of the Witwatersrand. The animals have bred freely and to date more than 50 infants have been born, approximately 75% of which have survived into adulthood. It is reliably estimated that this represents a survival rate considerably higher than that in the wild.

Study conditions

The semi-natural conditions in which our bushbabies have been housed and studied have been designed to duplicate as far as is practically possible the natural habitat to which the bushbaby has successfully adapted. While they fall far short of this goal the conditions are, nevertheless, vastly superior to small standard laboratory cages.

The animals under observation are housed in natural groups of one adult male, one or two adult females and their young in four large cages approximately 5′ x 6′ x 7′. The centres of all cages contain branching systems, and ledges and glass-sided nest boxes have been randomly placed on the walls. The floors are covered with a thick pile of wood-shavings. Temperature is allowed to fluctuate between 15°C and 30°C, which is well within the seasonal range characteristic of the natural habitat. The day/night cycle has been changed to suit the observers. The animals are active during 11 hours of red light (between 12.00 and 23.00 hrs.) which alternate automatically with 13 hours of white light when animals retire to the nest boxes to sleep. Daily diet consists of finely chopped-up fruits and vegetables in season, mealworms, milk supplemented with a commercially available vitamin B-complex syrup, and a commercially available food supplement in porridge form containing balanced proportions

Figure 1. Lesser bushbabies (*Galago senagalensis moholi*) photographed at night under natural conditions.

of proteins, vitamins, minerals and salts. Approximately every third day locusts are added to the diet. Animals will eat a variety of other foodstuffs, including nuts, dried fruit, bread and honey, but will not eat dead insects or raw meat. The cages are cleaned and food is placed on the ledges between 09.00 and 12.00 hrs. when the animals are asleep. A detailed description of the Animal Laboratory and methods of care and maintenance have been given elsewhere.[1]

Aim of research

The aim of the present series of studies is to obtain quantifiable data, on a time-sampling basis over long periods of time and under conditions of minimal interference, on various aspects of normal behaviour of the type that cannot be readily and completely obtained in the wild. No *a priori* theoretical assumptions have underlined our studies. Few experimental manipulations have been introduced and the only interferences to which our animals have normally been subjected are the daily removal of infants for weighing, the weighing of adults at regular intervals, the removal of sick animals for treatment and occasional changes in the composition of cage groups. Data in the form of activity counts and time scores are recorded on recording instruments and then entered after each observation session on data sheets which provide for the recording of descriptive data which cannot be quantified. To date approximately 5000 hours of laboratory observations and 500 hours of field observations have been completed.

The ultimate aim then of this series of studies is to provide a detailed and objective ethogram of the lesser bushbaby, and it is based on the assumption that no information of any value on the natural behaviour of a species is yielded by short studies of very few animals under highly artificial conditions. There are, of course, exceptions to this dictum in the form of highly specific studies such as those of Andrew,[2,3] Jolly,[4,5] Bishop, [6,7] and others.

This paper will be devoted to a broad, objective overview of the behaviour and other relevant aspects of the biology of the lesser bushbaby, based largely on quantitative and descriptive data from our own studies. Because of limitations of space, minimal reference is made to those of our own studies which have already been fully reported,[8,9,10] to the studies of others which are implicitly acknowledged, or to theoretical considerations.

General observations

1. The natural habitat

Galago senegalensis has adapted to a rather more generalised habitat than most prosimians, thriving throughout Africa in the drier, wooded bushland and savannah type country from as far north as Senegal, Nigeria, Sudan and Ethiopia, through Central and East Africa and down as far as Rhodesia, South-West Africa, Botswana and the northern Transvaal in the South.[11,12] Like all prosimians it is arboreal, but descends frequently to the ground to cross open spaces, to search for food and possibly to escape from predators.[13] *G. senegalensis* may be found at all heights and is equally at home on the heavier branches and trunks of large trees as it is in the fine branch zone. It thrives in a wide range of temperatures, for instance in South-West Africa where the seasonal variation is from -8°C to +47°C, and in the northern Transvaal where seasonal variation is from -5°C to +45°C.[13,14]

2. *Nesting*

In the laboratory *G. senegalensis* will use all available nest boxes, but shows a distinct preference for one. Very rarely is nesting material ever carried into the nest box except by the female prior to parturition. No recognisable nest ever results. In the wild *G. senegalensis* sleeps in dense thorny trees, on nests, forks, branches, in the midst of clumps of foliage and occasionally in tree hollows. In general, nests are difficult to locate from the ground. They may be occupied for long periods of time except in the case of females with young, when nesting sites may be changed every 10 to 14 days. Nests are invariably constructed from flat, fresh, soft, broad leaves and leafed branches and placed in thorn trees to form open-topped platforms. Occasionally they are built on a platform of twigs formed by old birds' nests and sometimes an old bird's nest is used without modification. With seasonal changes in leaf cover and temperature there is a gradual change in sleeping places throughout the year as some trees become more or less suitable. In cold weather *G. senegalensis* might expose itself to the direct rays of the sun at the ends of branches, but rarely does so in summer. Variations in type of nest, nest site and location of nest reported for other sub-species in different parts of America probably represent adaptations to localised habitats.[15]

3. *Activity rhythms*

Like all nocturnal prosimians, *G. senegalensis* begins its activities at dusk and settles down to sleep at about dawn. Both in the laboratory and in the field there is a period of very high activity in the first and second hours after waking, followed by irregular periods of activity and inactivity with a second period of high activity just before sunrise. *G. senegalensis* alternates between periods of rapid movement, sleep, rest, exploration, feeding, toilet activities and social activity independent of any pattern or rhythm. Heavy rain or strong wind make little difference to normal activity.

Regularity and bimodality of waking activity is much more marked in the social prosimians such as *Lemur*, where members of a troop rest, move and feed more or less as a unit,[16,17] than it is in the more solitary species where one animal may be active while another rests, as in *Microcebus* or *G. senegalensis.*[13,18]

4. *Ingestive behaviour*

Various insects as well as *Acacia* gum are the main food sources, and *G. senegalensis* has never been seen to display any interest in vegetable matter nor in small vertebrates in the wild although a variety of vegetable and animal matter is taken in the laboratory.

Even in the laboratory the most preferred food is insects. When *G. senegalensis* catches live prey with its hands, it shuts its eyes at the moment of impact and puts back its ears as a protection against insects with flapping wings and spurred legs (e.g. locusts). Once the fingers have closed round the prey and the head has been bitten, the eyes open and the ears come forward. Prey is first observed very carefully; the whole prey-catching gesture is very stereotyped and cannot be followed by the naked eye. Bearder confirms this behaviour for *G. senegalensis* in the wild.[13] Food is very rarely taken directly by the mouth and insects never. One hand is almost invariably used both to seize food and to hold it while being eaten. Food too large to be held or picked up with one hand is held in or down by both hands.

G. senegalensis does not use the lower procumbent incisors for eating. In biting or shearing food the canines or premolars are used, as Buettner-Janusch and Andrew[19] and Andrew[20] have shown for both *Galago* and *Lemur.* Premolars and incisors are used in detaching pieces of food only when the food is too large to be tackled directly by the molars.

When feeding in the wild, *G. senegalensis* moves rapidly from branch to branch and from tree to tree. *Acacia* gum is licked or eaten in small pieces and *G. senegalensis* may change its position several times while eating at the same spot. Bark may be chewed to expose the gum.

G. senegalensis drinks by lapping like a cat. It has never been seen to strike the surface of water or milk with the hand and lick the fingers. It will drink both water and milk, the former in small quantities in captivity, though it has never been seen to drink in the wild nor to lick the condensation from the surface of leaves, even in dry weather.

Food-stealing is common in *G. senegalensis* in captivity. Animals spend a great deal of time pursuing each other and all observers agree that food-stealing never leads to aggression.[15] The animal from which food is stolen may try to get it back or simply look for some other food. Food is usually stolen with the hand from the mouth or hand of the other.

Many insects found in the stomach of *G. senegalensis* are ground-living forms,[21] and Bearder[13] reports individuals spending as much as 15 minutes at a time on the ground foraging for insects. *Acacia* gum is found in the forks of trees, on the undersides of branches and near the bases of tree trunks, where it may be licked or eaten for up to 15 minutes at a time. As much as 2¼ hours may be spent in a single tree catching insects when conditions are favourable, while at other times as much as three hours may elapse without feeding. In general, the maximum amount of feeding occurs in the first few hours after dark. However, eating occurs throughout the night on an average of 6 times per hour in the wild. In captivity *G.*

senegalensis also feeds intermittently through the night. Diet and feeding habits vary seasonally. *Acacia* gum is available throughout the year and more time is spent eating gum and looking for insects in the winter than in the summer. During winter there is a noticeable change in the condition of the bushbabies.

Locomotion and posture

Detailed studies of the manipulative behaviour of a number of prosimians, including *G. senegalensis*, in respect of such activities as feeding, grooming, play, agonistic behaviour and general exploration have been reported by Bishop[6] and will not be repeated here.

1. *Locomotion*

G. senegalensis is a great hopper and leaper. Laboratory studies show that, except for rare occasions when it walks quadrupedally, its preferred mode of locomotion is by hops and leaps and it will jump to and from vertical supports with the same frequency and facility that it will use horizontal supports. Bearder[13] confirms this for *G. senegalensis* in the wild, where the natural habitat does not require exclusive adaptation to a vertical branch zone. In many thorny trees, too dense for jumping, *G. senegalensis* moves carefully and quadrupedally along the smaller branches. In locomotion across large open spaces, any small trees, fence posts or fallen branches may be used as vantage points. In order to progress from one tree to another without going to the ground *G. senegalensis* may make a series of jumps from one fence post to another spaced 10 feet apart. On one occasion *G. senegalensis* was seen walking quadrupedally along 70 yards of fencing wire, the accompanying tail movements appearing to subserve balance. When it has to descend to the ground *G. senegalensis* checks the environment carefully in all directions for as long as 20 minutes before leaving a tree. When on the ground it assumes an upright posture and takes long rapid bipedal jumps. Occasionally, movement along the ground is performed in a broken series of short hops covering distances of up to 60 yards. In the confined space of the laboratory, locomotion on the ground is always by a series of short hops, kangaroo fashion.

Normal standing jumps are performed from a crouched position, propulsion being achieved exclusively by the hind legs, which are brought forward to take the force of the impact on landing. During the jump the hands are usually held against the chest and are then extended to assist in gripping the substrate on landing. A variation of this occurs during very long jumps of 15 feet or more in the wild, when the arms are held forward and above the head to provide extra momentum. In the field as well as in the laboratory *G. senegalensis* carefully assesses the situation before making a long horizontal jump, by moving its head from side to side, checking all directions from a bipedal standing posture and gauging the distance from a number of vantage points. The ears move independently to and fro and the head may rotate owl-like through 180°. However, when a series of quick jumps is made in rapid flight, as in chasing, a succession of long jumps may be made without hesitation and with complete accuracy. *G. senegalensis* may also climb in any position that the substrate requires, upside-down on a horizontal branch, upwards or downwards on vertical surfaces.

2. *Sleeping and resting postures*

Sleeping postures vary greatly in *G. senegalensis*, particularly when they sleep in groups. They may sleep on the side with the head covered with hands and tail, or crouched in a saggital plane with the head bent and resting on the hands, the tail curled round the head. Occasionally animals sleep or rest in a sitting, vertical position or on their backs. During sleep the ears are retracted by means of the transverse folds. In stretching *G. senegalensis* bends the body almost to a lordosis, extending the limbs very much like a domestic cat. It does this frequently, particularly after sleeping, with many variations, one of which consists in hanging by the feet with hands extended downwards.

Maternal behaviour

Where observations of maternal behaviour have been made in the wild they fully support the detailed laboratory data that have already been reported.[10] In respect of the behaviour of the mother in changing infants from one nest to another, particularly in the first two weeks of life, some additional data are provided: Bearder[13] reported that an infant may be carried through several trees to a point some distance from the nest site, where it may be briefly groomed and left to cling, while the mother returns to get the other infant. Infants may be placed together or they may be left as much as 40 yards apart for short periods and sometimes for the greater part of the night. If left alone they may each be visited sporadically for short periods. This phenomenon, known as ‘parking’, is a common feature of maternal behaviour in the African lorisines.[15] The mother never fails to retrieve both infants back to the same nest site before dawn.

Infant development

G.s. moholi doubles its birth weight in from seven to nine days and a five-fold increase takes place over four weeks. Eyes are open at birth and within an hour infants can crawl on all fours awkwardly and with the belly on the substrate and can cling tightly to the mother's fur with the hands. By the fourth day infants can climb on wire mesh and Bishop notes that the full complement of adult prehensive patterns is present.[7] Before one week of age infants groom themselves and Andersson reports infant twins grooming each other on the second day after birth, in contrast to infant monkeys in which allogrooming does not occur.[22,23] By 10 days of age infants are extremely active in the nest, climb over adults, play and wrestle with each other, take small jumps of up to 6 inches and may stand bipedally with their hands against a vertical support and spend a good deal of time peering out of the nest box opening.

In the laboratory, infants emerge for the first time of their own accord from the nest box at between ten and fourteen days and Bearder notes that it is at about this age that they are first seen on their own in the wild.[13] The age at which the infants first emerge on their own does not seem to be related to the ease with which they are able to get out of the nest box, as Martin noted for *Tupaia.*[24] Infant *G. senegalensis* may get out of the nest box as early as even days, but they immediately get back in. It appears that not until they are at least ten days of age are they ready to face the outside world.

The threat posture has been seen well before two weeks of age. Urine-washing in its complete pattern, following ano-genital smelling and grooming, appears by three weeks and reciprocal ano-genital smelling has been seen at one month. Eating solids and drinking appears before three weeks. Insects are being caught and eaten by four weeks, but Bearder notes that in the wild gum is found and licked before insects are caught.[13] Suckling continues, but by about five weeks solids are being eaten more regularly and hands are employed more frequently in the eating process. By two to three weeks of age, infants run fast quadrupedally on branches and bipedal leaps are becoming longer and more frequent. Although they may frequently lose their balance at this age they seldom lose their grip.

Although at three to four weeks they can actively explore their environment, jump and take solid food, they are still dependent on their mothers for transport from tree to tree in the wild. At four to six weeks they may follow their mother, jump as far as three feet and descend to the ground for the first time, taking 1-foot hops to cross open spaces. After following their mothers from the sleeping trees in the evening, they may spend the rest of the night alone, feeding, playing, resting and moving from one tree to another over short distances. By this age play and locomotor ability have become quite elaborate.

In the wild, infants are probably self-sufficient and independent earlier than in the laboratory, where small cages keep animals together. After six weeks infants leave the sleeping site of their own accord and begin independent movement over an increasing range; but the family still comes together in the morning at the sleeping tree where infants probably still suckle.

Andersson reports that not until a year of age do males display the full adult aggressive pattern, emit the male call for the first time and begin to urine-wash with typical adult frequençy.[22] Females reach maturity earlier than males. The first onset of oestrus, mating and conception at about 200 days appears to be the rule in our laboratory.

Social behaviour

1. Olfactory communication: methods of scent deposition

The highly ritualistic method of depositing urine characteristic of the Lorisidae has been termed urine-washing by Boulenger.[25] In *G. senegalensis* foot and hand on the same side are raised simultaneously from the substratum; the hand is cupped under the uro-genital opening and a few drops of urine discharged into it. The hand then rubs the sole of the foot or grasps it firmly from one to several times; both hand and foot are then replaced on the substrate. The procedure may be repeated on the opposite side. The pattern is identical in both sexes, but it is performed far more frequently by the male than by the female. Under laboratory conditions, where animals are never isolated from each other, the male urine-washes from 3 to 22 times and the female from 1 to 3 times in a 50-minute observation period.

A second form of ritualised urine deposition, employed less frequently by the Galaginae than the Lorisinae, has been called rhythmic micturation.[26] Andersson describes it for *G. senegalensis* as follows:[22] the hindquarters are lowered to the substratum, a few drops of urine are released and at the same time the animal moves forward in a wriggling motion. It is more frequently seen in females than in males. *G. senegalensis* may also on occasion release a copious flow of urine on branches and may then wipe its feet in the urine or faeces thus deposited. This may not be a form of scent deposition, since other animals have not been observed to display any interest in it.

No concentration of cutaneous glands suitable for scent production is reported in *G. senegalensis*,[27] despite what appears to be glandular marking behaviour in this species. The adult male has a conspicuous bare patch on his chest which he rubs against the substratum or against branches. The chest is lowered against the substratum and the animal pushes first with one leg and then the other, sliding the chest over the surface. In chest-rubbing against a branch the arms are clasped tightly around the branch, hugging it closely. The animal then pushes forward with the hind legs, the head turning from side to side as the chest, side of the chin, and inner surfaces of the arms are rubbed along the branch. In the process the lower lip is dragged back and a tail of saliva is deposited. The sides of the mouth and chin may be rubbed against thin branches and ledges. Yasuda et al. report that the lips of *G. senegalensis* are well supplied with sebaceous glands.[27] After chest-rubbing against a branch the animal may bite it several times before repeating the movement. The same process may occasionally be performed by females and by juveniles. Vertical surfaces may also be marked in the same way, and while chest-rubbing against a flat surface the head may be rubbed when the animal comes up against a vertical surface.

Functions of scent-marking

The association between scent-marking, particularly urine-washing, and courtship is particularly obvious. The male will invariably urine-wash after genital-smelling or grooming others, particularly females, or after smelling a spot where a female has just urinated. When the female comes into oestrus, there is an approximate 30% increase in the frequency of male urine-washing and much of the male's urine is deposited directly on to the female. After genital-smelling a female, the male may vigorously chest-rub along her flank and back while adopting a mount-like position and this may be accompanied by urine-washing. In most and probably all prosimians, olfactory cues emananting from the female change and become stronger when the normally sealed vulva opens during oestrus.[11] There is an obvious increase in interest on the part of the male in females in neighbouring cages when they come into oestrus.

Urine-washing is more frequent in the dominant males, and chest-rubbing as well as urine-washing decrease in the male as dominance is lost and increase in the female as she becomes more dominant. The male urinates on strange females and on familiar females returned to the cage after a period of absence. Males generally urine-wash after chasing a subordinate or after staring at a subordinate prior to chasing.

In the nocturnal prosimians, particularly in the absence of sexual dimorphism, individual recognition is more likely to be olfactory than in the diurnal Lemuroidea, as Marler[28] has suggested and as Andersson[22] and Bearder[13] suggest for *G. senegalensis.* In *G. senegalensis* maternal recognition of infants from birth is certainly by olfaction. Repeated observations have shown that mothers are able to recognise their own infants from birth. When infants are removed by the observer for weighing they are returned in the hand for the mother to retrieve. All mothers will approach the hand and smell the infant, but only the real mother will pick it up, irrespective of which gets there first. The same occurs when mothers retrieve infants in distress.

A stranger placed in the cage increases the incidence of ano-genital smelling on the part of those present. Adult males in adjacent cages become aware of each other's presence and try to get at one another even when they cannot see one another. When strange animals avoid one another at first they will smell the places where each has been. A male may attack a female who has been with another male, though juveniles are never attacked under the same circumstances. Bearder suggests that mutual recognition is primarily by scent, which may carry information regarding age as well as sex.[13]

Hediger[29] pointed out that one of the aims of scent-marking in mammals appeared to be to keep the living area smelling of the animal, and Andersson[22] reports that *G. senegalensis*, particularly the male, thoroughly re-marks its cage by urine-washing after it has been cleaned. When placed in a new cage, the male may spend as much as an hour at a stretch chest-rubbing. New objects placed in the area are marked by urine-washing. Scent-marking is not restricted to making novel objects familiar, but also to keeping the familiar smelling strongly of the animal. For instance, a favourite site of scent-marking is the nest box and whenever an animal, but particularly the male, enters the nest box it invariably urine-washes, whether the nest box is empty or not. In the wild, scent-marking is concentrated in the sleeping tree and immediately adjacent area. Most scent-marking occurs after leaving the nest at sunset, which is presumably the time when the environment smells least of the animal. Scent-marking may reinforce territorial attachment as Martin[24] suggests, but it does not appear to act as a deterrent to rival conspecifics.[15]

It has been suggested for a number of prosimians that a function of scent-marking is the establishing of olfactory trails.[15] However, Bearder produces evidence for *G. senegalensis* which shows that scent-marking plays little or no role in establishing olfactory trails in the wild.[13] Most scent-marking occurs before the start of the night's activity in the small area with which the animal is most familiar, and thereafter it occurs only once or twice per hour, during which time *G. senegalensis* may move through as many as 45 to 50 trees over a distance of 200 yards. He reports too that *G. senegalensis* does not always use the same exact route, nor, with the exception of the sleeping area, does it mark at particular places. Finally, he reports that there is an absence of obvious sniffing during its movements in the wild except during social interactions and when investigating certain restricted areas within its home range. He suggests that a thorough knowledge of the home-range or territory is primarily visual.

It has also been suggested that a function of urine-washing is to facilitate the grasping of branches during movement through the environment. However, while male and female are equally active, the male urine-washes a great deal more than the female. Also, jumping or moving through the environment is not as a rule preceded by urine-washing. The evidence noted above that movement over large areas is seldom interrupted by urine-washing is yet another fact inconsistent with this hypothesis.

2. *Vocal communication*

Andersson was able to record a repertoire of 25 different vocalisations in *G. senegalensis* under laboratory conditions, but notes that if only discrete basic sounds are considered, while ignoring intermediate variations and combinations, then the range is reduced to 10 or 11 sounds.[22] Bearder reports that all the loud and many of the soft calls of *G. senegalensis* heard in the laboratory are also heard in the wild.[13] A number of authors have noted that the infant *G. senegalensis* emits tiny high-pitched cries which attract the mother and which cause her to go to the infant, pick it up and return it to the nest box, particularly if it is in difficulty.[15] Andersson reports that the small infant utters intense high-pitched clicks and crackles on loss of contact with the mother and later when it finds itself in difficulty or when it falls.[22] The incidence of clicking decreases as the infants gain confidence and become more independent. Juveniles utter clicks and crackles when introduced to a novel environment and, when frightened or disturbed, they may utter most adult calls except for the high intensity alarm call. In the wild, the female utters a soft maternal call on approaching the infants and the infants reply. These calls may be called contact calls uttered by mother or infant when separated and when they wish to restore contact. When the mother returns to the nest site just before daybreak, she utters a soft coo and three to four week old infants approach making frequent clicks. The infants follow the slowly-moving mother, continuing to make clicks which are not heard in her absence. Infants also utter soft notes while suckling and being groomed and when wriggling into close contact with the mother in search of the nipple.

Faint grunts are given when eating favourite food, a behaviour pattern found in many primates including human infants. Clicks and crackles are elicited by strange objects and by the smell of a strange fellow. A low moan preceded by a steeply rising then falling soft bark is given in warning in *G. senegalensis*. The range of alarm calls is correlated with the degree of intensity of the threatening situation. A typical situation is one in which one or two males start to call in a long-lasting series of increasingly penetrating sounds. Females may.

join in with the males. While calling the animals sit still, but they may periodically orient in different directions.

Some animals may move around and seek food while calling, but the call may change to a more penetrating-sounding chatter if a male discovers an intruder in his territory. The call appears to function to alert other animals. Activity stops, the animals jump higher in the branches and remain absolutely still. The call may continue for an hour or longer. No mobbing call has been reported for *G. senegalensis* similar to that reported for *G. crassicaudatus* by Jolly.[30] The majority of variants in the vocalisations of *G. senegalensis* are emitted during hostile encounters and mostly by the more submissive animal. The sound spectrograph shows that they are shortened versions of discrete calls. It is thought that they occur when the animal under stress is uttering various units of these discrete calls in such a rapidly changing series that it does not have time to give the full unit. The problems encountered in breathing control necessary to cope with strenuous exercise and simultaneous vocalising may account for the variations heard.

3. *Vision and visual communication*

Bearder showed that of 20 *G. senegalensis* none made any jumps over a period of two hours in complete darkness, although considerable movement took place if the environment was familiar.[13] If forced, *G. senegalensis* jumped randomly and landed on the floor. The long accurate jumps in an unfamiliar environment testify to the keenness of its vision. In the natural habitat *G. senegalensis* must be able to see at least over the distance of its maximum jumps (about 15 feet), but probably sees at least 30 to 40 yards. It moves about its habitat as actively and with the same facility in bright light and in almost total darkness. Sight and visual memory coupled with a good kinaesthetic sense enable *G. senegalensis* to move around its natural habitat with minimal reliance on olfactory cues. The status of strangers too is probably determined by visual as well as olfactory cues.

Andersson points out that in *G. senegalensis* the mouth, together with the eyes and various movements of the mobile external ears, give rise to several facial expressions, despite the physical handicap of a simple facial musculature, which, together with body movement and posture, provide the basis for a variety of visual signals.[2,22] Jolly suggests that the highly ritualised scent deposition gestures of *Lemur* almost certainly function as visual signals, both in those species which have scent glands, like *L. catta*, and in those which do not, like *L. macaco.*[16] The same could well be true in respect of the highly ritualised urine-washing displays of the Lorisidae.

4. Territory and home-range

In 15 *G. senegalensis* groups Bearder reports a mean home-range of 2.8 hectares (7 acres) varying from 1.25 to 3.95 hectares (3.0 to 9.6 acres).[13] He attributes this relatively large home-range for such a small animal to its great activity and the nature of the vegetation. Natural boundaries are formed by open spaces, roads, etc., and there is considerable overlap of home-ranges. They do not, however, appear to be as stable as in *Lemur*,[16] the size, shape and position of the home-range changing during the year. Like many prosimians and higher primates, *G. senegalensis* advertises its location vocally. It does not appear to occupy a territory in the sense of a core area defended against encroachment by conspecifics as in many other prosimians.[15]

In general, territorial fights in primates are uncommon, belligerence rarely going beyond various types of threat or agonistic display.[31] Conflicts may increase under crowded conditions, as in *Macaca mulatta* in India,[32] and fighting in male *G. senegalensis* is far more common in the laboratory than in the field. In the field, avoidance behaviour and, in particular, the typical male call, ensure separation between potential belligerents. Since the adult male is the dominant figure in each group and since, as Boulière pointed out,[33] territoriality is usually a male dominated activity, rivalry between adult males is probably the most important factor resulting in territorial spacing out.

In *G. senegalensis* the average path length during the night is about 2.0 km. (1.25 miles), representing movement through a mean number of 500 trees. It uses the same general pathways and although its route from one point to another is highly variable it frequently makes the same jumps and crosses the same open spaces. Although it is possible for the animal to traverse its home range in two hours it rarely does so. Field studies indicate that *G. senegalensis* has a thorough knowledge of its environment.

5. Social groups

The literature indicates that *G. senegalensis* is frequently seen in the wild singly or in pairs, occasionally in larger groups, and sleeping groups may be as large as seven.[15] Bearder found 238 animals in 119 sleeping groups.[13] Of these 75% were found in groups of two to four, the most common being an adult pair with young, an adult female with young, or a pair alone. Groups or individuals frequently use the same nests or sleeping sites, which form a home base to which they regularly return at the end of the night. The sleeping group does not function as a unit at night, but splits up for individual foraging; 70% of the waking time is spent alone and movement of two or more animals together for any length of time is rare. Before dawn, members of a group meet and move about together before arriving at the sleeping site. Each group lives in a definable area and the composition of the sleeping group varies during the year, probably due to increasing aggressiveness between adults as the young mature. Evidence from both the wild and the laboratory strongly suggests that the natural group in *G. senegalensis* is the family group and that the social organisation is relatively simple. In the laboratory, male and female with young and/or juveniles will live together peacefully. Provided that there is sufficient space, additional females and sub-adult males do not increase friction and females may live together peacefully and produce infants in the same nest.

6. Contact behaviour

In *G. senegalensis*, as in other Lorisidae, the need for bodily contact in adults does not appear to be as strong as that reported in some of the Lemuroidea,[16] although it may be equally strong in infancy. Adults display little contact behaviour except in courtship, mating, grooming, aggression, and very occasional play, although they always sleep in close contact even when there is ample space. Social contacts in the wild are brief and follow the same pattern as those observed in the laboratory, enabling the observer to tell the age and sex of the animals involved.

(*a*) *Greeting.* Naso-nasal contact is a common form of greeting in *G. senegalensis.* One animal will approach another slowly with the head stretched forward, ears slightly back and body held low. The individual being approached may remain quiet at first with head held slightly raised and may then respond by sniffing in turn or jumping away. Naso-nasal sniffing may be followed by naso-nasal contact and brief allogrooming of the head and especially chin area. This may be mutual or reciprocal. It may also be followed by high-intensity agonistic behaviour. Naso-nasal greeting is especially common between unacquainted individuals.

(*b*) *Grooming.* After waking and before leaving the nesting area, *G. senegalensis* goes through a sequence of toilet activities, yawning, stretching and intense self-grooming with the toothcomb, during which most accessible areas of the body are covered with much attention to, and much manipulation of, the tail. Evening toilet usually ends with a scratch with the toilet claw. *G. senegalensis* has never been observed to wash its face with its hands after the manner of a cat either in the laboratory or in the field.

G. senegalensis hold one another while grooming with all five digits flexed towards each other but rarely hold with both hands. The groomer usually puts one hand on the back or between the shoulder blades of the fellow. Males may hold the back of the female with both hands and groom it up the back with the toothcomb between the hand-holds. Allogrooming occurs frequently in a wide variety of situations and is not restricted to sexual situations; nor does it serve simply to keep the skin and fur in good condition, since with the aid of the toilet claw *G. senegalensis* can self-groom almost every part of its own body.

(*c*) *Agonistic behaviour.* When the tendency for offensive attack in *G. senegalensis* is very high, the ears are upright, the eyes are rounded, wide open and staring, the mouth is wide open and the tongue and upper canines are partly visible. In defensive threat, on the other hand, the open mouth becomes rounded, the lips are not tense but the teeth remain covered, only the tongue being visible in the mouth, and the head is raised. The eyes are wide open and the ears fully extended at first, becoming flatter as the threatening source approaches. If the threat intensity increases, and particularly if the animal is cornered, a defensive attack posture may be assumed, in which the animal rears up and takes a bipedal stance with the arms held up on either side of the head and fists clenched ready to cuff. This may be followed by a downward pounce accompanied by biting. When threatened by the approach of an aggressive superior, a submissive cringing posture is adopted in which the body is very tense and held low against the substratum, the head retracted against hunched shoulders, eyes narrowed and ears back and flattened against the head. The open-mouth, teeth-covered facial expression in defensive threat is often accompanied by swaying in highly nervous animals in which there is a strong tendency to flee. Andersson notes that this may have been ritualised from an intention movement for jumping, particularly in unfamiliar surroundings.[22] *G. senegalensis* narrows or closes its eyes and flattens its ears when approaching to face a fellow to groom or take food, indicating quite clearly that no aggression is intended.

Overt aggression often takes the form of chasing, particularly if the animals are not equally dominant. Chasing usually occurs between adult females and between an adult male and a maturing male, and it may also occur between an adult male and an adult female where the female is the aggressor. Aggression between females is, however, not as marked as aggression between males, and adult males are not normally aggressive towards sub-adult males. Minor disputes between relatively evenly matched animals may involve sparring matches in which the two animals stand on hind legs, cuff at one another, and grab at each other's hands and limbs. If one gets a hold, the dispute turns into a wrestling match, in which the animals hold and kick one another, rolling over together. This does not involve biting, never leads to serious injury and mainly involves juveniles and opposite sex adults. This type of behaviour is very frequent and may be more in the nature of play.

In the laboratory, relative dominance in a group of *G. senegalensis* is fixed; each animal knows its rank position, behaves accordingly, and the unit is stable. If an animal is removed and, more particularly, if a stranger is introduced, a marked change in the behaviour of the group may occur in which the social structure of the group may be re-organised. In the wild, *G. senegalensis* frequently come into contact with others. The resulting behaviour varies considerably and Bearder investigated this variability in the laboratory through diadic analysis of 130 paired presentations under a variety of conditions.[13] Dominance becomes obvious within the first few minutes of a 40-minute session. Aggression is usually uni-directional, the submissive animal retreating while emitting the full range of submissive vocalisations. Complete submissiveness is usually indicated by the subordinate animal going to the floor. The male is dominant in his own territory and is almost invariably dominant over all females. Sub-adult males are antagonistic towards each other and are usually dominant over females. Dominance in the male is usually benign, whereas when the female is dominant she is highly aggressive. Laboratory and field data suggest that the major factors determining dominance status are sex, age (the greater the age differential, the clearer the dominance status and the less the aggression), and familiarity with the environment (in the laboratory an individual may be sub-dominant in one cage but dominant over the same individual in another cage).

However, little significance can be attached to observations on linear hierarchies under laboratory conditions, since they could well be artefacts of confined conditions in normally solitary animals, as Martin has suggested in respect of laboratory observations on the Tupaioidea.[24]

Serious fighting in which the animals are completely silent only occurs in *G. senegalensis* between more-or-less evenly matched males and very rarely between adult females. Under laboratory conditions, the harmony of the group can only be maintained if it includes no more than one adult male. If adult males are brought together by the removal of a partition between adjacent home cages, vicious fighting will occur, in which severe damage is inflicted. In typical fighting the mouth is open, exposing both upper and lower teeth; ears are folded back and drawn back tightly so that the skin over the forehead and the top of the skull is also drawn back, giving the head a flattened appearance; the skin above the nose between the eyes is deeply wrinkled horizontally and the eyes are nearly closed in a pattern of protective responses similar to those when catching prey. No chasing occurs; animals spring at one another, grapple and fall to the ground, each attempting to gain a foothold on the other's ankles. Locked together, each attempts to throw itself on top of the other, while striking, pulling at fur, and biting at the other's face. Open wounds occur on the feet, hands and face. Loss of fur occurs on the head, neck and shoulders, and fur-plucking may be so violent that the skin is torn in places. Unless separated, the probability of one animal being killed is high. Overt aggression also occurs in the wild, but it is extremely rare for reasons already stated.

(*d*) *Play.* Play wrestling in twin infants appears from the sixth day of age. Like other prosimians, *G. senegalensis* engage in complex locomotor play by themselves at least while still young. They also spend a great deal of time in social play in the laboratory and have been observed play-wrestling in the wild. Adults have only occasionally been observed to play, and then only with juveniles. Social play is frequently associated with mutual grooming; play-wrestling often precedes or is followed by mutual grooming. As with most primates, the behavioural elements of fighting are seen in play-kicking, pushing, pulling at one another in apparent attempts to grasp one another around the ankles or to obtain a hold on one another's ears and hands. The sexual elements seen in the play of monkeys are, however, absent in *G. senegalensis.*[34] Most play contact is ventro-ventral. Wrestling bouts may occur with the animals hanging upside down from a branch suspended by the feet, cuffing and wrestling with the hands or hanging by the hands and kicking with the feet. *G. senegalensis* displays much curiosity towards strange inanimate objects placed in the cage, and, if small enough, will manipulate them playfully with the hands.

Acknowledgments

This research was supported by grants from the University Development Foundation and the University Council, University of the Witwatersrand, as well as by grants from the National Geographic Society, Washington D.C., and the Human Sciences Research Council of the Republic of South Africa.

NOTES

1 Doyle, G.A. and Bekker, T. (1967) 'A facility for naturalistic studies of the Lesser Bushbaby (*Galago senegalensis moholi*), *Folia primat.* 7, 161-8.

2 Andrew, R.J. (1964), 'The displays of primates', in Buettner-Janusch, J. (ed.), *Evolutionary and Genetic Biology of the Primates*, vol. 2, New York.

3 Andrew, R.J. (1962), 'Evolution of intelligence and vocal mimicking', *Science* 585-9.

4 Jolly, A. (1964*a*), 'Prosimians' manipulation of simple object problems', *Anim. Behav.* 12, 560-70.

5 Jolly, A. (1964*b*), 'Choice of cues in prosimian learning', *Anim. Behav.* 12, 571-7.

6 Bishop, A. (1962), 'Control of the hand in lower primates', *Ann. N.Y. Acad. Sci.* 102, 316-37.

7 Bishop, A. (1964), 'Use of the hand in lower primates', in Buettner-Janusch, J. (ed.), *Evolutionary and genetic biology of the primates*, vol. 2, New York.

8 Doyle, G.A., Pelletier, A. and Bekker, T. (1967), 'Courtship, mating and parturition in the lesser bushbaby (*Galago senegalensis moholi*) under semi-

natural conditions', *Folia primat.* 7, 169-97.

9 Doyle, G.A., Andersson, A. and Bearder, S.K. (1969), 'Maternal behaviour in the lesser bushbaby (*Galago senegalensis moholi*) under semi-natural conditions', *Folia primat.* 11, 215-38.

10 Doyle, G.A., Andersson, A. and Bearder, S.K. (1971), 'Reproduction in the lesser bushbaby (*Galago senegalensis moholi*) under semi-natural conditions', *Folia primat.* 14, 15-22.

11 Hill, W.C.O. (1953), *Primates: Comparative Anatomy and Taxonomy*, vol. I: *Strepsirhini*, Edinburgh.

12 Shortridge, G.C. (1934), *The Mammals of South West Africa*, London.

13 Bearder, S.K. (1969), 'Territorial and intergroup behaviour of the lesser bushbaby (*Galago senegalensis moholi*, A. Smith) in semi-natural conditions and in the field', unpublished MSc Thesis, University of the Witwatersrand.

14 Hoesch, W. and Niethammer, G. (1940), 'Die Vögelwelt Deutsch-Sudwest-Afrikas', *J.f. Ornithologie* 88, Sonderheft, 404.

15 Doyle, G.A. (1974), 'The behavior of prosimians', in Schrier, A.M. and Stollnitz, F. (eds.), *Behavior of Non-Human Primates*, vol. 5, New York.

16 Jolly, A. (1966*a*), *Lemur Behavior: a Madagascar Field Study*, Chicago.

17 Petter-Rousseaux, A. (1964), 'Reproductive physiology and behavior of the Lemuroidea', in Buettner-Janusch, J. (ed.), *Evolutionary and Genetic Biology of the Primates*, vol. 2, New York.

18 Martin, R.D. (1972), 'A preliminary study of the Lesser Mouse Lemur (*Microcebus murinus*, J.F. Miller 1777)', *Z.f. Tierpsychol.* Beiheft 9, 43-89.

19 Buettner-Janusch, J. and Andrew, R.J. (1962), 'The use of the incisors by primates in grooming', *Am. J. Phys. Anthrop.* 20, 127-9.

20 Andrew, R.J. (1964), 'The displays of primates', in Buettner-Janusch, J. (ed.), *Evolutionary and Genetic Biology of the Primates*, vol. 2, New York.

21 Haddow, A.J. and Ellice, J.M. (1964), 'Studies on bushbabies (*Galago* spp.) with special reference to the epidemiology of yellow fever', *Trans. roy. Soc. trop. Med. Hyg.* 58, 521-38.

22 Andersson, A. (1969), 'Communication in the lesser bushbaby (*Galago senegalensis moholi*)', Unpublished MSc Thesis, University of the Witwatersrand.

23 Sparks, J. (1967), 'Allogrooming in primates: a review', in Morris, D. (ed.), *Primate Ethology*, London.

24 Martin, R.D. (1968), 'Reproduction and ontogeny in tree-shrews (*Tupaia belangeri*) with reference to their general behaviour and taxonomic relationships', *Z.f. Tierpsychol.* 25, 409-532.

25 Boulenger, E.G. (1936), *Apes and Monkeys*, London.

26 Ilse, D.R. (1955), 'Olfactory marking of territory in two young male loris, *Loris tardigradus lydekkerianus*, kept in captivity in Poona', *Brit. J. Anim. Behav.* 3, 118-20.

27 Yasuda, K., Aoki, T. and Montagna, W. (1961), 'The skin of primates: IV. The skin of the lesser bushbaby (*Galago senegalensis*)', *Am. J. Phys. Anthrop.* 19, 23-33.

28 Marler, P. (1965), 'Communication in monkeys and apes', in DeVore, I. (ed.), *Primate Behavior*. New York.

29 Hediger, H. (1950), *Wild Animals in Captivity*, London.

30 Jolly, A. (1966*b*), 'Observations on *Galago crassicaudatus*', unpublished paper sent to Dr Sauer.

31 Bates, B.C. (1970), 'Territorial behaviour in primates: a review of recent field studies', *Primates* 11, 271-84.

32 Southwick, C.H., Beg, M.A. and Siddiqi, M.R. (1965), 'Rhesus monkeys in North India, in DeVore, I. (ed.), *Primate Behavior*, New York.

33 Bourlière, F. (1964), *The Natural History of Mammals*, New York.

34 Loizos, C. (1967), 'Play behaviour in higher primates: a review', in Morris, D. (ed.), *Primate Ethology*, London.

J. EPPS

Social interactions of Perodicticus potto *kept in captivity in Kampala, Uganda*

Introduction

Although the potto is relatively common within the limits of its distribution, little is known concerning its social behaviour, mainly because of the difficulties attendant on the study of a nocturnal animal that usually inhabits thick forest. It was therefore hoped that a study of pottos kept in outdoor cages in the area of origin, and so subjected to normal fluctuations of rainfall and temperature, might prove useful.

Observations were made of the social interactions of pottos caged at Makerere University, Kampala between February 1970 and May 1971. A total of thirteen pottos, five males and seven females, were kept at Makerere during this time, but one male and one female died of old age, and two females escaped during the first two months after capture. Seven of the pottos, given by Alan Walker, had been in captivity at Makerere Medical School for some time before the beginning of this study. The other six were brought to the Zoology Department by local people, who caught them in small plantations around Kampala. All the pottos used in the study had been caught within 100 miles of Kampala, either in Buganda or Bunyoro districts.

Housing and experimental procedure

The majority of the pottos were kept in a system of home cages, each 4½′ x 10′ x 9′. Each cage was provided with a system of branches and a sleeping box. The cages were well shaded with papyrus matting; the temperature never rose above 27° C, usually cycling between 17.5°C and 26°C. For observations of social interactions, the occupants of the home cages were allowed into a large central observation cage by way of swing doors, which were opened by a pulley system operated from the observation slit at one end of the cage. Light for observation was provided by two 60-watt red bulbs. The home cages were also lighted by red bulbs so that all the pottos could become habituated to red light. They did not appear to be affected by it in any way and were often seen licking insects off the glass around the bulbs. Automatic records of the nightly activity of an animal subjected to red illumination showed an initial decrease, but the animal soon became habituated and activity returned to normal.

For observations of social interactions, any two pottos were allowed into the observation cage together and their actions noted on a check sheet at three-minute intervals. In addition, details of any social behaviour that was seen between check sheet records were recorded on a tape recorder. Interactions between pairs of males, pairs of females and male-female pairs were watched during 2½-hour observation periods. Care was taken to ensure that the animals which met in the observation cage had not previously been in contact with each other. Additional observations were made of one female pair, one male pair and one male-female pair, each of which were transferred to a reversed daylight house for periods of two to three months.

Animals in the reversed daylight house were kept in a cage 6′ x 5½′ x 5′, which contained a sleeping box for each animal, a shelf for the food bowls, and was provided with a system of branches. The room containing the cage was kept dark during the day and was lighted with a 60-watt white bulb between 19.00 hrs. and 07.00 hrs. The hours of day and night were thus almost exactly reversed. The animals kept in this cage took only a week to ten days to become accustomed to the new regime. Light for observation during the artificial 'night' was provided by sunlight filtered through two layers of red cellophane covering a window measuring 4′ x 1′. Using a 12-channel activity recorder connected to a switchboard, continuous records of the behaviour of the animals were made during 1½-hour observation periods. The observer sat behind a curtain three feet from the cage and watched the animals through a peephole. Although the pottos were aware of the presence of the observer, they became habituated to it in two to three weeks.

In addition to direct observations of the behaviour of the pottos, records of nocturnal activity of single pottos were made on an automatic activity recorder. The animal was placed in a cylindrical cage 6′ high and 2½′ in diameter, which was hung on a frame balanced on a knife edge. A thread, attached to the opposite end of the frame from the cage, moved a lever recording on a kymograph drum revolving at 2 mm./sec. (Fig. 1). Any movements of the animal in the cage were thus recorded on the drum. It was found that all the pottos were active throughout the period of darkness, though there was a higher level of activity in the second half of the night when

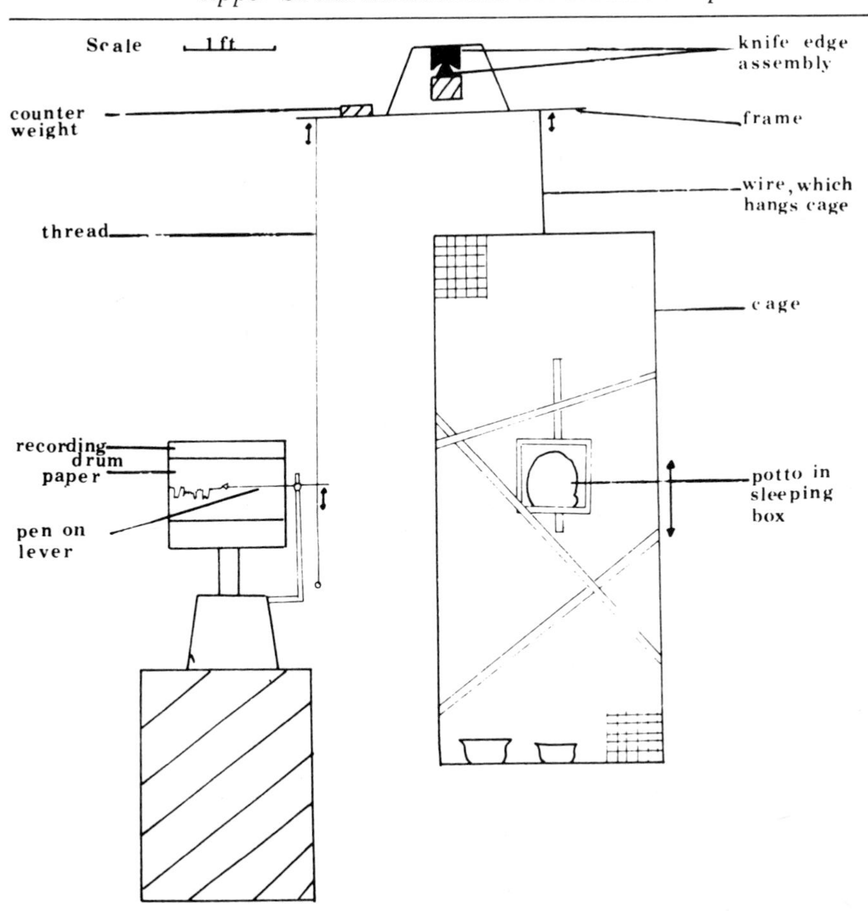

Figure 1. Diagram of the activity recording cage.

temperatures were lower. Records were made of the nightly activity of a single male and of a single female. An animal of the opposite sex to the experimental animal was then placed in a cage close to the activity cage in order to determine whether there was any effect on activity. The activity of a female was unchanged in the presence of a male, but percentage activity time of the male was increased from a mean of 67% to a mean of 90% when a female was caged next to him.

Communication

The pottos had a simple vocal repertoire of six sounds. These were usually heard accompanying aggressive postures, or when one potto was seeking contact with another. Four of these sounds have already been described by Andrew.[1] Grunts and shrieks were heard accompanying fights. Crackles appeared to indicate low intensity threat, and a chattering sound (described by Andrew as a series of screams) usually accompanied defensive threat postures. The spit-click, which sounded similar to the human 'ttt ttt' indicating scolding, may have been what Andrew described as 'loud sharp calls'.[1] The spit-click, together with a metallic clicking sound, appeared to act as a signal of the location of the caller. The metallic click, which was always preceded by a spit-click, was a high-frequency sound (fundamental frequency 15 KHz) with a distinct tonal quality.

Spit-clicks and metallic clicks were frequently heard from a male approaching a female. They were also heard on two occasions when a potto had been moved to a new cage from which the previous occupant had just been removed. The new occupant moved round the cage several times, clicking at intervals, before it entered its sleeping box. These clicks were also heard from a pair of females that were being observed for social interactions in the large observation cage. Before either animal moved out of its home cage into the observation cage, it would hesitate in the doorway and spit-click once or twice. If there was an answering spit-click from an animal already in the observation cage, the caller usually moved back into her home cage; if not, she moved out, still clicking. As she moved out, the spit-clicks were accompanied by metallic clicks, which were answered by the second animal still in her home cage. It could not be

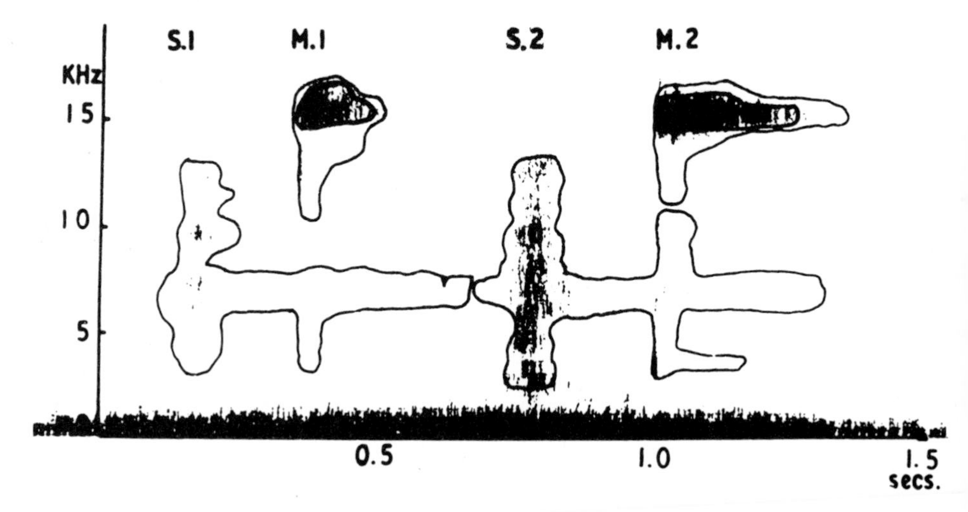

Figure 2. Sonogram of a click call sequence. The thick contour lines outline the fundamental frequency of each click. The thinner lines outline the whole frequency spectrum.
SI = spit-click from the animal furthest from the microphone
MI = metallic click from the same animal
S2 = spit-click from the animal nearest to the microphone
M2 = metallic click from the same animal

determined whether this was a call and answer sequence, but it appeared to be so; one animal would call and put its head on one side in the listening posture; if there was no answering click from the other animal it would click again, and this was usually 'answered' (Fig. 2). Metallic clicks are probably the same as a call, described by Walker, produced by an infant potto which had lost contact with its cloth surrogate mother.[2]

During spit-clicks, metallic clicks, chatters and crackles, the mouth was retracted at the corners. Otherwise no change of expression accompanied vocalisations.

It seems probable that a large amount of potto communication is olfactory. Any potto moving into a new area would urinate on the hardboard along the edge of the cage and on the poles. In addition 'key junctions' (near the entrance to the home cage, or leading from one level in the pole system to another) were marked by perineal rubbing.[3] After they had marked an area, the pottos frequently stopped and sniffed at junctions before moving onto a new pole. Pottos moving into an area that had previously been occupied by unknown animals showed considerable interest in marked areas, sniffing at them vigorously and leaving their own mark on top.

Social interaction

There was relatively little contact amongst any of the pairs of pottos that were observed. The check sheet records revealed that in male-female pairs in the outside observation cage only 12.0% to 19.4% of the activities recorded were of contact behaviour. No contacts at all were recorded for pairs of animals of the same sex meeting in the outside observation cage. In the reversed daylight house, where the animals were caged together continuously, the male-female pairs were in contact for only 15% of the 'night', the two females for only 13%, and no contacts at all were observed between the two males.

The majority of interactions observed between male and female pottos were amicable, though the female usually chattered in threat at the male when they first met. Most of the contacts between males and females were initiated by the male, who would move up to the female, sniff at her, and then (if she did not threaten him) groom her. Grooming was usually reciprocal rather than mutual, each animal in turn standing to groom the other, who sat with head extended towards the groomer. When the male groomed the female, he usually stood on his hind legs at a distance from her (Fig. 3); but when the female groomed the male, though she began by standing away from the male, he would gradually edge forwards until he was sniffing at her genital area (Fig. 4). At this point the female usually

Figure 3. Male grooming female.

Figure 4. Female grooming male.

Figure 5. Urine marking during allogrooming.

chattered in threat and moved off. Allogrooming was frequently accompanied by urine marking (Fig. 5). The groomer would wipe a drop of urine onto his hand and then rub it on the area being groomed.[4] This would result in some mixing of the smells of the two animals.

When a male and female had become familiar with each other, periods of grooming were longer and led to grappling (Fig. 6). This has already been described[5] and appears to be a type of friendly wrestling match; no aggressive vocalisations accompanied grappling.

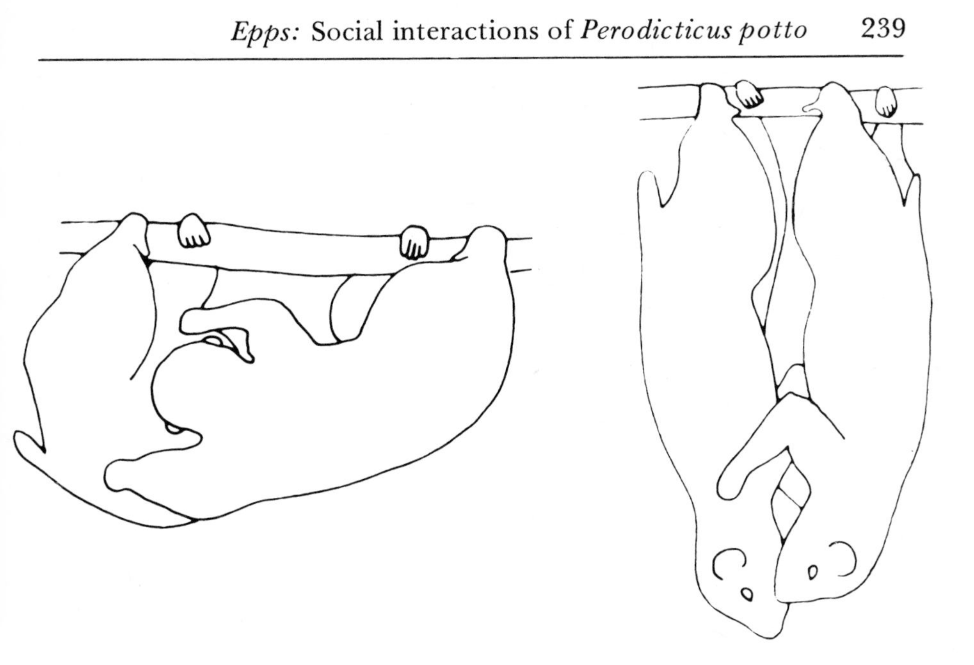

Figure 6. Grappling between two pottos.

Copulation was observed only between one pair; on each of the five occasions that it was seen, it was preceded by grappling, the male finally mounting the female from behind whilst the two animals were suspended from the ceiling (Fig. 7).

Although the male was the prime initiator of contacts, it was the female that determined their number and duration. When a male approached a female, he was frequently repulsed by vocal threats; allogrooming was often terminated in the same fashion, especially if the male sniffed at the female's genital area. The amount of aggression displayed by the female appeared to vary in relation to the oestrous cycle. In the one pair between which copulation was

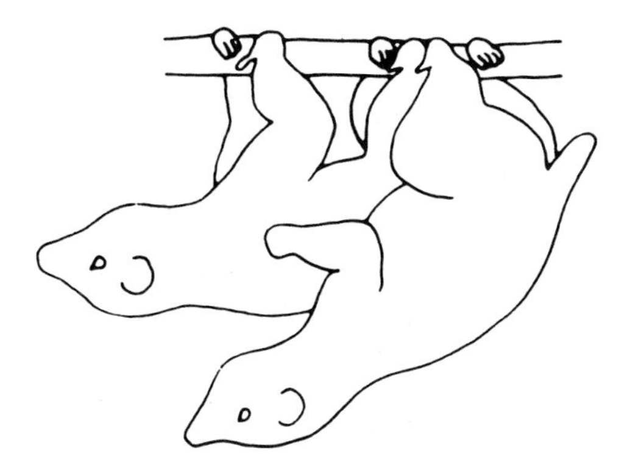

Figure 7. Copulation.

observed, there was a definite reduction in female aggression until just before they copulated, when the female actually sought the male. Two days after copulation had been observed, female aggression rose abruptly, and four days later she would allow very little allogrooming.

Figure 8. Fighting.

Interactions between animals of the same sex that had not met before were always aggressive. If the two animals came into contact, they would fight, wrestling together fiercely and rolling onto the floor (Fig. 8). Fighting was always accompanied by loud squeaks and chatters, and thus could be differentiated from the amicable grappling encounters observed after periods of grooming between a male and a female. A dominant-subordinate relationship was established between all the pairs of the same sex that were observed, usually after the first fight. The subordinate appeared to avoid contacts with the dominant whenever possible; this was frequently achieved by the subordinate remaining motionless under piles of shavings on the floor for a large part of each observation period. Pottos observed by Charles-Dominique[6] in forests in Gabon also avoided contacts with other pottos by remaining motionless, hidden in the vegetation. If a potto became aware of another in the area it was about to enter, it remained completely immobile until the other had moved away. Fights between animals with overlapping territories in the forests were rarely observed. In captivity, however, it is impossible for the animals to avoid contacts and fights do occur. Descriptions of the fights between captive pottos show extreme variety in the form of the fight.[7-10] It is possible that this is due to the fact that fights are rare in the wild state, and therefore no firm pattern has evolved.

Pottos have been observed to hold territories in the forest.[6] Those kept at Makerere exhibited strong territoriality. Even in the relatively small cage in the nocturnal house, it was found that the animals in two out of the three pairs that were kept there established separate areas of movement and rarely trespassed. The male and female met only for allogrooming and at the food bowl. The two males usually remained in their own areas of the cage. If one did move into the other's area, he was immediately threatened by the owner and retreated rapidly. However, the female pair kept in the nocturnal house did not develop separate areas of movement, but their behaviour was probably affected by the fact that they had already been caged together for two years.

An attempt was made to determine whether the olfactory marks left by the pottos served as a deterrent to intruders. It seemed possible that subordinate animals would avoid areas marked by dominant animals; therefore subordinates were allowed into an area that had been marked by a dominant animal, but from which the dominant animal had been removed. Two females that were observed under these conditions showed definite avoidance of the marked area; each moved into the observation cage only three times during a 2½-hour observation period. Each time an animal moved out, it only went as far as a 'key' junction that had been marked frequently by the dominant animal, sniffed this thoroughly and then ran back into its home cage. In a control experiment, in which one of the females used in the previous experiment was allowed access to the observation cage after it had been marked by an unknown female, the intruder spent 70% of the time in the observation cage, sniffing at the marked areas and leaving its mark on top. It therefore seems that pottos are able to recognise the marks of animals with which they are familiar.

Territory experiments with males were less conclusive than those using females. The first subordinate male used had not been in contact with the dominant animal for four weeks. This animal moved out into the observation cage for 54% of the 2½-hour observation period, but he never spent more than three minutes at a time in the observation cage, returning to his home cage at frequent intervals. The second experimental animal moved out into the observation cage only twice after the latter had been marked by the dominant animal. This animal spent 96% of the observation period in its home cage. However, when the same animal was used in the control experiment, it still spent the majority of the observation period in the home cage, only moving into the observation cage for 40% of the observation period after the cage had been marked by an animal with whom it had not previously been in contact. Nevertheless, in the control experiment, the intruder did not exhibit any of the usual signs of stress (holding itself stiffly and moving in jerks) that had been apparent when it was moving around an area marked by the dominant animal. It is possible that the avoidance of areas marked by known animals is less complete amongst males, which, Charles-Dominique has suggested,[6] may have a more rapid turnover of territories in the forest, with juveniles, as they mature, taking over the territories occupied by weak males. Unfortunately, shortage of males prevented the repetition of these experiments.

Discussion

Charles-Dominique's observations of pottos in a forest in Gabon showed them to be solitary territorial animals.[6,11] Encounters between pottos of the same sex were rare, and it seemed that such encounters were actually avoided by the intruder, who remained motionless until the other animal had moved away. In captivity, meetings between animals of the same sex result in fierce fights. It is therefore necessary for the potto to have some mechanism in its behavioural repertoire by which aggression may be overcome when a male and female meet, so that a pair-bond may be built up in order that mating may take place. Allogrooming appears to play a strong part in the formation of the pair-bond. It seems to be pleasurable both for the groomer and for the groomee; pottos that had been groomed were frequently observed soliciting further grooming by holding out their heads towards animals moving off after grooming encounters. Animals that were grooming frequently prevented the groomee from moving off by chattering at it and holding the head down so that grooming could be continued. It is interesting that on the few occasions that a male was heard threatening a female vocally, she immediately responded by holding her head out towards the male, who then ceased threatening and groomed her.

One interesting example, supporting the hypothesis that allogrooming serves as a means by which the pair-bond is built up between pottos, was observed between a male and a female that had been allowed together for seven 2½-hour observation periods. The female was by this time allowing the male to groom her for a short while each time they met. On the eighth night, the female moved up and sniffed at the male whilst he was chattering in threat at a male in an adjacent cage. He immediately turned round and attacked the female; they wrestled together for two minutes, the female shrieking loudly. The female eventually ran away, chattering. After this redirected aggression from the male, the female threatened him whenever he approached and would not allow him to groom her again. The urine marking that frequently accompanied allogrooming must have served a role in the formation of the pair-bond, making the marked animal more familiar and therefore less likely to elicit an aggressive reaction. Further smell sharing took place during the friendly grappling matches that frequently followed allogrooming. Grappling must cement the pair-bond, since all aggressive responses must be suppressed in what is, in essence, a play fight.

Charles-Dominique observed a similar pattern of social interaction between male and female pottos with adjacent territories in the forest.[6] A male potto would visit a female each night, and as she became used to his presence, she allowed him to groom her for gradually increasing periods. Grooming eventually led to grappling. This presumably led finally to copulation, as the female produced infants six months later. However, females in the forest took longer to accept the male. Limitations of space in captivity probably accelerated the pace of courtship and it is possible that, though the cages were kept separate, there had been some olfactory communication between the animals before they actually met.

There has been a great deal of speculation on the role of the nuchal spines of the cervical vertebrae of the potto, which are elongated and project $\frac{1}{8}''$ above the body suface.[12,13,14] These spines are covered with a layer of highly innervated skin;[15] stimulation of this area is, therefore, likely to be gratifying. Since allogrooming is concentrated on the area of the spines, it is probable that the hypothesis that they now function as a focus for amicable intraspecific interactions is correct.[14,16,17] It seems likely, however, that the sensory function of the spines is a secondary development and that the primary enlargement was associated with the development of the extensive cervical musculature.[11]

It was encouraging to find that the behaviour of the captive pottos corresponded quite well to that of the pottos observed by Charles-Dominique in a forest in Gabon.[6,11] It seems likely that cage studies of the behaviour of nocturnal prosimians may play a useful role in supplementing field studies in providing information on details of behaviour not easily obtainable in the wild, provided that the captive situation is carefully planned.

Acknowledgments

While conducting the research reported here the author was supported by the O.D.M. 'Study and Serve' scheme. Makerere University Research Grants Committee financed the housing and feeding of the pottos. Lance Tickell supervised the research and supplied help and criticism throughout. Indebtedness to Alan Walker is also gratefully acknowledged, not only for the original pottos, but also for help and criticism throughout the study and especially in the final stages of 'writing up' when the author was living in Nairobi.

NOTES

1 Andrew, R.J. (1963), 'The origin and evolution of the calls and facial expressions of the primates', *Behaviour* 20, 1-109.

2 Walker, A.C. (1968), 'A note on hand-rearing a potto', *Int. Zoo. Yb.* 8, 110-11.

3 Andrew, R.J. (1964), 'The displays of the primates', in J. Buettner-Janusch (ed.), *Evolutionary and Genetic Biology of the Primates*, vol. 2, London.

4 Urine marking of the groomee by the groomer was observed for the first time when the animals were in the full beam of a torch only 1½ feet from the observer. On this occasion, a drop of urine was seen to be deposited on the hand of the groomer each time it was passed back to the urino-genital area. All subsequent observations of marking during allogrooming were from a minimum distance of seven feet, with dim red illumination. It was assumed that this marking behaviour was the same as that observed previously, but in the light of Manley's[18] observations it is possible that at least some of the marking was genital scratch marking.

5 Bishop, A. (1962), 'Control of the hand in lower primates', *Ann. N.Y. Acad. Sci.* 102, 316-37.

6 Charles-Dominique, P. (1971), 'Eco-éthologie et vie sociale des prosimiens du Gabon', PhD thesis, Paris.

7 Bishop, A. (1964), 'Use of the hand by lower primates', in J. Buettner-Janusch (ed.), *Evolutionary and Genetic Biology of the Primates*, vol. 2, London.

8 Blackwell, K.F. and Menzies, J.I. (1968), 'Observations on the biology of the potto (*Perodicticus potto*), *Mammalia* 32, 447-51.

9 Vincent, F. (1969), 'Contributions à l'étude des prosimiens africains; le Galago de demidoff', PhD thesis, Paris.

10 Walker, A.C., personal communication.

11 Charles-Dominique, P. (1971), 'Eco-éthologie des prosimiens du Gabon', *Biol. Gabon.* 7, 121-228.

12 Sanderson, I.T. (1937), *Animal Treasures*, London.

13 Coon, C.S. (1962), *The Origin of Races*, New York.

14 Walker, A.C. (1968), 'A note on the spines on the neck of the potto', *Uganda J.* 32, 221-2.

15 Montagna, A.W. and Yun, J.S. (1962), 'The skin of the primates: 14. Further observations on *Perodicticus potto*', *Amer. J. Phys. Anthrop.* 20, 441-50.

16 Walker, A.C. (1970), 'Nuchal adaptations of *Perodicticus potto*', *Primates* 11, 135-44.

17 Kingdon, J. (1971), *An Atlas of East African Mammals*, 1, London.

18 Manley, G.H., this volume.

J. M. TANDY

Behaviour and social structure of a laboratory colony of Galago crassicaudatus

Introduction

Galagos are probably solitary or live in very small groups – perhaps only families. Sauer and Sauer found groups of two to nine individuals (*Galago senegalensis bradfieldi*) together.[1] Haddow and Ellice found groups of two to four animals (*G. senegalensis*) at a locality in Northern Karamoja, north eastern Uganda, and two to six animals in Tanzania.[2] Vincent found *G. demidovii* in the Congo (Brazzaville) in numbers of one to five.[3] He also found their nests in clusters: 'Adult pairs can be found in the wild outside the breeding season, and adult males and females can be accompanied by subadults of the same sex.'[3] Doyle, Andersson and Bearder, while searching for nests of *G.s. moholi*, found that 'but for the presence of mother and infants on them, they would, in most cases, not be recognized as nests'.[4] This may imply that the males do not stay with females and young. The work of Bearder and Doyle verifies this assumption.[5] Struhsaker found two to four *G. demidovii* in nests during a systematic study of numbers in nests and duration of occupancy.[6] The present author observed two *G. crassicaudatus* together at Athi River, Kenya, and single individuals at several localities in Tanzania during 1970-71.

There have been few behavioural studies of galagos in the wild: (1) *G.s. bradfieldi* in South West Africa by Sauer and Sauer;[1] *(2) G. demidovii* in the Congo (Brazzaville) by Vincent;[3] (3) *G. crassicaudatus garnetti, G.c. lonnbergi, G.c. umbrosus* and *G. senegalensis* in South Africa by Bearder and Doyle;[5] (4) *G. alleni* and *G. demidovii* near Kumba, West Cameroun, by Miller;[7] and (5) *G. alleni* and *G. demidovii* in Gabon by Charles-Dominique.[8] Several publications concern group behaviour of captive animals. Sauer and Sauer studied *G.s. bradfieldi*;[1] Lowther,[9] Doyle, Andersson and Bearder,[4] and Doyle, Pelletier and Bekker[10] studied *G.s. moholi*; Roberts observed

G. crassicaudatus;[11] and various species have been studied by other authors.

The purpose of the present study was to investigate what kind of social structure, if any, would develop among a group of feral *G. crassicaudatus* in captivity. The investigation was intended to be a general one from which could be developed a more intensive study to analyse their communication system. Behavioural data used as indicators of social structure were selected because of their ability to demonstrate positive or negative bonds between individuals.

Materials and methods

This study was conducted at the Physical Anthropology Laboratory of the University of Texas at Austin. The facilities included a thermostatically heated room, 5.4 x 8.3 x 3.1 m., in which were provided seven nest boxes and a variety of tree limbs and other climbing apparatus. The animals were on a natural day/night cycle from September until October 1971, at which time a light timer was installed. Their cycle was then reversed. 'Daylight' began at 18.30 hrs. when white fluorescent tubes switched on, and 'night' began at 06.30. Six 60-watt red bulbs burned continuously and provided enough light during dark periods for observation.

On 1 September 1971, eight feral *G. crassicaudatus* (three females, five males) were released into this room. The animals had probably been in contact with each other only since capture. On 10 September (nine days after importation), B died of pneumonia. Table 1 gives data on localities and dates of capture for the animals. Table 2 gives the animals' weights.

The taxonomy of *G. crassicaudatus* is complicated by a profusion of sub-species. The distribution of the 'sub-species' is continuous in eastern Africa from the Juba River to Natal.[12] It is possible that the

Table 1 Tag identification, sex, locality data, and date of capture for animals used in this study. Animal names represent colour of tags worn around the neck

Name	*Sex*	*Locality*	*Capture Date*
Purple (P)	male	Zigi Amani, Tanzania	14-17 August 1971
Red (R)	male	Zigi Kisiwani, Tanzania	2 August 1971
Black (Bl)	male	Fanusi, Tanzania	20 July 1971
Yellow (Y)	male	Gonja, Tanzania	31 March 1971
Blue (B)	male	Zigi Amani, Tanzania	14-17 August 1971
White metal (Wm)	female	Zigi Amani, Tanzania	14-17 August 1971
White plastic (Wp)	female	Zigi Amani, Tanzania	14-17 August 1971
Green (G)	female	Zigi Amani, Tanzania	14-17 August 1971

Table 2 The weights of the animals in grams

Name	*Weight (gm.)*
♂P	1134
♂R	1474
♂Bl	907
♂Y	1304
♀Wm	1134
♀Wp	1077
♀G	1049

subspecific differences represent a cline of continuously varying morphological characteristics. According to Doyle, Pelletier and Bekker,[10] *G. senegalensis* subspecific variations include fairly constant differences in gestation period and number of offspring. The animals used in this study are probably a mixture of what are considered to be *G.c. lasiotis* and *G.c. panganiensis.*[12]

The animals were fed mealworms, fruit, infant formula, peanut butter, Monkey Chow (commercial primate food) and vitamin supplements.

Observation periods varied from five-minute proximity checks to one hour or longer intervals. They were begun with a diary format and later switched to a modified checklist in which sequence and direction of events could be recorded. This paper concerns the period from 1 September 1971 to 23 April 1972. Because tabulated data include only days when all animals were identified, observations for most of the first part of September are excluded. Total data presented represent approximately 70 hours of observation (field hours not included). The galagos exhibited considerable investigative behaviour toward the observer sitting in their cage.

Results

Proximity

Each animal was scored as being in proximity to another if it was sleeping with or sitting by another at the time the observer entered the room. These were random samples which were not taken every time other observations were made. See Table 3 for proximity scores.

♂Y, ♂R and ♀Wm had proximity scores significantly less than their solitary scores. Probability is less than .0001 ($\chi^2 = 30.316$), .0001 ($\chi^2 = 21.053$) and .0012 ($\chi^2 = 11.250$) respectively for their scores representing equal frequencies of being alone versus with other animals. All animals were alone more often than with others. Proximity data indicate that the group as a whole does not associate randomly ($P < .0024$). When each animal's pattern of association with all other members of the colony is considered, proximity scores

Table 3 **Proximity data for about half of the observation sessions. The diagonal row represents the number of times the animal was alone. Side row totals represent the number of times the animal was with other animals**

	♂P	♂R	♂Bl	♂Y	♀Wm	♀Wp	♀G	TOTALS
P	50	5	1	1	8	5	12	32
R	–	62	1	1	3	2	2	14
Bl	–	–	47	5	6	13	7	33
Y	–	–	–	58	2	6	3	18
Wm	–	–	–	–	55	2	4	25
Wp	–	–	–	–	–	49	5	33
G	–	–	–	–	–	–	47	33

between ♂P and ♀G and between ♂B1 and ♀Wp are significantly greater than expected by chance. These pairs do not appear, however, when other interactions are considered. For example, ♂P grooms and is groomed by ♀Wm more often than any other animal.

Agonistic behaviour

Spats, 'spat-calls', cuffs, avoids, displaces, bites, chases, pulls, bipedal stances with or without the mouth open, and open mouth threats were considered agonistic actions. A spat is a combination of the bipedal stance with cuffs and bites (this looks much like a boxing match) and is accompanied by spat-calls. Spat-calls are noise-like sounds somewhat resembling the spitting sounds of young kittens when alarmed. Sounds resembling screams may be produced within such calls. A cuff is a slap at an animal with the hand. When approached by an animal in an agonistic situation, the individual approached may rise from its quadrupedal or sitting position to a bipedal one, the mouth may be held slightly open without emission of sound, and the arms may be raised; or the animal may remain sitting, with the mouth open displaying its teeth.

The following diary extracts illustrate short agonistic episodes:

11 November 1971: 16.15 R muzzles B1 and B1 avoids him. Then R follows B1; B1 spat-calls and partially stands bipedally with his mouth open.

16.20 R is licking himself and P is watching him. P approaches R and muzzles him. R partially rises bipedally, raises his left hand as in a cuff and gives a spat-call. P retreats.

16.25 Wm jumps to the box where P is. He genital-sniffs her. She spat-calls and cuffs him and then jumps away. He foot-rubs.

The highest numbers of agonistic encounters occurred when the animals were first released in the room and during the periods of female oestrus.

Non-agonistic behaviour

Muzzles (touching noses and sniffing), ano-genital muzzles, sniffs (smelling the area around an animal), grooming, 'crack-calls', mounts, follows and copulations were scored as non-agonistic actions. Crack-calls are low-frequency grating sounds given only by males. These were usually directed toward females. Such vocalisation often began when a male was following a female. It also occurred when a male was grooming a female and was not confined to periods of oestrus.

Sexual behaviour, such as crack-calls, copulations and mounts, was lumped with non-agonistic behaviour for convenience in data tabulation. Although a follow or crack-call was initiated sexually, the response from the recipient was usually agonistic. Out of 299 recipients, 293 were females. Since the act was apparently not initiated agonistically, there seems to be justification for including it with non-agonistic behaviour.

The following diary extracts illustrate a sexual episode:

30 October 1971: 16.15 Male P licks female G's genitalia and head. One of them makes a 'crackly' call (consisting of three rasps). P tries to manoeuvre into copulatory position.

16.20 P has a very large erection and thrusts at G a couple of times, but he cannot get into proper position because the ledge they are on is not wide enough. His hind feet are holding onto the wire below the ledge.

16.25 P and G have separated. P jumps to the ledge where R is sitting and then climbs back to G. He rapidly licks her head and gives several groups of the 'crack-call'. R urine-washes several times. The calls become more rapid and P has G in copulatory position. R is still urine-washing.

An excellent photograph of this species in the above-mentioned copulatory position can be found in Buettner-Janusch.[13]

Two oestrus periods occurred during this study. One lasted from 21 October to 1 November 1971, and another from 22 March to

27 March 1972 (length based on observed copulations). One infant was born to ♀G on 28 February 1972, but it was killed.

Table 4 summarises total interaction data recorded for each animal. Play encounters as observed by Roberts and Newell were never seen in these animals.[11,14]

Table 4 Total interaction data recorded for each animal

	♂P	♂R	♂Bl	♂Y	♀Wm	♀Wp	♀G
Groom	151	29	49	1	151	17	38
Muzzle	19	14	27	0	24	19	6
Sniff	119	128	184	16	119	167	37
Ano-genital muzzle	62	55	60	0	75	68	6
'Crack-call'	179	116	10	0	126	149	18
Follow	85	75	37	0	113	76	2
Mount	57	30	10	0	70	5	22
Copulate	21	13	0	0	29	0	5
Open mouth threat	2	2	1	0	0	2	0
Bipedal threat	24	23	12	1	8	9	1
Bipedal OM threat	6	4	3	0	3	0	0
Spat	232	141	34	13	93	54	21
'Spat-call'	205	113	30	10	104	96	26
Cuff	61	60	11	3	35	67	7
Bite	10	7	3	0	8	4	1
Avoid	219	106	65	2	254	155	9
Chase	20	16	5	0	7	6	0
Displace	13	5	5	0	4	5	0
Pull	1	2	1	0	0	0	0
Crouch	13	1	0	0	11	2	0
TOTAL	1,499	940	547	46	1,234	901	199
Total Agonistic	806	480	170	29	527	400	65
% Agonistic	54%	51%	31%	63%	43%	44%	33%
Total Non-agonistic	693	460	377	17	707	501	134
% Non-agonistic	46%	49%	69%	37%	57%	56%	67%

Other behaviour

'Loud calls', urine-washing and foot-rubbing are included as 'other behaviour' because their functions are not fully understood by the author. One loud call was heard approximately ten minutes after the animals were released into the enclosure. This call was not heard again until after a month in captivity. These calls occur much more frequently than observations indicate (128 recorded). They are often heard when the author is out of the room working elsewhere. I have never heard this call in the wild. It resembles that of a large bird giving a series of notes first rising rapidly and then falling gradually in frequency and amplitude. It is not possible to describe this call or others precisely until sonograms are available. The animal's body shakes violently during this call and the mouth is held only slightly open. The only animals seen making this call are ♂R and ♂P. It has been recorded eighteen times for the former and only three times for the latter. Occasionally, two occur almost simultaneously, with one male beginning and the other starting to call halfway through the first's call. The barking call frequently given in the wild to intruders has not been heard in our laboratory.

During urine-washing, an animal deposits urine on its hand and then rubs it on the sole of the foot of the same side of the body. Its possible functions are discussed by Sauer and Sauer, Andrew and Klopman, Doyle, and Quick.[1,15,16,17] It occurs in a variety of contexts. Individual animals exhibited significantly different frequencies of this behaviour, as the following table indicates:

♂P	♂R	♂Bl	♂Y	♀Wm	♀Wp	♀G
134	34	68	2	37	53	0

The probability that the distribution of frequencies of this action is the same as the distribution of frequencies of all actions of each individual is less than .0001 ($\chi^2 = 90.481$). Individual frequencies showing positive significant deviations at the .05 level from their frequencies predicted by total actions are underlined; negative significant deviations at the .05 level are circled.

Foot-rubbing consists of rapid rubbing of first one foot and then the other against a branch, box top, floor, etc. This makes a very loud scraping noise. The behaviour may be accompanied by urination or urine-washing. Probability that the following distribution of foot-rubbing is the same as the distribution of frequencies of all actions of each individual is less than .0001 ($\chi^2 = 318.450$):

♂P	♂R	♂Bl	♂Y	♀Wm	♀Wp	♀G
3	82	2	15	149	11	1

♀Wm has been seen foot-rubbing in the same spot where a few seconds previously ♂R had foot-rubbed. Twice, ♂Y moved from his usual place near the air conditioner in the corner onto the top of a nest box. Instead of moving farther, he spent approximately ten minuts foot-rubbing, and then quickly returned to the corner ledge.

Self-grooming was tabulated as a general indicator of the animals' activity levels away from other animals. The higher scores for males may be due to the fact that they are groomed by other animals less than are the females. Probability is less than .0001 ($\chi^2 = 53.135$) that the following distribution of the frequencies of self-grooming are the same as the distribution of frequencies of all actions of each individual:

♂P	♂P	♂Bl	♂Y	♀Wm	♀Wp	♀G
109	80	27	0	44	27	1

Discussion

Proximity

♂Y, ♂R and ♀Wm showed significant differences in solitariness versus association. All were more solitary than social. Y was an isolate or peripheral male characterised by almost total non-interaction with other colony members. ♂R was a subordinate animal which carried its tail in the curled-under submissive posture described by Sauer and Sauer,[1] and his ears were usually carried in a flattened position also indicative of sub-dominant status. ♀Wm was a hyperactive highly aggressive female. She was particularly agonistic toward ♀Wp. Lack of proximity is especially good for indicating these types of negative relationship.

Vincent appears to interpret his Congo data and the data of Haddow and Ellice on numbers of animals captured sleeping together as indicating relatively stable social groups.[2,3] This seems to be supported by Haddow and Ellice's yellow fever immunity tests.[2] In the laboratory, animals often sleep in groups, and these groups show some stability. Proximity scores of ♂P and ♀G indicate that they associate at random with all other individuals, but positively non-randomly with each other. The same type of relationship occurs between ♂Bl and ♀Wp. The former pair consists of a highly dominant male and a very subordinate, almost non-interacting female. The latter pair consists of a young adult male and a female which is subordinate to ♀Wm. The change in ♂Bl's proximity scores to greater solitariness (see below) does not seem to have affected his relationship with ♀Wp.

♂Bl's proximity scores changed markedly from February to April. From September to February he was found alone 24 times versus 31 times with others. From February to April he was seen alone 23 times but only twice with others. This change in proximity scores is statistically significant at the .0002 level. The change might be related to his sexual maturation and resulting difference in social standing in the colony.

Sauer and Sauer have suggested that agonistic dyads will tolerate each other during sleeping periods in order to stay in the same nest.[1] This type of occurrence was very infrequent in our colony.

Agonistic behaviour

During the present study, female antagonisms were high and directed toward the males during oestrus. ♂P, ♂R, ♀Wm and ♀Wp exhibited the highest frequencies of agonistic encounters. The highest scores for ♂R and ♀Wm were with ♂P, and ♀Wp's highest was with ♂R. By contrast, ♂Bl, ♂Y and ♀G had very low scores. ♂Y and ♀G's highest were with ♂P, and ♂Bl's was almost equally divided between ♂P and ♀Wm. Agonistic encounters always took place between two individuals without involvement of others.

Grooming

Grooming sessions were not timed, but were grouped with other actions into a general category of non-agonistic interactions. The increase in frequency of grooming among the animals over time may reflect closer personal bonds developing between animals as fear of the surroundings and of other animals lessened. More than half of the grooming recorded was directed by ♂P to ♀Wm and ♀G, who rarely reciprocated. What little grooming ♂R and ♂B1 did was mainly directed towards the females.

A grooming session often seems to be a stressful situation interspersed with spat-calls from the animal being groomed. But, while calling, an animal may be presenting an arm or bending its head to be groomed. Whether grooming functions as a group cohesive force in galagos, as it does in other primate groups, is uncertain.

Sexual behaviour

When females were not in oestrus, the scrota of the males were inconspicuous and fur-covered. During the oestrus period, the male's scrotum enlarged slightly and a pink-coloured patch of skin with a granular appearance was observed centred in the groove between the two scrotal sacs. This is very similar to that described by Manley for *Perodicticus* and *Arctocebus.*[18] In the females, the clitoris is conspicuous between oestrus periods, but the vulva opens only during oestrus.

During the first oestrus period, which occurred in October, copulations were restricted to ♂P and the females. The number of copulations recorded was low, only one for ♀Wm and five for ♀G. None was recorded for ♀Wp, although she underwent a large amount of investigation by the males and was mounted several times. Usually ♂R was able only to ano-genital muzzle as a female passed. It may be significant that ♂Y exhibited much scrotal swelling and pink colouration, but did not investigate the females. He may have been so intimidated by one or several of the others that his sexual behaviour was suppressed. Little sexual interaction occurred between ♂Bl and the females. That ♂Bl had no scrotal swelling or colouration, combined with his small body size, indicated that he was a subadult. Because of his subadult status, like the subadults of Roberts' group, he may have been more easily tolerated by the adult females than were the adult males.[11] This would explain his high non-aggressive scores and low aggressive ones between September and February.

There were probably no two females at the peak of oestrus at the same time, giving no opportunity to the observer to check for preference on ♂P's part. The males remained sexually excited after the females' vulval opening had closed and they were no longer receptive. This led to many agonistic encounters.

During the night (artificial light 'daytime' for the galagos) of February 28, ♀G gave birth to an infant. It was found dead the next morning, the brain and lower extremities having been eaten. Killing of infants in captive colonies has been thought to be related to over-crowding.[13,19] Females probably separate themselves from other animals during birth of infants.[9,19] Overcrowded conditions were not considered characteristic of the present study; ♀G was, therefore, left to give birth in the enclosure with the other animals. Either the colony was overcrowded (possibly indicated by the high percentage of agonistic interactions), or some other factor was involved. ♀G did not come into a post-partum oestrus as does *G.s. moholi.*[20]

On 22 March, ♀Wm came into oestrus. This precipitated both a sharp increase in fighting between ♂R and ♂P, and what appeared to be a shift in dominance status between them. The formerly submissive ♂R, which had not been allowed to copulate during the October oestrus period, became very aggressive. The aggression took the form of attacks and chases of ♂P and ♂Bl by ♂R, especially when Bl ventured onto the floor. ♂R successfully copulated with ♀Wm several times. ♂Bl shared typical scrotal swelling and pink colouration during ♀Wm's oestrus period. He mounted her several times, but he did not copulate while under observation.

The number of copulations is not considered a reliable means by which to judge dominance or group structure in other social species, because subordinate animals may copulate with females when a dominant animal is elsewhere. This was not observed in my group, possibly because the dominant male could never be any great distance from the other animals. Perhaps females discriminate against subordinates as mates.

Social structure

Field data for *G. senegalensis* and *G. demidovii* indicate that sometimes individuals form small groups. Although these species sleep together in groups, the individuals may separate for nightly foraging, but stay within hearing range of each other. They may regroup at daybreak. Haddow and Ellice used evidence of yellow fever immunity to support such regrouping.[2] They found four out of eleven groups which exhibited immunity. This might suggest little intermixture of groups. Observations in Cameroun by the present author in 1971 agree with such observations for *G. demidovii* and *G. alleni.* When lights were flashed into closely spaced trees, several pairs of red eyes could be seen. Closer inspection sometimes revealed two to four animals in the same tree. Observations by the present author in East Africa seem to indicate a more solitary way of life for *G. crassicaudatus* than is found in the other galagos. Observations by Bearder and Doyle on *G. crassicaudatus* in South Africa have shown one to three to be the most common number found together.[5]

Laboratory data from the present study suggest that 'activity centres' may exist rather than a cohesive group or groups. Flow diagrams are helpful for illustrating these types of relationships for behaviour scored (see Fig. 1). Whether these animals are capable of organising structurally beyond the mother-offspring level is a

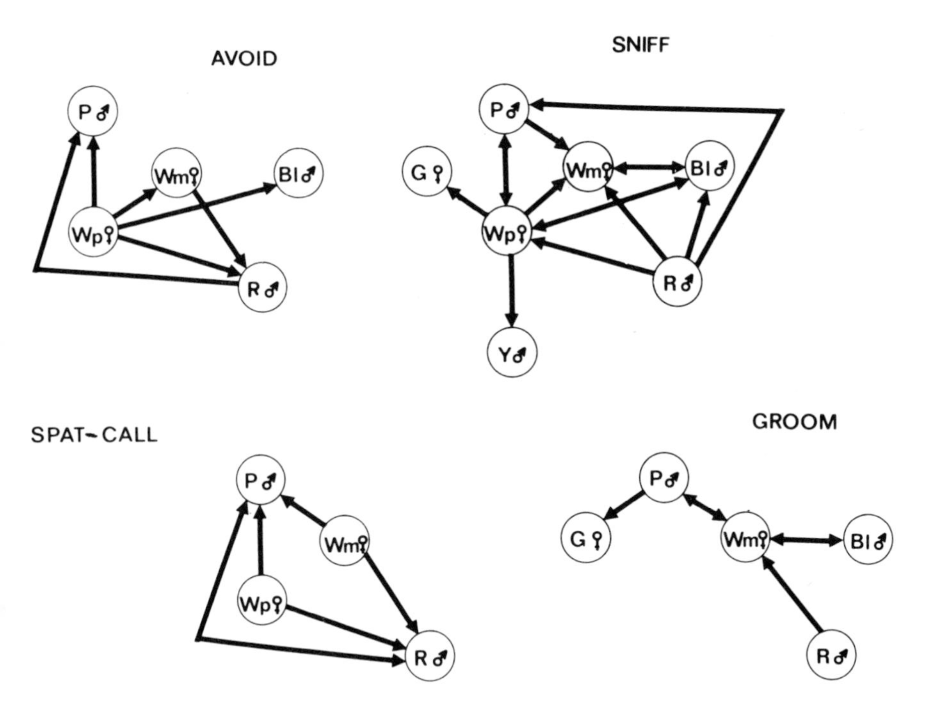

Figure 1. Flow diagrams illustrating four of the behaviour patterns scored. Arrows indicate ten or more interactions, and their directions.

question that needs to be answered. There are strong indications that most interactions between animals are sexual. The frequent sniffs of females by males instead of by other females, grooming of females by males instead of by other females (and vice-versa), and agonistic action of females toward the males are examples of this. A thorough understanding of social structure in *G. crassicaudatus* will require many additional hours of observation.

Urine-washing and foot-rubbing data suggest that more dominant animals are more territorial and that sub-dominant animals have a higher tendency to perform submissive and displacement activities. ♂P and ♂Bl were significantly high ($P < .05$) for urine-washing but significantly low ($P < .05$) for foot-rubbing. Just the opposite is true for ♂R and ♀Wm. ♀G was significantly low ($P < .05$) on both counts, while ♂Y was not significantly high or low for urine-washing but significantly high ($P < .05$) for foot-rubbing. ♀Wp was not significantly high or low for urine-washing but was significantly low for foot-rubbing.

The scent of the animal left with urine-washing might be a good territorial marker (but see Doyle).[16] The loud sound of foot-rubbing, however, might have an immediate effect. The contexts in which the latter has sometimes been seen, such as when a male is rejected by a female or when the isolated male ventures toward the other animals, suggest its role as a displacement activity or as a submissive signal.

Within the group of females, the differences in rate of urine-washing versus foot-rubbing do not seem to indicate dominance relationships as they do among males. For example, ♀Wm is dominant over ♀Wp and ♀G according to other data, but ♀Wm shows high levels of foot-rubbing and low levels of urine-washing. These actions by females may be directed toward males instead of other females. The females may also not be as strongly territorial as the males.

Within the group of males, the differences in rate of urine-washing versus foot-rubbing do seem to agree with other data regarding dominance relationships. ♂R and ♂Y are sub-dominant and ♂P and ♂Bl are dominant. ♂R is dominant over ♂Y and ♂P over ♂Bl. Strong territoriality in males might be adaptive, so that each male has optimum chances of reproducing.

Limitations of present laboratory studies

One must consider the problem of alteration of behaviour in the laboratory compared to that in nature. Overcrowding might cause abnormal levels of agonistic interactions. Possibilities for social stress might be greater in the laboratory than in nature, but physical stress due to parasites and disease may be less in captivity. Some differences of the colony in captivity versus nature are: 1. smaller animal/space ratio than in the wild; 2. opportunities for natural foraging behaviour extremely limited; 3. light cycle altered and animals moved to a different hemisphere; 4. frequent interactions with cage cleaner and observers; 5. probably abnormal age/sex ratio – groups contained neither infants nor young juveniles; and 6. the laboratory provides a more homogeneous environment. Inhibition of animals and alteration of stimuli may cause the absence of behaviour as well as creation of new behaviour (see Kaufmann and Kaufmann regarding this phenomenon in coatis, *Nasua narica*).[21]

The age/sex ratio may have an effect on the type of social structure observed. Such effects should become evident after several years of births and colony growth (which will require the provision of a larger living space).

Communicative behaviour

Few vocalisations have been observed except in a sexual or agonistic context. The spat-call is agonistic and the crack-call has been suggested to have a sexual function because it is usually heard only in male/female interactions. The loud call may function as a spatial indicator much as the morning call of the gibbon.[22] The acoustic qualities of this call are ideal for transmission over long distances and through dense vegetation.

Communication between animals in non-agonistic contexts may be primarily through olfaction. The animals in the present colony continually smell the substrate by lowering the nose to a tree limb or the floor as they walk. This is accompanied by loud sniffing sounds. Initial meeting of another animal usually initiates some sort of sniffing behaviour.

Ear positions are good situation indicators. The ears orient differently during alertness, fights, submissive gestures, etc.[1] Not all possible situations have been categorised in the present study, since this would require more detailed study of this behaviour. Ear positions most probably do function as communication signals. Perhaps in galagos, which seldom change facial expressions, they are analogous to the facial expressions of higher primates.

Terrestrial versus arboreal behaviour

Galagos hop well and are able to cover great distances quickly with this method of locomotion. During periods of extreme stress in the lab, the galagos leave the safety of high ledges and hop about on the floor, even though plenty of limbs are available. This appears to be strange behaviour for an arboreal animal. ♂R, which was very subordinate to ♂P for most of the first part of the study, spent most of his time on the floor. This is a behavioural pattern which developed during several months. ♂R did not do this when the colony was first established. He did not appear to be sick nor injured and could climb and jump well. He is becoming less agile than the others, perhaps because he spends so much time on the floor. Possibly a lack of exercise has resulted in his present 'heavyweight' condition. Considering this behaviour in captivity, it would be interesting to know how much time galagos in the wild spend on the ground. They may be more terrestrial than previously thought. In the farm-bush zone where *G. crassicaudatus* sometimes live, trees are fairly widely separated. This would necessitate locomoting on the ground at least during movement between trees, a situation which would seldom arise in riverine or other forested areas. Bearder and Doyle report on this aspect of behaviour in the field.[5]

Summary

Seven *Galago crassicaudatus* (four males and three females) were observed in early stages of contact and interaction after having been released into a laboratory enclosure. Animals were scored on proximity, agonistic, non-agonistic and other behaviour as possible indicators of social structure. It appears that there are personal preferences among the animals regarding other animals, which are apparent in the interaction data and proximity scores. At present small groupings may be forming, but the tendency seems to be toward solitariness. The animals interact with one another individually rather than as groups. Births of young in the colony may be sufficient stimulus to form family groups. Field data on several galago species indicate solitary and small group habits.

Various communicative methods are discussed. These include an agonistic spat-call, a crack-call probably related to sexual activity and a loud call of unknown function. Olfaction is important in communication as indicated by various types of sniffing and their frequencies, and ear positions may also be important in communication.

The enclosure was constructed to simulate a natural environment, but all stimulus possibilities available in nature were not present and several new ones were introduced. It is not known how such differences affected the animals' behaviour.

Acknowledgments

I would like to thank Y.G. Matola and P. Wegesa of the East African Institute for Malaria and Vector Born Diseases, Amani, Tanzania, for providing the animals used in this study. Thanks also go to Claude Bramblett, without whose help this study could not have been possible, for his aid in obtaining the animals, his suggestions and stimulating discussions, and for critically reading the manuscript. His wife Sharon very generously typed the manuscript. My indebtedness to Larry Quick is also acknowledged. Finally, I would also like to thank my husband Mills, who very patiently read the manuscript and made many constructive criticisms. This study was aided by an Undergraduate Research Grant from the University of Texas.

NOTES

1 Sauer, G.F. and Sauer, E.M. (1963), 'The South West African bushbaby of the *Galago senegalensis* group', *J. S.W. Afr. scient. Soc. Windhoek* 16, 5-36.

2 Haddow, A.J. and Ellice, J.M. (1964), 'Studies on bushbabies (*Galago* sp.) with special reference to the epidemiology of yellow fever', *Trans. Roy. Soc. trop. Med. Hyg.* 58, 521-38.

3 Vincent, F. (1968), 'La sociabilité du Galago de Demidoff', *Terre et Vie* 22, 51-6.

4 Doyle, G.A., Andersson, A. and Bearder, S.K. (1969), 'Maternal behaviour in the lesser bushbaby (*Galago senegalensis moholi*) under semi-natural conditions', *Folia primat.* 11, 215-38.

5 Bearder, S.K. and Doyle, G.A., this volume.

6 Struhsaker, T.T. (1970), 'Notes on *Galagoides demidovii* in Cameroun', *Extrait de Mammalia* 24, 207-11.

7 Miller, C.D. III, personal communication.

8 Charles-Dominique, P. (1971), 'Eco-éthologie des prosimiens du Gabon', *Biol. Gabon.* 7, 121-228.

9 Lowther, F. de (1940), 'A study of the activities of a pair of *Galago senegalensis moholi* in captivity, including the birth and postnatal development of twins', *Zoologica* 25, 433-59.

10 Doyle, G.A., Pelletier, A. and Bekker, T. (1967), 'Courtship, mating and parturition in the lesser bushbaby (*Galago senegalensis moholi*) under semi-natural conditions', *Folia primat.* 7, 169-97.

11 Roberts, P. (1971), 'Social interactions of *Galago crassicaudatus*', *Folia primat.* 14, 171-81.

12 Hill, W.C.O. (1953), *Primates: Comparative Anatomy and Taxonomy*, vol. 1. *Strepsirhini*, Edinburgh.

13 Buettner-Janusch, J. (1964), 'The breeding of galagos in captivity and some notes on their behavior', *Folia primat.* 2, 93-110.

14 Newell, T.G. (1971), 'Social encounters in two prosimian species: *Galago crassicaudatus* and *Nycticebus coucang*', *Psychon. Sci.* 24, 128-30.

15 Andrew, R.J. and Klopman, R.B., this volume.

16 Doyle, G.A., this volume.

17 Quick, L., unpublished data.

18 Manley, G.H., this volume.

19 Sauer, E.G.F. (1967), 'Mother-infant relationship in galagos and the oral child-transport among primates', *Folia primat.* 7, 127-49.

20 Doyle, G.A., Andersson, A. and Bearder, S.K. (1971), 'Reproduction in the lesser bushbaby (*Galago senegalensis moholi*) under semi-natural conditions', *Folia primat.* 14, 15-22.

21 Kaufmann, J.H. and Kaufmann, A. (1963), 'Some comments on the relationship between field and laboratory studies of behaviour, with special reference to coatis', *Anim. Behav.* 11, 464-9.

22 Carpenter, C.R. (1940), 'A field study in Siam of the behavior and social relations of the gibbon (*Hylobates lar*)', *Comp. Psychol. Monog.* 5, 1-212.

U. M. COWGILL

Co-operative behaviour in Perodicticus

Introduction

This study brings together observations concerning a variety of behaviour patterns of four *Perodicticus potto* (2 males, 2 females) gathered over the past eight years. These patterns are co-operative in character and may be divided into two general categories: those that involve the nuclear family only and those that occur between the family and the group. The individual that is most mobile between these two categories is the dominant 'alpha' male potto. His dominance is entirely expressed in either protective types of behaviour or co-operative ones. Despite the fact that some of his behaviour operates to ensure his own survival, it is nevertheless, as far as the family and group of which he is a part are concerned, protective and co-operative.

The death of Genet

The living conditions and type of food supply under which three *P. potto* have been maintained in captivity for the past eight years have been described earlier.[1] On the morning of 9 April 1971, the 'beta' male named M. Genet was clearly not feeling well. He was being groomed by the 'alpha' male and his mate on the shelf. He died sometime in the late afternoon of that day from organic causes of a non-infectious nature. In 1964, when his mate died as a result of pregnancy, it was noticed that the food consumption of the then remaining three animals was rather meagre. As a result of the more recent death (leaving two animals), this observation was again apparent and it appeared necessary to weigh the food before and after consumption to obtain some idea of how much they were eating. It has previously been noted that pottos consume food on a definite bidiurnal cycle.[2] This is confined to the bulk of food they eat, which is primarily bananas. At that time no such cycle was noted in the consumption of other foods. As will be seen later in this paper, foods other than liquids, if offered daily, are consumed on the same cycle. The data given below are all corrected for natural loss.

Initially, the two surviving animals, one 'alpha' male and his mate, were given the same weight of bananas as they had been when three were being maintained. The amount consumed per animal at this time was less than it had been prior to the death of the third. After two weeks, the quantity the pair had been eating was all they were served. It was noticed that they consistently left enough bananas for the absent third. After four weeks, the 200 gm. they were fed daily were reduced to 100 gm. for two weeks, but the pottos continued to leave enough food for M. Genet. The data are set out in Table 1,

Table 1 Average weight (gm.) of bananas eaten by *P. potto* on high and low consumption days. All data have been corrected for natural loss. With the exception of the first line, concerning three animals, the data for the 'length of time studied' category are consecutive

No. of Animals	*Bananas consumed*		*χ^2 periodicity*	*Amount fed per day*	*Length of time studied*	*Expected consumption*		*χ^2*
	High days	*Low days*				*High days*	*Low days*	
3	329	237	14.96***	450	7 weeks	–	–	
2	198	127	15.51***	450	2 weeks	220	158	8.28***
2	130	69	18.70***	200	4 weeks	198	127	49.84***
2	71	49	4.03*	100	2 weeks	100	70	14.71***
2	350	288	6.02**	450	3 days	220	158	183.78***
2	200	130	14.84***	450	1 week	220	158	6.77***

*** P = .01, ** P = .02, * P = .05

where the first χ^2 given is based on the hypothesis that there is no difference between every 48 hours, the high days, and the intervening low-consumption 24-hour period. The periodicity appears to be quite stable even when the food supply is low, though it is not as spectacular as it is when the supply is adequate. The same table shows what two pottos would be expected to consume on both days. The χ^2 between the observed and expected values is enormously significant, suggesting the persistence of food saving even when the supply is low. All during this ten-week period fruits, hard-boiled eggs, raw meat and insect larvae were offered to the animals. With the exception of a few larvae and some papaya, all of these foods were rejected. Papaya, a fruit highly valued by these animals, was occasionally offered, but only a half a fruit at a time, which is about 50 per cent less than their normal ration. An average animal portion was always left. To avoid starving further the already depressed and emaciated animals, the original amount of bananas was again supplied. For about three days, they consumed far more than was usual for them, but at the moment they are continuing to consume less per animal than they did when there were three.

Since this behaviour has been noted in two instances following death, though in the first case no quantitative data were collected, it would appear that the tendency to save food for an absent animal is an old and well established one.

The birth of twin pottos

On the evening of 24 September 1971, the two adult pottos left their tree-trunk dwelling and moved to the extreme end of the cage. At 21.00 hrs. they began to copulate, which on the basis of previous data is an unusually early hour for such an activity. All previous births have been preceded by copulatory behaviour initiated in the early evening.[3] On the following morning, dizygotic twin females were born. Data concerning the two babies are shown in Table 2. The

Table 2 Statistics on dizygotic twin pottos

Birth date	*Death date*	*Sex*	*Weight*	*Length*	*Respiration*
25/9/71 07.30 hrs.		♀	34.7 gm.	9 cm.	68
	26/9/71 13.30 hrs.		36.2 gm.	9 cm.	
25/9/71 08.00 hrs.		♀	40.0 gm.	11 cm.	72
	29/10/71 20.00 hrs.		81.7 gm.	13 cm.	

first baby was born outside of the tree trunk on the floor of the sawdust covered cage. The mother used the base of the tree trunk to assist her in her labour. She left the placenta of the first baby at the base of the tree trunk after severing the umbilical cord and moved back inside the trunk. The second baby was born immediately after this move. The mother was quite unconcerned about the second baby. For two hours after her birth, the placenta was partly inside the mother and partly out and the baby was still connected to it. The baby was frantically trying to induce the mother to accept her. By 10.30 that morning the mother had cleaned her off, consumed the placenta, and was attempting to feed two babies. During all this activity, the adult male was moving around at the end of the cage where the births were taking place. During the early evening (17.00 hrs.), the male began to carry bananas and mealworms to the adult female. While the adult female sat up to eat the bananas and thereby left the babies unprotected, the adult male groomed both babies, presumably to keep them warm. The following morning, the lighter of the babies was rejected. It was quite apparent that the adult female could not feed two babies simultaneously. Although all attempts were made to feed the rejected baby, she died one day after birth in convulsions, possibly due to some nutritive deficiency.

Initially, the remaining baby appeared to feed continuously. Twenty-four days after birth she moved back and forth between the two parents and wandered around the cage by herself at night. She was intermittently groomed by one or both parents. When the baby was four-and-a-half weeks old, the mother ceased to lactate. The baby was found still and cold on the floor of the cage. She was fed with a sterile baby doll bottle, containing 15 cc. of cow's milk, 0.5 cc. of a vitamin supplement (ABDEC), 2 gm. of meritine, a concentrated milk supplement, 30 mg. of sucrose and 30 cc. of distilled water. She was fed on demand. Table 3 shows the times of

Table 3 Feeding times and amount consumed by a 4½-week old potto, during hand-rearing

Date	*Amount consumed (cc.)*							*Total*
	03.00 hrs.	*07.00 hrs.*	*10.30 hrs.*	*13.00 hrs.*	*16.00 hrs.*	*20.00 hrs.*	*22.30 hrs.*	
10/26				15.0	14.6	2.1	2.1	33.8
10/27	4.20	1.1	1.1	1.1	1.1	4.2	2.1	14.90
10/28	9.45	1.5	1.5	1.5	2.1	3.3	3.3	22.65
10/29	3.80	1.1	1.1	0.5	1.1			7.60

her requests as well as the amount consumed. It is interesting to note that, even though the data are scanty, the presence of a bidiurnal cycle is apparent. During her life away from her mother, the peritoneal area had to be massaged to encourage urination. This necessity had been noted previously.[4] Both urination and defaecation were normal. On the evening of 29 October, her temperature began to increase. By 20.00 hrs. it had reached 39°C and she died half an hour later, presumably from a respiratory infection. Attempts were made to revive her, but to no avail.

Malnutrition of adult pottos

After she had given birth to twins, it was noticed that the female adult potto was suffering from a large number of deficiencies, presumably as a result of pregnancy. The male potto was also beginning to show signs of a calcium deficiency but not nearly as acute. The protein deficiency, which was exhibited only in the female, was apparent from cracked and bleeding skin and broken nails. A calcium deficiency was exhibited by muscular tremors. The latter were continuous in the female and intermittent in the male.

There was no doubt that some solution to the problem of finding acceptable food had to be found. Initially, an attempt was made to supplement the diet with calcium by adding calcium carbonate to mealworms and bananas. This caused the pottos respiratory difficulties, and so such treated food was rejected. The problem was finally solved by adding 5 gm. of calcium lactate daily to the drinking water. The pottos refuse to drink Pittsburgh tapwater, partly because of its high sodium concentration (60 ppm most of the time) and partly because of organic compounds such as phenols, which cannot be distilled off. For some time they had been drinking spring-water, which was close to the composition of distilled water.

The protein and vitamin deficiency problem was solved in several ways. The two pottos were hand-fed mealworms. On high consumption days they would eat 15 gm. each, but on the intervening low ones they consumed only 7 gm. each. This helped but did not entirely alleviate the deficiency symptoms. Finally, a mixture known as Nutri-Cal, the composition of which may be found in Table 4, was

Table 4 Composition of dietary supplement (Nutri-Cal)

Substance	*Amount*
Protein	1.0%
Carbohydrates	45.0%
Fats	35.0%
Calories/28.3 gm.	190
28.3 gm. contains	
Vitamin A	5,000 USP units
Vitamin D	250 USP units
Vitamin E	30 IU
Vitamin B_1	10 mg
Vitamin B_2	1 mg
Vitamin B_6	5 mg
Vitamin B_{12}	10 mcg
Nicotinamide	10 mg
Ca pantothenate	10 mg
Folic acid	1 mg
Iron (from Iron peptonised)	2.5 mg
Mn (from $MnSO_4$)	5.0 mg
Mg (from $MgSO_4$)	2.0 mg
I (from KI)	2.5 mg

hand-fed to the pottos daily. It is consumed on a bidiurnal rhythm. The mixture is sold in a tube and has a very thick consistency. It is put on the finger and the animals lick it off. On the average, the female consumed 15 gm. per day and the male about 7 gm.

Despite all efforts to eradicate the malnutrition problem, the female ceased to lactate. By 7 November, eight days after the death of the baby, she began to lactate again. Lactation persisted for about three months, with the male suckling regularly from the female throughout this time (see below).

One interesting observation should be noted here. Namely, the adult 'alpha' male always tastes all the food first. This pattern was apparent when there were a total of four pottos and is still the case with only two. If he decides it is consumable, he transmits that information in an unobvious fashion to the female, who then and only then will try the food in question. For example, Nutri-Cal is offered to the animals at a variety of times during the day. The male wakes up, tastes it, usually eats a portion, wakes up the female by walking around her, who will then also eat. If he decides that he does not want any food, he will still go through the tasting ceremony.

Discussion

The saving of food for an absent individual is such a curious phenomenon that even though it is an example of co-operative behaviour within the group, it is felt that the discussion of this aspect of behaviour should be considered separately. The remaining animals, one 'alpha' male and his mate, were apparently leaving food for a dead, less dominant male. This behaviour persisted in the captive *Perodicticus* even when the food supply was limited. Although there appear to be no comparable data on the subject, it has been noted in chimpanzees that food caught by more dominant individuals is shared with less dominant ones as a result of the latter's 'begging gestures'.[5,6] Of course, less dominant chimpanzees also share their food with more dominant individuals. In the baboon, a highly dominance-oriented species, such sharing has not been noted in the wild.[7] This suggests that the potto may be less dominance-oriented than the baboon.

This type of behaviour might tentatively be considered altruistic. Wynne-Edwards has proposed that altruistic behaviour is that which benefits a group or ensures its fitness and which is relatively independent of the individual exhibiting the behaviour.[8] Klopfer, on the other hand, is critical of this hypothesis as he states that, if the behaviour is disadvantageous to the individual, the genes responsible for it die when the individual fails to survive.[9] However, if the behaviour is advantageous, such action is irrelevant in terms of the group. Klopfer states that any advantage gained by the group is explicable in terms of natural selection alone. Mayr, who is concerned with the question of the widespread existence of altruistic behaviour, suggests that it may operate in populations containing genetically related individuals, i.e. a parent that sacrifices himself for his progeny ensures that safety of his genotype by such action.[10]

The behaviour of the pottos appears to be more similar to that suggested in the hypothesis of Wynne-Edwards where, though the animals go hungry in order to save food for an absent one, they nevertheless by so doing act to ensure the survival of the group. It may be that the ideas of Wynne-Edwards are more applicable to the higher vertebrates. No comment can be offered on the thoughts of Mayr, since the possible existence of kinship between the pottos is unknown. The suggestions of Klopfer, though applicable to other animals, do not appear to apply to pottos. Clearly, sharing of food in captive *Perodicticus* is a social phenomenon that is rather unexpected in a prosimian.

It should be considered that behaviour of this type in primates may be more developed in forest dwellers than in animals that live in a savannah, primarily for the reason that the former do not exhibit such a complicated social structure to ensure protection, since their environment already offers it.

The persistence of the bidiurnal cycle is a rather strange phenomenon. When it was first observed, the pottos had only been in contact with each other for a short period of time.[2] It is interesting to note that even when the food supply is limited, it persists. Similarly, if the cycle becomes disrupted, when the periodicity is restored it becomes so in the rhythm of the already established cycle. The other point to consider here is that this feeding pattern is already apparent in the young, or, at least, by the time the baby is a month old. All these observations tend to suggest that the cycle is not a reflection of prolonged captivity but one that is inherent and probably social in origin and possibly of survival value. Were it a reflection of a variable food supply in the wild, captivity of eleven years with ample available nourishment would tend to eradicate such a periodicity. Though pottos are basically insectivorous,[11] they will consume all stages of the insect life-cycle, so that presumably some stage would always be available in the wild. The other possibility is a low metabolic rate. The body temperature is similar to that of man,[12] but recent data show the metabolic rate to be normal.[13] It would appear that the periodicity in feeding habits may be territorial in origin and reflect some type of social phenomenon, since animals together seem to show the same cycle.

The baby female when fed on demand appeared to consume on a bidiurnal cycle. Possibly therefore this is an inherited phenomenon or perhaps it reflects the distribution of food supplied by the mother. If the latter is the case, then a new-born baby would be taught to feed on the cycle, since that is how food would be available. If this is the case, the bidiurnal cycle could possibly be a built-in population control. Namely, were a baby to be born on a low food supply day, it would have less of a chance of survival than had it been born on a high supply day. By the same token, it is possible to postulate that a baby's birth is arranged in such a fashion so that they are born on high supply days to ensure survival. New-born pottos begin to feed shortly after birth. The second baby that was born was found to urinate about four times a day and, on the second day after birth, produced about 2 cc. of urine at each time. This observation would tend to suggest that, though the baby was suckling, it was also drinking at the same time.

Co-operative behaviour may be divided into that type which is exhibited within the nuclear family and that which extends outside the family but within the group. Since the behaviour within the family appears to be somewhat different from that within the group, co-operative behaviour might well be separated into two general categories.

Within the nuclear family, much of the co-operative behaviour, though not all, is confined to the general activity of infant-rearing. It is also apparent that parturition is initiated by copulatory activity. During labour and the actual process of birth, the 'alpha' male rushes around 'trying to be useful'. Once birth has occurred, he grooms the baby and examines it extensively. Any time the mother found it necessary to raise herself on her perch, the male would groom the baby. During the first few days after birth, the 'alpha' male would carry food to the mother. After the baby was two weeks old, the mother would sometimes leave the tree trunk to drink fluid and often would leave the baby with the male potto. Other times she took the baby with her. Many of these observations have also been noted by Anderson.[14]

After the baby died and lactation was resumed, as a result of improved nutrition, the male suckled from the female. Initially, the suckling occurred six times a day. By mid-January the frequency had declined to four times daily. A similar observation has been noted by Anderson.[15] In the latter case the male was occasionally noted to suckle from the female in the presence of a live baby.

An explanation for the nursing adult male may be found in an examination of the nutrition of *P. potto*. An analysis of food eaten either by the pottos or the insects they eventually eat is given in Table 5. Comparative elemental data on the composition of insects may be found in Spector and Bowen.[17,18] In the wild, pottos live in a low-calcium region.[19] The types of food they are known to consume are all high in potassium. Although the varieties of insects they eat in the wild are not known, those they prefer in captivity are all high in potassium. Since zinc is known to accumulate in the mammalian prostate gland,[20] and in its absence spermatogenesis is adversely affected, it was thought wise to check that aspect of potto nutrition. Interestingly, zinc is accumulated by all the insects examined, and the concentration of zinc in the insects is higher than that of copper. The analyses clearly show that a diet consisting primarily of bananas will not be adequate. It is also possible that insects living in a low-calcium region will contain less of that element than the analyses show. It would appear that the types of insects *P.*

Table 5 The quantity of some elements in insects and food consumed either by *P. potto* or by the insects *

Dry Weight mg/100 gm.

Food	*Ca*	*Mg*	*P*	*K*	*Na*	*Fe*	*Cu*	*Zn*
Bran	120.0	62.5	1187.5	980.0	15.0	8.5	3.1	1.20
Bananas	36.8	87.5	9.0	1201.0	5.0	3.0	2.0	0.53
Pellets	955.7	237.5	1350.0	762.0	12.5	20.6	3.0	1.08
Insects								
Periplaneta americana L.	116.6	100.0	822.5	590.0	7.5	4.8	3.8	11.0
Blaberus giganteus L.	138.8	87.5	615.0	501.8	10.0	9.7	3.4	13.3
Blatta orientalis L.	214.2	125.0	1230.0	1055.6	12.5	9.8	3.4	14.2
Blattella germanica L. (male only)	359.8	112.5	1265.0	997.9	7.5	17.8	3.0	15.7
Tenebrio molitor L. (adults)	71.3	225.0	1632.5	879.2	1.5	7.6	3.9	12.8
T. molitor (mealworms)	62.7	200.0	1167.5	672.6	2.5	6.3	4.0	14.3

* Bran (crude) is consumed by *T. molitor*. Pellets are eaten by roaches. Insect data are averages obtained by drying (110°C, 1 week) 100 individuals with the exception of *B. giganteus* where only 15 were available. The chemical procedure employed is that described by Cowgill.[16]

potto is most likely to value are those high in potassium and low in sodium. This is rather an interesting point, since recently it has been suggested that the set-point for body temperature of a primate is maintained by the physiological ratio of sodium to calcium within the brain stem.[21] An excess of sodium in the absence of calcium perfused into the brain stem causes a rise in temperature, while an excess of calcium in the absence of sodium results in a fall in temperature. Pottos will not eat food high in sodium. They have, however, recently developed an appetite for milk, thus partly alleviating the problem of supplying them with adequate calcium. However, since the availability of calcium is low in the region where they live, such suckling behaviour would supply a higher calcium diet to the male.

Moreover, continuous suckling of this order may ensure the survival of future progeny in the sense of developing 'maternal behaviour' in the absence of the baby or ensuring its continuation in the presence of an infant by occasional suckling.[22] In the event of the death of a progeny, such behaviour may also ensure the survival of the female by preventing the difficulties that may in time develop when the results of lactation are not utilised. Some zoo animals which are permitted to lactate in the absence of their young may die of mammary tumours at a later time.[23,24] It has also been suggested that long-term nursing of infants in man contributes to a lower incidence of breast cancer.[25] More recent data tend to dispute this suggestion.[26] However, there does not seem to be any available information in man on the relationship between non-utilised lactation and the subsequent incidence of mammary tumours. It is therefore possible that suckling of an adult may have some survival value.

Co-operative behaviour within the group really is exhibited in quite a different form from that observed within the family. Initially, when the pottos first came, the 'beta' male lost his mate and became depressed and so isolated himself from the rest of the group. The 'alpha' male and occasionally his mate visited him each evening for a period of seven years.[1] When the 'beta' male became ill, he was found in that state with the other two pottos on the shelf, his head in the lap of the 'alpha' male and the rest of him in the lap of the female. When he finally died and his body was removed from the cage, the remaining two spent at least an hour each evening searching for him. This searching behaviour persisted until the birth of the twins. At the same time they saved food for him even to their own detriment.

Apart from the above patterns, which appear to act to make the captive group (whatever its size) cohesive, there are other forms of behaviour that are exhibited entirely by the 'alpha' male which serve to ensure the survival of the group. For an example, when the pottos first arrived, each was in a separate compartment in a carrying case. When the cases were placed inside the cage and the individual doors to the compartments were opened, the 'alpha' male came out by himself and investigated the entire cage. Then he visited each of the other animals, first his female, then the 'beta' male and then the 'beta' male's mate, at which point they all left their compartments. Any time the pottos have been moved over the past years, the same procedure has been repeated. Each evening since they arrived, the 'alpha' male potto has left the tree trunk first, examined the food, tasted it, and then returned to the trunk. This procedure, if the food supply were limited, would ensure the survival of the 'alpha' male. If, on the other hand, the food was in some way unhealthy to consume, this behaviour would ensure the survival of the group and not necessarily that of the 'alpha' male. When any of the animals have been ill, the 'alpha' male has tasted the food and then made all attempts to induce the sick animal to partake of the nourishment. Although these observations tend to indicate the behaviour of the 'alpha' male towards the rest of the group, they are nevertheless co-operative in character.

At this point in the discussion it might be well to consider the relevance of behavioural data obtained from captive animals in relation to that in the wild. Visiting in *Perodicticus*, for example, has been noted in captivity as well as in the wild,[1,27] though in the latter case this type of social activity has been confined to male-female contacts exclusively, while in captivity it has involved male-male as well as male-female visiting. It is not impossible to imagine that most of the types of behaviour mentioned here could be observed in the wild. The point that needs to be made is that this kind of behaviour is possible in prosimians that are this low in the evolutionary scale. How food saving would be noticeable to the observer in the field remains to be seen.

Acknowledgments

I gratefully acknowledge the support of the National Science Foundation which as a result of support of other work made the present study possible. I am also indebted to the Bio-Medical Sciences Support Grant awarded to the University of Pittsburgh by the National Institute of Health for travel aid to attend the Research Seminar in Prosimian Biology. I wish to thank Gulf Research and Development Company of Pittsburgh for supplying the cockroaches. J. Iben gave freely of her time and knowledge in trying to solve the problem of malnutrition among the pottos.

Editor's note

The results presented in this study and the conclusions drawn by the author, with respect to altruistic behaviour in P. potto, *if accepted, would place* P. potto *in a position quite unique amongst the primates and particularly amongst prosimians. At the conference, no-one was able to offer even a tentatively acceptable alternative explanation to the one offered by the author and, in fact, very little discussion took place. On reflection it is felt that results and conclusions of this nature are so controversial that the conventions of scientific research would require that, before they are accepted, the study should be repeated at least twice, carefully, using larger samples and extended over a longer period of time. It should also be duplicated exactly in a closely related species. Only when these requirements have been met can any definitive statement on the subject be added to the literature. This comment in no way reflects on the integrity of the author nor on the value of her research.*

NOTES

1 Cowgill, U.M. (1964), 'Visiting in *Perodicticus*', *Science* 146, 183-4.
2 Cowgill, U.M. (1965), 'A bidiurnal cycle in the feeding habit of *Perodicticus potto*', *Proc. nat. Acad. Sci. Wash.* 54, 420-1.

3 Cowgill, U.M. (1969), 'Some observations on the prosimian *Perodicticus potto*', *Folia primat.*' 11, 144-50.

4 Walker, A. (1968), 'A note on hand-rearing a potto', *Int. Zool. Yb.* 8, 110-11.

5 Van Lawick-Goodall, J. (1968), 'A preliminary report on expressive movements and communications in the Gombe Stream chimpanzees', in P. Jay (ed.), *Primates*, New York, 313-74.

6 Van Lawick-Goodall, J. (1971), *In the Shadow of Man*, Boston, 200-7.

7 Hall, K.R.L. and DeVore, I. (1965), 'Baboon social behaviour', in I. DeVore (ed.), *Primate Behavior*, New York, 53-110.

8 Wynne-Edwards, V.C. (1962), *Animal Dispersion in Relation to Social Behavior*, New York, 1-23.

9 Klopfer, P.H. (1969), *Habitats and Territories*, New York, 96-102.

10 Mayr, E. (1970), *Populations, Species and Evolution*, Cambridge, Mass., 114-7.

11 Jewell, P.A. and Oates, J.F. (1969), 'Ecological observations on the lorisoid primates of African lowland forests', *Zool. Afr.* 4, 231-48.

12 Wislocki, G.B. (1933), 'Location of the testes and body temperature in mammals', *Quart. Rev. Biol.* 8, 385-96.

13 Sucklin, J. (1969), 'The retia mirabilia of *Perodicticus potto*: a functional study', Unpublished PhD thesis, University of East Africa.

14 Anderson, M. (1971), 'The potto family', *Yale Alumni Mag.* 35, 24-7.

15 Anderson, M. (1971), 'A watched potto never grows: a chronicle of the prenatal and first months of *Perodicticus potto*', *Discovery* 6, 89-98.

16 Cowgill, U.M. (1966), 'Use of X-ray emission spectroscopy in the chemical analyses of lake sediments determining 41 elements', *Developments in Applied Spectroscopy*, 5, 2-23.

17 Spector, W.S. (ed.) (1965), *Handbook of Biological Data,* London and Philadelphia.

18 Bowen, H.J.M. (1966), *Trace Elements in Biochemistry*, New York, 70-2.

19 Papadakis, J. (1969), *Soils of the World*, New York, 146-50.

20 Mann, T. (1964), *The Biochemistry of Semen and of the Male Reproductive Tract*, London, 44-8.

21 Myers, R.D., Veale, W.L. and Yaksh, T.L. (1971), 'Changes in body temperature of the unanaesthetized monkey produced by sodium and calcium ions perfused through the cerebral ventricles', *J. Physiol.* 217, 381-92.

22 Cowie, A.T. and Tindal, J.S. (1971), 127 and 272. *The Physiology of Lactation*, Physiological Society Monograph 22, London, 127, 272.

23 Personal communication from J. Iben, a Pittsburgh veterinarian, who stated that a female lion in the local zoo died of mammary tumours some time after giving birth to young she was not allowed to nurse.

24 Heidrich, H.J. and Renk, W. (1967), *Diseases of the Mammary Glands of Domestic Animals*, Philadelphia, 312. Neoplasms may occur in female dogs who have never given birth or who have not done so for some time or who have repeatedly produced stillborn infants or who have undergone pseudo-pregnancy or false lactation.

25 Wynder, E.L., Bross, I.J. and Hirayama, T. (1960), 'A study of the epidemiology of cancer of the breast', *Cancer* 13, 559-601.

26 MacMahon, B. et al. (1970), 'Lactation and cancer of the breast', *Bull. Wld. Hlth. Org.* 42, 185-94.

27 Charles-Dominique, P. (1971), 'Eco-éthologie et vie sociale des prosimiens du Gabon', *Biol. Gabon.* 7, 121-228.

P. H. KLOPFER

Mother - young relations in lemurs

Introduction

Lemurs are lovely creatures, but this, of course, is rarely considered a sufficient reason for their study. As is the case with other primates, the justification for study is believed to be in the light they shed upon human behaviour, and particularly those kinds of human behaviour which some men would manipulate or control. However, there are countless problems in extrapolating from monkey to man. Most of these have been adumbrated often enough. Yet there remains a persistent inclination in each of us to indulge in extrapolations which would have us first identify an animal's activity as akin to ours, and then interpret the evolutionary origins of our acts in terms of the animal's performance. A robin indulges in a display that *we* (not the robin) label 'territorial defence'. The emotional state which we associate therewith is imputed to the robin; the robin's behaviour then becomes an analogue or even homologue of our own.[1,2] Hinde points to a related aspect of this problem when he underscores the difference between the statements, 'the rat ran to the goal box' and 'the rat ran to box X'.[3] More recently, Lehrman has traced the history of the belief of many U.S. Congressmen in the biological justification for defeating bills to establish child care centres.[4] Motherless rhesus monkeys display a depressed countenance that to psychiatrists is reminiscent of a condition in human foundlings, and is labelled anaclitic depression. The emotional concomitants of separation are assumed to be similar in monkey and man, as are the long term effects. *Ergo*, mothers being vital to the well-being of rhesus monkey infants in the same way as they are to human infants, surrogate mothering (or child care centres) are contrary to sound biological principles. Lehrman asks how the Congress would have responded had studies of bonnet macaques been presented to them rather than those of rhesus. In contrast to rhesus, bonnet monkeys accept surrogate mothers fairly readily.

The point, surely, is that we study monkeys not to understand man but in order to generate 'the rules of the game'. The game is the ecological game, staged in the evolutionary theatre.[5] What factors determine the origin, spread and preservation (or loss) of particular patterns of behaviour? Can one predict which factors will be significant for any particular species?

Consider, for example, the question of the exclusiveness of the mother-young relationship. Among goats, a newly parturient doe will generally accept all of her own kids, licking them and permitting them to nurse. Normally, she will violently repel, by butting and biting, any kids not her own, even though the alien kids were born simultaneously with her own, and at a distance of but two metres.[6,7] On the other hand, domestic milch cows can be relatively easily induced to accept alien calves. Lest it be thought this is but an artefact of domestication, the roe deer (*Capreolus capreolus*) has been reported to accept alien fawns, and moose cows (*Alces alces*) even attempt to lure alien calves to themselves.[6,7,8] Why these differences? One possibility lies in the fact that if the species in question is organised into a herd with high variability in the nearest-neighbour distances, young animals that wander randomly through the herd will be unlikely to encounter all the adults. If the young form small clusters within the herd, feeding at intervals throughout the course of their random wanderings, some mothers may not be nursed at all in the course of a day or two. The long-term effect of this on the gross milk production of the herd, to say nothing of the effect on the individual's udder, is surely detrimental. That would be a poor play, from the standpoint of every member of the species. By maintaining an exclusive nursing relationship, a constraint would be applied to the randomness of the youngsters' walk. Periodically the kids would have to take 'time out', and return to their own mothers, each of whom would then be assured of being nursed at sufficiently regular intervals. The sole disadvantage to this scheme lies in the inevitable fatal end of a youngster whose mother dies. If it cannot yet feed itself, it will starve. This apparently happens with insufficient frequency to counterbalance the benefits of the exclusive relationship.

If this kind of cost-benefit consideration is indeed one of the rules of the game, we should be able to predict the circumstances under which exclusive mothering would not occur. Surely this must be so in the most extremely different case of a stationary herd, with low variability in nearest-neighbour distances, and a higher rate of maternal mortality. Such a situation is posed by Northern elephant seals (*Mirounga angustirostris*), which form dense aggregations ('pods') of a few dozens of animals on certain Pacific beaches.[9,10] Baby seals do in fact nurse from mothers other than their own, and lactating seals are as tolerant of older aliens as of their own newborn.[9,10,11] I would expect that computer simulation would allow us to predict more precisely the break-even point, where the cost of exclusivity no longer exceeds the benefit. Other variables might also be important, but the process of identifying these is part of the rule-generation game.

Lemurs have particular advantages as players in this game. Their confinement to a rather restricted part of the globe, similar history, physical and ecological similarities, all combine to provide an array whose dimensions are probably fewer than would be the case for a comparable number of cercopithecine monkey species. As stated in an earlier report, 'the purposes of our studies are to uncover the factors that determine whether a mother will care for her young and whether that care is restricted exclusively to her own or extends to alien youngsters'. From the standpoint of the infant, we are interested in knowing how a separation from the mother affects its development under circumstances where a surrogate mother may or may not be available. Finally, we want to examine the long-term results of different rearing conditions. All this is to be studied against the different social backgrounds provided by three particular species of *Lemur; L. catta, L. fulvus* and *L. variegatus.* These three, while similar in their morphology, and, we assume, their physiology, are neither entirely sympatric nor identical in their patterns of maternal care. In *L. catta*, animals other than the mother may hold the infant before it is a week old and independent exploration begins before it has aged three weeks. In *L. fulvus*, however, other animals are generally prevented by the mother from taking over the infant until four weeks after its birth, a time which coincides with the commencement of independent explorations by the infant. The *L. variegatus* infant, finally, is denied early contact of any kind, except with the mother, until it is fully mobile, at about seven or eight weeks. Before this, it receives only intermittent attention even from its mother.

What is the significance of these different patterns of care? Are the more communally reared *L. catta* infants better buffered from the trauma of parental separation? Are the 'independent' *L. variegatus* better able to invade and colonise new areas? I do consider knowledge of the answers to be interesting *per se*, but the agencies that support this work can take comfort in the fact that by learning the answers we can bring the rules of the game to light. Those rules, surely, apply to all productions appearing in the evolutionary theatre, including the production we call man.

Summary of previous findings

Our first report merely described the pattern of events in captive groups of *Lemur fulvus* and *L. catta*, and in one group of *L. variegatus*. Those findings provided an indication of the character and degree of difference between the three species. In general, they allow the *L. fulvus* infant to be characterised as dependent for its care on a single individual, its mother, and relatively shielded from direct contact with other animals. *L. catta* infants, on the other hand, experience considerably more handling by other animals from their earliest days, while our single *L. variegatus* infant lived a solitary life until it was able to perform effective locomotion.

We subsequently sought assurance that our conditions of captivity were not unduly distorting the behaviour we were studying. To this end, several hundreds of hours of *Lemur catta* activity were logged under natural conditions, on the same protocol charts as those used for the captive groups.[12] The comparison of data from wild *L. catta* at Berenty, Madagascar, and captive *L. catta* at the Duke University Primate Facility showed the following: the start of a consistent increase in the proportion of time the infant spent off the mother (when in captivity) occurred at about 30 days of age. Specifically, the percentage of the observation time when either of two infants was not in contact with its mother rose from less than 10% by day 28, to over 15% by day 32 and over 30% by day 48 (one of the two reached the 30% mark on day 40). Wakeful contacts with other adults, simultaneously with maternal contact, occurred from the start. Contacts with others when separated from the mother were not noted until day 8, and then only infrequently. This corresponds closely with the situation at Berenty, where a significant number of wakeful contacts with animals other than the mother was also noted only during the September-October period. The first occasion on which a captive infant moved from the ventral surface to the back of its mother was day 3; this corresponds to the earliest age noted both by Jolly and by us at Berenty.[17] In the colony, the first independent exploration was noted as early as day 18, though this did not occur regularly until about day 30 (day 27 and 37, respectively). At Berenty, too, only two occasions of total separation were noted before day 30 (approximately), and several hundred thereafter. Thus, with respect to the specific indices of development which we used, the captive and wild animals appeared strikingly similar. While subjective reports must be treated sceptically, let me nonetheless record that the observers did indeed agree that this was the case, even with regard to other, less carefully measured attributes.

In order to assess the degree of buffering afforded infants of the different species, we examined in captivity the effects of mother-young separation in different social constellations (mother and infant alone; mother, father and infant; mother, father, infant, plus other adults and juveniles). At the time each infant attained 150 days of age, and again at 200 days, its mother was removed for seven days. A full series of observations has until now only been reported for *L. fulvus*. Where the infants were reared alone with their mothers, the infants remained close to their mothers for their first 80 days. The first separation of 1 m. distance did not occur until day 50; the maximum separation was permitted in their chamber on about the 110th day. The removal of their mothers at the 150th day led the infants to alternate between stupor and frenzy, but within a week following the reunion the animals had resumed their previous patterns of activity. The infants took the initiative in the re-establishment of relations, their mothers seeming relatively uninterested. For *L. fulvus* infants reared with both parents it appeared that there was a slight advance in the age by which the infant departed from its mother's back or lap. In other respects there were no differences from the mother-infant pairs. On removal of the mother, the infant increased the number of its grooming contacts with the male, but otherwise gave no indication of trauma. The infants shifted their attention back to their mothers on the latter's return. The infant kept with its parents and other animals was generally similar in behaviour. Again, following a week-long separation from the mother at 150 days, the infant seemed more affected by the reunion than the mother.

A final methodological consideration involved the effects of varying the number of animals in a group on the space available to any one of them. (In separate studies, at the Duke University Primate Facility, Miss Laura Vick reported that removal of one animal from a group immediately changed the status of the most subservient female from that of a peripheral to central animal, and similar phenomena have been noted by C. Chandler, personal communication). Perhaps even more importantly, one alters the character of the available space. Wherever any degree of dominance or attraction exists between two or more animals, the area surrounding them must be viewed as charged, analagous to the surround of atomic nuclei. The shape of the space available to other animals will then depend on their valence and will be altered as they move about. Complexities of the substratum, partitions, overhanging ledges, holes, may function as insulators, distorting the fields of charge and permitting distributions that would otherwise be impossible. Hence, the experiments on the infant separation trauma and its amelioration require an assessment of the role of the substratum on social spacing.[13]

Hence, the activities and distributions of four juvenile *L. fulvus* in a chamber with many artificial trees, and in the same chamber with fewer artificial trees, were compared. The number of active (and resting) periods did not differ significantly between the two environments (53 and 54, for the 'impoverished' and 'enriched' rooms, respectively). There were small differences between rooms in the amount of clumping, however. All the animals were alone 13 times in the impoverished room, compared with 28 in the enriched room (confidence intervals of these proportions are separate with confidence coefficients between .90 and .95). Clusters of 2 animals were seen 72 (impoverished) versus 53 (enriched) times (confidence intervals of these proportions are separate with confidence coefficients between .90 and .95). Large clusters were infrequent, and did not differ significantly, though during resting the enriched room seemed to promote more clumping (groups of 3 and 4: 12 *vs.* 5; difference too small to show significance, if any). The total number of animals involved in any size clump in all observations was 192 (impoverished) *vs.* 169 (enriched), which is not significant. Since the frequency of activity was unchanged between environments, differences in cluster frequency are not likely to be an artefact of the animals' speed or frequency of movement. The impoverished room did have fewer perches and rest areas, of course, though there were a sufficient number to allow total separation (3 shelves, a chair, a feeding ledge). Separation did not occur as often; under impoverished conditions, the animals were more often paired (whether resting or active) than under enriched conditions. The large clumps, on the other hand, seemed to occur more often among resting, 'enriched' animals. Hence, the effect of the artificial forest is not merely to provide a greater array of potential perches. Were that the case, one would expect fewer, rather than an equal or greater number of the 3 and 4 animal clumps. Nor did the 'forest' alter activity frequency, and thus, indirectly, the groupings. The observations themselves reveal no qualitative differences in the nature of the activities in which the animals engaged in the two environments. In a subtle, as yet undefined manner, the 'forest' did cause small differences in the groupings to emerge. Less dispersion of individuals in the impoverished room, and a predominance of pairs over other clumps of another size.[13]

In summary, observations of *Lemur catta* mother-infant pairs in a forest preserve in Madagascar provided assurance that the conditions of captivity in the Duke University colony were not seriously distorting the behavioural relations between mothers and their infants. Intensive observations of captive, individual mother-infant pairs of *L. fulvus* suggested that the frequency and distance of the infant's excursions from its mother were increased with the presence of other lemurs. The temporary absence of the mother upon the infant's attaining an age of 150 days was readily tolerated if another familiar animal was present, but otherwise proved traumatic. However, no subsequent effect of a separation of one week's duration was noted. On reunion, the infants seemed more eager to re-establish contact than their mothers.[13]

Answers to a number of other more specific questions are being sought. These include queries as to the proximate factors determining whether a mother will recognise neonatal young; whether she will accept them; whether she will discriminate between her own and alien young; how differences in maternal responses influence the infant's response to separation from its mother; and the basis for the success of surrogate mothers.

Subsidiary questions include the roles of other adults and peers in the development of young, and the importance of sex and individual differences in development. Obviously, the data collected thus far and presented below do not yet suffice to provide other than speculative answers.

Method

The observation chambers of the Primate Facility, within which the animals lived, were hexagonal rooms with a volume of about 82 cubic metres. The ceiling tapered to a height of 5 m.; the maximum distance from corner to corner across the floor was 5.3 m. Corner shelves and a single central post with several horizontal branches provided perches. Observations were made through one-way glass either from an overhead mezzanine or at ground level.[13] The rooms are not fully soundproof, but only especially loud sounds distract the animals.

The animals were maintained on a fixed schedule of 12 hours of light, 12 hours of dark (dim red light). While nocturnal observations were made to establish activity patterns, the maternal data were collected entirely during daylight hours. Observations were evenly spaced throughout those periods of the day when the animals were most likely to be active. (Nocturnal activities were observed with a light amplification device). These observations were supplemented by videotape records, which served to provide confirmation for periods of presumed inactivity. Each observation period, scheduled for a specified and different time each day, was of 60 minutes duration (in contrast to the 30-minute periods in the field). Data were compiled by noting the position of the infant at half-minute intervals and recording whether and with whom it was in contact, distance from the mother, and manner of behaviour, particularly whether grooming or being groomed. These data themselves were grouped into 30-minute data periods that facilitated the various comparisons.

The animals were generally introduced into their chambers several weeks prior to the birth of an infant. Formal recording of data commenced upon the discovery of a newly-born infant, usually within hours, always within 24 hours after birth. Upon attaining 150 days of age, the infant was subjected to the First Separation, which involved removal of its mother to separate quarters for a seven day period. At 200 days, there was a Second Separation, also of seven days' duration. Observations were made just before and just after the removal and return of the mother. At 215 days the observations were terminated.

Data

The data presented here are a supplement to those summarised above. They are based on observations of the following two groups:

(*a*) *L. catta*
♀ 534F Bilbo, infant ♂ 584M Cato
♀ 546F Balin, infant ♀ 585F Calpurnia
♂ 527M Socrates, and ♂ 548M Fredegar

(*b*) *L. fulvus*
♀ 554F Bathsheba, infant ♀ 572F Miriam
♀ 552F Shulamite, ♂ 553M David and
♂ 551M Ishmael

Mother, father, infant and others in L. catta

A group of four adult lemurs, two of each sex, was placed in the chamber on 10 March 1971. Bilbo's infant, a male, Cato, was born on 15 March; Balin's infant, the female Calpurnia, on 16 March. The graphs, Figs. 1*a, b* and 2*a, b* reveal the changes in the maximum distance between mother and infant at various ages and the proportion of the observation times the infant was not in contact with its mother.*

The two infants did differ in the age at which they first began to separate themselves from their mothers. Calpurnia was over 40 days old before she regularly moved more than an arm's length away, while Cato's explorations began fully 10 days earlier. However, both animals were showing maximal utilisation of the space at their disposal by the same age (approx 100 days). No change in this measure was evident subsequent to the separation and return of their mothers.

The graphs showing the % of time spent off their mothers reflect parallel behaviour. Calpurnia's rate of becoming 'independent' did not begin a rapid rise until around the 50th day; Cato preceded her by about 15 days. However, both achieved a plateau at about the

* In preparing graphs from the data sheets certain conventions were adopted for convenience' sake; observation hours with less than five minutes of activity were discarded; each graph point was made to represent a 4-day average, with dashed lines connecting points for which there were 2 or less observations in 4 days; during the 7-day separations, the averages were based on 4 and 3 days, respectively; distances or times during separations are based on next nearest animal to infant.

same time, around the 100th day.

Other major events are recorded on Table 1. In general, this captive, confined group proved similar to both those confined groups examined previously[11] and to wild groups at Berenty. It is clear, however, that there is considerably more variation than originally expected between infants (and mothers) in the ages at which certain activities first appear. Some of the differences may be more apparent than real, the result of sampling errors for instance. These are of particular significance in the case of infrequently occurring behaviour, especially where a group is being observed for only two to four hours out of 24. As the sample size grows, however, it will be possible to provide estimates of the intra-specific variations. Some other miscellaneous observations that were noted include the following (arranged chronologically).

25 March to 1 April 1971: Each mother frequently attempts to seize the other infant, attempting to keep both simultaneously. Even when there are no overt attempts at kidnapping, the mothers do much grooming of one another's infants.

13 May 1971: Kidnap attempts are no longer always resisted. Mothers occasionally nip their own infants when these seek their lap and make no protest when the other mother carries their own infant away. There is much *reciprocal nursing*, Bilbo even nursing Calpurnia while the latter is still in her mother's – Balin's – lap! Bilbo was the more aggressive kidnapper of the two mothers, but Balin actually had possession of both infants more often. She was also the more protective mother, inhibiting her infant's attempts at exploration. In fact, her infant, Calpurnia, was much later than Cato to move away from her mother.

8 May 1971: First noted wrist-marking from the axillary gland (by Cato).

8 June 1971: First noted mounting attempts by Cato (on Calpurnia) with pelvic thrusting.

10 June 1971: Calpurnia, the female infant, appears to dominate Cato. She is the one that initiates play and is always more active.

10 July 1971: First noted overhead tail flicking by Cato towards Calpurnia.

Effects of separation: there was an initial reduction in the infants' activity upon the removal of their mothers, but by day 3 their normal patterns had reappeared. Upon the mothers' return the

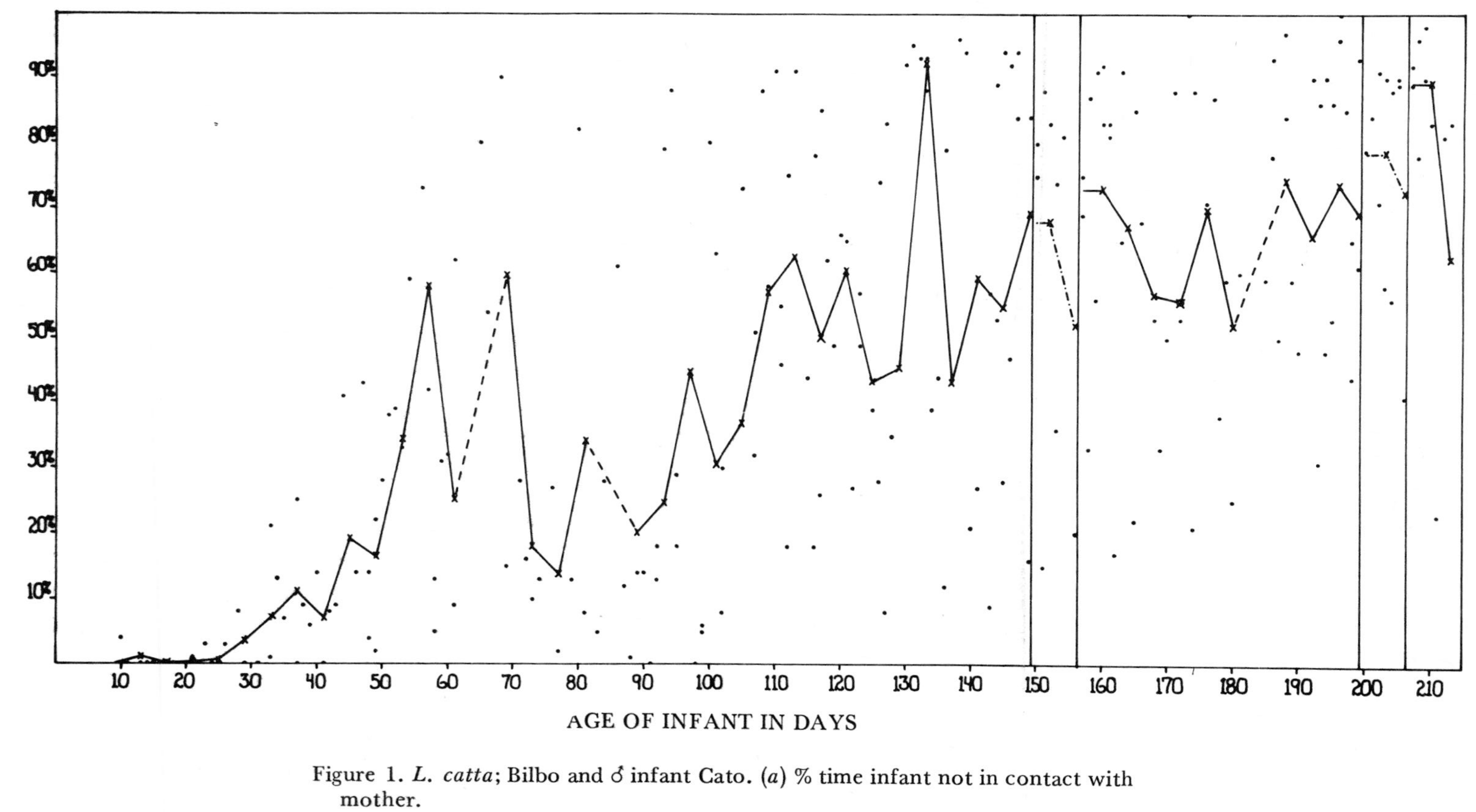

Figure 1. *L. catta*; Bilbo and ♂ infant Cato. (*a*) % time infant not in contact with mother.

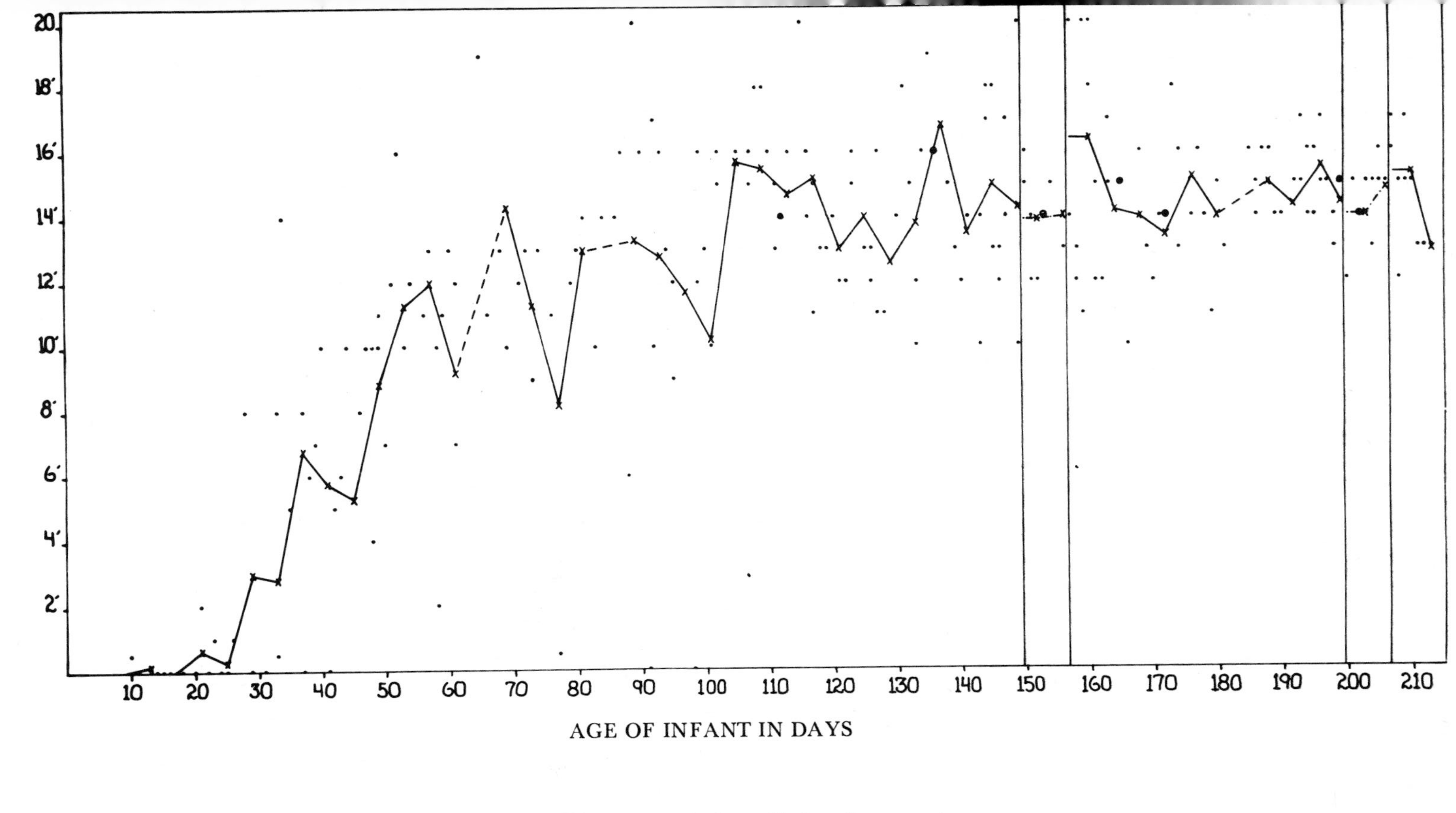

Figure 1 (*b*). Distance in feet of infant from mother.

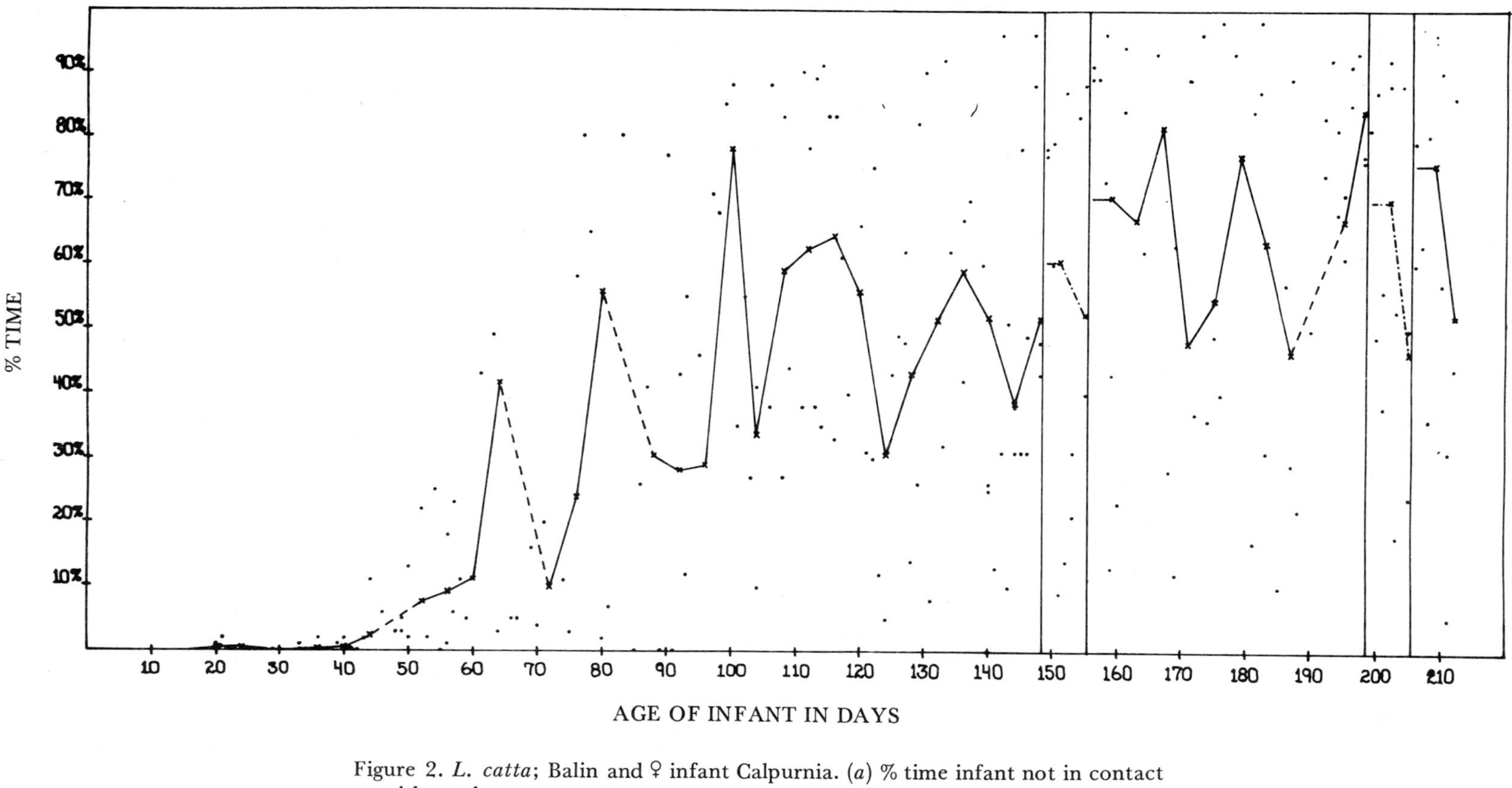

Figure 2. *L. catta*; Balin and ♀ infant Calpurnia. (*a*) % time infant not in contact with mother.

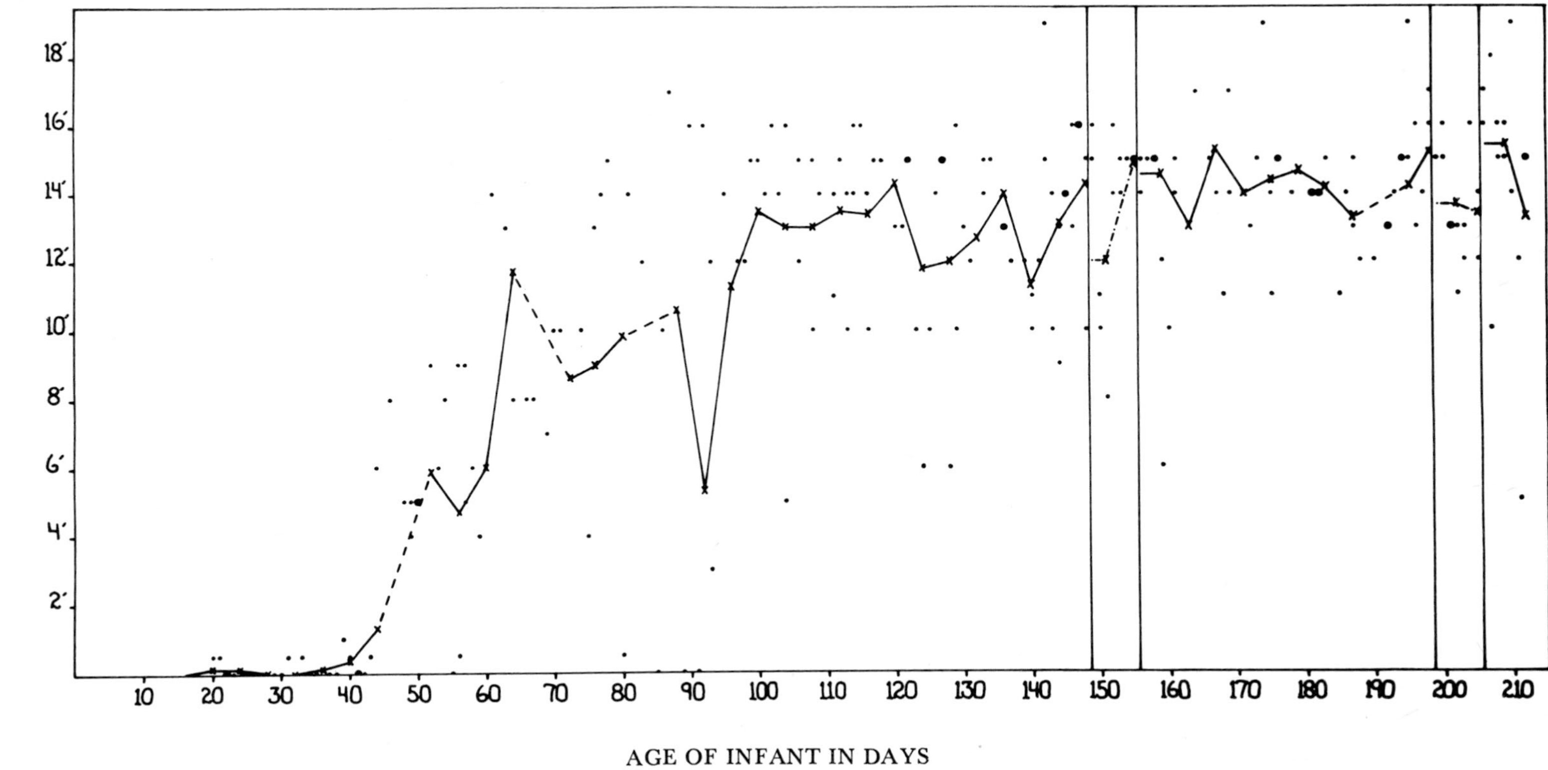

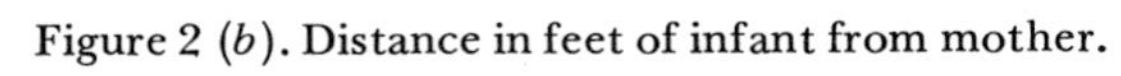

Figure 2 (*b*). Distance in feet of infant from mother.

Table 1 Major events in infant development

		in L. catta *(group)*		*in* L. fulvus *(group)*		*in* L. fulvus *(solitary)*		*in* L. fulvus *(pairs)*	
Sex of infant:		♂	♀	♂	♀	♀	♀	♂	♀
Age of infant when 1st groomed or handled by another animal (not its mother)		3 days;	2 days	7 days;	3 days	8;	2 days	?	?
Age of infant when 1st held by another animal (not its mother)		10 days;	20 days	46 days;	28 days	44;	30 days	?	?
Age of infant when 1st without contact with any other animal		22 days	33 days	49 days;	28 days	44;	30 days	49 days;	49 days
Age of infant when 1st off its mother for at least 30% of an observation hour		44 days;	61 days	? ;	82 days	77;	97 days	87 days;	72 days
Infant-mother and mother-infant grooming frequencies for 1 week before and 1 week after 1st separation	I-M	1.%*-2.1%	1.5%-1.5%	? ;	.8%-4.9%				
	M-I	4.8%-7.8%;	6.5%-3.2%	? ;	10. %-6.9%				
Infant-mother and mother-infant grooming frequencies for 1 week before and 1 week after 2nd separation.	I-M	2.5%-2. %;	4. %-3.7%	? ;	1.2%-5.1%				
	M-I	6.1%-3.3%;	3. %-7.1%	? ;	2.4%-4.9%				

*Mean % of 60-second intervals during observation hour when grooming occurred.

infants first fled from them, but after several minutes sought to climb onto their backs. In this, they were actively repulsed by their mothers. By the end of the first day following the return of the mothers, the pre-separation patterns were fully re-established. On 5 September both mothers were seen to be in full oestrus.

Relations to males: the infants repeatedly teased the males, bouncing on them, batting and cuffing them, or even biting. The males were very gentle in their responses, usually backing away or feinting in response. Their occasional bites never seemed to injure the infants, even to the extent of eliciting a scream.

Mother, father, infant and others in L. fulvus

A group of 4 adults, two of each sex, was moved into the chamber on the day of the infant's birth, 21/11/71. The graphs, Figs. 3 *a, b*, reveal changes in the distance between mother and infant as a function of age, and her growing independence. They indicate a slower initial rate of development than in *L. catta.*

The second adult female of the group, Shulamite, sat pressed close to Bathsheba, the mother, from the time of the infant's birth. She very probably groomed the infant from the day of its birth, though this was first observed only when the infant was 3 days old. At 11 days of age, the infant briefly climbed into the female Shulamite's lap, this occurring when she and Bathsheba were grooming one another. After a few minutes, Bathsheba took her back. Shulamite made repeated efforts to groom or seize the infant, but usually these were repulsed by Bathsheba. Once, when the infant was in contact with Shulamite, she was seen trying to nurse from her. Bathsheba seems to have been more tolerant of Ishmael's than Shulamite's attentions.

The age at which the infant was first alone, that is, without physical contact with others, was at 28 days, but it was over 80 days before this degree of independence was manifested for as much as one third of an observation hour.

The two separations (at 150 and 200 days) did lead to a subsequent slight increase in mother-infant grooming interactions, but this effect was transitory and no other changes were noted.

When the infant was 50 days old, Shulamite gave birth to twins. After 24 hours, one of the twins was found to have been badly bitten. Shulamite and the surviving twin were thereupon removed from the group. No change in the infant's behaviour was detected as a consequence of Shulamite's removal.

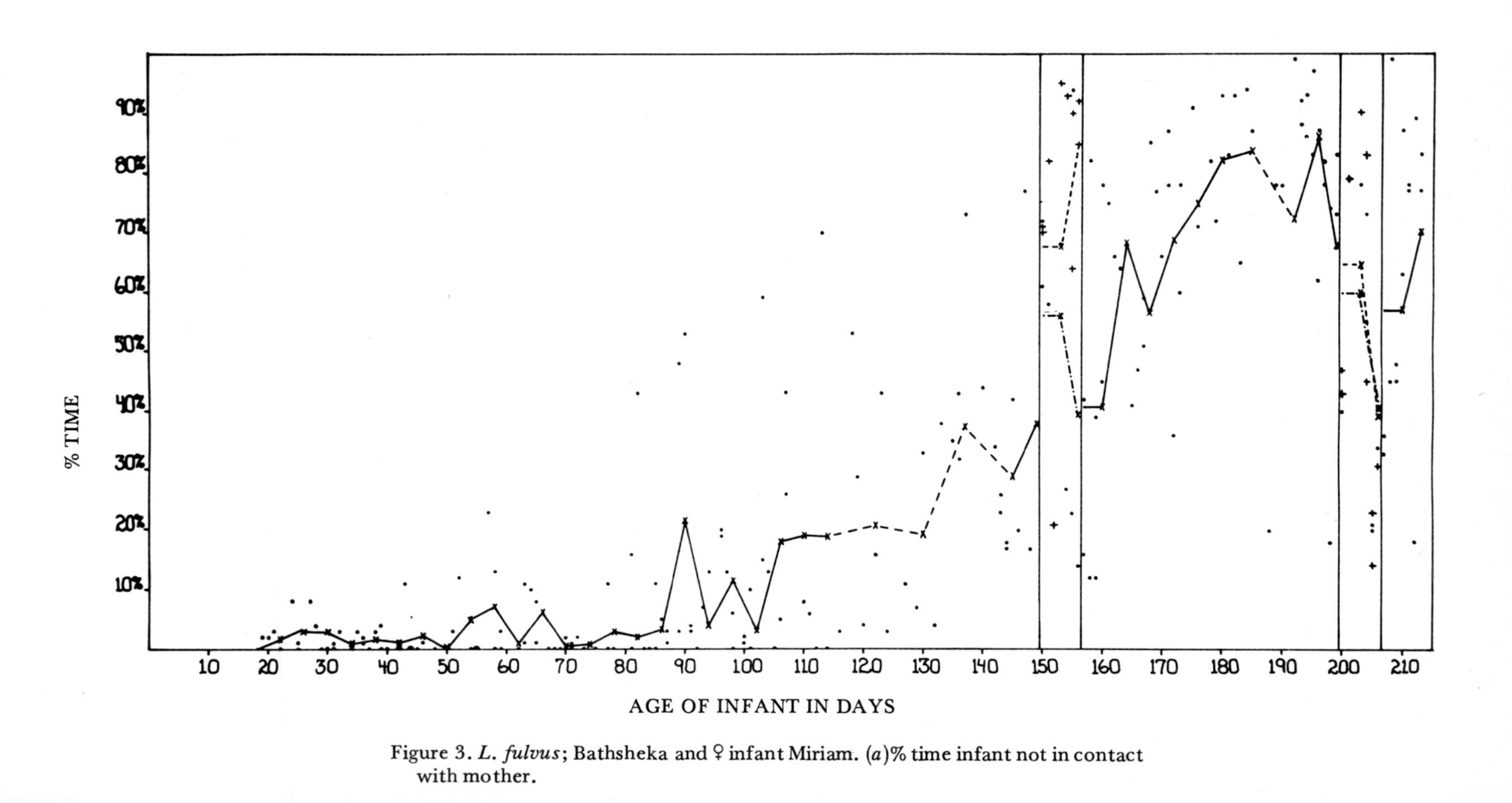

Figure 3. *L. fulvus*; Bathsheka and ♀ infant Miriam. (*a*)% time infant not in contact with mother.

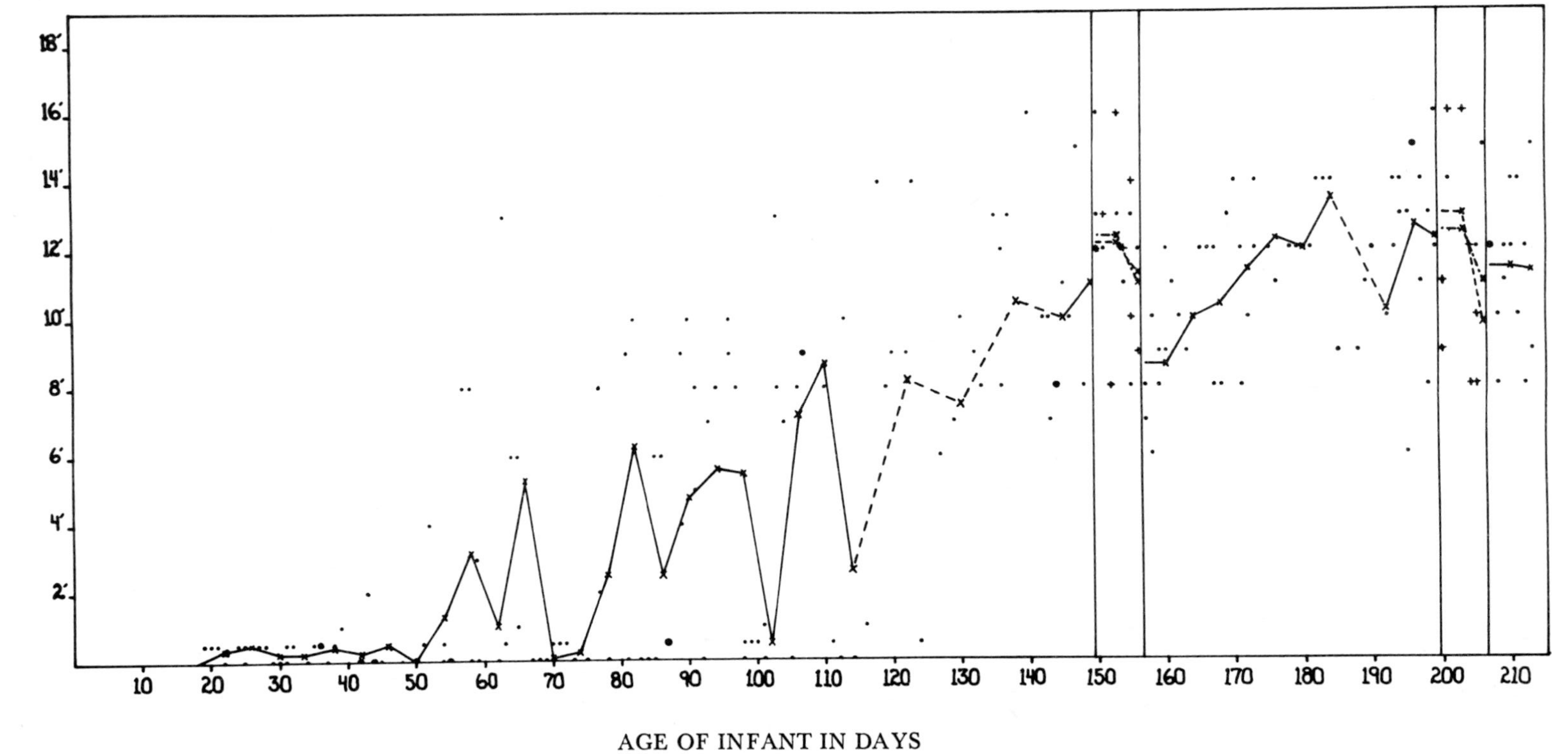

Figure 3 (*b*). Distance in feet of infant from mother.

Comparisons

With sample sizes as small as are these, and individual differences looming ever larger, generalisations must necessarily be tentative. The following relations appear to stand out: The age at which particular events first occur does not seem to differ between *L. catta* and *L. fulvus.* What does differ is the rate of increase in the recurrence of the behaviour in question. Thus, the first time an infant sits alone seems not to vary specifically, at least within pairs or groups (22 and 33 days for *L. catta*, 49, 28, 44 and 30 for *L. fulvus*), but infant *L. fulvus* consistently take longer before they are independent of the mother for 30% of their activity time.

Taking both species together, differences related to sex, as reported for some other primates,[14] have not yet been detected, nor is there any striking change in developmental rates as a consequence of rearing the *L. fulvus* infant in the company of its mother alone. (No data are yet available for solitary *L. catta.*) Solitary rearing produces, as its principal result, a reduction in activity during the separation period, a result that is hardly cause for surprise.

General conclusions

What can then be said about the functional significance of the differences that are seen? What in fact are the 'rules of the game'?

The ecology of *L. fulvus* and *L. catta* has been investigated by Petter and Sussman,[15,16] with particular attention to differences in their use of the habitat in areas where both species occur. Not unlike certain woodland warblers of the north-eastern United States, the two species have subdivided a common area. *L. catta* spends much time on the forest floor (the exact proportions of the active day apparently varying with season and weather).[17] *L. fulvus* is exclusively arboreal, slower moving, and with a more restricted vertical range within the trees as well. What the ultimate and proximate reasons for these, and other, differences may be remains a subject of investigation. Sussman's data do indicate that these differences are not a case of character displacement and hence susceptible to rapid change. That is, in areas where *L. fulvus* or *L. catta* occur alone, their respective behaviour *vis-a-vis* the habitat is apparently indistinguishable from their respective behaviour where the two overlap. We thus must suspect that whatever constraints limit their behavioural repertoires, these are relatively inflexible. It is, therefore, reasonable to expect that there would be concomitant effects upon maternal care patterns.

A leaping, vertically wide-ranging, arboreal-terrestrial animal might be far more likely to lose an infant, or, if on the ground, have greater need to spurt into the treetops than a more phlegmatic branch-walker. Would there then not be an advantage to promoting motor precocity or the acceptance of substitute mothers, or both? An actively-moving mother, as of the species *L. catta*, must profit more than a slow-ranging *L. fulvus* from not having continually to carry her infant. At the same time, the sudden appearance of a predator could pose an unacceptable risk if the more independent infant were not readily snatched into the bosom of the nearest adult, whether its mother or not. High rate of activity and widely ranging movements, in short, promote infant precocity; precocious infants have greater requirements for surrogate mothers for protection than do less precocious babes. That, in short, is the suggestion, tentatively put, as to the rules of this particular game.

Acknowledgments

This work was done at the Primate Facility of Duke University's Field Station for Animal Behaviour Studies, supported by NSF GB4000 and NIH FR 00388 and HD02319, as well as a Research Scientist Award. I also record my appreciation to Linda Hobbet, Martha Klopfer, Lee McGeorge, Charles Chandler, and June Bronfenbrenner for their many hours of observation and compilation. The good health of our colony is largely due to the ministrations of Dr Jan Bergeron and his staff.

Editor's note

Due to the author's unavoidable inability to attend the conference, this paper did not have the benefit of being discussed and, excepting minor editorial changes, is included in the form in which it was originally submitted.

NOTES

1 Leach, E. (1966), 'Don't say "boo" to a goose', *N.Y. Review of Books*, 7, no.10.

2 Klopfer, P.H. (1969), *Habitats and Territories: A Study of the Use of Space by Animals*, New York.

3 Hinde, R.A. (1966), *Animal Behaviour*, New Jersey.

4 Lehrman, D.S. (1971), Unpublished Lecture, Duke University.

5 Hutchinson, G.E. (1965), *The Ecological Theatre and the Evolutionary Play*, New Haven.

6 Klopfer, P.H. (1971), 'Mother love: what turns it on?', *Amer. Sci.* 59, 404-7.

7 Klopfer, P.H. and Klopfer, M.S. (1968), 'Maternal "imprinting" in goats: fostering of alien young', *Z.f. Tierpsychol.* 25, 862-6.

8 Altman, M. (1958), 'Social integration of the moose calf', *Anim. Behav.* 6, 155-9.

9 Bartholomew, G.A. and Collias, N.E. (1962), 'The role of vocalizations in the social behaviour of the northern elephant seal', *Anim. Behav.* 10, 7-14.

10 Klopfer, P.H. and Gilbert, B.K. (1967), 'A note on retrieval and recognition of young in the elephant seal, *Mirounga angustirostris*', *Z.f. Tierpsychol.* 23, 755-760.

11 Lluch-Belda, D., Irving, L. and Pilson, M. (1964), 'Algunas observaciones sobre mamíferos acuáticos', *Instituto Nacional de Investigaciones Biológico-Pesqueras, Comisión Nacional Consultiva de Pesca y Industrias.* 10, 1-23. Mexico, Conexas.

12 Klopfer, P.H. (1972), 'Patterns of maternal care in three species of *Lemur*: 2. Effects of early separation', *Z.f. Tierpsychol.* 30, 277-96.

13 Klopfer, P.H. and Klopfer, M.S. (1970), 'Patterns of maternal care in three species of *Lemur*: 1. Normative description,' *Z.f. Tierpsychol.* 27, 984-96.

14 Hinde, R.A. (1971), 'Development of social behavior', in Schrier, A.M. and Stollnitz, F. (eds.), *Behavior of Nonhuman Primates*, vol. 3, New York.

15 Petter, J.J. (1962), 'Recherches sur l'écologie et l'éthologie des lémuriens malgaches', *Mém. Mus. nat. Hist. nat.* 27, 1-146.

16 Sussman, R. (1972), Unpublished PhD Thesis, Duke University. (See also this volume.)

17 Jolly, A. (1966), *Lemur Behavior*, Chicago.

H. COOPER

Learning set in Lemur macaco

Although prosimians have been described as curious and manipulative,[1,2] little work has been done on assessing their learning skills in a carefully controlled situation, as has been done with the higher primates.[3,4] Part of this is due to the difficulty involved in obtaining and working with these reputedly nervous animals.

Lemurs can, however, be fruitful subjects for laboratory experiments in learning and problem-solving if they are carefully adapted to the experimenter and the experimental situation. The present study is a preliminary note on learning by one species of Malagasy lemur, *Lemur macaco*.

The learning set procedure, first introduced by Harlow,[5] has been shown to be a convenient inter- and intraspecific measure of learning ability. Learning set refers to progressive interproblem improvement between problems of the same type. Typically, the subject is given a large number of problems in which the stimulus elements change from problem to problem, but which are of equal difficulty and which all involve the same principle, such as object discrimination, discrimination reversal, or oddity. Performance is measured in terms of percentage correct responses, on individual trials or a series of trials, and the acquisition of learning set is indicated by progressive reduction in errors in successive problems.

This method has been shown to be fairly sensitive to interspecies differences within the primate order.[3,4,6] The ability to form a learning set is not restricted to the primates, however, as cats,[7] rats[8] and squirrels[9] can also develop this type of learning, though only to a limited extent. Learning set is also a function of age, as in macaques where the ability does not appear until about 200 days of age.[10]

Studies in discrimination ability of lemurs have been few in number.[11,12] A more recent study comparing learning through many discrimination reversals on a single problem[13] showed *Lemur* to be inferior to *Cercopithecus* in this particular problem. Similar results were found when these two species were compared for extinction.[14] The actual learning set procedure, however, allows great possibilities for analysis of response tendencies.

Methods

1. Subjects

The animals used in this study consisted of 2 *Lemur macaco albifrons* and 1 *Lemur macaco albifrons x fulvus* hybrid. All these animals were tame, and were each given 2-3 weeks of habituation to the transport and test procedure before the first problem was begun. Animals were housed in individual cages and were trained to enter a transport cage which had been adapted to the test apparatus. All subjects were experimentally naive before testing.

2. Apparatus

The test apparatus consisted of a Wisconsin General Test Apparatus,[5] suitably modified for use with lemurs (see Figs. 1 and 2). The animal compartment measured 50 cm. x 40 cm. x 40 cm., just large enough to accept the transport cage. Bars were mounted horizontally on the transport cage, 4 cm. apart. Two sliding screens, one opaque and one transparent, separated the animal from the test compartment, which measured 40 cm. x 40 cm. x 30 cm. A third screen containing a one-way mirror was located at the end of the test compartment to allow observation. A sliding plexiglass tray, 40 cm. x 30 cm., was used to present the stimulus objects. Two food wells, 4 cm. in diameter, were located on the tray 20 cm. apart, 1 cm. from the edge of the tray. The food wells were 1 cm. deep watch-glass shaped depressions, allowing easy retrieval of the small food rewards. In addition, the test tray protruded into the animal cage by 3 cm., allowing the animal to respond with the snout or the hand. Lighting was provided by a 100-watt bulb overhead, and a ventilation system was installed at the rear of the apparatus. All surfaces were painted grey.

The stimulus objects consisted of 500 randomly collected pieces of junk (beads, toys, bottle tops, etc.), mounted on plastic plaques measuring 6cm. x 6cm. and painted to match the test tray. The plaques were attached to strings to prevent the subjects from taking the stimulus objects into the cage. The response was a simple displacement of the object in any direction.

Objects were randomly paired, and then ordered in sequence randomly for each animal.

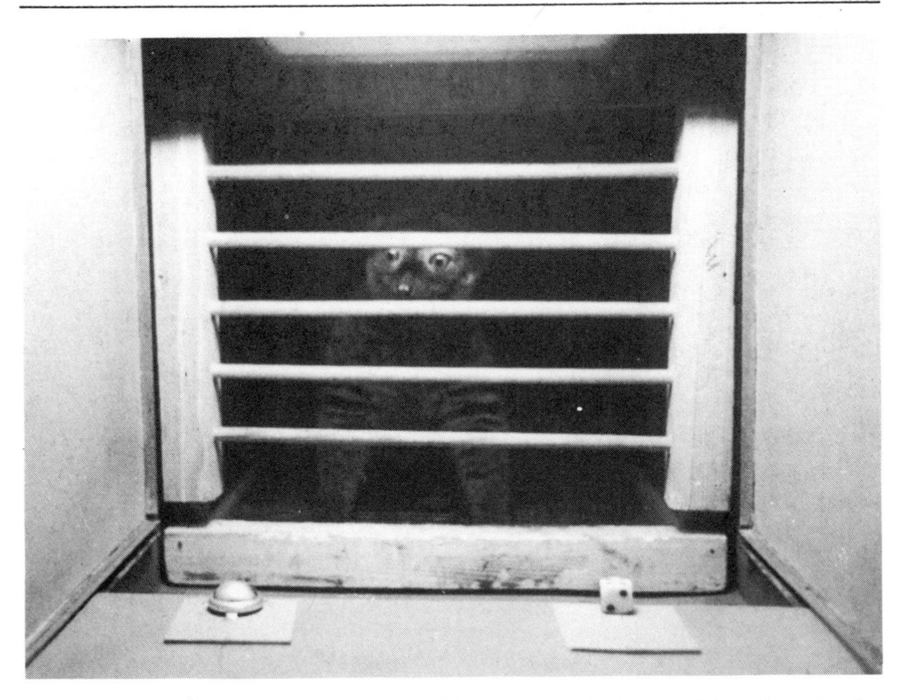

Figure 1. A female *Lemur macaco albifrons* attentively watching the stimulus objects before the test tray is pushed forward.

Figure 2. *Lemur macaco albifrons* displacing the stimulus object with the snout.

Procedure

Subjects were deprived of food for 16-20 hours before testing. Immediately following the test session, animals were allowed to eat their usual diet, consisting of bananas, apples, oranges etc., for 2-3 hours. Water was available *ad libitum.*

The reward consisted of dried raisins, a favourite food of the subjects. Adaptation to the apparatus was accomplished by first allowing the animals to find the food in the food wells, and then gradually covering the wells with neutral objects until the animals displaced the object within 20 seconds, 30 times in succession.

The form of response varied with each subject, some animals preferring to displace the object mainly with the hand, others mainly with the snout. Retrieval of the raisin was almost always accomplished with the mouth, only occasionally being transferred to the mouth by hand.

The subjects gave up to 100 responses a day, sometimes divided into two test sessions per day.

Testing was divided into three phases. The first 10 problems were presented for 100 trials each, for a total of 1000 trials. Between the 5th and 6th problems, each subject was given up to 200 trials in which the raisin was placed under one of two identical objects, to determine if the animal was utilising olfactory cues. Response was at chance level throughout.

The second phase consisted of 64 problems, each learned to a criterion of 17 correct responses in 25 or less consecutive trials. The third phase consisted of 192 additional problems presented for a fixed number of 6 trials each. Throughout testing, left-right position of the rewarded object was balanced both for trials within each problem, and for equivalent trials between problems.

The actual procedure in presenting the stimulus objects involved lowering of the opaque screen so that the experimenter was hidden from the animal's view. The tray was baited with a raisin associated with the positive object, and the stimulus objects were placed over the food wells. The screen containing the one-way mirror was then lowered, the opaque screen raised, and after a delay of two seconds the tray was advanced towards the animal to allow a response. A non-correction procedure was used throughout, only one response per trial being permitted. Time between trials was 15 seconds, and between problems 40 seconds. In the inter-problem interval, a 'free' raisin was given to the subject.

Results

Intraproblem learning curves are illustrated in Fig. 3. It indicates a gradual progressive improvement between successive blocks of problems. In the first problem block which consists of 10 problems run for 100 trials each, performance remains about chance level. This

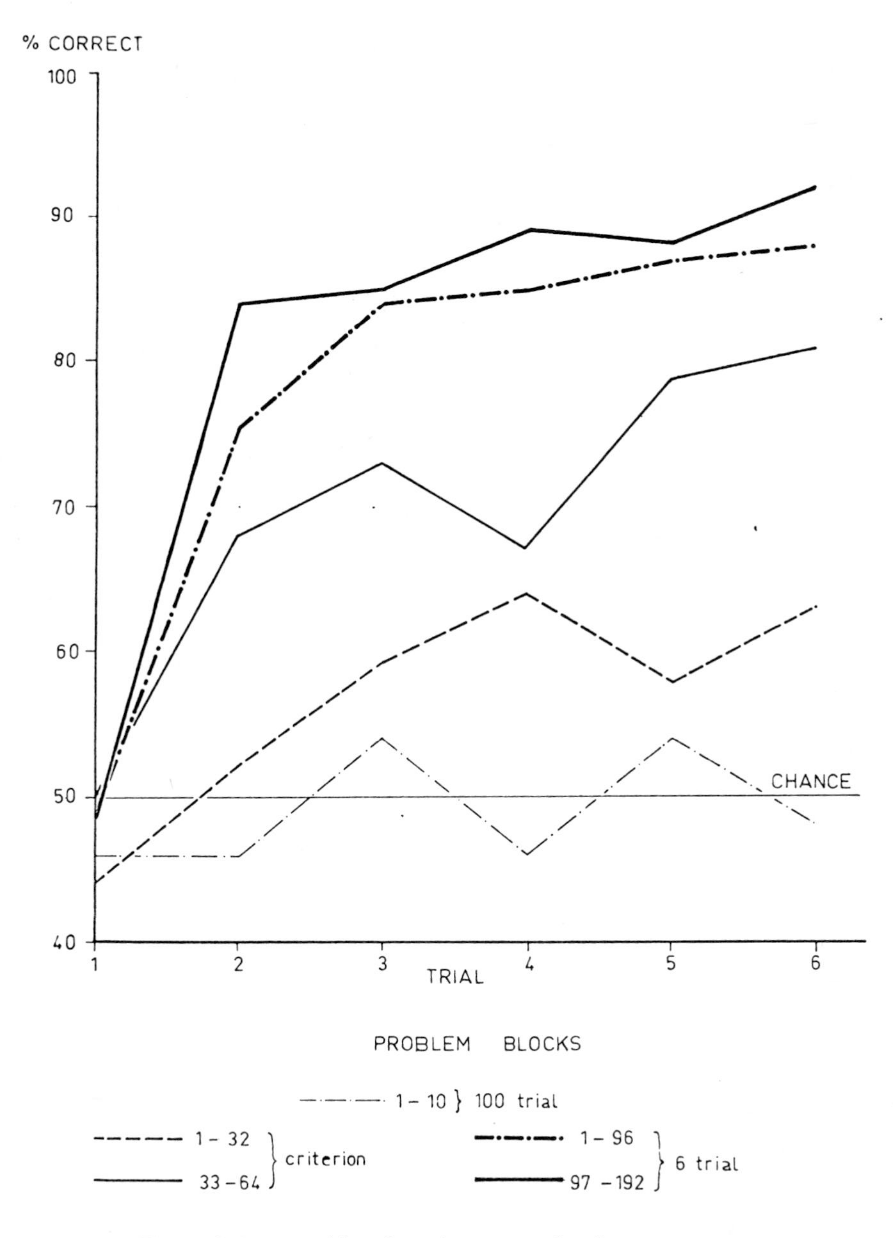

Figure 3. Intraproblem learning curves for *Lemur macaco*.

block of problems was designed thoroughly to adapt the animals to the experimental procedure, and to reduce frustration by ensuring a minimum of errors once a given problem was solved.

The second series of problems consists of 64 problems run to a criterion of 17 correct responses in 25 or less consecutive trials. This rather light criterion assured a minimum of learning on each problem (68%), while allowing problems which were solved rapidly to be completed without unnecessary overtraining. Problems were limited to a maximum of 50 trials. Dividing into two blocks of 32 problems each, improvement occurs over the initial problem blocks, and within the two blocks of criterion problems. Trial 2 performance for the first 32 problems is still at chance level, 52% correct, and for the second 32 problems has progressed to 68% correct. The average number of trials each animal required to solve the criterion problems was 23.5, 27.0 and 23.2 for the first 32 problems, and 19.3, 20.8, 19.6 for the second 32 problems, respectively.

The third phase consisted of 192 problems presented for 6 trials each. This gradual reduction in problem length, between each phase, from initially long problems to problems of shorter length, is considered to be the most effective for learning,[6] although learning has also been shown to be independent of problem length.[15,16] Improvement continues, especially at the trial 2 level, where 76% correct responses were made on the first block of 96, 6 trial problems, and 84% correct responses were made on the second block of 96 problems, of this phase.

Discussion

These curves illustrate that *Lemur* has a fairly efficient ability to form a learning set. Initial problems, where performance remains at chance level for the first six trials, are in contrast to later problem blocks, where 84% correct responses are being made on the second trial. On the early problem blocks, the learning curves are continuous S-shaped curves, and can be described by a single function. On later problems, however, learning curves are discontinuous, and can best be described as two functions; a rapid gain in performance between trials 1 and 2, followed by the gradual improvement between trials 2 to 6. The ability to solve problems in one trial, as measured by trial 2 performance, is typical of primates, but not non-primates,[4] and apparently also exists in *Lemur.*

Although the intraproblem curves are discontinuous, the inter-problem learning curves (Fig. 4) are continuous. Performance on any individual trial improved in a gradual and continuous manner, through each series of problem blocks. Learning set develops in an orderly fashion with experience, even though this learning is manifested as a discontinuous curve within individual problems.

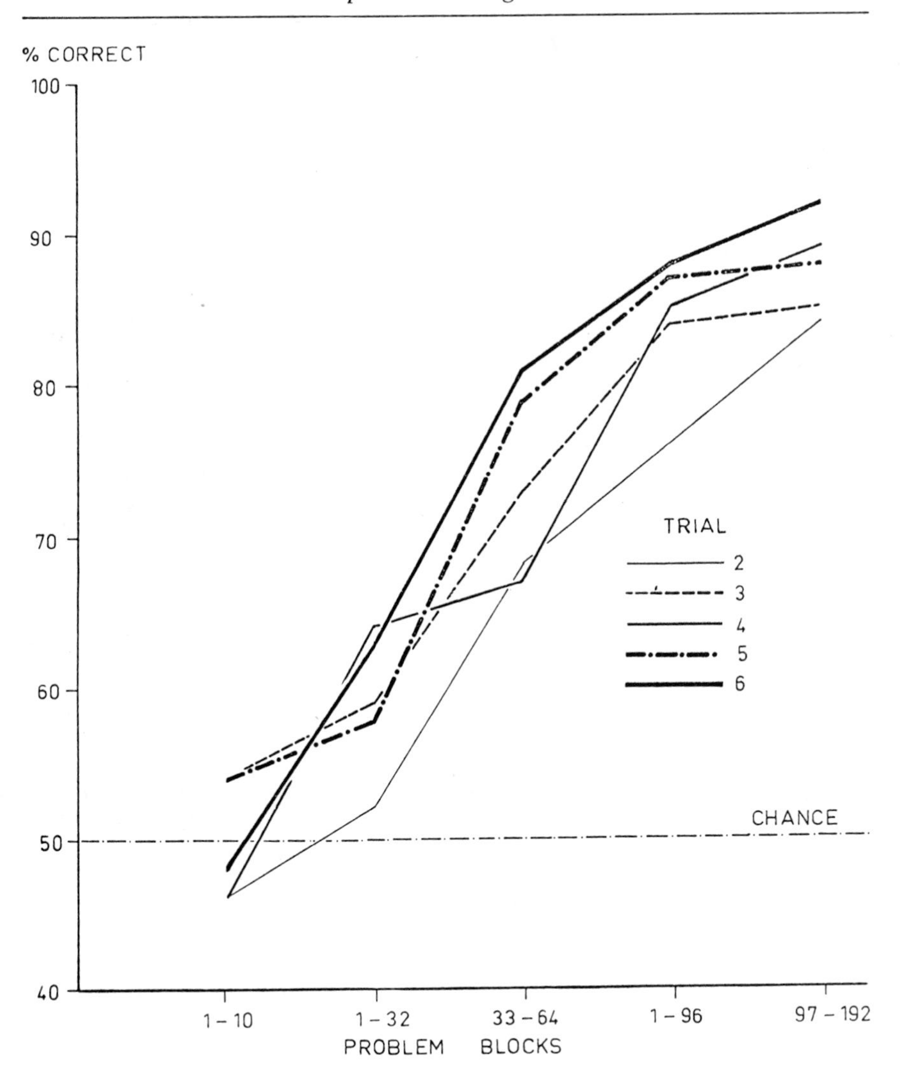

Figure 4. Interproblem learning curves for *Lemur macaco.*

Although one of the objectives of comparative psychology is assessing differences in learning ability between different species, this study of learning in *Lemur* is not directly comparable to similar studies with other primates in other laboratories, and therefore under different conditions. For the moment it is important simply to point out that *Lemur* shows the ability to form a learning set that follows the typical primate pattern.

NOTES

1 Jolly, A. (1964), Prosimians' manipulation of simple object problems', *Anim. Behav.*, 12, 560-570.

2 Ehrlich, A. (1970), 'Response to novel objects in three lower primates: greater galago, slow loris and owl monkey,' *Behaviour*, 37, 55-63.

3 Warren, J.M. (1965), 'The comparative psychology of learning', *Ann. Rev. Psychol.*, 16, 95-118.

4 Miles, R.C. (1965), 'Discrimination learning sets', pp. 51-95. In: A.M. Schrier, H.F. Harlow and F. Stollnitz (eds.), *Behaviour of non-human primates.* Vol. 1. New York: Academic Press, 51-95.

5 Harlow, H.F. (1949), 'The formation of learning sets', *Psychol. Rev.*, 56, 51-65.

6 Harlow, H.F. (1959), 'Learning set and error factor theory', pp. 492-537. In: S. Hoch (ed.), *Psychology: a study of a science.* Vol. 2. *General systematic formulations, learning, and special processes.* New York.

7 Warren, J.M. and Baron, A. (1956), 'The formation of learning sets by cats', *J. comp. physiol. Psychol.*, 49, 227-229.

8 Koronakos, C. and Arnold, W.J. (1957), 'The formation of learning sets in rats', *J. comp. physiol. Psychol.*, 50, 11-16.

9 Rolling, A.R. (1963), 'Successive object discrimination reversals by squirrels', Paper read at Eastern Psychol. Ass., New York.

10 Harlow, H.F., Harlow, M.K., Rueping, R.R. and Mason, W.A. (1960), 'Performance of infant rhesus monkeys on discrimination learning, delayed response, and discrimination learning sets', *J. comp. physiol. Psychol.*, 53, 113-118.

11 Jolly, A. (1964), 'Choice of cue in prosimian learning', *Anim. Behav.*, 12, 571-577.

12 Glickman, S.E., Clayton, K., Schiff, B., Guritz, D. and Messe, L. (1965), 'Discrimination learning in some primitive mammals', *J. gen. Psychol.*, 106, 325-335.

13 Rumbaugh, D.M. and Arnold, R.C. (1971), 'Learning: a comparative study of *Lemur* and *Cercopithecus*', *Folia primat.*, 14, 154-160.

14 Arnold, R.C. and Rumbaugh, D.M. (1971), 'Extinction: a comparative study of *Lemur* and *Cercopithecus*', *Folia primat.*, 14, 161-170.

15 Sterritt, G.M., Goodenough, E. and Harlow, H.F. (1963), 'Learning set development: trial to criterion vs. six trials per problem', *Psychol. Rep.*, 13, 267-71.

16 Rumbaugh, D.M. and Prim, M.M. (1963), 'A comparison of learning set training, methods and discrimination reversal training methods with the squirrel monkey', *Amer. Psychol.*, 18, 408.

PART I SECTION C
Olfactory Communication in Prosimians

R. J. ANDREW and R. B. KLOPMAN

Urine-washing: comparative notes

Urine-washing is unusual amongst mammalian fixed action patterns in that its complexity and peculiarity allow it to be recognised with near certainty as homologous in a restricted number of primate genera. The odd distribution of the behaviour and its possible systematic implications were first pointed out by Hill.[1] It is the purpose of this communication to consider to what extent urine-washing has the same form and causation in the various genera which show it, and to examine some broader questions. The first author has seen the behaviour in all of the species mentioned, which may make cross-species comparisons more reliable; in the case of *Galago demidovii, Loris* and *Perodicticus* fuller observations were made by the second author.

Data relate to captive animals in all cases, and are further restricted in *Loris, Aotus, Saimiri* and *Cebus* by the small numbers available for study (see Appendix). This is most serious in *Cebus*, where only one male was watched intensively. However, published sources independently confirm the existence of urine-washing in all these genera.

The movements of urine-washing may be most conveniently described in *Galago senegalensis* and *G. crassicaudatus*, where hand-reared individuals were watched extensively and at very close quarters, and the movements are relatively leisurely. The foot and hand of the same side are invariably involved, except in very incomplete intention movements, when one hand alone may be raised a short way from the substrate, sometimes repeatedly. The original description by Hill of both hands being rubbed together is almost certainly an error.[1] Hand and foot are raised almost or quite simultaneously. The foot passes under the genital region, with the sole turned inward and the big toe upwards; usually it appears not quite to touch the genital region. The hand is approximated to the foot. It may not quite reach the foot, but more usually the foot is grasped one or more times, with the palmar surface of the hand pressed against it. Urination begins as the hand and foot arrive under the belly, and probably ceases from the time when they are lowered to the time when the hand and foot of the other side have been raised in their turn.

The main variations in the form of the movements are in their degree of completeness; the addition of foot-rubbing (below) may perhaps be considered as the highest intensity form of the pattern. Raising the hand alone has already been noted, and urine-washing may also stop after the hand and foot of only one side have been raised. However, variation is also caused by interaction with other patterns. Thus urine-washing is often preceded by side-to-side swaying of the forebody as a part of parallax movement. As urine-washing develops, the two hands are alternately and repeatedly raised from the ground as part of the swaying, without any movement of the feet.

Foot-rubbing occasionally follows urine-washing. However, it also may follow rhythmic micturition (below) or ordinary micturition without urine-washing, so that it is probably best considered to be a separate, although closely related pattern. In its fullest form, several rubbing strokes of one hind foot over the substrate in the sagittal plane of the body are followed by a similar series of strokes from the other foot.

Urine-washing is basically similar in all the Lorisoidea studied, except that foot-rubbing has so far been seen only in *G. crassicaudatus*. *G. demidovii* urine-washes very frequently, and performs it rapidly, usually with at least one alteration between the sides. Finger flexing movements are unusually vigorous, as the hand grasps at the foot. Urine-washing in *Loris tardigradus* has been described by Ilse,[2] and no deviation from the pattern as already described was observed in the present study, except that incomplete movements might involve the foot alone. *Perodicticus potto* was not seen to urine-wash. This negative finding is confirmed by more extensive observations on another colony (G. Manley, unpublished data).

In the Pithecoidea, *Aotus trivirgatus* urine-washes in a manner almost identical to *Galago*, including the characteristic movement of rubbing and pressing the hand against the raised foot of the same side. *Saimiri sciureus* also rubs the hand against the ipsilateral foot in urine-washing, with the usual alternation between the limbs of each side.

The form of the pattern in *Cebus* is of particular interest. Even *Saimiri* has a limited repertoire of hand movements: the same two-handed grasp, in which the hands tend to rotate outwards, opposing each other, is used to hold food steady, to break it and to turn it in the hand. *Cebus* is thus the only genus, of those in which urine-washing has been examined, capable of varied and highly dexterous hand and arm movements. The male observed intensively was tentatively assigned to *C. albifrons*, but the behaviour has also been seen in *C. apella.* Urine-washing involved the usual basic movements with hand and foot raising, first on one side and then the other, but was considerably more variable than in any other species studied. Either the hand or the foot might be the first to be raised. At low intensities, it was common for the foot alone to be raised and placed with the sole under the penis and the thumb directed up the flank. Occasionally the same movement might be followed by scratching rather than urine-washing. More rarely, the hand alone was thrust under the genital region, the animal at the same time lying down. The foot might be rubbed repeatedly over the penis area, or be held relatively or quite still whilst the hand grasped at it with repeated flexings of the fingers. Both foot and hand sometimes appeared to move together. A very common variant was to bring the forearm, elbow or even the upper arm below the foot, so that one of these regions, rather than the hand, was wetted. The movement was a clasping one and was sometimes accomplished lying down; in general legs and arms were flexed at the beginning of urine-washing. Variation in foot movements also resulted in the spreading of urine widely over the body. Lateral rubbing movements of the foot might be replaced by forward strokes of the palm up over the flank, sometimes as high as the shoulder. Bouts of each type of stroke might alternate, and the hand was sometimes rubbed on the foot between bouts. Arm position was also adjusted to spread the urine widely: thus the arm might be progressively drawn out so that first the elbow, then the forearm, and finally the hand was wetted. Nolte has suggested that the behaviour of rubbing the body with pungent or acid substances shown by a *Cebus apella* might be related to urine-washing.[3] The present animal showed great interest in the areas damp with urine, often sniffing or licking at them. Functionally, the changes in form result in the presentation of scent on the body surface of the animal rather than a particular part of the substrate, which is likely to be more appropriate in a species where social interactions are of great importance. However, the most interesting aspect is the way in which an originally stereotyped movement has acquired variability during evolution, apparently as part of a general increase in the range of possible limb movements.

A number of other components were also commonly added to the basic pattern. After the initial crouch (if it occurred), and the commencement of urination, the tail was often coiled more tightly (sometimes twice), and shaken laterally; sometimes small amplitude lateral shaking of the body or even violent shuddering followed. Similar shuddering in association with urination in both horse and man has been observed by the first author, but its causation remains unknown.

The second topic on which comparative data may be of interest is that of *causation.* The main types of situation in which urine-washing occurs will be examined, but no attempt will be made to discuss motivational models. In the Lorisoidea, the situations may be classified as follows:

1. *Mobbing*

In *G. crassicaudatus, G. senegalensis* and *G. demidovii* urine-washing is common whilst gazing at a strange distant object; (its evocation by the introduction of a strange object into the home cage may be regarded as a variant of this situation). The gaze is usually intent and quite uninterrupted by the urine-washing. It is often accompanied by lateral swinging of head and forebody such as would yield parallax information. The animal may be silent or (particularly in the case of *G. senegalensis*) may give mobbing calls.

2. *Exploration of a strange area*

Rhythmic micturition (below) is more frequent than in 1, probably because it can be combined with forward locomotion in a way in which urine-washing cannot. Urine-washing is also common in all lorisoids observed in this situation.

3. *Defence and attack*

Urine-washing when approached by a superior of which the animal is nervous has been seen in *G. crassicaudatus* and *Loris*. Urine-washing may also immediately precede a leap to attack another male in *G. crassicaudatus* (although foot-rubbing is more usual in this situation; L. Rosenson, personal communication), or it may follow defence or a fight.

4. *Social interactions*

A hand-reared male *G. senegalensis* commonly urine-washed in the pauses between unusually vigorous grooming of the first author's hand. The association was very clear: two or three seconds of scraping with the lower incisors on first arrival might be immediately followed by urine-washing, after which scraping immediately resumed. No sexual responses were directed to the first author by the male at any time, but he sought body contact freely. No comparable association was seen in other species. This association has now been much more extensively documented by Doyle,[4] who has shown that male *G. senegalensis* commonly urine-wash after genital-smelling or grooming a conspecific, in particular a female in oestrus.

5. *Evocation by stimuli associated with urine-washing*

The female *Loris* often urine-washed immediately after the male had done so. The scent of urine may be involved, as it may also be in the concentration of general urination at a particular point in the cage (a folded towel). Ilse observed that captive *Loris* usually urine-washed on the same horizontal pole,[2] so that performance of the behaviour at particular scent posts is perhaps especially characteristic of this species.

6. As part of normal urination

It is possible in *G. demidovii*, where urine-washing was very frequent indeed, and in *G. senegalensis*, where urine-washing was sometimes seen in a very tame individual in the absence of any obvious cause except a full bladder, that the need to urinate may sometimes induce urine-washing.

In *Cebus* urine-washing was an almost invariable response to the appearance of the first author after some hours absence. It was never closely associated with threat (frown and *orbicularis oris* contraction) or grins, but no general conclusions as to causation are possible. In *Saimiri* the behaviour was typical when exploring a strange area, or when abandoned and giving contact calls. Latta et al. found urine-washing to be associated with external disturbances (cf. mobbing, above) and excitement in a colony of *Saimiri.*[5] It rose in frequency in both male and female during periods of copulation, but there was no direct association with copulation, and Latta et al. attribute the increase to general excitement.

Two other behaviour patterns, which have sometimes been classed as 'scent-marking' along with urine-washing, are widespread and apparently ancient in the mammals. The situations in which they occur provide useful material for comparison with urine-washing. The first pattern involves movements of rubbing the sides of the face and flanks on the ground or on fellows. It seems to have been primitively associated with copulation. Roberts has described the appearance of such movements in a generalised marsupial (*Didelphis virginiana*) as an invariable immediate precursor of copulation attempts induced by stimulation of the anterior hypothalamus.[6] The movements are specifically directed to furry surfaces, so a particular tactile sensation may be sought. Very similar behaviour occurs in equally generalised Eutherian mammals, *Solenodon* and *Tupaia.*[7] In *Tupaia glis* a male will often rub flanks and/or belly on the ground as it approaches an oestrous female to copulate, or, even more commonly, after smelling or licking her anogenital region. At full intensity, these rubbing movements pass into rolling on the ground. When similar movements are directed at the female rather than at the ground during copulation periods, they begin with wiping first the chin and throat, and then the side of the face on her, and end (much as in the opossum) with flank-rubbing. The indication that the behaviour is in part motivated by scent signals is confirmed by the fact that a hand-reared female would perform such rubbing after sniffing at a pool of urine (not necessarily from a *Tupaia*). Andrew[8] suggested that rubbing chin and throat should be considered quite a separate movement, used chiefly in marking scent posts; it is also as common as the first phase of flank rubbing.[8] It should be noted that flank rubbing may be inconspicuous, or even absent, in some other species of *Tupaia*. Sorenson[9] describes what appears to be flank-rubbing only in *T. longipes*.

Flank-rubbing motivated by scents is familiar in the rolling of the domestic cat at the scent of catnip and of the dog at scents associated with carrion and dung. In the female cat, the association of rolling and flank-rubbing with copulation remains clear. An increased tendency to seek rubbing contact over the body surface also occurs in greeting behaviour in these two species, and very generally through the mammals, both in social greeting and in association with copulation.

A second pattern of importance is the movement of drawing the anogenital region forward over the substrate. When rubbing on a vertical object, the movement may be first up and then down. *Tupaia glis* shows such 'anal rubbing' very clearly. The movement may follow defaecation, and may well have evolved from a purely defaecatory movement: a number of insectivores deposit faeces by such movements on vertical objects.[8] However, in *T. glis* it is likely that glandular secretion is being deposited. Anal rubbing may be directed to branches which appear to serve as scent posts, or (at least in a hand-reared female) to social companions. The association with copulation appears to be absent, although anal rubbing in the hand-reared animal was associated with grooming the hand or foot of the human rearer. A new *G. senegalensis* was very persistently and vigorously rubbed at first introduction. Rubbing could also be reliably elicited, along with the vertical leaps and body shakes given by many mammals in great excitement (e.g. rabbit, horse), by tickling her belly; novel stimuli may act in part, then, through their arousing properties. Neither flank nor anal rubbing occurs in the conspicuous mobbing displays of *Tupaia*.

Within the prosimians, *Lemur* shows characteristic anal rubbing remarkably like that of *Tupaia*, both in form and in the way in which it is directed to strange conspecifics (*L. fulvus*).[8] Causally it overlaps with urine-washing in that it is also given at the approach of a frightening stimulus object.[8] *L. fulvus* also shows what is probably the remains of flank-rubbing, in that males may rub the sides of the face (and top of the head) in female urine. Other than the above example, this pattern does not appear amongst the prosimians.

When compared with the above two patterns, urine-washing differs most obviously in its evocation (at least in most species in which it appears) by distant disturbing visual stimuli. It also differs in that it is not usually given in association with bodily contact with a conspecific. *G. senegalensis* is an exception here: not only is urine-washing common following genital smelling or grooming but it may be associated with scent marking by chest rubbing-against the female.[4]

Piloerection along back and tail in cat and dog (cf. also Von Holst, for *Tupaia*)[10] provides interesting resemblances in causation to urine-washing. It is also evoked by distant strange objects, and in particular by 'strange' animals. It also occurs in the preliminaries of defensive threat, but disappears when threat proper develops (Prescott, personal communication). In dogs, back hair erection is particularly prominent when a distant source of movement or sound is difficult to locate. Prescott has pointed out that such piloerection along the tail of the cat probably serves to release scent, as well as providing a visual cue.[11]

The origin and evolution of urine-washing remains mysterious. Simple urination (and defaecation) in a strange environment or in response to a strange object is probably widespread in mammals. In the Lorisoidea a slight elaboration of this has led to 'rhythmic micturition',[2] in which the anogenital region is repeatedly touched to the substrate, often as the animal moves slowly forward in exploration of a strange area. This has been seen in the present study in *Galago crassicaudatus, G. senegalensis, Loris* and *Perodicticus*, where it is slightly more elaborate (back legs are spread, for the pelvic lowering, and the short tail is flicked during micturition). However, comparable movements of deposition appear to occur outside Lorisoidea in species which are not known to urine-wash. A tarsier, for example, was several times seen to touch its genital area to the vertical post to which it was clinging and deposit drops of urine, whilst staring at a distant disturbing object. Sprankel[12] has suggested that urine-washing may derive from stepping movements following urination such as he saw in *Tupaia.* The first author has not seen behaviour definitely of this type in *T. glis*, and it appears that neither has Sorenson,[9] in all species of *Tupaia* watched by him. (A few drops of urine were often produced during anal marking but no obvious stepping occurred). One difficulty for such a derivation is that the simultaneous raising of foot and hand on the same side of the body is unusual in locomotion (although it does occur in the 'amble'). A movement such as the repeated rubbing of first one hind foot and then the other on the substrate, which has been described in *G. crassicaudatus*, would provide an alternative origin for the foot movement. Hand raising may be compared with the raising of one forepaw, held flexed backwards at the wrist, which can be seen in animals as widely separated as dog ('pointing') and *Tupaia* (where it is sometimes associated with urination), when trying to localise a particular somewhat distant stimulus. Finally, the side-to-side swaying of the body in parallax may have helped to constrain foot and hand movements to one side of the body at a time.

The final question to be considered is the systematic distribution of species which urine-wash. In discussion at the conference it proved possible to extend the number of species for which there is good evidence of the presence or absence of urine-washing, and this information has been summarised in what follows. The first point of importance is that in the Lorisoidea it is likely that urine-washing has been lost in certain lorisine species such as *Perodicticus*, and probably *Nycticebus* and *Arctocebus* (Manley, discussion). The pattern is present not only in the three species of Galaginae already cited (*G. crassicaudatus, G. demidovii* and *G. senegalensis*), but also in *G. alleni.*[13] Its presence in *Loris* shows that it is characteristic of Lorisinae as well as Galaginae. The possibility that urine-washing may have originally been present in the ancestors of *Perodicticus*, which is suggested by the wide distribution which has just been described, is strengthened by the description of genital scratching[13] in male pottos. This movement is very like the first phase of urine-washing and could easily have been derived from it. Urine-washing by *G. senegalensis*[4] after genital-smelling or grooming other animals, particularly oestrous females, provides an example of the elicitation of urine-washing in the same situation as that in which genital scratching occurs. Charles-Dominique (discussion) suggests that lorisoids which move chiefly on broad branches (e.g. *Perodicticus*) may not need to use palm and sole to spread urine on the substrate in the way which is necessary in animals which move amongst slender twigs, and adds *Euoticus* as a possible example of another lorisid in which urine-washing is absent for this reason.

Clearly, if urine-washing has been lost in some members of the Lorisoidea, similar loss might explain the patchy distribution in other groups. However, the possibility that the presence or absence of the pattern may have systematic implications deserves examination. In the Lemuroidea, urine-washing is probably present in *Microcebus murinus.* The pattern is certainly much rarer in this species than in comparable Lorisoids: the movement was never seen in the present colony, nor in a hand-reared male who was observed intensively by the second author. However, Charles-Dominique and Martin record urine-washing for *Microcebus*,[14] and Manley has shown that the characteristic movements of rubbing palm and sole together, first on one side of the body and then on the other, are involved.[15] It is possible that some components of the pattern are present in *Cheirogaleus*, in which the belly may be rubbed in urine (Perret, discussion). It is not clear why the pattern should be absent in other Lemuroidea, if it occurs in the Cheirogaleinae: *Cebus* (Pithecoidea) shows that it can be retained even by species with a monkey-like *facies.* Such evidence should be borne in mind in any consideration of the possibility of a polyphyletic origin of the Lemuroidea. Polyphylety is suggested much more strongly in the case of the

Ceboidea. Urine-washing is present in those of the Cebidae whose behaviour is well known (*Saimiri, Cebus* and *Aotes*) and could therefore be general in the family. Its apparent absence in *Lagothrix* and *Ateles* might be associated with their advanced type of locomotion, but this would not apply to marmosets (Hapalidae), in which it has not been seen by the first author, nor are there any published reports of it in the literature. Early independence of Hapalidae and Cebidae would be quite consistent with their marked differences in structure.

Appendix : species studied

Aotus trivirgatus: two wild-caught, female; *Cebus albifrons:* one male and one female, wild-caught; *Galago crassicaudatus:* 30-40, including both wild-caught and laboratory bred – three hand-reared animals in particular; *G. demidovii:* 8, wild-caught, male and female; *G. senegalensis:* 12, wild-caught and several laboratory bred – one hand-reared male; *Loris tardigradus*, 4, wild-caught, male and female; *Microcebus murinus:* 6, wild-caught, male and female; *Perodicticus potto:* 5, wild-caught, male and female; *Saimiri sciureus:* 2, wild-caught, female; *Tarsius:* 1, wild-caught, male; *Tupaia glis:* 8, wild-caught, male and female – one hand-reared female.

Acknowledgments

The authors wish to express their thanks to Professor D. Poulson and to the Department of Biology at Yale University for provision of the animal quarters which made these observations possible.

NOTES

1 Hill, W.C.O. (1938), 'A curious habit common to lorisoids and platyrrhine monkeys', *Ceylon J. Sci. B* 21, 66.

2 Ilse, D.R. (1955), 'Olfactory marking of territory in two young male loris, *Loris tardigradus lydekkerianus*, kept in captivity at Poona', *Brit. J. Anim. Behav.* 3, 118-20.

3 Nolte, A. (1958), 'Beobachtungen über das Instinktverhalten von Kapuzineraffen (*Cebus apella L.*) in der Gefangenschaft', *Behaviour* 12, 182-207.

4 Doyle, G.A., this volume.

5 Latta, J., Hopf, S. and Ploog, D. (1967), 'Mating behaviour and sexual play in the squirrel monkey', *Primates* 8, 229-46.

6 Roberts, W.W., Steinberg, M.L. and Means, L.W. (1967), 'Hypothalamic mechanisms for sexual, aggressive and other motivational behaviors in the opossum, *Didelphis virginiana*', *J. comp. Physiol. Psychol.* 64, 1-15.

7 Herter, K. (1957), 'Das Verhalten der Insektivoren', *Handb. Zool.* 8, 10.

8 Andrew, R.J. (1964), 'The displays of the primates', in J. Buettner-Janusch (ed.), *Evolutionary and Genetic Biology of Primates*, vol. 2, New York.

9 Sorenson, M.W. (1970), 'Behavior of tree shrews', in L.A. Rosenblum (ed.), *Primate Behavior*, vol. 1, New York.

10 Von Holst, D., this volume.

11 Prescott, R.G.W., personal communication.

12 Sprankel, H. (1961), 'Uber Verhaltensweise und Zucht von *Tupaia glis* (Diard 1820) in Gefangenschaft', *Z. wiss. Zool.* 165, 186-220.

13 Manley, G.H., this volume.

14 Charles-Dominique, P. and Martin, R.D. (1970), 'Evolution of lorises and lemurs', *Nature* 227, 257-60.

15 Manley, G.H. (1966), 'Prosimians as laboratory animals', *Symp. Zool. Soc. Lond.* 17, 11-39.

G.H. MANLEY

Functions of the external genital glands of Perodicticus *and* Arctocebus

Introduction

The existence of highly specialised concentrations of glands on discrete areas of the body-surface of lorisid prosimians has been recognised for some time. In the lorises of South-East Asia and India, the slow loris *Nycticebus coucang* and slender loris *Loris tardigradus*, the principal glandular fields in both sexes are located on the medial side of the upper arm (the flexor surface) and have been termed the brachial organs. In the two African members of the group, the potto *Perodicticus potto* and angwantibo *Arctocebus calabarensis*, they are found in the genital region, forming part of the scrotal surface of the male and lying closely adjacent to the vulva of the female.

In the case of the potto and angwantibo, the subjects of the present communication, some information is available concerning the morphology, histology and cytochemistry of the genital glands, particularly for the first of these two species, the potto. On the matter of function, however, our knowledge is far from adequate. We know little or nothing of the roles these specialised structures play in the lives of the animals bearing them. Speculation has been couched in general terms or, in a few instances, specific functions have been suggested; but, as far as the author can ascertain, no one has yet described patterns of behaviour directly and consistently associated with these organs. The topic was very briefly touched upon in an early work-in-progress report.[1]

The material presented here comes from a comparative study of the behaviour of the Lorisidae, carried out over a number of years on animals housed in relatively large cages in a nocturnal room, i.e. under a reversed-lighting regime. Details of housing conditions, cages, diet, etc. will be found in earlier papers.[2,3] The number of individuals of each species available for observation and manipulation had to be small, and this drawback, and certain others attendant upon captive conditions, is clearly limiting. It follows that conclusions drawn in this paper may be neither definitive nor exhaustive and should be viewed in that light.

Perodicticus potto

Observations on pottos covered a period of 19 months between April 1962 and November 1963. All the animals were of the western subspecies *potto* (see Hill).[4] The principal study animals comprised an adult male, and an adult female and her female offspring, which became full-grown in the course of the study. Initially the male, and the female with her infant, were housed separately; after four months the young female was isolated and the two adults housed together; six months later the young female joined them, being introduced to the male for the first time. Additional limited information came from two more adult females acquired late in the study period.

1. The external genital glands

Descriptions and illustrations of the external genitalia of the potto may be found in a number of publications[2,4–10] and histological and cytochemical studies of the glands have been conducted by Montagna and his colleagues.[7–9,11] A summary description may be appropriate here.

The major glandular areas lie on the scrotum of the male and the pseudoscrotum of the female. In the male, the scrotum has an extensive central area of closely applied tessellae, largely naked, the intersecting fissures being somewhat pigmented and, under normal conditions, narrow. Laterally and to some extent posteriorly the scrotal surface loses its sculpturing and becomes haired. In the female of this species a pseudoscrotum is present, i.e. the rima pudendum bisects a substantial mound with the clitoris situated at its anterior end, and the appearance of the glandular organ, although smaller, is extremely similar to that of the male. It has been shown that the tessellated areas in both sexes have rich fields of large apocrine glands surrounded by nerves containing abundant acetylcholinesterase.

In the female alone, an additional pair of discrete flask-shaped glands is found, flanking the vulva and opening into it[8] or close to the base of the clitoris.[10,12] These glands produce a thick, strongly-smelling secretion of semi-fluid keratinous composition. The observations recorded in the present paper almost certainly refer to the superficial glandular organs of both sexes and not to the female structures just mentioned.

Certain short-term changes in the conformation of the genitalia which are relevant here have been observed. These concern the male;

equivalent changes in the female genitalia were not noticed (compare changes accompanying oestrous cycles).[2] Under certain conditions and then only briefly, the male's testes project ventrally, the valley between them deepens and the fissures between the tessellae widen appreciably to as much as 2 mm. or more. Such changes have been noted (i) immediately after rousing from sleep, when the phenomenon is probably an effect of sustained warmth, (ii) during sexual behaviour, and (iii) during genital-scratching grooming (GSG) (see below).

2. *Genital-scratching grooming (GSG)*

Mutual grooming is extremely common in pottos living amicably together in captivity. Typically, the grooming individual grips the fur of the passive groomee and, using the characteristic licking and raking (dental comb use) actions of prosimians, attends to the fur lying between the clenched hands. Whereas most allogrooming is of this relatively uncomplicated type, the potto may incorporate an additional element of a quite unexpected and specialised nature to perform a behaviour pattern called by the author 'genital-scratching grooming' (GSG). In this, as ordinary grooming proceeds, the groomer is seen to move alternate hands caudad towards the genitalia, specifically the scrotum or pseudoscrotum. With the tips of the fingers applied to the glandular surface, the animal then scratches it in a typically human and completely unprosimian manner, i.e. the fingers are clenched and unclenched so that their tips are repeatedly drawn across the scratched surface (Fig. 1). The hand is then returned forward to regrasp the fur of the groomee, the other hand moves back, and so on. Regrasping appears to be simple and without special movements, but, even in the absence of these, one must suppose that the secretions of the scrotum or pseudoscrotum are transferred to the groomee's fur. The grooming animal's attention, in terms of nose and head orientation and gaze, is concentrated on the area being groomed and never diverted towards the genitalia during these scratching operations. Immediately after a GSG bout, however, the groomer often bends the head down to lick the scrotal or pseudoscrotal surface.

It should be emphasised that scratching actions of the kind described here occur in no other situation in the potto, nor have they been observed under any circumstances in the angwantibo, the lorises, or other prosimians such as *Galago senegalensis*. Normal scratching, of course, employs the 'toilet claw' of the foot.

The number of backward hand movements in a GSG bout varies from one to very many. Regular left-right hand alternation occurs in the majority of cases (88%; $n = 48$); otherwise one hand only may be used, or the movements of each hand grouped into blocks. With the fingertips touching the glandular area, 1-5 clenching-unclenching

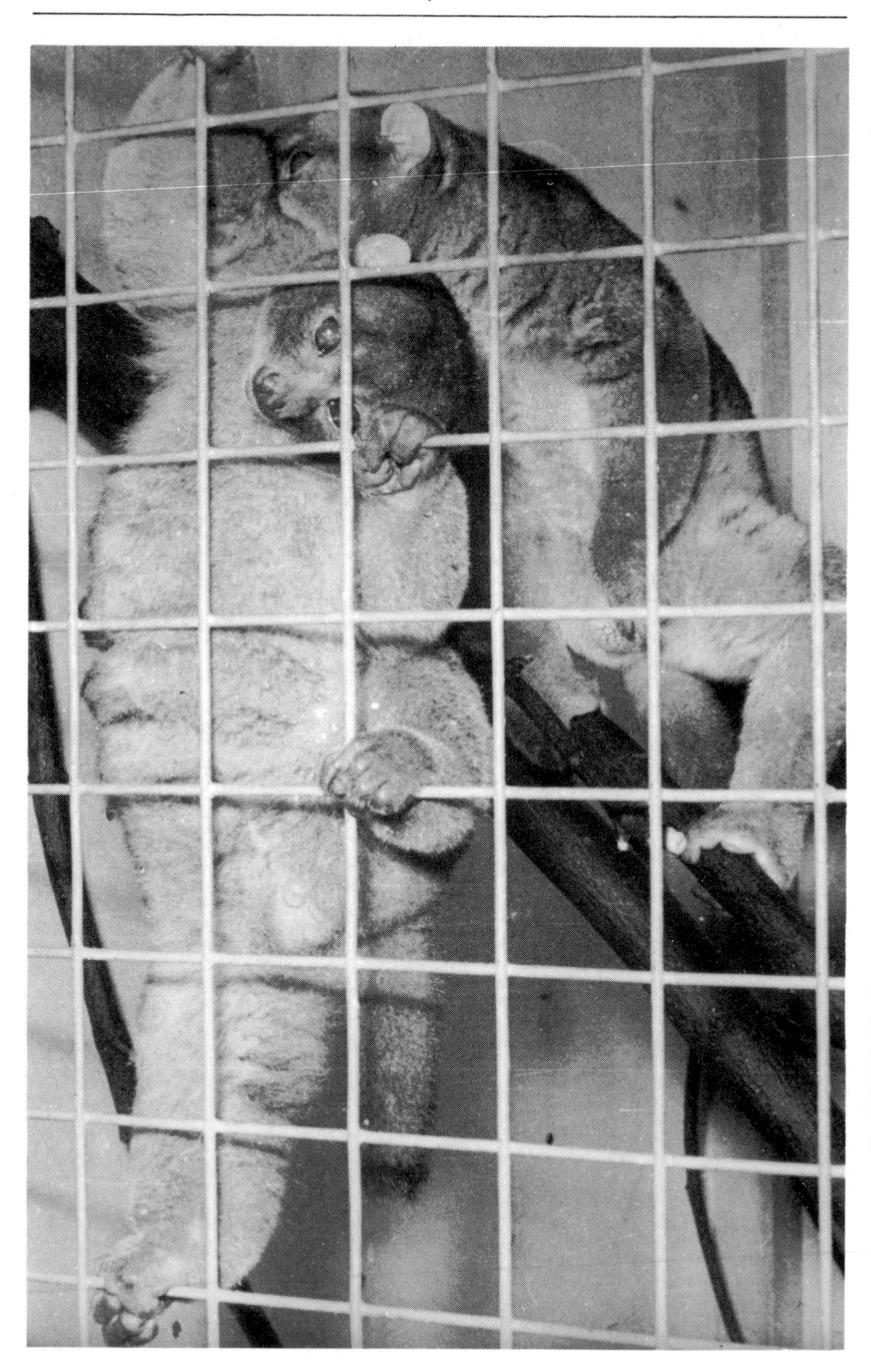

Figure 1. *Perodicticus potto potto.* Adult male (upper right) performing 'genital-scratching grooming'.

scratching actions are performed. In the majority of bouts observed the genitalia are scratched at least once and often many times (80%; $n = 61$), but sometimes no scratching at all is effected. This is because any hand movement may be incomplete, varying from no more than lifting the hand in an intention movement of genital-scratching to making scratching actions in the air immediately above the genitalia; not uncommonly, incompleteness takes the form of scratching a leg which may block the hand's path to the genitalia.

Licking of the scrotum or pseudoscrotum after GSG occurs more often than not (62%; $n = 50$) and, at this point, the groomer has been seen to lick its own *fingers* as well. After a GSG bout the groomee may lick or closely smell the groomer's genitalia. Attention to the genitalia of the groomer is occasionally also observed following apparently ordinary grooming bouts.

Both protagonists in a reciprocal grooming session may be seen performing the genital-scratching pattern, but only the individual actively grooming at the time does so. Like ordinary grooming, GSG takes place with the animals in virtually any position and the groomer may make caudad scratching movements even when fully suspended by feet alone.

With the somewhat limited data available, the parts of the body groomed during GSG are not clearly different from those covered by ordinary allogrooming (Table 1). Although the male seemed to concentrate upon grooming the other's back, little or no such focussing could be discerned in the female.

Table 1 Body areas groomed by two adult pottos during ordinary allogrooming (OG) and genital-scratching grooming (GSG). Percentages shown are of the totals indicated in brackets at the feet of the columns

Region groomed	*OG* ♂	*GSG* ♂	*OG* ♀	*GSG* ♀
Head and face	19.2	10.3	10.4	0.0
Neck	25.6	7.7	21.7	39.4
Back	16.8	58.9	24.5	9.1
Ventrum and flanks	12.0	7.7	9.4	27.3
Armpits	16.8	12.8	19.8	24.2
Limbs	4.0	2.6	12.3	0.0
Genitalia	5.6	0.0	1.9	0.0
	(125)	(39)	(106)	(33)

It will be evident from what has gone before that the GSG pattern is not confined to one sex but is performed by both males and females. However, the impression was gained that the male in the present study showed this behaviour more often than did the female and that this was not related to the frequency distribution between the sexes of allogrooming as such (about equal). In the captive conditions of the present study, at least, GSG was by no means rare and could be observed daily for many days on end.

GSG is not a component of sexual behaviour in this species, nor has any obvious relationship between its occurrence and the regularly recurring oestrous periods of the adult female been detected:[2] no exact counts were made but male and female certainly performed the scratching pattern both during the female's oestrus and throughout the dioestrous interval.

Table 2 summarises the GSG relationships developed among the three principal study animals in the different phases of the study. It reveals that mother and growing infant lacked such behaviour and that, later, the young but full-grown female was not seen to mark her mother; otherwise all other GSG relationships were established.

Table 2 GSG relationships among three pottos in different study phases
1. Adult female with growing infant (adult male isolated)
2. Adult male with adult female (young female isolated)
3. Young female, now full-grown, joins adult pair

Groomer showing GSG		*Groomee (recipient)* Adult ♂	Adult ♀	Young ♀
Phase 1	Adult♀	absent		–
	Young ♀		–	
Phase 2	Adult ♂		+	absent
	Adult ♀	+		
Phase 3	Adult ♂		+	+
	Adult ♀	+		+
	Young ♀	+	–	

The initial establishment of GSG relationships is also of significance here and could be traced in the behaviour of the animals following introductions. Although few introductions were possible in the present study, their results are highly suggestive.

In the case of the adult male and adult female, the full, permanent introduction was effected after a series of seven brief 'trial sessions' during the preceding three months. In these, aggressive interactions predominated but declined over time to a low level on the occasion of the full introduction. Then, a reciprocal grooming relationship (but without GSG) was established on the day following introduction, a full play relationship was established on the sixth day, and only after this, on the sixth day, did the male start marking the female with the GSG pattern. The female was first observed marking the male in this way several days later. In short, it would seem that the GSG relationship became established once the initial agonistic barriers had been fully overcome by mutual grooming and play.

When the full-grown young female was first introduced to the adult male and rejoined her mother, the process was accelerated apparently by the lack of aggression and hesistancy on the part of the young female. In this case the latter initiated grooming immediately and play occurred very soon afterwards, so that these relationships between the animals were established within the first half hour of the introduction. The male then GSG-marked the new female 35 minutes after the time of introduction and repeated the pattern at least 12 times in the following two hours. The young female first marked the male on the following day and was herself marked by her mother from the sixth day. The unusually high frequency of GSG marking shown by the male towards the new female should be noted.

3. *Angstgeruch*

Besides the GSG and ordinary allogrooming contexts, in which there is no evidence of any motivational conflict in the groomer, certain conflict situations in the potto seem to activate the genital glandular fields. In most instances this has been inferred from the attention given to the scrotum or pseudoscrotum by the animal itself (licking) or its cage-mates (licking or closely smelling); in some, secretion from the glands has been clearly visible. Unlike GSG, however, special associated behaviour is lacking.

No single interpretation is sufficient to account for the various occurrences but they tend to fall into a number of groups. The intention here is to concentrate on one group only, after dealing summarily with the others. Thus, on the criteria given above, stimulation of the scrotum or pseudoscrotum was observed with:

1. Physical thwarting of normally available locomotor patterns, e.g. preventing the performance of a cage stereotype in the adult female by simply blocking off a hand-hold; temporary leg malfunction in the same female making locomotion difficult and apparently painful. Neither overt fear nor social factors are evident here.
2. Thwarting of an aroused social tendency by inappropriate behaviour on the part of the other animal, e.g. the adult female behaving sexually at oestrus but eliciting only play from the male. Overt fear is again absent.
3. Conflict between two aroused social tendencies, e.g. attack-escape conflict as in both male and female during their early 'trial' introductions; escape inhibiting sex as in the final introduction, where the female was in oestrus but retained some fear of the male.
4. Fright or thwarting of a strongly aroused escape tendency, e.g. when the principal females were handled for cage transfer or

medical treatment; when the two additional study females, newly arrived and extremely timid, were even approached.

In the first three categories particularly, it is uncertain whether or not the undoubted stimulation of the scrotum or pseudoscrotum has a communication function, informing others of the motivational state of the animal, especially as the evidence of stimulation comes mainly from the animal itself rather than its congeners. The very variability of eliciting situations militates against this interpretation and the author is not convinced that cagemates displayed appropriate responses.

In the final category, however, a glandular secretion was patently visible and/or was clearly shown to be present by the behaviour of other animals. Two examples demonstrate these points. The first concerns the two very wild and timid adult females whose behaviour when approached was characterised by fleeing and protracted freezing; when touched or handled, besides pronounced defensive behaviour, these females would produce a copious secretion from the genital glandular surfaces, at times so plentiful as to flood them and drip to the floor as a clear brownish-yellow fluid. This was not urine.

The second example is provided by two occasions on which the principal adult female was handled in order to swab a head wound with antiseptic: when the animal was returned to the shared cage, the male and young female showed pronounced olfactory interest in her pseudoscrotum and this despite the competition of the strongly-smelling antiseptic, investigated only much later.

The tentative suggestion made here is that the potto may possess an *Angstgeruch* (fear scent) secreted by the external genital glands under these conditions and functioning as a warning signal to others of potential or actual danger. The matter is further considered when discussing a comparable phenomenon in the angwantibo and in the final section of this paper. The term *Angstgeruch* itself is taken from a publication demonstrating the existence of such a substance in the house mouse, *Mus musculus.*[13]

Arctocebus calabarensis

Observations on angwantibos of the subspecies *calabarensis* (see Hill)[4] (Fig. 2) centred on a male and two females and covered a period of 28 months between July 1964 and October 1966. A female infant studied during its first five months provided some further relevant information, as did a third adult female. The adult male was housed initially with one of the principal adult females for seventeen months, then spent five months with the second principal adult female before finally rejoining the first. The male fathered three infants in the course of this whole period.[14]

Figure 2. *Arctocebus calabarensis calabarensis.* Mother and infant.

1. The external genital glands

The external genitalia and associated glands of this species have been considered in several publications[4–6,11,15] but neither always entirely accurately nor completely, according to the author's own observations. For example, the erected penis is perfectly straight and not curved[5] and a tessellated area adjacent to the vulva of the female appears to have been overlooked. Rather surprisingly, the histological studies say little about the glandular areas and a distinct organ is not mentioned; apocrine glands in the skin of the genitalia are simply described as medium-sized[11] or as larger and more numerous than elsewhere, but not clustered.[15]

The genitalia and their glandular areas in *Arctocebus* are in some important respects quite unlike those of the better-known potto and deserve some attention here. In the male, the major difference concerns the size and position of the naked glandular area of the scrotum: in the angwantibo this is restricted to a small triangle, its apex pointing cephalad, on the posterior slope. Under normal conditions, the surrounding hair of the scrotum so encroaches on the naked triangle that only the closely-packed tessellae are visible, the boundaries of the glandular area being indeterminable.

No mention is found in the literature of a glandular area in the female angwantibo corresponding to that of the male, despite the prominence of this feature in the female potto. However, such an area is in fact present in the angwantibo, though relatively poorly defined. It lies immediately posterior to the posterior labium of the latero-laterally oriented vulva and comprises only a small number of tessellae in very low relief, barely discernible except along the anterior border and midline of the area. The area is most clearly seen when the female is in oestrus, at which time the labia and posterior slope become turgid and the surrounding hair is spread away from the general vulval region.

As in the potto, distinct short-term changes in the conformation of the external genitalia of the male angwantibo have been observed. Under certain circumstances the whole scrotal mass projects appreciably ventrally and caudally as the testes bulge downwards and rearwards. Concomitantly the area of the naked triangle is vastly increased and occupies most of the posterior hemisphere of the scrotal mass; the glandular tessellae become spread out and now lie in the centre of this expanded bare region, surrounded by a border of undifferentiated scrotal skin. This state has been observed (i) immediately after rousing from sleep, when it is probably an effect of sustained warmth, (ii) occasionally when being groomed by the female, (iii) when the male grooms his own genitalia vigorously, (iv) on one occasion soon after the arrival in the nocturnal room of two new adult females, (v) during sexual behaviour on the part of the male, and (vi) during episodes in which passing-over occurs (see below).

2. Passing-over

'Passing-over' is a male behaviour pattern directed towards the female. The male, moving along the upper surface of a horizontal branch on which the female has paused or is moving slowly, literally passes over her: employing three limbs for locomotion and keeping one leg cocked, the male moves forward over the entire length of the female's back in such a manner that his underparts are drawn over the fur of her back. The protuberant scrotal mass brushes or gently rubs against the fur as passing-over occurs. At times the pattern includes a distinct depression of the male's rear body and occasionally he may pause briefly in mid-passage along the female's back.

Passing-over may be performed either singly or several times in quick succession. The movement may be in either direction, starting from the rear end or the head end of the female, though the former is more usual. In some performances of the pattern the whole length of the female's back is not traversed but only some part of it, the male crossing obliquely over her.

It should be emphasised that, simple though the pattern might appear to be, it is a quite distinct and deliberate action and not just the 'accidental' outcome of one animal attempting to pass another on the same branch.

As *Arctocebus* of both sexes possess a urination pattern in which the rear body is depressed to deposit traces of urine on branch surfaces, it sometimes happens that the male, in passing-over, leaves a urine smear on the female's fur. The author is convinced, however, that this is no more than a by-product of passing-over and that the primary purpose of the latter behaviour is the deposition of olfactory substances emanating from the male's specialised scrotal glandular area.

Although not a constant feature of male sexuality, passing-over is most commonly seen when the male is engaged in sexual behaviour; it is thus associated with a number of patterns interpreted as having this motivation, e.g. persistent trailing after the female, repeated sniffing at her genitalia, a special courtship call, and another distinctive display that the author calls the 'chin rest'. When passing-over does occur here it appears during lower intensity sexual behaviour rather than the high intensity activities immediately preceding attempted or actual copulation. In the present study, passing-over tended to be observed when the two principal adult females came into oestrus. However, the adult male also occasionally displayed some sexual interest in the females during their dioestrous intervals and when one of them was pregnant and, at such times, passing-over could again be seen.

That passing-over is not entirely linked with male sexuality is indicated by the male's performance of the pattern in the several hours following his first introduction to the second principal adult female. No sexual behaviour could be recognised on that occasion (a week before the female's oestrus), yet passing-over was recorded twice and the male's genitalia revealed the conspicuous scrotal distension described above.

It would seem quite significant, too, that the male's first three close olfactory investigatory approaches to the female should have involved his sniffing closely at the fur of her *back* in preference to other parts including the genitalia; genital inspection only arose and became frequent later. This suggests the possibility that, on first meeting a strange female, the male may very soon ascertain whether or not she has been marked on the back by the passing-over patterns of another male. Some supporting evidence is provided by the male's behaviour upon eventual reintroduction to his original female: within seconds of being released into the cage he had checked the odour of her back.

When two adult females strange to one another were first introduced, the fur of the back was again the focus of much olfactory investigation, starting from the first two close approaches. The possibility thus arises that females too will check each other for scent-marks on meeting.

In short, the provisional picture that emerges is that males check females, females check each other, but females do not check males, all of which fits in well with the fact that passing-over is a masculine pattern and females have never been seen passing-over the backs of males. Clearly the observations recounted here are very few and, however suggestive, call for verification by planned series of introductions using greater numbers of animals.

3. *Angstgeruch*

A number of episodes witnessed during the course of these studies of captive angwantibos lead the author to suspect the existence of an *Angstgeruch* or fear scent such as has already been suggested might exist in the potto. Moreover, this would appear to be produced by the external genital glands of both sexes (cf. passing-over), so that the glandular area of the female, though ill-defined, would seem to be functional in this context at least.

As in the potto, indications of fear scent production were seen when an animal gave evidence of being more than usually frightened. Sources of such alarm were various but included interactions with conspecifics, unusual events in the nocturnal room and unusual noises in the building. At such times, although no visible secretion could be made out, the behaviour of the cage-mate unequivocally demonstrated the presence of an odoriferous substance on the tessellated glandular structure: the cage-mate would be drawn, often from some distance away, to smell closely that part of the alarmed individual's genitalia.

In both the potto and the angwantibo, the examples given reveal one aspect of these situations that might appear to go against the idea of a fear scent signal, namely, the response of the other animal. In most cases this has been to investigate the scent-producing individual rather than to show flight, freezing, etc. – the responses which might be expected on the alarm signal hypothesis. However, it is not inconceivable that what we are seeing is really a reflection of the captive state: animals long habituated to others in a relatively safe environment might be less responsive to signals of this kind than they would be in the wild.

A clue to what is possibly the true context in which the *Angstgeruch* operates in the two species comes from certain observations of mother-infant interactions in the angwantibo. On a number of occasions the author has been struck by the instant communication of alarm from mother to infant (the infant either on the mother or 'parked'), causing the latter instantly to suspend all activity or immediately attach itself to the mother. Unfortunately, on these occasions it has not been possible completely to exclude other factors such as visual cues, but the already-demonstrated fact that alarmed animals do produce an odoriferous secretion makes it at least possible that the infant was reacting to a specific olfactory warning signal.

Conclusions and discussion

It is not within the scope of this paper to consider mammalian scent-marking in general; marking of one individual by another such as has been reported here is by no means unknown and the reader is referred to a recent review.[16] Nor can the other marking systems (e.g. urinary) present in the Lorisidae be dealt with.

Concerning *Perodicticus* and *Arctocebus*, hypotheses regarding the functions of the external genital glands have usually been based on pure speculation rather than the observation of behaviour, and they range from the general to the specific and from the not improbable to the bizarre.[17]

Thus, in the case of the potto it has been suggested that the glands may be related to social and reproductive activities,[7,9] or may enable the sexes to locate each other at night;[6] a defensive function, the secretion having a repellent effect, has also been advocated.[18] One theory that has been advanced is that the glandular areas function as insect attractants,[19] and another that they serve as 'friction pads' and 'to some extent a clasping organ' in copulation.[5] (In the only reference to the angwantibo, the same author again suggests that the tessellated area of the male's scrotum acts as a 'friction pad' during mating.)[5] Several writers have reported observations leading them to believe that the potto's genital glands are used to mark branches or other objects in the environment.[20–22] Only two of these various hypotheses call for comment at this point.

Concerning a possible anti-predator defensive function, the present author's observations of glandular secretion in severe alarm might, at first sight, appear to be confirmatory; moreover, as has been noted,[18] this would tally with one function of the brachial organ of *Nycticebus*. However, unlike the slow loris, in which in most instances there can be no mistaking the nauseating, repellent nature of the fright secretion's odour, the potto's secretion does not recommend itself for inclusion in that class: in the author's experience the odour either escapes notice or simply lacks the overwhelming quality evident in the loris's discharge.

The author has himself seen both pottos and angwantibos apparently scent-marking branches with their genital glands, but so rarely as to make it unlikely that this could be the glands' function. It is possible to confuse such rubbing with rear body depression during urination and the anus-wiping that sometimes follows defaecation.

In the present paper a distinct behaviour pattern in each of the two species has been described, which results, in the author's view, in the scent-marking of another individual. In *Perodicticus*, the sexes mark one another and females may mark other females with genital-scratching grooming; in *Arctocebus*, only the male marks the female with his passing-over performance. What then is the significance of such olfactory marking in the natural life of these animals?

The evidence of recent field-studies confirms earlier reports that members of both species are solitary, i.e. are usually encountered singly in the wild.[22–25] However, such a designation has implications of a simplicity of social structure which is probably far from real; at the very least one has yet to consider precisely the general relationships between neighbouring individuals of the same or different sex, and the nature of the relationship between male and female at and around the time of mating. Associated with such relationships would be the ranging behaviour of individuals: whether home-ranges are exclusively occupied or overlapping and the significance of overlapping should it occur.

The very existence of scent-marking patterns of the kind described above would seem to indicate a social structure more complex than might at first be supposed and, moreover, that the social structures are different in the two species.

Charles-Dominique indeed has already presented evidence that consort pairing in the potto may be for a matter of months,[23,25] and although females do not form small social groups (cf. *Galago demidovii*),[25,26] that daughters may remain in home-ranges adjacent to those of their mothers. It may be the case in fact that the relationships between such animals are closer and even longer-lasting than these data indicate. It is not difficult to see how mutual scent-marking might be advantageous in such a social structure; not only would individual recognition be facilitated, but the exchanged scent-marks could provide the respective animals with immediate olfactory indicators of their already-established harmonious relationship.

In the absence of more field data concerning the social structure of the angwantibo, it is perhaps unwise to speculate too much on the manner in which it differs from that of the potto. However, evidence to be presented elsewhere does suggest that consort pairing at least is dissimilar and, here, the fact that males alone mark females certainly points to the conclusion that such differences exist.

As passing-over occurs mainly when the female is in oestrus it must be especially important that she bears the male's scent-mark at this time, and a tentative suggestion may be made of one reason why this should be so. Although the females of both species produce oestrous urine and become more restless at oestrus, observations reveal that in *Arctocebus*, additionally, the state is signalled by incredibly piercing, loud whistle calls. The calling behaviour especially might be expected to attract males from surrounding areas to the female, and it is then conceivable that if she carries the newly-applied scent-mark of one particular male it acts in some measure as a stamp or seal of ownership, dissuading others. Once copulation is effected, further mating by other males is mechanically frustrated, by the presence in the vagina of the successful male's copulatory plug.[14]

On the evidence currently available it is with some hesitancy that the *Angstgeruch* signal hypothesis is advanced: it is one thing to show that the genital glandular areas of the two species register alarm, but another to prove a real communication function in nature. However, as indicated earlier, the mother-infant relationship is one context in which such a signal is manifestly appropriate. And it is worth noting that (unlike the rapidly-moving, saltatorial and more social galagos) all the slow-moving lorisid species, in keeping with their stealthy and largely solitary way of life, lack alarm calls in the vocal repertoire.

Summary

The principal function of the external genital glands of *Perodicticus* and *Arctocebus* appears to be the scent-marking of other individuals. In the former species, marking is carried out by both sexes during the performance of 'genital-scratching grooming'; in the latter, only the male marks the female by 'passing-over' her back. The significance of such marking in the natural social life of the two species is discussed.

It is possible, though by no means certain, that an additional function of the glands of both sexes of both species is the secretion of an *Angstgeruch* or fear scent signal substance in appropriate circumstances.

Acknowledgments

This work was carried out while the author was Research Fellow at the Wellcome Institute of Comparative Physiology, Zoological Society of London, Regent's Park, London.

NOTES

1 Manley, G.H. (1967), 'Reproduction in lorisoid primates (Progress report)', *Sci. Report Zool. Soc. Lond.* 1966-67, 21-22.

2 Manley, G.H. (1966), 'Reproduction in lorisoid primates', *Symp. Zool. Soc. Lond.* 15, 493-509.

3 Manley, G.H. (1966), 'Prosimians as laboratory animals', *Symp. Zool. Soc. Lond.* 17, 11-39.

4 Hill, W.C.O. (1953), *Primates:* vol. 1, *Strepsirhini*, Edinburgh.

5 Sanderson, I.T. (1940), 'The mammals of the North Cameroons forest area', *Trans. Zool. Soc. Lond.* 24, 623-747.

6 Hill, W.C.O. (1957), 'Reproductive organs: external genitalia', in Hofer, H., Schultz, A.H. and Starck, D. (eds.), *Primatologia: Handbook of Primatology*, Basel, 630-704.

7 Montagna, W. and Ellis, R.A. (1959), 'The skin of primates: 1. The skin of the potto (*Perodicticus potto*)', *Am. J. Phys. Anthrop.* 17, 137-61.

8 Montagna, W. and Yun, J.S. (1962), 'The skin of primates: 14. Further observations on *Perodicticus potto*', *Am. J. Phys. Anthrop.* 20, 441-9.

9 Ellis, R.A. and Montagna, W. (1963), 'The sweat glands of the Lorisidae', in Buettner-Janusch, J. (ed.), *Evolutionary and Genetic Biology of Primates*, vol. 1, New York, 197-228.

10 Charles-Dominique, P. (1966), 'Glandes préclitoridiennes de *Perodicticus potto*', *Biol. Gabon.* 2, 355-9.

11 Machida, H. and Giacometti, L. (1967), 'The anatomical and histochemical properties of the skin of the external genitalia of the primates', *Folia primat.* 6, 48-69.

12 Manley, G.H., unpublished observations.

13 Müller-Velten, H. (1966), 'Über den Angstgeruch bei der Hausmaus', *Z. vergl. Physiol.* 52, 401-29.

14 Manley, G.H. (1967), 'Gestation periods in the Lorisidae', *Int. Zoo Yb.* 7, 80-1.

15 Montagna, W., Machida, H. and Perkins, E.M. (1966), 'The skin of primates: 33. The skin of the angwantibo (*Arctocebus calabarensis*)', *Am. J. Phys. Anthrop.* 25, 277-90.

16 Ralls, K. (1971), 'Mammalian scent marking', *Science* 171, 443-9.

17 Stümpke, H. (1962), *Bau und Leben der Rhinogradentia*, Stuttgart.

18 Seitz, E. (1969), 'Die Bedeutung geruchlicher Orientierung beim Plumplori *Nycticebus coucang* Boddaert 1785 (Prosimii, Lorisidae)', *Z.f. Tierpsychol.* 26, 73-103.

19 Cowgill, U.M. (1966), '*Perodicticus potto* and some insects', *J. Mammal.* 47, 156-7.

20 Bishop, A. (1964), 'Use of the hand in lower primates', in Buettner-Janusch, J. (ed.), *Evolutionary and Genetic Biology of Primates*, vol. 2, New York, 133-225.

21 Andrew, R.J. (1964), 'The displays of the primates', in Buettner-Janusch, J. (ed.), *Evolutionary and Genetic Biology of Primates*, vol. 2, New York, 227-309.

22 Jewell, P.A. and Oates, J.F. (1969), 'Ecological observations on the lorisoid primates of African lowland forest', *Zool. Afric.* 4, 231-48.

23 Charles-Dominique, P. (1968), 'Réproduction des lorisidés africains', in *Cycles génitaux saisonniers de mammifères sauvages*, 1. *Entretien de Chizé*, Paris, 1-9.

24 Charles-Dominique, P. (1971), 'Eco-éthologie des prosimiens du Gabon', *Biol. Gabon.* 7, 121-228.

25 Charles-Dominique, P. (1971), 'Sociologie chez les lémuriens', *Recherche* 2, 780-1.

26 Charles-Dominique, P. (1972), 'Ecologie et vie sociale de *Galago demidovii* (Fischer 1808; Prosimii)', *Z.f. Tierpsychol. Suppl.* 9, 7-41.

J. HARRINGTON

Olfactory communication in Lemur fulvus

Introduction

Communication by scent seems to be very important to most mammals. Except for aquatic mammals and the anthropoid primates, most mammals are macrosmatic, have specialised cutaneous scent glands, and deposit skin gland secretions, urine or faeces on objects or on other animals, often using specialised body movements. However, little is known about just what is communicated among mammals by scent. There is evidence that mammalian scents function as signals in alarm, territorial defence, breeding, orientation, agonistic behaviour, maternal behaviour, and the identification of individuals, sexes, social groups, and taxa. A great deal of new evidence on these points has appeared in the last five years. But our understanding of the communicative significance of most mammalian scents is still unclear, perhaps mainly because of the experimental difficulties of studying the responses of animals to olfactory signals.

Lemur fulvus, like other prosimians, is a representative mammal in the high development of its olfactory communication. Its sebaceous and apocrine sweat glands are numerous and highly developed, and are concentrated into specialised glandular fields.[1] *L. fulvus* is macrosmatic, and it frequently sniffs other animals and objects in the environment. It has conspicuous scent-marking behaviour. For these reasons *L. fulvus* is a suitable animal for a study of the important and little-understood subject of mammalian olfactory communication.

There is another reason for studying olfactory communication in *L. fulvus*, or in any other prosimian, and that is the particular importance of prosimian behaviour for the understanding of the evolution of behaviour in the primates. Prosimians preserve a number of primitive behavioural as well as anatomical characteristics. Furthermore, in the genera *Lemur* and *Propithecus*, primitive characteristics such as strictly seasonal breeding and well-developed olfactory communication coexist with some characteristics apparently evolved in parallel with the anthropoid primates, persistent bisexual social groups, diurnality, and considerable development of visual and vocal communication. A consideration of the olfactory communication of prosimians in relation to their communication in other sensory modalities and to their general biology, therefore, may throw light on the parallel evolution of behaviour in the anthropoid primates.

The following is a report of a field and laboratory study of olfactory communication in *Lemur fulvus.*

Field observations

L.f. fulvus was observed for 300 hours between February and July 1969 at the Forest of Ankarafantsika in northwestern Madagascar. Most of the observations were made under good conditions (2-50 m. away in deciduous forest) on two groups in which all individuals were known. Both groups had twelve members; their composition is given in Table 1. The breeding season occurred during the course of observations, but it was not marked by an increase in aggression such as occurs in *L. catta.*[2] Most intra-group interactions throughout the period of observations consisted of sitting peacefully in close contact with other animals and grooming each other. The two groups behaved territorially toward each other.

Special note of behaviour patterns which seemed to involve

Table 1 Composition of the *Lemur fulvus fulvus* groups observed

	Males	*Females*
	LE Group	
Adult	Russ (RU)	Yellow Eyes (YE)
	Yellow Eyed Male (YEM)	Pseudo Yellow Eyes (PYE)
	Light Eyes (LE)	Bear (BE)
	Rat (RA)	
Subadult	Young Male (YM)	Russela (RLA)
Juvenile	Patch (PA)	Light Face (LF)
	Mouse (MO)	
	PF Group	
Adult	Light Faced Male (LFM)	Left Slit (LS)
	Ear Gouge (EG)	Foxy (FO)
	Orange Eyes (OE)	Yellow Eyed Female (YEF)
	Burly (BU)	Mary (MA)
Subadult		Pale Face (PF)
Juvenile	Junior (JR)	Black Faced Female (BFF)
		Cleo (CL)

olfactory communication was taken. These consisted mainly of anogenital sniffing and three behaviour patterns which may function as scent-marking: marking with the anogenital region, with the forehead, and with the palms of the hands, although anogenital marking is the only one of the three for which there is direct evidence that an olfactory signal is deposited. Animals sometimes sniff places where others have anogenital marked and may then mark there themselves. Furthermore, both sexes have a prominent patch of naked perianal skin in which apocrine and sebaceous glands are concentrated,[1] and, in addition, males have a copious strong-smelling secretion, easily visible in the field, on the bottom of the scrotum. Urine and faeces are other possible sources of anogenital scent. Places which have been only head-marked or hand-marked are not sniffed by other animals. However, hand and head marking are classified provisionally as scent-marking because they are conspicuous rubbing movements which occur at the same time and in the same situations as anogenital marking, and there are some glands on nearly every part of the body surface.

Nearly all scent-marking occurred in one of four fairly well-defined situations: sexual behaviour, alarm response to a man on the

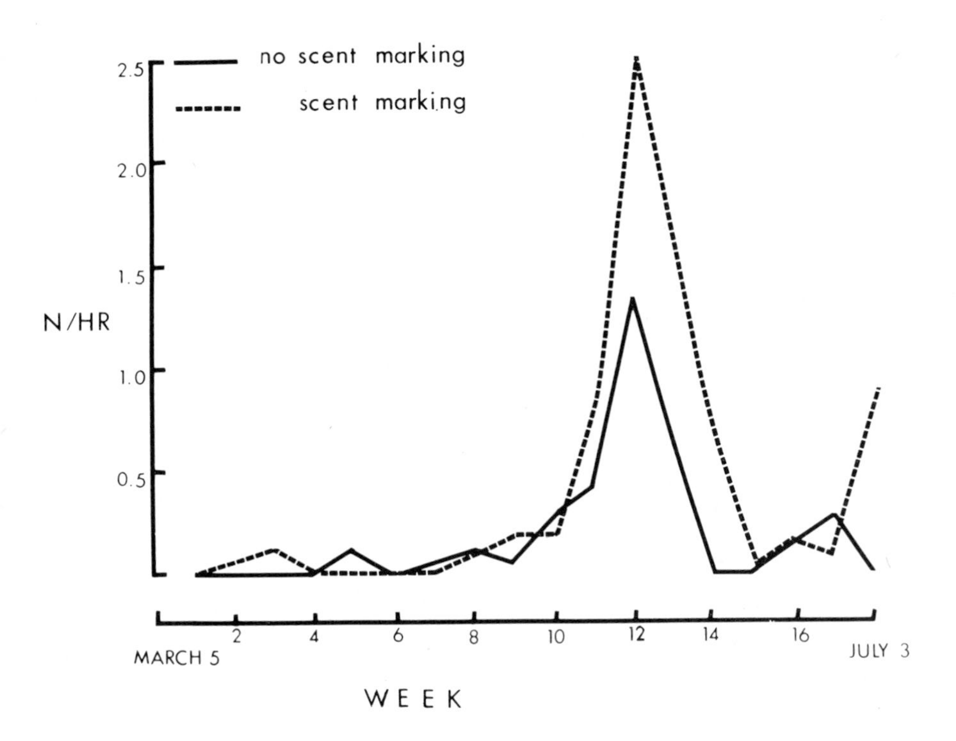

Figure 1. Male sniffs female anogenital scent (LE group).
Broken line: sniffing followed by scent-marking by male.
Solid line: sniffing not followed by scent-marking by male.

Table 2 Scent-marking by different individuals in various situations, for groups LE and PF

	Sex	*Alarm*	*Moving*	*Territorial*	*Play*	*Total*
			Anogenital marking			
Male	59	23	82	2	–	166
Female	18	100	85	6	–	209
Juvenile	–	–	–	–	1	1
Unident.	–	3	18	5	1	27
Total	77	126	185	13	2	403
			Head-marking			
Male	35	6	30	2	–	73
Female	–	1	–	–	–	1
Juvenile	–	–	2	–	–	2
Unident.	–	–	–	–	–	–
Total	35	7	32	2	–	76
			Hand-marking			
Male	10	–	6	–	–	16
Female	–	–	–	–	–	–
Juvenile	–	–	1	–	–	1
Unident.	–	–	–	–	–	–
Total	10	–	7	–	–	17
			Total scent-marking			
Male	104	29	118	4	–	255
Female	18	101	85	6	–	210
Juvenile	–	–	3	–	1	4
Unident.	–	3	18	5	1	27
Total	122	133	224	15	2	496

ground, territorial defence, and undisturbed moving around (Table 2). Most anogenital sniffing was accounted for by one situation – sexual behaviour (Table 3).

As the breeding season approached, males began to approach females and to try to clasp them around the waist, groom them, and sniff their anogenital regions. Usually the female would go away, stopping at several places as she went to anogenital mark or just to sit down. The male would follow and stop to sniff and scent-mark at the same places where the female had anogenital marked or sat. Before mounting, the male sniffed the female's anogenital region intensively and scent-marked the female and nearby objects. Fig. 1 shows that the frequency of sniffing of female anogenital scent by males, and of subsequent scent-marking by males, peaked during week 12 of the study. This was also the week in which other indicators of sexual activity reached a peak.

When the group was alarmed by a man on the ground, they moved

Table 3 Sniffing of anogenital scent in groups LE and PF

1. Sniffs anogenital scent, without scent-marking

	Sniffed				
Sniffers	*Male*	*Female*	*Juvenile*	*Unident.*	*Total*
Male	–	49	4	–	53
Female	11	1	1	–	13
Juvenile	5	2	1	–	8
Unident.	–	–	2	–	2
Total	16	52	8	–	76

2. Sniffs anogenital scent, then scent-marks

	Sniffed				
Sniffers	*Male*	*Female*	*Juvenile*	*Unident.*	*Total*
Male	7	100	–	1	108
Female	5	4	–	–	9
Juvenile	–	1	–	–	1
Unident.	–	–	–	–	–
Total	12	105	–	1	118

around quickly in the trees, swung their tails back and forth, gave loud grunting and shrieking vocalisations, and scent-marked. Sometimes the group mobbed the observer as well, the nearest animal approaching as close as 2 m.

The behaviour of the groups toward each other during territorial encounters was similar to the behaviour of a group toward a man on the ground. The two groups approached each other to within a few metres and gave tail swings, loud grunts and shrieks, and scent-markings.

Finally, many scent-markings were given when the group was not obviously alarmed at anything, but was just moving around in the trees, often just before beginning a long-distance progression.

These observations suggest some things that *might* be communicated by scent in *L. fulvus*. Scents might function for orientation, as alarm signals, as territorial markers, as primer or releaser pheromones in breeding, or for identification of individuals, sexes, social groups, or taxa. However, none of these communicative functions can be established from field observations alone. Field studies of olfactory communication must of necessity rely on observations of sniffing and scent-marking behaviour, which usually give only a suggestion of what is being communicated. Over-interpretation of such observations has resulted in concepts such as that of territorial scent-marking, critically reviewed by Schenkel,[3] for which definitive evidence is lacking. In most cases an experimental approach seems necessary to determine what is communicated by an olfactory signal.

Experiments

In these experiments an attempt has been made to determine whether the scent of *L. fulvus* conveys certain identifying information to other conspecifics; specifically, whether an animal's scent identifies it as an individual, as a male or a female, and as a member of a particular subspecies of *L. fulvus.*

The basic method in these experiments was: (1) to collect scents from animals caged individually, by leaving gauze pads in their cages overnight; (2) to present these pads to other animals caged individually; and (3) to record the responses of these animals to the scented pads. The method of collecting scent did not allow much control over the anatomical source of the scents. Presumably, the pads picked up a mixture of urine, faeces, and various skin-gland secretions. The animals were captives kept at Duke University, North Carolina. Most of them had been born in the United States.

1. *Identification of individuals*

To determine whether scents of two individuals of the same sex and subspecies were discriminated by another individual, the following method was used: a scent receiver was presented with a series of pads, one pad after another, all scented by the same animal, until the scent receiver became habituated to that animal's scent. Then the scent of a second individual was presented. It was anticipated that, if a scent receiver could discriminate between the two scent donors, its response would increase when it was presented with the scent of the second donor after habituation to the scent of the first.

Fig. 2 shows the results of six series of trials in each of which ♂ Oedipus was habituated to the scent of ♀ Sal, then presented with the scent of ♀ Frodo for ten more trials; and six control series in which ♂ Oedipus was habituated to the scent of ♀ Sal, then presented with the scent of ♀ Sal again for ten more trials. The criterion of habituation was five successive trials in each of which the pad was sniffed for 10 sec. or less. The figure shows that there is more sniffing in the ten post-habituation trials when a new scent is presented than when the same scent as before is presented ($p < .05$; 1-tailed Wilcoxon t-test). Table 4 gives the results of all the experiments, representing eleven different combinations of scent receiver and pair of scent donors. In all eleven, there is more sniffing in the post-habituation trials when a new scent is presented than when the same scent as before is presented. In eight of the eleven, the difference is significant at the .05 level. Fig. 3 shows the averaged results of all these experiments. The conclusion is that scents of *L. fulvus* of the same sex and subspecies can be discriminated by other *L. fulvus.*

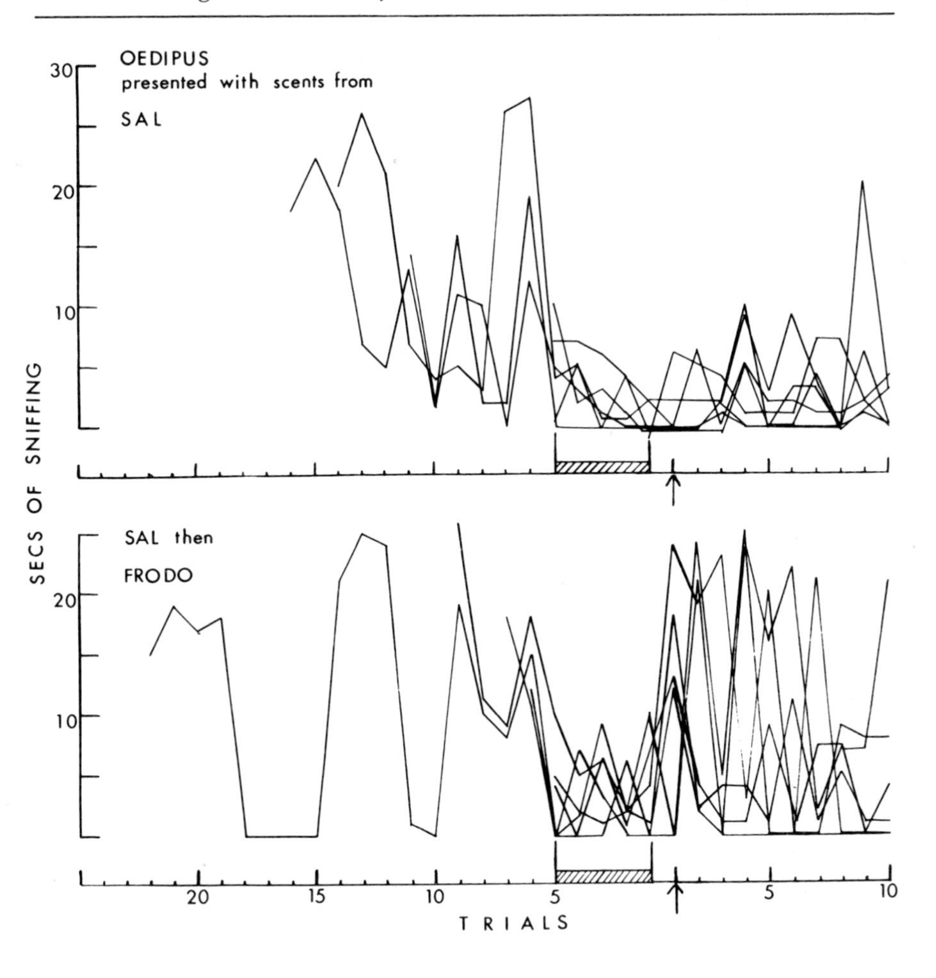

Figure 2. Discrimination of scents of ♀ Sal and ♀ Frodo by ♂ Oedipus. Upper graph shows 6 control series in which ♂ Oedipus was habituated to scent of ♀ Sal, then presented with scent of ♀ Sal for 10 more trials. Lower graph shows 6 experimental series in which ♂ Oedipus was habituated to scent of ♀ Sal, then presented with scent of ♀ Frodo for 10 more trials. Shaded area indicates 5 trials during which the criterion of habituation was reached. Arrow indicates first trial after habituation. Abscissa is numbered from right to left, starting to the left of the arrow, because different series took different numbers of trials to reach habituation.

2. *Identification of sexes*

To determine whether an animal's scent identifies it by sex, two males caged separately were presented with scents of a number of other individuals of both sexes, one scent after another. In addition, one of these males was presented with the scents of the same animals

Table 4 Discrimination of individuals: mean differences in amount of post-habituation sniffing between experimental and control series

Scent receiver	*Pair of scent donors*	*Difference (exptl-cont) (sec)*
Juvenal ♂ r	Thurber ♂ r Kingman ♂ r	*31.7
	Kingman ♂ r Fairchild ♂ r	*53.0
Oedipus ♂ f	Juvenal ♂ r Fairchild ♂ r	46.9
	Hermione ♀ r Calo ♀ r	*52.5
	Sal ♀ f Frodo ♀ f	*46.7
Calo ♀ r	Kingman ♂ r Thurber ♂ r	*69.0
	Kingman ♂ r Fairchild ♂ r	*77.5
	Sal ♀ f Frodo ♀ f	*35.5
Sal ♀ f	Juvenal ♂ r Fairchild ♂ r	22.7
	Hermione ♀ r Calo ♀ r	*82.0
Frodo ♀ f	Hermione ♀ r Calo ♀ r	31.2

N.B. Scent receiver was habituated to scent of donor named first.
* $p < .05$, 1-tailed Wilcoxon test.
r = *L.f. rufus*
f = *L.f. fulvus*

in pairs consisting of a male scent and a female scent of the same subspecies, presented at the same time. Both males were presented with each scent or pair of scents a total of ten times, spread over several days. The experiment was done in March, outside the usual breeding season in the northern hemisphere.

Fig. 4 shows the results for scents presented one at a time. Both males sniffed male scents significantly more than female scents ($p < .05$, Wilcoxon t-test). When the scents were presented in pairs to ♂ Juvenal, the male scent was sniffed more than the female scent in all six pairs (Table 5), and in three of these the difference was significant ($p < .02$, randomisation test). Only ♂ Oedipus showed any appreciable amount of marking in response to the scents; male and female scents did not elicit significantly different amounts of

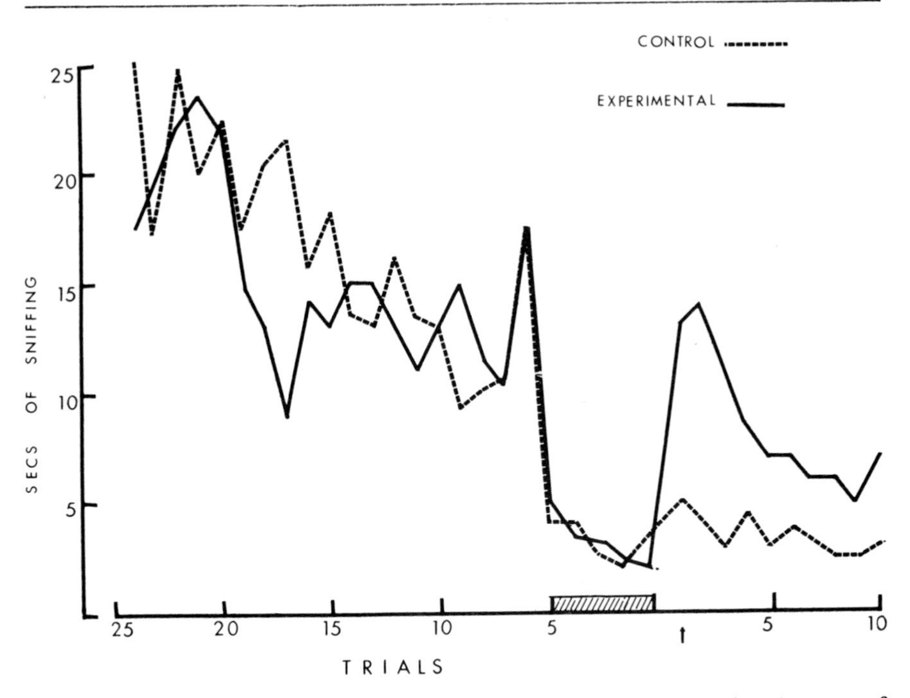

Figure 3. Discrimination of individuals: averaged results of all series. Average of 11 different pairs of graphs like those in Figure 2, representing 11 different combinations of scent receiver and pair of scent donors. Total of 71 experimental series and 71 control series.

marking. The conclusion is that the scent of an adult *L. fulvus* identifies it by sex to conspecifics. This finding was not surprising because, to the human nose, adult male *L. fulvus* have a strong smell, originating from the scrotal secretion, which is very different from the smell of a female.

3. *Identification of subspecies*

Neither of the preceding experiments showed any effect of subspecies on the response to scents. In the experiment on individual identification, scents of pairs of animals of the same sex and subspecies were discriminated by other animals of the same or different subspecies. Furthermore, there were no obvious differences in response to scents of different subspecies. This experiment involved two subspecies, *L.f. fulvus* and *L.f. rufus.*

In the experiment on identification of sexes, male scents were sniffed on the average more than female scents, regardless of subspecies. The marking responses of ♂ Oedipus in this experiment also did not show any effect of subspecies. This experiment involved three subspecies, *L.f. fulvus, L.f. rufus,* and *L.f. albifrons.*

An experiment was done to test further whether there was any

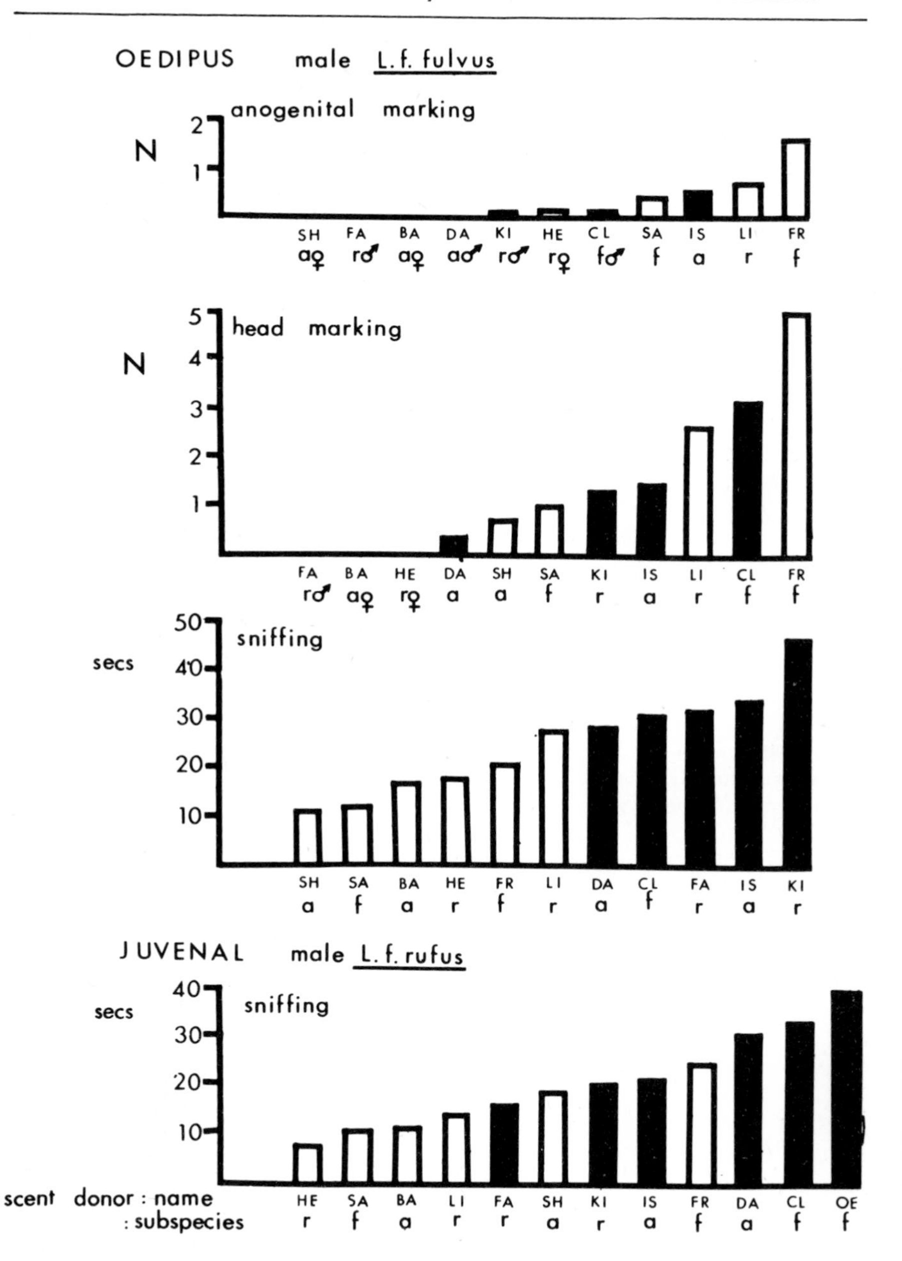

Figure 4. Discrimination of sexes: responses of ♂ Oedipus and ♂ Juvenal to scents presented consecutively. Dark bars = male scents. Abbreviations: HE = ♀ Hermione; SA = ♀ Sal; BA = ♀ Bathsheba; LI = ♀ Lisa; FA = ♂ Fairchild; SH = ♀ Shulamite; KI = ♂ Kingman; IS = ♂ Ishmael; FR = ♀ Frodo; DA = ♂ David; CL = ♂ Clarence; OE = ♂ Oedipus.

f = *L.f. fulvus*
r = *L.f. rufus*
a = *L.f. albifrons*

Table 5 Mean difference in amount of sniffing of paired male and female scents by ♂ Juvenal

Scent donors	*Difference (♂-♀) (sec)*
Oedipus ♂ f Sal ♀ f	3.0
Ishmael ♂ a Shulamite ♀ a	*22.0
Fairchild ♂ r Hermione ♀ r	0.2
David ♂ a Bathsheba ♀ a	*22.4
Kingman ♂ r Lisa ♀ r	6.9
Clarence ♂ f Frodo ♀ f	*22.1

* $p < .02$, 2-tailed randomization test for matched pairs.
f = *L.f. fulvus*
r = *L.f. rufus*
a = *L.f. albifrons*

effect of subspecies on the response to the scents. Three males were presented with paired scents of other animals. The two scents in each pair were from members of the same sex. One scent was from an animal of the same subspecies as the scent receiver, and the other scent was from an animal of a different subspecies. Each pair of scents was presented to each male twenty times (except for one pair which was presented to one male only fifteen times, because the animal swallowed a gauze pad on the fifteenth trial). Three subspecies were involved: *L.f. fulvus, L.f. rufus*, and *L.f. albifrons*. The object was to see if the animals responded differently to scents of their own and of other subspecies.

Table 6 shows the results. There were ten different combinations of scent receiver and pair of scent donors. In seven of these, one member of the pair elicited significantly more responses than the other, either sniffing, anogenital marking, or head marking ($p < .05$, signed ranks test). In three of these, the scent receiver responded more to the scent of his own subspecies and in the other four he responded more to the scent of the other subspecies. Thus the receiver may show a significant preference for the scent of one of a pair of animals of the same sex and different subspecies, but the direction of the preference is not determined by subspecies. This experiment, therefore, using three subspecies of *L. fulvus*, does not provide any evidence that the subspecies are identified by their scent.

Table 6 Mean difference in amount of sniffing, anogenital marking, and head-marking given to paired scents of own and other subspecies

		Differences (own/other subspecies)		
Scent receiver	*Scent donors*	*Sniff (sec)*	*Anogenital mark (N)*	*Head-mark (N)*
Juvenal ♂ r	Kingman ♂ r David ♂ a	*-12.4	0.1	-2.5
	Fairchild ♂ r Clarence ♂ f	11.8	0.6	-1.0
	Hermione ♀ r Sal ♀ f	*-7.2	-0.8	-1.5
	Red Child ♀ r Bathsheba ♀ a	4.9	*0.8	6.0
Oedipus ♂ f	Clarence ♂ f Fairchild ♂ r	*-21.5	0.5	3.5
	Sal ♀ f Hermione ♀ r	*-8.8	*-1.1	-1.6
	Frodo ♀ f Shulamite ♀ a	1.3	*1.5	1.4
Ishmael ♂ a	David ♂ a Kingman ♂ r	*28.0		
	Shulamite ♀ a Frodo ♀ f	-2.3		
	Bathsheba ♀ a Red Child ♀ r	-1.9		

* $p < .05$, 2-tailed signed ranks test.
Blank: no response.
f = *L.f. fulvus*
r = *L.f. rufus*
a = *L.f. albifrons*

Discussion

In summary, the scent of *L. fulvus* identifies it to other conspecifics by sex and individual identity. No evidence was found that the scent of *L. fulvus* identifies it by subspecies, among the subspecies *L.f. fulvus*, *L.f. rufus* and *L.f. albifrons*. Since *L. fulvus*, like *L. catta* and *Propithecus verreauxi*, live in social groups containing several individuals of both sexes, sexually and individually identifying information could be useful to it in a variety of social interactions. The significance of the apparent lack of discrimination among scents of the three subspecies is not clear, because of the uncertain condition of the taxonomy of the *L. fulvus* group, and the lack of knowledge of the geographical distribution of these three subspecies.

The subspecies of *L. fulvus* are distinguished from each other mainly by pelage, and they seem to be very similar in their morphology and behaviour.[4] Scent, then, seems to be another characteristic in which these forms do not differ from each other very much.

The social lemurs have highly developed olfactory communication, but they have considerable development of visual communication as well. They lack the elaborate facial communication of the monkeys and apes, but considerable increase in facial mobility has occurred as compared with the insectivores and the non-social prosimians.[5] There are also several conspicuous and stereotyped body movements, such as tail-swinging, cuffing, head jerking, and scent-marking, which may function as visual signals. The bright colours of these animals also imply some development of visual communication.

The social lemurs, then, are mostly 'olfactory' animals which have gone part of the way toward becoming 'visual' animals. Since the same transition has occurred during the evolution of the monkeys and apes, a consideration of the relationship between olfactory and visual signals in lemurs may provide suggestions as to how this transition might have occurred during the evolution of the anthropoid primates.

Both of the kinds of identifying information that were shown experimentally to be conveyed by the scent of *L. fulvus* are also conveyed quite conspicuously by the visual appearance of the animals. The sexes differ from each other in pelage, very strikingly in some subspecies; and the individual differences in pelage and other characteristics were sufficient for the observer to be able to identify easily nearly all of the individual animals studied in the field. The conveying of the same information by visual and olfactory means may be related to the activity cycles of the social lemurs. They are predominantly diurnal, but are also quite often active at night.

The production of olfactory signals often involves conspicuous body movements, i.e. scent-marking, which may themselves function as visual signals. *L. fulvus* obviously respond to anogenital marking as a visual signal when they go over to an animal which is anogenital marking from up to 5 m. away, and sniff and mark where the other animal has marked. Jolly also remarked on the large visual component in the stink fights and territorial battles of *L. catta.*[2] In these cases, scent-marking functions as both a visual and an olfactory signal. It is possible that scent-marking movements could be further selected for their signal function and evolve as visual signals on their own, without the production of an olfactory signal, as Andrew[5] suggested for presenting in monkeys and apes and Wickler[6,7] suggested for penis displays. If this is so, the scent-marking of lemurs may show how this process may have begun in the ancestors of the anthropoid primates.

A consideration of the general behaviour and ecology of the living prosimians may also suggest some reasons for the general decline in importance of olfaction in the anthropoid primates. The arboreal theory of primate evolution attempts to explain this decline as a result of an arboreal way of life.[8,9] However, prosimians and several other mammalian groups combine arboreality with highly developed olfaction, so arboreality alone does not necessarily result in the regression of olfaction.[10]

The prosimians, however, do have certain behavioural and ecological characteristics by which they differ from at least some of the anthropoids and whose loss in the anthropoids may have been associated with decreased selection for olfactory communication:

1. *Small home-ranges.* The prosimian data are for Lemuriformes, as well as *Galago* spp., which generally have group ranges of 10 hectares or less.[2,11,12] Some monkeys and apes, particularly the more terrestrial forms, have home-ranges one or two orders of magnitude greater. Larger ranges should decrease the usefulness of scent as a means of communication, because there must be a limit to the size of the area a group can effectively mark.
2. *Nocturnality.* Most prosimians are mainly or wholly nocturnal, and even the members of the diurnal genera *Lemur* and *Propithecus* are often active at night. This contrasts with the quite strict diurnality of nearly all monkeys and apes. Scent has obvious advantages as a means of communication for a wholly or partly nocturnal animal.
3. *Seasonal breeding.* The strict seasonal breeding of many prosimians contrasts with the less seasonal breeding of many anthropoid primates. A great deal of olfactory communication in mammals seems to be concerned with the timing of breeding, and therefore a loss of strictly seasonal breeding may be associated with decreased selection for olfactory communication.
4. *Relatively simple agonistic behaviour.* Most of the evidence on this point comes from *L. catta*, the only prosimian for which there are many field observations of agonistic behaviour. Jolly pointed out that *L. catta*, and perhaps the other social Lemuriformes, have a different system for controlling aggression from such primates as baboons and macaques.[2] Agonistic interactions in *L. catta* are almost entirely confined to the short breeding season, are often violent, usually involve only two animals, and involve much olfactory communication. In such primates as chimpanzees, macaques, and baboons, a more complex system is found, in which many agonistic interactions occur in non-sexual situations, elaborate signalling mechanisms have been developed to reduce aggression, more than two animals are often involved in interactions, and in which most signals are visual and acoustic.[13] It seems likely that olfactory

signals, whose coding possibilities in fast social interactions are more limited than those of visual and acoustic signals, are more appropriate for the relatively simple, all-or-none agonistic behaviour of *L. catta* than for the more complex agonistic behaviour of some monkeys and apes. This might also help explain 'why the highly social *Lemur* have not proceeded further with the evolution of facial expression',[5] since a great deal of the facial communication of anthropoid primates is concerned with subtle agonistic interactions.

5. *Oral grooming.* Mutual grooming in prosimians, as in the monkeys and apes, is an important kind of social interaction which seems to function to maintain social bonds. In addition, however, prosimian grooming may be a means of receiving or sending chemical signals, since all lemuriform and lorisiform and some tupaiiform prosimians groom with the tongue and the teeth.[14] The partial replacement of oral by manual grooming in the monkeys and apes may have eliminated an important means of exchanging chemical signals.

Acknowledgments

This paper is based on a PhD dissertation submitted to the Department of Zoology, Duke University. The author wishes to thank the following people for their help during the course of this research: P. Klopfer, J. Buettner-Janusch, J. Bergeron, N. Budnitz, R. Sussman, C. White, and L. Harrington. P. Marler and T. Struhsaker read and commented on the manuscript of this paper. The field work in Madagascar was made possible by the generous co-operation of la Direction des Eaux et Fôrets and le Ministre de l'Agriculture, Gouvernement Malgache; especially MM. Ramanantsoavina, Ramalanjoana, Andriamampianina, Razanajatovo, and Natai. The author also wishes to thank R. Albignac, G. Randrianasolo, and J.-A. Randrianarivelo for valuable advice and help in Madagascar.

Financial support was provided by NIH Grants HD02319 and RR00388, a Sigma Xi grant, a NASA Predoctoral Traineeship, and an NSF Predoctoral Fellowship.

NOTES

1 Montagna, W. (1962), 'The skin of lemurs', *Ann. N.Y. Acad. Sci.* 102, 190-209.

2 Jolly, A. (1966), *Lemur Behavior: A Madagascar Field Study*, Chicago.

3 Schenkel, R. (1966), 'Zum Problem der Territorialität und des Markierens bei Säugern – am Beispiel des Schwarzen Nashorns und des Löwens', *Z.f. Tierpsychol.* 23, 593-626.

4 Nute, P.E. and Buettner-Janusch, J. (1969), 'Genetics of polymorphic transferrins in the genus *Lemur*', *Folia primat.* 10, 181-94.

5 Andrew, R.J. (1964), 'The displays of the primates', in Buettner-Janusch, J. (ed.), *Evolutionary and Genetic Biology of Primates*, vol. 2, New York.

6 Wickler, W. (1966), 'Ursprung und biologische Deutung des Genitalpräsentierens männlicher Primaten', *Z.f. Tierpsychol.* 23, 422-37.

7 Wickler, W. (1967), 'Socio-sexual signals and their intra-specific imitation among primates', in Morris, D. (ed.), *Primate Ethology*, Chicago.

8 Smith, G.E. (1927), *The Evolution of Man*, 2nd ed., London.

9 Jones, F.W. (1926), *Arboreal Man*, London.

10 Cartmill, M. (1970), The Orbits of Arboreal Mammals: a Reassessment of the Arboreal Theory of Primate Evolution, Unpublished PhD Dissertation, University of Chicago.

11 Petter, J.-J. (1962), 'Recherches sur l'écologie et l'éthologie des Lémuriens malgaches', *Mém. Mus. nat. Hist. nat. Paris.*, Sér. A., 27, (1), 1-146.

12 Bearder, S.K. and Doyle, G.A., this volume.

13 Hamburg, D.A. (1971), 'Psychobiological studies of aggressive behaviour', *Nature* 230, 19-23.

14 Simons, E.L. (1962), 'Fossil evidence relating to the early evolution of primate behavior', *Ann. N.Y. Acad. Sci.* 102, 282-94.

A. SCHILLING

A study of marking behaviour in Lemur catta

Introduction

It is generally considered that three forms of marking are practised by the ringtailed lemur[1] (*Lemur catta*), a social prosimian inhabiting the south of Madagascar:

1. Genital marking, performed by both sexes, consisting of rubbing the genital organs on suitable supports and thereby depositing glandular secretions (Fig. 1*a*).
2. Brachial marking, performed only by the male, consisting of rubbing the inner surfaces of the forearms on suitable supports in a rhythmic movement, depositing secretions of the forearm and possibly the brachial gland (Fig. 1*b*).
3. Tail marking, performed by the male, involving rubbing the tail in such a way so as to impregnate it with scent from both arm glands (Fig. 1*c*).

In fact, the latter two forms may be considered analogous, in that the tail may become the object of brachial marking just like a branch. The structures used for marking, and the secretions deposited, are similar in both sexes. In the latter two types of marking behaviour, although there is an apparent difference in sequence and context, the components are generally common to both.

Marking may be considered as a rather stereotyped response to a complex of stimuli which are themselves modulated by ecological and ethological factors:

1. External physiological stimuli: *Lemur catta* possesses good vision and yet at the same time is a macrosmatic animal.
2. Internal physiological and psychological stimuli: phase of

(a)

Figure 1. The three different kinds of marking characteristic of *L. catta*. (*a*) genital marking; (*b*) brachial marking; (*c*) tail marking.

oestrous cycle, pregnancy or, in a more general sense, overall excitation level in both sexes.

3. Environmental factors: availability and nature of marking supports, climatic factors, etc.
4. Ethological factors: *Lemur catta* is one of the most social prosimians; marking behaviour could thus be involved in inter-individual and inter-group relationships.

This study was carried out in the gallery forest reserve at Berenty, Madagascar, in which *Lemur catta* is protected. Most of the time was spent following Troops I and II, previously described by Jolly and Sussman.[1,2] However, insufficient effective observations were made to consider this present work more than preliminary.

Morphological structures involved in marking behaviour

1. Genital marking

In males, the bare posterior surface of the scrotum is glandular in nature, a precise description of which has been given by Evans and Goy.[3] In females accessory glands are found on the labiae of the vulva, according to Jolly and Evans and Goy, or on the clitoris, according to Petter.[1,3,4]

(c)

2. *Brachial marking*

Histological studies of epidermal and glandular formations, which are developed or functional only in the males, have recently been carried out (Rumpler, Montagna and Yun, Kneeland).[5,6,7] On the upper part of the internal side of the arm, close to the shoulder, there is a multilobular sebaceous gland which reaches its maximum size (that of a walnut) in July (Petter).[4] On the antero-internal face of the forearm there is a bare cutaneous oval zone, covering a gland of similar size made up of sudoriporic tubules and particular interstitial cells. In females, there is a similar but smaller gland in the same place, in which adipocytes take the place of interstitial cells. In addition, but only in males, the overall zone is internally bordered by an epidermal horn spur. Andriamiandra and Rumpler[8] have shown that the sudoriporic antebrachial structures and sebaceous brachial structures are under testosterone control, a finding which permits the conclusion that they are secondary sexual characteristics.

Marking behaviour

1. *Genital marking* (Fig. 1*a*)

As has been previously noted by Jolly,[1] genital marking is identical in form in the two sexes. The area to be marked is selected, the animal turns about, balances on its forehands, and lifts itself to grab hold of the support above with the hind legs. The tail is curved forward, and the support is marked by the genital glands with an up and down movement, for 2-10 seconds. The trace thus left is strongly

(*a*) (*b*

scented, and produces a dark area several centimetres in length. According to Evans and Goy both brachial and genital secretion deposits remain attractive for several days.[3]

2. *Brachial marking of a branch* (Fig. 1*b*)

The animal, standing on its hind legs, grasps a suitable branch with one hand, and several centimetres above applies the oval area of the other arm. In a rapid movement, the forearm is pulled back towards the animal. This movement is then usually followed by the other arm, though there may occasionally be repetition, usually only once, on the same side. Up to 15 alternative movements by one male during one bout of marking have been observed; however, the average is about 4 movements. The behaviour has two results: first, the forearm spur cuts into the branch a scar in the form of a comma, the concavity of which indicates the forearm used. This scar is easily visible and about 2 centimetres in size (Figs. 2*a*, 2*b*). Secondly, the scar thus formed contains the scent left by the scraping of the antebrachial gland located near the spur. The secretion of the brachial gland is applied irregularly to the spur, with 'anointing behaviour' described by Jolly.[1] Indeed, by folding the forearms one over the other, the animal brings into contact the brachial gland and the antebrachial structures (the spur itself, or the oval zone). 'Anointing' may or may not be bilateral; it occurs only rarely (0 to 6 times per hour; only once per hour on the average), appearing before

Figure 2. Scars left on marked branches by the forearm spur of the male *L. catta*. (*a*) enlargement of the scar showing its comma form; (*b*) a typical aspect of a marked branch; (*c*) concentration of marked branches in the peripheral 'marking area' (the lengths of scars on the branches are underlined).

brachial marking, seldom before genital marking, or in isolation, without interrupting progression.

3. *Tail marking* (Fig. 1*c*)

This is brachial marking in which the support is the animal's own tail. 'The *Lemur* stands on his hindlegs, tail drawn forward between his legs, and up between his forearms. The spurs enter the fur deeply, and much of the tail must be in contact with the antebrachial gland.'[1]

4. *Associated marking*

About once in every 15 times, the two types of marking are performed successively on the same support, brachial marking almost always preceding genital marking.

Topography and study of marking on external supports

Marking, whether it be genital or brachial, occurs most frequently in the lower zones of a vertical stratification of the forest, on the ground, and on low branches (zones 1 and 2 of Sussman).[2] On the one hand, *Lemur catta* frequently descends to the ground (36% of the time according to Sussman);[2] on the other, it finds numerous fine tender-barked branches, more or less vertical, at low levels, which it can mark more easily than others. It is, therefore, at the lower levels of the gallery forest that most marking behaviour is observed, even though the animal traverses the higher branches, particularly those of the tamarind trees, equally often. It appears that the branches chosen for genital marking can be less vertical and thicker than those used in brachial marking. The few large branches marked were dealt with by genital marking. In fact, the actual nature of the support seems to be less decisive, as a *Lemur catta* has been seen to mark a thin weed with its genitalia, and on another occasion the pad of a cactus.

The type of support most often selected for brachial marking is thin and vertical, or at least upright (compatible with the posture of the animal). The thinness of the branches chosen (2.5 cm. average for 100 measured) may be due less to adaptation for movement of the support than to the nature of such branches: the bark is much more tender when the bush is young. This type of marking has sometimes been observed even on branches with a diameter of 0.5 cm. Fig. 3 gives the results of precise measurements taken on 100 branches marked in different zones of the study area.

It can be seen that marking is more frequent on the thinner branches (approx. 2 cm. diameter) but more extensive on branches of a large diameter (approx. 4 cm. diameter). The number of marks is extremely variable: from a few to several hundreds (3 times 450 marks, once 550). Although these marks become less easily visible upon drying, there remains the stain that forms, either all around or merely on one particular side of the branches, which is visible at several metres and which has an average length of 77 cm. (Fig. 2*b*). This is particularly striking in certain 'marking areas' (zones which are ecologically favourable to marking behaviour because of numerous vertical supports), where marking is frequent and usually occurs above the ground. In this case, the lemur begins marking from just above the ground (50 cm. on the average, never less than 10 cm.), then continues while climbing, or begins after having leaped down to a low position on the bush concerned. The average number of cuts made with the spur, in the course of 75 observed brachial markings, was 4. In the case of stains still visible, it may be inferred that certain branches must have been marked more than a hundred times. There were two important 'marking areas' frequently visited by troop I (home-range = 6 hectares: number of members = 15), which included 4 males, 2 of which were peripheral, 5 females, 2 juveniles, and 4 infants.

The first area (Fig. 2*c*), in the graduated zone which borders the territory of the troop to the north, was a rectangle of about 400 square m., limited by paths and, in the west, by a denser zone of vegetation. It was planted with a variety of fruit trees, and many bushes, rarely exceeding a height of 3 metres. Practically all of these

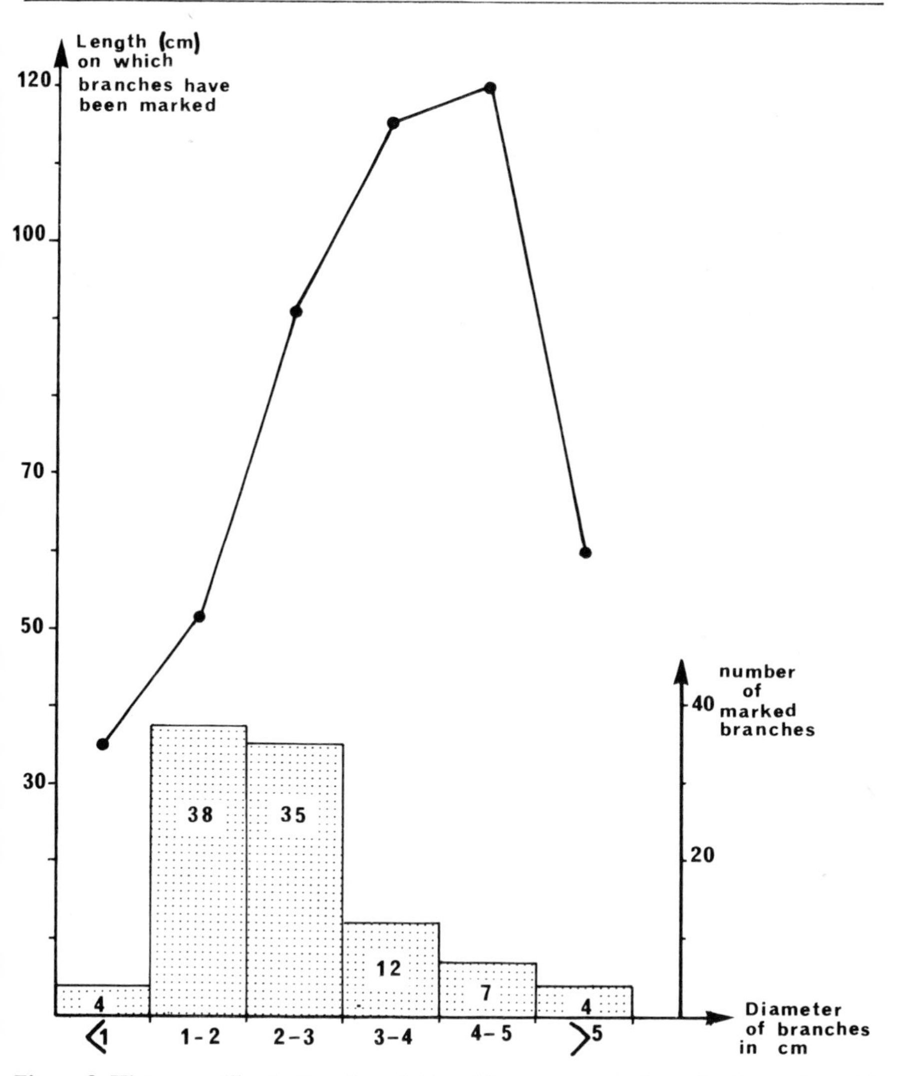

Figure 3. Histogram illustrating the relationship between the length of scars (graph) and the typical diameter of the branches marked (histogram).

trees were marked. The total length of the main stains represents an extent of 33 metres; and as the average density of these is estimated to be around 150 per metre of branch, more than 5000 of them could have been counted on this meagre surface.

The second 'marking area' (see Fig. 4) was located at the centre of the troop territory bordered by the dense gallery forest and a zone planted with cactus. It was an area of sparse undergrowth beneath the cover of large trees, especially tamarinds, where marking also occurred on the ground on low and isolated shrub trees. This zone, which did not need to be defended, was visited daily by the troop, which made incursions into the neighbouring cactus field to feed,

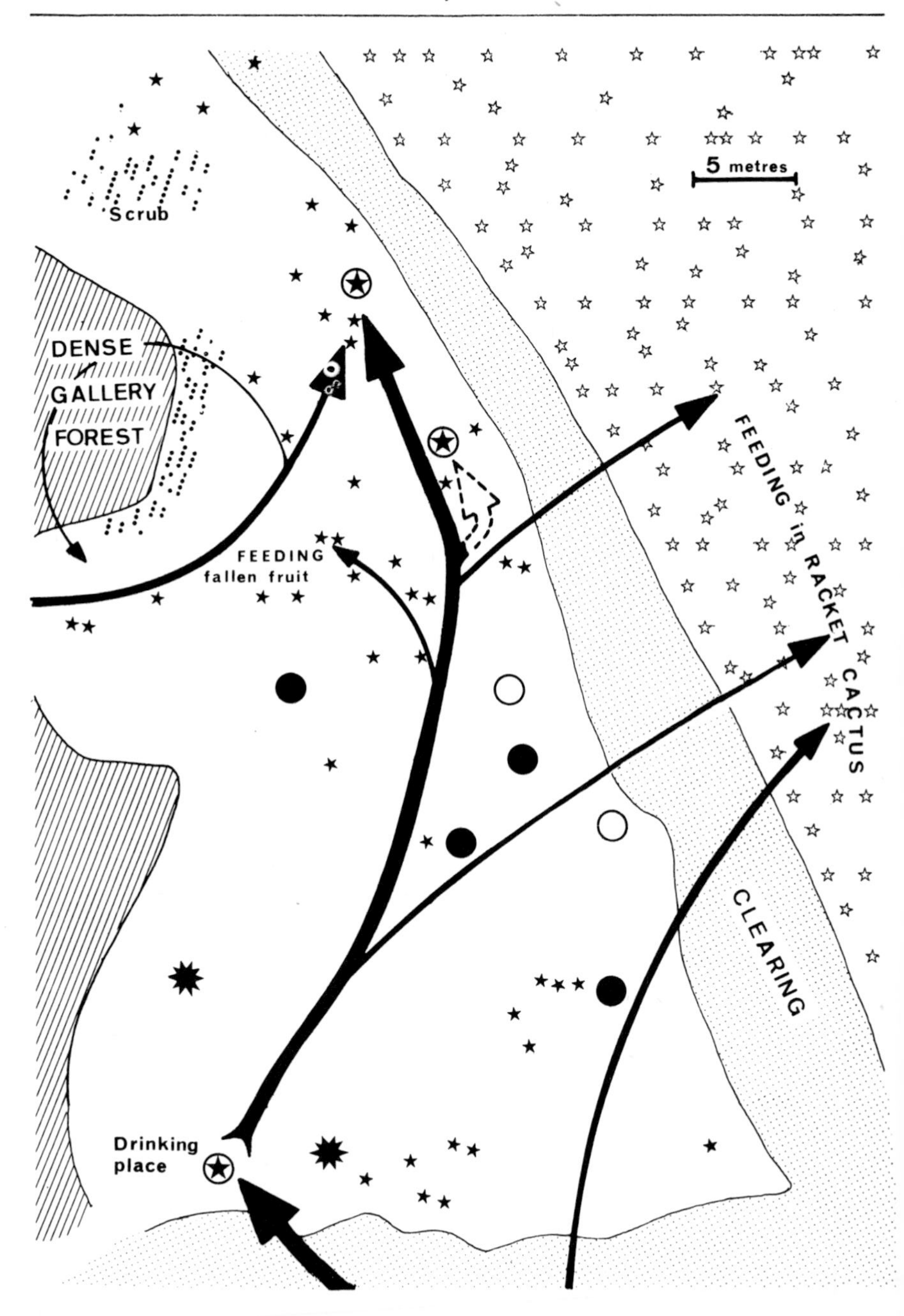

Figure 4. Map of the central 'marking area' of troop I in relation to its daily progression.
small black star = location of marked branches
open circle = tamarind trees
black circle = other large trees
star in circle = used for resting tamarind trees
large asterisk = dead trunk

and utilised certain tamarinds as resting trees for the morning siesta. Very often, at the end of the morning, the lemurs descended from the large trees, or went directly from the treeless boundary area, to play or eat fruit, fallen from the tamarinds, as well as to mark the bushes in this zone. It was not possible to observe a particular orientation of the marks or even a concentration in a precise area, except perhaps in the proximity of a tree in which the trunk contained a cavity filled with water where the troop came regularly to drink, or in an area where the animals found many fallen fruit under one of the resting tamarinds.

Associated behaviour patterns

1. Olfaction

It is evident that marking is associated with olfaction. As in all lemurs, *Lemur catta* possesses well-developed olfactory organs; but olfactory behaviour is far from being systematically associated with that of marking. For example, during a period of two-and-a-half hours of active and continuous marking by a male of troop I, the following was observed: 33 brachial markings preceded by smelling in one case; 60 genital markings preceded by smelling in 22 cases; 2 long smelling episodes without marking. It is probable that the animal is informed of the nature of the mark left by itself or by a conspecific, without stopping its progression or showing any behaviour detectable by the observer. In certain cases, nevertheless, the role of olfaction is evident:

(*a*) Verification of a former brachial mark before marking, either by a male or by a female.
(*b*) Verification of a mark left by a female before 'over-marking' by a male.
(*c*) Verification of marks left by the same animal, sometimes with re-marking.
(*d*) Processing with the nose on branches during a 'marking session' (see below).

2. Other behaviour patterns associated with marking of a support

(*a*) Licking, biting or even scratching of the bark of a branch before brachial marking.
(*b*) One case of faecal ingestion was followed by genital marking.

3. *Behaviour pattern not observed*

Marking of one individual by another, as has been seen in other lemurs, was not observed with *Lemur catta.*

Instances of marking

In a troop of *Lemur catta*, marking occurs either occasionally during progression, when the animal encounters a suitable branch and is stimulated, or more systematically during rest periods. In the latter case, the animal leaves the troop for a 'marking session' lasting two to twenty minutes, with frequent smelling and 'anointing', during which the neighbouring branches are actively marked, before coming back to rest with the troop. Fig. 5 gives an idea of the variability in the frequency of marking by a dominant male (see also Table 1).

Ethological context

1. *Age*

Infants and juveniles of less than one year of age do not mark.

2. *Genital marking by females*

Dominant females, or females carrying young, mark more frequently than other adult females.

3. *Marking and dominance in the males*

It was interesting to compare the frequency of marking by the dominant male of troop I with that of one of the two subordinate males, which were excluded from the troop, but which followed it at a variable distance (from fairly close to 200 metres away). Unfortunately, over several days of observations, only 10 hours of active marking were observed. For this reason the results summarised in the following table (Table 1) are only tentative.

4. *Significance of tail-marking*

As has been observed by Jolly,[1] tail-marking contains a definite aggressive element. All observations of the present study, like those of Sussman,[2] confirm the fact that this type of marking occurs only in an agonistic situation. This may be a simple confrontation between two animals, a fight between two troops, or a conflict situation with other species.

Marking generally takes place between two males, but it can also occur between a male and a female. Marking of the tail is a behaviour

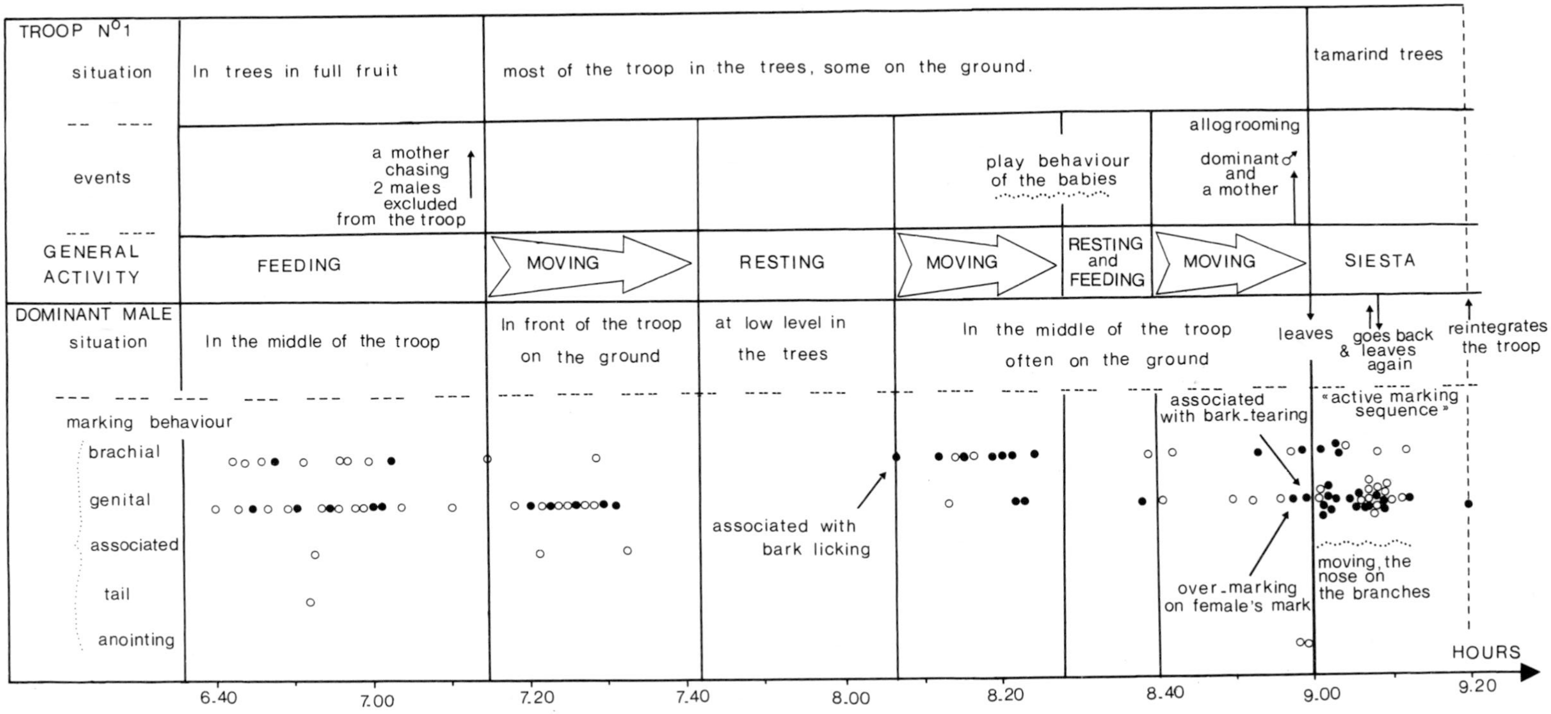

TROOP N°1
situation
In trees in full fruit
most of the troop in the trees, some on the ground.
tamarind trees
events
a mother chasing 2 males excluded from the troop
play behaviour of the babies
allogrooming
dominant ♂ and a mother
GENERAL ACTIVITY
FEEDING
MOVING
RESTING
MOVING
RESTING and FEEDING
MOVING
SIESTA
DOMINANT MALE
situation
In the middle of the troop
In front of the troop on the ground
at low level in the trees
In the middle of the troop often on the ground
leaves
goes back & leaves again
reintegrates the troop
marking behaviour
brachial
genital
associated
tail
anointing
associated with bark licking
associated with bark-tearing
over-marking on female's mark
«active marking sequence»
moving, the nose on the branches
HOURS
6.40
7.00
7.20
7.40
8.00
8.20
8.40
9.00
9.20

Figure 5. Chart of the marking behaviour of a dominant male, observed over 2½ hours, in relation to the general activity of the rest of the troop.
○ = occurrence of marking behaviour
● = occurrence of marking behaviour after preliminary sniffing

Table 1 Comparison of the frequency of marking between a dominant male and a subordinate male of the same troop

		Marking behaviour				*Anointing behaviour*
		Genital	*Brachial*	*Associated*	*Tail*	
Dominant Male	Number of markings observed in 4½ hours	74	39	8	1	5
	Average no. of markings per hour	16.5	8.7	1.7	0.2	1.1
Subordinate Male	Number of markings observed in 5 hours	43	50	10	2	13
	Average no. of markings per hour	8.6	10.0	1.1	0.3	1.7

pattern which should be included in the range of threats which a *Lemur catta* male, in all probability, uses for assuring its dominance over a conspecific. Ralls[9] has recently provided evidence illustrating the role that marking plays in dominance relationships for numerous mammals. In *Lemur catta*, other forms of behaviour accompanying this threat marking are a sort of grunting, a special high frequency whistling occurring only in this situation, and a characteristic waving of the tail, which is curved in front of the animal and towards the adversary. If the basis of the behaviour is the individual olfactory message which would be made up of the secretions of the brachial and antebrachial glands, Jolly is correct in thinking that 'tail-waving' represents an effective means of dispersing these odours.[1] The waving occurs more rarely than marking of the tail, which is usually sufficient to cause the adversary to flee. It may be a complex agonistic behaviour pattern, olfactory in nature, in which the first sequence involves the second, but is generally sufficient without it.

Marking as an olfactory message

1. Genital secretions of the females

There is no definite proof that the female odour informs the *Lemur catta* male of the stage of the oestrous cycle. However, this hypothesis may be considered, since a high frequency of marking by females in oestrus was observed. Conversely, marking of a support by a male is very often provoked by that of a female. This is exhibited either by over-marking of the female's trace, or by marking close by, either with the genitalia or with the arm.

2. *Brachial secretions*

Since development of the glands of the forelimbs used in marking is a secondary sexual character, it is logical to suppose that brachial marking plays a role in individual recognition, perhaps with respect to the individual dominance hierarchy. The olfactory message could be transmitted directly from one individual to another in the case of tail-marking and tail-waving. It could also be transmitted by means of the marks left on the bark of trees. Naturally, the brachial mark cannot be the only basis for inter-individual relationships, since in the *Lemur catta* troop females are sometimes dominant. Besides, even a dominant male has been observed to retreat before a female (usually a mother) in spite of vigorous marking of the forearm and then of the tail.

3. *Territorial marking*

The role of marking in territorial defence has been demonstrated in numerous mammal species,[10,11] and in particular in certain prosimians.[12] Thiessen and Lindzey have recently demonstrated the hormonal dependency of such behaviour.[13] In *Lemur catta*, all observers underline the abundance of marking in territorial defence. The behaviour, however, does not prove that the result (the odour mark) has a determining role in this defence. What must nevertheless be pointed out is the extraordinary abundance of marks left at the 'marking area' located at the periphery of the territory of troop I, and very often the site of territorial disputes with troop II. A territorial encounter between two troops is associated with much marking behaviour, both sexes participating and using all forms of marking (see Fig. 6). It appears to be primarily a means of impressing the opposite troop with the vigour and persistence of the behavioural act itself.

4. *Marking and orientation*

The role of marking in orientation seems to be doubtful, and learning of very precise routes is probably of visual origin. A young male *Lemur catta*, separated from its troop, did not sniff more than normally, but frequently stopped in its progression and uttered a sort of distress mewing, listened attentively to the group's answer, and then guided itself by the sound. Nevertheless, this did not prevent it from marking, mainly with the genitalia, after having carefully sniffed some point on the route where the other members of the troop had previously marked.

Figure 6. Photograph of a 'stink fight' between two troops on the periphery of their territories (courtesy of R.W. Sussman).

Marking without a communicative role

In this case, the marking stimuli are less specific, the behaviour reflecting a general state of excitation of the animal. It is in this light that an attempt is made to explain the 'stink-fight'.[1] At a less intensive level, any state of excitation could be translated into a marking response, being either of endocrine or nervous origin. Marking may be due to a conflict situation, to fear, or it may signify 'friendship' (as Petter suggested it did among tame animals in the Tananarive zoo).[4] Marking behaviour is primarily linked to olfaction, but it must not be forgotten that close connections exist between olfactory structures and those responsible for affective and emotional responses (Papez circuit, hippocampus, amygdaloid body). Marking may be a displacement activity for *Lemur*, as is suggested by some observations, for instance, after fleeing from an observer or an opponent in an agonistic situation, upon hearing a loud noise, falling from a tree after a badly calculated leap, or upon breaking of a branch, etc.

Conclusion

If, in certain cases, marking may seem to be explained as a response to specific external sensory stimuli where the motivation involved and the social role of the message are clear, in other cases the behavioural sequences may be one of the external manifestations of a complex psychophysiological situation which often appears to be the result of a conflict.

There is no proof that certain instances of marking, which appear to be provoked by specific stimuli, would not appear except as the result of a state of conflict or simple excitation which exists within the animal. Lastly, it is perhaps of interest to speculate on the scars left on the branches by *Lemur catta* as being visual signals, as no other primate is known to exhibit such a signal.

Acknowledgments

The author wishes to thank Henri de Heaulme for allowing this study to be carried out at his protected reserve in Berenty, South Madagascar.

NOTES

1 Jolly, A. (1966), *Lemur behavior*, Chicago.

2 Sussman, R.W. (1971), Unpublished doctorate thesis, Duke University. See also this volume.

3 Evans, C.S. and Goy, R.W. (1968), 'Social behaviour and reproductive cycles in captive ring-tailed lemurs (*Lemur catta*)', *J. Zool. Lond.* 156, 181-97.

4 Petter, J.-J. (1962), 'Recherches sur l'écologie et l'éthologie des lémuriens malgaches', *Mém. Mus. nat. Hist. nat. Paris*, sér. A., 27, 1-146.

5 Rumpler, Y. and Andriamiandra, A. (1968), 'Sur l'étude de glandes brachiales et antebrachiales du *Lemur catta*', *C.R.S. Soc. Biol.* 162, 1430; Rumpler, Y. and Andriamiandra, A. (1969), 'Les glandes brachiales et antebrachiales du *Lemur catta*', *Ann. Univ. Madag. Med. et Biol.* 11, 57-66.

6 Montagna, W. and Yun, J.S. (1962), 'The skin of primates: 10. The skin of the ring-tailed lemur (*Lemur catta*)', *Am. J. Phys. Anthrop.* 20, 95-118.

7 Kneeland, J.E. (1966), 'Fine structure of the sweat glands of the antebrachial organ of *Lemur catta*', *Z. Zellforsch. Mikr. Anat.* 73, 521-53.

8 Andriamiandra, A. and Rumpler, Y. (1968), 'Rôle de la testosterone sur le determinisme des glandes brachiales et antebrachiales chez le *Lemur catta*', *C.R.S. Soc. Biol.* 162, 1651.

9 Ralls, K. (1971), 'Mammalian scent marking', *Science* 171, 443-9.

10 Schultze-Westrum, T. (1965), 'Innerartliche Verständigung durch Düfte beim Gleitbeutler *Petaurus breviceps papuanus* Thomas (Marsupialia, Phalangeridae)', *Z. vergl. Physiol.* 50, 151.

11 Mykytowycz, R. (1965), 'Further observations on the territorial function and histology of the submandibular cutaneous (chin) glands in the rabbit *Oryctolagus cuniculus (L.)*', *Anim. Behav.* 13, 400-12; Mykytowycz, R. (1966), 'Observations on odiferous and other glands in the Australian wild rabbit,

Oryctolagus cuniculus (L.), and the hare *Lepus europaeus P.*: 1. The anal gland', *C.S.I.R.O. Wildl. Res.* 11, 11-29; Mykytowycz, R. (1968), 'Territorial marking by rabbits', *Sci. Amer.* 218, 116-26.

12 Ilse, D.S. (1955), 'Olfactory marking of territory in the young male Loris', *Brit. J. Anim. Behav.* 3, 118-20.

13 Thiessen, D.D. and Lindzey, G. (1970*a*), 'Territorial marking in the female Mongolian Gerbil: short-term reactions to hormones', *Hormones and Behav.* 1, 157-60; Thiessen, D.D. and Lindzey, G. (1970*b*), 'The effects of olfactory deprivation and hormones on territorial marking in the male Mongolian Gerbil', *Hormones and Behav.* 1, 315-25.

PART I SECTION D
Physiology of Behaviour in Captivity

A. PETTER-ROUSSEAUX

Photoperiod, sexual activity and body weight variations of Microcebus murinus *(Miller 1777)*

Introduction

Madagascan lemurs breed seasonally.[1] It has often been pointed out that transfer to Europe reverses the breeding season: members of the genus *Lemur* give birth in Madagascar in October, and in zoological gardens of the northern hemisphere in April. The Cheirogaleinae, which are small nocturnal species and include the lesser mouse lemur, *Microcebus murinus*, start mating in September in Madagascar and in April in Paris, under natural local light conditions.

Breeding condition is thus associated with the photoperiod: temperature, humidity and nutrition can be maintained constant in captivity, and these factors consequently have less influence on the observed variations in breeding activity.

It was observed a long time ago that certain cheirogalines, notably those of the genus *Cheirogaleus*, can accumulate fat reserves (under the skin, in the tail and around the viscera) and become torpid during the non-breeding season (viz. the dry season, which is the winter in Madagascar). There is evidence that these animals exhibit an annual cycle affecting the entire body metabolism, and not just reproduction.

In captivity, *Microcebus murinus* exhibit a comparable annual cycle, with fat-accumulation and semi-torpidity alternating with the breeding period. Over the course of several years, we have carried out various experiments to investigate the effects of photoperiodic variations on the physiology of *Microcebus*.

Housing conditions

The animals are housed individually or in groups of 2-3 in cages measuring 60 x 60 cm., each provided with small wooden nest boxes. The animals are fed with fruits, milk, honey, spice-cake, mealworms and fresh meat. Temperature is maintained constant at about 25°C.

Mating condition of females is recognisable from oestrous developments involving swelling of the sexual skin and opening of the vulva, which is otherwise closed.

In the following experiments, the animals were housed in 2 separate rooms:

A. room exposed to the natural daylight conditions of the Paris region.

B. room artificially lit with 'daylight' fluorescent tubes (40 W). In this room the amount of light received was 50-150 lux. The transformation from light to dark and *vice versa* occurred abruptly, without a 'twilight' phase.

Annual variations in body weight

Captive mouse lemurs exhibit regular annual variations in body-weight,[2] which is maximal during the non-breeding season, and minimal during the breeding season. Weight-loss begins 3-4 months before the onset of oestrus. Weight-loss and oestrus occur during the period of increasing day-length, and weight-gain occurs during the period of decreasing day-length; the changeover is abrupt and may be complete within 2 months.

The same phenomenon occurs in animals subjected to the natural daylight conditions of Paris or to an artificial lighting schedule following the Madagascar rhythm; but, as a result of the six-month difference in phase, when animals in one room are in breeding conditions, those in the other are not (Fig. 1).

The magnitude of annual variation in day-length is about 8 hours in natural daylight, but only 2 hours in artificial lighting (Madagascar rhythm). This difference in day-length fluctuation seems to have no effect on the reproductive processes, however.

The increase in weight affects all animals. It is of interest to note that a castrated male exhibited the same annual variation in body-weight, during the first year at least: weight-loss occurred in spring, with a minimum in July, followed by a sudden increase in September/October (Fig. 2).

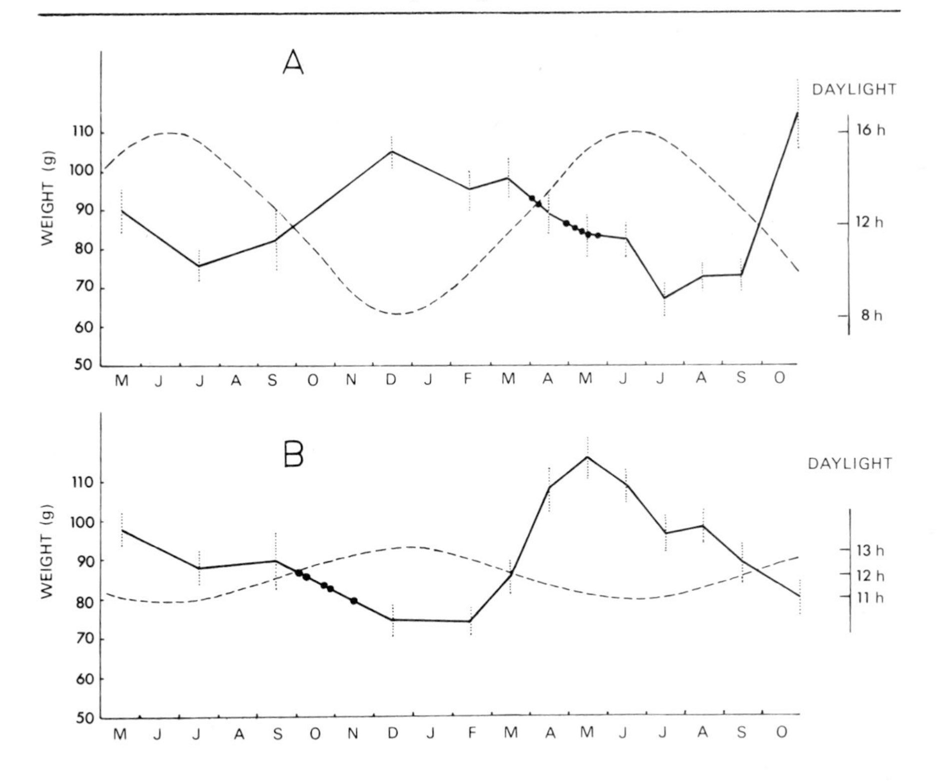

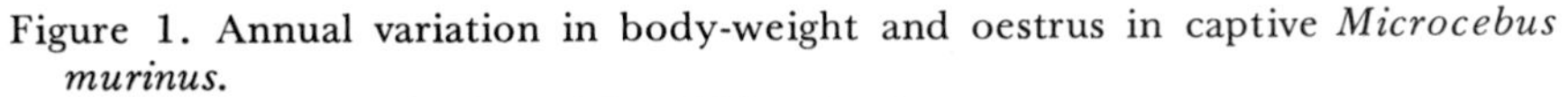
Figure 1. Annual variation in body-weight and oestrus in captive *Microcebus murinus.*

Solid line = mean body-weight of 15 animals.
Dotted line = annual variation in day-length.
Black circles represent the dates of onset of oestrus.
A – natural daylight in Paris
B – artificial Madagascan light regime

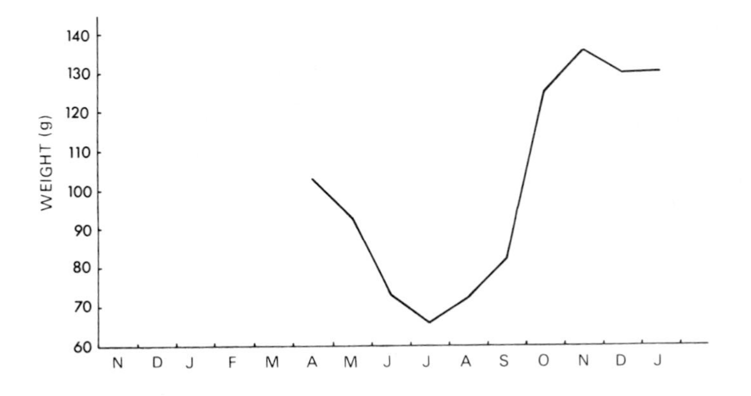

Figure 2. Body-weight variation in a castrated male, from 6 months after operation onwards (natural daylight conditions in Paris).

Adaptation to alteration in the light regime

Transfer from the Madagascan light regime to that of Paris

This transfer generally induces a rapid alteration in the breeding season. The timing of the alteration depends upon the timing of the change.

1. Change at the end of December

Fig. 3 shows the result of moving 7 females from the Madagascan light regime to that of Paris on 21 December. These animals were still in breeding condition in Madagascar, where maximum day-length conditions prevailed. They were suddenly subjected to the very short days of winter in Paris, followed by very rapidly increasing day-lengths. During January and February, their weights increased considerably as a result of the short days; thus, they entered into the non-breeding season earlier than animals under a Madagascan light regime. It was only after March/April that increasing day-length induced first weight-loss and then oestrus, which occurred at the end of April in the most precocious females; that is, only about a fortnight later than in those females under the Paris light regime. All females in this group came into oestrus between the end of April and the end of May.

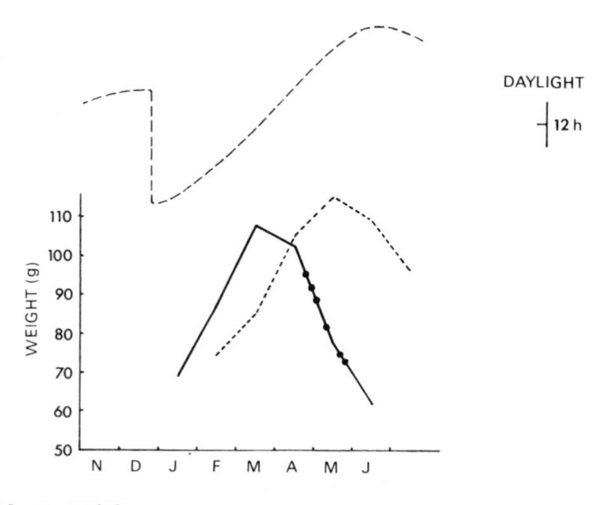

Figure 3. Mean weight and oestrus of 7 mouse lemurs transferred on 21 December, from the Madagascan light regime to natural daylight conditions in Paris (solid line). Black circles represent oestrus onset.
Lower dotted line = body-weight variation of mouse lemurs living under the Madagascan light regime at the same time of the year.
Upper dotted line = variation in day-length during the experiment. (In this and in subsequent figures, the scale for day-length variation is the same as in Fig. 1.)

2. *Change at the beginning of February*

Fig. 4 shows the transfer of 6 females on 7 February from the artificial Madagascan light regime to natural daylight conditions in Paris; none of these females had experienced pregnancy during the previous months. In this case, the effect of short days reinforced the fat-accumulation which would have occurred naturally at this time. The first oestrus occurred at the end of April, as with the preceding group; one female had still not come into oestrus by the end of July.

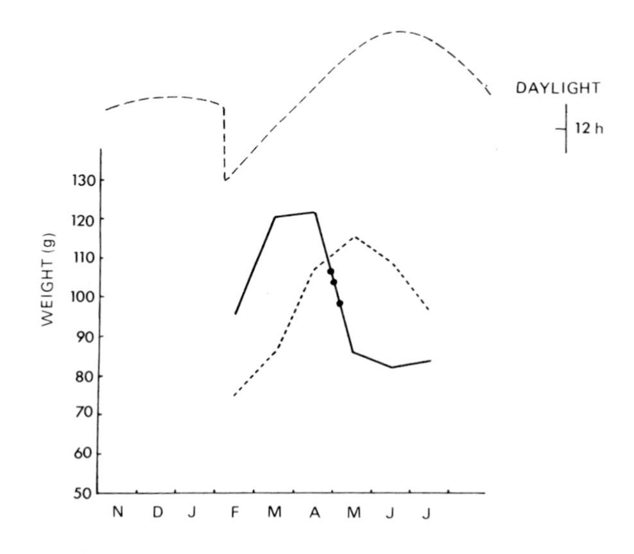

Figure 4. Mean weight and oestrus of 6 mouse lemurs transferred on 7 February from the artificial Madagascan light regime to natural daylight in Paris (solid line). Black circles represent oestrus onset.
Lower dotted line = body weight-variation of mouse lemurs living under the Madagascan light regime at the same time of the year.
Upper dotted line = variation in day-length during the experiment.

3. *Change at the end of April*

Fig. 5 shows the transfer of 6 females on 24 April from the artificial Madagascan light regime to natural daylight conditions in Paris. None of these females had experienced pregnancy during the preceding months. They had already been in the non-breeding season for 2 months. The abrupt change to long and increasing days hastened the weight-loss and induced oestrus, starting in June. The last female came into oestrus on 1 August, that is during the period of decreasing day-lengths, and one female did not come into oestrus at all. Thereafter, the effect of decreasing day-lengths accelerated fat-accumulation, and the breeding season was reduced to 3 months. Thus, in these three cases of transfer (from December to April) almost all of the animals came into oestrus a few months later.

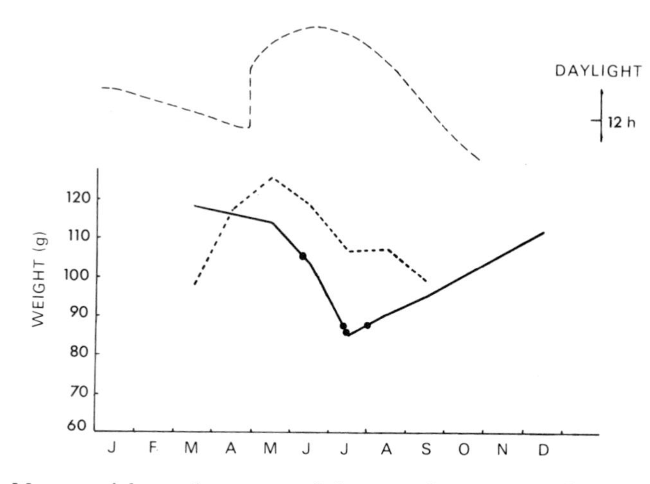

Figure 5. Mean weight and oestrus of 6 mouse lemurs transferred on 24 April from the artificial Madagascan light regime to natural daylight in Paris (solid line).
Black circles represent oestrus onset.
Lower dotted line = body-weight variation of mouse lemurs living under the Madagascan light regime at the same time of the year.
Upper dotted line = variation in day-length during the experiment.

4. *Change at the end of May*

If the transfer is too late, the increasing day-length sequence is too short to induce reproductive activity. Figure 6 shows the effects of actual importation to Paris of 4 females from Madagascar on 27 May. The history of these animals is unknown. Their weight diminished at once under the influence of long day-lengths, but they did not come into oestrus before the onset of the non-breeding season.

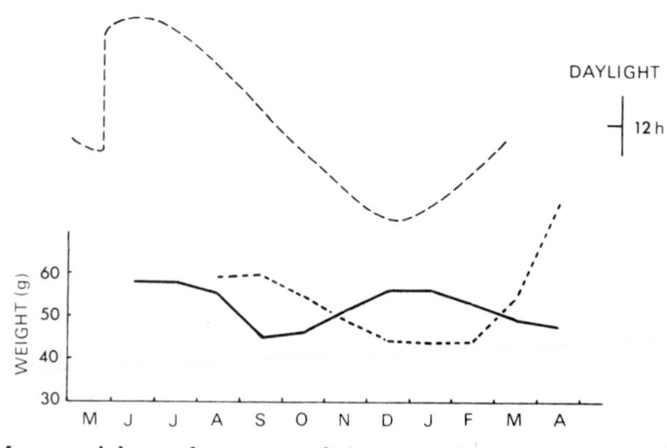

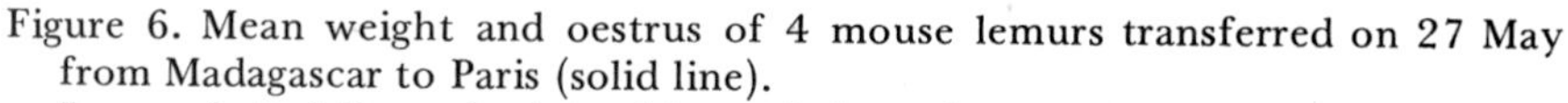

Figure 6. Mean weight and oestrus of 4 mouse lemurs transferred on 27 May from Madagascar to Paris (solid line).
Lower dotted line = body-weight variation of mouse lemurs living under the Madagascan light regime at the same time of the year.
Upper dotted line = variation in day-length during the experiment.

Transportation from the light regime of Paris to that of Madagascar

When transfer is made in the opposite direction, from the light regime of Paris to that of Madagascar, a different problem arises. The decreasing day-lengths of late summer in Paris are longer than the longest day of the Madagascan summer: at the end of the period of increasing day-lengths under the Madagascan regime, the animal will still be subjected to shorter day-lengths than those prevailing prior to transfer. Fig. 7 shows the transfer of 6 females on 3 August; the animals entered earlier into non-breeding condition, increased their body-weights, and were not stimulated to enter breeding condition before November; 3 of them came into oestrus in January. The remainder exhibited a similar weight-loss, but no reproductive activity.

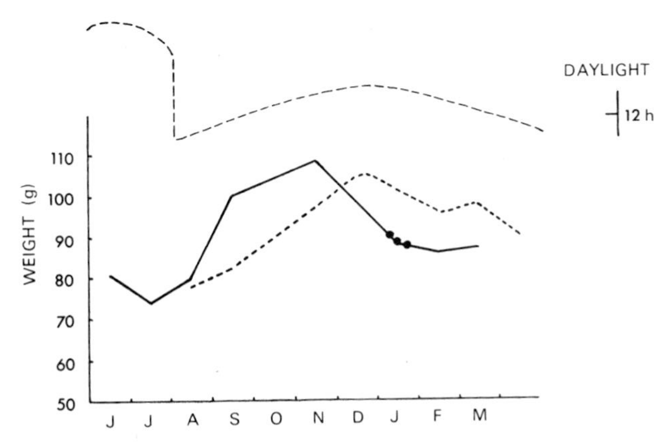

Figure 7. Mean body weight and oestrus of 6 mouse lemurs transferred on 3 August from natural daylight conditions in Paris to the artificial Madagascan light regime (solid line). Black circles represent oestrus onset.
Lower dotted line = body-weight variation of mouse lemurs living in Paris at the same time of the year.
Upper dotted line = variation in day-length during the experiment.

Applications

The transportation to Paris of mouse lemurs trapped in Madagascar during the breeding season results in abrupt, but brief, cessation of activity, and gives rise to the capacity to undergo another breeding period only a few months later. The weight-loss induced by this novel stimulation could have serious effects in cases where the animals do not possess adequate reserves.

It is consequently possible to control the body-weight of the animal by subjecting it to an appropriate lighting regime. Fig. 8 shows an example of such an experiment: a male, imported in December, suffered very severe weight-loss in spring. He was exposed in April to the Madagascan light regime, which resulted in an increase in weight, and in August a reverse transfer subjected him to the natural light regime of Paris. The resultant effect was that of brief breeding season stimulation, followed by renewed increase in weight.

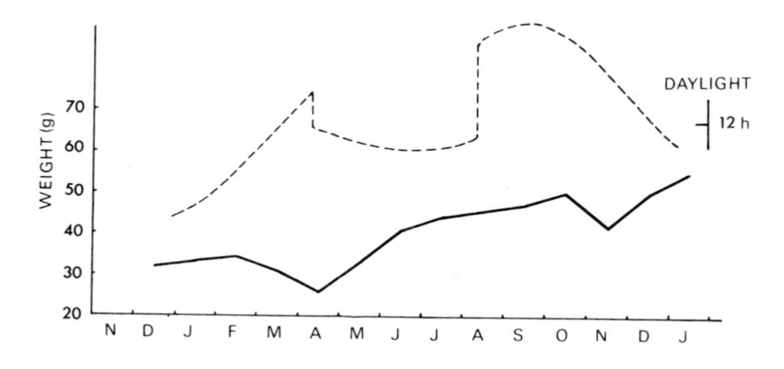

Figure 8. Body-weight of a mouse lemur transferred from the Madagascan light regime to that of Paris on 10 April, followed by the reverse transfer on 10 April (solid line).
Dotted line = variation in day-length during the experiment.

Discussion

These experiments demonstrate that in captive mouse lemurs reproductive activity and body-weight are correlated with variations in day-length.

The body-weights of captive mouse lemurs tend to increase considerably during the non-breeding period. Similar weights are unknown in the wild state, and this considerable increase may possibly be an effect of overfeeding in captivity. Lack of activity could also have this result; but the same phenomenon was observed with mouse lemurs kept in groups of 4 in large cages of 1.50 x 2 m. There is, perhaps, an alteration of nervous control of body-weight related to conditions in captivity.

Seasonal fattening related to the photoperiod is common in hibernating mammals, but it has never been demonstrated in lemurs other than the Cheirogaleinae. However, it is possible that an analogous situation exists in other Madagascan lemurs whose breeding season is similarly related to photoperiod variations. In lorisids, no evidence has yet been produced to demonstrate any effect of photoperiod variation on breeding condition or general metabolism.

Our observations suggest that the induction of a breeding season by an appropriate light regime is not dependent on the magnitude of the variation in day-length. It depends on the fact that days are increasing or decreasing in length; but, in a sudden transfer, increasing days do not have a stimulating effect if they are very much shorter than the days prior to the transfer. If the transfer occurs before the end of the active period, mouse lemurs experience an early non-breeding condition and accumulate fat; some 2 to 3 months later, they are ready to respond to new stimulation. If the transfer occurs later, the reaction is more irregular: some animals react very late, some do not respond at all.

In these experiments, adaptation to a change in light regime is rapid and generally free of difficulty. However, in one case of actual importation (on 1 February) a complete lack of response was observed: of 13 females, only one came into oestrus, and then only after 5 months. The history of these animals was unknown; possibly pregnancy or lactation during the preceding months could have delayed the effect of the stimulation.

NOTES

1 Petter-Rousseaux, A. (1968), 'Cycles génitaux saisonniers des lémuriens malgaches', in Canivenc, R. (ed.), *Cycles génitaux saisonniers des Mammifères sauvages*, Paris.

2 Petter-Rousseaux, A. (1972), 'Application d'un système semestriel de variation de la photopériode chez *Microcebus murinus* (Miller, 1777)', *Ann. Biol. Anim. Bioch. Bioph.* 12, 3.

M. PERRET

Variation of endocrine glands in the lesser mouse lemur, Microcebus murinus

Introduction

Studies of the reproductive and sexual cycles of mammals which show marked seasonal activity permit one to distinguish certain endocrine factors and to obtain a better understanding of their complexity. Little research of this type has been carried out with the Primates, owing to a lack of knowledge of their natural cycles and of the modifications produced by conditions in captivity. The Madagascan lemurs exhibit a seasonal pattern of activity which is maintained in captivity and is determined by the photoperiod.[1] We have therefore tried to determine the endocrine cycles of the lesser mouse lemur, *Microcebus murinus* (Cheirogaleinae), kept in captivity at the Laboratory of General Ecology in Brunoy.

Materials and methods

In captivity, *M. murinus* show a precise annual cycle of activity: during the period of short day-length (from September to February at our Paris latitude) sexual quiescence is complete in both sexes; a reduction in general activity occurs, and this is accompanied by a lowering of the internal body temperature and a gain in body-weight. In the springtime, general activity is stimulated, body temperature rises, animals lose weight and the sexual functions begin. In Paris, under natural daylight conditions, this period of activity lasts from February to August. All the animals used in this study were kept in captivity in Paris, except for six which were sacrificed in Madagascar. The conditions in captivity are constant with respect to temperature and food, but differ with respect both to available space and to the seasonal light-regime (which is either that of Paris or that of Madagascar). However, all dates given in this paper refer to the cycle exhibited by *Microcebus murinus* under the Paris light regime.

The animals were anaesthetised with ether for organ-collection. The endocrine glands (pituitary, thyroid, adrenals and gonads) were collected from different animals throughout the year, together with various other organs (kidneys, lungs, heart, liver, etc.). Fixation was carried out with Elftman's formol sublimate (1957). The glands were embedded in paraffin and cut at 3 to 5 μ. The annual development of the various endocrine glands was quantified by using simple histological procedures (standard staining techniques, measurements of nuclear and cellular diameters, measurements of the different components of each gland). In addition, the technique of immunofluorescence was employed to study the pituitary gland. Sections of the pituitary gland were treated with antibodies anti-STH, anti-TSH and anti-LH bovine, anti-HCG and anti-β (1-24) corticotropin. Five categories of cells were identified in the pituitary of *Microcebus murinus*: Cell-types STH, TSH, LTH and LH correspond to the classic descriptions; the corticotropic type, however, contrasts with the majority of observations in other mammals: it is basophilic and PAS-positive, and has been identified as equivalent to the category recognised in almost all other species as secreting gonadotropic hormone.[2] In addition to cytological studies and estimation of cell populations, the variations of chromophil reactivity of the cytoplasm and the intensity of fluorescence permitted estimation of the activity of cell types in the pituitary gland.

Pituitary cycles and target glands

Although the somatotropic cells in the pituitary gland (specifically stained green with luxol blue, using Shanklin's technique, 1959) remain numerous and active all the year, the other categories exhibit annual variations. These modifications have been studied in association with observations made on target glands.

1. Thyroid function

The pituitary thyrotropic cells of *M. murinus* are typically cyanophil cells; they are finely granulated and specifically stained by alcian blue, used as prescribed by Herlant (pH 1.5, after permanganate oxydation). They remain few in number throughout the year, but size and chromophily vary markedly: in winter the cells are small and strongly stained, while in spring they become enlarged and the secretory granules are more dispersed. Maximum size is attained in May-June. A slow involution begins in July or August, depending on the individual (decrease in cell size and increase in granulation).

Correlated with these pituitary variations are changes in the thyroid: there is a winter resting stage characterised by flattening of the epithelium and retraction of the colloid; followed by stimulation during springtime, which is indicated by growth of the epithelium and the appearance of absorption vacuoles in the colloid. As for the thyrotropic cells, activity is at its maximum in May-June and then decreases slowly, beginning in July (see Fig. 1).

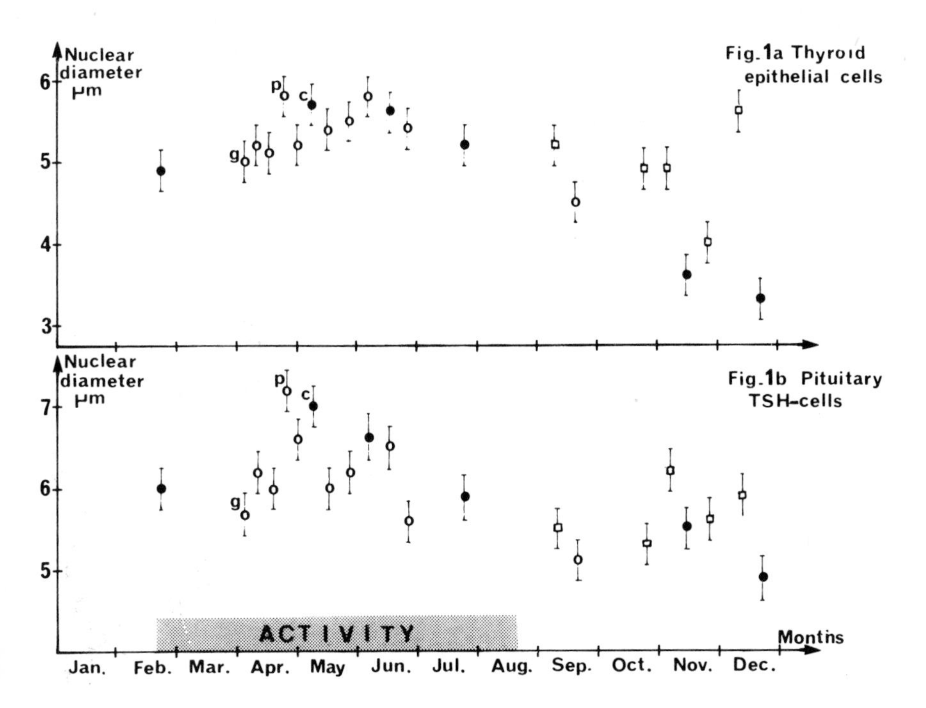

Figure 1. Annual variations of thyroid function.
● = male; o = female; □ = animal with chronic disease; *p* = pro-oestrus; c = castrated; g = pregnant; standard errors are indicated by vertical lines.
Correlation between thyroid activity and TSH-cell activity is highly significant (p=0.01). Thyroid activation occurs in early spring and is maximum in May-June. The resting condition begins in late July and is almost complete in September.

Castration, pregnancy (to a lesser degree), the ovulatory period, or a state of chronic illness are all accompanied by modifications of thyroid activity. It is probable that the steroid hormones (adrenal and gonadal) intervene in the regulation of the thyrotropic hormone, as is the case in most mammals. Observations concerning the thyroids of various hibernating animal species indicate that there is hyper-activity of the thyroid during the period of increased activity in spring. This active state continues in summer throughout the entire reproductive season of these animals (slow-worm, hamster, hedge-hog).

2. *Adrenocortical function*

The ACTH-cells have been identified by immunofluorescence with antibody anti-β (1-24) corticotropin. They are stained bright purple by the PAS-Orange G staining technique. Important modifications are evident in the cytoplasm and especially in their number. Observations reveal that there is a relationship between the physiological state of the adrenals and the histological state of the cell-type PAS positive ++. Development of the adrenals is accompanied by a state of pronounced activity of these cells.

During the winter, the ACTH-cells are rare and densely granulated, indicating weak secretory activity; in the adrenal, the cortex is reduced (zona fasciculata unorganised, cells small and heavily stained). Corticotropic function is stimulated during the month of February and is shown by multiplication and degranulation of ACTH-cells at the pituitary level and an increase in volume of the adrenal cortex, the organisation of cords in the zona fasciculata and the appearance of spongiocytes. The zona glomerula does not vary in size during the year and remains about 55 μ in thickness.

During the period of long day-length, there is pronounced corticotropic activity (ACTH-cells are numerous and active; the cortex is well developed, with spongiocytes present). Following this, during the month of July, the adrenal cortex decreases in size to reach a minimum in November. The spongiocytes disappear relatively early, and the size of the nucleus of fascicular cells is reduced from August onwards (see Fig. 2).

Certain animals have been observed to show hyperplasy and hypertrophy of the ACTH-cells, some of which exhibit a state of marked vacuolar degranulation. These animals were a castrated male (contrasting with normal males during the same period); females in oestrus (contrasting with females in di-oestrus), particularly females near the time of ovulation compared to females far from ovulation; and animals which died following a period of more or less extended exhaustion, compared to healthy animals during the same period. Accompanying such hyperactivity of the ACTH-cells (see Fig. 3), there is evident growth of the adrenal cortex.

Study of the adrenals during the oestrous cycle in females indicates that there is an accompanying cycle of corticotropic function. In fact, the adrenal cortex expands very rapidly at the beginning of oestrus and develops to its maximum size by the seventh day of vaginal opening, that is 5 days after ovulation. During this period, the pituitary is almost entirely filled by ACTH-cells showing signs of intense secretory activity (see Fig. 3*c*). The relationships between the adrenal cortex and the physiology of oestrus are certainly very pronounced in *M. murinus*, though they are perhaps accentuated in captivity. In addition, a female treated for 10 days after pro-oestrus

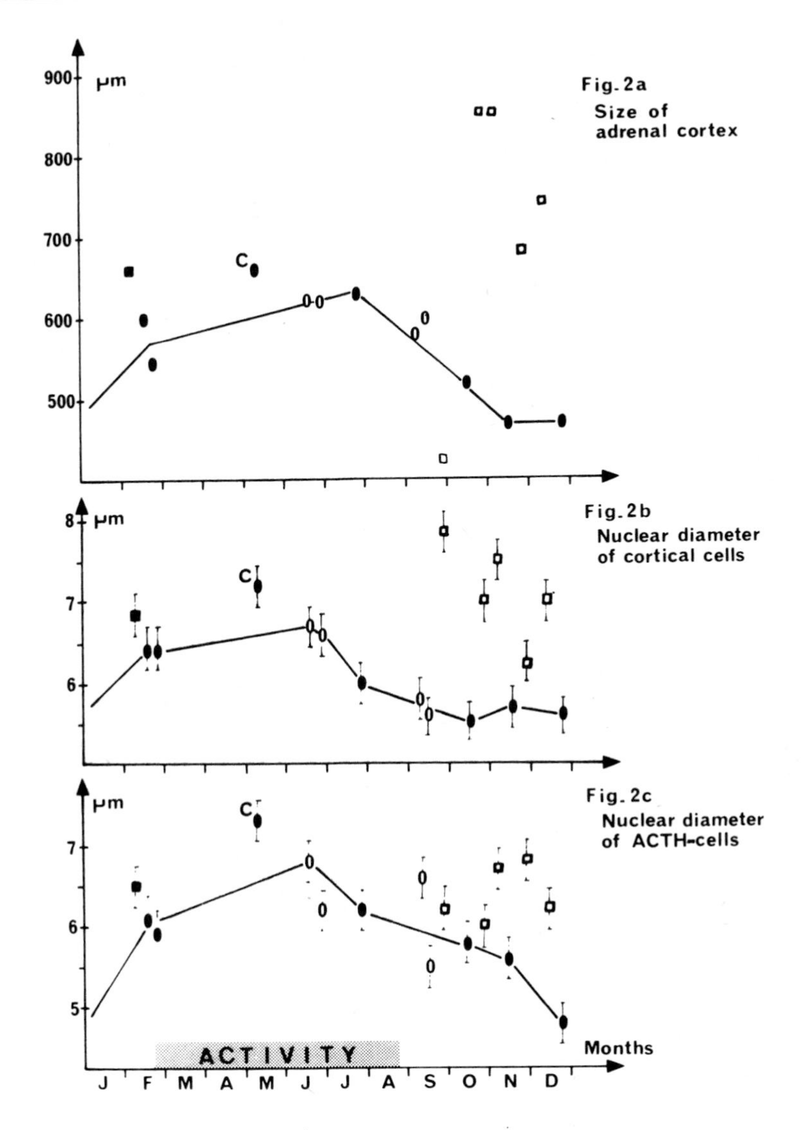

Figure 2. Annual variations of adrenocortical function.
● = male; o = female; □ = animal with chronic disease; C = castrated; standard errors are indicated by vertical lines.
Animals with chronic disease and castrated males show a clear increase of all morphological parameters of adrenocortical function.

with 1 mg./day of hydrocortisone ovulated 5 days late. The ACTH-cells of this female's pituitary were numerous, but inactive, and the adrenal was completely involuted. These observations, in association with the occurrence of adrenal hyperactivity after castration, suggest that corticosteroids play an active role in the sexual physiology of *Microcebus*.

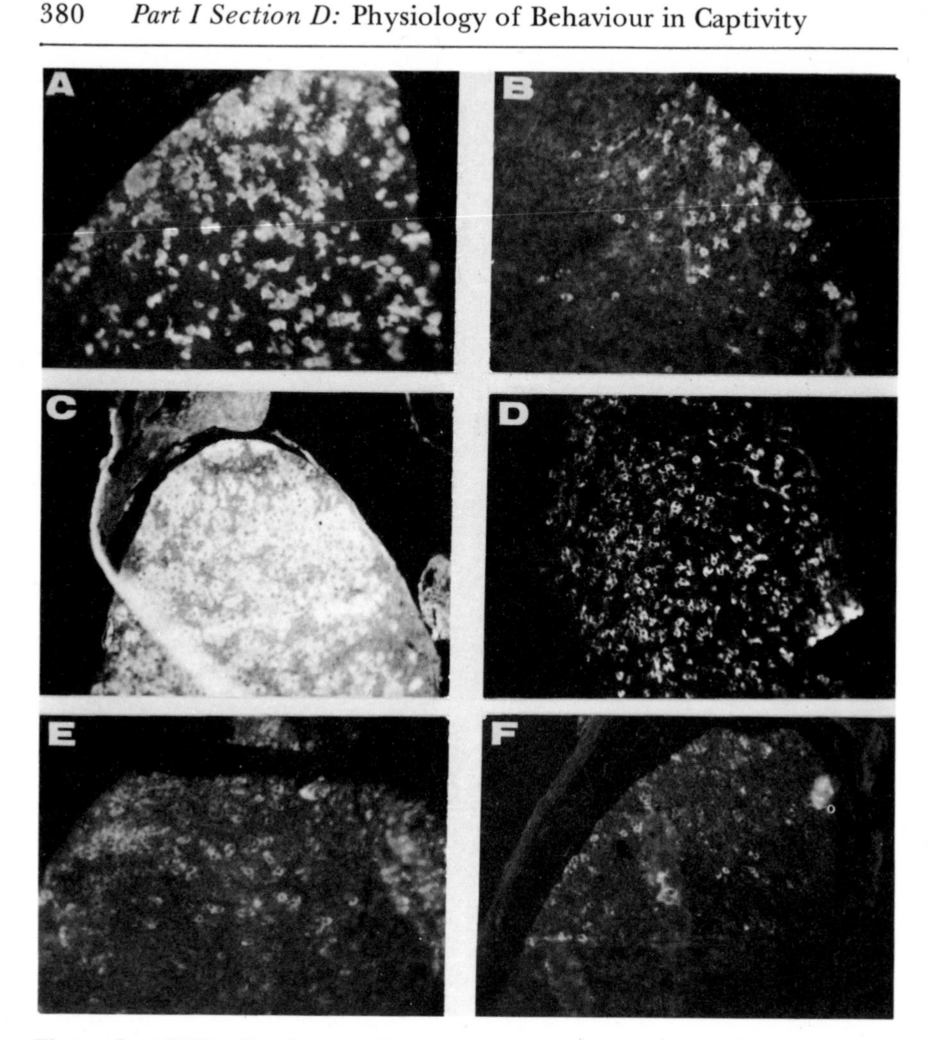

Figure 3. ACTH-cells: immunofluorescence with antibody anti-β (1-24) corticotropin. (*A*) Castrated male during May. (*B*) Normal male during May. (*C*) Female, 3 days after vaginal opening. (*D*) Female in di-oestrus. (*E*) Animal with chronic disease in the winter period. (*F*) Animal in good health in winter period. Hyperactivity is seen with animals *A*, *C* and *E*.

3. *Sexual cycles*

(*a*) *The male.* Annual variations in testis size are very marked in male *M. murinus*. During the winter the testes are just perceptible under the scrotal skin, although they cannot be measured, and during the breeding season they may reach a size of 2 cm. in length and almost as much in width.

In winter, the testes are completely involuted; activation (as for most northern species showing a seasonal sexual cycle) begins in February.[3] During the spring all the germinal elements develop in the seminiferous tubules, excepting the spermatozoa which appear later. Activity is at its maximum in May-June, and the epididymis canal is distended with spermatozoa. Spermatogenesis begins to decrease in activity in July, and regression is complete in September (see Fig. 4).

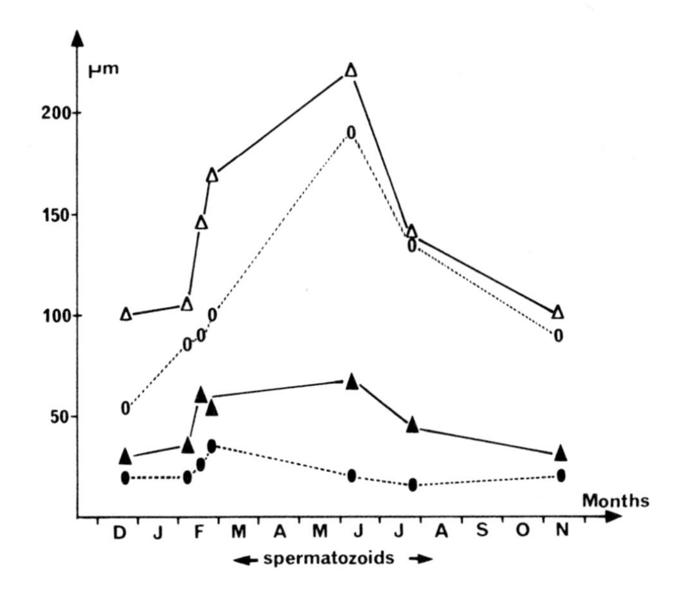

Figure 4. Changes in morphological parameters of testicular and epididymal functions during the year.
△ = diameter of seminiferous tubules;
o = diameter of epididymal tubules;
▲ = height of germinal epithelium;
● = height of epididymal epithelium.

The Leydig cells follow the same cycle of development, activity being maximal in spring. A slight disparity exists between maximum testicular activity and maximum interstitial activity. This separation is also apparent in certain other species showing a seasonal sexual cycle.[4] The curve of development of the gonadotropic LH-cells is, as expected, parallel to that of the Leydig cells (see Fig. 5). Another category of pituitary cells, the prolactin cells, also show annual variations. Easily distinguished by their red colour following treatment with Herlant's tetrachrome, these cells appear at the beginning of the period of activity and attain a maximal activity at the end of the breeding season. This hyperactivity (see Fig. 5) at the end of July is perhaps one of the factors causing reduction of spermatogenesis during this period.

We have not observed hyperactivity of the LH-cells in the

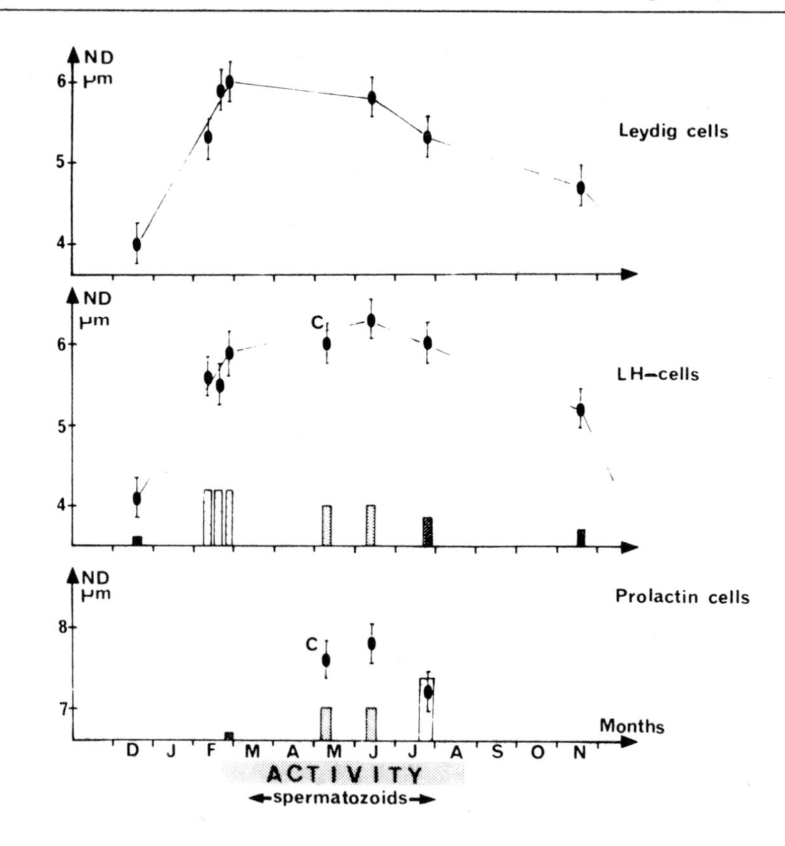

Figure 5. Annual variations of endocrine sexual function in the male. C = castrated; ND = nuclear diameter; standard errors are indicated by vertical lines. The surface of each rectangle is proportional to estimated cell population density; density of cytoplasmic granulation is indicated by the degree of darkness of each rectangle.

Maximum activity is seen in late February for Leydig cells, in May-June of LH-cells and in July for prolactin cells.

castrated male. This male was castrated 2 years ago, and it is possible that the adrenal cortex produces androgens to compensate for castration, and thus plays a normal feed-back role. Castration is accompanied by corticotropic hyperactivity,[5] possibly involving a suppression of the inhibition of ACTH synthesis due to the absence of testosterone, which has been shown to have an inhibitive effect on ACTH-cells and/or corticosteroid metabolism.[6]

(*b*) *The female.* The structure of the ovary and the accessory sex organs varies according to the season and the phase of oestrous cycle. The resting period begins in September and is characterised by degeneration of secondary follicles. Atresia is well advanced by November, and the ovary is invaded by dense conjunctive tissue.

During the oestrous cycle, development of the ovary follows the normal mammalian pattern: at pro-oestrus, each ovary develops one follicle and accessory sex organs enlarge. Follicular development in captive females does not seem to be very pronounced, many follicles showing signs of atresia. The vagina, normally closed, opens for about 10 days at the time of oestrus. Vaginal smears and histological studies have shown that ovulation occurs between the 2nd and 3rd day after vaginal opening. After ovulation, the corpus luteum develops in each ovary and no necrosis is observed. The absence of necrosis in captive females seems to be correlated with the appearance of the corpus luteum. The secretions of the corpus luteum may compensate for the possible lack of gonadotropins in these females.

Interpretation of the activity of the pituitary LH-cells during the oestrous cycle is difficult (see Fig. 6); there does not seem to be a well-defined cycle, but the cells do show variations both in their number and in chromophil reactivity. As in most mammals, the developing follicle would stimulate the formation of LH, and there would be a degranulation of these cells at the time of ovulation, followed by progressive granulation after the formation of the corpus luteum.[7]

The pronounced growth and abundance of ACTH-cells at the time

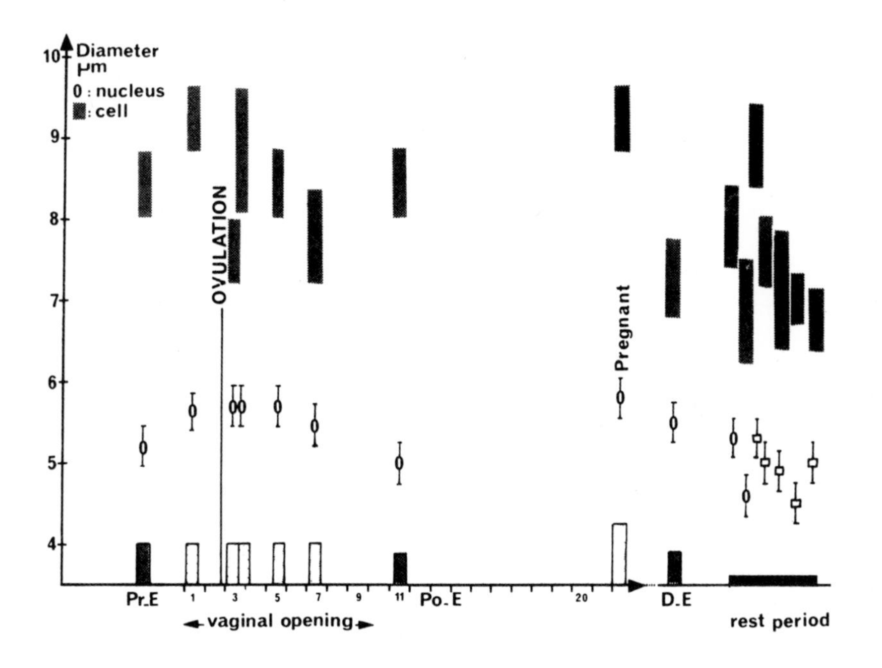

Figure 6. Variations in LH-cells in the female. (□ = animal with chronic disease; Pr.E = pro-oestrus; Po.E = post-oestrus; D.E = di-oestrus; standard errors are indicated by vertical lines.) The rectangles on the abscissa indicate cell population density and density of cytoplasmic granulation, as in Fig. 5. LH-cells show degranulation during ovulation, after which they enter gradually the phase of accumulation of their products.

of ovulation (see Fig. 3*C*) suggest that the cortico-gonadic relationships are particularly important in captive females. It is possible that, for these females, corticosteroid hyperactivity due to stressful conditions of captivity could explain the low rate of reproduction in captive animals. For instance, it is known that an excess of corticosteroids results in ovarian atresia. In addition, a female treated with hydrocortisone shows signs of follicular atresia, in spite of the presence of the corpus luteum.

As for the prolactin cells, they develop at the same time that the corpus luteum appears. The fact that these cells enlarge or regress according to the presence or absence of the corpus luteum could indicate a luteotropic action of prolactin in *M. murinus*, as is the case for mice, rats or hamsters.

Animals with chronic disease

It seemed interesting to include in this paper observations made on animals which died during the winter season (and accordingly exhibited no sexual activity), following a period of extended exhaustion. This is characterised by progressive loss in body-weight, followed by loss of appetite.

Several features are always evident in these animals:

1. *Hyperactivity of corticotropic function*

In the pituitary of these animals, ACTH-cells are numerous (see Fig. 3*E*) with a large nucleus (see Fig. 2*c*) and exhibit signs of great secretory activity (degranulation with cytoplasmic vacuolisation, which may lead to complete disappearance of granulations). Such activity is unusual at this time of the year.

The adrenals are significantly hypertrophied (see Fig. 2*a*) and the cortex accordingly exhibits a characteristic form. Spongiocytes with a large nucleus (see Fig. 2*b*) are visible, but organisation of cords in the zona fasciculata is defective, due to the presence of large PAS-positive regions. On the interior aspect of the Cortex, disorganisation is complete, and the cells remain small and easily stained.

2. *Increase in thyroid function*

In sick animals, greater activity of TSH-cells is observed in the pituitary (see Fig. 1). The thyroid is heterogenous and shows an intermediate state of activity. The follicles are more or less developed, and the colloid (though adhering to the epithelium) has few or no vacuoles. It seems justifiable to attribute this greater thyroid activity to the hypercorticosteroid state which exists in these animals.

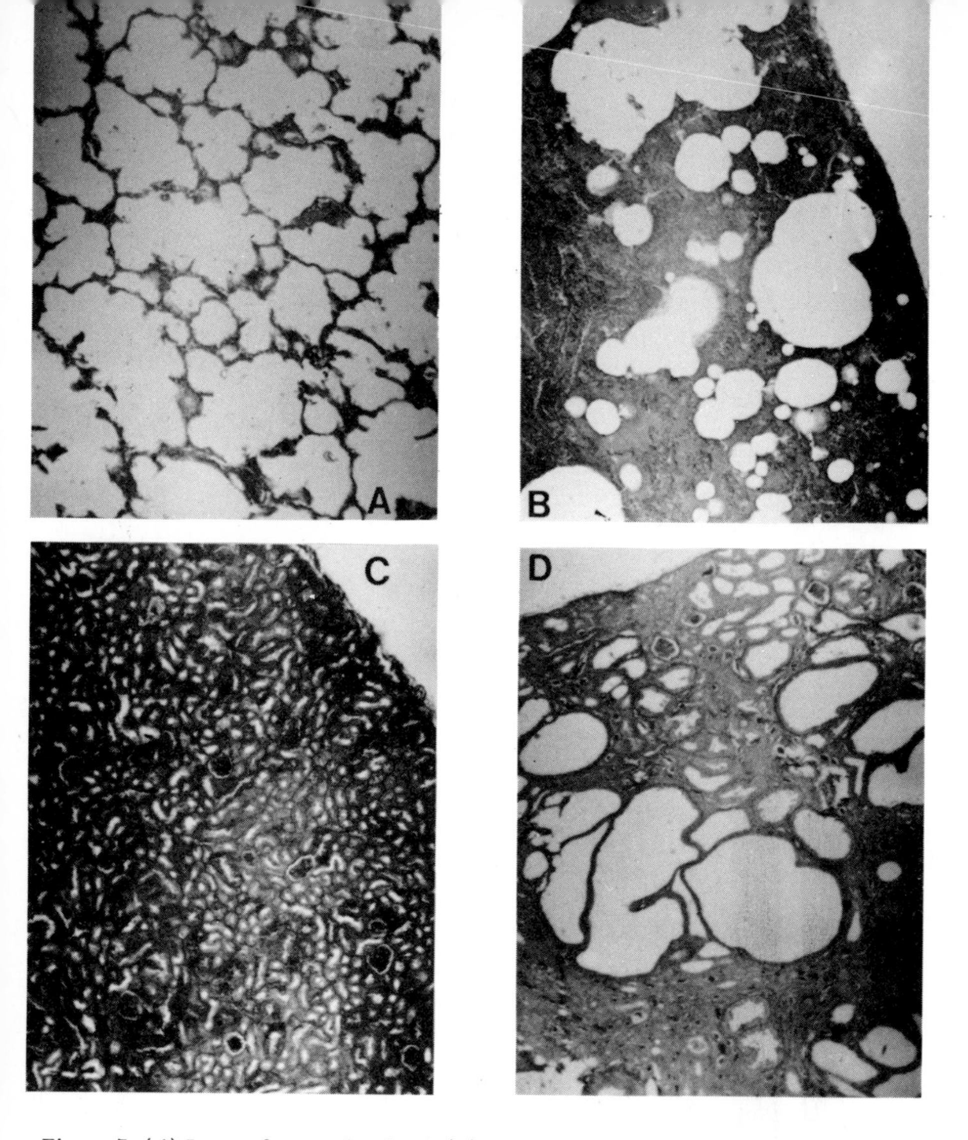

Figure 7. (*A*) Lung of normal animal. (*B*) Lung of animal with chronic disease. (*C*) Kidney of normal animal. (*D*) Kidney of animal with chronic disease.

3. Pulmonary lesions

At the end of a period of extended exhaustion in these animals, signs of more or less accentuated asphyxia have been observed. At necropsy, the lungs show blackish spots or large swellings. Histologically, they are characterised by enlargement of respiratory epithelium, which leads to disappearance of alveolar spaces in the majority of cases (see Fig. 7*B*). There is also exudate-formation in the alveoli, which often leads to haemorrhagic lung oedema. These phenomena (congestion; atelectasis; transudate and exudate formation in alveoli) are undoubtedly the result of a primary state of illness.

4. Kidney lesions

Animals which died after illness, and, to a lesser extent, animals which were in good health but had been in captivity for a long time, had smaller kidneys which exhibited scattered, clear blotches. In such animals, renal structure has more or less disintegrated, particularly at the cortical level. The cortical part of the kidney is invaded by large cavities, which may or may not be filled with PAS-positive mucus, corresponding to dilatated tubules (see Fig. 7*D*).

Other lesions have been observed irregularly in the gastro-intestinal tract and in the genital organs.

These fatal lesions would correspond to a greater secretion of corticosteroids. Studies made by Selye have demonstrated that long-term hyperproduction of adreno-cortical hormones induces a disturbance in physiological states and brings about lesions, particularly at the renal level.[8]

It is probable that such an increase of adrenal activity would occur as a response to stressful conditions of captivity. In any case, the fact that apparently healthy animals in captivity have more developed adrenals than animals captured in the wild indicates that *Microcebus murinus* are submitted to stress under the present conditions of captivity. Indeed, population density in the breeding colony is clearly far greater than that which is observed in the field (20 animals in a room of 6 square m.). Christian has demonstrated that there is an endocrine response to population increase which involves increased activity of the adrenals and a decrease in reproduction.[9] Thus, in captivity, stress (rather than nutritional factors) would be a major factor responsible for the low rate of reproduction and the death of certain animals.

NOTES

1 Petter-Rousseaux, A. (1970), 'Observations sur l'influence de la photopériode sur l'activité sexuelle chez *Microcebus murinus* en captivité', *Ann. Biol. Anim. Bioch. Biophys.* 10, 203-8.

2 Perret, M., Dubois, M.P. and Petter-Rousseaux, A. (1971), 'Les cellules corticotropes dans l'hypophyse d'un lémurien malgache: *Microcebus murinus* (Miller, 1777). Identification par immunofluorescence', *C.R. Acad. Sci.* 272, 2804-7.

3 Canivenc, R. (1968), 'Cycles génitaux des mammifères sauvages', *Entretiens de Chizé* 1, Paris.

4 Girod, C. and Cure, M. (1965), 'Etudes des corrélations hypophyso-testiculaires au cours du cycle annuel chez le hérisson (*Erinaceus europaeus* L.)', *C.R. Acad. Sci.* 261, 257-60.

5 Chester Jones, I. (1957), 'Adrenal gonad relationships', in *The Adrenal Cortex*, London.

6 Kitay, J.L. (1963), 'Pituitary adrenal function in the rat after gonadectomy and gonadal hormones replacement', *Endocrinology* 73, 253-60.

7 Herlant, M. and Ectors, F. (1969), 'Les cellules gonadotropes de l'hypophyse chez le porc', *Z. Zellforsch. Mikrosk. Anat.* 101, 212-31.
8 Selye, H. (1950), *Stress: A treatise based on the concepts of the General-Adaptation-Syndrome and the diseases of adaptation*, Montreal.
9 Christian, J.J. (1955), 'Effect of population size on the adrenal glands and reproductive organs of male mice in populations of fixed size', *Amer. J. Physiol.* 182, 291-300.

D. von HOLST

Social stress in the tree-shrew: its causes and physiological and ethological consequences

Introduction

Darwin based his theory of natural selection on two facts which were already well known in his time:

1. The number of individuals of a species population remains more or less constant over successive generations.
2. Any species is capable of producing many more progeny than are necessary to maintain a given population size.

As possible causes underlying this constancy in numbers of individuals, various factors have been proposed, such as: climate, food shortage, predators, diseases. Without doubt, all these factors can reduce population size; however, they cannot be regarded as the underlying causes of population regulation, because the number of individuals will remain constant even in their absence, at least in some species.[1]

Christian,[2] in 1950, propounded the hypothesis of self-regulation of mammalian populations: increasing density of individuals, according to this concept, leads to an increase in stress, to which animals adapt with hormonal changes. These adaptation reactions can be explained on the basis of the stress concept developed by Selye.[3] They are primarily characterised by increased activation of the hypophyseal-adrenal system and a corresponding reduction of hypophyseal-gonadal function. As a consequence of these endocrine changes, the fertility and vitality of the animals are sharply affected, thus counteracting the increase in population density. Excessive social stress, with subsequent exhaustion of the adrenal system, is even thought to be responsible for the death of animals, as in the spectacular population crashes in rodents (see Christian's summary).[4] Endocrine stress reactions in various social situations are well documented for a great number of mammalian species. However, until now there has been little clarification of the particular stimuli emitted from conspecifics which lead to these adaptation reactions, i.e. whether they are brought about by fighting, encounter frequency (density *per se*), or represent the consequences of dominance relationships.[4,5,6] This lack of clarity results largely from the difficulty of determining the level of activation of the adrenal system and, with it, the stressing effect of a given situation.

The following is a report on laboratory experiments with tree-shrews (*Tupaia belangeri*), which could help to clarify this question. Tree-shrews are small, diurnal mammals, which are distributed throughout South-East Asia. In nature they live singly or in pairs, apparently occupying territories which they defend fiercely against intruders of their own kind.[7,8] They are particularly suited for an investigation into the causes of social stress, because the activation of their sympathetic system can be observed directly.

Results

1. *Tail-ruffling (SST)*

Normally, the hair on the tail of tree-shrews lies flat. With any disturbance, however (such as an unexpected noise, approach of a strange person, fighting with a conspecific), the *Musculi arrectores pilorum* are activated by sympathetic fibres and cause the hair to stand almost vertically, thus giving the tail a bushy appearance (Fig. 1). This ruffling of the tail (SST = *Schwanzsträuben*) enables one to determine solely through visual observation how often and how long a tree-shrew is aroused in a certain situation.

The observer only distinguished between 'ruffled' (angle of hair spreading $> 45°$) and 'not ruffled' (angle $< 45°$) hair, although the *Musculi arrectores pilorum* can react differentially according to the degree of their activation by sympathetic fibres. It was not necessary, however, to distinguish tail ruffling further, since there was normally a clear-cut difference between 'no SST' (Fig. 1*a*) and 'SST' (Fig. 1*b*). This leads to the conclusion that during those social interactions which result in SST, the sympathetic nervous system is always strongly (or even maximally) activated. Without doubt, this results in a simultaneous output of catecholamines by the adrenal medulla, which will maintain or even increase the constriction of the pilomotor muscles. The lasting effect of catecholamines could possibly explain why, according to the disturbing situation, SST always persists for some time after cessation of the releasing stimulus. For example, a fight of few seconds' duration results in many minutes of SST.

Figure 1. The tail of *T. belangeri.* (*a*) Normal appearance. (*b*) Appearance during stress, when sympathetic fibres give the tail a distinctly bushy appearance. This ruffling of the tail is referred to by the author as *Schwanzsträuben* (tail ruffling), abbreviated as SST.

2. *SST-value*

The sum of episodes with SST recorded from an animal has been expressed as percentage of the 12-hour observation day (Fig. 2). This SST-value (% SST) can be taken as an indicator for the total duration of sympathetic arousal of the animal during the day. Observations of animals during their 12-hour activity period have led to the following results:

1. The SST-value may vary in any tree-shrew from 0-100%, depending on the environmental conditions of the animal.
2. The SST-value remains quite constant for months in a constant environment. Any change in the surroundings of an animal results in a corresponding change of the SST-level.
3. The SST-values are always correlated with specific ethological and physiological phenomena, which must be based on corresponding hormonal changes (summarised in Fig. 3).[9] This correlation is the same whether SST is triggered by members of the same species or is attributable to other causes.

3. *Direct correlation between SST-values and physiological changes*

Female fertility: A female tree-shrew which lives with a male in a large cage (floor area more than 5 square metres) normally shows very little (if any) SST per day. She gives birth every 45 days to 1-4 young (normally two). The young are fed by the mother directly after birth (Fig. 4), thereby increasing their body weight by up to 50%.[10] The mother then leaves the young and returns only once every 48 hours for feeding. The male (or other conspecifics) will

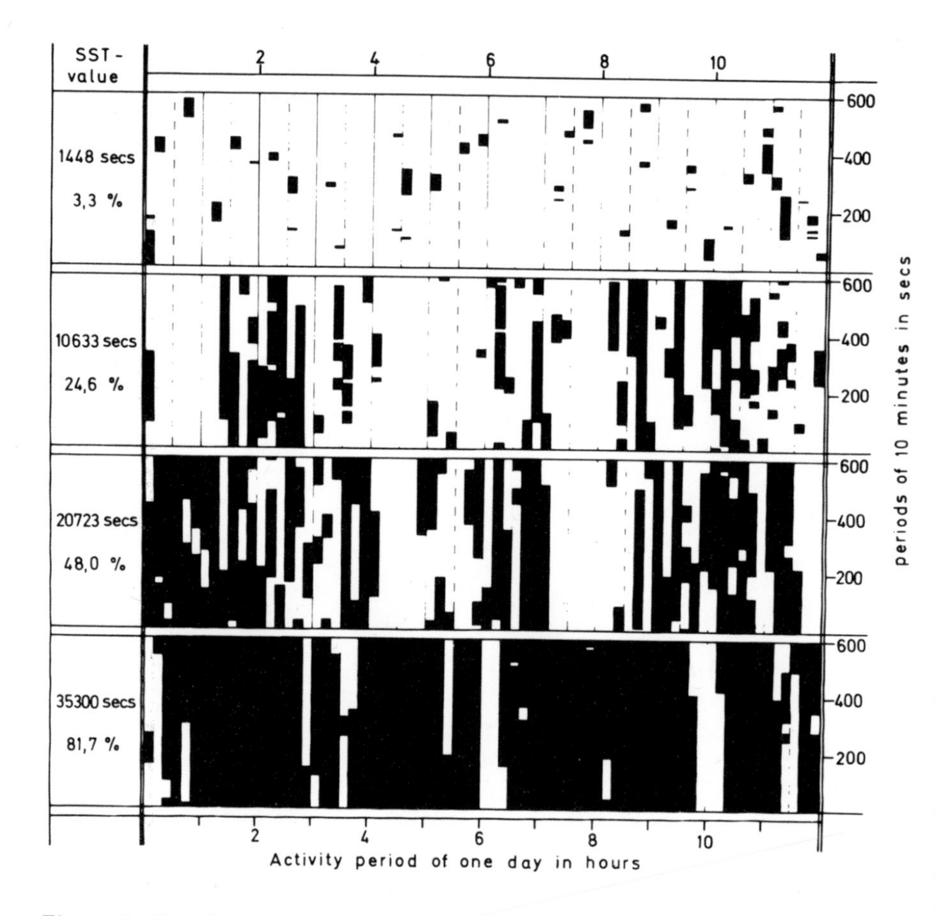

Figure 2. Distribution of times with (black) or without (white) tail-ruffling (SST) in 4 tree-shrews during their periods of activity. Each 12-hour observation day is divided into successive periods of 10 minutes, which are recorded side by side. (To assist interpretation, a vertical line has been drawn for every half-hour interval.) The ordinate indicates the time, in seconds, within each 10-minute period. The episodes of SST are recorded at the appropriate places within the individual periods of 10 minutes. As can be seen, times of SST are distributed more-or-less randomly over the day. The sum of episodes of SST gives the SST value in secs. (or in %) for the 12-hour observation period.

never enter the nest. If for any reason the SST-level of the mother rises above 20% per day, she continues to give birth regularly every 45 days; after the birth, however, any tree-shrew (often the mother) which enters the nest will remove the young and eat them ('cannibalism'; Figs. 5 and 6). This effect is caused by lack of sternal gland secretion with which the female normally marks the young at

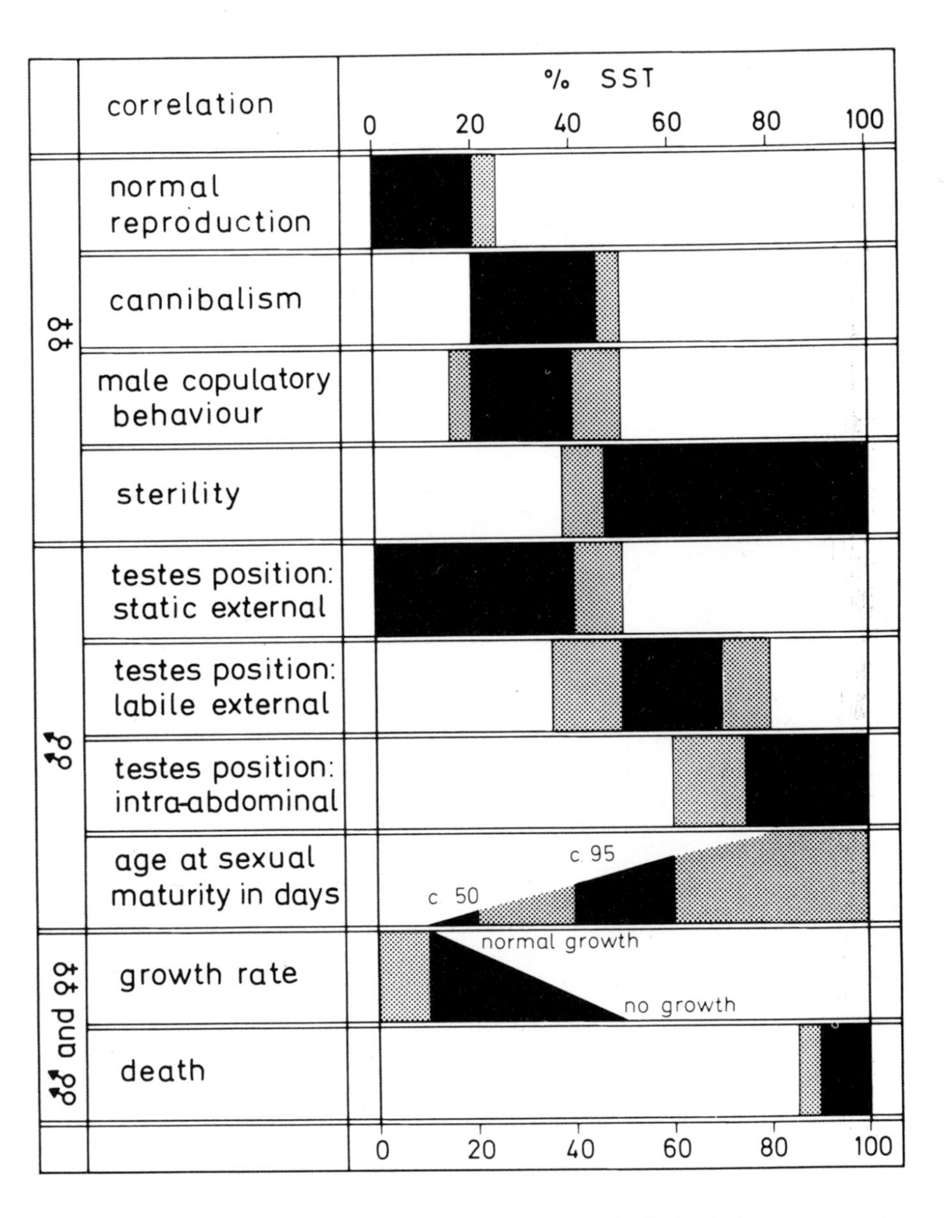

Figure 3. Tabular summary of some physiological and ethological phenomena in tree-shrews, showing their correlation with observed (black) or probable (grey) SST values.

Figure 4. Female *T. belangeri* during the birth of 4 infants.

Figure 5. Tree-shrew mother exhibiting cannibalism. The infant had been well fed just after birth, only 4 hours earlier.

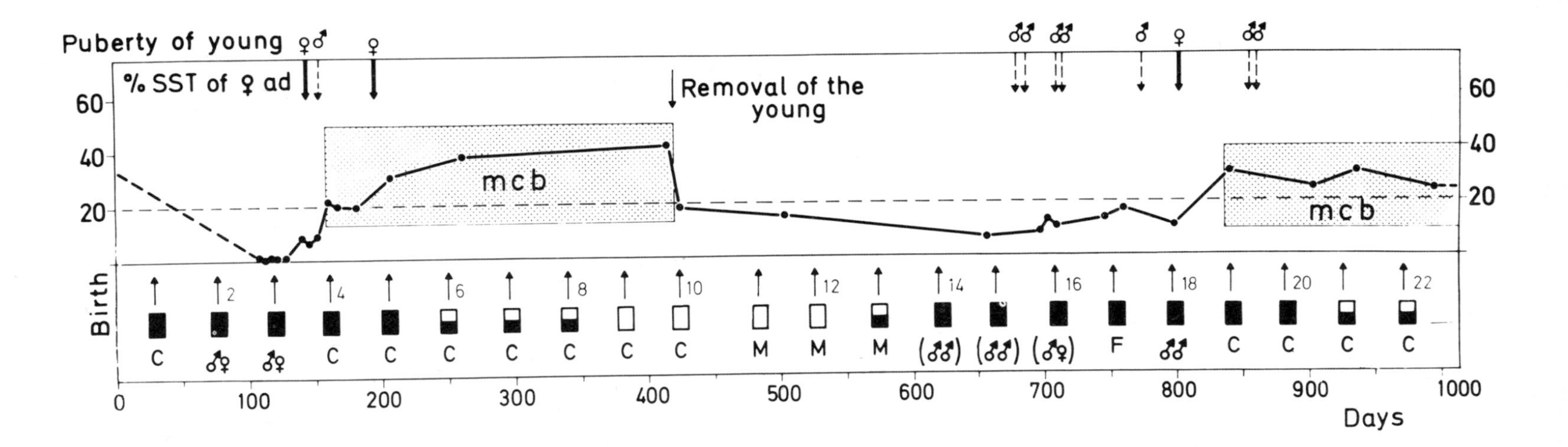

Figure 6. Correlation between the SST-value of a female, her reproduction and the occurrence of male copulatory behaviour (mcb), exhibited by the female. Each arrow on the lower scale indicates a birth. The nourishment of the infants after birth is indicated by the rectangles (black = normal, half-black = reduced, white = no milk given). The sexes of the young which reached sexual maturity are shown. The other litters did not survive because of cannibalism (C) or starvation, which was caused either by reduced maternal milk-production (M) or by disturbance of the maternal 48-hour feeding rhythm (F). The litters 14-16 (in brackets) would have died, as did litter no. 17, if the mother had not been forced to suckle by the experimenter (see text for details). It can be seen that an increase in the maternal SST-value is only brought about when female offspring attain puberty.

birth. At the same SST-level the female shows male copulatory behaviour, which is not seen below this SST-value (Fig. 6). When SST exceeds 50%, it results in sterility, arising from degeneration of ripening follicles in the ovaries.

Male fertility: With males, an increase in SST-values is equally correlated with a decrease in fertility (Figs. 7-9). This can be recognised externally by the state of development of the scrotum and the location of the testes:

(*a*) *Static external position of the testes* (Fig. 7*a*). When SST is below 50%, the testes lie outside the abdomen in a dark-pigmented scrotum, and cannot be squeezed into the body through the inguinal canal. Active spermatogenesis exists (Fig. 8*a*); correspondingly, the epididymes are filled with sperm (Fig. 9*a*).

(*b*) *Labile external position of the testes.* Above 50% SST, the scrotum loses its original compactness and the testes can easily be squeezed into the abdomen. As a consequence of any disturbance (fighting with conspecifics, capture by the experimenter), the testes are withdrawn instantly into the abdomen. This is impossible below 50% SST. When the disturbance ends, the testes return into the scrotum within a few seconds, possibly due to relaxation of the abdominal wall. In this physiological state, males are fertile; their spermatogenesis, however, is markedly lower than in animals with static external position of the testes.

(*c*) *Static intra-abdominal position of the testes* (Fig. 7*b*). If males live in a situation with SST exceeding 70%, the testes are withdrawn into the abdomen within 8-10 days, and remain there. At the same time, the testes and epididymes lose up to 80% of their initial weight. Spermatogenesis ceases completely (Figs. 8*b*, 9*b*). The scrotum is highly reduced, and after about two weeks it is essentially absent (Fig. 7*b*).

Also, as SST rises, the growth rate in young animals is progressively lowered and puberty delayed. All these changes are reversible (see also Fig. 6).

In a situation leading to more than 90% sympathetic arousal per day, within a few days the animal develops muscular twitching, convulsions, coma and dies.

4. *After-effects of situations leading to higher SST-values*

In the instances mentioned so far, there is a direct correlation between SST-values and changes in physiology and behaviour.

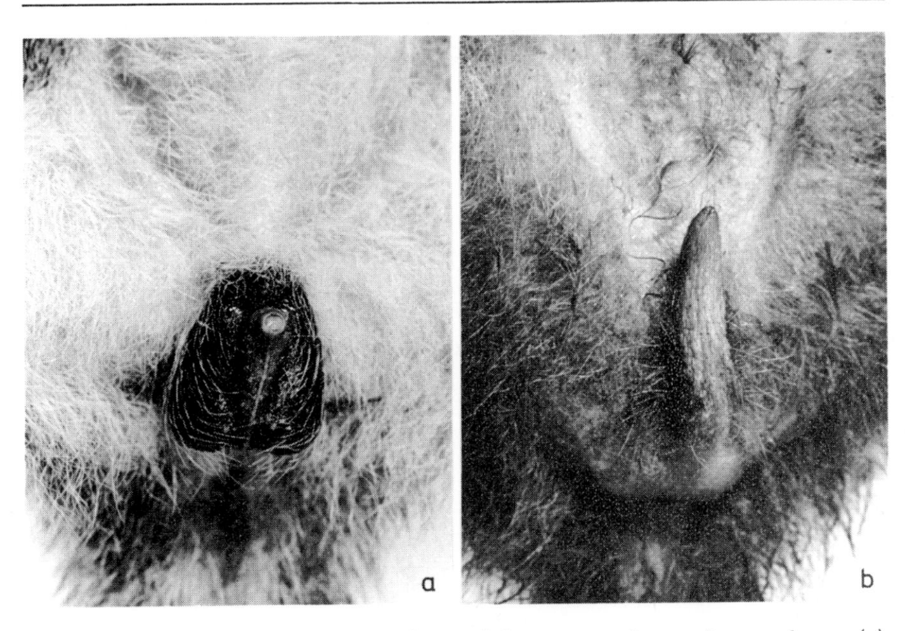

Figure 7. Position of the testes and form of the scrotum in a male tree-shrew. (*a*) Normal scrotum and static external position of the testes. (*b*) A male after 4 weeks of stress exposure leading to more than 70% SST. The scrotum is greatly reduced and the testes are in a static intra-abdominal position.

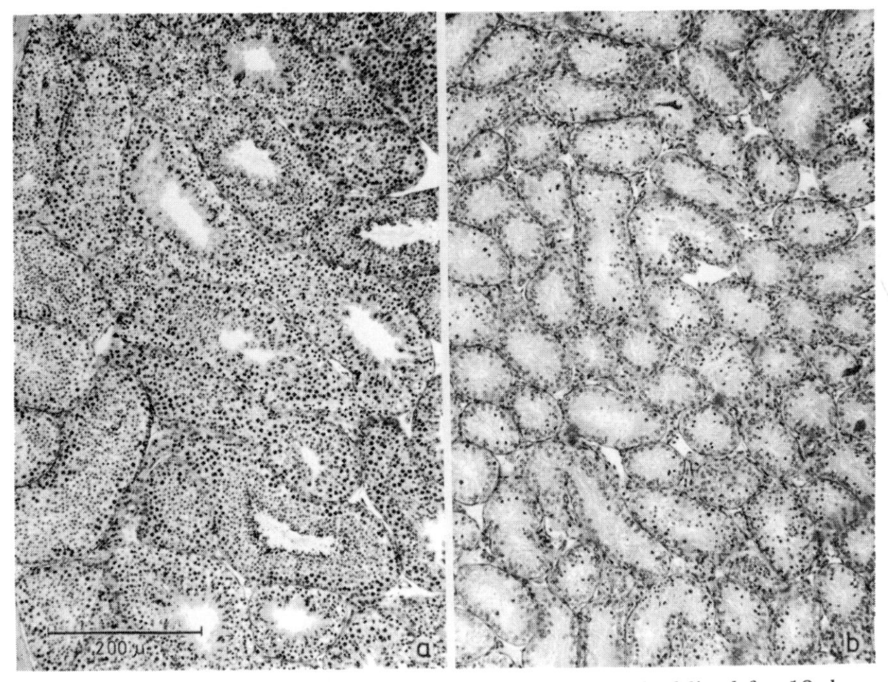

Figure 8. Testes of a control animal (*a*) and a male that had lived for 12 days with SST-values exceeding 70% (*b*). Sections 4 μ thick; staining with haematoxylin/eosin.

However, some disturbances of behavioural patterns and physiological processes can be observed in tree-shrews in the absence of any correlation with the social situation, or the SST-values present at that moment. Rather, they tend to be the result of relatively harmless (low SST-values), but long-lasting stressful situations. This continuous stress can result in after-effects lasting for months, or even lead to persistent physiological and morphological changes.

The following briefly covers some of these consequences, since they show how important the knowledge of an animal's history can be for the determination of the 'normal' behaviour or physiological state of a species.[9]

(*a*) *Body weight of adult males.* Males growing under undisturbed conditions achieve a body weight of more than 260 gms. as adults (at an age of about 6 months). If offspring are exposed to stress stimuli during their development, their growth rate is depressed. As adults, these animals are smaller (and lighter) than controls; even under optimal conditions they weigh only 150-250 gms. Their body weights depend on the duration and intensity of stress during their growth phases. At the same time, these animals differ distinctively in their physiological state from tree-shrews that have grown up undisturbed; their adrenal weight, relative to body weight, is

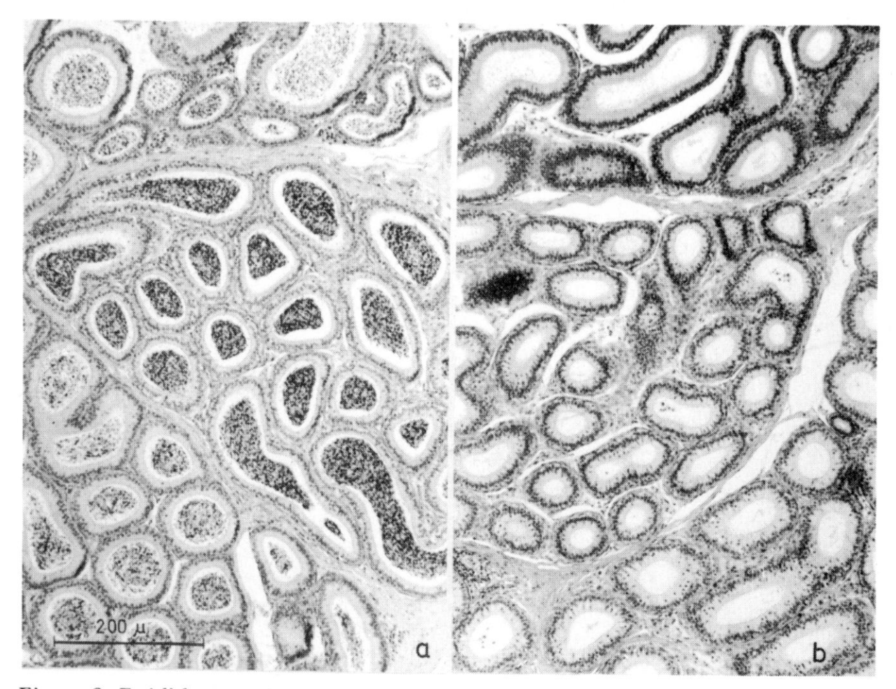

Figure 9. Epididymes of the males whose testes are illustrated in Fig. 8. (See text for details).

much higher than in controls. This indicates a higher basal activity of the adrenal system. Correspondingly, the concentrations of glucocorticoids and glucose in the blood are higher than in controls. Moreover, there are differences in liver glycogen, adrenal ascorbic acid concentration, and androgenic activity, i.e. stress during development results in animals that are morphologically and physiologically different from controls with undisturbed growth.[11]

(*b*) *Mammogenesis.* As mentioned earlier, a female displaying 20-50% SST can give birth to young every 45 days, these being eaten within a few hours after birth. While the first 2-3 litters are fed normally at birth (before they are eaten), subsequent litters are suckled only moderately, and after 4-6 litters, not at all. As far as we know, this is due to a failure of mammogenesis in the mother. When SST of the female falls below 20%, subsequent litters are not eaten, but they die of starvation nevertheless due to a lack of maternal milk production. Not before 2-4 litters have been produced (3-6 months) are the offspring suckled normally after birth (see also Fig. 6). The same effect on milk production is observed when a female has previously lived for many months with more than 50% SST, which induces sterility.

(*c*) *Suckling rhythm of females.* Even when the offspring of a female which had lived with high SST-levels for some time are fed normally again, i.e. after several months (see above), the subsequent litters usually die of starvation, because the normal 48-hour feeding rhythm of the mother is disturbed. Sometimes the mother suckles her young once to several times daily, sometimes normally – every 48 hours; in most cases, however, she does so after 3-4 days. The latter always results in the death of the young due to starvation during the first week, unless they are given an additional meal by the experimenter in the meantime. This 'omitting' of feeding is not due to inadequate milk production and associated tension of the mammary gland, because normal feeding of the young can be initiated regularly if the mother is locked in the nesting box with its young for some time (this was done, for example, with the litters 14-16 in Fig. 6). In experiments with mild stress of 6-9 months duration, the normal 48-hour feeding rhythm of the mother did not appear until 10-13 months after the end of the stressing period (*e.g.* litter no. 18 in Fig. 6).

These persistent after-effects of slight stress could certainly provide an explanation for the bad breeding results of several authors, and for the contradictory data on the 'normal' feeding rhythm in tree-shrews.

Conclusions

There is as yet no plausible physiological explanation of the after-effects mentioned above. The correlation between SST-value and growth, body weight, fertility and health of an animal, however, can be interpreted through a corresponding correlation between SST-value and the hormonal adaptation reactions of an organism to a stressor, as delineated by Selye in the 'general adaptation syndrome'.[3] This assumption is supported by data on adrenal function, i.e. adrenal weight and histology and corticoid levels in the blood. The externally measurable SST can, therefore, be taken as an indicator of the stress of an animal in a certain situation. This makes it possible to recognise through direct observation those social interactions, or stimuli arising from conspecifics, which cause SST and, with them, the detrimental ethological and physiological consequences.

Social causes of stress

Social stress is brought about almost exclusively by the following:

1. Density-effect: its extent depends solely on the number of mature animals of the same sex in the cage (without aggression).
2. Dominance-effect: its extent is based on dominance relationships that require at least one fight between the animals.

1. Density-effect

As mentioned earlier, a female tree-shrew can give birth to young every 45 days. After weaning, at an age of about 30 days, the young live with their parents in a close family group (without possibility of emigration in the laboratory). At night, family members usually sleep in the same nest. During the day as well, there is a strong tendency to rest with conspecifics; this might result in huddling-groups of up to 9 animals, lying in three layers one upon the other (Fig. 10). When meeting, the animals very often show mouth-licking (up to 15 min./day per animal; Fig. 11). Even after puberty, at about 50 days of age, there is no change in this close cohesion between parents and offspring for several months. Agonistic behaviour is completely absent and there are no dominance relationships of any kind. If, however, the young are placed with strange tree-shrews of the same sex, they will be attacked at once; this is not the case before they reach sexual maturity.

Social contact between parents and immatures never, or only very seldom, leads to arousal: therefore, SST-values of the parents remain

Figure 10. Huddling of adult tree-shrews in a family group.

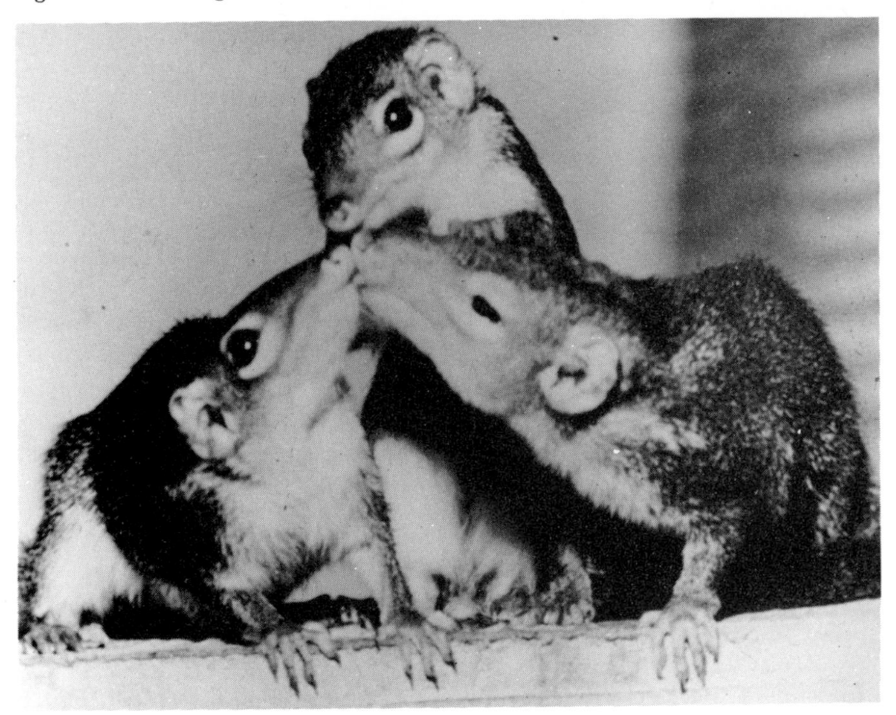

Figure 11. Mouth-licking (greeting) between adult male tree-shrews of a family group.

constant in spite of the increase in number of individuals in the cage. With puberty of the young, however, the SST-values of the parents increase distinctly to a new level: this happens in males only when a young male reaches maturity, and correspondingly in females only at sexual maturity of a young female (Figs. 6, 12).

In females, the addition of only one female to the cage results in a rise of SST above 20%, thereby suppressing further increase of the group because of cannibalism (cage size about 5-7 sq. m.). If a female gives birth only to males, their acquisition of maturity does not influence her. She continues to give birth regularly and raises the young normally until birth, or rather sexual maturity, of a female offspring (e.g. Fig. 6). This has resulted in groups of up to 9 animals. The mere presence of conspecifics, without aggression and dominance relationships, does not, however, lead to SST-values high enough to produce sterility either in males or in females.

2. *Olfactory marking*

The increase in the SST-values in parents does not result from observable visual, acoustic, or tactile stimuli. Instead, the arousing influence is based on some kind of olfactory marking of the young, independent of their immediate presence. Thus, after removal of the young, it takes several days before the SST-values of the parents decrease to the original levels. A corresponding increase in the SST-value also results from marking by strange tree-shrews of the same sex. This becomes apparent when the cage of an animal is inhabited for a short time by a strange conspecific with direct contact between the animals prevented, or when an animal is placed in the empty cage of a strange tree-shrew.

The increase in the SST-levels of the parents at the time of sexual maturity of their young indicates an influence of gonadal hormones. Support for this comes from experiments in which young males, castrated before puberty, do not cause an increase in the SST-value of their father, while castration after puberty results in a decrease in the SST-value of the father to the initial level. When castrated males are injected with testosterone, the SST-value of the father rises (Fig. 12).

The importance of gonadal hormones in the effectiveness of the olfactory marking can also be seen in their influence on the marking behaviour of male tree-shrews (females have yet to be investigated).[12] If a male is introduced into a thoroughly-cleaned experimental cage (PVC material; 70 x 50 x 50 cm. in size) for a fixed period of time (10 minutes in the present experiments), it will mark this cage several times with its sternal gland (Fig. 13). If the experiment is repeated, the number of sternal-markings is more or less the same for any tree-shrew, but there are considerable differences in mean values between the individuals, varying from 3 to

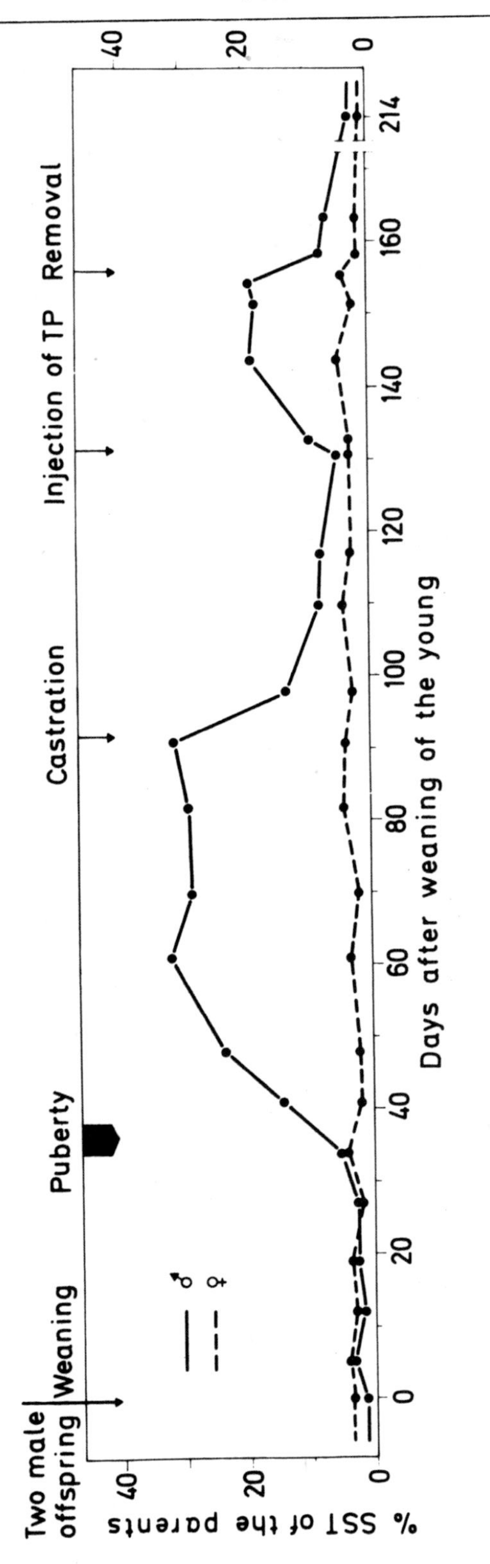

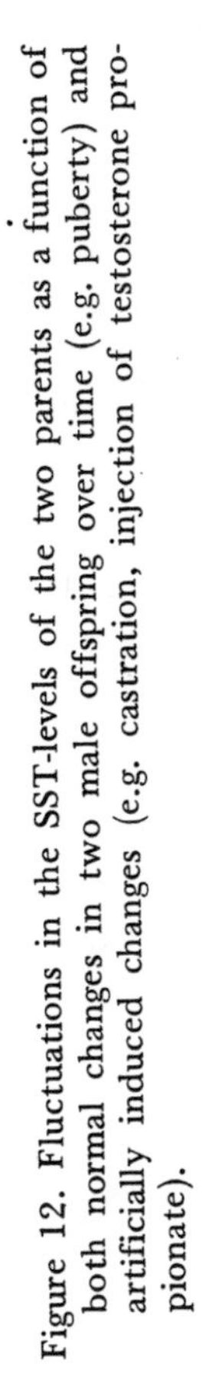
Figure 12. Fluctuations in the SST-levels of the two parents as a function of both normal changes in two male offspring over time (e.g. puberty) and artificially induced changes (e.g. castration, injection of testosterone propionate).

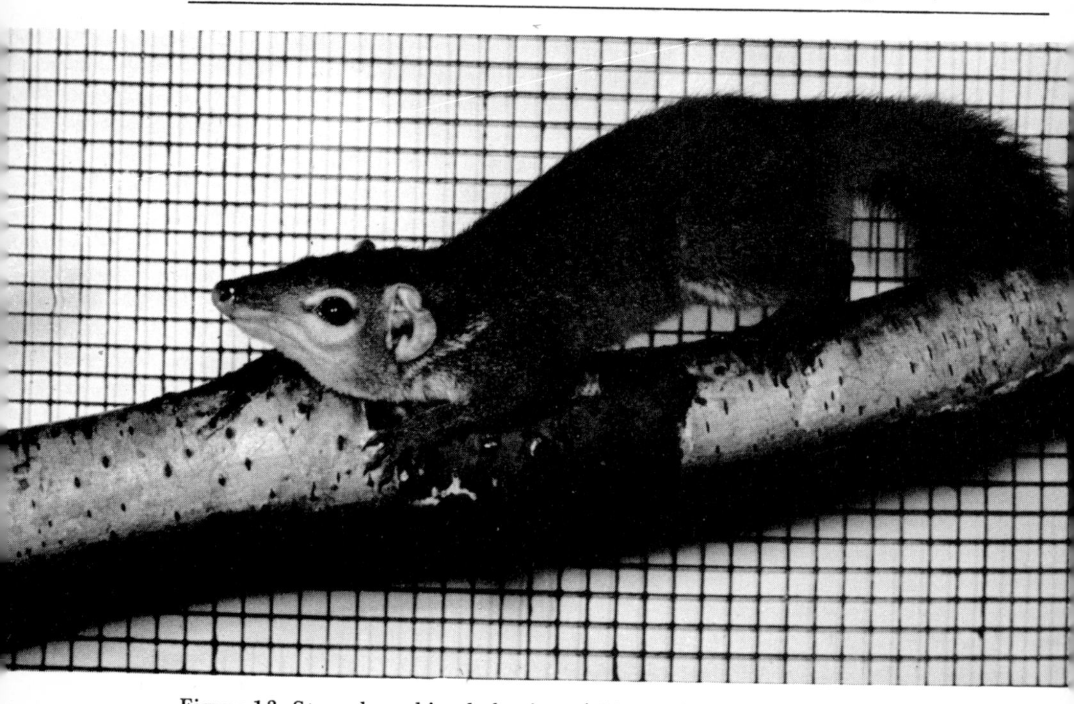

Figure 13. Sternal marking behaviour (chinning) of a male tree-shrew.

26 markings per 10 minute period (see also Fig. 14). These differences are at least to some extent due to corresponding differences in the blood levels of androgens. However, if the experimental cage is contaminated with urine, faeces and secretions of cutaneous glands of a strange mature male, the experimental animal is noticeably aroused, and marks significantly more than in an uncontaminated cage ($p < 0.05$-0.001). The increase in the mean values can be between 50% and 300% in different animals. The markings of immature or castrated males, as well as those of females or members of other species, have no or only very little effect on marking behaviour of males (Fig. 14). However, this does not mean that the secretions or excretions of the latter cannot be recognised or are odourless; males can accurately discriminate between the markings (secretions of sternal glands and urine) of mature, immature and castrated males as well as those of female tree-shrews, and from markings of other species, as has been shown by training experiments.[13]

The substances responsible for the arousing effect of the olfactory markings of mature males (and only of these) are not yet known. They could be the gonadal hormones themselves or their metabolites, excreted with urine or faeces, and/or secretions of certain glands.

The secretion of the male sternal gland, for example, has a highly arousing effect on other males. The production of this secretion is controlled by androgens, as is the marking behaviour which leads to

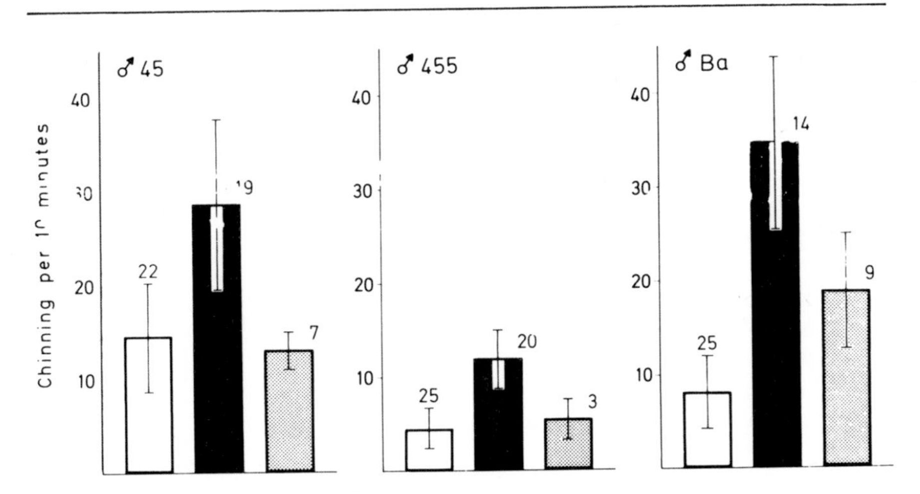

Figure 14. Chinning of adult male tree-shrews and its influence on conspecifics. Each bar indicates the mean value of chinning, with standard deviation and number of experiments, in the following experimental situations: cage uncontaminated (white bar), contaminated by a mature adult male tree-shrew (black bar), or contaminated by a castrated adult male tree-shrew (grey bar).

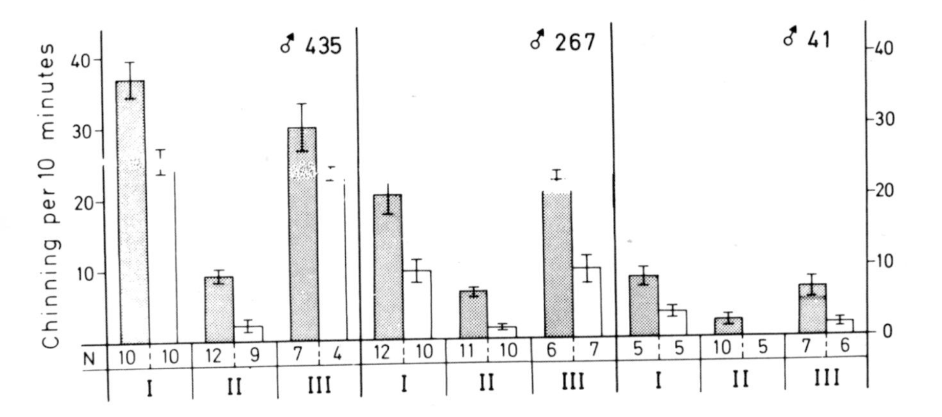

Figure 15. Chinning of male tree-shrews before (I) and after castration, either without (II) or with (III) testosterone-therapy. Experimental cage uncontaminated (white bar) and contaminated by a strange mature male (grey bar). Each bar indicates the mean value with its standard deviation. The number of experiments (N) is shown below each bar.

its distribution (Fig. 15). This does not mean, however, that only the secretion of the male sternal gland is involved. Rather, it is probable that various different (androgen-dependent) odorous substances work together in this arousing effect.

3. *Dominance-effect*

In the laboratory, as well as in nature, adult tree-shrews of both sexes immediately attack strange conspecifics of the same sex. Even in family groups, sooner or later (after 4-10 months), fighting will break out, usually terminating in less than a few minutes with the subjugation of one animal by the other, especially with males. During fighting, the tail hair of both animals is always maximally erected. Within 10 minutes after fighting has ended, the victor shows no further sign of physiological arousal (no SST) and pays virtually no attention to the submissive animal. The subordinate tree-shrew, however, cowers in a corner, which it leaves only to feed and drink. The hair of its tail is constantly erected. It hardly moves at all, spending more than 90% of the daily activity phase lying motionless in a corner and following the movements of the victor with its head. Approach of the dominant animal very often leads to 'fear squealing'.

During the following days, agonistic encounters between the two animals are very rare (less than a few minutes per day) or completely lacking in larger cages. The submissive animal, nevertheless, remains constantly aroused by the presence of the victor. This permanent arousal leads to a distinct loss in body weight and to death, even without any wounds. In contrast to the density-effect, the dominance-effect depends largely on vision. The olfactory markings or the smell of the victor alone never result in death. If the victor is removed, the submissive animal recovers in a few days, provided that the stressful situation has not persisted too long.

4. *Physiological cause of death*

The death of subordinate animals, even without wounds, leads to the question of the cause of death. The physiological and ethological changes under milder stress (below 80% SST) can be interpreted, as mentioned earlier, on the basis of Selye's 'general adaptation syndrome' and correspond with results from other species. Death under extreme social stress, however, has not yet been clarified for any species. Christian's postulation of exhaustion of the hypophyseal-adrenal system is based exclusively on adrenal weight and histology or histochemistry (mainly in rodents). It seems doubtful that these data allow a realistic statement as to the function of the adrenals.[14] Also, death either from hypoglycaemic shock or from haemolytic anaemia (snowshoe hare, vole) does not seem likely after more detailed investigation of the data.[15]

Therefore, the cause of death in subordinate tree-shrews was investigated in the following experiment. A cage (floor area 100 x 100 cm.; height 60 cm.) which could be divided by a vertical partition into two compartments of equal size was continuously occupied by a male 'fighter'. A strange male was introduced to the fighter, and it was usually attacked at once and subjugated in less than 3 minutes. The two tree-shrews were then separated from one another by the partition. Every 1 to 2 days, the fight and subsequent submission were repeated by removing the partition.

When the animals were separated by an opaque partition, the subordinate animal recovered as quickly as did the victor (no SST after 10 minutes). Even when a subordinate was defeated several times a day over a week or more, in this situation he lost little weight and never died.

If, however, both animals were separated from each other with a cage-wire partition, so that the subordinate animal was in continuous visual contact with the dominant rival, though immune from actual attack, it showed SST continuously and always died in less than 20 days.

The cause of the permanent arousal (SST), and with it the cause of death, is not the physical parameter of 'fighting', but the continued presence of the victor. The situation, therefore, does not operate as a source of stress through observable social interactions and their physiological results, but through central nervous processes based on experience and learning which take place in the subordinate animal. Correspondingly, there are distinctive differences in the serotonin metabolism in different brain areas between subordinate animals with or without presence of the victor, and controls.[16] Thus, tree-shrews die as a result of a type of stress which we can refer to as 'psycho-social'.

The physiological reactions of tree-shrews under this form of permanent social stress was examined. Some of the subordinates were kept in the stressful situation until death occurred after 2-16 days, but most experimental animals were removed from their cages at various periods after the first defeat, sacrificed and examined for physiological changes.[14,15]

Body weight of all animals decreased markedly from the time of first submission to death. The daily weight-loss was more or less constant for any individual, but there were considerable differences between animals (1.5-6.5% body weight loss per day). The more weight an animal lost per day, the sooner it died ($r = 0.89$; $p < 0.001$). However, death was not caused by exhaustion of available energy reserves of the body, as would occur with hypoglycaemia. In general, the glycogen content of the livers of subordinates was lower ($p < 0.003$; $n = 35$) than the mean control level, but even in dying animals the glycogen content was never exhausted. Also, the blood sugar content of stressed animals was as high as that of controls (90-105 mg./100 ml.).

The adrenal weight of the subordinates increased within 4 days from its initial value (18.0 ± 0.5 mg.; $n = 18$) to a new level (31.7 ± 0.9 mg.; $n = 21$), which was maintained virtually constant until death (Fig. 16). Correspondingly, the concentration of the glucocorticoid hormones in the blood was considerably above the control level. In the course of stress exposure, the haemoglobin content of the blood decreased ($r = 0.75$; $p < 0.001$) due to haemolysis and decrease in erythropoiesis. After 14 days, the mean value was only 50% of the initial value of 19.4 ± 0.2 *g.*/100 ml. blood ($n = 20$). Nevertheless anaemia was not the cause of death.

Non-protein nitrogen content of the blood increased in all tree-shrews during stress-exposure. The increase varied considerably from animal to animal, but all dying tree-shrews had values above 150 mg. per 100 ml. blood. This, together with the symptoms of dying animals (muscular twitching, convulsions, coma, decrease in body temperature) indicates *uraemia* as cause of death. The increase of non-protein nitrogen resulted from a marked decrease in renal blood flow. Thus, the glomeruli in the kidneys of subordinates were distended in comparison with control animals, and contained very few erythrocytes or none at all (Fig. 17). In the Bowman's capsule there was usually accumulated material, indicating increased permeability of the basement membranes. The lumina of the renal

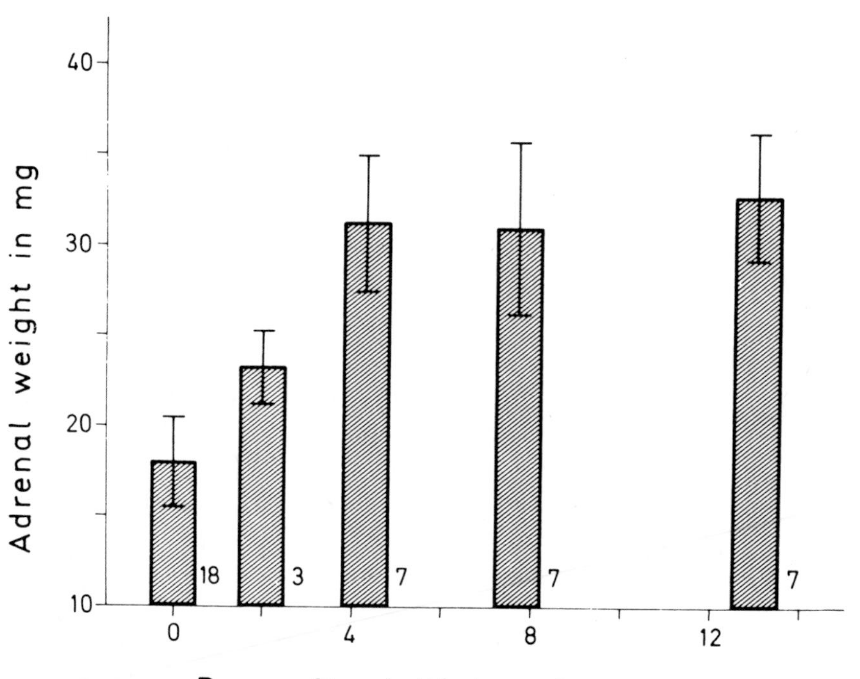

Figure 16. Adrenal weights in male tree-shrews under stress in the form of persistent visual exposure to a dominant conspecific following defeat. Each category (bar) shows the mean value with its standard error and the number of experimental animals. Categories: controls; subordinates after 2, 3-6, 7-10 and 11-16 days of stress, respectively.

tubules were more or less distended in all animals exposed to the stressing situation, while they had a closed appearance in controls (Fig. 17). The extent of decrease of renal blood flow varied from animal to animal, as can be seen histologically; there was a corresponding variation in the increase of non-protein nitrogen in the blood. In extreme cases, the blood urea-N rose nearly as quickly as in controls with surgically impaired renal function. Correspondingly, these subordinates died after 2 days of stress, as do animals without kidneys. This indicates almost complete arrest of renal blood flow due to psychosocial stress.

The exact cause of renal failure is not yet known. Primarily, there is probably constriction of the renal vessels through sympathetic fibres and catecholamines. Their influence alone, however, would not suffice; complementary mechanisms (renin-angiotensin, vasopressin, aldosterone) may be necessary to maintain or even increase the vasoconstriction.

The current concept of adrenal failure under strong social stress does not apply to tree-shrews; here death is brought about by uraemia which results from ischaemia or even anaemia of the kidneys. Whether this also applies to other species is unknown.

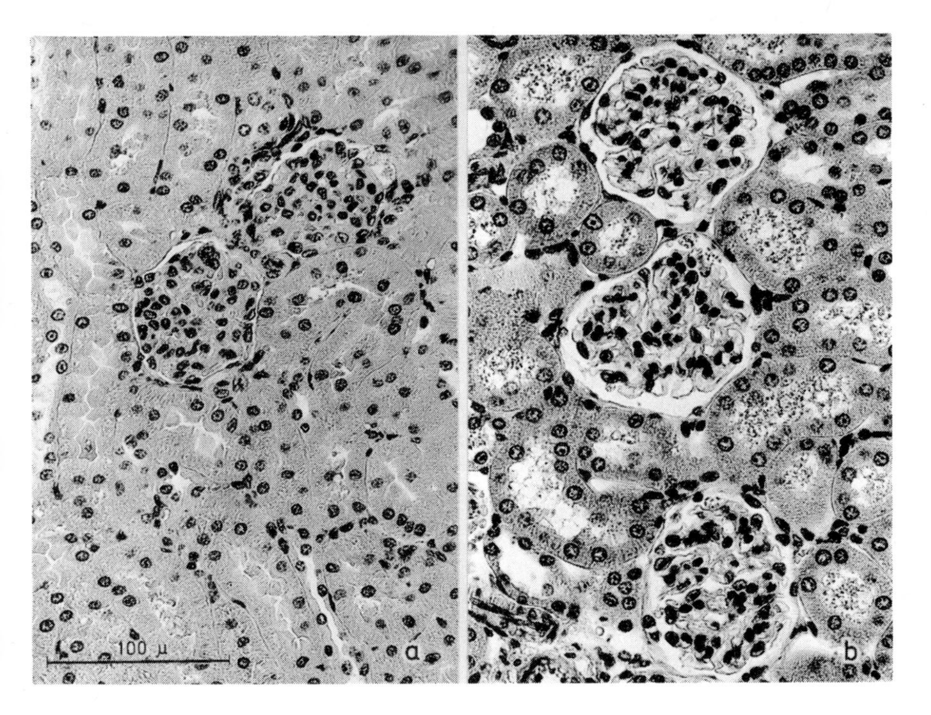

Figure 17. Kidneys of male tree-shrews. Sections 4 μ thick, staining with haematoxylin/eosin. (*a*) Control condition. (*b*) After 2 days of stress exposure: glomerular capillaries are distended, and contain very few erythrocytes or none at all. The lumina of the renal tubules are distended in comparison with controls.

Corresponding investigations of renal function have yet to be carried out in other animal species. However, the possibility seems to be present, since in all species so far studied (rats, rabbits, dogs, cats, monkeys, human beings) it is extremely easy to bring about renal ischaemia through stimulation of renal nerves and certain areas in the central nervous system, or through injection of catecholamines. Furthermore, various indications from the literature suggest that in natural populations too, it is renal function which is disrupted at times of high population densities and/or pronounced social stress (e.g. snowshoe hares, voles, lemmings, mice, prairie dogs, Sika-deer, human beings).[15]

Conclusion

This paper attempts to outline the social situations that lead to stress in tree-shrews, and to give some examples of the physiological consequences of social interactions. Whether and to what extent social stress may be effective in regulating numbers of tree-shrews in their natural habitat, as well as those of other species, cannot yet be discussed without detailed studies in nature. Without doubt, the behavioural situations leading to stress-reactions differ according to the different species and their social structures; the physiological reactions of organisms, however, are basically the same in all mammalian species. Thus, the present findings with tree-shrews show the significance which social interactions, and the resulting social or psychosocial stress, may have in the origin of many ethological and physiological disturbances and diseases (from changes in maternal behaviour to renal disease), possibly in human beings as well as in other mammals.

NOTES

1 Elton, C. (1942), *Voles, Mice and Lemmings*, Oxford.

2 Christian, J.J. (1950), 'The adreno-pituitary system and population cycles in mammals', *J. Mammal.* 31, 247-59.

3 Selye, H. (1950), *The Physiology and Pathology of Exposure to Stress*, Montreal.

4 Christian, J.J. (1963), 'Endocrine adaptive mechanisms and the physiological regulation of population growth', in Mayer, W.V. and van Gelder, R.G. (eds.), *Physiological Mammalogy*, vol. 1, *Mammalian populations*, New York.

5 Barnett, S.A. (1964), 'Social stress: The concept of stress', in Carthy, J.D. and Duddington, C.L. (eds.), *Viewpoints in Biology*, vol. 3, London.

6 Wynne-Edwards, V.C. (1962), *Animal Dispersion in Relation to Social Behaviour*, Edinburgh.

7 Cantor, T. (1846), 'Catalogue of Mammalia inhabiting the Malayan peninsula and islands', *J. Asiat. Soc. Beng.* 15, 171-279.

8 von Holst, D. and Raab, A. (1970), unpublished observations from Thailand (near Bangkok).

9 von Holst, D. (1969), 'Sozialer Stress bei Tupajas (*Tupaia belangeri*). Die

Aktivierung des sympathischen Nervensystems und ihre Beziehung zu hormonal ausgelösten ethologischen und physiologischen Veränderungen', *Z. vergl. Physiol.* 63, 1-58.

10 Martin, R.D. (1968), 'Reproduction and ontogeny in tree-shrews (*Tupaia belangeri*), with reference to their general behaviour and taxonomic relationships', *Z.f. Tierpsychol.* 25, 409-95 and 505-32.

11 Holst, D. von, (1972), 'Die Nebenniere von *Tupaia belangeri*', *J. comp. Physiol.* 78, 274-88.

12 Traum, S. (1972), Das Markierverhalten männlicher *Tupaia belangeri*. Diplomarbeit, Zool. Inst. Univ. München, unpublished.

13 Stralendorff, F.V. (1972), Das Sternaldrüsensekret männlicher *Tupaia belangeri*. Untersuchungen zu seinem olfaktorischen Informationsgehalt und dessen stofflichen Äquivalent. Diplomarbeit, Zool, Inst. Univ. München, unpublished.

14 Holst, D. von (1972), 'Die Funktion der Nebennieren männlicher *Tupaia belangeri*: Nebennierengewicht, Ascorbinsäure und Glucocorticoide bei kurzem und bei andauerndem soziopsychischem Stress', *J. comp. Physiol.* 78, 289-306.

15 Holst, D. von (1972), 'Renal failure as the cause of death in *Tupaia belangeri* exposed to persistent social stress', *J. comp. Physiol.* 78, 236-73.

16 Raab, A. (1971), 'Der Serotoninstoffwechsel in einzelnen Hirnteilen vom Tupaja (*Tupaia belangeri*) bei soziopsychischem Stress', *Z. vergl. Physiol.* 72, 54-66.

PART II
PROSIMIAN ANATOMY, BIOCHEMISTRY AND EVOLUTION

SECTION A
General Studies of Prosimian Evolution

E. L. SIMONS

Notes on Early Tertiary prosimians

Introduction

Several of my prior papers constitute reviews of Early Tertiary primate groups and more recent studies are incorporated in my book *Primate Evolution.*[1] What follows will be a series of notes and observations supplementary to the conclusions of that book. I will comment also on the most recent publications dealing with Early Tertiary prosimians. Because the scope of early primate history is so broad, and since more recent conclusions as to the relationships of evolving prosimians are stated at length in my book, I will take up only a limited series of topics in this contribution. Initially, I should say that discussions of prosimian evolution have, increasingly, been punctuated in recent years by sterile debates about higher category arrangements. Before discussing sequentially the Early Tertiary prosimians, some remarks on this subject seem warranted.

Use and misuse of higher categories

There has been some impetus given recently to revival of the allocation of Tarsiiformes to Anthropoidea, not Prosimii. This ranking was first seriously suggested by Hubrecht,[2] but actually even he was only making the point that *Tarsius* was more like Anthropoidea than are the lemurs. Such ranking was later espoused by Wood Jones, whose principal motivation for so doing seems to have been that he considered tarsiers more aesthetically pleasing as animals closely allied to man's ancestry than the monkeys and apes. Pocock proposed a suborder Haplorhini to include Mivart's Anthropoidea plus *Tarsius* and, by implication, its fossil allies.[3,4] Pocock, however, substantiated the transfer almost entirely on the basis of a few soft tissue characters. These were the dry rhinarium, the structure of the upper lip and nose and the nature of placentation. The latter had been urged by Gadow as resembling that of Anthropoidea.[5] Pocock gave only the post-orbital partition (or closure) of *Tarsius* as an osteological character allying it with monkeys, apes and man. Nevertheless, Simons and Russell[6] pointed out some time ago that the manner of bone growth accomplishing the partial postorbital closure of *Tarsius* is different from that of Anthropoidea. Thus partial postorbital closure and metopic fusion in it are probably consequences of the extraordinary orbital enlargement of that genus, and not of relatedness to higher primates. Starck reviewed the significance of the manner of placentation in *Tarsius* and concluded that the resemblance of its placenta in late stages of development to those of monkeys is based on convergence, since earlier in development and in implantation the two systems are quite different.[7] He concluded that the manner of placentation can only be used to bolster taxonomic systems that are founded on more secure data than that provided by the study of placentae, and should never be made the basis of classification on its own. Pocock's use of Haplorhini was more recently supported by Hill, but he founded his definition entirely on the nasal characters.[8] In any case, most authors have preferred to sustain the definition of Anthropoidea and Prosimii accepted by Simpson.[9]

A new transfer of Tarsiiformes into Anthropoidea is unwarranted for a whole series of cogent reasons. Some of these will perhaps not be taken as seriously by the neontologist as by students of primate history. Simply stated, this transfer of tarsioids to Anthropoidea would cancel out most of the familiar characteristics of higher primates that are to be found in their osteological and dental anatomy. If assignment to either suborder is to be considered possible for any extinct form, the bony parts (being the only portions of the animal concerned that are available) must include the defining characters of the two suborders. With tarsioids ranked among Anthropoidea, subordinal assignment of many extinct primate species would immediately be uncertain. It would be necessary to judge in each case whether the species concerned was significantly affined to *Tarsius* or not. Placing of *Tarsius* in Anthropoidea would thus raise the question with every single genus and species of Early Tertiary primates: whether it should or should not be carried along with tarsiers into Higher Primates. The futility of such procedures is well demonstrated by Hill, who assigned 31 extinct genera to Haplorhini.[8] My investigations show that with 23 of these genera there is no reason whatever to believe that there was (within Prosimii) a special relationship to *Tarsius*. Our quite inadequate knowledge of many Early Tertiary primate species and even genera would leave a very large number of them in an *incertae sedis* position between the two suborders. Retention of Tarsiiformes within Prosimii eliminates this problem. There is not even one species of past primates (with the possible exception of *Amphipithecus mogaungensis*) which would be in doubt as to its subordinal position if Simpson's usage is maintained.

That *Tarsius* possesses several interesting approximations toward Anthropoidea, not seen in other living prosimians, has been known at least since Burmeister published in 1846.[10] Nevertheless, these 'advanced features' are nearly all to be found only in the embryology and soft anatomy of *Tarsius* and are thus not applicable to fossils. It is to be expected that whenever an order of animals is divided up into two or more subsets (in this case suborders) each containing many members, that one or more members of one such subset will approximate more closely to the other subset (suborder) than does any other. Such consequences are inevitable whenever a broad spectrum of animals is ranked.

Transfer of Tarsiiformes to Anthropoidea would render non-diagnostic most characteristics that we consider today essential to the definition of Higher Primates. Characteristics that would have to be dropped include such features as enlarged, convoluted brains. The brain of *Tarsius*, although large in proportion to body size, is simple in construction and lacks well-defined convolutions. Postorbital closure and symphyseal fusion in the mandible would have to be discarded as distinctive of Higher Primates, since these advances are not fully expressed in *Tarsius.* The dental construction otherwise characteristic of Anthropoidea, such as presence of hypocones, would have to be dropped since *Tarsius* possesses teeth that are extraordinarily simple and primitively constructed and could hardly be less like that of living higher primates and still belong to a primate. The frontal fusion of *Tarsius*, resembling Anthropoidea, is most probably a parallelism or coincidence since it develops relatively late in juvenile life, not early as in Anthropoidea (defined excluding tarsiers). Moreover, as already stated,[6] the manner by which the parietal postorbital plates grow out in *Tarsius* differs from such growth in monkeys, apes and men. Frontal fusion in *Tarsius* occurs together with the marked flanging of the orbital margins and the full growth of the comparatively enormous orbits. The fusion is as easily interpreted as a result of the unique extreme of visual development seen in this animal as it is interpreted as a character of common heritage linking it with Anthropoidea. As will be discussed later, the character of possession of an ossified tubular auditory meatus in tarsiers and catarrhines can also be attributed to parallelism. Earliest catarrhines (such as the Yale skull of *Aegyptopithecus*) do not possess such a meatus. Plesiadapids, which on dental grounds alone cannot be related to catarrhines, do have an ossified tubular auditory meatus. Quite apart from the foregoing osteological consideration is the powerful evidence that derives from all recent biochemical investigation.[11] This shows that *Tarsius* is about as distinct from any Anthropoidea as is any other prosimian. Thus there is no longer any value in retaining the term Haplorhini in usage.

Such considerations as the above lead me to the plea that scholars in the future try to devote more time to the actualities of palaeontological discovery: the attempt to unravel, describe and explain the course and processes of the history of life. Such investigations are far more useful than the continued pedantic rearrangements of higher categories, the delimitation of which is ultimately a matter of taste. An example of the puerile nature of such efforts is the recent attempt by two or three authors to introduce the taxonomic rank of tribes into the literature on primates, although these have been little used before. Their widespread introduction into taxonomic discussions of the order now would serve no useful purpose. There is another reason, however, for *now* coining tribal group names among primates: it is the only rank where a present-day author can originate a higher category name and then refer frequently to himself as the author of such a new higher rank of primates. It is unlikely that arguments now in support of *new* families, superfamilies, infraorders or suborders for primates would gain wide acceptance. The practice of coining and discussing higher category names is rendered even more futile by Article 36 of the Code of Zoological Nomenclature, which gives authorship of all categories from subfamily through superfamily to whoever was the first proposer of any of the three. With primates, this means that authorship of a new higher category by a living author is now almost impossible to achieve, since so many prior alternative names are already available.

Szalay has recently argued that the primitive condition of the tympanic in primates is 'external to the bulla' and that therefore fossil forms like *Pronycticebus*, or the living lorisids, in which the ectotympanic is at or just inside the auditory opening are not derived from forms with a free ring inside the bulla like the living lemurs, but from forms like *Plesiadapis* which had a tube located largely outside the petrosal bulla.[12,13] Whatever the primitive position of the ectotympanic (which, developmentally, arises lateral to the bulla) was among primates, we know that *Pronycticebus* is ranked by all as close to species of the genera *Adapis, Notharctus* and *Smilodectes.* All three of these genera share with the Malagasy lemurs the condition of having a free ring within a petrosal bulla.

Discussions as to whether the ectotympanic was primitively located 'inside' the bulla as in lepticitid insectivores, tree shrews and lemurs, or 'outside' as postulated for other early insectivores, are unresolved, and are plagued by semantic problems, lack of direct observation and of precise definition. The ear drum and its supportive osseous ring functionally close the bulla. In a sense, the ring is no more inside nor outside the bulla than a window or door is inside or outside a house. Known fossils prove that, among primates, ectotympanic tubes have arisen independently, as among plesiadapids and hominoids; that in many forms the tympanic ring lies in the aperture of the bulla (which may thus be the primitive condition); and that extinct and living species that there is reason to consider primitive, such as adapids and lemurs, have a free *annulus tympanicus* located inside the bulla. We know that developmentally in animals as different as man and tarsier (which share an osseous, tubular ectotympanic) the tympanic begins development as a ring. I doubt that there are enough fossils documenting the ear structure, and enough understanding of the meaning of structure in the ear region, on which to base sound or far-reaching phylogenetic interpretations at present. Thus, I doubt the effectiveness of all such considerations as those of Szalay on the derivation of the tympanic in primates.

Palaeocene primates

There are four families of Palaeocene primates which have been dealt with in the recent literature: Plesiadapidae, Carpolestidae, Paromomyidae, and Picrodontidae. The first of these is the only group in which we know the postcranial anatomy well. *Contra* Szalay[13] no one, certainly not Napier and Walker,[14] has ever thought that this group, Plesiadapidae, represented vertical clinging and leaping primates. Rather, they have always been represented as heavily built, perhaps terrestrial, clawed, and rodent-like. What is important to remember is that we know that almost all of the members of the above listed families died out without descendants. Thus, when Napier and Walker discussed vertical clinging and leaping locomotion as being primitive for primates they, of course, had reference to the second, Eocene, radiation of primates of modern aspect. It is this second radiation and not the Palaeocene first radiation which is basal or 'primitive' to all the living primates.

The criticism of Charles-Dominique and Martin[15] that the position of Plesiadapidae as primates needs review has been dealt with adequately by Gingerich (this volume) and will not be taken up further here. Plesiadapids are of course primates. The striking similarities between molars of *Elphidotarsius* (Carpolestidae), *Pronothodectes* (Plesiadapidae), and *Pelycodus* (Adapidae, s.l.) justifies allocation of all to one order. The cranial and postcranial anatomy of the Adapidae in turn undoubtedly rank them as primates – a consideration that is quite apart from any similarities to modern forms in their dental structures.

Contra the discussion of the dental occlusion of *Plesiadapis* by Szalay,[12] the figure provided by that author of the lower incisor occlusion (when the molars are fully contacted) is not new and the figure not significantly improved over that illustrated by Simons.[16] Moreover, my 1960 drawing was made directly from the original materials in Paris, and Russell and I spent some time judging the issues, newly raised as novel by Szalay. Gingerich has fully demonstrated the facile nature of Szalay's argument that he had something new to report about the skull and dentition of this animal.[17] I am mystified that Szalay, in a more recent review,[13] entirely fails to take into account, or even mention, Gingerich's discussion. Instead, he lets stand his whole set of wrong assertions that prior workers never showed proper occlusion in *Plesiadapis.*

Justification of the picrodontids as primates recently attempted by Szalay and by others, nevertheless remains uncertain.[13,18] This is particularly true if groups with far more primate-like dental patterns, such as Microsyopidae, are excluded from the order by Szalay. Picrodontids are very small Palaeocene mammals of the genera *Picrodus* (middle) and *Zanycteris* (late) Palaeocene. Matthew originally considered *Zanycteris* a bat related to modern *Phyllonycteris* and the stenodermine and sturnirine bats.[19] Szalay carried out the type of one-sided transfer of these animals often now made, but which is outmoded. That is, he presented arguments purporting to show resemblance to and derivation from archaic primates, but he presented no complementary arguments to show why these genera are not bats. Instead, he deferred any discussion of why picrodontids are not bats to the prior comment of McGrew and Patterson.[20] Nevertheless, he neglected the cautions of these authors and of Simpson that there is very little to recommend transfer to the primates.[21,22] The derivation of the dental pattern of the teeth of *Picrodus* from primitive primate molars discussed by Szalay is topologically possible, but the lack of fossils documenting such a transition renders the derivation highly conjectural. What if the ear region of picrondonts, when discovered, proves to be even less like that of primates than the ear region of *Microsyops* (a non-primate for Szalay) is supposed to be? The final allocation of picrodonts depends on discovery of their now unknown basicranial structure.

Analyses currently being made at Yale by Brown, Gingerich, and myself on the Middle Palaeocene primate *Plesiolestes* suggest that *Plesiolestes* and *Palaechthon* are closely related (see Simons, 1963),[23] and that both are related to the ancestry of the family Microsyopidae (including *Microsyops, Niptomomys*, and *Uintasorex*). To the extent that this view is correct, all the above mammals may be considered primates.

Studies of the Carpolestidae, carried out recently under my direction by Rose, will lead to the publication of a definitive work on these poorly known animals. His investigations indicate that Jepson (personal communication) was correct in proposing that the three genera of these animals are all valid, *contra* Szalay,[13] and that they represent a time-successive series documenting exceptionally well a small and intriguing phylogenetic branch of Primates in extreme detail. Interestingly, there is high individual variability in the quarry samples of these primates. The genera concerned in this family are successively: *Elphidotarsius*, *Carpodaptes*, and *Carpolestes*. In Rose's judgement these appear to contain six valid, named species; two of *Elphidotarsius*, three of *Carpodaptes* and one of *Carpolestes*. The total number of specimens representing individuals of this primate family is approximately 225, including isolated teeth – of these only about a dozen specimens have been reported on in previous publications. All will be included in Rose's monograph. Because they document microevolution in action and ancient population variability, these animals, which left no descendants, are nevertheless of interest. The main trend in their evolution was toward extreme specialisation of the P_4 and P^{3-4}. This development placed the lineage as the principal primates to acquire the plagiaulacoid type of dentition, working on the assumption that its slight development in the German Palaeocene *Saxonella* is convergent. Interestingly, the molars, save for M_1, do not specialise away from the primitive primate condition, these teeth in all three genera being particularly close to those of plesiadapids such as *Pronothodectes*. The P^{3-4} of *Elphidotarsius*, not previously mentioned in print, are also close to those of *Pronothodectes* – the earliest plesiadapid. The resemblances confirm the view that carpolestids are primates, not *incertae sedis* as they were considered by Matthew and Granger,[24] and Saban.[25] They could have had a North American Puercan (early Palaeocene) common ancestry with plesiadapids. As mentioned, these animals make an outstanding example of evolution in action because of the documentation they provide of successive stages in the development of their basic dental adaptation. Trends are: (1) Reduction in size and number of the anterior premolars: P_2 is small in *Elphidotarsius* and is lost in the latter two genera; (2) P_3 decreases in size as P_4 enlarges; (3) The multiplication of cuspules on P_4 from about four in *Elphidotarsius* to eight or nine in *Carpolestes*; (4) The heel of P_4 is distinctly expressed in *Elphidotarsius*, but by the stage represented by *Carpolestes* entirely merges with the blade; (5) Trigonid of M_1 opens out and becomes blade-like in successive stages, thus extending the blade of P_4 posteriorly: by the time of *Carpolestes* this blade is doubled in length; (6) In P^{3-4} there is enlargement and relative complication of structure: the external row of cusps rises from two cuspules in *Elphidotarsius* to four or five in the later species.

There is little doubt that these fossils document the actual details of the successive evolutionary elaboration of the carpolestid dental mechanism. What this mechanism evolved to do is less certain. Direct attribution of the feeding habits of extant plagiaulacoid mammals (*Bettongia*, *Hypsiprymnodon*, *Burramys* or *Epiprymnis*) to carpolestids may not be relevant, since they share resemblances only in the lower blade and the teeth which oppose this above in these modern marsupials are nothing like those of carpolestids. Carpolestids were quite small, mouse-sized mammals. The smallest mandibles are little more than a centimetre and a half long – the largest, of *Carpolestes*, ranging perhaps up to 2.5 cm. long. Because of small size, their fossilisation and recovery could be considered less likely than with larger mammals. Nevertheless, in those few Palaeocene quarries where careful collection of small mammals has been conducted they occur in considerable numbers.

Finally, the feeding adaptations and diet of Palaeocene Primates have recently been speculated about by Szalay.[13] I doubt his conclusion (p. 27) that: 'It appears (but of course it cannot be proven) that the various Palaeocene families of primates fed predominantly on plant materials.' The Palaeocene primates exhibit a range of incisor, premolar, and molar configurations – and these configurations apparently represent a wide range of feeding adaptations. No living primates have procumbent incisors morphologically or functionally similar to those seen in specialised Palaeocene primates, though the Australian marsupial radiation provides close morphological parallels. Although some of the specialised Palaeocene primates, particularly certain Plesiadapidae, such as *Chiromyoides*, have often been analogised with the living Madagascan aye-aye, *Daubentonia*, their incisor function is clearly not the same. Daubentonia, unlike Palaeocene primates, has rodent-like, ever-growing incisors with chisel-like, sharpened tips. These front teeth function very differently from the occlusion plesiadapid incisors achieved. Kangaroos (Macropodinae) have a cropping incisor mechanism in which the lower incisors shear within a U-shaped arc formed by the upper incisors. They have no prominent projecting premolars and are herbivorous, feeding mainly by grazing and browsing. The Palaeocene primate *Plesiadapis* has a functionally similar incisor cropping mechanism, has no projecting premolars, and (based on analogy with macropodines) is apparently adapted to a predominantly herbivorous grazing feeding mode (see Gingerich, this volume).

The remaining Palaeocene primates, however, almost certainly did not feed predominantly on plant materials. The dentitions of Palaeocene and earliest Eocene primates of the genus *Phenacolemur*, for example, closely resemble the marsupial Sugar Glider *Petaurus* (Gingerich, in preparation), which feeds on insects, insect larvae, and small birds, as well as buds, blossoms, nectar, sap, and juices. The arrangement of cusps and crests of upper and lower molars of many of the Palaeocene primates (*Palenochtha, Elphidotarsius, Paromomys, Navajovius*) are closer to those of the living *Tarsius* than to those of any modern lemurs and lorises (some of which are mainly herbivorous). *Tarsius*, of course, is a predaceous insectivore with practically no plant component in its diet (see Fogden, this volume). Moreover, it is clear from all palaeo-ecological evidence that the regions from which these primitive primates come were warmer and wetter in the Palaeocene than they were again subsequent to the middle Eocene. This would ensure a year-round insect supply. The same tarsier-like molar conformation also characterises earliest anaptomorphids, adapids and microchoerines. Some of them, such as *Tetonius, Pseudoloris*, and *Nannopithex* are so tarsier-like as to strongly suggest that they had closely similar feeding habits; that is, they were mainly insectivorous. It is only with the evolution of larger body size, and much more broadened, flattened, and cuspidate cheek teeth in such genera as *Microchoerus* and *Notharctus* or *Smilodectes* that there is a suggestion of a shift toward plant-feeding in some Eocene primates. *Contra* Szalay,[13] these Palaeocene primates (and many of the early Eocene primates as well) *do*, by their clear-cut cheek tooth resemblance to *Tarsius*, substantiate the postulate made by Crook and Gartlan[26] that primitive primates were primarily insectivorous and carnivorous, with diets probably resembling those of *Microcebus, Galago demidovii* and *Tarsius.* Szalay was also incorrect in implying that long insect-stabbing upper front incisors, such as *Tarsius* has, need be a necessary correlate of insect-feeding preferences. Such evidence as is preserved in broken and fragile specimens shows that in many early Tertiary primates the upper front teeth must have been adequate for piercing or grasping insects. The long, pointed upper central incisors of *Tarsius* are unique among primates and may be a relatively recent acquisition. Upper incisor elongation is certainly not present in the smaller galagos nor in *Microcebus*, yet these animals are largely insectivorous (see Charles-Dominique, this volume).

Eocene prosimians

It has long been recognised that the various ancient families of primates have been split up too much into subgroups. This came about because of the wide adaptive range of their dental mechanisms, and because so little is known of their skulls and postcrania whereby group resemblances might be more easily demonstrated. Without such cranial and postcranial knowledge, any non-dental similarities between early groups of primates (apparently distinctive as super-families, families, or subfamilies) of course cannot be discussed. More postcranial and cranial remains are described from time to time, but stronger associations between groups of primates, particularly Eocene groups, have been made possible by recent improvements in understanding the topology of teeth among these early primates and by the reinterpretation of antemolar dental formulae in several different groups. Just as the new studies of Rose show that carpolestids and plesiadapids are much more closely related than had been thought, many of the smaller primates of Eocene times can now be drawn closer together. They share tarsioid dental features.

It is clear that there has always been a preferable case for ranking the mainly North American notharctine primates along with the European adapine primates in one family: Adapidae. Gazin chose to separate these two groups into two families Notharctidae and Adapidae and at one time I followed this, but do so no longer.[23, 27] Gazin essentially took the position adopted by Stehlin.[28,29] However, Gregory had already given cogent reasons for grouping the two as subfamilies within a single family.[30] Apart from the anatomical basis for this grouping, these subfamilies are now known to overlap geographically (in part) since *Pelycodus*, a notharctine, has been discovered in Europe. Moreover, the latest thinking on early Tertiary zoo-geographic relationships strongly weighs in favour of a direct North Atlantic connection between Europe and North America when *Pelycodus* existed (early Eocene).

Another slight change in grouping will be discussed later: that of the anaptomorphid (or tarsioid) prosimians and their relationship to two associations of genera which have often been treated as distinct families: the Omomyidae and the Microchoeridae (= Necrolemuridae). My present view is that, in the case of the above-mentioned five families (which have sometimes been recognised for various Eocene primates) a more probable ranking would recognise only three families. Two of these are extinct, while the third, Microchoerinae, is a group which has already been assigned to the family Tarsiidae.[31] *Contra* the statement in Szalay that their relationship to tarsiids has been 'alleged',[13] it is quite easy to demonstrate such an allocation for this particular Eocene subfamily. The assignment is founded on two principal sets of facts: 1. All details of dental and palatal morphology of the early microchoerine *Pseudoloris* are extraordinarily like those of *Tarsius*, save for its smaller size (see Simons, Fig. 66).[1] The upper teeth of another microchoerine, *Nannopithex*, are also extraordinarily like those of *Tarsius*, as are also those of at least one species of *Necrolemur*. 2. Craniological similarities shared by microchoerines and tarsiines are far too extensive to have their taxonomic assignment termed an allegation (see Simons, Table 1).[31]

In view of the astonishingly close dental similarities between *Pseudoloris* and *Tarsius*, any attempt to show that the groups are not clearly interrelated would merely be an attempt to create the illusion that there is something new to say about the relationships of these animals, and would certainly waste the time of students and scholars alike. No significant new microchoerine fossil material has been discovered during the past forty-five years, and in fact almost all the points and issues which could be raised in regard to this taxonomic matter are already dealt with in Stehlin,[28] Simpson,[32] Hürzeler,[33] and Simons.[31] In addition to this, a casual reading of the foregoing papers might fail to disclose a curious piece of motivation lying behind the strongly contrasting assessment of the microchoerines (or

Necrolemuridae as they termed them) held by Stehlin and Hürzeler. Stehlin had presented evidence for a relationship between *Necrolemur* and *Tarsius.* Later on, Hürzeler, who had studied with him, differed catagorically with the elder scholar, and argued that *Necrolemur* and its allies were not like tarsiers but resembled lemurs. All his positions for so arguing were taken in opposition to those of Stehlin, but when I investigated them, Stehlin always proved correct, Hürzeler wrong. As I have pointed out before, the science of palaeoprimatology is likely to operate at a level below the standards of competing sciences as long as hasty analyses, half-truths, and quotes out of context are the stock-in-trade of some of the workers in this field. This problem extends far beyond the analysis of *Necrolemur* and its allies. There have always been cogent reasons for sustaining close taxonomic ranking between tarsiines and microchoerines, and although representations to the contrary have been made, they can be sustained only by sophistry, not by facts. To some extent it may be a matter of taste whether one ranks the microchoerines in modern Tarsiidae or reserves a separate early Tertiary family rank for them, but their resemblances to *Tarsius* are real, not alleged.

It has been common practice since 1958, when Gazin raised the omomyines to a family rank separate from Anaptomorphidae, to deal with these families as distinct entities.[23] More recently Russell, Louis and Savage[34] pointed out that four of the five subfamilies included by Simpson[32] in the Anaptomorphidae were subsequently raised to family rank. Countering this trend, they again placed anaptomorphines and omomyines together in Anaptomorphidae, but gave little reason for this other than 'they overlap in many features'. In Simons, the view is also adopted that both should be considered subfamilies of one family, Anaptomorphidae.[1] Reasons for this are that the type and referred species of *Anaptomorphus* do not even exhibit some of the few characters, such as distinctly enlarged P_4, commonly accepted as distinguishing anaptomorphines. In addition, the lower dental formula for *Tetonius* of 0.1.2.3. proposed by Matthew and Granger was wrong.[24] This was revised by Gazin to an interpretation of $\frac{?}{1 \text{ or } 2}, \frac{1}{1}, \frac{2 \text{ or } 3}{2}, \frac{3}{3}$ for at least some *Tetonius.*[27] Recent studies by Szalay and at Yale (Simons; Bown and Gingerich)[1,35] show that in at least some *Tetonius* the position is $\frac{2}{2}, \frac{1}{1}, \frac{3}{3}, \frac{3}{3}$ and is therefore exactly the same as in the omomyines. There is thus clearly no wide divergence in dental formula, at least between the best-known members of the two subfamilies. Moreover, anaptomorphines are more closely similar to microchoerines in their dental formula than has been thought. Specimens of *Absarokius* represent a time sequence of three successive species. The earliest of these, *Absarokius abbotti*, is known from about 70 mandibles and maxillae recently found by my expeditions in the Bighorn Basin, Wyoming and about 60 more in other collections. This species first appears in the Lysite faunal zone, early Late Wasatchian stage, North America. It was apparently derived from a species of *Tetonius* of the earlier Wasatchian. Members of the latter are represented by about 75 new jaws in the Yale collection from the Bighorn Basin. These include half a dozen or so specimens preserving anterior teeth or teeth sockets. From all these new materials it is clear that the lower dental formula of *Absarokius noctivagus nocerai* (2.1.2.3.) arose from that of *Absarokius abbotti* (2.1.3.3.). This in turn is the same as is found in relevant Yale *Tetonius* specimens. Presumably the dental formula of the earliest anaptomorphids, s.l., was 2.1.4.3, as in *Teilhardina belgica.*[1] The dental formula derivation of *Absarokius noctivagus nocerai* shows the most plausible interpretation of the lower dental formula of the European microchoerine genera is the same as in advanced *Absarokius.* That is, *Pseudoloris, Nannopithex, Necrolemur,* and *Microchoerus* may well all have had the lower dental formula of 2.1.2.3., *contra* that proposed by Stehlin and supported by Simons.[29,31] This change in dental formula interpretation, however, does not cast doubt on a relationship of these animals to *Tarsius.* During the time represented by the late Wasatchian and Bridgerian North American provincial ages, the ancestors of microchoerines are not known. Nevertheless such forms, if derived from something like *Teilhardina*, would require an intermediate ancestor with a dental formula of 2.1.3.3. From such a stage the lower dental formula of *Tarsius* could be derived through loss of the lateral incisor. Clearly, the known microchoerines need not be in the direct lineage leading to *Tarsius* in order to be placed in the same family. In fact, *Pseudoloris* may yet prove to be a direct ancestor of *Tarsius*,[1] but this cannot be fully demonstrated without better knowledge of its anterior dentition.

The term Anaptomorphidae was coined by Cope, but without diagnosing it.[36] Therefore the first work which gave the group considerable discussion was that of Wortman, based on the Yale fossil prosimian collections.[37] He treated these animals as a family also including the omomyines and *Necrolemur.* Simpson indicated a similar concept unifying the three groups represented by *Anaptomorphus, Omomys*, and *Necrolemur* as subfamilies under Anaptomorphidae.[32] He also placed there the mainly Palaeocene paromomyines, together with another subfamily coined by him: Pseudolorisinae. In the latter he placed only *Pseudoloris.* Simpson was the only author to rank *Pseudoloris* separately in this manner. Simons demonstrated that *Pseudoloris* belongs in Microchoerinae (= Necrolemurinae).[31] Paromomyidae was raised in rank by Simons and has subsequently been considered as including Phenacolemuridae[38] by Simons.[1] As mentioned already, Gazin removed the omomyines to a separate family.[27] With these various removals the

Anaptomorphidae became a very much restricted taxon containing only the genera *Anaptomorphus, Tetonius, Absarokius*, and *Uintanius*, along with a few less-well-known, more doubtfully referred, or possibly synonymous genera, such as *Trogolemur, Uintalacus, Anemorhysis*, and *Tetonoides*. Simons, following Szalay in part, has suggested that *Tetonoides, Anemorhysis*, and *Chlororhysis* are probably all synonyms of *Tetonius*.[1] Sustaining the separation of anaptomorphines from omomyines are the following characteristics, most of which occur in species of the short list of anaptomorphine genera, *sensu stricto*, given above: Anaptomorphines have, relative to omomyines, cheek tooth cusps that are rounder or more inflated at the base, with talonids on the lower molars that are relatively less broad. Trigonids are typically shorter and narrower than in omomyines. As a result of these differences in anaptomorphines, molar cusps are somewhat more centrally shifted. In anaptomorphines, *s.s.*, such as *Tetonius*, the shape of the trigonid changes from front to back. The paraconid is progressively de-emphasised posteriorly, but in a manner different from that most often seen in omomyines. In *Tetonius* the M_1 trigonid cusps are distinctly formed and the angle formed by the crests running from paraconid and metaconid to the protoconid flares widely; in M_2 the paraconid and metaconid are brought closer together but are in line with each other; in M_3 they are even closer or may be coalescent. Omomyines differ from this in that the posterior de-emphasis of the paraconid in the successive molars comes about because this cusp is usually shifted more towards the midline, and decreases in size in the series from M_1-M_3. The P_4 in most anaptomorphines is typically much larger than are adjacent teeth, but *Anaptomorphus* (the type of the group) forms an awkward exception to this. The anaptomorphine genera *Tetonius* and *Absarokius* exhibit the *Nannopithex*-fold, which is also retained in microchoerines and is seen as well in the adapid *Pelycodus*, but not normally in omomyines. This unusual feature seems to bind all these groups together. Knowledge of anaptomorphine incisors is based mainly on the lower dentition, but where this is known the calibre of the central pair is usually great relative to the lateral pair of incisors. The same holds true of omomyines, but the disparity between central and lateral incisors is less great. Since anaptomorphines such as *Tetonius* are clearly related to modern *Tarsius* it is interesting that their lateral lower incisors are small compared to those of omomyines. This could be relevant to their complete loss in the modern genus. Interestingly, in the earliest Anthropoidea where lower incisor structure can be documented (*Apidium, Parapithecus, Aeolopithecus*), the lateral incisor root clearly has a far larger calibre than does the central root pair. This, then, is inconsistent with derivation of higher Primates from known Eocene prosimians, unless lower incisor size relationships somehow reversed in one or more lineages.

That some member of the 'Omomyidae' might have given rise to at least the platyrrhines was suggested by Gazin,[27] but he in turn was following or echoing the presentation of *Omomys* in this role by Wortman.[37] Now that assignment of the omomyines to Anaptomorphidae has been better justified, the question of whether or not omomyines were at least as tarsier-like as anaptomorphines is raised again. The attempt, on Simpson's part, to show that the postcranial structure failed to prove that the animal had ties with *Tarsius*, was probably well taken.[32] Nevertheless, his effort was slightly misleading inasmuch as close parallels to the unique adaptive features of the tarsier hind-limb may never be found in any early Tertiary form. This would certainly be the case if tarsier ancestors went through some of their final refinements in postcranial adaptation (such as development of tibiofibula) in the late Tertiary. Probably more relevant to this question is the unique skull of *Rooneyia* from the earliest Oligocene of West Texas. This animal is generally classified as an omomyine. Although the bullae are large and there is an ossified meatus, the pterygoid alae are long and the foramen magnum is so backwardly directed as to be quite unlike its orientation in *Tarsius.* In sum, there seems to be little in the craniology of *Rooneyia* that could be urged as being more tarsier-like than like galagines.

Recently Szalay (personal communication) has suggested that there are no Eurasian omomyines and that *Teilhardina* from the early Eocene of Belgium and France, long considered as a very similar, if not congeneric, relative of *Omomys*, is instead an anaptomorphine close to *Tetonius.* My studies of *Teilhardina* indicate that the situation is not so simple as that.[1] Although Teilhard originally referred this species to *Omomys*,[39] Simpson recognised its generic distinctiveness and coined the name *Teilhardina* for it.[32] However, he confused a mandible, edentulous save for P_4, of an insectivore with the hypodigm and so determined the wrong lower dental formula, 3.1.3 or 4.3. for *Teilhardina.* My study of the extensive new material, discovered since Simpson wrote, showed that the lower dental formula is 2?.1.4.3. The presence of four premolars, the relatively unenlarged P_4^4, and other unspecialised features, makes this species the most generalised anaptomorphid, *s. l.* known. The assessment of *Teilhardina* has been varied. Hürzeler[33] suggested that it was conceivably in or near the basal ancestry of the European Eocene microchoerines (Necrolemurinae). Later Quinet[40] proposed that *Teilhardina belgica* was at the base of the whole radiation of Anthropoidea. Although it seems to me too generalised for definite placement among anaptomorphines, *Teilhardina* may well be near the base of the Anaptomorphidae, *s. l.*

There are several other Eurasian Eocene prosimians, as well as a possible member of Anthropoidea, *Pondaungia,* from the late Eocene of Burma, which are currently being investigated at Yale; but there is not sufficient space to go into a discussion of all of these here. Valid

Eocene genera of adapids, anaptomorphids and microchoerines are reviewed in Simons.[1] It should be pointed out, however, that the pongid-like features of M^{1-2} of *Pondaungia cotteri*, Pilgrim,[41] tend to reinforce the original suggestion of Colbert that *Amphipithecus*, from the same general region and time in Burma, can be considered a higher primate.[42] One cannot explain away one of these as an anthropoid without discounting the other.

Oligocene primates

At the close of the Eocene, prosimian primates decline rapidly in variety and number in the northern hemisphere. The occurrence of some individuals of *Microchoerus ornatus* from Europe appears to extend the range of this genus into the early Oligocene. In North America only two genera and species, each known only from a single individual specimen, have been shown to exist in Oligocene times. These are a cranium of *Rooneyia viejaensis* from earliest Oligocene deposits of Presidio county in west Texas, and a single mandible of *Macrotarsius montanus* from a locality at Pipestone Springs, Jefferson county, Montana, again of earliest Oligocene age. Long after these had disappeared, in the Miocene, one additional primate *Ekgmowechashala* occurs in North America. Apart from these very few fossils from the northern hemisphere, all that we know about primates between about 35 and about 15 million years ago comes from the southern continents of South America and Africa.

At the outset of the record only higher primates, not prosimians, occur in either of these continents. In consequence, my concluding remarks will depart from the subject of this volume. Nevertheless, it seems advisable to stress to students of prosimians that the early Oligocene primates of the southern continents all show that they have clearly reached the grade we call Anthropoidea. Moreover, they exhibit far fewer features that could be taken as linking the two suborders than one might theoretically expect in view of their intermediate temporal position. In addition, as far as the African Oligocene primates are concerned, ample material from our recent expeditions to the Egyptian Oligocene deposits of the Fayum, Egypt now proves that the literature that has grown up around *Apidum* and *Parapithecus* is in many points wrong. Major errors about them have been published as recently as the *lapsus* of Hürzeler[43] suggesting that species of both of these genera were not even primates. As has been briefly mentioned elsewhere,[1,44–47] new *Parapithecus* material from the Fayum of Egypt shows that its initial describer, Schlosser, wrongly concluded that the type and only specimen was a juvenile with an unfused symphysis and a lower dental formula of 1.1.3.3.[48] As he then noted, the latter two features also typify modern tarsiers. The new Yale specimens of *Parapithecus* and *Apidium* show that the two are most closely related to each other, although differing distinctly in molar structure. Specimens of both show symphyseal fusion at an early age and have sockets indicating that large lateral incisors were present. Therefore, the dental formula in both is actually $\frac{2?}{2}, \frac{1}{1}, \frac{3}{3}, \frac{3}{3}$. These facts make it clear that, as far as the type of *Parapithecus* is concerned, breakage before or during collection shattered the symphysis and considerable bone from the incisor region was lost. When the horizontal rami were subsequently glued together, the remaining crack between the horizontal rami (resembling an unfused symphysis) and the high angle of posterior divergence of the tooth rows (exaggerated by the loss of symphyseal bone) mistakenly give it the dental formula, the high posterior mandibular divergence, and the unfused symphysis seen in *Tarsius.* These false evidences were the foundation for the conclusion of Schlosser, and later of Kälin,[49] that *Parapithecus* was an animal somehow intermediate between tarsioids and anthropoids. Actually, *Parapithecus* (at least in such parts as can be compared) is as different from any prosimians in its dental and mandibular structures as are many South American monkeys. In a longer discussion than can be detailed here (see Simons)[47] I have given the evidence that *Parapithecus* is neither a prosimian nor a hominoid, but is a monkey. It could be in or near the ancestry of later Old World monkeys. This conclusion is supplemented by a second account[50] which shows that the dental eruption sequence of *Apidium* and apparently also of *Parapithecus* is the same as in monkeys, not prosimians. In addition to this, Kay has been able to demonstrate that there are extensions of certain wear facets of both upper and lower molars in *Parapithecus* which are to be found elsewhere only among Old World monkeys.[51]

In view of the importance of the construction of the mandibular symphysis (whether fused or unfused) as one of the features used in segregating the catarrhine and platyrrhine primates, on one side, from the Prosimii on the other, it is surprising to read in Szalay[13] that:

> It is a well-known fact that living primates have solidly fused mandibular symphyses, yet the earliest members of the order, like all primitive mammals – with the possible exception of *Chiromyoides* – had a mobile joint at the symphysis. In all later lineages, however, including genera of the Madagascan radiation, the symphysis fuses early in ontogeny. It appears that among the primates, symphyseal fusion is an adaptive response to withstand great, horizontally directed stresses during chewing of plant materials, as opposed to the primitive placental conditions where the stresses of orthal shearing occlusion were primarily vertically directed.[13]

The foregoing observation conveys a misunderstanding of one of the fundamental generalisations of primatology: that all Anthropoidea, living and fossil, have fused symphyses, but that no living prosimians and few fossil ones show such fusion. Contrary to Szalay's remark that *Chiromyoides* is the only one of the earliest primates which might show fusion – that is, might lack the mobile, unfused symphysis – even casual inspection would show that *Chiromyoides* lacks symphyseal fusion. The actual early exceptions, that is, real cases of ancient prosimians with fused symphyses, are not mentioned. There *are* certain Eocene adapid prosimians where fusion occurs. That this is a convergent development, only parallel to the case in Higher Primates, is confirmed because earliest adapids, *Pelycodus*, and *Protoadapis*, lack symphyseal fusion. Again some few, certainly not all, of the Malagasy extinct prosimians show fused symphyses, but none of the living forms does. These large, extinct Malagasy prosimians are convergent in their dental mechanisms either to certain higher primates or towards other types of herbivores. Thus divergent, insular prosimians, like certain adapids, constitute an understandable (and equally independent) exception to the general rule that prosimians typically have unfused symphyses. To put this situation in numerical terms, there are approximately 190 species of Anthropoidea, living and fossil, and none show unfused symphyses. Enumerating prosimians on the other hand (and setting aside the less than twenty species of adapids and extinct Malagasy lemurs which do show fusion) there are no known species of living prosimians with such fusion. Moreover, there is a combined total of the remainder, some 140 living and fossil species, which have the primitive, unfused condition.

NOTES

1 Simons, E.L. (1972), *Primate Evolution: An Introduction to Man's Place in Nature*, New York.

2 Hubrecht, A.A.W. (1971), *The Descent of the Primates*, New York.

3 Pocock, R.I. (1918), 'On the external characters of the lemurs and of *Tarsius*', *Proc. Zool. Soc. Lond.* 55, 19-53.

4 Mivart, St. G. (1964), 'Notes on the crania and dentition of the Lemuridae', *Proc. Zool. Soc. Lond.* 1, 611-48.

5 Gadow, H. (1898), *A Classification of Vertebrata, Recent and Extinct*, London.

6 Simons, E.L., and Russell, D.E. (1960), 'Notes on the cranial anatomy of *Necrolemur*', *Breviora* 127, 1-14.

7 Starck, D. (1956), 'Primitiventwicklung und Plazentation der Primaten', *Primatologia* 1, 723-886.

8 Hill, W.C.O. (1955), *Primates, Comparative Anatomy and Taxonomy: 2. Haplorhini: Tarsioidea.* Edinburgh.

9 Simpson, G.G. (1945), 'The principles of classification and a classification of mammals', *Bull. Amer. Mus. Nat. Hist.* 85, 1-350.

10 Burmeister, H. (1846), *Beiträge zur Kenntnis der Gattung Tarsius*, Berlin.

11 Sarich, V. (1970), 'Primate systematics with special reference to Old World monkeys: A protein perspective', in Napier, J.R., and Napier, P.H. (eds.), *Old World Monkeys*, New York.

12 Szalay, F.S. (1971), 'Cranium of the Late Palaeocene *Plesiadapis tricuspidens*', *Nature* 230, 324-5.

13 Szalay, F.S. (1972), 'Paleobiology of the earliest primates', in Tuttle, R. (ed.), *The Functional and Evolutionary Biology of Primates*, Chicago and New York.

14 Napier, J.R. and Walker, A.C. (1967), 'Vertical clinging and leaping: A newly recognized category of locomotor behaviour of primates', *Folia primat.* 6, 204-19.

15 Charles-Dominique, P. and Martin, R.D. (1970), 'Evolution of Lorises and Lemurs', *Nature* 227, 257-60.

16 Simons, E.L. (1960), 'New fossil primates: a review of the past decade', *Amer. Sci.* 48, 179-92.

17 Gingerich, P.D. (1971), 'Cranium of *Plesiadapis*', *Nature* 232, 566.

18 Szalay, F.S. (1968), 'The Picrodontidae, a family of early primates', *Amer. Mus. Novitates* 2329, 1-55.

19 Matthew, W.D. (1917), 'A Paleocene bat', *Bull. Amer. Mus. Nat. Hist.* 37, 569-71.

20 McGrew, P.O. and Patterson, B. (1962), 'A picrodontid insectivore(?) from the Palaeocene of Wyoming', *Breviora* 175, 1-9.

21 Simpson, G.G. (1935), 'The Tiffany fauna, upper Palaeocene: I. Multituberculata, Marsupialia, Insectivora and ?Chiroptera', *Amer. Mus. Novitates* 795, 1-19.

22 Simpson, G.G. (1937), 'The Fort Union of the Crazy Mountain Field, Montana, and its mammalian faunas', *U.S. Nat. Mus. Bull.* 169, 1-287.

23 Simons, E.L. (1963), 'A critical reappraisal of Tertiary primates', in Buettner-Janusch, J. (ed.), *Genetic and Evolutionary Biology of the Primates*, New York.

24 Matthew, W.D. and W. Granger (1921), 'New genera of Palaeocene mammals', *Amer. Mus. Novit.* 13, 1-7.

25 Saban, R. (1961), 'Carpolestidae Simpson 1935', in Piveteau, J. (ed.), *Traité de Paleontologie*, Paris.

26 Crook, J.H. and Gartlan, J.S. (1966), 'Evolution of primate societies', *Nature* 210, 1200-3.

27 Gazin, C.L. (1958), 'A review of the middle and upper Eocene primates of North America', *Smithson. Misc. Coll.* 136, 1-112.

28 Stehlin, H.G. (1912), 'Die Säugetiere des schweizerischen Eocäns', *Verhdl. schweiz. paläont. Gesell.* 38, 1165-298.

29 Stehlin, H.G. (1916), 'Die Säugetiere des schweizerischen Eocäns', *Verhdl. schweiz. paläont. Gesell.* 41, 1299-552.

30 Gregory, W.K. (1920), 'On the structure and relations of *Notharctus*, an American Eocene primate', *Mem. Amer. Mus. Nat. Hist.* 3, 53-243.

31 Simons, E.L. (1961), 'Notes on Eocene tarsioids and a revision of some Necrolemurinae', *Bull. Brit. Mus. Nat. Hist.* (Geol. Ser.) 5, 45-69.

32 Simpson, G.G. (1940), 'Studies on the earliest primates', *Bull. Amer. Mus. Nat. Hist.* 77, 185-212.

33 Hürzeler, J. (1948), 'Zur Stammesgeschichte der Necrolemuriden', *Schweizer. paläont. Abhand.* 66, 1-46.

34 Russell, D.E., Louis, R. and Savage, D.E. (1967), 'Primates of the French Early Eocene', *Univ. Calif. Publ. Geol. Sci.* 73, 1-46.

35 Bown, T.M. and Gingerich, P.D. (1972), 'The dentition of the Early Eocene primates *Niptomomys* and *Absarokius*', in press.

36 Cope, E.D. (1883), 'On the mutual relations of the bunotherian Mammalia', *Proc. Acad. Nat. Sci. Phila.* 35, 77-83.

37 Wortman, J.L. (1903, 1904), 'Studies of Eocene mammalia in the Marsh collection, Peabody Museum', *Amer. J. Sci.* (1903), 15, 163-76, 399-414, 419-36; 16, 345-68; (1904), 17, 23-33, 133-40, 203-14.
38 Simpson, G.G. (1955), 'The Phenacolemuridae, new family of early primates', *Bull. Amer. Mus. Nat. Hist.* 105, 417-41.
39 Teilhard de Chardin, P. (1927), 'Les mammifères de l'Eocene inférieur de la Belgique', *Mém. Mus. roy. Hist. nat. Belgique* 36, 1-33.
40 Quinet, G.E. (1966), '*Teilhardina belgica*, ancêtre des Anthropoidea de l'ancien monde', *Bull. Inst. roy. Sci. nat. Belgique* 42, 1-14.
41 Pilgrim, G.E. (1927), 'A *Sivapithecus* palate and other primate fossils from India', *Mem. Geol. Surv. Ind.* 14, 2-26.
42 Colbert, E.H. (1937), 'A new primate from the Upper Eocene Pondaung Formation of Burma', *Amer. Mus. Nov.* 951, 1-6.
43 Hürzeler, J. (1969), 'Questions et reflexions sur l'histoire des Anthropomorphes, *Ann. Paléont.* 54, 13-41.
44 Simons, E.L. (1967), 'The significance of primate paleontology for anthropological studies', *Am. J. Phys. Anthrop.* 27, 307-31.
45 Simons, E.L. (1969), 'The origin and radiation of the primates', *Ann. New York. Acad. Sci.* 167, 319-31.
46 Simons, E.L. (1971), 'A current review of the interrelationships of Oligocene and Miocene Catarrhini', in Dahlberg, A.A. (ed.), *Dental Morphology and Evolution*, Chicago.
47 Simons, E.L. (1974), 'A new species of *Parapithecus* from the Oligocene of Egypt, with notes on the initial differentiation of Cercopithecoidea', in press.
48 Schlosser, M. (1911), 'Beiträge zur Kenntnis der oligozänen Landsäugetiere aus dem Fayum (Ägypten)', *Beitr. Paläont. Österreich-Ungarns* 24, 51-167.
49 Kälin, J. (1961), 'Sur les primates de l'Oligocène inférieur d'Egypte', *Ann. Paléont.* 47, 1-48.
50 Conroy, G.C. (1974), 'Dental eruption patterns in earliest Cercopithecoidea', in press.
51 Kay, R. (1974), 'Dental wear patterns in *Parapithecus*', in press.

A.C. WALKER

A review of the Miocene Lorisidae of East Africa

Introduction

The Miocene lorisids of East Africa are the only known African fossil prosimians sampled from populations older than the Pleistocene. As such, they offer the only palaeontological clues to prosimian evolution on the African mainland and may help in understanding the separate evolution of the lemurs of Madagascar. The African Pleistocene galagines seem so close, morphologically, to living species of *Galago* that they only serve to demonstrate that a species very close to, or the direct ancestral species of, *Galago senegalensis* was established 2 million years ago in northern Tanzania.[1] The subfossil Madagascan lemuroids are so recently extinct[2] that they must be considered as biologically contemporaneous with modern species and their importance as evolutionary indicators is at the same level as modern lemuroid skeletal remains. Although Miocene lorisids are our only knowledge of earlier African prosimians, it must be stressed that they are sampled from very local populations from a very few sites peripheral to early central volcanos and that each site has only sampled from populations living over a fairly short period of time. Strictly speaking, we are dealing with parts of isolated prosimian populations in a very limited area of East Africa, which were living at separate times over a fairly limited period. We are fortunate in that the fossil sites have steadily yielded new specimens and the number of isolated parts has now increased considerably since the first specimen was described by MacInnes. The deductions that can be made from the fossils have increased with the specimens. The evolutionary story has not, by the same token, become simpler.

Historical

The first, and for some time only, specimen was described by MacInnes as the holotype of *Progalago dorae.*[3] More material was added to the hypodigm of this species by Le Gros Clark and Thomas,[4] who created at the same time two new species of the genus, *P. robustus* and *P. minor*. Le Gros Clark later described what is still the most complete cranium, but refrained from designating it specifically.[5] Leakey made *Mioeutocus bishopi* as a new taxon to accept a facial skeleton from Napak.[6] Simpson, in his revision, divided the material available to him into two species of *Progalago, P. dorae* and *P. songhorensis*, and created a new genus, *Komba*, to receive *P. robustus* and *minor.*[7] At the same time, he proposed a new genus and species *Propotto leakeyi*, but the specimens concerned were later removed from the order.[8] The first postcranial remains were described in 1970.[9] Since Simpson's revision, more specimens have been discovered and full descriptions of these will be published shortly, only brief notes on the most important ones being given here.

The problems of interpretation

One of the basic problems of interpretation involves the dating of the different fossiliferous deposits. Simpson, while discussing what was then known about the localities, pooled all his available specimens.[7] The differences he could see in his samples could be accounted for, he suggested, by differences in facies or chances of discovery. In fact, now that larger samples are known, there do seem to be meaningful differences between the Songhor, Napak and Rusinga assemblages, and some of these differences may be accounted for on the suggested difference in age. The radioisotopic evidence has been summarised by Bishop,[10] and Bishop et al.,[11] and it appears likely that the Napak and Rusinga assemblages come from deposits roughly 18-19 million years old, while the available evidence points to the Songhor deposits being perhaps as much as a million years older. Differences in sampling in two contemporaneous deposits can produce the effect of mild assemblage differences that need not reflect the living assemblages. Similarly, even minor palaeoecological differences may reflect themselves in slightly different mammalian assemblages. The unknown factors are so many that, without detailed knowledge of the sedimentological and taphonomical features of the individual fossil sites, practically any statement that purports to deal with the 'palaeoecology' is meaningless.

Another basic problem is the fragmentary nature of the material.

There are the remains of only two individuals that probably have parts of the associated upper and lower dentitions. Equally, apart from these two cases, there is only one other in which dental and postcranial remains are associated. In some cases, therefore, the upper and lower dentitions of a species have to be matched by morphology and 'fit' between different individuals and this has in the past led to difficulties. The postcranial remains are assigned by size using modern species as a guide, and as chance would have it not one of the commoner postcranial elements is matched by any of the associated material.

On the comparative side, when dealing with the modern Lorisidae, there is a high proportion of monotypic genera (four out of five). These are distributed over enormous geographical ranges and hence representative samples across the ranges show fairly high variability, a complicating factor when comparing them with fossil samples from successive populations in small areas.

Taxonomy

The taxonomy of the Miocene lorisids is based mostly on dental and cranial remains. The scheme used here differs slightly from that of Simpson, since use is made of associated partial upper and lower dentitions that were not available to him.[7] Simpson felt that without postcranial remains he could not assign any species to either subfamily. I have attempted to do so, with the postcranial evidence now available, although the reasons for placement in a subfamily do not rest entirely on postcranial evidence.

Family Lorisidae Gregory 1915
- Subfamily Galaginae Mivart 1864
  - *Progalago* MacInnes 1943
    - *Progalago dorae* MacInnes 1943
    - *Progalago songhorensis* Simpson 1967
  - *Komba* Simpson 1967
    - *Komba robustus* (Le Gros Clark and Thomas 1952)
    - *Komba minor* (Le Gros Clark and Thomas 1952)
- Subfamily Lorisinae Flower and Lydekker 1891
  - *Mioeuoticus* Leakey 1962
    - *Mioeuoticus bishopi* Leakey 1962
    - *Mioeuoticus* sp. nov.

The size range of the fossil species, as far as can be judged from the known parts, is as great as seen today in the genus *Galago*. *K. minor* is the size of *G. demidovii*, *K. robustus* the size of *G. alleni*, while *P. songhorensis* is a little smaller and *P. dorae* a little larger than *G. crassicaudatus. M. bishopi* is roughly the size of *Arctocebus calabarensis*, and *Mioeuoticus* sp. nov. is a somewhat larger species, about the size of *Perodicticus potto.*

Because *Mioeuoticus bishopi* is the only species with a type based on upper dentition, difficulties of comparison arise since it can be compared only with referred upper dentitions of *Progalago dorae*, the only species similar to it in size. Simpson believed that the cranium described by Le Gros Clark belonged to *Progalago dorae*,[5,7] but I cannot accept this in view of new material from Songhor. I do accept the close similarity of the type of *M. bishopi* with the cranium and think that the minor differences are at a specific level. For this reason, whereas Simpson almost (but not quite) removed the genus *Mioeuoticus* in favour of *Progalago*, I have kept the genus and include the Rusinga cranium as a new species of it. This species will be named and described elsewhere.

An isolated tooth (R.649'49) was placed as a P_2 in *P. dorae* by Le Gros Clark and Thomas,[4] and later referred by Simpson to lorisid indet., *not P. dorae*, on the grounds of size.[7] Reading Simpson's note on this specimen, it is implied that there is present, represented by one tooth, another species of lorisid larger than *P. dorae.* This tooth is, in fact, a hominoid deciduous canine, probably a left lower deciduous canine of *Limnopithecus macinnesi.*[12]

Cranial and dental material

Twenty-seven specimens are known that include part or all of the mandibular body. As far as is known, no pair of these come from one individual. Of this number, roughly half preserve some part of the symphyseal region. Clark and Thomas believed that the degree of procumbency of the tooth-comb was less than in modern lorisids.[4] Simpson, using the same specimens, thought that the degree of procumbency was probably the same as in modern species.[7] Walker put forward new evidence from Uganda to support Simpson's view that the tooth-comb was fully developed in the Miocene forms.[13] Several new specimens are now known that give a picture of symphyseal morphology and the roots of the six anterior teeth for *Progalago dorae* and *songhorensis* and *Komba robustus* and *minor.*

All specimens that include the symphysis show that the two bodies were not fused. The shape of the symphyseal surface is nearly elliptical in all specimens. An inferior symphyseal tubercle is developed in all specimens, the strength of the tubercle being proportionate to the size of the mandible – *P. dorae* having the greatest tubercle and *K. minor* the smallest. This tubercle is for the insertion of the anterior belly of *m. digastricus* and is also developed in modern lorisids to a greater or lesser extent depending on size. The symphyseal surface of the larger species is also more rugose than the

Table 1 Depth of mandible at various levels in fossil lorisids

(Ranges of measurements in millimetres, numbers of observations in brackets)

	at P_3	*at P_4*	*at M_1*	*at M_2*	*at M_3*
K. minor	–	3.8-4.0(2)	2.8-4.0(5)	2.6-3.0(4)	3.0-3.5(2)
K. robustus (Songhor)	4.3-4.8(2)	3.8-4.5(3)	3.8-4.7(5)	4.0-4.3(3)	4.0-4.5(2)
K. robustus (Rusinga, Napak and Mwafanganu)	5.6(1)	5.3(1)	5.0-5.1(3)	4.4-5.3(6)	4.4-5.5(6)
P. songhorensis (Songhor)	–	6.0(1)	6.4(1)	6.2-6.5(2)	6.7(1)
P. songhorensis (Rusinga)	–	6.0(1)	6.1-6.6(2)	7.3(1)	7.5(1)
P. dorae	–	6.7-7.2(2)	7.1(1)	7.9(1)	8.7(1)
Mioeuoticus sp. nov.	–	8.5 (1)	8.4 (1)	9.0 (1)	9.0 (1)

smaller ones. This region does not differ in any meaningful way between *Progalago* and *Komba.* The mental foramina of *Progalago* and *Komba minor* are single and relatively large, whereas those of *K. robustus* are double and each is relatively small.

The form and dimensions of the mandibular bodies of *Progalago* and *Komba* species are different. Le Gros Clark and Thomas did not consider those features to be of generic significance, but Simpson did.[4,7] The mandibles of *Progalago* species deepen posteriorly to gain their maximum depth posterior to M_3, but the thickness of the mandibles is not increased proportionately so that the posterior part of the body is thin and plate-like. In species of *Komba*, the posterior deepening is not seen and roughly the same thickness of body is maintained along the length. Measurements are given in Table 1. A mandible that I have referred to *Mioeuoticus* sp., that was previously included in *P. dorae*, is extremely deep and of roughly constant depth along the length. It has a double mental foramen with each foramen of fair size.

Alveoli, or alveoli with broken roots, of the anterior six teeth are seen in ten specimens. In all, the canine and incisor roots were laterally compressed and packed tightly together. The angle of inclination of the alveoli or roots is within the range of modern species. Although no complete grooming comb tooth is known, the evidence strongly suggests that the comb in galagines was as well developed in the Miocene as at the present time.

The rest of the mandibular dentition is now known for the species in the following parts: *P. dorae* P_3-M_2 ; *P. songhorensis* P_3-M_3 ; *K. robustus* P_3-M_3 ; *K. minor* P_4-M_3 ; cf. *Mioeuoticus* sp. M_2-M_3. Scale drawings of reconstructed mandibles and dentition are given in Fig. 1. The general aspect of the lower dentition is very much like that found in modern galagines, except that the P_4 is not molariform in the fossils. The differences between the two *Progalago* and two *Komba* species lie mostly in the more bunodont character of the former against the relatively more acrodont character of the latter. There is, however, the same difference (albeit slightly less marked) between the small and large members of the modern genus *Galago* and the possibility that the differences are largely due to size cannot be excluded. Specimens of *K. robustus* from Rusinga and Mwafanganu that are larger than the Songhor ones have wider trigonids and talonids as well as less acrodont cusps. The relative size of the trigonid also decreases with increasing size of tooth in *K. robustus*, and in the Rusinga and Mwafanganu specimens, which are increased in size to nearly as large as the Songhor *P. songhorensis*, the

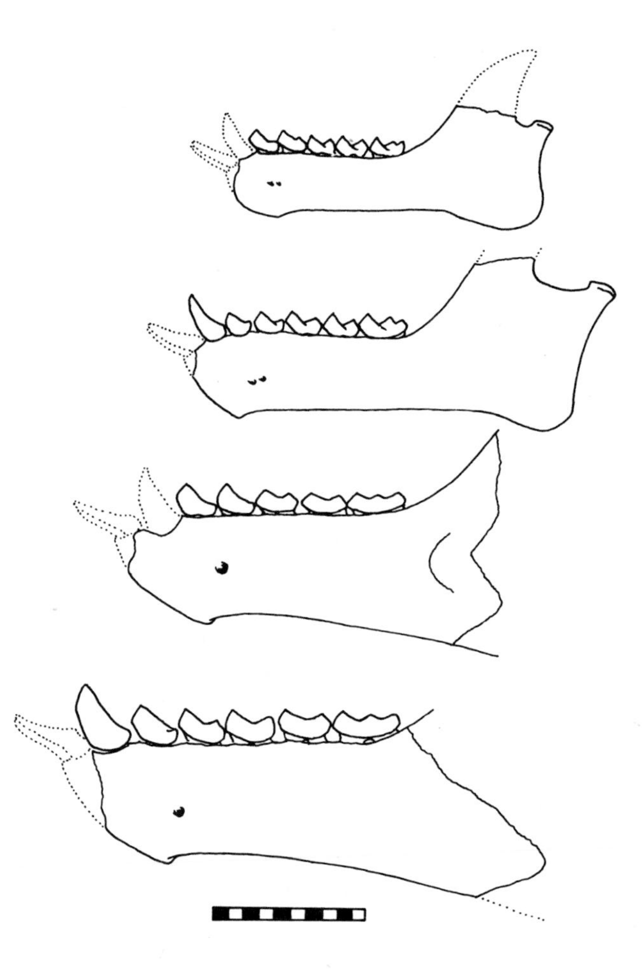

Figure 1. Lateral views of mandibles and dentition of Miocene lorisids. Species illustrated are, from above downwards, *Komba minor, Komba robustus, Progalago songhorensis* and *Progalago dorae*. The scale is in millimetres.

proportions of the talonids to trigonids are almost identical to those of that species. This may be taken as evidence that the dental differences between *Komba* and *Progalago* are not striking enough to warrant generic separation of the forms.

The lower dentition referred to *Mioeuoticus* has teeth larger than *P. dorae* with very expanded trigonids and talonids that indicate a quadrate form of upper molar.

The number of specimens of upper dentition has been increased since Simpson's revision and the finding of associated upper and lower teeth of *P. songhorensis* and *K. robustus* from Kathwanga, Rusinga Island leads to clearer associations of the known upper and lower dentitions. The upper dentitions are known from parts that include: *P. dorae* P^3-M^3; *P. songhorensis* P^2, P^4-M^3; *K. robustus* P^4-M^3; *K. minor* – nil; *Mioeuoticus bishopi* P^2, P^4-M^3; *M.* sp. nov. P^4-M^3. Information can be gathered about the anterior dentition of *P. dorae* and species of *Mioeuoticus*, and one deciduous premolar of *P. dorae* is known. The two most complete upper dentitions and facial skeletons are both of *Mioeuoticus*, a fact that is of some importance. Knowledge of the palate and facial skeleton of other species is limited to fragments of bone still in place around the tooth roots. *Progalago* and *Komba* species had, evidently, lightly built facial skeletons that have fragmented quickly prior to fossilisation or when weathering out of the sediments.

The upper dentition of *Progalago* and *Komba* shows the general situation as found in modern galagines, with the following exceptions: the P^4 is not as molariform, and the development of the molar hypocones is not as extreme. The degree of reduction of the third molar is less in the smaller species and in all forms is a little less than in modern galagines; a feature that is correlated, of course, with the fact that the M_3 in the fossils are relatively larger than their modern counterparts. Of the upper premolars, only P^2 is two-rooted in *P. dorae* and *K. robustus*, the remainder are three-rooted. In *Mioeuoticus bishopi* the same is true, but in *Mioeuoticus* sp. nov. both P^3 and P^4 are two-rooted as Le Gros Clark noted, not as Simpson recorded.[5,7] Simpson noted that the P^2 and P^3 preserved in one specimen assigned to *P. dorae* were 'clearly similar to *G. crassicaudatus* and markedly unlike the shorter, more transverse teeth of all Recent lorisines'. *Mioeuoticus bishopi, M.* sp. nov. and *Progalago dorae* have a depression in the palate just posterior to the canine that would have accepted the caniniform P_2. P^4 of *P. dorae* and *K. robustus* are basically bicuspid teeth and do not show the molarisation seen in Recent galagines. The P^4 of *Mioeuoticus* is also bicuspid but differs from the others in being lower crowned and more rectangular in outline due to the strong development of a distolingual cingulum. This cingulum is seen in *P. dorae* and *K. robustus*, but is very weakly developed.

The upper molars of *P. dorae* and *P. songhorensis* as they are known now are quite like those of *Galago* spp., except that the hypocone, though distinct, is more closely applied to the protocone and that the teeth are less wide transversely. The few known upper teeth of *P. songhorensis* are a little more transversely widened than in *P. dorae.* Upper molars of *K. robustus* are more acrodont with a more isolated hypocone than in *Progalago* spp. These teeth are also wide transversely. The M^3 is less reduced than in *P. dorae.* Upper molars of *Mioeuoticus* are strikingly different from those of *Progalago* and *Komba.* M^1 and M^2 are large and almost square in outline, with well developed buccal and distolingual cingula. M^3 is more triangular in outline due to the absence of the hypocone, but the distobuccal cingulum is still present. All the molars are low-crowned, bunodont teeth. *Mioeuoticus* molars show a striking resemblance to those of *Arctocebus*, but the latter are more acrodont. The resemblance to *Perodicticus* that Le Gros Clark pointed out is, I think, overstressed.[5]

Facial and palatal parts of *Progalago* and *Komba* are limited to small fragments supported by tooth roots. A fragment of maxilla shows that in *P. dorae* the canine root made a moderately strong jugum on the lateral surface, fairly similar to that of *G. crassicaudatus.* As far as can be judged, the palate of *P. dorae* was flat and not strongly arched. The orbits of *Progalago* and *Komba* were seemingly as large as in modern species of *Galago*, with the exception of *G. elegantulus,* but the orbital floor approaches the molars more in *P. dorae* than in the other species. The facial skeleton of *Mioeuoticus* is strongly built, with the canine root making a strong maxillary jugum. The orbits were large and their floors very thin. The frontal bones have a depressed area either side of a gentle sagittal swelling at the level of the zygomatic process. The ethmoid contributes to the medial wall of the orbit.

The rest of the cranium is known from two specimens only, a partial cranium with natural endocast and a cranium of *Mioeuoticus*, both from Rusinga. These two are strikingly different and are certainly from different genera. Now that the nearly complete cranium is seen to be of *Mioeuoticus*, it seems that the partial cranium is most probably either *K. robustus* or *P. songhorensis.* The two species are of roughly the same size and it is impossible to say, at present, to which species this belongs. Simpson thought that this specimen was probably of *K. robustus*, but he was using his assignation of the nearly complete cranium as *P. dorae* at the time. This no longer applies and the cranium could be of either *K. robustus* or *P. songhorensis.* The implications of recognising that the nearly complete cranium is of *Mioeuoticus* are considerable, for features of that cranium have an unmistakeable lorisine stamp, while the other cranium has, as Le Gros Clark and Thomas suggested, a definitely galagine aspect. Among the lorisine characters of the cranium of *Mioeuoticus* are the following:

1. Cranium strongly constructed.
2. Temporal ridges raised and very distinct.
3. Orbits more upwardly directed.
4. Bulla and mastoid weakly inflated.
5. Internal nares wide.
6. Basicranial flexion slight.

Among the galagine characters of the endocast specimen are the following:

1. Cranium lightly constructed.
2. Temporal lines present but not developed as strong ridges.
3. Bulla strongly inflated.
4. Basicranial flexion moderate (as in Recent *Galago* of same size).

The ear regions of both specimens have been cleaned on one side to expose the periotic. In all features except for the degree of bulla inflation they resemble each other as well as modern lorisids. Several features of the ear region show that these two specimens have unequivocal lorisid affinities. The details of the round and oval windows are exactly the same as in modern lorisids. The promontorial artery lay in a groove that runs across the surface of the periotic and was not encased in bone as in lemuroids. There is a clearly developed bony division of the tympanic cavity into hypotympanic and tympanic ones, unlike the lemurid condition where only one large tympanic cavity is found. On the cranial base there is in both specimens a well developed *foramen lacerum medium*, as in lorisids, but unlike the lemurid condition (with the exception of the cheirogaleines). The ectotympanic ring in the larger specimen lies in a position very similar to that seen in *G. crassicaudatus*, but the ring itself is thicker and less closely moulded to the bulla wall. The smaller specimen seems to have a thinner, more closely applied, ectotympanic.

Postcranial remains

Since my note on the first limb bone specimens, others have been found to give a more complete picture.[9] As far as can be told, no specimens are known of the postcranial remains of *Mioeuoticus*. Two sets of fragments are associated with dentition of *P. songhorensis* and *K. robustus* and one pelvic fragment with teeth of *P. dorae*. The following parts are now known, using the associations available and the size analysis that I applied to the original fragments.[9]

P. dorae Left acetabulum and ischium.

cf. *P. dorae*	Proximal part humerus, proximal and distal parts femur, distal hallucal phalanx, calcaneum.
P. songhorensis	Distal end tibia, capitate, a terminal phalanx, talus.
cf. *P. songhorensis*	Proximal and distal parts humeri, proximal part ti
K. robustus	Two vertebrae.
cf. *K. robustus*	Distal part humerus, proximal parts femora, calcaneum.
cf. *K. minor*	Distal part humerus, proximal and distal parts femora, calcaneum.

The new associated material confirms the usefulness of the size analysis that I used earlier. However, I made the size analysis using Simpson's dental references and the reassigning of dental material and the finding of new specimens has caused some minor changes in my placement of the limb bones. I no longer believe that we have any limb bone material of *Mioeuoticus* (I had, following Simpson's hints, dropped that genus in favour of *Progalago*) and the three fragments assigned to *M. bishopi* probably belong to other species of either *Progalago* or *Komba*, both of which genera are represented at Napak.

The major features of the limb bones of *Progalago* and *Komba* are galagine and even in some minute details the bones are remarkably like those of Recent species. In only one major respect do they differ and that is in the lesser elongation of the calcaneum. It may be, as I suggested before, that all tarsal elements were elongated and that the total foot length might have been approximately the same as in living species. However, the test of that hypothesis, the finding of an elongated cuboid, has not yet come about.

The major limb bone epiphyses in galagines do not fuse until after the eruption of the third molars. The pelvic specimen associated with teeth of *P. dorae* shows that the same was true in that species. I have been unable to confirm whether the same is true of the lorisines.

At all events the basic galagine locomotion was established by the early Miocene in the East African species and, as far as I can judge, no fossil material exists that can give us any indication as to whether or not the lorisines had by then developed their locomotor propensities, although the chances of their having done so are quite high.

Summary

The new specimens of fossil lorisid from Kenya help to show that both the Lorisinae and Galaginae were established by lower Miocene times. The lorisines are represented by two species of one genus separated in time and space, and although the sample is extremely small the specimens are the best we have of all the lorisids. The galagines are represented by four species and their fossil remains are more common but more fragmentary. Charles-Dominique reports mean densities of 10 lorisines of two species and 80 galagines of three species per square km. in the forests of Gabon.[14] On the basis of minimum numbers of individuals, and using all fossils, the ratio of lorisines to galagines in Miocene collections is close to 1:9. The close agreement of these two ratios may be due to chance or differences in preservability between the two subfamilies, but the small number of lorisine specimens remains a fact and it could possibly be that this represents a density situation such as is present today. A size range is seen among the galagines as great as is seen among living East African species.

There is a strong possibility that the ancestors of at least some modern *Galago* species are present in the Miocene assemblage. *K. minor*, as Le Gros Clark and Thomas noted, is especially close dentally to *G. demidovii* and *P. dorae* seems very close to *G. crassicaudatus.* Postcranially, all four galagine species appear to be very similar, despite the substantial size differences, and all closely resemble modern species except for the shortness of the calcaneum. It is very likely that the locomotion of the extinct forms was extremely similar to that of modern *Galago* species. Because the two smallest species are placed in *Komba* and the two largest in *Progalago*, and because of the possibility that some distinguishing features are merely those of size, the wisest step might prove to be to remove Simpson's genus *Komba* and leave all these Miocene galagines in one genus. The taxonomy as it stands implies that species of one or the other of the Miocene genera cannot be ancestral to *Galago* species, or, conversely, that the genus *Galago* is not a natural genus and that to reflect phylogeny it should be divided. Removing the taxon *Komba* would solve this difficulty.

Simpson's criteria for creating *Komba* as a distinct genus have a basis in his believing *Euoticus* and *Galago* to be distinct genera, and were further complicated by his acceptance of the Rusinga cranium as being *P. dorae.* It is very likely, given the apparently stable condition of the equatorial forest belt of Africa since before the Miocene, that the ancestors of East African species are to be found in East Africa and, similarly, of West African species in West Africa. Adaptation to a more arid environment, such as is seen in *Galago senegalensis,* could have been caused by a forest species colonising the peripheral areas or, alternatively, by adaptation of one species to the changing conditions of the peripheral areas as the forests contracted. *Galago inustus*, in this regard, can be viewed as representing either something like the *G. senegalensis* forest ancestor or as a forest species developed from *G. senegalensis. G. inustus* is not, as far as I can determine from the morphology, closely related to *G. elegantulus*, and those who favour a subgeneric rank for *Euoticus* should not place *G. inustus* therein.

Mioeuoticus might be considered as closest to *Arctocebus* among the lorisines, but equally plausibly it could be part of the ancestral African lorisine populations, or for that matter the ancestral lorisine one. Why lorisines and not galagines crossed into Asia is a vexing question. I can think of no reasonable explanation except that the galagine types of niches were occupied by tarsiids of some sort. In the absence of any fossil evidence the question remains unanswered.

Table 2 Sites from which Miocene lorisids have been recorded

	Songhor	*Rusinga*	*Napak*	*Moroto I*	*Mwafanganu*	*Koru*
K. minor	+	+	+	+	–	–
K. robustus	+	+	+	–	+	+
P. songhorensis	+	+	–	–	–	–
P. dorae	+	–	+	–	–	+
M. bishopi	–	–	+	–	–	–
Mioeuoticus sp. nov.	–	+	–	–	–	–

Acknowledgments

I thank the Director and Trustees of the National Museums of Kenya for permission to examine specimens in the National Museum, Nairobi. The late Dr L.S.B. Leakey and P.J. Andrews kindly gave me permission to study specimens collected by them.

NOTES

1 Simpson, G.G. (1965), 'Family Galagidae', in Leakey, L.S.B. (ed.), *Olduvai Gorge 1951-1961*, London.

2 Walker, A. (1967), 'Patterns of extinction among the subfossil Madagascan lemuroids', in Martin, P.S. and Wright, H.E. (eds.), *Pleistocene Extinctions*, New Haven, Conn.

3 MacInnes, D.G. (1943), 'Notes on the East African Miocene primates', *J. East. Afr. Uganda Nat. Hist. Soc.* 17, 141-81.

4 Le Gros Clark, W.E. and Thomas, D.P. (1952), 'The Miocene lemuroids of East Africa', *Fossil Mammals of Africa* 5, London.

5 Le Gros Clark, W.E. (1956), 'A Miocene lemuroid skull from East Africa', *Fossil Mammals of Africa* 9, London.

6 Leakey, L.S.B. (1962), in Bishop, W.W. (ed.), 'The mammalian fauna and geomorphological relations of the Napak volcanics, Karamoja, *Rec. Geol. Surv. Uganda (Entebbe)*, 1-18.

7 Simpson, G.G. (1967), 'The Tertiary lorisiform primates of Africa', *Bull. Mus. Comp. Zool.* 136, 39-61.

8 Walker, A. (1969), 'True affinities of *Propotto leakeyi* Simpson 1967', *Nature* 223, 647-8.

9 Walker, A. (1970), 'Post-cranial remains of the Miocene Lorisidae of East

Africa', *Am. J. Phys. Anthrop* 33, 249-61.
10 Bishop, W.W. (1971), 'The late Cenozoic history of East Africa in relation to hominoid evolution', in Turekian, K.K. (ed.), *The Late Cenozoic Glacial Ages*, New Haven, Conn.
11 Bishop, W.W., Miller, J.A. and Fitch, F.J. (1969), 'New potassium-argon age determinations relevant to the Miocene mammal sequence in East Africa', *Amer. J. Sci.* 267, 669-99.
12 Andrews, P.J., personal communication.
13 Walker, A. (1969), 'New evidence from Uganda regarding the dentition of Miocene Lorisidae', *Uganda Journal* 33, 90-1.
14 Charles-Dominique, P. (1971), 'Eco-éthologie des prosimiens du Gabon', Biol. Gabon. 7, 121-228.

C. P. GROVES

Taxonomy and phylogeny of prosimians

Phylogenetic systematics

The aim of the present paper is to present gaps in our knowledge as much as facts. The unsatisfactory state of primate taxonomy today is due, at least in part, to the explosion of research on primates (especially on prosimians) in recent years; the old certainties are gone and a host of new information is available, still largely undigested and still incomplete, but sufficient to point the way to radically new assessments. New areas of research are indicated in their turn, and it is the role of taxonomy to provide the synthesis and indicate possible avenues of useful future investigation.

It is necessary at the outset to ask oneself the question: what, precisely, is the function of taxonomy? The phenetic/phyletic and grade/clade controversies show no signs of abating; the theoretical approaches of the different schools are as wide apart as ever, but strange to say the effect of this in practical terms in minimal. Taking the mammals as an example, it is probably fair to say that the internal arrangement of most orders is widely accepted, only the Rodentia, Insectivora and Primates being sources of major disagreement. In the first case the phylogeny itself, and the roles of convergence and parallelism, are at stake; in the other two cases, there are radical divergences of opinion over how the agreed phylogenetic data should be employed to yield the most useful taxonomic scheme.

It is widely agreed that phylogenetically the lemurs (*sensu lato*) are more divergent from the tarsiers, monkeys, apes and man than these latter are from each other, and this theme recurred throughout the Research Seminar. In spite of this general agreement, taxonomic schemes vary widely; three in common use are presented in Table 1. Simpson's is a frankly grade-type of classification: for all its phyletic association with Anthropoidea, the tarsier is 'primitive', so it is ranked alongside the other 'primitive' primates in Prosimii.[1] Hill's scheme is strictly clade (phylogenetic):[2] the tarsier is related to the Anthropoidea (Pithecoidea) more closely than to the lemurs, so it must be placed nearer them in a classification. Romer's is essentially a compromise between these two opposing viewpoints.[3]

In the last analysis, a grade classification like Simpson's is essentially anthropocentric. When one asks *how* lemurs and tarsiers are primitive, the only possible answer is that they have small brains: they are less advanced up the ladder to man. The other 'primate trends'[4,5] – enhancement of digital mobility, orbital frontality, olfactory reduction and so on – which are also said to lead up the ladder to man, are consequently as anthropocentric as brain size; but all of them are subject to exceptions (frontal orbits of *Loris*, pseudo-opposable thumbs of many lemurs, hypocones present in many lemurs but absent in marmosets), so that we are left with brain size alone as the basis for subordinal division. Such a situation is of course an occupational hazard of grade taxonomy. If one attempts to characterise living species as 'primitive' or 'advanced' overall and then to make taxonomic groupings on this basis, there is always a strong risk that the members of one's resultant taxa may in fact be 'primitive' or 'advanced' in different ways, leaving the taxa undefinable and very probably polyphyletic.

It is here insisted that a phylogenetic ('clade') type of classification is more meaningful in all respects than a grade one. It avoids subjective assessments of overall evolutionary 'advancement'; its degree of objectivity varies only with the state of knowledge of the group's evolution; it excludes polyphyletic taxa; and perhaps most important, it maximises the amount of information content in a classification, enabling it to be used as a kind of one-dimensional phylogeny. The drawback to such a scheme is that its validity is totally dependent on the validity of the evolutionary scheme on which it is founded. It might at times be useful to retain an old classification in a case where phylogenetic views were in a state of flux: within the primates, for example, it is probably preferable for the moment to retain a grouping – Platyrrhini, Ceboidea or whatever – for the New World primates, although their homogeneity is by no means certain; but such an arrangement should only be an interim measure, and as far as suborders are concerned the relevant phylogenetic data are perfectly clear, and the advantages of a strict clade taxonomy should not be lost.

The study of fossil primates is progressing rapidly, but there are still so many gaps, uncertainties and disputes – not to say polemics – that comparative anatomy must still be the prime basis for phylogenetic reconstruction. For example, the detailed similarity of the foetal membranes of *Tarsius* and monkeys is so remarkable,[6] and is supported by such a wealth of other resemblances,[7,8] that to complain that the fossil record is inadequate to demonstrate the

Table 1 Three currently used classifications of Primates

Simpson, 1945[3] (grade)			Hill, 1953[4] (clade or phylogenetic)		Romer, 1967[5] (compromise scheme)	
SUBORDER PROSIMII	Infraorder Lemuriformes	Superfamily Tupaiodea	(not Primates)		(not Primates)	
		Superfamily Daubentonioidea	GRADE STREPSIRHINI	Suborder Lemuroidea	SUBORDER LEMUROIDEA	Family Daubentoniidae
		Superfamily Lemuroidea				
	Infraorder Lorisiformes			Suborder Lorisoidea		Family Lorisidae
	Infraorder Tarsiiformes		GRADE HAPLORHINI	Suborder Tarsioidea	SUBORDER TARSIOIDEA	Family Tarsiidae
SUBORDER ANTHROPOIDEA	Superfamily Ceboidea			Suborder Pithecoidea	SUBORDER PLATYRRHINI	Families Callithricidae, Cebidae
	Superfamily Cercopithecoidea				SUBORDER CATARRHINI	Family Cercopithecidae
	Superfamily Hominoidea					Families Pongidae, Hominidae

origin of monkeys and apes from a tarsier-like group is irrelevant;[9] nor is it really feasible to envisage separate derivation for Catarrhines and Platyrrhines from paromomyiform ancestors.[10]

There remains the problem of deciding the validity of shared characters. Detailed homology, as in the placentation example, can generally be relied upon to exclude convergence especially if backed up by other similarities. The same can probably be said of parallelism. The question of symplesiomorphy v. synapomorphy ('characters of common inheritance' and 'characters of independent acquisition') demands some knowledge of animals outside the group one is considering, as has become all too clear in the case of the tree-shrews, and their inclusion in, or exclusion from, the Primates.[11,12]

The species concept

The critical question for deciding the specific status of two extant forms is whether or not they form hybrid populations in a state of nature (if they share portions of their range). This is the meaning of reproductive isolation. Obviously, if the two are allopatric and never meet, then other means must be found to decide their status; but if they are sympatric and show no signs of forming a hybrid population between them, i.e. they do not form a morphological continuum when museum specimens or living examples are examined, then reproductive isolation is indicated. It is worth stressing this point, since many reports in the literature on primates give the impression that the two parent types are shown to be conspecific by the mere fact that they can hybridise under captive conditions.[13] Nor is the production of occasional wild hybrids indicative. The following cases can be used to indicate respectively the hybridisation of forms in captivity which are sympatric but reproductively isolated in the wild, and the production of hybrids in the wild without destroying reproductive isolation:

1. A recent checklist records the production, in captivity, of hybrids between taxa whose coexistence in the wild indicates their status as good, separate species;[14] among the Primates, these include:

   Hylobates lar x *H. pileatus* (two gibbons sympatric in Thailand)
   Cercocebus torquatus x *Papio sphinx* (redcap mangabey and mandrill, sympatric in Río Muni and Gabon)
   Cercopithecus cephus x *C. mona* (two guenons, sympatric in Gabon)
   Cercopithecus hamlyni x *C. lhoesti* (two guenons, sympatric in Kivu)

Cercopithecus mona x *C. nictitans* (two guenons, sympatric in Cameroun)
Macaca arctoides x *M. assamensis* (two macaques, sympatric in N. Burma)
Macaca nemestrina x *M. fascicularis* (two macaques widely sympatric in South-East Asia)
Papio hamadryas x *Theropithecus gelada* (sacred baboon and gelada, sympatric in Ethiopia)
Cebus albifrons x *C. apella* (two capuchin species, sympatric in Colombia and N.W. Brazil)

Especially noteworthy is the fact that several of these hybrids were capable of producing viable offspring when backcrossed with one or other of the parental species: in other words, the reason for their failure to form hybrid populations in the wild is purely ethological, not because of karyotypic incompatibility or for any other morphological cause. Instructive examples from outside the Primates can be found among the Cervidae: hybrids between *Cervus elaphus* (Red deer) and *Elaphurus davidianus* (Père David's deer), two species formerly sympatric in China and belonging to distinctly different genera by any standards, are not only obtained with great ease in captivity, but are 'endlessly fertile'!

2. There are, on the other hand, several cases of the production of occasional hybrids in the wild without the formation of hybrid populations, i.e. without actual gene-flow between the two parental species. Three such cases are known from primates:[14] *Cercopithecus mona* x *C. pogonias* in Cameroun, *C. mitis* x *C. ascanius* in Uganda, and *Macaca nemestrina* x *M. fascicularis* in Malaya. In at least two of these cases, habitat interference by man was suspected; it is likely that, in the macaque instance, all the males in a *nemestrina* troop had been shot, and the females joined a *fascicularis* troop. There is also at least one case, from the Bovidae this time, of regular hybridisation between two sympatric species which nonetheless manage to retain their separateness: it is claimed that the two ibex species *Capra caucasica* and *C. cylindricornis* frequently hybridise where their ranges overlap in the central Caucasus;[15] but the hybridisation goes no further than F_1, so that there is no gene-flow between the two species.

The major divisions of the Strepsirhini

There has been a tacit assumption that, because Madagascar is an island, and because its fauna is by and large very unlike that of the

African mainland, the Malagasy lemurs must be a homogeneous group: that there can have been only one invasion of Madagascar per mammalian order, and that this must have been in the far distant past. Hence, the aye-aye forms with the Lemuridae and Indriidae a single infraorder, Lemuriformes, while the lorises and galagos are set apart as Lorisiformes. Starting from this assumption, primatologists have searched for differences between the two groups and claimed to discover them; on this insecure basis, zöogeographers have felt at liberty to use the lemurs as illustrative of the distinctness of the Malagasy fauna, thereby completing the circularity of the argument![16]

Lorisiform-Lemuriform differences

The differences of the Lorisiform lemurs from the Lemuriform ones were first emphasised by Osman Hill,[17] who listed 39 characters said to distinguish them. Many of these however are at once qualified by such statements as 'except *Galago*', while many (perhaps most!) of the others certainly should have been. The most important of these characters, in that they have been given prominence in later works such as Le Gros Clark's great textbook,[4] would seem to be:

1. Tympanic ring external to bulla and fused with it.
2. Os planum (ethmoid plate) on orbital wall.
3. Large stapedial, small promontory artery, branching external to bulla and only stapedial entering bulla.
4. Microstructure of tooth enamel.
5. First ethmoturbinal enlarged, overlapping maxilloturbinal.
6. Vibrissae reduced.

Of these features, nos. 4 and 5 are of somewhat uncertain value since they have been investigated only in relatively few species. The first three are the 'key' features, quoted time and again in this regard. Table 2 shows however that, even if they are taken at their face value, only the tympanic region distinguishes the Malagasy and non-Malagasy groups as a whole. The characters are all more complicated than the list would indicate, however, and should be examined in detail.

Table 2: 'Lorisiform' characters in the Strepsirhini

	Lemuridae, Indriidae, Daubentoniidae	*Cheirogaleidae (Phanerinae)*	*Cheirogaleidae (Cheirogaleinae)*	*Lorisidae*
Tympanic ring	–	–	–	x
Os planum	–	–	x	x
Carotid arteries	–	x	x	x

1. The position of the tympanic ring, and its relation to the bulla, has been widely misunderstood or misinterpreted. Alan Walker has discussed the matter with the author, and permits the preliminary results of his researches to be mentioned here:

   (*a*) In the Malagasy lemurs the bulla is enlarged and inflated so that it overgrows the tympanic ring. The ring is therefore inside the bulla, and its inferior edge is separated from the bulla by a recess (except in *Microcebus*, according to Saban;[18] but Walker suggests that this latter condition may be an artifact of postmortem shrinkage of the extra-tympanic membrane).
   (*b*) In the Lorisidae the bulla is small and fails to overgrow the tympanic ring, coming into contact with it; and the latter grows laterally, forming the inferior margin of the mouth of the bulla.
   (*c*) In *'Galago' crassicaudatus* (probably forming genus *Otolemur*, see below), there is an intermediate condition: the bulla does not extend appreciably lateral to the tympanic ring, so there is little tympanic component in the bulla. This condition seems to occur in the *Mioeuoticus* cranium too.[19]

   This character therefore, though given great weight in most taxonomic schemes, is seen to have a very simple basis, being largely dependant on the size of the bulla itself. There is, however, a rather more fundamental difference, which Walker has demonstrated to me: in the Lorisidae, alone among the Strepsirhini, there is a partition within the bulla separating the hypotympanic cavity from the tympanic cavity proper; this occurs in the African Miocene crania as well.

2. The presence or absence of an ethmoid component (*os planum*) in the orbital wall is largely a matter of orbital frontality.[20] The Lemuridae and Indriidae lack it; the Lorisidae possess it; while the Cheirogaleidae are divided, *Phaner* lacking it and the other three genera possessing it. (Its evident existence in *Allocebus* has been ascertained from the holotype in the British Museum by Cartmill and myself.)

3. The most complex of the three characters, the disposition of carotid branches in the ear region, divides not the Lemuriformes from the Lorisiformes but the Cheirogaleidae and Lorisidae from the Lemuridae and Indriidae.[18] In the latter two families the common carotid artery divides into two major branches, the external and internal carotids (with the occipital artery arising from the same junction); the latter then, after giving off a 'rameau tubaire', enters the bulla via the carotid foramen, and on the cochlea it divides into a large stapedial and a small promontory branch. (The stapedial feeds the middle meningeal

arteries, while the promontory joins the *circulus arteriosus.*)

In the Cheirogaleidae and Lorisidae, on the other hand, the first junction gives off an additional artery, called the anterior carotid by Saban,[18] which runs forwards outside the bulla and enters the cranial cavity via a large foramen in the position of the human *foramen lacerum*, but undoubtedly not homologous with it (it is called the promontory foramen by Szalay).[21] The internal carotid enters the bulla and, as in the Lemuridae and Indriidae, divides on the cochlea into stapedial and promontory branches; but the stapedial is this time a very small artery – in fact it is absent in *Phaner* – while the promontory is large. The promontory does not enter the braincase right away, but leaves the bulla and goes through the promontory foramen along with the anterior carotid. The Lorisid-Cheirogaleid pattern therefore involves not an earlier stapedial-promontory branching as Le Gros Clark and others have stated (for this occurs in the usual place, *inside* the bulla), but the existence of an entirely new artery! It is tempting to identify this with the medial entocarotid – a primitive mammalian artery not otherwise recorded in primates – as do Van Valen and Szalay;[21, 22] but since its existence (as well as the changed pathway of the promontory) is closely bound to that of a promontory foramen, which is not present in fossil primates in which this region is discernable (*Plesiadapis, Phenacolemur, Adapis, Notharctus* etc.), this seems unlikely. It is best interpreted as a specialisation, probably derived from a modification of the 'rameau tubaire', which in the Lemurid type is a similar extrabullar branch of the internal carotid but in the Lorisid type arises from the anterior carotid.

It should finally be mentioned that the existence of the promontory foramen is associated also with a specialisation of the venous system, serving as the exit for veins draining the cavernous sinus into the external jugular vein.[18]

4. The final differentiating feature, vibrissal reduction, also falls down on closer examination.[8] The full complement of vibrissae – supraorbital, genal, mystacial, interramal and carpal – occurs in the Cheirogaleidae and in the lemurid genus *Varecia. Lemur* and *Hapalemur* lack only the interramal group. *Lepilemur*, the Indriidae and the Galaginae lack interramals and carpals. Finally the Lorisinae lack all except the mystacial group. There is therefore no clear-cut division that can be made.

The upshot of all this is that there is only one feature – bullar construction – in which a clear-cut Lemuriform/Lorisiform division exists; and no common specialisation uniting all the Malagasy lemurs can be found. However the Afro-Asian Lorisidae and the Malagasy

Cheirogaleidae do share a specialisation in the carotid system. Szalay underplays this resemblance, and argues that for mammals as a whole a tympanic external to the bulla is primitive;[21] but for lemurs, if they are truly derived from the Adapidae, it would have to be the other way around.

Zöo-geographic considerations

If, then, one can detect probable specialisations in common (synapomorphy) between two groups separated by the Mozambique channel, it is worth asking whether multiple invasions of Madagascar, or perhaps re-invasions of Africa from Madagascar, would have been possible. The mammalian fauna of Madagascar belongs to six orders: one, the Primates, is the order in question: another, the Chiroptera, is represented on the island only by worldwide genera. The other four are:

1. *Carnivora.* All Malagasy carnivores belong to the family Viverridae; they are assigned to seven genera, belonging to three subfamilies,[23] of which two (Cryptoproctinae and Galidictinae) are indigenous, the third (Hemigalinae) having members outside Madagascar – not indeed in Africa, but in South-East Asia! There is a suggestion of a close relationship between the Malagasy *Eupleres* and the Indochinese *Chrotogale* (Owston's civet). Later studies have extended this picture somewhat. Karyotype data[24] suggest that the Malagasy subfamilies are in fact closely interrelated and belong in a group with another Oriental subfamily, Paradoxurinae, contrasting with the true civets, Viverrinae, on the one hand, and the mongooses, Herpestinae, on the other. The most recent research finds that the Malagasy viverrids do in fact all form a homogeneous group, even with regard to the Oriental Hemigalines, and that one of the supposed hemigalines, *Fossa* (the Fanalouc) is the most primitive living form.[25]

2. *Rodentia.* After reviewing monophyletic and polyphyletic theories of the origin of Malagasy rodents, F. Petter concludes that they, like the carnivores, do derive from a close-knit common stock (probably among the Cricetidae), but have reached Madagascar via multiple invasions.[26]

3. *Lipotyphla.* Except for the ubiquitous *Suncus*, clearly of recent origin on Madagascar and probably introduced by man, all Malagasy insectivores belong to a single family, Tenrecidae, and to a single indigenous subfamily, Tenrecinae. The other subfamily, Potamogalinae, occurs on the mainland of Africa; a

morphological intermediate, *Protenrec*, is known from the early Miocene of East Africa.[27]

4. *Artiodactyla.* The African bush-pig is represented on Madagascar by a form currently classified as a distinct species, *Potamochoerus larvatus*, but probably conspecific with the East African *P. koiropotamus.*[28] It has evidently been there long enough to have formed subspecies, there being a large form (*edwardsi*) in the mountains, and two small ones (*larvatus* and *hova*) on the west and east coasts respectively. A hippopotamus, *H. lemerlei*, decidedly distinct from the African species, formerly lived on the island but is now extinct.

It would seem from the above survey that multiple invasions of Madagascar would have been possible, but at the same time all groups seem to have come from rather limited stocks. This would indicate that the island was populated from the nearest mainland – probably Africa – at a time when the latter was itself rather isolated, and had a restricted fauna. The absence of catarrhines, bovids, felids and so on may simply indicate that none were present in eastern Africa at the time; or else some factor such as size may have placed a limit on the fauna that was available for dispersal. The pre-Miocene fauna of Africa is poorly known, with the exception of the Oligocene Jebel-el-Qatrani formation of the Fayum, Egypt, whence remains are available of both Primates (primitive catarrhines, not strepsirhines) and rodents (family Phiomyidae, not closely related to the Cricetidae), but no viverrids; we have no means of knowing, of course, whether this was a pan-African fauna and the lemurs, cricetids and viverrids had not yet arrived, or whether it was in some way atypical of the more southern parts of Africa. Currently, however, the entry of viverrids and cricetids into Africa is provisionally dated to the end of the Oligocene.[29] A detailed consideration of the separation of Madagascar has recently appeared,[30] and an early date for separation is favoured, although a possible avenue of exchange may have remained open until the Oligocene.

Since the early Miocene *Mioeuoticus* had a lorisid bulla construction (see above), it would appear that whichever of the two types of bulla is derived from the other (probably the extra-bullar tympanic from the intra-bullar, see above) had already begun to evolve by then. This would put the cheirogaleid-lorisid divergence prior to 20 million years before present. The cheirogaleid-lorisid carotid type, and therefore the joint divergence of these two families from the lemurid-indriid stock, must therefore have been established even earlier.

Daubentonia

There is still one Malagasy 'lemur' which fits uneasily with the others, and indeed with the lemurs as a whole. This is *Daubentonia*. Many of its unique features can be explained as a consequence of its extraordinary incisor specialisation – high facial skeleton, large premaxillae, lack of interorbital depression[31] – and probably its brain form can be similarly explained;[32] but there are other features which certainly cannot be explained in this way:

1. In the orbit, there is a big maxillary plate which makes contact with the frontal wing, and no fronto-palatine suture or ethmoid exposure. This condition is similar to most of the Lipotyphla as well as to the Plesiadapidae and Adapidae, where this region has been preserved. It seems therefore to be a symplesiomorph primate feature. All living primates other than the aye-aye have either a long fronto-palatine suture, or else an ethmoid plate.

2. In the nasal cavity, *Daubentonia* has five endoturbinals and at least three ectoturbinals, as well as the single naso- and maxilloturbinals. The other lemurs all have four endo- and two ectoturbinals; while the Haplorhini have only two endo-, one ectoturbinal. In this morphological series, the aye-aye stands at the base of the primate stem and resembles the Lipotyphla.

3. In the claws (which occur on all digits except the hallux) the deep stratum comprises 46% of the thickness near the tip, and the terminal matrix giving rise to its comprises 40% of the total germinal matrix.[33] These proportions approach those of non-primates (marsupials, tree-shrews) at 75-79% and 73-81% respectively; lemurs (which have nails on all digits except the second toe) have neither deep stratum nor terminal matrix, not even on the 'toilet claw', the whole of the claw or nail being composed of superficial stratum, and the whole of the germinal matrix being basal matrix. The aye-aye claws cannot therefore be derived from the nails or claws of lemurs. (It is interesting that the lemurs resemble catarrhines in this, while tarsiers and platyrrhines retain at least a remnant of deep stratum and germinal matrix).[33,34]

4. In the muscular system there are numerous anomalies;[35] some are connected with the uniqueness of the third manual digit, but others cannot be so explained:

   (*a*) *M. trapezius* inserts on the occiput, and overlies the 'Acromio-trachelian' (the *M. levator claviculae* of apes, an anomaly in man), as it does in higher primates, but unlike any other lemur.

(*b*) *M. coraco-brachialis medius* originates from the coracoid itself as in higher primates, not from the tendon of of the short head of the *M. biceps brachii* as in other lemurs.
(*c*) *M. flexor digitorum sublimis* has a double origin like higher primates, with an ulnar head as well as the epitrochlear one found in lemurs.
(*d*) *M. obturator externus* is completely divided into two (except at the tendon of insertion), the two portions being separated by the obturator nerve. The overlying portion, essentially a separate muscle *(M. obturator intermedius),* occurs in rodents and insectivores but is quite absent from all other primates.
(*e*) In at least some specimens of the aye-aye, there are two *Mm. peroneo-tibiales*, an anomaly not seen in any other primates.

5. The carotid pattern is simpler than that of other lemurs, with an initial bifurcation into two branches, internal and external; the occipital artery then arises from the latter. This pattern is the one seen in tree-shrews, also in the orang utan and man; in other 'higher primates' the external carotid gives rise first of all to the posterior auricular, which in turn gives rise to the occipital.[18]
6. The mammae, two in number, are inguinal in position: unique among primates.
7. There is persistent oögenesis in the adult, as in the Lorisidae, but unlike any other Malagasy form.[36]

There are in this list a number of features in which the aye-aye resembles higher primates, and others in which it resembles a pre-primate mammalian condition. Sometimes, indeed, these features are the same, implying the existence of characters in which the 'Anthropoidea' have remained primitive while the lemurs as a whole (i.e. except *Daubentonia*) have acquired a specialised condition – a salutory antidote to the view of lemurs as 'lower primates'.

It is in fact none too easy to find specialisations that the aye-aye shares with other lemurs. The common possession of an epitheliochorial placenta may be a specialisation or it may not;[6,12] the M-loop of the colon ('ansa coli') is probably a specialisation;[4] and there appear to be common specialisations in the foot skeleton.[10] On the other hand there are features in common between the 'higher primates' and the remaining lemurs, which the aye-aye does not share: reduced deep stratum in nails and claws; olfactory reduction; pectoral mammae; simple form of *M. obturator externus*; and so on. The evidence is perfectly clear, in any case, that the aye-aye is the earliest lineage to split off from the Strepsirhine stem; it could even conceivably have been the first line of split off the primate stem as a whole, although it is still most probable that the initial division is the Haplorhine-Strepsirhine one.

The infraorders of Strepsirhini should therefore not be Lemuriformes and Lorisiformes, but Lemuriformes (including lorises) and Daubentoniiformes (Vallois, 1955 = Chiromyiformes Anthony and Coupin, 1931 = Chiromyoidea Pocock, 1918). Both are strongly specialised groups, the former with dental comb and toilet claw, the latter with its woodpecker-like functional complex.[31] Within the former group the superfamilies should be Lemuroidea (to include Lemuridae and Indriidae) and Lorisoidea (including Lorisidae and Cheirogaleidae), with the abandonment of any grouping that attempts to separate the Lorisidae from the Malagasy lemurs as a whole.

The phylogenetic scheme on which this classification is based is figured here in Fig. 1. It will be seen that Madagascar is shown to have been invaded three times by strepsirhine primates. An alternative would be to suppose that Madagascar was the original home of the Strepsirhini – their presence there and that of the rodents and viverrids dating perhaps from its original connection with India[37] – and that the presence of Lorisidae in Africa and Asia is due to one single dispersal event from Madagascar to Africa.

Taxonomic problems within families

1. Lorisidae

Traditionally this family has been divided into Galaginae (for the galagos) and Lorisinae (for the slow climbers). Not only, however, are there large differences between the African and Asiatic sections of the Lorisinae,[38] but there are even some indications of specialisations shared by the pottos and galagos but lacking in the lorises.[39] It is possible that the two lorisine groups have converged in their locomotor anatomy.

Within the Galaginae, studies in progress show that the genus *Euoticus* is highly distinctive and should be retained: although not unique in possessing 'needle-claws' (*Galago inustus* also has them), its dentition, skull form, body build, colour pattern and mammae (1 pair, pectoral, as opposed to 3 pairs, the norm in all other galagos and in pottos) set it well apart. Though restricted to a relatively small part of the Central African forest bloc, there are two very well-marked geographical forms (north-west and south-east of Mt. Cameroun), which are distinguishable by dental characters as well as by size and sharply distinct colour-patterns.

The next most distinctive form is *Galago crassicaudatus*, which can be separated as the genus or subgenus *Otolemur* (see Table 3). It appears to share features with the pottos (karyotype, serological reactions, skin structure) which the other galagos lack, and may perhaps be related to the stock which gave rise to them;[40] in this context its 'intermediate locomotion patterns' and the possible existence of a glandular area on the scrotum, like pottos, are particularly interesting.[40,41] Among its own unique features its peculiar nails, with their concave free edges, are especially interesting

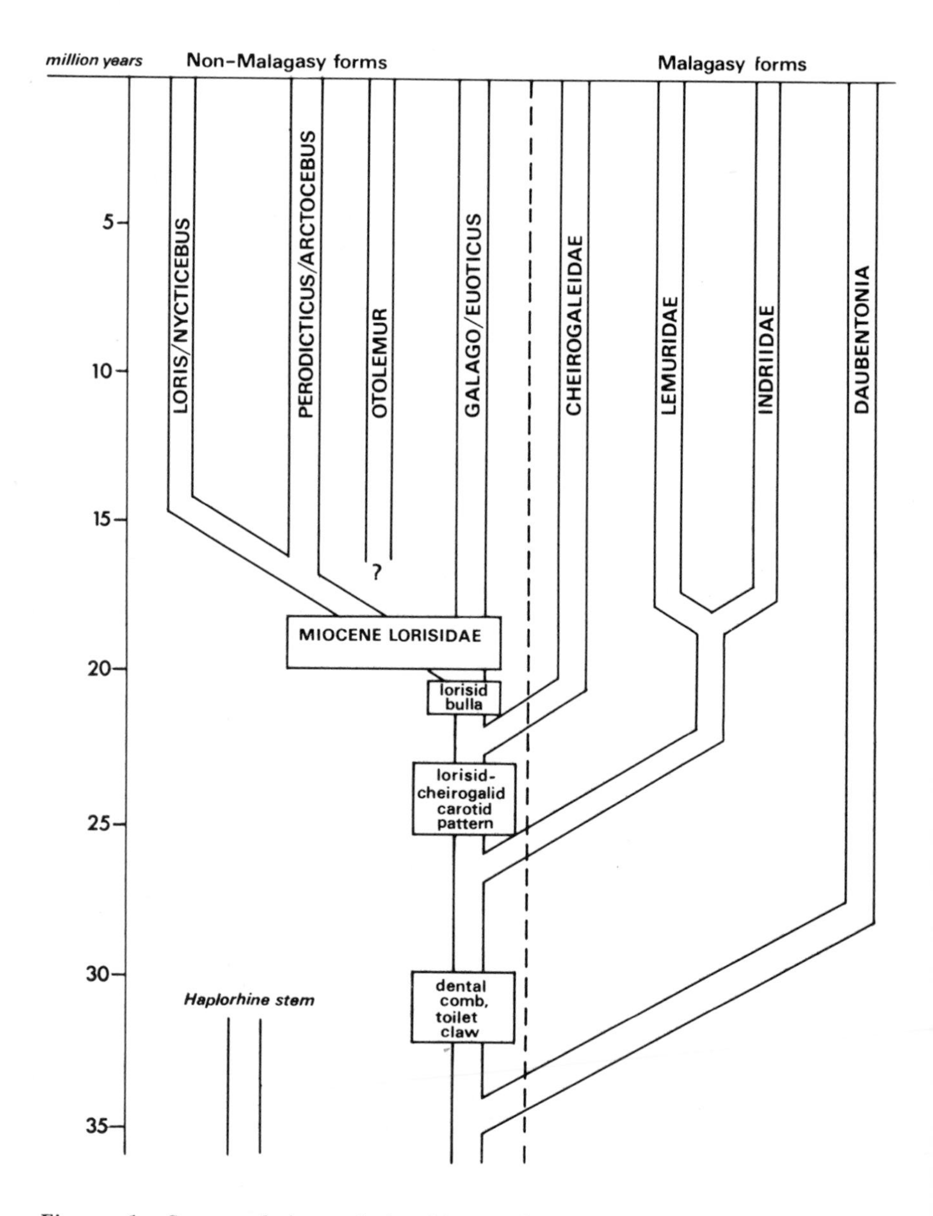

Figure 1. Suggested interrelationships and times of divergence of major strepsirhine groups. Taxa and lineages to the left of the dotted line are Afro-Asian; those to the right of it, Madagascan.

and are as yet unexplained. It is possible that there are more than one species in the *Otolemur* group: the small East African coastal races (*panganiensis, lasiotis* etc.) are very different animals from the much bigger south-eastern *crassicaudatus* and its relatives with their very long ears, while the beautiful silvery *monteiri* (= *argentatus*), also very large, from Angola and Zambia and the Western Rift lakes region as far north as Lake Victoria, tending to melanism in some parts of its range, is different yet again.

The other galagos form a close-knit group of at least 5 species:

(*a*) *G. demidovii* occurs, like the potto, throughout the forest bloc of Africa from Senegal to Uganda. From Senegal east to about Mt. Cameroun occur small races of about 115-130 mm. head and body length (N.B. these are approximately 95% limits, not observed ranges), in two colour-phases, greyish and red. From Cameroun to south of the Congo River is found a larger type, about 130-145 mm., more of a grey-brown colour but with occasional foxy-red individuals. North and east of the Congo and in Uganda is found the distinctive light brownish form *thomasi*, which goes up to 160 mm. in length.

(*b*) *G. senegalensis* replaces *demidovii* in the bush-savannah belt of Africa. It is larger on average, greyish in colour, with lighter (yellow to white) limbs and underparts. The biggest subspecies (head and body 160-190 mm.) are the northern series, found from Senegal to Ethiopia, Kenya and northern Tanzania. South of about Dodoma a very different type is found, perhaps representing a different species from which the prior name is *moholi*: much smaller (130-160 mm., no bigger than *G. demidovii thomasi*), grey with pinkish hues, the face white with striking dark eye-rings. There are also slight differences in the skull and teeth.

(*c*) *G. zanzibaricus.* The apparent coexistence in the same small forest in the Ulugurus (Tanzania) of the most Senegal-like race of Demidoff's bushbaby (*G.d. orinus*) with the most Demidoff-like race of the Senegal bushbaby (*G.s. zanzibaricus*)[42] has been recently called into question by Kingdon,[43] who suggests that they are actually the young and the adult respectively of a distinct species, *G. zanzibaricus*, whose range interdigitates with that of *senegalensis* which it replaces in forested coastal areas. It is about the same size as *G.s. moholi* or *G.d. thomasi* – 130-160 mm. – with cranial and dental characters intermediate between the two, but nearer to Demidoff's; it is yellow-brown to olive in colour with lighter coloured limbs and a brown tail; the hindlimbs are not as strongly elongated as those of *G. senegalensis*, the intermembral index being about 65 like *demidovii* and *alleni*, not 50-55 like *senegalensis.* In Fig. 2 the apparent distribution is mapped, on the basis of specimens in the British Museum and of Kingdon's records; the Mozambique *granti* seems likely to be a southerly representative of *zanzibaricus*, and has been so mapped. It will be seen that the distribution is more or less continuous down the coast, with inland outliers: Mt. Marsabit, Ugogo, and Ukinga. (Lawrence and Washburn record three specimens of *G.d. thomasi* from Ukinga,[42] while a dot in approximately this position is marked on Kingdon's map for *zanzibaricus*; whether Kingdon is thereby implying that the three '*thomasi*', like the type of *orinus*, are young *zanzibaricus*, is uncertain. Provisionally, both species are mapped here in Fig. 2).

(*d*) *G. inustus* is not, as claimed by some authors[8,42] an eastern representative of *Euoticus*, but despite its needle-claws is a forest version of *senegalensis.* From the latter it differs (apart from the shape of its nails) in its very dark colour, shorter hindlimbs and larger size (head and body 190-200 mm.).

(*e*) *G. alleni* is another, but rather more distinctive, forest version of *senegalensis.* Like *Euoticus* and *Arctocebus* it is restricted to the forest between the Niger and the Congo; unlike them it appears to be a more recent immigrant and does not show such striking geographic variation north-west and south-east of Mt. Cameroun, varying only in size (165-192 mm. in Nigeria, 205-230 mm. in Gabon).

2. *Cheirogaleidae*

That the dwarf lemurs have nothing in common with the Lemuridae has long been recognised,[8,44] and it is pleasant to find that this situation has at last been formally recognised by the recognition of a full family for them,[45] quite independently of the present contribution.

A new genus, *Allocebus*, has recently been added to the family.[46] Within the genus *Microcebus*, the species *M. murinus* apparently contains at least two species (a larger grey one weighing 75 g., and a smaller red one weighing 55 g.) which are sympatric in some areas;[47] for the vastly bigger *M. coquereli* (weighing 385 g.) the name *Mirza* Gray, 1870, is available (Table 3) if subgeneric division is required, although according to Petter (personal communication) it is really quite close to the small species.

The genus *Phaner* is somewhat isolated from the rest of the family: it lacks an os planum in the orbit,[20] has no stapedial artery,[18] and differs greatly in its foot structure and in its behaviour.[47] Subfamilial separation seems perfectly justified.[45]

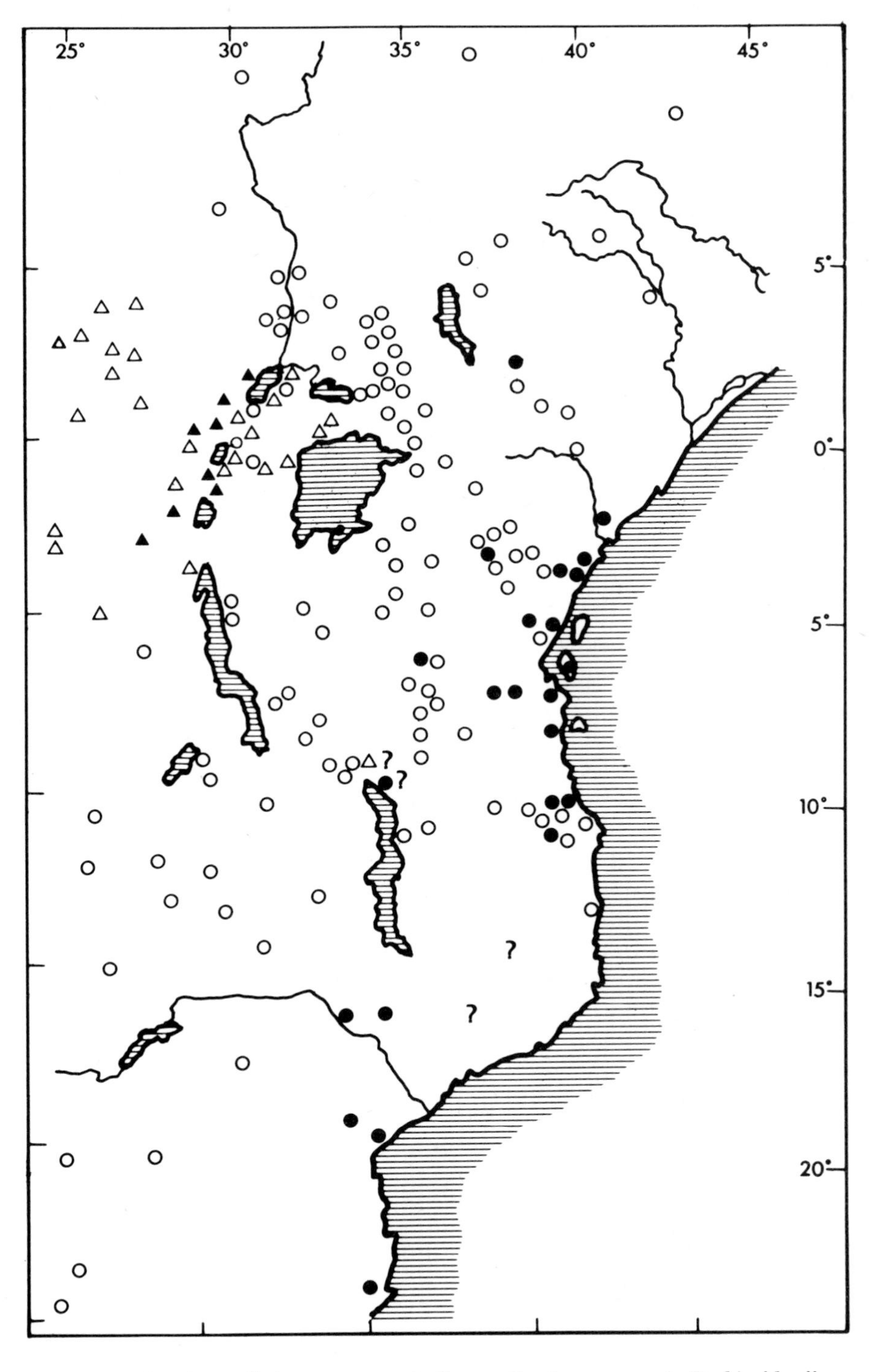

○ *G. senegalensis* ▲ *G. inustus* ● *G. zanzibaricus* △ *G. demidovii*

Figure 2. Distribution of members of the subgenus *Galago* in East Africa. Note that the identification of certain specimens from Ukinga (northern shore of Lake Malawi) is equivocal (see text).

Table 3: Generic names available in polytypic genera (as defined in Napier and Napier),[5] in the Strepsirhini

GALAGO	
G. alleni	*Sciurocheirus* Gray, 1872
G. senegalensis	*Galago* E. Geoffroy, 1796; *Otolicnus* Illiger, 1811; etc.
G. demidovii	*Galagoides* A. Smith, 1833; *Hemigalago* Dahlbom, 1857
G. crassicaudatus	*Otolemur* Coquerel, 1859; *Callotus* and *Otogale* Gray, 1863
G. elegantulus	*Euoticus* Gray, 1863
CHEIROGALEUS	
C. major	*Cheirogaleus* E. Geoffroy, 1812; *Cebugale* Lesson, 1840; etc.
C. medius	*Opolemur* Gray, 1872; *Altililemur* Elliot, 1912
C. trichotis	*Allocebus* Petter-Rousseaux and Petter, 1967
MICROCEBUS	
M.m. rufus	*Microcebus* E. Geoffroy, 1828; *Azema* Gray, 1870
M.m. murinus	*Scartes* Swainson, 1835; *Myscebus* Lesson, 1840
M. coquereli	*Mirza* Gray, 1870
HAPALEMUR	
H. griseus	*Hapalemur* I. Geoffroy, 1851
H. simus	*Prolemur* Gray, 1871
LEMUR	
L. catta	*Lemur* Linnaeus, 1758; *Procebus* Storr, 1780; etc.; *Odorlemur* Bolwig, 1958
L. mongoz	*Prosimia* Boddaert, 1785 (here designated as genotype)
L. variegatus	*Varecia* Gray, 1863

3. *Lemuridae*

As in the case of full family status for the Cheirogaleidae, the apartness of *Lepilemur* has long been hinted at,[44] but quite how extraordinary it is has only recently become apparent.[48] It is aptly removed to a subfamily of its own.[45] The remaining members of the subfamily are currently classed in the two genera *Lemur* and *Hapalemur*; to these at least one more, *Varecia*, should be added.[44] Additionally, there is the question of the relationships of *Lemur catta* to be considered: it shares characters of both skin and karyotype with *Hapalemur griseus*,[49,50] differing from other species of *Lemur*. Formerly it was thought that the little-known *Hapalemur simus* differed strongly from *H. griseus*, especially in lacking brachial and antebrachial glands; but it has recently been discovered[51] that it does in fact possess the glands, though very small and much nearer the elbow. (A diligent search on skins in the British Museum did successfully reveal the brachial gland; so inconspicuous is it that until my attention was directed to it (Petter, personal communication) I had not suspected its existence although perfectly familiar with the skins!)

Future studies may reveal whether these glands, and the other features that are in common, are primitive lemurine features that the *macaco*-group have lost, or whether they are common specialisations of *L. catta* and *Hapalemur*. If the latter, then the animals in question should be associated taxonomically at a closer level than to the *macaco*-group.

Within *Hapalemur*, the Lake Alaotra form is distinct and awaits formal diagnosis;[52] within *Varecia*, the problem of the red form, whether colour phase, subspecies or species, is still unsolved.[44]

The name *Prosimia* is available for the *mongoz-macaco-rubriventer* group of the genus *Lemur*, dating from Boddaert, 1785 (*Elenchus Animalium*, 43 and 65). It is first mentioned on p.43; further on, on p.65, the following species are listed within it: *mongoz, macaco, catta, volans* (i.e. *Galeopithecus*!), *spectrum* (i.e. *Tarsius*) and *minima* (i.e. *Microcebus*). The action of Gray[53] in excluding *L. catta* from this genus precludes the selection of the latter as genotype, and leaves the name available in the sense in which it is used by Osman Hill[8] and in the present survey, with *mongoz* as genotype. Provisionally, it will be ranked as a subgenus. The number of species that may be recognised is arguable. Study of the skins in the British Museum (Natural History), American Museum of Natural History, and Smithsonian Institution yields the following results:

(*a*) *Lemur (Prosimia) mongoz* group. These are distinguished primarily by the white facial 'mask', but are not easy to define by comparison with the next group. The recognisable forms, probably subspecies, are:

1. *mongoz*. Found north-east and south-west from Maroantsetra. Male reddish on head and cheeks; foreparts reddish, hindparts grey. Female light grey foreparts, yellow-red hind parts; cheeks not red. Both sexes have yellow belly, white chest.
2. *albimanus*. Comoro Is. Cheeks maroon-red; body maroon-grey, underparts yellower.
3. *coronatus*. Found north from Vohémar. Male grey, often grizzled or mottled, with a V-shaped red mark on crown, which is black behind; the red extends to the cheeks. Female is like *mongoz*, but with less contrast between grey foreparts and red hindparts, and has male's red chevron but no black on crown. Underside is all yellowish white.

(*b*) *Lemur (Prosimia) macaco* group. These generally have a black facial mask. The map (Fig. 3) is compiled from records of specimens on museum labels, and from the literature, especially Jolly and Schwarz;[54,55] it shows some apparent distributional overlaps, suggesting, as Hill opines,[8] that more than one species may be involved – a conclusion already hinted at by chromosome studies.[56] It should be noted, however, that if one follows Hill in separating *macaco* from the rest, then the name *fulvus* cannot be used for the composite species, as it is by no means the earliest available name. The valid names (as far as I can ascertain) in this group are, in order of priority: *macaco* Linnaeus, 1766; *albifrons* E. Geoffroy, 1796; *rufus* Audebert, 1799; *fulvus* and *collaris* E. Geoffroy, 1812; *mayottensis* Schlegel, 1866; *cinereiceps* Grandider and Milne-Edwards, 1890; *sanfordi* Archbold, 1932. It will be seen from the map that the forms *macaco-sanfordi-albifrons-rufus-cinereiceps-collaris* replace each other around the island starting at Majunga on the north-west coast and going north-east, then down the east coast. But *fulvus* appears to overlap with *rufus* in the west (Majunga district) and with *macaco* in the east (Tamatave district), with no sign of intermediacy. Moreover, *rufus* appears to approach *macaco* at Tamatave, and there may well be a ring-species effect here. On the other hand, specimens in the American Museum of Natural History appear to represent intergrades in certain cases: between *albifrons* and *rufus* at Maroansetra, and between *rufus* and *cinereiceps* at Ankazoabo; so some of the forms, at least, are conspecific.

1. *macaco.* Male black with maroon tints; female reddish-yellow becoming greyer with age, facial mask red. The typical form with long ear-tufts is found over most of the range marked on the map, while the form described as *flavifrons* (female) or *nigerrimus* (male), lacking ear-tufts, 'inserts' into this range at Maromandia.
2. *sanfordi.* Greyish, browner on back; head darker with a light grey zone on forehead. Sexes alike, except that male has ear-tufts. This form is unlike either of its geographical neighbours, most closely resembling *fulvus* in colour.
3. *albifrons.* Wood-brown, the tail becoming black towards the tip; hair round ears and on cheeks slightly elongated. Male is darker grey-brown, nearly black in midline of back; whole head, except facial mask, whitish; grey-white below; crown of head darker than body. Female is lighter, chocolate-grey-brown; crown darker, nearly black; supra-orbital region dark grey; light brown below, or orange; interramal space white.
4. *rufus.* Like the last, sexually dichromatic – but each sex has two varieties which do not seem to be geographic. In male, Variety A is grey, yellow below; Variety B is lighter, browner, and off-white below. Both have big white supraorbital spots, and the crown may be reddish or

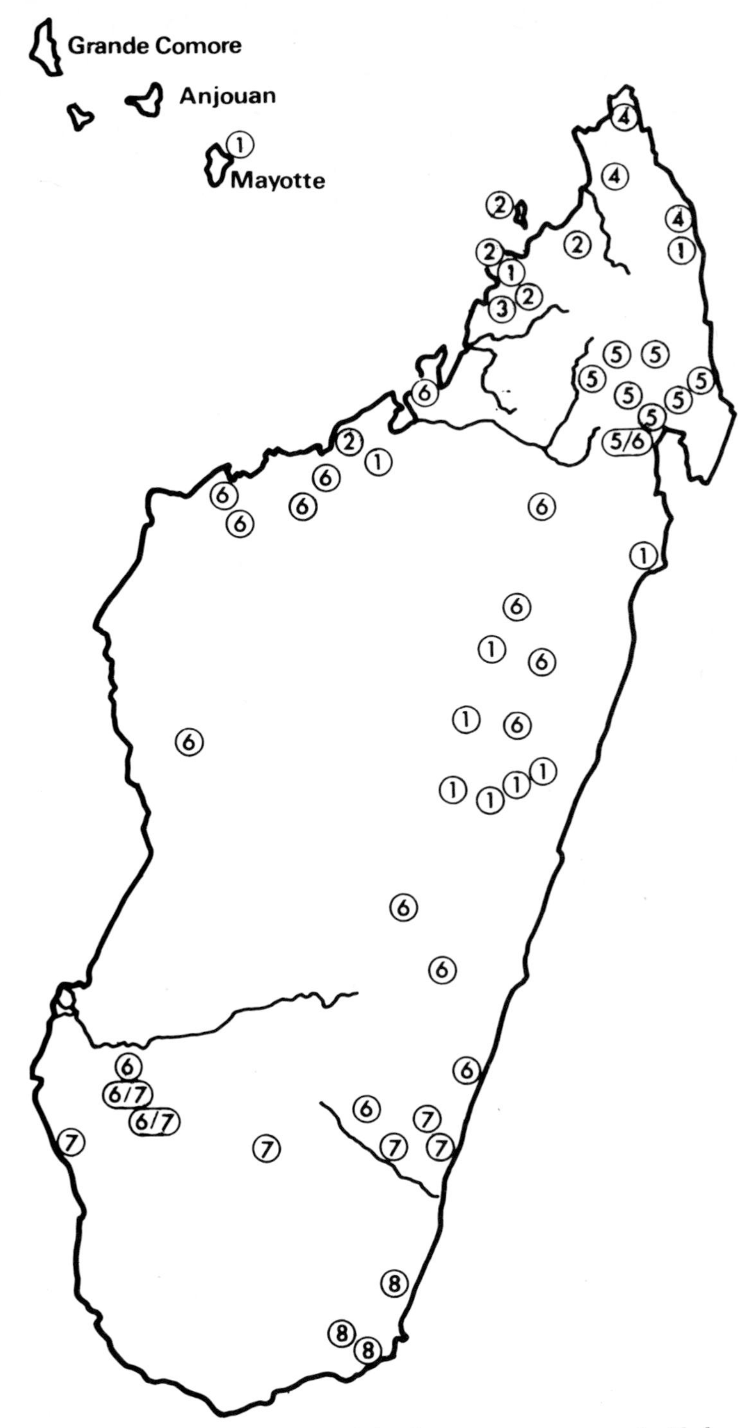

Figure 3. Distribution of members of the *Lemur macaco* group in Madagascar.

1. *fulvus* (? = *mayottensis*)
2. *macaco*
3. *flavifrons* (a localised variant of *macaco*)
4. *sanfordi*
5. *albifrons*

5/6. intergrade between *albifrons* and *rufus*

6. *rufus*

6/7. intermediate between *rufus* and *cinereiceps*

7. *cinereiceps* (the white-cheeked variant of *collaris*)
8. *collaris*

blackish. In female, Variety A is grey-red, becoming very red on underside and rump, with some red on cheeks; Variety B is foxy-red, yellower below, greyish crown.

5. *cinereiceps.* This has the supraorbital spots in the male like *rufus.* The male is yellow-grey or yellow-brown, with hindparts redder; the crown blackish; underside light red-brown; cheeks white. Female is greyish, becoming suffused with yellow on the foreparts and redder towards the hindparts; hindlimbs are red-yellow; underparts yellow; crown dark grey; face light grey.

6. *collaris.* Like the last, but with cheeks orange-red instead of white. The two have not usually been separated, but museum skins show the geographic separation. According to Petter (personal communication) the relatively slight differences between this race and the last may well reflect the situation elsewhere, where neighbouring populations thus far included under one head in fact can be distinguished by small colour attributes.

7. *fulvus.* Grey-brown, slightly redder in midback region. Ochery underparts; tail-root black; a black cap on crown; grey supraorbital spots. Not sexually dichromatic. Jolly says there are two isolated populations of this type, an eastern one (round Perinet) which are darker, with lighter white 'beards', and a western (around Ankarafantsika) which are grizzled, lighter grey, with tan or creamy beards.[54] All specimens in museums seen by me are of the eastern type, and agree approximately with Jolly's description. She comments that the existence of these two isolated populations may account for the reported chromosome polymorphism and also for the so-called *Lemur* sp. nov. of Chu and Swomley;[56,57] but Egozcue (personal communication) has informed me that this latter animal was in fact *collaris.*

8. *mayottensis.* This appears, though on the basis of only two specimens, to be but feebly differentiated from the last. One of the two falls in the range of variation of *fulvus*, the other is more intensely brown in colour.

(*c*) *Lemur (Prosimia) rubriventer.* The most distinctive of the three species-groups; it has a wide distribution, but no evident geographical variation.

The above notes may perhaps prove of use in working towards a total reassessment of the interrelationships of all lemurs. It will be seen that, while special studies on individual species and genera are still urgently needed, wide-ranging studies across the whole strepsirhine suborder are vital to a proper understanding of relationships.

Acknowledgments

I would like to acknowledge some very valuable discussions held during the course of the conference. Particular thanks for discussion and information are due to J.-J. Petter; additionally, to F.-K. Jouffroy, J. Egozcue, F.S. Szalay and W.W. Bishop. I would also like to acknowledge fruitful discussions on this and many other occasions with Robert Martin and Alan Walker.

NOTES

1 Simpson, G.G. (1945), 'The principles of classification and a classification of mammals', *Bull. Amer. Mus. Nat. Hist.* 85, 1-350.

2 Hill, W.C.O. (1953), *Primates: Comparative Anatomy and Taxonomy*, 1. *Strepsirhini*, Edinburgh.

3 Romer, A.S. (1967), *Vertebrate Paleontology*, 3rd ed., Chicago.

4 Le Gros Clark, W.E. (1959), *Antecedents of Man*, London.

5 Napier, J.R. and Napier, P.H. (1967), *Handbook of Living Primates*, London.

6 Luckett, W.P., this volume.

7 Pocock, R.I. (1918), 'On the external characters of lemurs and of *Tarsius*', *Proc. Zool. Soc. Lond.* 19-53.

8 Hill, W.C.O. (1955), *Primates: Comparative Anatomy and Taxonomy*, 2. *Tarsioidea*, Edinburgh.

9 Simons, E.L., this volume.

10 Szalay, F.S., in discussion.

11 Martin, R.D. (1968), 'Towards a new definition of Primates', *Man*, n.s., 3, 377-401.

12 Martin, R.D. (1968), 'Reproduction and ontogeny in tree-shrews (*Tupaia belangeri*) with reference to their general behaviour and taxonomic relationships', *Z.f. Tierpsychol.* 25, 409-95.

13 Albignac, R., Rumpler, Y. and Petter, J.J. (1971), 'L'hybridation des lémuriens de Madagascar', *Mammalia* 35, 358-68.

14 Gray, A.P. (1972), *Mammalian Hybrids: a Checklist with Bibliography*, Slough.

15 Heptner, V.G., Nasimowitch, A.A. and Bannikov, A.G. (1966), *Die Säugetiere der Sowiet-Union*, 1: *Artiodactyla and Perissodactyla*, Jena.

16 Darlington, P.J. (1957), *Zoogeography*, New York.

17 Hill, W.C.O. (1936), 'The affinities of the Lorisoids', *Ceylon J. Sci.*, B, 19, 287-314. See also Charles-Dominique, P. and Martin, R.D. (1970), 'Evolution of Lorises and Lemurs', *Nature* 227, 257-60.

18 Saban, R. (1963), 'Contribution à l'étude de l'os temporal des Primates', *Mem. Mus. Nat. Hist. Nat. Paris*, ser. A, 29, 1-378.

19 Walker, A., this volume.

20 Cartmill, M. (1971), 'Ethmoid component in the orbit of primates', *Nature* 232, 566-7.

21 Szalay, F.S. (1972), 'Cranial morphology of the Early Tertiary *Phenacolemur*, and its bearing on Primate phylogeny', *Amer. J. Phys. Anthrop.* 36, 59-75.

22 Van Valen, L. (1965), 'Treeshrews, primates and fossils'. *Evolution*, 19, 137-151.

23 Petter, G. (1961), 'Le peuplement en carnivores de Madagascar'. In 'Problèmes actuels de paléontologie (évolution des vertbrés)', *Coll. Int.. CNRS* 104, 331-42.

24 Wurster, D.H., and Benirschke, K. (1968), 'Comparative cytogenetic studies in the order Carnivora', *Chromosoma (Berlin)* 24, 336-82.

25 Albignac, R. (1972), 'Ethoécologie des Carnivores malgaches', in press.

26 Petter, F. (1961), 'Monophylétisme ou polyphylétisme des Rongeurs malgaches: problémes actuels de paléontologie', *Coll. Int. CNRS* 104, 301-10.

27 Butler, P.M. (1969), 'Insectivores and bats from the Miocene of East Africa, in Leakey, L.S.B. (ed.), *Fossil Vertebrates of Africa*, 1, London.

28 Mohr, E. (1960), *Wilde Schweine*, Wittenberg-Lutherstadt, 83-4.

29 Cooke, H.B.S. (1968), 'Evolution on Mammals on Southern Continents: 2. The fossil mammal fauna of Africa', *Quart. Rev. Biol.* 43, 234-64.

30 Walker, A. (1972), 'The dissemination and segregation of early primates in relation to continental configuration', in Bishop, W.W. and Miller, J.A. (eds.), *The Calibration of Hominoid Evolution*, Edinburgh.

31 Cartmill, M., this volume.

32 Radinsky, L., this volume.

33 Le Gros Clark, (1936), 'The problem of the claw in primates', *Proc. Zool. Soc. Lond.* 2, 1-24.

34 Thorndike, E.E. (1968), 'A microscopic study of the marmoset claw and nail', *Amer. J. Phys. Anthrop*, 28, 247-53.

35 Jouffroy, F.K. (1962), 'La musculature des membres chez les lémuriens de Madagascar', *Mammalia*, A, 3860.

36 Petter-Rousseaux, A. and Bourlière, F. (1965), 'Persistence du phénomène d'ovogénèse chez l'adulte de *Daubentonia madagascariensis*', *Folia primat.* 3, 241-4.

37 McElhinny, M.W. (1970), 'Formation of the Indian Ocean', *Nature* 228, 977-9.

38 Groves, C.P. (1971), 'Systematics of the genus *Nycticebus*', *Proc. 3rd. Int. Congr. Primat. Zürich 1970*, 1, 44-53.

39 Series: 'The skin of primates', see in particular Montagna, W. and Ellis, R.A., 'The skin of the potto', *Amer. J. Phys. Anthrop.* 17, 137-62; Montagna, W., Yasuda, K. and Ellis, R.A., 'The skin of the slow loris', *ibid.* 19, 1-22; Montagna, W. and Yun, J.S., 'The skin of the great bushbaby, *ibid.* 20, 149-66.

40 Bearder, S.K. and Doyle, G.A., this volume.

41 Tandy, J., this volume.

42 Lawrence, B. and Washburn, S.L. (1936), 'A new eastern race of *Galago demidovii*', *Occas. Papers Boston Soc. N.H.* 8, 255-66.

43 Kingdon, J. (1971), *Atlas of East African Mammals*, 1, London.

44 Petter, J.J., (1962), 'Recherches sur l'écologie et l'éthologie des lémuriens malgaches', *Mém. Mus. nat. Hist. nat. Paris,* sér. A, 27, 1-146.

45 Rumpler, Y., this volume.

46 Petter-Rousseaux, A. and Petter, J.J. (1967), 'Contribution à la systématique des Cheirogaleinae (Lémuriens malgaches): *Allocebus*, gen. nov., pour *Cheirogaleus trichotis* Günther 1875', *Mammalia* 31, 574-82.

47 Petter, J.J., Schilling, A. and Pariente, G. (1971), 'Observations éco-éthologiques sur deux Lémuriens malgaches nocturnes: *Phaner furcifer* et *Microcebus coquereli*', *Terre et Vie* 3, 287-327.

48 Charles-Dominique, P. and Hladik, C.M. (1971), 'Le *Lepilemur* du Sud de Madagascar: écologie, alimentation et vie sociale', *Terre et Vie* 3, 3-66.

49 Montagna, W. and Yun, J.S. (1962), 'The skin of primates: 10. The skin of the ring-tailed lemur', *Amer. J. Phys. Anthrop.* 20, 95-118.

50 Egozcue, J. (1967), 'Chromosome variability in the Lemuridae', *Amer. J. Phys. Anthrop.* 26, 341-8.

51 Petter, J.J., in press.

52 Petter, J.J. and Peyriéras, A. (1970), 'Observations éco-éthologique sur les lemuriens malgaches du genre *Hapalemur*', *Terre et Vie* 3, 356-82.

53 Gray, J.E. (1870), *Catalogue of Monkeys, Lemurs and Fruit-eating Bats in*

the Collection of the British Museum, London.
54 Jolly, A. (1967), *Lemur behaviour*, Chicago.
55 Schwarz, E. (1931), 'A revision of the genera and species of Madagascan Lemuridae', *Proc. Zool. Soc. London.* 399-428.
56 Chiarelli, B., this volume.
57 Chu, E.H.Y. and Swomley, B.A. (1961), 'Chromosomes of lemurine lemurs', *Science* 133, 1925-6.

W.P. LUCKETT

The phylogenetic relationships of the prosimian primates: evidence from the morphogenesis of the placenta and foetal membranes

Introduction

Comparative studies have demonstrated that the developmental pattern of the mammalian foetal membranes and placenta can provide considerable insight into the phylogenetic relationships among higher taxonomic categories.[1] Certain foetal membrane characteristics have undergone relatively little change during the evolutionary history of mammals, as suggested by their great similarity within the higher categories. Such conservative characters are the best kind for revealing phylogenetic relationships, because they are easier to trace across the wide gaps in the evolutionary record.[2] The conservative nature of the foetal membrane characteristics may be due in part to their relative isolation from the selective effects of the external environment,[1] but it is also likely that intense centripetal selective forces are acting upon them to maintain a relatively constant pattern.[3] There is no clear evidence that body size, locomotor activity or feeding habits have had any pronounced effect on the type of placentation that develops within any order of eutherian mammals.

The fossil record alone provides the essential time dimension for directly observing evolutionary changes, despite its limited nature. However, broad comparative analyses of characteristics in living animals provide additional insight into possible phylogenetic relations, particularly when an attempt is made to determine the primitive and specialised (derived) features of these characters. Convergence is less likely to occur in those characters or character complexes that are the product of the interaction of an extensive amount of genetic information.[4] The mammalian foetal membranes form such an interacting complex; they are derived from all of the embryonic germ layers and their interrelationships can be followed throughout their entire life history.[5] The terms 'primitive' and 'specialised' are relative; characteristics of common ancestry are primitive within a particular taxon, whereas others are more or less specialised in proportion to their departure from the ancestral condition.[6] Speculations concerning the primitive and specialised features of the developmental pattern of the foetal membranes in eutherian mammals are based upon a consideration of their nature in reptiles, birds, prototherians, marsupials, and all the orders of eutherian mammals.

It should be emphasised that mammalian classifications ought to be based upon a consideration of all of the available evidence, and the foetal membrane characters discussed here comprise only a portion of this evidence. The comparative development of the foetal membranes and placenta in primates has recently been discussed in detail,[3] and the present report will be limited to a summary of the main features of the foetal membranes in prosimian primates, and to a discussion of the possible relationships among the three major living taxa that have been included by many authors within the Prosimii: the Tupaiidae, Strepsirhini (=Lemuroidea and Lorisoidea), and the Tarsioidea.

Implantation

The early uterine blastocyst is fundamentally similar in all eutherian mammals prior to implantation; it consists of a peripheral epithelial layer of *trophoblast* surrounding a fluid-filled cavity, and a small cluster of *inner cell mass* attached to the inner surface of the trophoblast at one pole of the blastocyst (Fig. 1). A flattened epithelial layer of primitive endodermal cells becomes segregated from the internal surface of the inner cell mass, and this layer subsequently spreads peripherally to line the internal surface of the blastocyst in pre-implantation stages of tupaiids, strepsirhines, and tarsiids. Among the strepsirhines, the initial attachment of the bilaminar blastocyst is non-invasive in the lorisines *Nycticebus* and *Loris* and in the lemurid *Microcebus*, whereas there is a limited and variable amount of invasive activity by the trophoblast during the early stages of attachment in the two galagine species that have been examined (Figs. 2, 3, 4). Implantation in *Galago senegalensis* is central, and the invasive giant cell trophoblast of the attachment site is localised and transitory.[7] The invasive giant cell trophoblast degenerates by the time of the early primitive streak stage, and there is a complete replacement of the uterine epithelium at the previously denuded attachment site. The further development of the foetal membranes and placenta in this species is essentially identical to that of the lorisines.

The localised invasive attachment of the blastocyst in *Galagoides*

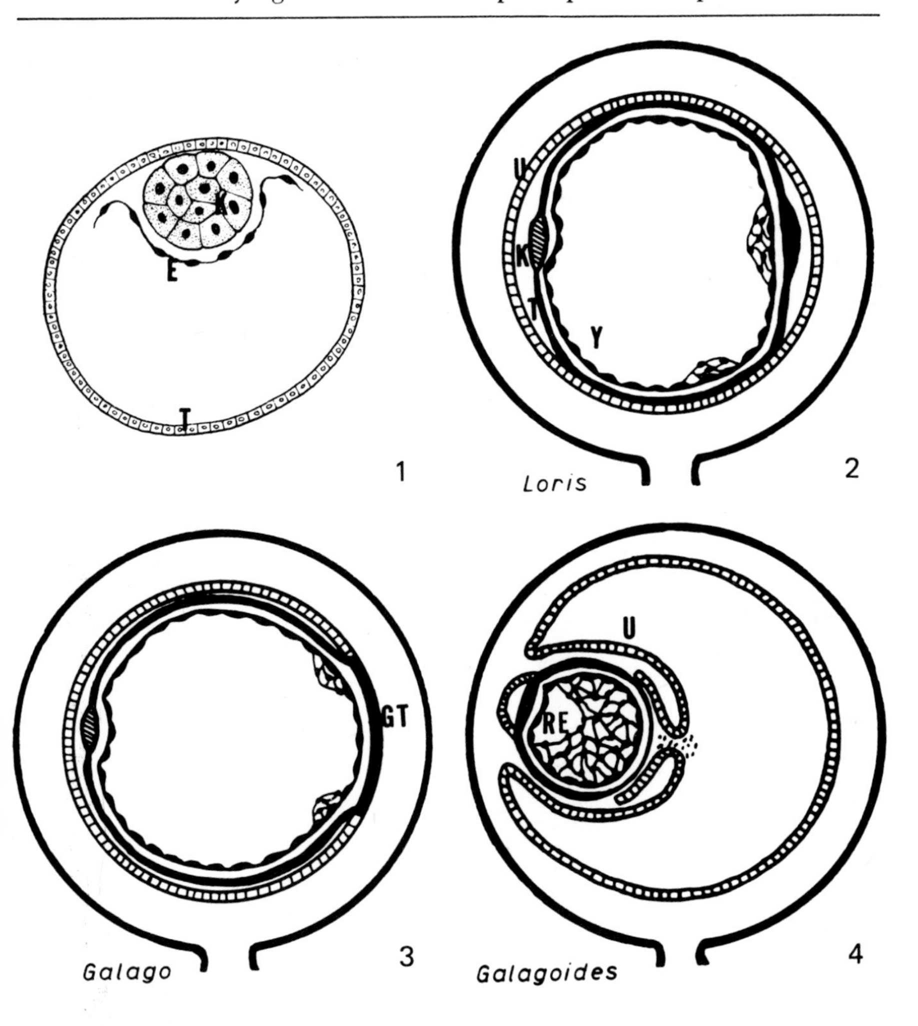

Figure 1. A schematic figure of a generalised eutherian blastocyst, prior to implantation, with peripheral trophoblast (T), inner cell mass (K), and endoderm (E).

Figure 2. Early implanted blastocyst of *Loris.* In this and all remaining figures the mesometrium of the uterus is oriented toward the bottom. Uterine epithelium (U), embryonic knot (K), trophoblast (T), yolk sac (Y).

Figure 3. Early implanted blastocyst of *Galago senegalensis*, with invasive giant cell trophoblast (GT).

Figure 4. Early implanted blastocyst of *Galagoides demidovii*, with reticulated endoderm (RE) in the yolk sac.

demidovii occurs within an implantation chamber that forms as the result of the closure of annular folds of the uterine mucosa about the blastocyst (Fig. 4). However, this condition is not strictly homologous to the interstitial implantation characteristic of the Hominoidea. The invasive activity of the para-embryonic trophoblast is more persistent than in *Galago senegalensis*, and it apparently contributes to the layer of giant cell trophoblast in the intimate zone of the definitive placenta,[8] although the critical intermediate stages are lacking. It seems likely that the implantation specialisations in *Galago senegalensis* and *Galagoides demidovii* have evolved independently from a non-invasive and centrally implanted blastocyst similar to that of *Nycticebus, Loris*, and *Microcebus*. However, further discussion of the significance of these specialisations is unwarranted without a knowledge of the mechanism of implantation in other species of the subfamily Galaginae.

Initial attachment of the blastocyst in *Tarsius* elicits a peculiar reaction in the necks of the underlying uterine glands, and this results in their proliferation and the formation of a unique nodular mass of modified epithelium (Fig. 5).[3] Subsequently, there is a proliferation of the trophoblast at the localised attachment site to form a trophoblastic preplacenta whose interstices soon become filled with maternal blood. This developmental relationship foreshadows the differentiation of the definitive haemochorial placenta in later stages, and this contrasts sharply with the condition in strepsirhines. Strepsirhines and *Tarsius* share a number of presumed primitive primate features, including the exposure of the embryonic disc to the contents of the uterine lumen by the rupture of the overlying layer of trophoblast, the participation of the para-embryonic trophoblast in the initial attachment, and the lateral orientation of the embryonic disc to the endometrium at the time of implantation (Figs. 2, 5). The first two of these features also appear to be primitive eutherian characteristics.

The tupaiid blastocyst implants superficially and bilaterally on specialised, gland-free endometrial pads, and the exposed embryonic disc is oriented antimesometrially (Fig. 6). Excluding the specialised nature of the endometrial pads, the relationships of the early implanted tupaiid blastocyst are quite similar to those of the insectivore families Soricidae and Talpidae. It is possible that these resemblances are due to primitive retentions in all three families.[9]

Development of the primitive streak

The initial differentiation of the primitive streak in strepsirhine primates and tupaiids follows the pattern characteristic for most other mammalian taxa. The earliest mesoderm derived from the primitive streak is axial (embryonic), and only later do cells of

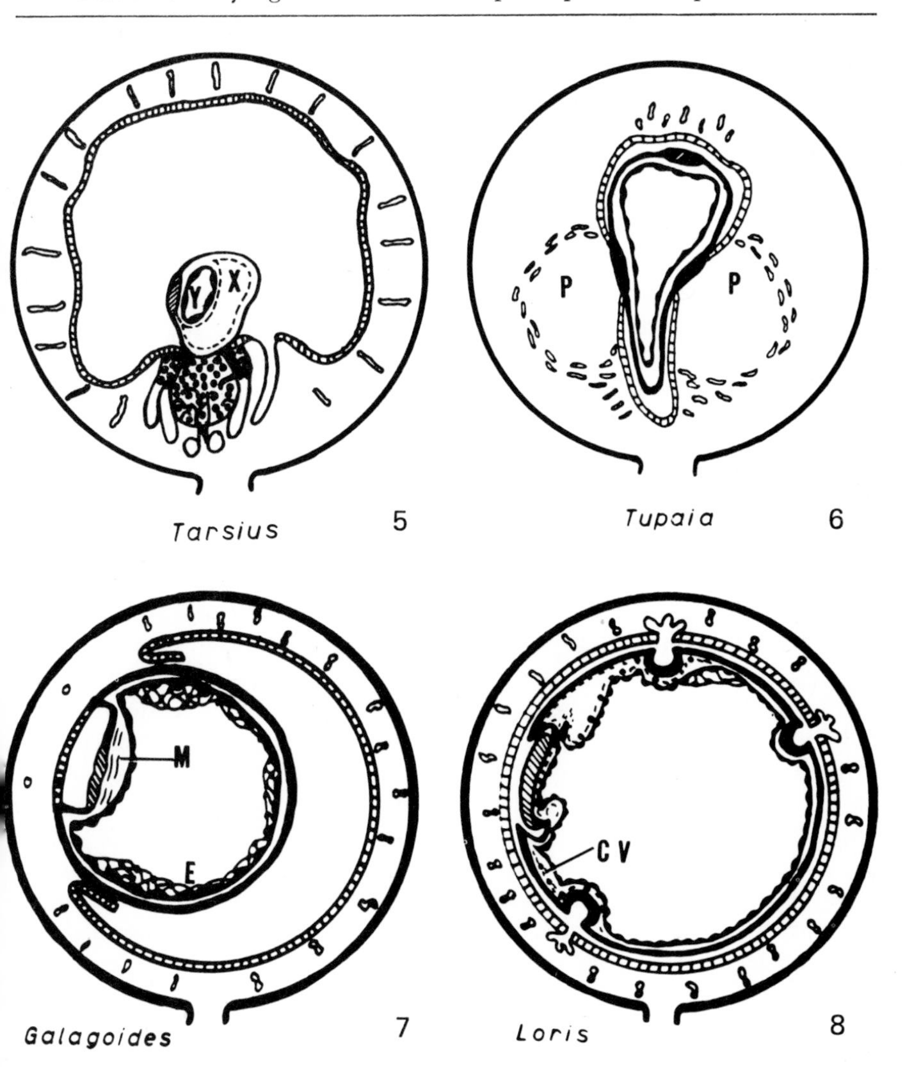

Figure 5. Early implanted blastocyst of *Tarsius*, with exocoelom (X), and nodular epithelial mass (N).

Figure 6. Early implanted blastocyst of *Tupaia*, with bilateral endometrial pads (P).

Figure 7. Early primitive streak stage of *Galagoides demidovii*, with embryonic mesoderm (M).

Figure 8. 12-14 somite stage of *Loris*, with amniotic folds and the development of a chorio-vitelline placenta (CV).

primitive streak origin spread beyond the margins of the embryonic disc to form the extra-embryonic mesoderm (Fig. 7). By contrast, there is a precocious differentiation of the caudal margin of the primitive streak in *Tarsius,*[10] so that extra-embryonic mesoderm is formed before the appearance of embryonic mesoderm (Fig. 5). This precociously differentiated extra-embryonic mesoderm gives origin to the mesodermal lining of the exocoelom and to the primordium of the body stalk, prior to the differentiation of more cranial portions of the streak. This specialised pattern of primitive streak formation is strikingly similar in *Tarsius* and the Anthropoidea (Platyrrhini and Catarrhini),[3] and it is correlated with the development of the mesodermal body stalk characteristic of all these taxa.

Development of the yolk sac and chorio - vitelline placenta

The migration of extra-embryonic mesoderm between the trophoblastic and endodermal layers of the strepsirhine and tupaiid blastocyst and the subsequent differentiation of blood vessels in this mesoderm result in the development of a chorio-vitelline placenta during the 12-14 somite stage (Figs. 8, 9). The chorio-vitelline placenta apparently functions as a site of foetal-maternal exchange until it is displaced topographically by the spread of the exocoelom and the allantoic mesoderm; this occurs at variable times after the 30 somite stage. The development of a vascular chorio-vitelline placenta at the equatorial pole of the blastocyst appears to be a primitive characteristic of therian mammals. It forms the definitive placenta for the great majority of marsupials that have been examined, as opposed to its transitory nature in eutherians.

The interior of the early implanted blastocyst of *Galagoides demidovii* is partly filled with a cellular meshwork of reticulated endoderm,[8] and the interstices of the meshwork are in continuity with a more distinct, eccentric primary yolk sac cavity (Fig. 4). The reticulated meshwork of endoderm becomes reduced as the expanding blastocyst increases in size. Most of the abembryonic endoderm is reduced to a simple squamous layer by the time of the early primitive streak stage, with only small isolated patches of persisting reticulated endoderm (Fig. 7). The continued expansion of the *Galagoides* blastocyst results in the rupture of the 'capsular' endometrium overlying its abembryonic pole. Thus, the abembryonic wall of the blastocyst secondarily acquires a relationship with the endometrium, similar to that seen in comparable stages of other strepsirhines. The subsequent development of the chorio-vitelline placenta in *Galagoides demidovii* appears to be identical to the general strepsirhine conditon.

The precocious development of extra-embryonic mesoderm and

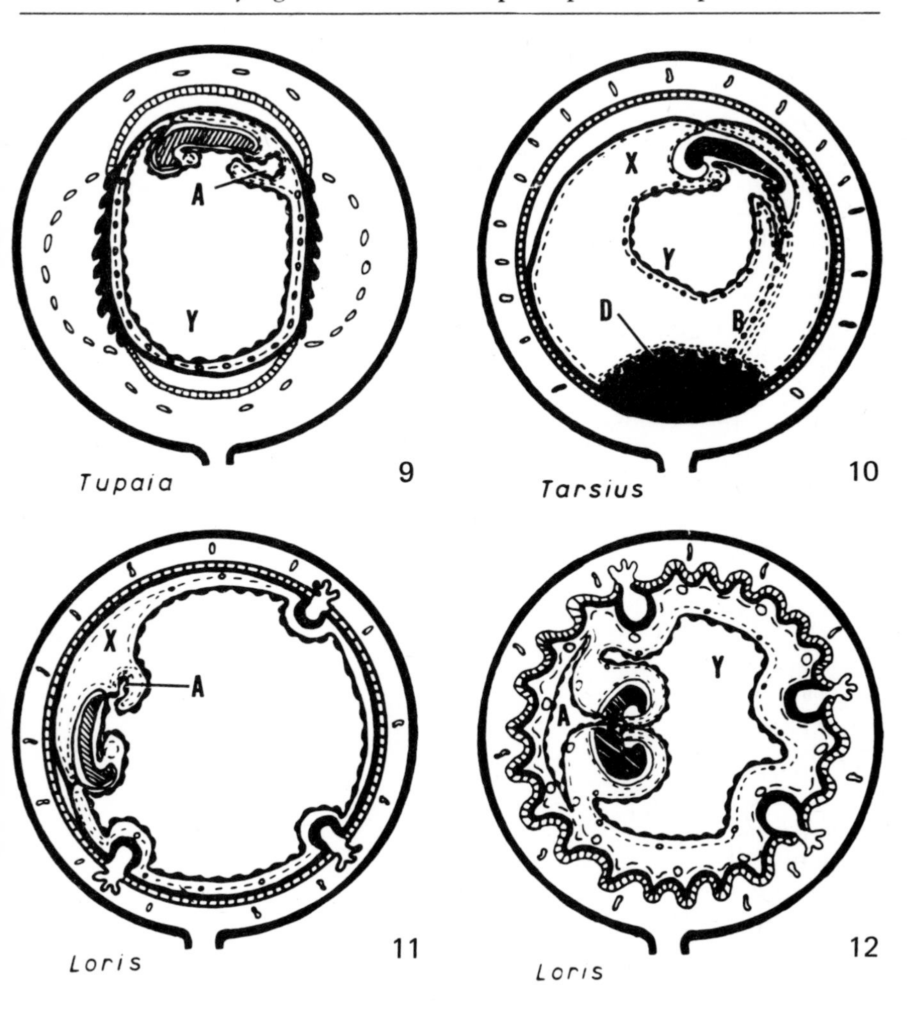

Figure 9. 23 somite stage of *Tupaia*, with chorio-vitelline placenta, recently fused amniotic folds, and allantoic diverticulum (A).

Figure 10. 17 somite stage of *Tarsius*, with free vascular yolk sac (Y), tubular allantoic diverticulum, recently fused amniotic folds, mesodermal body stalk (B), and early stage in the development of the definitive placental disc (D).

Figure 11. 22 somite stage of *Loris*, with chorio-vitelline placenta, recently fused amniotic folds, and allantoic diverticulum.

Figure 12. 39 somite stage of *Loris*, with free vascular yolk sac, allantoic vesicle, and the complete vascularisation of the chorion by allantoic mesoderm.

the exocoelom in *Tarsius* results in the separation of the endodermal yolk sac from its contact with the adjacent chorion prior to the development of blood islands in the splanchnic mesoderm of the yolk sac (Fig. 5). Therefore, a vascular chorio-vitelline placenta (that is, the fusion of the vascular yolk sac with the chorion) does not occur in *Tarsius.* When vitelline vessels subsequently develop in later stages, the vascular splanchnopleure of the yolk sac lies in contact with the large fluid-filled exocoelom (Fig. 10). It seems likely that the free vascular yolk sac of *Tarsius* plays an important role in foetal nutrition by its absorption of exocoelomic fluid, prior to the establishment of the chorio-allantoic placenta. The relationships of the yolk sac and exocoelom in *Tarsius* thus differ sharply from the condition in strepsirhine Primates, whereas the tarsiid condition is essentially identical with the pattern in platyrrhines and catarrhines.[3]

Allantoic diverticulum

The allantoic primordium arises as an endodermal diverticulum from the caudal end of the developing hindgut in strepsirhine and tupaiid embryos that bear about 13-20 pair of somites (Figs. 9, 11). The allantoic diverticulum, surrounded by a globular mass of highly vascularised mesoderm, projects freely into the exocoelom at this stage. The continued expansion of the allantoic vesicle and the exocoelom results in the displacement of the vascular splanchnopleure of the yolk sac from its contact with the chorion. The fusion of the vascular allantois with the chorion results in the replacement of the chorio-vitelline placenta, both topographically and functionally, by the developing chorio-allantoic placenta. Following the fusion of the allantois with the chorion in strepsirhines, the allantoic mesoderm rapidly spreads beyond the margins of the allantoic vesicle to vascularise the entire inner surface of the chorion (Fig. 12). The further expansion of the allantoic vesicle and its subsequent lobulation appear to be secondary phenomena.

The precocious spread of allantoic mesoderm has been interpreted as the initial step in the evolution of the mesodermal body stalk in haplorhine Primates (Tarsioidea, Platyrrhini, and Catarrhini).[11] However, it is equally plausible that this pattern of chorionic vascularisation is a secondary specialisation, because the chorio-allantoic placenta of strepsirhines is not established at a relatively earlier developmental stage than it is in tupaiids.[3] The expansion of the allantoic vesicle in tupaiids is directly involved with the vascularisation of the chorio-allantoic placenta; this appears to represent a primitive retention of the ancestral eutherian condition that is shared with carnivores, ungulates, and some insectivores.

The precocious development of a mesodermal body stalk between the caudal margin of the embryonic shield and the chorion in early primitive streak embryos of *Tarsius* occurs prior to the differentiation of the allantoic diverticulum. Nevertheless, the precociously developed extraembryonic mesoderm of the body stalk is clearly homologous to the allantoic mesoderm of strepsirhines and most other mammals.[11] Both are derived from the most caudal portion of the primitive streak and are ultimately associated with the allantoic diverticulum, and both regions serve as the pathway for the distribution of allantoic vessels to the chorio-allantoic placenta. The allantoic diverticulum of *Tarsius* invaginates the proximal portion of the body stalk in presomite stages; it retains a thin tubular appearance and never reaches the surface of the developing placental disc (Fig. 10).

The specialised pattern of body stalk and allantoic development in *Tarsius* is strikingly similar to the condition in the Anthropoidea, and differences in the length of the body stalk in the two groups are believed to be due to basic differences in the method of implantation.[3] The evolution of the haplorhine body stalk from the vesicular allantoic condition is considered to be the result of developmental acceleration in order to facilitate the vascularisation of the chorion at the earliest possible moment.[11] This concept is supported by a comparison of the initial vascularisation of the chorion in strepsirhines and *Tarsius*. The chorion is vascularised by the yolk sac mesoderm during the 12-14 somite stage in strepsirhines (as in tupaiids), prior to the differentiation of the allantois, whereas the initial vascularisation of the chorion in *Tarsius* is effected by umbilical vessels in the body stalk (allantoic) mesoderm during the 9-10 somite stage.[3] Thus, the process of chorionic vascularisation in *Tarsius* is clearly accelerated, as it is in other haplorhines, and a chorio-vitelline stage is bypassed.

Chorio-allantoic placenta

The fusion of the allantoic vesicle with the chorion and the vascularisation of the chorion by allantoic mesoderm initiate the differentiation of a non-invasive, diffuse, epitheliochorial placenta in all families of strepsirhine Primates. Much of the surface of the chorion differentiates into simple and branched villi that interdigitate with endometrial septa. There are two primary mechanisms involved in foetal-maternal exchange in strepsirhines: (1) exchange between the foetal and maternal capillaries across the intervening layers of the epitheliochorial placenta; and (2) absorption of uterine gland secretions by specialised chorionic vesicles or by less specialised 'bare areas' of the chorion. Histochemical evidence indicates that mucopolysaccharides and iron are absorbed by the trophoblastic lining of the chorionic vesicles in *Galago senegalensis*,[12] and this is doubtless true for strepsirhines in general.

In addition to the diffuse, epitheliochorial portion of the placenta in *Galagoides demidovii*, there is a small specialised zone overlying the site of the initial fusion of the allantoic vesicle with the chorion. Foetal and maternal tissues adhere more intimately in this zone, and a layer of columnar, pale-staining giant cells lies between the cuboidal layer of chorionic cytotrophoblast and the maternal stroma. Although the critical intermediate stages in the formation of this intimate zone are lacking, it seems most probable that the layer of giant cells is of trophoblastic origin.[3,8] If this is true, then the specialised intimate zone in *Galagoides* exhibits an endotheliochorial relationship, because maternal capillaries are in intimate apposition with the giant cell layer. Other authors have suggested that the layer of giant cells may represent hypertrophied uterine epithelium, and this region of the placenta would then be only a specialised epitheliochorial zone.[12,13]

A localised, discoidal, haemochorial placenta develops in *Tarsius*, as it does in all haplorhine primates, in contrast to the diffuse epitheliochorial condition in strepsirhines. Mesoderm derived from the body stalk invaginates the foetal surface of the trophoblastic preplacenta, and the appearance of blood cells within the allantoic vessels of these mesodermal villi during the 9-10 somite stage initiates the functional stage of chorio-allantoic placentation in *Tarsius*. Further growth and rearrangement result in the development of a continuous system of trophoblastic-covered foetal trabeculae which surround a lacunar system of circulating maternal blood. The definitive placenta of *Tarsius* is therefore labyrinthine and haemochorial.

The definitive chorio-allantoic placenta of tupaiids develops bilaterally and utilises the specialised gland-free sites of the initial attachment and chorio-vitelline placentation. The uterine epithelium and some of the underlying maternal stroma is destroyed by the trophoblast, but the endothelium of maternal vessels remains intact where it comes into contact with the advancing trophoblast. Recent re-examinations demonstrated that the maternal endothelium of tupaiids remains intact during the differentiation of the definitive chorio-allantoic placenta, and the latter is classified as bidiscoidal, labyrinthine, and endotheliochorial.[13,14]

Amniogenesis

In all prosimian primates and tupaiids the amnion is formed by the elevation of somatopleuric folds at the margins of the embryo proper and by the subsequent fusion of these folds over the dorsal aspect of the embryo (Figs. 8, 9, 10). Following this fusion, the amnion and chorion became separated by the exocoelom, although the site of fusion may persist for some time as a chorio-amniotic raphe.

Amniogenesis by folding appears to be the primitive condition in mammals and in other amniotes. The more specialised pattern of amniogenesis by cavitation, characteristic of Anthropoidea, is probably related to the attachment of the blastocyst to the endometrium by the trophoblast that persists over its embryonic pole.[3]

Phylogenetic considerations

The evidence for including the Tupaiidae within the order Primates has been based primarily upon certain resemblances in their brain, visual system, auditory bulla, and placentation.[15] However, resemblances may be due to homology, parallelism or convergence, and recent re-examinations of all of these tupaiid features suggest that their resemblances to Primates are the result of convergence or the retention of primitive eutherian characters.[16] The only foetal membrane characteristics shared by tupaiids and strepsirhines are those that are retentions of the primitive eutherian condition; these features are also shared with carnivores, some insectivores, and ungulates.[9] Immunological comparisons of serum proteins also support the removal of the Tupaiidae from any special relationship with the Strepsirhini.[17]

Many of the similarities between strepsirhine and tarsioid Primates may be due to the retention of primitive primate or eutherian features, and it is possible that these resemblances reflect a general prosimian *grade* of organisation when compared with a more advanced simian grade. These grade levels of organisation in Primates are generally considered to also represent a basic cladistic division into the suborders Prosimii and Anthropoidea.[6] However, the characteristics that have been used to evaluate primate subordinal relationships do not necessarily reflect the most conservative features, and an examination of other characteristics within the prosimian grade reveals that the Tarsioidea show closer affinities with the Anthropoidea than with the Strepsirhini.[3]

Table 1 summarises the major features in the development of the foetal membranes and placenta among the higher taxonomic categories of Primates. The developmental pattern of the Tarsioidea is most similar to that of the Anthropoidea, particularly in the specialised interrelationship between the yolk sac and allantois (including body stalk). The vascularisation of the chorion by either the yolk sac or the allantois is essential for the functional development of the mammalian placenta, and there is probably a fundamental interrelationship in their developmental patterns. Intense centripetal selective forces doubtlessly act to maintain a delicate balance of functional interrelationship between the yolk sac and the allantois during ontogeny.[3] The nature of the yolk sac and the allantois are two of the most conservative features during the evolution of the mammalian foetal membranes,[1] and the specialised nature of their developmental pattern in the Tarsioidea and Anthropoidea provides good evidence of affinities between these taxa. On the other hand, the similarities between *Tarsius* and the strepsirhines with respect to the initial attachment pole of the blastocyst and the mechanism of amniogenesis appear to be primitive primate and eutherian features; in this respect they have no bearing on the taxonomic relationships of these two groups.

The evidence from the developmental pattern of the foetal membranes supports the subordinal division of the Primates into Strepsirhini and Haplorhini. The foetal membrane evidence also supports the concept,[18] based primarily on dental similarities, that certain Eocene tarsiiforms may have been ancestral to the Anthropoidea. Immunodiffusion studies utilising an antiserum prepared against the serum proteins of *Tarsius* also suggest that *Tarsius* shared a more recent common ancestry with the Anthropoidea than with either the Lemuroidea or Lorisoidea.[19]

Table 1 Major features of the foetal membranes and placenta of Primates

	Strepsirhini	*Tarsioidea*	*Anthropoidea*
Implantation:			
Lateral Orientation of the Disc	x	x	x
Attachment Trophoblast:			
Para-embryonic Pole	x	x*	
Embryonic Pole			x
Amniogenesis:			
Folding	x	x*	
Cavitation			x
Yolk Sac:			
Large, Free, Reduced Later	x		
Small, Free		x	x
Chorio-vitelline Placenta:			
Present	x		
Absent		x	x
Allantoic Vesicle:			
Large, Permanent	x		
Small, Rudimentary		x	x
Chorio-Allantoic Placenta:			
Diffuse, Epitheliochorial	x		
Discoidal, Haemochorial		x	x

* It has been suggested that the differences in initial attachment and amniogenesis between Tarsioidea and Anthropoidea are the result of the development of a simplex uterus in Anthropoidea.[3]

Meaningful evolutionary classifications cannot be based upon the nature of any single feature or character complex, and it is essential that features of the dentition, nervous system, skeleton and other systems be examined on a broad comparative basis in order to determine their primitive and specialised features. Such determinations would contribute considerably to our understanding of the interrelationships among Primates at the higher taxonomic levels.

NOTES

1 Mossman, H.W. (1937), 'Comparative morphogenesis of the fetal membranes and accessory uterine structures', *Contrib. Embryol.* 26, 129-246; Mossman, H.W. (1953), 'The genital system and the fetal membrances as criteria for mammalian phylogeny and taxonomy', *J. Mammal.* 34, 289-98.

2 Farris, J.S. (1966), 'Estimation of conservatism of characters by constancy within biological populations', *Evolution* 20, 587-91.

3 Luckett, W.P. (in press), 'Comparative development and evolution of the placenta in Primates', in Luckett, W.P. (ed.), *Reproductive Biology of the Primates: Contributions to Primatology*, vol. 3, Basel.

4 Simpson, G.G. (1964), 'Organisms and molecules in evolution', *Science* 146, 1535-8.

5 Mossman, H.W. (1967), 'Comparative biology of the placenta and fetal membranes', *in* Wynn, R.M. (ed.), *Fetal Homeostasis*, vol. 2, New York.

6 Simpson, G.G. (1961), *Principles of Animal Taxonomy*, New York.

7 Butler, H. (1964), 'The reproductive biology of a strepsirhine *(Galago senegalensis senegalensis)*', *Int. Rev. Gen. and Exp. Zool.* 1, 241-96.

8 Gérard, P. (1932), 'Etudes sur l'ovogénèse et l'ontogénèse chez les lémuriens du genre *Galago*', *Archs. Biol. Paris* 43, 93-151.

9 Luckett, W.P. (1969), 'Evidence for the phylogenetic relationships of tree shrews (family Tupaiidae) based on the placenta and foetal membranes', *J. Reprod. Fert. Suppl.* 6, 419-33.

10 Hubrecht, A.A.W. (1902), 'Furchung und Keimblattbildung bei *Tarsius spectrum*', *Verh. K. Akad. Wet.* 2, (8), 1-113; Hill, J.P. and Florian, J. (1963), 'The development of the primitive streak, head-process and annular zone in *Tarsius*, with comparative notes on *Loris*', *Bibl. Primat.* 2, 1-90.

11 Hill, J.P. (1932), 'The developmental history of the primates', *Phil. Trans. Roy. Soc.* 221, 45-178.

12 Butler, H. and Adam, K.R. (1964), 'The structure of the allantoic placenta of the Senegal bush baby (*Galago senegalensis senegalensis*)', *Folia Primat.* 2, 22-49.

13 Hill, J.P. (1965), 'On the placentation of *Tupaia*', *J. Zool.* 146, 278-304.

14 Luckett, W.P. (1968), 'Morphogenesis of the placenta and fetal membranes of the tree shrews (family Tupaiidae)', *Am. J. Anat.* 123, 385-428.

15 Le Gros Clark, W.E. (1962), *The Antecedents of Man*, 2nd ed., Edinburgh.

16 van Valen, L. (1965), 'Treeshrews, primates, and fossils', *Evolution* 19, 137-51; Campbell, C.B.G. (1966), 'The relationships of the tree shrews: The evidence of the nervous system', *Evolution*, 20, 276-81; Noback, C.R. and Shriver, J.E. (1966), 'Phylogenetic and ontogenetic aspects of the lemniscal systems and the pyramidal system', in Hassler, R. and Stephan, H. (eds.), *Evolution of the Forebrain*, Stuttgart; Campbell, C.B.G. (1969), 'The visual system of insectivores and primates', *Ann. N.Y. Acad. Sci.* 167, 388-403.

17 Goodman, M. (1967), 'Deciphering primate phylogeny from macromolecular specificities', *Am. J. Phys. Anthrop.* 26, 255-76.

18 Simons, E.L. (1963), 'A critical reappraisal of Tertiary Primates', in

Buettner-Janusch, J. (ed.), *Evolutionary and Genetic Biology of Primates*, vol. 1, New York; Simons, E.L. (1969), 'The origin and radiation of the Primates', *Ann. N.Y. Acad. Sci.* 167, 319-31.

19 Goodman, M. et al., this volume.

L. R. WOLIN

What can the eye tell us about behaviour and evolution? or: *The aye-ayes have it, but what is it?*

Introduction

Primates are predominantly visual animals in that they appear to rely more heavily on vision than on the other sense modalities in guidance of behaviour. While the primate visual system may not be superior in all aspects to that of other vertebrate orders, the overall functional capacity of this system, in combination with the advanced development of the primate brain, results in a highly effective system.

In spite of the primacy of vision in primate behaviour, only relatively few species have been studied in any detail, and extensive study has been limited to only a few selected species of higher primates. With regard to the evolutionary aspect of vision in primates, virtually nothing is known, and the several aberrant genera which have aroused particular interest are more sources of contention than of enlightenment. The sub-order Prosimii is even less understood. To my knowledge, only *Tupaia* has been subject to anatomical, physiological and behavioural study with regard to vision.

The purpose of this paper will be to summarise what is known about certain characteristics of the prosimian eye and to attempt to relate this information to what is known about certain aspects of the behaviour of prosimians. Some comment will also be made with respect to our knowledge of the eye and behavioural patterns, as they may bear on our understanding of prosimian evolution. Specifically, with respect to the eye, I will deal with the following: pupil, tapetum, pigment layer, neuro-retinal structure and fundus patterns. In so far as behaviour is concerned, light sensitivity (nocturnal vs. diurnal habits), acuity and colour vision will be considered. Finally, some comments will be made regarding our need for information in each of these areas, so that those who attempt to repeat this task in the future may meet with somewhat less frustration than I have.

The pupil

Prince dealt with the relationship of pupillary patterns and visual function.[1] He notes that 'there are a surprising number of variations' of pupil shapes 'the reasons for which are not always immediately obvious'.

Among the prosimians, we find two major divisions with respect to pupil shape; circular and elongated. Within the second group there are variations which may be described as oval and elliptical, with variations in horizontal vs. vertical orientation as well. The following observations are based on my own examination of the eyes of a number of prosimian species and a survey of photographs of many species published in a variety of texts on vision, primatology and mammalian taxonomy.[2-8]

It has sometimes been assumed that a circular pupil is associated with a diurnal life style, whereas elongated or slit pupils are associated with nocturnal habits. There appears to be little relationship, however, between pupil shape and activity patterns. Among the prosimians, *Tupaia*, which is diurnal in habit, has an elongated (oval) pupil which is essentially horizontal in orientation. This is probably functional in terms of the lateral setting of the eyes and the temporal location of the central retinal area. Most of the Lorisiformes have elongated pupils, and most are vertically oriented. *Nycticebus coucang*, however, has a round pupil which is capable of contracting almost to pinpoint size. All of the Lorisiformes are reported as nocturnal. Most of the Lemuriformes about which information is available appear to have round pupils. Prince describes the crown lemur (= *L.mongoz coronatus*) as having an oval pupil with vertical orientation. *L. mongoz* is diurnal and probably also crepuscular in habit. However, *Indri*, which is diurnal in habit, has a round pupil. *Tarsius syrichta* and *Tarsius spectrum*, both of which are nocturnal, have horizontally oriented oval pupils, while *Microcebus*, which is also nocturnal, has a round pupil.

While it is obvious that information regarding pupil shapes among prosimians is scanty, it nevertheless appears reasonable to conclude that (as with other vertebrate orders) pupil shape shows no systematic relationship to observed living habits or behaviour. I am also inclined to suggest that pupil shape will not be found to be of any major significance in attempting to describe prosimian evolution.

Tapetum and pigment epithelium

The tapetum and pigment layers are both associated with light sensitivity. The tapetum is regarded as increasing light sensitivity by reflecting light back through the receptors, while pigments located just behind the neuroretina presumably absorb light. Many nocturnal species and some species categorised as crepuscular in habit have a tapetum.[2,9,10] Most diurnal primates do not have a tapetum. The diurnal Tupaiiformes appear not to have any tapetum, while the Lorisiformes, all of which are reported as being nocturnal, appear to have a tapetum. There are, however, some interesting exceptions. The tarsiers appear to have no tapetum,[10] although they are nocturnal. Among the Lemuriformes, *Propithecus* and *Lemur catta*, both reportedly diurnal in habit, have tapeta.[10] I should add that the tapetum of *L. catta* is *at least* as highly reflective as that of any Lorisiform which I have examined.

Information regarding retinal pigmentations is perhaps slightly less confusing, but data is also somewhat sparse. In general, it may be said that the pigment deposit tends to be somewhat greater in the diurnal than in the nocturnal species and that there is usually relatively less pigmentation in the centre than in the periphery of the retina.[9,10] Where a tapetum is found, no pigment or only very slight pigmentation occurs over the tapetum, but there may be relatively dense pigmentation at the periphery.

It thus appears that ocular characteristics such as the existence of a tapetum and the occurrence of retinal pigment cannot tell us, with any certainty, about behaviour. It may well be that the apparent discrepancies between anatomy and behaviour noted above may have some evolutionary significance in terms of changing patterns of adaptation; but in the absence of any information regarding the ancestors of living species, interpretation of these data may be extremely difficult.

Neuroretina

The neuroretina, containing the light-sensitive receptors and various transmitting and integrating elements, should show some more direct relationship to visual capacity and to visual guided behaviour. Definitive statement of such relationships will, however, be dependent on more detailed study of anatomy, physiology and behaviour, than are presently available. Certain areas of investigation, such as the detailed study of synaptic relationships within the reticular layers of the retina have only recently begun.

1. Receptors

The duplicity theory of vision based on the distinction between rods as low light level, achromatic receptors and cones as higher intensity chromatic receptors, appears to be well established. Walls, however, states that 'as a guide to the interpretation of habits in terms of 'structure', the Duplicity Theory, faced with transmutation, has come to the end of its usefulness'.[11] This statement is based upon the conclusion that many species have become nocturnal secondarily and that accordingly cones have in effect been 'transmuted' into functional rods.

Among the prosimians, rods are predominant. Only *Tupaia* and *Urogale* are reported as having essentially a pure cone retina.[10,12] Both genera are diurnal,[2,6] and *Tupaia glis* is reported to have colour vision.[13] No reports on the retina of *Ptilocerus* appear to be available, though Polyak states that this genus is nocturnal.[6] The tarsiers appear to have pure rod retinas,[10,14] and are nocturnal in habit. No information was uncovered regarding acuity or colour vision of this group. The uniformly nocturnal Lorisformes have predominantly rod retinas, but Kolmer reports some cones in *Nycticebus coucang,*[9] and Rohen reported some 'cone-like structures' in the retina of *Loris tardigradus.*[15] The Lemuriformes include species which are nocturnal, species which are diurnal and some that are crepuscular.[2,4,5,8,16] The eyes of several species have been studied and are reported to be primarily rod retinas, but with some cones.[9,10,17] The ratio of rods to cones varies from 1000:1 in *Cheirogaleus* to 5:1 in *Lemur catta.*[9] Appropriately enough, *Lemur catta* (and *L. mongoz*) have been reported to have colour vision. While not unusual, it is interesting to note that the nocturnal *Cheirogaleus* (species unspecified) apparently has some cones or cone-like receptors.[9] Where cones are found (except in Tupaiaformes), they appear to be localised in the central retinal area, but there is no report of a 'rod-free area' in any prosimian.

Receptor type and/or structure also has some relationship to visual acuity. While the overall structure of the retina also determines minimum acuity, absolute size of the receptors and their density per unit area, or centre-to-centre spacing, is the first limiting factor in determining acuity. Most prosimian species studied show some structural variation in receptors between the central and peripheral retina. There is frequently an elongation of the receptors in the central retina,[9,10,17] and almost always the density of receptors in the central retina is much greater than in the periphery.[10] As more behavioural studies of prosimian species become available, it will be interesting to see to what extent these anatomical reports correlate with behavioural results.

2. Cellular layers of the retina

A great deal of information processing apparently occurs within the retina. The relative numbers of receptors, bipolar cells and ganglion cells in various species has been regarded as intimately related to the limits of visual acuity and light sensitivity. While this is undoubtedly true, it is only a part of the story, as more recent studies of fine structure of the reticular layers of the retina have indicated.[18-21]

I shall not attempt to restate here all of the detailed information about each species which has been studied. Most of this work has been done by Detweiler,[3] Kolmer,[9] Rohen,[10,15,22] and Woollard.[17,23] The most recent and detailed work has been presented by Rohen and Castenholz,[10] while I have previously attempted a general survey of anatomic studies of the retina of primates.[14] On the basis of their excellent studies, Rohen and Castenholz have proposed four basic types of retinal organisation.[10] The classification is based on the ratio of receptor to bipolar to ganglion cells and the variation in these relationships between different portions of the retina. Group I, which includes *Tupaia glis* and *Urogale everetti*, would appear to be structured for high visual acuity. There is some differentiation between the central area and peripheral retina, with the central area better designed for high acuity. The second group is exemplified by *Microcebus murinus, Cheirogaleus medius* and *Lemur fulvus.* No pattern is illustrated for a central retinal area, and the general structure is one apparently designed for enhancement of light sensitivity. The third type of retina, exemplified by *Galago senegalensis* and *Indri indri*, reveals a central area which should manifest higher acuity and lesser light sensitivity than the more peripheral portions. The fourth type includes all the higher primates (except possibly *Aotes*), and is divided into three subtypes, the first of which is represented by *Tarsius bancanus.* This type of retina shows a distinct three-part organisation with structural organisation for acuity decreasing in a central to peripheral direction, while light sensitivity would be expected to be increased toward the periphery. The significance of these anatomical patterns will be more apparent, if and when single cell studies of the retinas of representative species are undertaken. Behavioural studies of visual acuity, light sensitivity and colour discrimination of a sample of species representative of different patterns of retinal organisation would of course be invaluable.

3. The reticular layers

These layers of the retina, which contain the synaptic structure of the retina, have only recently been subject to detailed study through the use of electron microscopy.[18-21] Although earlier work suggested that the horizontal and amacrine cells of the inner nuclear layer should have some relationship to the processing of visual information,[6] only in the last several years have such relationships in fact been demonstrated.[19] These synaptic relationships, and the cellular constituents of the inner nuclear layer from which they stem, indicate that the interpretation of retinal function based upon ratios of the elements of the cellular layers may be greatly modified by more detailed understanding of synaptic patterns. The horizontal cell synapses in the outer reticular layer and the amacrine cell synapses of the inner reticular layer have been implicated in various integrative processes including colour and brightness contrast, contour enhancement and motion sensitivity (including the directionally selective patterns revealed by unit studies). Dowling has also described three different types of amacrine cells with different areal distributions, the functional significance of which is as yet speculative.[19]

I must note at this point that I have not attempted any comprehensive survey of this particular topic. I will defer to those with expertise in this field for more detailed analysis and critical evaluation of this work. In spite of (or perhaps because of) my lack of expertise in these fields, I cannot but feel that analysis of the fine structure of the reticular layers of the retinas of prosimians will contribute substantially to our knowledge, and may make less obscure some of the relationships between ocular anatomy, behaviour and, perhaps, even evolution.

4. *Fundus patterns*

I will comment on this topic only briefly, since most of what I have to say has been said already.[14,24] The fundus pattern is a very gross index of structure and as such contains only very limited information. Even what appears obvious may not, in fact, be what it appears.

The eye of *Lemur catta* would appear from funduscopic examination to be that of a truly nocturnal animal, since it is but sparsely pigmented and has a highly reflective tapetum; but this is contradicted by published observations. *Tarsius syrichta*, on the other hand, has a moderately pigmented retina with a well developed macula and fovea and thus should be a diurnal animal with excellent acuity; but we know that the first point is untrue, while the second (to my knowledge) remains unknown.

I have never yet examined the eye of any primate which did not show an area which appeared to be a central visual area, this being defined by either the vascular pattern, the pigmentation, or both. Nevertheless, histological studies apparently fail to reveal a specialised central retinal area in some prosimian species.[10] Perhaps more detailed anatomical study or studies of receptive fields by single cell methods will change this picture. In any event, the correlations between fundus patterns and life style of prosimians are far from impressive, and with regard to any interpretation of the evolution of prosimians, I sorrowfully conclude that my limited studies of the ocular fundus have nothing to offer.

Some general comments

The subtitle of this paper may have aroused some curiosity on the part of the reader, and I feel at this point that an explanation is in order. I, like millions of other people, have never faced *Daubentonia* eyeball to eyeball, so to speak, and may never have the opportunity to do so. Information on this prosimian is scant. Nevertheless, I often get the feeling that these rare and exotic and little-known animals can provide a wealth of knowledge for us.

The survey of literature and the interpretations which have been made are not very positive. For each generalisation about possible relationships between anatomy and behaviour, there appears to be *at least* one significant contradiction. In attempting to find why this should be so, it occurred to me that I was dealing with categories, both anatomical and behavioural, which are not so absolute as they appear.

Reference has already been made to Walls' statement that rods and cones should perhaps be functional classifications, rather than purely structural ones.[11] The same approach is obviously applicable to behavioural categories. Animals may be nocturnal or diurnal or crepuscular as a preferred mode of life. Most of them would not be totally incapacitated if forced to operate under other than the preferred conditions. Some (e.g. *Lemur catta*) might be found, in fact, to be quite versatile. Colour vision is another area where greater flexibility of classification is in order. In the case of humans, we no longer think in terms of two or three types of colour blindness, but rather a very complex system of degrees and mechanisms of colour deficiency. I think it will be the exceptional species which shows no adaptability with respect to living habits, just as it will be the rare species with reveals total achromatopsia. Field studies as well as laboratory studies of animal behaviour should be designed to provide us with greater understanding of both the flexibility and the limitations of the subjects under study.

I have indicated in various places throughout this paper kinds of information which were lacking. Anatomical data (except for electron microscopic studies) on some species is fairly good, but the eyes of many species have been studied not at all. Behavioural data on many species is sketchy and experimental studies of visual function in prosimians is the exception rather than the rule. Most lacking (perhaps for valid reasons) are physiological studies of vision of prosimians. Within the limitations imposed by the decreasing populations of many prosimian genera, I hope that some of the newer techniques in each of these areas of study may be employed. If this can be done, I think that some of the questions regarding the relationships between the structure of the eye and behaviour may be answered. With regard to the other question of what the study of the eye can tell us about evolution, I again quote Walls, who said: 'The structural variations of the eyes are largely expressions of habit differences rather than stages in a majestic phylogenetic progress of increasing complexity and perfection, such as one sees in the heart, the brain, the ear or indeed in almost any other organ.'[2 5]

NOTES

1 Prince, J.H. (1956), *Comparative Anatomy of the Eye*, Springfield, Illinois.

2 Buettner-Janusch, J. (1966), *Origins of Man*, New York.

3 Detwiler, S.R. (1939), 'Comparative studies upon the eyes of nocturnal lemuroids, monkeys and man', *Anat. Rec.* 74, 129-45.

4 Morris, D. (1965), *The Mammals*, New York.

5 Napier, J.R. and Napier, P.H. (1967), *A Handbook of Living Primates*, New York.

6 Polyak, S. (1957), *The Vertebrate Visual System*, Chicago.

7 See *Primate News* (1971), Lyon, C.E., Ito, J., and West, E.S. (eds.), 9, 3-9, Oregon Regional Primate Research Center, Beaverton.

8 Walker, E.P. (1964), *Mammals of the World*, 1. Baltimore.

9 Kolmer, W. (1930), 'Zur Kenntnis des Auges der Primaten', *Z. Anat. Entwicklungsgesch.* 93, 679-722.

10 Rohen, J.W. and Castenholtz, A. (1967). 'Über die Zentralisation der Retina bei Primaten', *Folia primat.* 5, 92-147.

11 Walls, G. (1934), 'The reptilian retina', *Am. J. Opthal.* 17, 892-915.

12 Samorajski, T., Ordy, J.M., and Keefe, J.R. (1966), 'Structural organization of the retina in the tree shrew (*Tupaia glis*)', *J. Cell Biol.* 28, 489-504.

13 Shriver, J.W., and Noback, C.R. (1967), 'Color vision in the tree shrew *(Tupaia glis)*', *Folia primat.* 6, 161-9.

14 Wolin, L.R. and Massopust, L.C. (1970), 'Morphology of the primate retina', in Noback, C.R., and Montagna, W., (eds.) *Advances in Primatology: The Primate Brain*, New York.

15 Rohen, J.W., (1962), 'Sehorgan', *Primatologia* 2 (6), 1-210.

16 Petter, J.J., Schilling, A. and Pariente, G. (1971), 'Observations éco-éthologiques sur deux lemuriens malgaches nocturnes: *Phaner furcifer* et *Microcebus coquereli*', *Terre et Vie* 3, 287-327.

17 Woollard, H.H. (1926), 'Notes on the retina and lateral geniculate body in *Tupaia, Tarsius, Nycticebus* and *Hapale*', *Brain* 49, 77-104.

18 Allen, R.A. (1969), 'The retinal bipolar cells and their synapses in the inner plexiform layer', in Straatsma, B.R., Hall, M.O., Allen, R.A. and Crescitelli, F. (eds.), *The Retina*, Los Angeles.

19 Dowling, J.E. (1970), 'Organization of vertebrate retinas', *Invest. Ophthal.*' 9, 655-80.

20 Dowling, J.E. and Boycott, B.B. (1969), 'Retinal ganglion cells: a correlation of anatomical and physiological approaches', in Straatsma, B.R., Hall, M.O., Allen, R.A. and Crescitelli, F. (eds.), *The Retina*, Los Angeles.

21 Sjöstrand, F.S. (1969), 'The outer plexiform layer and the neural organization of the retina', in Straatsma, B.R., Hall, M.O., Allen, R.A., and Crescitelli, F. (eds.), *The Retina*, Los Angeles.

22 Rohen, J.W. (1966), 'Zur Histologie des *Tarsiusauges*', *Graefe Arch. Klin.*

Exp. Ophthal. 169, 299-317.
23 Woollard, H.H. (1927), 'The differentiation of the retina in Primates, *Proc. Zool. Soc. Lond.* 1, 1-17.
24 Wolin, L.R. and Massopust, L.C. (1967), 'Characteristics of the ocular fundus in primates', *J. Anat.* 101, 693-9.
25 Walls, G. (1939), 'Origin of the vertebrate eye', *Arch. Ophthal.* 22, 452-84.

PART II SECTION B
Anatomy and Function of Skull and Teeth

R. F. KAY and K. M. HIIEMAE

Mastication in Galago crassicaudatus: *a cinefluorographic and occlusal study*

Introduction

The extant genus *Galago* is a lorisiform prosimian of somewhat uncertain lineage. It does, however, have molars with sufficient affinity to some Eocene prosimians for it to serve as a reasonable functional analogue for the masticatory apparatus of those forms.

The ancestry of *Galago* is by no means clear. The oldest unequivocally lorisiform fossils known are the two genera *Komba* and *Progalago* from the lower Miocene of East Africa. *Propotto*, described by Simpson,[1] is now considered to be a fruit bat.[2] A single isolated second lower molar, the type specimen of *Indraloris* found in Siwalik deposits of Nagri age,[3] is poorly documented and is possibly an adapid.[4] On present evidence, it appears that the lorises stem from a family which had become clearly differentiated by the beginning of the Miocene.[5]

There is a considerable literature on the behaviour of the modern Galaginae but, as far as the authors are aware, no detailed investigation of their occlusion or masticatory behaviour has been undertaken. Mills has briefly described the wear facets produced on *Galago* molars and attributed their distribution to the presence of a 'balanced occlusion'.[6] The normal diet of *G. crassicaudatus*, as for *G. senegalensis*, is 'especially fruit and insects, but (the animal) is more or less omnivorous'.[7] The use of the hands in feeding has been discussed by Bishop.[8]

Description and discussion of feeding behaviour and the movements of the skull, lower jaw, tongue and food involved, have been complicated by the multiplicity of terms used to describe various aspects of the process. To clarify the report of the findings obtained in this investigation, the terms most commonly used, and their functional and morphological implications, must be discussed.

There are three stages in feeding: *ingestion* in which food is transferred from an extra-oral position; *mastication* in which 'the bite' is triturated and *deglutition* in which the reduced food is transferred from the mouth across the pharynx and into the oesophagus. Movements of the lower jaw and tongue are always rhythmic and cyclical in mastication and may be so in certain types of ingestion such as licking and lapping.[9–14]

Each masticatory cycle consists of three types of movement or *strokes* which follow each other in a smooth sequence:[14]

(*a*) A *preparatory stroke* or 'closing stroke',[9] in which the mouth is closed, either by upwards movement of the lower jaw alone or in combination with cranial movement at the atlanto-occipital joint (usually depression, but extension can occur). The effect is to bring the lower molars of the *active side* below the uppers and into a vertical relationship such that their buccal surfaces (external or lateral) are aligned, or such that those of the lower molars are lateral to the uppers. (It follows that in animals such as *Didelphis* and *Tupaia*, where the post-canine tooth row is linear, the anterior teeth may be well lateral to the corresponding upper teeth). The preparatory stroke is completed when tooth-food-tooth contact is reached; this point is not always possible to determine accurately from cineradiographs.

(*b*) A *power stroke* in which the food is triturated. The methods by which the food is reduced can be divided into two broad categories: food may be *crushed* between the teeth by forces acting more or less at right angles to their occlusal surfaces. In this case, the stroke begins at the point of initial tooth-food-tooth contact and is completed when that contact ceases. Both points may be determinable cinefluorographically and can occur within a wide range of gapes. The net effect of this type of power stroke on the teeth is the blunting of the tips of cusps. Alternatively, food may be *sheared, cut* or *ground* by the movement of the lower tooth across the upper which is produced by forces acting more or less parallel with the direction of movement. This type of power stroke normally begins with food between the opposing tooth surfaces, but throughout the contact between them is sufficiently close for its direction to be governed by their morphology. Shearing mastication produces wear facets on the slopes of cusps or ridges and sometimes in the basins between them, which are striated in a direction parallel to that of the movement producing them. (These movements are the 'contact' or 'glide' movements forming the upper limit of the envelope of motion.)[15] The stroke commences at the point of initial occlusal or tooth-food-tooth contact and is completed when the teeth come out of occlusion. The former point may be determinable cinefluorographically, the latter is not normally visible. However, dorso-ventral or antero-posterior projection cineradiographs may allow horizontal or transverse movements in this stroke to be analysed. The power stroke may or may not involve occlusal contact on both active and balancing sides of the jaw.

(*c*) A *recovery stroke* ('opening stroke') in which the mouth is opened to the maximum gape for the cycle. As in the preparatory stroke, this may result from predominantly vertical movement of the lower jaw only, or from such movement combined with cranial extension (usually) or depression.[12] The downward movement of the mandible may involve some lateral, protrusive or retrusive excursion. The stroke begins when occlusal or tooth-food-tooth contact is observed and completed when the upwards movement of the next preparatory stroke begins. The latter point is readily determinable from cineradiographs, the former not necessarily so.

A *stroke* may have one or more *phases* distinguished by at least one feature such as a significant change of direction of jaw movement. This situation normally occurs only in the power stroke, and is most clearly seen in a chewing stroke where the first upward and medial movement of the lower teeth on the active side brings them into centric occlusion (protocone in talonid basin). This is followed by a second downward and medial movement taking the lower molars out of centric occlusion and further towards the midline. The first of these movements may be regarded as 'phase I' and the second as 'phase II'. It does not follow that such phases always occur, since one or other may be suppressed depending on the architecture of the teeth and jaws.

At the completion of any masticatory activity, the lower jaw normally returns to the *rest position.*[6,12] This involves the curtailment of the recovery stroke of the terminal cycle. Similarly, when the animal begins to feed, the jaw is depressed from the rest position to maximum gape for the ensuing ingestive cycle.

The terminology used to describe the power stroke and derived from wear facet analysis has introduced two terms, the 'buccal phase' and the 'lingual phase', into the literature.[6,16,17] The former describes the movement of the jaw from the buccal position to centric occlusion on the active side and from the lingual position to centric occlusion on the balancing side. Mills then describes a 'lingual phase' in which the lower molars on the active side move from centric occlusion towards the midline whilst those on the balancing side are moving buccally. In other words a 'buccal phase' on one side of the jaw may occur simultaneously with a 'lingual phase' on the other.[16] This is a 'balanced occlusion'. Since the conclusions on which the terminology is based have been disputed,[13,14] and given that the terms may be justifiable only if applied to the active side, with no implications for balancing side occlusion, they are not used in this study. Instead two phases in the power stroke on the active side are recognised.

It follows from the above that any investigation into mastication must utilise two techniques; the movements of the cranium, lower jaw and to some extent the tongue and food can be recorded cinefluorographically; the occlusal relationships of the teeth during a chewing stroke have to be determined from the morphology of the molar crowns and their pattern of wear. This study has used both these methods to establish the pattern of feeding and mastication in *Galago*.

Materials and methods

1. *Cinefluorography.* The specimen of *Galago crassicaudatus* (a male) had radio-opaque markers inserted in upper and lower jaws (see Fig. 1) under general anaesthesia. Cinefluorographic recordings were made in the lateral, dorso-ventral and antero-posterior projections at 60 frames per sec. on 16 mm. Eastman Kodak Plus X negative film no. 7231, at 55-70 kilovolts and 14-18 milli-amperes, depending on the projection. The animal was fed on apple, 'Purina' monkey biscuits,[18] either plain or previously soaked in water and dipped in honey to give a softened but still crunchy consistency. The recordings were analysed by the methods described by Crompton and Hiiemae,[14] and Hiiemae and Crompton,[12] and as indicated in the text.

2. *Tooth morphology and occlusion.* Specimens of *Galago crassicaudatus* from the osteological collection at the Museum of Comparative Zoology, Harvard University, have been examined to determine the morphology of the teeth, the extent and definition of the wear facets and the geometrics of occlusal relations in the power stroke. This last method is based on a geological technique described by Lahee.[19] The terms used are defined in the following text. The method has been adapted for the determination of the direction of movement between occluding upper and lower teeth during tooth-tooth contact. Three points were chosen to define the occlusal plane, selected from the tips of the hypoconids of the left and right M_1, and of the hypoconids of the left and right M_3. (In the text, the letters I, C, PM or M refer to Incisors, Canines, Premolars or Molars, respectively. The superscript numbers refer to upper teeth, the subscript numbers to lower teeth.) A separate mesiodistal axis was chosen for the body of the dentary: a line drawn between the hypoconids of M_1 and M_3. The dentary was oriented normal to the occlusal plane and examined under a stereomicroscope with a camera lucida. The strike of each wear plane examined was estimated as a line drawn between two points on each wear plane parallel to the occlusal plane. The dentary was then oriented laterally so that the wear plane was parallel to the line of vision and the angle of dip could be measured. These procedures were repeated ten times for each wear plane studied on a sample of four specimens of *Galago*

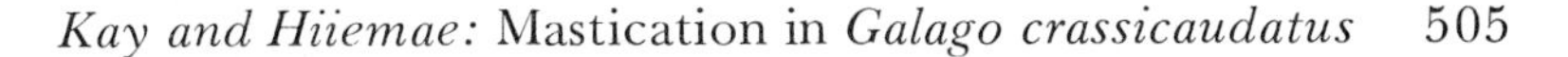

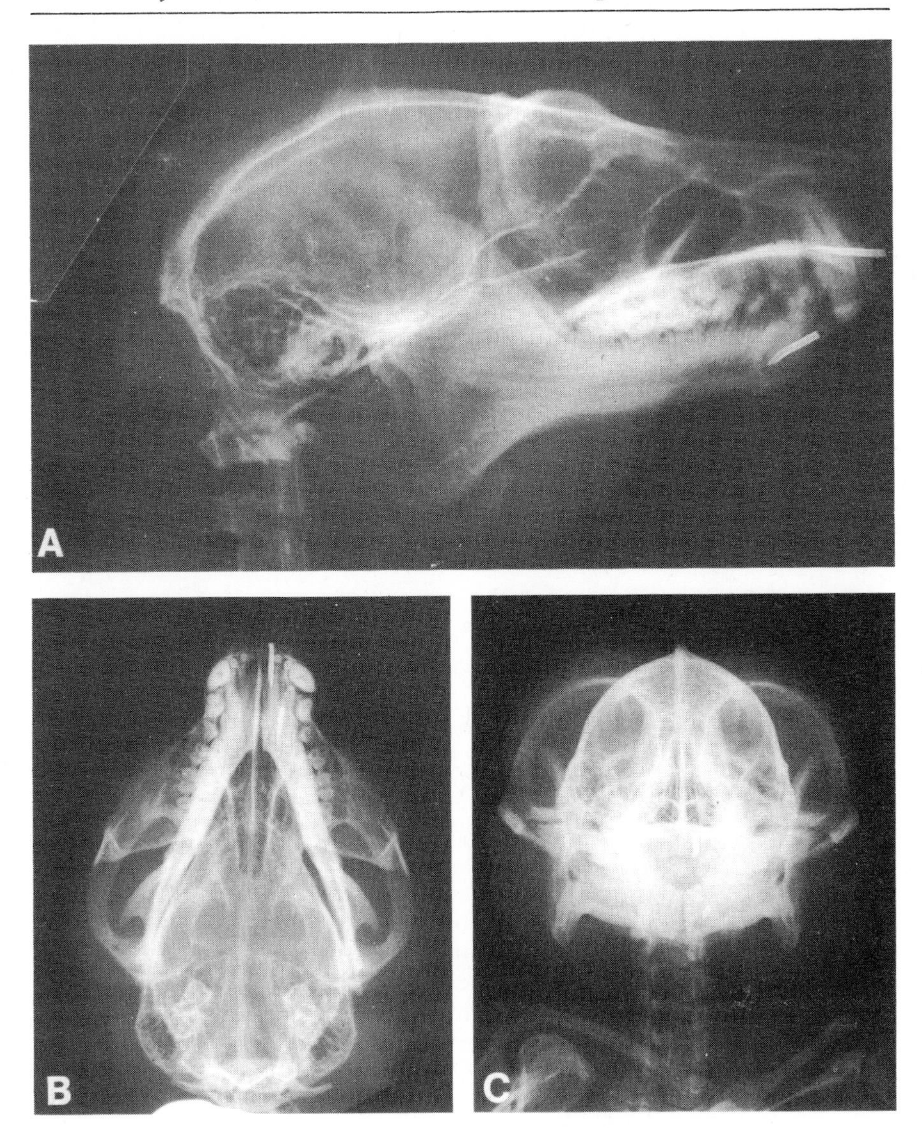

Figure 1. Laterial (*A*), dorso-ventral (*B*), and antero-posterior (frontal) (*C*) radiographs showing the appearance of the head of *Galago* in these projections and the positions of the markers. The radiographs are at the following approximate magnifications: *A* x 2.5, *B* and *C* x 2.

crassicaudatus. Given this raw data, the orientation and plunge for each facet were estimated by the geometrical method described by Lahee (Tables 1 and 2: see fig. 8).

Upper and lower molar dentitions of *Palenochtha* from the Middle Palaeocene, Gidley Quarry, Montana were used for comparative purposes.[20]

Table 1 **The dip and strike of facets 1 and 3. The figures represent the average of the mean of ten readings for each facet on each of four specimens**

	Wear Facet 1		*Wear Facet 3*	
	Dip°	*Strike*°	*Dip*°	*Strike*°
M_1	60	61	72	43
M_2	59	65	76	32
M_3	57	67	73	25

Table 2 **Movement during phase I of the power stroke in *Galago*, based on averaged data from four specimens**

	Tooth Position		
	M_1	M_2	M_3
Orientation°	66	59	50
Plunge°	54	52	55

Results

The experimental and analytical results obtained are described, using the terminology defined in the introduction, under three main headings: *Feeding Behaviour; The Masticatory Cycle*; and *Occlusion.*

The jaw apparatus in *Galago* is distinguished by the presence of a dental comb with a concomitant caninisation of the P_2 (P_1 is not present). The upper incisors are small and separated by a median diastema. The upper canine and the P_2 are large and protrude beyond the occlusal plane. P_2 and P^3_3 are somewhat sectorial, P^4_4 are semi-molariform. The post-canine teeth from $P^4 - M^2$ increase in their transverse width so that the outer profile of the tooth row bulges laterally (Fig. 1*B*). There is a less marked lateral bulge in the lower jaw. The mandible in *Galago* consists of two parts (dentaries). The mandibular symphysis is not ossified and fibrous material can be seen between the two dentaries in osteological material, suggesting that intra-mandibular movement is possible. The body of each dentary is stout, although shallow, whilst the ramus is much more slender with well developed and somewhat hooked coronoid and angular processes. The mandibular condyles have their longest axis at right angles to the midline of the skull and are sharply convex antero-posteriorly. There is a well-developed post-glenoid flange on the articular fossa. The radiographic appearance of the hard tissues is shown in each projection in Fig. 1.

Feeding behaviour

Relatively few animals have been studied using cinefluorography to record mastication: *Tupaia*,[21] *Didelphis*,[12] *Saimiri*,[13] *Ateles*,[22] and *Rattus*.[11] By comparison with the first two, *Galago* is a slow feeder, since a single chewing cycle takes longer, but it is more 'efficient' in that fewer cycles are needed to reduce the food sufficiently for deglutition.

1. *Ingestion*

Ingestion is normally preceded by inspection: the food is sniffed and/or licked. The actual method by which the bite is transferred to the mouth depends on the consistency of the food: liquids or semi-liquid 'mush' are ingested by licking or lapping; hard foods are ingested through the immediate post-canine region rather than through the front of the mouth. The incisors have never been seen to function as a biting organ during ingestion. Small bite-sized pieces of hard food are often directly picked up from the ground by rotation and tilting of the head to bring the canine and premolars of one side into position to pick up the bite; the latter is then 'tossed' distally onto the molars by a rapid movement of the head. Alternatively, one or other hand is used to manipulate the food into position for ingestion. A bite of hard food is separated from the remainder by a series of 'puncture-crushing' chewing cycles: this is 'ingestion by mastication'.[12] Throughout this procedure the bulk of the food is held in the ipsilateral hand. Tentative 'efficiency counts' indicate that three puncture-crushing cycles were required to separate a bite from the biscuit supplied in this study. At no time during these cycles do the teeth come into contact. Once the bite has been separated, the hand is dropped and mastication follows.

2. *Mastication*

As in *Didelphis* and *Tupaia*, mastication takes place in two stages for hard food – an initial group of puncture-crush cycles followed by normal chewing. Soft mush is only chewed. Frequently, towards the end of any chewing sequence, the tongue is protruded and the tip flipped up over the snout. At the completion of each cycle sequence, a bolus is passed. For hard food there appears to be a 1:1 relationship between bite and deglutition; i.e. one bite is ingested, completely reduced and the bolus passed before ingestion recommences. This is not necessarily true for soft food.

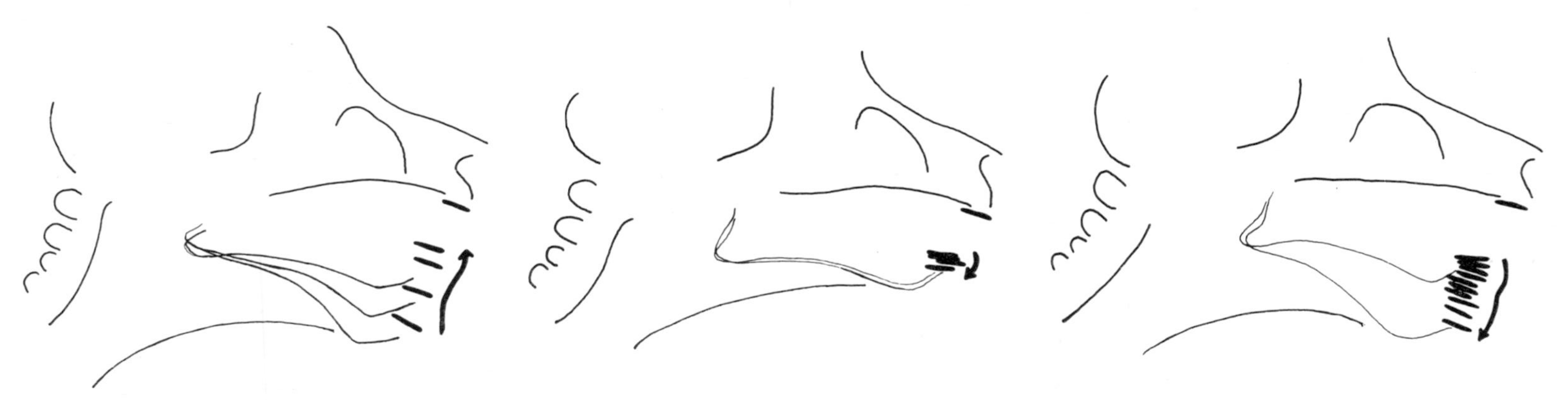

Preparatory Stroke

Power Stroke

Recovery Stroke

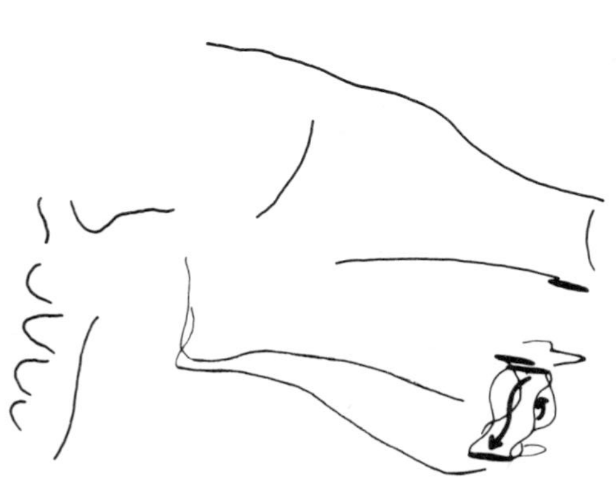

The masticatory cycle

The 'efficiency' with which *Galago* feeds means that the number of consecutive chewing cycles required to reduce a bite of soft food is small, and even for the hard biscuit used in this study rarely exceeds ten. As might be expected, fresh apple constitutes a greater problem, the number of cycles required rising towards twenty (see Figs. 4 and 5). Chewing proceeds rhythmically until mastication is completed and is followed by deglutition.

The amplitude of movement of the lower jaw and the degree to which the mouth is opened at the completion of the recovery stroke depends on the type of mastication in progress (see Figs. 4 and 5). The widest gapes are recorded during the ingestion of hard food, followed by those for puncture-crushing. Small gapes are associated with chewing and lapping.

The mandibular symphysis of *Galago* is very clearly mobile. It mediates rotation of either or both dentaries along their long axes so that they can evert or invert relative to each other. In addition, the symphysis may allow either dentary to slip further forward than the other, so that although the paths of movement of the dentaries on the active and balancing sides tend to parallel each other, this is not always the case. In fact there is some cinefluorographic evidence that independent movement on active and balancing sides does occur during the power stroke.

1. The preparatory stroke

During this stroke, the lower jaw moves upwards until tooth-food-tooth contact is reached. This movement, particularly from large gapes, normally involves a combination of cranial depression and mandibular elevation. As can be seen from Figs. 4 and 5, the amplitude of the vertical movement involved depends on the type of masticatory activity in progress. This movement is predominantly vertical, with at the most a very small lateral component (Fig. 3) and a slight retraction occurring while the mouth is still wide open (Fig. 2). At the completion of the preparatory stroke, the dentary and its post-canine teeth on the active side are below the corresponding upper teeth, with their lateral surfaces vertically aligned.

Figure 2. Compound profiles from cinefluorographic recordings of mastication taken in the lateral projection. Top row: superimposed tracings showing the position of the marker in sequential frames of a single chewing cycle; bottom left, to show the path of movement of the marker during a puncture-crushing cycle; and bottom right, in a chewing cycle. (The position of the marker at the beginning of each stroke is shown.)

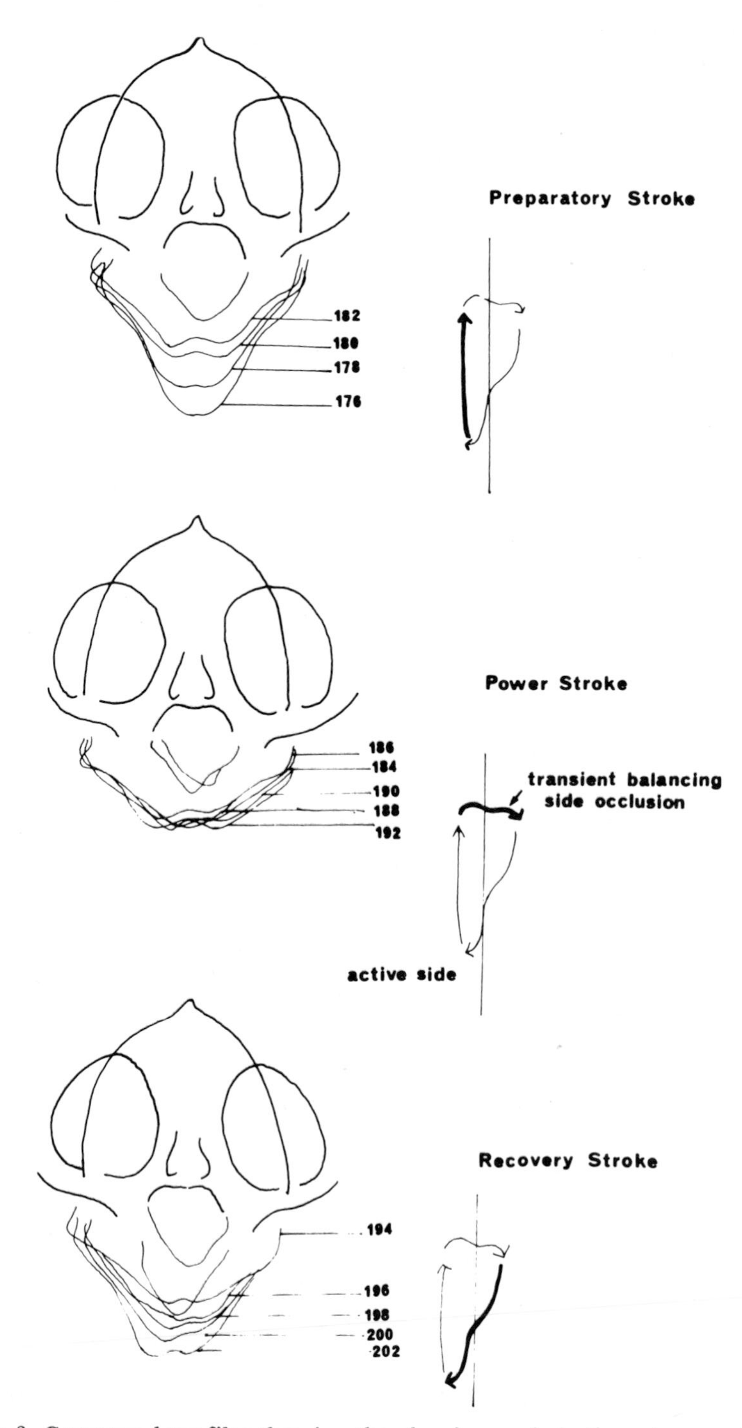

Figure 3. Compound profiles showing the chewing cycle in the antero-posterior (frontal) projection. Alternate frames have been used (as indicated by the numbering). The 'profile' of symphyseal and marker movement is shown on the right, somewhat enlarged.

2. *The power stroke*

The exact path of the movement of the dentary on the active depends on the type of mastication occurring.

In *ingestion and puncture-crushing* cycles, following separation of the bite, the power stroke continues as an extension of the preparatory stroke, but at a much reduced rate: there is a very slow upwards movement (Fig. 4) in which the food is compressed between the teeth. This is followed by a short period in which the jaw is 'held' at the minimum gape for that cycle. It can be argued that the food is being compressed during this time by largely isometric contraction of the jaw muscles. The dorso-ventral and antero-posterior recordings show that there is some medial movement of the dentary on the active side to bring it into, or just beyond, centric relation before the recovery stroke commences.

The movements of the dentary on the active side in the power strokes of *chewing* cycles are more complex. Viewed in lateral projection, the dentary is seen to come into occlusion, to move upwards and occasionally very slightly forwards and then to move slowly downwards until a clear freeway space is visible. In dorso-ventral and antero-posterior cinefluorographs, the movement of both dentaries in the horizontal and transverse planes can be seen. Initially the dentary on the active side moves slowly upwards, medially and slightly forwards until centric occlusion is reached. This is phase I. The dentary on the active side then continues to move medially through phase II, the symphysis crossing the midline (Fig. 3) to bring the balancing side into what appears from the cineradiographs to be a transistory occlusion. Both dentaries then drop and move into the recovery stroke. This long medial excursion of the active side dentary towards the balancing side lengthens the apparent duration of the power stroke.

3. *The recovery stroke*

The recovery stroke begins very slowly with the mandibular symphysis to the balancing side of the midline. When mandibular depression has reached a gape of about 8-10° (a large freeway space) the downward movement stops for a short time and the lower jaw is protruded (see Figs. 2 and 3). From the protruded position the lower jaw continues its downward and lateral swing towards the active side. The symphysis appears to cross the midline from active to balancing sides about halfway through the stroke (Fig. 3). The stroke is completed when the jaw reaches the maximum gape. At this point the dentary on the active side is normally well lateral to the midline.

The movements of the mandibular condyles on the glenoid fossae and of the symphysis cannot be clearly seen in all the radiographs.

However, the following movements appear to be occurring:

(*a*) *Preparatory stroke.* The jaw is rotating smoothly upwards with a little retraction (posterior translation) in the first stage. If the condyle on the active side has not reached a lateral position during the last phases of the recovery stroke, it moves laterally on the glenoid fossa during the early part of the preparatory stroke.

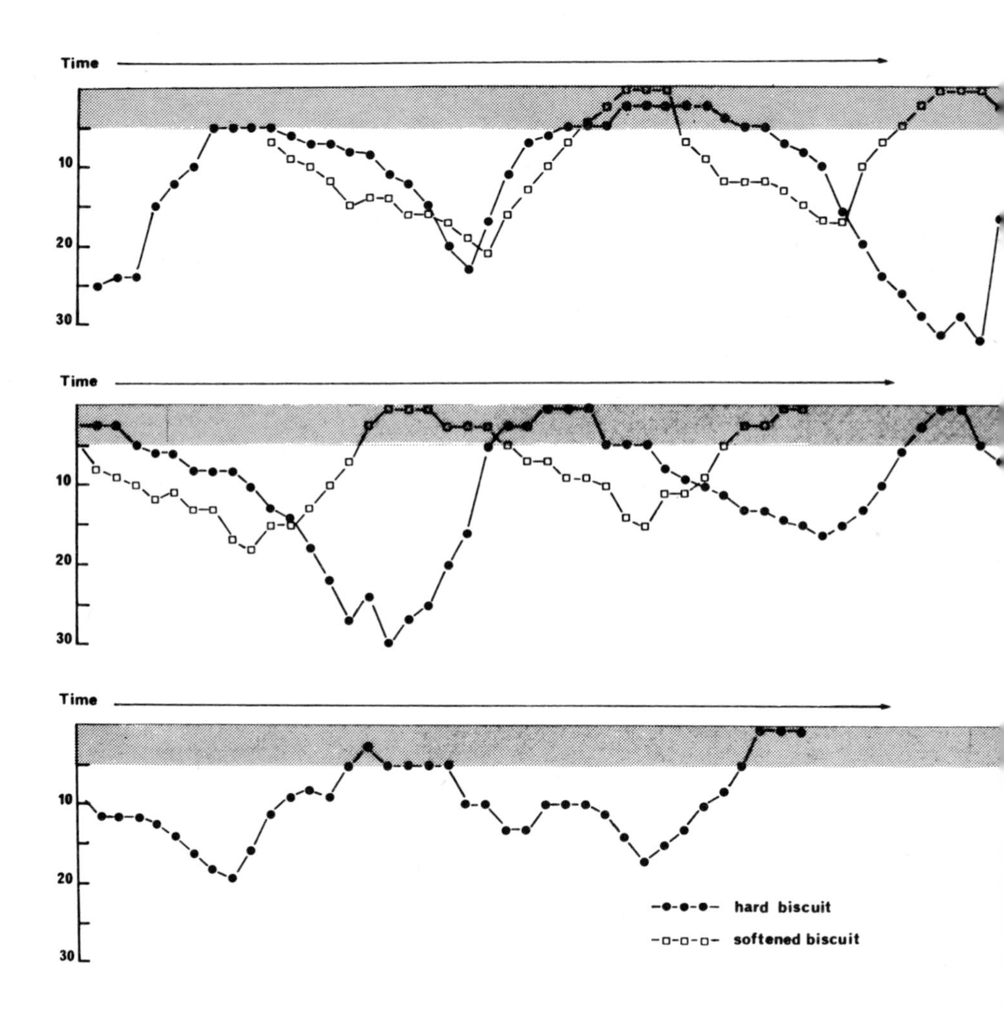

Figure 4. Plot showing the *gape* (in degrees) in lateral projection plotted against time for successive single frames during mastication on hard and very soft biscuit. Each horizontal line represents a total elapsed time of 0.83 secs. The shaded area represents the range of gapes, defined as 'wide freeway space', 'narrow freeway space' and into occlusion (i.e. equivalent to 5, 2.5, and 0 degrees of gape). The first three cycles for hard biscuit show the amplitude of movement associated with puncture-crushing mastication, and the second three that for chewing. Softened biscuit is not puncture-crushed and the maximum gape for the cycles shown is similar to that for the chewing cycles for the hard biscuit.

(*b*) *Power stroke.* During the last stages of the power stroke the condyle slips medially on the glenoid fossa, a Bennett movement. At the same time some slip occurs at the symphysis, the active dentary apparently moving slightly further forwards than the one on the

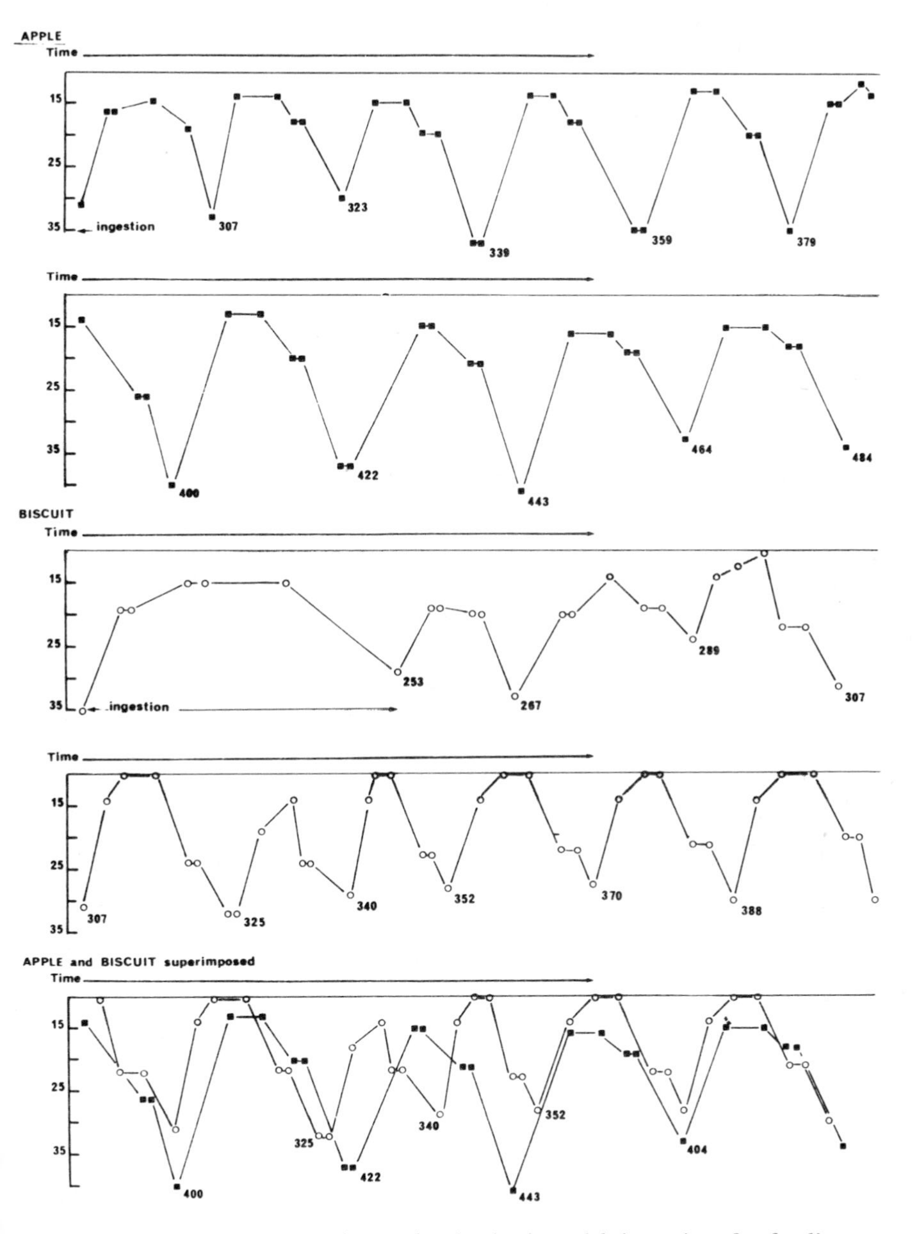

Figure 5. Plot showing successive cycles, beginning with ingestion, for feeding on apple and hard biscuit. Frames at the beginning of each stroke only have been traced and measured. The time elapsed for each line is 1.66 secs. Frame numbers are included for comparison of durations and to demonstrate rhythmicity.

balancing side. The behaviour of the mandibular condyles during the two phases of the power stroke is discussed below.

(*c*) *Recovery stroke.* After the initial slow opening phase, the jaw is then protruded (a forward translation of the condyle) and then depressed by condylar rotation on the glenoid fossa. The lateral movements of the jaw from the balancing to the active side will involve some condylar rotation about a vertical axis. The relative duration and the great regularity of the various types of masticatory cycle recorded is indicated in Figs. 4 and 5. The details of the time course for mastication in *Galago* will be discussed elsewhere,[2 3] but it is clear from the figures that a puncture-crushing cycle takes about 10% longer than a cycle involving chewing or lapping.

Occlusion

The cinefluorographic study confirms that *Galago* uses two different actions to triturate its food: puncture-crushing and chewing. Each produces a different type of wear on the teeth. In puncture-crushing the teeth do not come into contact and are used in a fashion analogous to a meat tenderiser; this blunts or flattens the tips of the cusps. In chewing the teeth contact in a regular fashion, their relative movement is controlled by the configuration of the cusps and ridges on their crowns. These movements produce clearly delineated matching wear facets on both upper and lower teeth. Examination of such wear facets allows the occlusal relationships between contacting teeth to be determined.

1. Molar occlusion in Galago

The upper and lower molars of *Galago* and *Palenochtha* (Fig. 6) are compared in Fig. 7. Fig. 6 illustrates the terminology used.[2] The molar morphology of *Palenochtha* from the middle Palaeocene is only slightly modified from the tribosphenic pattern which is likely to have characterised the Cretaceous ancestors of primates. A comparison between the molars of the two genera places the functional interpretation of *Galago* mastication in an evolutionary perspective. In both *Galago* and *Palenochtha* the protocone, which is displaced posteriorly, the paracone and metacone are all well developed on the upper molars. The stylar area, with para- and metastyles present in *Palenochtha*, is absent in *Galago*. The crests which primitively run antero- and postero-externally from the paracone and metacone respectively (paracrista and metacrista) are much abbreviated and unsupported by stylar cusps in the latter. The paraconule, metaconule and the crests (pre- and post-paraconule crista, premetaconule and post-metaconule crista) are also very much reduced or absent. The large hypocone on the postero-internal

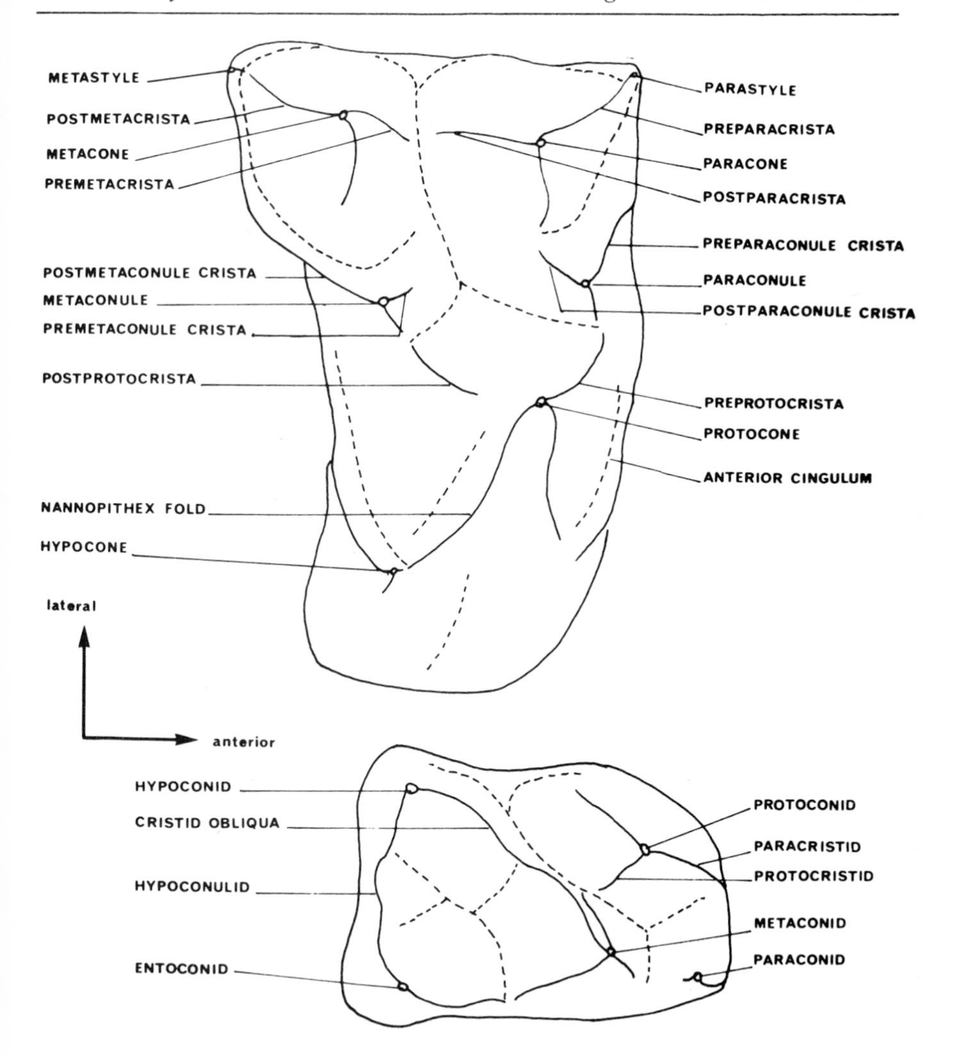

Figure 6. Outline drawings of upper and lower first molars of *Palenochtha minor* (AMNH 35443 and AMNH 35450 respectively) to show the nomenclature of the major features on their crowns.

cingulum is separated from the metacone by a distinctive notch in *Galago*. A cuspule is present in the same position on the first and second molars of *Palenochtha*.

The lower molars of *Galago* are modified by comparison with those of *Palenochtha*: the trigonid is reduced with the loss of the paraconid and is only slightly raised above the talonid. The metaconid has moved slightly posterior to the protoconid. The expanded and fully enclosed talonid basin is bordered by an enlarged hypoconid and a slightly smaller entoconid, whilst that of *Palen-*

ochtha is slightly open internally. The hypoconulid is reduced and virtually absent.

Consequent on these changes in gross morphology, the wear facets seen in *Galago* (although they are found, with two exceptions, in *Palenochtha*) have altered in size and relative importance. Those associated with transversely orientated embrasure shear in the primitive tooth (Facets 1*a*, 1*b*, 2*a*, 2*b* – see Crompton,[24] and also for the numerology and terminology used) are reduced from the sharply curvilinear and duplicated facets seen in the upper molars of

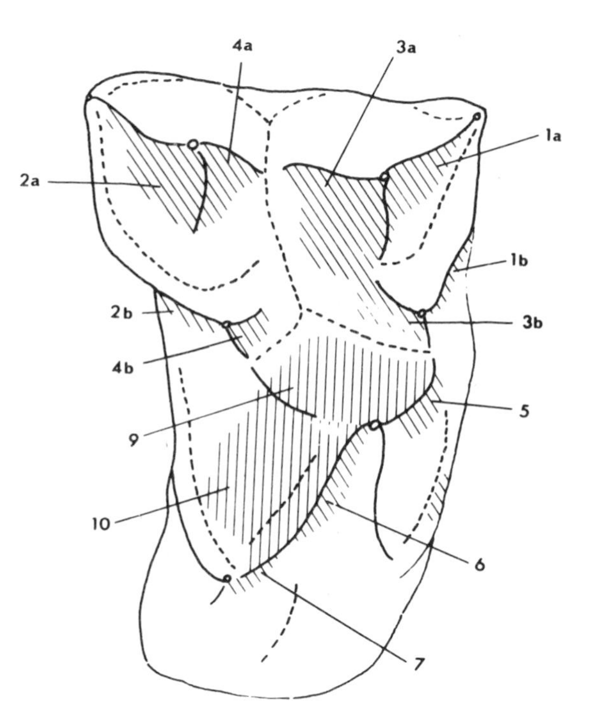

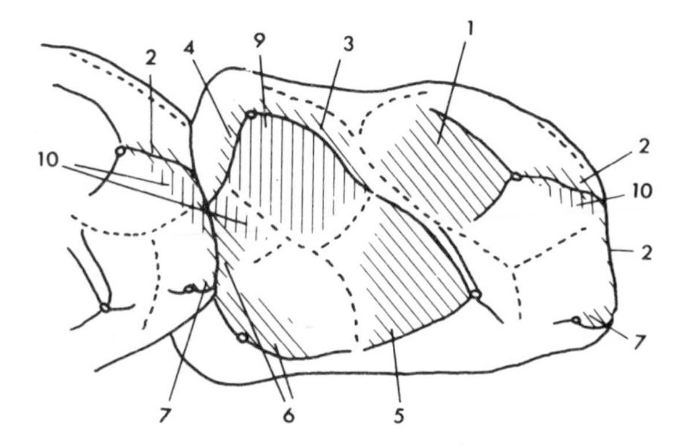

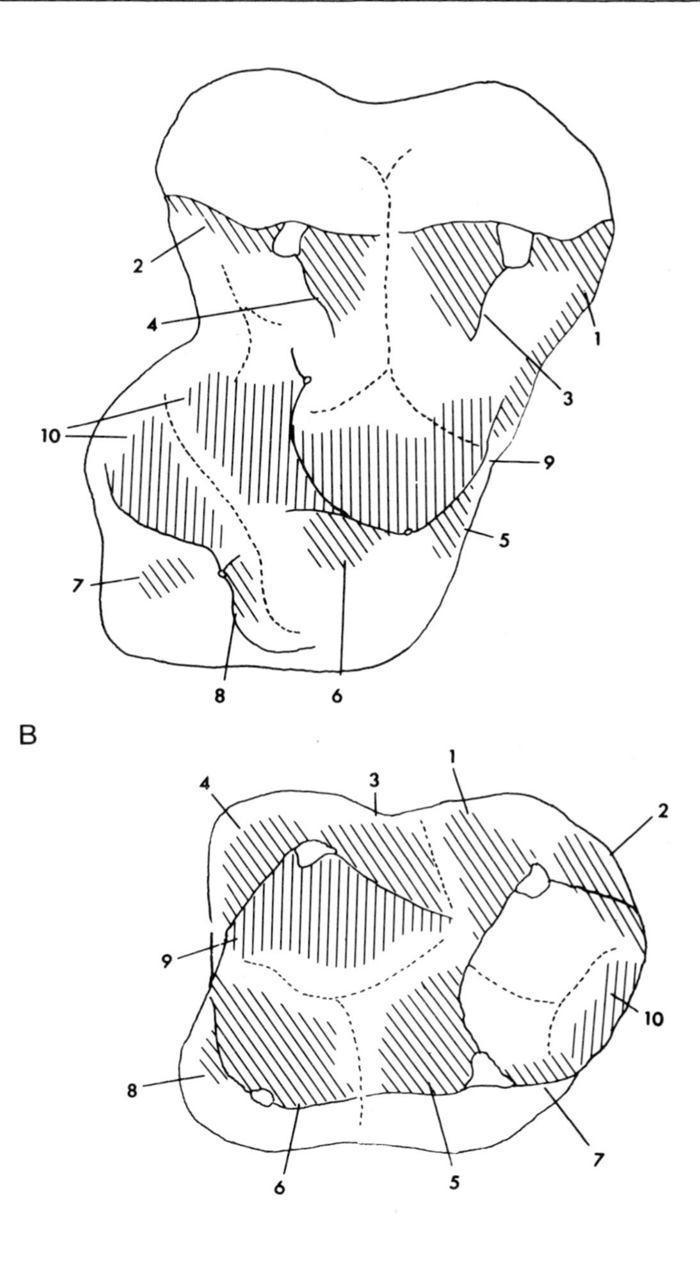

Figure 7. Upper and lower second molars of (*A*) *Palenochtha* and (*B*) *Galago* orientated as in Fig. 6 and shaded to show the positions and areas of the wear facets. The numbers show corresponding facets, but the cross-hatching does not indicate the direction of the striations. The molars of *Palenochtha* are from the Gidley Quarry, Montana; the specimen of *Galago crassicaudatus lasiotis* (MCZ 39414) was collected on the 'Tanganyika' Amboni Estate near Tanga. The teeth are shown at magnifications of approximately x 25 and x 15, respectively.

Palenochtha to comparatively small and much less transverse areas in *Galago*. The corresponding facets on the lower molars of *Galago* are greatly abbreviated: facet 1 no longer extends medially to the tip of the metaconid; facet 2 is rotated antero-externally, reducing its transverse orientation, and no longer extends far towards the metaconid; facets 3 and 4 are present in *Galago* on the postparacrista and the premetacrista as in *Palenochtha*, but facets 3*b* and 4*b* on the internal slopes of the postparaconulecrista and premetaconulecrista are absent in *Galago*. The wear facets found on the internal slopes of the talonid basin (facets 5, 6 and 9), an area associated with grinding, are much more extensive in *Galago*. Facets 5 and 6 are produced by contact between the preprotocrista and postprotocrista with the internal surfaces of the metaconid and entoconid. Facet 5 has so expanded in *Galago* that it reaches laterally almost to the protoconid. In *Palenochtha* facet 6 extends upwards and posteriorly along the nannopithex fold, supported at one end by a hypocone-like cingulum cusp. Wear facet 6 on the lower molars extends along the crest running anteriorly from the entoconid and posteriorly from that cusp onto the paraconid of the tooth behind it. The posterior cingulum cuspule occludes onto the paraconid. The corresponding facet in *Galago* does not reach the hypocone. The enlarged facet 9 seen in *Galago* results from contact between the outer surface of the protocone and the internal slope of the hypoconid. A partially enclosed basin has developed between adjacent teeth in *Galago* and is bordered by the entoconid of the anterior molar with the protoconid and metaconid of the posterior teeth. Wear facets 7, 8 and 10 are formed on the slopes of this basin by the hypocone. This arrangement, absent in *Palenochtha*, reflects an increased emphasis on crushing and grinding during the power stroke.

The major differences between the wear facets on the molars of *Palenochtha* and *Galago* demonstrate a shift in the emphasis placed on the leading edges of the shearing blades, as opposed to the wear surfaces supported by the cusps used for grinding the food. *Palenochtha* molars are designed to cut the food between a succession of shearing blades involving accentuated vertical and transverse movements in the first stage of the power stroke, whereas *Galago* molars are designed initially to cut and then to grind the food between a cusp and its related basin by a combination of shearing and compression. That this is the case in *Galago* can be readily demonstrated by considering the order in which the facets come into contact.

As the lower molar of *Galago* moves upwards and medially in phase I of the power stroke, facets 1, 2, 3 and 4 all contact simultaneously, followed successively by facets 5 and then 6. In *Palenochtha* the sequence is: facets 1*a*, 2*a*; 3*a* and 4*a*; 1*b*, 2*b*, 3*b*, and 4*b*, 5 and 6 to bring the teeth into centric occlusion (*sensu*

Crompton and Hiiemae).[14] As centric occlusion is reached in *Galago*, facets 7 and 8 (absent in *Palenochtha*) come into contact on the postero-medial slope of the hypocone and the posterior surface of the entoconid. The movement from the lateral position through a series of shearing contacts and into centric occlusion (phase 1) effectively completes the power stroke in a primitive mammal such as *Didelphis* or the Late Cretaceous *Didelphodus.*[24] In contrast, phase 1 in both *Galago* and *Palenochtha* is followed by a second movement in which the lower molars continue to move anteriorly and medially towards the midline and in doing so facets 9 and 10 move past each other. This is phase II of the power stroke. The relative degree of development of these facets in *Palenochtha* and *Galago* illustrates the greater functional significance of phase II in the latter. (See also Figs. 9 & 10.)

2. *Geometrical analysis of occlusion*

A geometrical construction has been developed to allow the direction of movement between any contacting pair of teeth to be estimated. This construction is based on the following: the wear facets on the molars of *Galago* are, to a first approximation, planes; more than one pair of non-parallel wear planes are in simultaneous contact in any pair of actively occluding teeth during tooth-to-tooth contact; for simultaneous contact to be maintained between these non-parallel matching wear planes, movement between the teeth must be confined to one, and only one, axis. Two angles, commonly used in structural geology, 'dip' and 'strike', completely characterise the position of any plane in space relative to a reference plane and an arbitrary axis in the reference plane (see Fig. 8). *Dip* is defined here as the angle of intersection between a wear plane and the occlusal plane. *Strike* is the angle between an approximate antero-posterior axis in the occlusal plane of the tooth and a line formed by the intersection of a wear plane with the occlusal plane. A line in space may be completely characterised by two angles which describe its relation to the occlusal plane and to the antero-posterior axis in that plane: its 'orientation' is the angle between the antero-posterior axis and the projection of the line on the occlusal plane; its 'plunge' is the angle between the projection of the line onto the occlusal plane and the line itself. The orientation and plunge of the line of intersection of two non-parallel wear facets are equal to the orientation and plunge of movement between the occluding teeth at that stage of mastication. Given that the dip and strike for each of the two wear facets is known, then the plunge and orientation of their line of intersection can be calculated.[19] From these figures the orientation and plunge of movement during phase I of the power stroke in *Galago* have been calculated.

As shown in Fig. 8, the upward movement of the lower jaw into

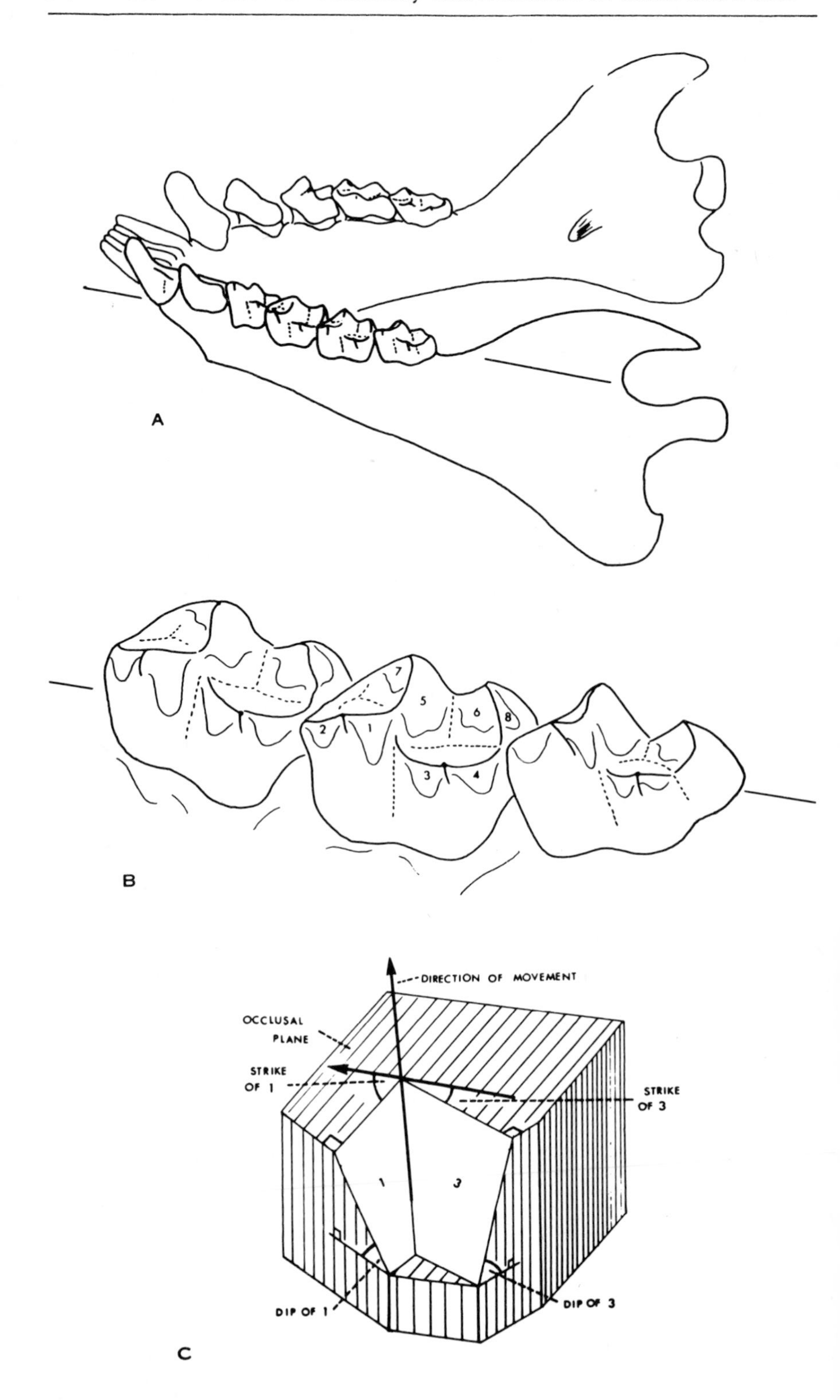
A
2
1
7
5
6
8
3
4
B
DIRECTION OF MOVEMENT
OCCLUSAL PLANE
STRIKE OF 1
STRIKE OF 3
1
3
DIP OF 1
DIP OF 3
C

centric occlusion during tooth-tooth contact may be described as a rotation of the dentary on the active side about a point located antero-medial to the active condyle, posteromedial to the third molar on that side and above the level of the occlusal plane. This results in an increase in the antero-posterior (mesiodistal) component of movement backwards along the post-canine tooth row and an increase in the transverse component of movement anteriorly along the tooth row. As the centre of rotation is above the tooth row, this will also augment the anterior component of movement posteriorly along the tooth row. (These observations and their relation to the concept of 'condylar rotation' will be discussed below).

Phase II of the power stroke may be described as an anteromedial translation of the active side of the lower jaw downwards and anteromedially out of centric occlusion during which contact is maintained between the protocone and talonid basin and the hypocone and trigonid basin. Since wear facets 9 and 10 on each lower molar are sub-parallel, an accurate estimate of orientation and plunge by the method using the intersection of planes is extremely difficult. However, visual observation suggests that the plunge of this movement is of the order of 50-55 degrees and is probably less steep than in *Palenochtha.* Kay shows that the movement of phase II in *Saimiri* is not a rotation about the condyle.[13] This appears also to be the case in *Galago*, a finding substantiated by the cinefluorographic observation that anteromedial condylar translation occurs in the later stages of the power stroke. The angle between the plunge of phase I and that of phase II is probably more oblique in *Galago* than in primitive primates.[23]

Fig. 10 is a scanning electron-micrograph showing the slope of wear facet 5 and the striations on its surface. At the bottom of this facet the striations continue to move without interruption into facet 9. This gradual change of direction supports the conclusion that phase I and phase II are parts of a single continuous power stroke on the active side and not the result of movement on the balancing side as suggested by Mills.[6] This conclusion is further substantiated by the observation that in *Tupaia* where phase II striations are seen, there is no medial excursion of the mandible such as to bring the balancing side into functional occlusion during the power stroke.[21] Moreover it has been shown that the orientations of phase II movements are not consistent with bilateral simultaneous contact in *Saimiri*, which has a fused symphysis.

A detailed view of facet 7 is shown in Fig. 9, demonstrating the

Figure 8. (*A*) Paraocclusal view of the lower jaws of *Galago* showing the arbitrary antero-posterior axis used. (*B*) The wear facets outlined in *A* enlarged. (*C*) Diagram showing the relationships of dip, strike and direction of movement to the occlusal plane, and the antero-posterior axis for wear planes 1 and 3.

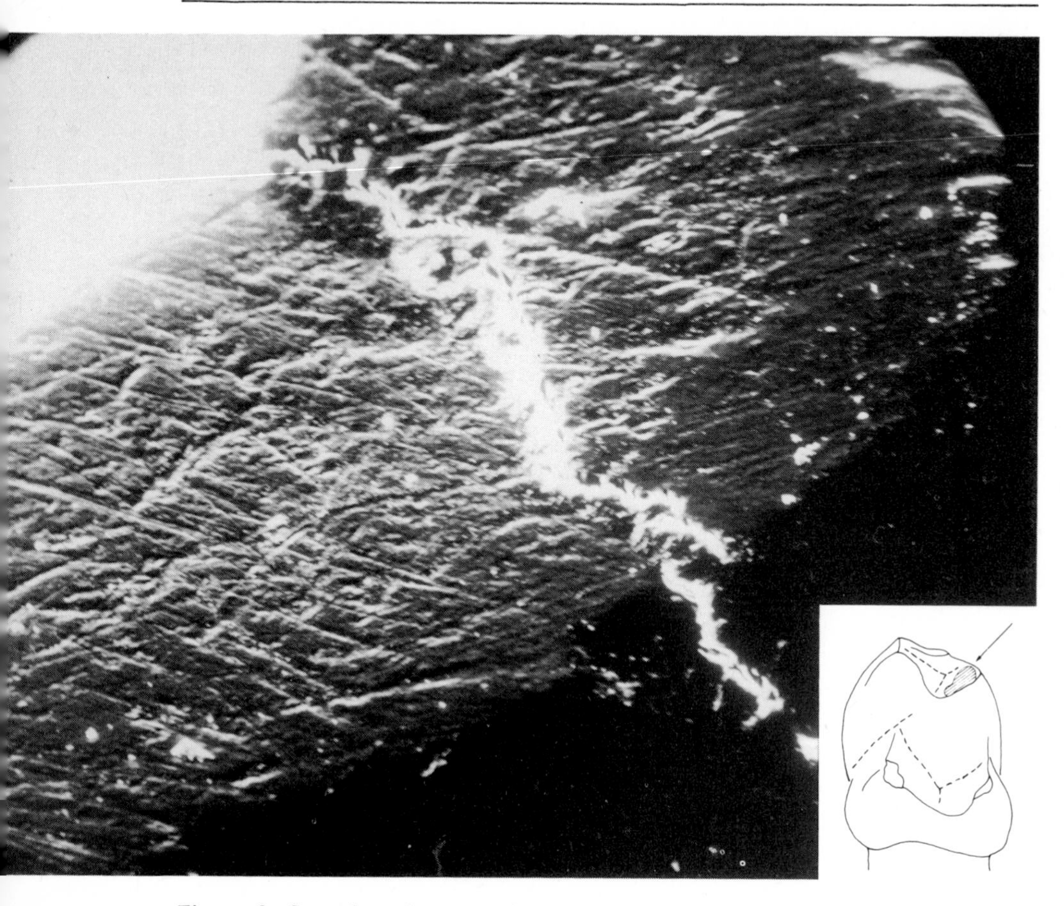

Figure 9. Scanning electron-micrograph of the lower second molar of *Galago* showing two sets of striations on the wear facet on the antero-internal surface of the metaconid (facet 7). Those running from upper right to lower left are produced during phase I of the power stroke, those from upper left to lower right by *casual* balancing side occlusion. The inset shows the position of the facet and the direction of phase I movement with the tooth viewed from above and behind. Magnification about x 800.

presence of two groups of striations running virtually at right angles to each other. The group crossing the field from upper right to lower left are parallel to the striations formed during phase I of the power stroke. The second group, passing from upper left to lower right, are only found on facet 7. Their orientation appears consistent with the effect of movement downwards and laterally by the molars on the balancing side following phase II contact on the active side. It can, of course, be argued that these striations are produced by upwards and backwards movement of the lower teeth on the active side as has been postulated from the position of wear facets and the orientation

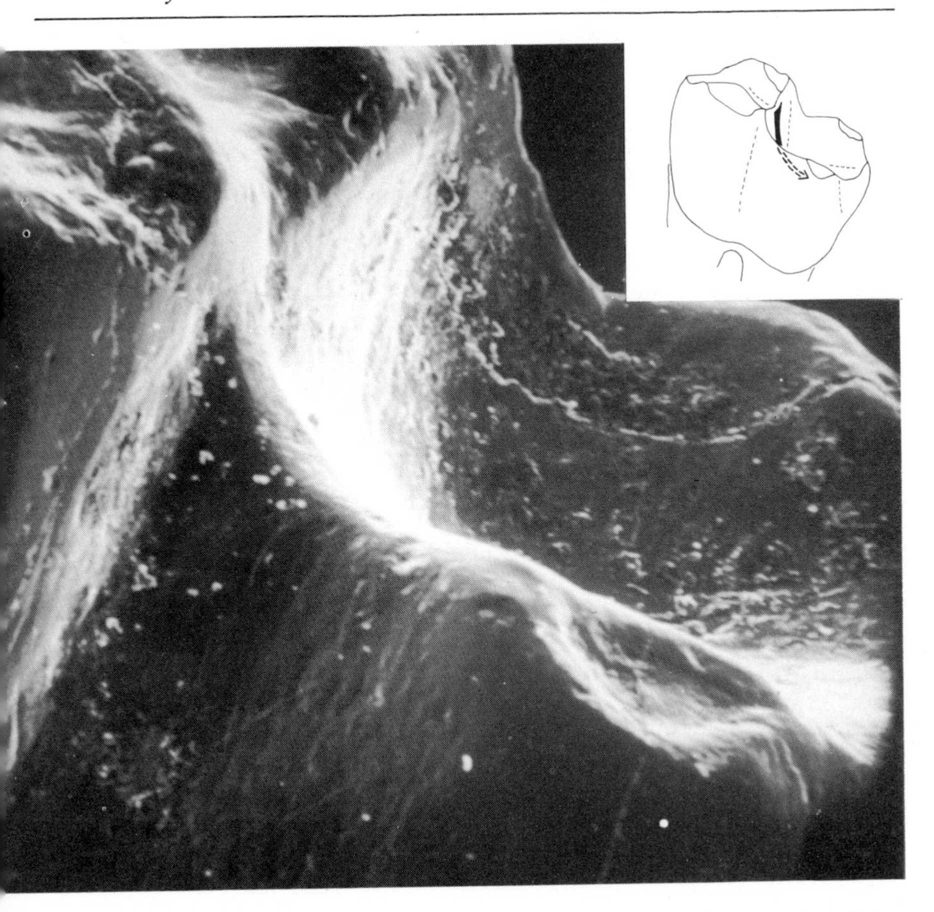

Figure 10. Scanning electron-micrograph showing the posterior margin of the trigonid and part of the talonid of the lower second molar of *Galago*. The protoconid (upper left) and the hypoconid (lower right) have been worn and dentine exposed during puncture-crushing mastication. The striations produced during shearing-grinding power strokes can be clearly seen on facet 3 (foreground) and on facet 5 (centre) where they curve continuously from phase I into phase II as indicated by the arrow in the inset. Magnification approximately x 100.

of the striations crossing them.[25] However, to date, in none of the assorted mammals studied has such a movement pattern been observed.

Discussion

This study has confirmed that in *Galago*, as in *Didelphis*,[12,14] only a very small proportion of the time spent triturating the food involves actual contact between upper and lower teeth, and that the movements of the whole lower jaw may not be predictable from the striations on the wear facets produced by such contact. The behaviour of active and balancing sides will not necessarily parallel each other where the mandibular symphysis is mobile and each dentary can rotate about its long (antero-posterior) axis. Although puncture-crushing and chewing power strokes utilise different parts of the teeth and produce different types of wear, the overall pattern of the masticatory cycle and the direction of the power stroke are similar. Since the shape of the teeth only guides and controls the relative movement of upper and lower teeth on a 'chewing' power stroke, this similarity of pattern suggests that the path of movement is almost certainly 'learned' at an early stage and is generally regulated by proprioceptive information emanating primarily from the muscles and the periodontal ligaments of the teeth, but also from the jaw joint. It follows that although direct extrapolation from one type to another, for example from *Galago* to the Miocene lorisiforms, is only wholly justifiable for the power stroke in chewing, it is likely that given occlusal and dietetic similarity, masticatory behaviour overall is likely to be comparable.

This view is to some extent substantiated by the observation that the incisors of *Galago* have a negligible masticatory function. Whilst this could be attributed to the presence of the 'dental comb' and therefore a consequence of this specialisation, behavioural studies of *Tupaia* and *Didelphis* have shown that in these animals, the incisors are rarely used and then mainly for grasping. It is suggested that since the incisors in primitive mammals have little functional importance, the adaptations for non-feeding purposes, such as grooming, do not mitigate against masticatory efficiency. If this is the case, then the spatulate 'cutting' incisors of the anthropoids are themselves a specialisation, reflecting a shift of 'biting' function from the canine-premolar series towards the incisors. This shift is not complete, since *Ateles*,[22] *Saimiri*,[13] and even man will use the premolars for tough food.

Puncture-crushing is often used in ingestion (ingestion by mastication) and involves all the cheek teeth but particularly the premolars. As might be expected, the tougher the food, the longer it is pulped in this fashion. In *Galago*, as in *Didelphis*, the gape between cycles is greatest during this stage of mastication. The explanation for the wide gape almost certainly lies in the behaviour of the muscles of mastication, as does the reason for the 'hold' seen in the recovery stroke. A suppression of downward movement while protrusion occurs, and at a gape slightly in excess of that seen in the rest position, is not unique to *Galago*. The same phenomenon occurs in *Tupaia, Didelphis, Ateles* and *Saimiri*. It is possible that protrusion following on limited opening, triggered by the external pterygoid, is necessary to give some greater mechanical advantage (in the widest sense) to the anterior suprahyoid muscles as they come into action as mandibular depressors.

The masticatory cycle in *Galago* conforms in its general pattern to that seen in all the other mammals studied using combined occlusal and cinefluorographic techniques. The essential difference between the primitive condition, as represented by *Tupaia* and *Didelphis*, is in the development of a consistent second phase in the power stroke. This is associated with the appearance of a 'grinding' type of trituration and is reflected in the changing distribution, extent and orientation of the wear facets. The facets seen in *Palenochtha* and *Galago* molars are generally very similar, but the gradual reduction and disappearance of facets 2*b*, 3*b* and 4*b*, the coalescence of facets 1*a* and 1*b*, together with the appearance of a functional hypocone with its associated facets 7 and 8, reflect a positive functional trend. The evolution of facets 1-6 in early mammalian molars has been traced from the Late Triassic to the Late Cretaceous by Crompton.[24] Their appearance and subsequent development can be related, at least in part, to the evolution of embrasure shearing as the mechanism of trituration. It involved a progressive increase in the number of transverse and obliquely orientated shearing blades coming into successive contact during a predominantly upwards and medially directed power stroke. The loss in *Galago* of many of the facets associated with this first phase of the power stroke, the reorientation of the remainder to give a more antero-posterior direction and the shift from sequential to near simultaneous contact reflects a reduction of simple embrasure shear with the new addition of a grinding action between areas contacting at centric occlusion and through the second phase of the power stroke.

In primitive mammals such as *Didelphis* there is little or no contact as the teeth move out of the power stroke, the recovery stroke involves a simple downwards movement from centric occlusion. In *Palenochtha*, as in all known Palaeocene Primates, a functional contact between the lateral slope of the protocone and the medial slope of the hypoconid and between the posterior slope of the protocone and the medial section of the paracristid can be observed. In these genera, the stylar areas, large in Late Cretaceous placentals, are much abbreviated, leading to a shorter transverse component of movement in phase I of the power stroke. This change, with the expansion of the contact areas of phase II, characterises the earliest primates, amongst other groups, appearing in the Caenozoic. *Galago* shows both these trends. The dip of many of the wear facets in *Galago* is much lower than for the corresponding facets in *Palenochtha* and the height difference between the trigonid and talonid basins is reduced. Both these features can be related to selection for an increasing component of occlusal force oriented normal to these surfaces, an adaptation for grinding, rather than parallel to the contacting blades as in shearing.

A two-stage power stroke was described by Mills for mammals, including primates: a 'buccal phase' producing wear facets with striations orientated transversely to the tooth row and a 'lingual phase' with wear facets striated obliquely.[6,16,17] Furthermore, he stated that the buccal phase on one side is coincident with the lingual phase on the other, producing a balanced occlusion, and suggested that this may help reduce the stress on the mandibular joint.[16] In the primitive mammal *Didelphis*, no lingual phase, and therefore no balanced occlusion were observed. The situation in *Galago* is somewhat different, since a phase II has developed. To that extent, phase I as described here can be equated with Mills' 'buccal phase' and phase II with his 'lingual'. However, this equation presents some difficulty in so far as the definitions given above carry the explicit statement that a phase on the active side has a simultaneous phase on the balancing side, whereas there is no evidence that phase I in *Galago* involves any tooth contact on the balancing side. Similarly, although the 'lingual phase' or phase II on the active side is, in *Galago*, followed by a delayed transitory occlusion on the balancing side, there is again no evidence that such contact is in any way functional.

Occlusion in *Galago* cannot therefore be regarded as 'balanced' in the sense Mills describes. Presumably a 'balanced' occlusion is one in which the force distribution about the jaws is such that no particular joint is unduly loaded. To achieve this in function would, in the extreme case, either require no food between the teeth on the active side, so eliminating any tendency for the lower jaw to 'rock' about a longitudinal axis, or comparable amounts of food on both sides and an effectively alternating power stroke. Neither of these conditions has been seen in *Galago*. Moreover there is a discrete and observable time lag between the completion of phase I with the onset of phase II and the transitory occlusion on the balancing side. The rapidity with which this contact occurs and is broken (assuming it to be true contact rather than simple vertical superimposition seen on the dorso-ventral projection radiographs) also mitigates against it having any real masticatory function. The terms 'buccal' and 'lingual' are therefore avoided in the present paper on the grounds that their connotations, as defined, cannot be justified on the evidence available.

In an earlier paper, Butler and Mills stated that 'the oblique scratches were produced (on certain wear facets) during rotation (of the mandible) about the contralateral condyle when the lower teeth were to the lingual side of the centric position'.[26] The transverse striations of the buccal phase were said to result from movement about the ipsilateral condyle. The free mobility of the mandibular condyles on the glenoid fossae has been observed in all the animals studied cinefluorographically and the actual range of movement occurring could not have been predicted from occlusal analysis.[13,14] Even so, a geometrical construction based on the occlusal analysis demonstrates that in *Galago*, as in *Saimiri*,[13] rotation is not occurring in the fashion described by Mills. Fig. 11 shows this construction: two lines, one from the lateral and one from the medial edge of the condyle, are drawn at a tangent to the arc of phase I movement to be expected at the second molar, given that this depends on condylar rotation. A third line is drawn showing the actual observed orientation of phase I movement. Expressing the orientation of each line as an angle between it and an arbitrary mesiodistal axis in the occlusal plane, it is clearly seen that the orientation of actual movement does not fall within the expected range for condylar rotation. This hypothesis may, therefore, be rejected. The smooth continuity between phase I and II on the active side and the medial shift of the condyle during the power stroke, suggests that the movements involved are smooth and unilateral. The mandibular joints may limit the possible range of movements in mammals including *Galago*, but do not appear to guide them.

It follows that since phase I movement is not a function of ipsilateral condylar rotation and that the angles of orientation and plunge of this movement change along the tooth row, no generalisations as to the direction and angulation of the power stroke in any mammal, let alone in fossil forms, can be made without certain precautions. Strict comparability in respect of the tooth or teeth used is essential. The movement described, since it represents a line in space, has to be characterised by relating that line to the occlusal plane or a prescribed axis in that plane. The power stroke in chewing in all mammals so far studied (including *Galago*) has upwards, medial and forwards components. It is the relative proportions of these three that delineate the distinctive nature of the power stroke in each animal. Evolutionary dental trends are normally reflected by the changing orientations of the power stroke. These cannot be adequately described in terms of either the vertical or horizontal plane alone.

This study has shown that in terms of general masticatory behaviour *Galago* conforms to the pattern found in primitive mammals such as *Didelphis* and *Tupaia* and in two other primates, *Saimiri* and *Ateles*. The implications of this observation will be discussed elsewhere.[23] However, *Galago* differs from the primitive mammals so far studied in that the power stroke has developed from a single movement carrying the lower teeth into centric occlusion (phase I) with the introduction of a crushing-grinding second phase (phase II). This appears to mark the first stage in a definite shift in primates and herbivores from simple embrasure shearing to a grinding type of power stroke.

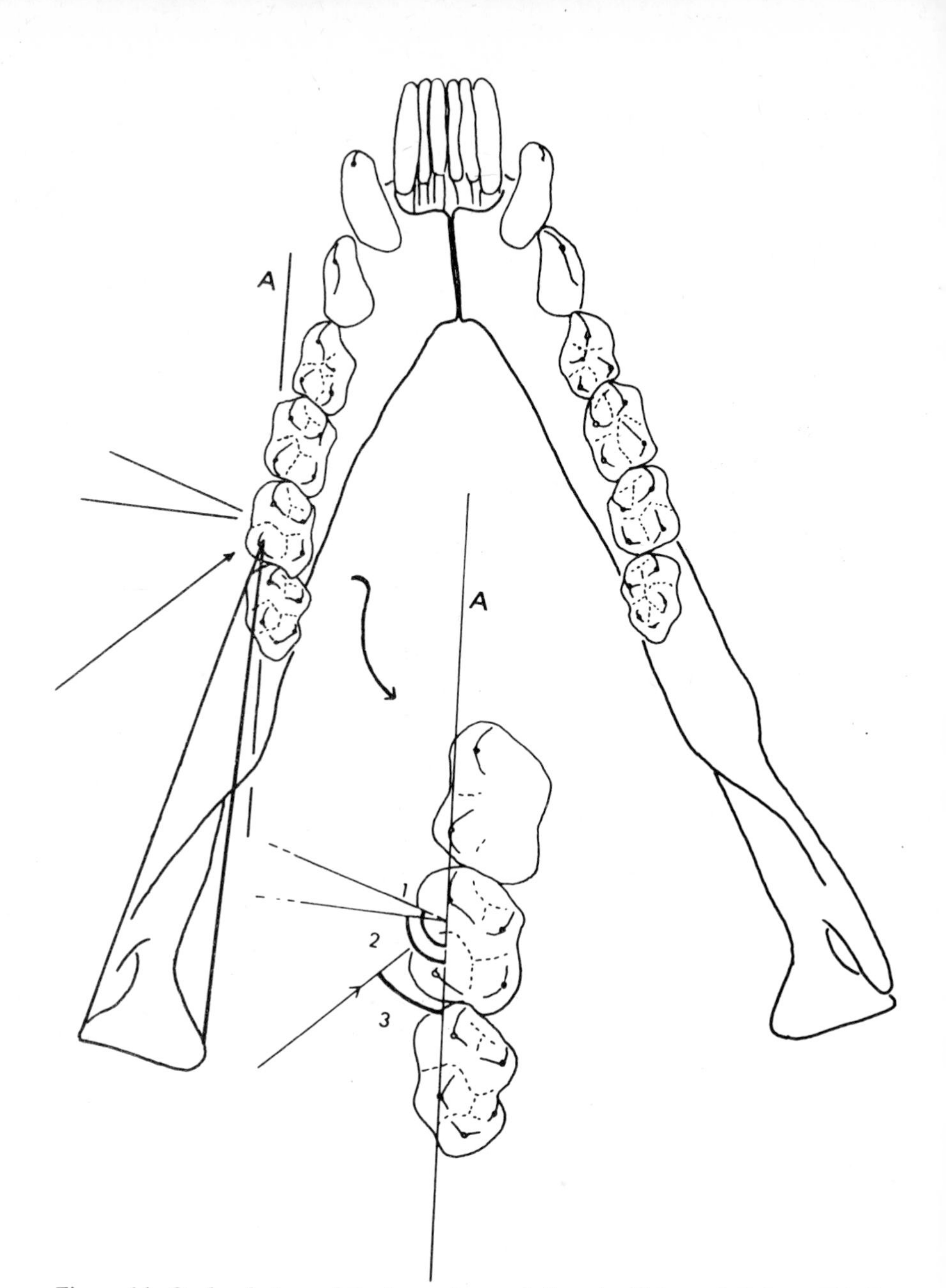

Figure 11. Occlusal view of the lower jaw and (inset) of M_{1-3} of *Galago*. The line A represents an arbitrary antero-posterior axis. Lines are shown projected from the lateral and medial borders of the condyle onto the hypoconid of M_2. On the assumption that the condyle is the centre of rotation for phase I of the power stroke, two lines (to the left of the complete tooth row) have been drawn at a tangent to the two axes of the condyle and represent the maximum range of orientations expected for the null hypothesis that condylar rotation occurs. The arrow to the left of the complete tooth row shows the observed orientations of the striations on the second molar produced in phase I. The angles 1 and 2 on the inset diagram show the angles between the antero-posterior axis and the range of angles subtended for condylar rotation. Angle 3 shows the observed angle between the orientation of phase I and the antero-posterior axis. The observed angle does not agree with that predicted for condylar rotation, confirming that such rotation does not occur during phase I of the power stroke.

Acknowledgments

The authors would like to express their great indebtedness to Dr B. Trum and especially to Dr F. Garcia of the New England Regional Primate Center for making available the specimen of *Galago crassicaudatus* used in this study, for their technical assistance and the use of laboratory facilities. One of us (R.F.K.) would like to acknowledge the support of the Wenner Gren Foundation for Anthropological Research Grant no. 2735, and the other (K.M.H.) USPHS Grant no. DE 02648. Mrs Gaby Dundon, Mrs Janet Lock and Mrs Pat Elson kindly typed and retyped the manuscript.

NOTES

1 Simpson, G.G. (1967), 'The Tertiary lorisiform primates of Africa', *Bull.Mus.Comp.Zool.* 136, 39-61.

2 Walker, A.C. (1969), 'True affinities of *Propotto leakeyi* Simpson 1967', *Nature* 223, 647-8.

3 Lewis, G.E. (1933), 'Preliminary notice of a new genus of lemuroid from the Siwaliks', *Amer.J.Sci.* 26, 134-8.

4 Simons, E.L., personal communication.

5 Walker, A.C., this volume.

6 Mills, J.R.E. (1955), 'Ideal dental occlusion in the Primates', *Dent.Pract.* 6, 47-61.

7 Ansell, W.H.F. (1960), *Mammals of Northern Rhodesia*, Lusaka.

8 Bishop, A. (1964), 'Use of the hand in lower primates', in Buettner-Janusch, J. (ed.), *Evolutionary and Genetic Biology of Primates*, vol. 2, New York and London.

9 Møller E. (1967), 'The chewing apparatus: an electromyograph study of the action of the muscles of mastication and its correlation with facial morphology', *Acta Physiol. Scand.* 69, suppl. 280.

10 Hiiemae, K.M. (1967), 'Masticatory function in the mammals', *J.Dent.Res.* 46, 883-93.

11 Hiiemae, K.M. and Ardran, G.M. (1968), 'A cinefluorographic study of mandibular movement during feeding in the rat (*Rattus norvegicus*)', *J.Zool.Lond.* 154, 139-54.

12 Hiiemae, K.M. and Crompton, A.W. (1971), 'A cinefluorographic study of feeding in the American oppossum, *Didelphis marsupialis*', in Dahlberg, A.A. (ed.), *Dental Morphology and Evolution*, Chicago.

13 Kay, R.F. (1972), 'Mastication in the squirrel monkey (*Saimiri sciureus*) and primate molar occlusion', unpublished MS.

14 Crompton, A.W. and Hiiemae, K.M. (1970), 'Molar occlusion and mandibular movements during occlusion in the American opossum *Didelphis marsupialis*', *Zool.J.Linn.Soc.* 49, 21-47.

15 Kraus, B.S., Jordan, R.E. and Abrams, L. (1969), *Dental Anatomy and Occlusion*, Baltimore.

16 Mills, J.R.E. (1963), 'Occlusion and malocclusion in the teeth of primates', in Brothwell, D.R. (ed.), *Dental Anthropology*, Oxford.

17 Mills, J.R.E. (1967), 'A comparison of lateral jaw movements in some mammals from wear facets on the teeth', *Archs.Oral.Biol.* 12, 645-61.

18 Manufactured by Ralston Purina Co., Saint Louis, MO 63188.

19 Lahee, F.H. (1957), *Field Geology*, 6th ed., New York.

20 *Palenochtha minor* was originally described by Gidley (1923) as '*Palaeochthon minor*', but Simpson (1937) transferred this species to a new genus *Palenochtha.* See Gidley, J.W. (1923), 'Palaeocene primates of the Fort Union, with discussion of relationships of Eocene primates', *Proc.U.S.Nat.Mus.* 63, 1-38; Simpson, G.G. (1937), 'The Fort Union of the Crazy Mountain Field, Montana and its mammalian faunas', *Bull.U.S.Nat.Mus.* 169, 1-287.

21 Hiiemae, K.M., unpublished observations.

22 Hiiemae, K.M. and Kay, R.F., unpublished observations.

23 Hiiemae, K.M. and Kay, R.F. (1973), 'Evolutionary trends in the dynamics of primate mastication', *Proc. Int. Congr. Primat.*, Vol. 3, pp. 28-69. Kay, R.F. and Hiiemae, K.M. (1974), 'Jaw movement and tooth use in recent and fossil primates', *Amer. J. Phys. Anthrop.*, 40, 227-56.

24 Crompton, A.W. (1971), 'The origin of the tribosphenic molar', in Kermack, D.M. and Kermack, K.A. (eds.), *Early Mammals*, London.

25 Gingerich, P., in press.

26 Butler, P.M. and Mills, J.R.E. (1959), 'A contribution to the odontology of *Oreopithecus*', *Bull.Brit.Mus.Nat.Hist.Geol.* 4, 1-26.

P. GINGERICH

Dental function in the Palaeocene primate Plesiadapis

Introduction

The main function of mammalian teeth is to reduce food matter to a size and consistency that can be swallowed and digested. The wide range of dental types seen in living mammals is the result of successive evolutionary radiations of animals adapted to masticating specific diets. In each radiation the dental morphology and adaptation are modified from those of an ancestral species to produce descendant species with a range of dental adaptations. Thus within any radiation it is usually possible to identify a number of dental types, each derived from the ancestral morphology and retaining features of it. For example, the Palaeocene primates *Plesiadapis, Phenacolemur, Carpolestes*, and *Palaechthon* exhibit four rather different adaptive modifications of the ancestral primate molar morphology. They also share many features which presumably were inherited from a common ancestor. The adaptive significance of the morphological differences seen in the teeth of these early primates can only be determined by a detailed consideration of how the teeth function.

Functional occlusion produces matching striated wear facets on upper and lower teeth. Much of the chewing behaviour of an animal can be reconstructed by studying these striated wear facets. Butler and Mills were the first systematically to map wear facets and classify them according to the direction of their striations.[1, 2] Mills distinguished two sets of wear facets on primate molars. One set is the result of an upward, medial, and slightly forward movement of the lower jaw into centric occlusion on the active side. This phase of occlusion Mills referred to as the 'buccal' phase. From centric occlusion the mandible moves forward, medially, and slightly downward on the active side, producing the second set of wear facets. This phase of occlusion Mills termed the 'lingual' phase.

Recent important cineradiographic studies of mastication in the opossum by Crompton and Hiiemae,[3] and in *Galago crassicaudatus* by Kay and Hiiemae,[4] have clarified several aspects of mandibular movement during molar occlusion. Fortunately, striated wear facets are as well preserved on the teeth of fossil mammals as on those of living animals. Thus it has been possible to identify both buccal and lingual phase facets on molars of the Eocene primate *Adapis*, as well as a third 'orthal retraction' set of facets indicating an upward and backward movement of the mandible during one stage of chewing.[5]

In this paper the wear facets on the molars and incisors of specimens of *Plesiadapis rex* (Gidley)[6] from the early Late Palaeocene Cedar Point quarry in northwestern Wyoming are described. This description will form the basis for a later comprehensive study of the evolution of dental function in the Plesiadapidae.

Molar morphology and function

Upper and lower molars of species of *Plesiadapis* have been illustrated and described in detail by Matthew,[7] Jepsen,[8] Simpson,[9] and Russell,[10] among others. The terminology used here in describing *Plesiadapis* molars is illustrated in Fig. 1. The lower molars are roughly rectangular; the protoconid and metaconid are joined by a strongly developed protocristid; the metaconid and paraconid are connate (as Matthew suggested, possibly what is here called the paraconid is really the metaconid, and what is here called the metaconid should really be considered a metastyle);[7] the paracristid runs forward to join the postcristid of the preceding molar; the cristid obliqua forms a strong shearing crest connected with the postcristid at the hypoconid; lingually the postcristid is supported by a well developed entoconid; a buccal cingulum is well developed on all molars, and a very weak lingual cingulum is suggested on some.

The upper molars are roughly triangular; the metacone and paracone support a continuous series of shearing crests: the paracrista, centrocrista, and metacrista; the paraconule and metaconule are joined to the protocone by the preprotocrista and postprotocrista respectively; the '*Nannopithex* fold'[11] is invariably a well developed crest running posteriorly from the protocone and turning abruptly buccally at the posterior margin of the tooth to form a post-cingulum; a cingulum borders the molars buccally, and a lingual cingulum is developed along the anteromedial border of the molars. Both upper and lower molars are low in profile and bear bulbous cusps.

Fig. 2 illustrates upper and lower second molars of *Plesiadapis rex* and indicates the wear facets formed during function. Each of the major crests connecting the cusps on the lower molars supports a buccal phase wear facet on its buccal surface. For descriptive

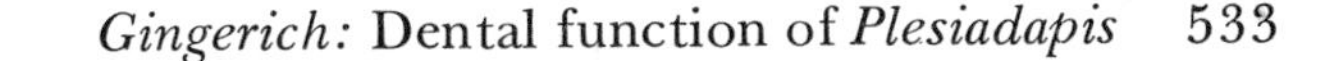

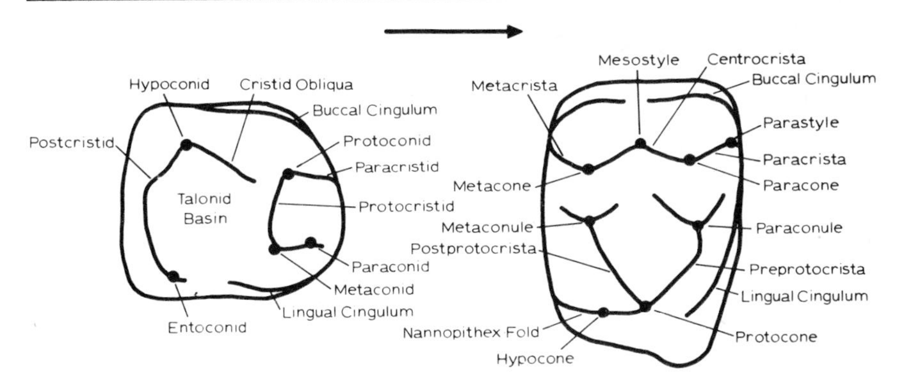

Figure 1. Terms used to describe molars of *Plesiadapis*. Lower molar is on the left, upper molar on the right. Heavy arrow is buccal to both molars and points anteriorly.

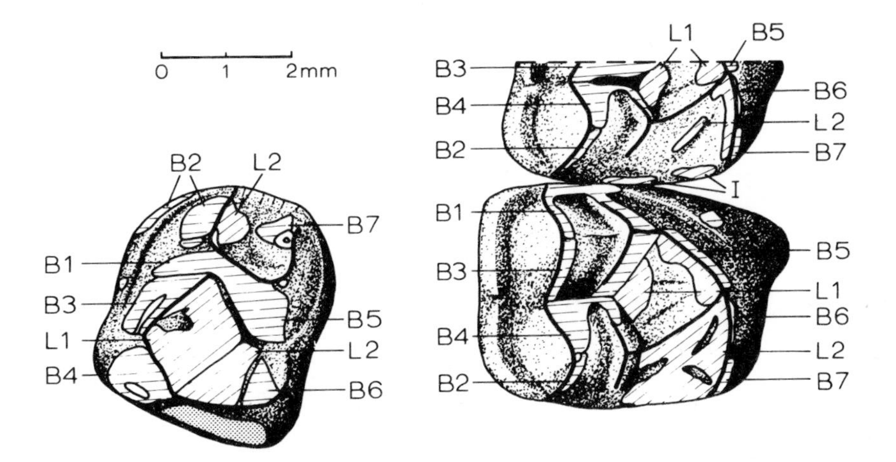

Figure 2. Wear facets on upper and lower molars of *Plesiadapis rex*, seen in occlusal view. On the left is a lower left second molar, on the right is part of an upper right first molar and an upper right second molar. Matching buccal (B) and lingual (L) phase facets are correspondingly numbered. Dotted area is an interstitial facet. Lines on the wear facets indicate the actual direction of striations and movement.

purposes, these are numbered B1-B6, following the numbering scheme recently introduced by Crompton.[12] An additional facet, B7, is present on the buccal side of the paraconid. Each buccal phase facet on the lower molars is matched by a buccal phase facet on the lingual side of the crests on the upper molars, correspondingly numbered B1-B7. The striations on all buccal phase facets are parallel, indicating that they were formed by a single jaw movement.

A second set of striated wear facets on the lower molars is formed mainly in the talonid basins. These are numbered L1 and L2 in

Fig. 2. The facet L2 generally continues posteriorly from the talonid of one molar onto the trigonid of the following molar. Thus the lower molar illustrated in Fig. 2 has a second L2 facet lingual to the protoconid, which is the posterior extension of the L2 facet on the preceeding molar. As the teeth become more heavily worn, this L2 facet on the trigonid may obliterate the neighbouring B7 facet. The L1 and L2 facets on the lower molars match facets developed on the upper molars, correspondingly numbered L1 and L2. Again, the striations on the lingual phase facets are parallel, indicating that they were formed during a single jaw movement. The two facets marked I on the upper first molar, and the corresponding area of L2 on the following molar have striation directions intermediate between those of the buccal and lingual phase facets. Matching intermediate facets are developed on the trigonid of the lower molars in some specimens. This confluency of buccal and lingual phase facets indicates that the buccal and lingual phases in *Plesiadapis* were consecutive components of a single transverse ectental jaw movement.

The buccal phase performs primarily a cutting function, with crests on the lower molars shearing past those on the upper molars. To better serve this function, the number of cutting edges on the molars was increased by two means. Cutting edges were added by developing new crests parallel to the previously existing ones. Thus the B1 protocristid cutting edge on the lower molars sheared past first the paracrista, then the preparaconule crista, and finally past the lingual cingulum on the corresponding upper molar. As the postmetaconule crista on the upper molars was limited in development by the L2 facet, a second B2 cutting edge was developed from the buccal cingulum of the lower molars. The second means of adding cutting edges is illustrated by both the B3 and B4 facets on the lower molars. As is shown in Fig. 2, here the enamel has been worn through relatively early, resulting in two resistant enamel cutting edges on a single crest. On heavily worn teeth almost all the crests bear two cutting edges as a result of wear having penetrated the enamel.

The molar teeth of *Plesiadapis* occluded in such a way that they produced on the active side a series of three compression chambers for expressing juices. At the beginning of the buccal phase the B3, B4, and L1 facets on the upper molars, and the L1, L2, B5, and B6 facets on the lower molars completely enclosed a space, the volume of which was progressively reduced as the teeth approached centric occlusion. The pulp remaining after this compression was then ground during the lingual phase.

The lingual phase performs primarily a grinding function. Food is ground between opposing planar areas of the matching lingual phase facets. In *Plesiadapis rex* the relatively rough texture of the enamel in the areas where lingual phase facets developed, illustrated in Fig. 2, caused windows in the facets until they became heavily worn. These windows functioned as cutting edges, suggesting that in *Plesiadapis* the lingual phase performed a cutting as well as a grinding function, at least in younger individuals.

Incisor morphology and function

The type species of *Plesiadapis, P. tricuspidens*, was so named because of its enlarged and distinctive tricuspid upper incisors.[13] The morphology of the lower incisors of *Plesiadapis* are also unique to the Plesiadapidae. These enlarged procumbent upper and lower incisors have been figured and described by Matthew and by Russell.[7,10] Both are illustrated in occlusal aspect in Fig. 3. The upper incisor (Fig. 3*a*) has a large anterior central cusp flanked by smaller medial and lateral cusps. Four ridges connect these three anterior cusps with a posterior cusp. These ridges became worn successively, beginning with the medial ridge, until the whole occlusal surface of the tooth anterior to the posterior cusp was excavated. A large interstitial facet developed where the right and left upper incisors contacted each other. With age, the large central anterior cusp and the medial and lateral cusps became heavily worn and blunted.

The lower incisor (Fig. 3*b*) has a slightly convex dorsal surface bordered laterally by a ridge of enamel running roughly parallel to the large interstitial facet on the medial side. The only cusp was small and developed at the posterior end of the lateral enamel ridge. With age, the tip of the incisor was gradually worn back. The tip was never sharply pointed. In some specimens the enamel is worn off on a large area of the central dorsal surface of the incisor crown; however, on other very heavily worn specimens this enamel covering is still complete.

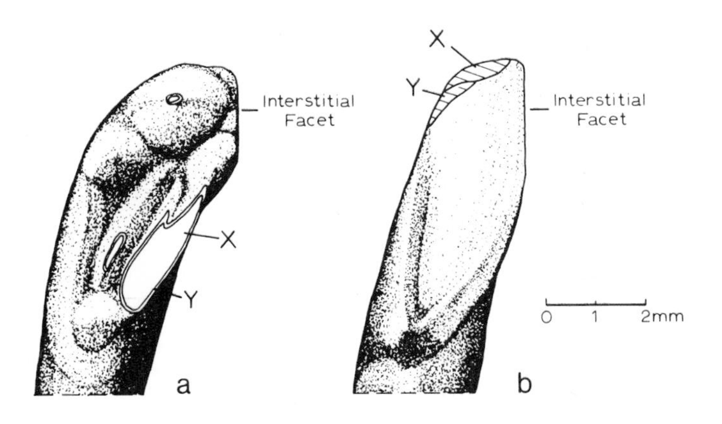

Figure 3. Wear facets on upper and lower incisors of *Plesiadapis rex*, seen in occlusal view. (*a*) right upper incisor; (*b*) left lower incisor. X and Y are matching facets on the upper and lower incisors.

The reconstructions of the skull of *Plesiadapis tricuspidens* by Simons,[14] Russell,[10] and Szalay[15,16] show that the function of these enlarged incisors was not fully understood. On the slightly worn specimens illustrated in Fig. 3 it is possible to distinguish two separate wear facets. The striated X facet on the lower incisors indicates an upwards, backwards, and slightly medial movement of the tip of the lower incisor across the ridges of the occlusal surface of the upper incisor. As the tip of the lower incisor approached the posterior cusp of the upper incisor it was forced more medially, and the leading edge of the Y facet on the lower incisor sheared past the enamel edge Y on the posteromedial side of the X facet on the upper incisor. The left and right lower incisors are just wide enough to have passed between the posterior cusps of the left and right upper incisors. From the wear on the incisors it is clear that the tips of the lower incisors functioned predominantly by shearing across the ridges (particularly the most medial ones) located just anterior to the posterior cusp on the upper incisors. The way the tip of the lower incisor closed the space in front of the posterior cusp on the upper incisor, shearing against the medial side of the posterior cusp, can best be seen in lateral view (Fig. 4). The result was a cutting device which would function most effectively in cutting stems of soft vegetation.

The tips of the upper and lower incisors could be opposed in *Plesiadapis*, but the tips of both the upper and the lower incisors became blunted with use. They therefore cannot have functioned as do the incisors of living rodents and lagomorphs, and would not have been effective in taking bites from a larger mass of food. Possibly the most similar incisor mechanism among living mammals is that found in the marsupial macropodines.[17] From striations on the interstitial facets of some of the lower incisors it is apparent that the

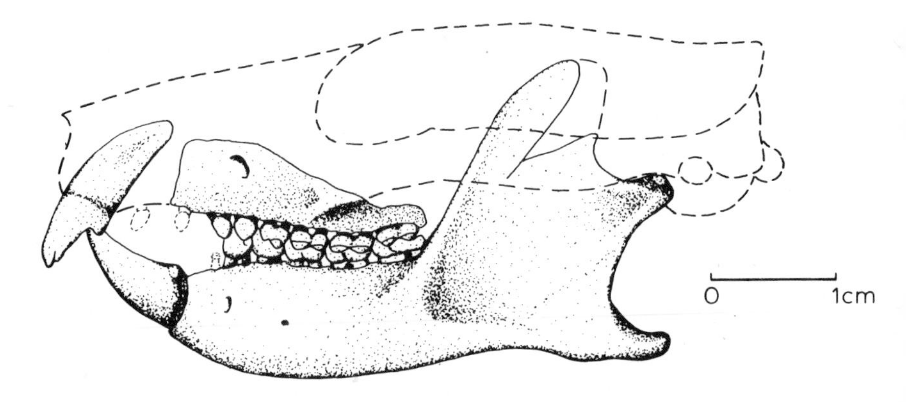

Figure 4. Reconstruction of centric occlusion in *Plesiadapis rex*, seen in lateral view. Stippled areas are actual specimens, dashed lines are hypothetical. Based on PU 21300, 21246, and an isolated upper incisor, all from the Cedar Point Quarry.

mandibular symphysis of *Plesiadapis* was highly mobile, as is true also of macropodines. The direction of movement of the lower incisors during function was upward and backward, the force for which was generated chiefly by the temporalis muscles.[18]

From this analysis of the function of the enlarged incisors of *Plesiadapis* it is apparent that when the molars were in occlusion, the incisors must have occupied approximately the position shown in Fig. 4. The lower incisors rested between and sufficiently far behind the upper incisors not to contact them during the buccal and lingual phases of molar occlusion. For the incisors to function it was necessary for the lower jaws to translate forward. With this forward translation, the mandibular condyle rode up on the articular tubercle and separated the upper and lower cheek teeth sufficiently for the incisors to function independently.

Discussion

Any attempt to determine the diet of an extinct mammal must consider the morphology of all the teeth: incisors, canines, premolars, and molars. The molars of *Plesiadapis* show relatively great development of buccal phase cutting edges, as do the molars of living grazing herbivores. The incisor mechanism of *Plesiadapis* indicates a cropping mode of food acquisition. The reduction and loss of canines and premolars in *Plesiadapis*, leaving a large diastema between the incisors and the remaining premolars and molars, is also consistent with a cropping incisor mechanism. The canine and premolar teeth of many mammals are used to shear and puncture large bites of food when they are first ingested. Reduction and loss of canines and premolars, with the resulting large diastema, is correlated with initial ingestion into the mouth of small pieces of food. Grazing mammals have a large diastema because they feed on food that is by its nature already in small pieces. Rodents have a large diastema because they gnaw a series of small pieces from a larger piece of food rather than ingesting the larger piece.[19] The great reduction of canines and premolars in *Plesiadapis*, compared for example with the large pointed premolars of *Phenacolemur*,[20] thus suggests that *Plesiadapis* also ingested food in small pieces. The shape of the cutting edges on the incisors of *Plesiadapis* further suggests that the original food mass was predominantly in the form of stems.

Teilhard de Chardin was the first to suggest that *Plesiadapis* may well have been a terrestrial animal, based on the relative abundance of *Plesiadapis* remains at Cernay.[21] In the Princeton collection from the Cedar Point quarry approximately 140 out of 400 catalogued placental mammal specimens are *Plesiadapis*, further supporting Teilhard's observation of their abundance. Simons suggested that some species of *Plesiadapis* most probably were terrestrial to account for the wide geographic distribution of the genus in North America and Europe.[22] The predominantly herbaceous diet and grazing feeding mode inferred above for *Plesiadapis* also suggest a terrestrial habitat. Among living gregarious terrestrial herbivores, the postcranial remains of *Plesiadapis tricuspidens* compare closely in size and many morphological features with the skeleton of a marmot (genus *Marmota*). The habits and appearance of *Plesiadapis* may have been similar to those of living marmots, although this hypothesis must be tested by more detailed comparative work on the postcranial anatomy of *Plesiadapis*.

Recently Charles-Dominique and Martin suggested that allocation of *Plesiadapis* to the order Primates should be reviewed.[23] *Plesiadapis* is included in Primates by palaeontologists for two reasons. The first reason was most clearly stated by Gidley and by Simpson.[6,9] The structure of the upper and lower molars of *Plesiadapis* is virtually identical to the structure of the molars of the undoubted primate *Pelycodus* of the early Eocene. To quote Simpson:

> As regards molar pattern, *Plesiadapis* resembles the primitive Notharctinae more closely than any other group. . . . The resemblance to *Pelycodus*, most primitive known notharctine, is really amazing and extends to the apparently most insignificant details. The upper molars are of almost identical structure throughout, differing only in details of the cingula and proportions such as may characterise species of one genus. In the lower molars, *Pelycodus* has the paraconids slightly more distinct, but the resemblance is equally striking and includes even such features as the minute grooving of the trigonid face of the metaconid and the exact structure of the complex grooving of the talonid face of the hypoconid and of the whole heel of M_3.[9]

As will be discussed below, molar morphology similar to that of *Plesiadapis* is known in living mammals only within the order Primates.

The second reason for including *Plesiadapis* in Primates concerns the presence and construction of the ossified auditory bulla. The bulla of *Plesiadapis* is apparently composed of an ossification completely continuous with the petrosal,[24] a characteristic of all known adult primates.[25]

Martin employed the term *synapomorph* to designate characters acquired during the separation of a new stock, distinguishing them from *symplesiomorph* characters which are present in the ancestral stock as well as the new stock.[26] Thus it is the synapomorph characters which distinguish and define any new stock. In his list of synapomorph characters of Primates, Martin included features of the orbit, auditory region, limbs, nervous system, and reproductive tract. No mention was made of the dentition, on which the palaeontologist must base much of his interpretation. In connection with the manner of elongation of M_3, the position and mode of disappearance of the paraconid, the structure of the paracristid, and the development of the '*Nannopithex* fold' to form a posterior basin on the upper molars (all of which are associated with increased propalinal mandibular movement and encroachment of the lingual phase facets from the talonid basin of one molar onto the trigonid of the following molar), Gidley stated:

> In fact, this peculiar development of the upper and lower cheek teeth apparently constitutes a distinctively primate characteristic, which while not found in all families of the order, seems to have been repeated over and over again, with slight variations, in several related or unrelated groups, and, so far as I am aware, is not found in any other order of mammals.[6]

As molar morphology similar to that of *Plesiadapis* and *Pelycodus* (in the manner of elongation of M_3, molar trigonid construction, and development of the '*Nannopithex* fold') is found only among living and fossil Primates, this morphology apparently constitutes an important synapomorph character of the order. Later radiations have modified this basic primate molar pattern, but it must, as it is present in all early members of the order, closely approximate the morphology developed by early primates during their separation from the ancestral mammalian stock. The fact that most Eocene tarsioid primates have enlarged central incisors similar to those of primitive plesiadapoids is additional evidence that the ancestry of Primates is to be sought in these archaic forms.

In conclusion, the Plesiadapidae are best considered an early, probably terrestrial, herbivorous radiation which shared little except common ancestry with the contemporary, presumably arboreal, primates that gave rise to Eocene and later radiations. The trends toward dental reduction and specialisation seen in plesiadapid lineages preclude their having had any close relationship with living prosimians, but the Plesiadapidae are nevertheless important in that further study should reveal much about the origin and evolution of the earliest primates.

Acknowledgments

I am indebted to Professor Glenn L. Jepsen, Princeton University, for my introduction to fossil primates, and the opportunity to work with him at the Cedar Point quarry for three summers. I thank Vincent Maglio and Donald Baird for allowing me to study specimens of *Plesiadapis* in the Princeton collections. David Parris of Princeton calculated the relative number of *Plesiadapis* specimens in the Cedar

Point collection and identified them as *P. rex.* I also thank Professor J.P. Lehman and D.E. Russell of the Institut de Paléontologie, Paris, for access to their collection of *Plesiadapis tricuspidens.* I have profited greatly from discussions with Elwyn Simons and David Pilbeam, Yale University. This work was supported by a travel grant from the US National Science Foundation and by a grant-in-aid of research from the Society of Sigma Xi.

NOTES

1 Butler, P.M. (1952), 'The milk molars of the Perissodactyla, with remarks on molar occlusion', *Proc. zool. Soc. Lond.* 121, 777-817.

2 Mills, J.R.E. (1955), 'Ideal dental occlusion in the Primates', *Dent. Pract. Bristol*, 6, 47-61.

3 Crompton, A.W. and Hiiemae, K. (1970), 'Molar occlusion and mandibular movements during occlusion in the American opossum, *Didelphis marsupialis* L.', *Zool. J. Lin. Soc.* 49, 21-47.

4 Kay, R.F. and Hiiemae, K.M., this volume.

5 Gingerich, P.D. (1972), 'Molar occlusion and jaw mechanics of the Eocene primate *Adapis*', *Am. J. phys. Anthrop.*, 36, 359-68.

6 Gidley, J.W. (1923), 'Paleocene primates of the Fort Union, with discussion of relationships of Eocene primates', *Proc. U.S. Nat. Mus.* 63, 1-35.

7 Matthew, W.D. (1917), 'The dentition of *Nothodectes*', *Bull. Am. Mus. nat. Hist.* 37, 831-9.

8 Jepsen, G.L. (1930), 'Stratigraphy and paleontology of northeastern Park County, Wyoming', *Proc. Am. Phil. Soc.* 69, 463-528.

9 Simpson, G.G. (1935), 'The Tiffany fauna, Upper Paleocene: 2. Structure and relationships of *Plesiadapis*', *Am. Mus. Novit.* 816, 1-30.

10 Russell, D.E. (1964), 'Les mammifères paléocènes d'Europe', *Mém. Mus. nat. Hist. nat. Paris*, sér. C, 13, 1-324.

11 See footnote in Simpson, G.G. (1955), 'The Phenacolemuridae, new family of early primates', *Bull. Am. Mus. Nat. Hist.* 105, 435. This crest or shelf is probably better termed a postprotocingulum.

12 Crompton, A.W. (1971), 'The origin of the tribosphenic molar', in Kermack, D.M. and Kermack, K.A. (eds.), *Early Mammals, Zool. J. Linn. Soc.* 50, Suppl. 1.

13 Gervais, P. (1877), 'Quelques ossements d'animaux vertébrés recueillis aux environs de Reims', *J. Zool. Paris*, 6, 74-9.

14 Simons, E.L. (1960), 'New fossil primates: a review of the past decade', *Am. Scient.* 48, 179-92.

15 Szalay, F.S. (1971), 'Cranium of the Late Paleocene primate *Plesiadapis tricuspidens*', *Nature* 230, 324-5.

16 Gingerich, P.D. (1971), 'Cranium of *Plesiadapis*', *Nature* 232, 566.

17 Ride, W.D.L. (1959), 'Mastication and taxonomy in the macropodine skull', *Systematics Association Publication* 3, 33-59.

18 Gingerich, P.D. (1971), 'Functional significance of mandibular translation in vertebrate jaw mechanics', *Postilla* 152, 1-10.

19 Hiiemae, K.M. and Ardran, G.M. (1968), 'A cine-fluorographic study of mandibular movement during feeding in the rat (*Rattus norvegicus*)', *J. Zool. Lond.* 154, 139-54.

20 A description of dental wear and function in *Phenacolemur* is in preparation.

21 Teilhard de Chardin, P. (1922), 'Les mammifères de l'éocène inférieure francais et leurs gisements', *Ann. Paléont.* 10, 169-76, and 11, 1-108.

22 Simons, E.L. (1967), 'Fossil primates and the evolution of some primate locomotor systems', *Am. J. phys. Anthrop.* 6, 241-54.
23 Charles-Dominique, P. and Martin, R.D. (1970), 'Evolution of lorises and lemurs', *Nature* 227, 257-60.
24 Russell, D.E. (1959), 'Le crâne de *Plesiadapis*', *Bull. Soc. Géol. Fr.*, sér. 7, 1, 312-14.
25 McKenna, M.C. (1966), 'Paleontology and the origin of the primates', *Folia primat.* 4, 1-25.
26 Martin, R.D. (1968), 'Towards a new definition of Primates', *Man* 3, 377-401.

D. SELIGSOHN and F. S. SZALAY

Dental occlusion and the masticatory apparatus in Lemur *and* Varecia*: their bearing on the systematics of living and fossil primates*

Introduction

Fossil primates are usually characterised by their dental and mandibular parts, primarily because teeth and jaw fragments are the only remnants in most cases. Diagnoses are based on these fragments, and more often than not it is the characters of the dentition, particularly of the molars, which serve to delineate the fossil taxa from one another or from living relatives.

Apart from the taxonomic importance of the dentition, the teeth, along with the jaws and skull, when preserved, provide the only basis for deciphering the role of each of the fossil species in their extinct ecosystem. Yet, without a clear appreciation of the dietary factors which molded the feeding mechanism of living species, and of their inherited morphology, there is little hope of understanding the palaeobiology of past radiations. Recent outstanding work by Charles-Dominique,[1,2] Hladik and Hladik,[3] Hladik et al.,[4] Petter and Hladik,[5] Petter and Peyrieras,[6] Sussman,[7] and others on the natural history of several primate species increasingly demonstrates that most living primates have a relatively species-specific dietary regime, in a feeding zone or area of the forest equally characteristic for most species. These fundamental works in primate natural history are of extreme value and supply the foundations upon which morphologists may begin causal analyses of the dental, mandibular, and cranial morphology. Most living primates are still very poorly understood in this respect, and consequently their extinct relatives fare no better.

We believe that the general practice of using dental criteria to separate fossil species and genera is well tested and sound. Yet in mammalian systematics the time is approaching when the diagnostic differences between species or at least genera should be causally explained.

The range of dental modifications in the Recent lemurs of Madagascar (the subfossils included) surpasses the spectrum of diversity shown by either the known catarrhines, platyrrhines, or the once very abundant tarsiiforms, the omomyids, microchoerids, and the extant tarsiids. They are therefore an ideal living group for the study of phylogenetic modifications of the masticatory system. The dentitions of the *Lemur-Varecia* group are highly derived ones, apart from the obvious specialisations of the tooth comb and associated modifications of second lower premolars. Because many features of their anatomy point to a closer common ancestry between *Lemur* spp. and *Varecia* spp. among known lemurids than any of these with either species of *Lepilemur* or *Hapalemur*, the divergence in their feeding mechanisms is of particular interest.

We have selected these two species groups of lemuriforms because they have been treated, as a result of no critical evaluation, under one genus, yet clearly fall into two very distinct genera according to the morphological criteria of the feeding mechanism, the dentition in particular. We have consequently analysed the differences in morphology, occlusion and other function and attempted to correlate those differences with what is known of the diet of these groups.

Although the extant *Varecia variegatus* and the extinct *V. jullyi* and *V. insignis* have been often categorically lumped in the genus *Lemur* with the *Lemur catta, L. rubriventer, L. mongoz,* and *L. macaco* group, the glaring differences in their dentition immediately signify an evolutionary divergence of generic level to students of fossil mammals. Based on what we know of the total morphology of *Lemur* and *Varecia*, it appears that they have evolved from a common ancestor. The divergence in the dentition and the musculoskeletal part of the feeding mechanism can be explained, we believe, by the differences of their dietary adaptations.

We hope that this study will achieve a better understanding of the phylogenetic modifications of the feeding mechanism in these genera, demonstrate the derived generic differences, point to some of the possible causes for these differences, as well as point out the necessity for rigorous research on as many living species as possible, to understand mammalian feeding mechanisms. Without scrutiny of diet, dental, mandibular and cranial morphology, and masticatory function in the living species, the morphology of fossils, rich with information, will supply us only with a small part of the precious knowledge of past adaptations.

Materials and methods

Numerous skulls of *Lemur catta, L. mongoz, L. macaco, L. rubriventer, Varecia variegatus, V. jullyi,* and *V. insignis* were examined primarily in the American Museum of Natural History, New York; the British Museum (Natural History), London; the Muséum National d'Histoire Naturelle, Paris; and the Academie Malagache, Tananarive. With the aid of a metric grid, line drawings were made of the upper left and lower right molars of *Lemur macaco* (AMNH no. 170764, Figs. 1 and 2) and *Varecia variegatus* (AMNH no. 77792, Figs. 3 and 4). Transparencies were then made of the line drawings of the lower right molars, and the masticatory cycle was simulated by moving the tracings of the lower molars across the upper ones. Figures were consequently prepared (see later – Fig. 6) from the superimposed drawings, controlled by rechecking the occluding dentitions.

The masticatory power strokes of both *Lemur macaco* and *Varecia variegatus* were then simulated by hand, using the mandibles and crania. Implicit in this simulation was the assumption that the lower jaws are naturally 'guided' through the masticatory stroke by virtue of the fact that the morphology of the upper and lower tooth rows complement each other.

For reference points, the tips of the protoconids of M_2 and M_3 were observed as they traversed the upper molars during numerous instances of simulated mastication (Fig. 6). These points on the transparencies, i.e. the tips of the protoconids of M_2 and M_3, were then plotted on the line drawings of the upper left molars indicating the successive points through which the protoconids passed during the initial (occlusal), centric occlusal and, where appropriate, terminal phases of the masticatory stroke. The main axes of the upper and lower tooth rows are also shown on Fig. 6. The axis of the upper tooth row is represented by a line connecting a point at the central base of the upper canine with a point at the centre of the glenoid fossa. The axis of the lower tooth row is represented by a line connecting a point at the central base of the lower P_2 with a point lying medially along the transverse axis of the mandibular condyle.

With the aid of these reference markers, the transparencies of the lower right molars were then superimposed on the line drawings of the upper left molars. The former were then moved over the latter with the appropriate dots on each coinciding in each successive phase of the simulated power stroke.

Additional information on dental, mandibular, and cranial morphology was gathered and analysed.

Molar morphology

1. *Lemur macaco*

The upper and lower molars of *Lemur macaco*, as in *L. catta* or *L. mongoz*, demonstrate relatively little relief in their crown structure (see Figs. 1 and 2). The upper molars possess low, sub-hemispherical paracones and metacones with edges, the former being more prominent than the latter. The protocones are broad, flattened, U-shaped shelves which tilt in a buccal and slightly posterior direction. This results in trigon basins (central foveae) which are relatively shallow and ill-defined. Projecting from the protocones of M^1 and M^2 are a pericone and a hypocone, mesio-lingually and disto-lingually located, respectively. The more mesial pericones are more lingually projecting and lie relatively flush with the palate. Viewing the palate ventrally, the hypocones are more prominently elevated above the palate than the pericones.

The lower molars are constructed in a complementary manner. The trigonid basins (anterior foveae) are but moderately elevated above the talonid basins (posterior foveae), while the relief of the cusps which surround these basins is minimised or eliminated. The trigonid and talonid basins are thus actually flattish planes, lingually down-tilted from the buccal shearing edges. The protoconids and metaconids of all three lower molars are separated by relatively restricted anterior foveae. There is a significant difference between the configurations of the anterior fovea of M_1 and M_2. The anterior

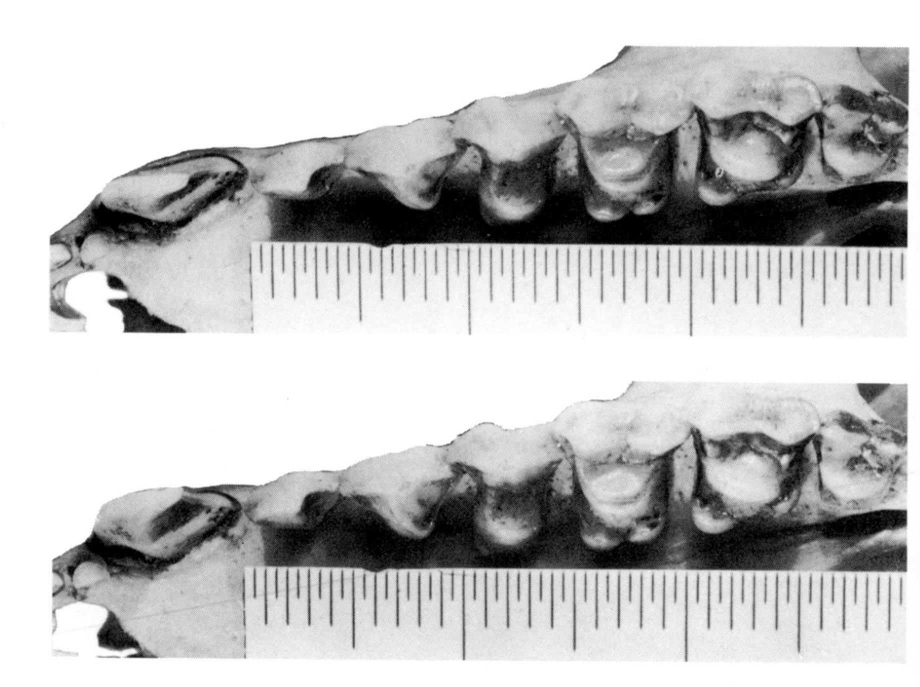

Figure 1. *Lemur macaco*, AMNH no. 41264, occlusal view of left upper dentition.

fovea of M_1 is 'closed off' anteriorly by a paracristid projecting antero-lingually from the protoconid. The anterior fovea of M_2, however, is open mesially. The lower molars are also distinctive in that the trigonid and talonid basins are displaced lingually on the occlusal surfaces of the crowns to a relatively great degree. The buccal faces of the lower molar crowns thus appear to slope downward in a markedly diagonal fashion from the rims of the trigonid and talonid basins. The talonid basin is completely closed off distally and lingually by a long postcristid.

Both M^1 and M^2 are oriented similarly with respect to the axis of the upper tooth row, with only a moderate rotation of M^3. This results in a relatively continuous system of ectolophic crests. With respect to M^1 and M^2, M^3 is only moderately reduced in size.

The lower molars are similarly oriented relative to the axis of the lower tooth row; there is relatively little tooth rotation.

2. *Varecia variegatus*

The molar teeth of *Varecia variegatus*, like those of *V. insignis* and *V. jullyi*, display more relief than those of *Lemur* spp. (see Figs. 3 and 4). The paracones and metacones of the upper molars are higher, steeper and mesio-distally more elongated than those of *Lemur*. As in

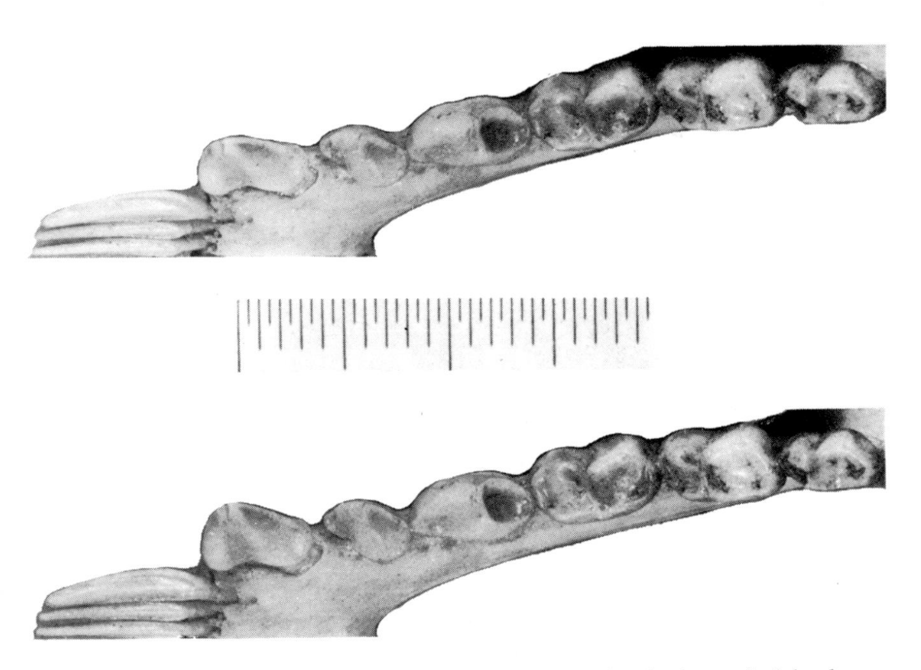

Figure 2. *Lemur macaco*, AMNH no. 41264, occlusal view of right lower dentition.

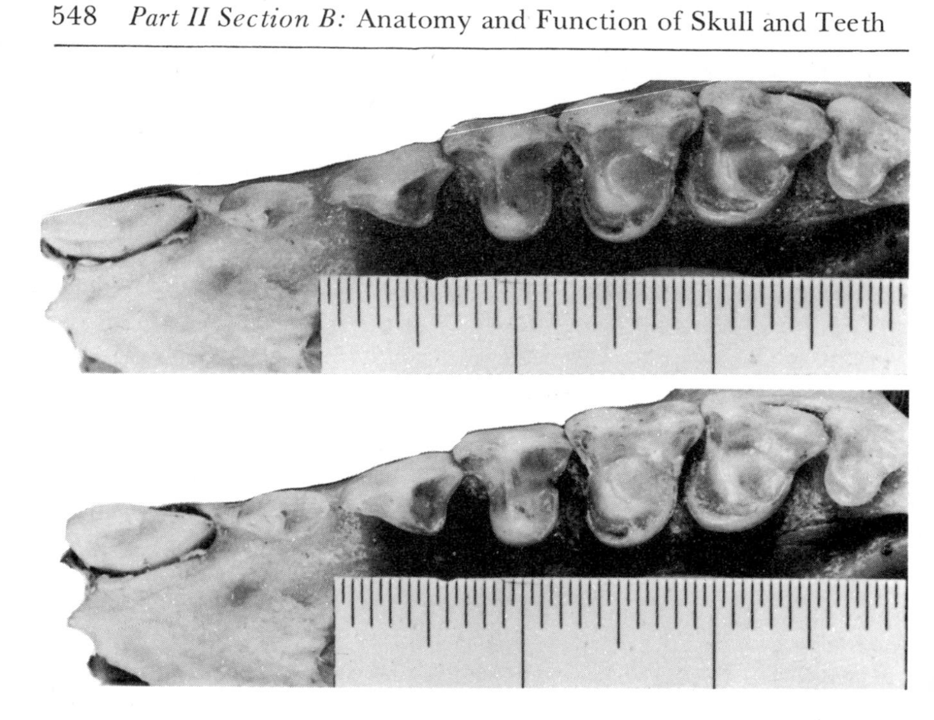

Figure 3. *Varecia variegatus*, AMNH no. 77792, occlusal view of left upper dentition.

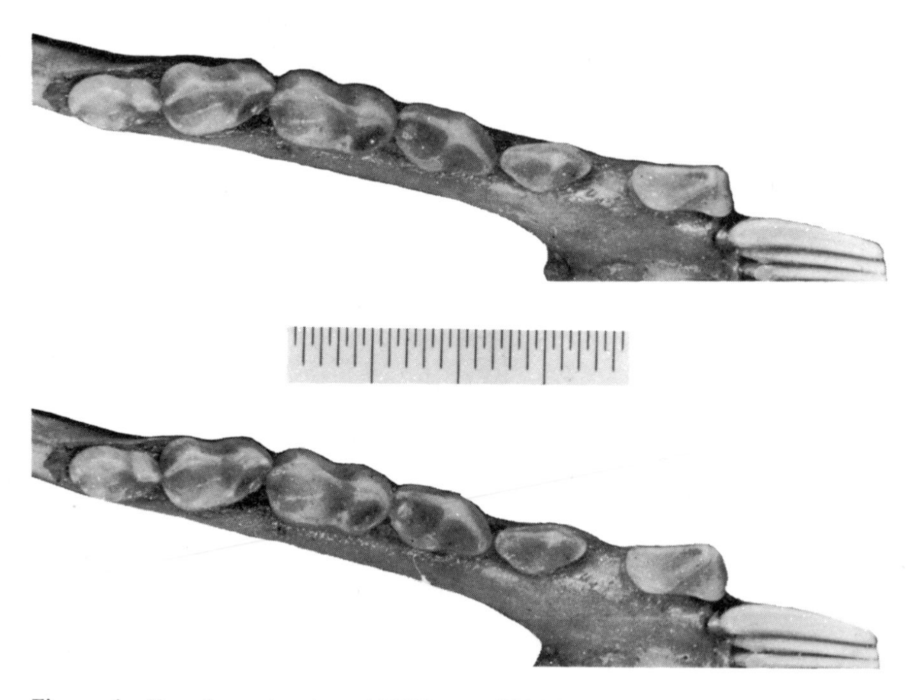

Figure 4. *Varecia variegatus*, AMNH no. 77792, occlusal view of left lower dentition.

Lemur, the paracones are more prominent than the metacones. The protocones of M^1 and M^2 are more restricted and conical in configuration, and are rotated greatly on the occlusal surface so as to dip virtually distally, and only slightly buccally. The basic bucco-lingual orientation of the protocone crests is thus different from the mesio-distal one of *Lemur*. The resulting trigon basins are relatively deeper and more defined, but are shifted from a central to a more distal position as compared to *Lemur*. There are extensive, shelf-like lingual cingula which project mesio-lingually from the base of the protocones of M^1 and M^2. These cingula 'dip' palatally as they extend mesially. The M^1 lingual cingulum usually possesses a small cuspule on the central portion of its lingual rim.

The lower molars of *Varecia variegatus*, *V. insignis*, and *V. jullyi* are similarly distinguished from those of *Lemur* spp. The trigonid and talonid basins are basically large excavations in the lower molar tooth crowns, enclosed by narrow, cusp-studded crests. The cusps and crests forming the buccal rims of the trigonid and talonid basins are far more prominent than those forming the lingual rims. The 'slopes' of the talonid basins thus complement the 'slopes' of the corresponding protocones. It should be noted that the buccal faces of the lower molar tooth crowns are more nearly vertically disposed.

The trigonid basin of M_2 tends to be more open anteriorly than that of M_1 and the talonid basins in general run in a mesio-buccal-disto-lingual direction. Because of their very low entoconids and postcristids, the talonid basins appear to extend through disto-lingually situated 'gaps' in their basin rims.

The upper molars, M^2 and M^3 in particular, tend to be rotated; M^3 is relatively and absolutely more diminutive than M^3 in *Lemur macaco*. Whenever rotation is present, it results in an interrupted ectolophic ridge system.

The second lower molar, as M_3, tends to be rotated, corresponding to the condition of M^2. The M_3, like M^3, is markedly reduced.

Molar wear

Because of the low relief of the lower molars of *Lemur macaco*, the cutting edges on them, resulting from tooth-on-tooth and tooth-food-tooth contact, are rather extensive and elaborate (Figs. 5*A* and *B*). The enamel cutting edges in the trigonid basins form large, modified figure 8s with extensive exposures of protoconid and metaconid dentine within the loops of these 8s. Occlusal wear around the talonid basins results in extensive exposures of dentine. The cutting edges generated around these exposures of denture run along the cristid obliqua and postcristid. On M_1 especially, the enamel cutting edges of both the trigonid and talonid regions form one continuous system.

In the upper molars extensive dentine exposures are evident along the crests of the ectolophs (the continuous crests of the paracone and metacone), especially on M^1, and along the crescentic rims of the protocones. Striated wear facets on the upper molars are evident, especially along the buccal rim of the crests of the protocones and along the centrocristas of M^1 and M^2 (see Fig. 5*A*). The striations on all these facets run in a lingual and moderately mesial direction. Striations on wear facets in the lower molars also run in a mesiolingual direction. These striated facets run along the lingual edge of the cristid obliqua, as well as along the buccal face of the crown below the protoconid and cristid obliqua.

Dentine exposures in the lower molars of *Varecia* (Fig. 5*D*) usually appear to be less extensive than those of *Lemur*. They are restricted to the narrow crests defining the anterior foveae and talonid basins and are best developed on the tips of the protoconids and hypoconids of M_1, M_2 and M_3. As in the lower molars, dentine exposures in the upper molars of *Varecia variegatus* (Fig.5*C*) are usually poorly developed, compared to those of *Lemur* spp., being essentially restricted in a comparable stage of wear to the very narrow ridges along the ectolophs and protocones.[8] The wear gradient in *Lemur macaco* appears to be drastically steeper than in *Varecia variegatus*. Faint striations, running essentially antero-posteriorly, were detected on M^1 and M^2 in the polished 'valleys' created at the junctures of the steep-sided paracones and preproto-cristas. The upper molar occlusal surfaces are, however, pre-ponderantly nondescript or mottled in appearance, with mottling being very evident within the trigon basins and along the ectoloph crests. The crest of the protocone of M^2 demonstrates a more pitted appearance.

Striations were detected on the buccal portions of the postcristids of M_1 and M_2, running in a markedly mesial and slightly lingual direction. As with the upper molars, however, the lower molar basins and crests are basically mottled or pitted, the pitting being most evident on the tops of the hypoconids of M_2 and M_3.

Occlusion

In studying the dentitions and the two types of wear on the molars of the two genera, two occlusal patterns became apparent. One pattern is associated with an inferred, primarily crushing-piercing type of chewing, as in *Varecia*, and the other with one in which cutting is of primary importance, as in *Lemur*. In the first type the cusps become abraded by the pounding-crushing contacts, whereas in the second type, in forms with extensive edge-to-edge shear, distinct and (depending on the diet) sometimes polished, sometimes grooved shear facets are especially characteristic of the molars. There

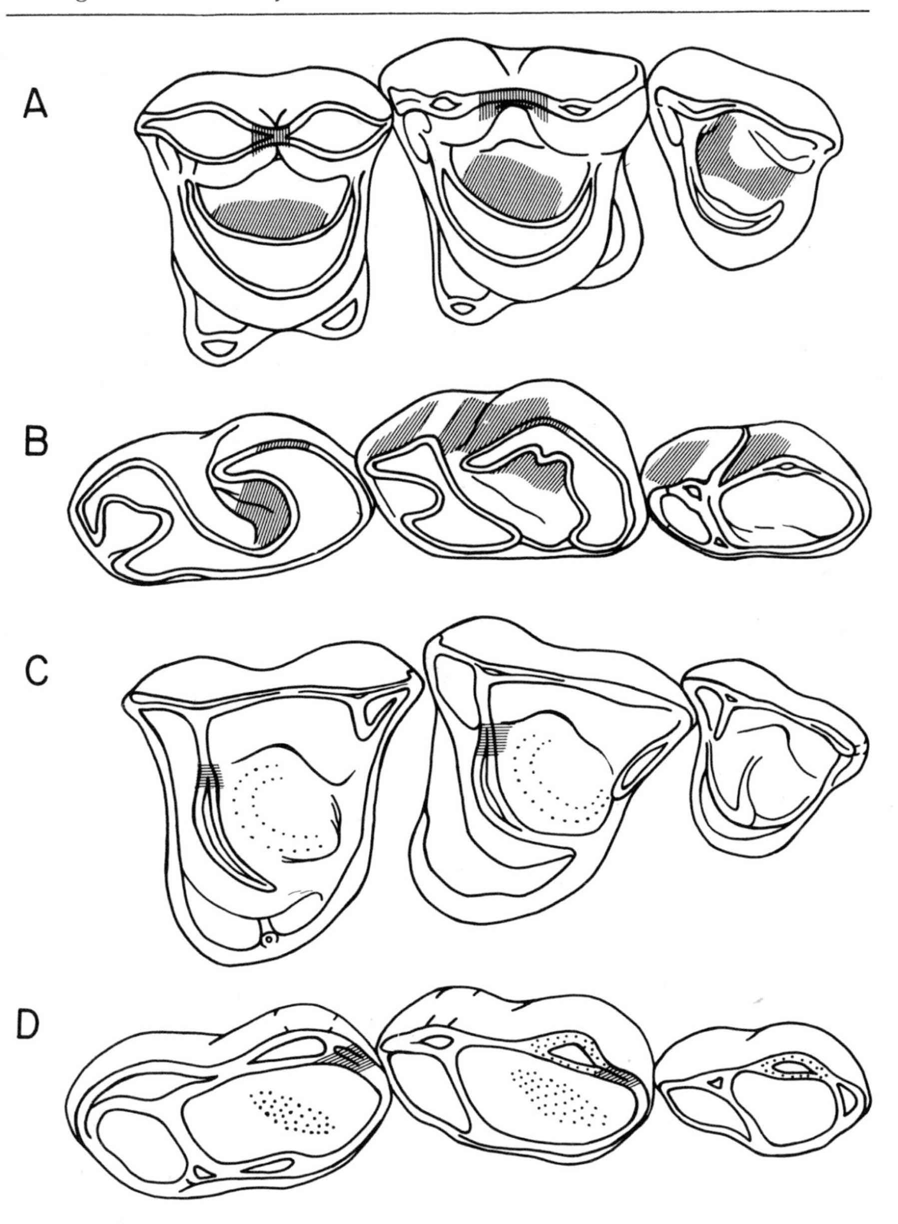

Figure 5. Contrast of molar wear in *Lemur macaco* and *Varecia variegatus*. *A* and *B* represent occlusal views of the upper and lower molars, respectively, of *Lemur macaco*, while *C* and *D* represent the occlusal views of the upper and lower molars, respectively, of *Varecia variegatus*.

The closely spaced parallel lines in *A*, *B*, *C*, and *D* represent patterns of striations in regions where they were observed. In *C* and *D*, dots in the trigon and talonid basins schematically indicate a mottled surface. Dots on the hypoconids of M_2 and M_3 indicate a more pitted surface.

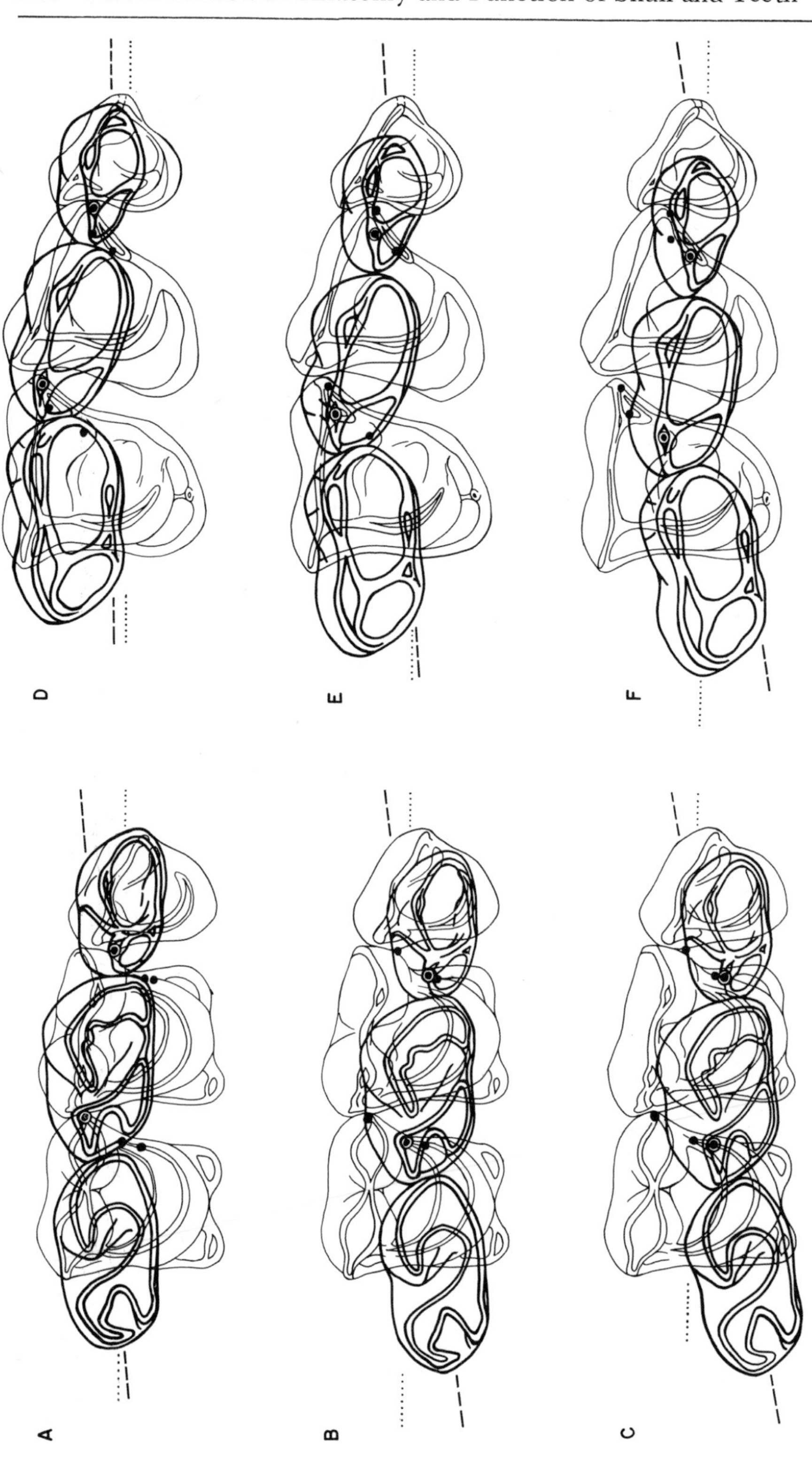
D
E
F
A
B
C

are some indications, according to the hitherto unappreciated discoveries of Every,[9,10] that cutting edges destroyed by the crushing-puncturing action are restored, and sharpened during non-chewing masticatory activities, though this is difficult to prove. The pioneering studies of Butler and subsequent works by Mills[11-16] have shown that grooves and striations, whenever present on contact facets, are particularly excellent indicators of the direction of jaw movements.

The patterns of molar occlusion for both *Lemur macaco* and *Varecia variegatus* are illustrated in occlusal view (Fig. 6). As mentioned previously, the protoconids of M_2 and M_3 are being used for reference. The figures indicate the paths taken by the lower molars as they move across the upper molars during the power stroke. In addition, the illustrations demonstrate the fact that the molar morphology of the *Lemur macaco* and *Varecia variegatus* differs considerably and that food is processed in distinctive ways.

The major 'theme' of the *Lemur* species group, as shown by the occlusion of *Lemur macaco*, appears to be primarily transverse cutting. The process is continuous but can be conveniently divided into two phases.

The first phase of the power stroke features the mesiolingual and moderately dorsal passage of the paracristid, cristid obliqua and postcristid of each lower molar across the appropriate ectoloph on the upper molars and premolars. Immediately following this, the diagonally sloping buccal faces of the lower molar crowns occlude with the ectolophs. This process is also mesiolingual and moderately dorsal. The entire first phase, when viewed in the coronal plane,

Figure 6. Occlusal diagrams of the right molars in *Lemur macaco* (*A*,*B*,*C*) and *Varecia variegatus* (*D*,*E*,*F*). The encircled dots indicate the tips of the protoconids of M_2 and M_3, while the three solid dots on M_1 and M_2 indicate the successive (and arbitrarily selected) positions through which the M_2 and M_3 protoconids pass during mastication.

The dotted lines represent the longitudinal axes of the upper tooth rows, which are defined here to connect a point at the central base of the upper canine with a point in the centre of the glenoid fossa. The axes of the upper tooth rows of both genera are aligned here in continuous and parallel fashion in *A* and *D*, *B* and *E*, *C* and *F*. The dashed lines represent the longitudinal axes of the lower tooth rows, which are defined here to connect a point at the central base of P_2 with a point medially situated on the transverse axis of the madibular condyle.

The three solid dots on M^1 and M^2 on *Lemur macaco* (*A*, *B* and *C*) indicate the positions of the M_2 and M_3 protoconids during the occlusal (*A*), centric occlusal (*B*), and terminal phases (*C*) of the masticatory stroke. The three solid dots on M^1 and M^2 in *Varecia variegatus* (*D*, *E* and *F*) represent the positions of the M_2 and M_3 protoconids as they move from the occlusal (*D*) to the centric occlusal phase (*F*) of the masticatory stroke.

A, *B* and *C* demonstrate that transverse motion is relatively greatly emphasised in *Lemur macaco*, while *D*, *E* and *F* indicate a relatively greater anterior component in the mandibular movement in *Varecia variegatus*.

describes a relatively low trajectory, being far more ectental (transverse) than orthal (vertical). This permits, as can be seen, the very extensive apposition and contact of cutting edges.

The conclusion of the first phase merges into the second phase, that of centric occlusion, followed by a final, short, mesio-ventro-lingual movement of the talonid basins across the protocones. Centric occlusion possibly results in the pulverisation of the previously sliced vegetable material as the plane-like talonid basins firmly engage with the shelf-like protocones. The terminal phase results in one last shearing movement: the cutting edges along the buccal rims of the talonid basins approach those along the protocone crests and probably rasp the food pressed in between. The striation patterns on the wear surfaces of the molars appear to reflect the primarily transverse cutting action of these teeth.

During the power stroke, the lower molars are displaced lingually to a relatively great extent.

The anterior fovea of M_2 seems to function, quite interestingly, as a sort of funnel or scoop. As the anterior fovea of M_2 lingually traverses the postprotocrista of M^1, it can conceivably catch food and channel it buccally toward the protoconid of M_2. This food would then be sheared as the cutting edges of the M_2 protoconid moved across the M^1 postprotocrista. The functional significance of the morphological differences seen in the M_1 and M_2 anterior foveae becomes readily apparent when one realises that the trigonid of the M_2 occludes with the postprotocrista of a preceding upper molar, whereas that of M_1 does not.

The function of the lingual cingulum and the pericones, so well developed in *Lemur* and many other primates, is difficult to explain. No pericones come into tooth-on-tooth contact with the lower molars until the crowns of the upper molars are well worn. Only then does the disto-lingual pericone of M^2 actually occlude with the M_2 postcristid, thus serving as an accessory cutting edge.

The basic functions of the cheek tooth row in *Varecia*, as displayed in *Varecia variegatus* (Fig. 6*D, E, F*), are significantly different from *Lemur* spp. The power stroke is comprised of two rather distinct phases. In the first phase, the lower molars are displaced upwards and mesially. The protoconids, hypoconids, and associated crests are thus 'slid' upward and mesially past the lingual rims of the upper molar ectolophs. In the second phase, the lower molars continue on their upward course, but are deflected lingually so that their movement in the occlusal plane is now mesio-lingual. With this movement, the molars ultimately achieve centric occlusion, as follows: (i) the protoconids of M_2 and M_3 engage with M^1 and M^2 lingual to the metacones; (ii) the protocones of M^1, M^2 and M^3 and the hypoconids of M_1, M_2 and M_3 are thrust respectively into the talonid basins of M_1, M_2 and M_3 and the trigon basins of M^1, M^2 and M^3.

The first phase of the masticatory stroke emphasises the sliding of sharp crests past each other. This situation seems well suited for rolling and cracking solid, particulate plant material. The second phase features the firm engagement of cusps with basins in mortar and pestle fashion. This occlusal pattern, it appears, excellently equips the molars to crush and mash stems or hard, discrete food objects once they have been cracked and splintered during the first phase. The wear of the teeth seems to corroborate this interpretation.

The movement of the lower jaw of *Varecia variegatus*, when viewed in the sagittal plane, emphasises both vertical and particularly mesial components, in contrast to the homologous jaw movement of *Lemur macaco*. The lower jaw of *Varecia variegatus* is thus displaced mesially to a relatively greater degree than it is displaced transversely, whereas the reverse is true for *Lemur macaco*.

The anterior foveae of both M_2 and M_3 appear to be quite functional. They are extensive and are posteriorly juxtaposed to the preceding talonid basins. During mastication, food is probably mashed within the talonid basins. Some of the processed food may be forced out of these basins, being carried outward primarily in a disto-lingual direction through the characteristic spout-like 'outlets' in the disto-lingual rim of the talonid basin described previously. A similar spout-like outlet characterises the lower molars of the Palaeocene picrodontids.[17] It would appear that the anterior foveae may be adapted to catch and retain this food-overflow. The retaining action of the anterior foveae and the extensive mesial travel of the cheek tooth row would insure that a maximum mass of food is available to the protocone-talonid and trigonid basin mashing apparatus. The anterior foveae of M_2 and M_3 can then be interpreted as functional extensions of the preceding talonid basins.

The extensive lingual cingula of M^1 and M^2 may partially function as broad retention areas for the masticated food carried mesio-lingually, although this is uncertain. We do not believe that extensive contralateral contact occurs or is responsible for the development of the lingual shelves in *Lemur* or *Varecia*. It is not necessary here to discuss the evidence for or against a 'lingual phase', *sensu* Mills.[15,16] It may be noted, however, that one of us (Szalay), after having re-examined the Palaeocene evidence, concluded that contralateral contact of molars is not responsible for several facets on the occluding hypoconid-protocone and the hypocone shelf-paracristid in Palaeocene primates. Every's studies should eventually reflect on this subject.

From the molar evidence, we have established a model for mastication in *Lemur macaco*, which emphasises transverse cutting. The premolars also seem to adhere to this model (see Figs. 1 and 2). The lingual elements of both P^3 and P^4 are relatively well developed, while the metaconid of P^4 is prominent. These features result in bucco-lingually elaborated cutting edges on both the mesial and distal slopes of these teeth. Running in parallel down the buccal and lingual margins of both the mesial and distal slopes of P_4 are two cutting edges. Because P_4 tightly interlocks with P^3 and P^4 during occlusion, it is naturally guided lingually between these teeth. The cutting edges on the mesial and distal slopes of P_4 thus make firm contact with the lingually extended cutting edges of P^3 and P^4. This situation seems to be conducive to efficient cutting and slicing, and appears to be consistent with the occlusal pattern of the molars outlined above.

The situation in *Varecia*, exemplified by the extant *V. variegatus*, is quite different (see Figs. 3 and 4). The lingual elements of P^3 and P^4 are weakly developed. The lingual portion of P^4 is distinctly more cuspate, while P^3 appears more caniniform. The lower fourth premolar has a low metaconid, with fossae replacing cutting edges on both the mesial and distal slopes. The lower fourth premolar is also rotated mesio-lingually. Occlusion, therefore, between it and P^3 and P^4 is altered. Because of its rotation, P_4 slides mesio-lingually between P^3 and P^4. This movement is facilitated by the fact that two relatively wide diastemata exist between the upper canine and P^2 and between P_2 and P_3, which permit the lower premolars to slide mesio-lingually between the upper premolars without interlocking. The morphology and occlusion of the premolars of *Varecia variegatus* thus minimise the possibility of efficient transverse cutting and slicing, and suggest instead a functional emphasis on the grasping, rolling and cracking of solid, particulate vegetable material. This reflects the functional theme seen in the occlusion of the molars.

Mandibular and cranial morphology

The lower jaw of *Varecia variegatus* displays marked thickening immediately beneath and medial to P_4-M_2. This thickening, when viewed ventrally, appears to align with the centrically occluded protocones and talonid basins. This thickening may possibly be a means by which the lower jaw is buttressed against stresses as they are generated ventrally through the medial half of the body of the mandible by the crushing function of the molars. *Lemur macaco* shows no such bucco-lingual thickening in the mandible.

The distinctive patterns of mastication in *Lemur* and *Varecia* seem also to be reflected in the osteology of the crania. In *Lemur macaco*, there is a tendency for the zygomatic arches (when viewed in a coronal plane) to be oriented ventro-medially, in some cases to a great extent (Fig. 7*A*). This may suggest a relatively more oblique line of action for the masseter in the coronal plane. At the same time, the pterygoid plates tend to be relatively more medially situated with respect to the upper tooth rows, with the lateral pterygoid plates projecting ventro-laterally from the basicranium. This might indicate that the medial pterygoid muscles in *L. macaco*, like the masseters, possess a relatively more oblique line of action when viewed in the coronal plane. Because the lateral pterygoid plates tend to be situated relatively more directly medial to the mandibular condyles, the medial pterygoid muscles, when viewed ventrally, would probably tend to have a predominatly medio-lateral orientation. The masseters and the medial pterygoids, then, would seem to be disposed so as to facilitate primarily transverse mandibular movement.

The cranium of *Varecia variegatus*, when similarly viewed, contrasts with that of *Lemur macaco*. In *Varecia variegatus* the zygomatic arches are not oriented ventro-medially, but are essentially vertical (Fig. 7*B*). A more vertical line of action is suggested for the masseter in *Varecia*. Similarly, the location of the pterygoid plates is relatively more lateral with respect to the upper tooth rows, while the pterygoid plates are also more vertically disposed. This suggests a more vertical line of action for the medial pterygoid muscles. In contrast to *Lemur macaco*, the medial pterygoid muscies may possibly facilitate more anterior mandibular movement, as the lateral pterygoid plates lie more anterior to the mandibular condyles.

In *Varecia variegatus*, then, the masseters and medial pterygoids appear to be better suited for producing relatively more orthal (vertical) movement in the lower jaw. The medial pterygoids also appear to have a greater ability to protract the mandible than in *Lemur macaco*. The evidence from the cranium apparently agrees with that of the molars in suggesting a more vertically and more anteriorly directed lower jaw.

Discussion and conclusions

Studies on tooth wear of extant and extinct taxa certainly contribute toward the understanding of functional significance of tooth morphology in a given species. On the other hand, meticulous studies on the natural history of living species, assessment of the dietary regime, and an understanding of the path of phylogeny and evolutionary dental morphology are of no less significance. It is only with a judicious combination of all these three approaches, and not by wear studies alone, that one can hope to understand the feeding mechanism (the teeth in particular) of any living or fossil species of mammal.

One of the greatest weaknesses of our pilot study is the still insufficient information about the dietary habits of the species in question in their natural habitats. It was the consensus at the conference among several students of prosimian ecology that most

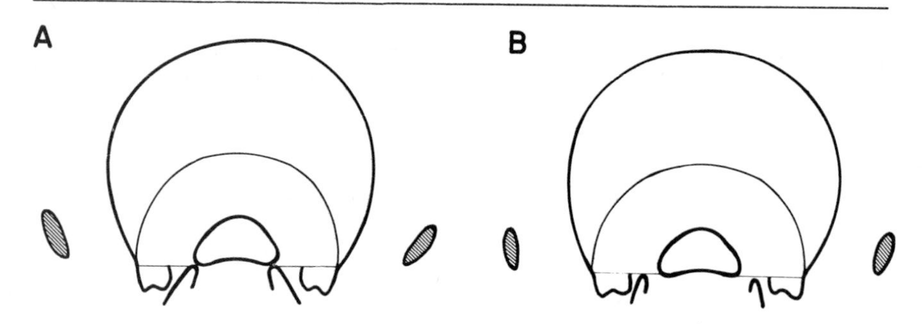

Figure 7. Comparative, schematic views of the posterior ends of the crania in *Lemur macaco* (*A*) and *Varecia variegatus* (*B*) to show the relative orientations of the zygomatic arches and pterygoid plates. In *L. macaco* the zygoma tends to be oriented ventro-medially, whereas in *V. variegatus* it is essentially dorso-ventral in orientation. In the former the pterygoid plates are distinctly more medial to the tooth row and closer to the articular condyles than in the latter. The cranial evidence indicates that in *L. macaco* both the masseters and pterygoid muscles were oriented in a relatively more diagonal fashion. In contrast, the masseter in *V. variegatus* lacks this relatively marked lateral component, and the pterygoid muscles appear disposed to pull more orthally and propalinally, rather than transversely.

of our knowledge on diet is derived from studies conducted during the dry seasons. This obviously has the drawback of not providing information on the diversity of the diet during the whole year. It should be remembered, however, that dry season studies might be indicative of the restrictive factors in the diet of a species, as food supply during the wet seasons is probably more abundant and diverse.

Petter and Jolly indicated that the natural diet of *Lemur macaco* consists chiefly of leaves and fruits.[18,19] Jolly lists *Harungana madagascariensis, Tristemma virusanum*, and *Trema orientalis* as species of 'lush green plants' which are particularly prominent in the diet of *Lemur macaco collaris.*[19]

The nature and consistency of most leaves and fruits are such as to warrant the dental morphology and masticatory jaw movements outlined above for *Lemur macaco*. As with those of *Lemur macaco, Lemur* spp. dentitions feature upper and lower tooth rows which: (*a*) tightly interlock during occlusion, (*b*) relatively evenly distribute cutting and shearing function throughout their premolar and molar regions, and (*c*) move transversely (ectentally) rather than mesiodistally (propalinally) with respect to one another during mastication. It would seem, then, that such dentitions have been selected to best process a diet of pliable, fibrous, vegetable material, such as leaves, given the hitherto unknown ancestral morphology.

The diet in the wild of *Varecia variegatus* is not well known. As J.-J. Petter remarked at this Research Seminar, study of this animal is extremely difficult as it is usually found at the very top of the canopy. R. Sussman (personal communication) has informed us of one report in which *Varecia variegatus* was observed in its habitat to have eaten large quantities of hard, spherical fruits or nuts, a specimen of which was collected by him. If this report is accurate, and does indeed reflect the food preference of *Varecia variegatus*, then some significant correlations can be made with the dental, mandibular and cranial morphology. It was noted above that mastication in *Varecia variegatus* seemed to stress relatively orthal puncturing, crushing and mashing. Such chewing would appear to be well adapted for processing stems as well as hard, spherical fruits, seed pods, or nuts. The relatively pointed and blade-like premolars, which occlude in the manner indicated above, combined with the extensive diastemata in the premolar region, all suggest that the premolars function primarily to grab and puncture rather than to cut or shear.

Similarly, some of the peculiar features of the molar region may be adaptations to process stems or solid, compact, tough-shelled nuts or fruits or seed-pods, presumably eaten by *Varecia variegatus.* Stems or fruits or nuts of the kind mentioned above would not require processing by an extensive portion of the upper and lower tooth rows, and could instead be efficiently cracked or mashed by just two adjacent pairs of upper and lower molars. It may be for this reason that in *Varecia variegatus*, M^3 and M_3 are reduced while M^2 and M_2 are often rotated. These rotations appear to have shifted the M^2 protocone and M_2 talonid basin mesially, thus functionally bringing the M^2 and M_2 mashing apparatus closer to that of M^1 and M_1. This phenomenon in the upper molars has resulted in the interuption of the primitively continuous M^1-M^2-M^3 ectolophs. For a leaf-eater, such a situation could be maladaptive. In *Varecia variegatus*, however, the sharply crested ectolophs, though oriented in a quasi-parallel rather than continuous fashion, can efficiently function to hold, squeeze, and splinter as the lower jaw is moved upward and anteriorly.

Summary

We have attempted to demonstrate the importance of evolutionary dental morphology and function in understanding the feeding mechanism in living and fossil species. We have indicated that for extant groups such studies facilitate interpretations of phylogenetic relationships, and that at the same time they shed light on the possible selective forces operating on the dentitions of closely related fossil taxa.

Because of the nature of the mammalian fossil record, students of mammalian palaeontology are very often constrained to resort to dental evidence in their efforts to elucidate the ecological role of extinct species. The study of dental function in extant groups (drawing as it does upon evidence which is not available in fossils) enables us to sharpen the interpretive tools that are needed to explain the (very often dental) fossil remains of extinct species. Two current taxonomically problematical groups of lemurines, *Lemur* and *Varecia*, were compared in this study with respect to dental morphology and dental function. We proceeded to demonstrate that the 'themes' of molar occlusion in each of these two groups differ dramatically, at a level expressed by generic separation; often mastication in *Lemur* emphasises transverse slicing and cutting, and that in *Varecia* features relatively orthal cracking and mashing. With this comparison, we hope to have contributed towards a clarification of the taxonomic status of these two groups, and towards some understanding of the forces of selection that produced them.

Acknowledgments

We are grateful to Dr Pierre Vérin of the Musée d'Art et Archéologie of Tananarive for his aid to Szalay while in Madagascar. The illustrations were skilfully prepared by Anita J. Cleary. The research was supported by NSF Grant GB 20085, and a grant for African studies by the Wenner-Gren Foundation for Anthropological Research, both to Szalay.

NOTES

1 Charles-Dominique, P. (1966), 'Analyse de contenus stomacaux d'*Arctocebus calabarensis, Galago alleni, Galago elegantulus, Galago demidoffi*', *Biol. Gabon.* 2, 347-53.

2 Charles-Dominique, P. (1971), 'Eco-éthologie des prosimiens du Gabon', *Biol. Gabon.* 7, 124-228.

3 Hladik, A. and Hladik, C.M. (1969), 'Rapports trophiques entre végétation et primates dans la forêt de Barro Colorado (Panama)', *Terre et Vie* 23, 25-117.

4 Hladik, C.M., Charles-Dominique, P., Valdebouze, P., Delort-Laval, J. and Flanzy, J. (1971), 'La coecotrophie chez un Primate phyllophage du genre *Lepilemur* et les corrélations avec les particularités de son appareil digestif', *C.R. Acad. Sci. Paris* 272, 3191-4.

5 Petter, J.-J. and Hladik, C.M. (1970), 'Observations sur le domaine vital, et la densité de population de *Loris tardigradus* dans les forêts de Ceylan', *Mammalia* 35, 394-409.

6 Petter, J.-J. and Peyriéras, A. (1970), 'Nouvelle contribution à l'etude d'un lémurien malagache, le aye-aye (*Daubentonia madagascariensis* E. Geoffroy)', *Mammalia* 35, 167-93.

7 Sussman, R., this volume.

8 Due to lack of data on the individual age of museum specimens and lack of information on the time of eruption of the molars, the statement 'comparable stage of wear' is a tenuous one indeed. Quite arbitrarily we assume that the third

molars erupted at comparable ages in these closely related taxa, and use the similarity in their wear as 'comparable stage' for the specimens compared.

9 Every, R.G. (1960), 'The significance of extreme mandibular movements', *Lancet* 2, 37-9.

10 Every, R.G. (1970), 'Sharpness of teeth in man and other primates', *Postilla* 143, 1-30.

11 Butler, P.M. (1937), 'Studies of the mammalian dentition: 1. The teeth of *Centetes ecaudatus* and its allies', *Proc. Zool. Soc. Lond.* 107, 103-32.

12 Butler, P.M. (1952), 'The milk-molars of Perissodactyla, with remarks on molar occlusion', *Proc. Zool. Soc. Lond.* 121, 777-817.

13 Butler, P.M. (1952), 'Molarization of the premolars in the Perissodactyla', *Proc. Zool. Soc. Lond.* 121, 819-43.

14 Butler, P.M. (1961), 'Relationships between upper and lower molar patterns', International Colloquium on the Evolution of Lower and Non-specialised Mammals, Brussels. *K. vlaamse Acad. Wetens. Belgie*, 117-126.

15 Mills, J.R.E. (1954), 'Ideal dental occlusion in the Primates', *Dent. Pract.* 6, 47-61.

16 Mills, J.R.E. (1966), 'The functional occlusion of the teeth of Insectivora', *J. Linn. Soc. Zool.* 47, 1-25.

17 Szalay, F.S. (1968), 'A family of early primates, the Picrodontidae', *Amer. Mus. Novitates* 2329, 1-55.

18 Petter, J.-J. (1962), 'Recherches sur l'écologie et l'éthologie des lémuriens malagaches', *Mém. Mus. nat. Hist. nat.*, sér. A, 27, 1-146.

19 Jolly, A. (1966), *Lemur Behavior*, Chicago.

I. TATTERSALL

Facial structure and mandibular mechanics in Archaeolemur

Introduction

Archaeolemurinae, the indriid subfamily containing the species *Archaeolemur majori* (Fig. 1), *A. edwardsi* (Fig. 2) and *Hadropithecus stenognathus*, has long been considered, particularly among English scholars, to be characterised by a variety of features linking it to the higher primates. This point of view is especially evident in the work of Major and Standing,[1,2] although, at about the same time as these two authors were writing, Grandidier expressed the opinion that 'Il est d'abord hors de doute qu'il faut les ranger parmi les lémuriens'.[3]

A recent reappraisal of the cranial anatomy of the archaeolemurines has fully confirmed Grandidier's views.[4] In all respects except, to a limited extent, their masticatory apparatus, the archaeolemurine lemurs remain extremely close to the living indriids; they were not advanced, for instance, over the indriine level in their neural or peripheral sensory organisation. Walker has recently shown that *Archaeolemur* possesses in its postcranial skeleton a variety of adaptations characteristic of the most terrestrial of the cercopithecoid monkeys,[5] but nonetheless from a phylogenetic point of view the postcranial anatomy of these animals bears an unmistakably lemuroid stamp.[6] Those features which led Major to the opinion that *Archaeolemur* might best be classified as a monkey are, in the words of Grandidier, 'les caractères acquis par une similitude de vie'.

The characters used by Major, and later by Standing, in suggesting that the archaeolemurines were more 'advanced' than the indriines reside primarily in the dentition, and, to a lesser extent, in the other components of their masticatory apparatus. This fact prompted the present investigation of the archaeolemurine masticatory system, and led to the generation of a model of masticatory function in these animals.

The mandible as a bent lever

Most previously proposed models of mandibular function have depended on the jaw's operation by means of independently-acting bent lever systems, usually comprising the temporalis and the masseter-internal pterygoid complexes.[7] Such models require that these levers produce rotation around the temporomandibular joint, which therefore acts as the fulcrum of the system.

The functioning of the temporomandibular joint as the fulcrum in a lever system is open to question on several counts. First, there is the inherent improbability of a system in which, for instance, depending on the position of the bite-point, only between about 30%-50% of the available power exerted by the temporalis and the masseter-internal pterygoid complexes is usefully expended at the dentition in *Archaeolemur edwardsi* (Fig. 3*a*). More important, however, is the fact that the remainder of the muscles' available effort goes to produce a reaction force at the jaw joint. Examination of the morphology, internal and external, of the condylar neck and

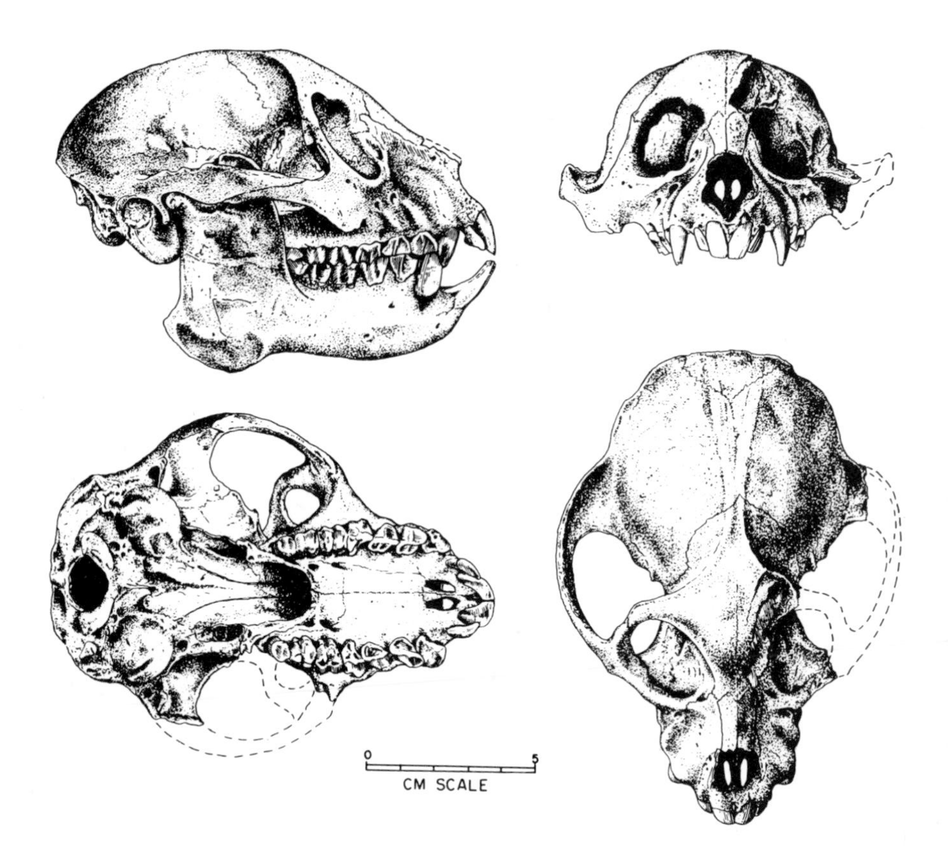

Figure 1. Cranium and mandible of *Archaeolemur majori* (BM M 7374).[21]

the glenoid fossa, of the low articular eminence and of the gracile postglenoid process in all the archaeolemurines, suggests very strongly that these structures could not have withstood the forces which would have been imposed upon them in a lever system. In addition, the functional surfaces of the temporomandibular joint in modern forms are lined with fibrous connective tissue ('fibro-cartilage'), rather than with the hyaline cartilage characteristic of the load-bearing joints of the body, such as the knee. Fibrocartilage, while providing an excellent sliding surface, is unsuited to withstand the large compressive forces which would inevitably be present if the jaw worked as a lever around the temporomandibular joint.

A further argument against a simple lever operation of the jaw is provided by the fact that in such an arrangement the resultant force at the dentition would be oriented tangential to the arc of adduction

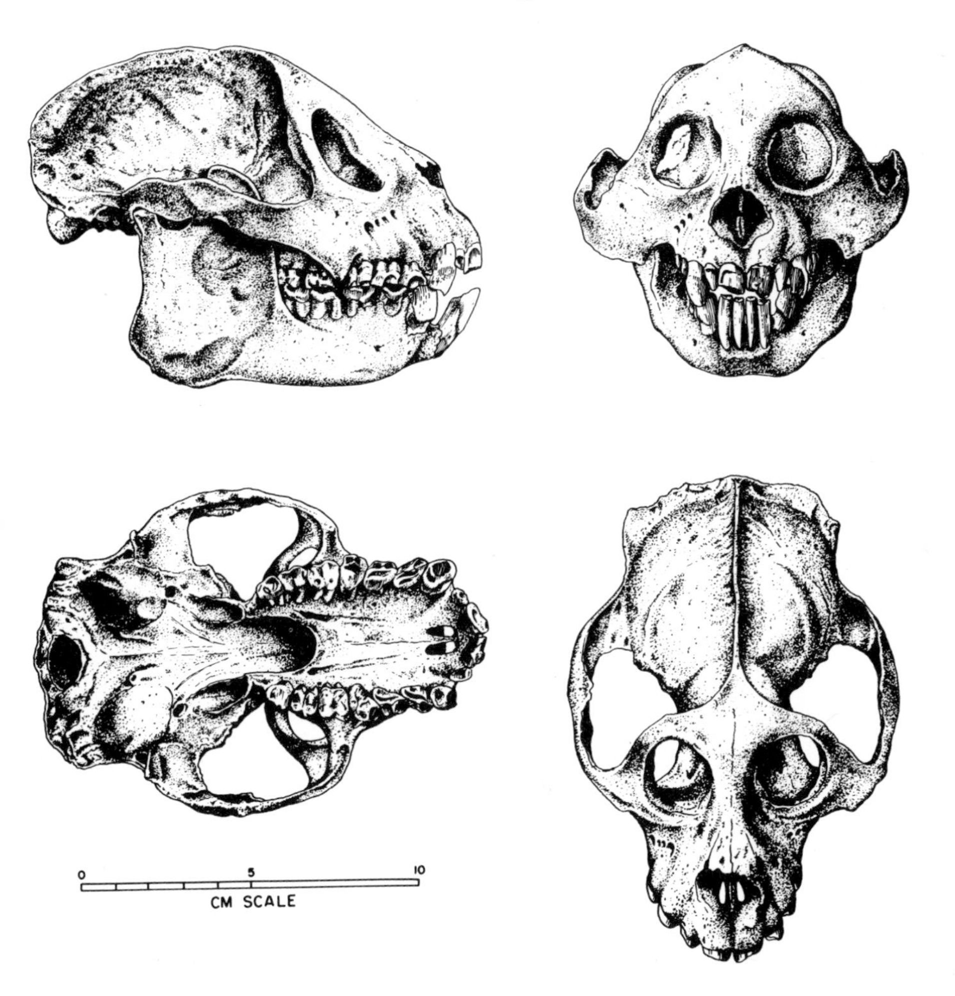

Figure 2. Cranium and mandible of *Archaeolemur edwardsi* (BM M 9909-9910).[21]

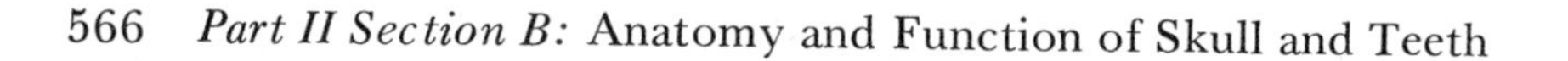

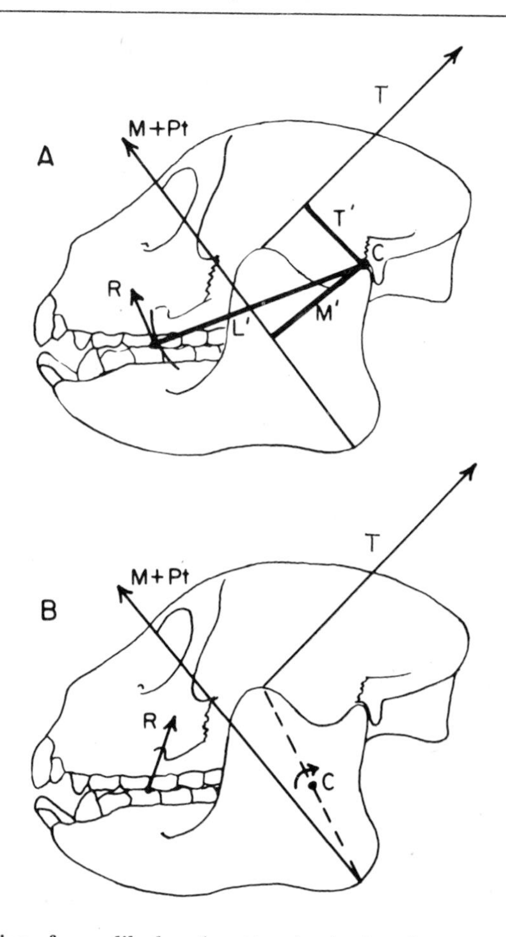

Figure 3. Mechanics of mandibular elevation in *Archaeolemur edwardsi*. (*A*) viewed as a bent lever system operating around the temporomandibular joint; (*B*) as a couple system producing torque around the mandibular foramen.

Abbreviations: C, point of rotation; M+Pt, vector of masseter-internal pterygoid complex; M', lever arm of masseter-internal pterygoid; T, mean vector of temporalis complex; T', lever arm of temporalis complex; R, resultant force at dentition; L', load arm.

of the jaw, i.e. orthogonal to the axis connecting the fulcrum and bite-point. (Some authors have indicated their belief that the occlusal force resulting from the rotation of the mandible about the jaw joint would be oriented vertically with respect to the tooth-row; this is emphatically not the case except where, as among Carnivora, the fulcrum and the bite-point lie in the same plane).[15] The architecture of the face in the archaeolemurines (and, indeed, in mammals in general) is totally inappropriate to the resolution of such anteriorly-directed resultant forces. A simple parallelogram of forces is sufficient to show that invariably a component vector will be directed out of the facial structure, placing the posterior part of the palate under considerable tension, and producing a bending movement in the face. Bone, while extremely strong in compression, is weaker in tension, and skeletal architecture is invariably designed to reduce, or even eliminate, tensile forces.[8]

All this is not to say that no forces ever occur at the temporomandibular joint; small locating forces necessarily occur, since the joint must, after all, function as a normal diarthrosis; but their magnitude is insignificant compared with those which would be present were the jaw joint a fulcrum. The fact that human patients with bilateral resection of the condyles are able to chew efficiently makes it sufficiently evident that no significantly large forces occur at this joint.[9]

Davis has proposed an alternative model of jaw function, based on the hypothesis that the temporalis and the masseter-internal pterygoid (anterior adductors) act not as separate lever forces but as a couple.[10] As Davis notes, this provides the advantage of reducing to zero the forces acting at the fulcrum. But Davis concludes that to the extent that the power exerted by the temporalis (an extremely large muscle in the bear, the animal under study) exceeds that exerted by the masseter, the jaw works as a bent (modified class I) lever. The misapprehension in Davis' scheme, even as it applies to carnivores, lies in the fact that the large size of the temporalis does not necessarily mean that it always exerts a force greater than that generated by the masseter, even though it may be capable of doing so. Moreover, and more importantly, the production of greater power by one muscle of the couple may in fact be *required* if a reaction force at the temporomandibular joint is to be eliminated.

More recently Turnbull, in a very useful work, has adopted Davis' scheme and generalised it to include mammals other than carnivores.[11] This is, however, unfortunate for a variety of reasons. For instance, although in some of the carnivores the condyle lies about midway between the mean points of insertion of the temporalis and anterior adductor complexes, with the result that it does approximate to (although it never precisely coincides with) the centre of rotation of the system,[12] in more herbivorous animals such as the primates, in which the temporomandibular articulation is elevated considerably above the plane of the tooth-row, the jaw joint and the centre of rotation of the jaw are far from coincident. It should also be noted that the couple action does not necessarily produce the vertical resultant force at the dentition that Davis illustrates, or the anterior one shown by Turnbull; one of the major advantages of a couple system lies in that the orientation of this force is variable along the tooth-row.

Facial structure

It has already been suggested that an anteriorly-directed resultant force would be inappropriate to the facial structure of *Archaeolemur* and other mammals.[13] What kind of resultant force, then, is the facial skeleton of *Archaeolemur* designed to resolve? Herein lies one of the keys to the operation of the archaeolemurine masticatory system. In essence, the facial region is designed as a bilateral tripod or pyramid, its base on either side formed by the robust maxillae. The apex of this structure lies at the most anterior and superior point on the frontal bone. The three legs of the tripod are formed by the facial profile, by the very strongly developed post-orbital bar/anterior zygomatic root complex, and by a strong trabecular structure, clearly visible in radiographs, running from the apex of the pyramid to the area of the palatine hamulus. The whole structure is reinforced by the bony palate and the medial septa of the nasal cavity and frontal sinus.

A structure of this sort is ideally designed to resolve forces directed towards its apex. The anteriorly-directed occlusal force in the face produced by a mandible rotating around the temporomandibular joint, then, would make functional anatomical nonsense of the facial architecture of *Archaeolemur.* Clearly, though, if masticatory stresses in the face of *Archaeolemur* were oriented towards the apex of the facial pyramid, the occlusal force generated by the masticatory musculature must have been variable in orientation according to the position of the bite-point along the tooth-row. Just such a vector is provided by the model of mandibular mechanics in *Archaeolemur* discussed below.

Mandibular movements

Several authors, particularly Mills,[14] have used the facets formed by wear on the occlusal surfaces of the teeth in reconstructing mandibular movements during mastication. This procedure is somewhat dubious in those cases where the dentaries are united in a mobile syndesmosis,[15] but is generally reliable when the mandibular symphysis is fused in an immovable synostosis, as in *Archaeolemur.*

Determination of the direction of jaw movements is normally made through examination of the orientation of the minute striae which are commonly found on the dental wear facets. In the case of *Archaeolemur*, although incisal and premolar wear is usually marked, the molars were subject to relatively little attrition and abrasion, and the molar wear facets in any single individual are inadequate for the complete reconstruction of mandibular movements. However, a composite pattern can be reconstructed from examination of molar wear in a variety of individuals. This question, and others related to it, are discussed at length elsewhere.[4]

In *Archaeolemur* there were three primary masticatory functions: incisal (nipping); premolar (shearing); and molar (crushing/grinding). The movement of the lower incisors against those of the upper jaw was the result of a direct posterior movement, combined with elevation. The condyles moved forward bilaterally as the mouth opened, and biting took place by means of a strong bilateral retraction of the condyles as the jaw moved upwards.

The premolar series of *Archaeolemur* is highly specialised as a continuous shearing blade. In young adults, at least, the shearing facets of the premolars, buccal in the lower series, lingual in the upper, are inclined at an angle of about 45° or less to the sagittal plane, and invariably bear clearly-marked striae which run almost directly transversely. The mandibular movements involved in this type of shear were probably as follows: as the jaw opened, both condyles slid forwards. After this, the ipsilateral condyle moved posteriorly on its articular surface at some point during elevation of the mandible to the point of tooth-food-tooth contact, or of dental approximation. A transverse movement then occurred, during which the ipsilateral condyle rotated medially; simultaneously, the contralateral condyle translated posteriorly. The mandibular movements involved in molar occlusion appear to have been substantially similar to those used to produce premolar shear.

Mechanics of mastication

That the basic mammalian pattern of jaw musculature is designed to produce an appropriate resultant force while eliminating any reaction at the temporomandibular articulation has been demonstrated in the case of *Archaeolemur majori* by construction of a simple two-dimensional static model. It is fully recognised that such a model of the jaw in elevation is highly oversimplified; however, it does serve to illustrate the possibilities inherent in a mechanism of this type.

The model consists of a plywood cutout of the mandible of *A. majori* suspended against a lateral outline of the cranium. The mandible is held in static equilibrium by means of springs of known stretching constant which represent the vectors of the major muscle groups (anterior adductors; anterior temporalis; posterior temporalis) and the resultant occlusal force. The model forcibly demonstrates that static equilibrium (as occurs during biting of a resistant object)[16] can only be achieved when the resultant force is of the type described earlier. Only in the presence of such a resultant do the available forces at the bite-point sum to zero. This is not to say that the resultant occlusal force equals all those muscle forces working to produce it; not only is a small amount of energy (variable, depending

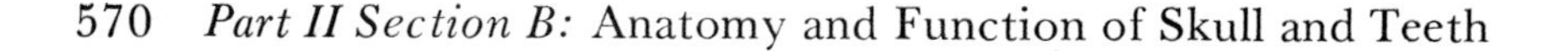

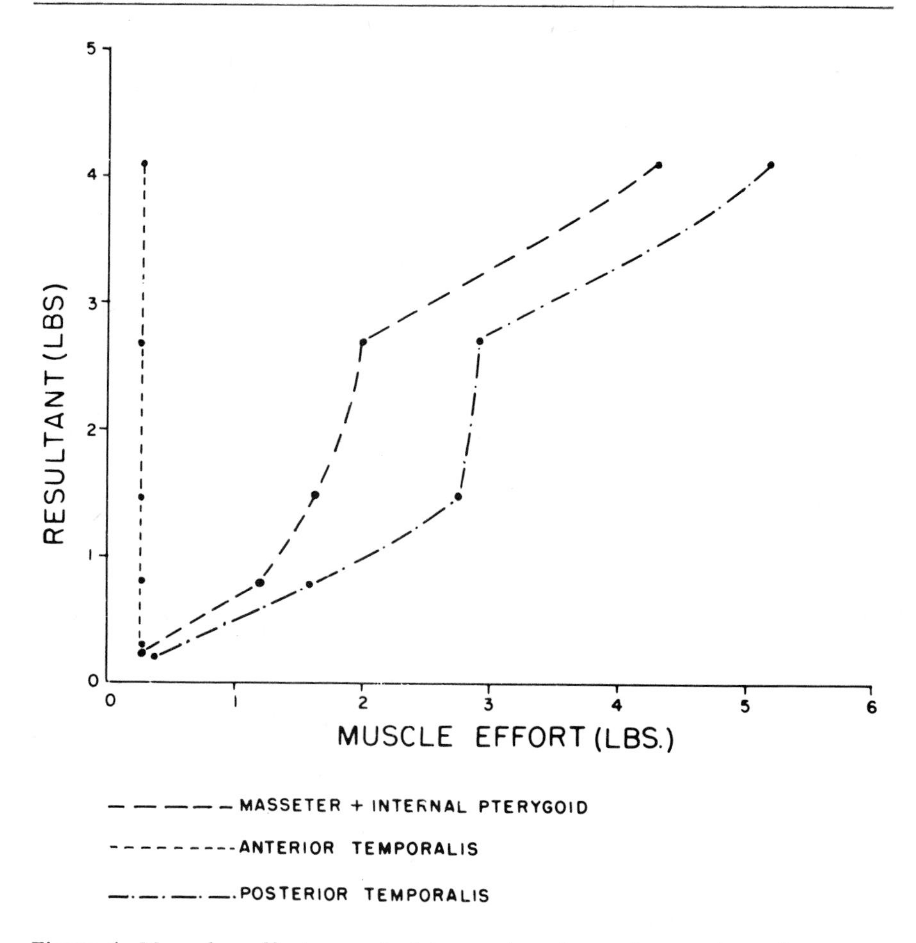

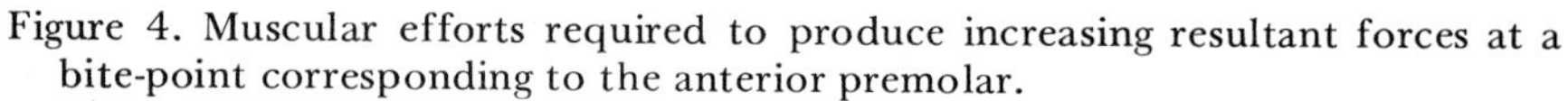

Figure 4. Muscular efforts required to produce increasing resultant forces at a bite-point corresponding to the anterior premolar.

on the bite-point) used in balancing the jaw, but, since muscles exert equal force at their origins and insertions, much effort is inevitably expended in producing forces on the neurocranium. This dome-like structure is well-adapted to resolve such forces; even those from the masseter are transferred from the zygomatic arch to the braincase by means of the temporalis fascia.[1 7]

Essentially, the mandible is held in equilibrium during biting by a couple action (Fig. 3) between the posterior temporalis and the anterior adductor muscles (particularly the superficial masseter, the dominant masseteric component, which originates very powerfully anteriorly via the anterior masseteric tendon). This posterior temporalis-anterior adductor couple produces an anti-clockwise torque around a point intermediate between the insertions of these two muscle groups; this coincides with the position of the mandibular

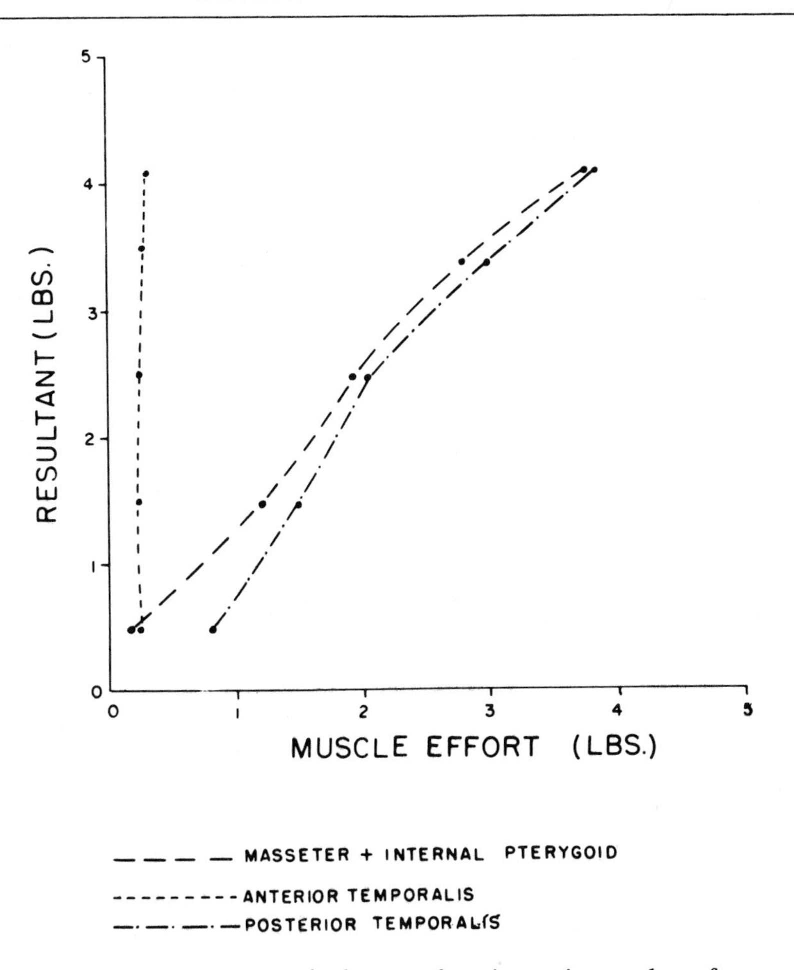

Figure 5. Muscular efforts required to produce increasing resultant forces at a bite-point corresponding to P4-M1.

lingula, the point of attachment of the sphenomandibular ligament. This ligament assists in the stabilisation of the jaw around this point during elevation. Since the condyle lies posterior to the lingula, the effect of the couple is to rotate the condyle ventrally. Such rotation is resisted primarily by the resultant occlusal force (i.e. the resistance at the teeth), and to a lesser extent by the antagonistic pull of the anterior temporalis.

Manipulation of the model described above can serve to provide an approximate idea as to the various muscular combinations which act to provide the appropriately-oriented masticatory force at different bite-points along the tooth row. Figs. 4 and 5 illustrate the muscle combinations which produce increasing resultant forces at two different bite-points; the anterior premolar and P4-MI, where wear indicates that much of the masticatory activity took place.

During incisal biting, at a 10° gape, the anterior temporalis contributes only a small, constant force to the system; the bulk of the effort is provided by the posterior temporalis working in the couple. Since the masseter produces what is essentially a balancing force at this bite-point, the effort of the posterior temporalis increases far faster than that of the masseter-internal pterygoid complex as bite-force is increased. At a wider gape, the anterior adductors play a greater role, although the posterior temporalis still predominates. The posterior temporalis must therefore produce a disproportionate amount of effort at this anterior bite-point, since it both balances the jaw and provides almost all the bite-force, while not in a particularly mechanical advantageous position to do the latter. This presumably explains the enormous size of the posterior temporalis and the diminution of the anterior moiety of this muscle in the extremely long-faced lemur *Megaladapis*, together with the fact that far more pressure may be exerted at the posterior than at the anterior teeth.

The retraction of the mandible during incisal biting results from additional pull by the posterior temporalis; the condyles translate posteriorly along a plane parallel to the major axis of action of the posterior temporalis, and therefore not resistant to their movement.

At a more posterior bite-point, corresponding to the middle of the premolar series, the anterior temporalis continues to play a minor role, while the anterior adductors are more strongly involved in the production of the resultant force, although their contribution is still less than that of the posterior temporalis. When biting at a 10° gape takes place at P^4_4-M^1_1, the anterior adductors and the posterior temporalis play more equal roles, although the effort expended by the anterior adductors only equals that of the posterior temporalis at high masticatory pressures. Under some conditions the contribution at this bite-point of the anterior temporalis may be rather greater than Fig. 5 indicates. During chewing at the most posterior teeth the couple muscles do little more than balance the jaw, i.e. their contribution, particularly that of the posterior temporalis, does not increase greatly as the bite force increases. Most of the occlusal force is provided by the anterior temporalis; indeed, at wide gapes, which bring the bite-point almost in line with the mean axis of action of this muscle, the force exerted by the anterior temporalis increases almost proportionately with the resultant force. Nonetheless, only a limited amount of masticatory activity took place at this point, as shown by the relatively low degree of wear on the teeth and by their small size; this was presumably due to the restricted amount of lateral motion possible.

Clinical evidence such as that advanced by Moss[18] substantiates the hypothesis of the mandible's rotation around the lingula. Although Moss' discussion is limited to the human temporomandibular joint, which possesses an articular eminence far more pronounced than that of *Archaeolemur*, the argument appears equally applicable in the case of the latter. During the opening of the jaw, the condyle rides forward on the articular eminence, while the angle of the mandible moves back. The functional surface of the temporal portion of the jaw joint is not comprised by the glenoid fossa, but by the more anteriorly positioned eminence, as can be seen from the distribution of fibrocartilage in the joint (Fig. 6). The glenoid fossa in man merely serves as a receptacle for the condyle when the jaw is at rest.

During motion of the jaw the mandibular foramen moves virtually not at all; this is necessarily so, since this foramen, adjacent to the lingula, is the point at which the inferior alveolar neurovascular bundle enters the mandible. Movement here during opening and closing of the jaw, as in a lever system, would tend to rupture the neurovascular triad, particularly the inferior alveolar nerve. It should also be pointed out that in the case of a mandible rotating around the temporomandibular joint, the amount of posterior movement of

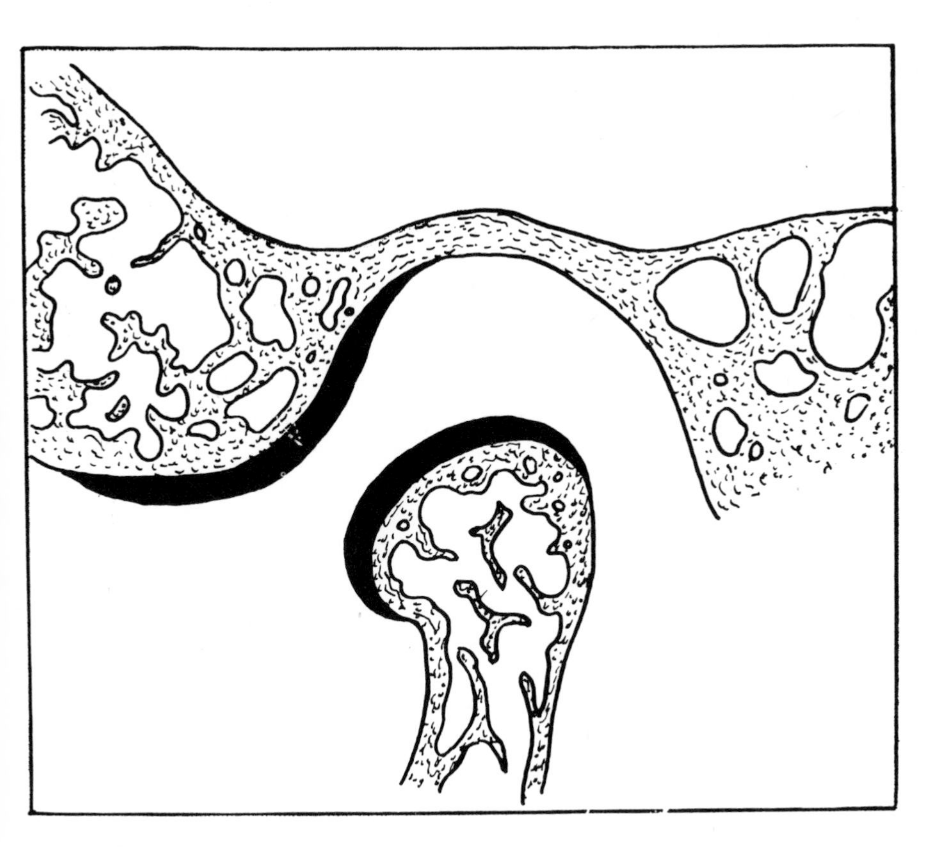

Figure 6. Diagram showing those areas (black) of the condylar head and the articular eminence in man which are covered with fibrocartilage; these represent the functional surfaces of the joint. Redrawn after Moss.[18]

the mandibular angle involved would be such as to endanger the pharyngeal structures of the throat.

In *Archaeolemur* lateral motion of the mandible during mastication was apparently achieved through contralateral rotation of the ipsilateral condyle; simultaneously the contralateral condyle translated posteriorly. Ipsilateral condylar rotation could have taken place at any phase of elevation, but presumably occurred generally during close approximation of the tooth-rows, when the ipsilateral condyle was in a posterior position. Translation of the contralateral condyle, and hence the prime component of lateral motion, was evidently largely effected by the almost directly posterior pull of the posterior temporalis; the movement would have been stabilised by the antagonistic action of the contralateral external pterygoid.

Since during lateral motion the ipsilateral condyle was effectively fixed, rotating, in this plane at least, around a point within itself, it would appear at first sight to have acted as the fulcrum of a lever. The geometry of the system, however, is such that the point of resistance, i.e. the food, would have tended to become the centre of rotation, causing the ipsilateral condyle to move anteriorly. The ipsilateral posterior temporalis would have acted to eliminate such movement. In the absence of a tendency of this sort, any reaction force in this plane would have been nullified by the anterior, antagonistic, pull of the ipsilateral lateral pterygoid.

Derivation of the masticatory apparatus

In most of its morphological details, as well as in its basic plan, the archaeolemurine masticatory apparatus is very close to that of the living indriines (Fig. 7). Apart from precise dental morphology, the most striking differences between the archaeolemurine and indriine chewing mechanisms lie in the form of the symphyseal region of the mandible, the robustness and proportions of the mandibular corpus, and in the manner in which the dentaries are joined. The lack of symphyseal fusion in the indriines is a particularly significant distinction; Crompton and Hiiemae have shown that a mobile symphysis implies a totally different pattern of mastication from that obtaining in mammals possessing synostosed dentaries: one in which the two dentaries may move independently of each other.[15]

In a living indriine such as *Propithecus*, given the presence of a mobile symphyseal syndesmosis, it is evident that the contralateral temporalis cannot function to produce internal movement of the ipsilateral dentary. Other, entirely ipsilateral, muscular combinations must be operative to produce jaw movements not dissimilar to those in *Archaeolemur*. As Smith and Savage have pointed out,[7] internal movement of the mandible by contraction of the temporalis of the

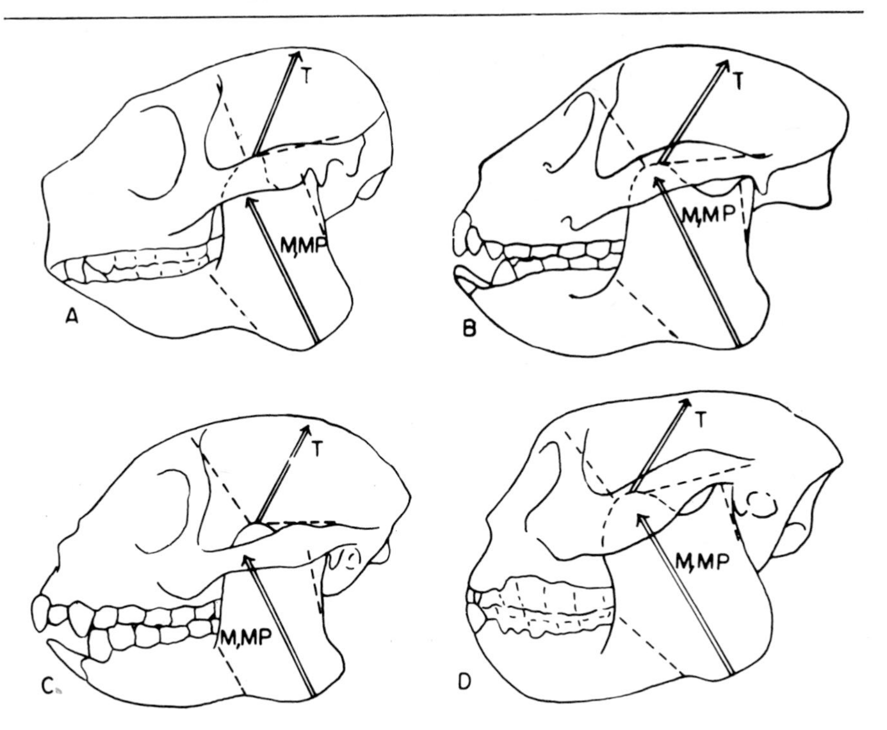

Figure 7. Comparison of the orientations in lateral view of the major masticatory muscles in four indriid species: *A, Propithecus verreauxi; B, Archaeolemur edwardsi; C, Archaeolemur majori; D, Hadropithecus stenognathus.* Abbreviations as in Fig. 3.

contralateral side would involve extreme stressing of the lower jaw; these stresses would be transmitted anteriorly, around the symphysis, and back again along the ipsilateral corpus. Despite this disadvantage, such a situation is precisely what we find among the archaeolemurines, together with massive jaws and an extremely strongly buttressed symphysis, characterised in particular by the possession of a strong inferior transverse torus.

Goodman has recently discussed the relation between form and function in the pongid symphysis.[19] By stressing both a photoelastic model and pongid mandibles coated with strain-sensitive lacquer, Goodman established that the inferior transverse torus of the pongid symphysis represents an adaptation for withstanding laterally-directed forces, i.e., those which would tend to spread the dentaries apart. This torus is well-developed in the mandible of *Archaeolemur*, while the forces developed in this area during contralateral temporalis action would have been very much of the sort Goodman describes.

It is clear, then, that the masticatory system of *Archaeolemur* could have been very simply derived from an ancestral pattern quite similar to that seen in the extant indriines, particularly *Propithecus*, by fusion of the mandibular symphysis. Other modifications, such as the dominance of the contralateral temporalis in lateral movements, change in symphyseal form and the increased robustness of the mandible, are all correlates of this fundamental shift, and presumably reflect a change towards a more powerful masticatory pattern, possibly accompanying the initial increase in size of the archaeolemurine ancestors. Presumably the change took the shape it did, with minimal modification of the pre-existing apparatus, because the common ancestor of Indriinae and Archaeolemurinae was already fairly highly specialised towards the modern indriine condition. In short, although the archaeolemurine masticatory apparatus resembles in some morphological respects (e.g. bilophodont molar teeth, possession of inferior transverse torus) that of various catarrhine primates, and was evidently the response to similar selective pressures,[4,20] in its general plan and in its origin it was radically different: an analogous variation on an altogether different theme.

Acknowledgments

The study of which this report is an outgrowth was supported by the Wenner-Gren Foundation for Anthropological Research Inc., by the Boise Fund of Oxford University and by various Fellowships of Yale University. Completion of this paper was assisted by a Grant-in-Aid of the Society of Sigma Xi. I should like to acknowledge here stimulating discussion of this topic with David Roberts, Department of Geology and Geophysics, Yale University. The illustrations were prepared by Mr Nicholas Amorosi, Department of Anthropology, American Museum of Natural History.

NOTES

1 Major, C.I.F. (1896), 'Preliminary notice on fossil monkeys from Madagascar', *Geol. Mag.* (decade 4), 3, 433-6; (1900), 'A summary of our present knowledge of extinct primates from Madagascar', *Geol. Mag.* (decade 4), 7, 492-9; (1901), 'Some characters of the skull in the lemurs and monkeys', *Proc. Zool. Soc. Lond.* 1901, 129-52.

2 Standing, H.F. (1908), 'On recently discovered subfossil primates from Madagascar', *Trans. Zool. Soc. Lond.* 18, 69-162.

3 Grandidier, G. (1905), 'Recherches sur les lémuriens disparus et en particulier ceux qui vivaient à Madagascar', *Nouv. Arch. Mus. Hist. Nat. Paris* 4 sér., 7, 1-142.

4 Tattersall, I. (in press), 'Cranial anatomy of Archaeolemurinae (Lemuroidea, Primates)', *Anthrop. Pap. Amer. Mus. Nat. Hist.*

5 Walker, A.C. (1967), 'Locomotor adaptation in recent and subfossil Madagascan lemurs'. PhD Thesis, University of London.

6 Jouffroy, F.-K. (1963), 'Contribution à la connaissance du genre *Archaeolemur* Filhol, 1895', Ann. Paléont. 49, 129-55.

7 e.g. Smith, J.M. and Savage, R.J.G. (1955), 'The mechanics of mammalian jaws', *School Sci. Rev.* 1955, 289-301; Ostrom, J.H. (1964), 'A functional analysis

of jaw mechanics in the dinosaur *Triceratops*', *Postilla* 88, 1-35; Crompton, A.W. and Hiiemae, K. (1969), 'How mammalian molar teeth work', *Discovery* 5, 23-4.

8 Oxnard, C.E. (1971), 'Tensile forces in skeletal structures', *J. Morphol.* 134, 425-35.

9 Moss, M.L. and Rankow, R.M. (1968), 'The role of the functional matrix in mandibular growth', *Angle Orthodontist* 38 (2), 95-103.

10 Davis, D. (1955), 'Masticatory apparatus in the spectacled bear *Tremarctos ornatus*', *Fieldiana: Zool.* 37, 25-46.

11 Turnbull, W.D. (1970), 'Mammalian masticatory apparatus', *Fieldiana: Geol.* 18 (2), 149-356.

12 The closeness of the jaw joint to the centre of rotation of the mandible may form part of the explanation for the tight, restricted condyle-glenoid relationship in carnivores, as opposed to the much looser temporomandibular articulation among herbivores; however, the point is far from clear, and cinefluorographic studies of movement at the carnivore jaw joint during chewing have yet to be made.

13 For the purposes of this discussion the dentitions and masticatory musculatures of the two species of *Archaeolemur* are virtually identical, the only major difference lying in the greater size of *A. edwardsi.* Unless otherwise stated, then, the discussion throughout applies equally to both species.

14 Mills, J.R.E. (1967). A comparison of lateral jaw movements in some mammals from wear facets on the teeth. *Archs. Oral. Biol.* 12, 645-61.

15 Crompton, A.W. and Hiiemae, K. (1969), 'Functional occlusion in tribosphenic molars', *Nature* 222, 678-9. Crompton, A.W. and Hiiemae, K. (1970), 'Molar occlusion and mandibular movement during occlusion in the American opossum', *Zool. J. Linn. Soc.* 49, 21-47.

16 Because, at least for the purposes of a simplified argument, the jaw in motion can be visualised as passing through an infinite series of static equilibria, it is permissible to generalise, within limits, from the static to the dynamic state.

17 Eisenberg, N.A. and Brodie, A.G. (1965), 'Antagonism of temporal fascia to masseteric contraction', *Anat. Rec.* 152, 185-92.

18 Moss, M.L. (1959), 'Functional anatomy of the temporomandibular joint', *New York J. Dentistry*, 29 (11), 315-19.

19 Goodman, M.C. (1968), 'Structure and function in the symphyseal region of the pongid mandible', Dissertation, Yale.

20 Jolly, C.J. (1970), '*Hadropithecus*, a lemuroid small-object feeder', *Man*, n.s., 5, 619-26.

21 B.M. = British Museum (Natural History), London.

R.G. EVERY

Thegosis in prosimians

Introduction

An analysis of thegosis in prosimians demonstrates how knowledge of this phenomenon has introduced a new dimension to the study of mammals. Many hitherto unsuspected questions can now be asked. In the area of primatology concerned with basic primate origins these questions are most pertinent. For, if basic theory proves fallacious, the implications are relevant to *Purgatorius ceratops*, the earliest known primate, and to *Homo sapiens*, the primate that concerns us most.

Knowledge of the origins of primates depends almost exclusively on teeth. Interpretation of the selective forces that produced teeth is therefore of crucial importance. The prime taxonomic feature for the earliest primates is the enlargement of the talonid and trigon basins, concurrent with the loss of the paraconid and reduction of the trigonid. The virtually universal interpretation of this phenomenon is that the evolution of primate teeth is dominated by a 'rubbing', 'grinding', 'mortar-and-pestle', 'surface-shearing' advantage of the enlarged basins.

From a study of thegosis this fundamental interpretation, basic to all primates, is now seen as fallacious.

Thegosis is a phylogenetically derived behaviour.[1] The term comes from Homer's use of the Greek *thego*: to whet, sharpen. In the *Iliad*, Homer accurately describes not only the boar's anatomical preparation of his formidable dental weapons (the lower canines) but also the animal's physiological preparation, the emotional state that basically determines his immediate behaviour. The term thegosis, therefore, includes the metaphorical connotation inherent in the original Greek *thego*: to excite, provoke.

Since all the author's basic concepts of thegosis emerged during the course of a psychiatrically-orientated dental practice, the initial experimental animal used was, of course, man himself. Most of these experiments, which began over thirty years ago, were possible only on man and cannot be duplicated on any other animal. In the study of any other animal one must, therefore, lean heavily on the evidence gained from man. Although this area of evidence is too extensive even to summarise here, it nonetheless forms the basis of this present study of prosimians.[2]

The earliest mammals represent another area of evidence that is also significant to this present study, but can only be briefly introduced.[3] When discussing this aspect of mammalian evolution, Simpson states in his introduction: 'The discovery of the ancestral type would introduce into the study of dentitions the principle of homology and would supply the background that would knit together knowledge of teeth and transform it from a mere morphological catalogue to a true and potent science. . . . The point of view adopted is principally functional and dynamic.'[4]

It seems that Simpson's prediction of a true and potent science may have been vindicated as far as general homology is concerned, but functionally it has remained virtually as he himself stated it in 1936. For example: 'The primitive upper molar did not consist simply of three cusps and nothing more. [Footnote] This fact, well understood by Cope and Osborn and by all their successors, has from time to time been hailed as a discovery and supposed to invalidate the Cope-Osborn theory.'

When one realises that Cope preceded Simpson by 53 years, it is sobering to read what was known and understood. Simpson says:

> The protocone, internal main tubercle, is crescentic and from it two sharp crests run externally (labially) along anterior and posterior sides. In occlusion these are shearing crests. . . . The main external cusps, paracone and metacone, are sub-equal, and it appears to have been a primitive character for them to bear crests descending anteriorly and posteriorly from each apex, to some degree uniting the two cusps. Aside from these crests, the cusps are sometimes nearly conical, sometimes compressed or lancelate, and sometimes strongly crescentic. The crescentic character was probably not generalised but was a modification that early arose within certain lines of descent . . . and greatly affected the subsequent development of their dentitions.[4]

Simpson was describing what he in that paper termed the 'tribosphenic molar':

> Its principal advance is the further development of opposition, first by the widening and basining of the talonid and second by the development of the hypoconid which occludes against the

central basin of the upper molar and between and internal to the main outer trigon cusps, now two in number, although not completely separated. In addition to this double opposition, the elements of alternation and shear are preserved.

It is undeniable that the advent of the reciprocally opposing, cusps-to-basins feature was an important evolutionary event. It allowed the division of exogenous substances that eluded the action of the cutting blades. It also had the advantage of squashing soft pulpy food and mixing it and other divided food with saliva, the first digestive juice of the alimentary canal. Odontologists, especially primatologists, have in a number of ways tended to be confused by this, however, and have postulated theories that now appear as fallacious. 'Preventing overclosure of the jaw' is one, 'grinding', 'mortar-and-pestle action', 'lingual phase of occlusion', 'lessening importance of shearing blades', and the 'shearing surface' as constituting the 'shearing' agent, are others. Since it is undeniable that the prime taxonomic feature diagnostic of the earliest primates is at present the enlargement and raising of the talonid relative to the trigonid, concomitant with corresponding changes in upper molars, whether such arguments are valid or not becomes of prime importance. For if they prove fallacious, Simpson's criterion for a 'true and potent science', that is, '. . . principally functional and dynamic', has not been vindicated; and, furthermore, knowledge of teeth, especially primate teeth, has remained virtually a 'morphological catalogue'.

In the hindsight knowledge of the phenomenon of thegosis, one may now detect some of the inherent difficulties. Apart from the misconceptions regarding phylogenetically derived behaviour and learning (the nature-nurture problem), which undoubtedly delayed the perception of thegosis as a fundamental mammalian characteristic, I believe functional analysis of teeth was, and still is, crippled by its functionless terminology, as the following examples show:—

'Crest' (crista) depicts a location, not a function; it means the summit of anything, as a comb or tuft on the head of a bird, roof-ridge, hill or wave; the only other anatomical use is a ridge along the surface of a bone.

'Selenodont' (Greek *selene*, a moon) and 'crescent' (Latin *crescere*, to grow) analogise a tooth-feature to a waxing moon.

'Trigon' is a triangle.

'Talon' (Latin *talus*, the heel) connotes, not a function, but a spatial relationship.

The suffix 'cone' has an inherent connotation of a solid figure with a circular base tapering to a point.

'Loph' is a crest (Greek *lophos*).

'Cingulum' is a girdle.

Apart from the apex of the cone, none of these terms connotes a function; none, especially not the cone, connotes a feature that has a functionally sharp cutting edge. 'Crest', in fact, merely serves to emphasise the (fallacious) importance of the so-called 'shearing surface' which it merely surmounts.

Simpson's own term, 'tribosphenic', is also suspect. 'Tribo-' (Greek *tribein*, to rub, grind, or pound), and '-sphenic' (Greek *sphenos*, a wedge) are, as Simpson explains, 'suggestive of the mortar-and-pestle, opposing action of protocone and talonid and of the wedge-like, alternating and shearing action of trigon and trigonid'.[4] The prefix depicting the talonid is functional; the suffix depicting the trigonid, unfortunately, is inconsistently morphological.

That Simpson's use of 'wedge-like' is merely morphological is clearly indicated in his discussion of symmetrodonts: 'These teeth have become triangular and in occlusion their opposed apices alternate and the teeth fit into each other like opposed series of blunt wedges. The crests have become sharp and pronounced and they add the important occlusal factor of shearing.'[4]

For Simpson to associate a wedge with a dividing action when the dividing agent is, in fact, not the wedge's thin edge but two blunt side edges, is particularly unfortunate. Although 'wedge' is used for anything more or less like a wedge, as a large piece of cake or the flying formation of geese, the basic connotation remains an instrument that has a function of splitting, forcing apart, or fixing tightly, etc. Its shape is characteristically thick at one end sloping to a thin (often sharp) edge at the other. In engineering, the inherent principle of any cutting instrument is described as the production of a crack that is propagated. Anyone who has used a wedge to split open a log is familiar with this action. If there is no existing crack to propagate, one is created. The wedge is hammered lightly into the grain of the wood until it grips; it is then hammered forcibly, propagating the crack. In this action, not only is the thin edge of the wedge functional, but also the sloping side surfaces. In fact, for a material that splits easily, the crack, once initiated, may be propagated by the sloping sides alone, i.e. surface against surface.

This, then, seems an appropriate place to discuss the common but fallacious concept of the so-called 'shearing surface', i.e., the idea that cutting and dividing exogenous material is achieved by grinding it between two opposed surfaces, especially in the manner of a mortar and pestle. A mortar-and-pestle unit is essentially two opposed surfaces, one concaved at a greater radius than the other's convexity. Since Butler's publication on the 'facet' and Mills' interpretations of 'buccal and lingual phases of occlusion',[5,6] the 'shearing surface' concept has, even more fallaciously, been argued to be the facet's inherent function. So much so that the facet is now identified as the 'shearing surface'; this has become the basic unit in the study of dentitions. It is the vogue to detect a 'shearing surface', to find its counterpart and its homology, to give it a number and analyse its efficiency and importance according to size. 'Shearing surfaces' are found to shear 'up' and 'down', 'across' and 'against' other 'shearing surfaces'. Occasionally (and often ambivalently) the functional significance of a 'crest' is noticed and mentioned.

The term 'shear', itself, needs analysis. Although 'shears' in engineering have come to apply to a larger instrument of similar kind to scissors, the term 'shearing', itself, is basically the dividing, cutting action of a single blade: a sickle, knife, chisel, lathe tool, bow or propeller of a ship, provide examples.

'Shearing', then, is a valid term for the initial actions of mastication – the action of Simpson's 'Opposition . . . of positive elements, cusps, crests, against other positive elements';[4] also for the first phase of the 'power stroke', described (for *Galago*) by Kay and Hiiemae in this volume,[7] by Crompton and Hiiemae (for *Didelphis*),[8] and commonly observed in man and other anthropoids. These constitute a number of relatively vertical strokes (made on a bolus of food) that terminate appreciably before the cusps and blades themselves come into apposition.

When the term 'shearing' is used for the action of 'shears' comparably to that of 'scissors', there is no confusion about the relationship of the blades. The efficiency of the cutting action requires the close apposition of the cutting edges. When a material is interposed between the blades' surfaces (i.e., between the facets of teeth-blades), the instrument is rendered useless.

Despite this overt mechanism, the identical term 'shearing' is used by some odontologists to apply precisely to the above situation that renders shears and scissors useless. Shearing is supposed to result when the material is interposed between two blade-surfaces (i.e., their facets) that move parallel but opposite to each other. This is the 'grinding', 'mortar-and-pestle', 'surface-shearing' action. To function at all, the surfaces would have to be considerably roughened in order to grip the substances being 'shorn', or the substance would have to possess considerable adhesiveness to grip the surfaces and, in addition, the forces perpendicular to the surfaces would have to be excessive. The reality is that the facet, compared to its adjacent tooth-surfaces, is characteristically polished, shiny and smooth.

'Shearing' is a term that not only may have different connotations for different individuals, but also may have one connotation for the same individual who uses it for a number of entirely different mechanisms.

Full discussion of this confusing term and its degradation of odontology is beyond the scope of this paper.[9] The situation, however, is so chaotic that I am suggesting a temporary – if not permament – suspension of its use.

Terminology

Should the phenomenon of thegosis eventually be generally accepted as a phylogenetically derived behaviour which was crucial to both the origin and the subsequent radiation of mammals, the magnitude of the inevitable change in our thinking can hardly be overrated. It would become apparent that the present discussion among odontologists about different notions of tooth function, and the terms used to label them, is highly significant. Even before thegotics has succeeded in making any great impact, there are signs of changes taking place. Odontology faces a considerable readjustment of its basic principles.

While odontology is in this state of flux, it would hardly be conducive to a flexible approach to argue that tradition justified the retention of confusing or inadequate terms, let alone those that are both morphologically and functionally erroneous. It is the very chronicity of a lesion that ultimately demands its excision. And it would hardly be conducive to the efficiency of the operation were the surgeon to discuss the functional properties of his lance in terms of 'crest' or 'cone'. When discussing the ancestral mammal, the theories and the theoreticans, Simpson comments: 'In this field there are and always have been workers whose opinions are not much influenced by fact.'[4]

Since I have the task of defining the phenomenon of thegosis and therefore the new science of thegotics, I am forced to use my own language if I am to facilitate the discussion. Fortunately, the changes are minimal and the introductions not extensive. In combination, however, they build to a complicated tooth-formula that at first sight appears rather formidable. However, the chemical formula for a complicated molecule appears to be even more complex at first sight. Yet there is no complaint that the chemical formula confuses our understanding rather than elucidates it. Nor is there any difficulty in devising an abbreviation, though this can only be introduced after its full components are known. This terminology is planned to suggest the selective forces in the evolutionary process. The aim is to simplify the conception of both the evolutionary process (the selective forces that produced the dentition) and the functional spatial relationships of the teeth themselves.

Although my professional as well as research experience has involved the articulation of artificial replacements for natural (human) teeth, and I have never delegated this aspect of the 'technical work' to others, and although I am familiar with working on a target viewed in a mirror, I nonetheless cannot achieve the feat, frequently demanded of me, of conceiving the spatial relationships of teeth as they are depicted in much of the current literature. The frequent practice of depicting right upper molar teeth alongside left lower teeth is particularly objectionable. In order to perceive the dynamics of function, I have first to convert (in my mind) the lower occlusal contours to their mirror image (i.e., to the right side) before I can procede to articulate in my mind the appropriate contours to their appropriate counterparts – as I am accustomed to do with the concrete objects. This defies my capacity.

The practice of depicting right upper molars with left lowers may have some advantage in presenting a 'radiographic profile' of a static occlusal relationship. Yet even this facility is doubtful; why, for example, was it not detected earlier that, in this view, the articulating blades of the 'tribosphenic' molars are mutually convex?[1,3]

If workers cannot perceive static relationships, how can they perceive dynamic ones? And if workers cannot perceive dynamic spatial relationships in one direction (transverse) how can they perceive these in two (transverse followed by oblique)?

The magnitude of this problem is exemplified by my colleagues in dentistry who, when fabricating artificial replacements, endeavour to articulate the teeth so that they achieve a so-called 'balanced occlusion'. This is an extremely difficult task to achieve visually, and is seldom even attempted. If a 'balanced occlusion' is ever achieved it is through the agency of a mechanical device: the 'grinding in' on an anatomical articulator using carborundum powder.

That scientists seldom appreciate the spatial relationships involved, despite their day-to-day involvement with the problem, is suggested by the consistent failure of authors to publish illustrations that depict the valid (frontal) cross-sectional relationship of teeth in lateral occlusion. Invariably the jaw (of a transverse masticating animal) is depicted, not as rotating about an axis near the postglenoid tubercle, but translating bodily in a transverse direction, which is fallacious.[10]

The new terminology not only suggests precise morphological and functional characteristics, but also retains the connotation of phylogenesis despite an adaptive variation where, for example, a transverse blade becomes virtually longitudinal, or a feature becomes amalgamated with another, or absorbed in a wall or base, or lost.

When the primate dentition is analysed in the light of thegosis, its striking characteristic is not, as has been generally supposed, the enlargement of the basins, but the modification and elaboration of cutting blades. Contrary to current opinion, bluntness, although elaborated, has not become the dominant characteristic. Sharpness (the cusp-blade unit) remains the prime advantage; its apparent depreciation is a result of the overt shortening of some of the blades that in the ancestral alpha dentition dominated the occlusal surface.[3] The remainder of the blades, however, are elongated. Apparently overlooked is the concomitant of a large basin: the elongation of its peripheral blades. Moreover, and significantly, a new cusp-blade unit has appeared.

Cingular blades, also, are elaborated. Not only are they raised higher on the crown, but new cingular blades and cusps are frequently present. Cingular blades are subsidiary (reserve) blades which function in the gerontic stage of the tooth's life-cycle. They are, therefore, appropriately oriented to the primary (occlusal surface) blades they eventually replace.[3]

The overall effect is that, because 'basined' surfaces have been enlarged and heightened on the crown, the cutting blades, instead of being spread out over the occlusal surface as in the ancestral condition, are now featured more on the periphery. This therefore ensures a greater ontogenetic continuity of cutting blades generally, an advantage – not a sacrifice – to the greater ontogenetic continuity of the 'basined' surfaces of which blades form an integral part.

These and other significant changes will be discussed later when using the facility of the new terminology.

The prime evolutionary advantage of teeth is their sharpness; the secondary advantage is bluntness. It is the sharp conical apex and blunt conical walls of our original aquatic ancestors' teeth that allowed their original grasping advantage. In the original mammalian molar it is the sharp thegosed apex of the cusp and the sharp thegosed blade that runs from it that allowed the additional advantage of cutting in scissor-like action. In the later mammalian molar it is the blunt, opposable, platform-surface bounded by sharp cusps and blades that allowed the additional advantage of cutting, cracking, splitting, crushing and mashing by striking action.

These three basic features of the characteristic (basic) mammalian dentition – thegosis, sharpness, and opposable sharpness and bluntness – can therefore be used to build a functional terminology.

Apart from *thegosis* itself, the terminology is virtually built up around four basic new terms:

1. *Scissorial:* Dividing exogenous material by point-cutting in a scissor-like action.
2. *Incusive* (Latin *incuso*, to hammer; *incus*, anvil): Dividing exogenous material by striking when held by a surface as an anvil or chopping-block.
3. *Akis* (Greek *akis*, small pointed object, splinter, arrow-head, point of a chisel): A cusp with a sharp edge running from the apex as an arrow-head or point of a chisel or sickle.
4. *Drepanon* (Greek *drepane*, a sickle): A concave blade with the sharpness on the edge of the concavity as a sickle.

There are further terms; some of these are mentioned below; all (including the above basic units) are presented in detail elsewhere.[3]

Types of dentitions

Dentitions can be categorised according to the location of the thegosis-facet relative to the direction of the enamel prisms (or rods). The cutting blades of primitive teeth before the Cretaceous and (for advanced teeth) those of typical insectivores and carnivores, for example, are thegosed on the surface of a plate of enamel, that is (predominantly) across the ends of the enamel rods. This is *alpha thegosis.*

The cutting blades of the typical herbivores, for example, are thegosed transversely across a plate of enamel including the amelo-dentinal junction (the line of demarcation between enamel and dentine), that is (predominantly) on the sides of the enamel rods. This is *beta thegosis.*

The present study of fossil and extant Primates shows clearly that this Order is characterised by cutting blades that function for a part (sometimes a considerable part) of the animal's lifetime with alpha thegosis before beta thegosis eventually appears.

Types of blades

1. Scissorial blades

'Scissorial' is a general term to include the basic action common to all paired blades, whether vertical or horizontal.

Scissorial action is virtually the first category of Simpson's definition: 'Shearing: (*a*) The parallel passage of crests sharing equally in shear.'[4] Although Simpson later explains that 'the crests shear past each other, a truly scissor-like, crest-past-crest shearing mechanism',[4] his use of 'parallel' in the definition is misleading, if not erroneous. Parts of the juxtaposing blade-surfaces may be regarded as parallel, but the blades' functional cutting edges are far from being so. Simpson's use of 'parallel' would be appreciated by those who, despite the defining 'crest', conceive of the parallel surfaces as the cutting agents.

The essence of scissorial action is that the diagonally opposing blades, while confining an interposed exogenous material to their cutting actions, concentrate the dividing force at a point, thus utilising the principle of producing a crack which is propagated: point-cutting. Were the blades in direct opposition, i.e., parallel, there would be even more efficient confining, but it would be at the cost of cutting along a line, i.e., the force required for such an action would be multiplied. There is, therefore, an optimum angle for efficient point-cutting compromised with the requirements for efficient confining. The difference between point-cutting and line-cutting actions is comparable to that between guillotine and axe.

In the evolution of early mammalian dentitions there was an interesting switch in the mechanism of cutting blades.[3] The first evidence we have of this is in the extinct dryolestids and paurodonts. Its occurrence in the ancestral mainstream, however, first appears in the Cretaceous therian *Pappotherium*, i.e. concurrent with the appearance of opposable basins and their peripheral blades and cusps.

In the primitive cusps-in-a-row and symmetrodontoid conditions there is a relatively simple action of blades with cusp-to-valley interdigitation. In the advanced condition, there is a cusp-to-cusp opposition with the blades vertically concave (i.e. the concavity of one blade opposes that of the other) and horizontally convexed (i.e. the convexity of one blade opposes that of the other). This complicated system occurred to the advantage of ensuring automatic confining of exogenous material to the dividing, cutting action of the blades. It is an essential characteristic of advanced alpha dentition.

In beta dentition, however, with the transfer from a predominantly vertical blade action to a predominantly horizontal one, the advantage of the opposing-convexity system has given place to another. The presence of blade-concavities perpendicular to the line of cutting action remains, however, very much an advantage. Alpha/beta dentition is intermediate.

Because primate dentition is intermediate, the horizontal-opposing-convexity system is somewhat depreciated. For this reason, and because the spatial relationships are somewhat difficult to describe, it is only mentioned here. Nonetheless, this feature has had important bearing on the characteristics of primate dentition.

In the primitive dentition with simple scissor-like (in some aspects more guillotine-like) blades, point-cutting begins at one end of the blade-pair and proceeds along the whole course of the cutting edges to terminate at the other end. The blades are zig-zagged (somewhat in the manner of tailors' pinking-shears), which aids, primitively, the confining of exogenous material to the cutting action. In the advanced alpha dentition (Simpson's 'tribosphenic' dentition, and the dentition basic to my own terminology) there is a profound change. Alphascissorial blades are now consistently drepanoid (sickle-like), and at both ends of the blade there is an *akis* (in an upper blade) or *akid* (in a lower),thus forming a *diakidrepanon* (upper) or *diakidrepanid* (lower). In function, each akid of the (dynamic) diakidrepanid approximates to and juxtaposes its counterpart akis of the (static) diakidrepanon; the concavity of the diakidrepanid, therefore, also approximates to and juxtaposes its counterpart concavity of the diakidrepanon. Such a morphological arrangement allows functional point-penetration at each juxtaposing akid-akis extremity, followed by point-cutting as the blades juxtapose. Interposed exogenous material is thereby confined to the dividing action of the blades. (It is to be emphasised that the horizontal-opposing-convexity feature has an important bearing on this confining action; it and other relevant features of the advanced condition are presented elsewhere.)[3]

In the basic (advanced) alpha dentition, each diakidrepanoid blade is combined with another, each blade sharing a cusp of the other; the combination therefore becomes a *triakididrepanon*, or *triakididrepanid.* In occlusal view the triakididrepanoid unit has the appearance of an asymmetrical Gothic arch; both blades are (horizontally) convexed, one is oriented (predominantly) transversely across the occlusal plane, the other obliquely. As Simpson has emphasised, Cope and Osborn were aware of the blades that ran from the three cusps of the 'trigon'.[4] Had they observed that there were two profoundly different blade systems present in early mammals (i.e. cusp interdigitation and cusp apposition) they might have realised that the blades at each cusp of their 'trigon' form a unit 'trigon' that, in fact, is homologous with the 'trigonid'.

These terms are abbreviated in the following way. With, for example, the prefix *para*:

Para-Triakis for Para-Triakididrepanon (PaT)
Paratransversis for Para-diakidrepanon-transversa (Pat)
Paraobliquis for Para-diakidrepanon-obliqua (Pao).

Because lower transverse blades articulate with upper transverse blades, and oblique with oblique, the terms allow easy appreciation of occlusal relations.

In the primate (alpha/beta) dentition, the angle between the triakididrepanoid blades progressively becomes obtuse. Despite such adaptive variation (discussed below), where both transverse and oblique blades progressively become longitudinal, the terminology, by retaining the phylogenetic connotation, also retains the connotation of occlusal relationships.

The Palaeocene primate *Plesiadapis rex* (Fig. 1*a*) illustrates the characteristic widening of the angle of the triakididrepanoid blades; the Paratransversis and Metatransversis have virtually become as obliquely orientated mesially as the Paraobliquis and Metaobliquis are distally. The Eocene primate *Adapis magnus* (Fig. 3*b*) illustrates further widening; the angle has now become obtuse.

In *Plesiadapis* the paraobliquakis and metatransversakis, although immediately adjacent, are virtually discrete entities; whereas in *Adapis* they are completely amalgamated. In *Plesiadapis* the Paraobliquis and Metatransversis each share an akis with their respective cingular blades; whereas in *Adapis* they are widely separated from all contact. The parastylo-Cingulis and metastylo-Cingulis themselves are now combined at a shared (pseudo-) akis.

Thegosis characteristically occurs as the teeth erupt and come into occlusion. The behaviour is adapted to facilitate the alignment of the scissorial blades which are thus brought into accurate functional apposition. A further advantage of this facility is the alignment that

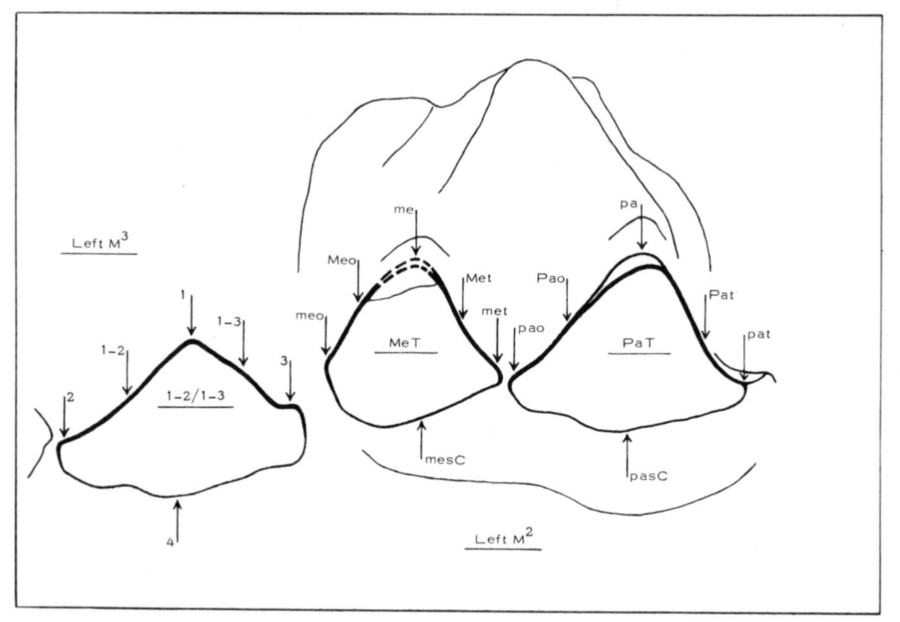

Figure 1(*a*). *Plesiadapis rex.* Left M^2 and part of M^3. Buccocclusal view. (Figs. 1*a*, *b* and *c* are from the same specimen.)
1, 2 & 3 Akis (a cusp with a sharp blade running from the apex).
1-2 Oblique diakidrepanon (a sickle-like blade with an akis at either end).
1-3 Transverse diakidrepanon (phylogenetically orientated transversely across the occlusal plane).
1-2/1-3 Triakididrepanon (a combination of two diakidrepana).

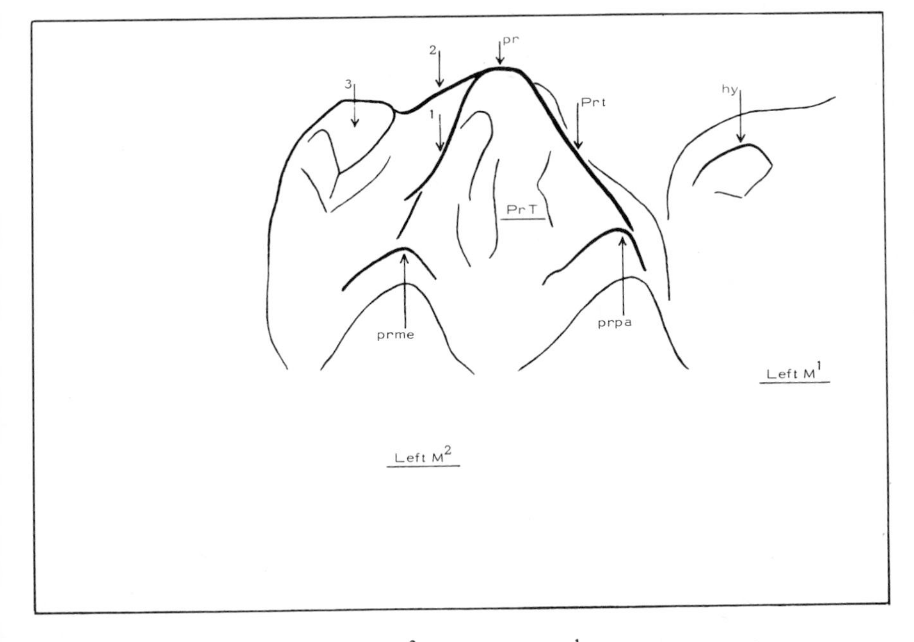

Figure 1(*b*). *Plesiadapis rex.* Left M^2 and part of M^1. Buccocclusal view.

1 The (relict) Protobliquis (Pro).

2 The (pseudo-) Protobliquis, i.e., functioning with the blades from the entoakid.

3 Facet on the (pseudo-) Hypo-Triakis, with abrasion scratches superimposed.

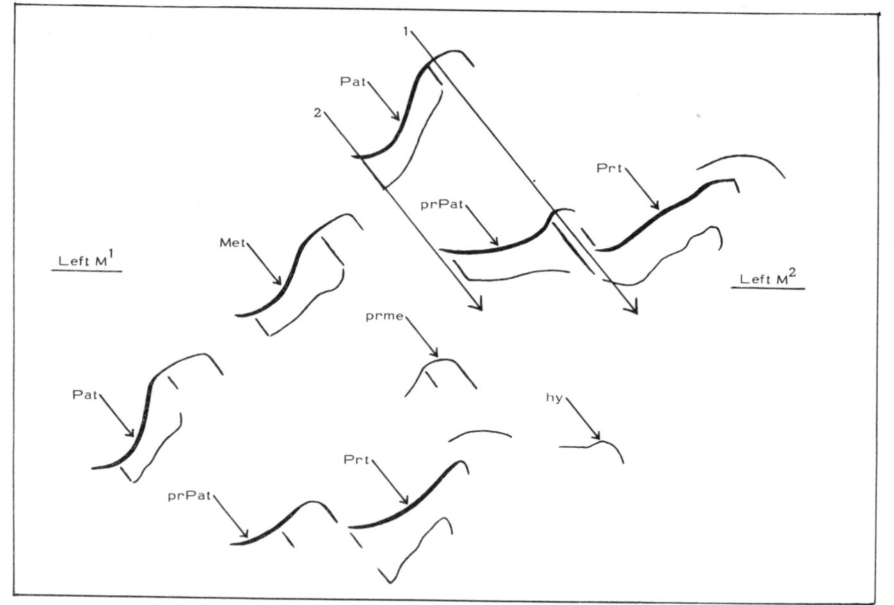

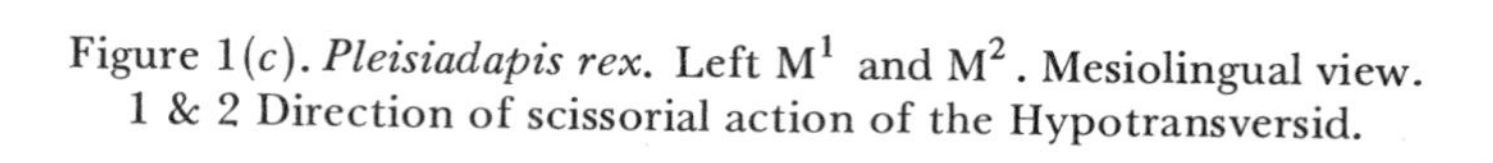

Figure 1(*c*). *Pleisiadapis rex.* Left M^1 and M^2. Mesiolingual view.
1 & 2 Direction of scissorial action of the Hypotransversid.

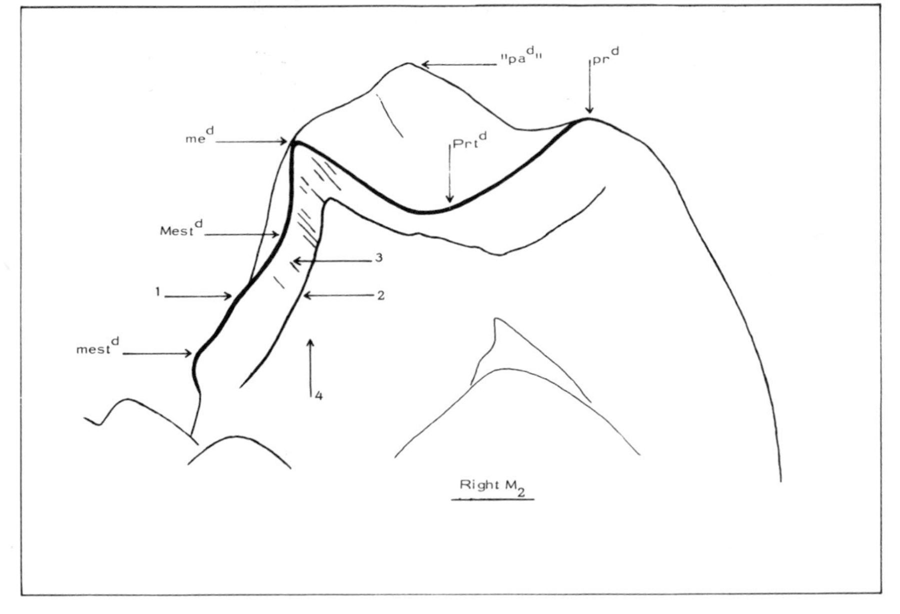

Figure 2 (*a*). *Plesiadapis rex*. Right M_2. Distal (slightly buccocclusal) view.
1 Leading edge of the Metastylotransversid.
2 Trailing edge of the $Mest^d$.
3 Thegosis-facet forming a narrow strip at the scissorial edge.
4 Primary (developmental) hollowed-out blade-surface.

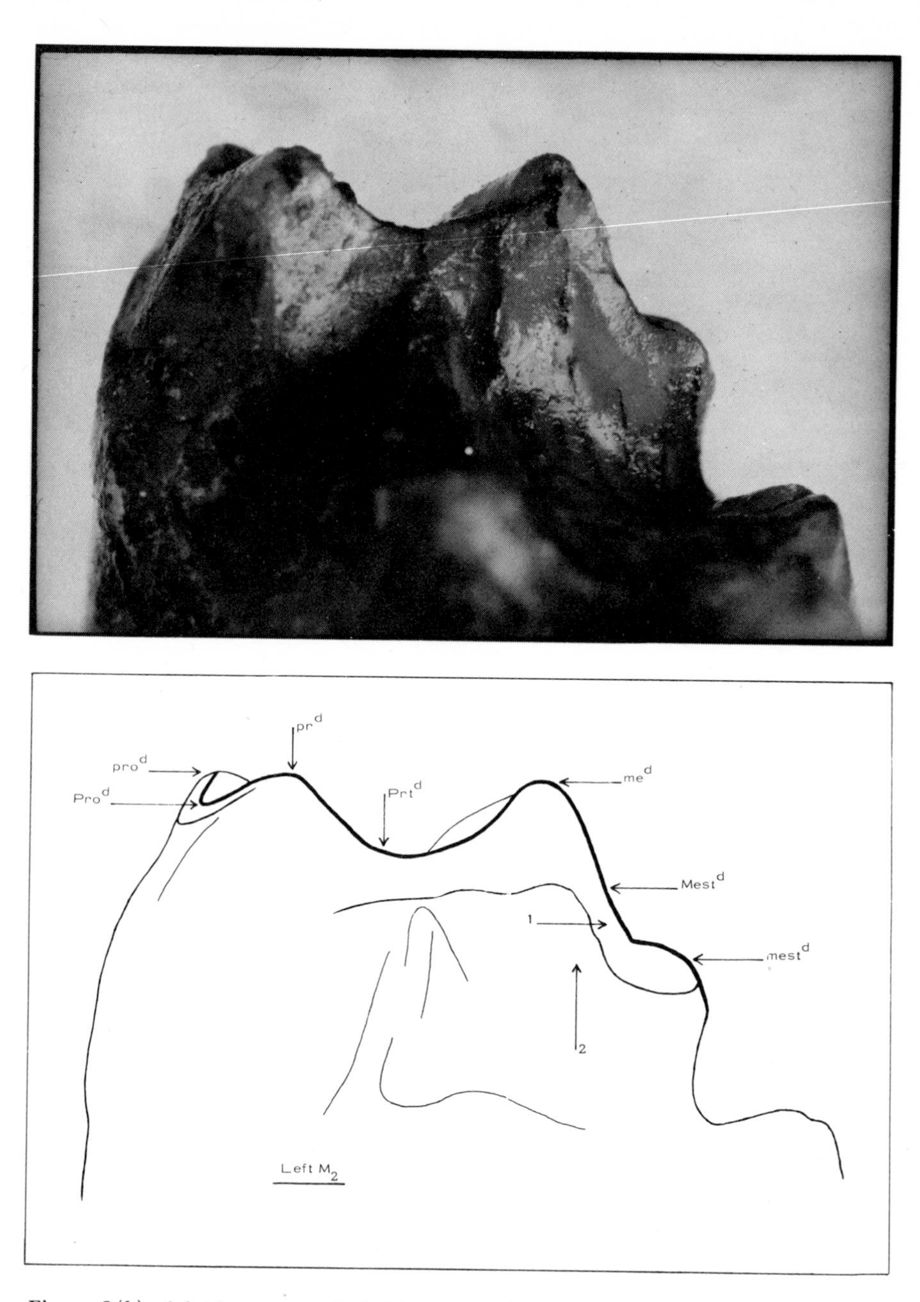

Figure 2(*b*). *Adapis magnus.* Left M_2. Distal (slightly buccal) view.
(Figs. 2 *b* and *c* are from the same tooth).
pro^d The protobliquakid, the pseudo-parakid, i.e. functionally homologous to the original parakid.
Pro^d The functional Protobliquid.

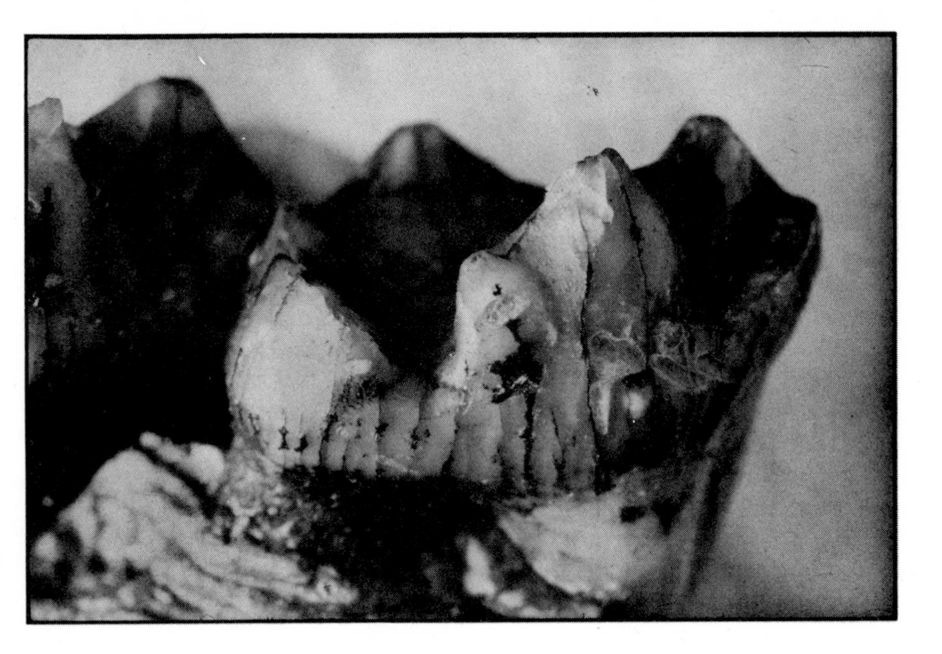

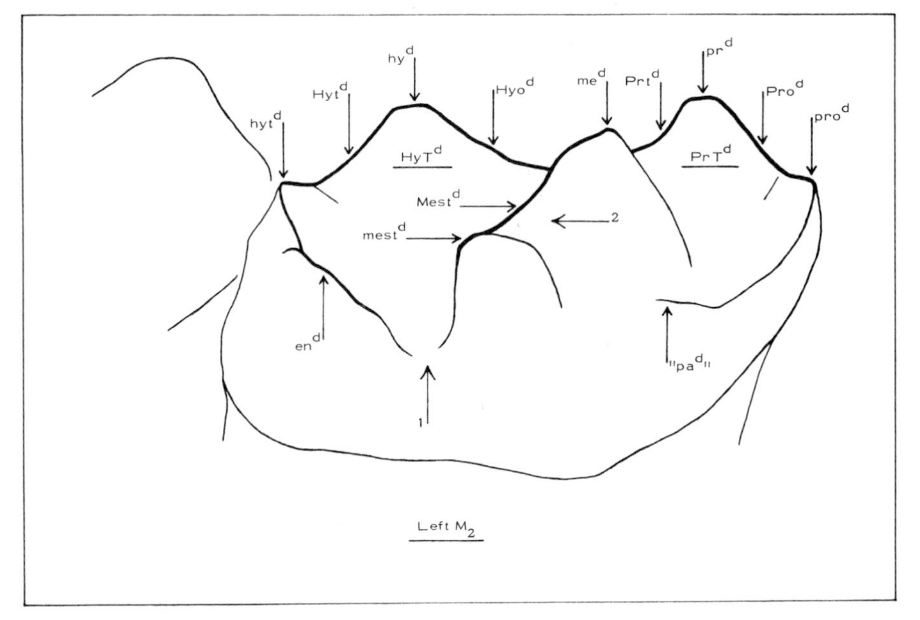

Figure 2(c). *Adaptis magnus*. Left M_2. Linguocclusal view.
(NB The lingual surface of this specimen has a number of preparation artifacts.)
1 Talonid escapement.
2 Metastylotransversid escapement.

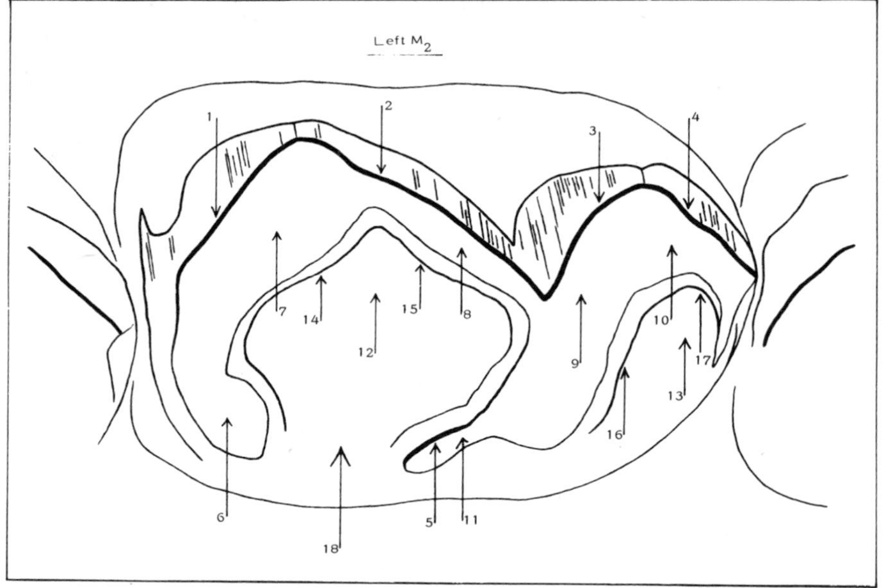

Figure 3(*a*). *Adapis magnus.* Left M_2. Occlusal view.

1-2-3-4 & 5 Leading edges of secondary, scissorio-incusive blades, now the leading blades of the tooth, i.e. Hypotransversid, Hypobliquid, Prototransversid, Protobliquid, and Metastylotransversid.

6-7-8-9-10-11 Secondary (abrasive) hollowing to the secondary blades. Secondary hollowing, also, now forms secondary (dentinal) incusive basins coalesced into one continuous basin.

12 & 13 Remnants of primary (enamel) incusive basins of Hypo-Triakid and Proto-Triakid.

14-15 & 16-17 Remnants of primary (scissorio-incusive) blades of HyT^d and PrT^d, now incusive only.

18 Talonid escapement.

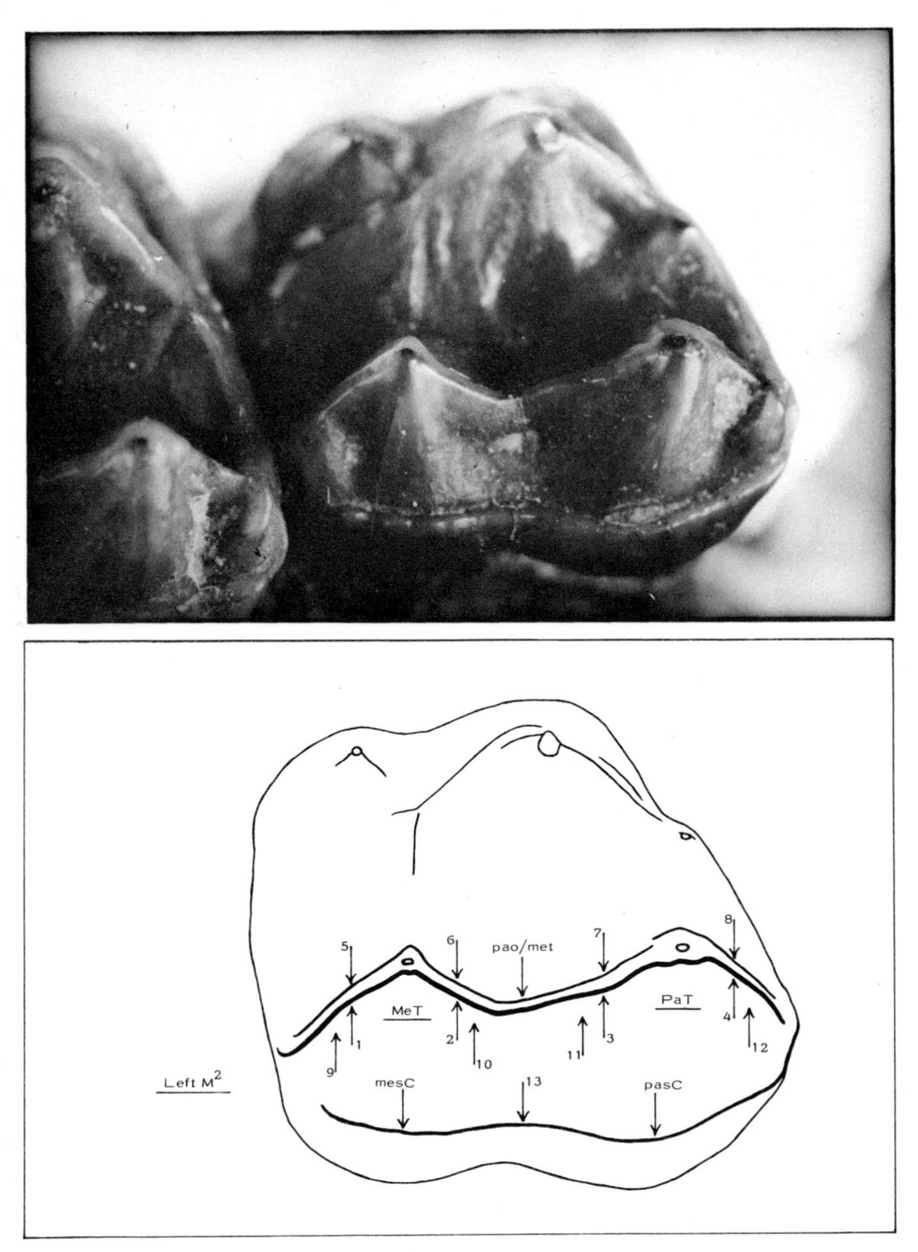

Figure 3(*b*). *Adapis magnus.* Left M^2. Buccocclusal view.
(Figs. 3, *b* and *c* and Figs. 4, *a*, *b* and *c* are from the same tooth.)
1-2 & 3-4 Leading edges of the blades of the Meta-Triakis and Para-Triakis.
5-6 & 7-8 Trailing edges of the blades of the MeT and PaT.
9 & 10 Primary hollowed-out surfaces to the leading edges of the MeT.
11 & 12 Similar surfaces to the PaT.
13 Shared (pseudo-) akis of the parastylo-Cingulis and metastylo-Cingulis.
pao/met Amalgamation of paraobliquakis and metatransversakis.

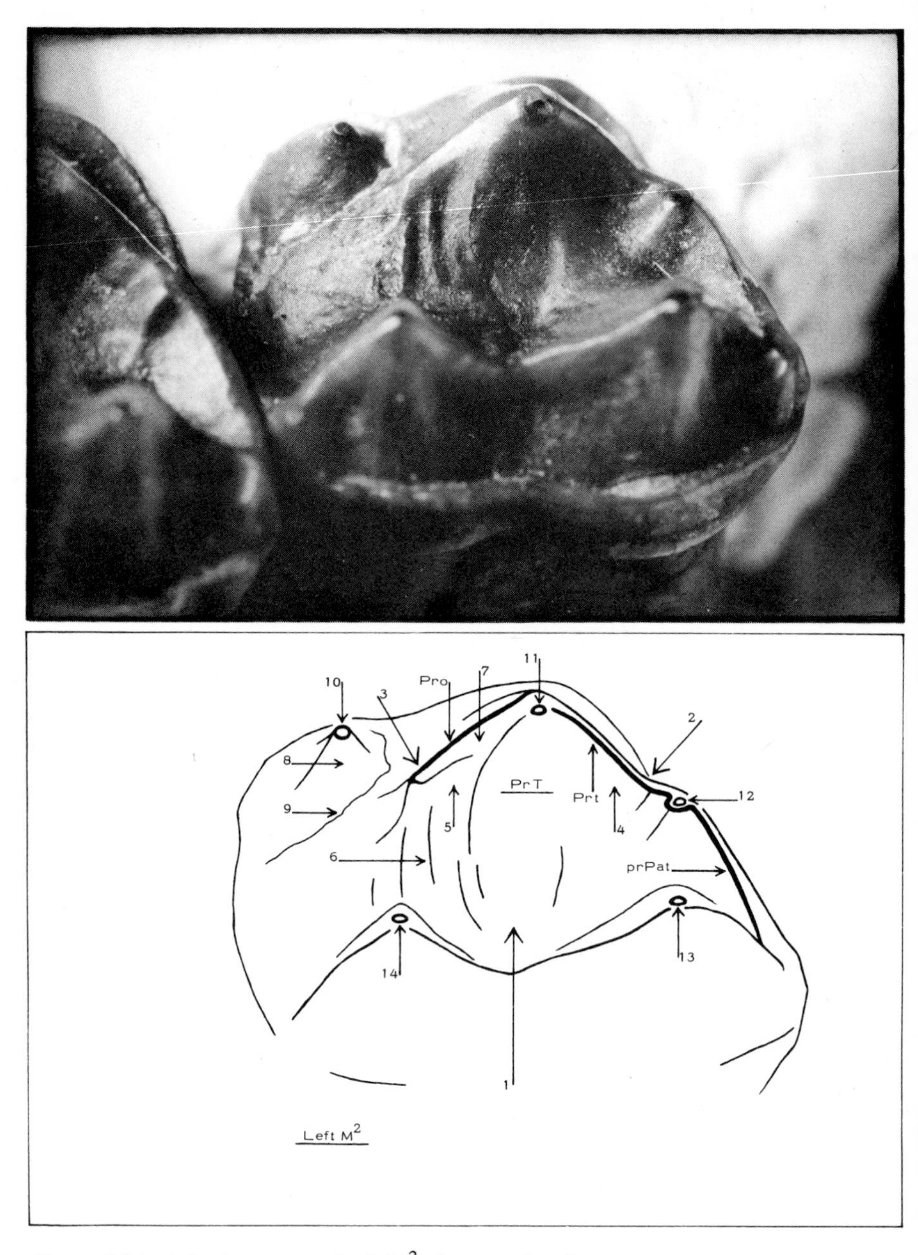

Figure 3(c). *Adapis magnus.* Left M^2. Buccocclusal view.

1 Direction of scissorio-incusive action; terminal occlusion of the hypoakid.

2 Point where the Hypobliquid in terminal occlusion crosses the Prototransversis.

3 Point where the Hypotransversid in terminal occlusion crosses the Protobliquis.

4 & 5 Primary hollowed-out incusive surfaces to the leading edges of the Proto-Triakis.

6 Ridges concaved towards the centre of the incusive action.

7 Facet caused by oblique thegosis of the Hypotransversid with the Protobliquis.

8 Exposed dentinal core of the (pseudo-) Hypo-Triakis; the enamel cap has fractured in preparation (9).

10 Functionally abraded apex of the dentinal core.

11, 12, 13 & 14 Corresponding dentinal abrasion of the protoakis, protoparakis, parakis and metakis.

can be achieved over an appreciable range of morphological variation. It is a feature, moreover, that offers considerable opportunity for biological progress.

To facilitate this and subsequent action, the juxtaposing surfaces of scissorial blades are hollowed out[3] (or are oriented to produce the equivalent effect). In the alpha stage (e.g. Fig. 3*b*; Fig. 4*a*) the hollowing is primary (developmental); in the beta stage (Fig. 5*c*) it is part primary and part secondary (abrasive). This therefore allows:

1. Close juxtapositioning of the cutting edges.

Advantage:

(a) Obviates the jamming of an interposed exogenous material between, and therefore separating, the cutting edges.
(b) Facilitates division of exogenous material.

2. The restriction of thegosis to a narrow strip at the cutting edge. This is particularly evident in the beta condition where the facet is limited by the width of the enamel plate itself.

Advantage:

(a) Facilitates the thegotic process through an economy of time; a minimum amount of tooth-substance has to be removed.
(b) Ensures the ontogenetic continuity of thegosis through an economy of tooth-material, especially the crucial edge-forming enamel.
(c) Further facilitates division.

The origin of the widened angles of primate triakididrepanoid units, both upper and lower, is a significant aspect of the primate dentition. To understand this, additional features of the ancestral alpha dentition need some discussion.

When describing the basic 'tribosphenic' lower molar, Simpson gives the morphology of an overt feature of this and subsequent alpha dentitions (where it often becomes even more overtly adapted to appear – additionally – in the upper molars), but he gives it no functional significance: 'The external cusp, protoconid, is crescentic, and from its apex run two sharp, shearing crests each of which sinks, or is distinctly notched, near the middle and then rises again to a distinct cusp.'[4]

Although the presence of this 'notch' on the ancestral Protobliquid and Prototransversid is occasionally mentioned in the literature, the absence of any attempt to give it functional significance is as conspicuous as is the absence of the feature in the descendent primate dentition.

One of the advantages of the scissorial diakidrepanoid blade is that it allows an extended length of point-cutting edge. It also capitalises on the horizontal-opposing-convexity principle mentioned above.

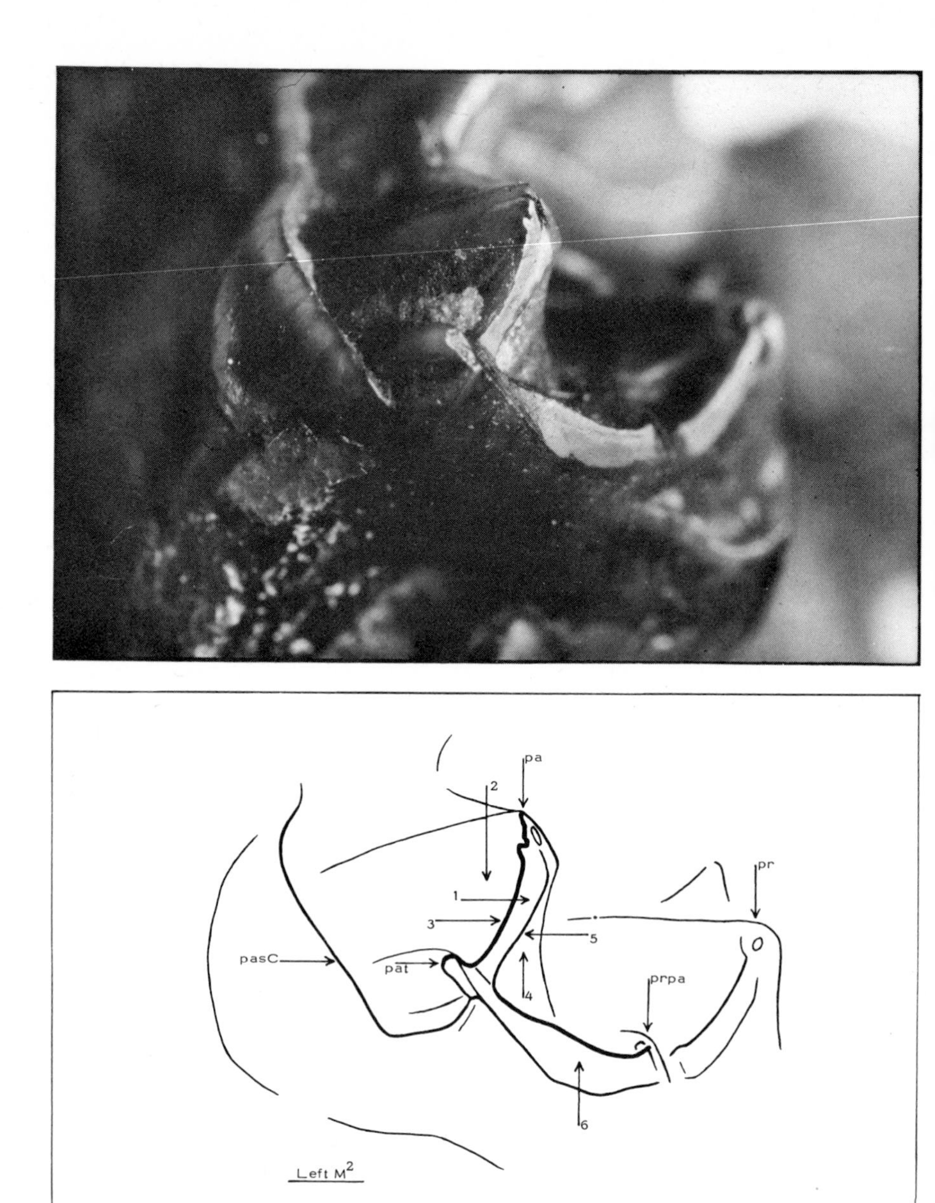

Figure 4(*a*). *Adapis magnus.* Left M^2. Mesiobuccocclusal view.

1 Alpha-thegosis of the Paratransversis; there is primary hollowing (2) to the leading edge (3), also primary hollowing (4) to the trailing edge (5).

6 Alpha-thegosis of the proto-Paratransversis, with corresponding characteristics.

pasC The parastylo-Cingulis, an ontogenetic, reserve, leading blade of the crown.

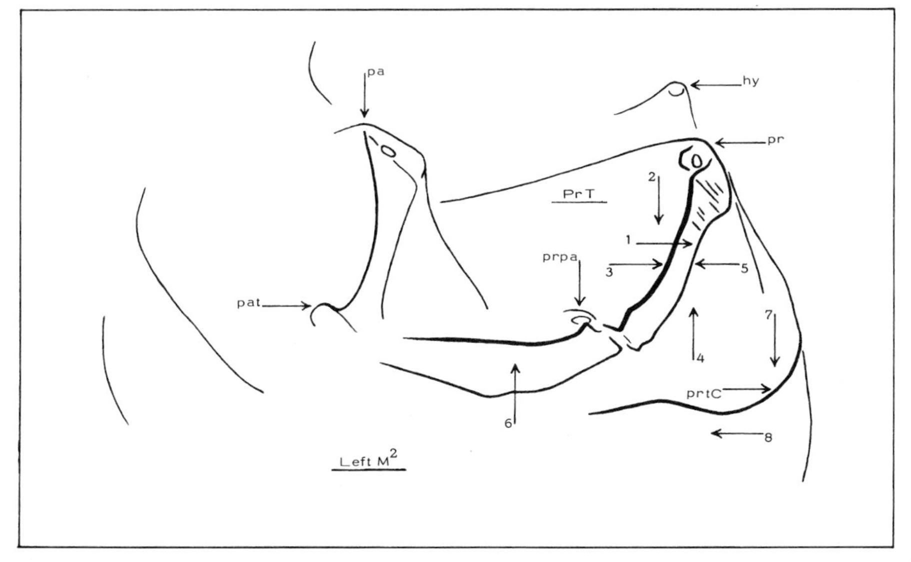

Figure 4(*b*). *Adapis magnus.* Left M^2. Mesiocclusal view.

1 Alpha-thegosis of the Prototransversis. The primary hollowing (2) to the leading edge (3) also forms part of the incusive basin of the Proto-Triakis; the primary hollowing (4) to the trailing edge (5) forms part of the mesiobuccal wall of the crown.

6 Alpha-thegosis of the proto-Paratransversis, with corresponding features.

7 & 8 Primary hollowing to the prototransverso-Cingulis.

prtC The prototransverso-Cingulis, an ontogenetic, reserve, trailing blade of the crown.

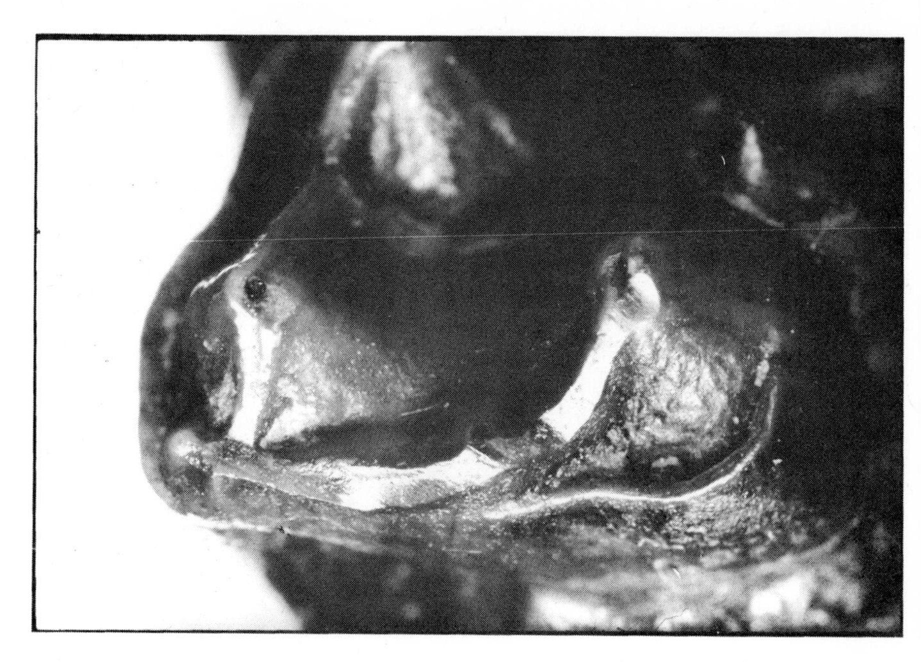

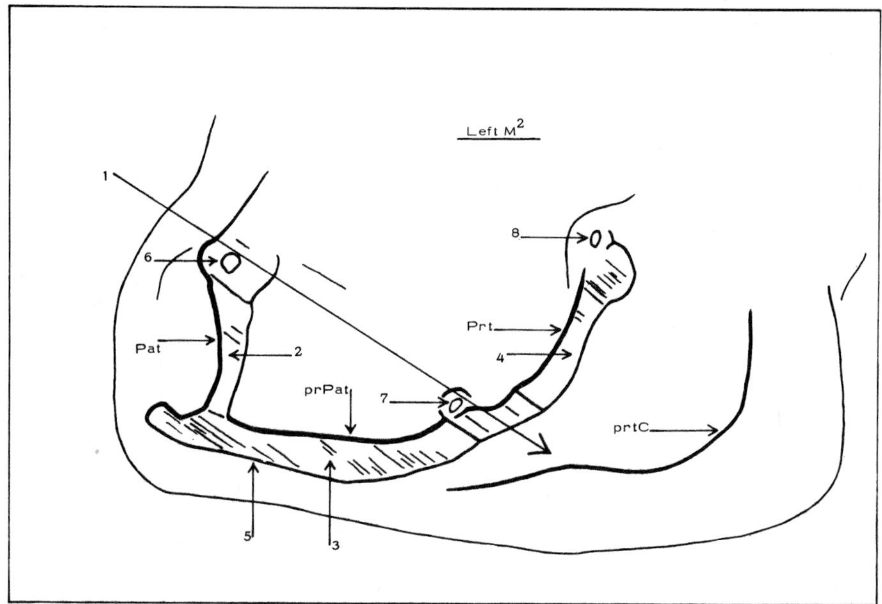

Figure 4(c). *Adapis magnus.* Left M^2. Mesiocclusal view.

1 Direction of scissorial action.

2, 3 & 4 Narrow strips of thegosis faceting of the Paratransversis, proto-Paratransversis and Prototransversis.

6, 7 & 8 Beginnings of secondary hollowing at the apices of the parakis, protoparakis and protoakis.

Simple diakidrepanoid blades have an inherent disadvantage that becomes especially significant when blades are elongated: when the central concavities of two blades approximate, point-cutting becomes line-cutting. It is a disadvantage, however, that is obviated by the morphological *fissuring* of one or both the blades;[3] point-cutting is thus maintained.

In the (advanced) alpha dentition the Protobliquid is the dominant blade of the lower molar and is characteristically fissured. Moreover, it overshadows the basin of the adjacent mesial tooth, on the distal margin of which is the Hypotransversid and entoakid. Inevitably, then, when the (adjacent) basin is progressively raised, the function of the Protobliquid is affected. When the two structures (i.e. the Protobliquid and the Hypotransversid) eventually become level, the portion of the original Protobliquid lingual to the interproximal contact-point is rendered functionless and is, in fact, vestigial or lost (Figs. 2*c*; 6*b*, *c*). Significantly, the characteristic scissorial, diakidrepanoid blade, although greatly shortened, has remained intact (Figs. 2*b*, *c*; 5*a*, *b*). In some primates the relict of the parakid is apparent, in others its former location is represented by a void.

The question arises, since there is still the characteristic akid at the free (lingual) end of the remaining blade: has the parakid in some primates 'migrated', or is the existent akid invariably a new evolutionary feature? Since in many species an obvious remnant parakid occurs, I therefore propose the functional term *protobliqakid.* The blade remains the Protobliquid.

The new blade, mentioned above, that occurs as a result of the enlargement of the incusive surfaces, runs distolingually from the metakid (Figs. 2*a*, *b*, *c*; 5*a*, *b*). A blade in this position also occurs in many dentitions other than those of the primates, particularly alpha/beta and beta dentitions. It even occurs, but in a primitive form, in alpha dentitions where there is a quasi-enlargement of the basin of the Proto-Triakis as a result of a secondary reduction of the Para-Triakis (e.g. *Didelphis*).

Crompton (whose system of facet numbering is increasingly used) depicts the facet of this blade as 'shearing surface 5'.[11] He seems more concerned, however, with the size and 'efficiency' of this and other surfaces than he is with their cutting edges. Although Crompton recognises my personal communications with him regarding such features,[11] he has yet to recognise the phenomenon of thegosis upon which all this is based. He recognises, for instance, that 'the food to be sheared was trapped in an ellipsoid space which decreased in size as the leading edges passed by one another. Once the leading edges had passed one another, no further shearing was possible'. Despite this, Crompton seems ambivalently preoccupied with the 'shearing surface', and his analysis of this feature becomes the dominant basis for his discussion of tooth function generally, for instance:

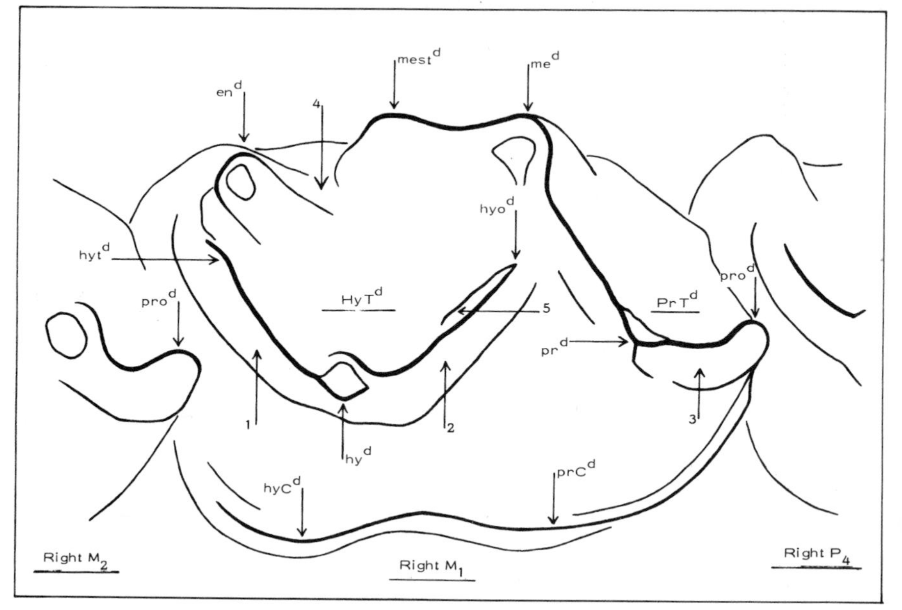

Figure 5(*a*). *Adapis parisiensis.* Right M$_1$ and parts of P$_4$, M$_2$. Buccoclusal (slightly distal) view.
(Figs. 5, *a*, *b* and *c* are from the same specimen.)
1, 2 & 3 Alpha-thegosis of the Hypotransversid, Hypobliquid and Protobliquid.
4 Talonid escapement.
5 Facet caused by oblique thegosis of the Hypobliquid with the Prototransversis.
hyo^d (Pseudo-) hypobliquakid, absorbed in the wall of the Prototransversid.

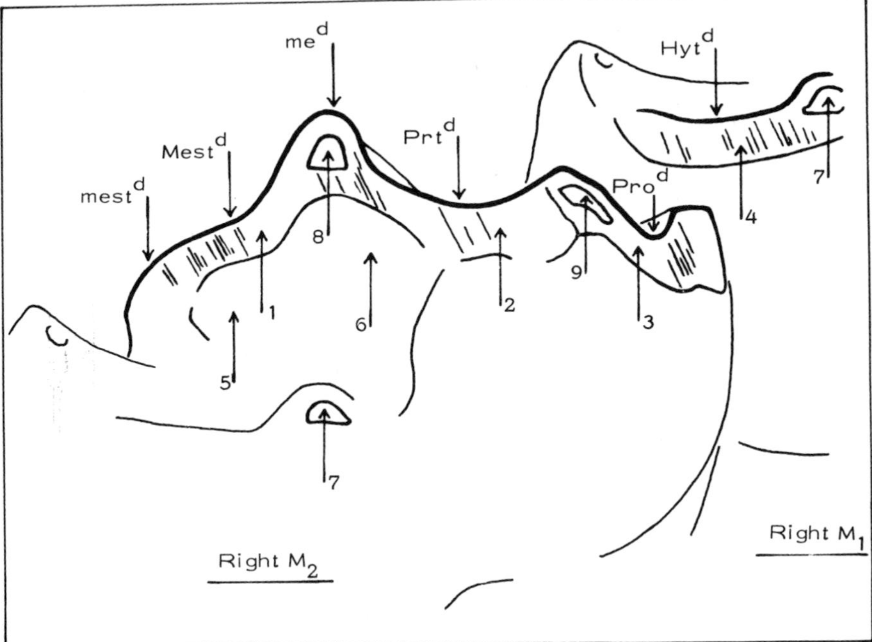

Figure 5(*b*). *Adapis parisiensis.* Right M_2 and part of M_1. Distobuccocclusal view.
1, 2, 3 & 4 Alpha-thegosis of the Metastylotransversid, Prototransversid, Protobliquid and Hypotransversid.
5 & 6 Primary hollowing of $Mest^d$ and Prt^d.
7, 8 & 9 Beginnings of secondary hollowing and beta-thegosis at the hypoakid, metakid and protoakid.

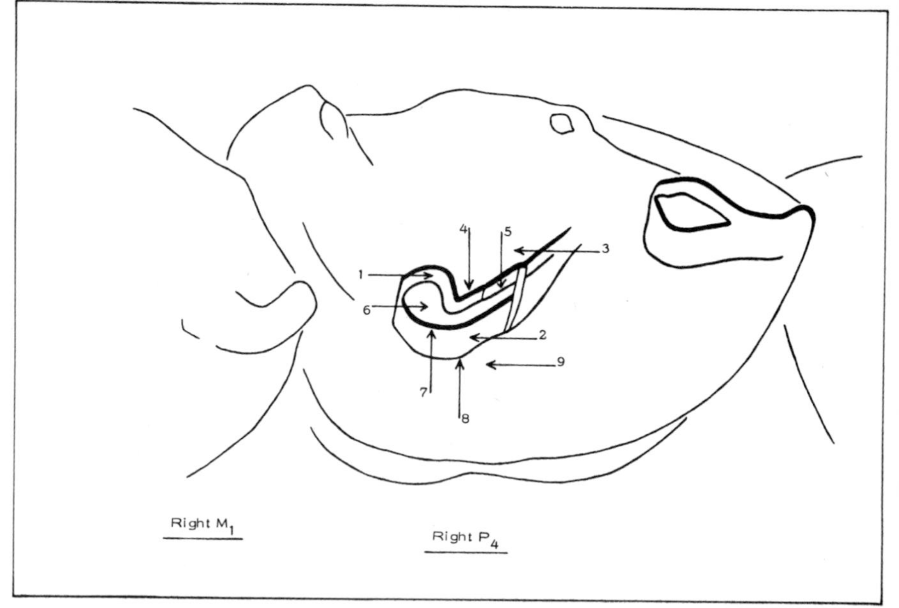

Figure 5(*c*). *Adapis parisiensis.* Right P_4. Buccocclusal (slightly distal) view.

1 Beta-thegosis of the primary Hypobliquid blade; there is primary hollowing (3) to the leading edge (4) and secondary hollowing (6) to the trailing edge (5).

2 Beta-thegosis of the secondary Hypobliquid blade; there is secondary hollowing (6) to the leading edge (7) and primary hollowing (9) to the trailing edge (8).

The position, orientation, and relative size of the shearing surfaces in *Didelphodus* were determined by the direction of mandibular movement and the type of food to be broken down. The teeth are constructed so that some of the shearing surfaces were used simultaneously while others were used successively. . . . The evolution of the tribosphenic molar from those of a form such as *Kuehneotherium* consisted essentially of increasing the transverse component of mandibular movement, increasing the size and efficiency of some of the shearing surfaces present in the early therians, and adding shearing surfaces that were used successively.

In prosimians generally the new blade is seen to have become highly developed. The blade is diakidrepanoid, it has a characteristically concave *escapement* area (Fig. 2*c*),[3] and, contrary to Crompton's suggestion of 'shearing efficiency' of a large facet, the juxtaposing surface of this scissorial blade is characteristically hollowed: the facet, therefore, characteristically forms a narrow strip at the cutting edge.

Its counterpart, the Prototransversis (with which the Prototransversid of the ancestral alpha dentition orginally articulated), also shows the scissorial diakidrepanoid characteristic (Figs. 1*c*; 4*b*, *c*). It has the narrow strip of thegosis-faceting; the hollowing (present on both sides of the blade) is particularly evident within the basin itself (Figs. 1*b*; 3*c*; 4*b*, *c*).

In some prosimians, e.g. *Palaeopropithecus* and *Lepilemur* (Fig. 6*a*, *b*, *c*), the blade is so extensive that the entoakid has entirely disappeared. That this is a loss and not an amalgamation is evident from the fact that the extremity of the blade (i.e. the metastylotransversakid) is separated from the extremity of the Hypotransversid (i.e. the hypotransversakid) by the escapement valley that phylogenetically separates it from the entoakid. Since this new scissorial blade is virtually an extension, at the metakid, of the transverse blade of the Proto-Triakid, I therefore propose the term *Metastylo-diakidrepanid-transversa*, abbreviated to *Metastylotransversid*.

With respect to two disputed primates *Tupaia* and *Plesiadapis*, the absence of the Metastylotransversid in the former and presence in the latter confirms that *Tupaia* should be placed clearly in the Insectivora and *Plesiadapis* equally clearly in the Primates (Fig. 2*a*).

2. *Scissorio-incusive blades*

Despite Simpson's statement, when discussing the origin of the basining of the talonid and central basin of the upper molar, that 'in addition to this double opposition, the elements of alternation and shear are preserved',[4] his preoccupation with the 'tribo-' function for the basins has emphasised the importance of the central bluntness rather than the peripheral sharpness of these structures. Simpson

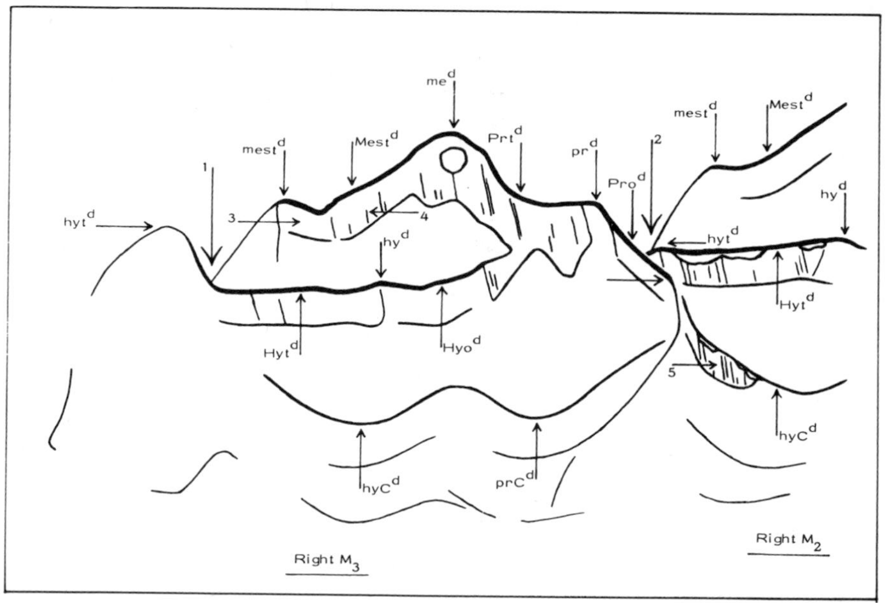

Figure 6(*a*). *Lepilemur mustelinus.* Right M3. Buccocclusal (slightly distal) view.
1 & 2 Talonid escapement, now between the metastylotransversakid and the hypotransversakid (phylogenetically between the metastylotransversakid and the entoakid), i.e. the entoakid is lost.
3 & 4 Alpha-thegosis of the Metastylotransversid (facet 3 is in the shade).
5 Alpha-thegosis of the hypo-Cingulid.

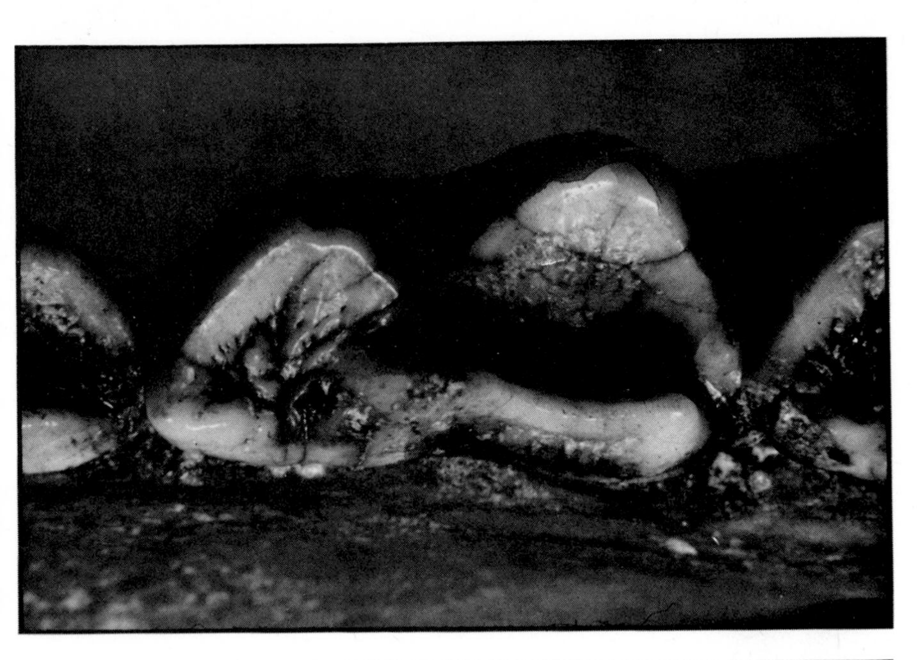

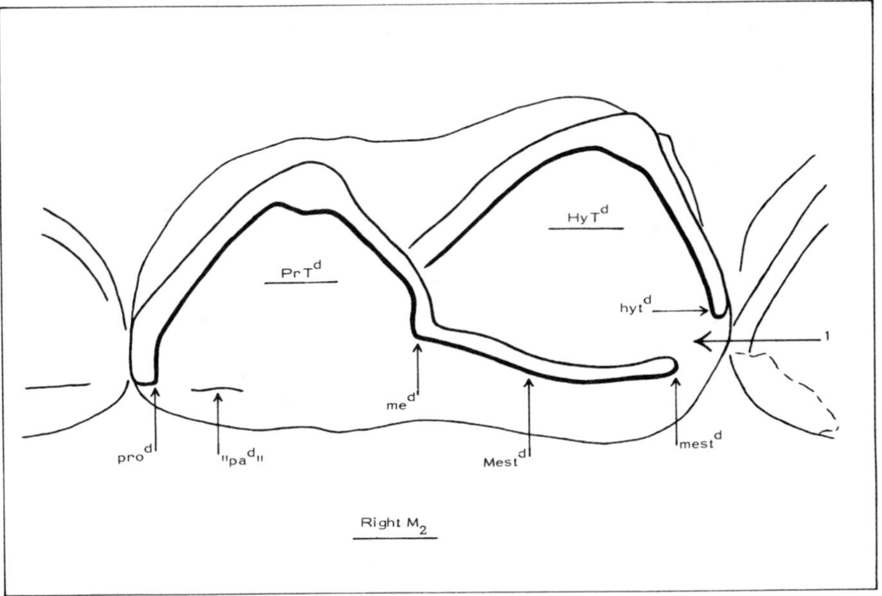

Figure 6(*b*). *Palaeopropithecus maximus.* Right M_2 . Occlusal view.
1 Talonid escapement, as in *Lepilemur.*

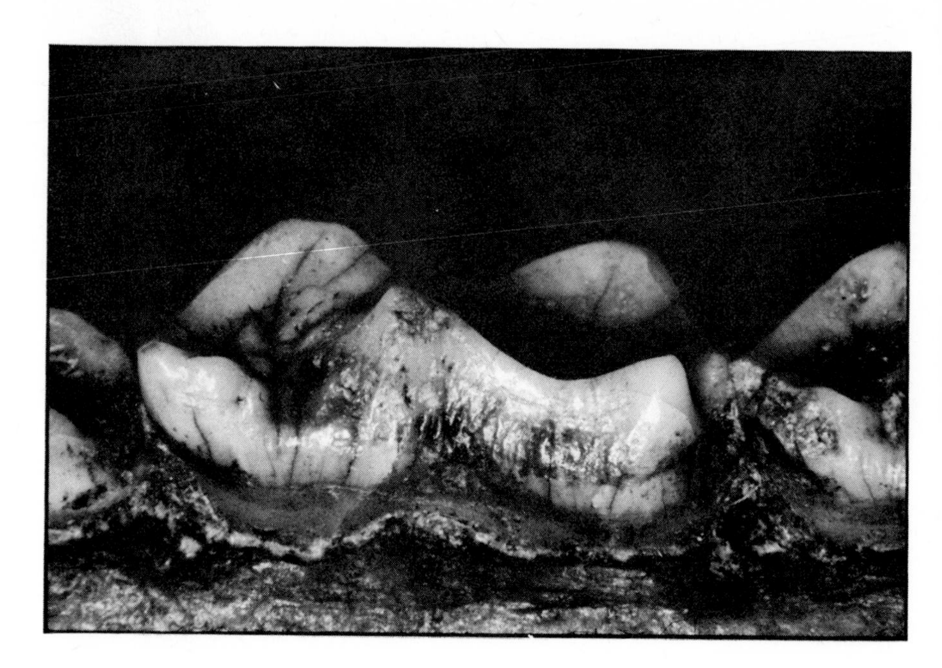

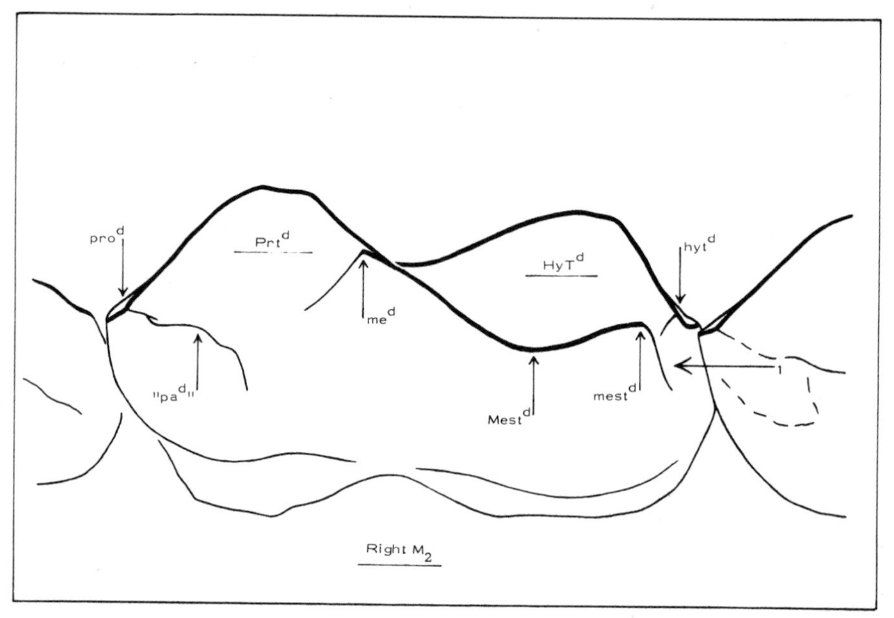

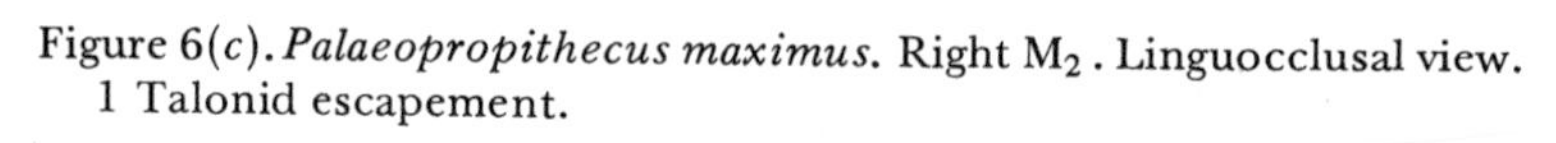

Figure 6(*c*). *Palaeopropithecus maximus.* Right M_2 . Linguocclusal view.
1 Talonid escapement.

relates the origin of these structures (including their peripheral sharpness, which he, ironically, did not entirely overlook) to a development from a 'protribosphenic' Middle Jurassic molar whose talonid showed 'simple opposition', i.e. an initial structure with bluntness but no sharpness.

This argument now proves fallacious. The talonid of the Middle Jurassic pantotheres functioned as a scissorial blade that protected the vulnerable interproximal mucous membrane from pathogenic impaction of exogenous material. There is no (blunt) opposability of the talonid.[3]

It can hardly be overemphasised that the prime evolutionary advantage of teeth is their sharpness, the secondary advantage is bluntness. For although most triakididrepanoid blades function purely scissorially, those of the talonid double in function, acting both scissorially and incusively.

The transverse blade of the Hypo-Triakid occludes first scissorially with the transverse blade of the Meta-Triakis, then incusively with the oblique blade of the Proto-Triakis; whereas the oblique blade occludes, first scissorially with the oblique blade of the Para-Triakis, then inclusively with the transverse blade of the Proto-Triakis.

There are, therefore, two fundamentally different occlusal relations:

1. Scissorial: the blades are in point-cutting relationship, i.e. transverse blade occludes with transverse blade, and oblique with oblique.
2. Incusive: the blades are crossed diagonally in striking relationship, i.e. transverse blade occludes with oblique blade, and oblique with transverse.

The dominant function of the incusive (basin) surface is to confine exogenous material to the striking (dividing) action of the incusive blade-and-cusp. Occasionally, however, exogenous material will elude the blade-and-cusp, yet be divided (cracked, split, mashed) by the mutual striking action of the incusive surfaces alone.

With a double (scissorio-incusive) function for the feature there was, therefore, selective advantage for a greater reserve bulk of tooth substance for this segment of the tooth; this allowed greater ontogenetic viability for the crucial scissorio-incusive blade-and-cusp. Hence the Hypo-Triakid/Proto-Triakid levelling, and the inevitable extension of incusive functions to the Proto-Triakid.

With a history of general acceptance of the dominant 'tribo-' function for the pre-primate basin it is hardly likely that the enlarged primate basin, or even the 'lingual-phase-of-occlusion' concept, should raise any question. From the knowledge of the phenomenon of thegosis a number of pertinent questions can now be asked. One of these is fundamental to an understanding of the whole primate dentition: is the increased bulk of the talonid, for example, primarily adapted to the increased functional wear of the basin or to that of its peripheral blades and cusps?

Perhaps the answer can be found in two characteristic features: the differential hardness and the differential morphology of enamel and dentine, the two basic tooth-structures.

Fig. 3*c* shows the dentinal core of the hypoakis of *Adapis magnus* left M^2; the enamel cap has fractured off in preparation. The apex of the dentinal core has been worn flat by functional abrasion, as has that of the protoakis, protoparakis, parakis and metakis; all are equally worn (see also Fig. 4*c*). Fig. 5*a, b* show the same worn dentinal apices of *Adapis parisiensis* right M_1. Fig. 5*c* (P_4 of the same specimen) shows the extension of dentinal abrasion along the (now) beta scissorial blades. With continued functional abrasion the dentinal areas widen, lengthen, and eventually coalesce. They thus form a strip along the entire blade system, i.e. the Protobliquid, Prototransversid, Metastylotransversid, Protobliquid, Hypotransversid, and entoakid (Fig. 3*a*).

At this stage, the primary blades progressively lose their thegotic contacts and, therefore, progressively lose their scissorial function; their incusive function remains.

The scissorial function of the secondary blades (now the leading scissorial blades) are, however, enhanced. This is a result of the widening and deepening of the secondary hollowing, which now becomes a secondary incusive basin.

Finally, i.e. in the merontic condition, the secondary (dentinal) basins take over from the primary (enamel) basins, which disappear. The occlusal surface now becomes one large, secondary, dentinal basin, encircled by peripheral (secondary) blades that are still subject to beta-thegosis, although now ontogenetically (i.e. adaptively) attenuated (Fig. 7*a, b, c*).

Even with a superficial observation of this phenomenon, it is surprising that the overt breaching of enamel at the cusp apex, the

Figure 7(*a*). *Galago crassicaudatus.* Left M^1, M^2, M^3. Occlusal view. Showing abrasion (scratch and pit marks) on the secondary (dentinal) incusive surfaces. The enamel in this specimen is clearly defined, being slightly decalcified in preparation.

1 & 2 Remnants of the primary (enamel) incusive surfaces of the Proto-Triakises of M^1 and M^2.

3, 4 & 5 Secondary (dentinal) incusive basins of the remnant Proto-Triakises of M^1, M^2 and M^3.

6, 7 & 8 Remnants of the primary incusive basins of the (pseudo-) Hypo-Triakises of M^1, M^2 and M^3.

9 & 10 Secondary incusive basins of the remnant (pseudo-) Hypo-Triakises of M^1 and M^2.

11 Coalescence of secondary incusive basins of M^1.

12 Tangent to the general direction of mandibular movement in the scissorio-incusive action, i.e. mastication. This is the direction of the so-called 'buccal phase of occlusion'.

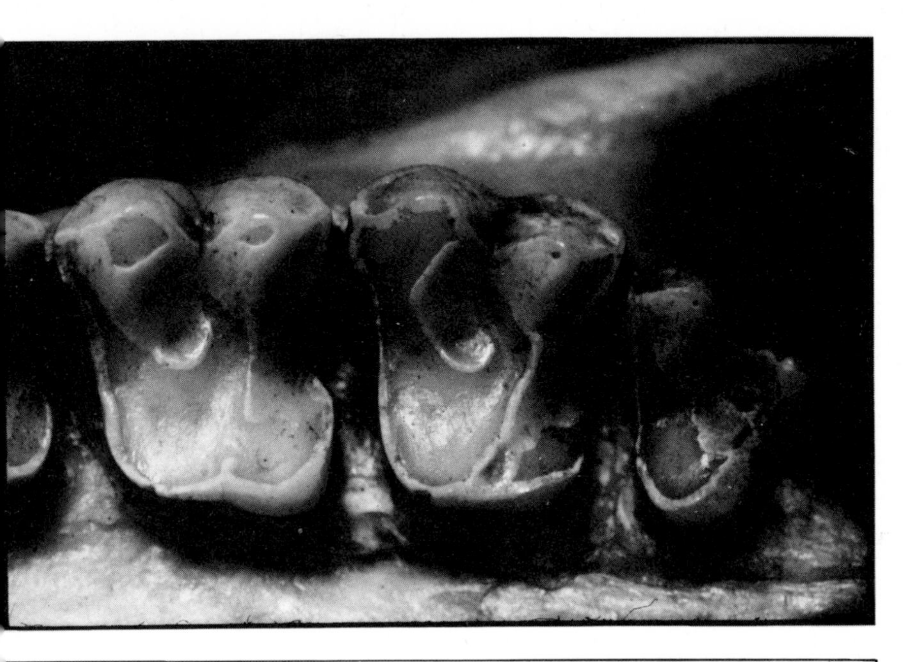

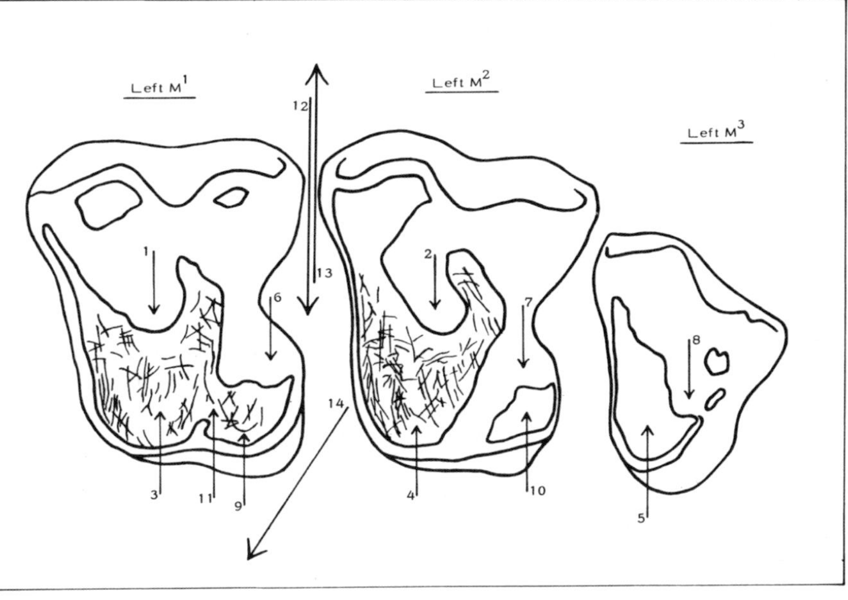

13 Tangent to the general direction of the (dominant) transverse thegotic action (as shown in young adult *Galago* specimens).
14 Tangent to the general direction of the (dominant) oblique thegotic action (from young adult specimens). This is the direction of the so-called 'lingual phase of occlusion'. The evidence of functional (masticatory) action, however, is not the matching parallel striations on facets but the curved, divergent, scratch, pit and gouge marks of abrasion. The evidence of multiple incusive basins and abrasion, alone, falsifies the concepts of a 'mortar-and-pestle' action and a 'lingual phase of occlusion'.

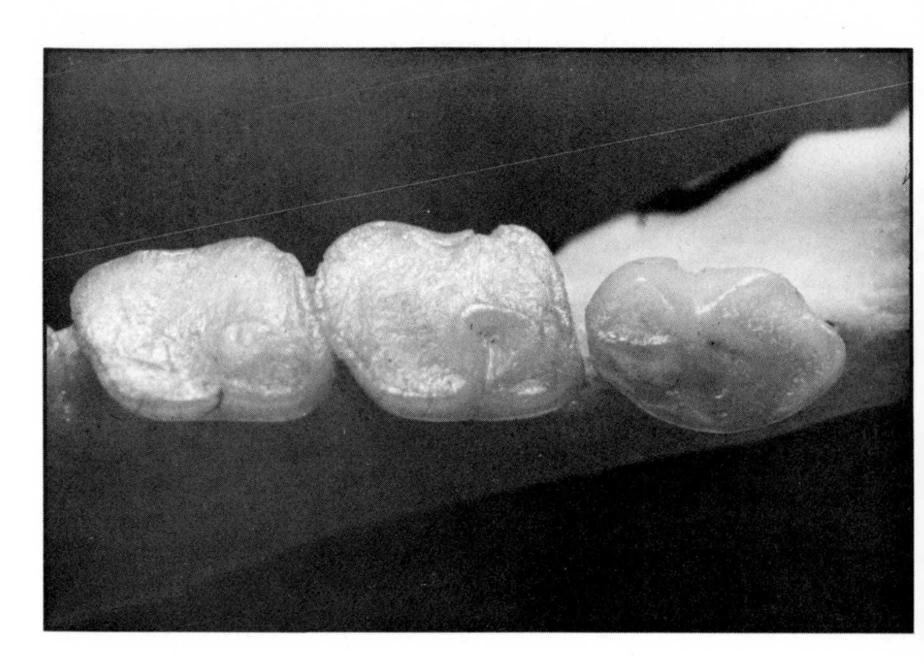

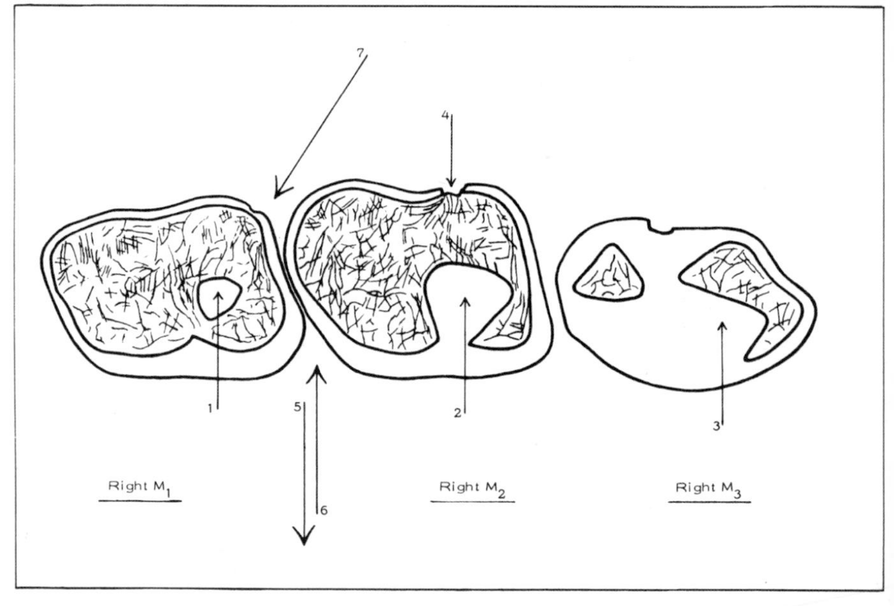

Figure 7(*b*). *Galago crassicaudatus.* Right M_1, M_2, M_3. Occlusal view. Showing abrasion on secondary incusive surfaces. All secondary incusive basins of M_1 and M_2 are coalesced.

(Figs. 7*b* and *c* are from the same animal.)

1, 2 & 3 Remnants of primary incusive basins of the Hypo-Triakises of M_1, M_2 and M_3.

4 Small fracture of enamel periphecy, with abrasion channelled into the opening.

5 Tangent to the general direction of scissorio-incusive action.

6 Tangent to the general direction of transverse thegotic action.

7 Tangent to the general direction of oblique thegotic action.

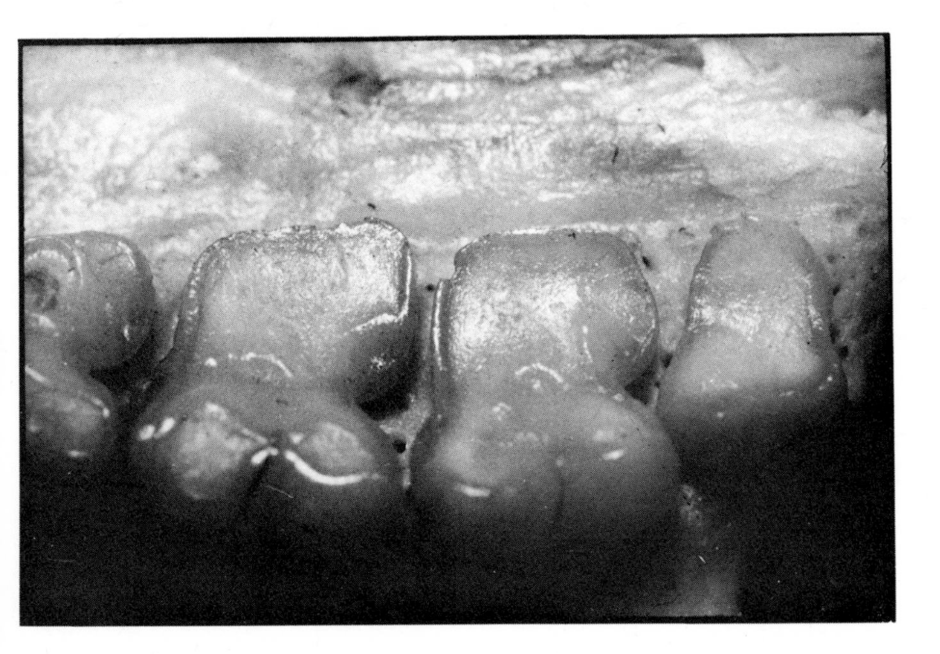

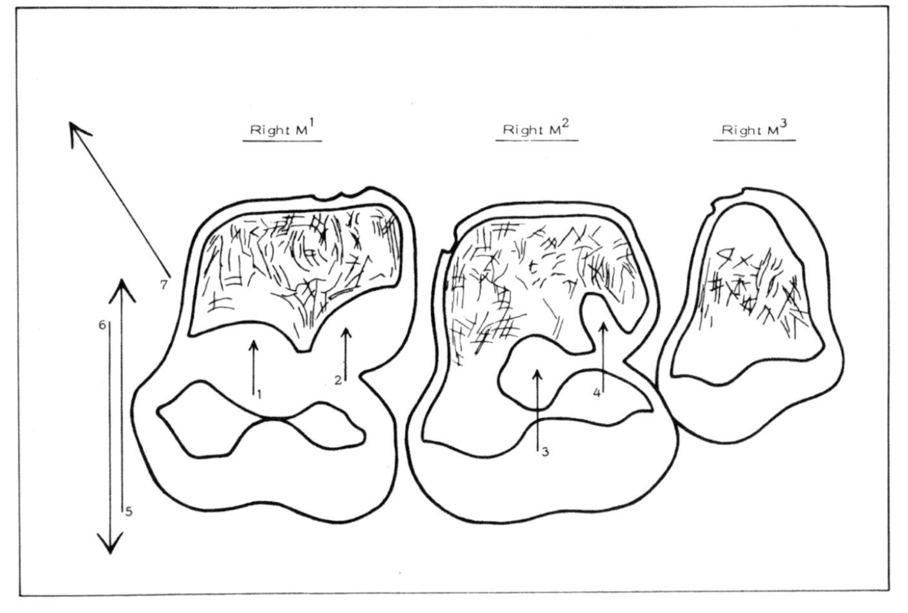

Figure 7(*c*). *Galago crássicaudatus.* Right M^1, M^2, M^3. Occusal view, showing abrasion on secondary incusive surfaces; all basins are now coalesced.
1, 2, 3 & 4 Remnants of primary incusive basins.
5 Tangent to the general direction of scissorio-incusive action.
6 Tangent to the general direction of transverse thegotic action.
7 Tangent to the general direction of oblique thegotic action.

exposure and rapid wear of the dentine along the blade system, has been interpreted as evidence of dominant 'tribo-' functions, or, as Kay and Hiiemae suggest in this volume, that 'puncture crushing' before the teeth 'come into contact' 'blunts or flattens the tips of the cusps'.[7]

Three further questions arise for the evolutionists:

1. What is the adaptive significance of a finely tapered dentinal core, especially that of a 'pestle' cusp, when its effect is merely to jeopardise the cusp's viability?
2. What is the adaptive significance of a 'pestle' cusp, especially when 'blunted', when its function is not only jeopardised by its shortening, i.e. removal of contact with the 'mortar' but by the creation of a second 'mortar' in addition to the (reduced) first, the second even deeper, moreover, than the first?
3. What is the adaptive significance of a 'lingual phase of occlusion' when no cutting blade is oriented in this direction and, in fact, the phase produces no evidence of (functional) abrasion that is in any way different from that produced by the buccal phase (Figs. 1*b*; 7*a, b, c*)?

As is to be expected, the morphological characteristics of dentine when analysed are seen to be adapted, as is its differential hardness, to facilitate the ontogenetic viability of the functional features of the enamel which the dentine supports. They are adaptations in terms of the prime evolutionary advantage of teeth: their sharpness.

Evolutionary control of tooth-wear

One feature of anthropoid thegosis that may prove crucial to our understanding of prosimian phylogenies is the suggestion that in primates the direction of the final phase of the (transverse) masticatory action is opposite to that of the dominant thegotic action. The two actions are discrete, not only directionally, but also temporally – occurring at quite different times.[1,2,10,13]

The phylogenesis of upper canine thegosis is an important aspect of this evidence and relevant to thegosis in prosimians, but it is excluded here, partly because my studies of this aspect of the evidence have been limited by the paucity of fossil canines, and partly because I have concentrated on molar thegosis.

As the idea gains further ground that the sharp boundary produced by faceting is the functional cutting agent, it can be anticipated that those who find the thegosis concept unacceptable will attempt alternative arguments of 'self-sharpening' mechanisms for various teeth (i.e. automatic sharpening, concomitant with the masticatory action itself).

The self-sharpening idea occurred to me some 25 years ago. My electric shaver at that time had rather soft stainless-steel blades that, too often, wore out and had to be replaced. Although the metal was soft, the blade edges were always sharp. It was immediately clear, however, that there was no parallel between this action and that of mastication. The action of an ordinary pair of scissors, where virtually each stroke is a cutting one, could more reasonably be considered as a parallel. Unlike the shaver, scissors characteristically become blunt, not sharp, with use. To resharpen a scissor blade effectively one has to remove a whole layer off one surface, evenly and precisely, so as to maintain the optimum angle of bevel to the leading cutting edge. It is better done by an expert who knows what is the optimum angle, what force to use in the action, and when to stop.

A fundamental characteristic of phylogenetically-derived behaviour is the inherent mechanism that starts (evokes) the behaviour and stops (attenuates, terminates) it. This mechanism is species-specific: it is as characteristic and stereotyped as the animal's morphology.

Considering the strong selective forces that must be operating on brachydont teeth to conserve their irreplaceable substances, evolutionary processes have doubtless operated to produce a mechanism to resharpen enamel blades in a controlled manner. With fine control of both the occurrence and the extent of the action there would be selective advantage; without it, there would be no way of balancing the need for blades to be resharpened with the need to avoid over-sharpening, i.e. unnecessary wastage of tooth-substance. The structures of the oral cavity, particularly the peridontal membranes, are richly supplied with proprioceptive pathways; it would be surprising were they unable to cope with a mechanism guiding and eventually terminating a masticatory stroke just short of collision of tooth-substances, a collision that would bring discomfort to the mouth and no advantage whatsoever to the division of exogenous material. The jarring discomfort of chewing on an undetected spicule of bone, even a grain of sand, illustrates the argument.

This paper is in a way a preliminary report as I have only recently had access to prosimian material. The exposure, however, has shown many avenues that I hope to pursue.

Acknowledgments

I thank Mrs Margaret Black of Kent, England, who not only financed the entire trip from London's antipodes but also most of several months' expenses (including the photography) while I studied the relevant material and wrote this paper. There were many others who contributed, chiefly Frank Canaday through the Explorers' Club, New York, and my own family. I also thank all those individuals and institutions who gave me access to their material.

NOTES

1 Every, R.G. (1970), 'Sharpness of teeth in man and other primates', *Postilla* (Yale) 143, 1-30.

2 Every, R.G. (1960), 'The significance of extreme mandibular movements', *Lancet* 2, 37-9.

3 Every, R.G. (1972), *A New Terminology for Mammalian Teeth: Founded on the Phenomenon of Thegosis*, Christchurch, N.Z.

4 Simpson, G.G. (1936), 'Studies of the earliest mammalian dentitions', *Dent. Cosmos* 78, 791-800.

5 Butler, P.M. (1952), 'The milk molars of the Perissodactyla, with remarks on molar occlusion', *Proc. Zool. Soc. Lond.* 121, 777-817.

6 Mills, J.R.E. (1955), 'Ideal dental occlusion in the Primates', *Dent. Pract.* 6, 47-61.

7 Kay, R.F. and Hiiemae, K.M., this volume.

8 Crompton, A.W. and Hiiemae, K.M. (1970), 'Molar occlusion and mandibular movements during occlusion in the American opossum, *Didelphis marsupialis*', *Zool. J. Linn. Soc.* 49, 21-47.

9 Every, R.G., in preparation.

10 Every, R.G. (1965), 'The teeth as weapons: their influence on behaviour', *Lancet* 1, 685-8.

11 Crompton, A.W. (1971), 'The origin of the tribosphenic molar', in Kermack, D.M., and Kermack, K.A. (eds.), *Early Mammals, Zool. J. Linn. Soc.* 50, Suppl. 1, 65-87.

12 Gingerich, P.D., this volume.

13 Every, R.G., in preparation.

List of abbreviations

en^d	entoakid
hy	hypoakis
hyC^d	hypo-Cingulid (Hypo-cingulo-diakidrepanid)
hy^d	hypoakid
hyo^d	hypobliquakid
Hyo^d	Hypobliquid (Hypo-diakidrepanid-obliqua)
hyt^d	hypotransversakid
Hyt^d	Hypotransversid (Hypo-diakidrepanid-transversa)
HyT^d	Hypo-Triakid (Hypo-Triakididrepanid)
me	metakis
me^d	metakid
meo	metaobliquakis
Meo	Metaobliquis (Meta-diakidrepanon-obliqua)
mesC	metastylo-Cingulis (Metastylo-cingulo-diakidrepanon)
mest^d	metastylotransversakid
Mest^d	Metastylotransversid (Metastylo-diakidrepanid-transversa)
met	metatransversakis
Met	Metatransversis (Meta-diakidrepanon-transversa)
MeT	Meta-Triakis (Meta-Triakididrepanon)
pa	parakis

'pa^{d}'	(relict) parakid
pao	paraobliquakis
Pao	Paraobliquis (Para-diakidrepanon-obliqua)
pao/met	amalgamation of paraobliquakis and metatransversakis
pasC	parastylo-Cingulis (Parastylo-cingulo-diakidrepanon)
pat	paratransversakis
Pat	Paratransversis (Para-diakidrepanon-transversa)
PaT	Para-Triakis (Para-Triakididrepanon)
pr	protoakis
prC^{d}	proto-Cingulid (Proto-cingulo-diakidrepanid)
pr^{d}	protoakid
prme	protometakis
Pro	Protobliquis (Proto-diakidrepanon-obliqua)
pro^{d}	protobliquakid (pseudo-parakid, i.e. functionally homologous to the original parakid)
Pro^{d}	Protobliquid (Proto-diakidrepanid-obliqua)
prpa	protoparakis
prPat	proto-Paratransversis (proto-Para-diakidrepanon-transversa)
Prt	Prototransversis (Proto-diakidrepanon-transversa)
PrT	Proto-Triakis (Proto-Triakididrepanon)
prtC	prototransverso-Cingulis (Proto-cingulo-diakidrepanon-transversa)
Prt^{d}	Prototransversid (Proto-diakidrepanid-transversa)
PrT^{d}	Proto-Triakid (Proto-Triakididrepanid)

J. H. SCHWARTZ

Premolar loss in the primates: a re-investigation

Introduction

From his study of fossil and living lemuroids, Gregory derived a theory of premolar loss during primate evolution:

> The indrisine lemurs (Indrisidae) afford an instructive example of the reduction in the number of the premolars from four to two on each side, by the loss of the anterior two, probably conditioned by the shortening of the face and by the crowding of the dentition due to the increase in the anteroposterior diameter of the remaining cheek-teeth.
>
> In the final stage of this evolution [referring to the differentiation of the premolars], the first and second premolar disappear entirely and the remaining premolars are homologous with the third and fourth of the primitive placental dentition.
>
> In all primates the elimination of the two anterior premolars is probably correlated with the shortening of the face, with the marked increase in size of P^3 and P^4, and often with the anteroposterior lengthening of the molars.[1]

It was these three statements (the only three which Gregory appears to have made in regard to premolar loss in the primates) which led the author to reconsider premolar loss. There did not seem to be a sufficient basis for such a generalised theory. Therefore, the purpose of this paper is to re-evaluate this theory, and to suggest that the second lower premolar lost during primate evolution may have been the homologue of the fourth premolar of the primitive placental dentition.

Palaeocene primates

Among the Palaeocene primates one finds evidence of various trends of specialisation. The succession in the Carpolestidae (*Elphidotarsius-Carpodaptes-Carpolestes*) demonstrates a reduction in size and number of the anterior premolars with a concomitant increase in size of a specialised posterior premolar.[2] Although this can serve as supportive evidence for Gregory's theory of premolar loss, it must be remembered that the Carpolestidae were a specialised evolutionary branch of primates which left no descendants,[2] and that this specific pattern of premolar loss in the family was the result of evolutionary pressures towards specialisation involving not only morphology and function of each individual tooth, but the interaction of morphology and function of the dentitions of both upper and lower jaws.

The Plesiadapidae (Fig. 1), further shown to be primates by Gingerich,[3] also display a trend for reduction in size and number of the anterior premolars as the posterior premolar increased in size and complexity.[4,5] Further specialisations in dentition, such as the lack of the lower canine, reduction of the upper canine, and the functional significance of the enlarged and procumbent upper and lower incisors,[3] contribute to the consideration that this family was an aberrant branch in the evolution of the Primates and cannot, therefore, be used as evidence for premolar loss.

The Paromomyidae (Paromomyinae[6]) were not as specialised in the premolar region as the Carpolestidae or Plesiadapidae; the posterior lower premolar was about as high as M_1 and either similarly or less molariform than the same tooth in the Eocene subfamily

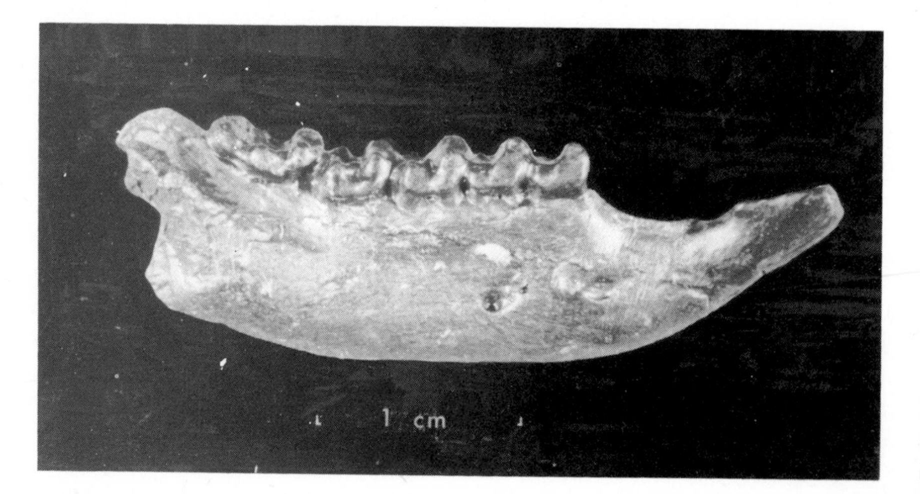

Figure 1. *Plesiadapis fodinatus*, B.M. (N.H.) 13922. Right dentary, showing complete tooth-row.

Omomyinae. The members of this family were specialised in the direction of greatly enlarged lower front teeth (either an incisor or a canine), a condition which, in comparison with succeeding Eocene primates, would exclude them from direct relationship to the evolution of later primates.

Eocene primates

The Adapidae, including the North American subfamily Notharctinae and the European subfamily Adapinae, retain the original placental mammal number of premolars.[7,8] The first premolar of both the upper and lower dentitions is greatly reduced in size and, as with the Palaeocene primates, can be regarded as evidence of eventual evolutionary loss of this tooth. Starting with the second premolar and proceeding posteriorly, there is a gradient of increasing size, and incorporation of gross morphological characteristics such that the spatulate shape seen in P^2_2 grades into the molariform P^4_4 and does not create a break in the premolar-molar morphological continuum (Figs. 2 and 3). Unlike the Palaeocene primates, the Adapidae do not display specialisations which would exclude them from relationship to the evolution of later primates. Both the Adapinae and Notharctinae have peg-like upper incisors with a definite gap between the central incisors. (Le Gros Clark describes the upper incisors of *Adapis* as being 'of moderate size and furnished with a chisel-like cutting margin', and says that 'in neither genus, however, are the upper incisors separated by a conspicuous median gap' (p.93).[4] This is not so, as can clearly be seen in specimens of *Adapis parisiensis* and *A. magnus* at the British Museum of Natural History (B.M.N.H., M.1633, M.1345, M.4487, M.7504) and *A. parisiensis* (M.583) in the Watson Collection, Cambridge (Fig. 4)). Although the Adapidae differ from extant primates in dental morphology, the presence of a median gap between peg-like upper incisors relates them to the strepsirhines (i.e. prosimians excluding *Tarsius*), which have similarly shaped upper incisors and a median gap for the passage of a rhinarium.[9] Further relationship of the Adapidae to strepsirhine evolution is seen in the similarity of the ectotympanic region of the basicranium.[10]

The family Anaptomorphidae is currently generally considered to contain the subfamilies Anaptomorphinae and Omomyinae,[2] as had been proposed by Simpson and *contra* Gazin.[6,11] Description of the differences in dental morphology between these two subfamilies is given by Simons and Simpson.[2,6] One particular characteristic of the Anaptomorphinae is that P^4_4 is generally very enlarged and is noticeably higher than the crowns of adjacent teeth, while that of the Omomyinae is generally not elevated above adjacent teeth, and P_4 is incipiently or nearly molariform, similar to that of the

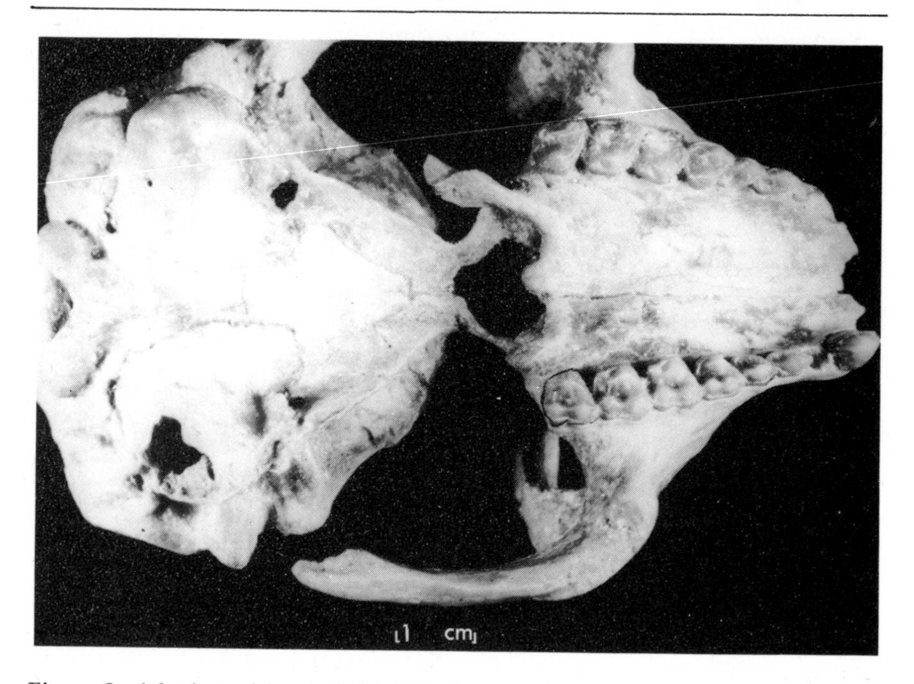

Figure 2. *Adapis parisiensis*, B.M. (N.H.) M.1345. Ventral view of skull, showing upper dentition. Note smooth transition from premolars to molars.

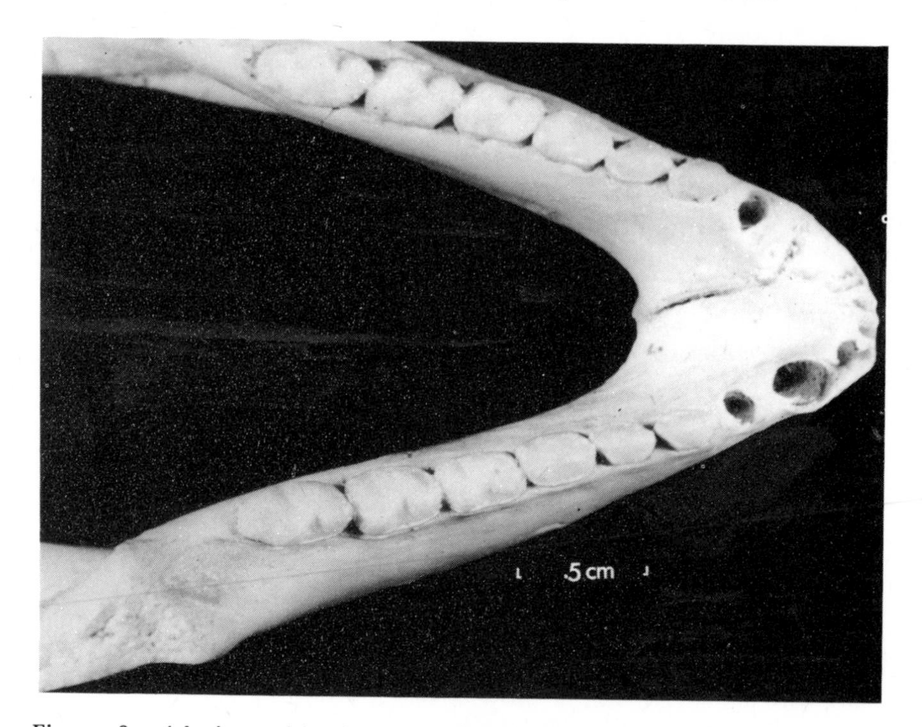

Figure 3. *Adapis parisiensis*, B.M. (N.H.) M.1634. Lower jaw, showing incomplete dentition. Note smooth transition from premolars to molars.

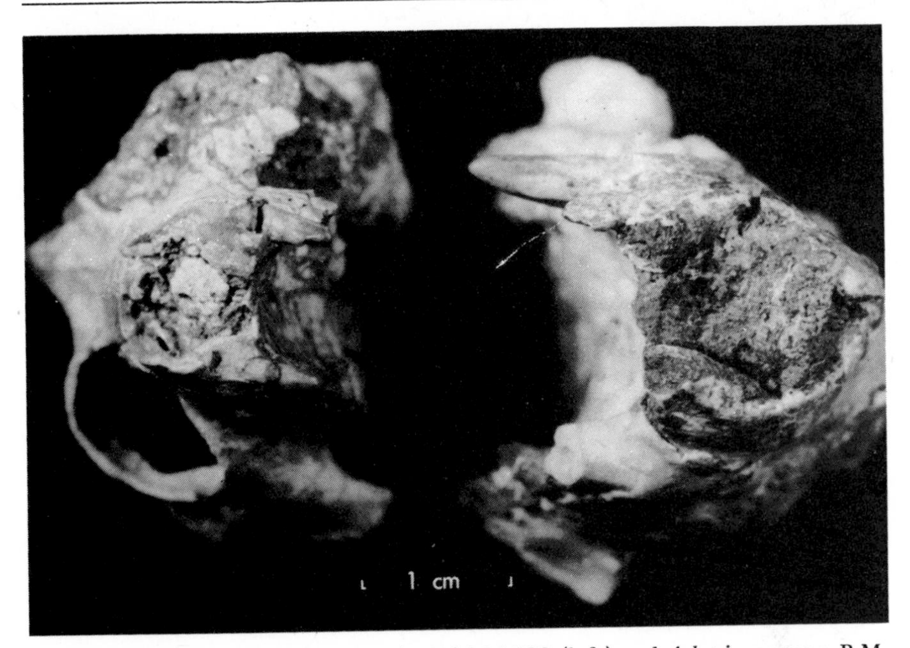

Figure 4. *Adapis parisiensis*, B.M. (N.H.) M.1633 (left) and *Adapis magnus*, B.M. (N.H.) M.4487 (right). Frontal view of snout, showing anterior upper dentition.

Microchoerinae (= Necrolemurinae). The lower dental formula of the Omomyinae is typically 2.1.3.3., but that of *Teilhardina belgica*, morphologically similar to *Omomys*, is 2?.1.4.3. It has been suggested that the Anaptomorphinae, Omomyinae, and Microchoerinae were derived from a form similar to or prior to *Teilhardina*, giving all three subfamilies common tarsioid affinities.[2,10,12] As with the Carpolestidae, reduction in the number of premolars can be seen in successive species of several genera of the Anaptomorphidae: (1) *Teilhardina*, the basal form, which has a greatly reduced P_1; and (2) the anaptomorphine genera *Tetonius* and *Absarokius*. Although anaptomorphid molar and upper and lower dental arcade morphology indicate tarsioid affinities, relationships to the evolution of *Tarsius* are not as close as are those of the Microchoerinae.

Although members of the microchoerine genera *Microchoerus* and *Necrolemur* show specialisations in their cusp morphology which remove them from direct relation to tarsioid evolution, members of the genera *Pseudoloris* and *Nannopithex* are generalised so as to indicate more direct tarsier affinities of the subfamily.[10] In general, the Microchoerinae are characterised by one enlarged lower anterior tooth; by a P_4 which is not elevated but nearly molariform, with a short heel; and by $P^{3,4}$ which are low and oval in shape (Figs. 5 and 6). (For molar morphology see Stehlin,[8] Simons,[10] and Simpson[6].)

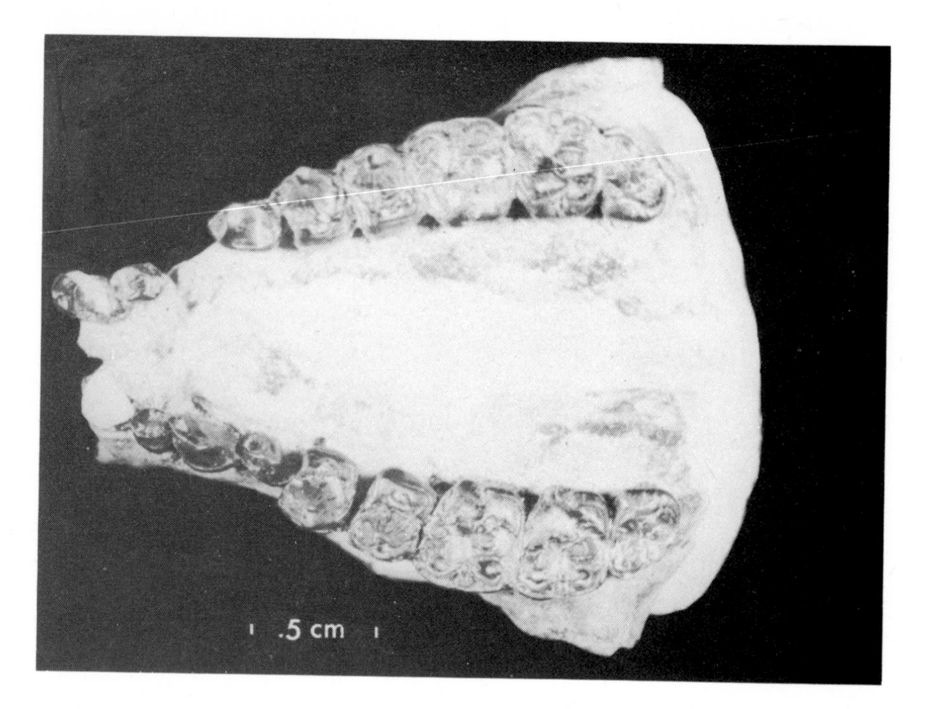

Figure 5. *Microchoerus erinaceus*, B.M. (N.H.) 25.229. View of palate, showing upper dentition. Note the relatively smooth transition from premolars to molars.

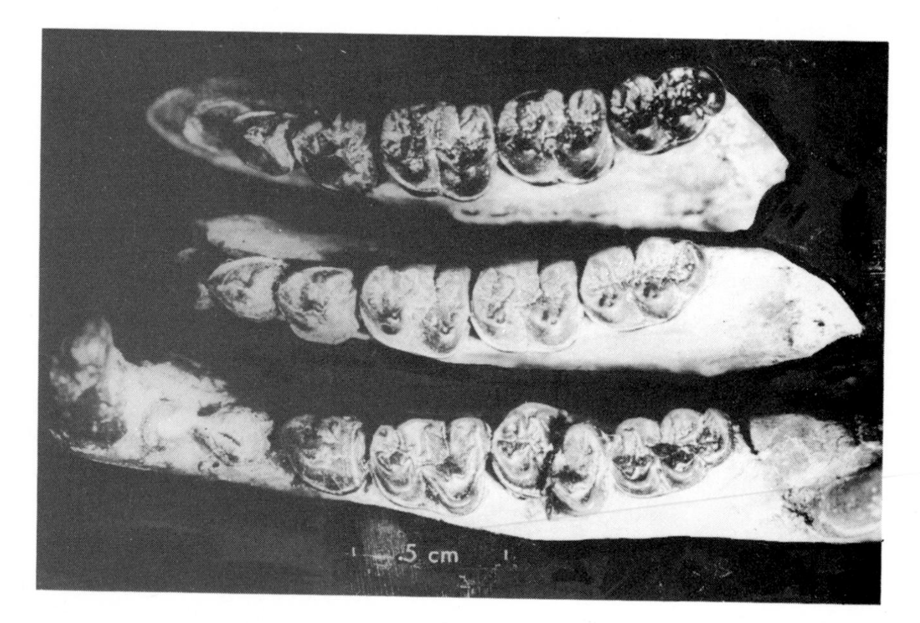

Figure 6. *Microchoerus erinaceus*; left mandibles, showing cheek-teeth.
Top: Sedgwick Museum, Cambridge C9670.
Middle: Sedgwick Museum, Cambridge C9681.
Bottom: B.M. (N.H.) 25.229.

The lower dental formula for all four genera is 0.1.4.3. and the upper is 2.1.3.3., except for *Nannopithex raabi* which is 2.1.3 *or* 4.3.[10] Recently it has been proposed that a new interpretation of the lower dentition of Microchoerines shows that the dental formula is 2.1.2.3.[2] Studies and x-rays made by the author on specimens from the British Museum (M.H.) and the Zoological Museum (Cambridge) indicate that the morphology of the tooth now proposed as a canine is more like that of a premolar; this is further indicated by comparison of x-rays of microchoerines and adapids, whose lower premolar regions are similar. As with the Adapidae and some of the Anaptomorphidae, the anterior premolar is greatly reduced in size.

Oligocene primates

A few members of the Eocene tarsioid families persist into the Oligocene (e.g. *Rooneyia*,[13] *Microchoerus, Macrotarsius*), but the majority of the fossils are catarrhine. (This does not include the Parapithecidae (*Parapithecus* and *Apidium*), which the author feels are not related to catarrhines, but represent, in comparison with other Oligocene primates, specialised and aberrant evolutionary forms.)[2,14,15] The Oligocene catarrhines have a lower dental formula of 2.1.2.3., and show at least incipient catarrhine characteristics of an enlarged canine and a sectorial anterior premolar. This is most definitely seen in the earliest catarrhine fossil, *Oligopithecus savagei*, where the enlarged anterior premolar even shows wear which would necessitate a 'sizeable upper canine',[16] indicating the typical catarrhine characteristic of a honing triad (i.e. large upper canine and lower canine and anterior premolars).[17] Subsequent forms, *Aegyptopithecus* and *Propliopithecus*, show the same condition. Furthermore, as opposed to the Adapidae, Omomyinae, Microchoerinae, and Parapithecidae, which show a gradient from the premolars through the molar set, the catarrhines of the Oligocene show a distinct break between the two sets of teeth. The primates of the Miocene and upwards, then, were to undergo further specialisations based on these Oligocene foundations.

Recent primates

Although the fossil record may be lacking in direct evidence of evolutionary divergence, major changes in the evolution of gross dental morphology had been established by the end of the Oligocene: strepsirhines and haplorhines can be differentiated on the basis of dental, palatal, cranial and post-cranial morphology and dental formulae.[9] (Unfortunately there is a dearth of platyrrhine material until the Miocene with *Homunculus, Cebupithecia*, and *Neosamiri*, which are very similar to modern types.[18] In dental formula and premolar-molar gradients, the platyrrhines are similar to many prosimians.) Since it would appear from the fossil evidence that haplorhines (more specifically, catarrhines) were evolving differently from the strepsirhines (except for the Indriidae) in the differentiation of the premolars from the molars, at least in the lower jaw, it should be possible to see this in the dentitions of extant primates.

The following notes are based on specimens studied at the British Museum (Natural History), London and supplemented by reference to prior publications:

The Lemuridae (e.g. *Lemur catta, Varecia variegatus, Hapalemur griseus, Cheirogaleus major, Microcebus murinus, Phaner furcifer*) Lorisidae (Lorisinae: e.g. *Loris tardigradus, Nycticebus coucang, Arctocebus calabarensis, Perodicticus potto*) and Galaginae (e.g. *Galago senegalensis, G. crassicaudatus, G. demidovii, Euoticus elegantulus*) all have the dental formula $\frac{2.1.3.3}{2.1.3.3}$, a tooth-comb (generally considered to contain the lower incisors and canines), and a caniniform lower anterior premolar. Both upper and lower jaws show a gradient from the premolar set, through the molariform last premolar, and on through the molars; however, the gradient in the lower jaw does not generally include the caniniform premolar (Fig. 7). *Lepilemur mustelinus* differs only in its dental formula, $\frac{0.1.3.3}{2.1.3.3}$. (Because of its unique specialisations, the aye-aye will not be considered in this discussion.) Thus, this characteristic of a dental gradient is generally maintained from the Eocene to modern strepsirhines.

Tarsius presents some difficulty in interpretation. The upper dental formula is clearly 2.1.3.3., but the anterior premolar tends to be relatively smaller than in other primates with three premolars, so that the dental gradient is generally seen to begin with the second premolar. Although the lower dental formula is commonly accepted as being 1.1.3.3., it can equally well be argued that it is 2.1.2.3. The last premolar may vary in its degree of molarisation from specimen to specimen, such that one might see a gradient in one animal, but some degree of differentiation between the premolar and molar sets in another. Qualitatively, it appeared that there was a correlation between the size of the larger of the anterior teeth and the degree of molarisation of the last premolar (i.e. the larger that tooth, the greater the differentiation between the premolar and molar sets.

The Cebidae (Fig. 8) (dental formula $\frac{2.1.3.3}{2.1.3.3}$) show wide intra-familial variation such that the post-canine dentition of the upper and lower jaws of *Aotus trivirgatus* and *Cacajao calvus* tends to be graded; that of *Chiropotes satanus, Pithecia monachus* and *Cebus*

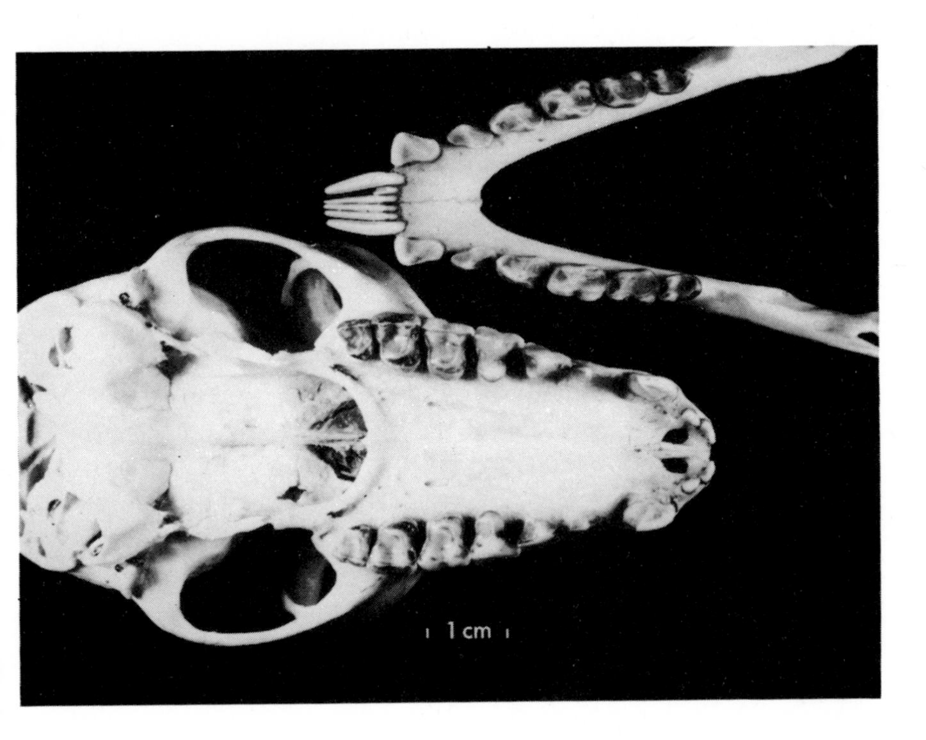

Figure 7. *Varecia variegata* (*'Lemur variegatus'*), B.M. (N.H.) 75.1.29.10 Skull and lower jaw, showing full dentition. Note the smooth transition from premolars to molars. (N.B. One of the incisors in the tooth-comb has been broken.)

apella less so; whereas that of *Callicebus torquatus, Alouatta seniculus, Saimiri sciureus, Ateles fusciceps, Brachyteles arachnoides, Lagothrix flavicauda* and *Callimico goeldii* presents a clearer distinction between the premolar and molar sets. Of the Callithricidae (dental formula $\frac{2.1.3.2}{2.1.3.2}$), *Saguinus midas* can be interpreted as having a dental gradient, but not *Callithrix flaviceps* or *Cebuella pygmaea.* In practically all specimens examined, if there was any degree of differentiation between the premolar and molar sets it was more so in the lower than upper jaw, the last premolar of which was relatively more molariform. As with the variability exhibited in the genus *Tarsius*, the Ceboids which showed less grading tended to have relatively larger upper and lower canines and slightly sectorial anterior lower premolars.

The catarrhines (Figs. 9 and 10) (dental formula $\frac{2.1.2.3}{2.1.2.3}$) differ from most extant primates not only in the number of premolars, but also in the anterior premolar of the lower jaw which – except in the Hominidae – is enlarged, elongated, and specialised sectorially so as

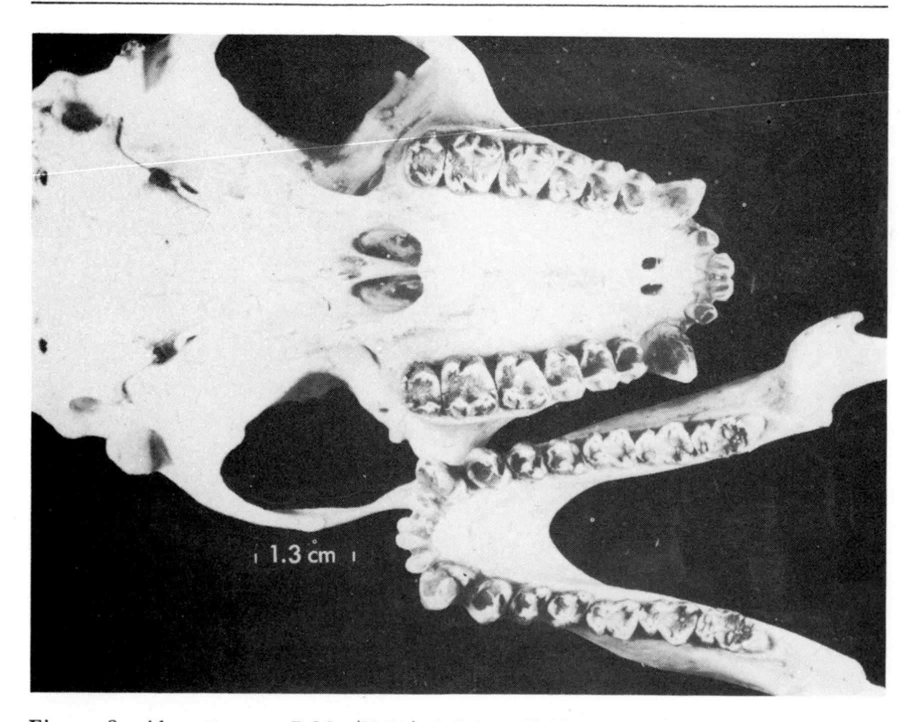

Figure 8. *Alouatta* sp., B.M. (N.H.) 3.7.11. Skull and lower jaw, showing full dentition. Note the fairly clear division between premolar and molar morphology, which is more obvious in the lower jaw.

to accommodate the large upper canine. Except for *Erythrocebus patas* and *Cercopithecus aethiops* (subfamily Cercopithecinae), in which the posterior premolar tends to be relatively larger and molariform compared with the other cercopithecines, the Cercopithecinae, Colobinae, Pongidae and Hylobatidae were seen to have relatively the same condition: clearly distinct sets of premolars and molars in the lower jaw, with the anterior premolar larger than the posterior and the upper jaw showing more of a tendency for grading between the two sets of cheek-teeth. Once again, the degree of differentiation or grading appeared to be correlated with the relative sizes of the upper and lower canines and lower anterior premolar, the components of the honing triad.

The Indriidae (Fig. 11) differ from the Lemuridae and Lorisidae in that they have fewer anterior teeth in the lower jaw and fewer premolars. The dental formula is $\frac{2.1.2.3}{1\ or\ 2.1\ or\ 0.2.3.}$,[19,20,21] but it is commonly accepted, by homology with the anterior dentitions of the lemurids and lorisids, that the lower dental formula is 1.1.2.3.[22] Indriids also differ from other prosimians in that the anterior lower premolar is not as much caniniform as it is sectorial, analogous in

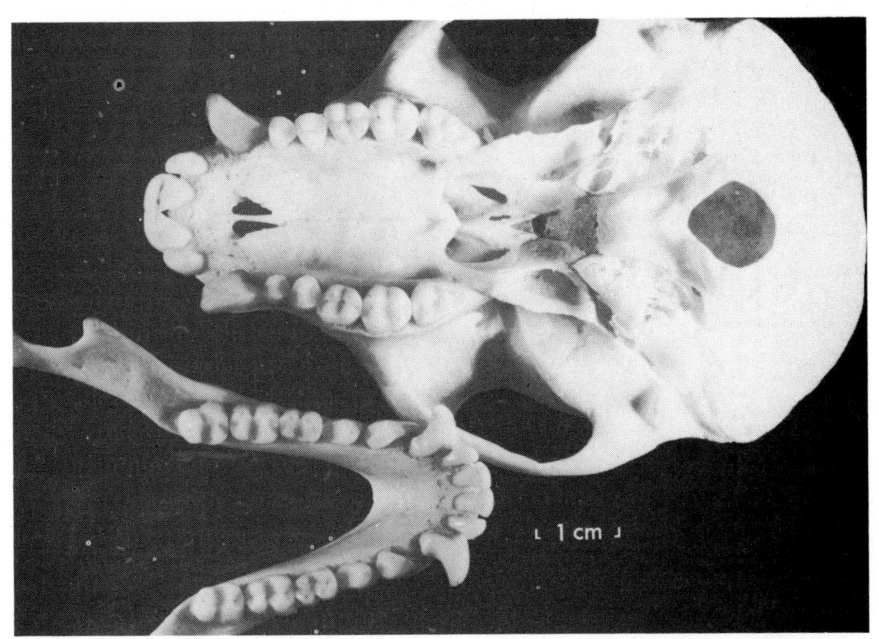

Figure 9. *Cercopithecus aethiops*, B.M. (N.H.) 66.504. Skull and lower jaw showing full dentition. Note the marked distinction in morphology between the last premolar and the first molar in both upper and lower tooth-rows.

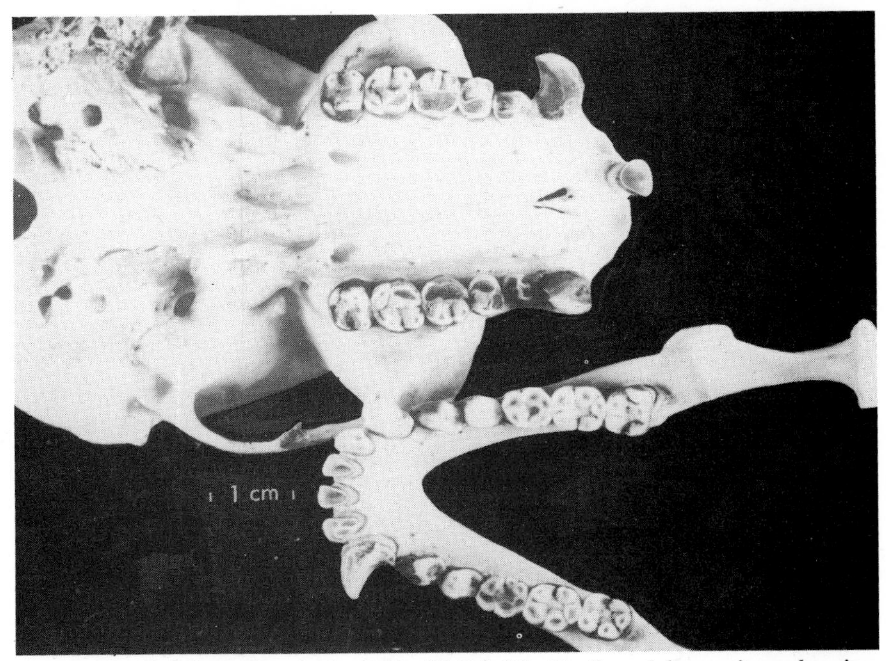

Figure 10. *Hylobates hoolock*, B.M. (N.H.) 33. Skull and lower jaw, showing most of the dentition (N.B. 3 incisors have been lost from the upper jaw). Note the marked distinction in morphology between the last premolar and the first molar in both upper and lower tooth-rows.

function and similar in shape to some catarrhines and those platyrrhines with relatively enlarged upper canines. The morphology of the lower jaw is similar to that of catarrhines, in that the anterior premolar is larger in height and length than the posterior premolar and there is distinct differentiation between the premolar and molar sets. This is more marked in *Propithecus* than in *Indri*, and is apparently correlated with *Propithecus* having a relatively larger upper canine than *Indri.* As with catarrhines and platyrrhines, there is a greater tendency for grading in the upper jaw. The Indriidae are unique among primates, though, in the fact that the number of deciduous molars is greater than the number of replacing premolars (Fig. 12): the formula for the deciduous dentition is said to be $\frac{2.1.3}{2.1.3}$, the same as lorisids and lemurids, except that *Lepilemur*[23] is $\frac{1.1.3}{2.1.3}$. Current research by the author indicates that the formula for the deciduous dentition of indriids is $\frac{2.1.2}{1.1.4}$ and that the replacing premolars in the lower jaw are not P_2 and P_4 or P_3 and P_4.[19,1]

Discussion

Comparison of the tooth rows and premolar regions of extant and fossil primates seems to support a separation of the strepsirhines and haplorhines. The strepsirhines, apart from the aye-aye and indriids, have lost one premolar on each side of the upper and lower jaws during their evolution and have maintained a dental gradient. The haplorhines, on the other hand, have gone through stages where one and then, in some cases, another premolar has been lost; the premolar region exhibits variability posteriorly and specialisation anteriorly, with the result that this region is, with few exceptions, distinguishable as a unit separate from that of the molars. The issue at hand is to try to understand what the premolar region represents and to what extent variability within it is significant.

Gregory interpreted the premolar region as 'intermediate in position and function between the canines and the molars'.[24] Butler adds agreement in the presentation of his Field Theory.[25,26] This theory states that there is a gradient of forms as one passes along the series of teeth indicating the '. . . presence in the embryonic jaw of morphogenetic fields that influence the tooth rudiments in different parts of the jaw to develop in divergent ways', and that the whole series of teeth is a '. . . range of forms into which a given type of tooth germ can develop when placed in the given range of environments'.[25]

The gradient is the 'field'; a tooth develops according to its position in the field; 'field genes' control regional differentiation; the

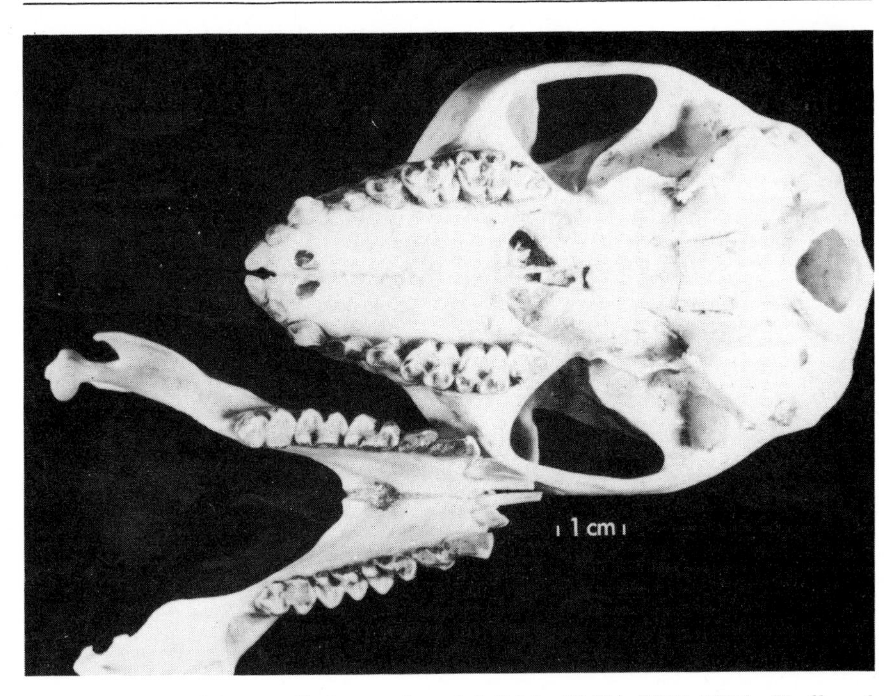

Figure 11. *Propithecus diadema edwardsi*, B.M. (N.H.) 1939.1204. Skull and lower jaw, showing the complete dentition (N.B. Most of the outer tooth of the tooth-comb has been lost from the right side of the lower jaw). Note that this indriid resembles the catarrhines in the marked distinction in morphology between the last premolar and the first molar in both upper and lower tooth-rows.

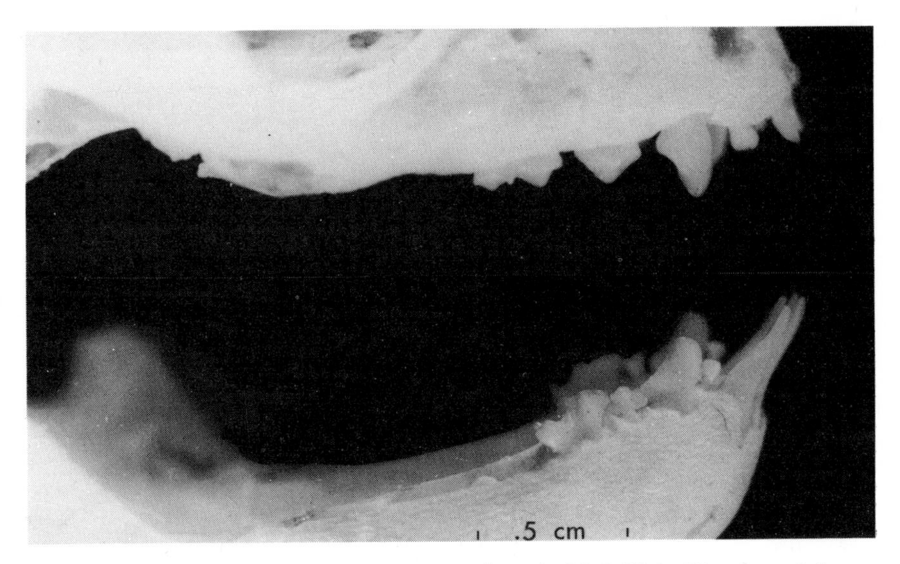

Figure 12. *Propithecus verreauxi*, B.M. (N.H.) 30.3.15.1. Upper and lower deciduous dentition (see text for explanation).

field is differentiated into incisor, canine, and molar regions; the premolar 'region' is the result of the combined affects of the canine and molar regions; and the morphology of each premolar is determined by the predominance of either the canine or molar region, or the equilibration of both (Fig. 13).[25,26] In terms of function, it would appear that the role of the premolar, whose morphology can range from being caniniform to molariform, is that of a potential resource which can be tapped by evolutionary pressures to enhance or expand the functional area of either the canine or the molars, or both. For example, the battery of cheek-teeth of *Equus* was expanded, during the course of evolution, by the molarisation of the premolars;[25] in lemur, the lower anterior premolar became extremely caniniform 'by assuming the function of the true canine, which . . . has been relegated to the

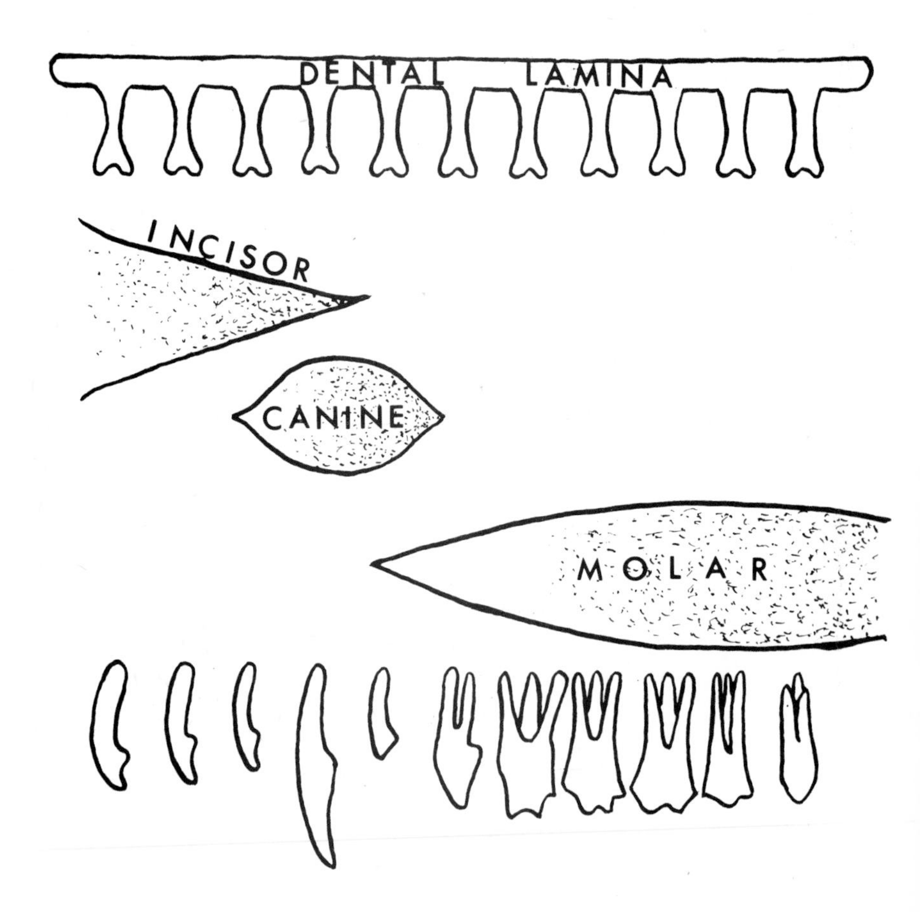

Figure 13. Diagrammatic representation of the morphogenetic fields as proposed by the Field Theory (Butler); [25,26] these three fields (incisor, canine and molar) are presented in relation to the dental lamina and the realised eutherian dentition. Note the overlapping influence of these fields in this instance.

incisor series' (Le Gros Clark, p.94).[4] Butler's Field Theory can be readily accepted at this level, considering only molarisation or caninisation; but the application of the concept of a morphogenetically patterned premolar region must also take into consideration the consequences of tooth loss.

According to the Field Theory, the characteristics of a premolar are determined by its position in the jaw in relation to the positions of the canine and molars. The area between the canine and molar fields is a gradient of the genetic potentials of each field. Evidence from the evolution of herbivores (e.g. the horse), where the length of the jaw and the number of molariform teeth were increased, indicates not only that the field can change so that there is a simultaneous modification of several teeth,[25] but also, more significantly, that the size, and thus genetic potential, of a field is influenced by the size of the jaw. If the potential of a field increases with the increase in the size of the jaw, it must also diminish in intensity with a decrease in the size of the jaw. Therefore, the reduction of the jaw and the process of tooth (i.e. premolar) loss, which occurred during the course of primate evolution, must have affected the magnitude of the potentials of both the canine and molar fields. In the light of this, the data presented earlier in this paper become critical in the interpretation of premolar loss in the primates, as to whether reduction has occurred consistently in an anteroposterior direction, or whether the second premolar to be lost was not, in fact, the numerically second premolar.

Subsequent to the Field Theory, other points of interpretation of dental gradients have been put forward. Patterson would have a premolar field as well as incisor, canine, and molar fields.[27] To postulate such a field seems to obscure the functional nature of premolars, which is potentially that of supplementing or approximating the function of the canine and/or molars. Van Valen proposed a model for a prepattern-gradient theory where 'all sites for prospective teeth have identical prepatterns, different parts of which are activated to different degrees by gradients of evocators', such that 'a prepattern is the competence to form a set of structures (a morphogenetic program for differentiating cusps, crests, etc. in a prescribed morphological relation to each other), the presence and relative development of each of these structures depending on the strength of an appropriate evocator'.[28] First of all, although Butler did not state it as such, it is obvious from the above quotations that the Field Theory incorporates this concept;[25] and, secondly, by eliminating morphogenetic fields, one is eliminating the control over the evocators, which would be analogous to eliminating the codon from the gene. In sum, the Field Theory, at least at this point, is the most direct means for interpreting differences in dental gradients.

The present consensus of opinion, that the upper and lower first premolar (as interpreted from its small size in the Adapidae and

Microchoerinae or its non-appearance in the evolution of the Carpolestidae, Plesiadapidae, and some Anaptomorphinae) was the first premolar lost in primate evolution, is in accordance with the Field Theory. The resultant reduction of the canine field did not eliminate the potential for influencing the formation of a caniniform premolar, as seen in the lemurids and lorisids, or of a sectorialised caniniform premolar, as seen in the indriids, some ceboids, and the catarrhines. However, it need not be the case that the anterior premolar is lost first, for 'in spite of the already reduced size of the first premolar in most eutherians . . . this position has not always been the initial one vacated in Tertiary therians', as is seen, for example, in the Palaearctic eutherian genus *Talpa*, where the upper and lower P2 were suppressed or lost, 'P_1 being retained because of its importance as a substitute canine'.[29] The diminution of the canine field in the primates must have been secondary in importance to the maintenance of the molar field and its molarising effect on neighbouring premolars, for the gradation of the premolar set into the molars, as seen in Eocene strepsirhines, to some degree similarly exists in the Lemuridae, the Lorisidae, *Tarsius* and, with some variation, the Ceboidea, the latter two groups having three premolars as well. As most of these living primates are herbivorous or frugivorous,[30,31,32] diet probably provided the pressure on early primates for the initial selection of the molar field potential over that of the canine field. The further process of reduction from three to two premolars is not so clear-cut, however.

It is not necessary that one should be able to observe the decrease in size and eventual loss of that tooth. The formation of the tooth follicle and subsequent development of the tooth germ before hard tissues are formed is greatly affected by the environment of the jaw, adjacent developing follicles, and already developed and erupted teeth.[33] Studies carried out by Osborn on the reptile *Lacerta vivipara* show that (1) a tooth germ may develop from the dental lamina but never fully develop and erupt; and (2) that the presence and proximity of young teeth can inhibit the development of a new tooth.[34,35] Although mammals differ from reptiles in the direction of budding of the anterior teeth and in the number of successions of teeth, the same principles of tooth formation or inhibition hold.[35] In most cases, the permanent premolars of non-hominid primates erupt after all but the canine and third molar of the permanent dentition have erupted.[36] Thus, the area in which the tooth germs of the premolars develop would be subjected to crowding and inhibition, which might even prevent a follicle from budding from the dental lamina, as would be the case with the Indriidae. Generally, during the course of the evolution of therian dentition, 'late eruption is directly correlated with late onset of ontogenetic development and, under certain circumstances, these manifestations may strongly indicate a trend toward evolutionary loss of the tooth involved', but this does not apply to the primate canine.[29]

Since the functional importance of the canine, and therefore the canine field, seems too great for this tooth to be lost, its late development would cause added crowding on the developing premolars. It is possible, then, that even though the deciduous molars had erupted, the permanent premolars might be subjected to such pressures that not only would one or more be distorted in shape, but eruption or tooth germ formation might not occur.[29] Again, one must consider the case of the Indriidae.

Oligopithecus is the first of the directly related fossil primates to exhibit a reduction in premolar number to two. As stated earlier, this haplorhine also shows at least an incipient catarrhine condition of specialisation of the lower anterior premolar towards being sectorial, as indicated by its relatively greater size and anteroposterior elongation. For example, the characteristics of a sectorial premolar (development of a robust (anterior) root and an increase in the height of the primary cusp) are characteristics which can be related to a canine rather than a molar field. The development of such a specialised tooth would increase its functional importance relative to the premolar, or premolars, posterior to it and, as a member of a late-developing group of teeth, would take precedence in the utilisation of the space available for development. It seems unlikely that the sectorial premolar could be the third premolar of the primitive placental dentition.

First of all, one must take into consideration the condition in living primates. Even assuming bias on the part of the author in his interpretation of the variability of grading versus distinctiveness of premolars and molars in the platyrrhines, one cannot deny the morphological differences which exist between the posterior premolar and first molar of the catarrhine and indriid dentitions. As outlined earlier in this paper, the strepsirhines excluding the indriids exhibit similar gradations of characteristics from the premolars through the molars. Since this gradation seems to be a function of the morphogenetic potential of the molar field, one would expect, with the reduction from three to two premolars, a relatively undisturbed premolar-molar gradient if the reduction had been the result of the loss of the second premolar. As shown in the above review, this is not the case with the catarrhines, fossil or extant. This would indicate that, in the early stages of catarrhine evolution, there was an alteration in the milieu of the posterior premolar; that there had been a diminution in the influence of the molar field over the development of a molariform premolar. Secondly, if the second premolar lost was the numerical second, the result would have been a further decrease in the morphogenetic potential of the canine field over the adjacent premolar. The characteristics of a sectorial premolar are too closely allied to those of a canine, with its strong, high primary cusp and robust root, to expect that there had been further reduction of the canine field. And, thirdly, with the shortening of the lower jaw and the selective pressures for a specialised and robust anterior premolar, the pressures for the non-eruption of a premolar would be directed toward the distal end of the premolar series, with the result that the area of development of the last premolar would be impinged upon, leading to the loss of that tooth and a decrease in the potential of the molar field. Therefore, it is suggested that, even though Gregory's interpretation of the lemuroids may be applicable to the loss of the first premolar of the lower jaw, the second premolar lost during the course of primate evolution was, in fact, the homologue of the fourth premolar of the primitive placental dentition – in other words, P_4. The pressures involving reduction in the upper jaw from three to two premolars were not the same.

The discussion of the interrelationships between the lower canine and the sectorial premolar must be expanded to include the function of these two teeth as a unit as well as the functional interactions of this unit with the upper dentition. Zingeser points out the basic mechanisms of the interaction between the lower canine and sectorial premolar and the upper canine: the mesiobuccal cusp of the sectorial premolar and the distal surface and cingulum of the lower canine form a V-shaped notch for the upper canine in the centric position, 'against which this tooth differentially wears to sharpen blade and cusp-tip during honing activities'. The upper canine, sectorial premolar and lower canine form a tongue-in-groove grinding and honing triad, whose primary function is to sharpen the upper canine.[17]

First of all, this adds weight to the previous discussion by emphasising the importance of maintaining the morphogenetic potential of the canine field during the loss of the second premolar as a mechanism for the eventual development of the lower unit of the honing triad. Secondly, it introduces the concept of the lower canine and sectorial premolar being a unit which must be responsive to the evolutionary changes occurring in the upper jaw, particularly those of the upper canine, thus implying a basis for a difference in the evolutionary processes affecting the upper and lower jaws.

Whereas in the lower jaw the combined shortening of the jaw and the increase in size of the anterior premolar would create pressures in a posterior direction, thus causing the loss of P_4, the pressures created in the upper jaw, by its shortening and by the increase in size of the canine would not be impeded by the added functional requirements of another tooth (i.e. an analogue to the sectorial premolar). Since it is the upper canine alone which is the basis of the honing triad of the catarrhines and the analogous condition of the Indriidae, there would not be a need to maintain its morphogenetic field, so that premolar loss could occur in an anteroposterior direction. As opposed to the lower jaw, where evolutionary pressures appear to have been acting to preserve the canine field, thereby differentiating the dental gradient, evolutionary pressures affecting the upper jaw might have been operating to preserve its molar field, as suggested by the higher incidence of gradation in the catarrhines and Indriidae; in which case P^2 would be lost.

In conclusion, it is suggested that differing demands on the canine field caused different modes of premolar loss in the upper and lower jaws during the course of primate evolution, resulting in the loss of the homologues of P^1 and P^2 and P_1 and P_4 of the primitive placental dentition.

Acknowledgements

I would like to thank R.D. Martin and J.W. Osborn, and Miss T. Molleson and members of the staff of the Department of Zoology in the British Museum (Natural History), London, without whose assistance and criticism this paper could not have been written.

NOTES

1 Gregory, W.K. (1922), *The Origin and Evolution of the Human Dentition*, Baltimore, pp. 146, 155.

2 Simons, E.L., this volume.

3 Gingerich, P., this volume.

4 Le Gros Clark, W.E. (1962), *The Antecedents of Man.* Edinburgh.

5 Simpson, G.G. (1935), 'The Tiffany fauna, Upper Palaeocene II: Structure and relationships of *Plesiadapis*', *Amer. Mus. Novit.* 816, 1-30.

6 Simpson, G.G. (1940), 'Studies on the earliest primates,' *Bull. Amer. Mus. Nat. Hist.* 77, 185-212.

7 Gregory, W.K. (1920), 'On the structure and relations of *Notharctus*, an American Eocene primate', *Mem. Amer. Mus. Nat. Hist. (n.s.)* 3, 49-243.

8 Stehlin, H.G. (1916), 'Die Säugetiere des schweizerischen Eocäns', Teil 7, 2 Hälfte, *Abhandl. schweiz. paläont. Ges.* 41, 1299-552.

9 Hill, W.C.O. (1953), *Primates*, vol. 1: *Strepsirhini*, Edinburgh.

10 Simons, E.L. (1961), 'Notes on Eocene tarsioids and a revision of some Necrolemurinae', *Bull. Brit. Mus. Nat. Hist. (Geol.)* 5, 45-69.

11 Gazin, G.L. (1958), 'A review of the middle and upper Eocene primates of North America', *Smithson. Misc. Coll.* 136, 1-112.

12 Simons, E.L. (1963), 'A critical reappraisal of Tertiary primates', in Buettner-Janusch, J. (ed.), *Evolutionary and Genetic Biology of Primates*, vol. 1, New York.

13 Wilson, J.A. (1966), 'A new primate from the earliest Oligocene, West Texas: preliminary report', *Folia primat.* 4, 227-48.

14 Simons, E.L. (1967), 'The earliest apes', *Sci. Amer.* 217, 28-35.

15 Simons, E.L. (1969), 'The origin and radiation of the primates', *Ann. N.Y. Acad. Sci.* 167, 319-31.

16 Simons, E.L. (1962), 'A new Eocene primate genus, *Cantius*, and a revision of some allied European lemuroids', *Bull. Brit. Mus. Nat. Hist. (Geol.)* 7, 1-36.

17 Zingeser, M.R. (1969), 'Cercopithecoid canine tooth honing mechanisms', *Amer. J. Phys. Anthrop.* 31, 205-13.

18 Stirton, R.A. (1951), 'Ceboid monkeys from the Miocene of Colombia', *Univ. Calif. Publ., Bull. Dep. Geol. Sci.* 28, 315-56.

19 Friant, M. (1948), 'Classification générique des lémuriens actuels, principalement basée sur l'ostéologie et la dentition', *Acta anat.* 6, 152-74.

20 James, W.W. (1960), *The Jaws and Teeth of Primates,* London.

21 Spreng, H. (1938), 'Zur Ontogenie des Indrisinengebisses', Dissertation, Bern University.

22 Mivart, St. G. (1867), 'On the skull of *Indris diadema*', *Proc. Zool. Soc. Lond.* 1867, 247-56.

23 Friant, M. (1947), 'L'état de la dentition d'un lémurien nouveau-né (*Lepilemur leucopus* F. Major)', *Rev. Stomat.* 48, 597-9.

24 Gregory, W.K. (1916), 'Studies on the evolution of the primates', *Bull. Amer. Mus. Nat. Hist.* 35, 239-355.

25 Butler, P.M. (1939), 'Studies of the mammalian dentition: differentiation of the post-canine dentition', *Proc. Zool. Soc. Lond.* 109, 1-36.

26 Butler, P.M. (1963), 'Tooth morphology and primate evolution', in Brothwell, D. (ed.), *Dental Anthropology,* Oxford.

27 Patterson, B. (1956), 'Early Cretaceous mammals and the evolution of mammalian molar teeth', *Fieldiana Geol.* 13, 1-105.

28 van Valen, L. (1970), 'An analysis of developmental fields', *Develop. Biol.* 23, 456-77.

29 Ziegler, A.C. (1971), 'A theory of the evolution of therian dental formulas and replacement patterns', *Quart. Rev. Biol.* 46, 226-49.

30 Chance, M.R.A. and Jolly, C.J. (1970), *Social Groups of Monkeys, Apes and Men,* London.

31 Jolly, A. (1966), *Lemur Behavior,* Chicago.

32 Napier, J.R. and Napier, P.H. (1970), *A Handbook of Living Primates,* London.

33 Butler, P.M. (1956), 'The ontogeny of molar pattern', *Biol. Rev.* 31, 30-70.

34 Osborn, J.W. (1970), 'New approach to *Zahnreihen*', *Nature* 225, 343-6.

35 Osborn, J.W. (1971), 'The ontogeny of tooth succession in *Lacerta vivipara* Jaquin (1787)', *Proc. Roy. Soc. Lond. (B)* 179, 261-89.

36 Schultz, A.H. (1935), 'Eruption and decay of the permanent teeth in primates', *Amer. J. Phys. Anthrop.* 19, 489-581.

A.C. BERRY

Non-metrical variation in the prosimian skull

Introduction

In the older descriptive works on anatomy, reference is frequently made to unusual ossicles, extra foramina and other minor variations in the bones of the skull. Initially these variations were regarded as nothing more than minor curiosities. Russel and Le Double noted the different incidences of these variants in different populations,[1,2,3] while Montagu used their varying incidences in human and other primates in evolutionary studies.[4,5]

Schultz reported variations in crania of several monkey species and carried out an intensive study on both metrical and non-metrical variation in large samples of *Alouatta, Cebus* and *Ateles.*[6,7,8] Wood Jones advanced human population studies,[9,10] based on the variation in incidence of a number of these non-metrical minor variants in the human skull (reviewed by Brothwell),[11] and this method has recently been considerably expanded (see below).

Before these variants can be critically harnessed for population comparisons, it is necessary to establish what factors control their occurrence. Genetical studies on mice by Grüneberg and co-workers (reviewed by Grüneberg),[12] discussed below, indicate that there is a considerable degree of genetical control, particularly if a wide range of variants is used.

Berry, using a range of skeletal variants, has been able to study genetical distances, based on differences in variant incidences, between different rodent populations.[13,14,15] He has also applied the method to samples of seals.[16] A range of 30 minor variants in the human cranium was described by Berry and Berry,[17] and genetical distances between human population samples have been calculated by Laughlin and Jørgensen for Eskimo,[18] by Kellock and Parsons for Australian Aborigines,[19,20] by De Villiers for South African negroes,[21] and by Berry for Northern Europeans.[22]

A pilot study on a number of non-human primates (Berry and Berry) showed considerable variation, both inter- and intra-specific, in the incidences of non-metrical variants.[23] Similar findings for *Aotus trivirgatus* were reported by Balan and Thorington.[24]

It appeared likely that variation of this type existed in the prosimian cranium. Blood samples for the investigation of serum protein and enzyme variation are hard to obtain for many prosimians, whereas a considerable amount of skeletal material is available in museums. Hence it seemed worthwhile to see whether non-metrical variants could be used as genetical indicators in the study of these species.

Materials

The present study has been limited to *Galago* spp. because of the good numbers available for study at the British Museum (Natural History). There is bound to be similar variation in other prosimian species. Although variants of the vertebral column are likely to exist and have been utilised in rodent studies (Berry and Searle),[25] the present investigation was confined to the skull simply because post-cranial material is not often preserved in collections. In general a sample of about 50 crania is desirable.

The following crania were examined.

G. crassicaudatus agisymbanus	31 from Zanzibar 28 from Pemba Island, about 40 miles away Total: 59 (mixed sexes)
G. senegalensis albipes	47 males 47 females from Uganda Total: 94
G. senegalensis moholi	51 (mixed sexes) from Zambia, S. Rhodesia, Malawi and Angola
G. demidovii	53 (mixed sexes) from various subspecies and localities

Method

1. Characters

Each skull was examined for the presence or absence of the following characters (many of which have been previously described and illustrated).[17,24,25] Some of these, common in man and higher primates, were absent in galagos or occurred in only one species.

1. *Asymmetry of nasal bones.* Usually the nasal bones articulate

with the frontal bones in a symmetrical manner. Occasionally, however, the articulation is asymmetrical. This variant was not scored in our previous primate study.

2. *Sutural bones.* Except for the variant below (No. 3), these were never seen in the present study.
3. *Bregmatic bones present.* Rarely an ossicle is found in the position of the anterior fontanelle. It occurs in man, larger primates and one of the *Galago* species studied.
4. *Asymmetrical bregma.* In larger primates, the suture between the frontal bones rarely persists into adult life. When it does the skull is said to be metopic. In the galagos studied the suture never fused, but usually joined the sagittal and coronal sutures to form a symmetrical junction at the bregma. Occasionally, however, the junction was asymmetrical. This variant is one in which scoring is to some extent subjective, as microscopic examination emphasises degrees of asymmetry unnoticed by the naked eye.
5. *Parietal foramina present.* The emissary foramina in the parietal bone that lie close to the sagittal suture were only seen in the larger *Galago crassicaudatus.*
6. *Stylomastoid foramen present.* There is a moderately large foramen lying in the groove between the bulbous mastoid part of the temporal bone and the occipital condyle. Posterior to this the much smaller stylomastoid foramen may be seen when present.
7. *Posterior condylar pits.* At the junction of the posterior part of the occipital condyles with the occipital bones there may be a small pit. This is difficult to score in galagos because of varying degrees of development in the pit, and its use was abandoned. However, it is a useful variant in larger primates.
8. *Anterior condylar canal double.* This canal, which carries the hypoglossal nerve, may be double. It is a widespread variant and occurs in rodents (where it is called *foramen hypoglossi*), man and other primates.
9. *Foramen paraovale divided.* This foramen lies close to the junction of the sphenoid with the temporal bone, immediately medial to the foramen ovale. Rarely it is divided by a bony septum.
10. *Pterygoid-sphenoid lamina incomplete.* The external pterygoid lamina stretches posteriorly and fuses with the posterolateral corner of the sphenoid, arching over the foramen ovale. This variant has been previously used only in *Aotus trivirgatus.* (Thorington refers to it as 'external pterygoid lamina joined posteriorly to sphenoid').[24]

11&12. *Lateral or median sphenoidal foramina present.* The basal aspect of the body of the sphenoid bone is sometimes pierced by small foramina. These may be either lateral, in which case they are frequently paired, or a single median foramen may occur, usually more anterior. Rarely, a single lateral foramen may occur with a median one. In this case, both have been scored. There is considerable variation in the expression of these foramina and those that are merely a pin-prick have been scored as ½.

Occasionally in *G. crassicaudatus agisympanus* a single midline foramen was noted immediately anterior to the suture between the anterior and posterior parts of the sphenoid bone. Such a foramen was not scored. These foramina have not been noticed in other primates, but have been described in mice.

13. *Accessory palatine foramina present.* The foramina at the postero-lateral corners of the palatine bone were counted, and any in excess of one considered to be accessory palatine foramina.
14. *Asymmetry of palatine suture.* The suture between the palatine bones and the palatine processes of the maxilla usually crosses the midline suture of the hard palate symmetrically. Occasionally, however, the junction is asymmetrical.
15. *Triple orbital foramina.* In *G. crassicaudatus agisymbanus* the

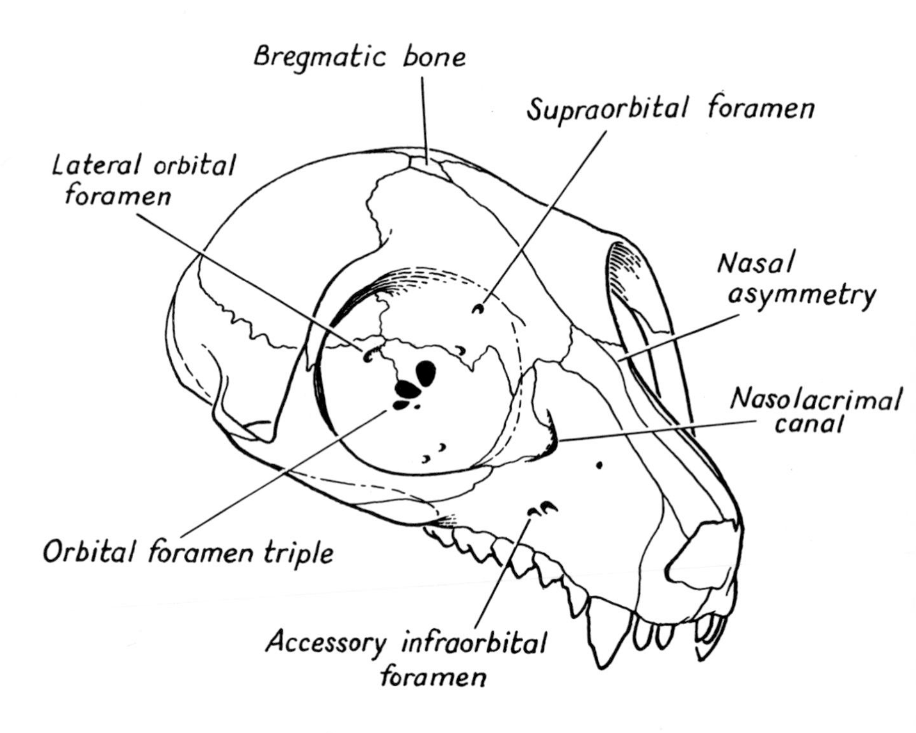

Figure 1. Anteriolateral view of a skull of *Galago* sp., to show the variants described in the text.

posterior wall of the orbit is perforated by two clearly defined foramina. Rarely, a third opening is clearly demarcated; but more frequently this third opening is only partially separated by an incomplete bony spicule from the other foramina (partial triple orbital foramen). This was scored as ½.

A tiny but distinct foramen sometimes occurred medial to the other foramina discussed above. This was not scored in the present study, but could well be used in future.

16. *Lateral orbital foramen absent.*

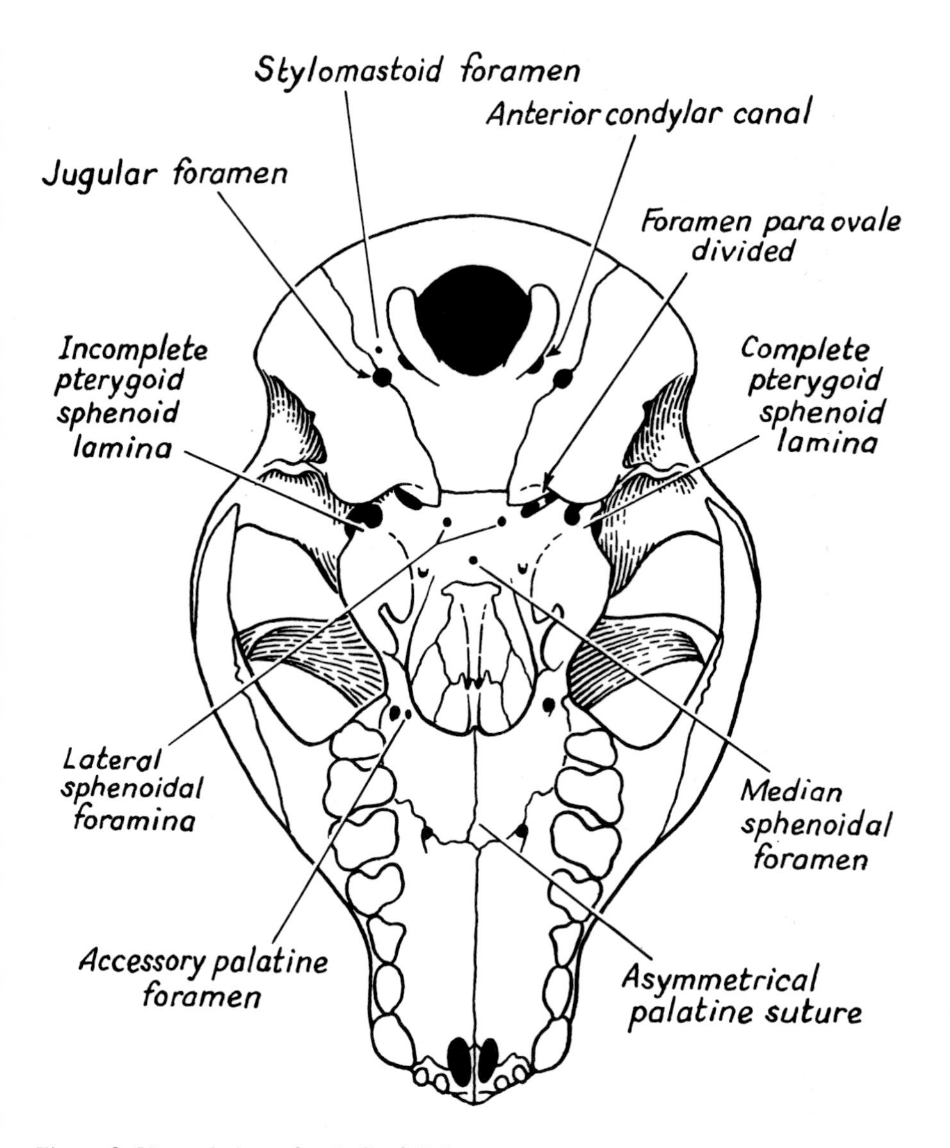

Figure 2. Ventral view of a skull of *Galago* sp.

17. *Lateral orbital foramen double.* A small foramen is usually found latero-superior to the main orbital foramina. The double foramen was only seen in *G. c. agisymbanus.*
18. *Supraorbital foramen absent.* The supraorbital foramen opens inside the orbit and is almost universally present. In rodents it has been named the frontal foramen.
19. *Accessory supraorbital foramen present.* Occasionally, further foramina were noted in the supraorbital region. These were scored as accessory supraorbital foramina.
20. *Accessory nasolacrimal foramen present.* Rarely the foramen immediately anterior to the orbit is double. This variant is called 'accessory pre-orbital foramen' in rodents.
21. *Accessory infraorbital foramen present.* The infraorbital canal runs from the orbit and emerges on the lateral surface of the maxilla. The emergent foramen may be double.
22. *Accessory mandibular foramen.* The anterior foramen on the external surface of the mandible may be double. In large primates, multiple foramina are frequently present.

Scoreable variation was found for 20 of these 22 characters in the three *Galago* species examined (see Figs. 1 and 2).

2. *Scoring difficulties*

Since the method entirely depends on accurate scoring of the variants described, it is important to consider the hindrances to accurate scoring.

(*a*) Observer subjectivity: inevitably, several of the variants will be sometimes scored as present by one observer and yet absent by another; ideally all material being compared should be scored by the same worker. Where this is not possible, careful agreement on the scoring methods and thresholds is essential.

(*b*) With smaller crania, the bone is very delicate and it is hard to know whether variants are truly present or merely artefacts from damage to the bone. The orbit is particularly vulnerable.

(*c*) Where the bone has been incompletely cleaned, foramina may be obscured and made unscoreable by fragments of soft tissue. Enzymatic cleaning as described by Searle is a great advantage, but is unpopular with museum curators since it results in disarticulation of some of the cranial elements in smaller species.[26]

(*d*) Foramina and sutures may be obscured by putty, glue, indian ink and other museum artefacts.

3. *Statistical procedure*

The incidences of the individual variants are calculated in each sample. As many of the variants are bilateral, the incidence for these are calculated for the number of sides scored. Certain variants are frequently not scoreable in some skulls, and consequently the total number scored differs for each variant. The incidences of the individual variants in different samples can then be compared with a χ^2 or similar test. A male sample of *G. senegalensis albipes* was compared in this way with a female sample from the same species.

For population comparisons, however, it is usually more profitable to calculate an 'estimate of divergence' between the samples. This calculation gives the root mean of the summed squared differences of transformed incidences of the individual variants. It is a measure of the genetical distance separating the two samples. The value of the method depends on the degree of correlation of variants with each other. Although no correlation study has been done on the *Galago* data, studies on other species have shown only very low levels of correlation.[17,27,28,29]

The calculation of the estimate of divergence, devised by Professor C.A.B. Smith, has been described in detail previously.[22,23] It is given by:

$$\text{Estimate of Divergence} = \sqrt{\frac{\Sigma\left[(\Theta_1 - \Theta_2)^2 - \left(\frac{1}{n_1} + \frac{1}{n_2}\right)\right]}{\text{number of variants}}}$$

– where Θ is the angular transformation in radians of the percentage frequency (P) of the character, such that $\Theta = \text{Sin}^{-1}\,(1\text{-}2p)$; n_1 and n_2 represent the number of individuals in the sample being compared.

The variance of this function can easily be calculated.

Estimates of Divergence were calculated between *G. crassicaudatus agisymbanus* samples from Zanzibar and Pemba Island, and between the two *G. senegalensis* subspecies.

Results

1. *Incidences of variants.* Table 1 shows the percentage incidences of 20 variants in 5 samples.
2. *Comparison of incidences of variants* in male and female. *G. senegalensis albipes:* The only variant to show a difference in incidence between the sexes with a probability of $<5\%$ was 'Asymmetrical nasal bones'. This occurs more frequently in males than in females ($p>0.01$).
3. *Estimates of Divergence:*
   (i) Between *G. crassicaudatus agisymbanus* from Zanzibar and the same subspecies from Pemba Island: 0.084.

Table 1 Incidence of 20 minor skull variants in 5 samples of *Galago*

		G. crassicaudatus agisymbanus		*G. senegalensis albipes*		*moholi*	*G. demidovii*
		Zanzibar	*Pemba Is.*	♂	♀		
	Number of crania in sample:	*31*	*28*	*47*	*47*	*51*	*57*
	Variant	%	%	%	%	%	%
1.	Nasal asymmetry present	6.7	7.4	22.2	2.2	3.9	0
3.	Bregmatic ossicle present	0	25	0	0	0	0
4.	Asymmetrical bregma present	35.0	38.9	23.0	22.0	33.0	34
5.	Parietal foramina present	5.2	5.4	0	0	1.0	0
6.	Stylomastoid foramen present	13.8	12.7	71.8	57.5	11.7	32.5
8.	Anterior condylar canal double	9.8	3.6	8.9	7.9	8.7	28.9
9.	Foramen paraovale divided	0	0	14.1	5.7	0	0
10.	Pterygoid-sphenoid lamina incomplete	15.5	17.8	8.7	5.6	13.2	14.3
11.	Lateral sphenoidal foramen present	9.2	18.8	52.0	48.0	11.7	3.8
12.	Median sphenoidal foramen present	15.0	17.8	0	2.2	2.1	19.6
13.	Accessory palatine foramen present	28.2	31.5	17.9	6.6	4.2	3.9
14.	Asymmetrical palatine suture present	14.3	14.8	13.2	13.0	0	4.2
15.	Orbital foramen triple	0	0.9	36.1	23.5	0	3.7
16.	Lateral orbital foramen absent	0	0	10.5	38.4	36.9	20.4
17.	Accessory lateral orbital foramen present	3.7	7.7	0	0	0	0
18.	Supraorbital foramen absent	0	0	0	0	0	4.2
19.	Accessory supraorbital foramen present	8.8	18.9	3.2	1.1	2.1	0
20.	Accessory nasolacrimal foramen present	0	0	0	0	0	0
21.	Accessory infraorbital foramen present	1.8	0	1.0	0	2.0	1.0
22.	Accessory mandibular foramen present	40.3	35.1	3.2	9.6	6.0	4.0

(ii) Between *G. senegalensis albipes* and *G. senegalensis moholi:* 0.473.

Discussion

1. *General conclusions*

The results show that non-metrical minor variants do occur with varying incidences in *Galago* skull samples. Apart from the exception noted, the variant occurrence is unaffected by sex. Why asymmetrical nasal bones should occur more frequently in males than in females is hard to explain. It could be a chance finding. The low estimate of divergence between separated samples of the same subspecies is to be expected, but the dramatic difference in the incidence of bregmatic bone (Pemba 25%, Zanzibar 0%) is the sort of difference often found between island populations.[15,30] The large estimate of divergence found when the subspecies of *G. senegalensis* are compared is in reasonable accord with expectation. The amount of variation seems small compared with that found in other species. Not only is there a low number of traits found to exhibit variation in the samples examined, but even when variation does occur it frequently does so only rather rarely. This is a disadvantage to studies of the present type, as population incidences below about 10% affect distance estimates disproportionately.

Why the galagos should have this tendency to invariability is not obvious. It would be interesting to know if it was characteristic of other prosimian species and also whether galagos have a similar tendency when other systems such as coat colour or blood biochemistry are considered. In general, man seems to exhibit much more variation, though to some extent this is due to the frequent occurrence of sutural bones and tori in human crania. The large primates such as *Gorilla* and *Pongo* show considerably higher incidences of multiple foramina compared to the galagos. It is tempting to suggest that the number of accessory foramina might be related to skull size, since accessory foramina are considerably commoner in the larger species. This does seem to apply when the primates are considered as a whole or when the galagos alone are considered, but in rodents the opposite is the case – multiple foramina being commoner in the tiny harvest mouse than in the rat.

2. *The validity of the method*

Earlier studies have shown that the method gives apparently meaningful genetical distances between samples within a species.[13,15,17] However, once comparisons are attempted between samples quite unrelated to one another (such as between clearly separated species) confusing results emerge.

The comparability of the genetical distances calculated between samples depends on the assumption:

(*a*) that the variants occur independently of one another,
(*b*) that their occurrence is largely genetically controlled.

Correlation studies in the mouse and man all show little correlation between the variants.[17,27–9] Although no such study has been done on the present material, it is likely that the above results hold good.

The *genetical control* of the variants has been established in mice by Grüneberg, who showed them to be controlled multigenically, with occurrence or non-occurrence of a variant resulting from a final threshold effect.

Searle,[26,31] Deol and Truslove,[32] and Howe and Parsons[33] showed that maternal physiology and diet could exert a considerable effect on individual variants. These effects were largely related to birth-weight. If a large number of variants are used together, however, these maternal and dietary effects seem to cancel one another out and variation is largely under genetical control.[33] In general, the factors affecting non-metrical skeletal variants are thought to be some way from the primary site of gene action, so that they are highly sensitive to minor fluctuations in the developing foetus.[12] Hence the term 'epigenetic' was used by Berry and Searle to describe the causation of these non-metrical minor variants.[25]

3. The validity of the method when compared with other methods

The tools usually used in population studies tend to be morphological measurement, coat colour variation and variations in blood biochemistry. Comparison of metrical and non-metrical methods on Indian rats led Berry and Smith (unpublished) to believe that their distances based on non-metrical methods were more realistic than those based on metrical methods. With human material, the two methods tend to show similar relationships,[21,22,34] though a negative correlation was found in one study.[35]

Coat colour is a classical taxonomic character (for obvious reasons). In some circumstances it may be adaptive, but direct evidence for this is surprisingly slight. Certainly much of intraspecific colour variation is a consequence of the genetical history of the population: it may represent clinal variation or random assortment of genes, as in Hebridean races of *Apodemus sylvaticus.*[15,36]

Comparison of the results of biochemical studies with non-metrical ones have produced unexpected results. Berry and Murphy, working on mice, found seasonal variations in both systems.[37] In humans, Berry found little correlation between distances calculated on blood groups and those from non-metrical variants;[22] but in general the latter distances tended to be larger than serological ones. The multi-factorial control of the minor skeletal variants means that a very large number of loci are represented when a wide range of variants is being considered. This should make the method a more delicate indicator of genetical variation than methods using serological markers which are represented (albeit very closely) by only a few loci.

Conclusion

This study was undertaken as a pilot project. It has established that variation does exist in prosimian crania and that it can be harnessed for use in population studies. Further variants could well be discovered by those more familiar with prosimian anatomy.

The present study has also entirely neglected the teeth. Minor dental variants occur in most animals and have been used in population studies of rodents,[38] gibbons,[39] and humans.[40] This would appear to be a potentially fruitful field.

Acknowledgment

My thanks are due to A.J. Lee for drawing the figures, and to the Keeper of Zoology, British Museum (Natural History), for giving me access to the material.

NOTES

1 Russel, F. (1900), 'Studies in cranial variation', *Am. Nat.* 34, 737-47.

2 Le Double, A.F. (1903), *Variations des os du crane*, Paris.

3 Le Double, A.F. (1906), *Variations des os de la face*, Paris.

4 Montagu, A.M.F. (1933), 'The anthropological significance of the pterion in primates', *Am. J. Phys. Anthrop.* 18, 159-336.

5 Montagu, A.M.F. (1937), 'The medio-frontal suture and the problem of metopism in the primates', *J.R. Anthrop. Inst.* 67, 157-201.

6 Schultz, A.H. (1923), 'Bregmatic fontanelle in mammals', *J. Mammal.* 4, 65-77.

7 Schultz, A.H. (1926), 'Studies on the variability of platyrrhine monkeys', *J. Mammal.* 7, 286-305.

8 Schultz, A.H. (1960), 'Age changes and variability in the skulls and teeth of the central American monkeys, *Alouatta, Cebus* and *Ateles*', *Proc. Zool. Soc. Lond.* 133, 337-90.

9 Wood-Jones, F. (1930-31), 'The non-metrical morphological characters of the skull as criteria for racial diagnosis', I, II and III, *J. Anat.* 65, 179-95, 368-78, 438-45.

10 Wood-Jones, F. (1933-34), 'The non-metrical morphological characters of the skull as criteria for racial diagnosis', IV, *J. Anat.* 68, 96-108.

11 Brothwell, D.R. (1958). The use of non-metrical characters in the skull in differentiating populations, *Dt. Ges. Anthrop.* 6, 103-9.

12 Grüneberg, H. (1963), *The Pathology of Development*, Oxford.

13 Berry, R.J. (1963), 'Epigenetic polymorphism in wild populations of *Mus musculus*', *Genet. Res.* 4, 193-220.

14 Berry, R.J. (1964), 'The evolution of an island population of the house mouse', *Evolution* 18, 468-83.

15 Berry, R.J. (1969), 'History in the evolution of *Apodemus sylvaticus* (Mammalia) at one edge of its range', *J. Zool. Soc. Lond.* 159, 311-28.

16 Berry, R.J. (1969), 'Non-metrical skull variation in two Scottish colonies of the Grey Seal', *J. Zool. Soc. Lond.* 157, 11-18.

17 Berry, A.C. and Berry, R.J. (1967), 'Epigenetic variation in the human cranium', *J. Anat.* 101, 361-79.

18 Laughlin, W.S. and Jørgensen, J.B. (1956), 'Isolate variation in Greenlandic Eskimo crania', *Acta Genet.* 6, 3-12.

19 Kellock, W.L. and Parsons, P.A. (1970*a*), 'Variations of minor non-metrical skeletal variants in Australian Aborigines', *Am. J. Phys. Anthrop.*, n.s. 32, 409-21.

20 Kellock, W.L. and Parsons, P.A. (1970*b*), 'A comparison of the incidence of non-metrical cranial variants in Australian Aborigines with those of Melanesia and Polynesia', *Am. J. Phys. Anthrop.*, n.s. 33, 235-9.

21 De Villiers, H. (1968), *The Skull of the South African Negro*, Johannesburg.

22 Berry, A.C. (1971), 'The use of minor skeletal variants in human population studies', PhD Thesis, University of London; and Berry, A.C. (1974), 'The use of non-metrical variations of the cranium in the study of Scandinavian population movements', *Am. J. Phys. Anthrop.*, in press.

23 Berry, A.C. and Berry, R.J. (1971), 'Epigenetic polymorphism in the primate skeleton', in Chiarelli, A.B. (ed), *Comparative Genetics in Monkeys, Apes and Man*, London and New York.

24 Balan, B.B. and Thorington, R.W. (1969), 'Discrete variation in *Aotus* skulls', Paper read at Annual Meeting of Am. Soc. Mammalogists.

25 Berry, R.J. and Searle, A.G. (1963), 'Epigenetic polymorphism of the rodent skeleton', *Proc. Zool. Soc. Lond.* 140, 577-615.

26 Searle, A.G. (1954*a*), 'Genetical studies on the skeleton of the mouse. IX: Causes of skeletal variation within pure lines', *J. Genet.* 52, 68-102.

27 Truslove, G.M. (1961), 'Genetical studies on the skeleton of the mouse. XXX: A search for correlations between some minor variants', *Genet. Res.* 2, 431-8.

28 Ossenberg, N.S. (1970), 'The influence of artificial cranial deformation on discontinuous, morphological traits', *Am. J. Phys. Anthrop.*, n.s. 33, 357-71.

29 Hertzog, K.P. (1968), 'Association between discontinuous cranial traits', *Am. J. Phys. Anthrop.*, n.s. 29, 397-404.

30 Berry, R.J. (1967), 'Genetical changes in mice and men', *Eug. Rev.* 59, 78-96.

31 Searle, A.G. (1954*b*), 'Genetical studies on the skeleton of the mouse. XI: The influence of diet on variation within pure lines', *J. Genet.* 52, 413-24.

32 Deol, M.S. and Truslove, G.M. (1957), 'Genetical studies of the skeleton of the mouse, XX: Maternal physiology and variation in the skeleton of C57BL mice', *J. Genet.* 55, 288-312.

33 Howe, W.L. and Parsons, P.A. (1967), 'Genotype and environment in the determination of minor skeletal variants and body weight in mice', *J. Embryol. Exp. Morph.* 17, 283-92.

34 Knip, A.S. (1970), 'Metrical and non-metrical measurements on the skeletal remains of Christian populations from two sites in Sudanese Nubia', I and II, *Proc. Kon. Ned. Akad. v. Wetensch. Amsterdam* C 73, 433-68.

35 Berry, A.C., Berry, R.J. and Ucko, P.J. (1967), 'Genetical change in ancient Egypt', *Man* 2, 551-68.

36 Delany, M.J., (1964), 'Variation in the long-tailed field-mouse (*Apodemus*

sylvaticus L.) in North West Scotland. I: Comparison of individual characters', *Proc. Roy. Soc. Lond. B* 161, 191-9.

37 Berry, R.J. and Murphy, H.M. (1970), 'The biochemical genetics of an island population of the house mouse', *Proc. Roy. Soc. Lond. B.* 176, 87-103.

38 Corbet, G.B. (1964), 'Regional variation in the bank vole *Clethrionomys glareolus* in the British Isles', *Proc. Zool. Soc. Lond.* 143, 191-219.

39 Frisch, J.E. (1963), 'Dental variation in a population of gibbons (*Hylobates lar*), in Brothwell, D.R., (ed.), *Dental Anthropology*, Oxford.

40 Hanihara, K. (1967), 'Racial characteristics of the dentition', *J. Dent. Res.* 46, 923-6.

M. CARTMILL

Daubentonia, Dactylopsila, *woodpeckers and klinorhynchy*

Introduction

Many insects, especially among the Lepidoptera and Coleoptera, feed as larvae on the cambium and woody tissues of the trunks and branches of trees. A great number of insect species have adopted this way of life; over 500 genera of wood-boring and cambium-eating insects, comprising many hundreds of species, infest the forests of the Congo region alone.[1] Wood-boring adaptations take many forms. Some larvae feed on cambium and phloem immediately under the bark; some can only feed on dead and partly decayed wood; some kill the tree or branch by tunnelling in the cambium and eat deeper into the underlying wood as it begins to decompose.[2,3] All these wood-boring adaptations have the effect of protecting the larvae and pupae from vertebrate predators to some degree.

The greater the specialisation of the larva for boring in wood, the greater the specialisation demanded of its predators. Larvae feeding directly under the bark of a tree may be pulled from crevices by the slim barbed tongues of titmice, nuthatches, woodhewers and other birds,[4,5] or fished out by the probing fingers of marmosets,[6] or of squirrel monkeys[7] and orangutans.[8] Insects boring in fallen branches or old tree stumps can be exposed and eaten by any predator strong enough to break open the rotten wood concealing them – by coatimundis, for example.[9] To recover insect larvae from their tunnels in the sound wood of living trees, however, requires special adaptations (*a*) for boring or cutting through the concealing wood and (*b*) for probing the exposed tunnel to catch the retreating insect.

In most continental forest communities, woodpeckers are the principal vertebrate predators on wood-boring insect larvae. Wood-pecking birds of the family Picidae inhabit forests in every part of the world except Madagascar, New Guinea and Australia, and various

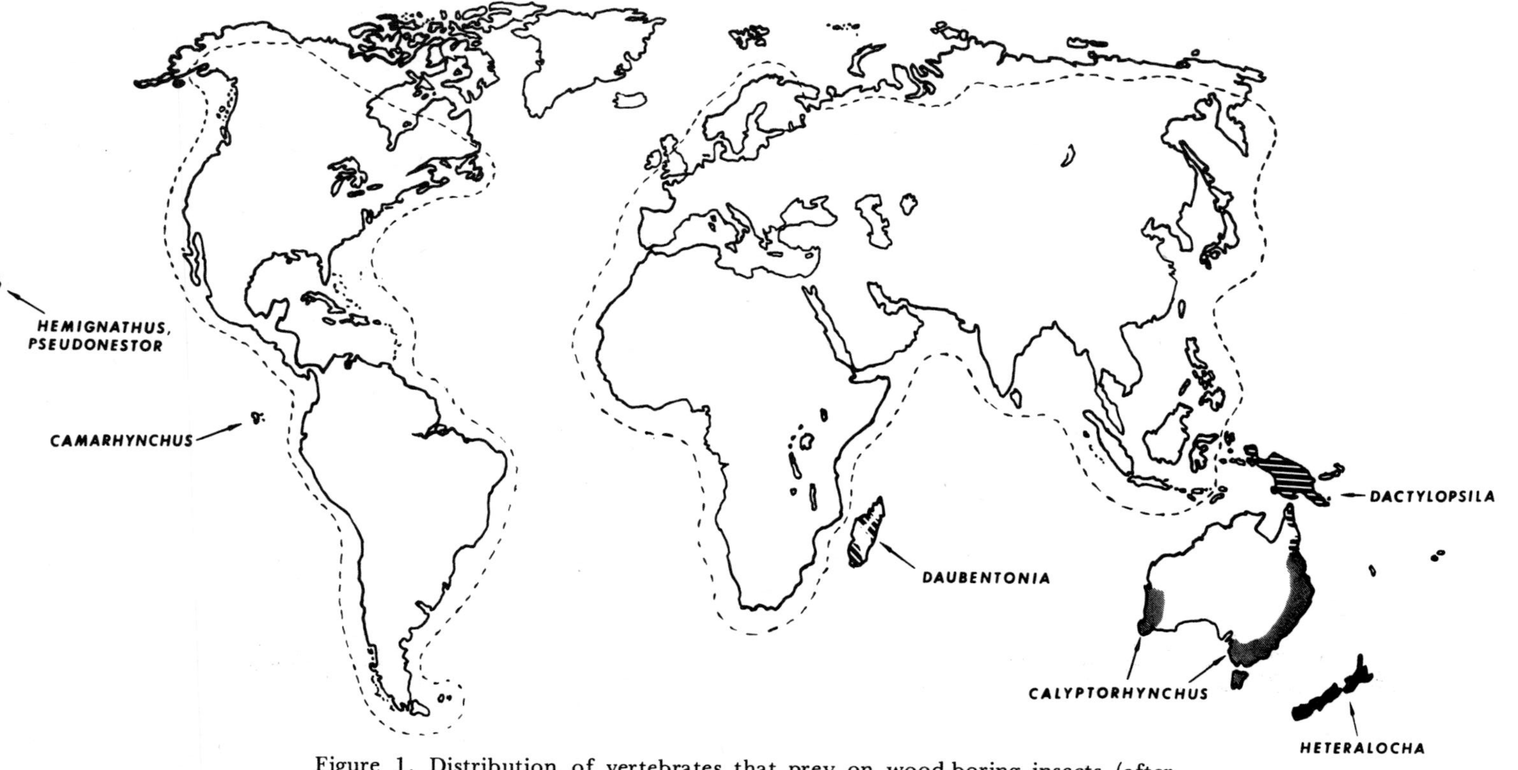

Figure 1. Distribution of vertebrates that prey on wood-boring insects (after various sources). The dashed line shows the limits of woodpecker distribution. The distribution shown for *Calyptorhynchus* includes only those species which feed on wood-boring insects. *Daubentonia* currently has a distribution more restricted than that shown here; the hatched areas represent all records of living or subfossil *Daubentonia*, including the extinct *D. robusta*.

oceanic islands (Fig. 1).[10] Although many woodpeckers get most of their food by picking insects from the surface of trees and from crevices in the bark,[11,12] a woodpecker can also expose a deeply imbedded insect by cutting into the wood with rapid blows of its chisel-like beak; the insect is then either picked up with the beak or extracted from its tunnel by the bird's long, barbed tongue.

On some of the oceanic islands never colonised by woodpeckers, other birds have developed adaptations analogous to the picid specialisations for tunnelling and probing in wood. On Hawaii, honeycreepers of the superspecies *Hemignathus lucidus* use their short, straight lower beaks to peck and excavate in rotting wood, and employ their long, slender and curved upper beaks to probe exposed tunnels.[13,14] The extinct huia *Heteralocha acutirostis* of New Zealand foraged for wood-boring insects using a sexual division of labour; the male bore a stout beak for excavating, and the female a long curving beak for probing.[15] The woodpecker finch *Camarhynchus pallidus* of the Galapagos Islands is well-known for its habit of using a cactus spine or twig as a tool for probing insect tunnels after they have been exposed by ripping off bark and wood with the beak.[16] In Australia, parrots of the genus *Calyptorhynchus*, especially *C. funereus* and *C. lathami*, bite and wrench chunks of wood from living trees to expose burrowing moth larvae; no probing adaptation seems to have been developed in these birds.[17,18] The Hawaiian honeycreeper *Pseudonestor* has a parrot-like beak and habits like those of *Calyptorhynchus.*[19]

New Guinea and Madagascar are the only extensive areas of the world which have not been colonised by woodpeckers or by any of the various woodpecker-like island birds. In these areas, the woodpecker niche is filled by specialised mammals, which have enlarged incisors that serve as wood-cutting organs and an elongated finger which is used as a probe.

The lemuriform primate *Daubentonia madagascariensis* forages for wood-boring grubs in the rain forest of the eastern coast of Madagascar. In feeding, it begins by tapping the surface of a branch, trunk, or fruit, turning its large ears toward the spot being investigated. Detecting an insect, it probes the crevices of the infested plant with a grotesquely attenuated third finger, exposing deep-burrowing larvae by biting and tearing at the wood with its rodent-like incisor teeth. Larvae are pulled out whole or piecemeal by the third finger and licked off.[20,21] Unlike the other Madagascar primates, *Daubentonia* possesses true claws on all its digits except the first toe.[22] These claws represent an adaptation for foraging on trunks, comparable to the specialised stiff tail which helps a foraging woodpecker to support its weight; they make movement and posturing on large trunks easier for *Daubentonia* than for a clawless animal of similar size.[23,24]

In New Guinea, similar morphology and foraging habits have been developed by diprotodont marsupials of the genus *Dactylopsila.* Much like *Daubentonia*, these animals feed by tapping and sniffing infested branches and trunks to locate prey, tearing off bark and wood with their enlarged incisors, and inserting an elongated finger – in this case, the fourth – into exposed tunnels to snare burrowing larvae.[25,26] Like *Daubentonia*, but unlike many of the other small arboreal diprotodonts, *Dactylopsila* has well-developed claws on all digits except the hallux; it differs from *Daubentonia* in being of much smaller size and in having rooted incisors. At least two species of *Dactylopsila, D. palpator* and *D. trivirgata*, inhabit New Guinea. *Dactylopsila palpator* has a longer and more attenuated fourth finger and larger wood-cutting incisors than *D. trivirgata*;[27,28] it is also distinguished from *D. trivirgata* by certain details of the basicranium,[29] and is placed by some in a separate genus *Dactylonax. Dactylopsila trivirgata* occurs in two areas on the eastern coast of the Cape York Peninsula, on the Australian mainland.[30] In the more southerly of these two areas, *D. trivirgata* is sympatric with *Calyptorhynchus funereus*;[17] there is presumably some competition between the two species in this small zone of sympatric occurrence.

Cranial morphology of Daubentonia *and* Dactylopsila

The skull of *Daubentonia* (Fig. 2) appears aberrant in a number of respects when compared to the other Madagascar lemurs. Its most obvious peculiarity is the great hypertrophy of its large, continually growing upper and lower incisors; these, together with the large post-incisor diastemata in both upper and lower jaws, give the skull a superficially rodent-like appearance. *Daubentonia*'s brain is globose and foreshortened, and its dorsal-ventral depth is unusually great relative to its size.[31–3] It is also unusually large; corrected for allometry, the brain/body-weight ratio of *Daubentonia* is greater than that of any other prosimian, and even exceeds that of many anthropoids.[34–6] Seen in midsagittal section, the facial skeleton of *Daubentonia* shows, like the brain, a pronounced decrease in anteroposterior length and increase in dorsal-ventral height by comparison with other prosimians.[32,37] The dental arcade is short and narrow. Klinorhynchy is marked in *Daubentonia*: that is, this lemur's facial skeleton is bent downward with respect to the cranial base, so that the obtuse angle between the floor of the nasal fossa and the clivus is about ten degrees less than in *Lemur.*[37]

The zygomatic arches of *Daubentonia* are exceptionally high, especially at their anterior ends,[37] and the zygomatic processes of the maxilla arise far rostrad, above P^4. In other extant lemuriforms, they ordinarily arise above M^2. Although *Daubentonia*'s orbits are of

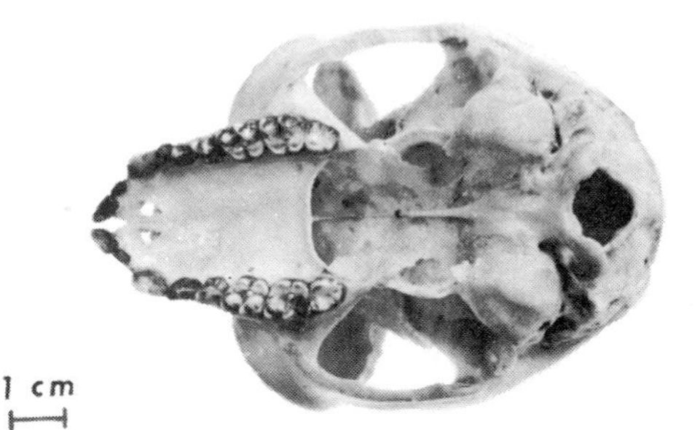

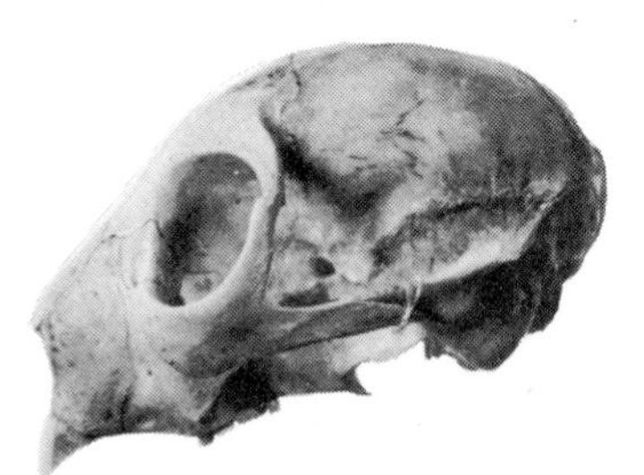

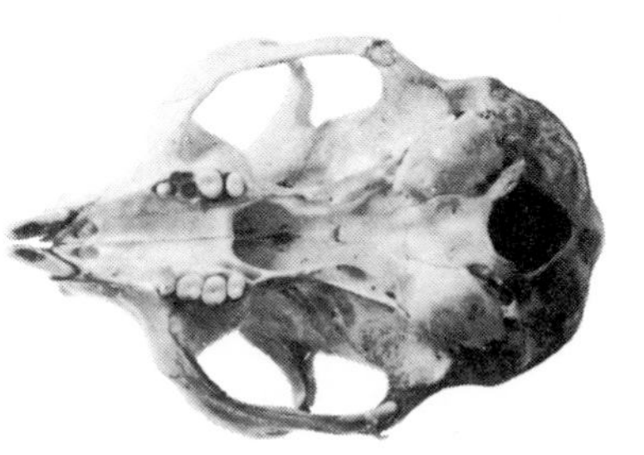

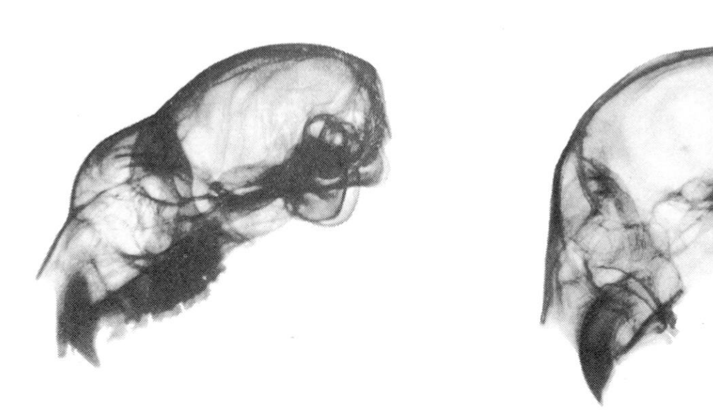

Figure 2. Skulls of prosimian primates. Top row, lateral and palatal views of *Lemur mongoz*, Am. Mus. Nat. Hist. no. 100539, showing relatively primitive configuration of zygomatic arch and pterygoid laminae. Second row, *Propithecus diadema*, A.M.N.H. no. 100557, showing indriid configuration. Third row, *Daubentonia madagascariensis*, A.M.N.H. no. 100632. Bottom row, lateral radiographs of *Propithecus verreauxi* (left), Field Mus. Nat. Hist. no. 8342, and *Daubentonia* (right), Yale Peabody Mus. no. 302, oriented on horizontal through clivus to show increased klinorhynchy in the latter; the *P. verreauxi* skull has been enlarged slightly for comparison with *P. diadema*. In the Yale specimen of *Daubentonia*, the occiput is damaged.

approximately the same diameter as the orbits of other lemuriforms of comparable body size, the relative inter-orbital distance is appreciably greater in *Daubentonia* (Table 1). The result of these three peculiarities is that, in effect, the medial margin of the orbit of *Daubentonia* is displaced laterally, the anterior margin is displaced still further forward, and the lateral margin is displaced dorsally. The orbits of *Daubentonia* are therefore as convergent as those of other Madagascar lemurs of similar size,[38] but directed much less frontally – that is, they face upwards toward the skull roof, like the orbits of *Nycticebus* or *Perodicticus.*

The basicranium of *Daubentonia* is also aberrant. The sphenoid is greatly inflated, the *sinus sphenoidalis* extending back through the entire basisphenoid.[31] The lateral pterygoid laminae extend from the posterior molar alveoli backward in a virtually straight line to the lateral side of the auditory bulla, where they have a broad articulation with the petrosal,[39] and fuse with the squamosal medial to the glenoid fossa.

Table 1 Relative interorbital breadth in lemuriform prosimians[51]

Species (and sample size)	*Interorbital breadth in orbital diameters*	
	Sample mean	*s.d.*
Lemur catta (6)	0.84	0.08
Lepilemur mustelinus (6)	0.51	0.03
Hapalemur griseus (8)	0.49	0.04
Cheirogaleus major (4)	0.56	0.04
Microcebus murinus (6)	0.31	0.02
Phaner furcifer (2)	0.56*, 0.54*	–
Propithecus verreauxi (6)	0.90	0.03
Avahi laniger (4)	0.62	0.05
Indri indri (6)	0.93	0.11
Daubentonia madagascariensis (3)	1.15*, 1.16*, 1.25*	–

* Values for single specimens.

Table 2 Klinorhynchy in phalangeroids[51]

Species (and sample size)	*Angle between clivus and palate*	
	Sample mean (degrees)	*s.d.*
Eudromicia caudata (6)	165.4	4.4
Trichosurus vulpecula (6)	167.3	1.3
Pseudocheirus lemuroides (7)	165.8	2.5
Phalanger orientalis (6)	163.6	5.4
Phalanger maculatus (6)	162.1	4.2
Phascolarctos cinereus (6)	183.9	4.5
Dactylopsila trivirgata (6)	160.0	4.8
Dactylopsila palpator (6)	154.5	6.0

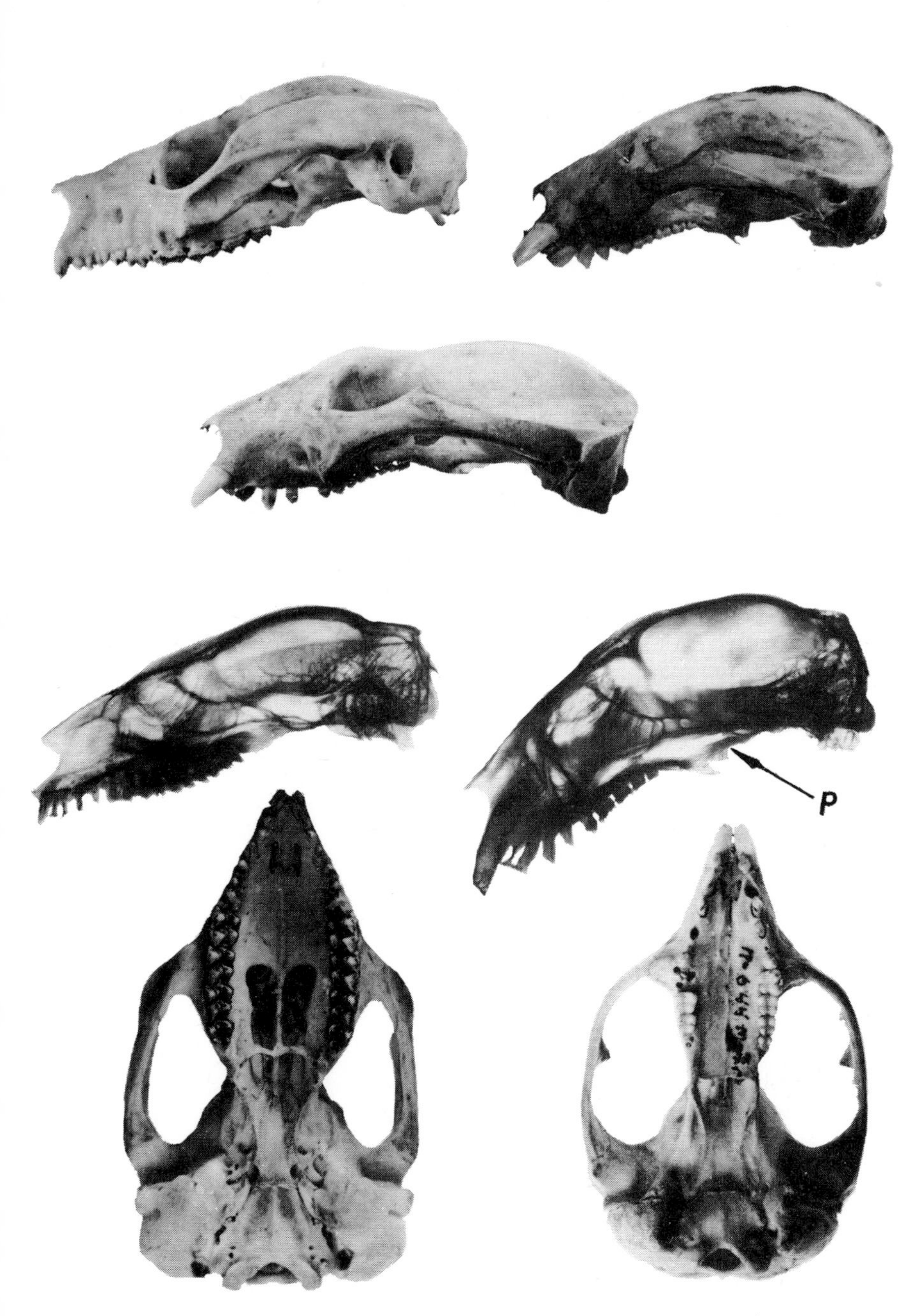

Figure 3. Skulls of phalangeroid marsupials (not to same scale). Top row, *Pseudocheirus* sp. (left), Duke University Dept. of Anatomy collection, and *Dactylopsila palpator* (right), Amer. Mus. Nat. Hist. no. 191043, showing differences in zygomatic arches. Second row, *Dactylopsila trivirgata*, A.M.N.H. no. 143852, a less specialised species of the genus. Third row, lateral radiographs of *Pseudocheirus lemuroides* (left), A.M.N.H. no. 60923, and *D. palpator* (right), A.M.N.H. no. 151964, oriented on clivus; increase in klinorhynchy, in size of olfactory bulbs and frontal cortex, and in size and density of lateral pterygoid processes (p) is evident in *Dactylopsila.* Bottom row, palatal views of same specimens seen in top row; *D. palpator* lacks palatine vacuities.

Most of the peculiarities of cranial morphology which distinguish *Daubentonia* from the other lemuriforms also distinguish *Dactylopsila* from the other phalangeroids, and are especially pronounced in the more specialised species *Dactylopsila palpator.* By comparison with a less specialised phalanger of comparable size (e.g. a small species of *Pseudocheirus*), *Dactylopsila palpator* displays marked klinorhynchy (Table 2), increased relative interorbital breadth (exceeded only by the equally but differently specialised *Phascolarctos* among the species listed in Table 2), and reduction in the length of the snout and augmentation of its vertical height. *Dactylopsila*'s brain is globular and exceptionally large. The dental arcade of *Dactylopsila* is relatively short and narrow. The anterior root of the zygomatic arch extends forward to a point above P^3 in *Dactylopsila*, whereas it lies above the M^{1-2} interspace in most other phalangers; the anterior zygomatic root is also unusually high in *Dactylopsila.* As a consequence, *Dactylopsila*, like *Daubentonia*, has markedly convergent but dorsally-directed orbits. In most phalangeroids, the medial pterygoid laminae are prolonged backward along the cranial base as distinct ridges springing from the basisphenoid; in *Dactylopsila*, the *lateral* pterygoid plates are also prolonged backward as low ridges of dense bone, which extend caudad almost to the glenoid fossa (Fig. 3).

Interpretation

Since *Daubentonia* and *Dactylopsila* have had no common ancestor since at least the mid-Cretaceous, it must be assumed that all the peculiar cranial traits which they share with each other but not with their closest relatives are habitus traits, functionally related to their other adaptations for preying on wood-boring insects. This assumption is further supported by the fact that comparable peculiarities – a shortened face, a foreshortened and rounded braincase, a wide and densely ossified interorbital region, moderate klinorhynchy, and so on – are characteristic of woodpeckers, and are most pronounced in those woodpecker species whose diets contain the largest proportions of wood-boring insects.[40,41] Evidently, all these traits of cranial morphology are primary or secondary adaptations for generating and dissipating the large forces needed to chisel or bite through wood.

When a mammal bites into wood with its incisors, the biting force exerted by the masticatory musculature is transmitted through the wood to the upper incisors. Since the mandible cannot as a rule be protruded far enough to direct the transmitted force along the dorsal surface of the snout (Fig. 4*A*), incisal biting tends to bend the end of the rostrum dorsally, inducing compressive stresses in the roof of the rostrum and tensile stresses in its floor. The amount of force that can be exerted in biting is limited by the capacity of the facial skeleton

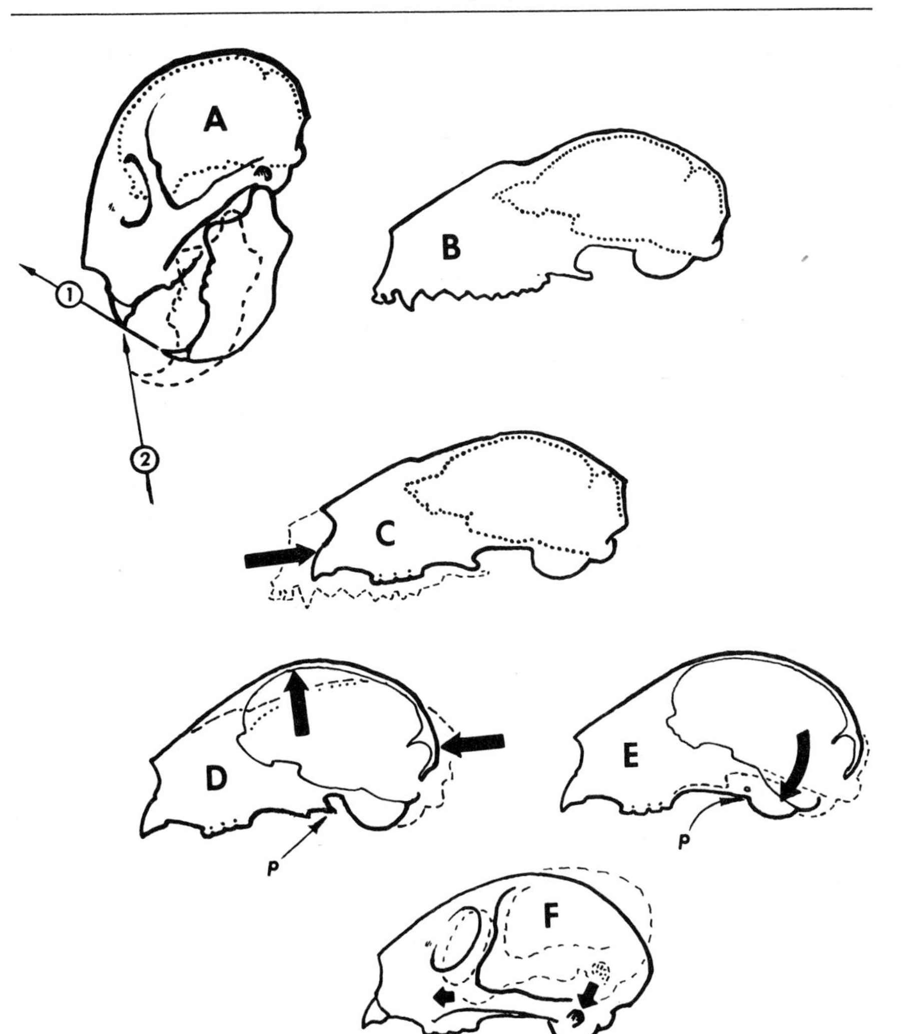

Figure 4. Generation and dissipation of biting stresses in the skull of *Daubentonia.* (*A*) Lateral view of skull. When the jaws are parted in biting into wood, the force transmitted to the rostrum is roughly normal to the rostral axis (arrow 1). To align this force with the rostrum (arrow 2) would require an impossible protrusion of the mandible (dashed line). (*B*) Diagrammatic outline of indriid skull with superimposed silhouette of brain (dotted line). (*C*) Recession of the rostrum (arrow) relative to the proportions seen in other lemurs of similar size (dashed line) reduces the bending moments acting on the rostrum. (*D*) Remodelling of the braincase (arrows) increases the vertical height of the interface between splanchnocranium and neurocranium. (*E*) Klinorhynchy (curved arrow) permits further expansion of the area through which the face is hafted onto the braincase, by presenting the cranial base for attachment of caudad projections from the lateral pterygoid processes (p). (*F*) Klinorhynchy and forward displacement of the anterior zygomatic root (arrows) allow the zygomatic arch to act under tension as a guy wire resisting upward displacement of the end of the rostrum.

to withstand these bending stresses.

Most of the shared cranial peculiarities of *Daubentonia* and *Dactylopsila* have the effect of reducing and countering bending forces in the snout, thus increasing the amount of force that these animals can exert when they tear into infested wood in search of wood-boring insects. Their most obvious adaptation to this end is a reduction in the relative length of the dental arcade. This has the effect of drawing the maxillary incisors back toward the braincase, therefore reducing all the bending moments of the force exerted on the facial skeleton through the maxillary incisors (Fig. 4*C*).

Most of the other cranial specialisations of *Daubentonia* and *Dactylopsila* have the effect of countering bending forces rather than reducing them. In less aberrant lemurs and phalangers, the facial skeleton resembles a long truncated cone affixed to the narrow end of the prolate braincase (Fig. 4*B*). In *Daubentonia,* the height of the braincase is increased, thus, in Radinsky's words, 'strengthening the skull against the biting stresses from the incisors' by increasing the diameter of the facial cone's base in the midsagittal plane (Fig. 4*D*).[33] The same adaption is evident in *Dactylopsila.* In both animals, the braincase is also rotated or tilted back and up with respect to the axis of the snout, bringing the occipital end of the cranial base forward and downward (Fig. 4*E*) and so further increasing the neurocranial surface available for attachment of the facial skeleton. This is accompanied by an extension of the lateral pterygoid laminae back along the ventral surface of the braincase toward the ear region.[42] As the face has rotated downward and shifted back with respect to the cranial base, the cribriform plate has rotated with the olfactory fossa, so that in *Daubentonia* and *Dactylopsila* the olfactory lobes of the brain face more nearly ventrally than in other lemurs or phalangers when the skull is oriented on the clivus. In *Daubentonia*, the vertical height of the facial cone's base is also augmented by pneumatisation of the frontal region and the sphenoid.

Since the biting force tends to bend the end of the rostrum upward, it places the frontal region under compression and the palate under tension. Relative interorbital breadth is increased somewhat in both *Daubentonia* and *Dactylopsila*, enabling this region to withstand more compressive stress. Tensile stresses in the floor of the snout are transmitted backward along the palate and alveolar bone to the pterygoid processes, which in *Daubentonia* and *Dactylopsila* are broadly fused with the cranial base back as far as the glenoid region (Fig. 4*D-E*).

In both *Daubentonia* and *Dactylopsila*, the anterior root of the zygomatic arch arises relatively further rostrad than it does in less specialised lemurs and phalangers. In *Daubentonia*, this has the effect of bringing the post-orbital bar more nearly into alignment with the long axis of the rostrum; and it might be argued that this serves to permit more efficient transmission of compressive stresses from the end of the rostrum backward and upward through the post-orbital bar.[43] But this analysis fails to account for the similar specialisation seen in *Dactylopsila*, which lacks a post-orbital bar. The forward shift of the anterior zygomatic root seen in these animals makes more sense when we note that the associated klinorhynchy involves a depression of the posterior zygomatic root with respect to the long axis of the snout (Fig. 4*F*). These changes in the relative positions of the two ends of the zygomatic arch permit the entire arch to act like a guy wire or tie beam connecting the anterior end of the snout to the depressed glenoid region, thus helping to counteract bending moments acting on the rostrum.

Increase in the vertical height of the anterior zygomatic root serves the same function. Elastic deformation of the rostrum by bending forces transmitted through the maxillary incisors would displace the lacrimal region backward and the molar alveoli forward, and therefore produce torsion in the zygomatic arches. In Fig. 5, two points (x, y) are diagrammed as alternative centres of rotation for this torsion. If the bite force transmitted through the upper incisors produced torsion around x, the moment arm of this force would be p_1. Rotation around x requires anterior displacement of y, and the forces resisting this displacement have a moment arm L equal to the distance from x to y. In effect, this system represents a bent lever. If the anterior zygomatic root is shifted forward, p_1 decreases accordingly; if the vertical height of the anterior zygomatic root is increased, x and y move further apart, and the load arm (L) of the bent lever system increases relative to the power arm p_1. The same analysis applies for a bending force with a moment arm of p_2 acting around y (Fig. 5); or around any point between x and y. The configuration seen in *Daubentonia* and *Dactylopsila* (Fig. 5*B*) therefore permits the arches to resist torsion more efficiently, and so helps to counter bending forces acting on the rostrum.

Several features of the skull of *Dactylopsila* suggest that this marsupial may at times use its rostrum as a chisel, after the fashion of a woodpecker. The enlarged upper incisor in this genus shows pronounced wear on the labial surface; the tip of the tooth is ordinarily worn off flat in a plane roughly perpendicular to the longitudinal axis of the palate. As Thomas pointed out, the lower incisor cannot occlude with this wear facet.[27] If *Dactylopsila* ever thrusts or pries into a piece of wood using its upper incisors alone, as this wear pattern suggests,[44] its entire rostrum will sometimes be placed under compression, and the palate and pterygoid processes must be adapted to carry compressive stresses as well as tensile ones. Two peculiarities of *Dactylopsila* look like adaptations to this end. First, its palate is narrow, dense, and thick, lacking the palatine vacuities seen in all other phalangerid genera; in *D. trivirgata*, the palate is thin and faintly translucent where vacuities might be

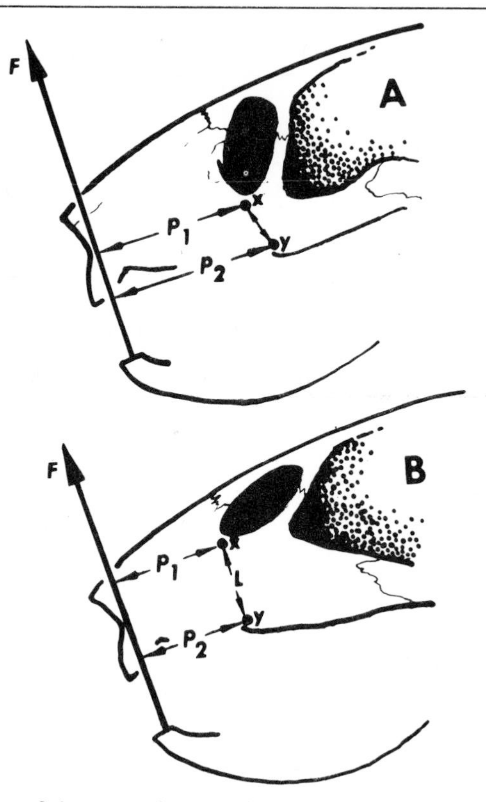

Figure 5. Effect of increase in anterior zygomatic height. (*A*) Unspecialised configuration. (*B*) Configuration seen in *Daubentonia* and *Dactylopsila*. The biting force (F) transmitted to the upper incisors tends to produce torsion in the zygomatic arch; the configuration seen in *Daubentonia* resists this torsion more effectively, and thus reduces strain in the rostrum during gnawing. See Explanation in text.

expected to occur, but in the more specialised *D. palpator* even this vestige of the primitive condition has disappeared (Fig. 3). Secondly, the lateral pterygoid processes or crests which extend backward toward the ear region of *Dactylopsila* are thickened struts which cast a dense x-ray shadow; by contrast, those of *Daubentonia* are relatively delicate bony plates, evidently better suited to carrying loads under tension (through periosteal fibre tracts?) than under compression.

Discussion: gnawing dentitions and skull form

Although a distinguished proponent of the notion that klinorhynchy reflects postural uprightness has been bold enough to assert that *Daubentonia* is 'nearly always bipedal' in its locomotion,[45] klinorhynchy clearly has nothing to do with upright posture in either *Daubentonia* or *Dactylopsila*. Radinsky suggests that klinorhynchy in

Daubentonia 'brings the axis of the rostrum more into line with the incisors, so that stresses resulting from biting are transmitted along the long axis of the rostrum'.[33] The difficulty with this interpretation becomes apparent when we translate it into the equivalent suggestion that depression of the occipital end of the braincase (Fig. 4*E*) somehow causes biting stresses to be transmitted along the long axis of the rostrum. In fact, biting stresses in *Daubentonia* cannot be aligned with the long axis of the rostrum, at least when the incisors are parted far enough to bite into wood; the mandible cannot be protruded so far (Fig. 4*A*).

The interpretation advanced here appears preferable: that deflection of the visceral cranium (or retroflexion of the braincase) serves to augment the vertical height of the interface between visceral and neural parts of the skull, thus helping to stabilise the facial skeleton against bending moments in the sagittal plane. How generally applicable is this interpretation? It would be a mistake to attempt to apply it to all instances of pronounced klinorhynchy. In *Tarsius*, klinorhynchy is a secondary effect of the hypertrophy of the eyes,[46] while in *Homo* it probably represents, at least in part, a postural adaptation.[47] But in the interpretation of klinorhynchy in rodents, lagomorphs, and other animals with enlarged anterior teeth used in cutting hard materials, the thesis advanced here should be tested before resorting to explanations involving presumed changes in posture. It seems possible, for instance, that *Lepus* is more klinorhynch than *Ochotona* because it is a larger animal capable of generating larger bending moments in its long snout, not because its facial skeleton has been 'subjected to readjustment by the adoption of upright posture'.[48]

Daubentonia's brain is unexpectedly large for a prosimian of its mass; in 13 of the 14 dimensions of various parts of the brain measured by Stephan et al.,[36] *Daubentonia* exceeds *Indri* in absolute size, though its body weight is less than half that of *Indri*. The suggestion that *Daubentonia* is a dwarf form[49] seems untestable. Moreover, a similar increase in size and height of the braincase is seen in *Dactylopsila*, and is at least as pronounced in *D. palpator* as in the somewhat smaller *D. trivirgata*. Dabelow proposed that central flexion of the cranial base ordinarily results in ('*verursacht*') enlargement of the brain, because the dorsal surface of the braincase must increase in length as the base becomes flexed.[50] In both *Daubentonia* and *Dactylopsila*, klinorhynchy has been accompanied by a slight increase in central flexion between the pre- and post-sellar parts of the cranial floor; more importantly, it has also been accompanied by an increase in the vertical height of the frontal end of the braincase (Fig. 4*D*). It seems possible that the unusual growth of the brain of *Daubentonia* is a secondary result of a change in the pattern of mechanical forces operating on the skull.[51]

Acknowledgments

The data in Tables 1 and 2 are taken from a study funded in part by grants-in-aid from the Wenner-Gren Foundation and from The Society of the Sigma Xi. My attendance at the Research Seminar on Prosimian Biology was made possible by funds provided by the Duke University Department of Anatomy. The skull of *Pseudocheirus* sp. reproduced in Fig. 3 was generously provided by Dr Nancy Bowers of the University of Auckland.

NOTES

1 Mayné, R. (1962), 'Hôtes entomologiques du bois. II: Distribution au Congo, au Rwanda et au Burundi. Observations éthologiques', *Publ. Inst. Nat. Etude Agron. Congo Belge*, sér. sci., 100, 1-514.

2 Graham, S.A. (1952), *Forest Entomology*, New York.

3 Kirkpatrick, T.W. (1957), *Insect Life in the Tropics*, London.

4 Gardner, L.L. (1925), 'The adaptive modifications and the taxonomic value of the tongue in birds', *Proc. U.S. Nat. Mus.* 67 (19), 1-49.

5 Skutch, A.F. (1945), 'Life history of the allied woodhewer', *Condor* 47, 85-94.

6 Hershkovitz, P. (1969), 'The recent mammals of the neotropical region: a zoogeographical and ecological review', *Quart. Rev. Biol.* 44, 1-70.

7 Allen, J.A. (1916), 'Mammals collected on the Roosevelt Brazilian Expedition, with field notes by Leo E. Miller', *Bull. Am. Mus. Nat. Hist.* 35, 559-610.

8 Harrisson, B. (1962), *Orang-Utan*, London.

9 Kaufmann, J.H. (1962), 'Ecology and social behaviour of the coati, *Nasua narica*, on Barro Colorado Island, Panama', *Univ. Calif. Publ. Zool.* 60, 95-222.

10 Peters, J.L. (1948), *Check-List of Birds of the World*, vol. 6, Cambridge, Massachusetts.

11 Short, L.L. (1970), 'Notes on the habits of some Argentine and Peruvian woodpeckers (Aves, Picidae)', *Am. Mus. Novit.* 2413, 1-37.

12 Short, L.L. (1971), 'Systematics and behavior of some North American woodpeckers, genus *Picoides* (Aves)', *Bull. Am. Mus. Nat. Hist.* 145, 1-118.

13 Amadon, D. (1950), 'The Hawaiian honeycreepers (Aves, Drepaniidae)', *Bull. Am. Mus. Nat. Hist.* 95, 151-262.

14 Munro, G.C. (1960), *Birds of Hawaii*, Rutland (Vermont) and Tokyo.

15 Buller, W.L. (1888), *History of the Birds of New Zealand*, 2nd ed., published by the author, London; rev. ed. by Turbott, E.G. Honolulu, 1967.

16 Lack, D. (1947), *Darwin's Finches*, London.

17 Eastman, W.R. Jr., and Hunt, A.C. (1966), *The Parrots of Australia*, Sydney.

18 Frith, H.J. (ed.) (1969), *Birds in the Australian High Country*, Sydney.

19 Perkins, R.C.L. (1903), 'Vertebrata', in Sharp, D. (ed.), *Fauna Hawaiiensis*, London, vol. 1, part 4.

20 Owen, R. (1866), 'On the aye-aye (*Chiromys* Cuvier)', *Trans. Zool. Soc. Lond.* 5, 33-101.

21 Petter, J.-J. and Petter, A. (1967), 'The aye-aye of Madagascar', in S.A. Altmann (ed.), *Social Communication among Primates*, Chicago, 195-205.

22 Le Gros Clark, W.E. (1936), 'The problem of the claw in primates', *Proc. Zool. Soc. Lond.* (1936), 1-24.

23 Petter, J.-J. and Peyrieras, A. (1970), 'Nouvelle contribution à l'étude

d'un lémurien malgache, le Aye-Aye (*Daubentonia madagascariensis* E. Geoffroy)', *Mammalia* 34, 167-93.
24 Cartmill, M. (in press), 'Pads and claws in arboreal locomotion', in Jenkins, F.A. Jr. (ed.,) *Primate Locomotion*, New York, 45-83.
25 Rand, A.L. (1937), 'Results of the Archbold expeditions, no. 17: Some original observations on the habits of *Dactylopsila trivirgata* Gray', *Am. Mus. Novit.* 957, 1-7.
26 Fleay, D. (1941), 'The remarkable striped possum', *Victorian Nat.* 58, 151-5.
27 Thomas, O. (1910), 'A new genus for *Dactylopsila palpator*', *Ann. Mag. Nat. Hist.*, ser. 8, 6, 610.
28 Tate, G.H.H. (1945), 'Results of the Archbold expeditions, no. 55: Notes on the squirrel-like and mouse-like possums (Marsupialia)', *Am. Mus. Novit.* 1305, 1-12.
29 Tate, G.H.H. and Archbold, R. (1937), 'Results of the Archbold expeditions, no. 16: Some marsupials of New Guinea and Celebes', *Bull. Am. Mus. Nat. Hist.* 73, 331-476.
30 Marlow, B. (1965), *Marsupials of Australia*, Brisbane.
31 Starck, D. (1954), 'Morphologische Untersuchungen am Kopf der Säugetiere, besonders der Prosimier, ein Beitrag zum Problem des Formwandels des Säugetierschädels', *Z. wiss. Zool.* 157, 169-219.
32 Hofer, H. (1958), 'Vergleichende Beobachtungen über die kraniocerebrale Topographie von *Daubentonia madagascariensis* (Gmelin, 1788)', *Morph. Jb.* 99, 26-64.
33 Radinsky, L.B. (1968), 'A new approach to mammalian cranial analysis, illustrated with examples of prosimian primates', *J. Morph.* 124, 167-80.
34 Bauchot, R. and Stephan, H. (1966), 'Données nouvelles sur l'encéphalisation des Insectivores et des Prosimiens', *Mammalia* 30, 160-96.
35 Bauchot, R. and Stephen, H. (1969), 'Encéphalisation et niveau évolutif chez les Simiens', *Mammalia* 33, 225-75.
36 Stephan, H., Bauchot, R. and Andy, O.J. (1970), 'Data on size of the brain and of various brain parts in insectivores and primates', in Noback, C.R. and Montagna, W. (eds.), *The Primate Brain*, New York, 289-97.
37 Biegert, J. (1957), 'Der Formwandel des Primatenschädels und seine Beziehungen zur ontogenetischen Entwicklung und den phylogenetischen Spezialisationen der Kopforgane'. *Morph. Jb.*, 98, 77-199.
38 Cartmill, M. (1971) 'Ethmoid component in the orbit of primates'. *Nature*, 232, 566-567.
39 Saban, R. (1963), 'Contribution à l'étude de l'os temporal des Primates'. *Mem. Mus. nat. Hist. nat., Paris*, n.s. 4, sér. A (Zool.), 29, 1-378.
40 Burt, W.H. (1930), 'Adaptive modifications in the woodpeckers'. *Univ. Calif. Publs. Zool.*, 32, 455-524.
41 Hofer, H. (1945), 'Untersuchungen über den Bau des Vogelschädels, besonders über den der Spechte und Steisshühner'. *Zool. Jb.*, 69, 1-158.
42 In this respect, as in several other points of cranial anatomy, *Daubentonia* resembles the indriids.
43 Cf. Tattersall's analysis of *Archaeolemur*. (This volume).
44 I am indebted to Dr. W.L. Hylander for this suggestion.
45 DuBrul, E.L. (1958), *Evolution of the Speech Apparatus*, Springfield (Ill.): Charles C. Thomas.
46 Spatz, W.B. (1968), 'Die Bedeutung der Augen für die sagittale Gestaltung des Schädels von *Tarsius* (Prosimiae, Tarsiiformes)'. *Folia primat.*, 9, 22-40.
47 Weidenreich, F. (1924), 'Die Sonderform des Menschenschädels als Anpassung an den aufrechten Gang'. *Z. Morph. Anthrop.*, 24, 157-189.
48 DuBrul, E.L. (1950), 'Posture, locomotion, and the skull in Lagomorpha', *Am. J. Anat.* 87, 277-313.

49 Stephan, H. (1972), 'Evolution of primate brains: a comparative anatomical investigation', in Tuttle, R.H. (ed.), *The Functional and Evolutionary Biology of Primates*, Chicago, 155-74.

50 Dabelow, A. (1929), 'Uber Korrelationen in der phylogenetischen Entwicklung der Schädelform. I. Die Beziehungen zwischen Rumpf und Schädelform', *Morph. Jb.* 63, 1-49.

51 These data are taken from Cartmill, M. (1970), 'The orbits of arboreal mammals: a reassessment of the arboreal theory of primate evolution', Ph.D. thesis, University of Chicago.

PART II SECTION C

Morphology of the Brain

D. E. HAINES, B. C. ALBRIGHT, G. E. GOODE and H. M. MURRAY

The external morphology of the brain of some Lorisidae

Studies on the evolution of the central nervous system can utilise two tools: palaeoneurology and comparative neurology. The former gives information on the relative size, shape and topography of the endocranium of animals in lineages, while the latter involves comparison of primitive and more advanced extant forms.[1] The members of the family Lorisidae represent, to a limited degree, a series in the development of the central nervous system. The animals used in the present study represent more primitive members of the family (*Galago senegalensis*) and some members that have an advanced degree of CNS development (*Perodicticus potto*). The present report on the external morphology of the brain of some Lorisidae is preparatory to studies in these animals of fibre connections of the cerebellum and red nucleus, ascending projections from the cord, and descending cortical pathways.

Descriptions of the external morphology of the brains of a variety of prosimians are found throughout the literature. Elliot Smith reviewed numerous earlier works.[2] More recently topography of the cortex in several prosimians has been reviewed by Connolly,[3] Friant,[4] and others.[5–8] Le Gros Clark has dealt not only with the cortex, but also with other areas of the CNS of tree shrews,[9–11] mouse lemur,[12] insectivores,[13,14] and tarsier.[15] The anatomy of the brain stem of the lesser galago and slow loris has recently been reported,[16,17] and the cytoarchitectonics of the thalamus have also received some attention.[18] The tarsier has received little attention, even though there is considerable debate as to its phylogenetic significance.[19–21] The phylogenetic changes in the olive, pons and pyramid have been postulated recently in a study that included several lorises and lemurs.[22]

As complete as many of these studies appear to be, many details are still not known. For example, there is disagreement as to the precise exit of cranial nerve rootlets. With the exception of a recent study,[23] the rhomboid fossa of all prosimians has been ignored. The gross relationships of the mesencephalon with the metathalamus also need clarification. These and other areas merit clearer description and illustration.

Materials and methods

The brains of the lesser bushbaby (*Galago senegalensis senegalensis*), greater bushbaby (*Galago crassicaudatus*), slow loris (*Nycticebus coucang*) and potto (*Perodicticus potto*) were used in this study. The animals were anaesthetised with either intraperitoneal or intravenous administration of Nembutal, then perfused with physiological saline followed by 10% formalin. After perfusion, the brain was removed with the dura intact. Special effort was exerted to leave the cranial nerve roots as long as possible. The dura and arachnoid were removed and most observations were made under a dissecting microscope with a zoom lens.

Scale drawings, at 1 mm. = 1 cm., were made and repeatedly checked from all angles for accuracy. A special effort was made when removing the arachnoid not to damage the rootlets of the cranial nerves so that their precise origin from the brainstem could be determined.

Histological sections at 30 to 40 microns were made at some brainstem levels to verify what deep nuclei were responsible for elevations and/or grooves or fissures. These sections were stained either with the luxol fast blue-cresylecht violet method,[24] or with a cresyl acetate violet method.[25]

Medulla oblongata and pons

1. Ventral and ventro-lateral surfaces

Owing to variation in the exit of the first cervical nerve, medulla length is measured on mid-sagittal sections. The measurement from the ponto-pyramidal groove to the last visible fibre bundle of the pyramidal decussation corresponds to the external length of the pyramids. *G. crassicaudatus* has the longest pyramidal length (10 mm.) followed by *N. coucang* (9.5 mm.) and *P. potto* (9 mm.) (Figs. 1-3). In *G. senegalensis* the pyramids are 7 mm. in length (Fig. 4). The width of the medulla is considerably greater than its dorso-ventral dimension. In *P. potto* and *G. crassicaudatus*, width measurements were nearly equal (11 mm.) and approximately 1 mm. greater than that of *N. coucang*. The width of the medulla in *G. senegalensis* is 8.3 mm. and of proportionate size to the other specimens.

List of abbreviations

AOT	Posterior accessory optic tract
AS	Aqueduct of Sylvius
A-SMV	Attachment of the superior medullary velum
BCI	Brachium of the inferior colliculus
BCS	Brachium of the superior colliculus

Cranial nerves

I	Olfactory (see O1. B, O1. T)
II	Optic (see ON, OC, OT)
III	Oculomotor
IV	Trochlear
Vm	Trigeminal, motor division
Vs	Trigeminal, sensory division
VI	Abducent
VII	Facial
VIIi	*Nervus intermedius*
VIIIc	Cochlear division
VIIIv	Vestibular division
IX	Glossopharyngeal
X	Vagus
XI	Spinal accessory
(XIb	bulbar rootlets)
(XIs	spinal rootlets)
XII	Hypoglossal

C-1	1st cervical nerve
CC	Corpus callosum
CI	Inferior colliculus
CP	Cerebral peduncle (crus cerebri in text)
CS	Superior colliculus
CT	*Corpus trapezoideum*
CuT	Cuneate tubercle
DCN	Dorsal cochlear nucleus
DCN-CS	Dorsal cochlear nucleus cut surface
FaC	Facial colliculus
FC	*Fasciculus cuneatus*
FG	*Fasciculus gracilis*
GT	Gracile tubercle
HT	Hypoglossal tubercle
ICP	Inferior cerebellar peduncle
LCT	Lateral cuneate tubercle
LF	Longitudinal fissure
LGB	Lateral geniculate body
LL	Lateral lemniscus
MCP	Middle cerebellar peduncle
ME	Median eminence of rhomboid fossa
MGB	Medial geniculate body
MI	*Massa intermedia*
MOR	Maxillo-ophthalmic ramus
O1	Olive
O1B	Olfactory bulb
O1T	Olfactory tract

OC	Optic chiasma
ON	Optic nerve
OT	Optic tract
P	Pituitary gland
PHR	Prehypoglossal region
PM	Ponto-medullary junction (Ponto-pyramidal + Ponto-trapezoidial grooves)
Pn	Pons
Pp	Palaeopallium
PrPn	Pre-trigeminal pons
PtPn	Post-trigeminal pons
Pul	Pulvinar
Py	Pyramids
RP	Reticular prominence

Sulci

c	Sulcus c
ce	Central sulcus
cg	Cingulate sulcus (intercalary sulcus)
col	Collateral sulcus
cre	Rectus communicating
d	Sulcus d
di	Diagonal sulcus
e	Sulcus e
g	Genual sulcus
ip	Intraparietal sulcus
l	Lunate sulcus
lc	Lateral calcarine
o	Orbital sulcus
oct	Occipito-temporal sulcus
pc	Paracalcarine
r	*Sulcus rectus*
ra	Rhinal fissure anterior
rc	Retrocalcarine
rp	Rhinal fissure posterior
tm	Middle temporal sulcus
ts	Superior temporal sulcus
SCP	Superior cerebellar peduncle
SF	Superior fovea
SL	*Sulcus limitans*
ST-V	Spinal tract of fifth nerve
SyC	Sylvian complex
TG	Trigeminal ganglion
TT	Transpeduncular tract
VA	Vestibular area
VCN	Ventral cochlear nucleus
VT	Vagal tubercle (dorsal motor nucl. of Vagus)

Figure 1. Ventral view of the medulla and pons of *Galago crassicaudatus.* Note the exit of the abducent (VI) nerve, the size of the *corpus trapezoideum* and the distinct ponto-trapezoidial groove, the exit of the rootlets of the hypoglossal (XII) nerve, and the location of the reticular prominence. Scale = 1 mm.

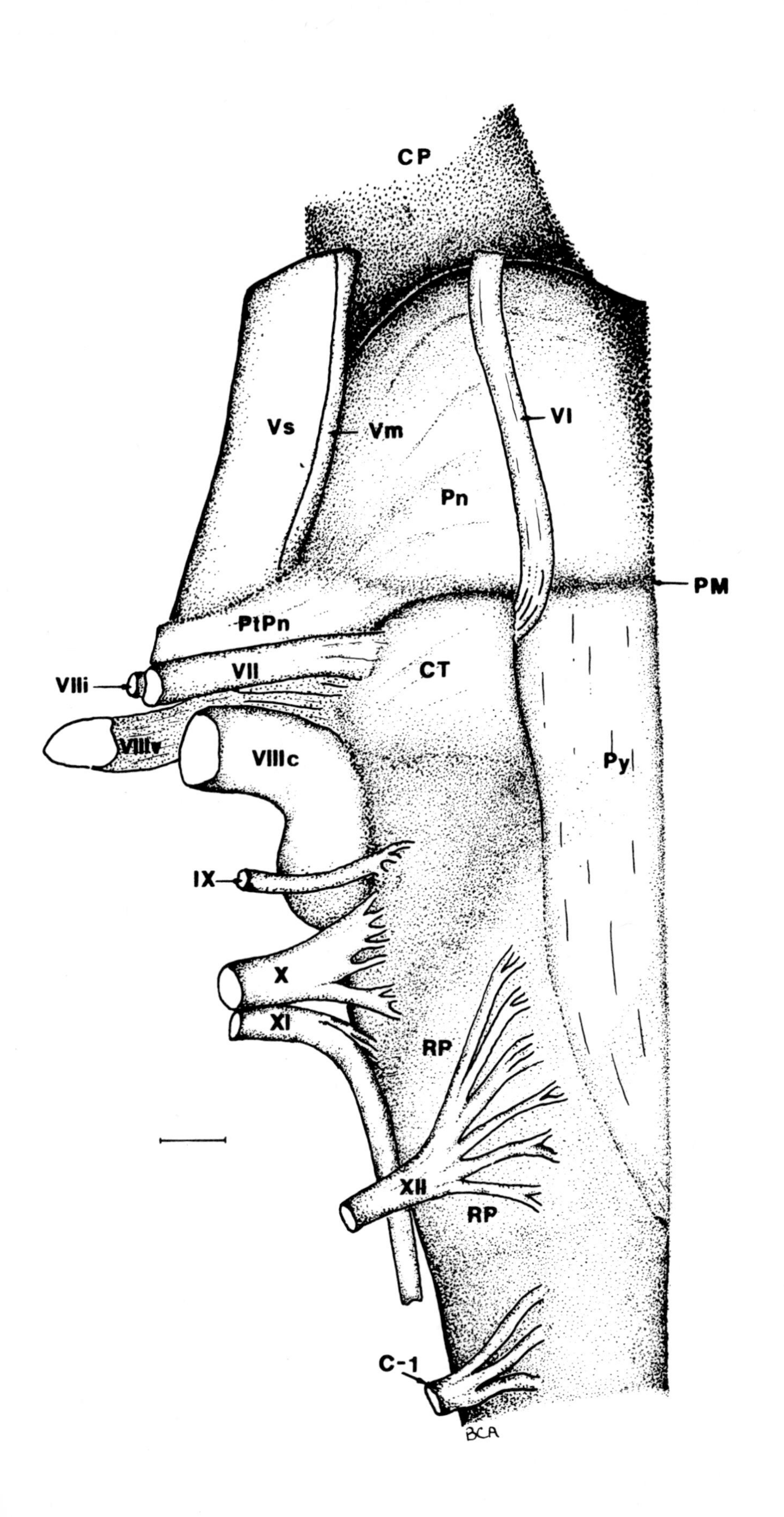

CP
Vs
Vm
VI
Pn
PM
PtPn
VII
VIIi
CT
VIIIv
VIIIc
Py
IX
X
XI
RP
XII
RP
C-1
BCA

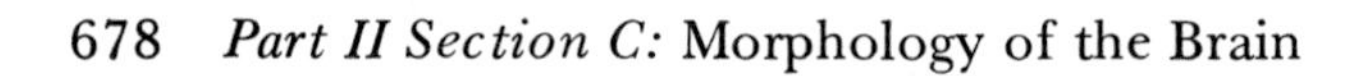

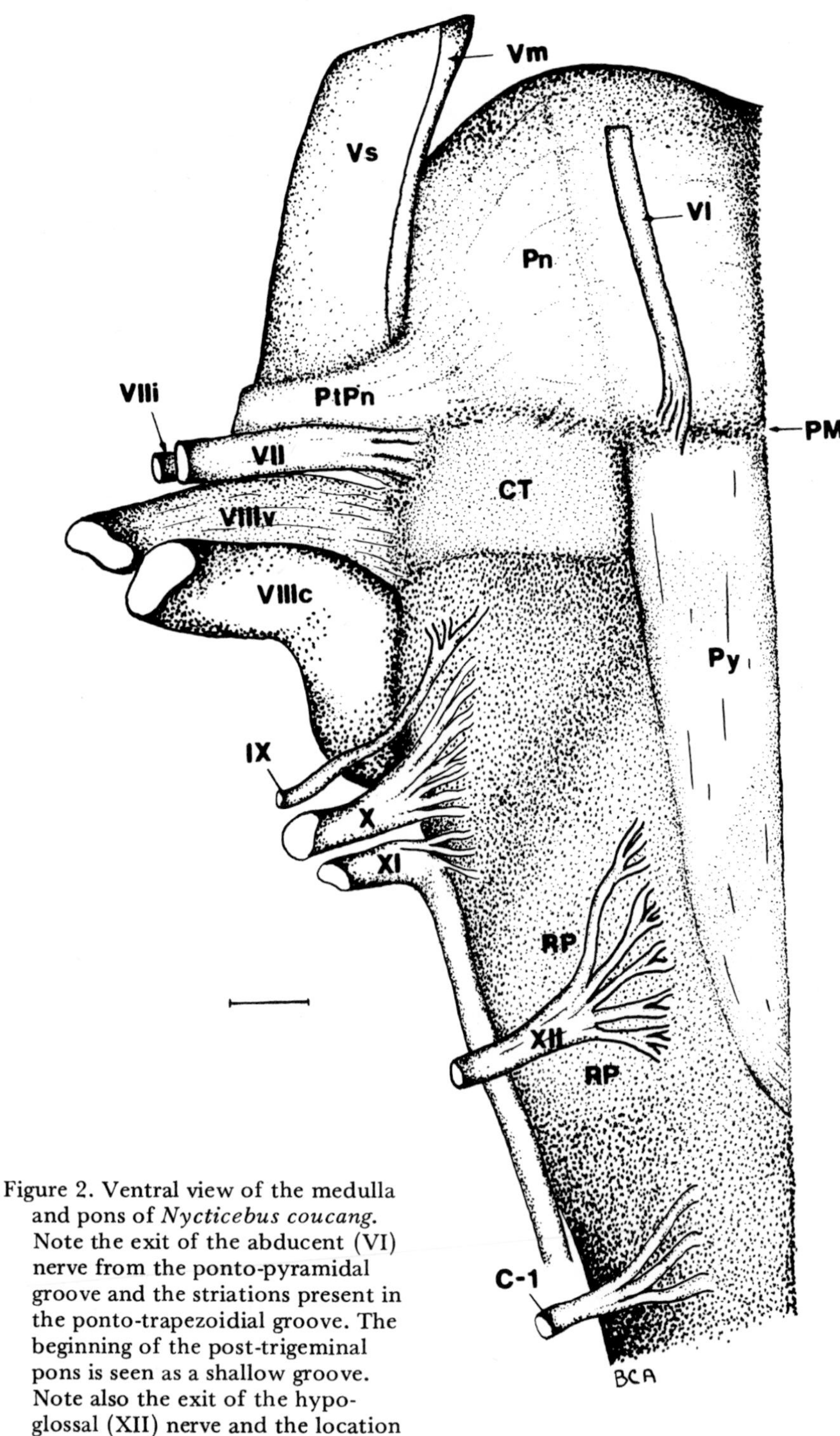

Figure 2. Ventral view of the medulla and pons of *Nycticebus coucang*. Note the exit of the abducent (VI) nerve from the ponto-pyramidal groove and the striations present in the ponto-trapezoidial groove. The beginning of the post-trigeminal pons is seen as a shallow groove. Note also the exit of the hypoglossal (XII) nerve and the location of the reticular prominence.

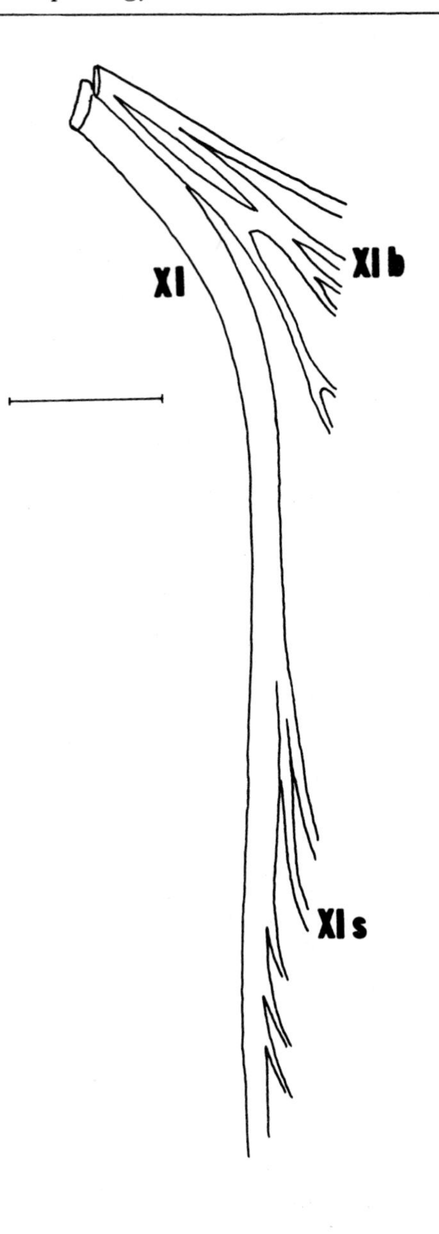

Figure 2(*a*). The spinal accessory (XI) nerve of *Nycticebus coucang* has distinct bulbar and spinal divisions. Scale = 1 mm.

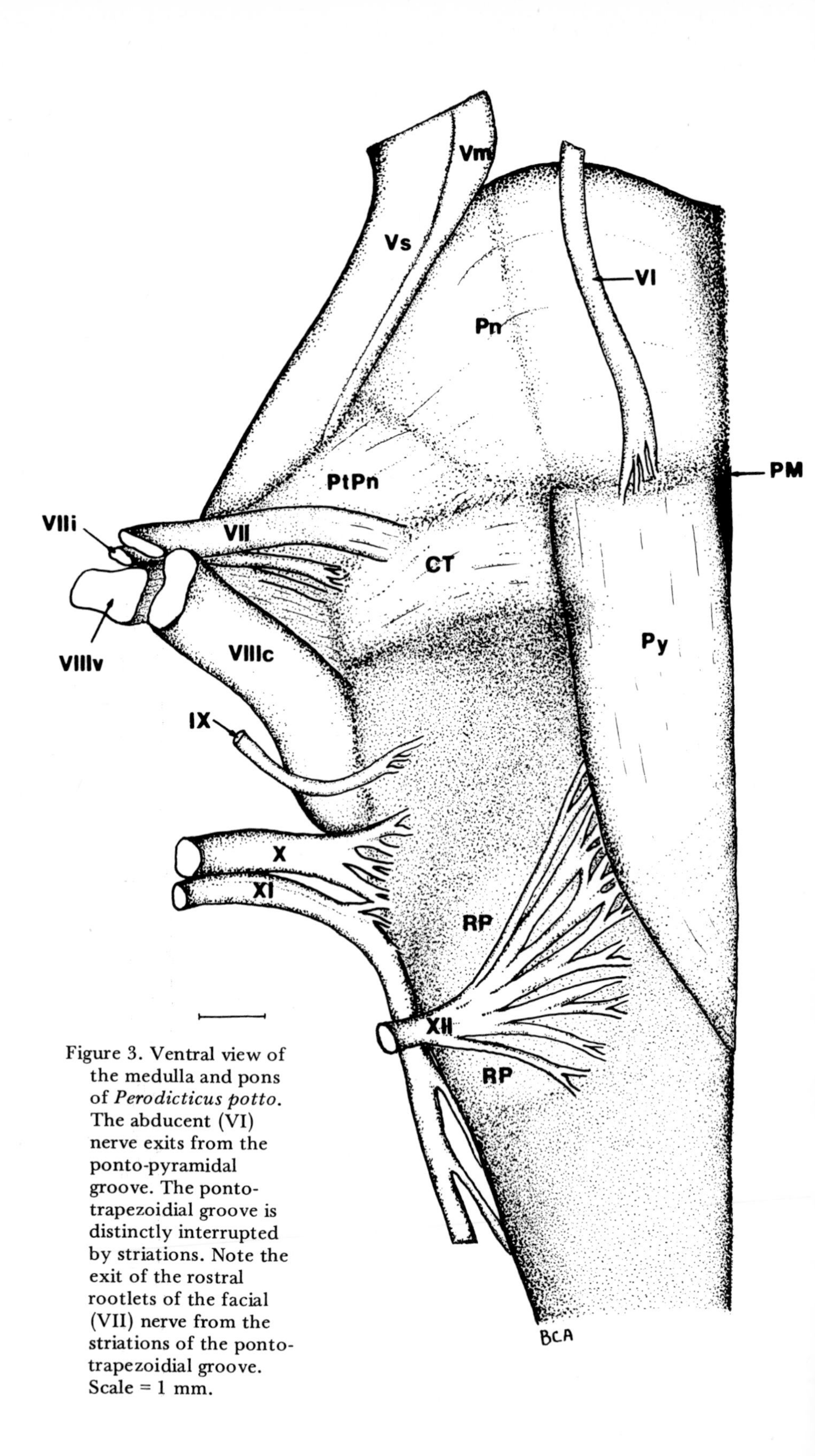

Figure 3. Ventral view of the medulla and pons of *Perodicticus potto*. The abducent (VI) nerve exits from the ponto-pyramidal groove. The ponto-trapezoidial groove is distinctly interrupted by striations. Note the exit of the rostral rootlets of the facial (VII) nerve from the striations of the ponto-trapezoidial groove. Scale = 1 mm.

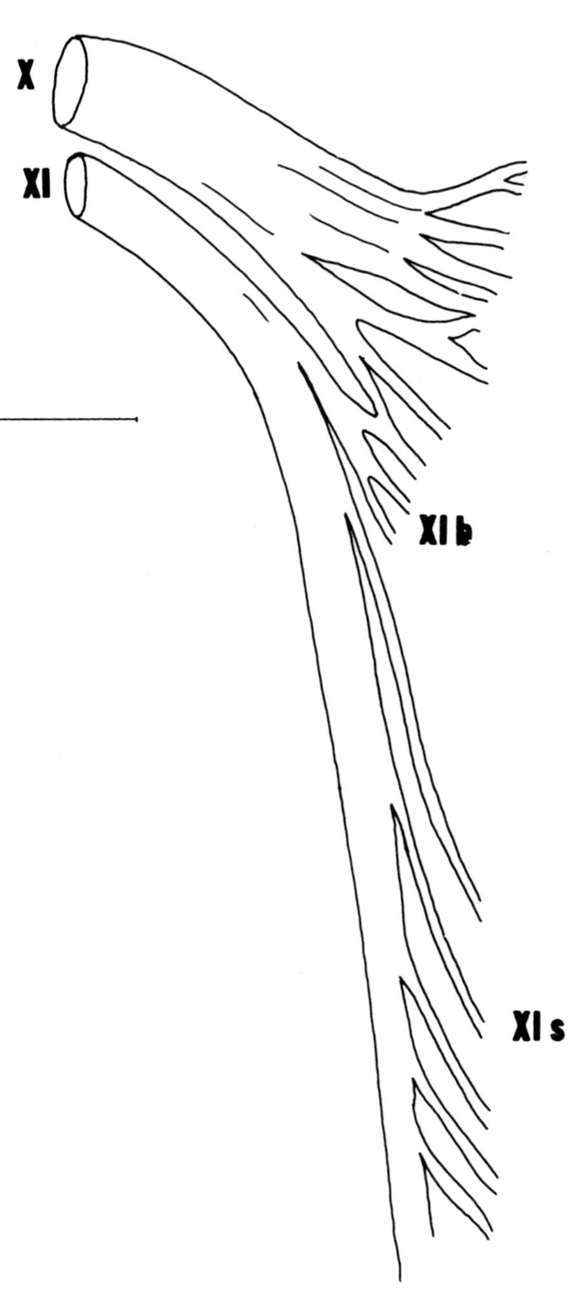

Figure 3(*a*). The spinal rootlets of the spinal accessory (XI) nerve of *Perodicticus potto* are distinct from the bulbar rootlets. Note the association of the bulbar rootlets of XI with the caudal rootlets of the vagus (X) nerve. Scale = 1 mm.

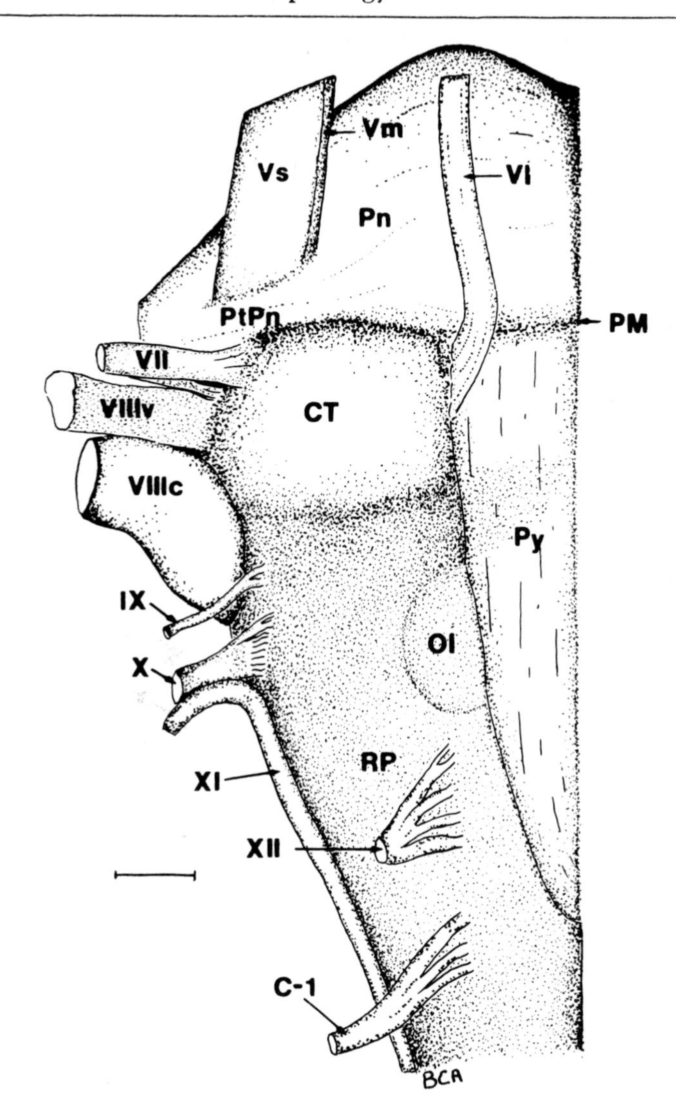

Figure 4. Ventral view of the medulla and pons of *Galago senegalensis.* Note the location of the olive and the size of the *corpus trapezoideum.* The rootlets of the abducent (VI) nerve exit caudal to the ponto-medullary groove and lateral to the pyramids. Scale = 1 mm.

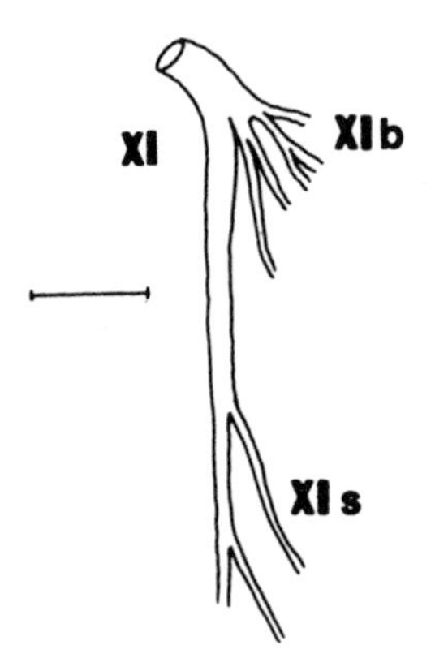

Figure 4(*a*). Rootlets of the spinal accessory (XI) nerve. Note the bulbar (XIb) and spinal (XIs) divisions. Scale = 1 mm.

Excluding *P. potto*, the pyramids are observed to have certain similarities. They are: a gradual uniform tapering, a less distinct lateral border in the region of the olive, and an occasional shallow depression caudal to the *corpus trapezoideum* (Figs. 1, 2, 4). The pyramids of *P. potto* are distinctly different in appearance. In addition to being shorter than *G. crassicaudatus* and *N. coucang*, they appear to be thicker and broader and to taper more abruptly near the region of the hypoglossal (XII) nerve (Fig. 3). The ventral elevation of the pyramids in *P. potto* (0.8 mm.) is 0.2 to 0.3 mm. greater than that measured in *G. crassicaudatus*. The lesser bushbaby and slow loris have similar pyramidal heights of approximately 0.4 mm. The pyramids in slow loris appear flatter and show a less definable ponto-pyramidal groove (Fig. 2). Just caudal to the ponto-pyramidal groove, the pyramids in *P. potto* are 2.5 mm. in width and approximately 0.5 to 0.6 mm. wider than in the greater bushbaby and slow loris respectively. In the lesser bushbaby, pyramidal width is 1.5 to 1.7 mm.

The abducent nerve (VI) is composed of 3-4 rootlets. In both galagos it exits from the lateral border of the pyramids caudal to the ponto-pyramidal groove (Figs. 1, 4). In *P. potto* and *N. coucang*, however, the rootlets exit through the pyramids at the level of the ponto-pyramidal groove (Figs. 2, 3).

The *corpus trapezoideum* is distinct and shows medial to lateral striations; its size and prominence vary considerably. It is similar in the galagos, with well-defined rostral and caudal borders, the rostral forming the ponto-trapezoidial groove. Of the prosimians studied, *G. senegalensis* demonstrates the largest *corpus trapezoideum*, measuring 2.5 mm. in rostrocaudal length (Fig. 4). This measurement is only 0.2 mm. less than that in *G. crassicaudatus* (Fig. 1). In *N. coucang* and *P. potto* it is considerably less prominent, and measures 2 mm. and 1.6 to 1.7 mm., respectively (Figs. 2, 3). In association with the decrease in width, striations appear in the ponto-trapezoidial groove.

The motor root of VII is formed by the union of 3 to 4 short rootlets. Except for *P. potto*, the nerve exits distinctly through the rostral region of the *corpus trapezoideum.* In *P. potto*, the rostral rootlets exit from the striations of the ponto-trapezoidal groove (Fig. 3). The smaller *nervus intermedius*, composed of 2 to 3 fine rootlets, exits through the *corpus trapezoideum* dorsal to the main motor root. The divisions of the vestibulo-cochlear nerve (VIII) are easily distinguished (Figs. 1-4). The vestibular division is striated and exits lateral to the *corpus trapezoideum.* This component is similar in width to that of the cochlear division, yet characteristically flatter. The truncated cochlear division is easily traced toward the cochlear nuclei. It is located caudal to the vestibular division and passes ventral to it. The glossopharyngeal (IX) nerve is small and exits as 2 to 3 fine rootlets ventral to the tuberculum quintae. These rootlets are oriented mediolaterally and in *P. potto* and *N. coucang* appear to be separated into one medial and two lateral rootlets. The vagus (X) nerve exits caudal to the glossopharyngeal nerve and is formed from easily observable rootlets. These rootlets vary from 5 to 6 in *G. senegalensis* (Fig. 4) to approximately 10 in *P. potto* (Figs. 3, 3*a*). In *P. potto* and *N. coucang*, the rootlets have a slightly longer course prior to forming the common trunk. In addition, the rootlets of X in *P. potto* are divisible into dorsal and ventral groups. The ventral roots are shorter and fewer in number. The bulbar and spinal divisions of the accessory nerve (XI) are distinct due to an interval between the two groups of rootlets. The bulbar division is composed of 3 to 6 rootlets, the spinal division 4 to 6 rootlets. In all genera, considerable variation in the number and arrangement of the bulbar rootlets is observed (Figs. 2*a*, 3*a*, 4*a*). These rootlets exit from the postero-lateral surface and are often difficult to observe by direct ventral observation.

In all genera a distinctly raised area, the reticular prominence (see Discussion), is observed on the ventro-lateral medullary surface (Figs. 1-3). In *P. potto*, *N. coucang*, and *G. crassicaudatus* the raised area is approximately 3.8 to 4.0 mm. in length and 2.0 mm. in width. In *G. senegalensis* the area is less definable and accurate measurements are not possible.

In *P. potto* and *G. crassicaudatus* the olives are difficult to identify. They appear to be oriented longitudinally and located adjacent to the pyramids, approximately 1.5 mm. rostrolateral to the pyramidal decussation (Figs. 5*b*, 5*c*). This slightly raised area, approximately 1 mm. in width, is distinguished from the above described reticular prominence by a shallow post-olivary groove. The

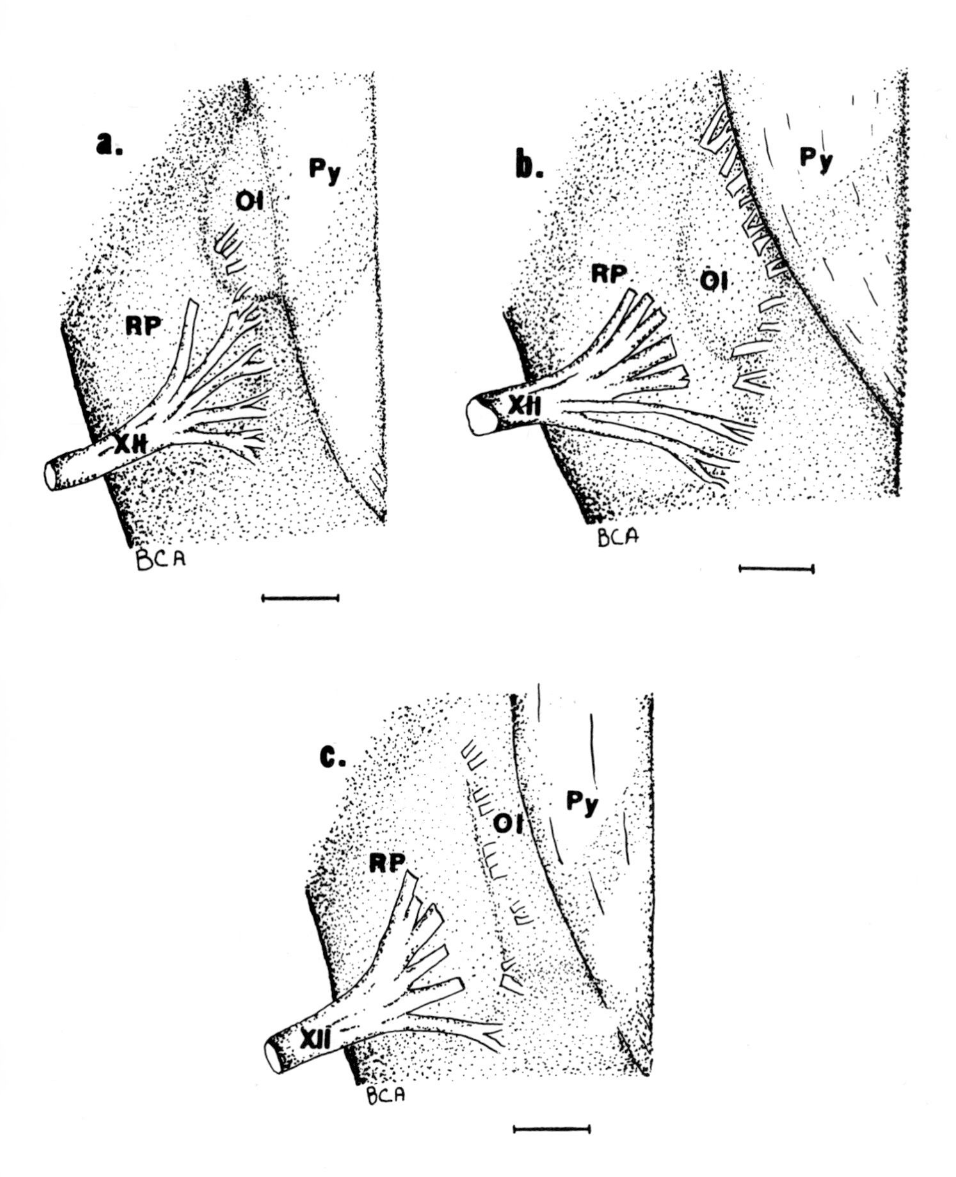

Figure 5. Ventral view of the caudal medulla of *Nycticebus coucang* (*a*), *Perodicticus potto* (*b*), and *Galago crassicaudatus* (*c*). Note the difference in location and shape of the olives. The olive (Ol) is distinguished from the reticular prominence (RP) by a shallow postolivary groove. Scale = 1 mm.

olives in *N. coucang* and *G. senegalensis* are similar in appearance (Figs. 4, 5*a*) and are located adjacent to the pyramids, but further rostral than in *P. potto* and *G. crassicaudatus.* Although equal in width (1 mm.), length measurements are 1.5 to 1.7 and 2.5 mm. in *G. senegalensis* and *N. coucang* respectively. The lateral borders of the pyramids are slightly raised and appear to cover the medial aspect of the olive.

P. potto shows an elaborate arrangement of hypoglossal nerve rootlets (Figs. 3, 5*b*). As many as 15 small primary rootlets are observed to join immediately in the formation of 7 to 8 secondary rootlets. The rostral rootlets exit from the lateral border of the pyramids from the preolivary groove. Caudally, the rootlets appear to follow the contour of the reticular prominence. Due to an ill-defined olivary eminence, it is not possible to determine the exact relationship between the exit of XII and the position of the olive. The number, arrangement and location of exit of hypoglossal rootlets in *G. crassicaudatus* and *N. coucang* are similar despite the difference in location of the olive (Figs. 1, 2, 5). In these Lorisidae, 12 to 13 primary rootlets join to form 6 to 7 secondary rootlets. In *G. crassicaudatus* they exit from the mid-longitudinal region of the olive (Fig. 5*c*). In *N. coucang* only the more rostral rootlets exit from the olive, the caudal rootlets follow the reticular prominence (Fig. 5*a*). In *G. senegalensis* the hypoglossal nerve is composed of 4 to 5 rootlets. All the rootlets exit caudal to the olive, from the ventrolateral surface of the medulla (Fig. 4).

Pons

In all genera, the pons appears to be well-developed, having the common features of: a distinct basilar groove, cross-pontine striations, and an obliquely-exiting trigeminal nerve root. The trigeminal nerve divides the lateral pontine surface (middle cerebellar peduncle) into pre- and post-trigeminal bundles (Figs. 1-4, 6). In those animals of comparable brain size, measurements of the rostrocaudal length are equal (5 mm.). In *P. potto* (Fig. 3), however, the pontine width (9 mm.) is approximately 1 mm. greater than that observed in *N. coucang* and *G. crassicaudatus.* In *P. potto* the pons appears thick and its anterior border is abrupt. In the other genera the pons is less pronounced and slopes rostral gradually. The ventrolateral surface of the pons in the galagos is smooth (Figs. 1, 6), except for a shallow groove running caudomedially from the exit of V in *G. crassicaudatus.* This groove, also seen in *N. coucang* and *P. potto*, marks the beginning of the post-trigeminal pons. In *P. potto* a definite ridge is observed on the pontine surface which appears to join the lateral border of the pyramids with that of the cerebral peduncles. There is only a slight indication of this ridge in *N. coucang.* None of the genera exhibited oblique striations that might be indicative of

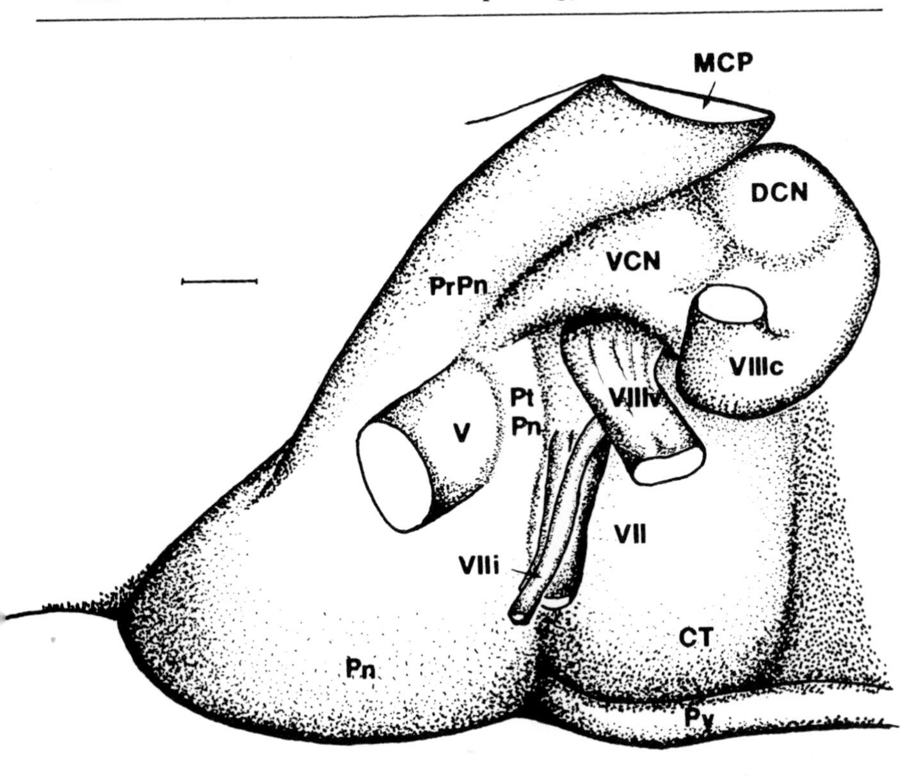

Figure 6. Lateral view of medulla and pons of *Galago senegalensis*. Note the exit of the vestibular division of the vestibulocochlear nerve (VIIIv) and the relative size of the pre-(PrPn) and post-(PtPn) trigeminal pons. Scale = 1 mm.

post-trigeminal pontine fibres originating from rostral pontine areas.

At the dorsal limits of the exit of the fifth nerve, the post-trigeminal pons is approximately half the width of the pre-trigeminal bundle. The post-trigeminal pons in *G. senegalensis* is relatively uniform in width throughout its course (Fig. 6). In contrast, the post-trigeminal pons in *P. potto* shows a greater amount of tapering, which corresponds to the broader width at its ventral origin.

The fifth nerve is relatively large, and both motor and sensory divisions are easily distinguishable. Considerable variation is noted in the number and origin of the rootlets which compose the smaller motor component. The trigeminal ganglion is associated with the ventro-caudal aspect of the pyriform lobe (see later: Figs. 32-4). The motor component of V passes laterally toward the mandibular division. Rostral to the ganglion, the maxillary and ophthalmic divisions occur as a single branch, the maxillo-ophthalmic ramus (Figs. 32-4). This large ramus courses medial to the pole of the pyriform lobe and divides before leaving the skull.

2. Rhomboid fossa

The rhomboid fossa is similar in appearance and proportionately equal in size (Figs. 7-10). The rostral and caudal extensions are separated by a distinct vestibular area. The approximate length and width of the fossa in *N. coucang, P. potto* or *G. crassicaudatus* is 7.0 to 7.5 mm. by 8.0 mm. In the rostral portion, the facial colliculus and median eminence are separated by a shallow groove which courses from the midline to the superior fovea (Figs. 7, 10). A faint *sulcus limitans* is present medial to the large vestibular area (Figs. 7, 9). The continuation of the sulcus cannot be accurately followed. The inferior fovea can only be identified as a region. The caudal region is characteristically bordered by the obex and two eminences which correspond to those formed by the hypoglossal and external motor nucleus of the vagus (Figs. 8, 10). In *N. coucang* and *P. potto* a pre-hypoglossal region is observed (Figs. 8, 9). The attachments of the superior and inferior medullary vela around the fossa are typical of higher primates.

3. Dorsal medulla

Fasciculi graciles and *cuneati* are easily distinguished. They are separated by the posterior intermediolateral groove (Figs. 7-10). The *fasciculus cuneatus* is bordered laterally by the posterolateral groove, which terminates rostrally at the caudal border of the inferior cerebellar peduncle. In upper cervical and lower medullary regions, the *fasciculus gracilis* is approximately 1/4 the size of *fasciculus cuneatus* (Figs. 7, 8). Both *fasciculi* expand in width as they approach their respective tubercles. This is more evident in the *fasciculus gracilis* (Fig. 8). The posterior intermedio-lateral groove broadens to separate gracilis and cuneate tubercles. This groove is less evident in *N. coucang*. The tubercle *gracilis* in *N. coucang, P. potto*, and *G. senegalensis* is approximately half the size of tubercle *cuneatus* (Figs. 8-10). In *G. crassicaudatus*, the tubercle *gracilis* appears similar in size to that of cuneate tubercle (Fig. 7). Except in *G. senegalensis*, a small eminence is observed rostral to the cuneate tubercle (Figs. 7-9). It may identify the location of the lateral cuneate nucleus. The rostral border of both tubercles serves for the attachment of the inferior medullary velum.

The cochlear nuclei are difficult to distinguish. The dorsal nucleus is usually observed extending to variable distances over the surface of the vestibular area. The inferior cerebellar peduncle courses rostrally from the lateral aspect of the medulla, where it passes deep to the cochlear nuclei (Figs. 8, 10). Apart from being easily traced and serving for the attachment of the superior medullary velum, the superior cerebellar peduncle is separated from the middle cerebellar peduncle by a groove.

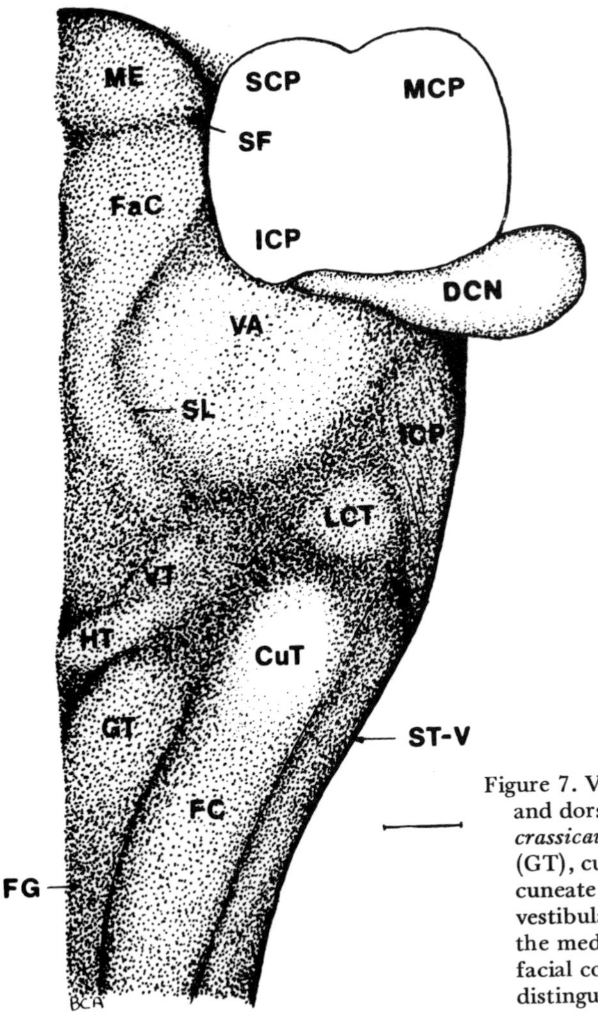

Figure 7. View of rhomboid fossa and dorsal medulla of *Galago crassicaudatus.* Note the gracile (GT), cuneate (CT) and lateral cuneate (LCT) tubercles. The vestibular area (VA) is large and the median eminence (ME) and facial colliculus (FaC) are easily distinguished. Scale = 1 mm.

Mesencephalon and diencephalon

1. Dorsal view

The inferior colliculi appear as dorsally directed structures whose apices are located more laterally than those of the superior colliculi. A series of shallow grooves appear on the caudal face of the inferior colliculus. They are formed by the folia of the anterior lobe of the cerebellum (Figs. 11-14). The trochlear (IV) nerve, after its exit from the medullary velum, traverses the caudolateral surface of the inferior colliculus in a dorsal oblique direction (Figs. 12, 14). Upon

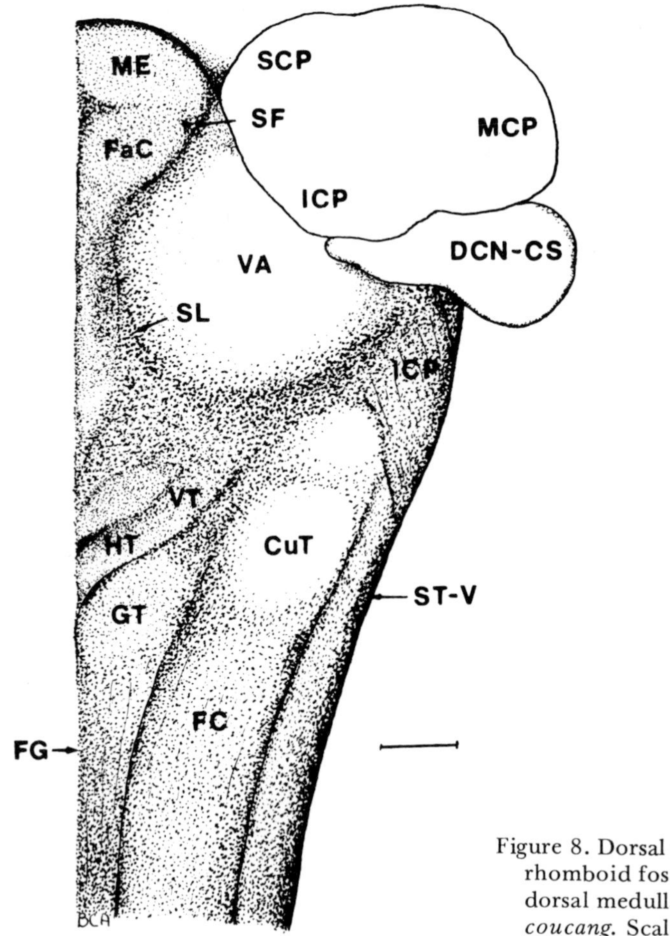

Figure 8. Dorsal view showing rhomboid fossa and tubercles of dorsal medulla in *Nycticebus coucang.* Scale = 1 mm.

reaching the lateral surface of the colliculus, the nerve turns ventrally to travel with the maxillo-ophthalmic ramus of the trigeminal nerve.

The superior colliculi are large, somewhat flattened, oval structures (Figs. 12, 13, 15). The inferior and superior colliculi are separated by the intercollicular groove (sulcus) which is narrow in the lesser galago (Figs. 14, 15),[20] the colliculi abutting along the greater part of their common borders. The groove appears somewhat wider in both the greater galago and slow loris (Figs. 11, 12), and reaches its greatest dimension, 2 to 3 mm., in the potto (Fig. 13).

2. *Lateral view*

The lateral lemniscus is observed to course in a slightly dorsal oblique

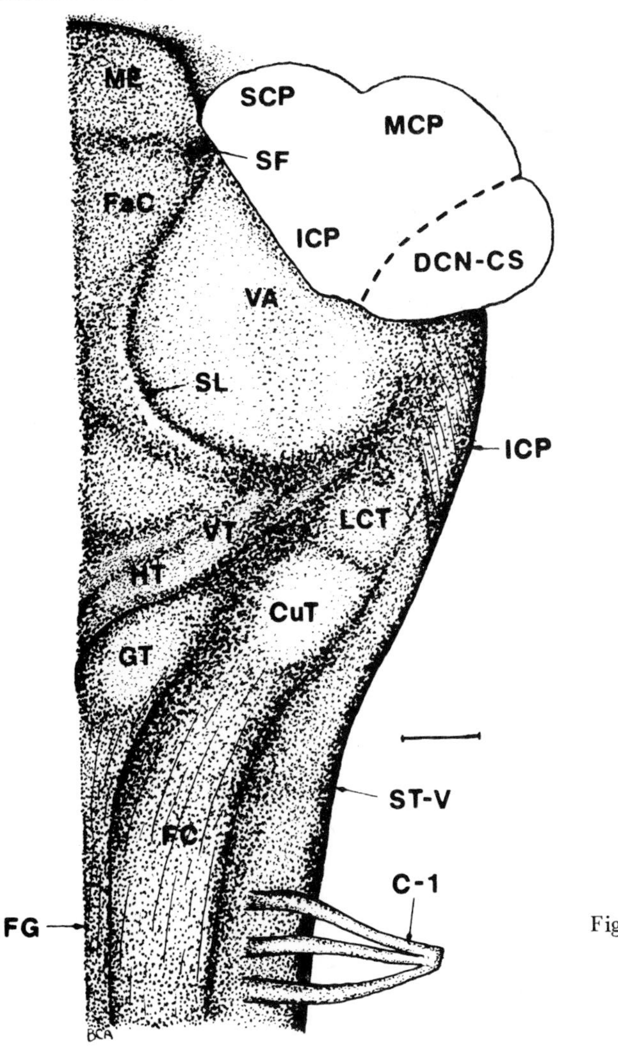

Figure 9. Dorsal view of rhomboid fossa and dorsolateral medulla in *Perodicticus potto*. Scale = 1 mm.

direction across the ventrolateral surface of the inferior colliculus (Figs. 11-14). The brachium of the inferior colliculus leaves the rostrolateral surface and passes obliquely in an anteroventral direction to join the medial geniculate body (Figs. 11, 12, 15).

The medial geniculate body is a distinct rounded tubercle located lateral to the superior colliculus. The medial-lateral diameter is greatest in the greater galago (Fig. 11), (2.8 mm.), followed by the potto, slow loris and lesser galago (2.6, 2.4, 2.3 mm. respectively) (Figs. 13-15). The dorsoventral diameter is greatest in the potto, 3.2 mm., while in the other animals it is 2.5 mm.

The lateral geniculate body appears as a somewhat flattened,

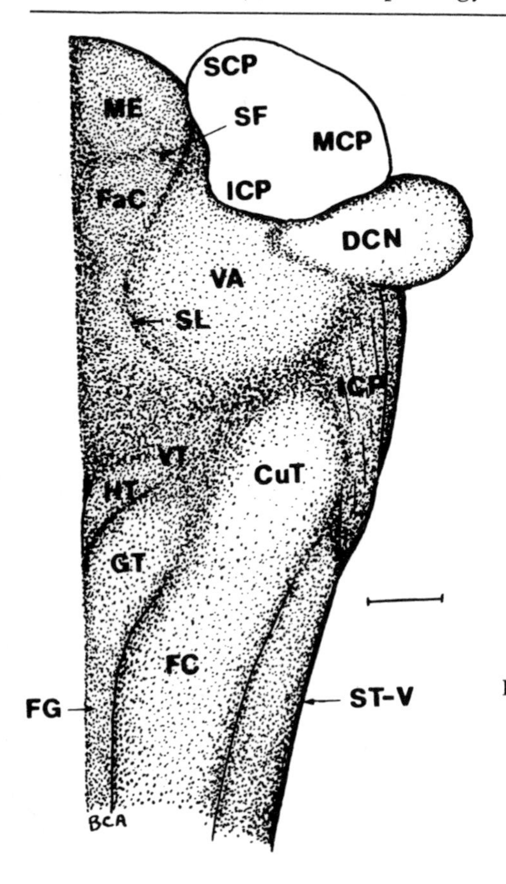

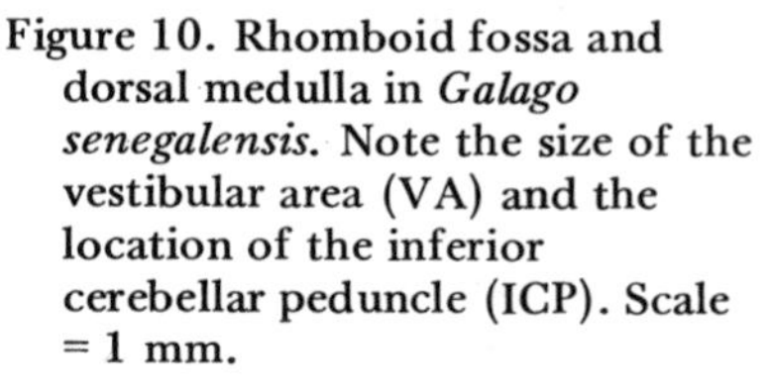
Figure 10. Rhomboid fossa and dorsal medulla in *Galago senegalensis.* Note the size of the vestibular area (VA) and the location of the inferior cerebellar peduncle (ICP). Scale = 1 mm.

crescentic structure located rostrolateral to the medial geniculate body. It attains its greatest medial-lateral width in the potto (Fig. 13), 2.8 mm., while in the greater galago (Fig. 11) it is 2.4 mm.; in the slow loris and lesser galago it is 2.2 mm. at its widest point (Figs. 12, 14, 15). The dorsal-ventral extent is 4.5 mm. in both the potto and lesser galago, while in the greater galago and slow loris it is 4.2 mm. and 3.0 mm. respectively.

Passing laterally from the superior colliculus to join the dorsomedial border of the lateral geniculate body is the striated brachium of the superior colliculus (Figs. 11-14). A deep groove is observed between the ventrolateral aspect of the superior colliculus and the brachium. There is no distinct margin noted between the brachium and the lateral geniculate body, nor at the ventral junction of the lateral geniculate body and optic tract.

3. *Ventral view*

Figs. 16-19 are ventral views of the mesencephalon and diencephalon of the animals used in this study. The interpeduncular fossa at the anterior border of the pons is widest (3.8 to 4.0 mm.) in the lesser galago (Fig. 19), being almost 1½ times to twice as wide as in the other animals (Figs. 16-18). The medial-lateral extent of the crus cerebri at this point is 2.0 to 2.1 mm. in the lesser galago, 3.2 mm. in the potto and slow loris, and 3.0 to 3.1 mm. in the greater galago. (The medial-lateral measurement did not take the curvature of the crura into account.)

The transpeduncular tract is observed as a dorso-ventrally oriented elevation on the crus cerebri. Fibres of the crus appear to pass either over or through the tract. It is 1.0 mm. in width in the greater galago and potto (Figs. 16, 18), and is slightly narrower (0.7 to 0.8 mm.) in the lesser galago and slow loris (Figs. 17, 19). At the dorsal limit of the crus, the tract continues as the posterior accessory optic tract onto the lateral surface of the mesencephalon, coursing caudally around the medial geniculate body (Figs. 16, 17). Using gross morphological techniques, it cannot be determined whether the tract joins the superior colliculus or the brachium of the superior colliculus. A discrete bundle of fibres is observed ventral to the lateral surface of the medial geniculate body. It is probable that this bundle represents portions of the accessory optic system. The optic tract flattens as it courses caudodorsally from the optic chiasma to terminate in the lateral geniculate body.

The 5 to 7 rootlets of the oculomotor (III) nerve exit from the interpeduncular fossa in all the Lorisidae studied except *N. coucang* (Figs. 16, 18, 19). The oculomotor nerve rootlets of individuals of this species exit from the interpeduncular fossa and from the crus cerebri (Fig. 17). At least 2 rootlets are observed to exit bilaterally from the crus in each animal.

4. *Midsagittal view*

The height of the colliculi above the roof of the aqueduct is greatest in *G. crassicaudatus* (inferior colliculus, 3.8 mm.; superior colliculus, 3.6 mm.). In the greater galago, the rostrocaudal length of the inferior colliculus is greatest, while this parameter of the superior colliculus is increased in *P. potto* (3.4 mm.) as compared to the other animals.[26]

The height of the cerebral aqueduct varies among the different Lorisidae examined, being greatest in the greater galago. A shallow groove is noted in the lateral wall of the aqueduct in both galagos.

The oval to circular massa intermedia (or commissure mollis) is large in all animals.[26] Its long axis is 3.7 mm. in *G. senegalensis*, 5.7 mm. in *G. crassicaudatus* and *N. coucang*, and 6.0 mm. in *P. potto*.

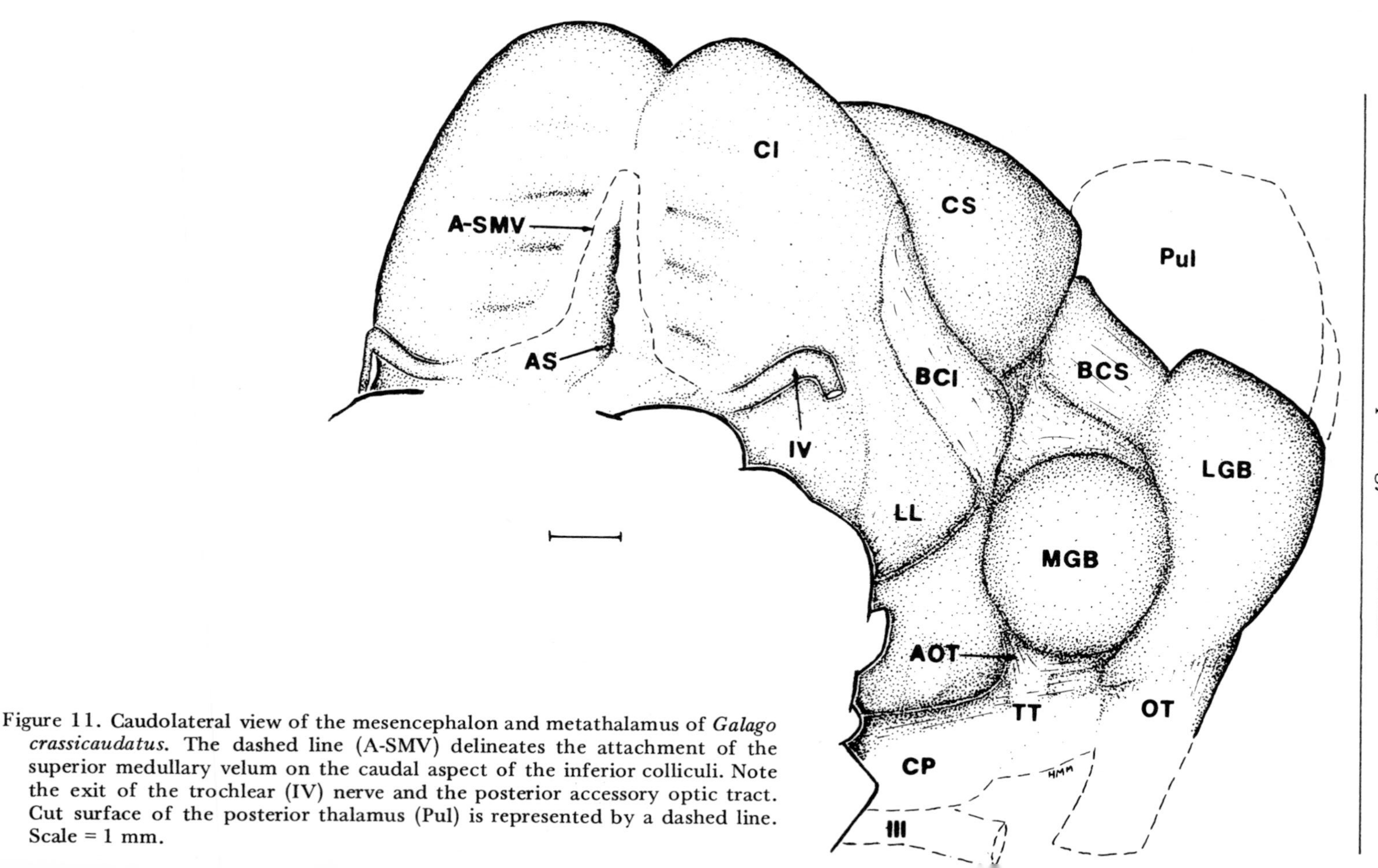

Figure 11. Caudolateral view of the mesencephalon and metathalamus of *Galago crassicaudatus*. The dashed line (A-SMV) delineates the attachment of the superior medullary velum on the caudal aspect of the inferior colliculi. Note the exit of the trochlear (IV) nerve and the posterior accessory optic tract. Cut surface of the posterior thalamus (Pul) is represented by a dashed line. Scale = 1 mm.

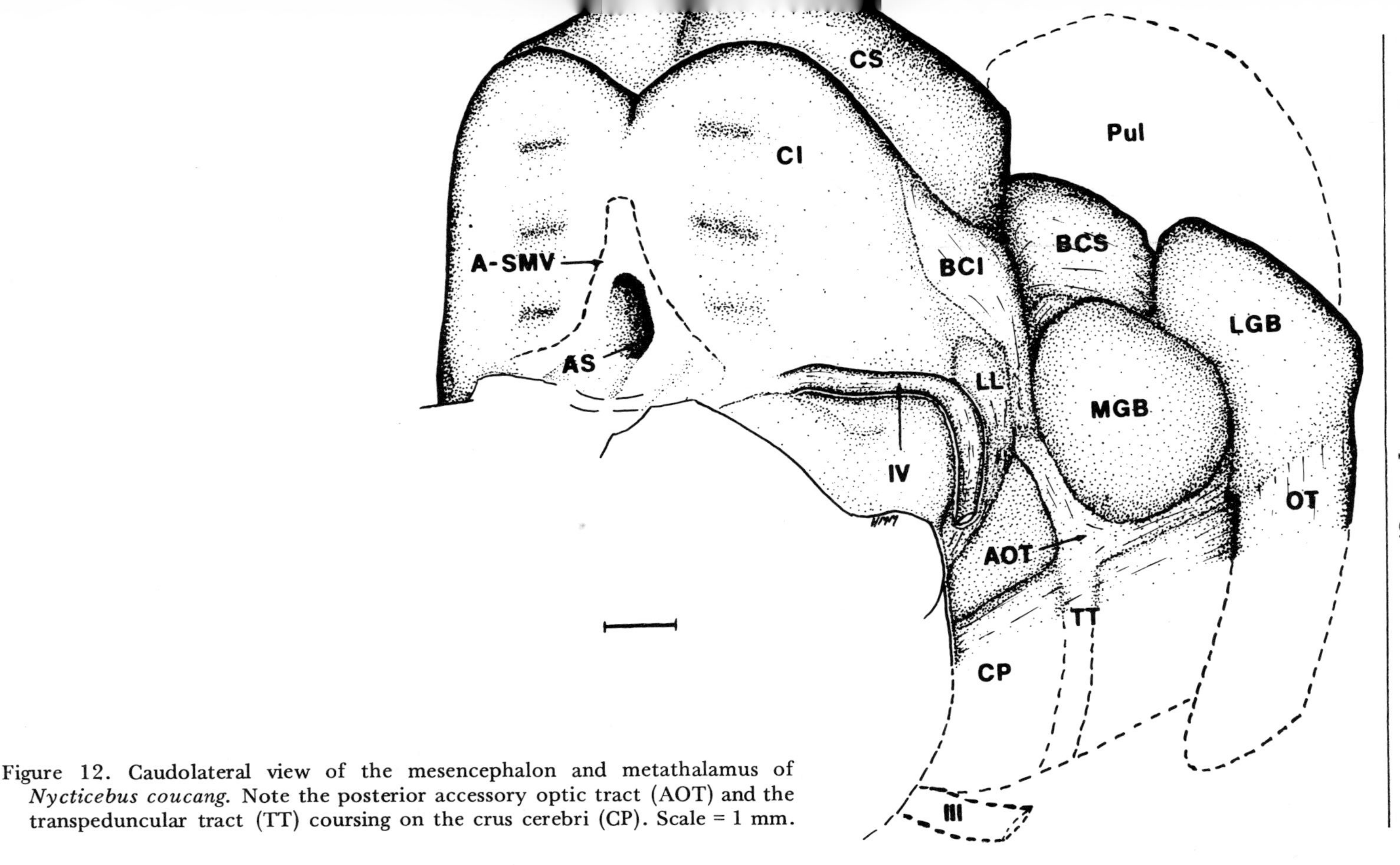

Figure 12. Caudolateral view of the mesencephalon and metathalamus of *Nycticebus coucang*. Note the posterior accessory optic tract (AOT) and the transpeduncular tract (TT) coursing on the crus cerebri (CP). Scale = 1 mm.

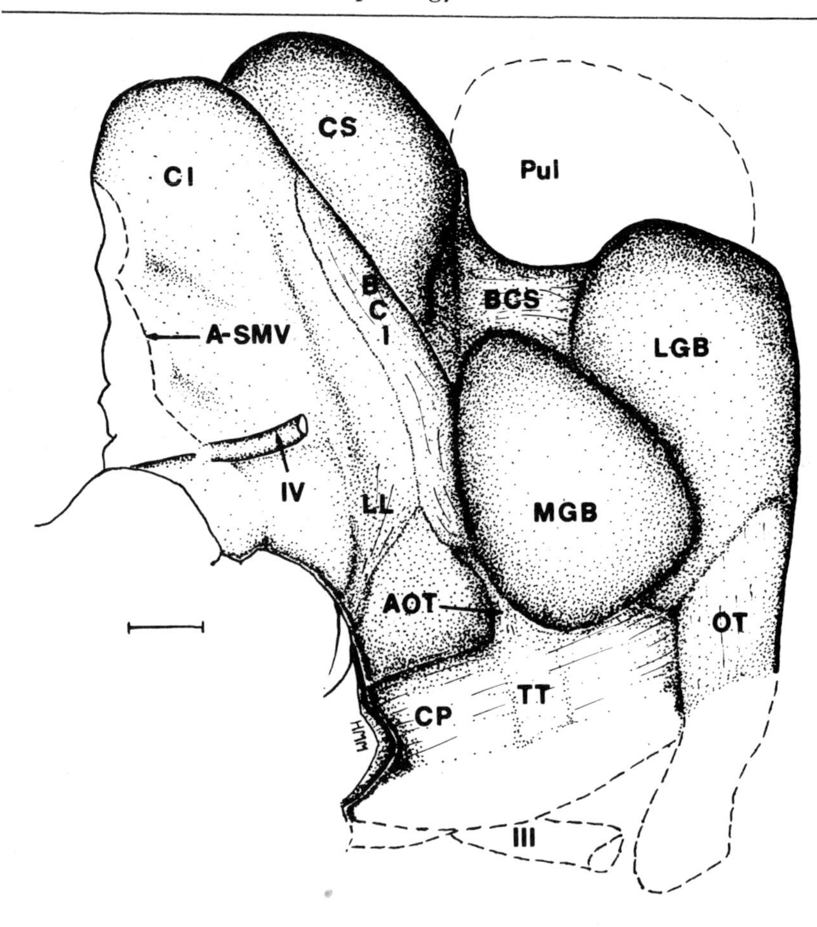

Figure 13. Lateral view of the mesencephalon and metathalamus of *Perodicticus potto* from a slightly caudal aspect. The medial geniculate body (MGB) and lateral geniculate (LGB) are distinct metathalamic structures located on the lateral aspect of the mesencephalon. Scale = 1 mm.

The pituitary gland (see later – Figs. 32, 33) is large in the greater galago, having a rostrocaudal measurement of 5.5 mm. and a dorsoventral measurement of 2.6 mm. The diameter of the gland at its widest point is 4.0 mm. The pituitary of the potto and slow loris is somewhat smaller, while that of the lesser galago is the smallest (rostrocaudally, 3.0 mm.; dorsoventrally, 2.0 mm.; diameter at widest point, 3.2 mm.).

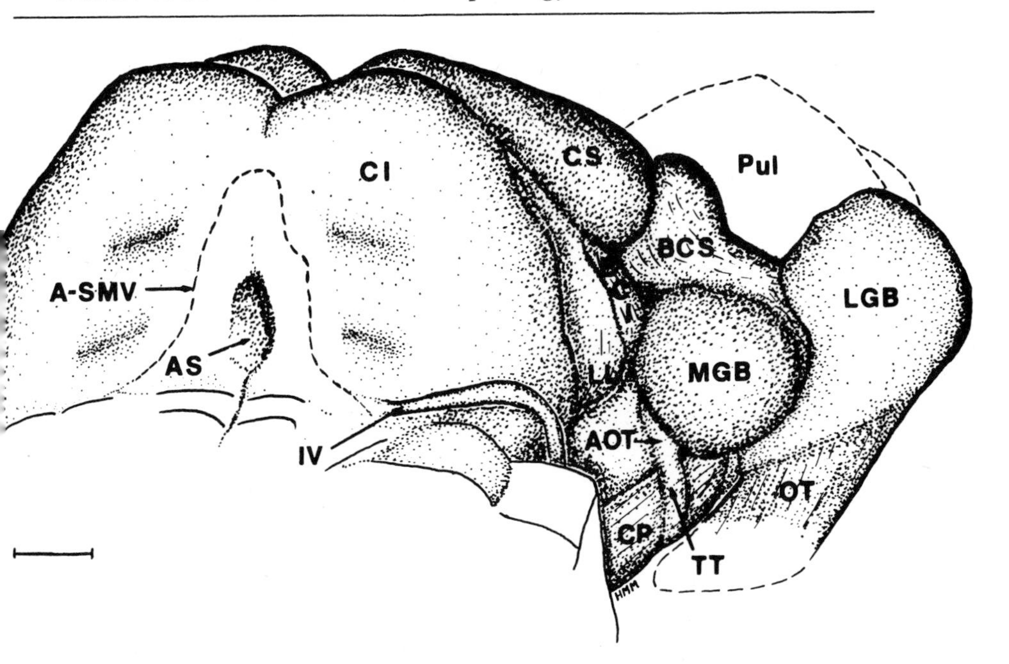

Figure 14. Caudolateral view of the mesencephalon and metathalamus of *Galago senegalensis.* Note the course of the trochlear (IV) nerve, the transpeduncular tract, and the relatively large size of the lateral geniculate body (LGB). Scale = 1 mm.

Telencephalon

1. Dorsolateral surface

The sulcal pattern on the dorsolateral surface of the specimens examined increased in number and complexity in the following order: *G. senegalensis, G. crassicaudatus, N. coucang*, and *P. potto* (Figs. 20-3). These patterns are discussed in relation to the sylvian complex, i.e. sulci that are dorsal, rostral and caudal to the complex.

An arcuate sulcus (intraparietal s.) is evident dorso-mesially from the termination of the sylvian complex. This indentation varies in the lesser galago from a pit to a sulcus not longer than 3 mm. In all specimens except the slow loris the sulcus curves around the termination of the sylvian complex (Figs. 20-3). The sylvian complex in slow loris appears to extend dorso-mesially to within 3 mm. of the longitudinal fissure. This extension is the intraparietal sulcus. Between this arcual sulcus and the midline (Fig. 26), a sagittal sulcus (sulcus *c*) begins approximately 12 mm. from the frontal pole and extends not more than 6 mm. toward the occipital pole. This sulcus is slightly concave toward the midline.

Rostral to the sylvian complex a variety of grooves and sulci are

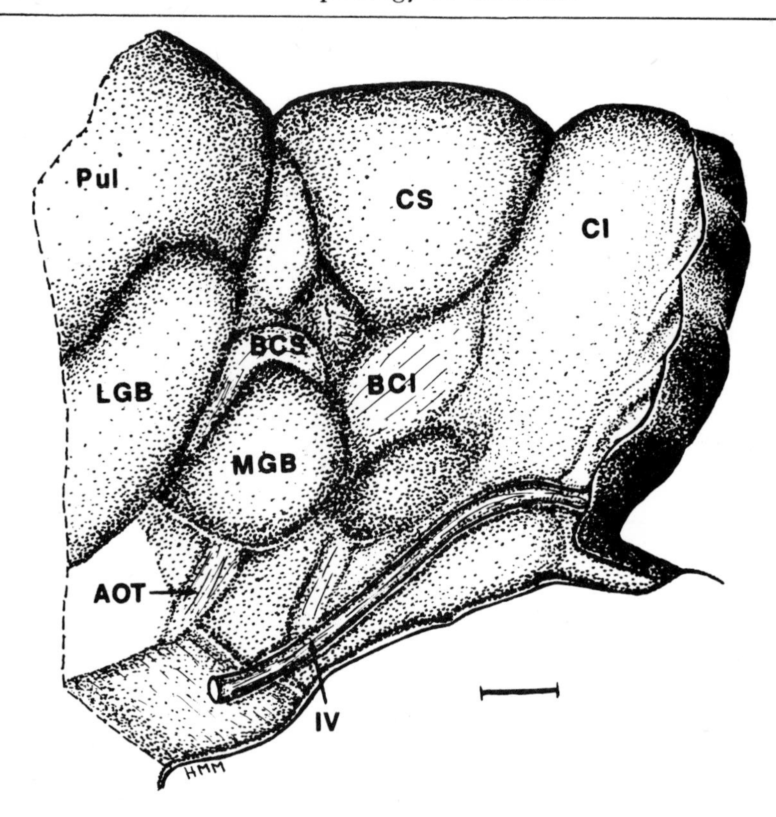

Figure 15. Lateral view of the mesencephalon and metathalamus of *Galago senegalensis.* Note the relative size of the superior colliculi (CS), the lateral geniculate body (LGB), and the indentations present on the caudal face of the inferior colliculi (CI). Dashed line represents cut surface. Scale = 1 mm.

seen. A pit is sometimes seen in the lesser galago, whereas definite sulci are present in the other Lorisidae. A straight sulcus (s. rectus) occurs caudal to the frontal pole. This sulcus is absent in lesser galago (Fig. 24) whereas variations are numerous in the slow loris, the sulcus being either straight, arcuate or triradiate. A sulcus just rostral to the sylvian complex is seen in the slow loris and greater galago. This sulcus (sulcus *e*) is large and arcuate in the slow loris, but small and variable in the greater galago (Figs. 25, 26). In the majority of our specimens of these animals, there is a communicating groove (rectus communicating s.) connecting the straight sulcus (s. rectus) with sulcus *e*. Rarely does the straight sulcus extend inferiorly to the orbital surface of the frontal pole. There is a central sulcus approximately midway between the frontal pole and sylvian complex in the potto (Figs. 23 and 27). It begins approximately 4 mm. from the midline and extends, in the coronal plane, inferolaterally for 11 mm. On the right hemisphere, the central sulcus continues as a

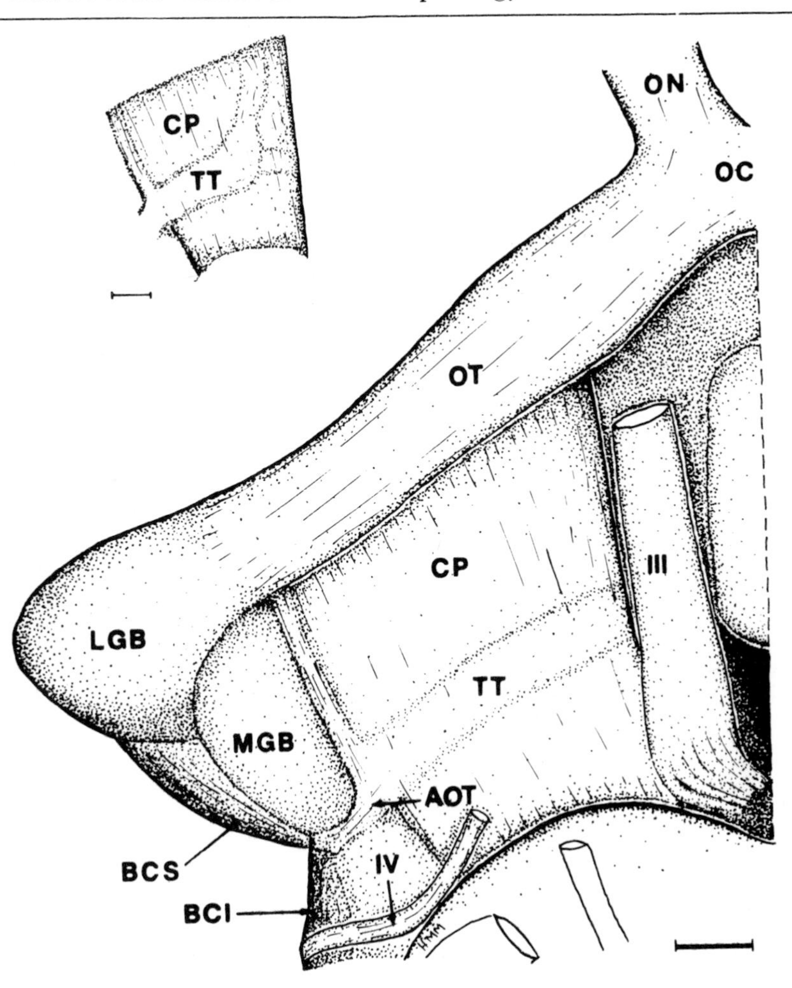

Figure 16. Ventral view of the mesencephalon and diencephalon of *Galago crassicaudatus*. The rootlets of the oculomotor (III) nerve exit through the interpeduncular fossa. Scale = 1 mm.

Inset: Variation in the course of the transpeduncular tract noted in one animal. Scale = 1 mm.

groove to join a caudal pit located approximately 3 mm. from the midline.

Caudal and parallel to the sylvian complex there is a groove or a sulcus (superior temporal s.). The lesser galago consistently shows a gyrus rostral and caudal to this groove, whereas slow loris and potto exhibit a more rounded temporal area (Figs. 20-3). The dorsal termination of the superior temporal sulcus is consistently more complex than its inferior terminus. In the slow loris, the dorsal terminus is a groove (sulcus *d*) which extends toward the sylvian complex (Fig. 26). In the potto and slow loris the area also exhibits a

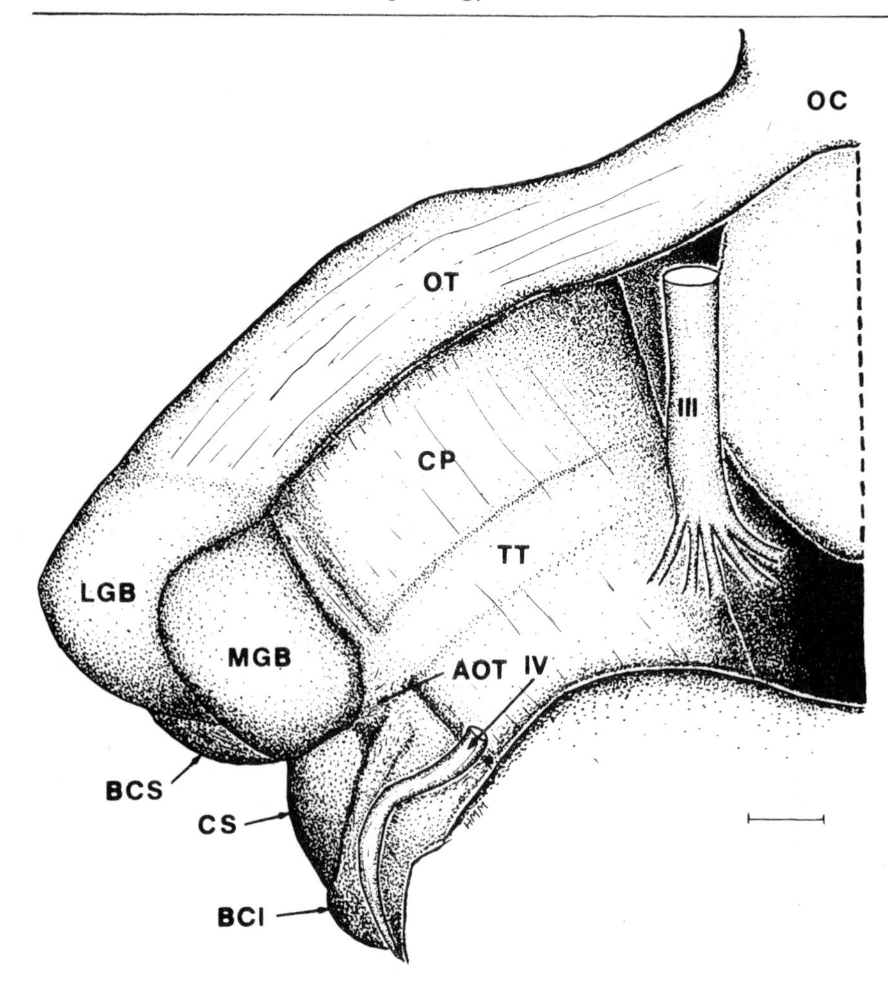

Figure 17. Ventral view of the mesencephalon and diencephalon of *Nycticebus coucang.* Note the rootlets of the oculomotor (III) nerve as they exit from the interpeduncular fossa and through the crus cerebri. Scale = 1 mm.

deep triradiate pit. In the right hemisphere of the potto, a groove arches rostrally to the sylvian complex. Caudally, this groove extends not more than 3 mm. toward the occipital pole (Fig. 23).

Sulci associated with the occipital region are consistently seen only in the slow loris. A sulcus in the coronal plane (lunate s.) indents the superomedial border and extends laterally from 3 to 7 mm. (Fig. 22). A similar groove is evident in both galagos 5 to 6 mm. rostral to the occipital pole (Fig. 32*b*). In one specimen of slow loris, this sulcus was found to be continuous with a groove which extends over the infero-lateral border on to the tentorial surface. A similar pattern is observed in the potto (Fig. 34*c*).

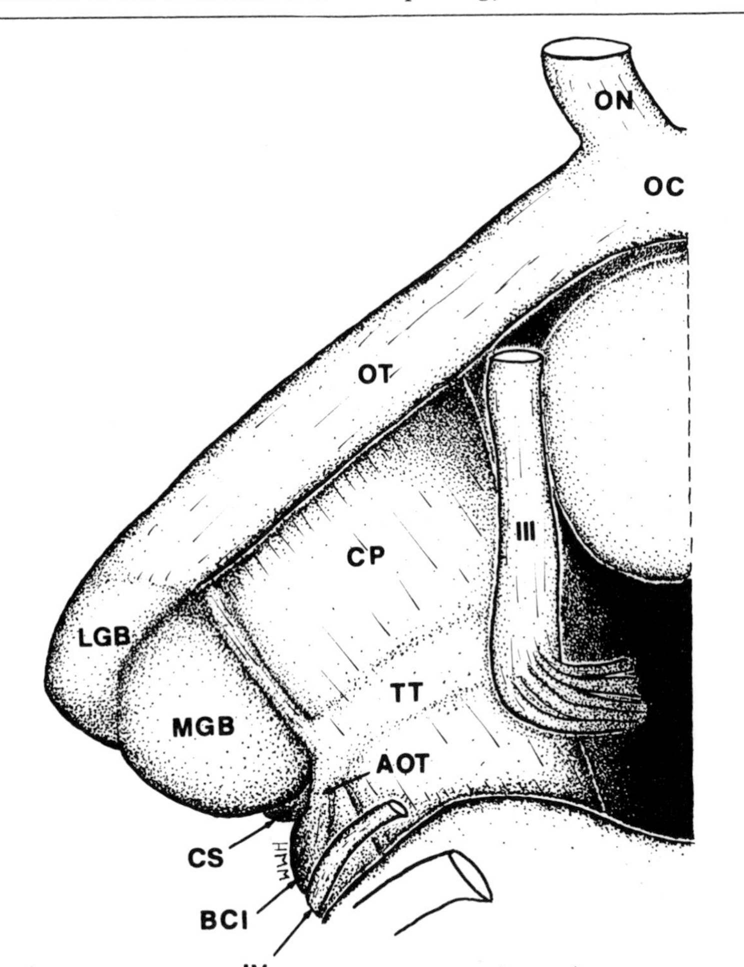

Figure 18. Ventral view of the mesencephalon and diencephalon of *Perodicticus potto*. Scale = 1 mm.

2. *Ventral and mesial surfaces*

The rhinal fissure exhibits an anterior and posterior division in these animals. It is shallow in the lesser galago, but more definite in the larger animals (Figs. 24-7 and 32-4). The posterior division is attenuated caudally. The olfactory bulbs project beyond the frontal pole approximately 3 mm. in the lesser galago and potto, and 5 mm. in the greater galago and slow loris. There is a distinct lateral olfactory tract. The pyriform lobe extends inferiorly and is visible below the posterior rhinal fissure from a lateral view.

On the orbital surface a straight sulcus (orbital s.) is seen (Figs. 32-4). In the lesser galago only a shallow groove is evident. Variations

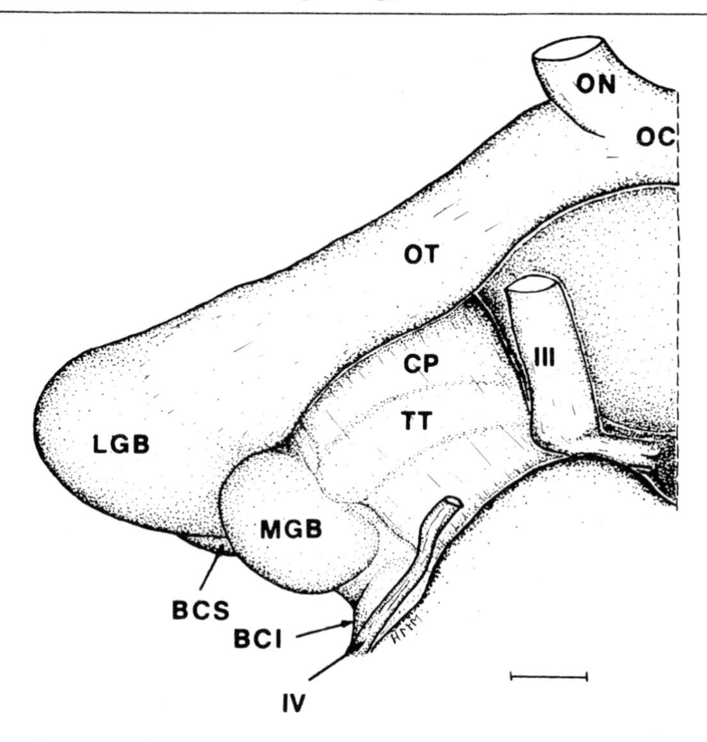

Figure 19. Ventral view of the mesencephalon and diencephalon of *Galago senegalensis.* Scale = 1 mm.

are seen in three of the lorises, the orbital sulcus being continuous with s. rectus (Fig. 33*c*).

Other than the calcarine complex, the sulci of the tentorial surface are consistent only in slow loris. Ventrolaterally, a straight sulcus emerges to indent the inferolateral border. In one specimen, three parallel sulci crossed the surface 7, 10 and 12 mm. from the occipital pole. There is no sulcus coursing parallel to the temporo-occipital polar axis.

On the mesial surface a long cingulate sulcus extends from the frontomesial to the occipitomesial surface. It branches rostrally in the potto; caudally, it continues as a groove to the paracalcarine sulcus in the greater galago and slow loris. In the lesser galago it joins the paracalcarine directly (Figs. 28-31). A short superiorly directed sulcus is interposed between the frontal pole and the origin of the cingulate s. in the greater galago. A similar sulcus is seen in the potto, but not in the slow loris.

The calcarine complex is similar in these animals. The retrocalcarine s. extends toward the occipital pole; the paracalcarine s. is directed superiorly (Figs. 28-31).

The rostrocaudal length of the corpus callosum is 11.5 mm. in the lesser galago, 13.6 mm. in the slow loris, 13.8 mm. in the potto, and

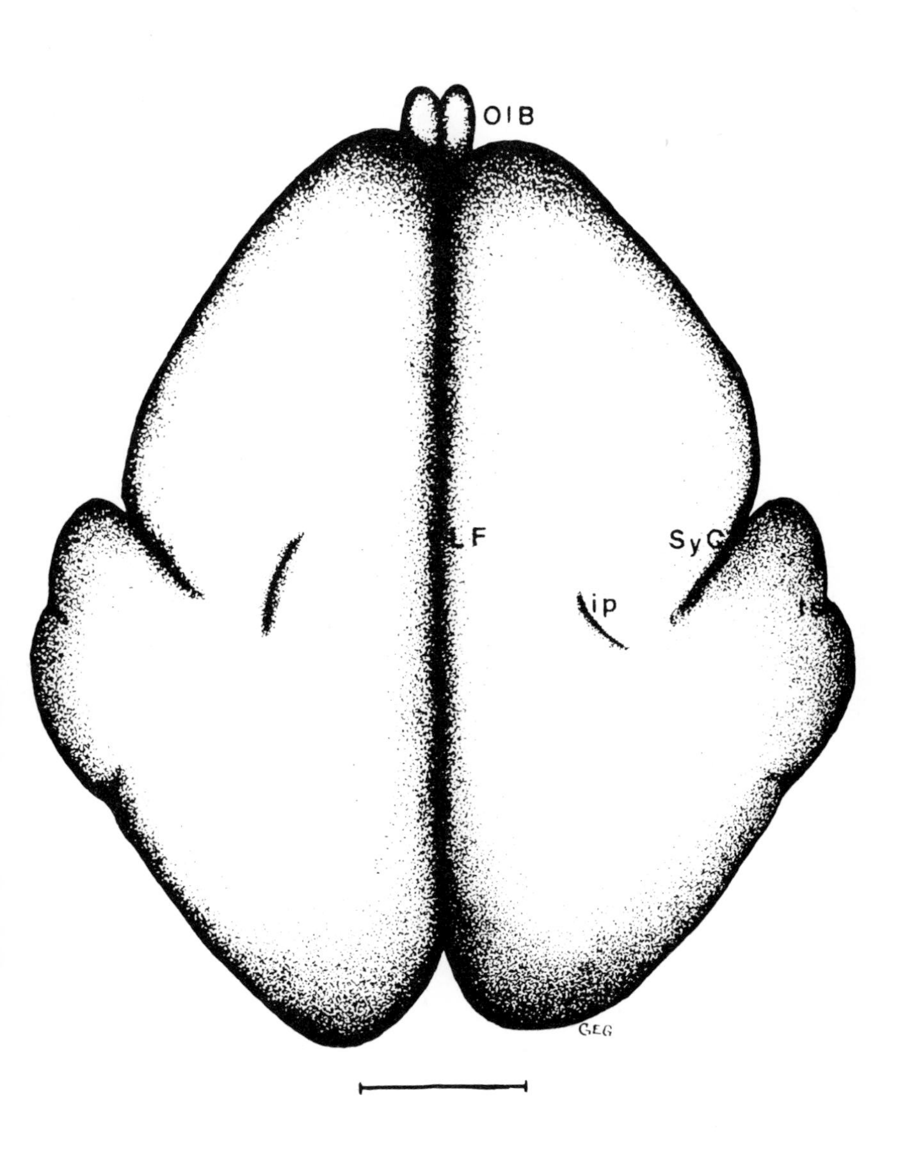

Figure 20. Dorsal view of the telencephalon of *Galago senegalensis*. The Sylvian complex (suprasylvian plus pseudosylvian) separates temporal cortex from frontal cortical areas. Scale = 0.5 cm.

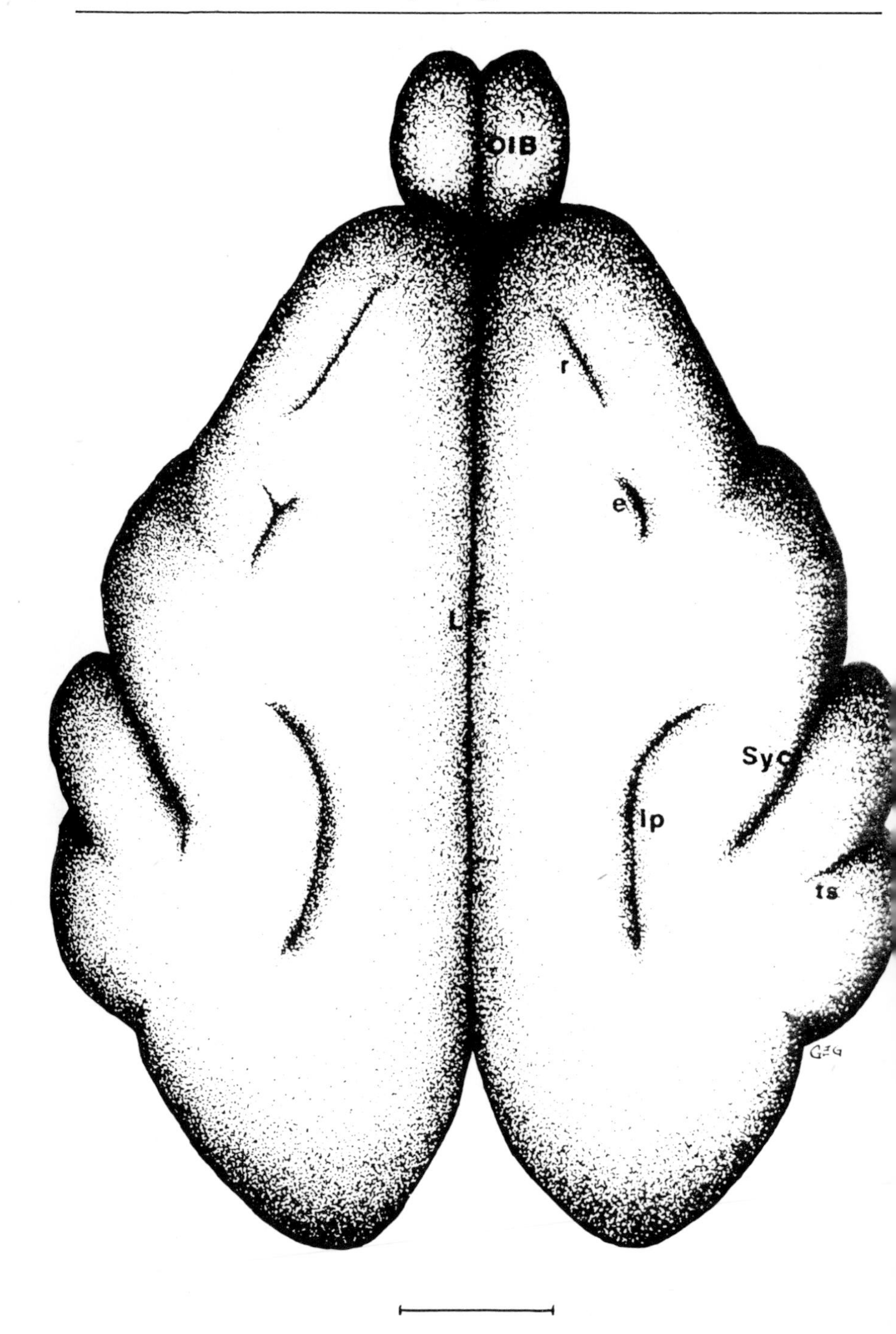

Figure 21. Dorsal telencephalon of *Galago crassicaudatus.* Note the variation of sulcus *e* on the two hemispheres as well as the development of the parieto-temporal cortex. Scale = 0.5 cm.

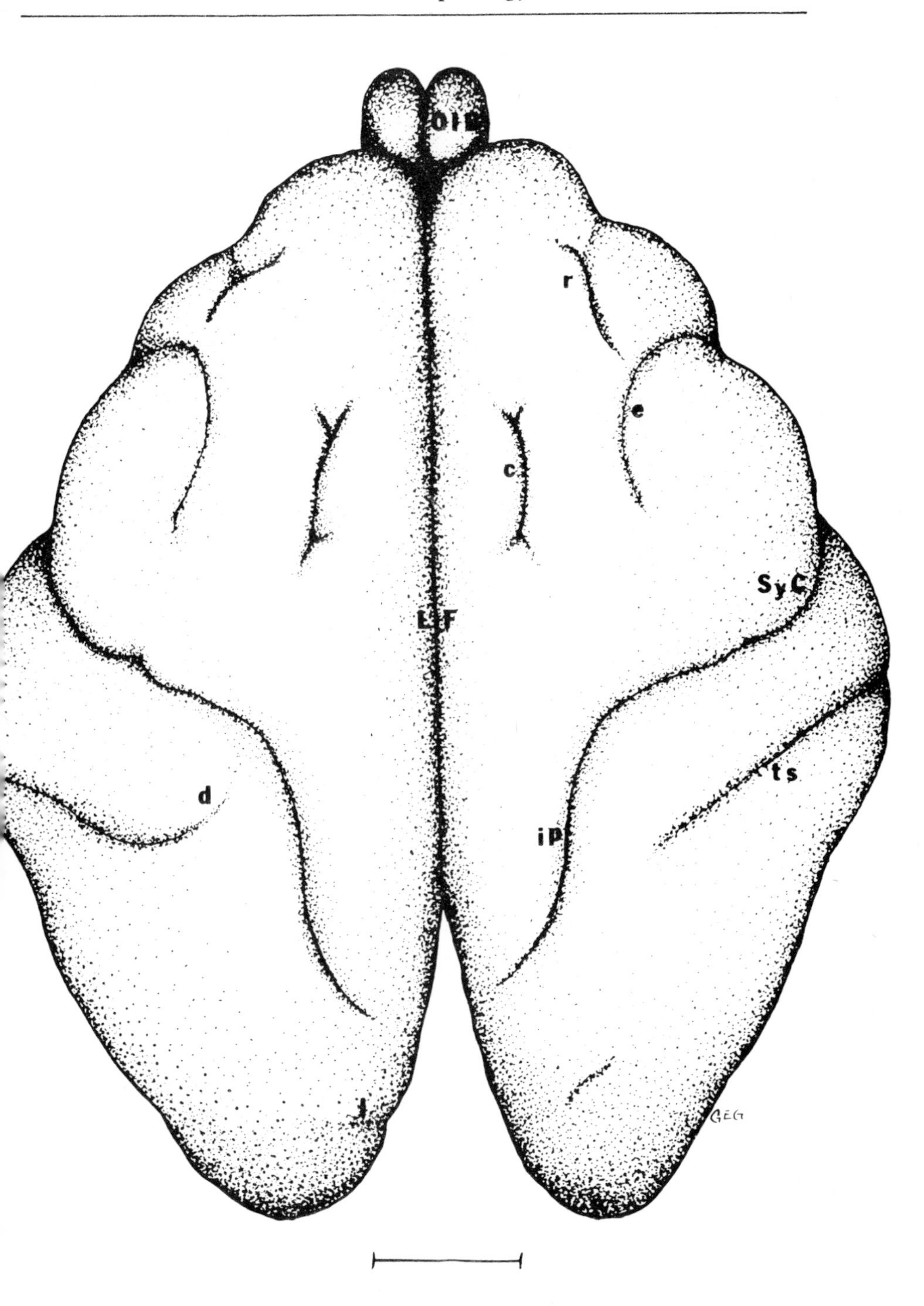

Figure 22. Telencephalon of *Nycticebus coucang.* Notice the indentation of the frontal margin by the sulcus rectus and the position of the lunate sulcus (1). Scale = 0.5 cm.

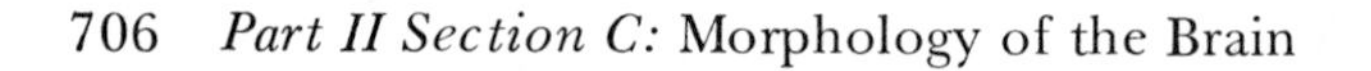

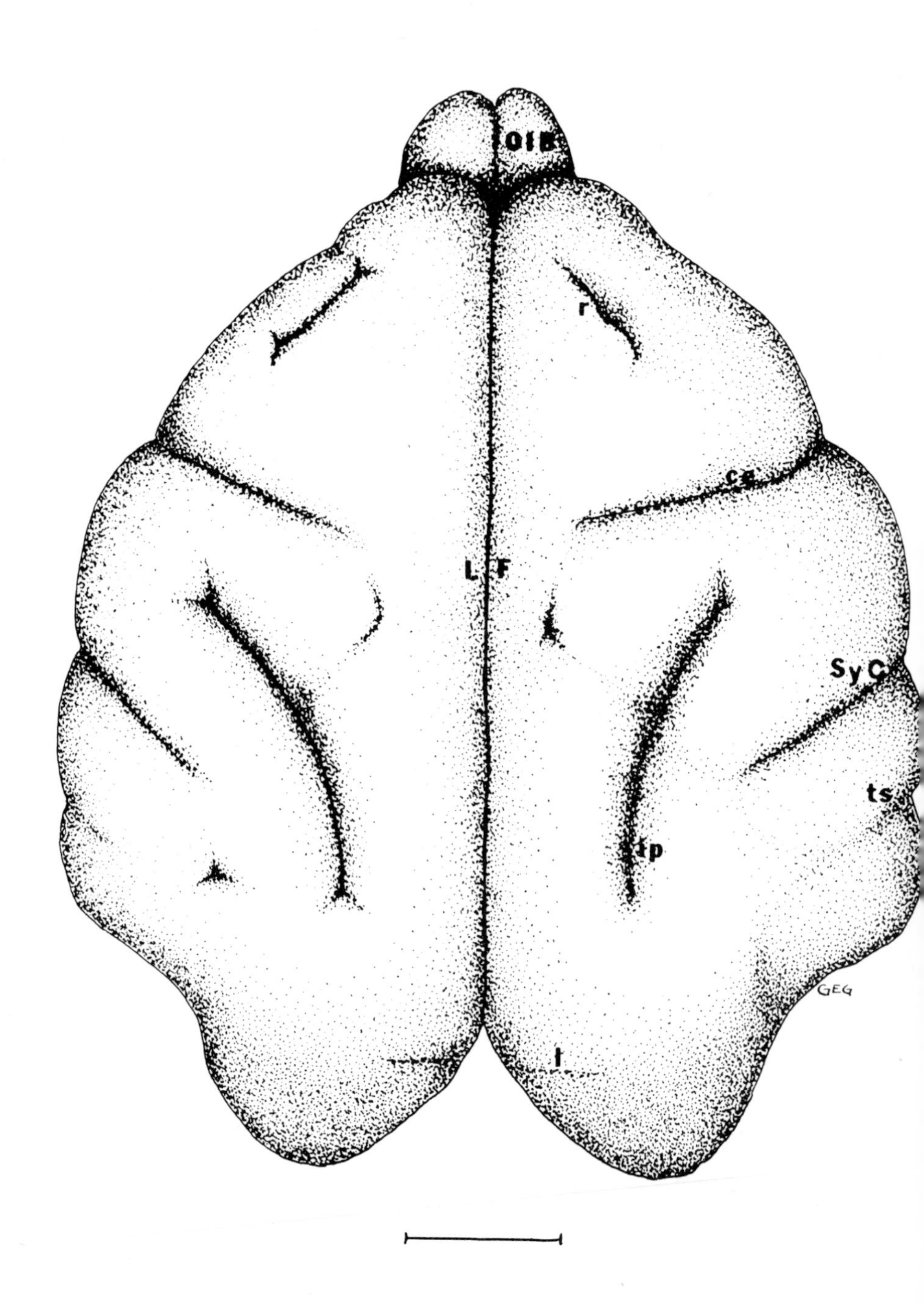

Figure 23. This dorsal view of the telencephalon of *Perodicticus potto* exemplifies the transition from sagittal sulci to coronal sulci in these prosimians. Scale = 0.5 cm.

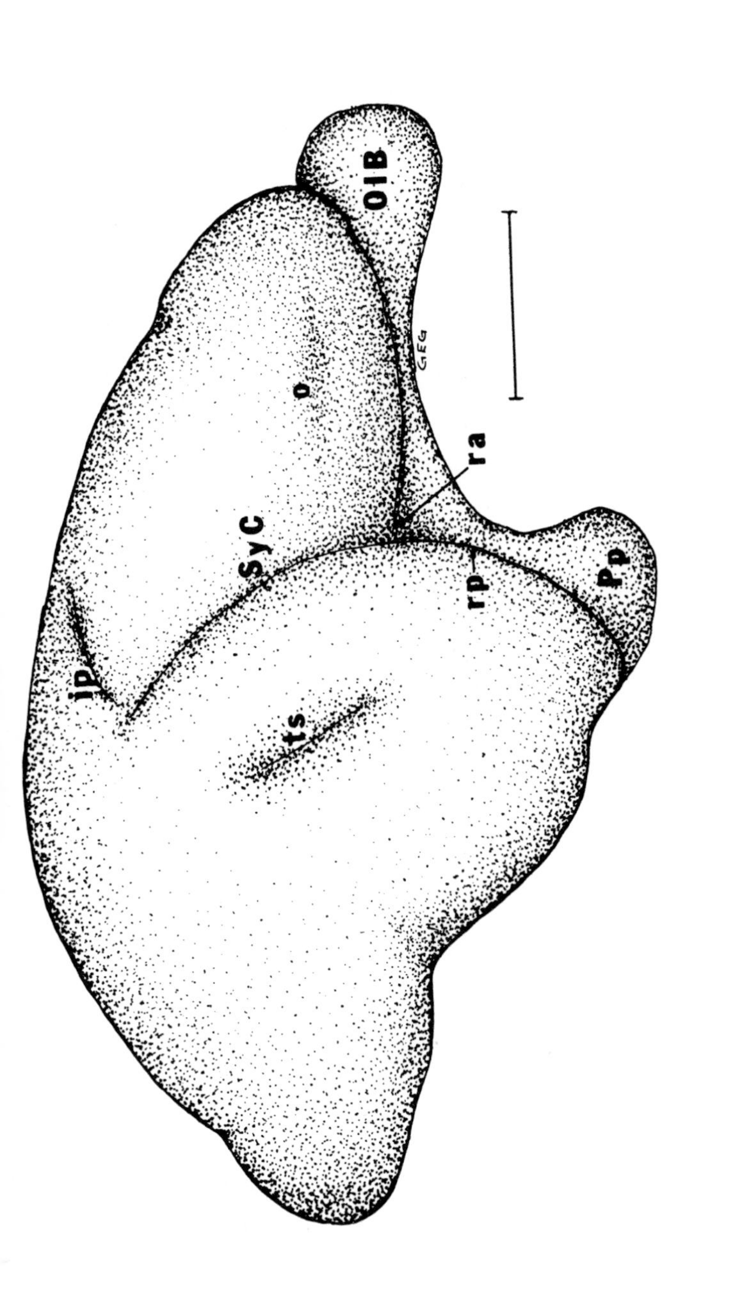

Figure 24. Lateral view of the telencephalon of *Galago senegalensis.* The majority of cortical infoldings are dimples or grooves, except the Sylvian complex (SyC) and intraparietal sulcus (ip). Scale = 0.5 cm.

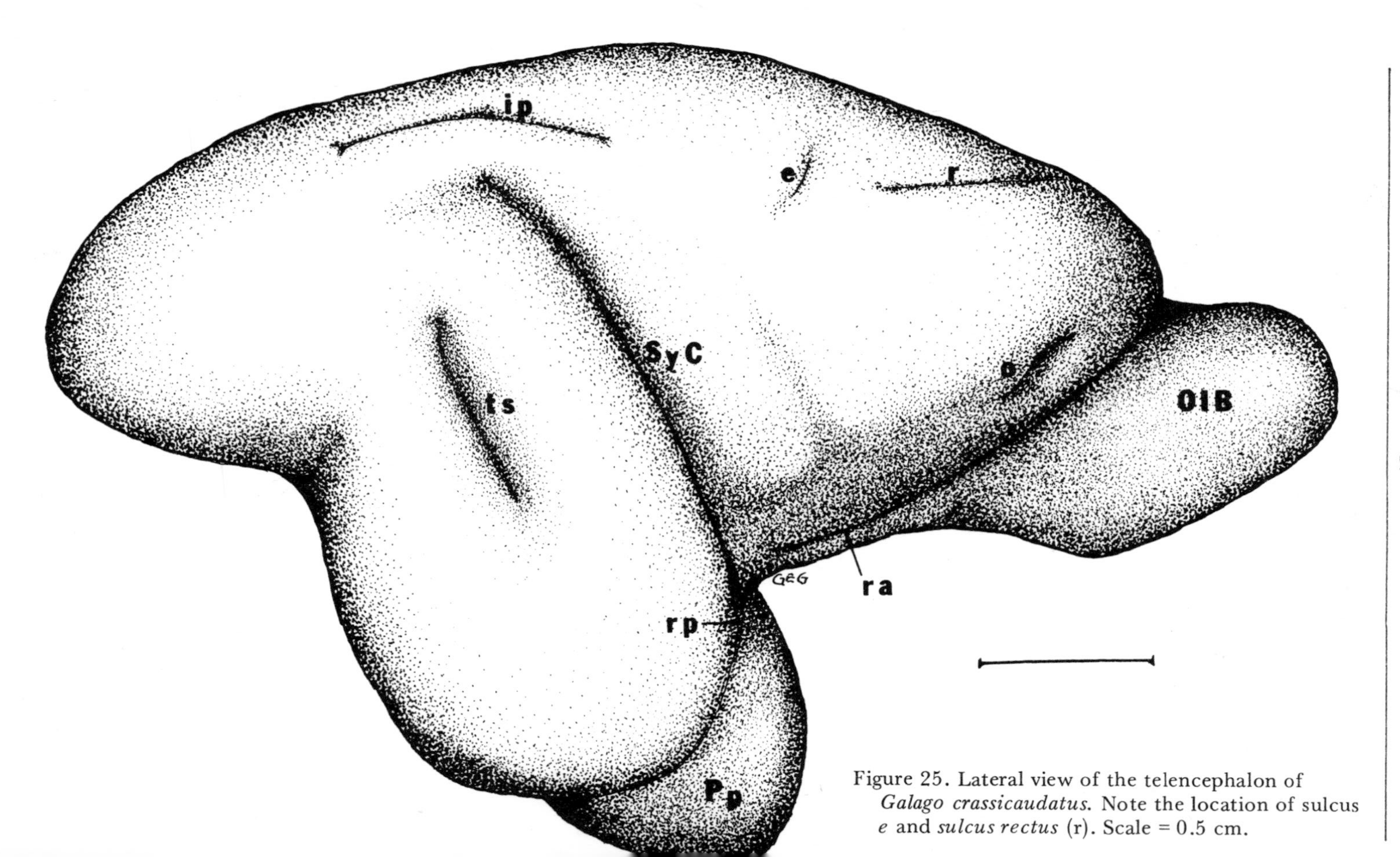

Figure 25. Lateral view of the telencephalon of *Galago crassicaudatus.* Note the location of sulcus *e* and *sulcus rectus* (r). Scale = 0.5 cm.

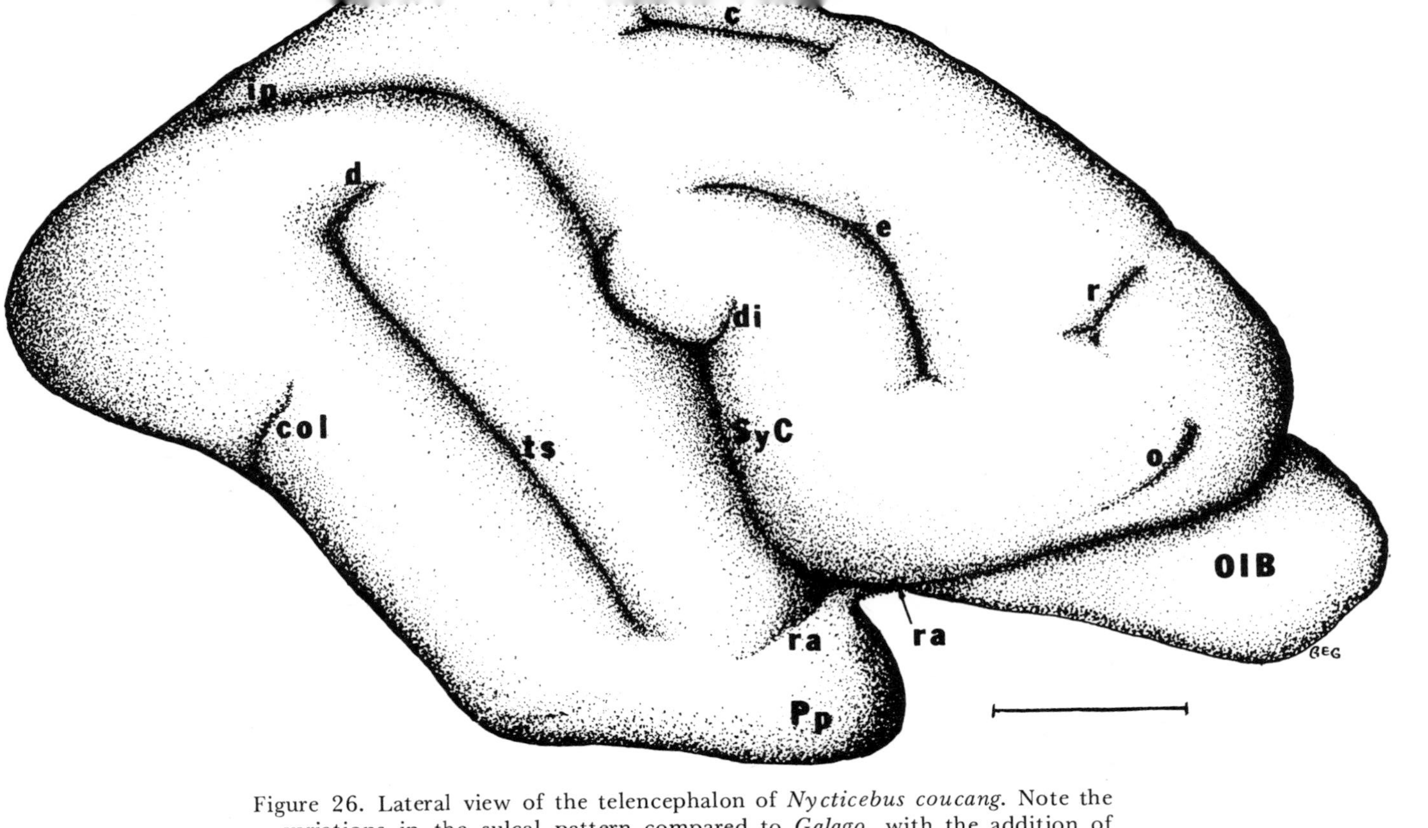

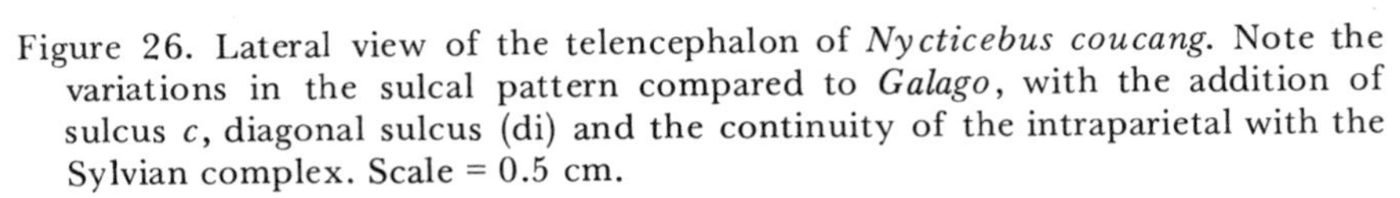

Figure 26. Lateral view of the telencephalon of *Nycticebus coucang.* Note the variations in the sulcal pattern compared to *Galago*, with the addition of sulcus *c*, diagonal sulcus (di) and the continuity of the intraparietal with the Sylvian complex. Scale = 0.5 cm.

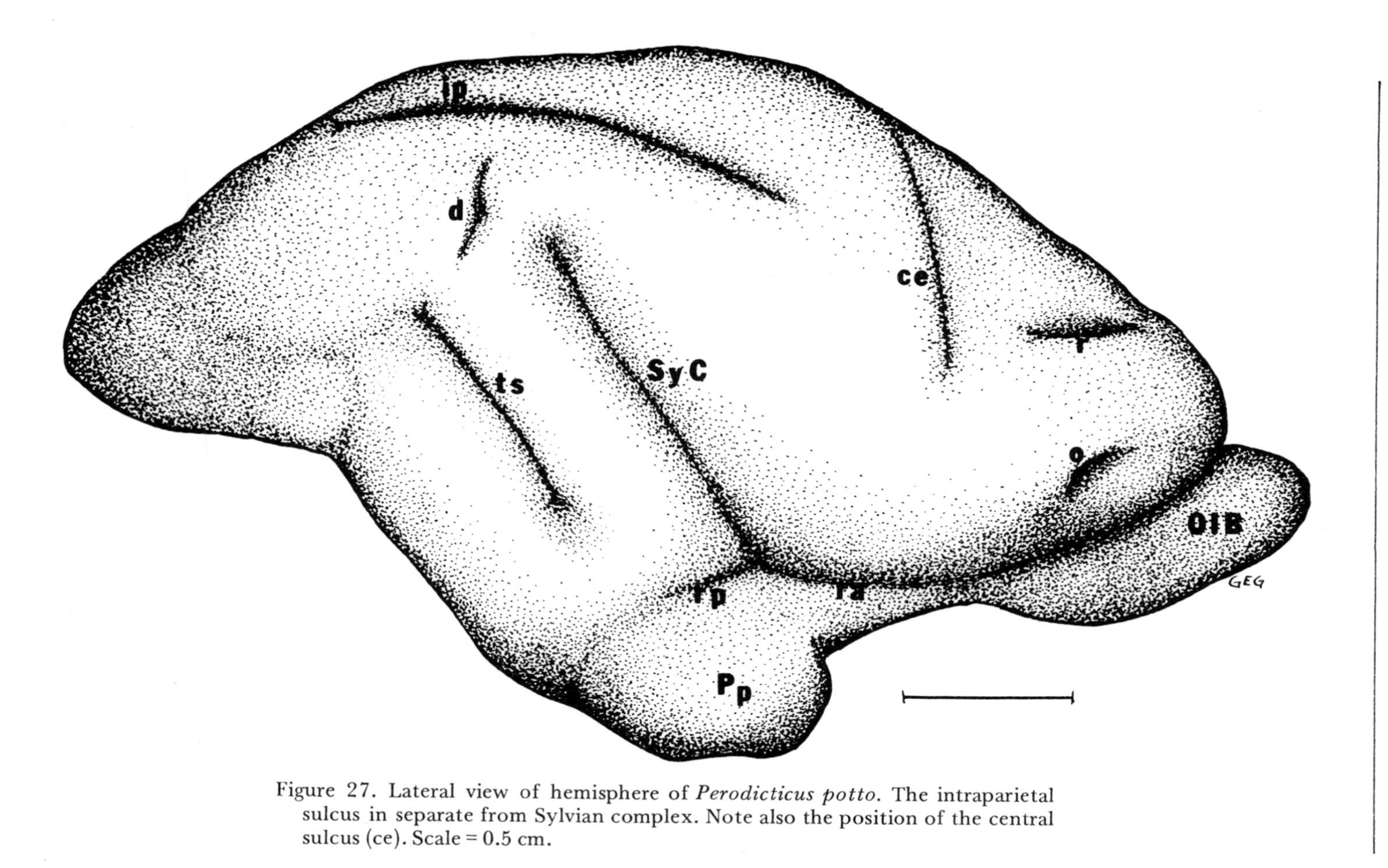

Figure 27. Lateral view of hemisphere of *Perodicticus potto*. The intraparietal sulcus in separate from Sylvian complex. Note also the position of the central sulcus (ce). Scale = 0.5 cm.

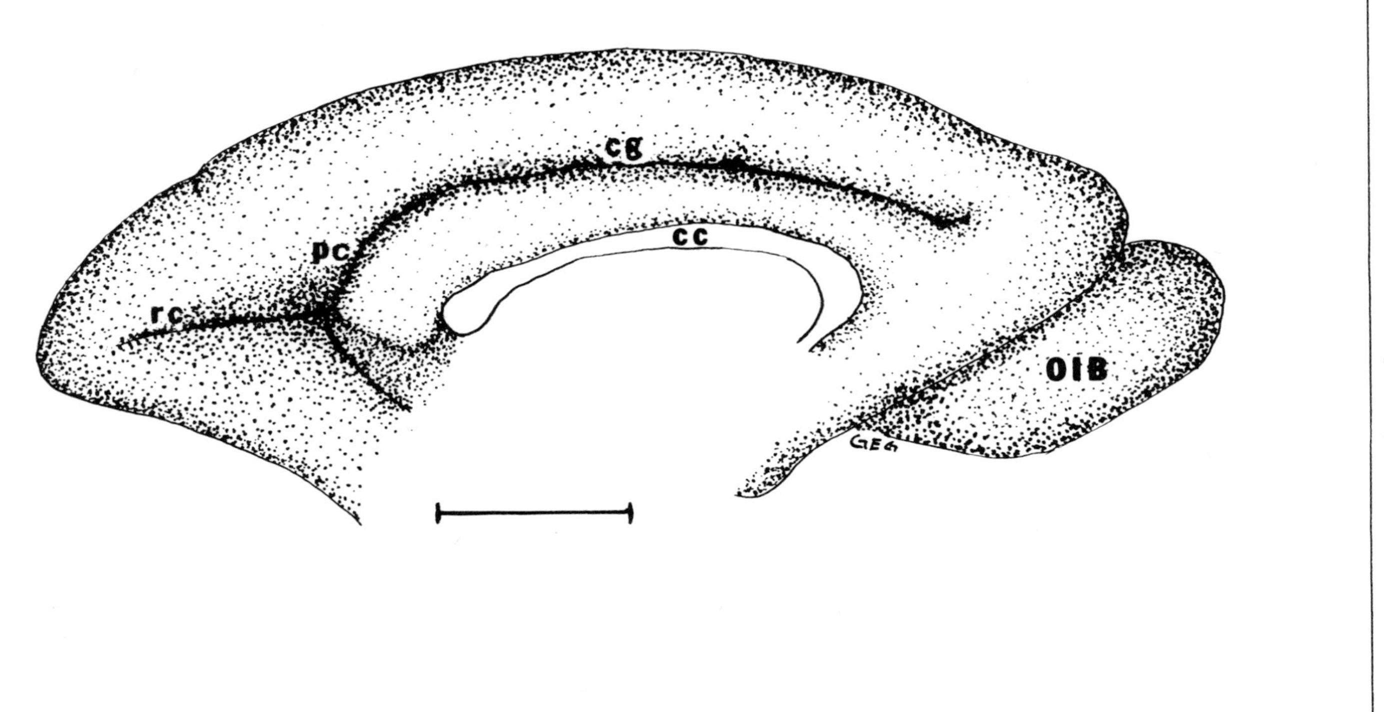

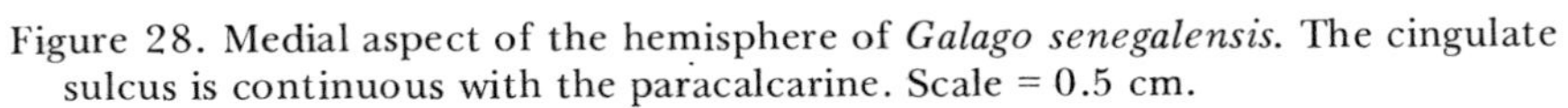

Figure 28. Medial aspect of the hemisphere of *Galago senegalensis*. The cingulate sulcus is continuous with the paracalcarine. Scale = 0.5 cm.

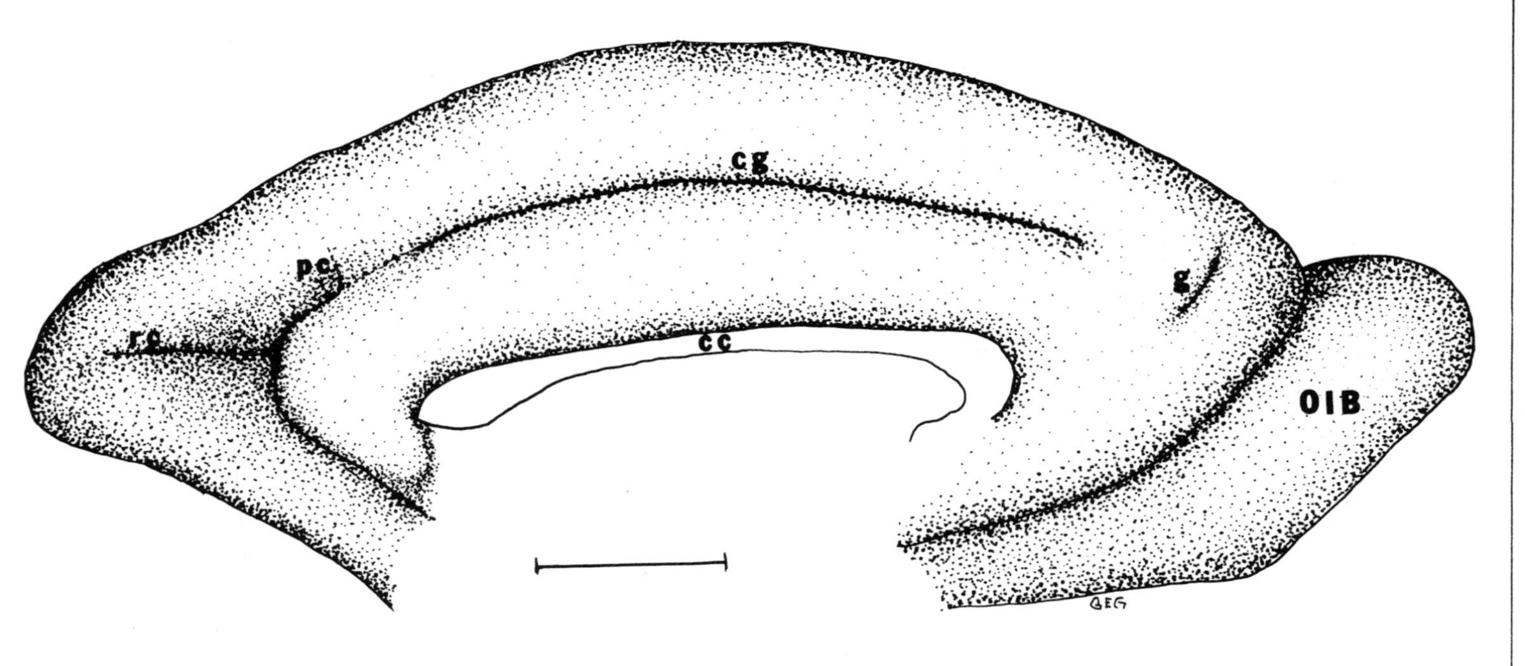

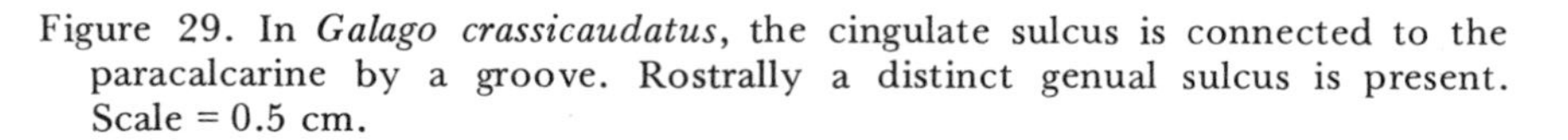

Figure 29. In *Galago crassicaudatus*, the cingulate sulcus is connected to the paracalcarine by a groove. Rostrally a distinct genual sulcus is present. Scale = 0.5 cm.

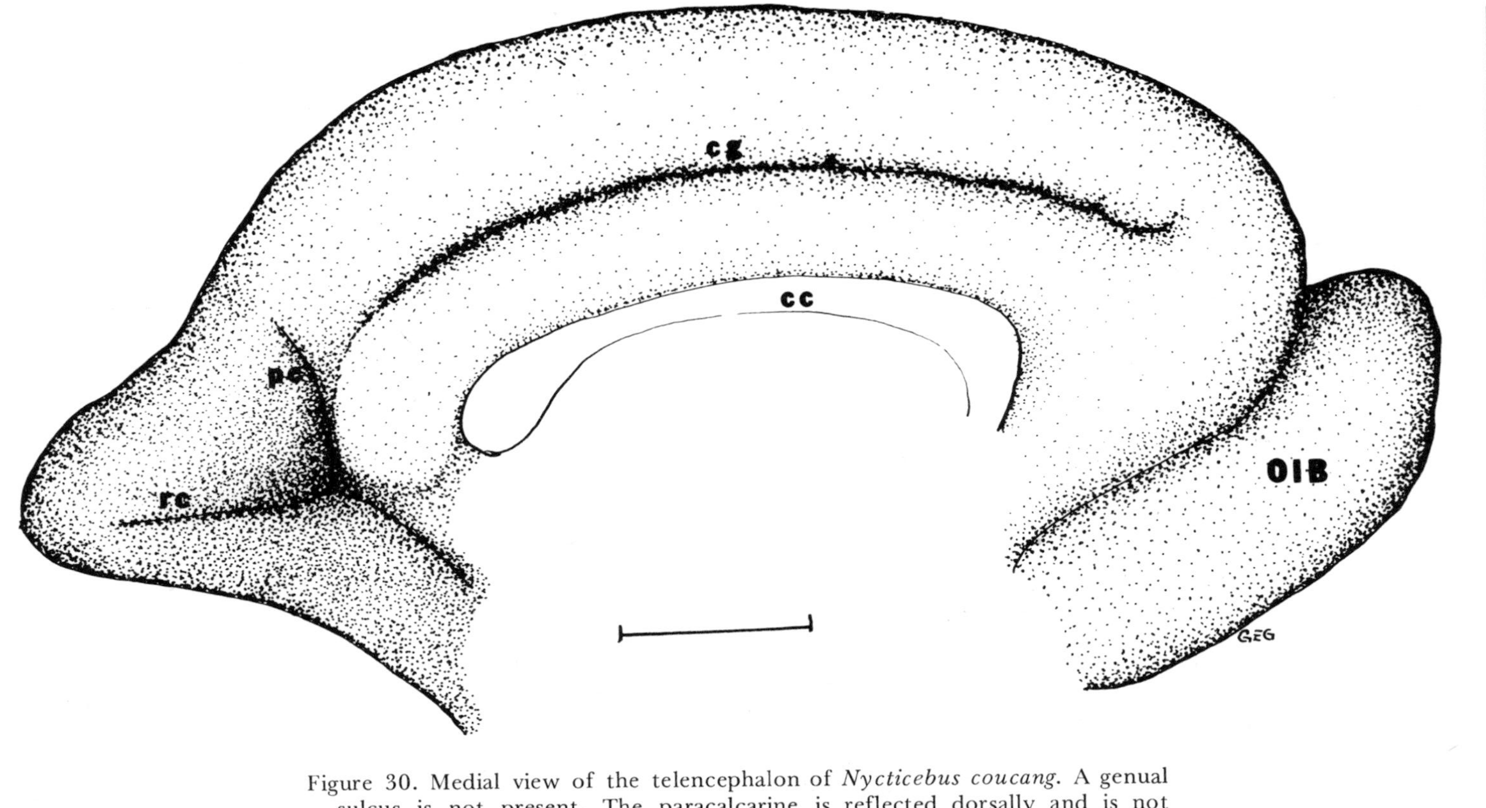

Figure 30. Medial view of the telencephalon of *Nycticebus coucang*. A genual sulcus is not present. The paracalcarine is reflected dorsally and is not associated with the cingulate sulcus as seen in *Galago*. Compare with Fig. 29. Scale = 0.5 cm.

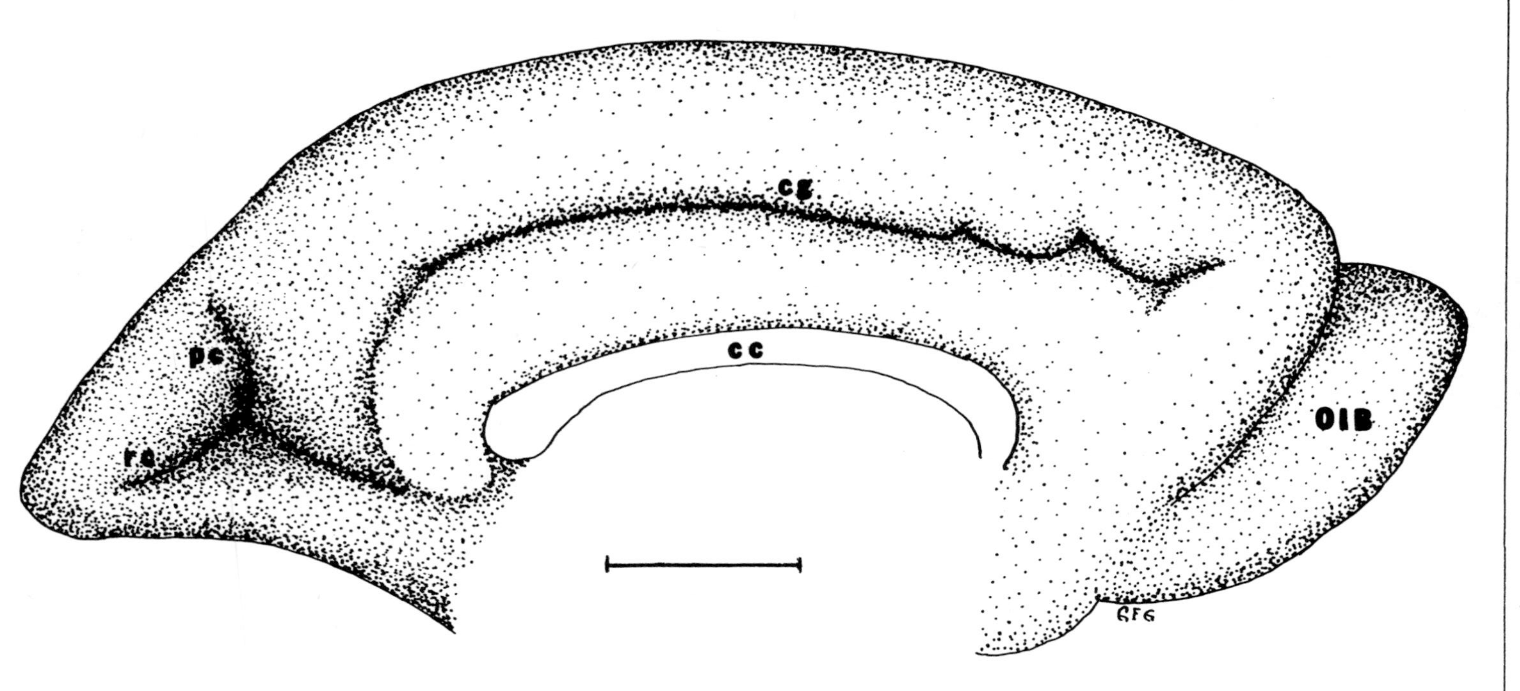

Figure 31. The medial aspect of the hemisphere of *Perodicticus potto* is similar to *N. coucang*, although variations of the rostral terminus of the cingulate sulcus are not seen in the slow loris. Scale = 0.5 cm.

15.0 mm. in the greater galago. The splenium of the corpus callosum overlies the rostral portion of the superior colliculus in all the animals except *G. senegalensis.*[26] The fornix is 0.8 mm. in thickness in the greater galago, while it is 0.5 to 0.6 mm. in the other animals.

Discussion

In Monotremata and Marsupialia, the pons, olive and pyramids are simple in structure and remain relatively unchanged throughout these groups.[22,28] The spiny anteater (*Tachyglossus aculeatus*) has a post-trigeminal pons; the *corpus trapezoideum*, olive and pyramids cannot be seen grossly. In the wallaby, the pons is entirely pre-trigeminal, the pyramids are thin and the olive cannot be seen.[22] The pontine and medullary components in the Insectivora are similar to that observed in the wallaby. In the elephant shrew, the pons and *corpus trapezoideum* are approximately equal in size and are separated by a transverse groove. The fifth nerve exits laterally from the groove.[13]

The pons in *Tupaia minor* measures 2.5 mm. in length and is located rostral to a broad *corpus trapezoideum.*[9] The fifth nerve exits from the lateral pons, and the pyramids are attenuated.[9] We found the pons and medulla of *Tupaia glis* and *Tupaia minor* to be similar in appearance. The pontine surface measures 3 mm. in length and is only 0.5 mm. greater than the *corpus trapezoideum*. The lateral pons shows a very small post-trigeminal bundle. The pyramids are uniform, 6.5 mm. in length and 0.8 mm. in width. The pons in *Tarsius* is notably smaller but more prominent than in *Tupaia.*[23]

In primates the pons, in particular, undergoes considerable increase in size. Phylogenetically, pontine growth occurs in all dimensions, but characteristically more caudally. This caudal increase is believed to be partially responsible for the decrease in the size of the *corpus trapezoideum*;[22] exceptions may be due to specialised sensory mechanisms such as audition. In general, the lower primates have a large *corpus trapezoideum.* In the higher primates (macaque), the *corpus trapezoideum* is covered rostrally by the pons.[22] In chimpanzee and man it is no longer visible.[22,29] It has been proposed that in primates the ponto-trapezoidial groove is obliterated by the caudal growth of the pons.[22]

The pons and medulla of the genera included in this study are typically primate in appearance. The striations observed in the ponto-trapezoidial groove of *N. coucang* and *P. potto* do not necessarily imply a more caudal-pontine growth. This could, in fact, be related to a decrease in the size of the *corpus trapezoideum* and other structures related to audition. Despite the location and shape of the olive, *P. potto* has a well-developed ventral brainstem.[23] This observation is based on the appearance of the pons, the pyramids and

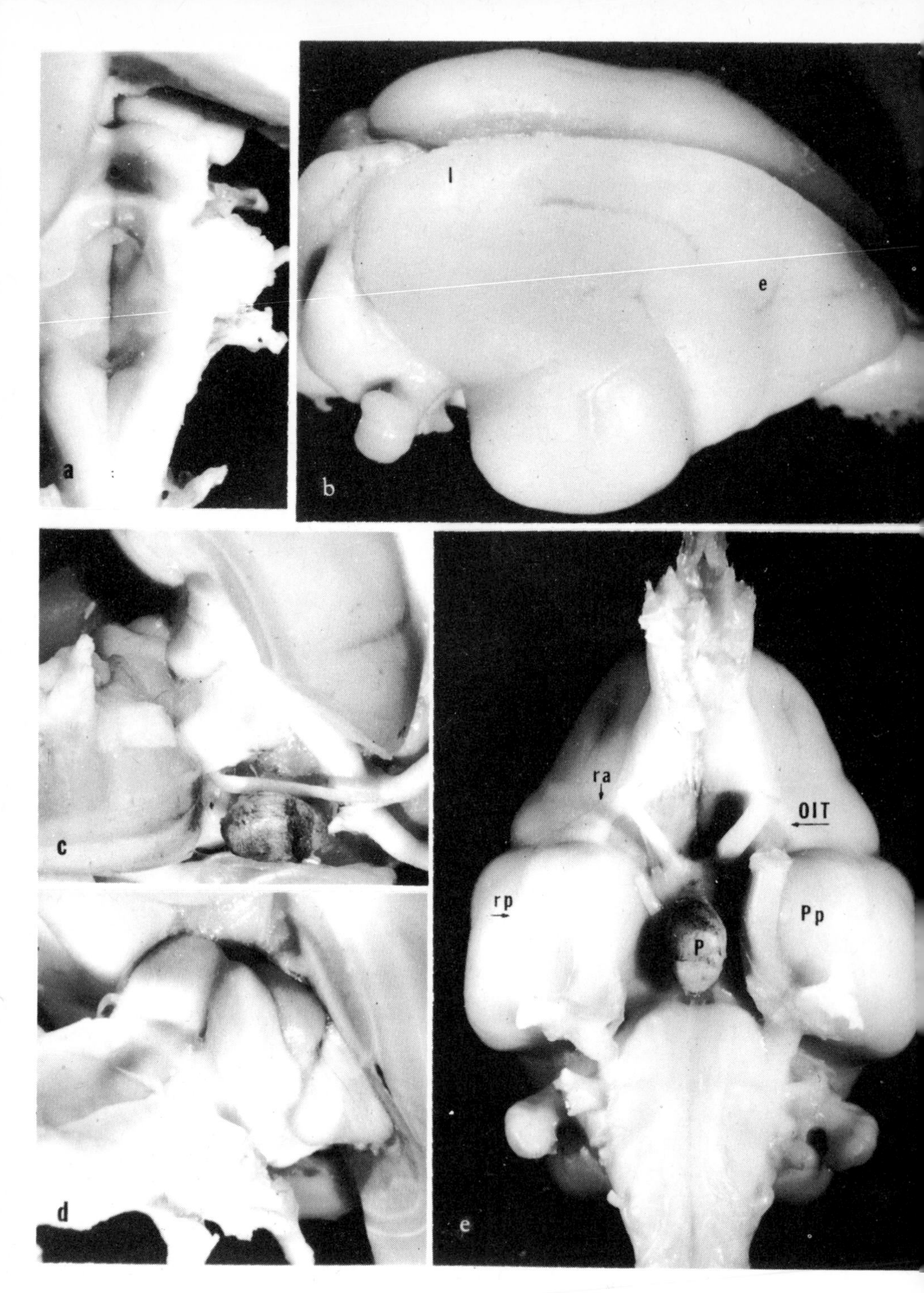

Figure 32. Photographs of the cerebral cortex and brain stem of *Galago crassicaudatus*.

(*a*) Dorsal view of the rhomboid fossa, dorsal and dorsolateral medulla and superior and inferior colliculi. Compare with Fig. 7.

(*b*) Dorsolateral view of the cerebral cortex. Compare with Fig. 25.

(*c*) Ventrolateral view of the myelencephalon and mesencephalon. Note the trapezoid body, pons, crus cerebri and optic tract. Compare with Figs. 1 and 16.

other brainstem structures when compared with the other Lorisidae.

In the lesser bushbaby and slow loris, the *corpus trapezoideum* was described as being located within the caudal portion of the pons.[16,17] Since the *corpus trapezoideum* was considered to be a portion of the pons, some considered the pyramids to be superficial to the caudal pons. However, this caudal area is not pons, but the *corpus trapezoideum*, and should not be included within the borders of the pons in lower mammalian forms. Our observations show that the pons has distinct rostral and caudal borders. The caudal border corresponds to the ponto-medullary groove. The *corpus trapezoideum* is located caudal to the pons and is covered medially by the pyramids. The ponto-medullary groove can therefore be described regionally as the ponto-trapezoidial groove and the ponto-pyramidal groove. The pyramids are located in a superficial position only in the medulla. Rostrally the pyramids pass into the basilar pons.

It is our experience that the small primary rootlets of cranial nerves were extremely difficult to observe without adequate removal of the arachnoid. This may account for the difference in the number of rootlets comprising the vagus and hypoglossal nerves reported in our study and the observations reported for *N. coucang.*[17] We noted in excess of six primary rootlets for the vagus nerve (versus three) and 12 to 13 primary rootlets (versus four) for the hypoglossal nerve.

Histological studies on the potto show a large *nucleus reticularis lateralis.*[23] This nucleus is responsible for the raised area on the ventrolateral surface of the medulla. It is similar in the greater bushbaby, slow loris, and potto, but less evident in the lesser bushbaby. We suggest that this elevation be called the reticular prominence.

The features of the prosimian mesencephalon that call for special note are the *corpora quadrigemina.*[2] The general features of these structures are remarkably constant in all the Metatheria and Eutheria.[2] Wood Jones described the dorsal surface of the midbrain of a generalised marsupial as having well-developed *corpora quadrigemina*, the inferior colliculi being larger and more conspicuous (his Fig. 95).[28]

In *Tupaia* the *corpora quadrigemina* are noteworthy for the large size of the anterior pair, which indent the occipital lobe of the cerebrum and the anterior surface of the cerebellum.[9] Large, anterior colliculi are reminiscent of the optic lobe of the reptilian brain.[10] On

---

(*d*) Caudolateral view of the mesencephalon. Note the anterior medullary velum, brachium of the inferior colliculus, medial and lateral geniculate bodies. Compare with Fig. 11.

(*e*) Ventral view of the brain and brain stem. Both divisions of the rhinal fissure can be seen as well as the lateral olfactory tract (OIT). Note the size of the pituitary gland.

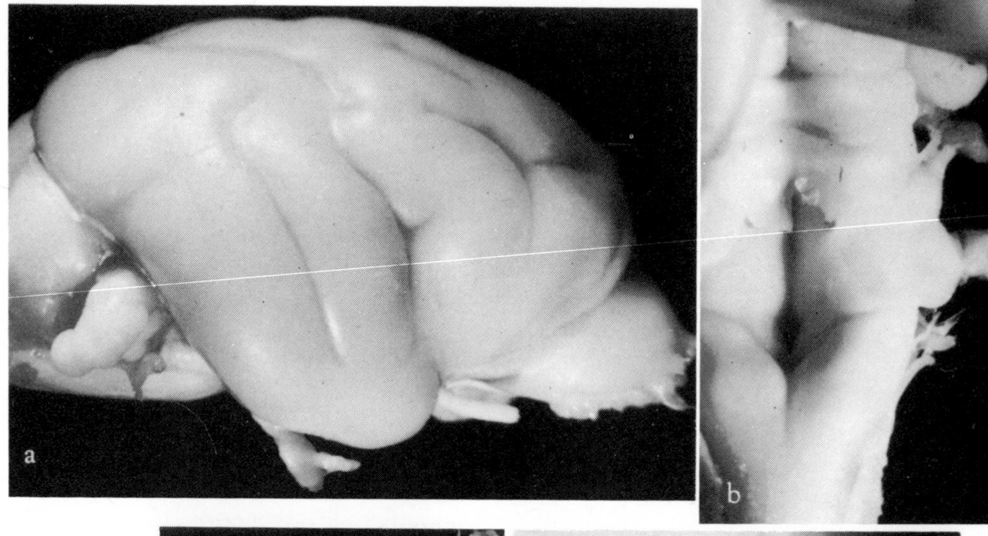

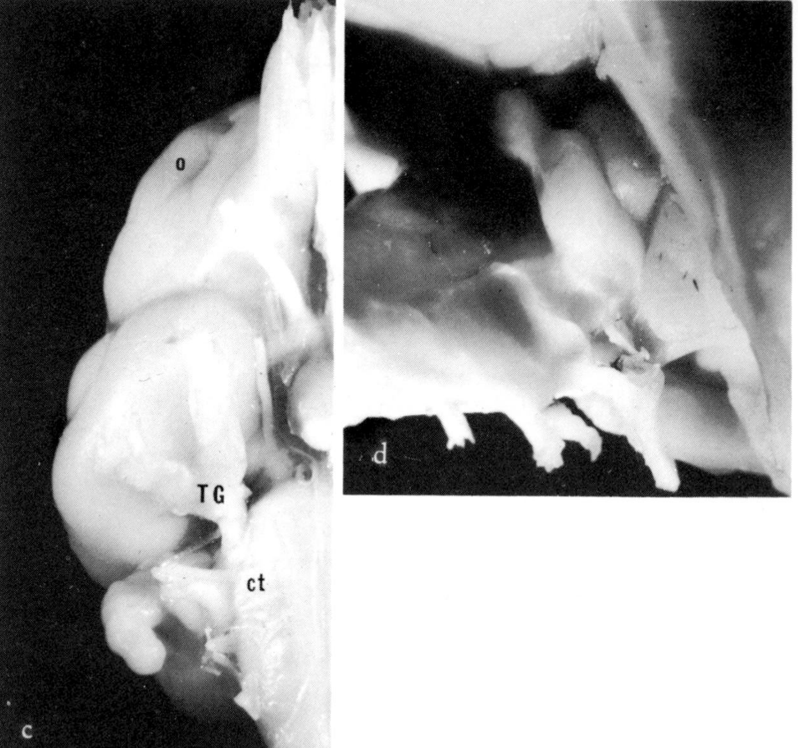

Figure 33. Photographs of the cerebral cortex and brain stem of *Nycticebus coucang.*

(*a*) Lateral view of the cerebral cortex. Compare with Fig. 26.

(*b*) Dorsal view of the myelencephalon and mesencephalon. Compare with Fig. 8.

(*c*) Ventral view of the cerebral cortex and brain stem. Note the trigeminal ganglion (TG) and trapezoid body (CT). Compare with Fig. 2.

(*d*) Caudolateral view of the mesencephalon. Compare with Fig. 12.

the other hand, the posterior colliculi are very inconspicuous in *Tupaia*,[9] forming a transversely elongated strip which lies under the posterior border of the anterior colliculi, again reminiscent of the reptilian condition.[10]

The brain of *Tupaia* exhibits a 'definite hypoplasia' of secondary auditory centres which can be correlated 'with the atrophy of the external ear'.[10] The anterior colliculus of *Ptilocercus* is described as being much smaller relative to the posterior colliculus than is the case in *Tupaia.*[11]

The dorsal surface of the mesencephalon in *Tarsius* exhibits well-developed inferior and superior colliculi.[21] The superior colliculus in this animal attains dimensions that almost warrant the designation of optic lobe.[19,20] While Woollard describes the inferior colliculus and the medial geniculate body as being of normal size,[21] Tilney writes that these two structures are larger and better defined than in all other primates.[20] Both collicular elevations of the mesencephalon are prominent in *Lemur mongoz.*[20]

The anterior quadrigeminal bodies are smaller in *Lemur* than in *Tarsius.*[2] The lateral geniculate body and the superior colliculus in *Galago* and the slow loris are described as distinct features of the brain stem.[16,17] Their prominence has suggested the importance of vision in these nocturnal animals.[16,17]

The inferior and superior colliculi are large elevations on the dorsal surface of the mesencephalon of the Lorisidae. The size of the inferior colliculus may reflect a retention of primordial or rudimentary control over audition.[20] In the present study the inferior colliculus is more prominent in the galagos than in the slow loris and potto. This parameter may be correlated with the size of the external ear. Both galagos have large pinnae, and as Osman Hill has noted, 'the external ear reaches its maximum expression in Galagidae. In addition to its larger size it differs from that of the lorises in its greatly expanded, spoon-shaped, transversely ribbed cymba.'[30] The size of the inferior colliculus and the pinna may also be correlated with that of the *corpus trapezoideum* of the medulla. As was noted above, the *corpus trapezoideum* in the galagos is larger than that in the potto or slow loris.

The superior colliculus is a large, oval somewhat flattened structure in all the animals studied, but in proportion to the size of its brain, the lesser galago has the largest superior colliculus. This correlates with the size of the eyes in proportion to the size of the head in this animal.

The accessory optic system consists of the transpeduncular tract, its nucleus and the accessory optic tract and its nuclei.[31] A distinct transpeduncular tract (*tractus peduncularis transversus*)[32] has been shown in *Tupaia*,[10] *Tarsius* and *Lemur.*[15,2] It was not described in the slow loris or lesser galago.[17,16] In the present study, a well-defined accessory optic system is seen. An account of reduction

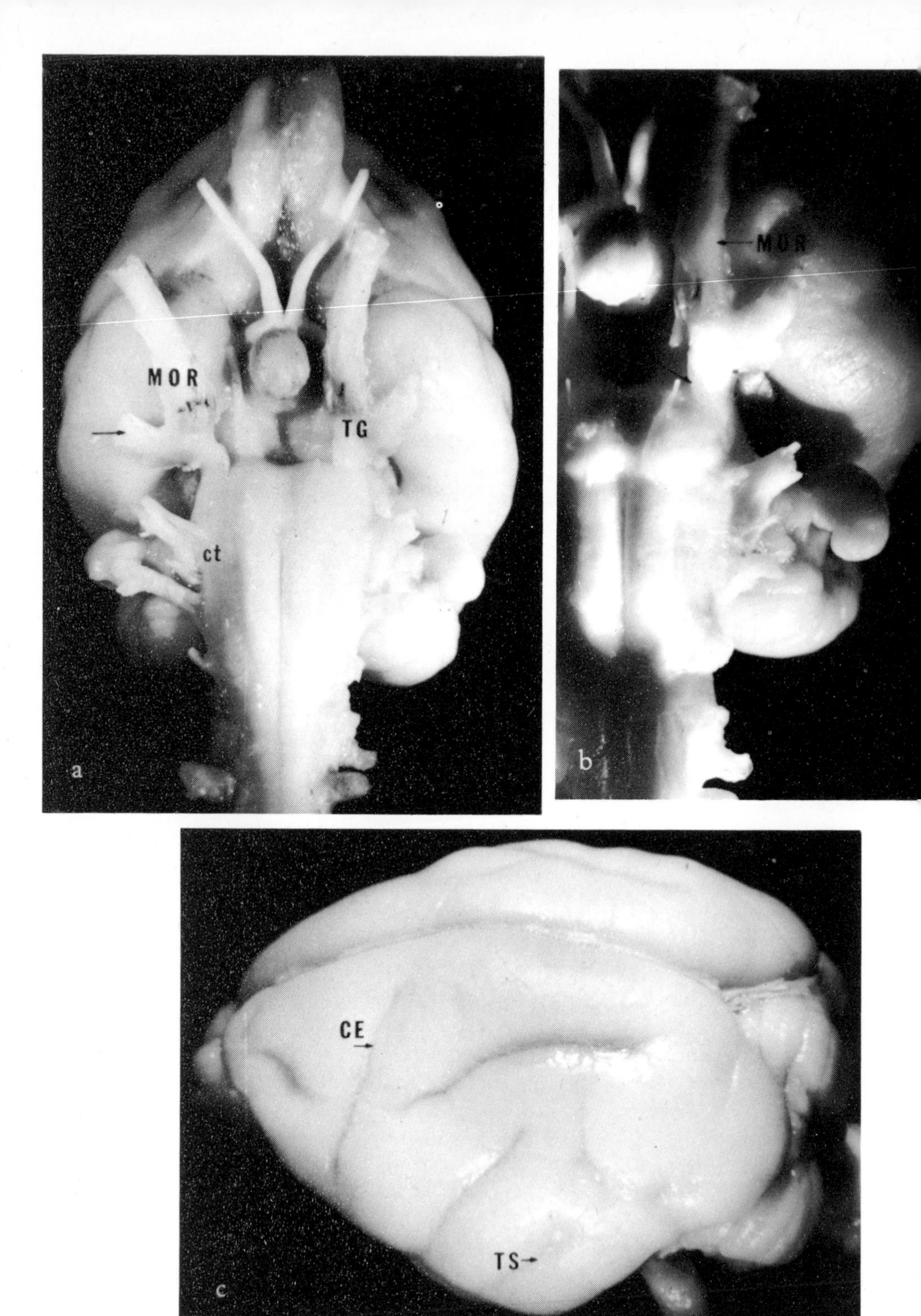

Figure 34. Photographs of the cerebral cortex and brain stem of *Perodicticus potto*.

(*a*) Ventral view of the cerebral cortex and brain stem. Note the maxillo-ophthalmic ramus (MOR) and the mandibular division (arrow) from the trigeminal ganglion (TG). Compare with Fig. 3.

(*b*) Ventral view of the brain stem. Note the small motor (arrow) and larger in the projections of the accessory optic system of some prosimians over that of *Tupaia* was recently published.[33]

It has been reported that in one of eighteen specimens of *N. coucang* the rootlets of the oculomotor nerve left the cerebral peduncle.[17] In all of the individuals of this species in our study, at least two rootlets of the nerve left the crus while the rest left from the interpeduncular fossa. We have seen oculomotor nerve rootlets leave the crus in only one human brainstem. This is considered to be an embryological variant in man, but it may describe the normal condition in the brain of *Nycticebus.*

Le Gros Clark discussed the hypothetical brain of a common progenitor of Primates.[12] This brain would have been lissencephalic except for calcarine, hippocampal and ectorhinal fissures and perhaps an indication of the suprasylvian fissure. Modern tree shrews have advanced little from this basic pattern.[9] *Tarsius spectrum* has only a small suprasylvian element on the orbital surface and a triradiate calcarine complex.[3] The brain of *Microcebus murinus*, although advanced in comparison to *Tupaia*, is still simple when compared to the other lemurs.[4,12] The brain of the lesser galago is advanced in its sulcal pattern when compared to the dwarf and mouse lemurs.[27]

Following the rule of gyrification,[6] the cortex of the animals in our study exhibit an increase in the variety and number of grooves and sulci. The lesser galago exhibits a more complex sulcal pattern than has previously been reported for the dwarf galago,[7,34] or lesser galago.[5,16]

By its sulcal pattern, the lesser galago shows differential cortical growth around the sylvian complex. This is seen in the development of the intraparietal sulcus and a deep superior temporal groove. A triradiate depression is present in the frontal region and a transverse groove noted in the occipital area. With the addition of sulcus *e* in the greater galago the same cortical pattern is seen, but the individual elements of the pattern are more distinct. This gives the impression that the smaller brain has been magnified. The slow loris and potto both have complex sulcal patterns. The majority of sulci are in the sagittal plane, although there are some located coronally (Figs. 32-4).

Compared to the general mammalian cortex, the lower primates may characterise trends of neocortical growth associated with modifications and development of visual and association cortex These modifications can be correlated with an increase in the number of coronally oriented sulci.[35]

The importance of a definitive description of cortical morphology, including dimples, fissurettes, and subzonal sulci is seen in the studies

sensory roots of the trigeminal nerve. The abducent nerves exit the pontopyramidal groove through the pyramids. Compare with Fig. 3.

(*c*) Dorsolateral view of the cerebral cortex. Note the central sulcus and the modifications dorsocaudal to the terminus of the superior temporal sulcus. Compare with Fig. 23.

of Welker and Campos,[36] Sanides and Krishnamurti,[8] and Le Gros Clark.[12] Although the approach of form analogy to suggest neocortical evolution has proved its merit, it relies too heavily on a strict orthogenetic interpretation. The interpretation of neocortical evolution initiated by Abbie,[37,38] and independently proposed by Sanides,[39] reveals a more fundamental correlate between gyrus formation and cortical cytoarchitecture in phylogeny. The expansion of the cortex may also be correlated with brainstem development. This may be associated with the size and complexity of the pons and cerebellum, and the development of the pyramids in these animals. Emphasis can now be placed on the evaluation of gross morphology in the light of combined comparative studies of fibre projections, cytoarchitecture and electrophysiology.

Acknowledgments

The authors wish to express their appreciation to Dr Y. Rumpler for generously supplying two brains of the mouse lemur. This paper was presented at the Research Seminar on Prosimian Biology at The Institute of Archaeology, April 14-17, 1972 by George E. Goode. Partial support for this project was provided by an A.D. Williams Grant (3558-550). The authors are especially indebted to Mrs M.P. Harrell for typing the preliminary and final drafts of this paper.

NOTES

1 Holloway, R.L. (1968), 'The evolution of the primate brain: some aspects of quantitative relations', *Brain Res.* 8, 1-52.

2 Elliott Smith, G. (1903), 'On the morphology of the brain in the Mammalia, with special reference to that of the lemurs recent and extinct', *Trans. Linn. Soc. Lond.* 8, 319-432.

3 Connolly, C.J. (1950), *External Morphology of the Primate Brain*, Springfield, Illinois.

4 Friant, M. (1970), 'Le cerveau des lémuriens de Madagascar', *Acta Neurol. Belg.* 70, 439-470.

5 Kanagasuntheram, R. and Leong, C.H. (1966), 'Observations on some cortical areas of the lesser bushbaby (*Galago senegalensis senegalensis*)', *J. Anat.* 100, 317-33.

6 Ariens Kappers, C.U., Huber, G.C. and Crosby, E.C. (1936), *The Comparative Anatomy of the Nervous System of Vertebrates, Including Man*, New York.

7 Zuckerman, S. and Fulton, J.F. (1941), 'The motor cortex in *Galago* and *Perodicticus*', *J. Anat.* 75, 447-56.

8 Sanides, F. and Krishnamurti, A. (1967), 'Cytoarchitectonic subdivisions of sensorimotor and prefrontal regions and of bordering insular and limbic fields in slow loris (*Nycticebus coucang coucang*)', *J. Hirnforschung* 9, 225-52.

9 Le Gros Clark, W.E. (1924), 'On the brain of the tree shrew (*Tupaia minor*)', *Proc. Zool. Soc. Lond.* 1053-74.

10 Le Gros Clark, W.E. (1929), 'The thalamus of *Tupaia minor*', *J. Anat.* 63, 177-206.

11 Le Gros Clark, W.E. (1926), 'The anatomy of the pen-tailed tree shrew (*Ptilocercus lowii*)', *Proc. Zool. Soc. Lond.* 1179-308.
12 Le Gros Clark, W.E. (1931), 'The brain of *Microcebus murinus*', *Proc. Zool. Soc. Lond.* 463-86.
13 Le Gros Clark, W.E. (1928), 'On the brain of *Macroscelides* and *Elephantulus*', *J. Anat.* 62, 245-75.
14 Le Gros Clark, W.E. (1932), 'The brain of the Insectivora', *Proc. Zool. Soc. Lond.* 975-1013.
15 Le Gros Clark, W.E. (1930), 'The thalamus of *Tarsius*', *J. Anat.* 64, 371-414.
16 Kanagasuntheram, R. and Mahran, Z.Y. (1960), 'Observations on the nervous system of the lesser bushbaby (*Galago senegalensis senegalensis*)', *J. Anat.* 94, 512-27.
17 Krishnamurti, A. (1966), 'The external morphology of the brain of the slow loris (*Nycticebus coucang coucang*)', *Folia primat.* 4, 361-80.
18 Kanagasuntheram, R., Wong, W.C. and Krishnamurti, A. (1968), 'Nuclear configuration of the diencephalon in some lorisoids', *J. Comp. Neurol.* 133, 241-68.
19 Tilney, F. (1927), 'The brain stem of *Tarsius*: A critical comparison with other primates', *J. Comp. Neurol.* 43, 371-432.
20 Tilney, F. (1928), *The Brain From Ape to Man*, New York.
21 Woollard, H.H. (1925), 'The anatomy of *Tarsius spectrum*', *Proc. Zool. Soc. Lond.* 1071-185.
22 Marsden, C.D. and Rowland, R. (1965), 'The mammalian pons, olive and pyramid', *J. Comp. Neurol.* 124, 175-88.
23 Gerhard, L. and Olszewski, J. (1969), 'Medulla oblongata and pons', in Hofer, H., Schultz, A.H. and Starck, D. (eds.) *Primatologia, Handbook of Primatology*, Basel.
24 Klüver, J. and Barrera, E. (1953), 'A method for the combined staining of cell and fibers in the nervous system', *J. Neuropath. Exp. Neurol.* 12, 400-3.
25 Hall, J. (1970), personal communication.
26 For midsagittal views, see Fig. 31 in Haines, D.E., this volume.
27 *Microcebus murinus* has been studied in our laboratory.
Affinities of Cheirogaleinae and Galaginae based on other characters have recently been discussed by Charles-Dominque, P. and Martin, R.D. (1970), 'Evolution of lorises and lemurs', *Nature* 227, 257-60.
28 Wood Jones, F. (1947-1950), 'The study of a generalized marsupial (*Dasycercus cristicauda*) (Krefft)', *Proc. Zool. Soc. Lond.* 26, 493-9.
29 Shantha, T.R. and Manocha, S.L. (1969),s'The brain of the chimpanzee (*Pan troglodytes*)', Basel.
30 Osman Hill, W.C. (1953), *Primates, Comparative Anatomy and Taxonomy*, I: *Strepsirhini*, Edinburgh.
31 Giolli, R.A. (1963), 'An experimental study of the accessory optic system in the cynomolgus monkey', *J. Comp. Neurol.* 121, 89-107.
32 Marburg, O. (1903), 'Basale Opticuswürzel und Tractus peduncularis transversus', *Arb. a.d. Neurol. Inst. a.d. Wien. Univ.* 10, 66.
33 Giolli, R.A. and Tigges, J. (1970), 'The primary optic pathways and nuclei of primates', in Noback, C.R. and Montagna, W. (eds.), *The Primate Brain*, New York.
34 von Bonin, G. (1945), *The Cortex of Galago*, Urbana, Illinois.
35 Le Gros Clark, W.E. (1947), 'Deformation patterns in the cerebral cortex', in *Essays on Growth and Form Presented to d'Arcy Wentworth Thompson*, Oxford.
36 Welker, W.I. and Campos, G.B. (1963), 'Physiological significance of sulci in somatic sensory cerebral cortex in mammals of the family Procyonidae', *J. Comp. Neurol.* 120, 19-36.

37 Abbie, A.A. (1940), 'Cortical lamination in the Monotremata', *J. Comp. Neurol.* 72, 428-67.

38 Abbie, A.A. (1942), 'Cortical lamination in a polyprodont marsupial, *Perameles natusa*', *J. Comp. Neurol.* 76, 509-36.

39 Sanides, F. (1942), 'Die Architektonik des menschlichen Stirnhirns', *Mon. Neurol. Psych.*, Berlin-Göttingen-Heidelberg, 98.

D.E. HAINES

The cerebellum of some Lorisidae

Introduction

Few portions of the mammalian central nervous system have undergone as obvious a development in the course of phylogeny as the cerebellum. Its enlargement in the higher mammals, especially the primates, is related to subsequent development of many important brain stem areas, for example the pons, ventrolateral nucleus of the thalamus, reticular formation, inferior olivary nucleus, red nucleus, and others.

Elliot Smith studied the cerebelli of *Lemur, Tarsius* and several other forms.[1] He proposed that the cerebellum be divided in the following manner: 'The mesial part of the cerebellum is divided into three regions by means of the deep *fissura prima* . . . and by the shallower *fissura secunda.* The region in front of the *fissura prima* (which is homologous with the so-called "preclival" fissure)[2] may be called the anterior lobe, that between the *fissura prima* and *fissura secunda* the central, or, better middle lobe; the third region being the posterior lobe.' Bolk studied the cerebelli of *Lemur, Nycticebus* and *Perodicticus*, and proposed a schema considerably different from that of Elliott Smith.[3] Subsequent investigators used the terminology of Elliott Smith.[1] Woollard and Tilney briefly discussed the cerebellum of *Tarsius*;[4,5] Le Gros Clark mentioned the cerebellum of *Tupaia, Ptilocercus* and *Microcebus*;[6–8] Kanagasuntheram and Mahran commented on the cerebellum of *Galago senegalensis*,[9] and Krishnamurti briefly mentioned the cerebellum of *Nycticebus coucang.*[10] With the exception of the work of Bolk,[3] the above-mentioned studies did not deal primarily with the cerebellum; consequently, this structure received little or no detailed attention. The above-mentioned papers have been reviewed elsewhere.[11]

Bradley, although he did not deal with the prosimian cerebellum, proposed a different scheme. He stated:[12,13]

> The simplest workable type of cerebellum was found to be one in which a mesial sagittal section shows five lobes separated from each other by four fissures. In order to avoid confusion, no pre-existing names were applied to these lobes and fissures, but the simplest method of designating them – that of letters and figures – was employed. The five lobes were called A, B, C, D, and E. . . . The fissures were similarly designated as I, II [*fissura prima*],[14] III, and IV.

This method did not receive wide support and was not adopted, to any extent, by subsequent investigators.

Over a period of years Larsell developed the method of dividing the mammalian cerebellum into three main lobes and ten lobules.[15,16] His technique was developed by studying the sequence of development of lobules and fissures in the embryo. The postlateral fissure (fissure IV of Bradley) is the first to appear in the embryo and it divides the developing cerebellum into the flocculonodular lobe and the *corpus cerebelli.* The primary fissure (*fissura prima* of Elliot Smith; fissure II of Bradley) is the second to develop and it divides the *corpus cerebelli* into anterior and posterior lobes. The anterior lobe is composed, in the adult, of lobules I to V; the posterior lobe of lobules VI to IX; the flocculonodular lobe is lobule X. A more detailed discussion of this method is given elsewhere.[11,15,16]

The method of Larsell has been applied to the cerebelli of early postnatal and adult specimens of *Galago senegalensis* and *Tupaia glis.*[11,17] In these studies it was concluded that the cerebellum of *G. senegalensis* has all the characters of higher primates, but at a basic level of organisation. The cerebellum of *Tupaia* has morphological characteristics which are seen both in non-primate mammals and in some basic primate forms.

The morphology of the deep cerebellar nuclei of prosimians has been virtually ignored. Tilney stated that the cerebellar nuclei of *Tarsius* 'emphasise their general resemblance to lower mammalian forms'.[5] The cerebellar nuclei of *Lemur catta, Avahi laniger, Loris tardigradus* and *Tarsius spectrum* have been divided into a *nucleus lateralis* (NL), *nucleus interpositus anterior* (NIA), *nucleus interpositus posterior* (NIP), and *nucleus medialis* (NM).[18] In *G. senegalensis* and *T. glis* the cerebellar nuclei are also divided into NL, NIA, NIP, and NM.[19] In *Galago* the nuclei are relatively easy to distinguish as separate cell masses, and the lateral nucleus (dentate) appears concave medially and shows a primitive hilus. However, in *Tupaia* the NL, NIA, and NIP are joined in caudal and ventral areas, and the lateral nucleus is merely an irregular mass of cells with no distinct hilus. In this respect, the morphology of the deep cerebellar nuclei of *Tupaia* closely resemble the basic pattern seen in the hedgehog and the mole.[18,19] The characteristic undulatory appearance of the lateral cerebellar nucleus of higher primates is exemplified at a basic level of organisation in the lesser *Galago*.

The cerebellar corticonuclear projections and cerebellar efferents of prosimians have not been investigated. The concept of medial, paravermal, and lateral functional zones, as originally proposed using the Marchii technique, has undergone some modification in deference to recent results of cerebellar corticonuclear studies using the newer silver degeneration techniques (see Discussion). The present paper will include a discussion of corticonuclear projections in the lesser *Galago* following lesions of the so-called lateral and medial functional zones.

Rationale

The cerebellum functions in the area of coordination and integration of unconscious proprioception with motor function. The animals most often used in a comparative study of cerebellar morphology or cerebellar corticonuclear connections represent divergent (and to a certain extent specialised) evolutionary forms such as the cat, rat and monkey. Not only are these animals divergent forms, but they have developed distinctly different locomotor patterns. There are undoubtedly neurological specialisations, coupled with osteological and myological specialisations, which are related to the evolution of the locomotor characteristics of each group.

The neurological basis for the evolution of bipedal locomotion has not been adequately explored. The family Lorisidae has members which illustrate different locomotor patterns, coupled with varying degrees of brain development. A study of the cerebellum as one centre related to postural orientation may provide clues relative to the neurological specialisations which enhanced the evolutionary development of bipedal locomotion. Some results of work on the cerebellum of the tree shrew (*Tupaia*) are included, since this animal may represent a living example of the primitive stock from which primates, and possibly all placental mammals, arose.

Materials and methods

The cerebelli of adult specimens of the greater bushbaby (*Galago crassicaudatus*), the lesser bushbaby (*Galago senegalensis senegalensis, G. s. moholi*), the slow loris (*Nycticebus coucang*), and the potto (*Perodicticus potto*) were used in this study. Several specimens of the tree shrew (*Tupaia glis*) were also used.[20] Gross examinations of the cerebelli were made under a dissecting microscope with a zoom lens. The gross drawings were made using calipers and metric rules, and repeatedly checked from all angles for accuracy.

The deep cerebellar nuclei of *G. senegalensis, G. crassicaudatus, N.*

coucang and *P. potto* were studied in transverse and horizontal serial sections. The method used in our laboratory for serial sectioning allows exact determination of all dimensions of the nuclei under consideration. The Klüver and Barrera method for combination staining,[21] and a cresyl acetate violet Nissl stain, were used to stain the deep nuclei and surrounding fibre bundles. All sections used in the present study were cut at 30 or 40 microns on a sliding microtome modified for frozen sections.

Lesions were placed in the cerebellar cortex of nine lesser bushbabies (*G. s. senegalensis, G. s. moholi*) and two tree shrews (*Tupaia glis*). The animals were anaesthetised intravenously with sodium pentobarbital. The nuchal musculature was retracted, a small area of occipital bone removed, the dura incised, and a small portion of cortex was aspirated. The animals were allowed to survive 3 to 5 days, then sacrificed by perfusion with saline followed with 10% formalin. The brain and spinal cord were removed *in toto.* The entire brainstem was serially sectioned and stained with the Fink and Heimer method for degenerating fibres.[22] Alternating sections were stained with a cresyl acetate violet method for cell bodies.[23] Projected drawings of the deep cerebellar nuclei of normal and lesioned animals were made on a Leitz Prado projector.

In the present study, horizontal sections are defined as those cut on a plane parallel to the floor of the fourth ventricle. A transverse section is cut on a plane perpendicular to the horizontal plane.

Results

1. Nomenclature

The nomenclature of the gross morphology of the cerebellum in this study is that of Larsell. The terminology of the cerebellar nuclei and their homologisation with the nuclei of other animals has been established by Rüdeberg.

2. Anterior lobe of the corpus cerebelli

The anterior lobe of the *corpus cerebelli* is composed of all portions rostral to the primary fissure. It can be divided into either four or five lobules; the merits of both approaches will be discussed below.

The anterior lobe of the cerebellum is well developed in all the Lorisidae examined in the present study. The anterior vermis is relatively large and, with the exception of a modest lobule I (lingula), the hemispheric portions of the lobules II through V are well developed.

(*a*) *Lobule I.* Vermian lobule I, the lingula, is small and composed of two or three folia. The lingula is larger in *G.*

crassicaudatus than in *P. potto, G. senegalensis*, and *N. coucang* (Fig. 1). In the latter three, it is composed of either a single lobule, or two sublobules. The principal pattern appears to be that of two sublobules. In the greater galago, the lingula is large and occasionally shows three sublobules. In greater and lesser galago, the lingula has no hemispheric extension (see later, Fig. 18).[11] However, the potto and slow loris have what appears to be a laterally attenuated portion of the lingula that may represent the hemisphere (HI) of the lingula in these animals. The precentral fissure separates the lingula from the ventral central lobule.

(*b*) *Lobules II and III.* Lobules II and III of the vermis represent the ventral and dorsal central lobules respectively. The ventral central lobule II in all specimens examined, with the exception of the greater galago, consists of a single folium (Figs. 3, 4; Haines[11] Fig. 1). In the greater galago, lobule II has two sublobules (II*a*, II*b*) with sublobule II*a* being small in most animals. The shallow fissure separating sublobules II*a* and II*b* does not extend to the cerebellar margin (see also Fig. 18). The predominant pattern appears to be that of a single folium for lobule II in the Lorisidae.

(*c*) *Lobule III*, the dorsal central lobule, is composed of sublobules III*a* and III*b* in slow loris, potto, and lesser bushbaby (Figs. 3, 4; Fig. 1 of Haines 1969).[11] In the greater galago, two or three sublobules may be present (Fig. 2). The intracentral fissure extends to the lateral cerebellar margin only in *G. crassicaudatus* (Fig. 2), while in the lesser galago, slow loris and potto it terminates before reaching the margin (Figs. 3, 4). The hemispheric portions of lobule II (HII) and III (HIII) are separated from each other by the intera-central fissure in greater galago. Since the intracentral fissure terminates prior to reaching the cerebellar margin, the hemisphere portions HII and HIII in potto, slow loris, and lesser bushbaby share a common area (HII + HIII).[24] The preculminate fissure extends to the lateral cerebellar margin separating the dorsal central lobule from the ventral culminate lobule.

(*d*) *Lobules IV and V.* The ventral and dorsal culminate lobules (IV and V respectively) are located between the preculminate and primary fissures and constitute the larger portions of the anterior lobe. In all genera studied, lobule IV consists of two sublobules and apposes the flocculus and/or middle cerebellar peduncle laterally (Figs. 2-4; Figs. 1, 2 of Haines).[11] The hemisphere area, HIV, is a single folium in all animals. Vermian lobule V is composed of two or three main sublobules in the slow loris, and in greater and lesser bushbabies (Figs. 2, 3). In the potto, lobule V had four sublobules and appears somewhat enlarged over the same area in the other genera. The intraculminate fissure extends to the lateral cerebellar margin, dividing a rather small HIV from the relatively well developed HV region. It is interesting to note that the hemispheric portions of the anterior lobe (HII-HV) of the Lorisidae of the present study are well developed, while in the tree shrew (Fig. 11 in Haines 1969)[11] and other basic forms this region is usually poorly developed.

List of abbreviations

Roman numerals associated with lower case letters, either in the text or in figures, indicate sublobules. Roman numerals in the discussion or in figures prefixed by an upper case H indicate the hemispheric portion of that specific vermian lobule.

Ac	anterior commissure
AMV	anterior medullary velum
ANS	ansiform lobule
APML	anterior paramedian lobule
CP	*copula pyramidis*
Cr I	crus I of the ansiform lobule
Cr II	crus II of the ansiform lobule
DPF	dorsal paraflocculus
F	flocculus
For	fornix
FB	fibre bundle (position of efferents of the lateral cerebellar nucleus)
FAP	ansoparamedian fissure
FIC	intracentral fissure
FICL	intraculminate fissure
FN-L	flocculonodular lobe
FNP	flocculonodular peduncle
FP	primary fissure
FPC	preculminate fissure
FPn	prenodular fissure (posterolateral fissure)
FPo	postpyramidal fissure
FPr	prepyramidal fissure
FPRC	precentral fissure
FPS	posterior superior fissure
h	hilus of the lateral cerebellar nucleus
Hab	habenula
ICP	inferior cerebellar peduncle
ICS	intercrural sulcus
LS	simple lobule (*lobulus simplex*)
MB	mammillary body
MCP	middle cerebellar peduncle
NIA	anterior interposed nucleus (*nucleus interpositus anterior*)
NIP	posterior interposed nucleus (*nucleus interpositus posterior*)
NL	lateral cerebellar nucleus (*nucleus lateralis*)
NM	medial cerebellar nucleus *(nucleus medialis)*
p	position of the parafloccular stalk
Pc	posterior commissure

Pi	pineal body
PL-CC	posterior lobe of the corpus cerebelli
PPML	posterior paramedian lobule
PS	parafloccular stalk
SCC	splenium of corpus callosum
SCP	superior cerebellar peduncle
VL	lateral vestibular nucleus
VM	medial vestibular nucleus
VPF	ventral paraflocculus
VS	superior vestibular nucleus
4th	fourth ventricle
5th	trigeminal nerve root
I	lingula
II	ventral central lobule
III	dorsal central lobule
IV	ventral culmen
V	dorsal culmen
VI	declive (*lobulus simplex*)
VII	folium tuber
VIII	pyramis
IX	uvula
X	nodulus

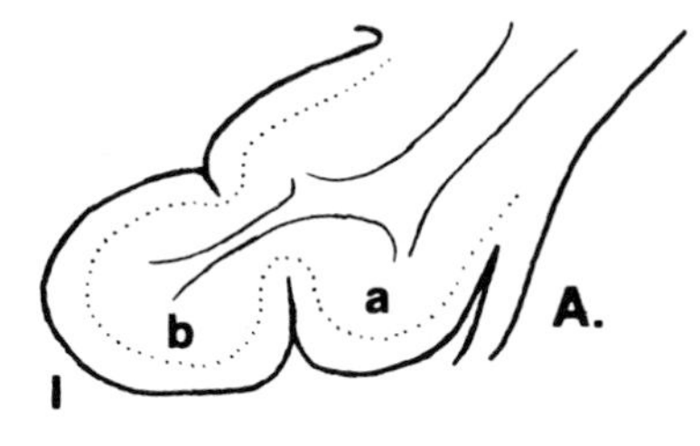

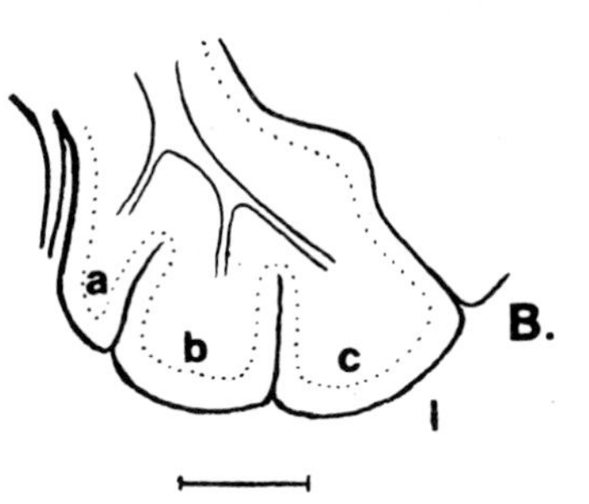

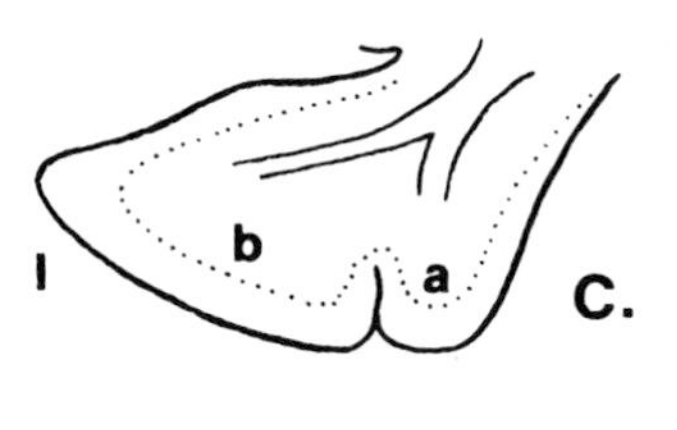

Figure 1. Mid-sagittal view of the lingula (lobule 1) of *P. potto* (*A*), *G. crassicaudatus* (*B*), and *N. coucang* (*C*). Scale = 1 mm.

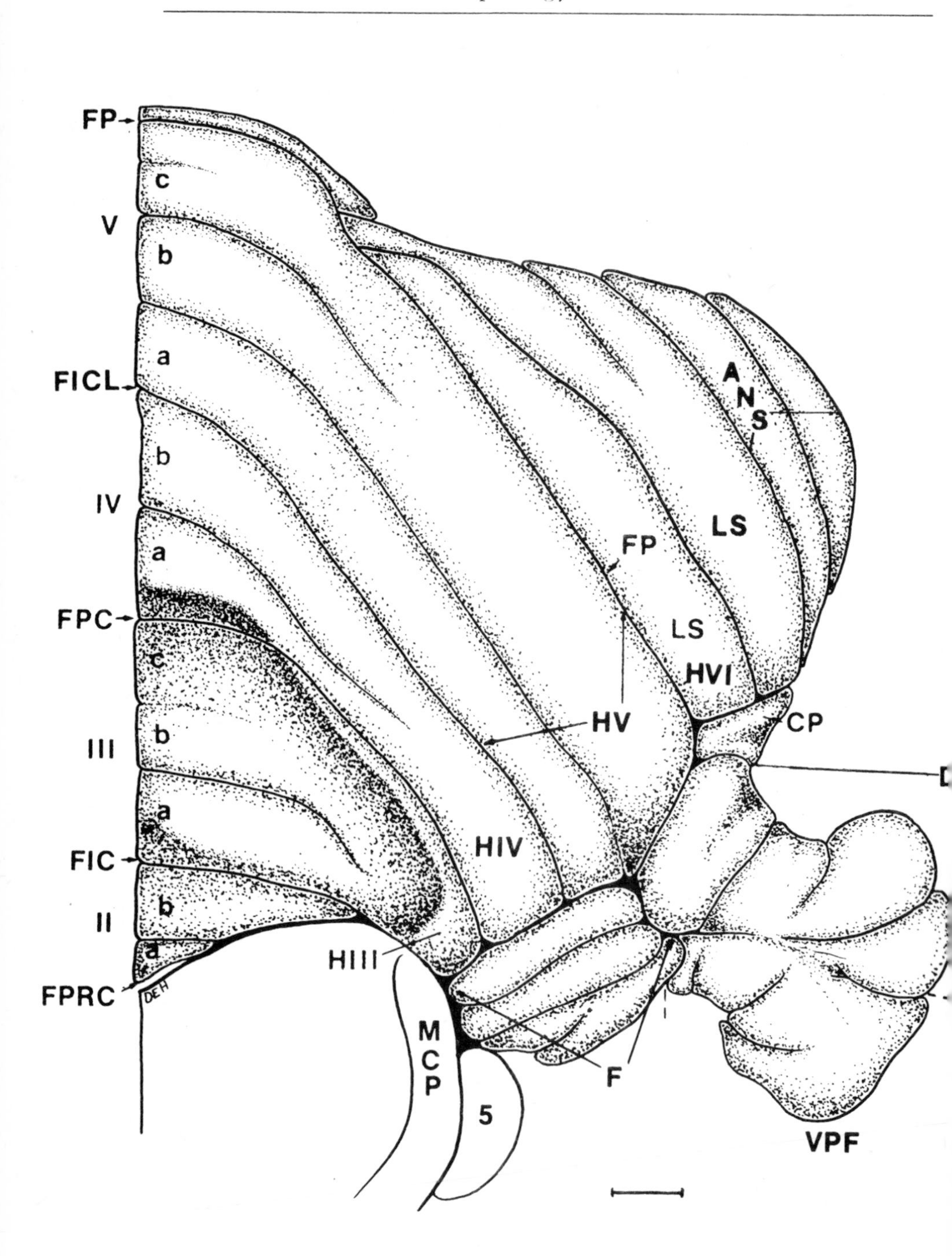

Figure 2. Rostral and slightly dorsal view of the anterior aspect of the cerebellum of *Galago crassicaudatus*. Note the indentation of the anterior lobe formed by the inferior colliculus. Scale = 1 mm.

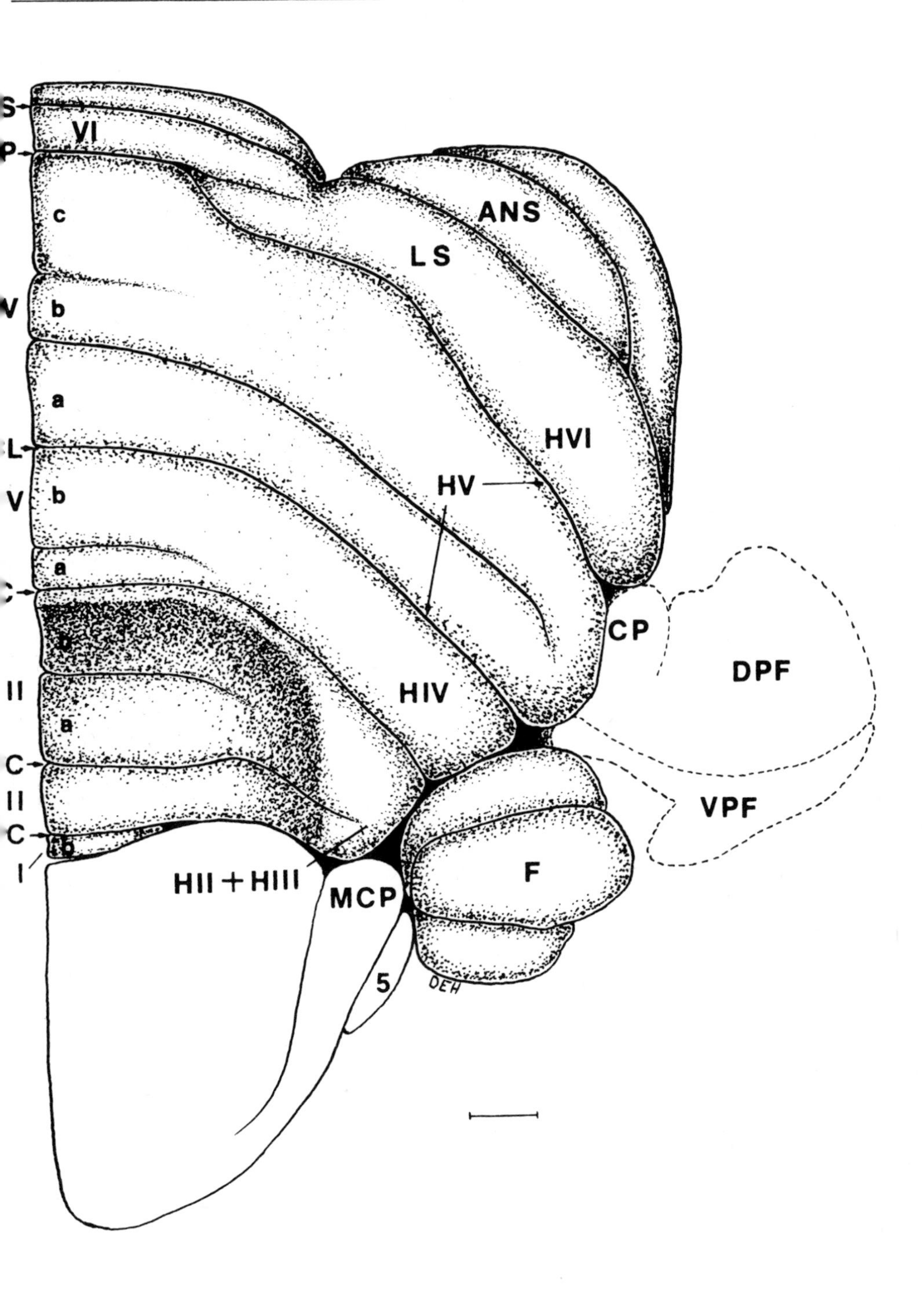

Figure 3. Rostral and slightly dorsal view of the anterior aspect of the cerebellum of *Nycticebus coucang*. Scale = 1 mm.

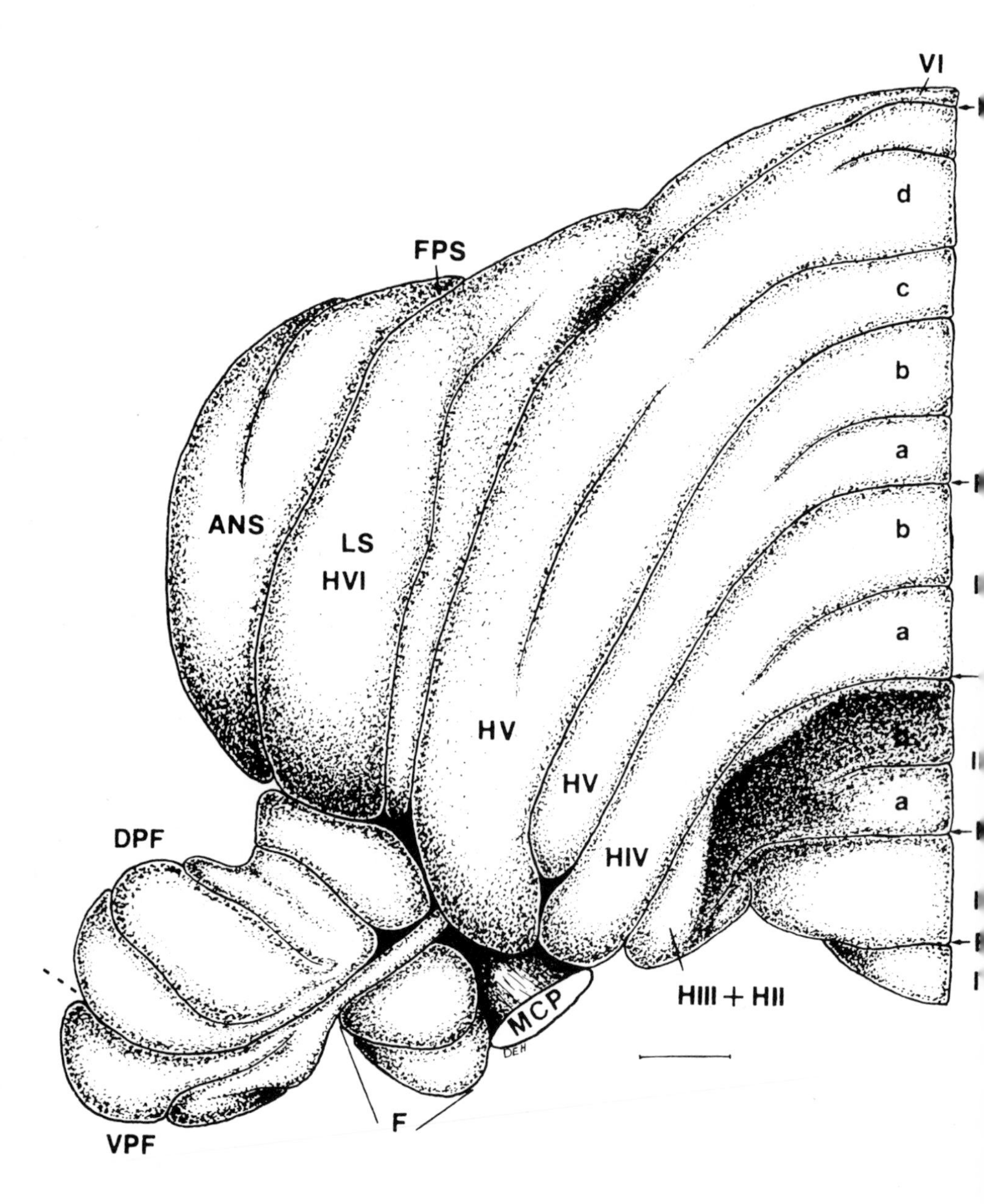

Figure 4. Rostral and slightly dorsal view of the anterior aspect of the cerebellum of *Perodicticus potto*. Scale = 1 mm.

3. *Posterior lobe of the corpus cerebelli*

The posterior lobe of the *corpus cerebelli* is located between the primary fissure and posterolateral fissure and it is the largest of the three principal lobes. The posterior lobe is divisible into a simplex lobule (VI), a folium-tuber region (VII), a pyramidal lobule (VIII) and the uvular lobule (IX).

(*a*) *Lobule VI.* The lobulus simplex in the vermis is a single folium, and its continuation with the hemisphere is distinct (Figs. 6, 8). Laterally, the HVI portion of the lobule is in contact with portions of the dorsal paraflocculus and *copula pyramidis.* The posterior superior fissure separates lobules VI and VII, and extends, almost uninterrupted, from the vermis to the lateral cerebellar margin. The hemisphere portion (HVI) of the lobulus simplex is usually a single folium of variable width in slow loris (Fig. 5), while in greater and lesser galago and potto this lobule is composed of two folia (Figs. 4, 6, 7). When two folia are present, the rostral folium is usually the narrowest (Figs. 4, 7).

Even though the vermis portion of the *lobulus simplex* appears relatively small in size, the HVI portion of this lobule in all genera is more extensive. In the slow loris, lobule HVI is not foliated on the surface; but several folia are present on the anterior aspect of the lobule (HVI) deep within the primary fissure (Fig. 9). A similar pattern to that shown in the loris (Fig. 9) is also observed on HVI within the primary fissure of the other genera studied.

(*b*) *Lobule VII.* Vermian lobule VII appears to be subdivided into two primary sublobules. These are divided by the ansoparamedian fissure into VIIA and VIIB. In its course from the vermis to the lateral cerebellar margin, the ansoparamedian fissure is discontinuous in the paravermal area in both greater and lesser bushbaby and potto (Figs. 6, 8, 12). In the slow loris it is a continuous fissure (Fig. 10). The hemisphere portions of VIIA and VIIB are the ansiform lobule and anterior paramedian lobule respectively. In potto and greater galago the ansiform lobule is made up of three folia, the rostral two folia interpreted as crus I and the caudal folium as crus II (Figs. 4, 6, 8). In the slow loris and lesser galago, the anisform lobule (HVII) is usually composed of two large folia, each interpreted as a crus (Figs. 7, 10, 11). In some specimens of greater galago, crus II of the ansiform lobule (HVII) is located on the caudal aspects of crus I deep within the ansoparamedian fissure (see inset, Fig. 6). Variations of the ansiform lobule in the lesser galago have been reported previously.[11]

The anterior paramedian lobule is a lateral continuation of vermian lobule VIIB. In the potto it appears to be composed of a wide folium subdivided by a shallow intrinsic sulcus. The continuity between vermian lobule VIIB and the anterior paramedian lobule is obvious (Fig. 8). The anterior paramedian lobule of the slow loris is slightly larger than that of the potto, and the continuation between this lobule and vermian lobule VIIB is relatively distinct (Fig. 10). The anterior paramedian lobule of the lesser galago is evaluated anew in light of the comparative data obtained in the present report on other Lorisidae. The ansoparamedian fissure extends to the cerebellar margin caudal to crus II of the ansiform lobule, and the prepyramidal fissure passes laterally around the hemisphere to join the former fissure. The

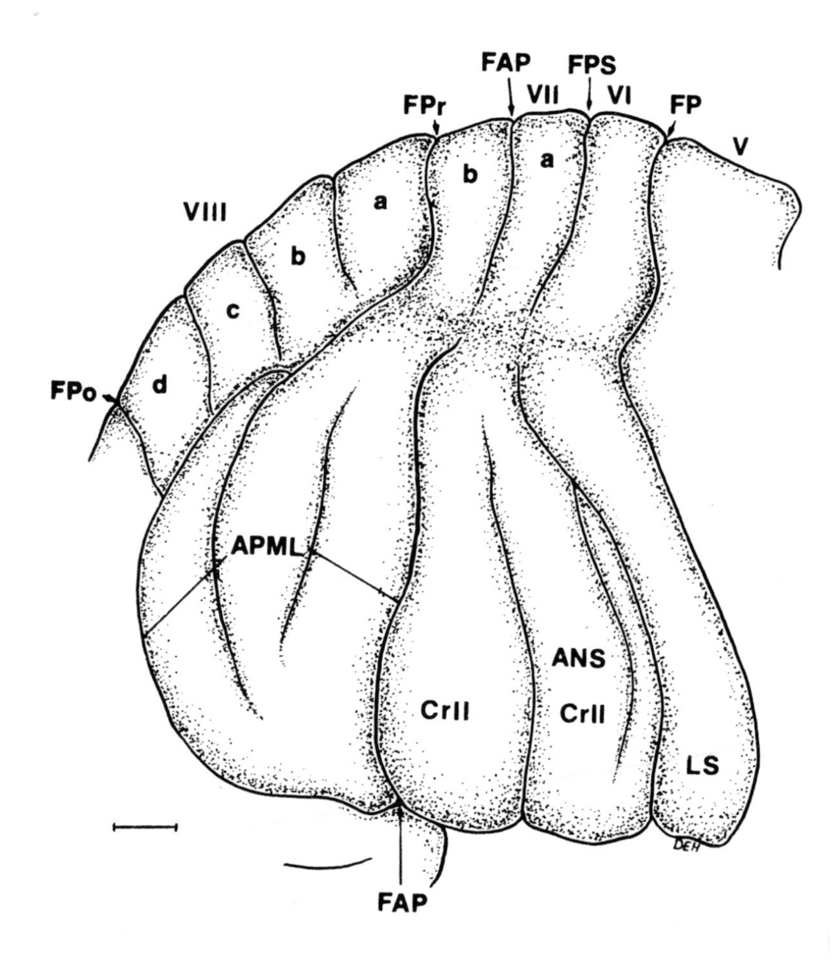

Figure 5. Dorso-lateral view of the hemisphere of the posterior lobe of the cerebellum of *Nycticebus coucang*. Note that some fissures (FAP and FPS) are discontinuous in the paravermal region. Scale = 1 mm.

anterior paramedian lobule is interpreted as the three folia between these fissures (Fig. 7). This would suggest that this lobule is slightly larger than previously interpreted.[11] In the greater galago the anterior paramedian lobule consists of four

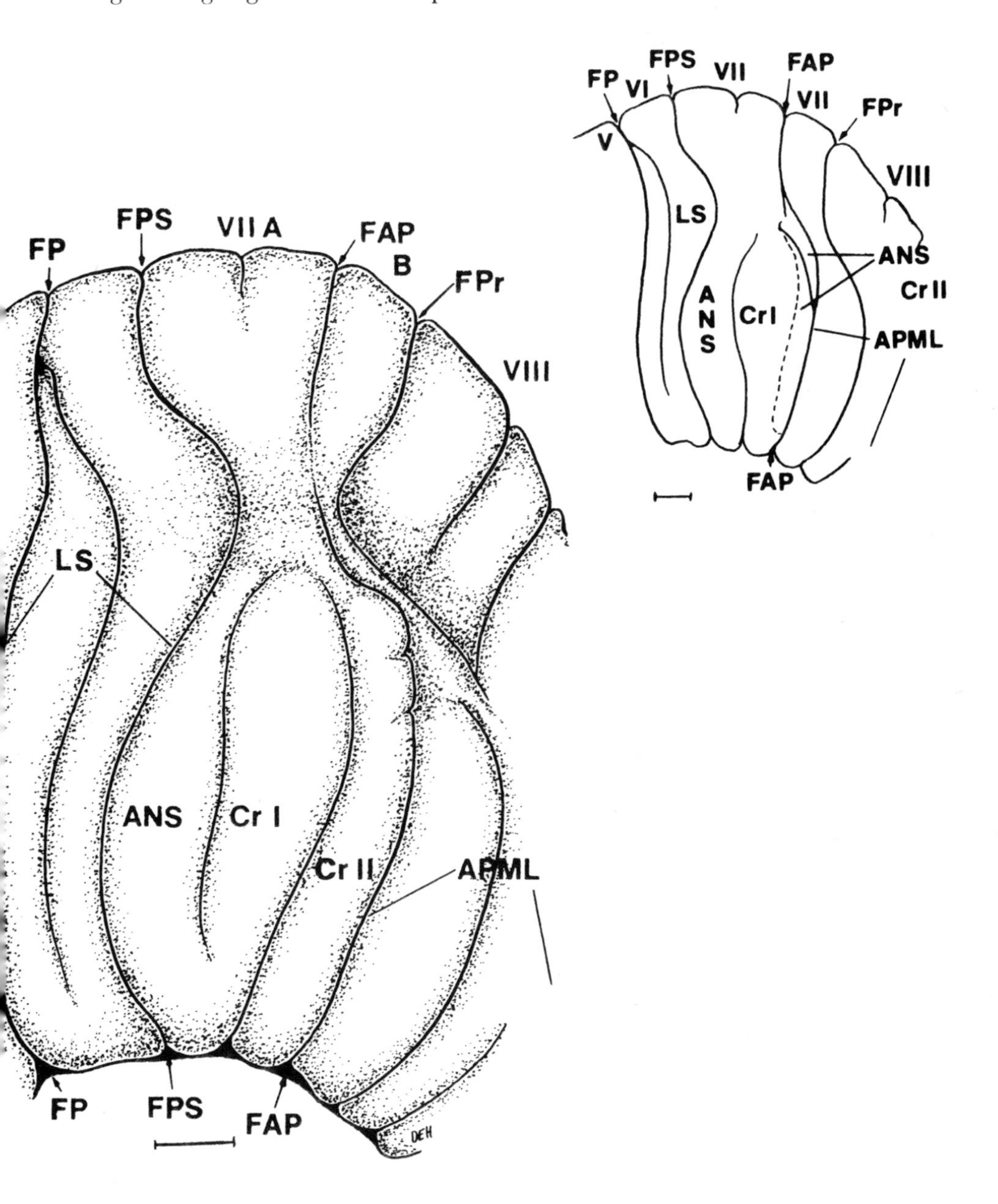

Figure 6. Dorso-lateral view of the hemisphere of the posterior lobe of the cerebellum of *Galago crassicaudatus*. In the inset, note that ansiform lobule crus II is located on the caudal aspect of crus I within the ansoparamedian fissure. Scale = 1 mm.

principal folia covering a large portion of the hemisphere of the posterior lobe (Figs. 12, 13). This conclusion is based on the fact that, when vermian lobule VIII is carefully dissected, no fibres from the pyramidal lobule seem to enter these folia. The observation is in accord with the embryological development of this area in other mammals.

(*c*) *Lobule VIII.* The prepyramidal fissure separates the *folium tuber* region from the pyramidal lobule, vermian lobule VIII. In the foetus, this fissure passes laterally and terminates by dividing the area of the future posterior paramedian lobule from the more rostrally located ansoparamedian lobule. The latter will subdivide into the ansiform and anterior paramedian lobule of the adult. Caudal to the prepyramidal

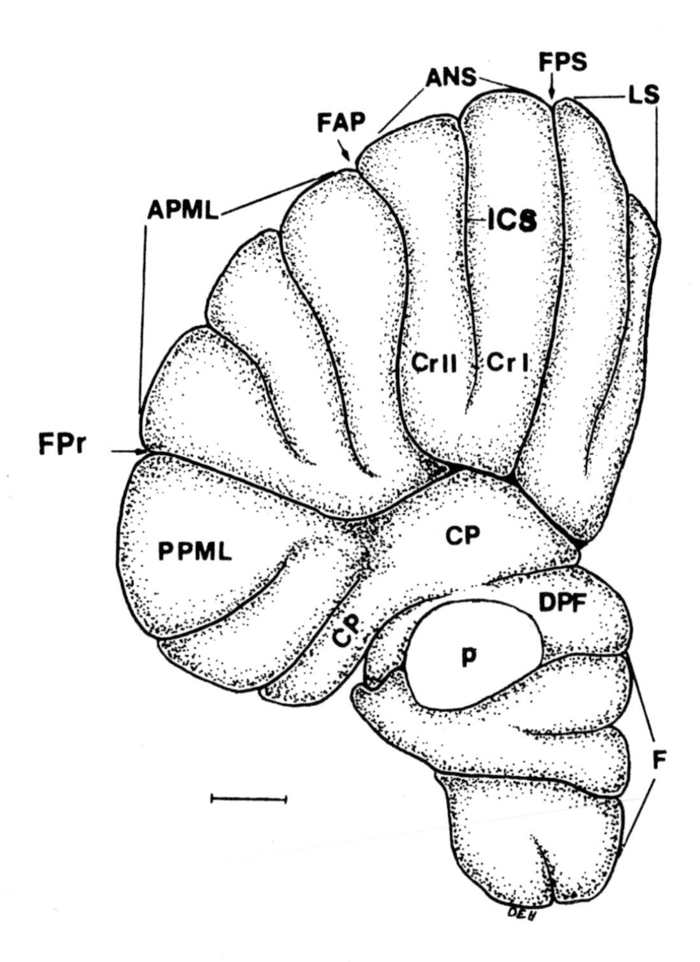

Figure 7. Lateral view of the cerebellar hemisphere of *Galago senegalensis.* The paraflocculus (p) has been removed. Note the anterior and posterior paramedian lobules and the continuation of the copula pyramidis with the dorsal paraflocculus. Scale = 1 mm.

fissure, in the foetus and adult, are the hemisphere portions (HVIII) of the pyramidal lobule. These are the posterior paramedian lobule, *copula pyramidis*, and the dorsal paraflocculus.

The prepyramidal fissure in the Lorisidae of the present

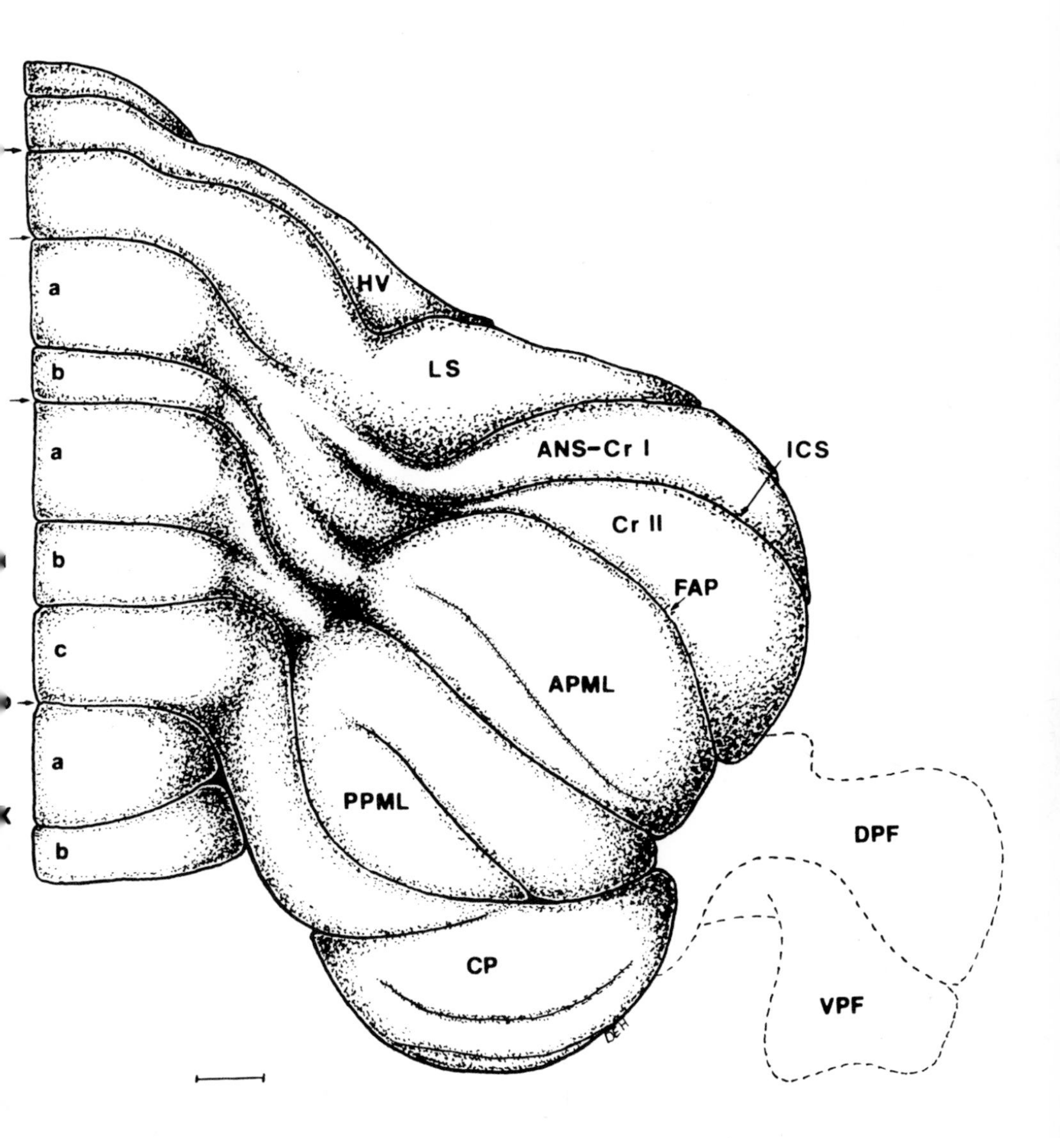

Figure 8. Caudal and slightly dorsal view of the posterior lobe of the cerebellum of *Perodicticus potto*. Note the obvious continuations of vermian lobules, V, VI, VII, and VIII with their respective portions of the hemisphere. The paraflocculus is outlined to scale. Scale = 1 mm.

study extends from the vermis to the lateral cerebellar margin. In the area of the paramedian fissure, it is somewhat obliterated since it travels from the vermis into the paramedian fissure, where it turns abruptly caudal before passing laterally between the anterior and posterior portions of the paramedian lobule.

The posterior paramedian lobule in the slow loris and potto is made up of two folia; in the lesser galago it may be one or two folia, but in the greater galago three folia are usually present (Figs. 7, 8, 10, 13). The posterior paramedian lobule is medially continuous with the base of vermian lobule VIII. The remaining portions of HVIII are the *copula pyramidis* and the dorsal paraflocculus. In the potto, the lower portion of lobule VIII is continuous into the lower to

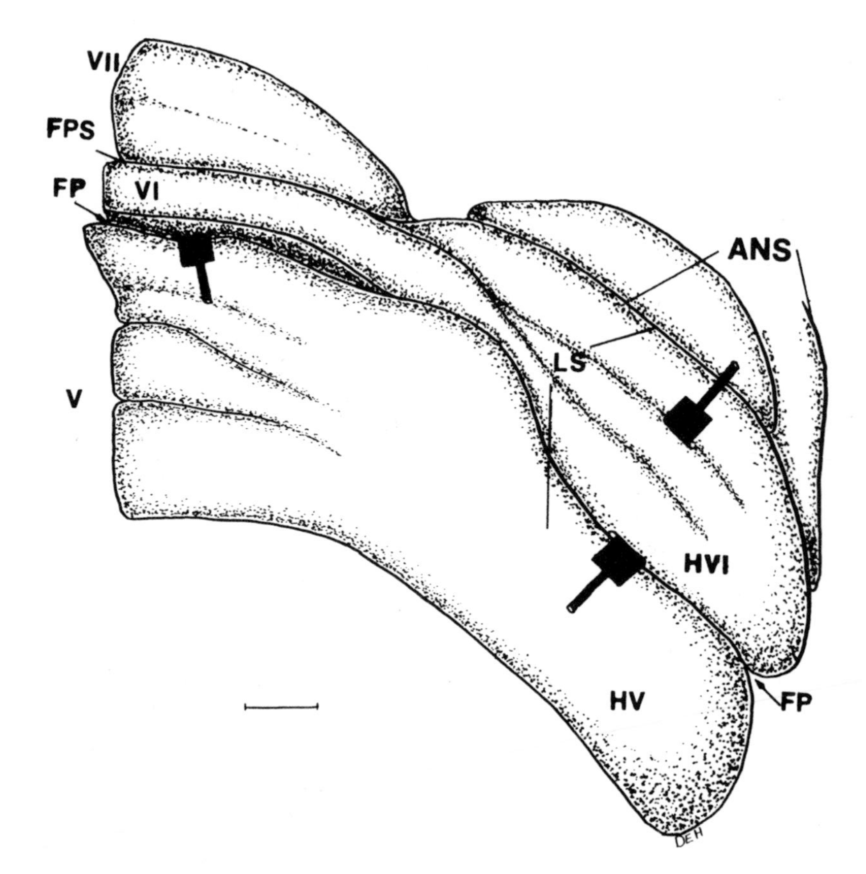

Figure 9. Anterior view of lobules V, VI, and VII of *Nycticebus coucang*. The caudal lip of lobule V is retracted so that the primary fissure is opened. Note the fissuration of the *lobulus simplex* within the fissure. See Results. Scale = 1 mm.

three folia which pass laterally toward the stalk of the dorsal paraflocculus (Figs. 8, 15). Because of its obvious continuation with the pyramidal lobule this area is interpreted as *copula pyramidis* in the potto (Fig. 8). The cortex portion representing the *copula pyramidis* in the slow loris and greater galago is somewhat reduced in size and located ventral to the posterior paramedian lobule (Figs. 10, 13). In both greater galago and slow loris, this cortical area consists of two

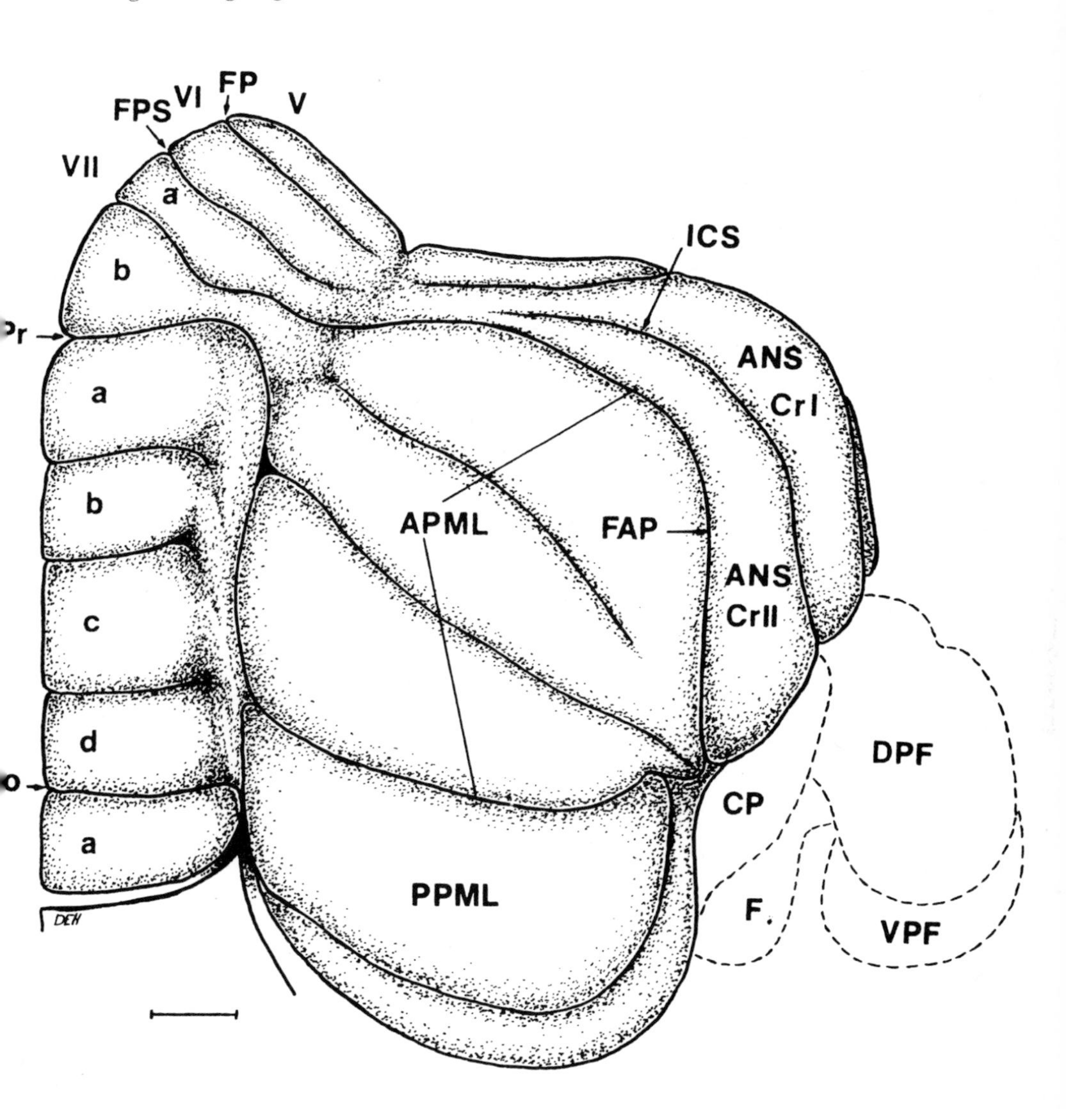

Figure 10. Caudal and slightly dorsal view of the posterior lobe of the cerebellum of *Nycticebus coucang*. Note the distinct continuation of vermian lobules VI and VII with their respective portions of the hemisphere. Scale = 1 mm.

folia (Figs. 11, 13, 18, 19). In the slow loris, the division of the copula is ventral with the lateral-most folium passing laterally to join with the dorsal paraflocculus (Figs. 16, 19). In the greater galago, a moderately-sized folium passes from ventral to lateral and a second flattened folium extends onto the parafloccular stalk in close association with the dorsal paraflocculus (Figs. 13, 18). In general, the appearance of the copula of greater galago and slow loris is similar, while the same area in potto is tentatively interpreted as being somewhat larger (Figs. 8, 15). These observations lead to a reconsideration of the same area in the lesser galago. The *copula pyramidis* had been interpreted as the large area on the ventral aspect of the posterior lobe hemisphere.[11] When the limits of the anterior paramedian lobule were reconsidered and subsequently modified, the area of the copula was also reconsidered. In view of the consistent appearance of this area in the greater galago and slow loris, it was concluded that the copula in the lesser galago may possibly be smaller. The pyramidal lobule and surrounding structures were dissected and fibre bundles followed into hemisphere folia. The posterior paramedian lobule is the lower two folia of the hemisphere in the lesser galago and the copula is the

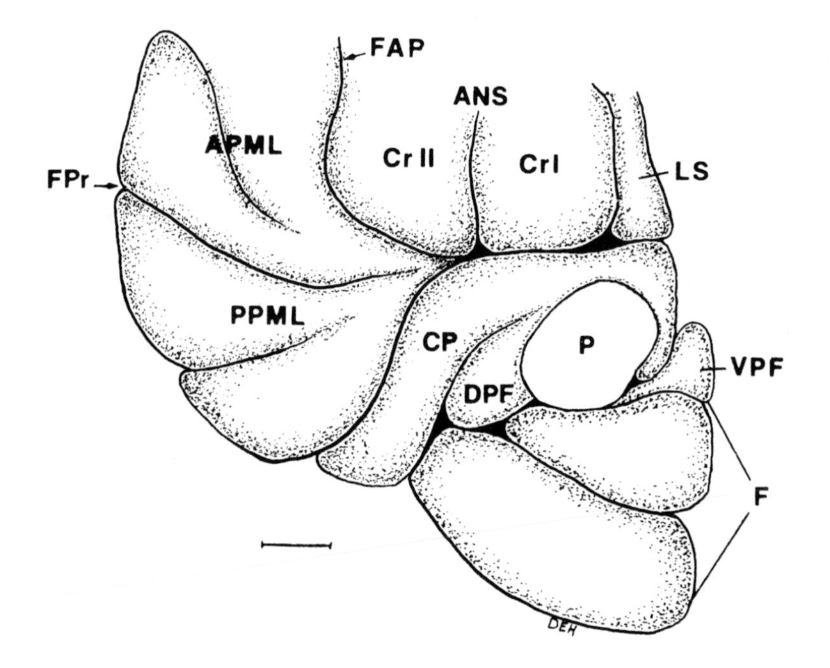

Figure 11. Lateral view of the ventrolateral aspect of the posterior lobe of the cerebellum of *Nycticebus coucang*. Note the continuation of the *copula pyramidis* onto the dorsal paraflocculus. The paraflocculus has been removed. Scale = 1 mm.

lowermost folium on the ventral aspect (Figs. 7, 21). This relatively large folium passes laterally into the caudodorsal aspect of the dorsal paraflocculus where it is broad and flattened (Fig. 7). In general, the configurations of the greater and lesser galagos resemble each other on the

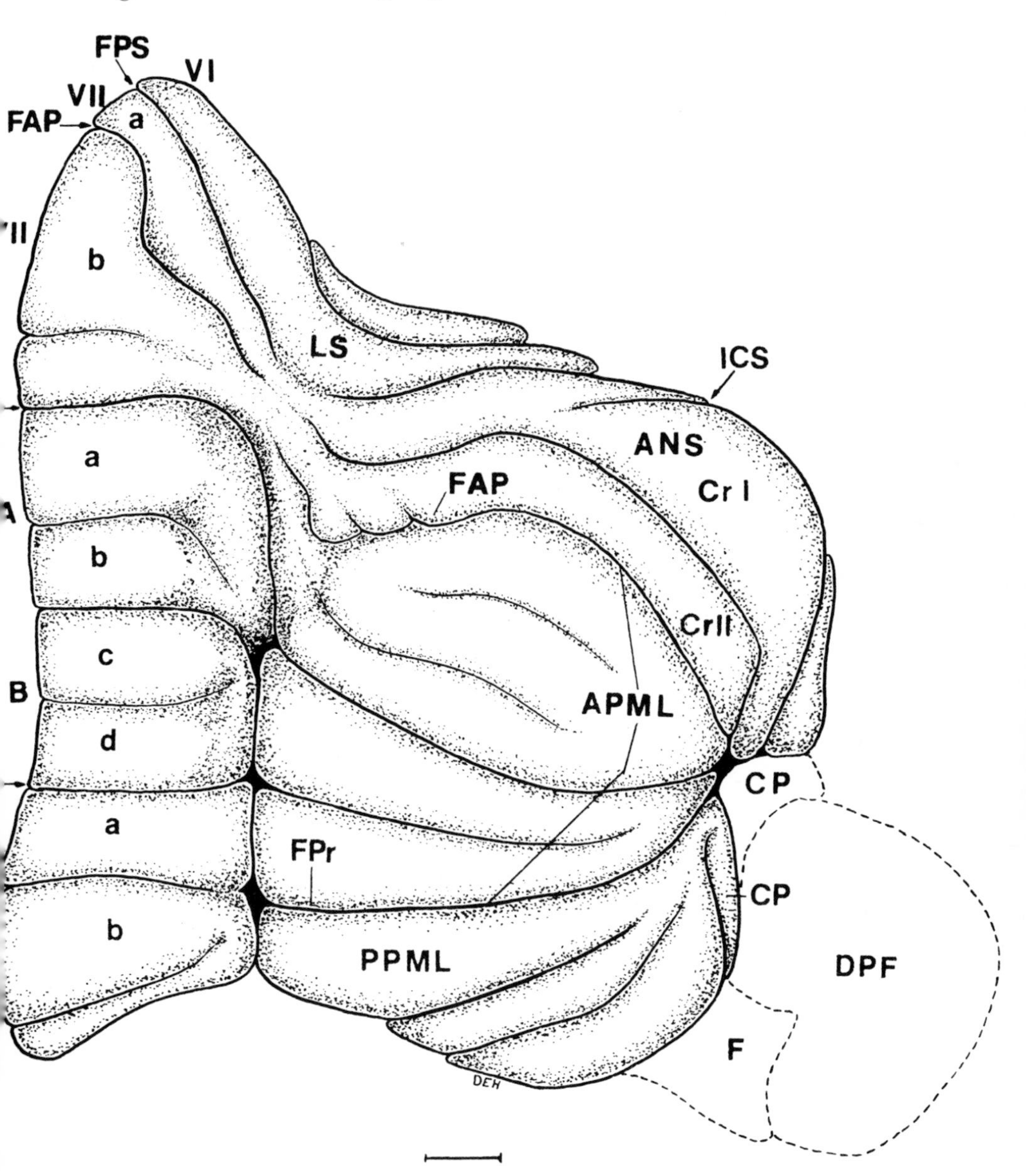

Figure 12. Caudal and slightly dorsal view of the posterial lobe of the cerebellum of *Galago crassicaudatus.* The continuations of vermian lobules V, VI, and VII with their respective areas of the hemisphere is distinct. Note the large anterior paramedian lobule. Scale = 1 mm.

ventrolateral aspect of the posterior lobe of the hemisphere (Figs. 7, 13).

The most lateral portion of HVIII is the dorsal paraflocculus. In all genera studied it is only modestly developed, being composed of two to three principal folia (Figs. 14, 16). The dorsal paraflocculus is slightly larger in the potto than in the other genera (Fig. 15). The division between dorsal and ventral portions is distinct on the rostral surface, but somewhat obliterated on the caudoventral surface. In the slow lorises of the present study a slight variation in the size and shape of the paraflocculi is noted (Fig. 17), although they did show a distinct general pattern.

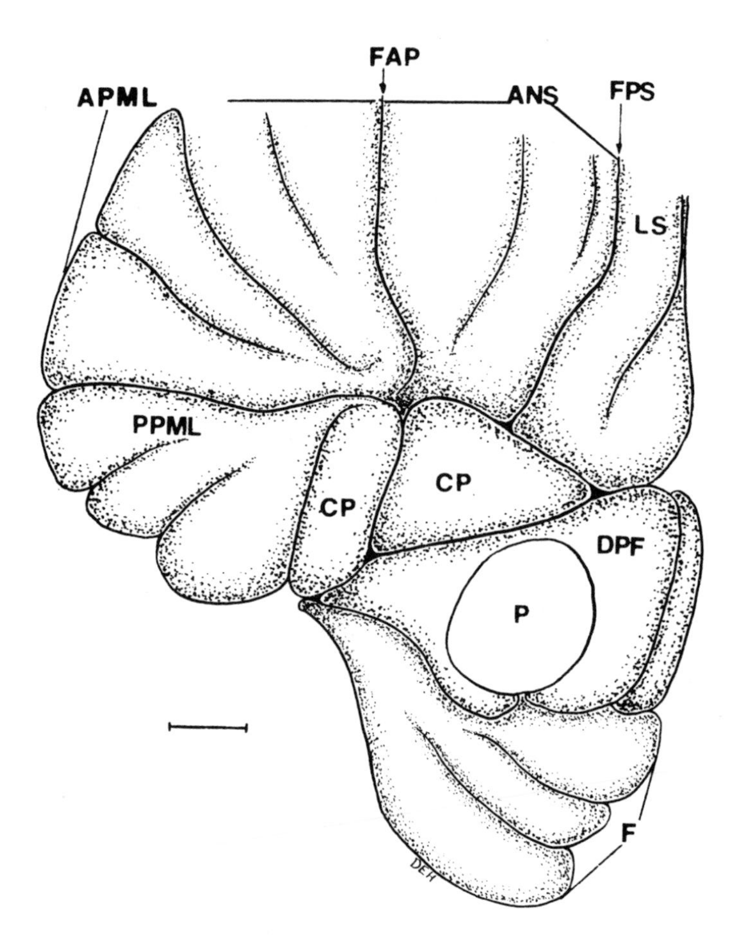

Figure 13. Lateral view of the ventrolateral aspect of the posterior lobe of the cerebellum of *Galago crassicaudatus*. Note the large anterior paramedian lobule, and the arrangement of the *copula pyramidis*. The paraflocculus has been removed. Scale = 1 mm.

(*d*) *Lobule IX.* The postpyramidal fissure extends laterally from the vermis to terminate by dividing the paraflocculus into dorsal and ventral portions. The fissure is continuous from the midline into the hemisphere, but its extreme lateral course is somewhat obliterated by the relation of the paraflocculus and development of the posterior lobe cortex during morphogenesis. Vermian lobule IX, the uvula, is

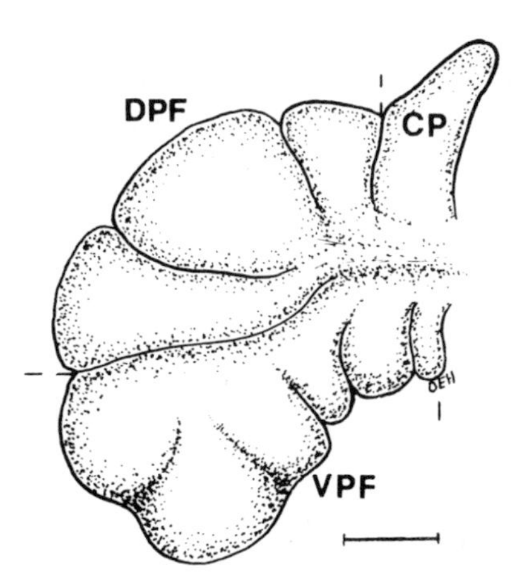

Figure 14. Dorsal view of the paraflocculus of *Galago crassicaudatus.* Note the dorsal and ventral portions, and the intimate association of the *copula pyramidis* with the dorsal paraflocculus. Scale = 1 mm.

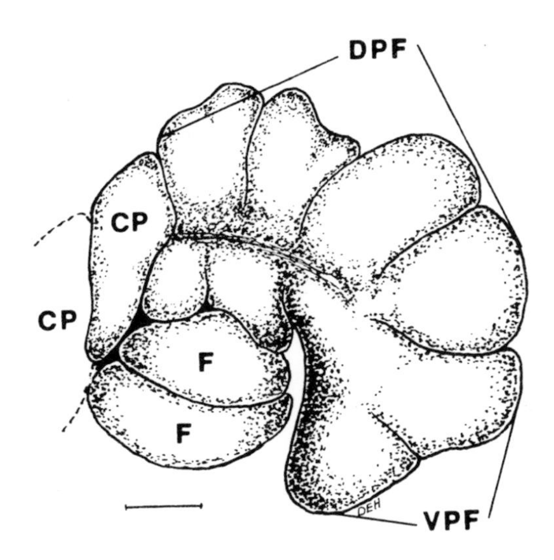

Figure 15. Dorsolateral view of the paraflocculus of *Perodicticus potto.* Note that the dorsal division is slightly larger in the potto than in the other animals of the present study. Scale = 1 mm.

composed of three main sublobules in all the Lorisidae studied, with the exception of the potto. In the latter, the uvula consists of two elongated sublobules.

The only hemispheric representation of the uvula is the ventral paraflocculus, lobule HIX, and in the adult specimens, the continuation between these two structures is reduced to a bundle of fibres located deep with respect to the flocculonodular bundle (Figs. 6, 14 in Haines).[11] The ventral paraflocculus is also only moderately developed. In the potto it consists of four folia (Fig. 15), while in the greater and lesser galago and slow loris it is somewhat smaller, being composed of two to three main folia (Figs. 14, 16, 17; Fig. 7 in Haines).[11]

4. *Flocculonodular lobe*

The prenodular or posterolateral fissure (of Larsell) separates the uvula from the nodulus in the vermis and laterally separates the flocculus from the ventral paraflocculus. The flocculus in the slow loris, potto, and greater galago is composed of two to four rostro-caudally elongated folia (Figs. 4, 11, 13, 18, 20). In the lesser bushbaby the folia are somewhat shortened but four are usually present (Fig. 7). The nodulus is relatively simple in all specimens. It appears to have two main sublobules in greater and lesser bushbabies and slow loris. In the potto it is slightly larger and has three sublobules.

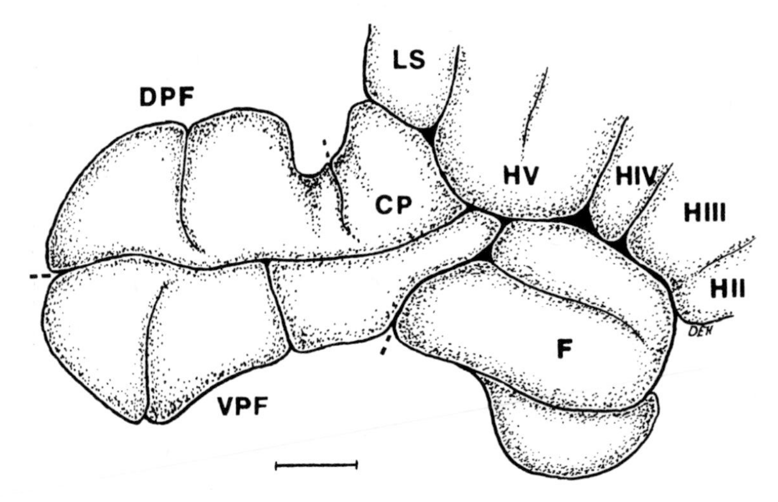

Figure 16. Rostral view of the paraflocculus, flocculus, and ventrolateral portions of the hemisphere of *Nycticebus coucang.* The separation between dorsal and ventral divisions of the paraflocculus is distinct. Note the continuation of the *copula pyramidis* onto the dorsal paraflocculus. Scale = 1 mm.

The flocculus is the hemispheric portion of lobule X, (HX), and it is connected with the nodulus by a medullated bundle of fibres, the flocculonodular peduncle (floccular peduncle of Larsell). This connection is a low ridge of tissue in the slow loris and greater galago, in which delicate striations are visible (Figs. 18, 19). In the potto, however, only a low ridge is evident and no striations can be seen (Fig. 20). In the lesser bushbaby the connections are similar to those seen in the greater galago and slow loris (Fig. 21). This striated ridge passes directly between the most ventral portions of the flocculus and the nodulus of the vermis. The gross separation of the fibres of the flocculonodular peduncle and the fibres passing from the uvula (IX) to the ventral paraflocculus (HIX) is obliterated in the adult.[11,17]

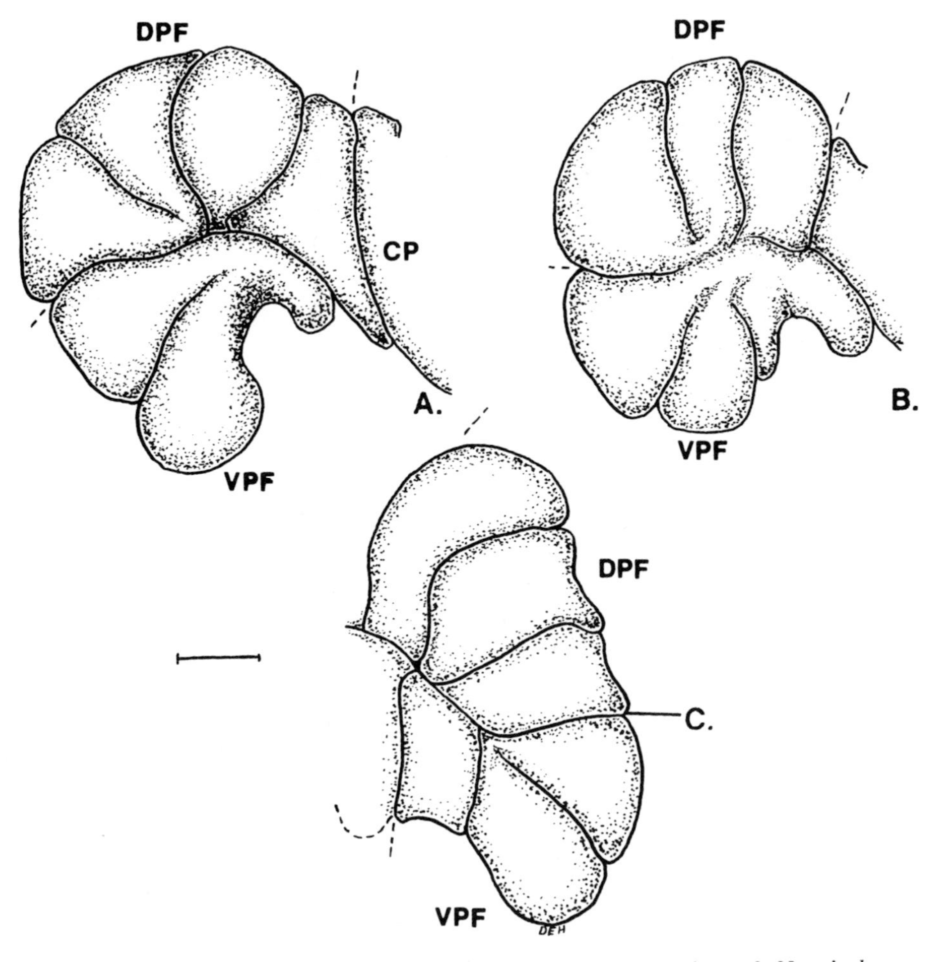

Figure 17. Variations in the structure of the paraflocculus of *Nycticebus coucang.* Dorsal views (*A, B*) to show the slight variation in the size of dorsal and ventral divisions. Ventral and slightly caudal view (*C*) of the paraflocculus. Scale = 1 mm.

5. *Cerebellar nuclei of Lorisidae*

The deep cerebellar nuclei are the principal centres for the origin of efferent information passing out of the cerebellum. In many mammalian forms, the deep nuclei are divisible either partially or totally into a *nucleus lateralis* (NL), a *nucleus interpositus anterior*

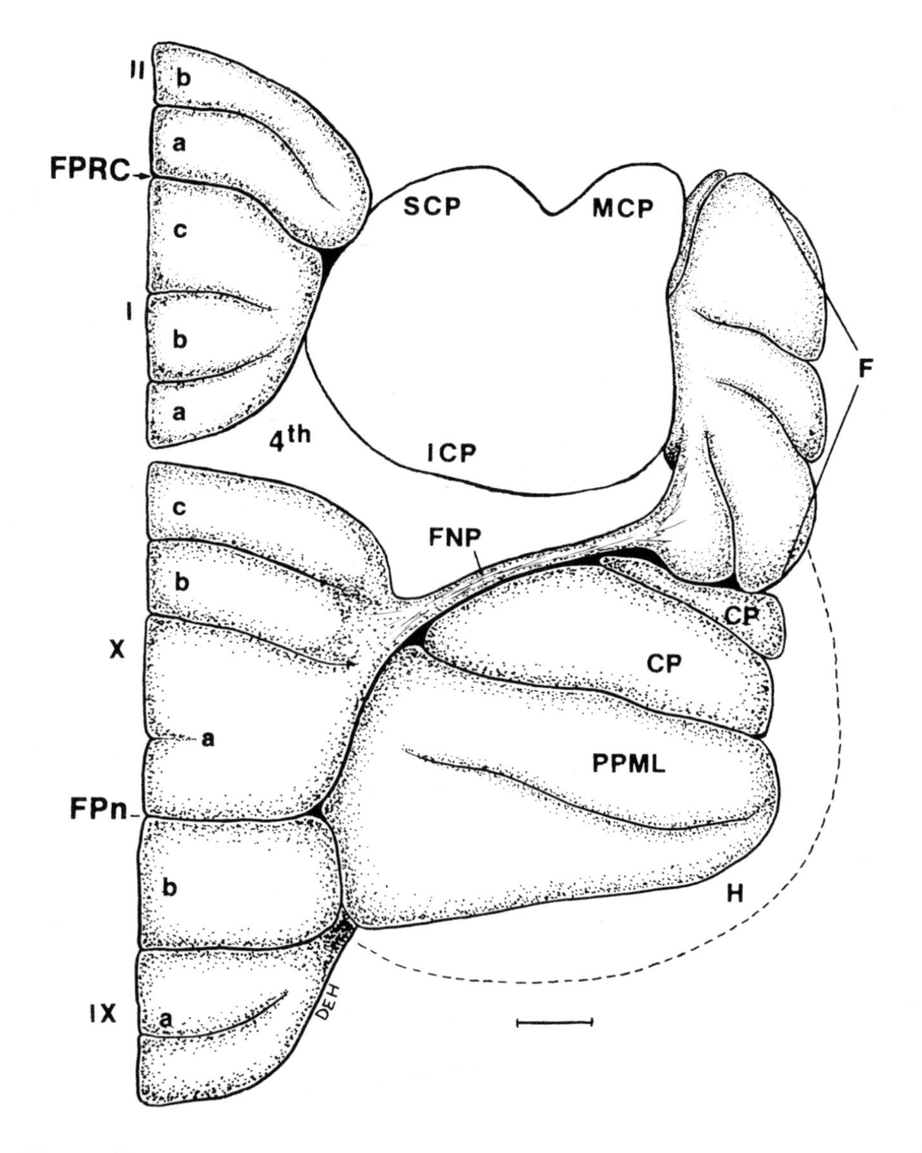

Figure 18. Ventral view of the cerebellum of *Galago crassicaudatus.* The cerebellar peduncles (SCP, MCP, ICP) have been cut. Note the nodulus (X), the flocculus (IX) and the flocculonodular peduncle. The lingula shows no hemisphere, and the hemisphere area of the ventral central lobule is limited. The posterior paramedian lobule and two divisions of the *copula pyramidis* are clearly evident. Scale = 1 mm.

(NIA), a *nucleus interpositus posterior* (NIP) and a medial cerebellar nucleus (NM), on each side.

(*a*) *Nucleus lateralis.* The lateral cerebellar nucleus of slow loris and potto closely resemble each other (Figs. 22, 23). The NL of the slow loris measures approximately 2500 by 1900 microns at its greatest diameter and about 2400 microns at

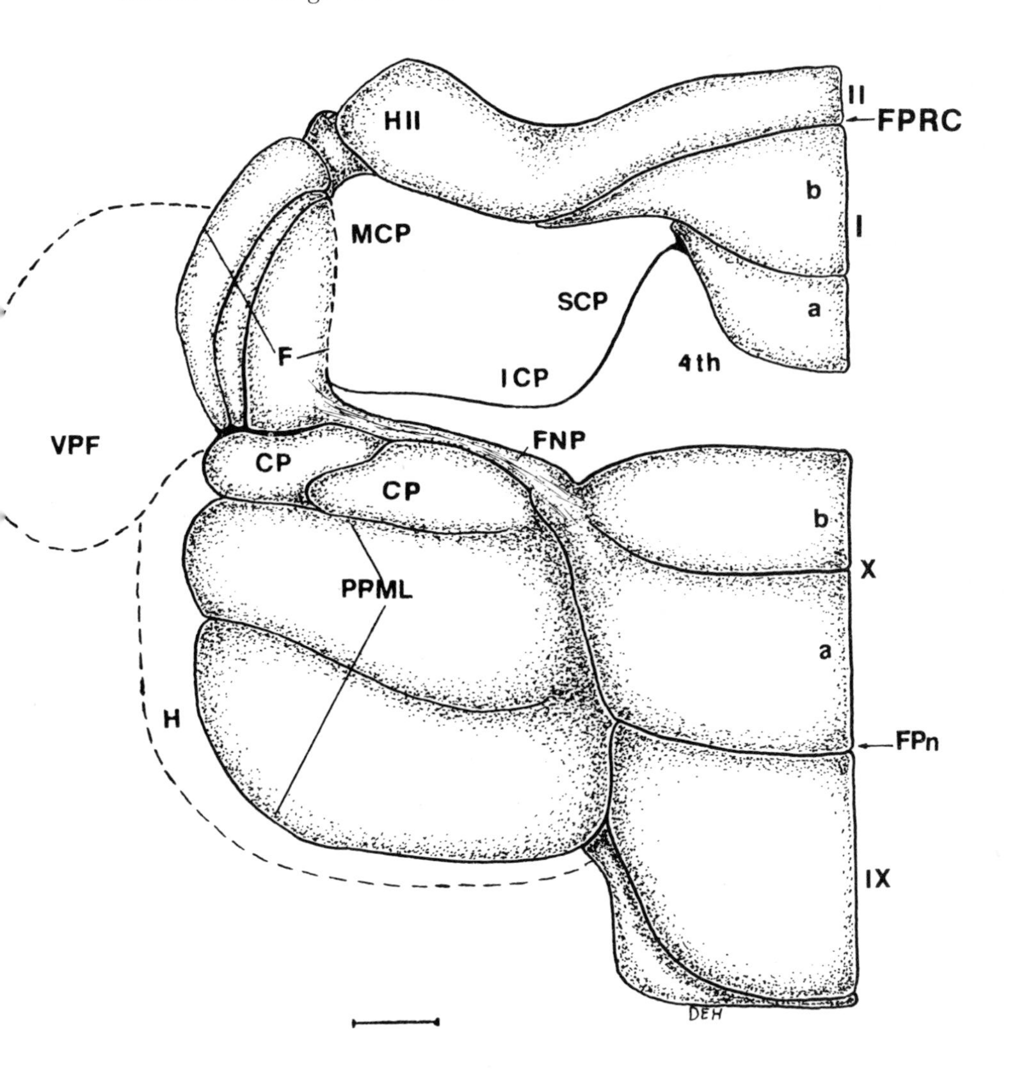

Figure 19. Ventral view of the cerebellum of *Nycticebus coucang*. The cerebellar peduncles (SCP, MCP, ICP) have been cut, exposing the entire ventral area. Note the orientation of the folia of the flocculus, the size of the nodulus, and the striations of the flocculonodular peduncle. The posterior paramedian lobule is large and the two divisions of the *copula pyramidis* are distinct. There is a well developed HII area, and lobule I has a small lateral projection (see Discussion). Scale = 1 mm.

its greatest rostro-caudal length. In the potto the NL is of similar size, measuring about 2600 by 1800 microns at its greatest diameter and approximately 2200 microns at its greatest rostro-caudal length. Throughout their dorsal area, the NL and NIA are joined so closely that it is not possible to define the borders of each nucleus exactly. At approximately mid-horizontal level, the NL and NIA are separated from each other by a delicate lamina of fibres in their middle third (Fig. 23). This separation becomes more evident toward the

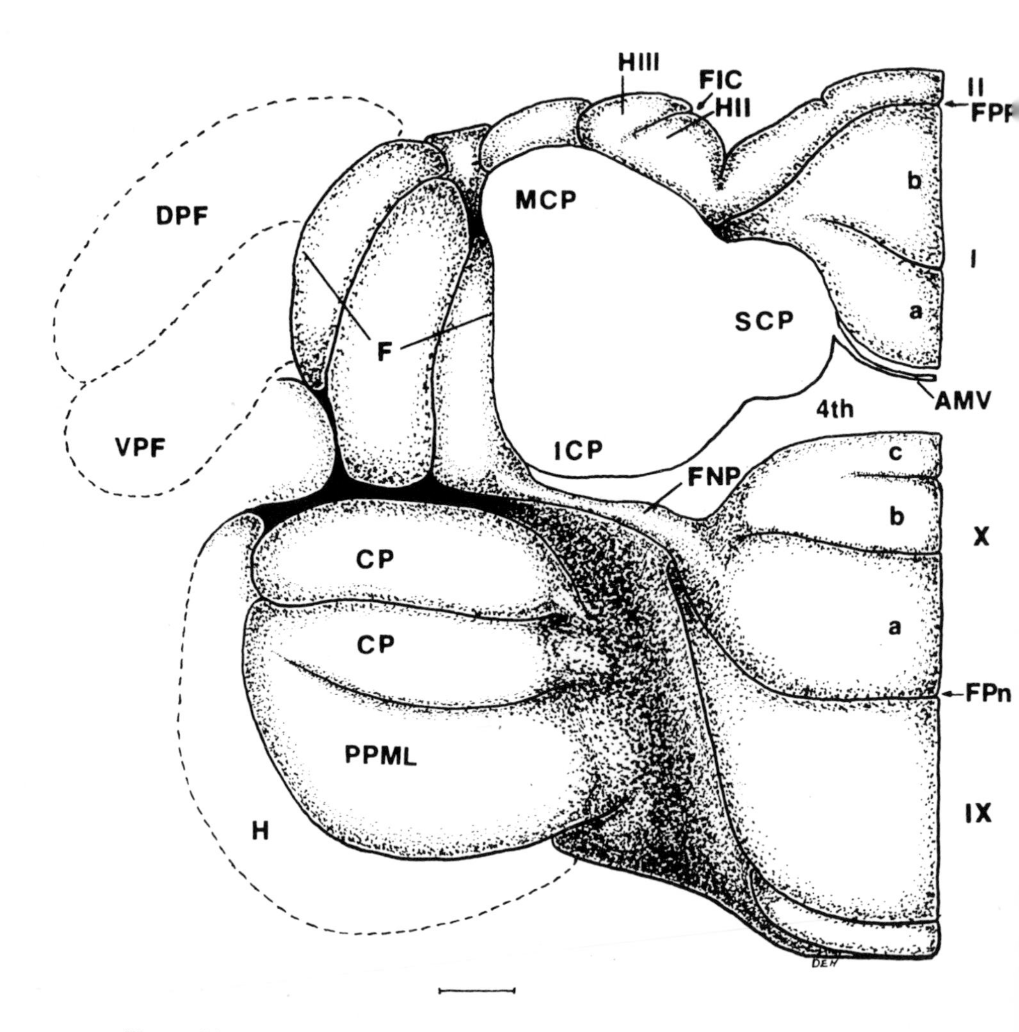

Figure 20. Ventral view of the cerebellum of *Perodicticus potto*. The cerebellar peduncles (SCP, MCP, ICP) have been cut, exposing the entire ventral surface. The nodulus is only moderate in size, the flocculus is distinct, and the flocculonodular peduncle is non-striated. The copula is well developed. Note the lateral extension of lobule I, and the moderate development of lobules HII and HIII. Scale = 1 mm.

rostral pole of the NL in the ventral area. The caudomedially directed pole of the NL is continuous with the NIA throughout the dorsoventral extent (Figs. 22, 23). This is especially evident in potto (Fig. 23*B-E*). In the slow loris and potto, clusters of cells are present between the NL and NIA in the ventro-caudal region (Figs. 22*F*, 23*E-F*).

In overall observation, the *nucleus lateralis* of the potto and slow loris is somewhat semilunar in shape, with a distinct ventromedially directed hilus (Fig. 32*c*, *d*). It is significant to note, however, that the NL and NIA are difficult to divide morphologically into distinct separate cell masses.

Identification of the hilus area is facilitated by the presence of fibre bundles presumed to constitute efferents of the NL. The cell population of the NL in potto and slow loris consists of small and medium-sized fusiform and stellate shaped cells (14-23 microns) dispersed throughout a population of medium and large fusiform and stellate shaped cells (24-35 microns). In the ventrocaudal region of the NL of the slow loris there appears to be a higher population of smaller cell bodies (14-21 microns); however, larger cells (22-28 microns) are also seen in this area. In the potto the range of cell size is essentially the same as in the slow loris, although

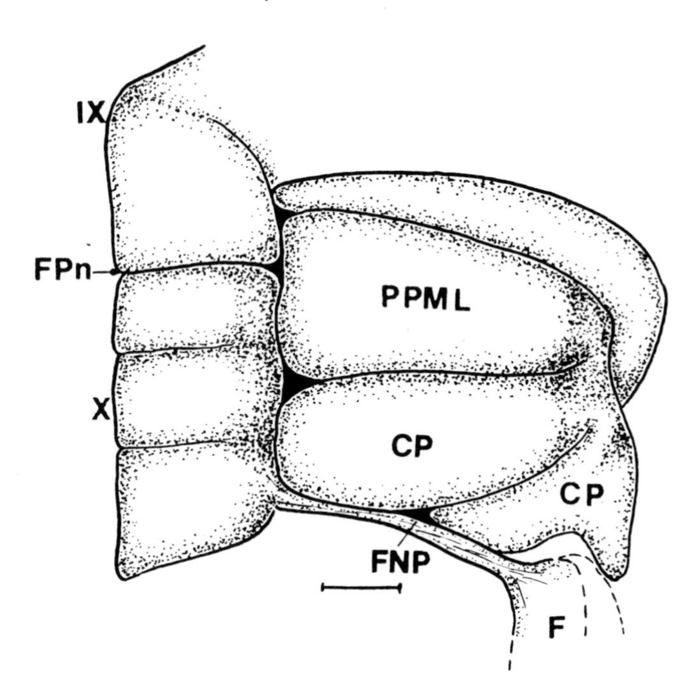

Figure 21. Ventral view of caudal area of the cerebellum of *Galago senegalensis*. Note the nodulus, flocculus, and the delicate flocculonodular peduncle. The *copula pyramidis* is large. Scale = 1 mm.

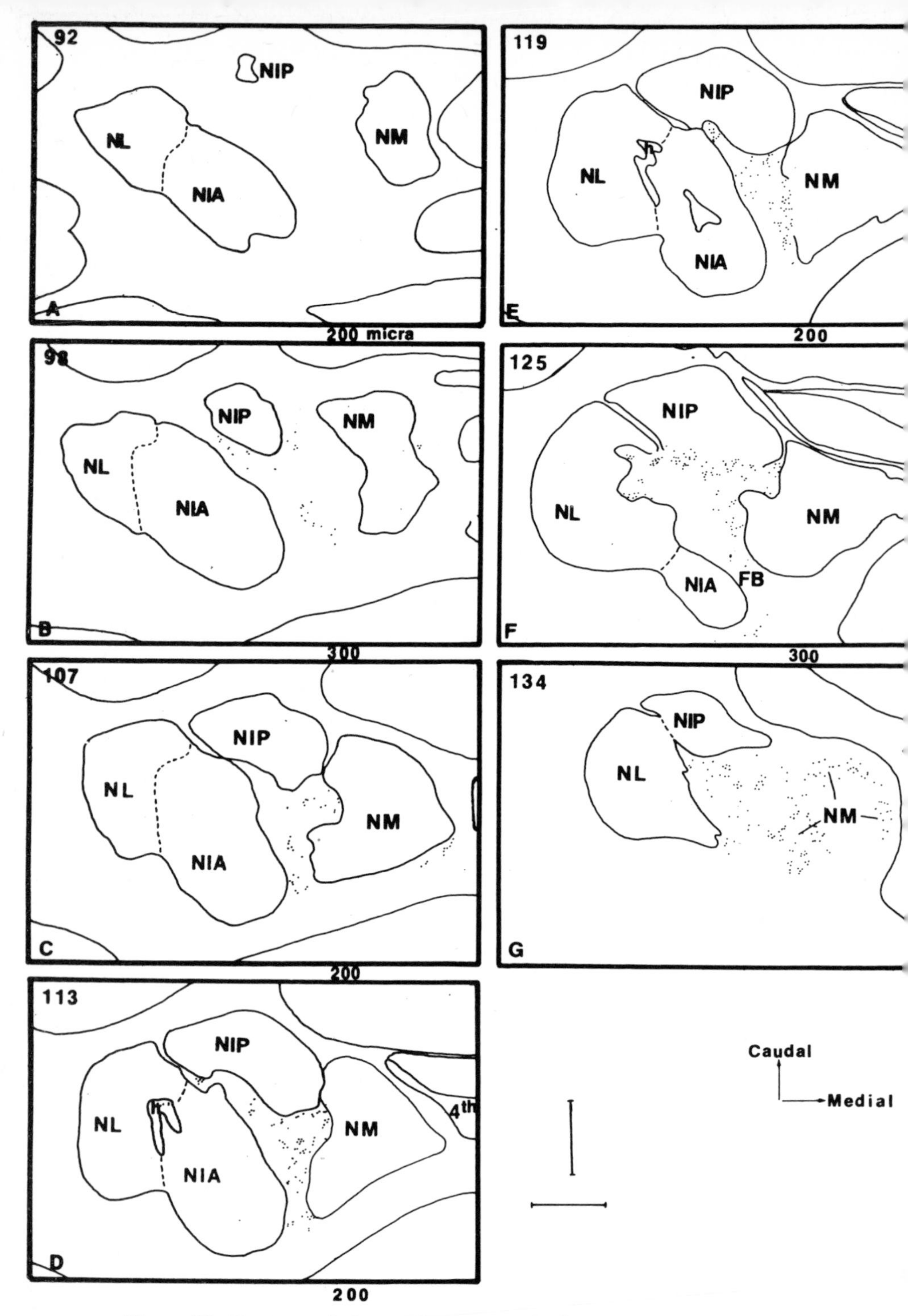

Figure 22. Tracings of the cerebellar nuclei of *Nycticebus coucang* in horizontal section. The upper left number in each tracing is the section number of that specific tracing, and the numbers between each tracing represent the depth in microns separating one from the next. The individual dots represent the positions of cell bodies seen in that particular 40 micron section. These cells did not appear to be located within the boundary of any individual nucleus, nor did they form a subnucleus. These tracings are from dorsal (*A*) to ventral (*G*). Scale = 1 mm.

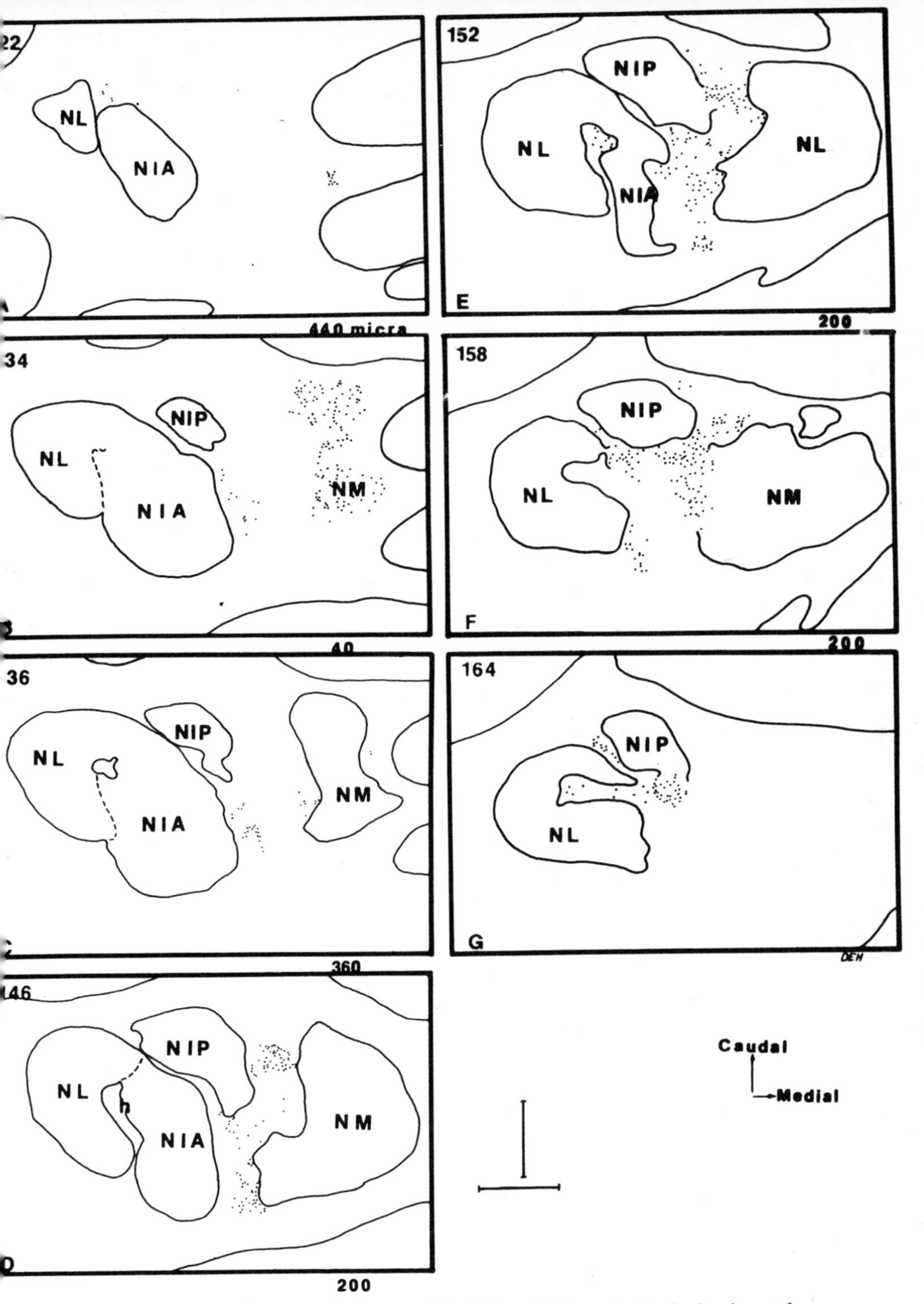

Figure 23. Tracings of the cerebellar nuclei of *Perodicticus potto* in horizontal section. The broken line in some drawings represents an arbitary division of nuclei. For additional explanation see Figure 22. Tracings are from dorsal (*A*) to ventral (*G*). Scale = 1 mm.

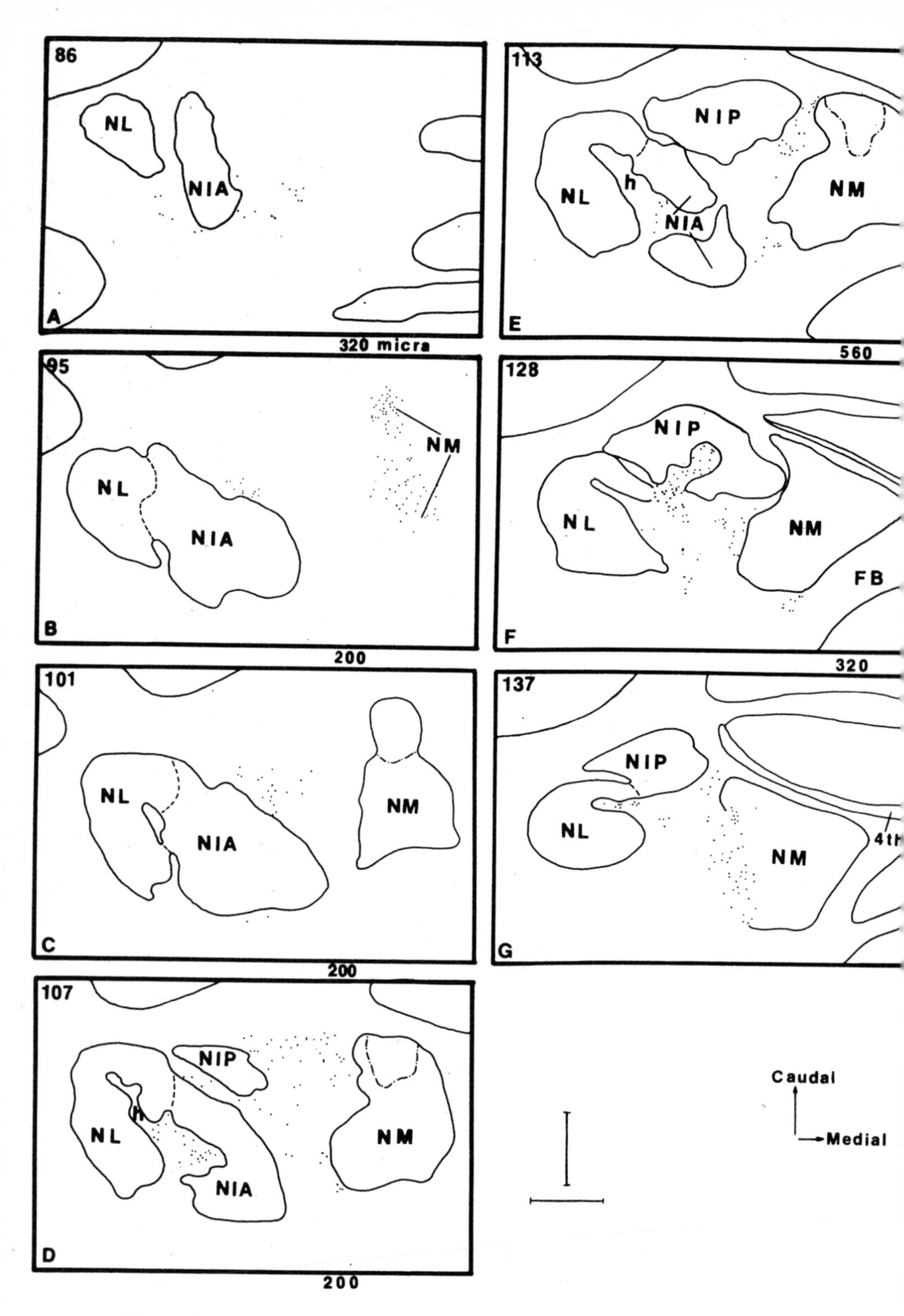

Figure 24. Tracings of the cerebellar nuclei of *Galago crassicaudatus* in horizontal section. Note the compact appearance of cells in the caudal area of NM (*B*). The area of compact cells is indicated by the dot-dash line (*C-E*). For additional explanation see Fig. 22. Tracings are from dorsal (*A*) to ventral (*G*). Scale = 1 mm.

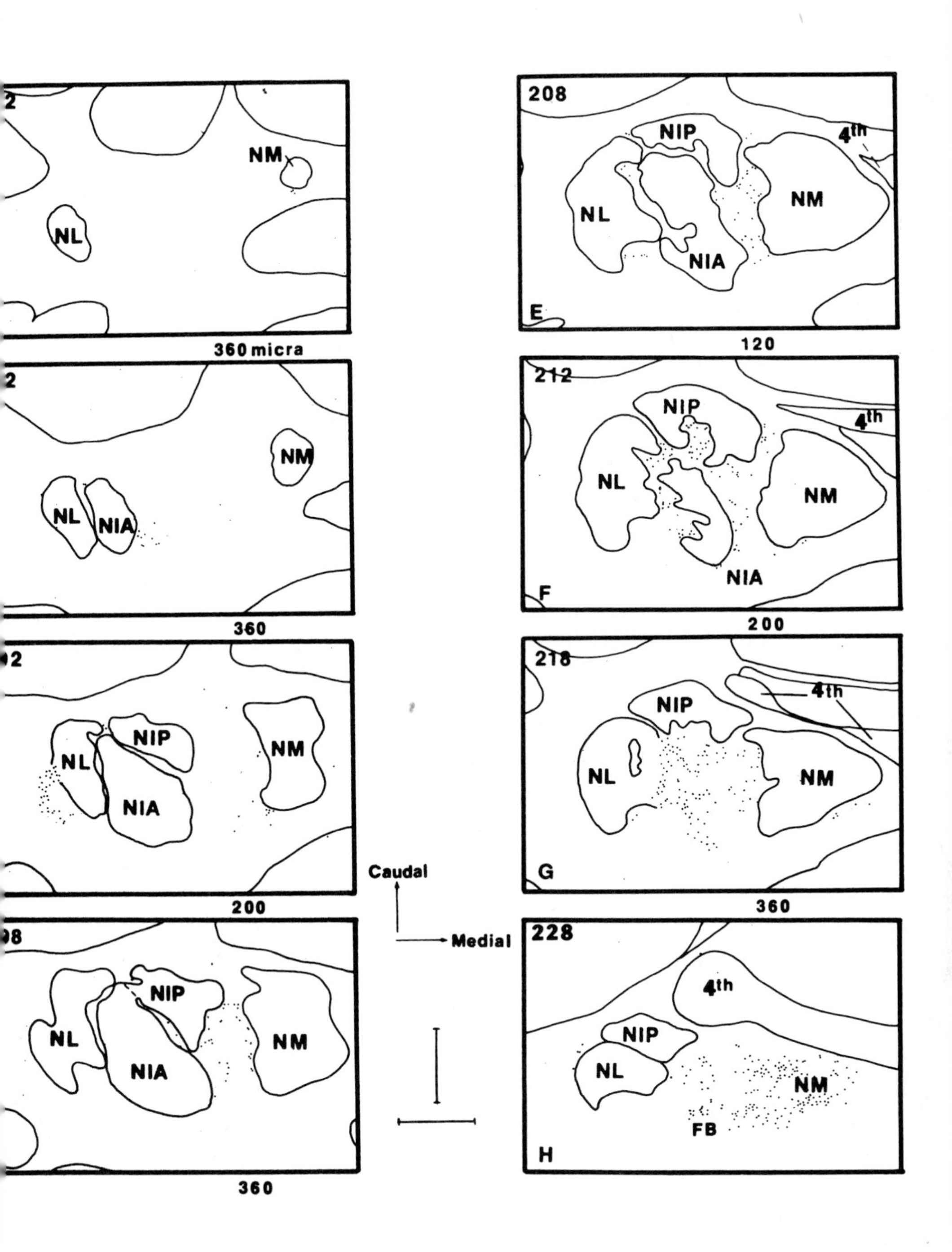

Figure 25. Tracings of the cerebellar nuclei of *Galago senegalensis* in horizontal section. Tracings are from dorsal (*A*) to ventral (*H*). For additional explanation see Fig. 22. Scale = 1 mm.

there is a predominant population of oval-shaped cell bodies (Fig. 33*e*). There is no regional concentration of large or small cell bodies, therefore the NL of the potto and slow loris does not appear to be subdivisible into subnuclei.

In the greater and lesser galago the NL is separated from the NIA throughout by a lamina of delicate medullated fibres. The NL of the greater bushbaby is approximately 2800 by 1700 microns at its greatest diameter and about 2300 microns at its greatest rostrocaudal length. The approximate size of the NL in the lesser bushbaby has been reported elsewhere.[19] The only point at which it is difficult to define the boundary of NL and NIA is at the caudomedially directed pole of the NL in the greater galago (Fig. 24*D, E*). The NL of both galagos is also in intimate contact with a small portion of the NIP in the most caudal and ventral area (Figs. 24*G*, 25*H*). In both the bushbabies the NL is medially concave, being more so in the greater than in the lesser bushbaby (Fig. 32*a*). A medially oriented hilus is apparent in the lesser galago (Fig. 25*D-F*). In the greater galago the NL is somewhat larger and the hilus is more distinct and oriented ventromedially (Fig. 24*D-F*). No large- or small-celled areas are identifiable; therefore, no subnuclei are definable. The size and shape of the cell bodies seen in the NL of the greater and lesser galago are within the ranges mentioned above for the slow loris.

(*b*) *Interposed nuclei.* The interposed nuclei in the Lorisidae studied can be divided into a *nucleus interpositus anterior* (NIA) and a *nucleus interpositus posterior* (NIP). The NIA extends most rostrally and the NIP extends most caudally in all genera studied (Figs. 21-25). As previously noted, the NIA is closely associated with the NL throughout the dorsal area in the slow loris and potto, but separated from NL, in dorsal and rostral areas, by a delicate lamina of fibres in both galagos. The NIA is strikingly similar in all genera studied; it has a rounded bulbous anterior pole, has its most intimate contact with the NL, disappears first in serial section passing dorso-ventrally, and in traverse section the NIA terminates caudolaterally in close association with the NL and NIP (Figs. 22-25, 32*d*). The NIA and NIP in the slow loris form a group of cells that measures approximately 2200 by 2100 microns at its greatest diameter and about 3400 microns at its greatest rostro-caudal length. These nuclei in the potto, also taken collectively, form a group of cells about 2000 by 1900 microns at its greatest diameter and about 2900 microns at its greatest rostro-caudal length.

The NIP also presents a rather consistent pattern in the greater and lesser bushbabies, slow loris and potto (Figs.

22-25). It appears in horizontal section (passing dorsal to ventral) at about the same time as the *nucleus medialis* (NM). The NIA and NIP collectively form a group of cells in the greater bushbaby about 2000 by 2200 microns at its greatest diameter and approximately 3000 microns at its greatest rostro-caudal length. These dimensions are somewhat smaller in the lesser bushbaby.[19] The NIP is located caudomedial in relation to the NIA, and in all specimens has a lateral portion extending caudal to the NL. The NIP is largest in the potto and slow loris, and it comes in intimate contact with the NM in caudal areas (Figs. 22, 23). In the greater and lesser galago, the NIP is slightly smaller (Figs. 24, 25). In all genera, the NIP is separated from the NL and NIA by a relatively consistent lamina of fibres. There appear to be no cyto-architectural subdivisions of either the NIA or NIP, and the predominent population is of medium (16-24 microns) and large (25-36 microns) fusiform and stellate-shaped cell bodies (Fig. 32*b*, *d*).

(*c*) *Medial nucleus.* The *nucleus medialis* (NM), or tectal nucleus, is large in all genera and located just lateral to the midline and dorsolateral to the fourth ventricle. The NM of the slow loris is an irregular mass of cells, somewhat flattened dorsoventrally, measuring about 2400 by 2000 microns at its greatest diameter and about 2200 microns at its greatest rostro-caudal length. The NM in potto is slightly larger, measuring about 2600 by 2100 microns at its greatest diameter and 2500 microns at its greatest rostro-caudal length. The NM in the greater bushbaby is an irregular mass of cells approximately 2200 by 2200 microns at its greatest diameter and about 2350 microns at its greatest rostro-caudal length. In the lesser bushbaby, the NM is considerably smaller.[19] The NM of all genera is relatively large and somewhat irregular in shape (Figs. 22-25). With the exception of the slow loris, the NM is separated from the NIP and NIA by small bundles of fibres. In the slow loris the NM and NIP are in close contact in the middle third at their caudal aspect (Fig. 22*C-F*). Compared in size with the lateral three nuclei, the NM is proportionately the same size in greater and lesser galagos and slow loris (Figs. 22, 24, 25). In the potto the NM is distinctly larger in comparison with the lateral three nuclei (Fig. 23). The general cell population is of small (14-18 microns) and medium-sized (19-23 microns) oval, fusiform, and stellate-shaped cell bodies. Intermixed with these small cells is a general population of medium and large (24-37 microns) oval, a few fusiform elongate, and stellate-shaped cell bodies. There is no organisation of cell bodies in a manner that would suggest the existence of a subnucleus of the NM in any of the Lorisidae studied. However, in the greater galago the cell bodies in the caudal third of the NM are distinctly compact in their organisation (Figs. 24*B*, 32*b*). This area is composed of small and medium-sized cell bodies with some large cells scattered throughout. Therefore it could not be considered as a subnucleus based strictly on cell size. The caudal portion of the NM of the potto also appears to have a more compact cell population than the rostral areas. However, this condition is not as accentuated as in the greater galago.

6. *Cerebellar corticonuclear projections*

The cerebellar cortex projects onto the deep cerebellar nuclei in a highly organised manner. The zonal concept (see Discussion) proposed that the cortex of the vermis projected onto the medial cerebellar nucleus, the paravermal cortex projected onto the interposed nucleus (NIA and NIP), and the lateral functional zone projected onto the lateral cerebellar nucleus. Recent observations on corticonuclear projections of the cat, a monkey and the rat (see Discussion) has shown that these projections are not necessarily limited to a single specific deep nucleus.

In the present study, lesions were placed in the cortex of the lateral and medial functional zones of the lesser galago (*G. s. senegalensis* and *G. s. moholi*) and the tree shrew (*Tupaia glis*). The animals used for this portion of the present study therefore represent a prosimian with a specialised locomotor pattern and an animal which may or may not be a primate.[59] It has been suggested by several investigators that the tree shrews represent either a living example of the stock from which modern primates evolved or the ancestral stock of all placental mammals. These considerations, coupled with the quadripedal locomotor pattern of tree shrews, have led to its use, on a comparative basis, in a long-term study of the phylogeny of cerebellar function as related to the evolutionary development of upright patterns of locomotion.

(*a*) *Lesions of the lateral functional zone.* The degeneration pattern following lesions of the lateral functional zone is reported (Figs. 26-9). A lesion placed in the lateral functional zone in the cortex of crus II of the ansiform lobule and the rostral folium of the anterior paramedian lobule shows degeneration in the lateral cerebellar nucleus (NL), and in both the anterior and posterior portions (NIA, NIP) of the interposed nuclei (Fig. 26). Fibres of passage enter the NL and the NIA from the caudal and caudolateral aspect (Fig. 26*A*), and fibres of passage extend into the lamina separating the NL and NIA (Fig. 26*A*, *B*). Most of these fibres appear to enter the extreme lateral portion of the NIA. There is terminal degeneration in the caudal half of the NL and in the lateral portion of NIA (Fig. 26*B-D*). A small amount of terminal degeneration is seen in the rostromedial areas of the NIA in dorsal levels (Fig. 26*A, B*) and in the rostral area of the NL in mid-horizontal levels (Fig. 26*C, D*). In more ventral areas, terminal degeneration in the NIA is extremely sparse (Fig. 26*E*). Fibres of passage are noted to course rostromedially through the anterior area of the NL. Some of these fibres terminate on cell bodies in the medial area of the NL and others appear to extend into the most lateral area of the NIA (Fig. 26*D, E*). A small bundle of fibres is noted passing towards, and entering, the caudolateral area of the NIP (Fig. 26*A-D*). This small discrete fasciculus of degenerating fibres terminates on cell bodies in the caudolateral area of the NIP (Figs. 26*D-F*, 35*c*, *e*).

A smaller lesion of crus I of the ansiform lobule in the cortex of the lateral functional zone shows a pattern of degeneration similar to that seen above (Fig. 27). Fibres of passage enter the caudolateral aspect of the NL and NIP, and a small bundle is noted entering the caudolateral area of the NIP (Fig. 27*D–F*). Terminal degeneration is dense in the lateral area of the NL (Fig. 34*a*) and sparse, but consistent in the lateral areas of the NIA and NIP (Fig. 27*B-F*, Fig. 35*d*,*f*). In horizontal and transverse section, the degeneration in the lateral nucleus is most dense in the dorsal 4/5, leaving the ventral area relatively free of terminal debris (Figs. 26, 27). In both animals illustrated (*G. senegalensis senegalensis*, Br. Ser. 6, and *G. senegalensis moholi*, Br. Ser. 12) it is also noted that the degeneration in the NL is most dense in the caudo-lateral portion of the nucleus (cf. Figs. 26 and 27).

As noted above, the dorsal and ventral paraflocculi are hemisphere areas HVIII and HIX, respectively. Due to their embrylogical development and anatomical position in the adult, they represent the most ventrolateral area of the cortex of the lateral functional zone of the posterior lobe. A lesion of the total dorsal and ventral paraflocculus results in dense degeneration in the ventrolateral caudal half of the NL (Figs. 28*A-E*, 34*a*). There is also a heavy amount of terminal degeneration in the medial area of the NL (Fig. 28*C*), and an extremely light amount of degeneration in the caudolateral area of the NIA (Figs. 28*C*, 34*d*, *e*). A small bundle of degenerated fibres extends caudorostrally ventral to the NIP and NIA (Figs. 27*B*, 34*C*). This bundle appears to give off a few fibres which terminate on cell bodies located slightly ventral to the NIA and NIP proper (Fig. 28*B, C*); however, its primary contibution appears to be to a small concentration

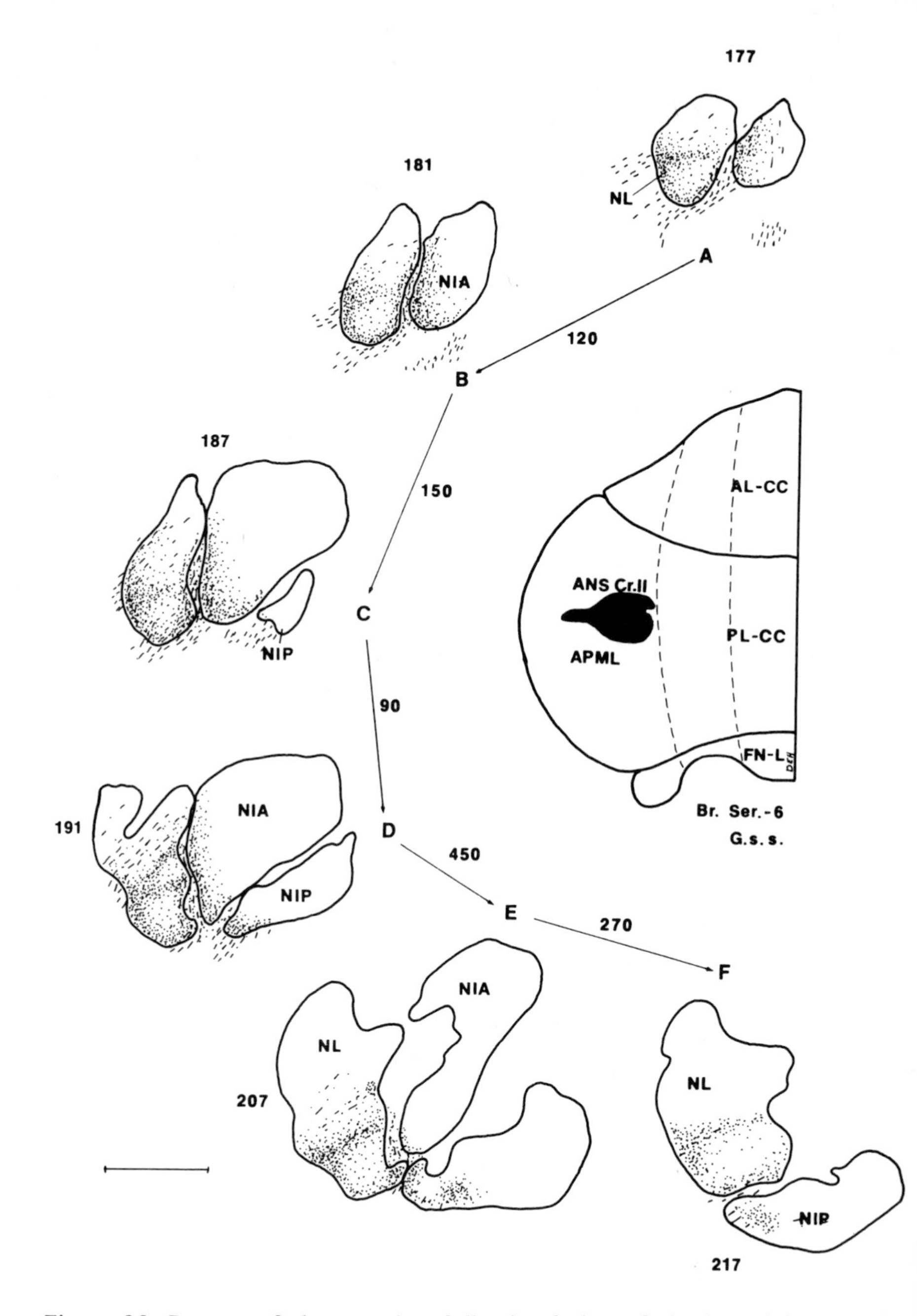

Figure 26. Pattern of degeneration following lesion of the lateral functional zones, which include crus II of the ansiform lobule and the anterior portion of the anterior paramedian lobule. Lesion of an adult *Galago senegalensis senegalensis.* The sections are on the horizontal plane and cut from dorsal (*A*) to ventral (*F*). The numbers between each section represent the distance in microns separating each drawing. The short lines represent fibres of passage and the dots represent areas of terminal degeneration. Note that the terminal degeneration is located in the lateral area of the NL, NIA, and NIP. Scale = 1 mm., in relation to the tracings of the nuclei.

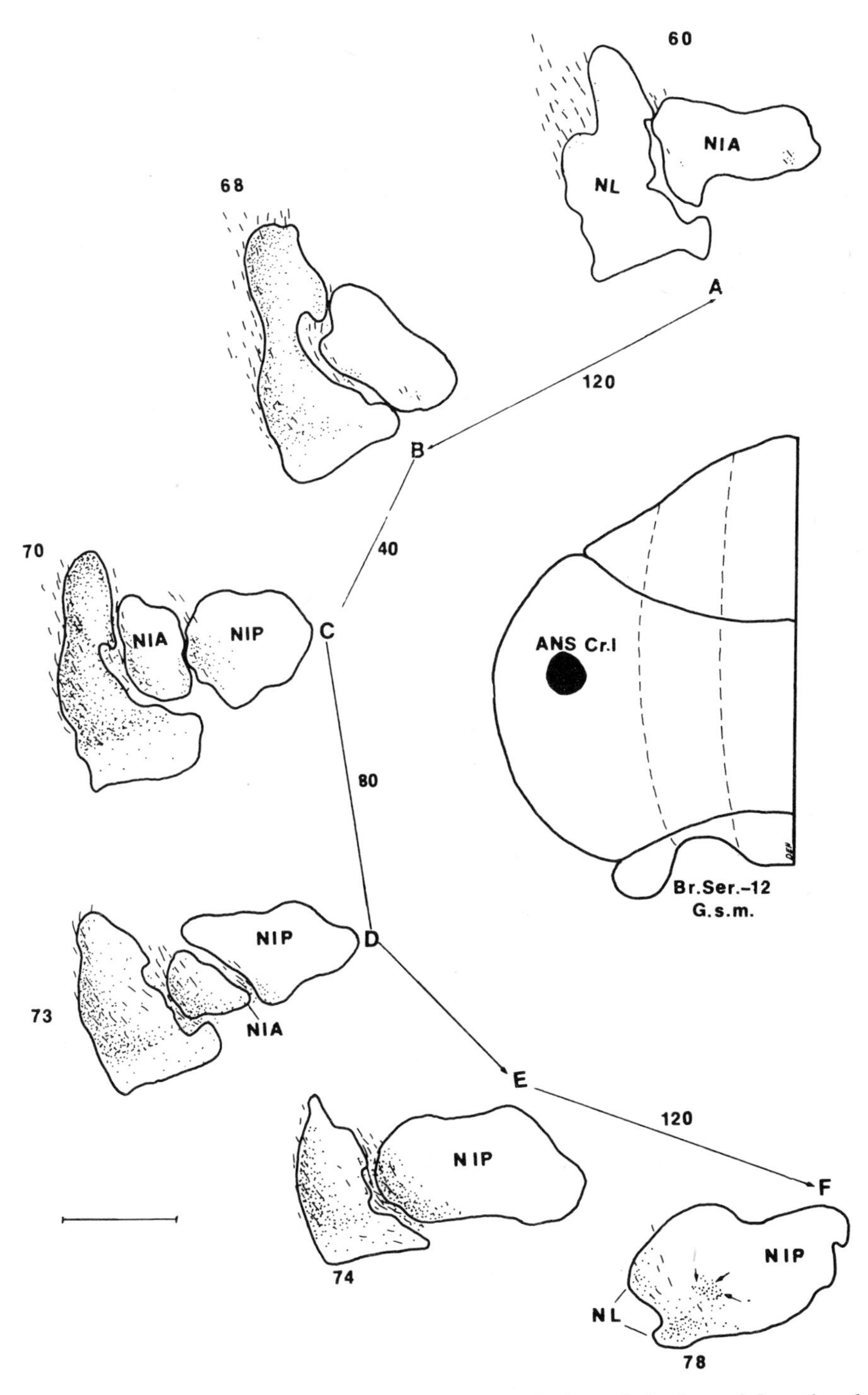

Figure 27. Pattern of degeneration following a lesion of the lateral functional zone, which was located in crus I of the ansiform lobule. Lesion of an adult *Galago senegalensis moholi.* The sections are cut on the transverse plane from rostral (*A*) to caudal (*F*). Note the pattern of degeneration and compare with Fig. 26. Terminal degeneration is seen in the lateral areas of the NL, NIA, and NIP. The ventral area of the NL shows only sparse degeneration. Scale = 1 mm., in relation to the tracings of the nuclei.

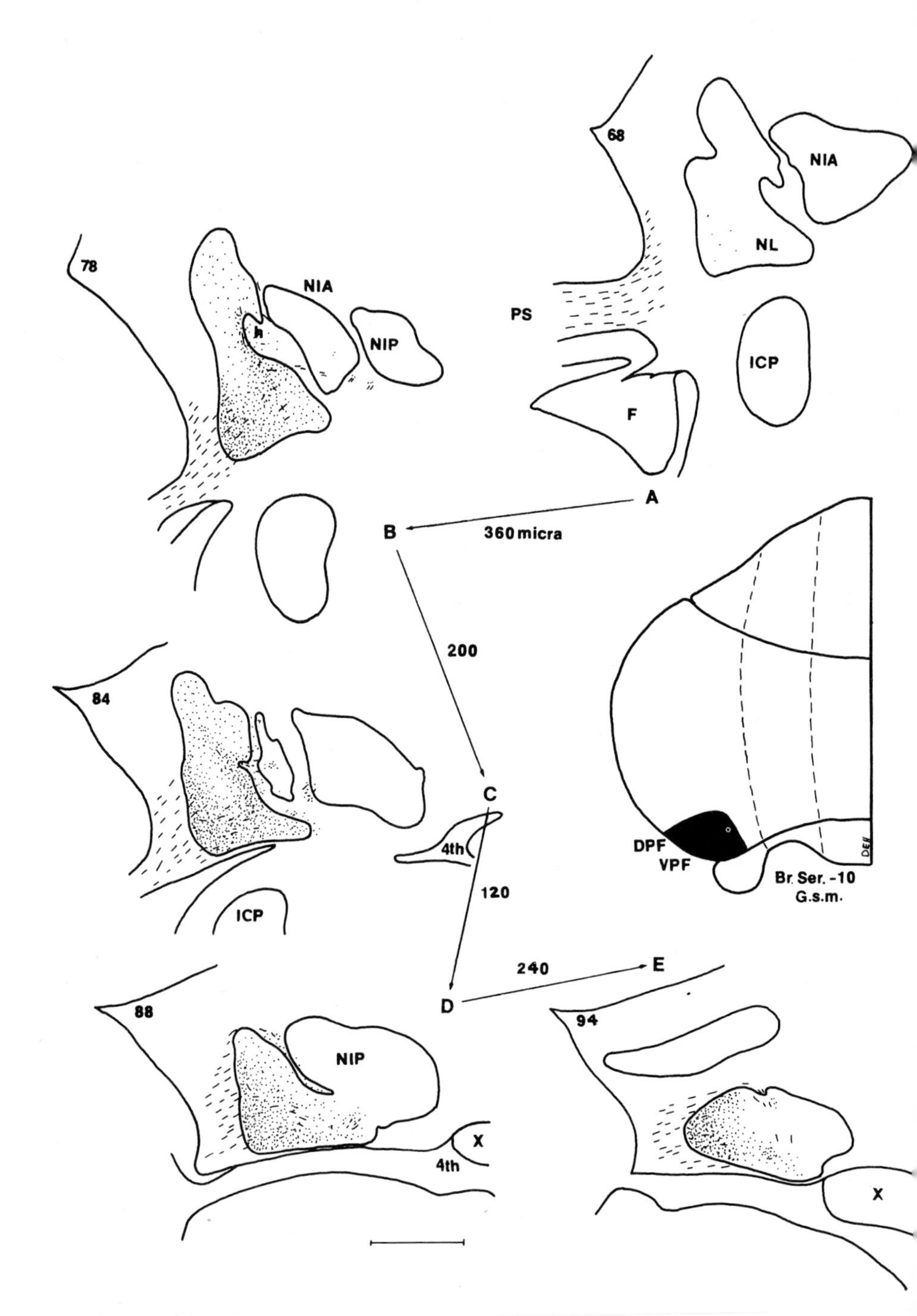

Figure 28. Pattern of degeneration following lesion of the lateral functional zone, which includes a total destruction of the dorsal and ventral paraflocculi. Fibres of passage are indicated by short lines and terminal degeneration by dots. Lesion of an adult specimen of *Galago senegalensis moholi.* Note the sparsity of degeneration in the dorsal area of the NL. The terminal degeneration and presence of fibres of passage is extremely sparse in the NIA and NIP. Scale = 1 mm., in relation to the tracings of the nuclei.

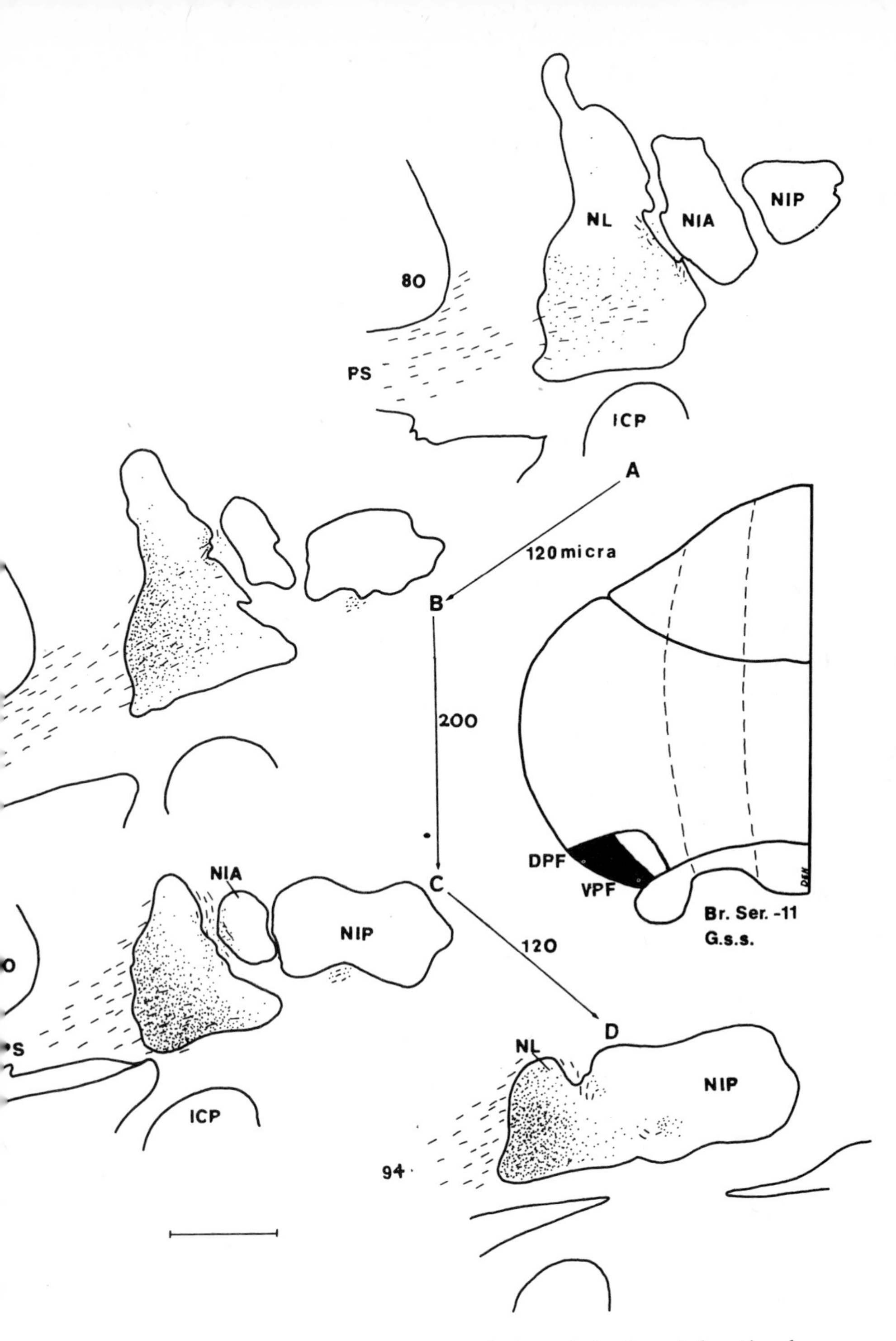

Figure 29. Pattern of degeneration following lesion of the lateral functional zone, which includes total destruction of the lateral two thirds of the dorsal and ventral paraflocculi. Lesion of an adult *Galago senegalensis senegalensis*. Note the sparse degeneration in the medial areas of the NL as compared with Fig. 28. Small areas of degeneration in the NIA and NIP are noted. Terminal degeneration is most dense in the lateral and ventral regions of the NL. (Compare with Fig. 28.) Scale = 1 mm., in relation to the tracings of the nuclei.

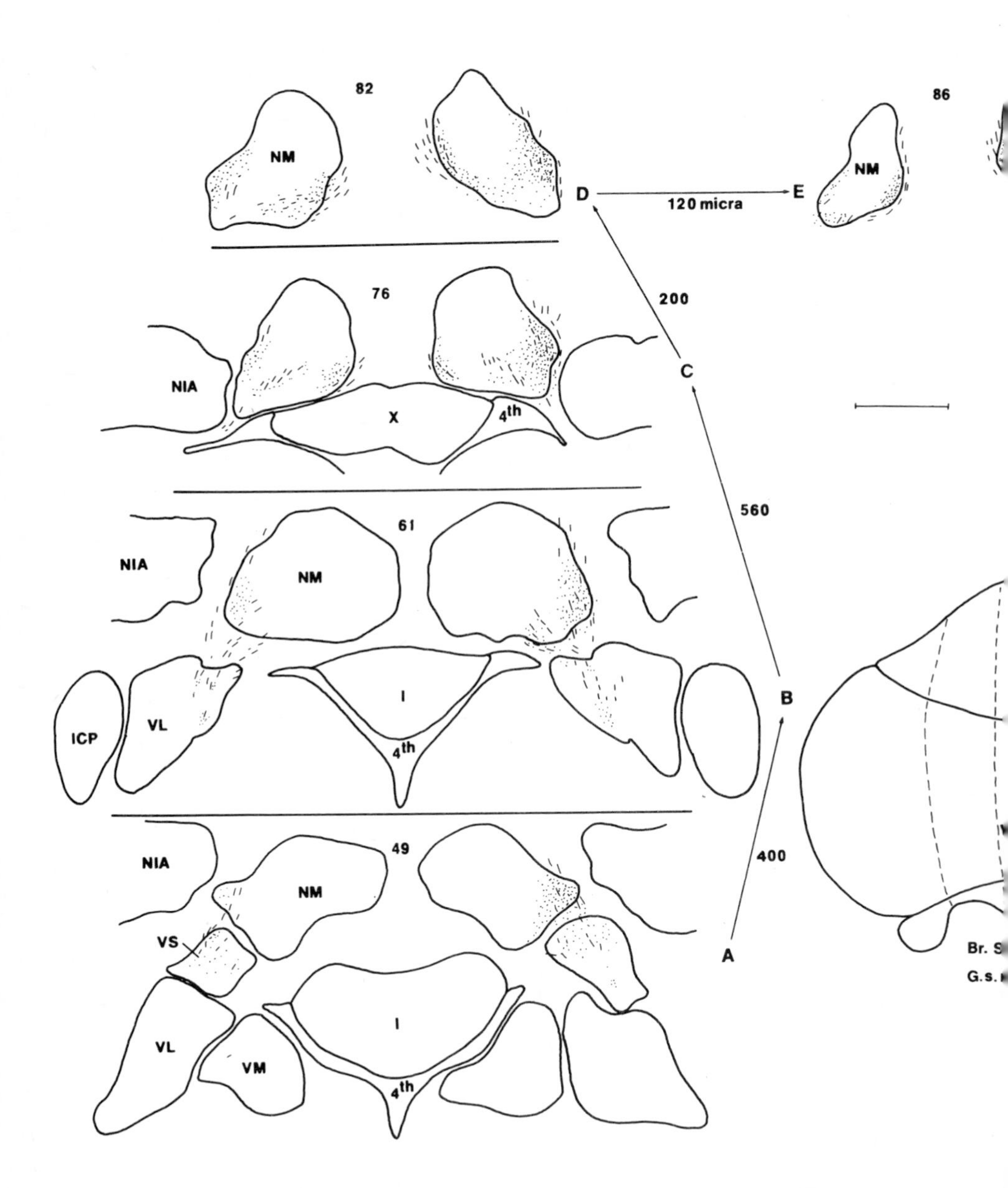

Figure 30. Pattern of degeneration following lesion of the medial functional zone, which includes lobule VIIIB and the upper lip of lobule IX. Lesion of an adult *Galago senegalensis moholi.* The projections are bilateral and pass to the NM, the lateral vestibular nucleus and the superior vestibular nucleus. The heavier degeneration to the nuclei on the right is due to the fact that the lesion is larger on the right side of the midline. Right and left sides are indicated by L. and R. Scale = 1 mm., in relation to the tracings of the nuclei.

of cells in the caudolateral area of the NIP (Fig. 28*D*).

Further substantiation of the sparse fibres to the NIA and NIP following parafloccular lesions was considered mandatory. Therefore, the lateral 2/3rds of both the dorsal and ventral paraflocculi was removed in *G. senegalensis senegalensis*, Br. Ser. 11 (Fig. 29). Fibres of passage entered the lateral aspect of the NL in the caudal half of its length, essentially the same as that noted in a lesion of the entire paraflocculus (Fig. 28). Heavy terminal degeneration is present in the lateral and ventral areas of the NL in its caudal half (Fig. 29*B-D*). Terminal degeneration is sparse in the dorsal area of the NL with complete removal of the paraflocculus (Fig. 28), or almost completely absent when only the lateral 2/3 of the paraflocculus is removed (Fig. 29). In both cases illustrated, there is a persistent amount of extremely sparse fibres of passage and terminal degeneration in caudal and lateral areas of the NIA and NIP (Figs. 28*B-E*, 29*B-D*, 34*c*, *e*). In these descriptions it is *stressed* that these projections to NIA and NIP from the cortex of the paraflocculus are *extremely* sparse.

(*b*) *Lesions of the medial functional zone.* The fibre projections from lesions of the cortex of the medial functional zone are shown in Figs. 30 and 36. In one animal, a lesion was placed in the cortex of lobules VIIIB and IX on the midline with slightly more of the lesion on the right hand side of the vermis (Fig. 30). Corticonuclear fibres pass rostrally toward the caudal pole of the NM. Fibres of passage are seen coursing bilaterally in a caudorostral direction on the medial and lateral sides of the NM (Fig. 30*e*). There is a small amount of terminal degeneration in the medial aspects of the NM and a moderate amount in the ventrolateral areas (Fig. 30*B-D*). The amount of terminal degeneration and fibres of passage is slightly greater in the NM on the right. This is due to the fact that the lesion involved more of the cortex on the right side of the midline than on the left.

Fibres of passage course rostrally on the lateral and medial aspects of the NM. Some of the fibres located medial to the NM terminate on small- and medium-sized cell bodies in the ventromedial area of the nucleus. Other fibres travelling in this area course rostrally, then turn laterally to travel through the ventral area of the NM (Figs. 30*C-D*, 35*a*, *b*). These fibres show a small amount of terminal degeneration throughout their course, although the majority of the fibres appear to pass out of the NM at its ventrolateral aspect. These fibres then enter, and terminate on cell bodies in, the medial portion of the lateral vestibular nucleus (Figs. 30*B*, 36*b*, *e*). A second distinct fasciculus of degenerating fibres courses rostrally through the lateral portion of the NM. Some fibres enter the centrolateral area of the NM to terminate, while others extend rostral to enter the superior vestibular nucleus. The fibres passing into and terminating on the superior vestibular nucleus are thin fibres and the pattern of terminal degeneration is sparse (Fig. 36*a*, *c*, *b*), and located in the caudodorsal area of the nucleus (Fig. 30*A*).

Lesions of the medial functional zone cortex of the posterior lobe which do not extend over the midline (not illustrated) show the same pattern of projection to the NM, the lateral and superior vestibular nuclei. The projections in cases of lesions restricted to one side of the midline are almost exclusively into the nuclei of the ipsilateral side. In the present study, no degenerated fibres were seen to cross the midline between the medial nuclei.

Discussion

The cerebelli of the Lorisidae considered in this study show many similarities, yet at the same time some interesting differences are noted. The anterior lobe of the *corpus cerebelli* is divided into the five lobules characteristic of many mammalian forms. The division of the anterior lobe into five lobules has been accomplished in a wide variety of mammals ranging from the rat to the higher primates.[16] The anterior lobe of the avian cerebellum also has five characteristic subdivisions which are indicated as folia instead of lobules.[16,25] It has been recently suggested that the anterior lobe may consist of four lobules.[26] The criterion for this is: 'In order to be considered a lobule, the morphological feature in question must be separated from its neighbours by deep fissures that, as seen in mid-saggital section, must approach that fastigial nucleus closely, and must not conjoin them laterally in surface aspect.' This technique, in the form stated,[26] can only be applied to the animals of the present study in a modified form. In the greater galago and slow loris, lobules IV and V of Larsell could correspond to lobule IV of Dillon and Brauer. However, if the above definition is applied to the potto and the lesser galago, lobules II and III of Larsell would correspond to lobule II of Dillon and Brauer.[16,26] Considerably more work must be carried out on other prosimian forms to evaluate which of these methods are applicable to the suborder as a whole. The method of Larsell is utilised in the present investigation since it is well established in a variety of vertebrates.[15,16]

Figure 31. Midsagittal views of the cerebellum and diencephalon of *Nycticebus coucang* (*a*), *Galago crassicaudatus* (*b*), and *Perodicticus potto* (*c*). For abbreviations not listed on p. 730, see Notes.[33]

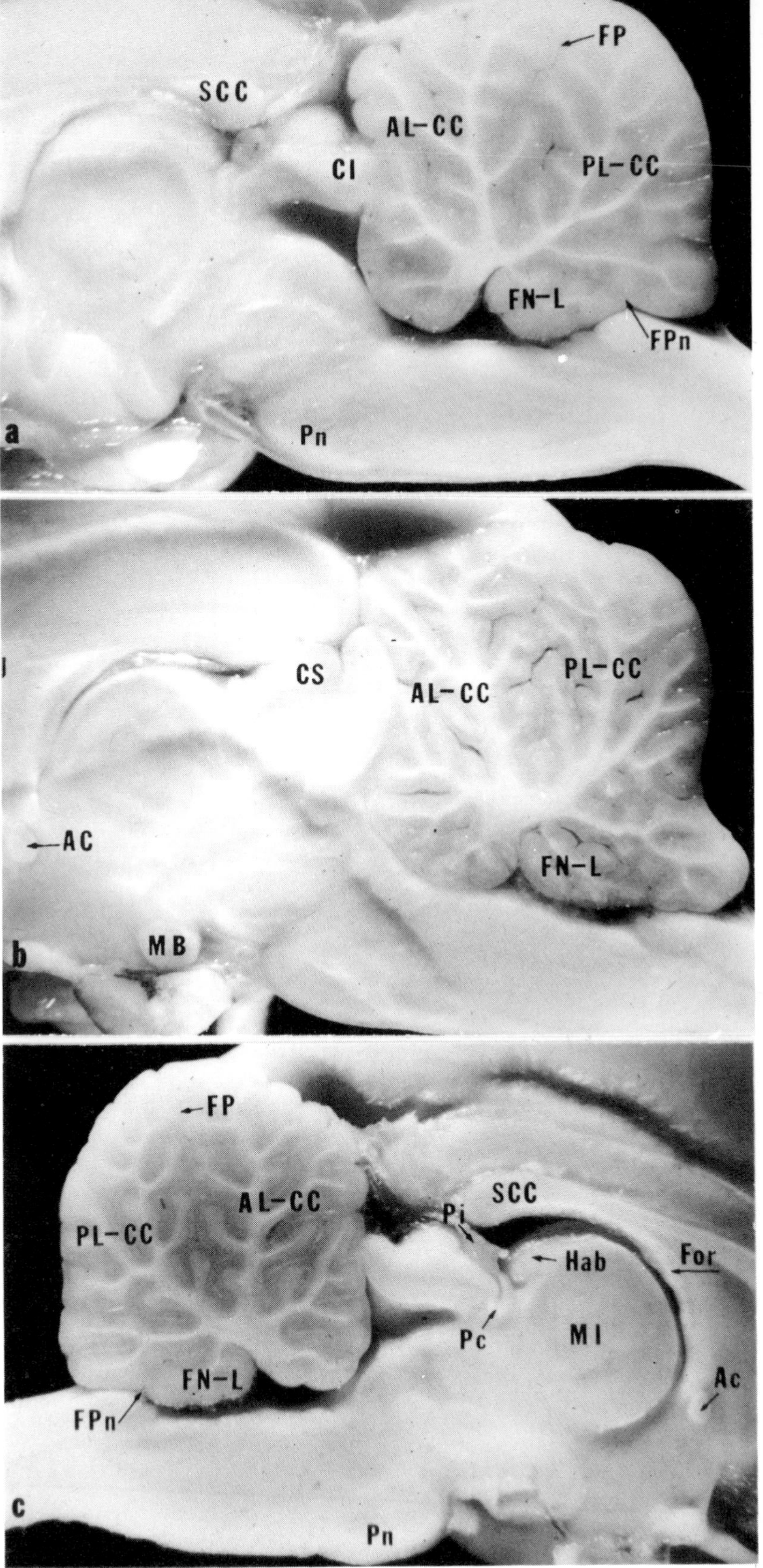
FP
SCC
AL-CC
CI
PL-CC
FN-L
FPn
a
Pn
CS
AL-CC
PL-CC
AC
FN-L
MB
b
FP
AL-CC
SCC
PL-CC
Pi
Hab
For
MI
Pc
FN-L
Ac
FPn
c
Pn

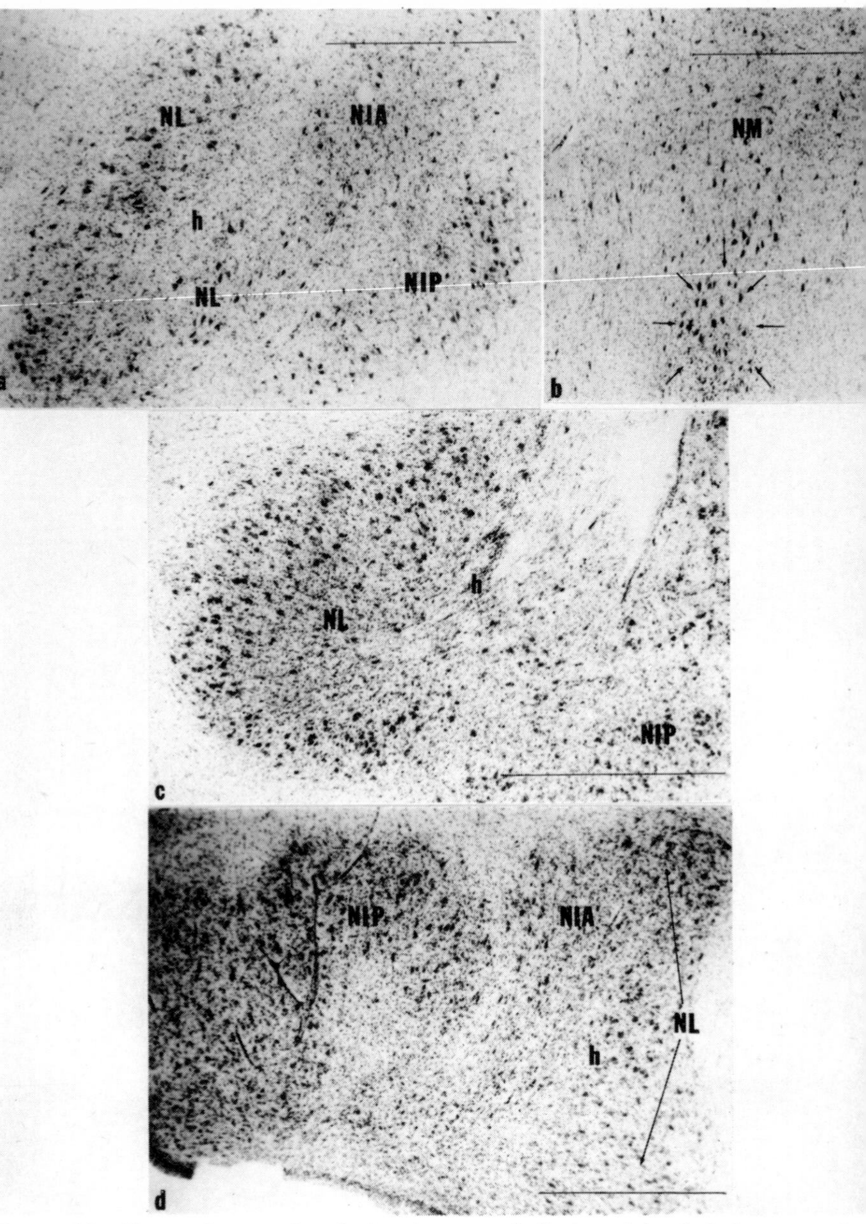

Figure 32. Photomicrograph of the deep cerebellar nuclei of *Galago crassicaudatus* (*a,b*), *Perodicticus potto* (*c*), and *Nycticebus coucang* (*d*). Scale = 1 mm.

(*a*) The NL, NIA, and NIP of *G. crassicaudatus* are sectioned in horizontal section. Note the undulatory appearance of the NL. This photomicrograph is approximately at the level of Figure 24*E-F*.

(*b*) The NM of *G. crassicaudatus*. Note the compact appearance of the cells in the caudal area, versus the reticulated appearance of the cells in the rostral areas. This photomicrograph is approximately at the level of Figure 24*B*.

(*c*) The NL, and NIP of *P. potto*. The NL lacks the undulatory appearance characteristic of the greater galago. Note the hilus. This photomicrograph is approximately at the level of Figure 23*F*.

(*d*) The NL, NIP, and the caudolateral portion of the NIA of *Nycticebus coucang* as seen in transverse section.

Figure 33. Photomicrographs of the typical appearance of cell bodies of the NL (*a*), the NIA (*b*), the NM (*c*), and the NIP (*d*). The more oval shaped cell bodies of the potto are exemplified by the photomicrograph of cells from the NL (*e*). Scale = 50 microns.

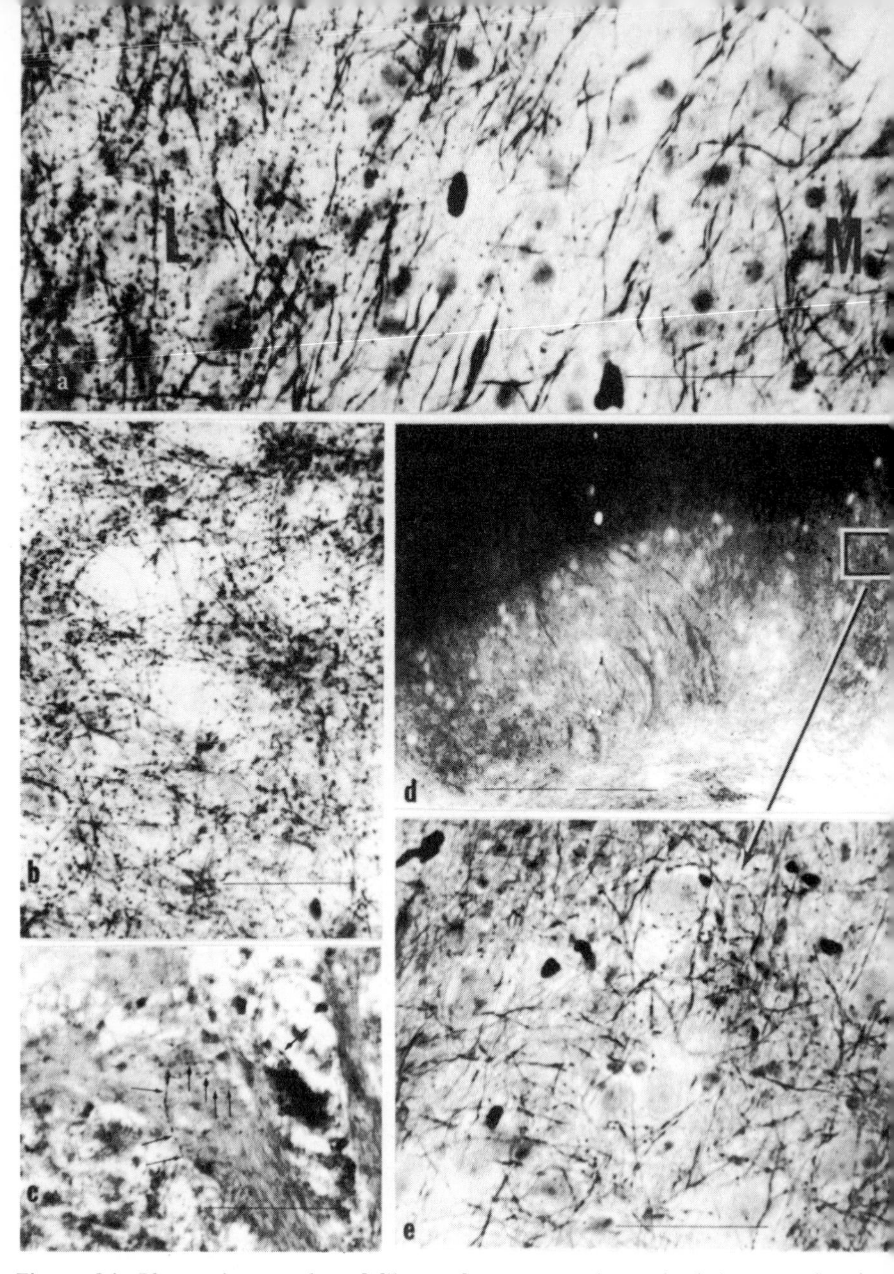

Figure 34. Photomicrographs of fibres of passage and terminal degeneration in the deep cerebellar nuclei.

(*a*) Photomicrograph of heavy degeneration in the lateral area of the NL (L) and sparse degeneration in the medial area of the NL (M). From *G.s. moholi* No.12. Scale = 50 microns.

(*b*) Photomicrograph of dense degeneration in the ventral area of the NL. From *G. s. senegalensis* No.10. Scale = 50 microns.

(*c*) Photomicrograph of degeneration of fibres of passage in a small bundle located ventral to the NIP. This bundle courses in a caudorostral direction. *G. s. senegalensis* No.10. Scale = 50 microns.

(*d*) Low power photomicrograph of the NIP, NIA, and NL. Taken from Fink-Heimer stain. From *G. s. senegalensis* No.10. Scale = 1 mm.

(*e*) Photomicrograph of the light degeneration in the caudolateral area of the NIA. This is high power of the boxed area of Figure 34*d*. Note the fibres of passage and sparse amount of terminal degeneration. *G. s. senegalensis* No.10. Scale = 50 microns.

The anterior lobe of the cerebellum is poorly developed in basic subprimate mammalian forms. In the insectivores and the elephant shrews, it is hidden by forward expansion of the posterior lobe vermis (middle lobe of Elliot Smith in the older literature), and has poorly developed hemisphere areas.[2,7-9] In *Ptilocercus lowii* the cerebellum is simpler than that of *Tupaia*.[7] In both of these animals, the anterior lobe is small with poorly developed hemispheres.[7,11,17] In a variety of basic forms, the anterior lobe of the *corpus cerebelli* shows a characteristically poor level of development.[15,16,30] It is equally significant to note that in rodents, carnivores, lagomorphs and other mammalian forms the anterior lobe is poorly developed, even though the total size of the cerebellum may equal or exceed that of some primates.[16] In the Lorisidae of the present study the hemisphere of the anterior lobe (HII-HV) is well developed. In a variety of Old and New World primates, regardless of the total size of the cerebellum, the hemisphere of the anterior lobe is well developed.[16,31] It appears that well developed hemispheres of the anterior lobe represent characteristics shared by all primates.

The lingula of the greater and lesser bushbabies was simple and restricted to the vermis. In the slow loris and potto the lingula appeared to have a lateral portion which may represent some development of a very primitive hemisphere. It has been suggested that the hemisphere of the lingula, when present, is related to the development of a prehensile tail.[16] Since the potto and slow loris obviously lack prehensile tails, the development of the lingula in these animals awaits another answer. It has recently been shown that lesions of the lingula in a monkey cause the animal to rotate in a clockwise direction when aroused.[32] This observation would support the concept of vestibulo-cerebellar fibres projecting to the lingula. Since the potto and slow loris do not have a tail that could assist them in maintaining balance, a modest development of the lingula may provide a closer functional relationship between the cerebellum and vestibular centres for maintainence of balance in an arboreal habitat. The greater and lesser bushbabies have long tails to assist them in maintenance of balance, therefore some adaptive specialisation of the nervous system would be unnecessary.

The posterior lobe of the corpus cerebelli is well developed in all genera of the present study. Even though the size and shape of the cerebelli, pons and cerebral cortices vary somewhat,[33] the relative complexity of the posterior lobe is similar in the slow loris, lesser galago and potto. A notable exception is the greater galago. The cerebellum of this animal is consistently more complex than that of the other three genera. The anterior and posterior paramedian lobules are most complex in this genus. The ansiform lobule, size and shape of the dorsal paraflocculus, and limitations of the *copula pyramidis* (the potto excepted) are similar in all groups. Enlargement of the posterior lobe, especially the hemispheres, is characteristic of

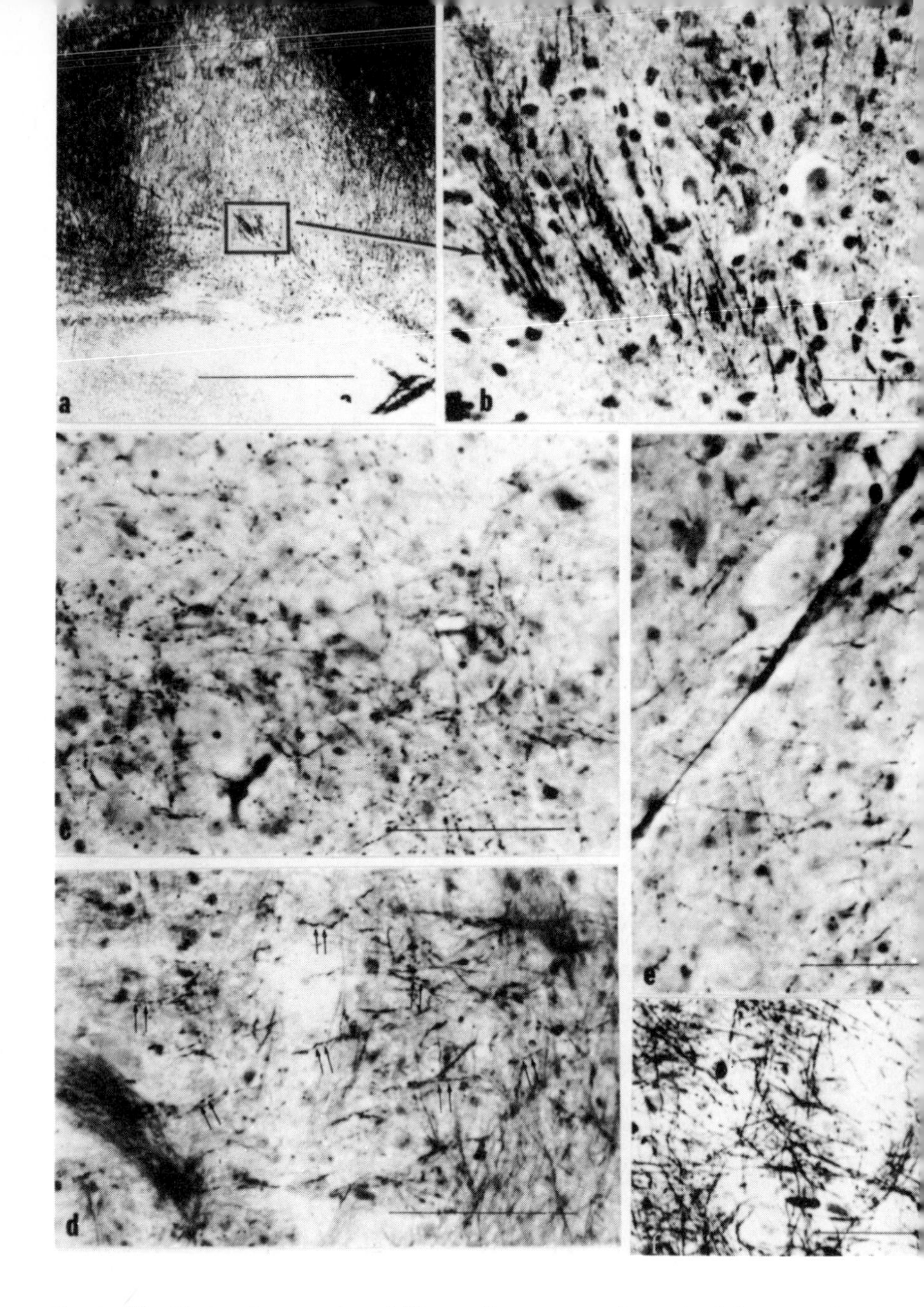

Figure 35. Photomicrographs of fibres of passage and terminal degeneration in deep cerebellar nuclei.

(*a*) Low power photomicrograph of the NM stained with the Fink-Heimer stain. From *G. s. moholi* No.9. Scale = 1 mm.

(*b*) Photomicrograph of the bundle of fibres coursing through the ventral area of the NM. This is a high power view of the approximate area indicated by the box in Figure 35*a*. Note the fibres of passage in the bundle and the terminal degeneration in the area lateral to it (upper right corner of photo). From *G. s. moholi* No.9. Scale = 50 microns.

(*c*) Photomicrograph of fibres of passage and terminal degeneration in the lateral area of the NIP. Note the fibres of passage and the moderate terminal degeneration. From *G. s. senegalensis*. No.6. Scale = 50 microns.

(*d*) Photomicrograph of degeneration of fibres of passage in the small

all primates.[16,34,35] The cerebelli of the Lorisidae are more complex than the same structure in many new world monkeys.[16,31]

The deep cerebellar nuclei of the Lorisidae of the present study can be divided into four cell masses on either side: *nucleus medialis, nucleus interpositus anterior, nucleus interpositus posterior*, and *nucleus lateralis.* The cerebellar nuclei of primates have been reported as four separate cell masses for a variety of different genera.[18,19,36,37] The most obvious change in the morphology of the cerebellar nuclei is the increasingly complex undulatory appearance of the lateral cerebellar nucleus as one ascends the primate scale. The greater and lesser galago show the most advanced degree of differentiation of the cerebellar nuclei in the present study. The potto and slow loris show a lesser degree of differentiation, as indicated by the intimate association of the NL with the NIA and NIP. The NIA and NIP are relatively large in all genera of the present study. In the potto, the NIP is slightly larger than in the other genera, but the NIA is essentially similar. The NIA is a slightly elongated cell mass that expands further rostral than any other cell group. In view of recent evidence of somatotopical projections from the NIA to the red nucleus,[38] the pathways between NIA and the mesencephalic red nucleus may be well developed in the Lorisidae. It has been suggested that the fusion of the cerebellar nuclei represents a basic level of organisation from which higher levels of differentiation arose.[39]

The cerebellar nuclei of a variety of sub-primate mammals have also been divided into four cell masses on each side.[39–42] It should be noted, however, that these animals are specialised mammals that have been evolving divergently from the primates for considerable time. If basic forms which may have a more direct relationship with extant primates are considered, it is noted that there is a progressive degree of phylogenetic development of the cerebellar nuclei from a single cell mass to two bilateral cell masses and eventually four bilateral cell masses.[19,30,43,44] The enlargement of the lateral cerebellar nucleus appears to be correlated with increase in size of the hemisphere of the anterior and posterior lobe. In the present study, the degree of differentiation of the cerebellar nuclei in ascending order of complexity would be: slow loris, potto, greater galago, lesser galago. The development of the NL is greater in *Galago*, especially *G. senegalensis*, and appears to be partially related to the

---

bundle of fibres travelling in a caudorostral direction in the ventral area of the NIA. From *G. s. moholi* No.12. Scale = 50 microns.

(*e*) Photomicrograph of fibres of passage and terminal degeneration in the lateral portion of the NIP. From *G. s. senegalensis* No.6. Scale = 50 microns.

(*f*) Photomicrograph of sparse fibres of passage and terminal degeneration in the ventromedial area of the NIP. From *G. s. moholi* No.12. Scale = 50 microns.

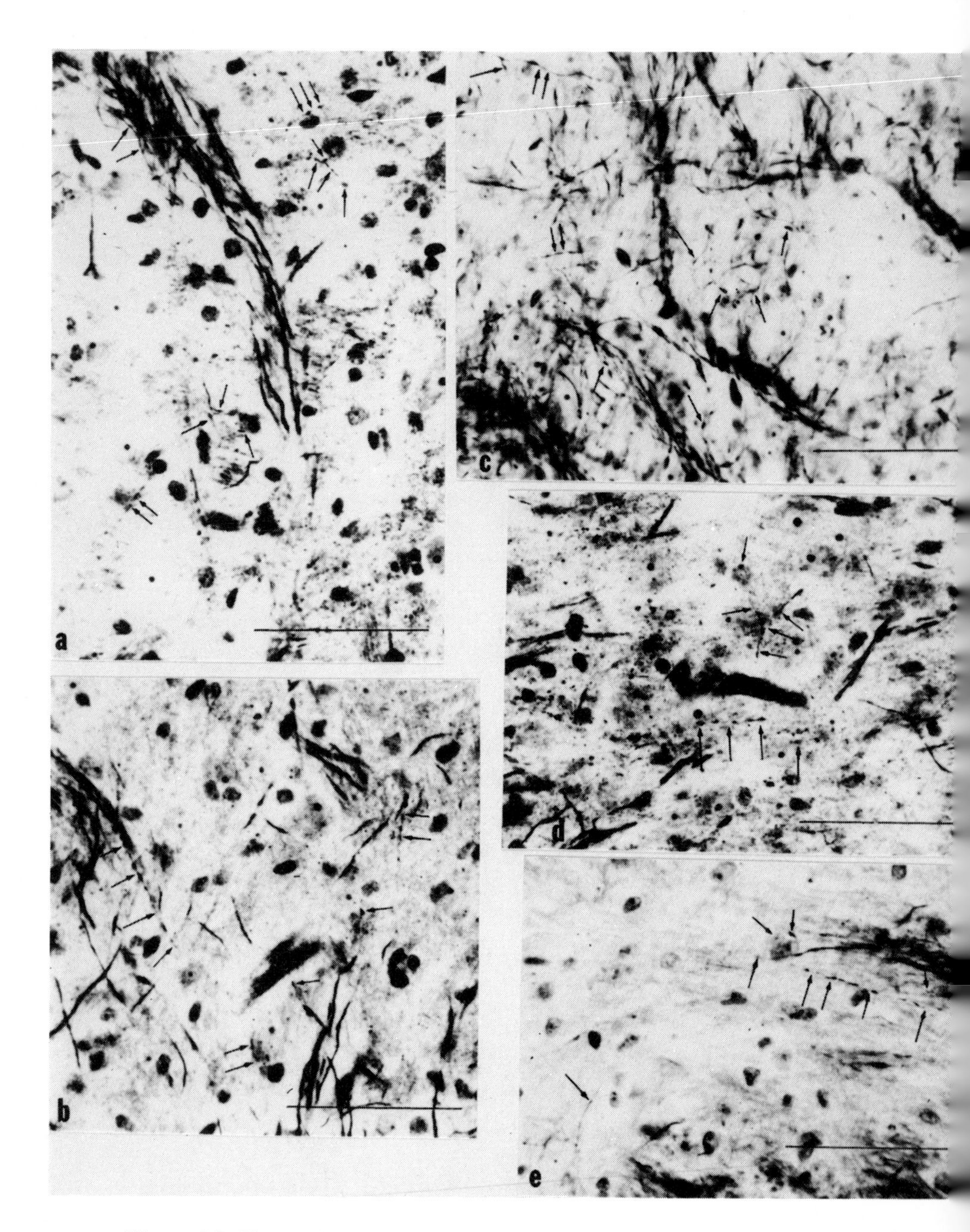

Figure 36. Photomicrograph of fibres of passage (double arrows) and terminal degeneration (single arrows) in the superior vestibular nucleus (*a,c,d*), and in the lateral vestibular nucleus (*b,e*). Note that the projection from the cortex of the medial functional zone passes to two of the vestibular nuclei. These are sparse yet consistent projections, and can be termed cerebellovestibular in nature. *G. s. moholi* No.9. Scale = 50 microns.

saltatory locomotor pattern of these animals. Results of recent work on the *nucleus dorsalis* of Clarke (spinal cord afferents to the cerebellum) further support the concept of neurological specialisations which may be related to locomotion.[45] Rüdeberg has established the homologous characteristics of the deep cerebellar nuclei in a wide variety of vertebrates including man, and his terminology is used in this report.[46]

The zonal concept of cerebellar corticonuclear projections suggested that the vermal cortex (medial functional zone) projects onto the NM, the paravermal cortex (paravermal functional zone) onto the NIA and NIP, and the lateral hemisphere (lateral functional zone) projecting on to the NL.[47] Recent evidence using newer degeneration techniques has not totally supported this concept.[48-53] The results of the present study also show that corticonuclear fibres from the cerebellar cortex project, in overlapping fields, on to the deep nuceli. In the present study, the lateral functional zone of the posterior lobe projects to the NL, NIA, and NIP. The heaviest projection is to the dorso-lateral half of the NL. Fibres passing to the NIA and NIP are, at best, moderate to sparse. These observations are in general agreement with other recent studies that have shown projections from the lateral functional zone on to portions of several of the deep cerebellar nuclei.[47-9,51] If the results of the present study are correlated with previous reports, it appears that the lateral functional zone projects onto the lateral portions of the interposed and lateral cerebellar nuclei. Lesions of the parafloccular cortex in the *Galago* projected to ventrolateral and caudal area of NL, and to extremely small areas of the NIA and NIP. It has been previously suggested, for the rat, that the parafloccular cortex projects to a single cerebellar nucleus.[48] All corticonuclear projections from the cortex of the posterior lobe passed into the caudal one-half to two-thirds of the deep nuclei, leaving rostral areas free of degeneration.

Corticonuclear projections from the vermis of the lesser *Galago* pass into the ventromedial and ventrolateral areas of the NM when the lesion is on the midline. If the lesion is restricted to one side of the midline, the fibres appear to pass principally to the ventrolateral areas of the NM. Vermian lobules VIII and IX in lesser *Galago* give cerebello-vestibular fibres which pass to the lateral and superior vestibular nuclei. In *Tupaia*, it tentatively appears that the vermis of the posterior lobe projects to the NM and lateral vestibular nucleus only.[53] Cerebello-vestibular fibres have been reported in the cat, but are apparently not present in the rat or in *Macaca.*[48,50] In *Galago*, the fibres passing to the superior and lateral vestibular nuclei are sparse. It is not surprising that the fibre projections from the cerebellar cortex are slightly different from those previously reported for other animals.[48-50] It has been acknowledged by previous investigators that true species differences do occur.[48] It is also suspected that the evolutionary development of the locomotor pattern of the particular animal under consideration will be reflected by specialisations within the nervous system.[45]

The question of the phylogeny and taxonomy of the tree shrews has been the subject of considerable interest. To some, they represent primates,[54-5] but to others, they have little or no primate affinities.[56-7] A recent review of neuroanatomical evidence would tend to support the concept that the tree shrews should more correctly be considered as a separate order.[58-9] Romer has suggested that tree shrews may represent the primitive stock from which all placental mammals arose, not just extant primates.[60] The tentative results of the present investigation, and some previous studies on the tree shrews,[58,61] indicate that valuable information can be obtained by comparative studies on these animals. If *Tupaia* is further compared with mammals to which it may be related phylogenetically (for example, the elephant shrews and various prosimian forms), new information relative to its position should be realised.[62]

Acknowledgments

The author wishes to express his sincere appreciation to David Cohen and Donald Wright, and to Betsy Harmeling and Jo Bishop for their technical help. The expert assistance of Janet Hall of Duke University is appreciated. The Department of Anatomy and School of Basic Sciences and Graduate Studies has enthusiastically supported this project, and grants from the MCV Foundation and the A.D. Williams Committee (3558-550) partially supported this research. The author is especially indebted to C.J. Gray for typing the preliminary draft of this paper and to M.P. Harrell for typing the final draft.

NOTES

1 Elliot Smith, G. (1903), 'On the morphology of the brain in the mammals, with special reference to that of the lemurs recent and extinct', *Trans. Linn. Soc. Lond.* 8, 319-432.

2 Parentheses are those of Elliot Smith (1903).

3 Bolk, L. (1906), *Das Cerebellum der Säugetiere: Eine vergleichende anatomische Untersuchung*, Haarlem.

4 Woollard, H.H. (1925), 'The anatomy of *Tarsius spectrum*', *Proc. Zool. Soc. Lond.* 1071-184.

5 Tilney, F. (1925), 'The brain stem of *Tarsius*: A critical comparison with other primates', *J. Comp. Neurol.* 43, 371-432.

6 Le Gros Clark, W.E. (1924), 'On the brain of the tree shrew (*Tupaia minor*)', *Proc. Zool. Soc. Lond.* 1053-74.

7 Le Gros Clark, W.E. (1926), 'The anatomy of the pen-tailed tree shrew (*Ptilocercus lowii*)', *Proc. Zool. Soc. Lond.* 1190-307.

8 Le Gros Clark, W.E. (1931), 'The brain of *Microcebus murinus*', *Proc. Zool. Soc. Lond.* 463-86.

9 Kanagasuntheram, R. and Mahran, Z.Y. (1960), 'Observations on the nervous system of the lesser bushbaby (*Galago senegalensis senegalensis*)', *J. Anat.* 94, 512-27.

10 Krishnamurti, A. (1966), 'The external morphology of the brain of the slow loris *Nycticebus coucang coucang*'. *Folia primat.* 4, 361-80.

11 Haines, D.E. (1969), 'The cerebellum of *Galago* and *Tupaia*. I: Corpus cerebelli and flocculonodular lobe', *Brain Behav. Evol.* 2, 377-414.

12 Bradley, O.C. (1904), 'The mammalian cerebellum: Its lobes and fissures', Part I', *J. Anat.* 38, 448-75.

13 Bradley, O.C. (1904-1905), 'The mammalian cerebellum: Its lobes and fissures, Part II', *J. Anat.* 39, 99-117.

14 Author's parentheses.

15 Larsell, O. and Jansen, J. (1967), *The Comparative Anatomy and Histology of the Cerebellum from Myxinoids through Birds*, Minnesota.

16 Larsell, O. and Jansen, J. (1970), *The Comparative Anatomy and Histology of the Cerebellum from Monotremes through Apes*, Minnesota. (All the earlier works of Larsell and others are reviewed in this volume.)

17 Haines, D.E. (1971), 'The cerebellum of *Galago* and *Tupaia*. II: The early postnatal development', *Brain Behav. Evol.* 4, 97-113.

18 Okawa, K. (1957), 'Comparative anatomical studies of cerebellar nuclei of mammals', *Arch. Hist. Jap.* 13, 21-58.

19 Haines, D.E. (1971), 'The morphology of the cerebellar nuclei of *Galago* and *Tupaia*', *Am. J. Phys. Anthrop.* 35, 27-42.

20 The author is fully aware that *Tupaia* is not a member of the Lorisidae, and is even excluded from the suborder Prosimii by some authorities.

21 Klüver, J. and Barrera, E. (1953), 'A method for the combined staining of cells and fibers in the nervous system', *J. Neuropath. Exp. Neurol.* 12, 400-3.

22 Fink, R.P. and Heimer, L. (1967), 'Two methods for selective silver impregnation of degenerating axons and their synaptic endings in the central nervous system', *Brain Res.* 4, 369-47.

23 The author expresses his sincere thanks to I.T. Diamond for allowing Janet Hall of his laboratory to coach the author in the proper methods for using the Fink and Heimer stain on *Galago* and *Tupaia*. Her expertise has been an invaluable aid.

24 The hemispheric portion of a vermian lobule is indicated by the prefix H. For example, the hemisphere of lobule II is indicated by HII.

25 Larsell, O. and Whitlock, D.G. (1952), 'Further observations on the cerebellum of birds', *J. Comp. Neurol.* 97, 545-66.

26 Dillon, L.S. and Brauer, K. (1970), 'A proposed method for establishing homologies among the lobules of the anterior lobe in the mammalian cerebellum', *J.f. Hirnforsch.* 21, 217-32.

27 Elliott Smith, G. (1902), 'Notes on the brain of *Macroscelides* and other Insectivora', *Linn. Soc. J. Zool.* 28, 443-8.

28 Le Gros Clark, W.E. (1928), 'On the brain of the Macroscelididae (*Macroscelides* and *Elephantulus*)', *J. Anat.* 63, 245-75.

29 Le Gros Clark, W.E. (1932), 'The brain of Insectivora', *Proc. Zool. Soc. Lond.* 975-1013.

30 Obenchain, J.B. (1925), 'The brains of the South American marsupials *Caenolestes* and *Orolestes*', *Chicago Field. Mus. Nat. Hist., Zool. Ser.* 14, 175-232.

31 Haines, D.E. and Leichnetz, G.R. (1970), 'The gross morphology of the cerebellum of the adult marmoset', *Anat. Rec.* 169, 332.

32 Crosby, E.C., Taren, A.J. and Davis, R. (1969), 'The anterior lobe and lingula of the cerebellum in monkeys and man', *Top. Probl. Pyschiat. Neurol.* 10, 22-39.

33 Haines, D.E., Albright, B.C., Goode, G.E. and Murray, H.M., this volume.

34 Larsell, O. (1953), 'The cerebellum of the cat and the monkey', *J. Comp. Neurol.* 99, 135-99.

35 Shantha, T.R. and Manocha, S.L. (1969), 'The brain of chimpanzee (*Pan troglodytes*). I: External morphology', *The Chimpanzee* I, 187-237.

36 Courville, J. and Cooper, C.W. (1970), 'The cerebellar nuclei of *Macaca mulatta*: A morphological study', *J. Comp. Neurol.* 140, 241-54.

37 Tamagaki, R. (1959), 'Comparative anatomical studies on the cerebellar nuclei of the primates', *Bull. Kob. Med. Col.* 17, 387-421.

38 Courville, J. (1966), 'Somatotopical organisation of the projection from the nucleus anterior of the cerebellum to the red nucleus: An experimental study in the cat with silver impregnation methods', *Exp. Brain Res.* 2, 191-215.

39 Ohkawa, K. (1957), 'Comparative anatomical studies of cerebellar nuclei of mammals', *Arch. Hist. Jap.* 13, 21-58.

40 Flood, S. and Jansen, J. (1961), 'On the cerebellar nuclei in the cat', *Acta Anat.* 46, 52-72.

41 Korneliussen, H.K. (1966), 'On the morphology and subdivision of the cerebellar nuclei of the rat', *J. f. Hirnforsch.* 10, 109-22.

42 Snider, R.S. (1940), 'Morphology of the cerebellar nuclei in the rabbit and cat', *J. Comp. Neurol.* 72, 399-415.

43 Hines, M. (1929), 'The brain of *Ornithorhynchus anatinus*', *Phil. Trans. Roy. Soc. Lond.* (*B*) 217, 155-287.

44 Abbie, A.A. (1934), 'The brain stem and cerebellum of *Echidna aculeata*', *Phil. Trans. Roy. Soc. Lond.* (*B*) 244, 1-74.

45 Albright, B.C. and Haines, D.E. (1972), 'Morphology of Clarke's column in *Galago senegalensis*', *Anat. Rec.* 172, 447.

46 Rüdeberg, S-I. (1961), 'Morphogenetic studies on the cerebellar nuclei and their homologization in different vertebrates including man', Dissertation, University of Lund.

47 Jansen, J. and Brodal, A. (1940), 'Experimental studies on the intrinsic fibers of the cerebellum. II: The corticonuclear projection', *J. Comp. Neurol.* 73, 267-321.

48 Goodman, D.C., Hallett, R.C. and Welch, R.B. (1963), 'Patterns of localization in the cerebellar cortico-nuclear projections of the albino rat', *J. Comp. Neurol.* 121, 51-67.

49 Eager, R.P. (1963), 'Efferent cortico-nuclear pathways in the cerebellum of the cat', *J. Comp. Neurol.* 120, 81-104.

50 Eager, R.P. (1966), 'Patterns and mode of termination of cerebellar cortico-nuclear pathways in the monkey (*Macaca mulatta*)', *J. Comp. Neurol.* 126, 551-66.

51 Dow, R.S. (1936), 'The fiber connections of the posterior parts of the cerebellum in the rat and cat', *J. Comp. Neurol.* 63, 527-48.

52 Walberg, F. and Jansen, J. (1964), 'Cerebellar corticonuclear projection studied experimentally with silver impregnation methods', *J.f. Hirnforsch.* 6, 338-54.

53 Unpublished observations on the tree shrew (*Tupaia*).

54 Sorenson, M.W. (1970), 'Behavior of Tree Shrews', in Rosenblum, L.A. (ed.), *Primate Behavior: Developments in Field and Laboratory Research*, New York.

55 Le Gros Clark, W.E. (1959), *The Antecedents of Man*, Edinburgh.

56 Campbell, C.B.G. (1966), 'The relationships of the tree shrews: The evidence of the nervous system', *Evolution* 20, 276-81.

57 van Valen, L. (1965), 'Tree shrews, primates and fossils', *Evolution* 19, 135-51.

58 Haines, D.E. and Swindler, D.R. (1972), 'Comparative neuroanatomical evidence and the taxonomy of the tree shrews (*Tupaia*)', *J. Human Evol.*, in press.

59 Straus, W.L. (1949), 'The riddle of man's ancestry', *Quart. Rev. Biol.* 24, 200-23.

60 Romer, A.S. (1969), 'Vertebrate history, with special reference to factors related to cerebellar evolution', in Llinas, L.L. (ed.), *Neurobiology of Cerebellar Evolution and Development*, AMA Education and Research Foundation.

61 Martin, R.D. (1968), 'Reproduction and ontogeny in tree-shrews (*Tupaia belangeri*), with reference to their general behaviour and taxonomic relationships', *Z.f. Tierpsychol.* 25, 409-95 and 505-32.

62 Since submission of this manuscript the author has become aware of additional papers by Achenbach and Goodman (*Brain Behav. Evol.* 1,43-57, 1968) and Eager (Symp. on the Role of Vestibular Organs in Space Exploration, Pensacola, Fla., 1967) which report cerebellar corticovestibular fibres in the rat and *Macaca.*

L. RADINSKY

Prosimian brain morphology: functional and phylogenetic implications

Introduction

Two sources of information are available for studies of prosimian brain evolution: the brains of modern prosimians and endocranial casts of fossils. The scanty fossil record of prosimian brains has been reviewed recently,[1] and not enough new information has become available since then to warrant another review. Studies of modern prosimian brains offer access to much more information from which evolution may be interpreted, but few broad comparative surveys have been carried out. A notable exception is the work of Stephan, Bauchot and Andy,[2,3,4] whose quantitative studies of major brain structures in insectivores and prosimians are excellent examples of an important but under-utilised approach.

Most comparative studies involving prosimian brains have been purely descriptive and have dealt with gross external morphology alone. The best comparative study of prosimian brain morphology, both for the number of species treated and for the analysis of details, is that of Smith,[5] which, however, covered only 11 of the 17 commonly recognised genera. Recently,[6] I published figures of endocasts of 12 genera of recent prosimians, with some functional interpretation. The present study was undertaken firstly to survey the entire range of diversity of external brain morphology of extant prosimians so that data at least comparable to that published for ceboid primates would be available for prosimian studies,[7,8] and secondly to interpret functionally morphological differences and consider their relevance to phylogenetic interpretations. The use of endocranial casts made it possible to examine external brain morphology of all living genera of prosimians.

Methods

Although the value of endocasts for illustrating prosimian brain morphology was demonstrated a century ago,[9] little use has been made of them since then. As has been amply demonstrated, endocasts of prosimians, as of most mammals, reproduce in fine detail most of the external morphology of the brain, including all of the cerebral sulci, as well as the pattern of blood vessels, cranial nerve stumps, and often braincase sutures.[6,10] In addition, endocasts reproduce the shape and proportions of the major parts of the brain as they existed in life, undistorted by the shrinkage that usually accompanies brain fixation.

Endocasts were made from skulls representing all genera of living prosimians, using a latex casting technique which does not damage the skull.[6] Interpretation of prosimian external brain morphology

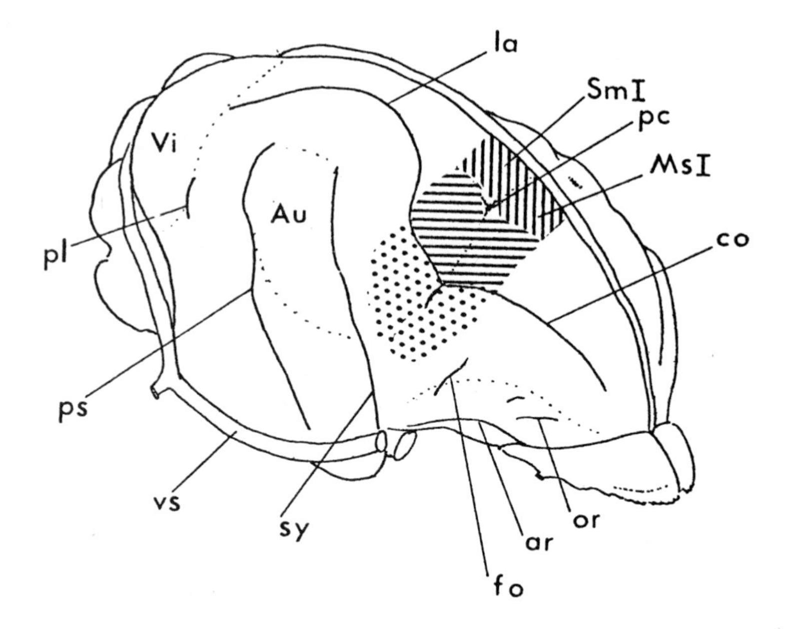

Figure 1. Endocast of *Lemur fulvus*, to show sulcal terminology used and to indicate cortical areas as inferred from references cited in text. Stippled area indicates head, horizontal hatching fore limb, and vertical hatching hind limb areas of somatic sensory and motor cortex. Abbreviations: ar, anterior rhinal fissure; Au, auditory cortex; co, coronal sulcus (? = sulcus rectus); fo, fronto-orbital sulcus; la, lateral sulcus (? = intraparietal); MsI, primary motor cortex; or, orbital sulcus; pc, postcruciate sulcus (transverse branch = central sulcus); pl, postlateral sulcus (= lunate sulcus); ps, postsylvian sulcus (= posterior suprasylvian, superior temporal or parallel sulcus); SmI, primary somatic sensory cortex; sy, sylvian sulcus (= anterior suprasylvian sulcus); Vi, visual cortex (Brodmann's area 17); vs, vascular sinus overlying posterior rhinal fissure.

was aided by the histological and physiological cortical mapping studies of Brodmann,[11] Campbell et al.,[12] Kanagasuntheram et al.,[13] Mott and Kelley,[14] Sanides and Krishnamurti,[15] and Zuckerman and Fulton.[16] Terminology for sulci used in this study, plus cortical areas as interpreted from the preceding references, are shown in Fig. 1. Examination of drawings of endocasts superimposed in proper position on skulls in many cases provided insights into the functional significance of differences in brain shape and proportions. To facilitate comparisons, all endocast-skull drawings were oriented with the floor of the medulla horizontal in lateral view and in the plane of the paper in dorsal view. Finally, the quantitative data provided by Stephan et al.[4] were used to check and supplement some of the ideas that resulted from study of the endocasts alone.

Observations

1. *Family Lemuridae*

(*a*) Subfamily Lemurinae: see Fig. 2. The three commonly recognised living genera of lemurines are *Lemur, Hapalemur* and *Lepilemur*, in order of decreasing size. Brains of the genus *Lemur* have the following sulci: coronal, postcruciate, lateral, postlateral, sylvian, postsylvian, orbital (usually represented by two dimples), and sometimes a fronto-orbital sulcus. A vascular sinus overlies the posterior rhinal fissure. The relationship between the caudal end of the sylvian and the postsylvian is variable. There is also considerable variation in the degree of overlap of the cerebellum and in the degree of convexity of the frontal lobes as seen in lateral view; the specimens of *Lemur variegatus* and *L. rubriventer* shown in Fig. 2 represent the extremes of variation of those features present in the sample available to me. My sample, even combined with figures from the literature, is too small to determine the extent to which species of *Lemur* may be distinguished by those or other neuro-anatomical features.

The brain of *Hapalemur* displays a sulcal pattern similar to that of *Lemur*, except that it lacks the postlateral sulcus. (The fronto-orbital sulcus is absent in the endocast I studied, but is present in a brain figured by Beddard.)[17] The one endocast available to me displays general proportions similar to those of *Lemur* species, except for being relatively narrower and in having the olfactory bulbs relatively more ventrally located than in most specimens of *Lemur*. (The *Lemur rubriventer* endocast figured here has the olfactory bulbs about as low as in *Hapalemur*, but differs in having a more convex frontal lobe).

Lepilemur, which is the smallest of the lemurines, has a sulcal pattern similar to that of *Lemur*, except that it lacks the postsylvian, postlateral and fronto-orbital sulci, and in having a less pronounced coronal sulcus. Brain shape and proportions fall within the range observed in *Lemur*.

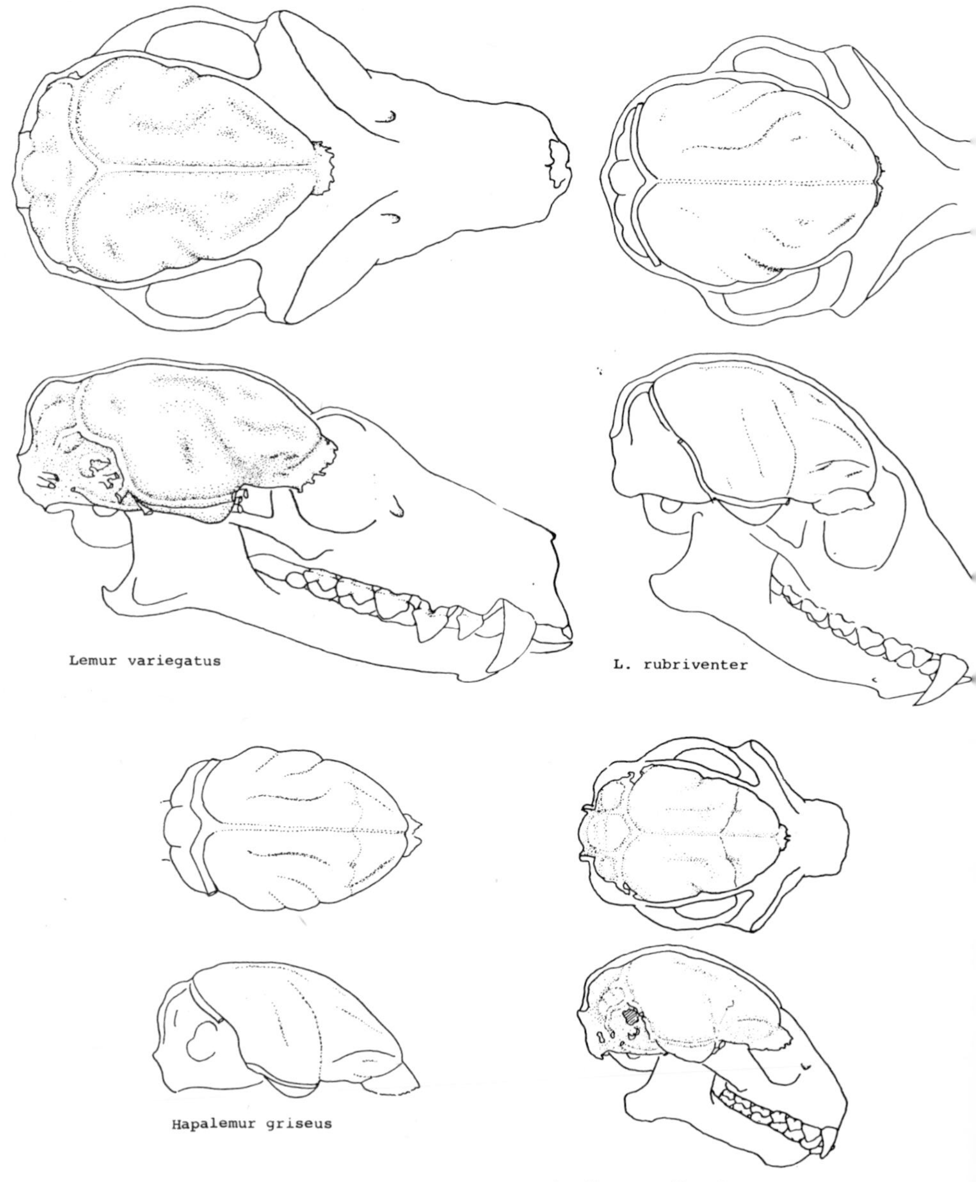

Figure 2. Endocasts of lemurines. All x 0.75.

(*b*) Subfamily Cheirogaleinae: see Fig. 3. The cheirogaleines include three genera, *Phaner, Cheirogaleus* and *Microcebus.* The brain of *Phaner*, the largest of the cheirogaleines, is similar in size to that of *Lepilemur*, the smallest lemurine, and has a similar sulcal pattern except that it lacks the postcruciate sulcus and possesses a short postsylvian. The brain of *Cheirogaleus* is similar to that of *Phaner* except that it is slightly smaller and lacks an orbital sulcus. *Microcebus*, the smallest of the lemurids, is one of the few prosimians for which the brain has been studied in any detail.[18] It differs from the preceding cheirogaleines in lacking all sulci except for the sylvian and a faint trace of a postsylvian. The olfactory bulbs are more compressed laterally than in the other lemurids, presumably a correlate of *Microcebus* being small and having relatively large eyes.

2. *Family Indriidae*

The family Indriidae (see Fig. 4) includes three living genera, *Indri, Propithecus* and *Lichanotus* (=*Avahi*), in order of decreasing body size. The brain of *Indri* is similar to that of comparable-sized *Lemur* specimens, except for having a more pronounced postlateral sulcus and broader frontal lobe, with the postcruciate sulcus better developed and, most striking, having a transverse branch that crosses the coronal sulcus. Related to the latter features, the motor area for the hand is larger and bulges out more in *Indri* than in *Lemur* (see

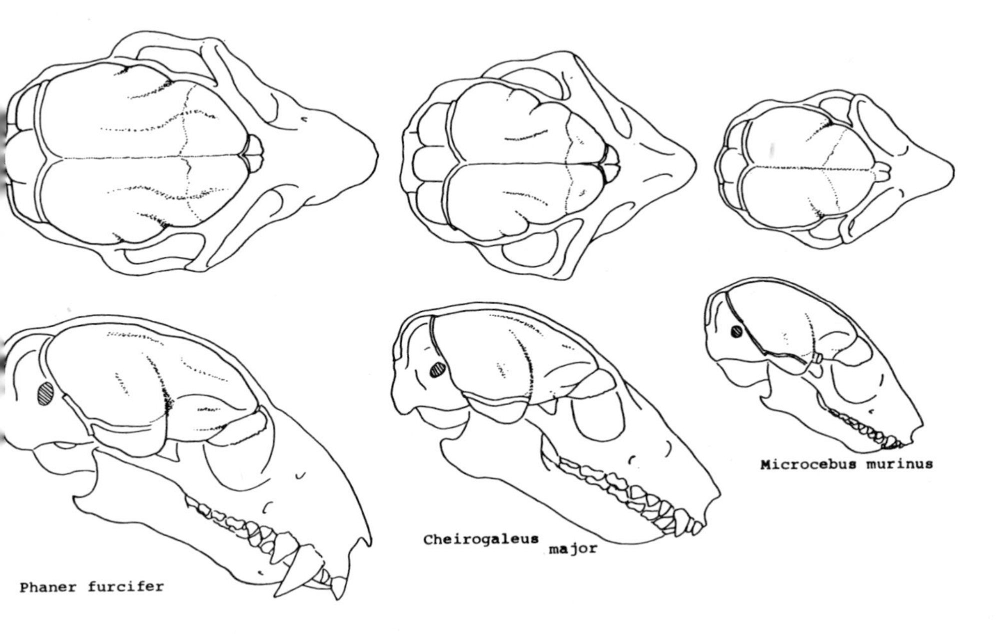

Figure 3. Endocasts of cheirogaleines. All natural size.

area marked 'x' in Fig. 4). The caudal end of the postsylvian is particularly variable in *Indri.*

The brain of *Propithecus* has a sulcal pattern similar to that of *Indri*, except that the sylvian sulcus extends further caudally than the lateral sulcus, usually terminating in a short caudomedially directed segment. Also, the transverse branch of the postcruciate sulcus, which is the homologue of the central sulcus, is more pronounced in *Propithecus* than in *Indri.* In shape, the brain of

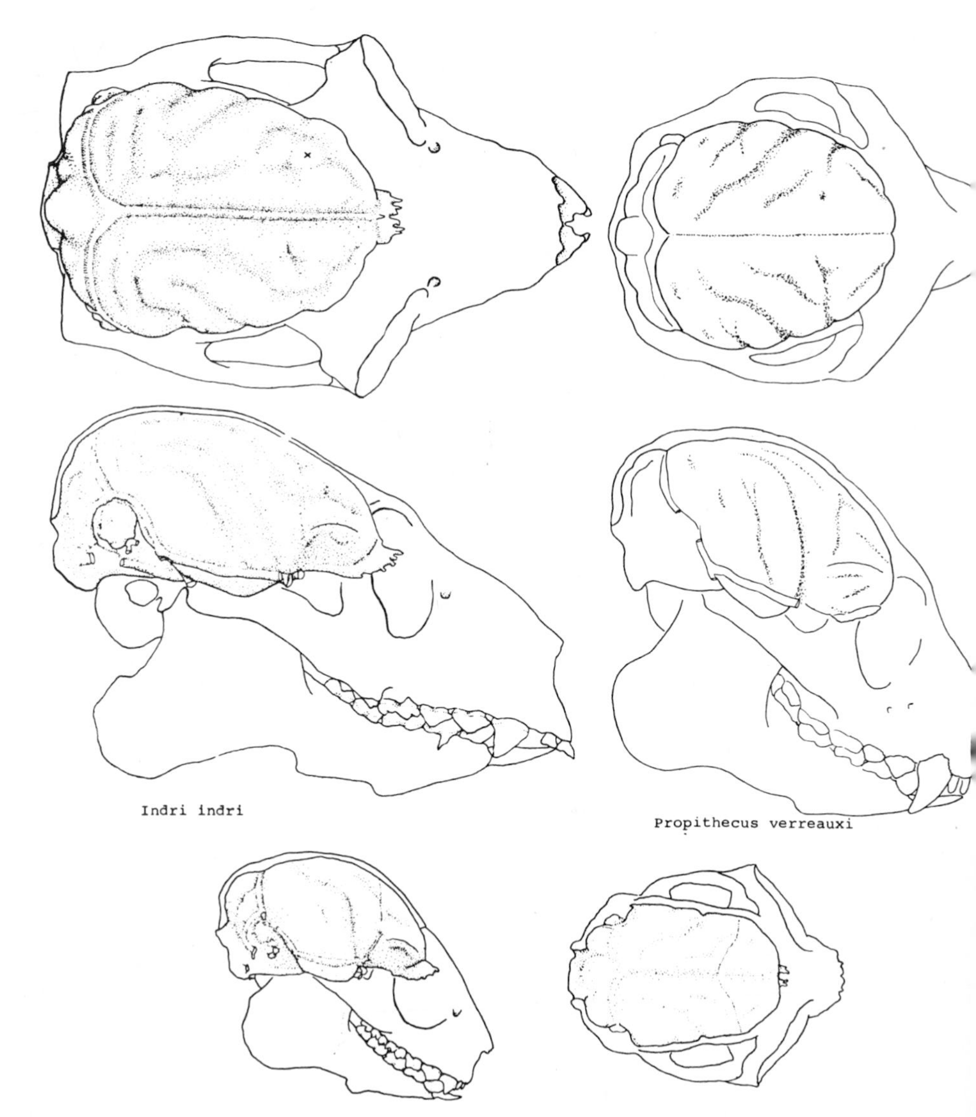

Figure 4. Endocasts of indriids. x (top left illustration) marks inferred hand area in primary motor cortex of *Indri.* All x 0.75.

Propithecus differs from that of *Indri* in having a more convex frontal lobe dorsal profile, with more downwardly rotated olfactory bulbs.

The brain of *Lichanotus* lacks the postcruciate and fronto-orbital sulci seen in its larger relatives, and the postsylvian and postlateral sulci are less developed. As in *Propithecus*, the sylvian extends far caudally, past the end of a shorter lateral sulcus. The frontal lobes are relatively broad, as in *Indri* and *Propithecus*, and intermediate in dorsal profile convexity between the latter two genera.

3. *Family Daubentoniidae*

The brain of *Daubentonia madagascariensis*, the sole surviving species in the family (see Fig. 5), is bizarre among prosimians and has been the subject of several investigations.[6,19–21] It is relatively high and narrow, and has a more convex and downwardly rotated frontal lobe than any comparable sized prosimian brain (compare with *Indri* and *Lemur*), and is further unique in lacking orbital impressions. The coronal sulcus is broken and variable in development, as is the postcruciate complex. The anterior suprasylvian is separated from the pseudosylvian, exposing an area of cortex that in all other prosimians lies buried in a sylvian sulcus. The postsylvian is variably developed and apparently there is no postlateral sulcus.

4. *Family Tarsiidae*

The brain of *Tarsius*, the sole living genus of this family (see Fig. 5), is relatively short and broad, and has relatively small frontal lobes, small olfactory bulbs and a wide occipital-temporal area (comprised of visual cortex), compared to other prosimians. The only sulcus present is the sylvian. The small frontal lobes are primitive and the large visual area and small olfactory bulbs are advanced characters.

5. *Family Lorisidae*

(*a*) Subfamily Galaginae: see Fig. 6. The genus *Galago* includes species differing more in size than do species of other prosimian genera, and with correspondingly greater differences in brain morphology. The largest species of *Galago*, *G. crassicaudatus*, has a brain close in size to that of the lemurine *Hapalemur*, and with a similar sulcal pattern, including coronal, lateral, orbital, sylvian, postsylvian and (occasionally present) a faint postcruciate sulcus. Six *G. crassicaudatus* brains examined by Haines et al. had a broken coronolateral sulcal complex, while in the two endocasts available to me that sulcal complex appeared to be continuous.[22] The olfactory bulbs are relatively larger in *Galago crassicaudatus* than in *Hapalemur*, and the frontal lobes are relatively narrower, although there is apparently considerable variation in that feature (see Fig. 6). The brain of *Galago senegalensis* is considerably smaller than that of *G. crassicaudatus*, with the sylvian and postsylvian the only clear sulci, and a faint trace of orbital and coronal sulci. The frontal lobes of *G. senegalensis* appear shorter and are downwardly rotated compared to those of *G. crassicaudatus*. An endocast of *Galago senegalensis* figured by Clark and Thomas,[23] is broader across the temporal lobes and has a more tapering frontal lobe profile in dorsal view than the endocast figured by me. An endocast of *Galago demidovii* reveals a brain somewhat smaller but otherwise similar to that of *G. senegalensis*.

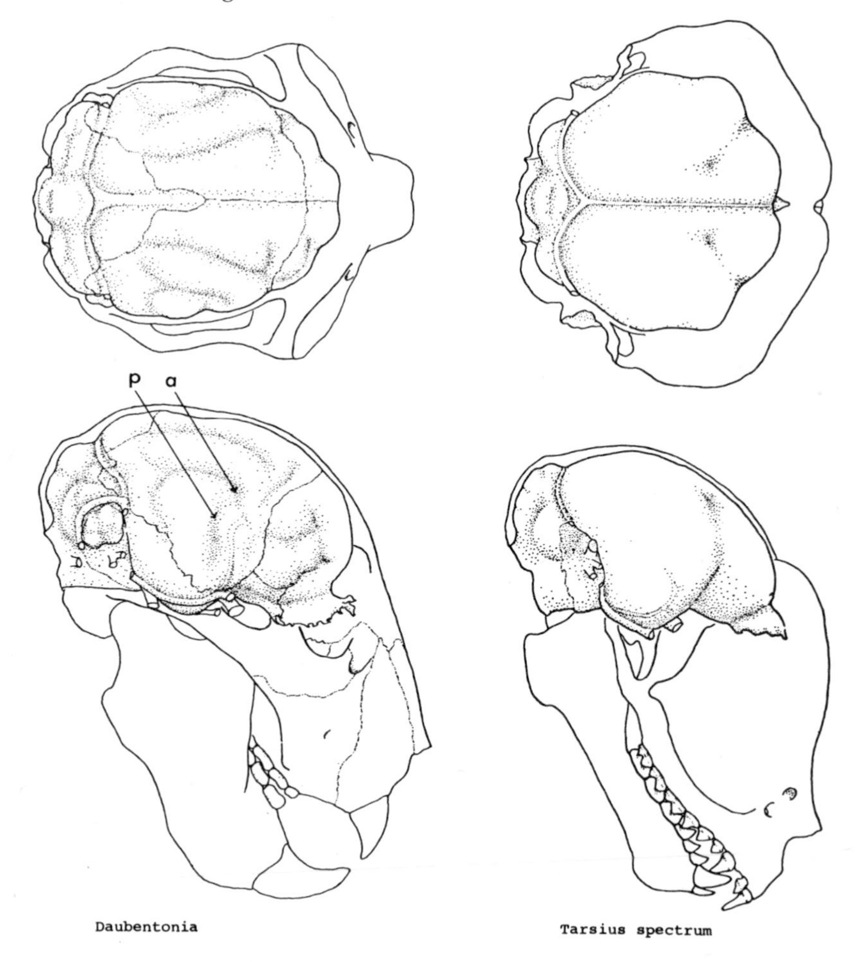

Figure 5. Endocasts of *Daubentonia* and *Tarsius*, the former x 0.75, the latter x 1.5. Abbreviations: a, anterior suprasylvian sulcus; p, pseudosylvian sulcus.

The brain of *Euoticus elegantulus* is slightly larger than that of *Galago senegalensis*, and has lateral, coronal and orbital sulci, in addition to the sylvian and postsylvian. The olfactory bulbs are less downwardly rotated than in *G. senegalensis.* The frontal lobes have the same narrow, tapering outline in dorsal view as in *Galago.*

(*b*) Subfamily Lorisinae: see Fig. 7. The lorisines include two African genera, *Perodicticus* and *Arctocebus*, and two genera confined to Asia, *Nycticebus* and *Loris.* The brain of *Perodicticus* is about the same size as that of *Galago crassicaudatus*, but with relatively broader frontal lobes and olfactory bulbs, smaller orbital impressions, and two major additions to the sulcal pattern: a short but prominent

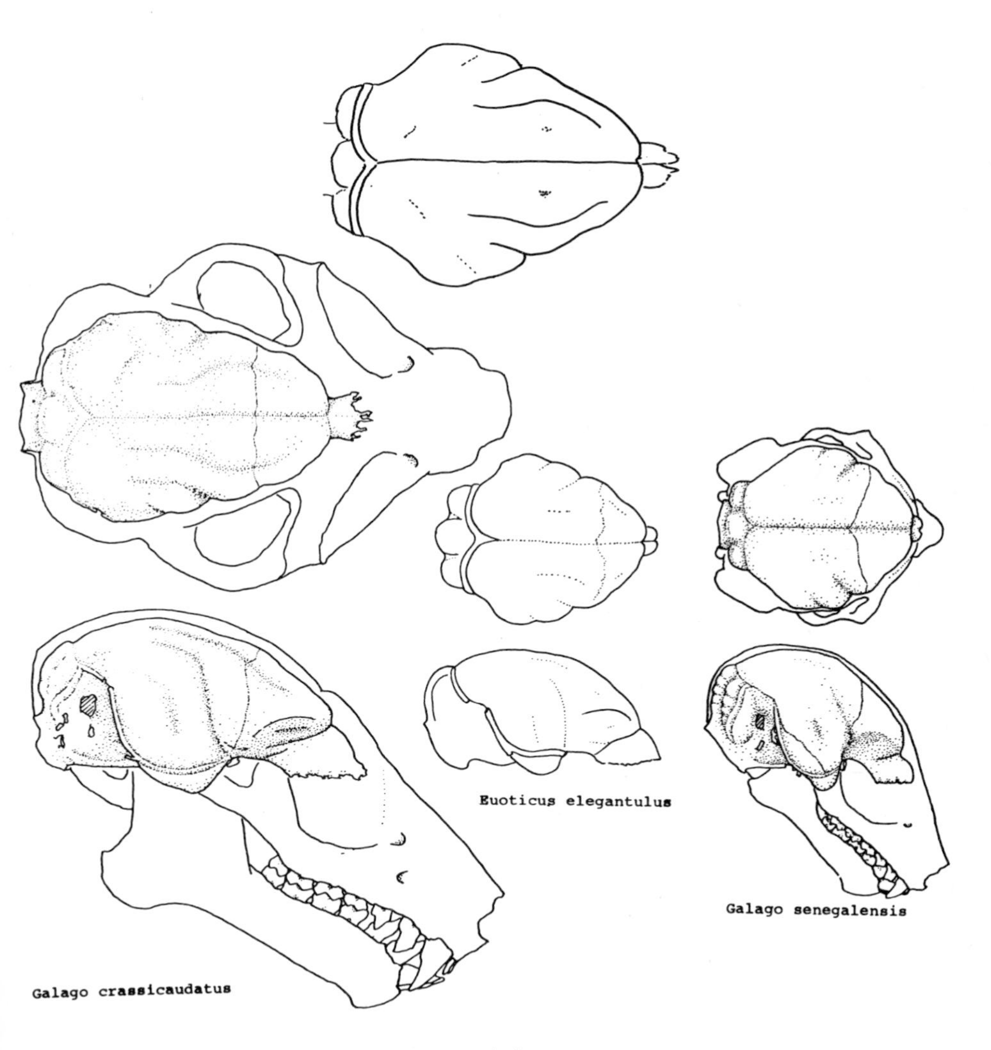

Figure 6. Endocasts of galagines. All natural size.

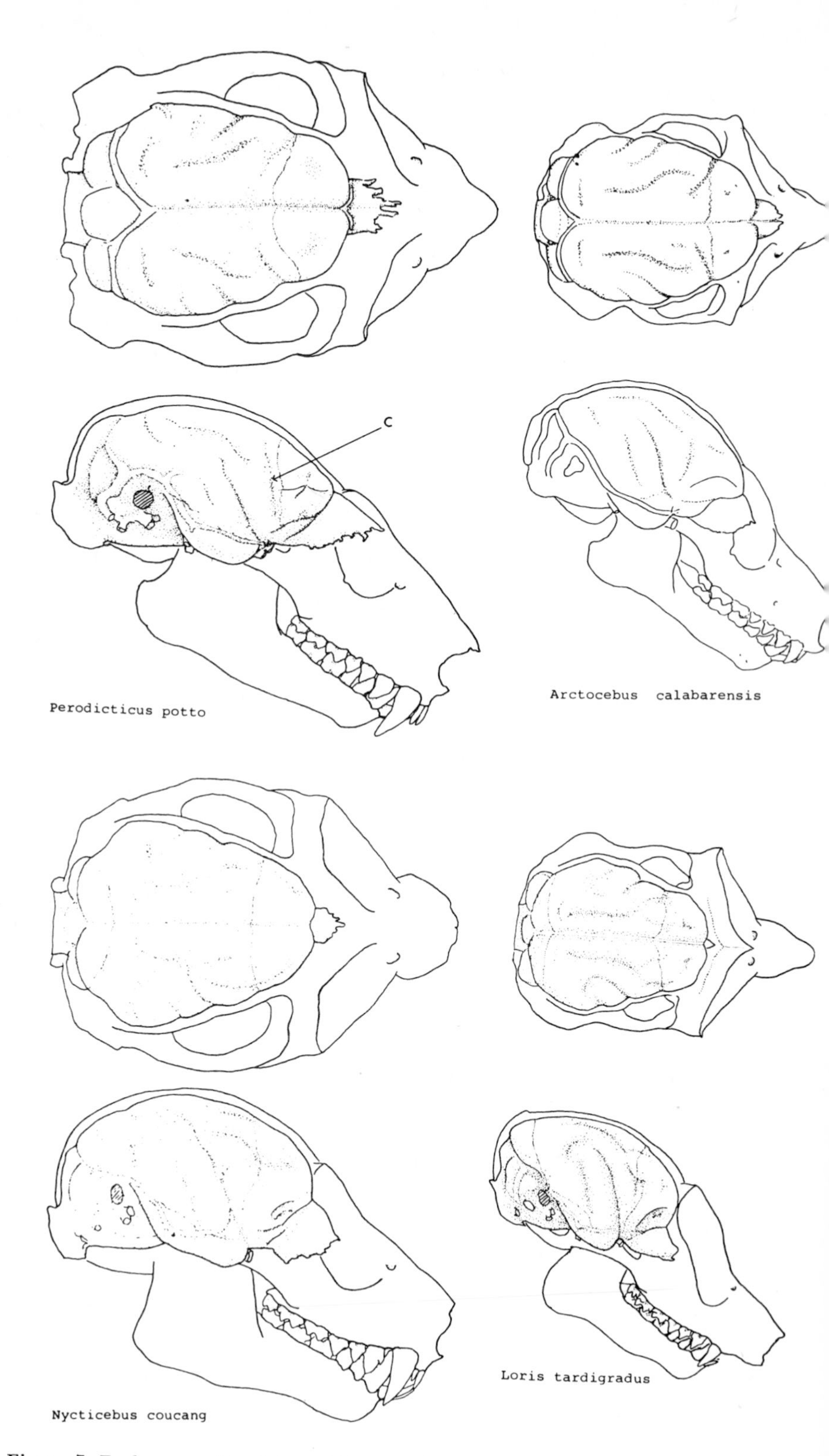

Figure 7. Endocasts of lorisines. All natural size. c = central sulcus.

postlateral sulcus and a long central sulcus (developed from the postcruciate complex) that cuts across the coronal sulcus and descends well down across the coronal gyrus. Those sulci are better developed in *Perodicticus* than in comparable-sized indriids or lemurids, and the central sulcus is better developed than in any other prosimian, regardless of size.

Arctocebus calabarensis has a brain smaller than that of *Perodicticus*, with a more convex dorsal profile and more downwardly rotated frontal pole, but with the same basic sulcal pattern, including the postlateral and central sulci. The posterior rhinal fissure is higher in *Arctocebus* than in the other prosimians examined.

Nycticebus coucang has a brain about the size of that of *Perodicticus potto*, but with a more convex dorsal profile. The sulcal pattern is unique in that there is a short portion of the central sulcus extending ventrally from the coronal sulcus, a longitudinally oriented postcruciate sulcus, and the sylvian sulcus rather than the coronal sulcus connects up with the lateral sulcus. The last is reminiscent of the condition in the indriids *Propithecus* and *Lichanotus*, but in these two genera the lateral sulcus is more extensive. Like *Perodicticus* the brain of *Nycticebus* has a short but marked postlateral sulcus. The olfactory bulbs are narrower in *Nycticebus* than in *Perodicticus*, being similar in relative width to those of the galagines. An endocast of *Nycticebus pygmaeus* is slightly smaller than that of *N. coucang* and displays the same sulcal pattern except for the apparent absence of a postlateral sulcus.

The brain of *Loris tardigradus* is smaller than that of *Nycticebus coucang*, with a more convex frontal lobe dorsal profile and more ventral and caudally displaced olfactory bulbs. The same sulci that are present in *Nycticebus* occur in *Loris*, including the orbital and postlateral sulci; the main difference is that in *Loris* the lateral joins the coronal in the normal prosimian manner, rather than the sylvian, as in *Nycticebus*.

Discussion

1. *Cortical folding*

A common kind of difference in external brain morphology observed between prosimian genera is in the number of neocortical sulci. Having a highly folded cortex is considered an advanced character, and in large part because of that the brain of *Microcebus*, for example, has been described as primitive.[8,24] However, it is clear from the above survey of prosimian brains that as a general phenomenon smaller brained forms have fewer sulci than their larger brained relatives. This is evident in examining brains of the lemurid genera arranged in order of size, in comparing the small indriid *Lichanotus* with *Propithecus* or *Indri*, or the small galagines *Galago senegalensis* and *G. demidovii* with the larger species *G. crassicaudatus*. The same allometric phenomenon may be observed in ceboid primates (see figures in Hershkovitz),[8] artiodactyl ungulates, and other groups of mammals, and was considered by Clark and others to result simply from the need to maintain an approximately constant proportion of gray cortex to internal white matter:[25] the thin sheet of cortex expands as a surface area, proportional to the square of a radius, while the white matter expands as a volume, proportional to the cube. Given that consideration, the lack of sulci in small prosimian brains cannot in itself be considered a primitive character; it merely reflects their small size. Only when brains of similar size are compared would fewer sulci suggest relatively less neocortex.

The quantitative data presented by Stephan and Andy, which must be interpreted with some caution owing to the small sample sizes and the fact that their indices are based on body weight, nevertheless provide the possibility of a check on inferences derived from endocast observations.[3] For example, the cerebral cortex of *Loris* is more highly folded than that of some similar sized or slightly larger-brained prosimians, such as *Cheirogaleus*, *Lepilemur* and *Lichanotus*, and the direct volumetric measurements of Stephan and Andy show *Loris* to have relatively more neocortex.[3] The same work shows the brain of *Microcebus* to have relatively more neocortex than that of *Cheirogaleus*, *Lepilemur*, *Hapalemur* and *Lichanotus*, all of which have more cortical sulci.

Details of sulcal pattern also differ between brains of various prosimian genera, and some of those differences can be interpreted by reference to the cortical mapping studies listed above and summarised in Fig. 1. For example, brains of *Indri* and *Propithecus* differ from *Lemur* brains in having a more prominent development of the transverse branch of the postcruciate sulcus and a larger area of cortex immediately rostral to that sulcus and medial to the coronal sulcus. That cortical area, marked 'x' in the *Indri* brain in Fig. 4, is the primary motor area for the fore limb; its relatively greater size in the large indriids suggests finer muscular control of the hand in *Indri* and *Propithecus* than in *Lemur*.

A postlateral sulcus delimits the rostral boundary of the visual cortex in all lorisine brains (except *Nycticebus pygmaeus*), but is not present in brains of comparable sized lemurids, indriids or galagines. By analogy with work that has been done on sulci of the somatosensory cortex, the presence of the postlateral sulcus in lorisine brains suggests expansion of the visual cortex (see Welker and

Campos for discussion of functional significance of sulci).[26] The volumetric measurements of visual cortex analysed by Stephan indicate that *Nycticebus coucang* and *Loris* do indeed have relatively considerably more visual cortex than those other prosimians.[2] However, the brain of *Perodicticus*, which is similar in size to that of *Nycticebus coucang* and has a similar development of the postlateral sulcus and occipital lobe, has considerably less visual cortex according to Stephan's data (measurements of the visual cortex of *Arctocebus* are not available).[2] This contradiction of an inference based on observation of sulcal pattern emphasises the need to interpret with caution and to check when possible, with direct measurements or experiments, hypotheses based on such observations.

A difference in sulcal pattern, the significance of which is not immediately obvious, is the presence of a well developed central sulcus in *Perodicticus* and *Arctocebus.* The dominance of that transverse sulcus, which separates primary motor cortex from primary somatic sensory cortex, over the longitudinally oriented coronal sulcus, which separates head from forelimb representations in both areas, distinguishes the brains of anthropoid primates from those of prosimians. Studies of the developing thalamus and cortex in *Perodicticus* and *Arctocebus* might yield insights into the question of why the central sulcus is so prominent in anthropoid primates.

Another difference in sulcal pattern which is equally puzzling is the caudal extension of the sylvian sulcus in *Propithecus, Lichanotus* and *Nycticebus.* A similar pattern occurs in some cebid monkeys, where the sylvian sulcus is confluent with the intraparietal.[27] In most but not all of the cebids that have that pattern, the overlap appears to have resulted from expansion of the tail representation area in somatic sensory cortex and consequent lateral displacement of the intraparietal sulcus. That particular explanation is not applicable to *Propithecus, Lichanotus* and *Nycticebus*, and since the cortical areas immediately involved are association areas, it is difficult to suggest a functional interpretation.

The peculiar sulcal pattern of *Daubentonia* appears to have resulted largely from changes in brain shape, and will be discussed below.

2. *Brain shape and proportions*

Another kind of difference between various prosimian genera is in shape and proportions of the brain. Endocasts are particularly well suited for revealing this kind of difference because they reproduce the shape and proportions of the major parts of the brain as they were in life. Examination of endocasts superimposed on the outline drawings of the skull may provide insight into the significance of differences in brain shape, since skull shape may affect brain shape.

A commonly occurring phenomenon is that the frontal lobe is more convex in smaller forms than in their larger relatives. This is seen in the following pairs: *Lemur rubriventer* and *L. variegatus; Propithecus* and *Indri; Galago senegalensis* and *G. crassicaudatus; Arctocebus* and *Perodicticus*; and *Loris* and *Nycticebus.* The more convex and downwardly rotated frontal pole, which is often correlated with a shorter and/or more ventrally flexed splanchnocranium, appears to be a general allometric phenomenon, and has been discussed in the literature (see Biegert and references cited therein).[28] In some cases, splanchnocranial deflection and the accompanying rotation of the frontal pole appears to arise from the need to accommodate larger orbits in a smaller face; *Tarsius* is an outstanding example of this.

Splanchnocranial deflection and consequent modification of brain shape may also occur in large forms, in response to other than allometric demands. The skull of *Daubentonia* appears to have been modified to resist stresses from biting into hard materials (the aye-aye bites into fruit stones and the wood of trees to get at boring insect larvae). Splanchnocranial rotation and rotation of the incisors has brought the long axis of the rostrum into line with the upper incisors, which allows more effective absorption and transmission of stresses than does the condition in other prosimians. The braincase is relatively high and narrow, which would also appear to facilitate resistance of stresses in the saggital plane.[29] The splanchnocranial flexion and changes in shape of the braincase, both in response to selective pressures relating to mastication, have resulted in the globose shape, with downwardly rotated frontal pole and olfactory bulbs, which characterises the brain of *Daubentonia.* Those modifications, perhaps coupled with what appear to be relatively smaller orbits in *Daubentonia* than in *Lemur* and *Indri*, probably account for the lack of orbital impressions in *Daubentonia.* The disrupted sulcal pattern of the frontal lobe appears to have resulted from those changes in frontal lobe shape, as may the separation of anterior suprasylvian and pseudosylvian sulci. Another suggestion to account for the latter is that *Daubentonia*, which uses hearing to locate its prey, may have an expanded auditory cortex (which lies between suprasylvian and pseudosylvian sulci).

The frontal lobes are relatively broader and less tapering in all indriids than in comparable-sized lemurids. Some of that difference might be the result of an expanded motor cortex (see above); but examination of endocasts suggests that motor cortex expansion alone is insufficient to account for the observed difference in frontal lobe shape. Examination of dorsal view drawings of endocasts superimposed on skulls reveals that indriids also have more forwardly facing orbits and that the anterolateral border of the frontal lobe approximately parallels the plane of the orbit in both groups (compare Figs. 2 and 4). This suggests that forward rotation of the orbits altered the shape of the frontal lobes in indriids, producing a relatively broader and less tapering outline.

The brain of *Nycticebus* is relatively globose for a prosimian of its size (compare with *Perodicticus, Galago crassicaudatus*, small indriids and lemurines), which may be correlated with its having relatively large and forwardly facing orbits. That effect on brain shape is even more marked in *Loris*, which has relatively even larger orbits. Similarly, the more forwardly facing orbits in all lorisines may be responsible for the smaller orbital impressions in those prosimians than in comparable sized galagines and lemuroids. Enormously expanded orbits probably account in part for the relatively short and wide brain shape of *Tarsius*, although expansion of the visual cortex probably also played a role in bringing about the very broad occipito-temporal region.

A greater degree of overlap of cerebrum over cerebellum has been noted by previous workers to distinguish brains of galagine and lorisine lorisids from those of comparable sized lemurids and indriids. This appears to be supported in the sample of endocasts I have studied, although the relatively great amount of intraspecific variation present in some taxa for which I have several specimens (*Lemur, Indri, Perodicticus*) suggests that it is a character that must be used cautiously. Quantitative data presented by Stephan and Andy on neocortical volumes reveal a relatively greater amount of neocortex in lorisines and galagines than in comparable-sized lemurids and indriids, which may be the explanation for the difference in cerebellar exposure.[3]

Another feature that should be examined carefully is relative size of olfactory bulbs. Appearances may be deceptive and, for example, measurements by Stephan and Andy of olfactory bulb size reveal that the olfactory bulbs of *Tarsius* are relatively larger than those of the indriids and *Hapalemur*, a difference that was not obvious from examination of endocasts.[3] The observation that lorisids have relatively larger olfactory bulbs than do lemuroids holds true for lemurines and indriines, but not for *Daubentonia, Cheirogaleus* and *Microcebus*, according to Stephan and Andy's data.[3,6]

3. *Phylogenetic implications*

Features of external brain morphology characterise various groups of prosimians; the families Lemuridae and Indriidae, the subfamilies Lorisinae and Galaginae, and the genera *Daubentonia* and *Tarsius*, represent six taxa that are clearly distinguishable on the basis of external neuroanatomy. While lorisines and galagines are currently placed in the same family (Lorisidae) by most primate taxonomists, the differences in external brain morphology between them are as great or greater than those that distinguish other families of prosimians, and in fact appear greater than the differences in brain morphology between galagines and lemurids. However, this is not to say that the latter two groups are related. Before neuro-anatomical features are used to support hypotheses of phylogenetic affinity, they should be carefully evaluated in light of their functional significance relative to that of other characters used to distinguish higher taxa of prosimians. At present I would only emphasise that neuro-anatomical data support evidence from other parts of the body that galagines and lorisines are more different from each other than are lemurids and indriids.

Simpson stated that *Loris* most closely resembles *Arctocebus*, while *Nycticebus* is closest to *Perodicticus*.[30] The brain morphology suggests, on the contrary, that *Loris* and *Nycticebus* are most closely related (the former appears to be a dwarfed version of the latter), while *Perodicticus* and *Arctocebus* are linked by the common possession of a central sulcus.

The differences in sulcal pattern that distinguish *Propithecus* from *Indri* and *Nycticebus* from *Loris* (continuity of sylvian and lateral sulci) should not, I believe, be given more than generic value, since in both cases they result from a change in the course of one sulcus in brains that are otherwise quite similar within each respective pair. Except for the differences in development of the central sulcus in lorisines, it appears that the features that distinguish brains of the prosimian genera within the major polygeneric groupings (Lemuridae, Indriidae, Galaginae, Lorisinae) can be attributed mainly to size differences.

It has been suggested by Le Gros Clark and others that similarities in external brain morphology indicate phylogenetic affinity between *Tarsius* and the Eocene prosimians *Tetonius* and *Necrolemur*.[24] However, as I have argued previously, I find the main source of special resemblance to arise from the failure of *Tarsius*, alone among the living prosimians, to have evolved frontal lobes larger than the Eocene condition, and such a similarity, i.e. retention of a primitive condition, cannot be taken as evidence of special phylogenetic affinity.[1] I see no features in the brains of any modern prosimians that suggest special relationships to any particular group of early Tertiary prosimians.

Advanced development of a central sulcus in *Arctocebus* and *Perodicticus* should not be taken to indicate special affinities to the anthropoid primates, for that sulcus also occurs, although less well developed, in the Asiatic lorisines and in indriids. It would not be surprising to find independent development of that sulcus in several groups of primates; parallel evolution of an equally prominent sulcus, the cruciate sulcus, has been demonstrated in the Carnivora.[31] Nor should the presence of a well developed coronal sulcus be considered to preclude from the ancestry of anthropoids any particular line of fossil prosimians, for the variable development of central and coronal sulci in various living prosimians suggests the possibility of the central sulcus replacing the coronal in prominence during brain evolution within a given lineage.

Acknowledgments

This work was supported in part by National Science Foundation Grant GB-31242 and a grant from the Abbott Memorial Fund, Anatomy Department, University of Chicago.

NOTES

1 Radinsky, L. (1970), 'The fossil evidence of prosimian brain evolution', in Noback, C.R. and Montagna, W. (eds.), *The Primate Brain: Advances in Primatology* 1, New York.

2 Stephan, H. (1969), 'Quantitative investigations on visual structures in primate brains', *Proc. 2nd. Int. Congr. Primat.* 3, 34-42.

3 Stephan, H. and Andy, O.J. (1969), 'Quantitative comparative neuro-anatomy of primates: an attempt at a phylogenetic interpretation', *Ann. N.Y. Acad. Sci.* 167, 370-87.

4 Stephan, H., Bauchot, R. and Andy, O.J. (1970), 'Data on size of the brain and of various brain parts in insectivores and primates', in Noback, C.R. and Montagna, W. (eds.), *The Primate Brain: Advances in Primatology* 1, New York.

5 Smith, G.E. (1903), 'On the morphology of the brain in the Mammalia, with special reference to that of the lemurs, recent and extinct', *Trans. Linn. Soc. Lond. (Zool.)* 8, 319-432.

6 Radinsky, L. (1968), 'A new approach to mammalian cranial analysis, illustrated by examples of prosimian primates', *J. Morph.* 124, 167-80.

7 Anthony, J. (1946), 'Morphologie externe du cerveau des singes platyr-rhiniens', *Ann. Sci. Nat. (Zool.)* 8, 1-150.

8 Hershkovitz, P. (1970), 'Cerebral fissural patterns in platyrrhine monkeys', *Folia primat.* 13, 213-40.

9 Gervais, P. (1872), 'Mémoire sur les formes cérébrales propres à l'orde des Lémures', *J. Zool.* 1, 5-27.

10 Bauchot, R. and Stephan, H. (1967), 'Encéphales et moulages endocraniens de quelques insectivores et primates actuels', *Colloq. Internat. C. N. R. S.* 163, 575-86.

11 Brodmann, K. (1908), 'Beiträge zur histologischen Lokalisation der Grosshirnrinde', 7: Die cytoarchitektonische Cortexgliederung der Halbaffen (Lemuriden)', *J. Psychol. Neurol.* 10, 287-334.

12 Campbell, C.B.G., Yashon, D. and Jane, J.A. (1966), 'The origin, course and termination of corticospinal fibres in the slow loris, *Nycticebus coucang* (Boddaert)', *J. Comp. Neurol.* 127, 101-12.

13 Kanagasuntheram, R., Leong, C.H. and Mahran, Z.Y. (1966), 'Observations of some cortical areas of the lesser bushbaby (*Galago senegalensis senegalensis*)', *J. Anat.* 100, 317-33.

14 Mott, F.W. and Kelley, A.M. (1908), 'Complete survey of the cell lamination of the cerebral cortex of the lemur', *Proc. Roy. Soc. Lond. B* 80, 488-506.

15 Sanides, F. and Krishnamurti, A. (1967), 'Cytoarchitectonic subdivisions of sensorimotor and limbic fields in slow loris (*Nycticebus coucang coucang*)', *J. f. Hirnforsch.* 9, 225-52.

16 Zuckerman, S. and Fulton, J.F. (1941), 'The motor cortex in *Galago* and *Perodicticus*', *J. Anat.* 75, 447-56.

17 Beddard, F.E. (1901), 'Notes on the broad-nosed lemur, *Hapalemur simus*', *Proc. Zool. Soc. Lond.* 1901, 121-9.

18 Le Gros Clark, W.E. (1931), 'The brain of *Microcebus murinus*', *Proc. Zool. Soc. Lond.* 1931, 463-86.

19 Hofer, H. (1956), 'Das Furchenbild der Hirnrinde von *Daubentonia madagascariensis* (Gmelin, 1788) und seine morphologische Bedeutung', *Zool. Anz.* 156, 177-94.

20 Hofer, H. (1958), 'Vergleichende Beobachtungen über die kraniocerebrale Topographie von *Daubentonia madagascariensis*', *Morph. Jb.* 99, 26-64.

21 Starck, D. (1954), 'Morphologische Untersuchungen am Kopf der Säugetiere, besonders der Prosimier, ein Beitrag zum Problem des Formwandels des Säugerschädels', *Z. wiss. zool.* 157, 169-219.

22 Haines, D.E. et al., this volume.

23 Le Gros Clark, W.E. and Thomas, D.P. (1952), 'The Miocene lemuroids of East Africa', *Fossil Mammals of Africa* No. 5, British Museum of Natural History, London.

24 Le Gros Clark, W.E. (1971), *The Antecedents of Man*, 3rd ed., Chicago.

25 Le Gros Clark, W.E. (1947), 'Deformation patterns in the cerebral cortex', in Le Gros Clark, W.E. and Medawar, P.B. (eds.), *Essays on Growth and Form Presented to d'Arcy Wentworth Thompson*, Oxford.

26 Welker, W.I. and Campos, G.B. (1963), 'Physiological significance of sulci in somatic sensory cerebral cortex in mammals of the family Procyonidae', *J. Comp. Neurol.* 120, 19-36.

27 Radinsky, L. (1972), 'Endocasts and studies of primate brain evolution', in Tuttle, R. (ed.), *Functional and Evolutionary Biology of Primates*, New York.

28 Biegert, J. (1963), 'The evaluation of characteristics of the skull, hands and feet for primate taxonomy', in Washburn, S.L. (ed.), *Classification and Human Evolution*, Chicago.

29 Cartmill, M., this volume.

30 Simpson, G.G. (1967), 'The Tertiary lorisform primates of Africa', *Bull. Mus. Comp. Zool.* 136, 39-62.

31 Radinsky, L. (1971), 'An example of parallelism in carnivore brain evolution', *Evol.* 25, 518-22.

PART II SECTION D
Prosimian Locomotion

C. J. JOLLY and A. T. GORTON

Proportions of the extrinsic foot muscles in some lorisid prosimians

Interpreting the relationship between structural and functional diversity in the living world was a respectable occupation for naturalists long before Darwin substituted a mechanistic for a theistic explanation of the existence of the phenomenon. Today, however, confirmation of the universal existence of adaptation (or Divine Providence) is perhaps the least of our reasons for investigating the concordance between structure and function in particular cases. Much more important is the prominent role played by such deductions in two central procedures of evolutionary biology; determination of the phyletic relationships between species, and elucidation of the adaptive trends that have led to the array of relationships and characteristics observed in extant and fossil organisms. Whether one is concerned with living or extinct forms, and whether the operation is performed intuitively or with the aid of mathematical procedures, the construction of phyletic trees involves a critical assessment of the significance, or 'phyletic weight', of observed resemblances and differences. In this judgment, the functional interpretation of characters plays a vital part, since only thus can the potentially misleading products of parallelism, convergence, and rapid adaptive shifts be distinguished from phyletically significant heritage characters.

This rationale is implicit in recent reconsiderations of lorisid classification which question the phyletic unity of a subfamilial or familial group which includes both the Asiatic and African 'slow climbers' while excluding all forms of *Galago* and *Euoticus*.[1] To the extent that the resemblances between Asiatic and African lorisines can be attributed to a single morphological-functional pattern associated with common locomotor habits, they must be regarded as poor indicators of special phyletic relationship, although it is not unlikely that they are indeed heritage characters. (This interpretation does not, of course, prohibit the establishment of a phenetically-defined lorisine grouping on the grounds of adaptive resemblance.)[2]

On the other hand, the monophyletic status of the Lorisidae (*sensu lato*) as opposed to other extant prosimians is supported by a variety of details of dental and cranial structure to which high phyletic weight may be accorded.[3]

In drawing out correlations between features of structure and function in extant forms we are able to make use of two sets of data which are logically independent of each other: the comparative anatomy of the system under examination, and information about the naturalistic behaviour of the living animal. Each of these two sets may be used to derive hypotheses testable against the other. Since correlation, not causality, is the object, this process may proceed in either direction.

By contrast, when attempting a similar procedure in forms known only as fossils, our *independent* knowledge of their natural history is virtually limited to ambiguous information derived from their geological and taphonomic setting. Hypotheses about the behaviour of fossil forms (their locomotor repertoires, dietary emphases, and so on) thus rest largely upon the extent to which analogies can be drawn with living forms in which behavioural-structural correlations have been satisfactorily established. (It is perhaps worth emphasising that the most telling analogies are often derived from forms relatively remote, in the phyletic sense, from the species under consideration.)[4,5,6]

Isolated foot bones, especially of the tarsus, are among the postcranial elements most frequently recovered as fossils. Furthermore, there are few of the more contentious issues in prosimian evolution (the occurrence of vertical clinging and leaping adaptations in Eocene prosimians,[4,7] and the nature of the distinctive adaptation of basal primates,[8,9] are cases that come to mind), which do not involve to some extent the functional interpretation of hindlimb anatomy. Interpretation of skeletal elements, whether fresh or fossil, is sterile if it does not view them as parts of a total mechanical system including, in the living, muscles, cartilage, ligaments, membranes and tendons as well as bones. The present paper is an examination of variability among four lorisid species (*Galago senegalensis, G. crassicaudatus, Nycticebus coucang*, and *Perodicticus potto*) with respect to one aspect of a single part of this total system – proportional relationships among the weights of the extrinsic muscles of the foot. Parallel work on joint structure and the relationships between the bones, tendons, muscles and ligaments of the leg and foot is currently in progress. Since these factors, and others, profoundly influence the nature of the force exerted by a particular muscle,[10,11] the mechanical interpretations of muscular proportions which are offered in the discussion must be regarded as tentative and provisional. Regardless of their interpretation, however, the results presented below do seem to suggest that certain muscular ratios vary among the four in proportion to the importance of particular locomotor activities in the repertoire of each species.

Materials and methods

Material examined consisted of the legs and feet of two *Galago crassicaudatus* (three limbs), five *G. senegalensis* (six limbs), and one *Perodicticus potto* (one limb). All but one subadult lesser bushbaby were apparently mature animals. The galagos came from south-eastern Kenya, the potto from the Mabira Forest, Uganda. The former were feral animals which had been experimentally infected with hookworm prior to sacrifice; the latter had been held captive for a brief period before its death from unknown causes. None of the subjects showed any obvious pathology affecting the skeleton or musculature. All had been preserved in neutralised formaldehyde solution for five years prior to dissection. The skeleton exhibited no signs of demineralisation, and the musculature was apparently neither macerated nor differentially dehydrated in any specimen.[1 2]

After dissection of the leg, each of the extrinsic foot muscles (Table 1) was carefully removed from its origin, and its tendon severed at a standard point, usually immediately proximal to its passage beneath its most proximal retinaculum, if any. Muscle bellies and attached portions of tendon were then immersed in 10% formaldehyde solution for at least a week prior to weighing.

Each muscle was removed from the solution immediately before weighing, blotted quickly and without pressure to remove surface moisture, and weighed on a chemical balance legible to 0.003 g. The whole weighing procedure was then repeated. If the discrepancy between the two determinations exceeded 0.01 g., reweighings were repeated until a 'run' of three consistent weights was obtained. Specimens were always returned to the fluid and subsequently blotted between determinations. Weights were rounded to the nearest 0.01 g. After weighing, the saturated muscles of a represent-

Table 1 List of muscles discussed and abbreviations used in Tables 2-6

Gn	*Gastrocnemius*	Fdt	*Flexor digitorum tibialis*
So	*Soleus*	Fdf	*Flexor digitorum fibularis*
Pt	*Plantaris*	Pl	*Peroneus longus*
Edl	*Extensor digitorum longus*	Pb	*Peroneus brevis*
Ehl	*Extensor hallucis longus*	$\overline{X}$	Sample Mean
Ta	*Tibialis anterior*	Range obs.	Observed range in the sample
Tp	*Tibialis posterior*	S.D.	Estimate of population Standard Deviation
		C.L.	Upper and lower 95% Confidence Limits of the Population Mean

ative of each species were teased and interspecific comparisons made in terms of the architectural arrangement of their fibres and tendons.

Weight data were punched onto cards and a computer was used to calculate a number of ratios in each specimen, as well as a series of statistics (Mean, Standard Deviation, 95% Confidence Limits of the Mean) partially describing the distribution of each of the ratios in each species as represented by our samples. Ratios calculated fall into three groups:

1. The weight of each individual muscle expressed as a percentage of the total extrinsic foot musculature (referred to as 'Relative Weight') (Table 2).
2. The weights of individual muscles expressed in terms of the total musculature acting upon a particular joint or set of joints (Tables 3, 4, 5).
3. Ratios comparing the weight of certain pairs of muscles with partially similar action (Table 6).

Grand includes in his account of the anatomy of the leg and foot of *Nycticebus coucang* expressions of the weights of the extrinsic foot muscles of eight animals (nine limbs).[13] Although Grand weighed only to the nearest 0.1 g., used fresh rather than preserved material, and did not state how much of the tendon of insertion is included in the weight, his data comprise a valuable series for comparison with the African lorisids. Grand's data are presented in the form of ratios of the total musculature operating upon each of three joints (as in Tables 3, 4 and 5); but Relative Weights and ratios between pairs of muscles could be extracted from them by simple arithmetic manipulation.

Results

A general, qualitative account of lorisid limb musculature is to be found in the work of Jouffroy, who summarised her own and previous findings, comparing them with the lemuroid condition.[14] In general, the origins, insertions and gross structure of individual muscles were quite similar in the three species dissected. The following differences were noted:

Gastrocnemius in *Galago* has a large fabella in each tendon of origin; these are absent in *Perodicticus* and *Nycticebus*, as previously noted by Jouffroy.[14] Grand reports an origin from the fascia of flexor digitorum tibialis in the slow loris.[13] The bellies of the muscle are more slender and cylindrical, and their insertion into the common triceps tendon more fibrous, and less aponeurotic, in *Perodicticus* than *Galago*.

Soleus in the potto is similar in size and shape to the two heads of

Table 2 Relative weights of the extrinsic musculature of the foot

+ In this and subsequent tables indicates cases where there is no overlap in 95% confidence limits between the two *Galago* species.

* In this and subsequent tables indicates cases where *Perodicticus* falls more than 2.5 S.D. from the Mean for *Nycticebus*.

	Galago senegalensis				*G. crassicaudatus*				*Nycticebus coucang*				*P.potto*
	$\overline{X}$	Range obs.	S.D.	C.L.	$\overline{X}$	Range obs.	S.D.	C.L.	$\overline{X}$	Range obs.	S.D.	C.L.	
Gn	24.8	23.7-26.6	1.2	23.6-26.1	+18.8	17.7-19.3	0.9	16.5-21.0	5.2	3.8-87.4	1.3	4.2- 6.3	11.9*
Pt	7.7	5.8- 8.9	1.1	6.5- 8.9	6.3	5.9- 6.9	0.6	4.9- 7.7	1.0	0.6- 1.6	0.4	0.4- 1.5	1.5
So	3.8	3.1- 4.4	0.6	3.2- 4.4	5.0	4.6- 5.6	0.5	3.7- 6.3	9.4	8.1-10.6	0.7	8.7-10.0	8.8
Edl	3.5	1.2- 4.6	1.2	2.2- 4.7	2.8	2.6- 3.0	0.2	2.2- 3.3	4.2	3.6- 5.2	0.5	3.8- 4.6	5.4
Ehl	3.7	3.3- 4.1	0.3	3.4- 4.0	3.3	3.1- 3.4	0.1	3.0- 3.5	2.9	2.1- 3.8	0.5	2.6- 3.3	3.7
Ta	13.0	12.2-15.0	1.1	11.9-14.2	15.1	14.2-15.9	0.9	13.0-17.2	12.4	10.5-15.3	1.2	11.5-13.3	10.9
Tp	3.5	3.1- 4.1	0.4	3.1- 3.9	3.2	3.0- 3.4	0.2	2.8- 3.6	3.8	3.0- 5.2	0.8	3.2- 4.5	4.5
Fdt	6.7	5.1- 7.9	1.1	5.6- 7.9	8.7	7.7-10.0	1.2	5.8-11.7	32.5	29.2-40.3	1.6	31.2-33.7	20.6*
Fdf	14.8	13.1-16.9	1.6	13.1-16.5	18.9	17.3-21.6	2.4	13.0-24.7	14.3	12.4-16.5	0.8	13.6-14.9	15.0
Pl	11.7	10.3-13.0	1.0	10.6-12.8	13.5	13.2-13.9	0.4	12.5-14.5	11.4	10.8-12.5	0.6	10.9-11.9	12.3
Pb	6.8	5.8- 8.4	1.1	5.6- 7.9	+4.6	4.5- 4.6	0.1	4.3- 4.8	3.0	2.5- 3.6	0.2	2.8- 3.2	5.5*

Table 3 Weights of the individual talocrural muscles expressed as a percentage of the total for the group

	Galago senegalensis				*G. crassicaudatus*				*Nycticebus coucang*				*P.potto*
	$\overline{X}$	Range obs.	S.D.	C.L.	$\overline{X}$	Range obs.	S.D.	C.L.	$\overline{X}$	Range obs.	S.D.	C.L.	
Gn	57.3	55.3-61.7	2.3	54.8-59.7	52.0	50.7-53.5	1.4	48.6-55.5	23.0	17.8-30.7	4.3	19.4-26.6	38.0*
Pt	17.6	14.8-19.6	1.7	15.9-19.3	17.4	16.1-19.8	2.1	12.2-22.6	4.5	2.7- 7.3	1.9	2.1- 6.9	4.9
So	8.7	7.1- 9.9	1.2	7.4- 9.9	+13.9	13.2-15.1	1.0	11.2-16.4	42.3	38.4-48.2	3.0	39.7-44.8	28.2*
Edl	7.9	3.2-10.8	2.7	5.1-10.7	7.7	7.3- 8.0	0.4	6.8- 8.6	18.7	15.4-22.8	2.3	16.9-20.5	17.1
Ehl	8.5	7.7-10.6	1.1	7.3- 9.7	9.0	8.9- 9.0	0.1	8.9- 9.1	12.9	9.6-15.3	1.7	11.6-14.2	11.8
Gn+Pt+So	83.6	81.5-86.2	1.8	81.7-85.5	83.3	82.3-84.3	0.4	83.0-83.7	68.9	64.3-71.8	4.0	65.8-72.0	71.0

Table 4 Weights of individual invertors and evertors of the foot expressed as a percentage of the total for the group

	Galago senegalensis				*G. crassicaudatus*				*Nycticebus coucang*				*P.potto*
	$\overline{X}$	Range obs.	S.D.	C.L.	$\overline{X}$	Range obs.	S.D.	C.L.	$\overline{X}$	Range obs.	S.D.	C.L.	
Ta	31.3	28.4-36.2	2.8	28.4-34.1	33.5	32.7-34.1	0.7	31.7-35.2	19.6	16.6-22.6	1.9	18.1-21.1	20.2
Tp	8.5	7.5- 9.4	0.8	7.6- 9.3	7.1	6.7- 7.5	0.4	6.1- 8.0	6.1	4.7- 8.2	1.3	5.1- 7.0	8.3
Fdt	16.1	12.8-18.9	2.6	13.4-18.8	19.3	17.6-21.5	2.0	14.5-24.2	51.3	46.3-53.5	2.3	49.6-53.8	38.3*
Pl	27.9	25.6-29.9	1.5	26.4-29.5	30.0	28.5-32.1	2.0	25.1-34.8	18.0	16.5-19.6	1.0	17.3-18.8	22.9*
Pb	16.2	13.8-19.8	2.7	13.4-19.0	+10.1	9.6-10.6	0.5	8.8-11.4	4.7	3.9- 5.1	0.4	4.4- 5.0	10.2*
Ta+Tp+Fdt	55.9	50.8-58.5	2.8	52.9-58.8	59.9	57.4-62.3	2.4	53.8-66.0	77.1	76.1-78.7	0.8	76.5-77.8	66.9*

Table 5 Weights of individual flexors and extensors of the digits expressed as a percentage of the total for the group

	Galago senegalensis				*G. crassicaudatus*				*Nycticebus coucang*				*P.potto*
	$\overline{X}$	Range obs.	S.D.	C.L.	$\overline{X}$	Range obs.	S.D.	C.L.	$\overline{X}$	Range obs.	S.D.	C.L.	
Fdt	23.2	20.0-26.0	2.2	21.0-25.6	56.1	21.9-30.7	4.0	15.9-36.1	60.2	56.9-63.0	1.8	58.8-61.6	46.1*
Fdf	51.6	47.7-56.2	2.8	48.7-54.6	56.1	51.8-61.8	5.2	43.2-68.9	26.4	23.3-28.0	1.3	25.4-27.4	33.5*
Edl	12.2	4.1-15.9	4.3	7.7-16.7	8.3	7.3- 9.2	0.9	6.0-10.6	7.8	6.6-10.1	1.0	7.0- 8.6	12.0*
Ehl	12.8	11.4-14.6	1.3	11.5-14.2	+9.7	9.0-10.3	0.7	8.0-11.4	5.4	3.8- 7.1	0.9	4.7- 6.0	8.3*
Fdf	75.0	70.9-82.2	4.0	70.8-79.1	82.0	80.5-83.7	1.6	78.1-86.0	86.7	84.5-88.3	1.4	85.7-87.8	79.7*

Table 6 Intra-group ratios

	Galago senegalensis				*G. crassicaudatus*				*Nycticebus coucang*				*P.potto*
	$\overline{X}$	Range obs.	S.D.	C.L.	$\overline{X}$	Range obs.	S.D.	C.L.	$\overline{X}$	Range obs.	S.D.	C.L.	
Gn/so	6.70	5.70-7.75	0.85	5.80-7.60	+3.77	3.44-4.00	0.29	3.04-4.49	0.56	0.42-0.75	0.12	0.45-0.66	1.35*
Fdt/Fdf	0.45	0.39.0.53	0.05	0.40-0.50	0.47	0.35-0.58	0.12	0.18-0.76	2.28	2.04-2.65	0.18	2.15-2.42	1.38*
Ta/Fdt	2.54	2.00-3.27	0.57	1.94-3.13	2.11	1.90-2.26	0.19	1.64-2.57	0.50	0.42-0.66	0.07	0.45-0.56	0.75*
Pl/Pb	1.75	1.29-2.06	0.28	1.45-2.04	+2.96	2.89-3.03	0.07	2.78-3.13	3.86	3.34-4.96	0.52	3.46-4.26	2.23*

gastrocnemius. Distally it develops numerous tendinous fibres on its deep surface which extend directly into the common tendon of *triceps surae.* In *Galago* most of the fibres of the muscle insert fleshily on the deep surface of the gastrocnemius tendon, direct contributions by tendons developed in soleus being few. In all specimens of our series, soleus originated by a slender flat tendon from the head of the fibula; none showed the origin from the fibular shaft reported by Nayak.[14] *Plantaris* in our potto was distinct from gastrocnemius, though weakly developed. Its origin was femoral, not fibular as reported by Nyak.[14]

Flexor digitorum tibialis of galagos resembles its homologue in the lemuroids, a single-headed muscle arising from the lateral condyle and posterior surface of the proximal third of the tibia. In *Nycticebus* and *Perodicticus* it has a more extensive origin and is almost completely divisible into two heads. In *Nycticebus* the deep head, which is presumably the homologue of the single head of other prosimians, is supplemented by a superficial head with an extensive origin on the medial surface of the tibia, the medial surface of the distal end of the femur, and the medial aspect of the knee.[13] In our *Perodicticus*, the accessory, superficial head originated only from the medial tibial surface.

In *Galago* the relation between *tibialis anterior* and *extensor hallucis longus* is especially close, the tendon of the hallucial extensor passing between the fibres of the larger muscle, and lying in a superficial groove in its tendon under the extensor rectinaculum. In the potto the two muscles and their tendons are merely adjacent.

Extensor digitorum longus supplies no tendon to the reduced second digit in *Perodicticus*;[14] digits II, III, IV and V are supplied in *Galago.* In *Nycticebus*, according to Grand, it supplies 'each toe'.

The tiny *peroneus digiti quinti* was absent in our potto, as in some *Nycticebus*,[13] but occurred in all *Galago* specimens. When present it was weighed with peroneus brevis, although its contribution to the weight of that muscle was negligible.

The results of analysis of the relative and proportional weights of the eleven extrinsic foot muscles are presented in full in Tables 2-6.

In the following exposition of the salient points of interspecific difference, comparisons are first made between *Galago senegalensis* and *Nycticebus coucang*, which are represented by larger samples of material. *G. senegalensis* and *P. potto* are then compared with both.

Fig. 1 contrasts the profiles of relative muscle weight in *Galago senegalensis* and *Nycticebus coucang*, and demonstrates the major difference between them, namely the dominance of gastrocnemius in the bushbaby, and flexor digitorum tibialis in the slow loris. Within the group whose primary action is plantarflexion of the foot at the talocrural joint (Gn, So, Pt), gastrocnemius is by far the largest component in the bushbaby, and is supplemented by a substantial plantaris, another two-joint muscle crossing both knee and ankle. In

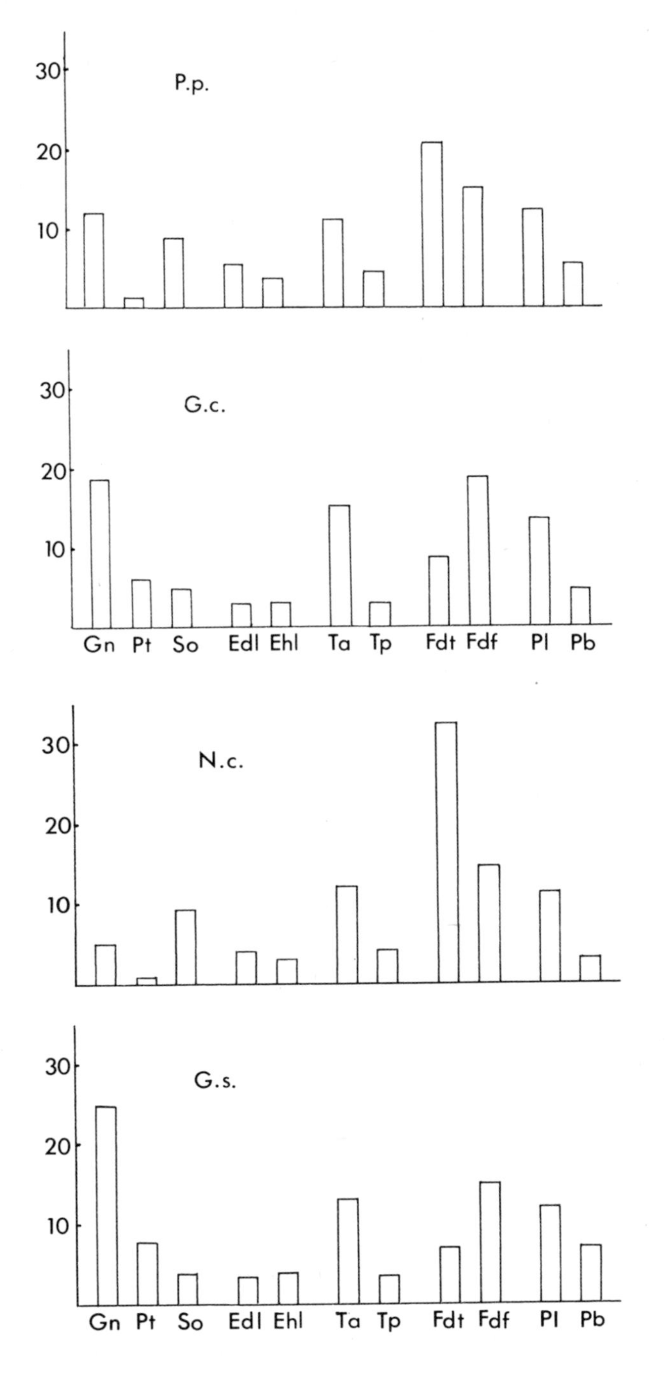

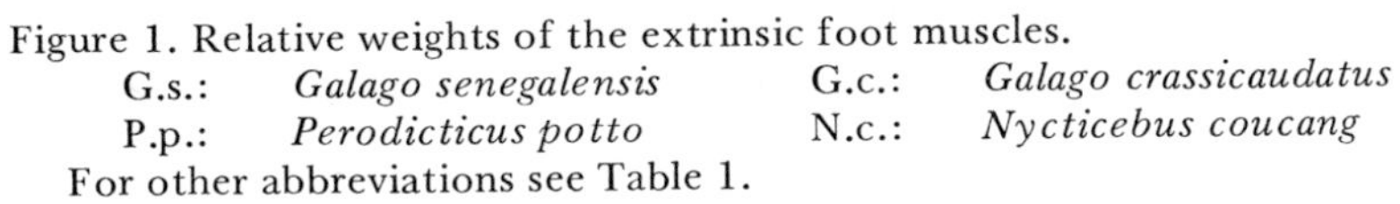

Figure 1. Relative weights of the extrinsic foot muscles.

G.s.: *Galago senegalensis* G.c.: *Galago crassicaudatus*

P.p.: *Perodicticus potto* N.c.: *Nycticebus coucang*

For other abbreviations see Table 1.

the slow loris both plantaris and gastrocnemius are very weak, and together weigh less than soleus. This contrast appears in the Gn/So ratio (Table 6), which averages 6.70 in the bushbaby, but only 0.56 in the loris. The long digital extensors, which are also dorsiflexors of the foot on the leg, are comparatively small and rather variably developed muscles which do not differ in relative size in the two forms. The plantarflexors therefore represent a higher proportion of the total talocrural musculature in the bushbaby (Table 3). Tibialis anterior and posterior, which are apparently primarily invertors of the foot, are rather similar in relative size in the two forms.

Flexor digitorum fibularis, which has little or no rotatory action on the foot, is very similar in relative size in the two forms, but the combined digital flexor and pedal invertor, flexor tibialis, is much larger by all standards in the loris, as well as exhibiting the specialised structure and long head of origin described above. With a relative weight averaging 32.5, it is much the largest of the extrinsic foot muscles, corresponding in importance to combined gastrocnemius and plantaris in the bushbaby. The importance of the combined flexor and invertor is very apparent when compared with that of the simple digital flexor (Fdt/Fdf, Table 6), or with that of the simple invertor (Ta/Fdt, Table 6), and produces the high ratio of invertors to evertors of the foot seen in the loris (Table 4).

The two major peronei are the principle evertors of the foot, and peroneus longus is, in addition, a powerful adductor of the mobile first metatarsal. In both species peroneus longus has a relative weight of about 11, but peroneus brevis is more than twice as large in the bushbaby (Table 2). The difference in relative size of the two muscles is further illustrated by the ratio Pl/Pb (Table 6).

The generalisations that can be made about the muscular proportions of *Perodicticus potto* on the basis of a single specimen are of course limited. Pending a study of more extensive material, attention may be drawn to those features in which the single specimen lies so far from the mean of the other series that a species-specific difference is clearly implied. In general, the profile of Relative Muscle Size in the potto differs from that of *Galago senegalensis* in the direction of *Nycticebus coucang* (Fig. 1). However in those characters in which the loris differs most strikingly from the galago, the potto is consistently less extreme in its deviation. Soleus, in the potto, is relatively as large a muscle as it is in the loris, but it is a much smaller part of the total talocrural musculature (Table 3) because of the relatively larger gastrocnemius. The difference between the two is brought out by the Gn/So ratio (Table 6). Since the relative weights of soleus and plantaris do not differ in the two lorisines (Table 2), the low Gn/So ratio in the loris is clearly due to a deficit in gastrocnemius. On the other hand, it is noteworthy that the talocrural plantarflexors are a similar proportion of the total talocrural musculature in the two lorisines (Table 3), both of which differ from the galago in this respect.

Flexor digitorum tibialis is large and bicipital in the potto. Although it exceeds *flexor fibularis* in weight, the disproportion between them is less pronounced than it is in the loris (Table 6). The more modest size of the muscle is reflected also in the lower invertor/evertor and digital flexor/extensor ratios in the potto (Tables 4 and 5).

Finally, peroneus brevis is relatively larger in our potto than in any of the loris series (Table 2), as is reflected in its lower Pl/Pb ratio (Table 6).

As with *Perodicticus*, our sample of three limbs of *Galago crassicaudatus* must be regarded with caution as representative of the species as a whole. They do, however, show some noteworthy differences from the other species, especially the lesser galago. The plantarflexors of the talocural joint comprise, as in the lesser bushbaby, about 83% of the total musculature acting upon the joint, but within the group the proportion contributed by gastrocnemius is much less (Table 3), giving a Gn/So ratio as close to that of the potto as to that of *G. senegalensis* (Table 6). No significant difference between the two galagos can be shown in the relative size of soleus or plantaris, so that (as in the case of the lorisines) the difference can be attributed to variation in the size of gastrocnemius. Again as in the lorisines, this reduction accounts fully for the smaller

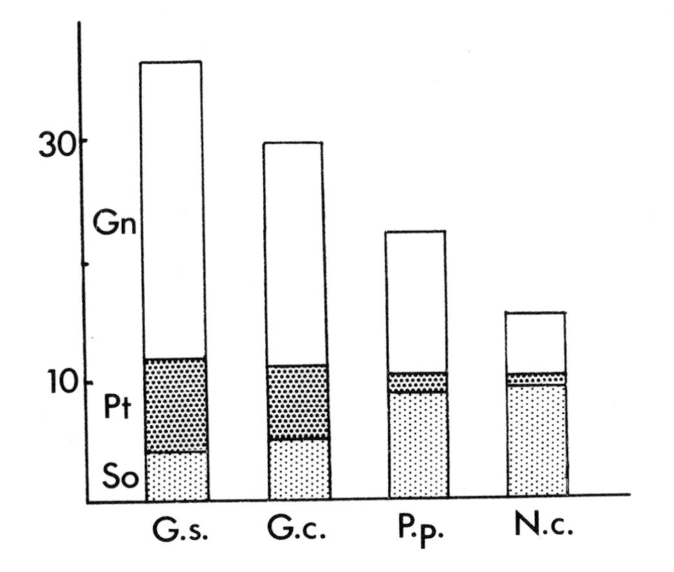

Figure 2. Relative weights of the talocrural plantarflexors.
Segments of each column represent the relative weights of soleus, plantaris and gastrocnemius.
Abbreviations as in Fig. 1.

relative size of the total plantarflexors (range 29.3-30.0) in *G. crassicaudatus* when compared with *G. senegalensis* (range 33.5-39.3) (Fig. 2).

Unlike the talocrural musculature, the proportions of the digital flexors of the greater bushbaby show no significant trend in a loris-like direction, the relative size of flexor digitorum tibialis and the ratio Fdt/Fdf showing values little different from those seen in *G. senegalensis*.

A significant difference does, however, appear in the peronei. Peroneus brevis is a smaller part of the total crural musculature, and of the total invertor-evertor group, than in the lesser galago, a difference that is also apparent in the ratio Pl/Pb (Table 6). In each of these expressions of the relative size of peroneus brevis *G. crassicaudatus* lies closer to the potto than it does to its smaller relative.

Discussion

The finding that gastrocnemius is a relatively large muscle in *Galago senegalensis* is not unexpected, in view of its importance, as a two-joint muscle, in the leap,[15] which is a common and characteristic gait in its repertoire.[16] Since leaping and hopping are reported to be less frequent in the behavioural repertoire of *G.s crassicaudatus*,[17-19] the reduced relative size of gastrocnemius in this form is explicable, as is its still smaller size in the lorisines. Clearly, the same explanation is insufficient to account for the difference between the two lorisines, neither of which has ever been observed to leap, hop or gallop.[20] It is possible that the femoral origin of flexor tibialis in the slow loris may partially replace the gastrocnemius as an accessory flexor of the knee, thus permitting its further reduction, but this action of the long head of flexor tibialis has not been demonstrated.

In contrast to gastrocnemius, the one-joint plantarflexor, soleus, is presumably able to exert its action at any position of the knee, which might account for its importance in the lorisines, whose mode of locomotion, especially on small branches, involves great variation in orientation of footholds.[13,21] The ability to plantarflex the foot with the knee fully extended may be an important factor since slow-climbing involves maintaining a grip with the trailing foot until the last possible moment. Plantaris, which varies almost exactly inversely to soleus in the four forms investigated here, is a two-joint flexor with, presumably, the additional function of tensing the strong plantar aponeurosis, and thus helping to support the elongated tarsus against the powerful internal forces of dorsiflexion which must be exerted upon it as the body weight is raised about a fulcrum at its distal end. This function is surely minimal in the short, hand-like foot of the lorisines, as is the development of the plantar aponeurosis itself.

The most striking muscular specialisation in the leg of the loris is the hypertrophy, bicipital structure and femoral origin of flexor digitorum tibialis. As a combined invertor and digital flexor this muscle performs two of the most important motions involved in deliberate climbing, especially when performed in a network of steeply-sloping, small-diameter footholds.[13] Its less specialised condition in the potto is less easy to interpret on the basis of present evidence about the natural history of the two forms. Perhaps future quantitative studies, modelled upon those of Charles-Dominique,[21] may indicate that the slow loris, while sharing the basic locomotor repertoire of the potto, may spend more of its time in the demanding small-branch milieu, and less on the larger boughs which, as footholds, do not require the same degree of pedal inversion. It is known that in primary forest 71% of the supports utilised by pottos exceeded 5 cm. in diameter, while in secondary forest 52% were in this size-range.[21]

Comparable data on *Nycticebus* are not available, although Grand[13] suggests that the form seems to prefer footholds about 3 cm. in diameter. Possibly, on the other hand, the femoral origin of flexor tibialis, and consequent reduction of gastrocnemius, represents a condition that would be adaptive for both species, but has arisen in the ancestors of *Nycticebus* by mutational events which by chance have not occurred in the *Perodicticus* stock. In any case, it would certainly be premature to interpret the apparently more primitive condition in the potto as evidence for the separate derivation of *Perodicticus* and *Nycticebus* from an ancestral stock whose locomotor repertoire did not include a predominant component of slow and deliberate climbing.

Variations in the peronei do not follow the galagine-lorisine dichotomy as do the other characters which vary in the four forms. However, the low Pl/Pb ratio seen in *G. senegalensis* may be associated with both its comparatively frequent descent to the ground,[16] and its small size, both of which are known to affect this ratio in cercopithecines.[10,22] The difference between the two lorisine species cannot be explained in terms of size differences, and perhaps may be regarded as an additional indication that the potto, though predominantly an arboreal animal, may not be as constantly dependent upon the pedal grip as the slow loris, either because it spends more time on large boughs, or because it descends more readily to the ground. Pottos in the wild are known to cross open ground between trees, especially where the forest is broken.[18,21] Their movement when placed on the ground is easy and as rapid as it is in the trees,[20] showing none of the discomfiture and difficulty in locomotion which Grand reports for *Nycticebus* forced to walk on a flat surface.[13]

It cannot be emphasised too strongly that the interpretations suggested in this section must be regarded as hypotheses to be tested against additional subsystems of the locomotor apparatus, further behavioural observations, and more species. Nevertheless they do support the view that despite the undoubted heuristic value of the concept of locomotor categories, the complex of adaptive characters distinctive of a particular species can be more fully elucidated by reference to its total locomotor repertoire, which is likely to comprise a unique combination of gaits and postures. They also illustrate the insight that may be gained by comparing groups of closely-related forms, and thus partly controlling the heritage component of inter-specific variation.[10,23]

NOTES

1 Groves, C.P., this volume.
2 Washburn, S.L. (1951), 'The new physical anthropology', *Trans. N.Y. Acad. Sci.*, ser. 2, 13, 198-304.
3 Le Gros Clark, W.E. (1959), *Antecedents of Man*, Edinburgh.
4 Cartmill, M. (1972), 'Arboreal adaptations and the origin of the order primates', in Tuttle, R. (ed.), *The Functional and Evolutionary Biology of Primates*, Chicago.
5 Cartmill, M., this volume.
6 Jolly, C.J. (1971), '*Hadropithecus*: A lemuroid small-object feeder?', *Man*, 5, 619-26.
7 Napier, J.R. and Walker, A.C. (1967), 'Vertical clinging and leaping – a newly recognized category of locomotor behaviour of primates', *Folia primat.* 6, 204-19.
8 Szalay, F.S. (1968), 'The beginnings of primates', *Evolution* 22, 19-36.
9 Martin, R.D. (1968), 'Towards a new definition of primates', *Man* 3, 376-401.
10 Tuttle, R. (1972), 'Relative mass of cheiridial muscles in catarrhine primates', in Tuttle, R. (ed.), *The Functional and Evolutionary Biology of Primates*, Chicago.
11 Hall-Craggs, E.C.B., this volume.
12 We are greatly indebted to Tom Miller and Alan Walker for the gift of cadavers used in this study, and to Fred Brett for his stalwart assistance with the computations.
13 Grand, T.I. (1967), 'The functional anatomy of the ankle and foot of the slow loris (*Nycticebus coucang*)', *Am. J. Phys. Anthrop.* 28, 168-82.
14 Jouffroy, F.K. (1962), 'La musculature des membres chez les lémuriens de Madagascar, étude descriptive et comparative', *Mammalia* 26, 1-386.
15 Hall-Craggs, E.C.B. (1965), 'An analysis of the jump of the lesser galago (*Galago senegalensis*)', *J. Zool.* 147, 20-9.
16 Doyle, G.A., this volume.
17 Walker, A.C. (1967), 'Locomotor adaptation in recent and fossil Madagascan lemurs', PhD thesis, University of London.
18 Kingdon, J. (1971), *East African Mammals*, Vol. 1, London.
19 Bearder, S.K. and Doyle, G.A., this volume.
20 Walker, A.C. (1969), 'The locomotion of the lorises, with special reference to the potto', *E. Afr. Wildl. J.* 7, 1-5.
21 Charles-Dominique, P. (1971), 'Eco-éthologie des prosimiens du Gabon', *Biol. Gabon.* 7, 121-228.

22 Jolly, C.J. (1967), 'The Evolution of the Baboons', in Vagtborg, H. (ed.), *The Baboon in Medical Research*, Austin, Texas.
23 Jolly, C.J. (1972), 'The large African monkeys as an adaptive array', in Napier, J.R. and Napier, P.H. (eds.), *Systematics of the Old World Monkeys*, New York.

F.K. JOUFFROY, J.P. GASC, M. DÉCOMBAS and S. OBLIN

Biomechanics of vertical leaping from the ground in Galago alleni *: a cineradiographic analysis*

General observations

It is a well-known fact that galagos display a remarkable jumping ability; in height, as well as length, they can jump a distance fourteen times their body size (tail excluded). It is a record challenged among primates only by the tarsier and among the other mammals by the rat-kangaroos (Potoroinae) and the kangaroo-rats (Zapodinae).

Compared to the other galagos, *Galago alleni* displays the most specialised type of locomotion: 'The animal moves almost exclusively by vertical jumps between small trunks and vines, moving from one vertical starting place to another one.'[1] If the animal descends to the ground for feeding, selecting insects or gathering fallen fruits, at the least startling noise it jumps into the nearest tree. This type of jump is very common in captivity. In captivity, *Galago alleni* shows remarkable adaptability to its new 'environment', especially in its locomotor abilities. In the absence of tree trunks or vines (lianas), or whatever other vertical supports might be supplied, it will move easily on plane, usually horizontal, surfaces. On horizontal surfaces the animal is observed running quadrupedally, or moving by little bipedal hops like a jerboa, or even leaping from one horizontal platform to another. As in its native forest, it leaves the platform for a higher position at the slightest noise. Even after gathering food from the cage floor it will make the same movement. This jump is extraordinarily quick. The preparatory and launching phases together do not last more than a few tenths of a second.

Interpretation of photographs, hand and foot prints

Before our cinematographic and cineradiographic recording we conducted some preliminary investigations on the animal's locomotion by means of photographs and footprint registration with the animal running on smoked paper tracks.[2] The analysis of these prints (Fig. 1) shows that before leaping the animal is in a plantigrade position with its heels pressing strongly on the ground (Fig. 2*B*), but that when it runs quadrupedally it does so in a semidigitigrade manner with the tuber of the calcaneum slightly above the ground (Fig. 2*A*). The absence of hand prints just before the leap shows that the animal raises its trunk and jumps in a bipedal attitude. The distance between the hallux and the other digits,[3] as well as the general orientation of the feet, is quite variable. In the few cases where the landing places have been registered it was observed that the hands touched first (making smudged prints).

Cinematography and cineradiography

The vertical jump from the ground of *Galago senegalensis* has been studied by Hall-Craggs,[4,5] who filmed at 1000 frames/sec. vertical jumps of up to 7′ 4¾″. We shall quote some of his conclusions: the fore feet take little part; effective use is made of the length of the synchronously acting hind limbs (as a single structure); the force is applied to the ground in the region of the tarso-metatarsal joint. In an attempt to study more precisely the movements of the bones, we have analysed the same type of jump, but over a shorter distance, in *Galago alleni*, using for this purpose the cineradiographic technique.[6]

In cineradiography, the image is produced on a small screen that becomes fluorescent when subjected to X-rays, and the image is retained on movie film. Use of an image intensification system reduces the radiation necessary to produce an acceptable image. The quality and amount of information thus obtained depends not only upon radiological factors (e.g. number and speed of electrons, sensitivity of the fluorescent screen), and upon characteristics of the photographic system, (e.g. quality of the optics, frames per second), but, in addition, requires an awareness of both the ecology and behaviour of the animal studied. It is essential to obtain an animal's 'co-operation' in conforming as closely as possible to its natural behaviour, or at least (for a captive animal) to its habitual behaviour. In this study we used a species with which we were familiar and an animal which one of us had personally raised for seven years.

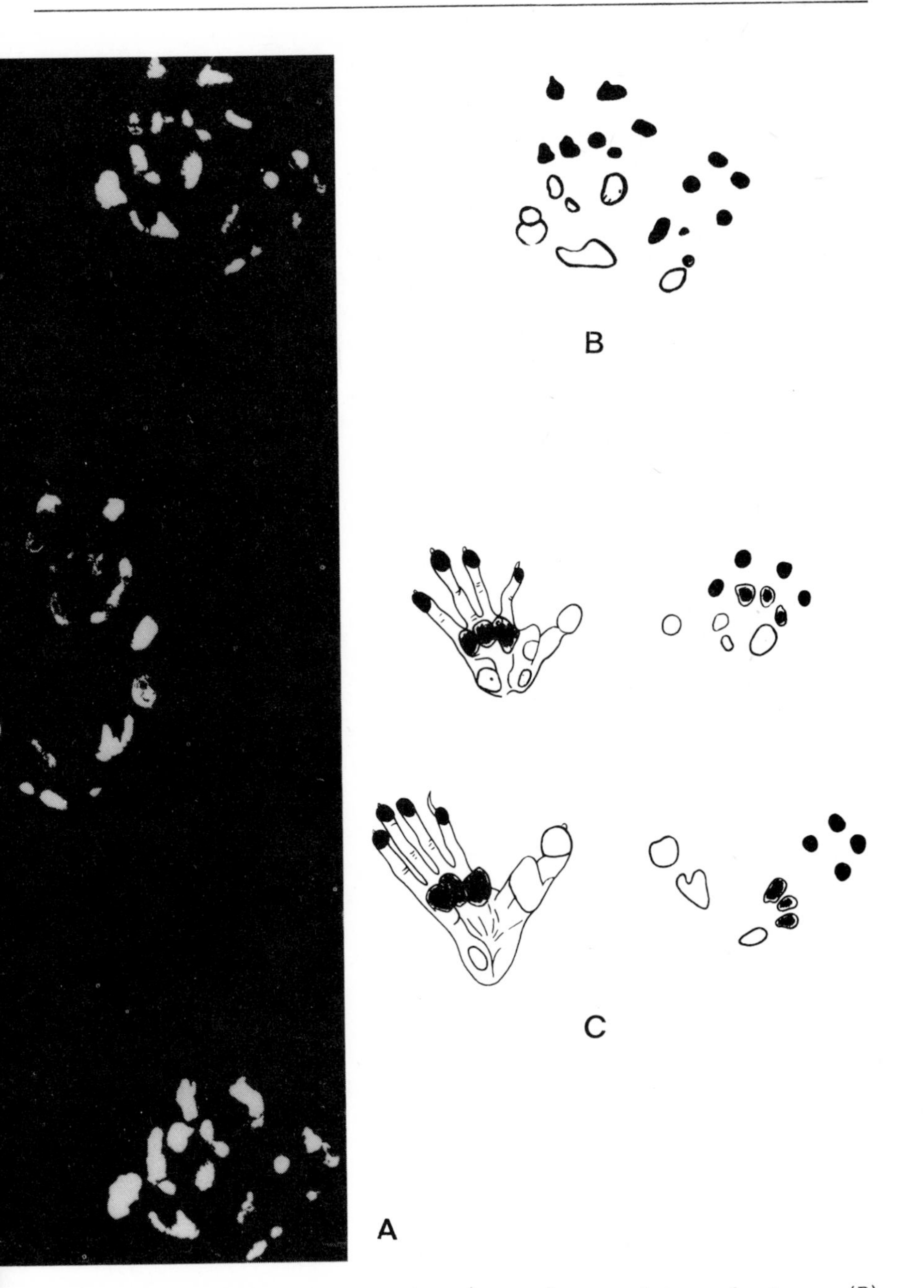

Figure 1.(*A*) Tracks of *Galago alleni* running on the ground at a moderate pace.(*B*) Schematic drawing of the top print.(*C*) Correspondence between the foot and hand prints (right) and the foot and hand pads (left).

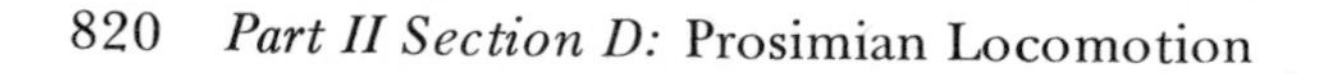

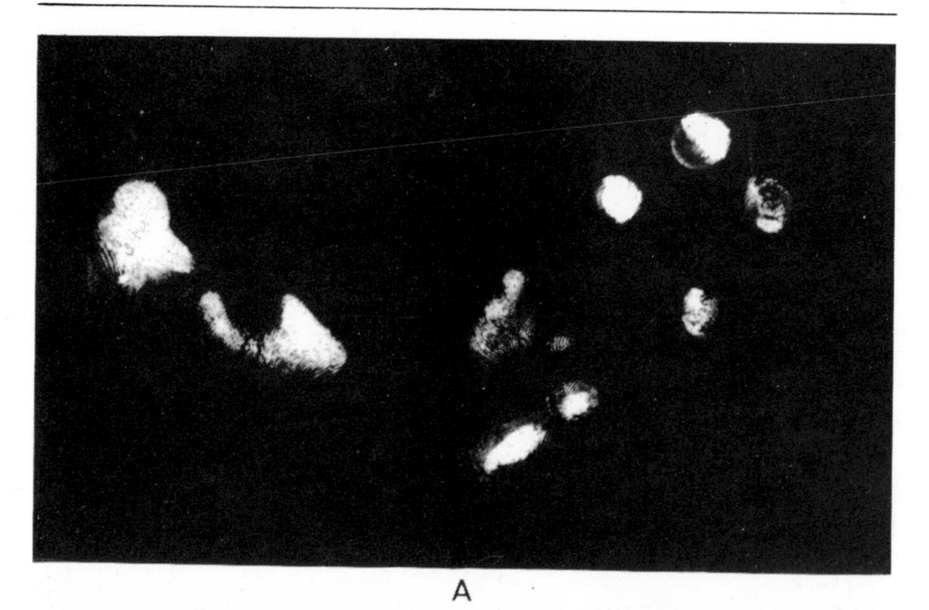

Figure 2. Right footprint: (*A*) while running in a quadrupedal posture on the ground; (*B*) a leap about to be launched.

Materials and methods

A Massiot Philips Medio 20 X-ray generator was employed at 60 KV. The milli-amperage was automatically regulated during the operation, but never exceeded 15 mA. The maximum length of uninterrupted recording under these conditions is 60 seconds. A pause of equal time is recommended before continuing with a new film sequence.

The operator, protected by a lead apron, follows the field on a fluoroscope reflected by a mirror. The cinefluoroscopic apparatus, mounted on a 'C' arm, is suspended by a motorised telescopic column which permits raising and lowering; this in turn is freely movable along the ceiling on a double system of perpendicular rails. Thus, although the diameter of the image intensifier is only 15 cm., the mobile suspension system makes it possible to centre the image rapidly on the movement of the animal.

For movements as rapid and sudden as the galago jump, the inertia of the apparatus, along with the time-lag between the observation of the event and the start and actual operation of the motors, makes centring of the apparatus a difficult task. Centring requires, besides rapid reflexes, sufficient familiarity with the tame animal to predict its direction of movement. In addition, the sharp noise made by the starting of the X-ray generator always startles the animal, interrupts its movement and, in most cases, provokes an escape reaction. For this reason we were unable to obtain any record of locomotion in another prosimian (*Cheirogaleus* sp.), which crouched on the ground in fright and attempted to hide itself in the smallest crevices. In the case of the galago, this escape reaction was used to advantage, since we were able to study the vertical or semi-vertical jump that the animal always employed when frightened, in an attempt to reach the security of its nestbox.

A 16 mm. Arriflex camera was run at 64 frames per second using Kodak Plus X, T.V. reversible film. When filming small animals the periphery of the field is brightest, obscuring the image definition on projection. For this reason we found it necessary to stop down the lens (Schneider-Kreznach Arriflex Cine Xenon 1:2/35) to a setting of f 5.6. The film was initially projected at 16 frames per second to permit observation of movements in slow motion. The most interesting sequences were then examined with the aid of a film editor. Two procedures were used to analyse these sequences frame by frame:

1. By printing all the frames on photographic paper with an enlarger. This extremely laborious process is only practicable for short sequences of the order of 40 frames.
2. By single frame projection onto a screen covered with tracing paper. In this way, the axes of various bones can be traced and successive images superimposed as required. Sequences of 160 frames (2.5 seconds duration) are readily analysed by this technique.

The animal

The subject of this study, a female *Galago alleni*, was a pet of one of the authors who captured it in primary forest in Gabon (May 1963: Mission biologique au Gabon, CNRS, France). It was brought back to France and raised first in the author's house and then, from 1968 until its death in 1971, in the laboratory.

The galago lived in a large-meshed, metallic cage measuring 30 x 45 cm. and 100 cm. high. A box, 18 x 16 x 15 cm., fastened 60 cm. above the floor, served as a site for sleeping, eating and retreat. In order to avoid upsetting the normal behaviour of the animal, it was placed during the recording sessions in a cage of similar dimensions to its own, but constructed with a plywood frame covered with netting and with the nestbox replaced by a wooden platform. The 'C' arm of the apparatus was positioned to deliver a horizontal X-ray beam.

Because of the restrictions of the technique, we analysed only the vertical or semi-vertical (within 30° of the vertical) jumps that the galago invariably made to reach its platform. For angular measurements we used sequences in which the animal faced the apparatus (frontal view) and also sequences in which the hindlimbs were seen in lateral view. Frontal views were most common because, by natural curiosity, the animal tended to watch the investigator and was therefore frontally positioned. Angular measurements thus recorded are not usually the actual angles between two bones, but instead represent a projection of the angles on to the frontal or parasagittal plane. Comparisons of angular variation are valid only during the time when the animal moves its limb segments in the same plane.

Slow motion analysis

Slow motion analysis at 16 frames per second yielded the following data: Of the seven jump sequences filmed, six were asymmetrical (i.e. all the body weight shifted to a single foot, the active one, before the jump), and one was symmetrical. In every case of an asymmetrical jump the active foot was always the left. The galago started from a quadrupedal posture, which in two cases was such that the animal was orientated laterally to the apparatus, and then, lifting its fore limbs, it was seen to be in a bipedal posture before pushing off with the active hind limb.

In the case of a symmetrical jump, three successive stages are distinguishable in slow motion: (1) straightening of the head and the beginning of realignment of the trunk; (2) lifting of the fore limbs with accompanying extension of the trunk, lowering of the pelvis and closure of the femoro-tibial angle; and (3) pushing off with both hindlimbs.

Single frame projection

1. *The symmetrical jump* (Figs. 3 and 5*A*). During this type of jump the different segments of the hindlimbs move alike and are strictly synchronised. The whole movement lasts 0.35 sec. (23 frames). It can be divided into a first, or preparatory phase (*phase d'appel*) of relatively long duration (0.26 sec.), and a very quick springing phase that lasts 0.09 sec.

   During the preparatory phase, the animal crouches with the feet in full plantigrade contact with the ground. The main movements occur at the femur and tibia levels. The femur undergoes a rotatory movement that diminishes the femoro-tibial angle in such a way that the pelvic/femoral articulations are lower than the knees. This movement covers the whole preparatory phase. About 0.05 sec. before the end of this phase, the tibia moves towards the ground, narrowing the tibio-tarsal angle. The outcome of this movement is that it narrows even more the femoro-tibial angle.

   The springing phase straightens, in a perfect alignment, all the hindlimb bone segments. 0.09 sec. before the launching, the heads of both the femurs begin the ascent; at 0.05 sec. before the launching, the tibiae follow the same ascending movement; and at 0.03 sec. before the launching the tarsus is raised, by dorsiflexion of the tarso-metatarsal articulation, before the digits finally leave the ground.

   This symmetrical type of jump, during which the propulsive strength is equally distributed between the two limbs, appears – at least in our film recordings – to be less frequent than the asymmetrical one.

2. *The asymmetrical jump* (Figs. 4 and 5*B*). In this type of jump, the right and left limbs play quite different roles, and this explains why no exact chronological correlations can be noted between the different bone segments of the two limbs.

   During the preparatory phase, the animal's weight is shifted on to one foot, while the other is lifted: the active foot (*pied d'appel*) and the inactive foot (*pied libre*), respectively. The left foot was always the active one during our film recording, probably as a result of the experimental conditions. The preparatory phase thus involved the shifting of the right foot, with all the articulations of the left limb flexing at the same time.

   Taking the right limb first, the foot is initially in contact with the ground and the tarso-metatarsal articulations are dorsiflexed. Just before lifting, successive oscillations of the tarsus are observed (measured by the tibio-tarsal angle). The position of the tibia, relative to the horizontal, remains almost constant.

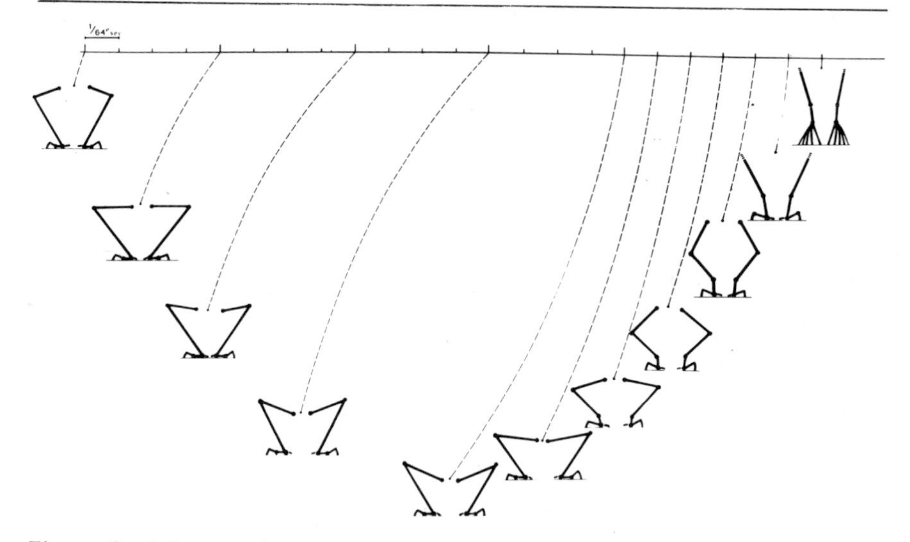

Figure 3. Schema of the movements of the hind limb segments during a symmetrical leap. Time elapsed: 1/64 sec.

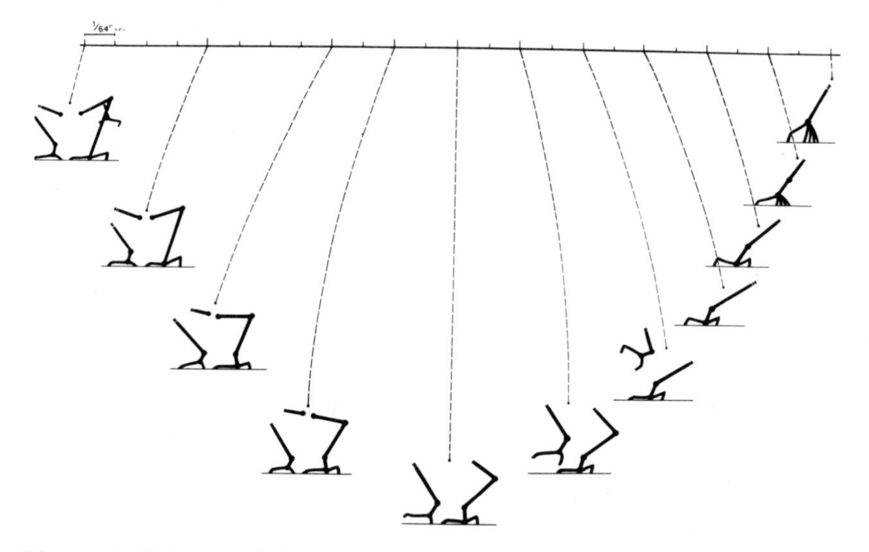

Figure 4. Schema of the movements of the hind limb segments during an asymmetrical leap. Time elapsed: 1/64 sec.

The oscillation amplitude is about 15°; its duration is from 0.04 to 0.08 sec. These oscillations last as long as the foot is not lifted. During this stage the pelvis is lowest, relative to the ground, and remains stable.

The left limb, meanwhile, remains immobile and flexed. This limb is entirely responsible for the launching phase. It supports the entire body weight. 0.34 sec. before launching, the springing phase begins simultaneously at two levels; the femur becomes

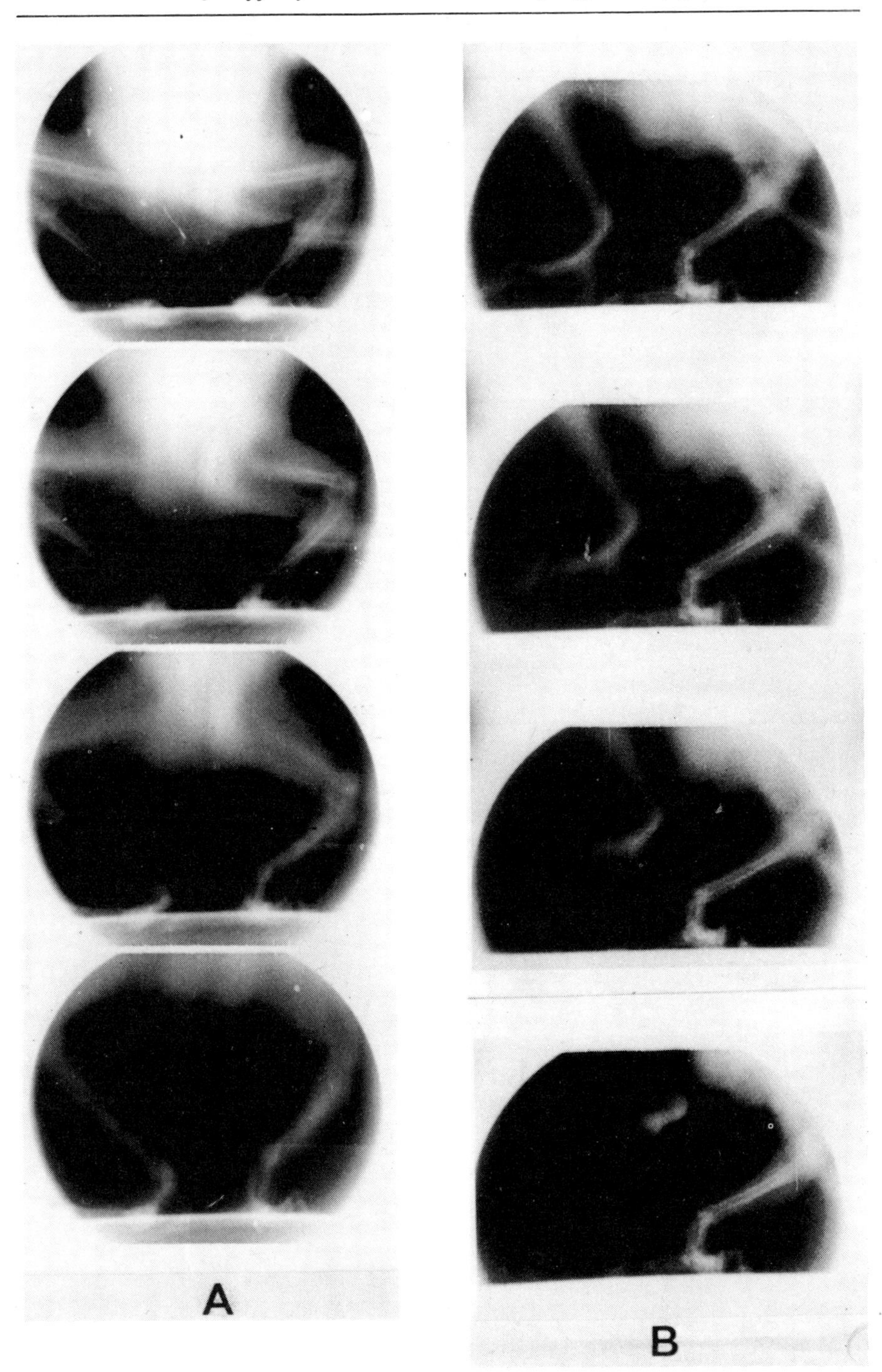

Figure 5. Cineradiographic frames showing two characteristic phases of a symmetrical (*A*) and an asymmetrical (*B*) launching. In each case, the time elapsed between two successive frames is 1/64 sec.

more vertical and the pelvis rises correspondingly; the tarsus rises and leaves the ground by tarso-metarsal dorsiflexion. The angle of the tibia relative to the horizontal remains constant during this stage. 0.28 sec. before launching, a preparatory movement of the tibia is observed: it reduces the tibio-tarsal angle by 40° (0.12 sec.) while the femur and the tarsus continue their upward motion. 0.09 sec. before the launch, the tarsus, which has previously become vertical in a parasagittal plane, inclines itself in the frontal plane (30°) to position itself in the direction of the jump. Finally, a slight push with the tip of the hallux takes place as the last contact with the ground.[7]

Discussion

The *Galago alleni* symmetrical jump appears to be exactly the same as the one recorded for *Galago senegalensis* by Hall-Craggs,[5] who concluded that 'each pair of limbs may be considered as a single structure'. The *G. alleni* specimen that we studied, however, used this type of jump far less than the asymmetrical one.

The high frequency of the asymmetrical jump is perhaps correlated with the ability of *G. alleni* to move in conditions that vary considerably, and to leap to any landing place. The symmetrical jump supposes a landing place situated vertically above, or directly in front of, the take-off point. The asymmetrical jump, on the other hand, enables the animal to choose a greater variety of landing places.

In both cases, a preliminary movement is observed that tends to flex the limbs before extension. This complex movement can be related to contractions of the muscle antagonists. The durations of the two phases, preparatory and launching, differ markedly. Under the conditions examined above, the launching phase (extension) began 0.34 sec. before the launch in the asymmetrical jump and only 0.09 sec. before the launch in the symmetrical jump.

In the case of the asymmetrical jump, the different bones follow each other harmoniously in their displacements, and the whole of this movement is easily analysed: firstly, the active limb femur becomes vertical, and secondly, the semi-plantigrade foot posture is attained before the tibia becomes more vertical. During the symmetrical jump, on the contrary, tibia and tarsus extend at the same time (0.03-0.04 sec. before the launch) and until this movement takes place the feet remain in a plantigrade position on the ground.

This preliminary analysis, which emphasises the precision of the data that can be recorded with the technique radiocinematography, should be completed by the study of other types of locomotion or movements such as grasping or mastication, for instance, which would give us a more thorough understanding and a more general knowledge of prosimian biology.

NOTES

1 Charles-Dominique, P. (1971), 'Eco-éthologie des prosimiens du Gabon', *Biol. Gabon.* 7, 121-228.

2 Décombas, M. (1970), 'Analyse comparée des pistes chez divers rongeurs et prosimiens', *Mém. D.E.A. Université de Paris*, 7, 26 (unpublished).

3 In more than 500 measurements of rodent and prosimian prints, the distance separating the hallux, or the thumb, from the other digits has always been found to be larger on the right side.

4 Hall-Craggs, E.C.B. (1964), 'The jump of the bushbaby, a photographic analysis', *Med. Biol. Ill.* 170-4.

5 Hall-Craggs, E.C.B. (1965), 'An analysis of the jump of the lesser galago (*Galago senegalensis*)', *J. Zool.* 147, 20-9.

6 Jouffroy, F.K. and Gasc, J.P. (1974), 'A cineradiographic analysis of jumping in an African prosimian (*Galago alleni*)', in Jenkins, F.A. (ed.), *Advances in Primatology*, 3, *Primate locomotion*, New York.

7 The film shown at the Research Seminar was produced by the Laboratory of Comparative Anatomy of the Muséum National d'Histoire Naturelle, Paris, and by the Audio-visual Department of the University of Paris.

E.C.B. HALL-CRAGGS

Physiological and histochemical parameters in comparative locomotor studies

Introduction

In previous communications the results of a study of the jump of *Galago senegalensis* have been published. In an initial paper an account was given of the method used to record the early stages of the jump using still and high-speed cinematography.[1] Later, osteometric data obtained from the hind limb skeleton of *Galago senegalensis* and a number of allied prosimian types were presented and their significance discussed.[2] Subsequently an attempt was made to combine the accumulated findings into an analysis of the jump as a whole.[3] Since this work, many additional and better studies have been made of locomotor patterns throughout the primate order including those of prosimians.[4] However, a number of these have been behavioural in nature or strictly concerned with musculoskeletal features that could be correlated with fossil evidence. No emphasis has been placed on the mechanical functions that individual muscles are required to perform, nor on the physiological characteristics that allow them to perform these functions. Over the last decade considerable advances have been made in the physiological and histochemical characterisation of limb muscles and their component fibres, and where possible these characteristics have been linked with the apparent functional calls made on the muscle. Because of this it is proposed to describe the results of some physiological experiments made on the locomotor muscles of *Galago senegalensis*, up to now only briefly reported,[5] and to discuss these in combination with a review of new knowledge of the differing contraction and metabolic properties of skeletal muscle.

Anatomical and physiological considerations

The Galaginae have a reputation for their striking saltatory mode of progression and this has been confirmed by observation of *Galago senegalensis* in its natural habitat. Despite the fact that it is a nocturnal creature, when disturbed in its arboreal habitat during the day it will make accurate jumps from branch to branch and if forced to the ground will proceed by a series of short hops in the upright position using only its hind limbs. When the jump is observed under controlled conditions it is found to be made only when a safe perch can be reached, and – using a chimney-like cage with smooth walls and an adjustable roof of wire netting – a maximum, almost vertical, jump of 7 feet 4¾ inches was recorded. In view of the fact that the animal's centre of gravity is probably only 1½ inches from the ground, this was considered by any standards a remarkable achievement. When the vertical and horizontal course of the jump were controlled by a shallow cage having a transparent front and a background ruled out in 5.0 cm. squares, it was possible to make a simultaneous photographic record of the height achieved in the jump, together with a slow motion study of the critical period when the animal had commenced the jump but still remained in contact with the ground. It is during this period that the height reached is determined, for it is related to only two variables – the angle and velocity of take off – while the weight of the animal and air resistance remain constant. Repeated observation of jumps made to differing heights under these conditions suggested that these variables were accurately gauged by the animal, which could only be persuaded to leap when the aiming point was kept in view. The take-off velocity, upon which so much depends, is attained by the animal while it is in contact with the ground and is related to the force applied to its centre of gravity either during this time or over the distance travelled by the animal during this time. A study of individual frames of the slow-motion film showed that the forelimbs contributed little or nothing to the jump and that the force used was equal and opposite to that applied by the hind-limbs against the ground. It was also possible to decide upon the moment at which the centre of gravity began to move, the moment at which the animal lost contact with the ground and the angle of take off. This information, combined with a knowledge of the frames exposed/sec. by the camera, allowed an estimate to be made both of the distance travelled by the centre of gravity and the time that had elapsed. Knowing the height reached in the jump and the angle of take-off, the take-off velocity could be calculated for each jump. Having established this, the acceleration required to reach this velocity over the observed time or distance could next be calculated, and this combined with the weight of the animal allowed a figure for the amount and direction of the reaction against the ground to be obtained from a force diagram. In one example a 69.0 cm. jump was found to require a force of 1328 g.wt. It is a fortunate fact that the two hind limbs are used in a symmetrical manner, for it could be further said that half this force was exerted by each limb.

A consideration of the anatomy of the hind limb shows that this force is largely developed by, and transmitted through, the two

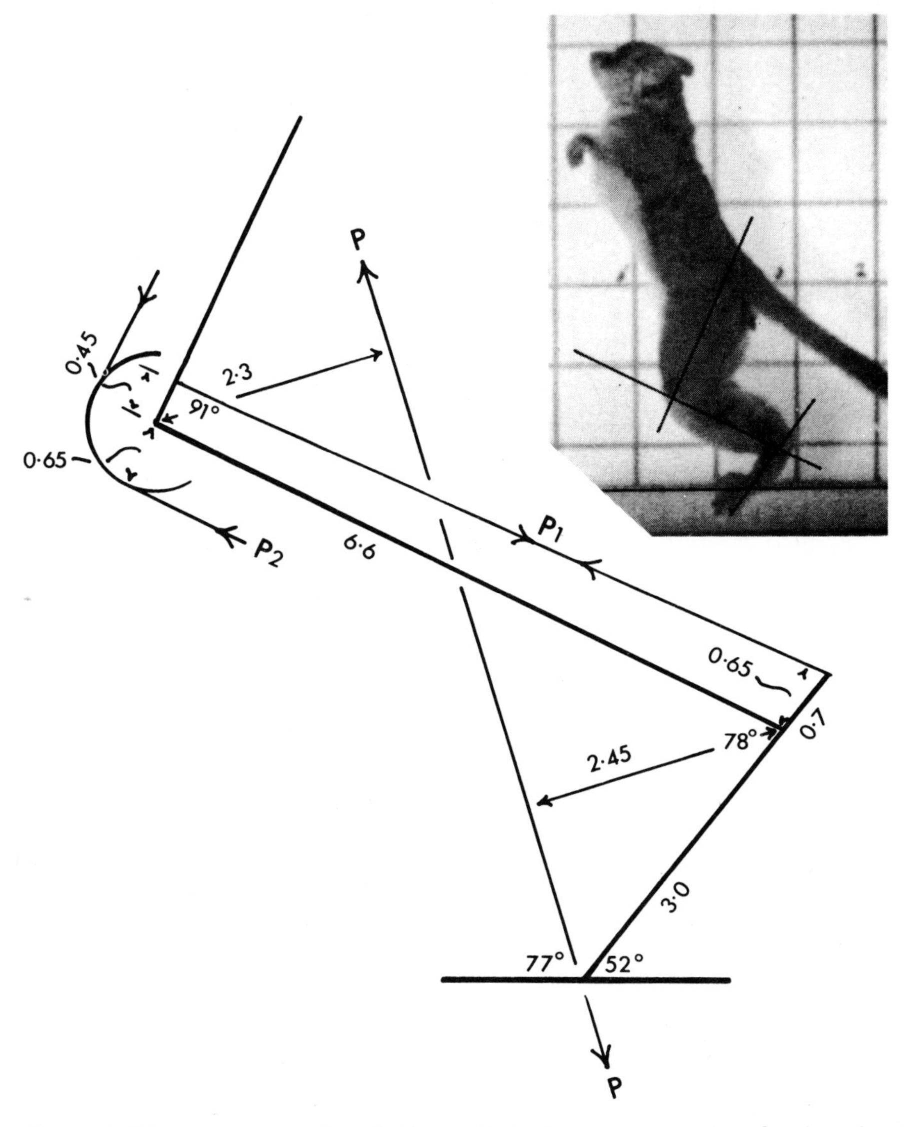

Figure 1. Diagram representing the lower limb skeleton as a series of rods and a pulley from which can be calculated the contribution of *triceps surae* (P1) and *quadriceps femoris* (P2) to the total force required to make the jump (P). Inset is a photograph from which the angles made by the rods were taken.

muscle groups *quadriceps femoris* and *triceps surae.* The *quadriceps femoris* extends the knee joint by acting around the axis of the joint over a pulley formed by the femoral condyles. In so doing, it is required to undergo considerable shortening. The *triceps surae* plantarflexes the foot at the ankle joint, acting as a lever whose fulcrum lies at the tarso-metatarsal joint. Its attachment in large part to the femur has, however, important consequences. Not only does it plantarflex the foot, but it also flexes the knee, and as a result the *quadriceps femoris* must counterbalance the force generated by the *triceps surae* before it is able to extend the knee. This 'two-joint' arrangement of *triceps surae* is not altogether as disadvantageous as it appears. From the scale drawing (Fig. 1) it can be seen that this muscle forms one side of a parallelogram, and that as the knee extends and the ankle plantarflexes the *triceps surae* needs to shorten little. The increased bulk necessary in the *quadriceps femoris* to compensate for this feature also has the effect of transferring the weight of the limb nearer to the centre of gravity of the animal, thus decreasing the inertia of the extremity, a feature observed in many 'fleet-footed' animals.

Turning again to the scale drawing (Fig. 1), which was prepared from a frame of the high-speed cine-film and measurements taken from X-rays of a comparable animal, it is possible, when the total force required to make the jump is known, to calculate estimates of the forces actually developed by these two groups of limb muscles. This was done at a number of points during the jump and the figures appear in Table 1. It can be seen that peak tensions of approximately 4.65 kg.wt. and 5.70 kg.wt. might be expected to be developed by the *triceps surae* and *quadriceps femoris* muscles respectively, and that, because of the changing arrangement of the limbs about the direction of the total force required to be applied to the centre of gravity, the estimated values fall off throughout the latter part of the jump. If these figures can be accepted, it becomes of interest to examine the anatomy of these two muscle groups and to know whether tensions of this order can be demonstrated to be developed in an experimental animal, especially in view of the considerable difference in size between the two. It would also be worthwhile to

Table 1 Estimates of tensions developed by *triceps surae* (P1) and *quadriceps femoris* (P2) at different intervals after the beginning of a jump of 69.0 cm.

Time elapsed *sec*	*P1* *g wt*	*P2* *g wt*
.04	4454	4576
.05	4648	5313
.06	4058	5709
.07	3652	5514
.08	2503	4083
.09	1598	1660

see how the changes in tension needed are related to their physiological properties.

Thompson has said that in similarly constructed animals the height to which they are able to jump would be constant whatever the size of the animal.[6] However, differences in construction may well lead to differing performances. Of the possible differences, those in the proportion of the body weight devoted to jumping musculature might be expected to have a considerable effect.

In a small number of specimens of *Galago senegalensis* the body weights and weights of *triceps surae* and *quadriceps femoris* were obtained and compared with similar figures obtained from *Perodicticus potto* (Table 2). It was found that in *Galago* the combined weights of the two muscles formed 8-12 percent of the body weight, but in *Perodicticus* the same groups formed little more than 2 per cent of the body weight. These figures show that a considerably greater proportion of the body weight is devoted to jumping musculature in *Galago* than in *Perodicticus.*

A comparison was also made of the proportion that the weight of

Table 2 Body weights and muscle group weights in specimens of *Galago senegalensis* and *Perodicticus potto*

Specimen	*Body weight*	*Weight of quadriceps femoris + triceps surae*	*% of body weight*
	g	*g*	*%*
Galago senegalensis	300	24.5	8.2
,,	230	27.1	11.8
,,	168	20.9	12.4
,,	275	31.1	11.3
Perodicticus potto	1429	19.4	1.4
,,	1025	21.9	2.1

Table 3 Weight of *quadriceps femoris* and *triceps surae* muscles in *Galago senegalensis* and *Perodicticus potto*

Specimen	*Weight of triceps surae*	*Weight of quadriceps femoris*	*Weight of triceps surae as % of quadriceps femoris*
	g	*g*	*%*
Galago senegalensis	2.73	21.8	12.5
,,	2.96	24.1	12.3
,,	3.0	17.9	16.8
,,	2.6	23.2	11.2
,,	3.5	27.6	12.7
Perodicticus potto	6.2	13.2	47.0
,,	5.0	16.9	29.6

the *triceps surae* muscle forms of the weight of *quadriceps femoris* (Table 3). The evidence confirms the suggestion that there is a tendency in agile animals for the limb weight to be nearer to the centre of gravity, for in *Galago* the weight of *triceps surae* is only about 15 per cent of *quadriceps femoris* whereas in *Perodicticus* it is 30 per cent or more.

In order to explore the tensions that these muscle groups were in fact able to develop, isometric and isotonic recordings were made under controlled laboratory conditions. Under general anaesthesia, the lower limb was securely fixed by steel pins passed through the upper and lower ends of the femur and by a clamp holding the foot. The patellar tendon and *tendo calcaneus* were exposed, as were the nerves to each muscle group. The tendons were transfixed and ligated to steel hooks, which were in turn linked to either an isometric or isotonic lever recording on a revolving smoked drum. The nerves were divided and prepared for stimulation. All incisions were kept irrigated with liquid paraffin and muscle temperature was maintained by radiant heat controlled by a thermistor.

Only one series of satisfactory isotonic recordings was made, but these allowed both force-velocity (Figs. 2 and 3) and power-velocity curves to be constructed. The force velocity curves for each muscle showed the shape already established for muscles from many different types of animals.[7] The power-velocity curves showed that the peak output occurs at a higher rate of shortening in *quadriceps femoris* than in *triceps surae* – a fact which may be correlated with the distance over which each muscle is required to shorten during the jump; approximately 17 per cent of its initial length in the case of

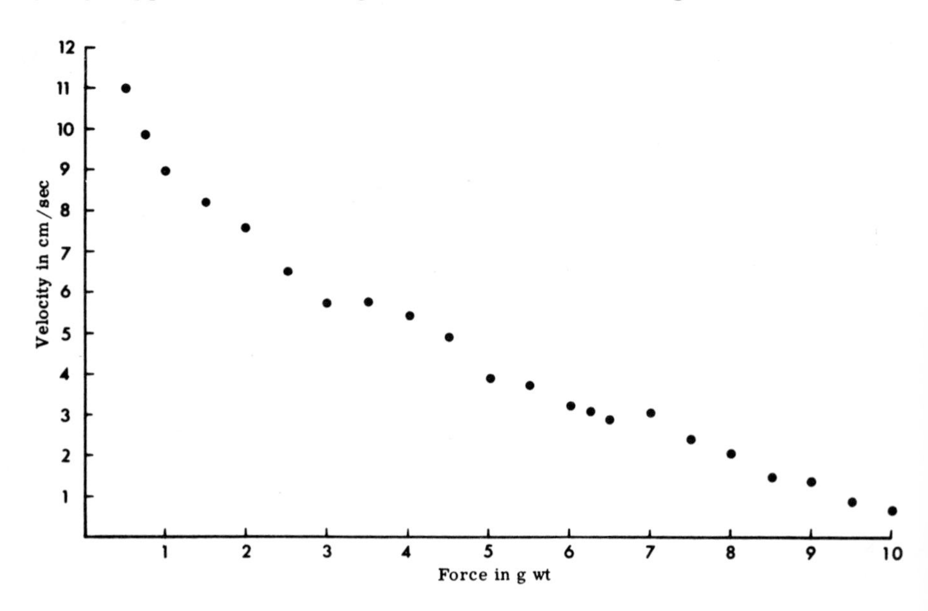

Figure 2. Force-velocity curve, *quadriceps femoris*.

quadriceps femoris and 5·6 per cent in the case of *triceps surae*. In terms of actual velocity of shortening and tensions developed, these results fell well below those that had been estimated as necessary to make the jump. This may have been for a variety of reasons, connected, for example, with the method of anaesthesia or inertia and friction in the mechanical method of making the recordings. The isotonic recordings were also made without any pre-loading, a situation which may in fact not be the ideal condition for the development of maximum tension (see later). The isometric recordings were, however, made over a range of muscle lengths more closely related to those found *in vivo*. This was achieved by taking an X-ray of each animal immediately before recordings were made. Before each tendon was divided, the position of a mark on the tendon was recorded when the limb was arranged in the fully extended position that it would take up at the end of the jump. The length of the muscle in this position was called the terminal length (see Fig. 4). The tendons were then divided and isometric recordings begun with the muscle at this length. These were repeated at increasingly greater muscle lengths up to a length just beyond that estimated from the X-rays to represent full flexion of the knee and full dorsiflexion of the ankle. This was called the initial length (Fig. 4). The difference

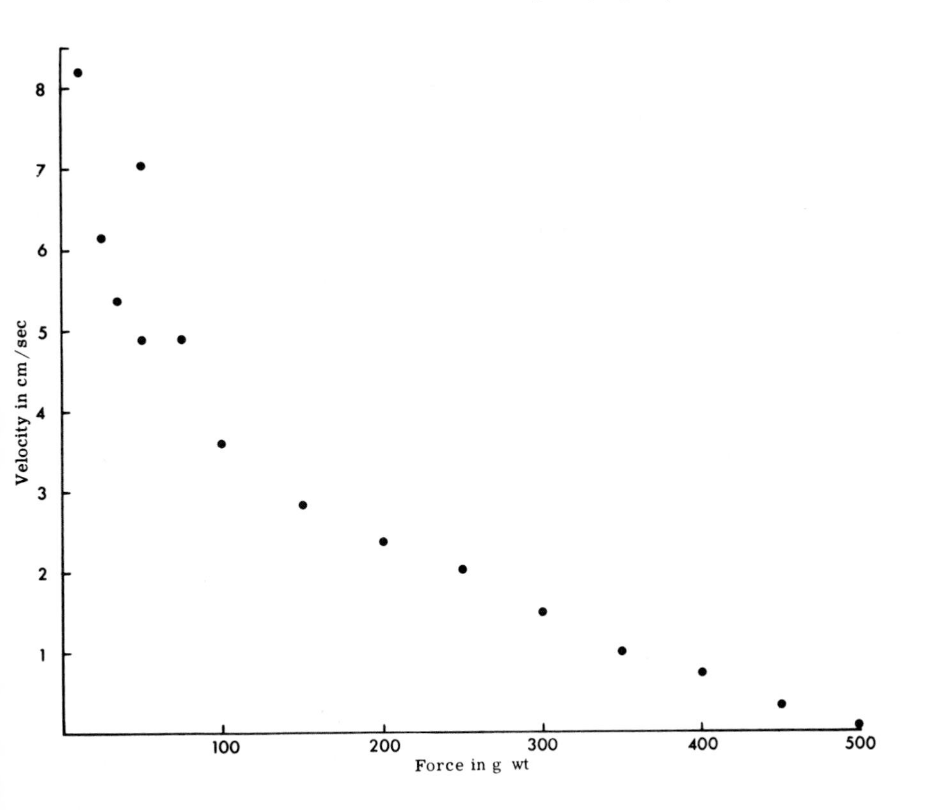

Figure 3. Force-velocity curve, *triceps surae*.

between these two lengths, as suspected from a consideration of the anatomy, proved to be small in the case of *triceps surae* – only 2.0 and 2.5 mm. in two animals – although the theoretical minimum and maximum lengths (Fig. 4) were 10.0 mm. above and below the range actually employed in jumping. It was found that stimulation at the terminal length produced a tension of just over 60 per cent of the maximum tension recorded (Fig. 5). This value was 3.3 and 3.375 kg.wt. in two experiments and was developed very close to a length estimated to exist at the beginning of the jump, i.e. the initial length.

The change in length of *quadriceps femoris* needed to execute the jump was calculated to be 1.25 cm. At the initial length the tension developed was approximately 4.75 kg.wt., but at this point the passive tension was 2.5 kg.wt. (Fig. 6). The distance over which *quadriceps femoris* must shorten is considerably greater than that required of *triceps surae* and as a result, while the maximum tension developed is greater than that developed by *triceps surae*, the tensions developed near the terminal length would appear to fall below those possible in *triceps surae*.

A further experiment was performed in which an attempt was made to reproduce as far as possible an *in vivo* situation. Using the isotonic lever, the mechanical leverage existing at the knee and ankle joints was more nearly simulated by attaching the muscle nearer to the fulcrum than the load. The load was also rigidly attached to the

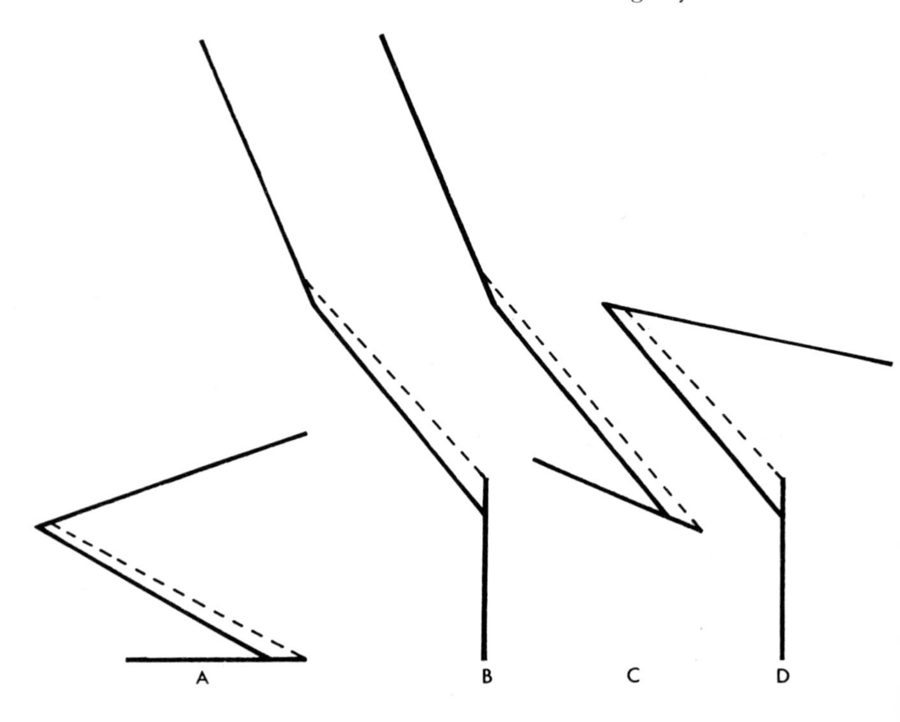

Figure 4. Diagram illustrating the 'initial length' of *triceps surae* (*A*) the 'terminal length' (*B*) and the theoretical 'maximum (*C*) and minimum (*D*) lengths'.

lever. The expected amount of shortening was again calculated from an X-ray and the muscles linked to the lever at their terminal lengths. From this position they were then extended by known amounts, and at each length stimulated while they were loaded with a range of weights. In this way, portions of their force-velocity curves were recorded at lengths believed to exist during the jump (Figs. 7 and 8). The results of this experiment showed, as might be expected, that at a given rate of shortening the force developed increased as the length of the muscle was increased. At its initial length *triceps surae* was now found to develop a force of 4.2 kg.wt. and the *quadriceps femoris* a force of 6.0 kg.wt. midway between its initial and terminal lengths. These tensions showed an increase over those previously recorded and would in fact have been sufficient to perform the jump originally described. However, at these peak tensions there remains a considerable discrepancy in the speed of shortening – a factor possibly related to the experimental conditions. Despite this, the geometry of the levers is such that the maximum force is required of each muscle when the speed of shortening would need to be least (Table 1) and falls off as the animal accelerates, thus allowing greater

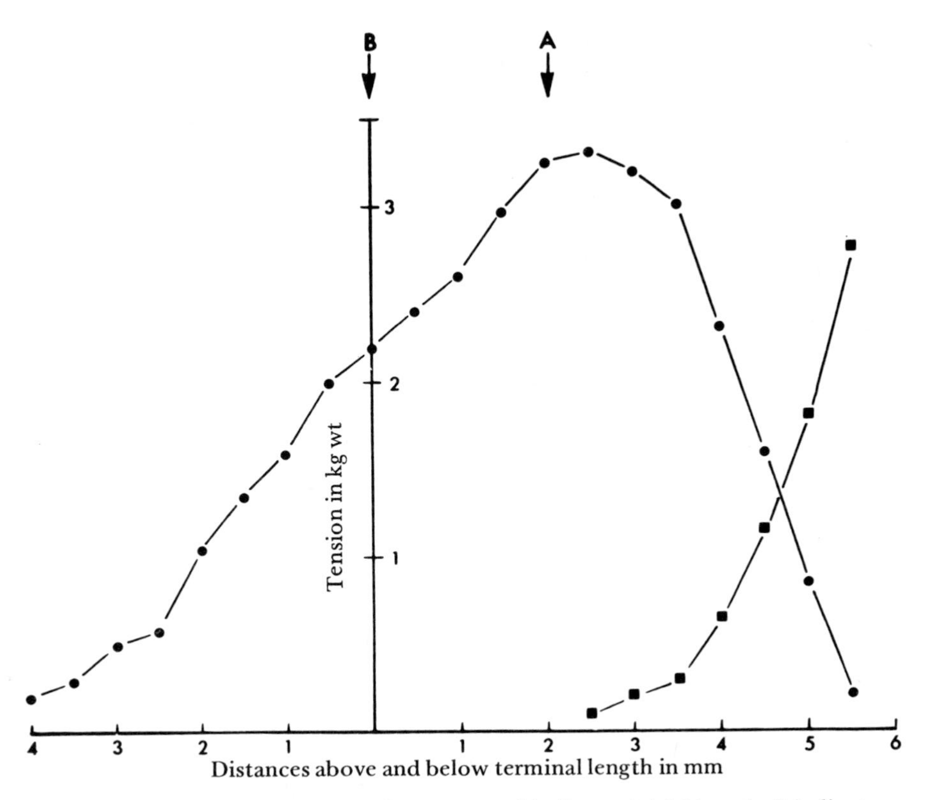

Figure 5. Length-tension curve, *triceps surae*. *A* indicates initial length, *B* indicates terminal length, ■ = passive tension, ● = developed tension.

speeds of shortening to be reached towards the end of the jump.

A further discrepancy also appears to exist between the maximum tensions developed by these muscles and their size. Following two of the experiments described, the two muscles were weighed and figures obtained for tension in kg.wt. per g. of muscle (Table 4). The figures for *quadriceps femoris* were in the same order as those obtained from rabbit *triceps surae* in two preliminary experiments and were well within the maximum figures of 0.88 kg.wt./g. quoted by Alder et al. for rabbit *tibialis anterior.*[8] The *triceps surae* of *Galago*, however, was found to develop between four and five times as much tension per g. as *quadriceps femoris.* A possible explanation for this lies in a differing fibre architecture, for the tension is related to the effective cross-sectional area of the fibres and (for muscles of the same weight) this will vary inversely with the length of their component fibres. Using a method described by Elliott and Crawford,[9] muscles were macerated and small fascicles containing perhaps half a dozen fibres were sorted into groups by length, dried and weighed. The weight of each group was divided by its mean length and the sum of the quotients gave an index of the relative fascicular cross-sectional area.

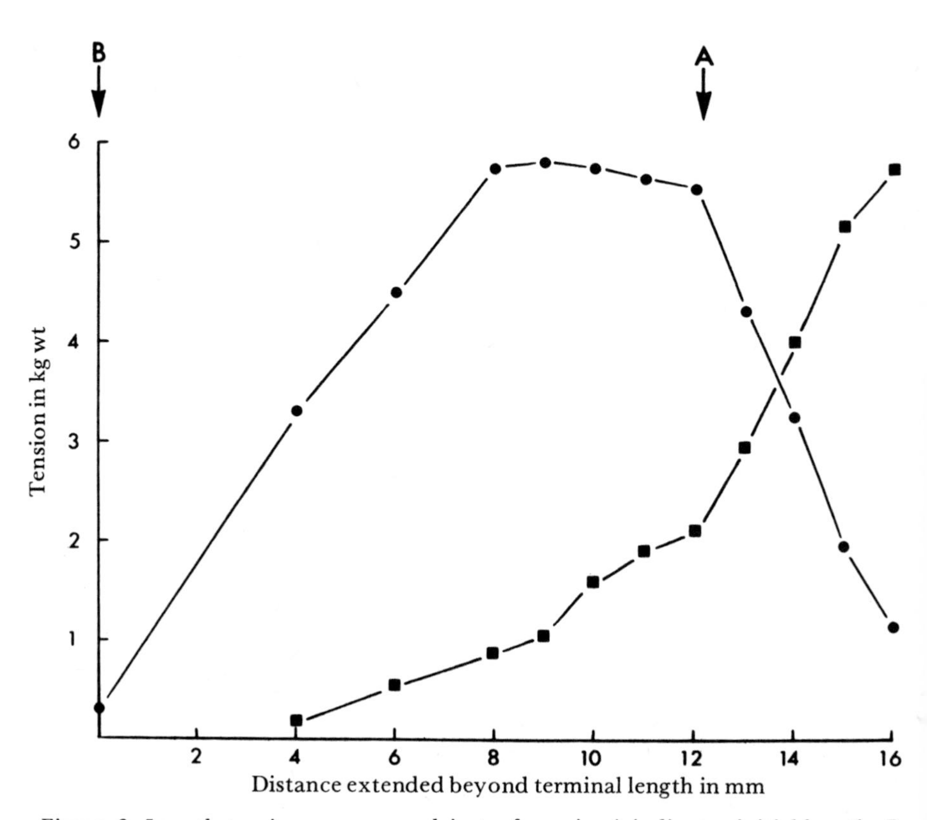

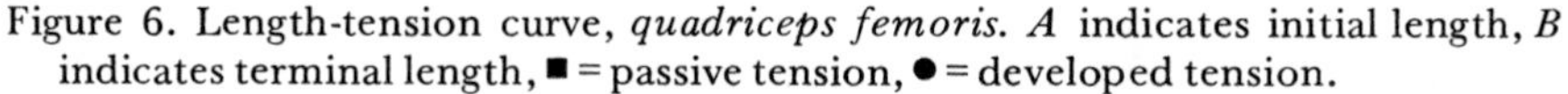
Figure 6. Length-tension curve, *quadriceps femoris. A* indicates initial length, *B* indicates terminal length, ■ = passive tension, ● = developed tension.

This figure was corrected for degree of extension by counting the number of sarcomeres per unit length in phase contrast microphotographs of specimen fibres. The final figure for relative fascicular area per gramme dry weight was found to be twice as great for *triceps surae* as for *quadriceps femoris* in both experiments. On these grounds, *triceps surae* could be expected to develop twice the tension per unit weight that can be developed by *quadriceps femoris*, thus accounting in part for the difference of four to five times that had been observed.

Table 4 Weights of muscle groups and maximum tensions recorded from them

Specimen	*Muscle*	*Weight*	*Max. tension developed*	*Tension developed per Kg of muscle*
		g	*Kg wt*	*Kg wt*
Galago	Quad. fem.	10.88	6.25	0.57
senegalensis	Tri. surae	1.37	3.49	2.56
,,	Quad. fem.	12.05	5.8	0.48
,,	Tri. surae	1.48	3.3	2.23
Rabbit	Quad. fem.	12.7	6.63	0.52
	Tri. surae	9.8	6.6	0.67

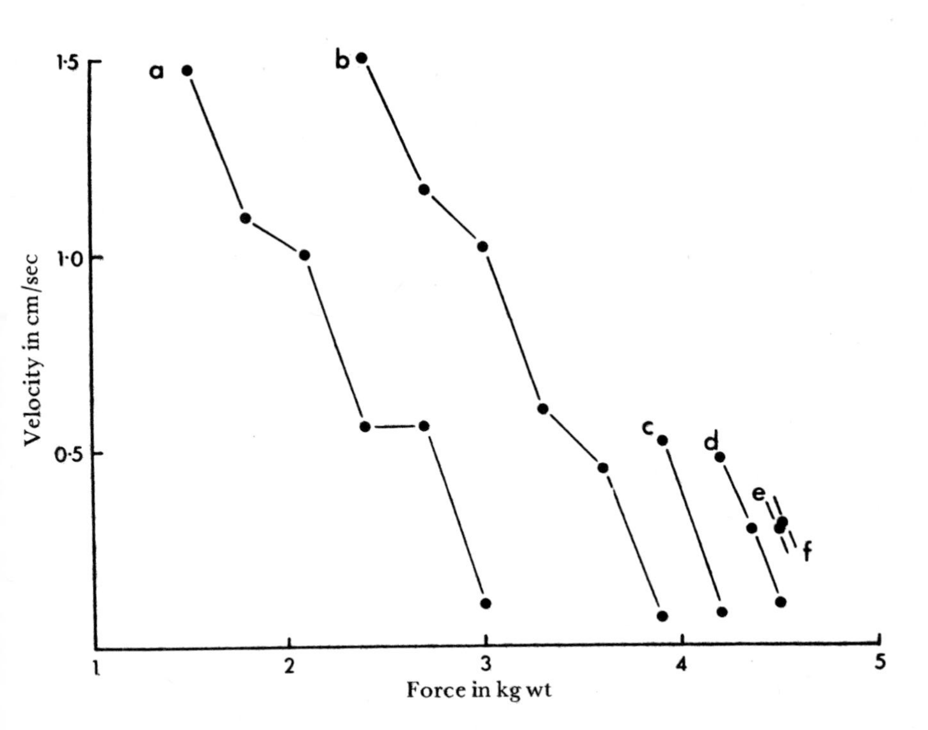

Figure 7. Portions of a series of force-velocity curves at different muscle lengths. The muscle was lengthened 1.0 mm. between *a* and *b* and 0.5 mm. between subsequent curves. The muscle length used at *c* corresponded to the initial length; *triceps surae.*

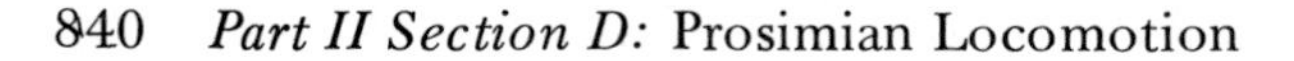

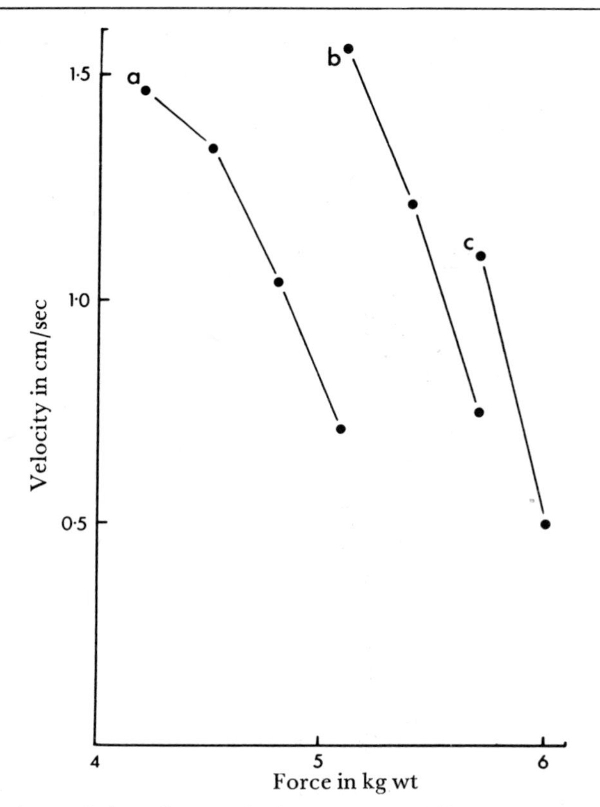

Figure 8. Portions of three force-velocity curves at different muscle lengths. The muscle was lengthened by 2 mm. between curves *a* and *b* and 1 mm. between curves *b* and *c*. The muscle length used at *c* lay midway between the initial and terminal lengths; *quadriceps femoris.*

Histochemical evidence

It has, of course, long been known that skeletal muscles differ from each other in both morphological and physiological detail. Ranvier was well aware of this when he demonstrated that a stimulus rate that produced tetanic fusion in a red muscle was insufficient to produce the same effect in a white muscle.[10] Over subsequent years the contraction characteristics of red and white limb muscles have been more fully analysed and a fairly wide spectrum of properties revealed.[11] Together with these studies, advances in histochemical techniques for demonstrating the nature of enzymes present in the individual fibres comprising these muscles have allowed some correlation to be made between metabolism and function. Stein and Padykula classified three distinct fibre types in the white *gastrocnemius* and red *soleus* of the rat.[12] This classification was based on the cytochemical distribution of the activity of succinic dehydrogenase (SDH). In the rat the white muscle was found to contain all three types (*a, b* and *c*, Fig. 9) and the *soleus* two (*b* and *c*, Fig. 10). A reciprocal relationship between the presence of oxidative enzymes and those of the glycolytic pathway has also been demonstrated (Figs. 11 and 12),[13] and differences revealed in the acid and alkaline stability of myofibrillar and intermyofibrillar adenosine triphosphatase (MATPase and IMATPase) have provided a further technique for characterising the fibres of fast and slow muscles (Fig. 13).[14]

Romanul applied a large number of histochemical tests to serial sections of the calf muscles of rats and was able to define eight fibre types.[15] From this more extensive study it emerged that there were: (1) a group of fibres with a high capacity to utilise glycogen, but with a low lipid and oxidative metabolism and a low myoglobin content; (2) an intermediate group with a moderate ability to metabolise glycogen and lipids and a high oxidative metabolism; and (3) a third type with a low capacity for glycogen breakdown, very active lipid and oxidative metabolism and a high myoglobin content.

The metabolic varieties of the fibres of limb muscles having been established, it remains to link these more closely with the physiology of the muscle and if possible with those of motor units. A number of workers have been able to record contraction properties of individual units either by using reflexly evoked stimuli or by the stimulation of single ventral root fibres.[16] Edström and Kugelberg found that repetitive stimulation of motor nerve fibres produced sufficient change in phosphosylase and glycogen levels to allow them to identify the fibres of the motor unit involved.[17] They were thus able to say that the fibre composition of units was largely uniform in fibre type and that the fatigability of units was closely correlated with their fibre types. More recently, by combining this method with a range of physiological determinations particularly aimed at the response of a fibre to fatigue, Burke et al. have described three varieties of motor unit in the cat *gastocnemius.*[18] These were a fast but rapidly fatigued unit, a fast unit resistant to fatigue and a slow and fatigue-resistant unit. Serial sections taken from the experimental muscles enabled them to show these units to be composed of fibres of low, moderate and high oxidative activity, respectively.

Little information appears to be available about fibre types in non-human primates. Beatty et al. have presented some histochemical and biochemical data from the Anthropoidea and in particular from the rhesus monkey.[19] As information about the Prosimii appears to be lacking, it seems worthwhile to report briefly on a number of preparations obtained post-mortem from members of this sub-order.[20] The quality of these is poor, owing to the time that elapsed between death and removal of the muscles, but in a number it is possible to identify the fibre type after incubation to demonstrate SDH.

Of the muscles examined in *Galago senegalensis*, the small *rectus femoris* showed the typical chequerboard pattern of a fast muscle

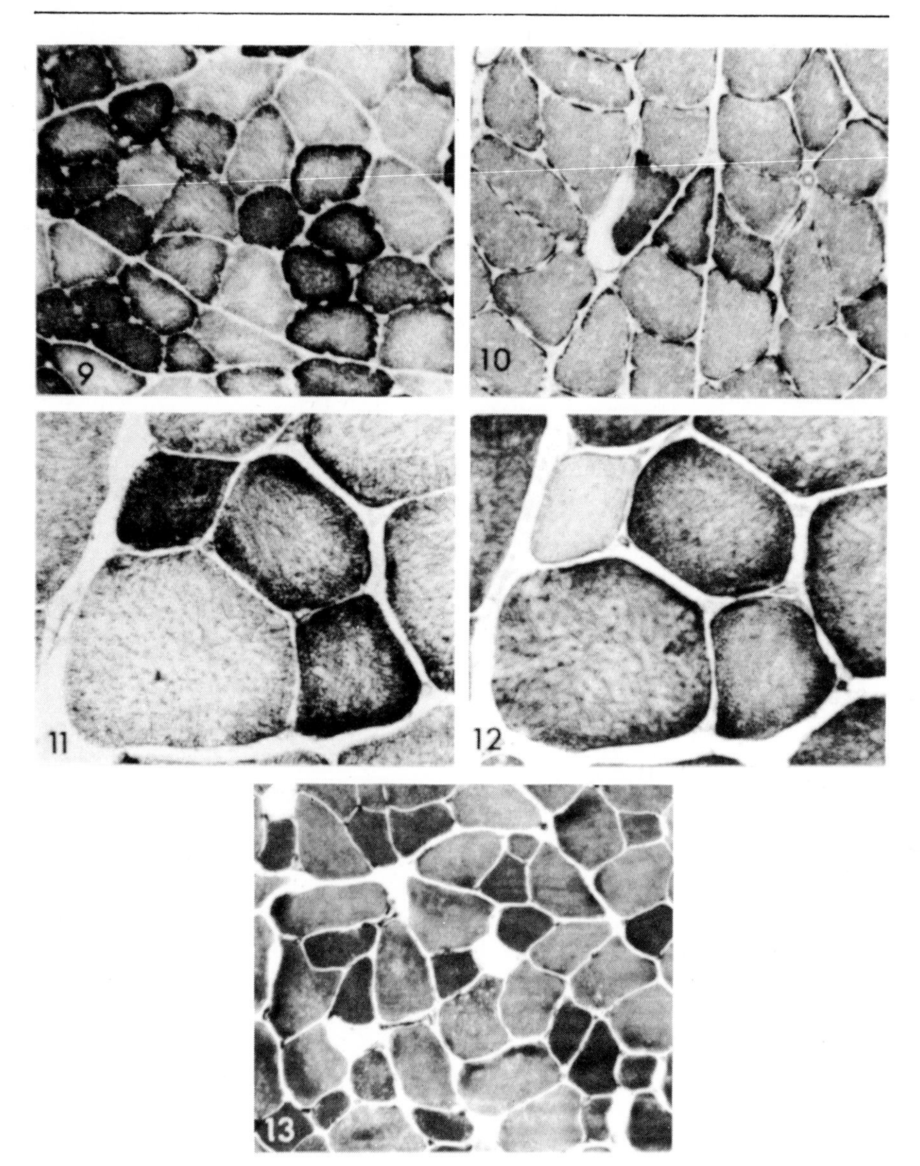

Figure 9. Rat *plantaris* muscle, SDH; *a, b* and *c* fibres are seen to be present.

Figure 10. Rat *soleus* muscle, SDH; only *b* and *c* fibres are present.

Figure 11 and 12. Serial sections of rat *plantaris* muscle. In Fig. 11 the oxidative enzyme reduced diphosphopyridinenucleotide-tetrazolium reductase is demonstrated, and in Fig. 12 the glycolytic enzyme α-glycerophosphate dehydrogenase. Comparison of the two figures shows the reciprocal relationship of these two enzymes.

Figure 13. Rat *plantaris* muscle, MATPase; using this reaction, three fibre types can also be distinguished: $\alpha\beta$ or *a* fibres (moderate staining), α or *c* fibres (dark) and β or *b* fibres (light).

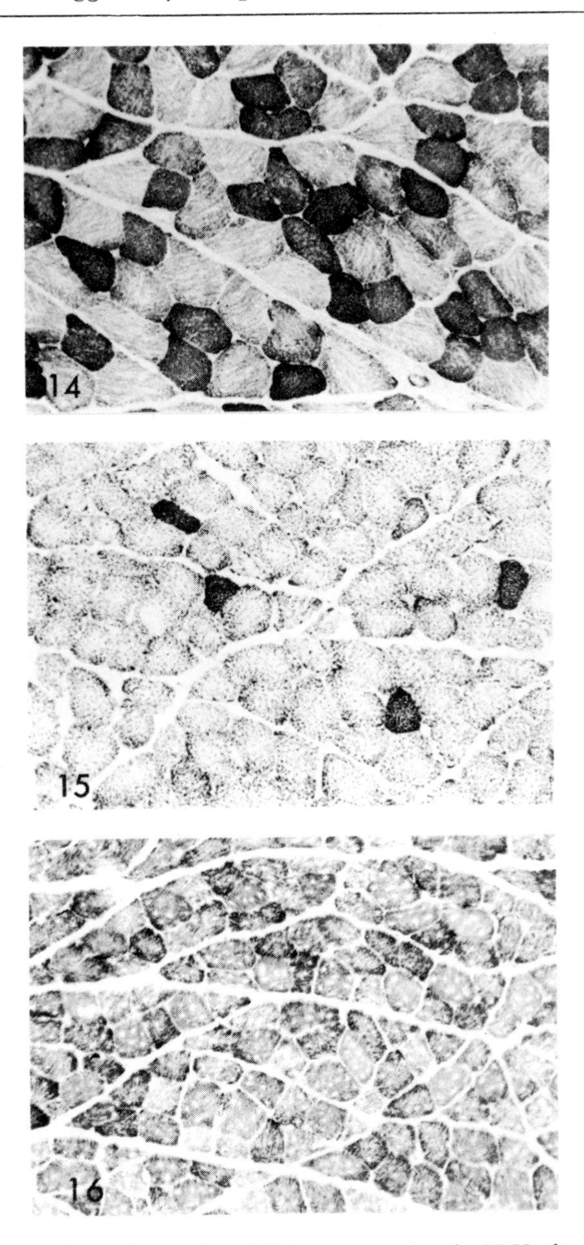

Figure 14. *Rectus femoris* muscle of *Galago senegalensis*, SDH; three fibre types are seen.

Figure 15. *Vastus medialis* muscle of *Galago senegalensis*, SDH; nearly all the fibres are Type *a*.

Figure 16. *Vastus medialis* muscle of *Loris tardigradus*, SDH; showing a high proportion of *b* and *c* fibres.

and all three types of fibre could be distinguished (Fig. 14). The two *vasti*, however, appeared to have a higher proportion of *a* fibres and in some regions of *vastus medialis* (Fig. 15) this was almost the only type seen. In contrast to this, the *vasti* of *Arctocebus* and *Loris tardigradus* showed a high proportion of *b* and *c* fibres (Fig. 16), the calibre of which often appeared to exceed that of the *a* fibres. The same picture was seen in the *gastrocnemii* and the *biceps* and *triceps brachii* of the slender loris.

It would be a mistake to reach any firm conclusions on this evidence, but one cannot avoid drawing attention to the presence of a high proportion of possibly fast *a* fibres in the *quadriceps femoris* of *Galago*, where speed of shortening is a paramount need, and the predominance of fatigue-resistant *b* and *c* fibres in the muscles of more slowly moving species that are known to maintain a posture for long periods without movement.

Summary

In summary it can be said that *Galago senegalensis* shows a number of features that appear to be positively advantageous to its saltatory mode of progression. Its long hind limb with a wide range of extension allows the maximum time or distance for the application of force against the ground. Against this must be placed the poor mechanical advantage of the leverage of the foot, which calls for great forces from the *triceps surae.* The geometry of the attachments of *triceps surae* have been seen to limit the shortening needed, thus increasing the possible cross-sectional area of the muscle fibres per unit weight when compared with the *quadriceps femoris.* This feature is reflected in the length-tension diagrams which show that *triceps surae* only uses the optimum portion of the curve when shortening during the jump. The attachment of this muscle to the femur also has the effect of transferring the weight of the limb nearer to the centre of gravity and reducing the inertia near the moment of take-off. It would appear that the *quadriceps femoris* may be largely composed of muscle fibres having a high speed of shortening, a fact that again is seen to be necessary from a study of its mechanical action.

Since the work described here was completed, major advances in the investigation of the physiological and histochemical characteristics of skeletal muscle have been made. Many of these new techniques could be applied to comparative locomotor studies in the Prosimii, which present such a wide range of locomotor patterns, and it is thought that if these were to be combined with the type of anatomical and mechanical data presented here a significant contribution to knowledge in both the field of primatology and muscle biology could be made.

NOTES

1 Hall-Craggs, E.C.B. (1964), 'The jump of the bushbaby – a photographic analysis', *Med. Biol. Ill.* 14, 170-4.

2 Hall-Craggs, E.C.B. (1965), 'An osteometric study of the hind limb of the Galagidae', *J. Anat.* 99, 119-26.

3 Hall-Craggs, E.C.B. (1965), 'An analysis of the jump of the lesser galago (*Galago senegalensis*)', *J. Zool.* 147, 20-9.

4 e.g. Grand, T.I. (1967), 'The functional anatomy of the ankle and foot of the slow loris (*Nycticebus coucang*)', *Am. J. Phys. Anthrop.* (n.s.) 26, 207-18. Napier, J.R. and Walker, A.C. (1967), 'Vertical clinging and leaping – a newly recognized category of locomotor behaviour of primates', *Folia primat.* 6, 204-19. Simons, E.L. (1967), 'Fossil primates and the evolution of some primate locomotor systems', *Am J. Phys. Anthrop.* (n.s.) 26, 241-54.

5 Hall-Craggs, E.C.B. (1966), 'Muscle tension relationships in *Galago senegalensis*', *J. Anat.* 100, 699-700.

6 Thompson, D'A.W. (1942), *On Growth and Form*, Cambridge.

7 Wilkie, D.R. (1956), 'The mechanical properties of muscle', *Brit. Med. Bull.* 12, 177-82.

8 Alder, A.B., Crawford, G.N.C. and Edwards, R.G. (1958), 'The growth of the muscle *tibialis anterior* in the normal rabbit in relation to the tension-length ratio', *Proc. Roy. Soc. B.* 148, 207-16.

9 Elliott, D.H. and Crawford, G.N.C. (1963), 'The relationship between the size of muscles and their tendons in the adult rabbit', *J. Anat.* 97, 139.

10 Ranvier, L. (1873), 'Propriétés et structures différentes des muscles rouges et des muscles blancs, chez les lapins et chez les raies', *C.R. Hebd. Séanc. Acad. Sc. Paris* 77, 1030-4.

11 Denny-Brown, D.E. (1929), 'The histological features of striped muscle in relation to its functional activity', *Proc. Roy. Soc. B* 104, 371-411. Cooper, S. and Eccles, J.C. (1930), 'The isometric responses of mammalian muscles', *J. Physiol.* 69, 377. Buller, A.J., Eccles, J.C. and Eccles, R.M. (1960), 'Differentiation of fast and slow muscles in the cat hind limb', *J. Physiol.* 150, 399-416. Close, R. (1967), 'Properties of motor units in fast and slow skeletal muscles of the rat', *J. Physiol.* 193, 45-55.

12 Stein, J.M. and Padykula, H.A. (1962), 'Histochemical classification of individual skeletal muscle fibres of the rat', *Am. J. Anat.* 110, 103-23.

13 Dubowitz, V. and Pearse, A.G.E. (1960), 'Reciprocal relationship of phosphorylase and oxidative enzymes in skeletal muscle', *Nature* 185, 701-2.

14 Guth, L. and Samaha, F.J. (1969), 'Qualitative differences between actomyosin ATPase of slow and fast mammalian muscle', *Exp. Neurol.* 25, 138-52.

15 Romanul, F.C.A. (1964), 'Enzymes in muscle. I: Histochemical studies of enzymes in individual muscle fibres', *Arch. Neurol.* 11, 355-68.

16 Henneman, E. and Olson, C.B. (1965), 'Relations between structure and function in the design of skeletal muscles', *J. Neurophysiol.* 28, 551-98. Close, R. (1967), 'Properties of motor units in fast and slow skeletal muscles of the rat', *J. Physiol.* 193, 45-55.

17 Edström, L. and Kugelberg, E. (1968), 'Histochemical composition, distribution and fatiguability of single motor units: Anterior tibial muscle of the rat', *J. Neurol. Neurosurg. Psychiat.* 31, 424-33.

18 Burke, R.E., Levine, D.N., Zajac, F.E., Tsaris, P. and Engel, W.K. (1971), 'Mammalian motor units: physiological-histochemical correlations in three types in cat gastrocnemius', *Science* 174, 709-12.

19 Beatty, C.H., Basinger, G.M., Dully, C.C. and Bockek, R.M. (1966), 'Comparison of red and white voluntary skeletal muscles of several species of primates', *J. Histochem. Cytochem.* 14, 590-600.

20 The author is indebted to the Zoological Society of London for making specimens available.

PART II SECTION E
Chromosomes, Proteins and Evolution

N. A. BARNICOT

The molecular and cytogenetic approaches to prosimian phylogeny

Although inherited variation is a cornerstone of the theory of natural selection very little was known about the mechanisms of heredity in 1859. Following the rise of Mendelian genetics a re-examination of evolutionary theory was initiated by mathematical geneticists in the 1920s. This led to clearer concepts in which gene mutation and gene frequency change were seen as the fundamental processes involved in evolutionary divergence. It follows that changes in the structure of the genome (i.e. the genetic material as a whole) should provide the most basic information about evolutionary divergence; but the chemical nature of the genome was not known and it was impossible to compare the genomes of distantly related species by classical breeding methods.

This impasse was resolved barely 20 years ago by developments in molecular biology. The crucial advances were the discovery of the structure of DNA and its role in protein synthesis, and the invention of methods for determining the complete amino acid sequences of protein chains. At about the same time work on the three-dimensional structure of certain proteins by X-ray crystallographers made it possible to understand how the amino acid sequence is related to the functional properties of the molecule.

These discoveries opened new prospects for evolutionary biology which were soon exploited, notably by Margoliash and his colleagues in their remarkable studies on the amino acid sequences of the enzyme cytochrome *c* in a very wide range of organisms.[1] It was no doubt reassuring to find that a scheme of phylogenetic relationships based on this single protein, which represents only a minute fraction of the whole genome, was reasonably consistent with prevailing taxonomic views. But it should be noted that there is little or no fossil evidence for many invertebrate groups and comparative anatomy is not much help; these parts of the cytochrome *c* phylogeny are therefore difficult to check against other types of evidence.

I do not wish (after this appalling example of potted history) to detail the biochemical facts on which the molecular approach to phylogeny is based. This is ably done by Dr Doolittle and other authors in the papers that follow. I believe that amino acid sequence data on several different proteins in a well-chosen spectrum of primates will go far to solve problems of phylogenetic relationship. But it is worth remembering that relationship is only one, and perhaps not even the most interesting, facet of the evolutionary puzzle. A diagram of the course of evolutionary divergence with a correct branching pattern and correct distances along the branches is certainly well worth having; but it is merely a description of what happened. The much more difficult problem of explaining why evolution took that particular course remains, and it will certainly need many other approaches to reach plausible solutions.

The study of proteins in an evolutionary context has generated some ideas that are disturbing to orthodox belief in the all-pervasive power of natural selection. The question is whether natural selection has in fact been responsible for most of the amino acid substitutions that have been incorporated in proteins over the course of time. Kimura pointed out that if the rates of change in the few proteins that had then been examined could be taken as typical of the thousands of proteins that must be coded by the genome, they would imply an intolerable load of genetic deaths if the changes were due to selection.[2] One way out of this difficulty is to suppose that many of these mutations are selectively neutral. A small fraction of those that occur then become fixed in the population by random processes (genetic drift). Selection still operates, but mainly in a negative sense, to weed out unfavourable mutants. The wide variation in the evolutionary rates of different proteins could be accounted for by differences in their tolerance of mutational change in structure. Various arguments for and against this theory of protein evolution by neutral mutations have been advanced, and the controversy continues.[3–5] Perhaps it should be remarked that neither Kimura nor his protagonists have denied the importance of natural selection at levels of organisation higher than protein molecules, so that there is no cause for general alarm.

Polymorphic variation of proteins is remarkably frequent in man and in a few other organisms that have been closely examined.[6] The neutral mutation theory sees this polymorphism as a stage in the passage of neutral alleles towards fixation by drift, and it enables one to predict the numbers and frequencies of such alleles to be expected in populations of given effective size. Studies on biochemical polymorphism in non-human primates, such as Hewett-Emmett and I have attempted (this volume), may therefore help to check the validity of this theory if they can be combined with field observations on population size and breeding structure. Although non-human primates are not ideal for work of this kind, they do have the advantage that considerable effort is now being devoted to studying their behaviour in the wild. Perhaps some of the smaller and more abundant prosimians that are not too difficult to locate and catch, such as galagos or mouse-lemurs, would provide favourable material.

One attraction of the neutral mutation theory is that it predicts constant rates of change in proteins provided that mutation rates remain constant. If this is the case, and if the theory is substantially right, then proteins can serve as evolutionary clocks. Approximately constant rates of change have in fact been claimed for some proteins, but Goodman et al. (1971) have postulated declining rates for anthropoid haemoglobin chains and there are various other facts in the literature which suggest that rates are not always constant.[7]

A striking example has been presented by Staehlin (1972), who studied the sequences of the hormone calcitonin in man, pig, ox, sheep and salmon.[8] He found 18 differences between man and pig in a chain of 32 residues and only 15 differences between man and salmon. He suggested that major changes in the function of this calcium-controlling hormone may have occurred early in the evolution of artiodactyls.

To estimate rates of change we need an acceptable phylogenetic tree, measures of distance (i.e. divergence) along the branches and reliable dates for the nodes. Some of the problems of constructing phylogenetic trees from protein sequence data are discussed by Cook and Hewett-Emmett in this volume. It is easy to forget the assumptions on which these procedures are based and to relax into the pleasant belief that one need only feed the data into a computer for infallible phylogenies to emerge. As a rule no measure of the reliability of a dendogram is given and it is tempting to attach undue weight to details of the branching pattern that may in fact be supported by very slight evidence.

Palaeontological dates for branching points in primate phylogeny are often approximate. This is not so much because dating of the geological deposits is uncertain, but because we are unlikely to find fossils that are true common ancestors (i.e. lie exactly on nodes in the phylogenetic tree). We therefore have to extrapolate to a greater or lesser extent; and, although I hesitate to mention it for fear of rousing either Professor Simons, Dr Szalay or both, it is not unknown for palaeontologists to disagree in their attributions of ancestors.

There is much evidence pointing to changes of evolutionary rates at the morphological level.[9] Adaptation to new ecological situations may at first be relatively rapid and later slow down. It would be surprising if major adjustments of habit are not reflected in structural and functional change of at least *some* proteins. But it may not always be possible to get evidence of rate changes in the evolution of a protein. If we have, say, only two living descendants (or two groups of closely related descendants) from some early common ancestor, we cannot tell whether the sequence differences between them that we see today accumulated at a steady rate, or whether divergence was more rapid at some restricted period. To investigate this we need data on the survivors of lines that branched from these lineages at some suitable stage in their divergence. Even if such branches ever existed it may be that extinction will have deprived us of the evidence.

If we are interested in the structural divergence of genes it would be best to analyse and compare the genes themselves, because most amino acid changes in proteins do not give unequivocal information about the precise nature of the gene change and because some gene changes are not manifested in proteins at all. In general it is still impossible to isolate and analyse particular genes, but it is worth mentioning that Min Jou et al. have now reported the complete base sequence of the gene coding for the coat-protein of phage MS 2.[10] Admittedly the genome of this virus is much smaller than that of higher organisms, and the genetic material is RNA, not DNA, but the work serves to show how rapidly molecular genetics is advancing. In any case it is already possible to make a *general* comparison of the genomes of higher organisms by means of DNA hybridisation methods. These techniques test the overall structural similarity of large segments of DNA containing perhaps many thousands of genes. Some work has been done on primates,[11] and it is a pity that we were unable to include a paper on this approach, since it is a valuable complement to studies on single proteins.

Despite technological advances amino acid sequence analysis is still a slow (and expensive) business and quicker, if less informative, methods still command attention. Immunological comparisons of primate proteins were initiated by Nuttall over fifty years ago and the methods have been refined and extended since then.[12] Insofar as amino acid substitutions alter the molecular suface to which antibodies stick, we should expect serological and chemical comparisons to give similar results. This seems to be the case, though very few proteins have been intensively studied by both methods. However, discrepancies sometimes occur if the same antigenically potent substitution happens to have taken place in two distantly related species (see Nisonoff et al. on cytochrome *c*).[13] Bauer gives us a clear account of the rationale of the serological approach to phylogeny and both he and Goodman present massive new data on primates. It is interesting to see that Bauer's estimate of the time since divergence of man and chimpanzee, based on a technique different from that of Sarich and Wilson,[14] and on an examination of many serum proteins rather than one, is substantially longer.

Lastly, with the contributions of Chiarelli, Egozcue and Rumpler we step from the molecular to the microscopic world of chromosomes. The distinction is of course arbitrary and essentially one of method, because re-arrangements of the genetic material also occur at submicroscopic levels. Comparisons of protein sequences have shown the importance of gene duplication for the genesis of novel proteins, and hybrid protein chains, such as those of the Lepore haemoglobins, testify to breaks and errors of reunion within the confines of structural genes.

The karyotypes of a substantial proportion of the Prosimii have now been described and some striking differences both within and between species have been recorded. But, as Chiarelli and Egozcue point out, chromosome numbers, lengths and centromere positions provide only limited clues to the cytological events that produced karyotype divergence. Small deletions and translocations, and even quite large inversions, may be virtually undetectable in somatic cells by the older methods of examination. Even if meiotic material is available, the amount of evolutionary change is sure to be underestimated.

In the last few years new staining techniques have revealed patterns of banding in the arms of mammalian chromosomes. These patterns are certainly very gross in comparison with the exquisite banding of Dipteran giant chromosomes, but they suffice to make the identification of each member of the chromosome set much more reliable. They also make it easier to detect structural deviations and to give a precise account of the changes that occurred in karyotype divergence. In a recent study using these procedures De Grouchy et al. concluded that at least five chromosome mutations must have been incorporated during the divergence of man and chimpanzee.[15] Curiously enough, four of them are pericentric inversions, a type of change which is liable to generate abnormal meiotic products. Pearson et al. have shown that the intense fluorescent staining which marks the distal part of the long arm of the human Y-chromosome is also present in the chimpanzee and gorilla, but not in the orang and gibbon.[16] Human Y-chromosomes show considerable individual variations in length, apparently due to loss of material in this fluorescent region, and in the chimpanzee it is conspicuously small. It seems that material from certain regions of chromosomes can be lost without causing deleterious effects. It may well be that these are regions in which particular sequences are repeated hundreds or thousands of times. There is also evidence that such repetitious DNA is concentrated near the centromeres of many chromosomes and this may be one reason why 'centric-fusions' between acrocentric chromosomes are such a conspicuous feature of karyotype evolution. Naturally the anthropoid apes have received preferential attention from workers using these new methods and prosimians will have to wait their turn in the queue. It need hardly be added that many prosimian karyotypes differ from one another far more than do these of man and chimpanzee and it will probably be correspondingly more difficult to unravel the changes that have taken place over such relatively long periods of divergence.

Obviously genes are ideally precise markers of regions of the chromosomes. If we had sufficiently good linkage-maps they might give us a much more detailed picture of the rearrangements of chromosomal material that occurred in evolution than we can get by microscopy. This may seem utopian, because fine linkage-mapping by classical methods is hopelessly tedious in slow-breeding species. But there are other possibilities. A good deal of work is now being done with cell cultures, though the long-cherished hope that somatic crossing-over can be used for gene-mapping has not yet been fulfilled. Nevertheless cell cultures (see Ruddle for a recent review)[17] are beginning to be useful in assigning particular genes to particular chromosomes. This approach depends on being able to detect the specific product of a gene in the cells of a tissue that is easy to culture, and biochemical genetics has given us numerous enzyme variants that are good markers for such experiments. The technique of fusing cells from different species (e.g. mouse and man), which results in the progressive elimination of some of the chromosomes from the hybrid cell lines, has been very important in this work. Attempts have also been made to use radio-labelled messenger RNA as a specific 'stain' to reveal the sites of particular genes, but the results are still controversial. I mention these developments simply to show that the prospects of comparative cytogenetics in the next few decades are distinctly attractive.

When we are dealing with protein sequences it is easy to derive quantitative measures of evolutionary divergence by simply counting amino acid substitutions. In the case of karyotypes there is no great difficulty, apart from human fatigue, in measuring chromosome arms. But to express karyotype differences effectively we also need to take into account re-associations of arms, or parts of them, due to translocations, as well as changes (i.e. paracentric inversions) that do not alter arm lengths. The best measure of divergence would seem to be the number of mutational events (i.e. translocations, deletions, inversions, etc.) accounting for the observed differences between two species. To make such estimates we must be able to homologise chromosome arms. The resolution of our technique will put an upper limit to the number of observable changes and we should have to assume that invisible changes are in proportion to those we can actually see. Since at present we can identify points of breakage only very crudely, there may be an element of doubt as to whether a given structural change present in two species is really due to the same mutational event.

As in the case of protein evolution we should obviously like to go beyond descriptions of what happened in evolution and to understand why karyotype divergence took a particular course in certain lineages. Presumably karyotypes diverge because, like gene mutations, breaks and errors in reunion of chromosome strands are going on all the time and sometimes the products spread through a population. Many of these products must surely be eliminated quite rapidly because they entail loss of essential material or lead to uncompensated meiotic troubles that lower fitness. We now have many examples of more or less deleterious chromosome changes in man. On the other hand we know from classical studies on *Drosophila* that some inversions persist in wild populations and it is supposed that they are favourable because they ensure tight linkage of co-adapted blocks of genes. At present our knowledge of the organisation of genes in chromosomes and its functional significance is somewhat rudimentary, at least for higher organisms. Advances in this area of cell biology will surely have much to offer to students of karyotype evolution.

NOTES

1 Fitch, W.M. and Margoliash, E. (1967), 'Construction of phylogenetic trees', *Science* 155, 279-84.

2 Kimura, M. (1968), 'Evolutionary rate at the molecular level', *Nature* 217, 624-6.

3 King, J.L. and Jukes, T.H. (1969), 'Non-Darwinian evolution', *Science* 164, 788-98.

4 Maynard-Smith, J. (1968), ' "Haldane's dilemma" and the rate of evolution', *Nature, Lond.* 219, 114-16.

5 Clarke, B. (1970), 'Darwinian evolution of proteins', *Science* 168, 1009-11.

6 Harris, H. (1969), 'Enzyme and protein polymorphism in human populations', *Brit. Med. Bull.* 25, 5-13.

7 Goodman, M., Barnabas, J., Matsuda, G. and Moore, G.W. (1971), 'Molecular evolution in the descent of man', *Nature* 233, 604-13.

8 Staehlin, M. (1972), 'The calcitonins: an example of unusual evolution', *J. Molec. Evol.* 1, 258-62.

9 Simpson, G.G. (1944), *Tempo and Mode in Evolution*, New York.

10 Min Jou, W., Haegeman, G., Ysebaert, M. and Fiers, W. (1972), 'Nucleotide sequence of the gene coding for the bacteriophage MS 2 coat protein', *Nature* 237, 81-8.

11 Kohne, D.E., Chiscon, J.A. and Hoyer, B.H. (1972), 'Evolution of Primate DNA sequences', *J. Hum. Evol.* 1, 627-44.

12 Nuttall, G.H.F. (1904), *Blood Immunity and Blood Relationship*, London.

13 Nisonoff, A., Reichlin, M. and Margoliash, E. (1970), 'Immunological activity of cytochrome *c*. II: Localisation of a major antigenic determinant of human cytochrome *c*, *J. Biol. Chem.* 245, 940-6.

14 Sarich, V.M. and Wilson, A.C. (1967), 'Immunological time scale of hominid evolution', *Science* 158, 1200-3.

15 De Grouchy, J., Turleau, C., Robin, M. and Klein, M. (1972), 'Evolutions caryotypiques de l'homme et du chimpanzé: Etude comparative du topographie des bandes après dénaturation ménagée', *Ann. Genet.* 15, 79-84.

16 Pearson, P.L., Bobrow, M., Vosa, C.G. and Barlow, P.W. (1971), 'Quinacrine fluorescence in mammalian chromosomes', *Nature* 231, 326-9.

17 Ruddle, F.H. (1973), 'Linkage analysis in man by somatic cell genetics', *Nature* 242, 165-9.

J. EGOZCUE

Chromosomal evolution in prosimians

Introduction

The study of primate chromosomes has experienced great expansion in the past few years. Unfortunately, data on prosimian chromosomes remain scarce. With the exception of the Lemurinae, which have been fairly well studied, the gaps in our knowledge of prosimian chromosomes are wider than in any other primate group. Since the publication of our review of primate chromosomes,[1] only a few papers have dealt with the problem of prosimian karyology; most of them (Rumpler and Albignac, Rumpler, Egozcue)[2-4,5,6] are devoted to the genus *Lemur*, and only two deal with the genera *Galago* and *Nycticebus* (Egozcue, Hayata et al.).[7,8] In spite of these shortcomings, the steps of chromosomal evolution can be traced in some prosimian genera. Following previous practice (Egozcue),[1] the charts of chromosomal evolution presented in this paper apply only to structural chromosome changes, and it is not pretended that they have phylogenetic implications. For comparison of chromosomal formulae, the chromosomes are classified into submetacentrics and acrocentrics; the sex chromosomes are always excluded. Karyotypes are compared only when numerical and morphological similarities warrant it.

Lemuroidea

While extensive data are available on the genus *Lemur*, chromosome studies have been almost entirely lacking for other Lemuroidea. Although the preliminary program of the Third Conference on Experimental Medicine and Surgery in Primates (Lyon, June 21-23, 1972) includes a report on the chromosomes of the Indridae (Rumpler and Albignac),[9] and a paper on the chromosomes of the

Table 1 Chromosomes of Lemuroidea

Species	*2N*	*S*	*A*	*X*	*Y*
Lemur mongoz	60	4	54	A	A
L. fulvus fulvus	60	4	54	A	A
L. f. rufus	60	4	54	A	A
L. f. albifrons	60	4	54	A	A
L. f. sanfordi	60	4	54	A	A
L. f. collaris	60	4	54	A	A
L. catta	56	8	46	M	A
L. catta	56	10	44	M	—
L. f. collaris	52	12	38	A	A
L. f. collaris	48	16	30	A	—
L. variegatus	46	18	26	M	A
L. macaco	44	20	22	A	A
Hapalemur griseus	54	10	42	A	A
H. griseus	58	6	52	—	—
Cheirogaleus major	66	—	64	A	A
Microcebus murinus	66	2	64	—	—
Propithecus verreauxi verreauxi	48	—	—	—	—
P. v. coquereli	48	—	—	—	—

Lemuridae, Indridae and Daubentoniidae has been presented by Rumpler in this volume,[10] no details of the karyotypes are known to me at the time of this writing, except for the incomplete data published by Chu and Bender on *Propithecus.*[11]

Recently, we have studied the chromosomes of two female *Microcebus* in blood cultures.[12] The diploid number is 2N = 66, and the karyotype is composed of one pair of submetacentric chromosomes and 32 pairs of acrocentrics.

Table 1 summarises the diploid numbers and chromosomal formulae of the Lemuroidea. While data on the different species of the genus *Lemur* have been confirmed by several authors, the chromosomal complements described so far in the genera *Hapalemur* (Chu and Swomley),[13] *Cheirogaleus* (Chu and Swomley, cited by Chu and Bender),[11] and *Propithecus* (Chu, cited by Chu and Bender)[11] have to be considered as tentative until Rumpler's data are published in full.[10]

The possible pathways of chromosomal evolution in the Lemurinae are shown in Fig. 1. Only confirmed data have been included. However, the diploid numbers and chromosomal formulae of *Hapalemur* also fit the pathways illustrated in Fig. 1, and correspond to intermediate steps in the chromosomal evolution of the Lemurinae (Fig. 2).

In the genus *Lemur*, hybrids have been obtained between species with very pronounced chromosomal differences. Gray lists some of these hybrids and qualifies the crosses as unlikely.[14] However, such hybrids do occur, and some of them have been studied by Rumpler and Albignac. The chromosomal complement of the hybrids

Table 2 Chromosomes of lemur hybrids

Species	*2N*	*S*	*A*	*X*	*Y*
L. f. fulvus x *L. f. albifrons* (1)	60	4	54	A	A
L. f. fulvus x *L. f. rufus* (2)	60	4	54	A	A
Hybrid 1 x Hybrid 2	60	4	54	A	—
L. macaco x *L. f. albifrons*	52	12	38	A	A
L. macaco x *L. f. rufus*	52	12	38	A	—
L. f. collaris[a] x *L. f. collaris*[b]	54	10	42	A	—

[a] *L. f. collaris* with 2N=60
[b] *L. f. collaris* with 2N=48

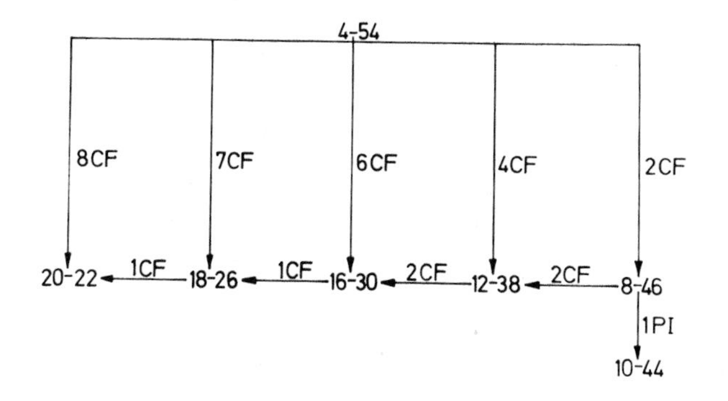

Figure 1. Chromosomal evolution in the genus *Lemur*. CF stands for Centric Fusion, PI for Pericentric Inversion. (Numbers indicate submetacentrics and acrocentrics, respectively – cf. Table 1.)

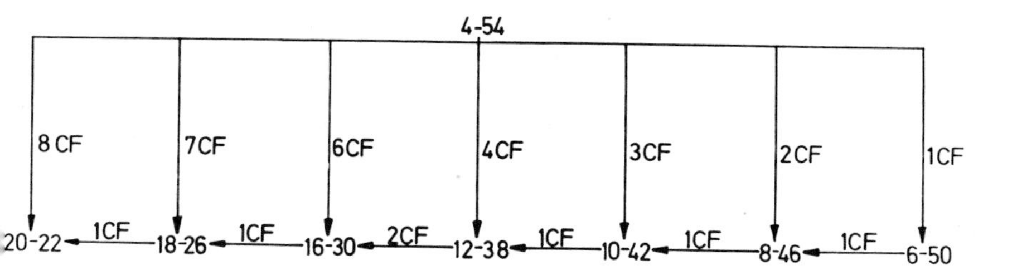

Figure 2. Chromosomal evolution in the genera *Lemur* and *Hapalemur* (see caption to Fig. 1).

(Table 2) is made up of the haploid set from each parent.[6] That such hybrids can be produced and are viable suggests that, in spite of the considerable structural changes that have taken place in the chromosomes of Lemuridae, a high degree of genetic homology must exist between the different species concerned.

Lorisoidea

Table 3 shows the chromosome numbers of the Lorisoidea. Some of the results have recently been confirmed by De Boer.[15] In the Lorisidae, a case of intraspecific polymorphism has been described in *Nycticebus coucang*, where diploid numbers of 2N = 50 and 2N = 52 are found. Of special interest are the striking similarities between the karyotypes of *Nycticebus coucang* and *Arctocebus calabarensis.* There is some discrepancy regarding the chromosomal formula of *Loris tardigradus* (Manna and Talukdar).[16] However, as we have stated before,[6] some differences may be due to the subjective criterion used by an author in classifying subterminal chromosomes as acrocentric or submetacentric, rather than to actual polymorphism in the species. Since all animals studied in this group belong to different genera, no attempt has been made to establish any pathways of chromosomal evolution, in spite of the fact that from a strictly morphological point of view it would not be difficult to establish a relationship among the chromosomal complements of the Lorisidae.

In the Galagidae, different results have been published for *Galago senegalensis* and *Galago crassicaudatus.* As far as *Galago senegalensis* is concerned, the only study conducted on suitable material revealed a karyotype composed of 30 submetacentrics and 6 acrocentrics.[7] In *Galago crassicaudatus*, these results have been confirmed by Hayata et al.[8] The chromosomal complement described by Chu and Bender for *Galago crassicaudatus* has not been confirmed, but it provides a chromosomal link between the greater and the lesser bushbabies.[17] The possible pathways of chromosomal evolution in the Galagidae are shown in Fig. 3.

Table 3 Chromosomes of Lorisoidea

Species	*2N*	*S*	*A*	*X*	*Y*
Lorisidae					
Perodicticus potto	62	24	36	S	A
Loris tardigradus	62	34	26	S	S
L. tardigradus	62	38	22	S	S
Arctocebus calabarensis	52	50	—	S	S
Nycticebus coucang	52	2	48	S	S
N. coucang	50	48	—	S	S
Galagidae					
Galago crassicaudatus	62	6	54	S	A
G. crassicaudatus	62	30	30	S	A
G. senegalensis	38	30	6	S	—

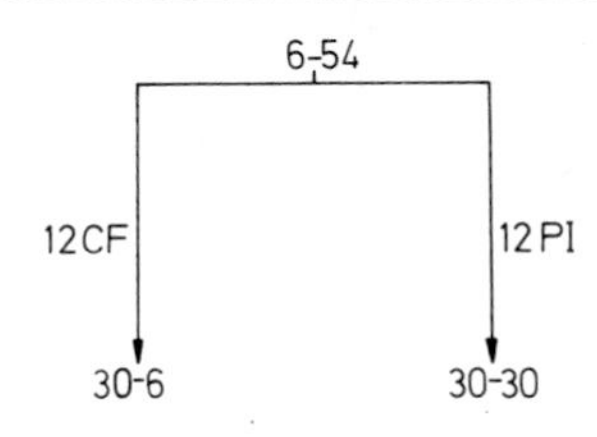

Figure 3. Chromosomal evolution in the Galagidae (see caption to Fig. 1).

Tarsioidea

Data on the chromosomes of Tarsioidea are extremely scarce. The diploid number of 2N = 80, described by Klinger in *Tarsius bancanus*,[18] was confirmed by Chiarelli and Egozcue in testicular material of *Tarsius syrichta*,[19] and is the highest diploid number of any primate. The karyotype of *Tarsius bancanus* is composed of 14 submetacentrics and 66 acrocentrics. Since the only animal studied was a female, no information is available on the sex chromosomes (Table 4).

Table 4 Chromosomes of Tarsioidea

Species	*2N*	*S*	*A*	*X*	*Y*
Tarsius bancanus	80	14	66	—	—
T. syrichta	80	—	—	—	—

Discussion

In spite of the scarcity of data on the chromosomes of prosimians, it is still possible to trace the structural changes that may have given rise to new species. This is permissible because of the fact that the chromosomal mechanisms that may lead to speciation are monotonously similar in mammals, and almost always correspond to centric fusion or pericentric inversion events. In other words, at the present time mammals seem to be in a process of chromosome stabilisation characterised by centric fusions and pericentric inversions.[21] Both mechanisms increase the number of submetacentric chromosomes in a given karyotype; furthermore, centric fusions increase chromosomal size while reducing the diploid number of a species. Since there is some evidence that smaller chromosomes and acrocentric chromosomes tend to have a lower number of chiasmata during meiosis I, and thus show a tendency to separate precociously into univalents and provoke meiotic problems,[20] it seems logical that any mechanisms that increase the two main morphological factors of chromosome stability in meiosis, i.e. chromosomal size and metacentricity, will have an important role in chromosome evolution. This is not to deny the importance of centric fission (Todd),[21] a mechanism that may have been predominant at some stages of mammalian evolution.

From a taxonomic point of view, chromosomal data may be of importance, as in the case of the proposed classification of *Lemur fulvus* in a single species with *L. macaco* (a suggestion not supported by chromosomal studies). In other cases, karyological studies are of no assistance, for example in deciding whether to classify *L. catta* in the genus *Hapalemur*, or whether to give *L. variegatus* separate generic status (*Varecia*), as would apparently be warranted by data from other sources.

Recently, the new techniques of quinacrine fluorescence and Giemsa banding of chromosomes have developed to provide an extraordinarily useful tool for comparative cytogeneticists. The fact that both the Q (fluorescent) and G (Giemsa) bands permit detection of chromosome homology will be useful for establishing interspecific relationships on a more secure basis. I have not had the opportunity to apply these techniques to prosimian chromosomes; but the studies carried out so far in our laboratory suggest that banding patterns are extremely stable, and that they thus provide an excellent means of establishing chromosome homologies between closely related species.

Acknowledgments

Most of our studies on prosimian chromosomes were conducted at the Oregon Regional Primate Research Center, with the skillful technical assistance of Fay Hagemenas. I thank R.D. Martin and N.A. Barnicot, University College, London, for supplying the samples of *Microcebus*.

NOTES

1 Egozcue, J. (1969), 'Primates', in Benirschke, K. (ed.), *Comparative Mammalian Cytogenetics*, New York.

2 Rumpler, Y. and Albignac, R. (1969), 'Etude cytogénétique de quelques hybrides intraspécifiques et interspécifiques de lémuriens', *Ann. Sci. Univ. Besançon* 6, 1.

3 Rumpler, Y. and Albignac, R. (1969), 'Etude cytogénétique de deux lémuriens, *Lemur macaco macaco*, Linné 1766, et *Lumur fulvus rufus* (Audebert 1800) et d'un hybride *macaco macaco/fulvus rufus*', *C. R. Soc. Biol.* (Paris) 163, 1247-50.

4 Rumpler, Y. and Albignac, R. (1969), 'Existence d'une variabilité chromosomique intraspécifique chez certains lémuriens', *C. R. Soc. Biol.* (Paris) 163, 1989-92.

5 Rumpler, Y. (1970), 'Etude cytogénétique du *Lemur catta*', *Cytogenetics* 9, 239.

6 Egozcue, J. (1972), 'The chromosomes of lemurs', *Folia primat.* 17, 171-6.

7 Egozcue, J. (1970), 'The chromosomes of the lesser bushbaby (*Galago senegalensis*) and the greater bushbaby (*Galago crassicaudatus*)', *Folia primat.* 12, 236-240.

8 Hayata, I., Sonta, S., Itoh, M. and Kondo, N. (1971), 'Notes on the karyotypes of some prosimians, *Lemur mongoz, Lemur catta, Nycticebus coucang* and *Galago crassicaudatus*', *Japan. J. Genet.* 1, 61.

9 Rumpler, Y. and Albignac, R. (1972), 'Cytogenetic study of the Indridae', *Third Conf. on Exptl. Med. and Surg. in Primates*, Lyon, June 21-23.

10 Rumpler, Y., this volume.

11 Chu, E.H.Y. and Bender, M.A. (1962), 'Cytogenetics and evolution of primates', *Ann. N.Y. Acad. Sci.* 102, 253-66.

12 Egozcue, J. (1972), unpublished data.

13 Chu, E.H.Y. and Bender, M.A. (1961), 'Chromosomes of lemurine lemurs', *Science* 133, 1925-6.

14 Gray, A.P. (1954), *Mammalian Hybrids*, Technical Communication No. 10 of the Commonwealth Bureau of Animal Breeding and Genetics, Alva.

15 De Boer, L.E.M. (1972), 'Chromosome studies in some primates', *Mammal. Chromos. Newsl.* 13, 4.

16 Manna, G.K. and Talukdar, M. (1968), 'An analysis of the somatic chromosome complements of both sexes of two primates, the slender loris, *Loris tardigradus*, and rhesus monkey, *Macaca mulatta*', *Mammalia* 32, 118.

17 Chu, E.H.Y. and Bender, M.A. (1961), 'Chromosome cytology and evolution in primates', *Science* 133, 1399-405.

18 Klinger, H.P. (1963), 'The somatic chromosomes of some primates (*Tupaia glis, Nycticebus coucang, Tarsius bancanus, Cercocebus aterrimus, Symphalangus syndactylus*)', *Cytogenetics* 2, 140-51.

19 Chiarelli, B. and Egozcue, J. (1968), 'The meiotic chromosomes of some primates', *Mammal. Chromos. Newsl.* 9, 85-6.

20 Egozcue, J. (1969), 'Aneuploidia cromosómica: nuevas observaciones sobre la separación precoz de bivalentes en meiosis I', *Sangre* (Barcelona) 14, 442.

21 Todd, N.B. (1967), 'A theory of karyotypic fissioning, genetic potentiation and eutherian evolution', *Mammal. Chromos. Newsl.* 8, 268.

Y. RUMPLER

Cytogenetic contributions to a new classification of lemurs

Introduction

The systematic relationships of the Malagasy lemurs are still a matter of controversy, as is testified by the successive different classifications that have been proposed. In recent years, however, new studies have led to a certain degree of convergence of the conclusions of various authors, permitting us to attempt a revision of the classification, and in this paper we propose a classification of lemurs based on a systematic cytogenetic study of almost all of the Malagasy lemur species. For this revision, we have utilised our cytogenetic data to modify one of the most recent classifications (Hill)[1] in cases where our results generally accord with the conclusions of recent publications.

Results

We propose to raise the existing subfamily Cheirogaleinae Gregory, 1915 to the rank of the family Cheirogaleidae, with type genus *Cheirogaleus* E. Geoffroy, 1812. This family includes the Cheirogaleinae Gregory, 1915 and a new subfamily Phanerinae, with type genus *Phaner* Gray, 1870. The family Lemuridae is modified to include two subfamilies: the Lemurinae Mivart, 1864 and the Lepilemurinae, with type genus *Lepilemur* I. Geoffroy, 1851, corresponding to the tribe Lepilemurini Stephan and Bauchot, 1965.

Discussion

Study of the chromosomes of species within the previously recognised subfamily Cheirogaleinae indicates that there are two separate groups: (*a*) *Cheirogaleus* and *Microcebus* species; (*b*) *Phaner* species. *Cheirogaleus* and *Microcebus* have the same diploid numbers (2N = 66) and fundamental numbers (FN = 66). The diploid number (2N = 48) and particularly the fundamental number (FN = 62) of *Phaner* are markedly different from those of *Cheirogaleus* and *Microcebus* (Rumpler and Albignac).[2] Other characters which have been reported recently, such as the dermatoglyphic patterns (Rakotosamimanana and Rumpler),[3] the presence of a specific marking gland on the anterior aspect of the neck in *Phaner* (Rumpler and Andriamiandra),[4] and the peculiar behaviour of *Phaner* (Petter et al.),[5] all support the chromosomal evidence in favour of separation of the two groups. On the basis of cranial and behavioural characters, Petter has already proposed that the subfamily Cheirogaleinae should be raised to family level,[6] and recent parasitological investigations similarly underline the specificity of the nematode parasites of members of this group,[7] which are quite distinct from those of species within the family Lemuridae. This point of view has been accepted in the present investigation, and the new family Cheirogaleidae has already been described as containing the 2 subfamilies Cheirogaleinae and Phanerinae (Rumpler and Albignac).[2] We have left *Allocebus* in the subfamily Cheirogaleinae along with *Cheirogaleus* and *Microcebus* (see Petter-Rousseaux and Petter),[8] although at the present time we have not been able to study the karyotype of this genus and therefore cannot provide evidence to support this provisional allocation.

The subfamily Lemurinae includes 3 genera: *Lemur, Hapalemur* and *Varecia.* The 3 genera have the same fundamental number (FN = 64), which indicates robertsonian evolution within this group. There are no cytogenetic characters indicating that the variegated lemur should be placed as a separate genus, rather than as a species within the genus *Lemur* (i.e. as *Lemur variegatus*). However, other recently described characteristics, such as the quite distinctive dermatoglyphic patterns (Rakotosamimanana and Rumpler),[9] the presence of a marking gland on the neck (Rumpler and Andriamiandra),[4] and various cranial features,[10] support J.-J. Petter's suggestion that this animal should be placed in the independent genus *Varecia.*

The subfamily Lepilemurinae contains quite distinctive animals which were previously placed in the genus *Lepilemur* and classified together with *Lemur* and *Hapalemur.* Stephan and Bauchot have already proposed that the sportive lemurs should be more clearly distinguished, by placing them in a separate tribe Lepilemurini (as distinct from the Lemurini).[11] The extremely distinctive karyotypes of the sportive lemurs, with diploid numbers varying between 2N = 20 and 2N = 38, and fundamental numbers ranging from FN = 44 to FN = 38 (Rumpler and Albignac),[19] have induced us to raise them to the level of a subfamily, Lepilemurinae. Other recently described characters of the sportive lemurs, in particular the dermatoglyphs (Rumpler and Rakotosamimanana),[13] haematological characters,[14,15] and parasitological features (Chabaud, Brygoo and Petter),[7] along with behavioural peculiarities (Petter, Charles-Dominique and Hladik),[16,17] also argue in favour of this allocation. These recently recognised characteristics, when viewed along with the very distinctive karyotypes of the *Lepilemur* group, might even indicate that they should in fact be classified in a separate family, Lepilemuridae.

The 3 genera of the family Indriidae are clearly distinguished both by their karyotypes and by their fundamental numbers.[18] In particular, the genus *Avahi* is clearly distinct from the other two genera, *Indri* and *Propithecus.* Other characters, such as the marking glands (Rumpler and Andriamiandra),[4] emphasise the distinctiveness of each of these 3 genera. Despite the differences which exist between the 3 genera in the palmar and plantar pads, however, there is marked homogeneity within the family Indriidae as a whole (Rakotosamimanana and Rumpler).[3]

The family Daubentoniidae is represented by the sole genus *Daubentonia.* The diploid number (2N = 30), the karyotype and the fundamental number (FN = 54) are all extremely distinct from those of other lemurs and would provide a further argument – if such were necessary – for classifying the aye-aye quite separately.

We therefore propose the following new classification:

Superfamily LEMUROIDEA (Mivart, 1864)
Family CHEIROGALEIDAE Rumpler & Albignac, 1972
1° – Subfamily CHEIROGALEINAE Gregory, 1915
genera: *Microcebus* E. Geoffroy, 1828
Cheirogaleus E. Geoffroy, 1812
Allocebus Petter-Rousseaux & Petter, 1967
2° – Subfamily PHANERINAE Rumpler & Albignac, 1972
genus: *Phaner* Gray, 1870

Family LEMURIDAE Gray, 1821
1° – Subfamily LEMURINAE Mivart, 1864
genera: *Lemur* Linnaeus, 1758
Hapalemur I. Geoffroy, 1851
Varecia Gray, 1863
2° – Subfamily LEPILEMURINAE Rumpler & Rakotosamimanana, 1972
genus: *Lepilemur* I. Geoffroy, 1851
Family INDRIIDAE Burnett, 1828
genera: *Avahi* Jourdan, 1834
Propithecus Bennett, 1832
Indri E. Geoffroy & Cuvier, 1795
Family DAUBENTONIIDAE Gray, 1870
genus: *Daubentonia* E. Geoffroy, 1795

NOTES

1 Hill, W.C.O. (1953), *Comparative Anatomy and Taxonomy of the Primates*, vol. 1: *Strepsirhini*, Edinburgh.

2 Rumpler, Y. and Albignac, R. (1972), 'Cytogenetic study of the endemic Malagasy lemurs subfamily Cheirogaleinae Gregory, 1915', *Am. J. Phys. Anthrop.* 38, 261-4.

3 Rakotosamimanana, B.R. and Rumpler, Y. (1972), 'Evolution of palm, sole pads and dermatoglyphic patterns in the Malagasy lemurs', in press.

4 Rumpler, Y. and Andriamiandra, A. (1971), 'Etude histologique des glandes de marquage de la face antérieure du cou des lémuriens malgaches', *C.R. Soc. Biol.* 165, 436-44.

5 Petter, J.-J., Schilling, A. and Pariente, G. (1971), 'Observations éco-éthologiques sur deux lémuriens malgaches nocturnes: *Phaner furcifer* et *Microcebus coquereli*', *Terre et Vie* 25, 287-327.

6 Petter, J.-J. (1962), 'Recherches sur l'écologie et l'éthologie des lémuriens malgaches', *Mém. Mus. Nat. Hist. Nat.* (n.s.) Sér. A, *Zoologie* 27, 1-146.

7 Chabaud, A.G., Brygoo, E. and Petter, A. (1965), 'Les nématodes parasites des lémuriens malgaches. VI: Description de six espèces nouvelles et conclusions générales', *Ann. Parasit. Hum. Comp.* 40, 181-214.

8 Petter-Rousseaux, A. and Petter, J.-J. (1967), 'Contribution à la systématique des Cheirogaleinae', *Mammalia* 31, 574-82.

9 Rakotosamimanana, B.R. and Rumpler, Y. (1970), 'Etude des dermatoglyphes et des coussinets palmaires et plantaires de quelques lémuriens malgaches', *Bull. Ass. Anat.* 148, 493-510.

10 Mahé, J., personal communication; thesis in preparation: 'Crâniométrie des lémuriens'.

11 Stephan, H. and Bauchot, R. (1965), 'Hirn-Körpergewichts-Beziehungen bei den Halbaffen (Prosimii)', *Acta. Zool.* 46, 1-23.

12 Rumpler, Y. and Albignac, R. (1972), 'Etude cytogénétique de *Varecia variegata* et du *Lemur rubriventer*', *C.R. Soc. Biol.* 165. 741-7.

13 Rumpler, Y. and Rakotosamimanana, B.R. (1972), 'Coussinets palmo-plantaires et dermatoglyphes des représentants des Lemuriformes malgaches', *Bull. Ass. Anat.* 154.

14 Richaud, J., Rumpler, Y. and Albignac, R. (1971), 'Quelques données hémobiologiques de la sous-famille des lémurinés malgaches', *Arch. Inst. Pasteur Madagasc.* 40, 137-44.

15 Buettner-Janusch, J., Washington, J.L. and Buettner-Janusch, V. (1971),

'Hemoglobins of Lemuriformes', *Arch. Inst. Pasteur. Madagasc.* 40, 127-36.
16 Petter, J.-J. (1960), 'Remarque sur la systématique du genre *Lepilemur*', *Mammalia* 24, 76-86.
17 Charles-Dominique, P. and Hladik, C.M., 'Le *Lépilemur* du sud de Madagascar: Ecologie, alimentation et vie sociale', *Terre et Vie* 25, 3-66.
18 Rumpler, Y. and Albignac, R. (1973), 'Etude cytogénétique des Indriidae', International Congress of Anthropological and Ethnological Sciences, Chicago, in press.
19 Rumpler, Y. and Albignac, R. 'The cytogenetic study of the subfamily Lepilemurinae', in preparation.

B. CHIARELLI

The chromosomes of the prosimians

In many different animal groups, recently acquired knowledge on number and morphology of the chromosomes has furnished important data for reconstruction of their phylogenies and has provided us with one more criterion for taxonomic organisation. The group of the prosimians, which are so heterogeneous in terms of their remote evolutionary relationships and their geographical isolation, seems particularly suitable for a karyological approach to their phylogeny (Fig. 1). The phylogenetic organisation of the species currently assigned to this group has in fact been the object of continuous discussion, notably with respect to classification above the family level.

Knowledge of chromosomes is important in the study of phylogeny and taxonomy because these structures are the direct carriers of the genetic information, which is itself organised in a stable arrangement on the chromosome. For each species of Eukaryota, the structural affinity between homologous chromosomes is consistently controlled by their pairing at meiosis in each individual. Incomplete chromosome pairing at meiosis suggests the presence of structural changes in the organisation of the genetic information which is linearly distributed along the chromosomes. If such structural changes are not restricted, they cease to be compatible with the functional organisation of the genetic information. The result of such incompatibility is the inability of homologous chromosomes to pair at meiosis, with consequent disruption of the meiotic process.

In all higher organisms, pairing of homologous chromosomes at meiosis plays the rôle of a filter through which only a functionally patterned genetic system can pass. Furthermore, in natural populations meiosis represents a barrier which prevents exchange between diverging genetic systems, and it should therefore be considered as a basic mechanism involved in the division of populations into new species.

For any given species, constancy of karyotype appearance (which we can appreciate at the level of each chromosome) is therefore a consequence of the meiotic filter. On the other hand, the existing differences between karyotypes of different species represent the product of variation in the somatic chromosomes of the germinal cell line which were able to filter through the meiotic sieve and become

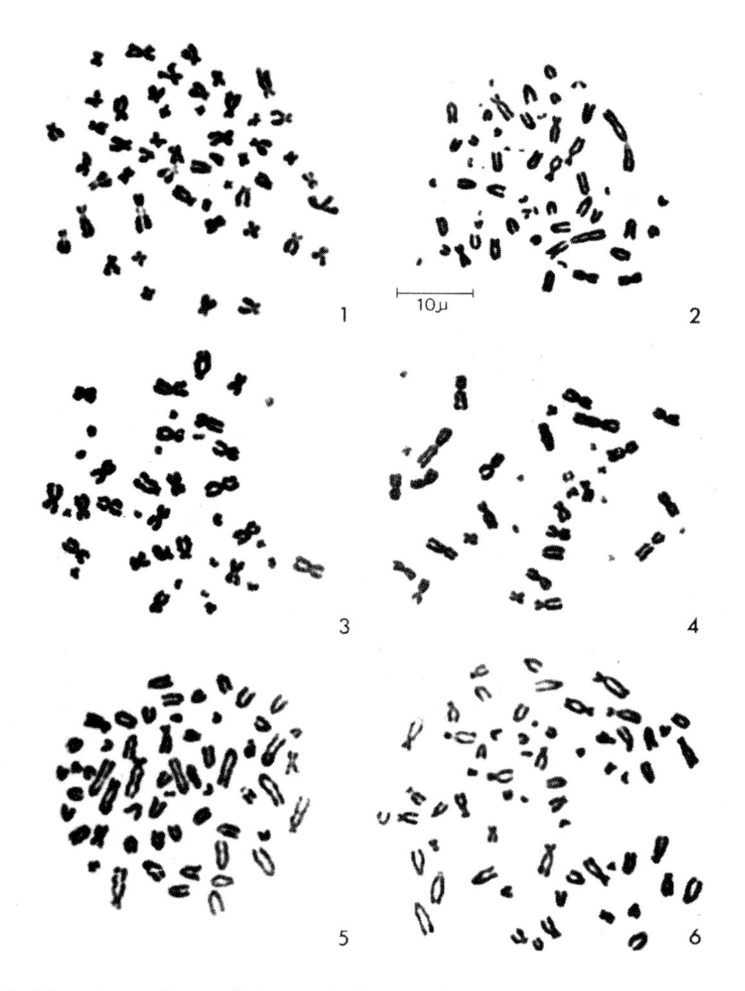

Figure 1. Metaphase plates of six prosimian species.
1 – *Nycticebus coucang* (female; 2N = 50)
2 – *Lemur catta* (male; 2N = 56)
3 – *Galago alleni* (female; 2N = 40)
4 – *Galago senegalensis* (female; 2N = 36)
5 – *Galago demidovii* (male; 2N = 58)
6 – *Galago crassicaudatus* (male; 2N = 62)
(These metaphase plate photographs were kindly placed at our disposal by L.E.M. de Boer, Institute of Genetics, University of Utrecht, The Netherlands.) The magnification is the same for all figures (1000 x), and the actual scale is indicated by the 10 μ line on photograph 2.

established because of some selective advantage.

On the basis of hypotheses about the possible steps by which the chromosomes of two related species became differentiated, one can construct a phyletic line connecting them and establish their taxonomic affinities.

However, before one can conclude that there is chromosome identity between individuals of different species it is important to be sure of their homologous genetic content. The only control one can have at the moment is the observation of the behaviour of the chromosomes during meiosis (expecially diakinesis) in individuals derived from cross-breeding of two related species (F_1). In cyto-taxonomic research, therefore, data on hybrids deserves particular consideration.

The purpose of the present study is to analyse how far the karyological information so far available for the prosimians is of value in arriving at a systematic and phylogenetic organisation of this group.

The taxonomic nomenclature and order followed here is the traditional one.[13] A synthesis of the information available for the chromosomes of those species which have been studied to date is presented in Table 1.

From this table it can be seen:

1. That there is a total lack of karyological information for many important genera: *Dendrogale, Ptilocercus, Avahi, Indri.*
2. That for many other genera only preliminary information is available (*Propithecus, Microcebus, Cheirogaleus, Phaner, Lepilemur, Daubentonia*).
3. That great variability exits in the number of chromosomes for all of those genera which have been more extensively studied (e.g. *Galago* and *Lemur*).

Disregarding, for the moment, the intrageneric chromosomal variation mentioned in point 3, (which will be discussed later) a more general question may be asked: Can the phylogenetic relationship between families be established on the basis of variations in the chromosome number between the genera belonging to each family within this group of Primates?

Mere statement of the diploid number of somatic chromosomes excludes one of the most easily occurring types of chromosome mutation; namely, centric fusion. This difficulty can be partially overcome by resorting to Mattheys's fundamental number (FN), since it takes into consideration solely the number of arms. The use of the fundamental number is, moreover, based on the assumption that centric fusion (or misdivision of the centromere) is one of the more successful types of mutation in passing the filter of meiosis. Such a mutation would not in fact interfere with the organisation of

Table 1 Numerical and morphological data on the chromosomes of Tupaioidea, Lorisoidea, Lemuroidea and Tarsioidea (2n = diploid chromosome number; M = metacentric, S = submetacentric, A = acrocentric; x and y = sex chromosomes)

Taxa	*2n*	*M-S*	*A*	*X*	*Y*	*References* (see Notes)
1. Tupaiidae						
Tupaia glis	60-62	14-12	44-48	M-S	A	1,2,7,17,27,34,35,32
Tupaia montana	52-68			M	A	1,32
Tupaia minor	66					1,31,32
Dendrogale sp.						
Urogale everetti	44					2,8,17,19
Ptilocercus lowii						
2. Lorisidae						
Loris tardigradus	62	34-38	26-22	S	S-A	28,39
Nycticebus coucang	50-(52)	48		S	S	3,17,30,33,35
Nycticebus pygmaeus	50					17
Arctocebus calabarensis	52	50		S		29
Perodicticus potto	62	24	36	S	A	2,8,16,23
3. Galagidae						
Galago senegalensis	36-38	22-24-30	14-12-6	S	S-A	4,16,22,23,24,38,40,49,50
Galago crassicaudatus	62	6-30	54-30	S	A-S	16,17,23,24,33,38
Galago alleni	40					4
Galago demidovii	58					5

4. Lemuridae						
Microcebus murinus	66		64	A	A	16,18,42,44,45,46
Cheirogaleus major	66		64	A	A	17,18,42,44,45,46
Phaner furcifer	48					42
Hapalemur griseus	54-58	10-6	42-50	A	A	16,18,42,44,45,46
Lemur catta	56	10-14	44-50	S-A	A	2,10,11,16,17,18,21,33,41,13,20,42,44,45,46
Lemur variegatus	46	18	26	S	A	10,16,17,18,21,20,42,44,45,46
Lemur macaco	44-48-52-58-60	20-16-4	22-30-52-54	A	A	2,10,11,16,18,21,23,43,12,9,14,20,42,44,45,46
Lemur mongoz	60	4	54	A	A	10,16,18,21,33,12,9,20
Lepilemur mustelinus	22-38					42
5. Indridae						
Propithecus verreauxi	48					17,42,44,45,46
Avahi laniger	—					
Indri indri	42					48
6. Daubentoniidae						
Daubentonia madagascariensis	30					42
7. Tarsiidae						
Tarsius syrichta	80					15
Tarsius bancanus	80	14	66			35

the genetic information on the chromosomes. The only change which would result would be a reduction or increase in randomnesss of the distribution of genetic information in the offspring. Reduction or increase in the number of chromosome units certainly has adaptive advantages for an organism, in that it reduces or increases the potential variability in a population. In addition, there must be a relation between chromosome morphology and chiasma frequencies, although at the moment we do not have enough information to definite this relationship exactly.

(N.B. Calculation of the fundamental number (FN) is made by counting a metacentric or submetacentric as 2 and an acrocentric or sub-acrocentrix as 1 in any general karyotype).

Data of this type are shown in the fourth column of Table 2. Matthey's fundamental number is 70 to 84 in Tupaioidea; 87 to 102 in Lorisidae; 61 to 94 in Galagidae; 62 to 70 in Lemuridae; 94 in Tarsiidae and 54 in Daubentoniidae.

Karyological data elaborated in this way lend themselves to more reliable analysis and yield more useful taxonomic information.

No problem exists with regard to the taxonomic, and hence

Table 2 Basic karyological information for the prosimians

Genera	*Known/Studied*	*2n*	*Fundamental number*	
Tupaia	11/3	52-60-62-66-68	70-72-74-76	70
Dendrogale	2/0			↓
Urogale	1/1	44	84-80	↓
Ptilocercus	1/0			84
Loris	1/1	62	98-101	87
Nycticebus	2/2	50	100	↓
Arctocebus	1/1	52	102	↓
Perodicticus	1/1	62	87	102
Galago	6/4	38-62	61-64-69-75-94	61→94
Microcebus	2/1	66	68	62
Cheirogaleus	2/1	66	68	↓
Phaner	1/1	48	62	↓
Hapalemur	2/2	54-58	64	↓
Lemur	6/6	44-46-48-52-56-58-60	62-64-66-70	↓
Lepilemur	1/1	22-38	42-36	70
Propithecus	2/1	48	?	
Avahi	1/0			
Indri	1/1	42	?	
Daubentonia	1/1	30	54	54
Tarsius	3/2	80	94	94

phylogenetic, separation of the Tupaioidea with respect to the Lorisoidea and Lemuroidea. However, difficulties do arise with these two latter superfamilies, whose relationships are more controversial. A ready interpretation is provided by the data on fundamental numbers available to us. The Lorisidae can, in fact, be sharply distinguished from Lemuridae because their fundamental numbers are different and do not in any way overlap (87-102 vs. 64-70). The taxonomic position of the Galagidae, whose fundamental numbers (FN : 61-94) provide a kind of bridge between the Lemuridae and Lorisidae, is particularly interesting.

The strikingly large difference in the diploid chromosome number between the two *Galago* species previously studied (*Galago senegalensis* with 2N = 38, and *G. crassicaudatus* eith 2N = 62) provides a tempting basis for speculation about a possible polyploid mechanism as the origin of this variation.

However this hypothetical indication of polyploidy has been definitely eliminated by the recent findings that:

(*a*) There is polymorphism in the chromosome number of *Galago senegalensis*, due to a mechanism of centric fusion (Ying and Butler).[50]

(*b*) There is a diploid chromosomal number of 58 in *Galago demidovii* and one of 40 in *Galago alleni* (De Boer, personal communication).

(*c*) There is an identical DNA content in the two *Galago* species (7.54 ± 0.09 a. u. in *G. senegalensis* and 7.26 ± 0.09 a. u. in *G. crassicaudaus*); Manfredi-Romanini et al.[38]

These findings provide specific support for a relatively close taxonomic relationship of the Galagidae to the Lorisidae, and at the same time open the field to extensive speculation about the adaptive advantages of such chromosomal polymorphism in the Galagidae. The fact that extensive polymorphism due to centric fusion also exists in the genus *Lemur* (see Table 3) with production of karyological subspecies, as has recently been underlined by Rumpler and Albignac,[46] enormously increases our interest in the adaptive advantage of these variations and in such a peculiar mechanism of speciation.

Another aspect of particular interest concerns the relationship of the living prosimians to other groups of primates, especially the South American monkeys. According to one classic theory, the South American monkeys originated from a group of now extinct North American prosimians (Omomyidae), representatives of which migrated from North to South America. However, recent re-evaluation of the theory of continental drift makes room for speculation about direct migration of some early stock of primates directly from Africa to South America before the end of the Eocene. This hypothesis was recently set out by Lavocat,[36] who conducted a

Table 3 Chromosome polymorphism in the genus *Lemur* (after Rumpler et al. 1971)

Species	*Chromosomes*						
	2n	*M*	*S*	*A*	*X*	*Y*	*FN*
L.m. macaco	44	12	8	22	A	A	64
L.f. fulvus	48	10	6	30	A	A	64
L.f. fulvus	58	–	4	52	A	A	62
L.f. fulvus	60	–	4	54	A	A	64
L.f. albifrons	60	–	4	54	A	A	64
L.f. sanfordi	60	–	4	54	A	A	64
L.f. collaris (barbe blanche)	52	8	4	38	M	A	66
L.f. collaris a (barbe rouge)	60	–	4	54	A	A	64
L.f. collaris b (barbe rouge)	48	8	8	30	A	–	64
L.f. collaris c (barbe rouge)	52	6	6	38	A	A	64
L.f. collaris a x *L.f. collaris* b	54	4	6	42	A	–	64
L.m. macaco x *L.f. rufus*	52	7	5	38	A	–	64
L.m. macaco x *L.f. albifrons*	52	7	5	38	A	A	64
L.m. macaco x *L.f. collaris* b	46	10	8	26	A	–	64
L.f. fulvus x *L.f. albifrons*	58	–	4	52	A	–	62
L.f. fulvus x *L.f. albifrons* (1)	60	–	4	54	A	A	64
L.f. fulvus x *L.f. rufus* (2)	60	–	4	54	A	–	64
Hybrid *F2* : (1) x (2)	60	–	4	54	A	–	64
Lemur variegatus	46	14	4	26	M	A	66
Lemur catta	56	2	6	46	M	A	66

comparative study of the African Miocene rodents and the South American caviomorphs and came to the conclusion that the close morphological relationship between these two groups can only be explained on the basis of such migration. A research programme with the aim of finding chromosomal evidence for a possible relationship between some group of living prosimians and South American monkeys is now being carried out in our Institute.

Apart from the general morphological detail of the chromosomes, another important parameter to take into consideration in developing hypotheses about phylogeny is, as has been mentioned above, the DNA content. Unfortunately only limited data are available so far (Manfredi-Romanini),[37] and the subject is mentioned here in the hope that this might lead to a cooperative effort in obtaining materials for such analysis.

NOTES

1 Arrighi, F.E., Sorenson, M.W. and Shirley, L.R. (1969), 'Chromosomes of the tree shrew (Tupaiidae)', *Cytogenetics* 8, 199-208.

2 Bender, M.A. and Chu, E.H.Y. (1963), 'The chromosomes of Primates', in

J. Buettner-Janusch (ed.), *Evolutionary and Genetic Biology of Primates* 1, 261-310, New York.

3 Bender, M.A. and Mettler, L.E. (1958), 'Chromosome studies of primates', *Science* 128, 186-90.

4 de Boer, L.E.M. (1972), 'The karyotype of *Galago alleni* Waterhouse 1837 (2N = 40) compared with that of *Galago senegalensis* Geoffroy 1796 (2N = 38) (Primates, Prosimii: Galagidae)', *Genetica* 43, 183-9.

5 de Boer, L.E.M. (1972), 'The karyotype of *Galago demidovii* Fischer 1808 (2N = 58) (Primates, Prosimii: Galagidae)', *Genen en Phaenen* 15, 19-22.

6 Borgaonkar, D.S. (1968), 'Additions to the lists of chromosomes in the Insectivores and Primates', *J. Hered.* 58, 211-13.

7 Borgaonkar, D.S. (1969), 'Insectivora cytogenetics', in Benirschke, K. (ed.), *Comparative Mammalian Cytogenetics*, New York.

8 Buettner-Janusch, J. (1963), 'An Introduction to the Primates', in Buettner-Janusch, J. (ed.), *Evolutionary and Genetic Biology of Primates*, vol. 1, New York.

9 Chiarelli, B. (1961), 'Ibridologia e sistematica in Primati. I: raccolta di dati', *Atti Assoc. Genetica Ital.* 6, 213-20.

10 Chiarelli, B. (1961), 'Some chromosome numbers in Primates', *Mammal. Chrom. Newsl.* 6, 3-5.

11 Chiarelli, B. (1963), 'Primi risultati di ricerche di genetica e cariologia comparata in Primati e loro interesse evolutivo', *Riv. Antropol.* 50, 87-124.

12 Chiarelli, B. (1966), 'Ibridologia e sistematica in Primati: deduzioni sul cariotipo degli ibridi', *Riv. Antropol.* 53, 113-17.

13 Chiarelli, B. (1972), 'New data on prosimian chromosomes', *J. Human Evol.* 1, 61-3.

14 Chiarelli, B. and L. Barberis (1966), 'Some data on the chromosomes of Prosimiae and of New World monkeys', *Mammal. Chrom. Newsl.* 22, 216.

15 Chiarelli, B. and Egozcue, J. (1968), 'The meiotic chromosomes of some Primates', *Mammal. Chrom. Newsl.* 9, 85-6.

16 Chu, E.H.Y. and Bender, M.A. (1961), 'Chromosome cytology and evolution in Primates', *Science* 133, 1399-405.

17 Chu, E.H.Y. and Bender, M.A. (1962), 'Cytogenetics and evolution of Primates', *Ann. N.Y. Acad. Sci.* 102, 253-66.

18 Chu, E.H.Y. and Swomley, B.A. (1961), 'Chromosomes of lemurine lemurs', *Science* 133, 1925-6.

19 Dodson, E.O. (1960), cited by Bender, M.A. and Chu, E.H.Y. (1963), 'The chromosomes of Primates', in Buettner-Janusch, J. (ed.), *Evolutionary and Genetic Biology of Primates*, vol. 1, New York.

20 Eckhardt, R.B. (1969), 'A chromosome arm number index and its application to the phylogeny and classification of lemurs', *Am. J. Phys. Anthrop.* 31, 85-8.

21 Egozcue, J. (1967), 'Chromosome variability in the Lemuridae', *Am. J. Phys. Anthrop.* 26, 341-8.

22 Egozcue, J. (1968), 'The meiotic chromosomes of the lesser bushbaby (*Galago senegalensis*)', *Mammal. Chrom. Newsl.* 9, 92-3.

23 Egozcue, J. (1969), 'Primates', in Benirschke, K. (ed.), *Comparative Mammalian Cytogenetics*, New York.

24 Egozcue, J. (1970), 'The chromosomes of the lesser bushbaby (*Galago senegalensis*) and the greater bushbaby (*Galago crassicaudatus*)', *Folia primat.* 12, 236-40.

25 Egozcue, J. (1972), 'Chromosomal evolution in prosimians', this volume.

26 Egozcue, J. (1972), 'The chromosomes of lemurs', *Folia primat.* 17, 171-6.

27 Egozcue, J., Chiarelli, B., Sarti-Chiarelli, M. and Hagemenas, F. (1968), 'Chromosome polymorphism in the tree shrew (*Tupaia glis*)'. *Folia primat.* 8, 150-8.

28 Egozcue, J., Ushijima, R.N. and Vilarasau de Egozcue, M. (1966), 'The chromosomes of the slender loris (*Loris tardigradus*)', *Mammal. Chrom. Newsl.* 22, 204.

29 Egozcue, J. and Vilarasau de Egozcue, M. (1966), 'The chromosomes of the angwantibo (*Arctocebus calabarensis*)', *Mammal. Chrom. Newsl.* 20, 53.

30 Egozcue, J. and Vilarasau de Egozcue, M. (1967), 'The chromosome complement of the slow loris (*Nycticebus coucang*)', *Primates* 7, 423-32.

31 Elliot, O.S. and Lisco, H. (1969), cited by Borgaonkar, D.S. (1968), in Benirschke, K. (ed.), *Comparative Mammalian Cytogenetics*, New York.

32 Elliot, O.S., Wong, M. and Borgaonkar, D.S. (1969), 'Karyological study of *Tupaia* from Thailand', *J. Hered.* 60, 153-7.

33 Hayata, I., Sonta, S., Itoh, M. and Kondo, N. (1971), 'Notes on the karyotype of some prosimians, *Lemur mongoz, Lemur catta, Nycticebus coucang* and *Galago crassicaudatus*', *Japan. J. Genet.* 1, 61-8.

34 Hsu, T.C. and Johnson, M.L. (1963), 'Karyotypes of two mammals from Malaya', *Amer. Nat.* 97, 127-9.

35 Klinger, H.P. (1963), 'The somatic chromosomes of some Primates: *Tupaia glis, Nycticebus coucang, Tarsius bancanus, Cercocebus aterimus, Symphalangus syndactylus*', *Cytogenetics* 2, 140-51.

36 Lavocat, R. (1969), 'La systématique des rongeurs hystricomorphes et la dérive des continents', *C.R. Acad. Sci.* (Paris) (Série D) 269, 1496-7.

37 Manfredi-Romanini, M.G. (1972), 'Nuclear DNA content and area of primate lymphocytes as a cytotaxonomical tool', *J. Human Evol.* 1, 23-40.

38 Manfredi-Romanini, M.G., de Boer, L.E.M., Chiarelli, B. and Tinozzi Massari, S. (1972), 'DNA content and cytotaxonomy of *Galago senegalensis* and *Galago crassicaudatus*', *J. Human Evol.* 1, 473-6.

39 Manna, G.K. and Talukdar, M. (1968), 'An analysis of the somatic chromosome complements of both sexes of two Primates, the slender loris, *Loris tardigradus*, and Rhesus monkey, *Macaca mulatta*', *Mammalia* 32, 118-24.

40 Matthey, R. (1955), 'Les chromosomes de *Galago senegalensis* Geoffroy (Prosimii – Lorisidae – Galaginae)', *Rév. Suisse Zool.* 62 (suppl.), 163-206.

41 Rumpler, Y. (1970), 'Etude cytogénétique de *Lemur catta*', *Cytogenetics* 9, 239.

42 Rumpler, Y., this volume.

43 Rumpler, Y. and Albignac, R. (1969), 'Etude cytogénétique de deux lémuriens, *Lemur macaco macaco*, Linné 1766, et *Lemur fulvus rufus* (Audebert 1800) et d'un hybride *macaco macaco/fulvus rufus*', *C.R. Soc. Biol.* 163, 1247-50.

44 Rumpler, Y. and Albignac, R. (1969), 'Etude cytogénétique de quelques hybrides intraspécifiques et interspécifiques de lémuriens', *Ann. Scien. Univ. Besancon* 6, 1-3.

45 Rumpler, Y. and Albignac, R. (1969), 'Existence d'une variabilité chromosomique intraspécifique chez certains lémuriens', *C.R. Soc. Biol.* 163, 1989-92.

46 Rumpler, Y. and Albignac, R. (1971), 'Etude cytogénétique des Cheirogaleidae', *IVe congrés international de Primatologie*, Beaverton, in press.

47 Rumpler, Y. and Albignac, R. (1972), 'Etude cytogénétique de *Varecia variegata* et de *Lemur rubriventer*', *C.R. Soc. Biol.* 165, 741-5.

48 Rumpler, Y. and Albignac, R. (1972), 'Cytogenetic study of the Indriidae', *Third Conf. on Exptl. Med. and Surgery in Primates*, Lyon, 21-23 June, in press.

49 Ying, K.L. and Butler, H. (1969), 'The somatic chromosomes of a lesser bushbaby (*Galago senegalensis*)', *Mammal. Chrom. News.* 10, 21.

50 Ying, K.L. and Butler, H. (1971), 'Chromosomal polymorphism in the lesser bushbabies (*Galago senegalensis*)', *Can. J. Genet. Cytol.* 13, 793-800.

M. GOODMAN, W. FARRIS Jr.,
W. MOORE, W. PRYCHODKO,
E. POULIK and M. SORENSON

Immunodiffusion systematics of the primates II: findings on Tarsius, *Lorisidae and Tupaiidae*

Introduction

The rabbit and chicken antisera used in the present study are listed in Table 1. These antisera were produced against plasma proteins from *Tarsius syrichta, Galago crassicaudatus, Nycticebus coucang, Loris tardigradus, Perodicticus potto, Tupaia chinensis*, and *Urogale everetti.*

Using these antisera, and taking the sera and plasmas from the species listed in Table 2 as test antigens, 800 trefoil Ouchterlony plate comparisons were carried out. A trefoil Ouchterlony plate consists of three wells surrounding a centre field of agar. Antiserum

Table 1 The antisera used in this study and the number of plate comparisons performed with each antiserum type

ANTISERUM					
Produced in	*Donor species*	*Antigenic preparation*	*No. of lines**	*No. of antisera*	*No. of plate comparisons*
Rabbit	*Tarsius syrichta*	plasma	8.3	1	56
Rabbit	*Galago crassicaudatus*	plasma	6.6	2	43
Rabbit	*Nycticebus coucang*	plasma	6.3	1	15
Rabbit	*Loris tardigradus*	plasma	6.8	3	160
Rabbit	*Urogale everetti*	plasma	6.9	3	171
Rabbit	*Tupaia chinensis*	plasma	7.4	2	202
Chicken	*Perodicticus potto*	plasma Fr 12	3.3	2	51
Chicken	*Perodicticus potto*	plasma Fr 14	3.1	2	42
Chicken	*Perodicticus potto*	plasma Fr 13	4.4	2	32
Chicken	*Perodicticus potto*	plasma	4.5	2	28

*Average number of precipitin lines which the antiserum developed in plate comparisons with the homologous antigenic preparation.

Table 2 Type and number of primate and other mammalian samples (individual or pooled sera and plasma) used as test antigens in this study

Scientific name	*No. of Samples*
HOMINOIDEA	
Homo sapiens	10
Pan troglodytes	3
Gorilla gorilla	2
Pongo pygmaeus	1
Hylobates lar	2
CERCOPITHECOIDEA	
Macaca mulatta	1
Macaca fuscata	6
Cercopithecus aethiops	1
Papio anubis	2
Papio hamadryas	1
Theropithecus gelada	1
Erythrocebus patas	1
Cercocebus torquatus	1
Presbytis eristatus	2
CEBOIDEA	
Ateles geoffroyi	1
Ateles panisus	1
Ateles sp.	1
Saimiri sp.	2
Saimiri sciureus	2
Cebus albifrons	1
Cebus sp.	2
Cacajao rubicundua	1
Aotes trivirgatus	3
TARSIOIDEA	
Tarsius syrichta	2
LORISOIDEA	
Galago crassicaudatus	6
Galago senegalensis	1
Galagoides demidovii	3
Nycticebus coucang	6
Loris tardigradus	3
Perodicticus potto	3
Arctocebus calabarensis	1
LEMUROIDEA	
Lemur fulvus	3
Lemur fulvus albifrons	1
Lemur catta	1
Lemur mongoz	2
TUPAIODEA	
Urogale everetti	1
Tupaia glis	4
Tupaia chinensis	3
Tupaia longipes	1
Tupaia montana	1
Tupaia minor	1
Tupaia palawanensis	1

NON-PRIMATES	
Rhynchocyon sp.	1
Petrodromus sultan	3
Nasilio brachyrhynchus	1
Atelerix sp.	1
Hemiechinus auritus	1
Tenrec ecaudatus	1
Scapanas aquaticus	1
Sorex cinereus	1
Eptesicus sp.	1
Pipistrellus hesperus	1
Dasypus novemcinctus	3
Bradypus tridactylus	1
Myrmecophaga tridactyla	1
Manis pentadactyla	1
Bos taurus	3
Loxodonta africana	1
Canis familiaris	2
Potos flavus	2
Ursus arctos	1
Eumetopias jabatus	1
Rattus sp.	1
Cavia sp.	1
Citellus mexicanus	1
Suncus murinus	1
Elephantulus myurus	1

is put into the bottom well and antigen preparations from two different species are placed in each of the two top wells. The reactants diffuse towards each other in the limited area of agar, producing precipitin lines. If antigen from the species against which the antiserum was produced (the homologous species) is in one of the top wells and antigen from a heterologous species in the other top well, the precipitin line of the homologous reaction will extend beyond the precipitin line of the heterologous reaction, provided that the two species do not share all of the antigenic sites against which there are anti-bodies. This extension is called a spur. The fewer the antigenic sites shared by the two species, the longer the spur. When one heterologous species is compared to another, the one sharing more antigenic sites with the homologous species yields the larger net spur. Spurs are numbered according to length and intensity, on a scale from *one*, which denotes a reaction of identity when antigen from both the right and left top wells show precipitin lines, to *five*, the largest spur possible. Photographs of several species comparisons (tarsier vs. man and man vs. galago) developed by rabbit antiserum to tarsier serum in 'Y' trefoil Ouchterlony plates are shown in Fig. 1, with the net spur sizes recorded for the man vs. galago comparison given in the legend to Fig. 1. The spur size results from the 800 plate comparisons gathered in the present study were recorded on IBM coding forms and then converted by a computer

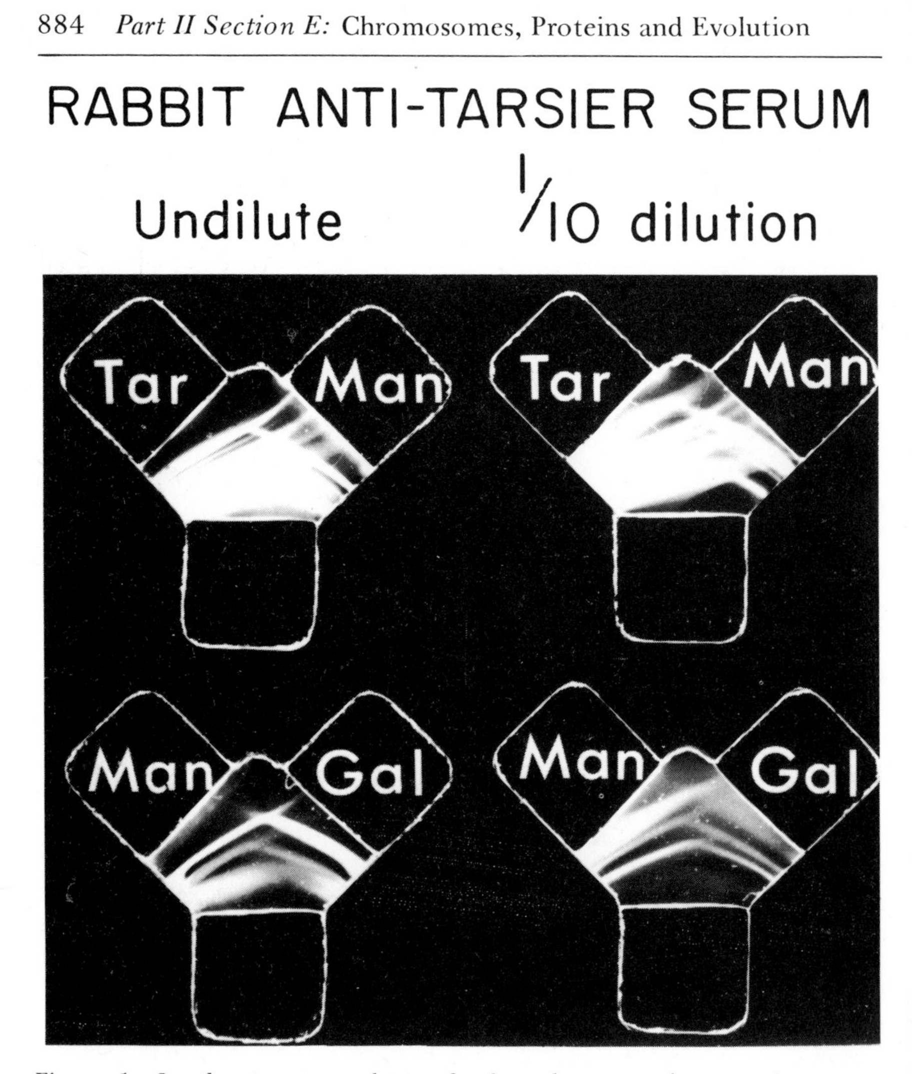

Figure 1. In the top two plates, the homologous antigen, tarsier serum, developed 8 distinct precipitin lines, each of which yielded strong intense spurs against the heterologous antigen, human serum. In the bottom two plates, in which there is comparison of two heterologous antigens, human and galago sera, the crossreacting serum proteins (diffusing into the agar from each antigen well) developed about three precipitin lines, with the human proteins developing positive net spurs against the galago proteins. In this heterologous to heterologous comparison, recording of the first two precipitin lines from the plate with 1/10 dilutions of human and galago serum, and of the third precipitin line from the plate with undilute human and galago serum, yielded the following net spur sizes:

	Human - Galago	*Net spur size*
Line 1	2 – 1	1
Line 2	3 – 1	2
Line 3	4 – 3	1

program IMMDF into the Taxonomic Distance Tables shown in Tables 3-12. The methods for producing the antisera, carrying out the trefoil Ouchterlony plate comparisons, and converting the plate results into Taxonomic Distance Tables are thoroughly described in a previous paper.[1]

These taxonomic distance data (Tables 3-12) provide evidence for the following conclusions about the cladistic relationships of different prosimian lineages.

Table 3 Taxonomic distance table: rabbit anti-*Tarsius syrichta* plasma

Species	*Antigenic distance*
Tarsius syrichta	0.000
Presbytis cristatus	4.128
Theropithecus gelada	4.188
Erythrocebus patas	4.249
Pan troglodytes	4.249
Gorilla gorilla	4.249
Pongo pygmaeus	4.249
Cercocebus torquatus	4.249
Hylobates lar	4.249
Cercopithecus aethiops	4.249
Papio hamadryas	4.249
Macaca fuscata; M. mulatta	4.533
Homo sapiens	4.650
Saimiri sp.	4.896
Aotes douroucouli	4.992
Cebus albifrons; C. sp.	5.037
Ateles sp.	5.158
Galago crassicaudatus	5.320
Tupaia chinensis	5.364
Cacajao rubicundua	5.400
Lemur fulvus	5.485
Perodicticus potto	5.509
Loris tardigradus	5.616
Nycticebus coucang	5.673
Citellus mexicanus	6.169
Dasypus novemcinctus	6.169
Bos taurus	6.411
Bradypus tridactylus	6.775
Petrodromus sultan	6.775
Atelerix sp.	6.956

Table 4 Taxonomic distance table: rabbit anti-*Galago crassicaudatus* plasma

Species	*Antigenic distance*
Galago crassicaudatus	0.000
Galago senegalensis	0.114
Galagoides demidovii	0.275
Nycticebus coucang	2.242
Perodicticus potto	2.400

Arctocebus calabarensis	2.626
Loris tardigradus	2.926
Lemur mongoz	6.236
Homo sapiens	6.800
Saimiri sciureus	6.951
Tupaia glis	7.515

Table 5 Taxonomic distance table: rabbit anti-*Nycticebus coucang* plasma

Species	*Antigenic distance*
Nycticebus coucang	0.000
Loris tardigradus	1.395
Galago crassicaudatus	2.727
Galago senegalensis	2.864
Arctocebus calabarensis	3.250
Perodicticus potto	3.422
Lemur mongoz	7.685
Homo sapiens	8.021
Saimiri sciureus	8.100
Tupaia glis	8.830

Table 6 Taxonomic distance table: rabbit anti-*Loris tardigradus* plasma

Species	*Antigenic distance*
Loris tardigradus	0.000
Nycticebus coucang	1.430
Galagoides demidovii	1.930
Galago crassicaudatus	1.996
Galago senegalensis	2.186
Perodicticus potto	2.441
Lemur fulvus	4.295
Lemur catta	4.576
Lemur sp.	4.743
Lemur mongoz	4.764
Ateles panisus	4.839
Homo sapiens	4.978
Macaca fuscata	5.027
Ateles geoffroyi	5.186
Pipistrellus hesperus	5.270
Gorilla gorilla	5.591
Papio anubis	5.594
Erythrocebus patas	5.627
Macaca fuscata	5.700
Pan troglodytes	5.810
Hylobates lar	5.957
Tupaia chinensis	6.156
Presbytis cristatus	6.206
Potos flavus	6.303
Aotes trivirgatus	6.350
Citellus mexicanus	6.505
Bos taurus	6.509
Atelerix sp.	6.895

Bradypus tridactylus	6.919
Dasypus novemcinctus	7.059

Table 7 Taxonomic distance table: rabbit anti-*Urogale everetti* plasma

Species	*Antigenic distance*
Urogale everetti	0.000
Tupaia minor	1.302
Tupaia longipes	1.392
Tupaia glis	1.399
Tupaia montana	1.434
Tupaia chinensis	1.441
Tupaia palawanensis	1.611
Lemur fulvus	3.453
Galago crassicaudatus	3.690
Loris tardigradus	3.690
Nycticebus coucang	3.987
Macaca fuscata	3.987
Potos flavus	4.038
Homo sapiens	4.058
Dasypus novemcinctus	4.080
Canis familiaris	4.228
Aotes trivirgatus	4.279
Petrodromus sultan	4.287
Tenrec ecaudatus	4.342
Perodicticus potto	4.352
Cavia sp.	4.494
Atelerix sp.	3.579
Scapanas aquaticus	4.652
Elephantulus myurus	4.895
Eptesicus sp.	4.980

Table 8 Taxonomic distance table: rabbit anti-*Tupaia chinensis* plasma

Species	*Antigenic distance*
Tupaia chinensis	0.000
Tupaia longipes	0.406
Tupaia glis	0.439
Tupaia minor	0.666
Tupaia montana	0.717
Tupaia palawanensis	1.141
Urogale everetti	1.331
Tarsius syrichta	5.882
Macaca fuscata fuscata	6.108
Lemur fulvus	6.114
Galago crassicaudatus	6.128
Aotes trivirgatus	6.222
Lemur fulvus albifrons	6.336
Bradypus tridactylus	6.368
Homo sapiens	6.369
Macaca fuscata	6.408
Loxodonta africana	6.423

Perodicticus potto	6.538
Ursus arctos	6.629
Potos flavus	6.726
Nycticebus coucang	6.819
Manis pentadactyla	6.901
Canis familiaris	6.961
Eumetopias jabatus	7.105
Dasypus novemcinctus	7.161
Petrodromus sultan	7.244
Rattus sp.	7.309
Tenrec ecaudatus	7.377
Cavia sp.	7.377
Rhynchocyon sp.	7.513
Myrmecophaga tridactyla	7.581
Nasilio brachyrhynchus	7.784
Scapanas aquaticus	7.852
Atelerix sp.	7.852
Eptesicus sp.	7.920
Suncus murinus	7.920
Sorex cinereus	8.192

Table 9 Taxonomic distance table: chicken anti-*Perodicticus potto* plasma Fr 12

Species	*Antigenic distance*
Perodicticus potto	0.000
Galago crassicaudatus	2.108
Arctocebus calabarensis	2.367
Nycticebus coucang	3.158
Loris tardigradus	3.275
Galago senegalensis	3.492
Lemur mongoz	5.158
Tupaia glis	5.758
Macaca irus	5.833
Cebus apella	5.908
Tarsius syrichta	6.058
Homo sapiens	6.583

Table 10 Taxonomic distance table: chicken anti-*Perodicticus potto* plasma Fr 14

Species	*Antigenic distance*
Perodicticus potto	0.000
Arctocebus calabarensis	0.995
Loris tardigradus	1.676
Nycticebus coucang	1.774
Galago crassicaudatus	1.891
Galago senegalensis	2.052
Lemur mongoz	4.242
Tarsius syrichta	4.306
Macaca irus	4.375
Homo sapiens	4.415
Cebus apella	5.210

Table 11 Taxonomic distance table: chicken anti-*Perodicticus potto* plasma Fr 13

Species	*Antigenic distance*
Perodicticus potto	0.000
Arctocebus calabarensis	1.250
Galago crassicaudatus	2.235
Galago senegalensis	2.500
Nycticebus coucang	2.500
Loris tardigradus	2.538
Lemur mongoz	5.346
Tarsius syrichta	5.790
Cebus apella	5.255
Macaca irus	6.385
Homo sapiens	6.645
Macaca mulatta	6.710

Table 12 Taxonomic distance table: chicken anti-*Perodicticus potto* plasma

Species	*Antigenic distance*
Perodicticus potto	0.000
Arctocebus calabarensis	1.704
Nycticebus coucang	2.574
Galago crassicaudatus	2.981
Galago senegalensis	2.981
Loris tardigradus	3.093
Lemur mongoz	5.241
Tupaia glis	6.796

Major ancestral branchings

At the base of the Primates, the tree shrew separates, or has already been separated, from the line to lemuroids, lorisoids, *Tarsius*, and Anthropoidea. (The question of whether or not the tree shrews should be treated as Primates is left unanswered by our present data.) In the early Primates, four major lines branch apart, separating the ancestors of lemuroids, lorisoids, *Tarsius* and Anthropoidea, with the lorisoids and lemuroids sharing a slightly more recent common ancestor with each other, and, similarly, *Tarsius* and Anthropoidea sharing a slightly more recent common ancestor before branching apart. (Taxonomic distance tables for anti-ceboid sera reveal the Ceboidea to be closer to catarrhines than to any living prosimian group.)

Subsequent branchings

1. Lorisidae

Three lines separate at the base of the family, one leading to *Galago* and *Galagoides*, another to *Nycticebus* and *Loris*, and the third to *Perodicticus* and *Arctocebus.* The *Nycticebus, Loris* and *Galago* lines may have slightly closer cladistic relationships with each other than with the *Perodicticus/Arctocebus* line. However, the latter line may simply have diverged more from the ancestral state than the former, thus accounting for the slightly smaller antigenic distances between the former lines. The splitting between *Galago* and *Galagoides* appears to be much more recent than either the *Nycticebus-Loris* split or the *Perodicticus/Arctocebus* split.

2. Tupaiidae

After the ancestral branching apart of *Urogale* and *Tupaia, T. palawanensis* appears to separate from the remaining *Tupaia* complex, then the line or lines to *T. montana* and *T. minor*, and finally quite small divergences separate *T. glis* and *T. longipes* from *T. chinensis.* Additional antisera to either *T. minor* or *T. montana* are needed to determine whether these two species represent a distinct monophyletic group within the genus *Tupaia.*

Acknowledgments

This research was supported by NSF grant GB7420. We thank Francis Vincent and Alan Walker for contributing blood samples from *Galagoides demidovii* and also John Buettner-Janusch and William Montagna for contributing blood samples from several lorisoids and lemuroids. The help of Mary Etta Hight in obtaining samples is also appreciated.

NOTES

1 Goodman, M. and Moore, G.W. (1971), 'Immunodiffusion systematics of the Primates. I: the Catarrhini', *Syst. Zool.* 20, 19-62.

N. A. BARNICOT and
D. HEWETT-EMMETT

Electrophoretic studies on prosimian blood proteins

Introduction

A simple way of comparing proteins from different species or individuals is to study their migration under the influence of an electric potential. Filter paper, cellulose acetate paper or agar gel can be used as supporting media, but better resolution is usually obtained with fine-pored media such as starch or acrilamide gels. If the protein is sufficiently concentrated its position in the gel at the end of the run can be revealed by applying some non-specific dye; but many proteins, notably enzymes, have to be located by special staining methods that depend on their characteristic activities. Many enzymes from blood or other tissues show more than one band when examined by standard methods. In some cases these *isoenzyme* bands are due to the presence of more than one form of the enzyme, each controlled by a different genetic locus, and sometimes to the presence of their interaction products as well. In other cases the reasons are less clear.

It is obvious that these electrophoretic procedures give much less information about the structure of the protein than do amino acid sequence analyses. As a rule, only amino acid changes that result in alteration of the electric charge of the molecule will be detected. Identical mobilities in the medium by no means prove identity of structure and differences of mobility may be due to a change in one amino acid or may be the resultant of many differences. Despite these limitations, the method does allow many specimens of many different proteins to be examined quite rapidly, and in interspecies comparisons regularities of taxonomic interest often emerge (see Tashian;[1] Barnicot and Cohen;[2] Barnicot and Hewett-Emmett).[3]

Electrophoresis is also very useful in detecting inherited variations of a given protein within a species; that is, biochemical polymorphism. It has been found that in man and a few other species that have been closely studied, about one third of the proteins examined are polymorphic. It is known that most of the numerous haemoglobin variants that have been described in man are due to mutations producing a change in a single amino acid in one or other of the two chains composing the molecule. The problem of why there is such widespread protein polymorphism has been touched on in the introduction to this section of the symposium.

We have been able to examine certain red cell and serum proteins in a fair range of prosimian species (Table 1) and we are very grateful to the workers, listed at the end of this paper, who have been good enough to send us material. In this paper we wish to draw attention to some of the main findings; a detailed report will be published elsewhere.

Table 1 Material examined

	Individual samples
Tarsius bancanus	1
Microcebus murinus	7
Cheirogaleus major	2
Lemur catta	5
Lemur mongoz	6 (parents and 4 offspring)
Lemur macaco	3 (parents and 1 offspring)
Hapalemur griseus	9 (includes 2 parents and 1 offspr
Galago senegalensis	8
Galago crassicaudatus	19
Galago demidovii	7
Galago alleni	2
Euoticus elegantulus	2
Arctocebus calabarensis	2 (one examined for Hb only)
Perodicticus potto	1
Nycticebus coucang	3
Tupaia tana	3
Tupaia belangeri	9 (4 of them sibs)
Tupaia minor	9

Since we are concerned with patterns of bands, varying in intensity and position, it would no doubt be possible to make more precise numerical descriptions and to combine the results from different proteins to give measures of overall resemblance by the multivariate methods that are now so much in vogue. So far we have not felt that the added labour would be worthwhile.

A key to the lettering of the diagrams is given in the caption to Fig. 1.

1. Haemoglobins (Hb)

In Fig. 1 the haemoglobins of prosimians, roughly arranged in taxonomic groups, are compared with the common human adult phenotype Hb-A. We did not detect any clear examples of haemoglobin polymorphism. It will be noted that the major Hb of *Tupaia* (three species) is conspicuously slow (i.e. cathodal or nearer the negative pole). This confirms a previous report on *T. glis.*[4] The major Hbs of both *Microcebus* and *Cheirogaleus* are somewhat anodal to those of the other Madagascan species examined, in agreement with Buettner-Janusch et al.[5] These authors also found that all specimens of *Lemur catta* and *Hapalemur griseus* had two haemoglobins in roughly equal concentrations, whereas *L. mongoz* and *L. macaco* had only one;[4,5] this was confirmed by our investigations.

The presence of two major haemoglobins in all specimens of *Galago senegalensis* and *G. crassicaudatus* has been reported by several workers.[4–6] We found this to be so in these two species, but only one major Hb was present in *G. demidovii* (N = 7) and *Euoticus elegantulus* (N = 2). The results for *G. alleni* are uncertain, since the specimens were very dilute. Nute et al.[7] reported two major Hbs in *G. demidovii* and it seems possible that there may be regional variation in this species. The presence of two major haemoglobins in

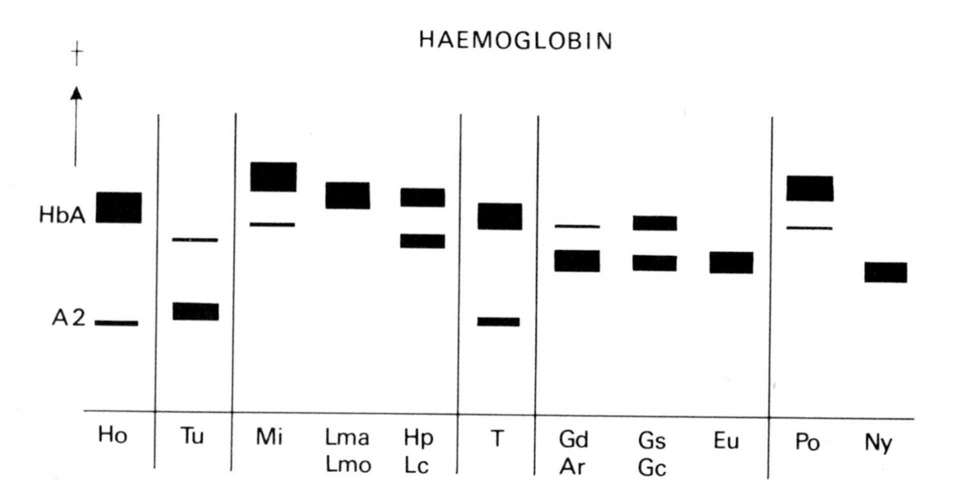

Figure 1. *Haemoglobins* (pH 8.6 tris-EDTA-borate buffer system).
Arrow indicates direction of the anode.
In this and subsequent figures: HO = human
Tu = *Tupaia*, 3 species Mi = *Microcebus*
Lma = *Lemur macaco* Lmo = *Lemur mongoz*
Lc = *Lemur catta* Hp = Hapalemur T = *Tarsius*
Gd = *Galago demidovii* Gs = *G. senegalensis*
Gc = *G. crassicaudatus* Eu = *Euoticus* Ar = *Arctocebus*
Po = *Perodicticus* Ny = *Nycticebus*
The horizontal line at the bottom in this and subsequent figures indicates starting point of run.

virtually every individual of certain prosimian species is perhaps best explained by postulating gene duplications at very high frequencies; similar situations are known in certain anthropoids (see Sullivan).[8]

A minor Hb component Hb-A_2 is shown in the diagram (Fig. 1) of a human specimen. This contains a chain (δ) similar to the β-chain and probably derived from it by duplication. A similar and probably homologous minor component occurs in pongids and gibbons. The only prosimian in which we found a minor Hb of comparable mobility to Hb-A_2 was *Tarsius*, but its concentration was considerably higher than that of Hb-A_2. The chain composition is not yet known and there is therefore no evidence as to whether it contains a homologue of the δ-chain.

A conspicuous minor component slightly cathodal to the major one is also seen in *Microcebus*, but its strength seems to vary in different individuals. Weak bands anodal to the major Hb, or in roughly the position of Hb-A_2, are sometimes seen in certain other prosimian species after o-tolidine/peroxide staining. It may be that some of these components are polymers or other secondary products of the major Hb, but this needs further study.

Buettner-Janusch and Twichell showed that in some prosimians (and also in certain other mammals) the haemoglobin is more resistant to denaturation by alkali than is human Hb-A.[9] We also found that *Galago crassicaudatus* Hb shows marked alkali resistance and that the two bands are not differentially affected. Resistance was also high for *Microcebus* and *Cheirogaleus* Hbs, but did not differ appreciably from human Hb-A in *Tarsius* and was only slightly raised in tree-shrews.

2. *Phosphoglucomutase (PGM)*

This red cell enzyme shows several isoenzyme bands. A common human type (PGM2) is shown on the left in Fig. 2. It is know that in man the more cathodal pair of bands and the anodal set are controlled by different gene loci.

The tree-shrews, represented in the diagram by *T. belangeri*, have very strong anodal bands. There is a clear polymorphism of these bands in this species; one of nine animals showed a four-banded pattern interpretable as a heterozygote for the two types shown in Fig. 2.

There was also polymorphism of the more cathodal bands in *Microcebus* (see Fig. 2) and probably in *Cheirogaleus*. The patterns in these two species are rather distinct from those found in the lemurine material. In some specimens of *Lemur mongoz* and *L. macaco* faint bands in a very anodal position were sometimes, but not consistently, detected.

The pattern in *Tarsius* is unusual and difficult to interpret. Since

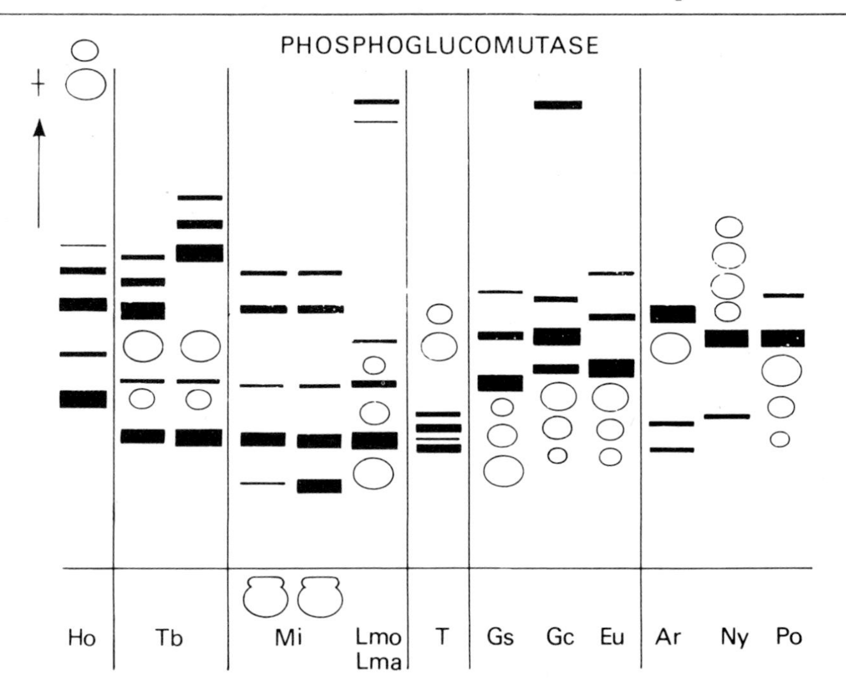

Figure 2. *Phosphoglucomutase* (black) and achromatic bands (white). Tb = *Tupaia belangeri.* Two phenotypes of *Microcebus* and *T. belangeri* are shown.

the specimen was taken some time after death the possibility of deterioration cannot be excluded.

It is not clear whether the set of bands in the galagines corresponds to the anodal set or the cathodal ones found in other species. Even very strong staining failed to show more cathodal bands, but the presence of achromatic bands (see below) in this region may interfere; this may also be true for *Perodicticus*, though further study is needed. The pattern shown for *G. crassicaudatus* suggests a heterozygous type, but it was seen in all the specimens. A very anodal band, which may sometimes be double, is consistently found in this species and faint bands rather less anodally situated were detected in some *G. senegalensis* specimens.

3. Achromatic bands

Fig. 2 also shows (white circles) the locations of an oxidase which prevents darkening of the gel background by the tetrazolium stain and therefore shows up as an unstained area. Wide variations in the mobility of this enzyme are seen in higher primates.[2,3] It will be noted that: (1) the cheirogaleines have more cathodal bands than the lemurines;[15] (2) the gradient of band strengths in galagines (except *G. senegalensis*) is the inverse of that in the lemurines; (3) the bands are rather more anodal in the lorisines than in the galagines, notably in *Nycticebus*. The pattern in this latter species suggests heterozygosity, but it was observed in all six animals.

4. Adenylate kinase (AK)

In the buffer system described by Fildes and Harris,[10] human homozygous types show a series of bands decreasing in strength anodally. The prosimian patterns are similar to this (Fig. 3), though in some species the set of bands is much more anodal and in some there are faint cathodal bands in addition. We may note again a clear difference between the two cheirogaleines and the lemurines. Clear examples of polymorphism were seen in *Galago senegalensis* and *G. crassicaudatus* (Gs, Gc in Fig. 3).

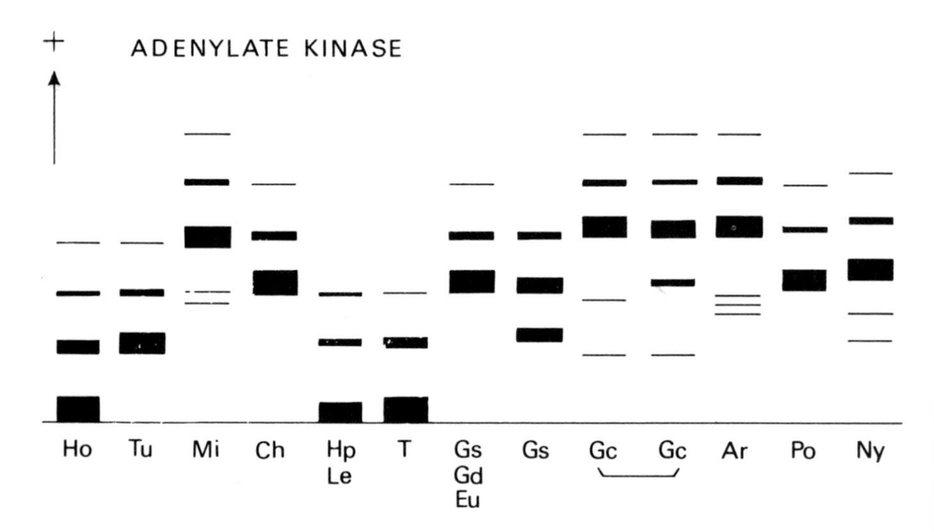

Figure 3. *Adenylate kinase.* Ch = *Cheirogaleus*, Le = *Lemur* (3 species). Note that two phenotypes are shown both for *Galago senegalensis* (Gs) and for *G. crassicaudatus* (Gc).

5. Acid phosphatase (AP)

The patterns show wide interspecific, and in some cases intraspecific, variations. Since many of the prosimian patterns lie well outside the range of the human control, cross-comparison of different gels is inexact and we show only (Figs. 4 and 5) comparisons based on individual gels.

All the tree-shrews showed a rather characteristic pattern of two widely separated bands. The more anodal band was clearly seen even in very fresh material and is therefore unlikely to be due to deterioration.

The bands in *Microcebus* and *Cheirogaleus* were substantially cathodal to those in the other Madagascan forms. Due to the small

ACID PHOSPHATASE

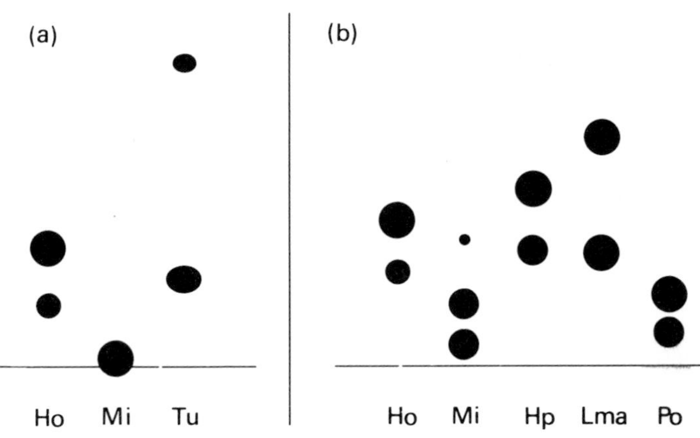

Figure 4. *Acid phosphatase.* Diagram of comparisons on two different gels (a) and (b). Gel (b) was run at increased voltage with cooling plates. The human control is an AB type.

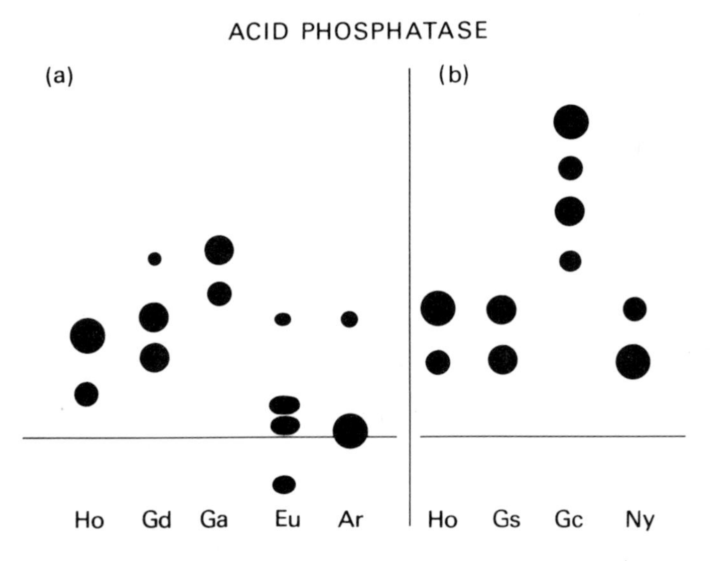

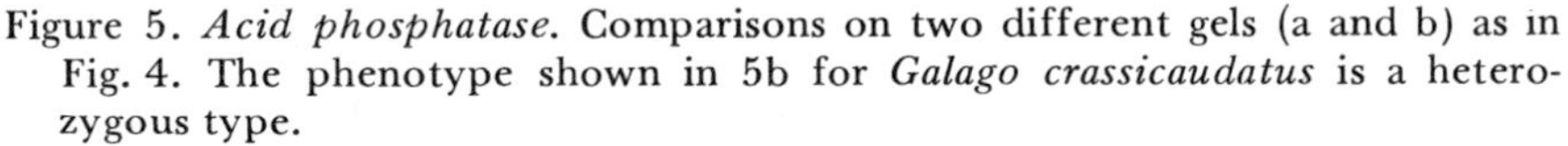
Figure 5. *Acid phosphatase.* Comparisons on two different gels (a and b) as in Fig. 4. The phenotype shown in 5b for *Galago crassicaudatus* is a heterozygous type.

migration distance, only one band is usually seen in the former; but it can be resolved into two by higher voltage runs using cooling-plates (Fig. 4*b*).

There is considerable variation between galagine species (Fig. 5*a*). Clear polymorphism was found in *G. crassicaudatus* (Fig. 5*b*), the types probably corresponding to those described in families of this species by Buettner-Janusch and Wiggins.[11]

6. *Malate dehydrogenase (MDH: cytoplasmic form)*

Tariverdian, Ritter and Schmitt have shown that this enzyme is more anodal in prosimians than in anthropoids.[12] We have examined some half dozen prosimian species (Fig. 6) and our results confirm this. It may be significant that the bands were relatively anodal in *Nycticebus* and *Perodicticus* and also in the single galagine examined so far.

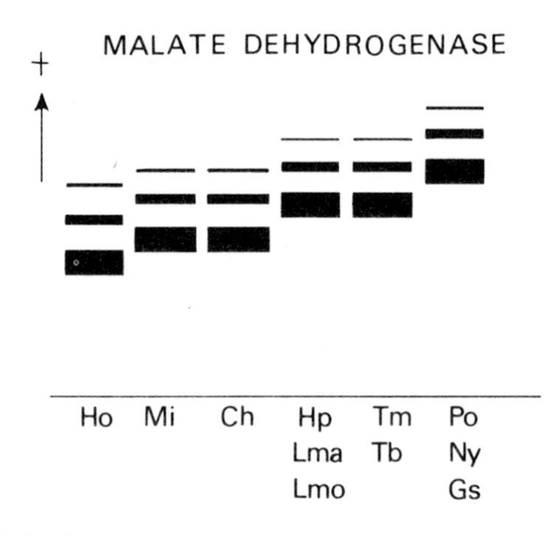

Figure 6. *Malate dehydrogenase.*

7. *Phosphohexose isomerase (PHI)*

This is an interesting enzyme since, as Tariverdian et al. have shown, there is a mobility gradient with hominoids at one end, prosimians at the other and monkeys in between.[13] However, we do not advocate a return to the concept of the Great Chain of Being on this account. We have not completed our studies, but it is already evident that there are wide mobility differences among prosimians. *Microcebus* and *Tupaia* show relatively cathodal bands and *Nycticebus* is at the opposite extreme. Mobility differences are also seen between

Hapalemur and *Lemur catta*, and between *Galago senegalensis* and *G. crassicaudatus*.

8. Serum proteins

(*a*) *Albumin:* The albumins of most of the prosimians are anodal to human albumin at pH 8.6, *Tarsius* being the most notable exception. There were some interspecies differences in the lemurines and tree shrews. An interesting feature is the presence of two albumin components of unequal strengths in all the galagines (*G. alleni* was excluded since the serum was in poor condition) and also in *Arctocebus* (Fig. 7). Simply inherited variations of albumin are known in man and various other mammals; but heterozygotes for such variants generally have two bands of roughly equal strength. Some experiments on the serum of one individual of *Galago crassicaudatus* showed that: (*a*) the albumin was single in runs on agar at pH 8.6, suggesting that the components do not differ in charge, (*b*) that a single band is also obtained if the pH of starch gels is lowered to 7.5, (*c*) that, at this pH, addition to the serum of mercaptoethanol (a disulphide bond-splitting reagent) results in two bands even at this lower pH, while at pH 8.6 the slower band is enhanced. Possibly the albumin in these species tends to undergo a conformational change, or to bind some relatively large molecule in the higher pH range; but the explanation is still unclear.

(*b*) *Transferrin* (Tf): The position of the iron-transport protein, transferrin, is also shown in Fig. 7. Some species show polymorphism; for simplicity we have illustrated only one phenotype in such cases, choosing a presumed heterozygote to show the mobility range of the protein.

The tree shrews and all the prosimians have transferrins anodal to the common human type TfC, and the lorisids tend to be most extreme in this respect.

Marked polymorphic variation was found in the lemurines, in agreement with Nute and Buettner-Janusch,[14] each of the four species showing three different transferrins. Three transferrins were also seen in our small sample of *Microcebus murinus*. There was clear polymorphism in *Tupaia belangeri, T. minor, Nycticebus* and *Galago demidovii*; but in these cases the phenotypes were less easy to interpret due to the presence of minor bands, which are clearly seen after autoradiography with radioactive iron (^{59}Fe). This complication was noted by other authors in *Lemur* species and was attributed to variations in the numbers of sialic acid residues attached to the protein.[14] Our

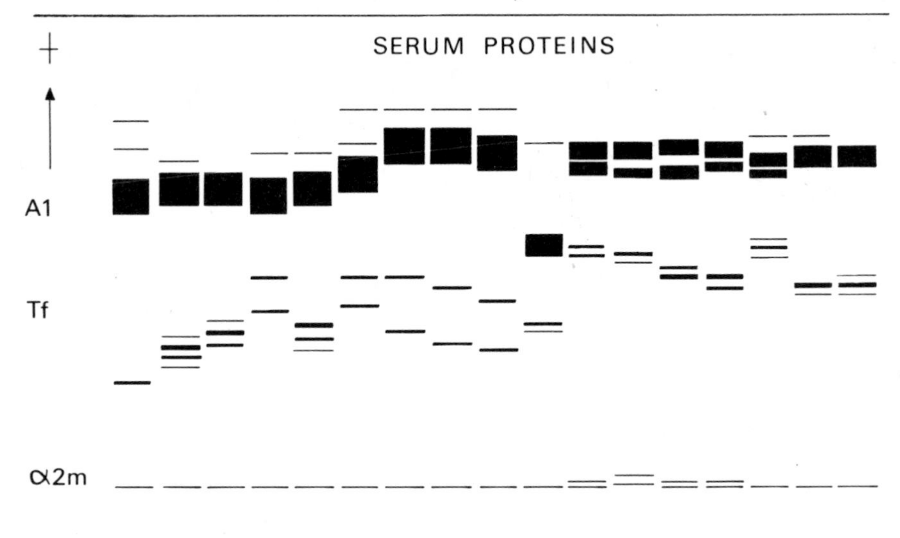

Figure 7. *Serum proteins*
Al = albumin (heavy black areas)
Tf = transferrin
α^2 m = α^2 macroglobulin
Thin lines anodal to albumin are prealbumin bands.
Tm = *Tupaia minor.*

own lemurine material took up the isotope only weakly, so that minor bands were not seen. It may be noted, however, that minor bands were not seen in *Microcebus* specimens, some of which took up ^{59}Fe strongly.

Discussion

Bearing in mind the methodological limitations mentioned in the introduction and the dangers of characterising a species from small samples, the following points are worth making.

The tree shrews, represented by three *Tupaia* species, differ from the prosimians most conspicuously in their Hb and AP patterns.

The two cheirogaleine species resemble one another quite closely but differ from the lemurines with regard to many proteins, notably in their Hb, AP, AK, achromatic band and PGM patterns. The comparison should be extended by the examination of the Indriidae.

The resemblances between *Tarsius* and man in Hb and AK patterns is interesting in view of morphological, and especially embryological, evidence of relationship between tarsioids and anthropoids. However these similarities are offset by marked differences in albumin and PGM. Since we had only one specimen of *Tarsius*, which had been dead for some hours before the blood was taken, the results clearly need confirmation.[16] At this level of analysis it would not be particularly surprising to find phenotypic resemblances for a few proteins even in forms that are in fact only distantly related.

Perhaps the most striking and consistent features of the galagines are the double albumin and transferrin bands. However, the former occurs in *Arctocebus*, and multiple transferrin bands, though somewhat different in pattern, are found in lorisines, Madagascan lemurs and tree shrews.

No useful generalisations can be made about polymorphism from such small samples, except that it certainly exists in several of the species. Variations of AK, AP, PGM and Tf were seen in some of the forms examined and our data confirm the impression from earlier work that transferrin may be very polymorphic in some species, especially lemurines. It is hoped that such variations may be put to good use in studies on the population genetics of prosimians.

Acknowledgments

We are very grateful to the following for providing us with material:

R. Albignac	*Hapalemur, L. catta*
P. Charles-Dominique	*G. alleni, G. demidovii, Euoticus, Arctocebus*
F.C. Jones and F.G. Garcia	*G. crassicaudatus, G. senegalensis, Nycticebus*
Ilaar Muul and Lim Boo Liat	*Nycticebus*
R.D. Martin	Tree-shrews, *Microcebus, Cheirogaleus, G. demidovii*
J.-J. Petter	*L. macaco, L. mongoz*
A. Petter-Rousseaux	*Microcebus, Perodicticus*
J. Rollinson	*Tarsius*
P.V. Tobias	*G. senegalensis*
D.A. Valerio	*G. crassicaudatus*
A. Voller	*G. crassicaudatus*

We wish also to acknowledge a grant from the Medical Research Council in support of this work, and to express our thanks for the technical assistance of Carol O'Reilly and Jackie Hillier.

NOTES

1 Tashian, R.E. (1965), 'Genetic variation and evolution of the carboxylic esterases and carbonic anhydrases of primate erythrocytes', *Am. J. Hum. Genet.* 17, 257-72.

2 Barnicot, N.A. and Cohen, P. (1970), 'Red cell enzymes of primates (Anthropoidea)', *Biochem. Genet.* 4, 41-57.

3 Barnicot, N.A. and Hewett-Emmett, D., (1971), 'Red cell and serum proteins of Talapoin, Patas and Vervet monkeys', *Folia primat.* 15, 65-76.

4 Buettner-Janusch, J. and Buettner-Janusch, V. (1964), 'Haemoglobins in Primates', in Buettner-Janusch, J. (ed.), *Evolutionary and Genetic Biology of Primates*, New York.

5 Buettner-Janusch, J., Washington, J.C. and Buettner-Janusch, V. (1971), 'Haemoglobins of Lemuriformes', *Arch. Inst. Pasteur Madagascar* 40, 127-36.

6 Hoffman, H.A. and Gottlieb, A.J. (1969), 'Investigations of non-human primate electrophoretic variation', *Ann. N.Y. Acad. Sci.* 162, 205.

7 Nute, P.E., Buettner-Janusch, V. and Buettner-Janusch, J. (1969), 'Genetic and biochemical studies of transferrins and haemoglobins of *Galago*', *Folia primat.* 10, 276.

8 Sullivan, B. (1971), 'Comparison of the haemoglobins in non-human Primates and their importance in the study of human haemoglobins', in Chiarelli, A.B. (ed.), *Comparative Genetics in Monkeys, Apes and Man*, London and New York.

9 Buettner-Janusch, J. and Twichell, J.B. (1961), 'Alkali resistant haemoglobins in prosimian primates', *Nature* 192, 669.

10 Fildes, R.A. and Harris, H. (1966), 'Genetically determined variation of adenylate kinase in man', *Nature* 209, 261.

11 Buettner-Janusch, J. and Wiggins, R.O. (1970), 'Haptoglobins and acid phosphatases of *Galago*', *Folia primat.* 13, 166.

12 Tariverdian, G., Ritter, H. and Schmitt, J. (1971*a*), 'Zur transspezifischen Variabilität der NAD-Malatdehydrogenase (E.C.:1.1.1.37) der Primaten', *Humangenet.* 11, 339.

13 Tariverdian, G., Ritter, H. and Schmitt, J. (1971*b*), 'Transpecific variability of phosphohexose isomerase (E.C.:5.3.1.9) in Primates', *Humangenet.* 12, 105.

14 Nute, P.E. and Buettner-Janusch, J. (1969), 'Genetics of polymorphic transferrins in the genus *Lemur*', *Folia primat.* 10, 181.

15 The specimens of *Microcebus* represented in Fig. 2 are both *M.m. murinus*. Further work subsequent to the conference has shown oxidase bands somewhat anodal to the origin in *Cheirogaleus* and in four individuals of *M.m. rufus*.

16 Subsequent to the Research Seminar meeting, we received an additional sample of *Tarsius bancanus* blood, by courtesy of Dr Carsten Niemitz. Once again we identified two haemoglobins, with the same mobilities as before; the minor band represented 18% of the total quantity. The unusually slow albumin was also confirmed.

F. M. BUSH, D. E. HAINES and
K. R. HOLMES

Haemoglobin of the lesser bushbaby, Galago senegalensis: *starch gel electrophoresis and alkali-resistance*

Introduction

Heterogeneity of haemoglobin has been found in the suborder Prosimii. In a survey of some Lorisiformes, Buettner-Janusch and Twichell mentioned the presence of at least two haemoglobin components in a number of species of Galaginae, including *Galago crassicaudatus crassicaudatus, G. c. agisymbanus*, and *G. senegalensis senegalensis.*[1] Later, Buettner-Janusch described the variability in electrophoretic patterns of haemoglobins for five specimens of *G. c. crassicaudatis.*[2] Among 19 specimens of *G. c. crassicaudatus*, ten had two haemoglobin components and nine had three components (Buettner-Janusch and Buettner-Janusch).[3,4] In another subspecies, *G. c. panganiensis*, electropherograms of 23 adult and 13 young (over two months old) exhibited the presence of three haemoglobin components, two major bands and one satellite band of variable intensity (Stormont and Cooper).[5] The significance of this variability among the species so far studied has been minimised by the discovery that aged haeomolysates pooled from six specimens of *G. senegalensis* and *G. crassicaudatus* had two major and one minor electrophoretic bands, and that the minor band was absent in fresh samples (Hoffman and Gottlieb).[6] This phenomenon exhibited by aged haemolysates has been confirmed by Nute, Buettner-Janusch and Buettner-Janusch,[7] who observed the occurrence of two major components in fresh samples and also satellite bands in aged samples when they compared the electrophoretic patterns from 57 specimens of *G. c. crassicaudatus* and *G. c. argentatus*, 12 of *G. demidovii*, and nine of *G. senegalensis.* The satellite bands formed from polymerisation which could be reversed by treatment of the aged samples with 2-mercaptoethanol.

The two major haemoglobin components have been isolated by

DEAE-Sephadex column chromatography from individual samples in 13 specimens of *G. crassicaudatus* (Nute et al.).[7] Subsequent separation of both components in urea gels revealed the occurrence of identical alpha-chains and different beta-chains. Similar patterns in urea gels for unfractionated haemolysate samples in single specimens of *G. senegalensis* and *G. demidovii* suggested a similar polypeptide substructure for *Galago* haemoglobins. Accordingly, Nute et al. proposed an $alpha_2 beta_2{}^1$ polypeptide substructure for the component migrating slowest toward the anode on starch gels and an $alpha_2 beta_2{}^2$ polypeptide substructure for the rapidly-migrating electrophoretic band.[7]

Measurement of alkali-resistant haemoglobin reveals significant differences between the *Galago* species studied. Mean levels were found to be considerably higher in *G. crassicaudatus* than in *G. senegalensis.* Five specimens of adult *G. c. agisymbanus* had a mean level of 36.0% (range, 32.0-46.8) and 18 *G. c. crassicaudatus*, 36.5% (25.3-59.1);[3] 19 *G. crassicaudatus* spp., 27.7% (19.2-36.3);[3] and four *G. s. senegalensis*, 8.4% (7.2-10.1).[1]

The paucity of information about the haemoglobin in other colonies of the lesser bushbaby, particularly during ontogeny, prompted the present investigation. This paper reports on the results of additional studies on neonate, juvenile, and adult *G. senegalensis* haemoglobin, including studies on alkali-resistance, polypeptide chain composition, and electrophoresis in alkaline and acid buffer systems.

Materials and methods

The adult bushbabies used in the present study have all been maintained in captivity either at the Medical College of Virginia or at Michigan State University for periods of two to four years. The neonates were offspring of adults in the respective colonies. All animals were on a 12-hour day/night light cycle and received water and a standard diet of bananas, monkey chow, and a vitamin supplement. Blood samples were obtained from 34 different animals and were drawn from a superficial vein on the caudal aspect of the lower leg, designated as the short saphenous vein in *Tarsius.*[8] All animals were sampled at least twice. Each sample was drawn in a heparinised syringe between 8.30 and 12.00 a.m.

The procedure for preparation of red cell haemolysates was the same as described by Bush and Townsend.[9] Plasmas were separated after centrifugation of the sample at 2,000 g for 10 minutes. Red cells were washed at least three times with physiological saline kept at 4°C. The cells were lysed with distilled water and toluene and then the solution was exposed to carbon monoxide to form the CO-haemoglobin derivative. Each sample was centrifuged at 10,000 g for one hour and the clear red haemoglobin layer was filtered to remove any traces of cell debris. Small quantities of each haemolysate were further diluted either by adding approximately three volumes of distilled water or by making up to 0.1 M in 2-mercaptoethanol, a procedure followed by Nute et al.[7]

Haemolysate samples were subjected to vertical starch gel electrophoresis according to the method of Smithies.[10] Gels were prepared with the Tris-EDTA-borate buffer, pH 8.5 (Manwell, Baker and Betz).[11] Electrophoresis was for seven hours at 4°C, with power of 280 V supplied by a Beckman Duostat. Gels were slit longitudinally; one half was stained for haemoglobin by the standard peroxidative procedures involving H_2O_2 and either benzidine or *o*-dianisidine (Chernoff and Pettit),[12] and the other half was stained with amido black 10B in methanol: distilled water: acetic acid (50:50:10) solution. Some haemolysate samples were subjected to electrophoresis in a 0.1 M formate buffer, pH 1.8 (Muller) for chain separation;[13] individual components of other haemolysates were cut from the Tris-EDTA-borate gels and either incubated with 6 M urea for 30 minutes and transferred into the sample slots of acid gels, or transferred directly to the acid gels. These samples were subjected to electrophoresis at 150 V for seven hours and then stained as described above.

Alkali-resistant haemoglobin was determined by the method of Szakacs.[14] The haemoglobin concentration for each sample was adjusted to 10 g haemoglobin per 100 ml. The modified Wong method (MacFate) was used to determine the iron content in haemoglobin.[15] For each sample, the haemoglobin concentration was:

$$\text{g haemoglobin} = \frac{\text{mg iron/100 ml}}{3.4}$$

All colorimetry was accomplished with a Beckman D. U. Spectrophotometer.

Results

Electrophoresis of freshly prepared haemolysates or aged haemolysates treated with 2-mercaptoethanol produced identical patterns, indicating that the erythrocytes of all of these bushbabies contained two major components (Figs. 1 and 2). We refer to these as the rapidly (R)- and slowly (S)-migrating haemoglobins. In most animals, the S haemoglobin had a slightly greater stain intensity than did the R haemoglobin. These same two-band patterns were found in haemoglobin samples of neonate and juvenile animals between the ages of five days (Fig. 3) and six months (Fig. 2), and for all adult samples. No sexual dimorphism in electrophoretic patterns was detected.

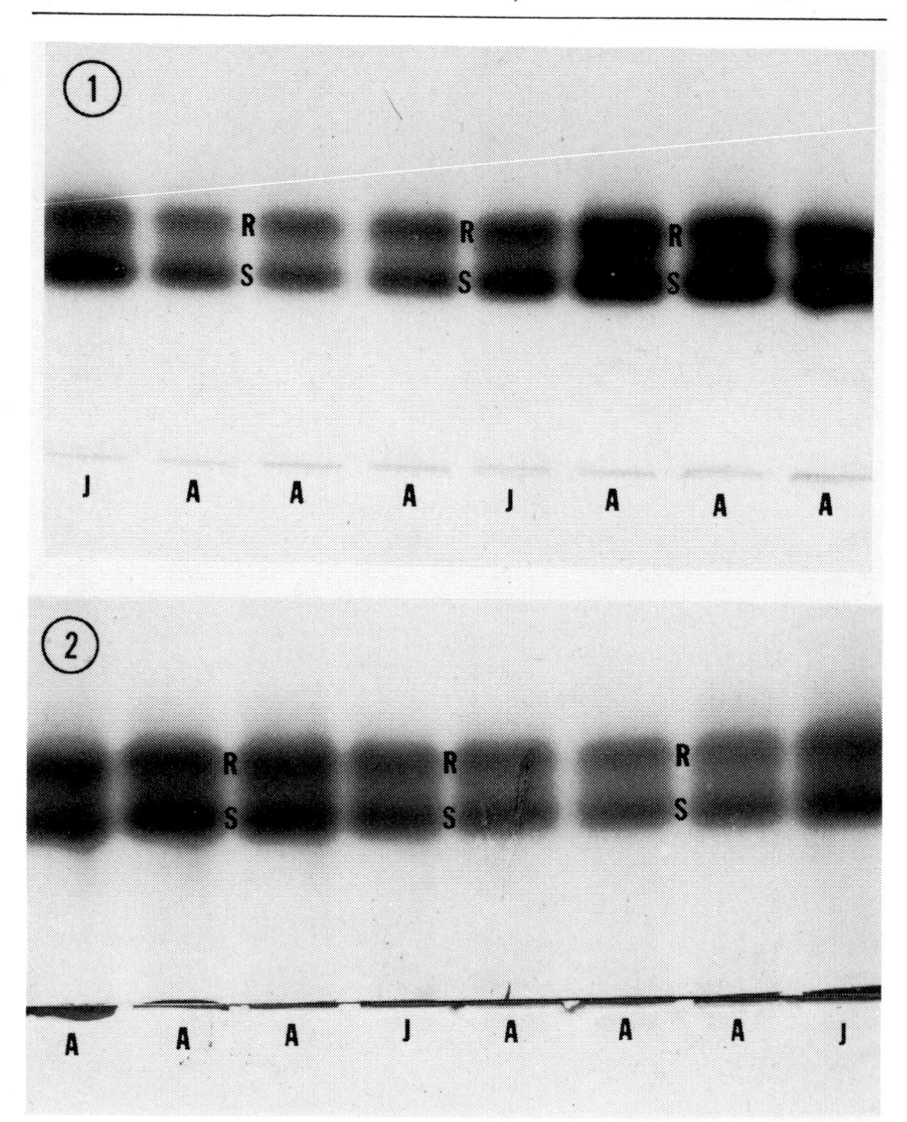

Figure 1. The electropherogram of haemoglobins prepared from fresh haemolysates in juvenile (J) and adult (A) *Galago senegalensis.* Tris-EDTA-borate gel stained with *o*-dianisidine and hydrogen peroxide. Anode is at the top. R is the rapidly-migrating band and S is the slowly-migrating band.

Figure 2. The other half of the electropherogram of *Galago senegalensis* haemoglobins, stained with amido black.

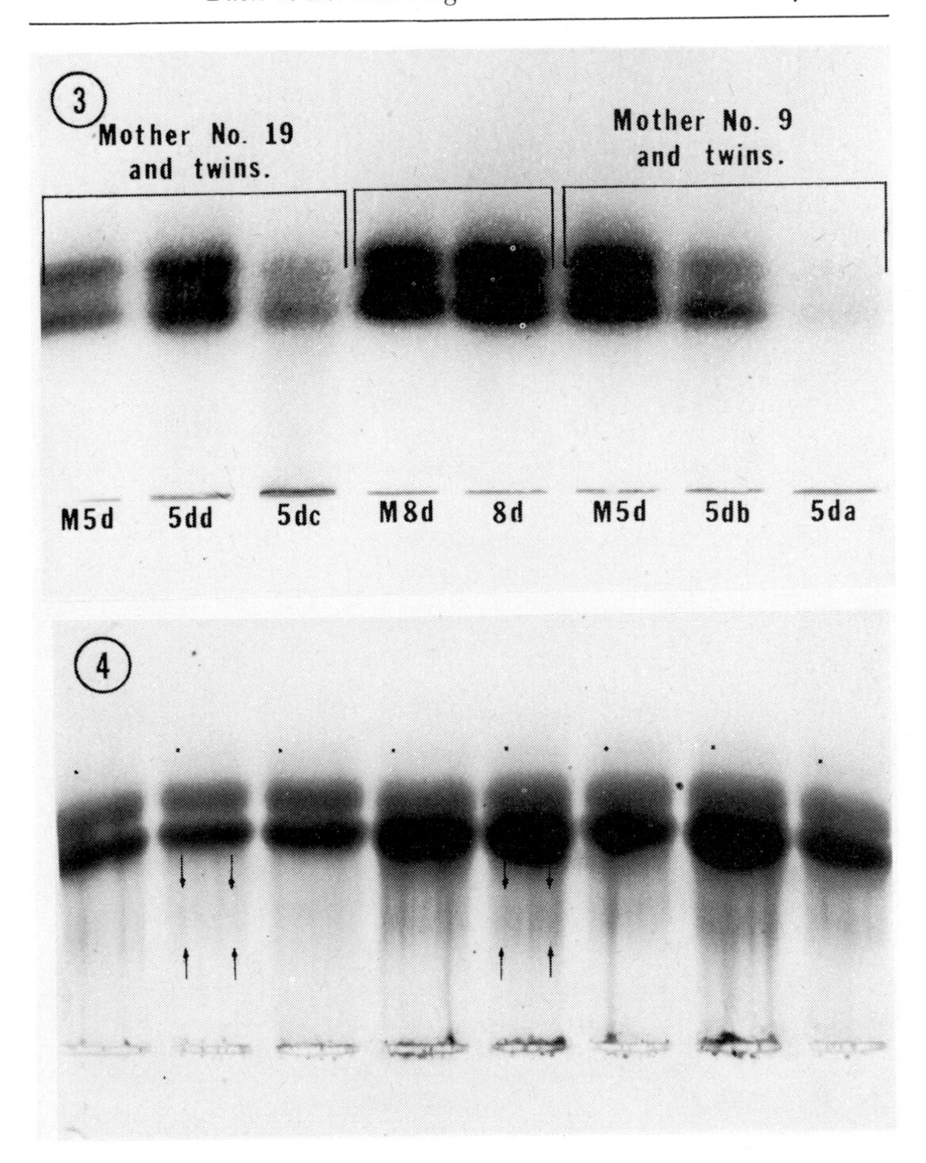

Figure 3. The electropherogram of *Galago senegalensis* haemoglobins prepared from haemolysates of neonates and their respective mothers. Patterns stained with *o*-dianisidine. Abbreviations: 5da and 5db = young of mother 9; 5dc and 5dd = young of mother 19; 8d = young of mother 8.

Figure 4. The electropherogram of *Galago senegalensis* haemoglobins resolved in formate buffer, pH 1.8, 1.2 M. Anode is at the bottom. Arrows denote diffuse zone in amido black patterns. Dots indicate non-haeme protein.

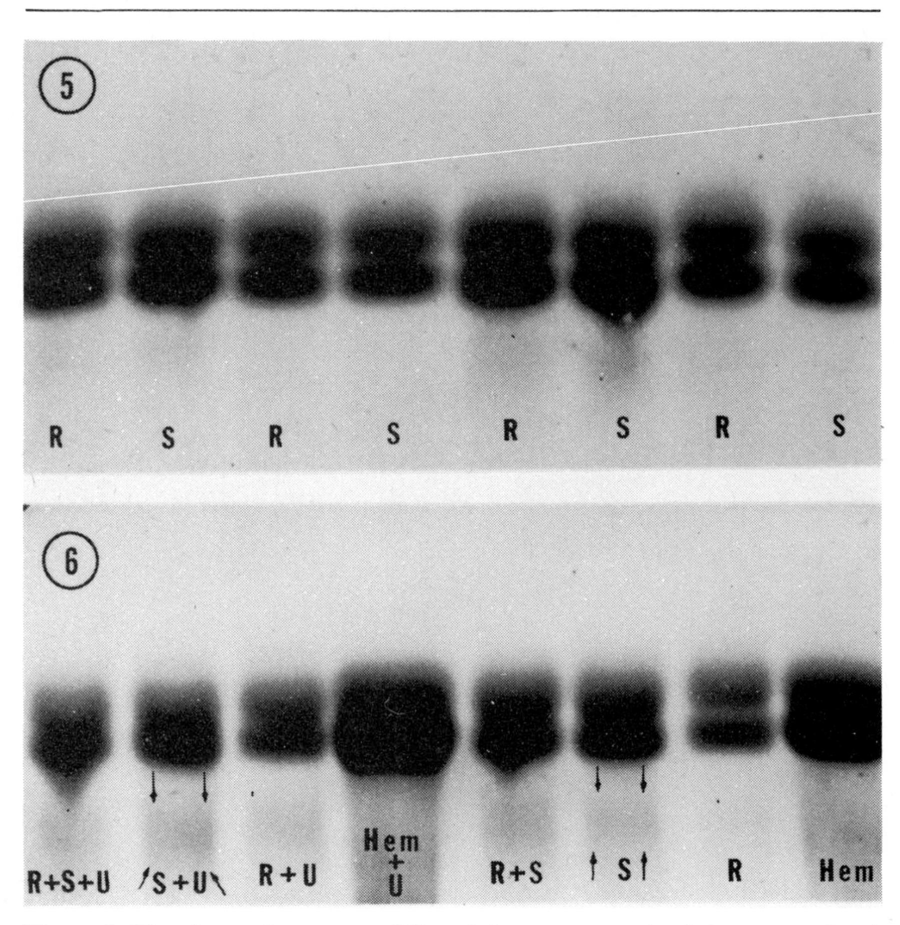

Figure 5. The electropherogram of R and S components in *Galago senegalensis* cut from the starch gels and subjected to electrophoresis in formate buffer. Anode is at the bottom of amido black pattern.

Figure 6. The electropherogram of the R and S components and haemolysates (Hem) subjected to electrophoresis in formate buffer, and the same components incubated with 6 M urea (U) and then subjected to electrophoresis. Anode is at the bottom of amido black pattern.

The formate gel patterns of unfractionated haemolysates treated with 2-mercaptoethanol were characterised by two major components: one anodic satellite band, and a more slowly migrating diffuse zone (Fig. 4). The satellite band proved to be a non-haeme protein, because it failed to stain with benzidine. When the two major components were cut from the starch gels and then subjected to electrophoresis in formate gels, the segments containing the proteins with the mobilities of the R and S haemoglobins resolved during electrophoresis into a four-band pattern of benzidine-positive components (Fig. 5). Each of the R and S haemoglobins separated into two bands. The more anodic band of the R haemoglobin had the same mobility as the more anodic band of the S haemoglobin. The second bands of each haemoglobin were also of similar mobility. Treatment of the haemolysate and the two bands separately or in combination with 6 M urea did not appreciably alter the patterns (Fig. 6).

Tests for the presence of haemoglobin resistant to denaturation by alkali are shown in Table 1. The mean level was 9.3% (range 6.6-13.0) for the 31 animals studied. Females had slightly, but not significantly (for *t*-test, $p>0.05$), higher levels than were found in

Table 1 Alkali-resistant haemoglobin in the lesser bushbaby, ***Galago senegalensis***

(a) MALES

Animal number	*Age*	*Per cent alkali-resistant*
4 Michigan		9.0
30	Up to 4 years in captivity	9.6
32		12.9
41 B		13.0
2 Virginia		7.1
3 A		9.5
36		10.4
38		7.6
42		7.7
46		8.9
Juvenile	6 weeks	6.6
	6 months	7.4
Neonate	8 to 10 days	7.3
		Mean ± S.E. = 9.2 ± 0.7

(b) FEMALES

Animal number	*Age*	*Per cent alkali-resistant*
5		9.4
25		9.8
37		12.0
41		7.3
41 C		10.9
1 Virginia		7.1
3		8.9
5		9.6
39		7.7
45		11.2
47		8.8
51		9.4
53		10.2
55		10.9
57		11.0
59		7.2
63		10.8
Juvenile	6 weeks	10.8
	6 months	11.4
Neonate	8 to 10 days	8.3
		Mean ± S.E. = 9.5 ± 1.2

males: females = 9.5 (7.1-12.0), males = 9.2 (6.6-13.0). Percentages for neonate and juvenile bushbabies were within the range of adult levels. Measurements of alkali-resistant haemoglobin taken at six weeks and at six months from the same juveniles revealed no significant change in the level of pigment. Although the number of young tested was small, family data failed to reveal any appreciable difference between the levels of parents and those of their offspring. For example, animals 41 B and 41 C, born at different intervals to female 41, had higher levels than their mother, but their levels fell close to the higher values found in some other adults.

Discussion

The starch gel patterns from haemolysates in our colony of bushbabies confirm the results previously reported by Nute et al. for the few specimens so far sampled in this species.[7] The reproducibility of the two-band pattern in our animals, the findings by Nute et al. in *G. senegalensis* and in over 100 specimens of *G. crassicaudatus*,[7] and the report by Barnicot and Hewett-Emmett in the present seminar of two major components in all of eight specimens of *G. senegalensis* and 19 specimens of *G. crassicaudatus*,[16] indicates a remarkable homogeneity of haemoglobin phenotype for these species of *Galago* from different colonies. Such homogeneity suggests homozygosity, although it is difficult to generalise from such small samples. The uniformity in patterns for neonate, juvenile, and adult haemoglobins in our animals agrees with the statement made by Hoffman and Gottlieb,[6] who mentioned the similarity in the electrophoretic mobilities at pH 9.1 of haemoglobin from adult *Galago* blood and from foetal *Galago* cord blood.

Our results of unfractionated haemolysates and segments cut from the starch gels and resolved in formate buffer do not refute the evidence by Nute et al. of the three-band pattern for the unfractionated sample of haemolysates in one specimen of *G. senegalensis* and the DEAE-Sephadex fractionated samples of several *G. crassicaudatus* separated in urea gels.[7] However, the separation of the R and S haemoglobins cut from the starch gels and resolved with the formate buffer into four benzidine-positive components gives rise to some problem in interpretation. The presence of a fast-migrating band in the R component with the same electrophoretic mobility as a fast-migrating band in the S component agrees with the concept of alpha-chains with the same mobilities. The presence of a slowly-migrating band in the S component with similar electrophoretic mobility as the slowly-migrating band of the R component, even in samples treated with urea, would not agree with the results of beta-chains with different mobilities. Such a discrepancy could arise by several means; for example, the charges on the beta-chains may be similar enough in the acid system even though the amino acid compositions of the two components are different, or there may be incomplete separation of both components that migrate so close together in alkaline gels, or buffer interaction with the haemoglobin. The presence of the non-haeme protein as well as the diffuse zone in amido black patterns that stains intensely with benzidine, the existence of satellite bands in aged haemolysates, and the findings of a two-band pattern, including a diffuse zone, in urea gels by Nute et al. for the single unfractionated haemolysate in *G. demidovii*, all show the technical problems in handling haemoglobins from these animals.[7] In man, Muller reported a slow band of undetermined nature for both fractionated and unfractionated globins in acid gels that possibly represented agglomeration of single chains.[13] Because most of these investigators stained the haemoglobins only with amido black, it is possible that this diffuse staining in acid gels represents the separated haeme that would be benzidine-positive.

The mean level of 9.3% for alkali-resistant haemoglobin in our animals is slightly higher than the mean level found previously by Buettner-Janusch and Twichell for four specimens of this species.[1] Our discovery of similarities in levels of alkali-resistant haemoglobin for neonates, juveniles, and adults is consistent with the similarities in electrophoretic patterns for these animals of different ages. This similarity in levels corresponds to the results found previously by Buettner-Janusch in the young Ringtail Lemur, *Lemur catta.*[2] In that species, the percentage of alkali-resistant haemoglobin was nearly the same in the four-month old animal as that found in the adult.

Knowledge of the amino acid sequence in *Galago* haemoglobins does not seem to account for the differences found between alkali-resistant haemoglobins from *G. senegalensis* and *G. crassicaudatus.* A partial sequence of both alpha- and beta-chains has been determined for the prosimians *G. crassicaudatus, Lemur fulvus*, and *Tupaia glis* (Hill; see Sober).[17] Both *G. crassicaudatus* and *L. fulvus* have fairly high levels of alkali-resistant haemoglobin, while *T. glis* has a lower level comparable to that found in *G. senegalensis.*[1] Haemoglobins of *G. crassicaudatus* and *T. glis* have a number of residues in common, particularly in the beta-chain; whereas, there are fewer corresponding residues in either chain of *G. crassicaudatus* and *L. fulvus*.

It appears unlikely that this difference in alkali-resistance in the two *Galago* species relates directly to the oxygen-binding affinities of the haemoglobins. Compared with anthropoid haemoglobins, prosimian haemoglobins have lower oxygen affinities in the physiological range, a large acid Bohr effect, and a larger alkaline Bohr effect (Sullivan).[18] Sullivan showed that the haemoglobin oxygen affinity in *G. senegalensis* was similar to that of most lorisiform haemoglobins, that a slight increase and shift of the alkaline Bohr effect occurred at higher pH values, and that a slightly greater increase in the base-level oxygen affinity was found than in *G. crassicaudatus.* The oxygen affinities, but not the alkaline Bohr effects, were disparate between *G. senegalensis* and *T. glis*, while the alkaline Bohr effects were different in the two species of bushbabies. Furthermore, the increased base-level oxygen affinity and decreased alkaline Bohr effect were found in *L. fulvus*, and neither was characteristic in *G. crassicaudatus.*

We found no association between the percentage of alkali-resistant haemoglobin and the stain intensity of electrophoretic bands. Buettner-Janusch and Buettner-Janusch reported that no obvious correlation existed between the level of alkali-resistant pigment in *Galago* and the number of bands on starch gels; they did not identify the alkali-resistant haemoglobin with either band.[3,4] No specific relationship was found between the proportion of alkali-resistant haemoglobin and haemoglobin types in six species of macaques (Ishimoto et al).[19] Dissimilarities in alkali-resistant haemoglobin and the type of placentation in prosimian and anthropoid species have been discussed by Buettner-Janusch and Buettner-Janusch, who hypothesised that the shift from the epitheliochorial placenta in lower mammals to the haemochorial placenta in higher Primates affected the haemoglobin advantageously in the latter species.[4] Interestingly, the mean levels of alkali-resistant haemoglobins in the lesser and greater bushbabies are far apart; but both species possess the diffuse, non-deciduate epitheliochorial type placenta (Gérard; see Hill).[20] A more valid comparison between this blood property and placental structure might be permissible if the level of alkali-resistant haemoglobin were known in *G. demidovii*, which has an endotheliochorial type placenta (Gérard, see Hill),[20] closer to the type characteristic of higher Primates. Thus, the present body of evidence does not appear to be sufficient to disclose whether or not either the presence of alkali-resistant haemoglobin or two major haemoglobin components in *Galago* species have any adaptive significance. Recently, Sullivan[18] has implied a possible difference in function between the two major bands in *G. crassicaudatus*, after study of the haemoglobin oxygen affinity in this species.

Acknowledgments

We thank H.C. Dessauer (Department of Biochemistry, Louisiana State University) E.S. Kline (Department of Biochemistry) and J.I. Townsend (Program in Genetics, Virginia Commonwealth University) for assistance with techniques or criticisms of the manuscript. M.F. Dolwick, G.E. Goode, and D. Wright, Virginia Commonwealth University, helped with the typing or provided useful comments during the course of this research.

NOTES

1 Buettner-Janusch, J. and Twichell, J.B. (1961), 'Alkali-resistant haemoglobins in prosimian primates', *Nature* 192, 669.
2 Buettner-Janusch, J. (1962), 'Biochemical genetics of the primate haemoglobins and transferrins', *Ann. N.Y. Acad. Sci.* 102, 235-48.
3 Buettner-Janusch, J. and Buettner-Janusch, V. (1963), 'Haemoglobins of *Galago crassicaudatus*', *Nature* 197, 1018-19.
4 Buettner-Janusch, J. and Buettner-Janusch, V. (1964), 'Haemoglobins in primates', in Buettner-Janusch, J. (ed.), *Evolutionary and Genetic Biology of Primates*, New York.
5 Stormont, C. and Cooper, R. (1965), 'Haemoglobin and serum proteins in *Galago crassicaudatus panganiensis*', *Fed. Proc.* 24, 532.
6 Hoffman, H.A. and Gottlieb, A.J. (1969), 'Discussion paper: Investigations of nonhuman primate haemoglobin–electrophoretic variation', *Ann. N.Y. Acad. Sci.* 162, 215-238.
7 Nute, P.E., Buettner-Janusch, V. and Buettner-Janusch, J. (1969), 'Genetic and biochemical studies of transferrins and haemoglobins of *Galago*', *Folia primat.* 10, 276-87.
8 Hill, W.C.O. (1955), *Primates, Comparative Anatomy and Taxonomy*, vol. 2: *Haplorhini: Tarsioidea*, Edinburgh.
9 Bush, F.M. and Townsend, J.I. (1971), 'Ontogeny of haemoglobin in the house sparrow', *J. Embryol. Exp. Morph.* 25, 33-45.
10 Smithies, O. (1959), 'An improved procedure for starch-gel electrophoresis: further variations in the serum proteins of normal individuals', *Biochem. J.* 71, 585-7.
11 Manwell, C., Baker, C.M.A. and Betz, T.W. (1966), 'Ontogeny of haemoglobin in the chicken', *J. Embryol. Exp. Morph.* 16, 65-81.
12 Chernoff, A.I., and Pettit, N.M., Jr. (1964), 'Some notes on the starch-gel electrophoresis of haemoglobins', *J. Lab. Clin. Med.* 63, 290-6.
13 Muller, C.J. (1960), 'Separation of the α- and β-chains of globins by starch gel electrophoresis', *Nature* 186, 643-4.
14 Szakacs, J.E. (1964), 'Measurement of alkali-resistant haemoglobin in blood (method of Singer et al.), in Sunderman, F.W. and Sunderman, F.W. Jr., *Haemoglobin: Its Precursors and Metabolites*, Philadelphia.
15 MacFate, R.P. (1964), 'Measurement of haemoglobin in blood. II: Determination of iron concentration in haemoglobin solution', in Sunderman, R.W. and Sunderman, F.W. Jr., *Haemoglobin: Its Precursors and Metabolites*, Philadelphia.
16 Barnicot, N.A. and Hewett-Emmett, D., this volume.
17 Sober, H.A. (1968), *Handbook of Biochemistry*, Cleveland.
18 Sullivan, B. (1971), 'Structure, function, and evolution of primate haemoglobins. II: A survey of the oxygen-binding properties', *Comp. Biochem. Physiol.* 40 B, 359-80.
19 Ishimoto, G., Tanaka, T., Nigi, H., and Prychodko, W. (1970), 'Haemoglobin variation in macaques', *Primates* 11, 229-41.
20 Hill, W.C.O. (1953), *Primates: Comparative Anatomy and Taxonomy*, vol. I: *Strepsirhini*, Edinburgh.

K. BAUER

Comparative analysis of protein determinants in primatological research

Introduction

There are two main ways of establishing relationships between different species. The first is to examine a number of morphological parameters comparatively and to draw conclusions from the resulting data. This approach can be termed the classical anatomical and palaeontological method. The second technique is to investigate the same relationships at the molecular level.

Since the genetic information which is expressed in the phenotype of an organism is coded for in the DNA sequence, this might at first sight appear to be the level most suited for this type of investigation. Unfortunately, technical difficulties have inhibited progress in this field, and it is not possible at present to determine the numbers of mutational changes separating homologous DNA samples isolated from different species.

The translations of structural genes in the form of amino acid sequences have been amenable to experimental analysis since Sanger's pioneering work:[1] most of our knowledge about basic mechanisms of information-transfer and modification (both from generation to generation and inside the cell) originate from work at this level.

Single proteins contain only very small fractions of the information coded in the whole genome, so it is not possible to employ them as adequate representatives. In other words: to draw phylogenetic conclusions regarding a whole species, a panel of various such proteins must be examined. However, actual elucidation of polypeptide sequences is only possible with rather sophisticated and time-consuming methods. Therefore, this approach poses technical problems if it is to be applied to phylogeny. A more rapid, simpler technique for comparing proteins is urgently needed for this field of research.

This paper describes an immunological method based upon a comparative analysis of the determinant patterns reacting in pairs of cross-reacting protein homologues. No isolation procedure is needed for the analysis proper, which uses unfractioned blood plasma or serum. The method permits detection of mutations separating protein homologues from various species.

We have applied this approach to primate phylogeny, deriving a time-scale of primate evolution from data on a minimum of 15 different plasma proteins per species. The resulting data on the ages of the last common ancestors for man and various subhuman primates are consistent with estimations based on classical palaeontological work.

The principle of comparative determinant analysis (CDA)

Genetic information, coded in a structural gene, is translated into an amino acid sequence, which thus contains only information from the structural gene. The information transfer from DNA to protein levels is characterised by partial loss due to the degeneracy of the genetic code. However, all information present in a polypeptide must come from the corresponding cistron. Moreover, any substitution through which two polypeptide chains differ must have its equivalent in an accepted mutation at the DNA level distinguishing the two genes. But it is important to remember that not all mutations established as the nucleotide base level are apparent at the level of amino acid sequence.

With respect to information transfer, there exists a very similar relationship between the polypeptide chain level and the level of antigenic structure of a protein. Of all the amino acids present, only a certain fraction takes part in the formation of the antigenic structure of the molecule. These amino acids must:

1. be situated at the surface of the native molecule, in order to be accessible to the antibody,
2. form part of an antibody-binding site.

The second condition depends on the antigen-producing species (in our experiments: man) in relation to the antibody-producing species (in our experiments: the rabbit). For example, if human transferrin is injected into a rabbit, antibodies will be synthesised against those parts of the human molecule which are different from the rabbit transferrin molecule. The immunisation procedure thus becomes, in a certain sense at least and with certain limitations, a process of comparison between the two (sets of) structural genes (in our example, human and rabbit genes) coding for transferrin. Those parts of the human molecule against which antibodies are produced and with which they are able to react are the *antigenic determinants.*

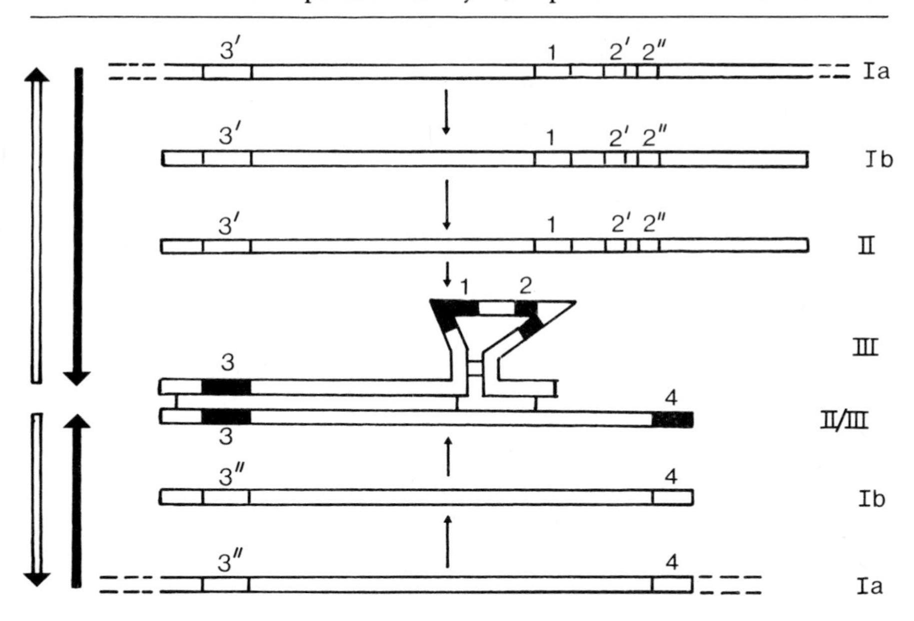

Figure 1. Diagram illustrating the three levels of potential investigation of genetic information; *nucleic acid* (DNA: I*a*; m-RNA: I*b*), *polypeptide chain* (II) and the *antigenically active polypeptide*, either folded (upper chain) or unfolded (lower chain). Immunogenetic comparative determinant analysis (CDA) utilises the third level (III). The flow of information is indicated by the black arrows, whilst the white arrows indicate the direction of the analytical approach. Black areas shown at level III show the determinants: 1 and 2 represent CDs, 3 is a CD coded for on two different structural genes, 4 is an SD. In most cases, no differentiation between the structural genes is possible through CDA, the main exception to the rule being immunoglobulin chains, for which specific antisera are available.

The antigen-antibody-reaction in such a system is therefore one between antigenic determinants, located on the surface of an otherwise inert residual molecule, and a mixture of determinant-specific antibodies; in short between a determinant carrier and a number of antibody populations.

A single determinant consists of a small number of amino acids which either represent a section from a polypeptide chain (*sequence determinant* – SD) or are only brought together by the molecule's conformation (*conformation determinant* – CD). It is apparent that both types of determinant are directly coded for in the corresponding structural gene, and that DNA sequence, polypeptide, and determinant pattern are colinear structures connected by information transfer from the genome. A model of this relationship is given in Fig. 1. Evidently, all determinants are coded in the structural gene itself, while in the actual definition of the single determinant – which takes part in the immunisation process – the genes from the immunised species also take part in the manner described.

If determinant patterns of a pair of homologues are analysed comparatively with the same antiserum, any mutations accepted in those parts of the structural genes corresponding to a determinant will be observable through techniques which allow us to state whether one of them carries more determinants than the other. Appropriate techniques will be described below. By their use for any pair of cross-reacting homologous proteins, one of the following statements can be made (Fig. 2):

1. homologue 1 is identical to homologue 2,
2. homologue 1 carries more determinants than does homologue 2,
3. homologue 1 carries less determinants than does homologue 2.

There are two limitations of such comparative determinant analysis (CDA):

1. Since any one determinant is formed by a *group* of amino acids, one is dealing with larger units in CDA than in sequence analyses, so that the structure examined only partly reflects the information involved.
2. Since only part of the whole genetic information coded in a gene takes part in the formation of the molecule's antigenic structure, only this fraction of the information is actually investigated. The residue, especially that corresponding to internal segments of the protein, is beyond the reach of CDA.

In spite of this loss of information, results obtained by CDA can be employed as an estimation of mutational differences between pairs of homologous proteins, because, among all changes occurring,

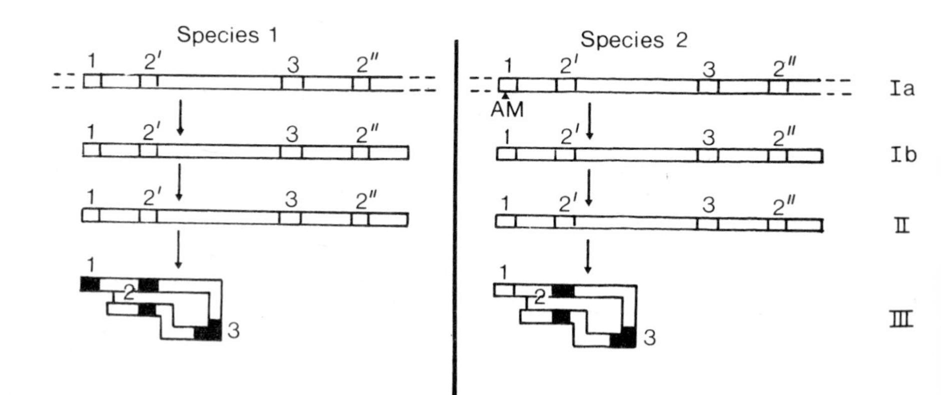

Figure 2. Diagrammatic representation of a CDA comparison between two structural genes (homologues) in species 1 and species 2. The antibody-binding capacity of the determinants in the final, native proteins, shown in the figure at the third level (III) is the property which is analysed. Because of the presence of an accepted mutation (AM) in the gene of species 2, determinant 1 loses its capacity to bind the antibody specific for the original determinant. If an Ouchterlony test of the two homologues is conducted, species 1 will 'spur over' species 2 (see Fig. 3*b*). For key, see Fig. 1.

there is no apparent departure from randomness in the selection of mutational events detected in the determinant structure. Since all species commonly used as antiserum producers (e.g. rabbit, sheep, horse, goat) are related to man via the eutherian mammal common ancestor, no difference would be expected if antiserum producers were exchanged. This expectation is borne out by our finding that antisera (to human proteins) from the species mentioned do not differ systematically in the range of animals with which they react.[2] There is one more possible limitation, which might greatly interfere with CDA and its application to primatological problems. Many proteins – among them the majority of plasma proteins on which our investigation is based – contain prosthethic groups, which are mainly carbohydrate. The whole principle of CDA, however, only applies to semantides, i.e. informational macro-molecules, connected directly with genetic information.[3] It is only for semantides, which include nucleic acids and polypeptides, that the relationship of *direct homology* can exist at all.[4] Episemantides, on the other hand, may be related by *indirect homology*, in cases where the enzymes taking part in their synthesis are homologous. In most cases, it is difficult to decide from the experimental findings alone whether two carbohydrate determinants occurring in two different species are – indirectly – homologous or analogous. This renders the use of episemantides in immunogenetic work very problematic whenever more than one species is involved. A very good example is provided by the classic carbohydrate determinants, ABO-blood group antigens. If experimental data on *these* determinants alone were used, certain intestinal bacteria would have to be considered as close relatives to primates (Table 1). This observation clearly demonstrates that phylogenetic immunogenetic research should be restricted to semantide determinants.

In order to exclude the use of episemantide determinants in our own analysis of proteins we have conducted the following experiments:

1. Denaturation by various agents, the effect of which should be confined to the polypeptide portion of the protein, destroyed the antibody-binding capacity of the plasma proteins in our test panel either totally or to a large degree. There were two exceptions to this rule, one containing only 0.8% and the other 38% carbohydrate (pre-albumin and acid α_1-glycoprotein, respectively).
2. The use of various glycolytic enzymes in no case interfered with antigenicity or destroyed any single determinant used for our study.[5]

While additional experiments, especially on the structure of carbohydrate moieties of plasma proteins, are desirable, these results indicate that the determinants of plasma proteins in our test panel

Table 1 Comparison of bacterial surface antigens and blood group determinants in various species of intestinal bacteria and of primates

The occurrence of these 'blood group' agents, which are carbohydrate in structure, is apparently independent of taxonomic relationships of species. The lack of such relationships is due to the episemantide nature of carbohydrates. Only semantides, such as polypeptides, are suited for this type of phylogenetic work. (Data compiled from ref. 28 – see Notes.)

Species (O-antigen)	*Location of determinant*	*Specificity*
E. coli (86)	O-antigen	B
E. coli (127)	O-antigen	H (O)
E. coli (128)	O-antigen	H (O)
S. poona (13, 22)	O-antigen	H (O)
S. grumpensis (13, 22)	O-antigen	H (O)
S. atlanta (13, 22)	O-antigen	H (O)
S. berkeley (43)	O-antigen	B
Arizona (9)	O-antigen	H (O)
Arizona (21)	O-antigen	B
H. sapiens	Erythrocytes	A, B, H (O)
P. troglodytes	Erythrocytes	A, H (O)
G. gorilla	Erythrocytes	B
P. pygmaeus	Erythrocytes	A, B
Hylobates lar	Erythrocytes	A, B
Papio	Erythrocytes	A, B
Erythrocebus	Erythrocytes	A

(see later: Table 5) are generally semantide in nature (i.e. are amino acid groupings).

It has been demonstrated that it is possible to compare two homologous proteins (viz. their structural genes) to determine the occurrence of established mutations by investigating their determinant structures. Homology can be demonstrated immunologically by the existence of cross-reaction itself, i.e. the reaction of a protein other than the one used for immunisation (the immunogen), if appropriate controls for the specificity of the reaction are conducted. Of course, single proteins comprise too small a fraction of the information present in the genome of higher organisms to permit any meaningful analysis. Buettner-Janusch states in this connection: 'Haemoglobin is . . . nice . . . to work with . . . but I would like to examine 30 proteins before I made a taxonomic decision.'[6] The minimal condition that we have employed was 15 proteins per primate species, and in most cases more proteins were included in the investigation. Such a panel of proteins is taken as a statistical sample of the total genetic information present.

The proteins investigated in our laboratory included only blood proteins (a few other proteins examined by other authors were also considered), and it may be argued that this selection is not random in the statistical sense. On the other hand, there are good reasons for our choice. If intracellular proteins are analysed, complicated isolation procedures must precede the actual investigation. These procedures may well interfere with the conformation of the proteins, so that certain determinants are altered or destroyed and escape detection, in spite of their presence in the native molecule. It is difficult to standardise the isolation procedure for all species examined to the extent that such possible deterioration effects are at least kept equal (assuming that the different homologues are equally sensible to denaturation). The same changes must be expected to a still higher degree if structural proteins, such as keratin or myosin or cell wall proteins, must be solubilised before the investigation. By contrast, plasma proteins comprise a wide variety of proteins as regards physicochemical properties (molecular weight, electrophoretic mobility) and functional activities (enzymatic, antigenic, immunoglobulin, transport, coagulation etc.). In all of these respects, blood proteins appear to be a good choice if a whole spectrum of properties is to be included in such an investigation. We therefore believe that deviation from a random sample of all proteins coded is not very large in the case of blood proteins. It should be mentioned, however, that an ideal genome information sample should include structural as well as other types of genes, and also those DNA sequences coding for classes of RNA other than m-RNA.

If the concept described here is accepted for the estimation of evolutionary rates from the CDA of a panel of plasma proteins, the next step (calculation of other parameters, especially times of divergence, leading to the different lines of descent) becomes feasible with familiar methods established in biochemical phylogenetical studies.[3] Various aspects of this calculation will be discussed below (see section on 'Evaluation').

The technique

From the range of immunological methods available, only those which *directly compare numbers of determinants* can be employed for CDA. This criterion is *not* fulfilled by quantitative techniques which measure the amount of precipitate formed in an antigen-antibody-reaction, or which are based on indirect quantitative determinations, such as the various modifications of complement fixation tests. The reason for this limitation is that determinants are to some degree interchangeable in the reaction, as far as the amount of reaction product is concerned.

The minimum requisite for a precipitation reaction is the presence of two determinants per molecule, regardless of their specificity.[7] If the assumption is made that a protein carries two determinants and the antibodies specific for them are abundant in the reaction medium, a maximum precipitation should occur even without marked participation of additional determinants. Therefore, in all immunological techniques measuring the amount of precipitate, great influence is exerted by one group of parameters which is normally unknown and not easy to estimate: the relative quantities of the various determinant-specific antibodies. Without a knowledge of the composition of the antiserum, which is practically very difficult to obtain, it is impossible to distinguish the various determinants and their participation in the total reaction. Moreover, the binding constants for each determinant-antibody-reaction should be known in order to estimate the composition of the reaction product. Therefore, this type of method is unsuited for elucidation of determinant structures, which are the only features of a protein molecule indicating substitutions at the base level with appropriate immunological methods. This fact is one reason for the failure of attempts to derive a time-scale of primate evolution with techniques not connected with determinant structures.[8] For further critical discussions of these authors' approach and results, we refer to papers by Simons,[9] Read and Lestril,[10] and Bauer.[11]

CDA was especially adapted from standard immunological methods for use in relationship to palaeontological studies, and it was first applied to the primate order.[11] It consists of the following steps:

1. *Comparative Ouchterlony analysis*

The gel diffusion technique developed by Ouchterlony[12] detects differences in determinant patterns of pairs of cross-reacting antigens. The three reactants, the antiserum and the two antigens, are poured into small reservoirs punched into an agarose gel covering a microscopic slide. The reactants are allowed to diffuse into the gel and to react with each other as and where they meet. If the antiserum contains appropriate antibodies, specific for at least two determinants per molecule of the two antigens, a precipitation line develops between each of the sample wells and the central trough containing the antiserum. Where these lines meet, various patterns can develop which are indicative of the relationship of the antigenic structures present in the two samples. Fig. 3 shows the main patterns. In general, we followed the technique described by Clausen;[13] our own modification is summarised below.

(*a*) *Gel*: 2.2 ml of a 1% solution made of agarose (Behringwerke, Marburg, Germany) in a pH 8.6 barbiturate buffer were poured on microscopic slides and allowed to cool. They were kept overnight at 4°C. The trough pattern shown in Fig. 3 was punched into the layer and the superfluous gel removed.

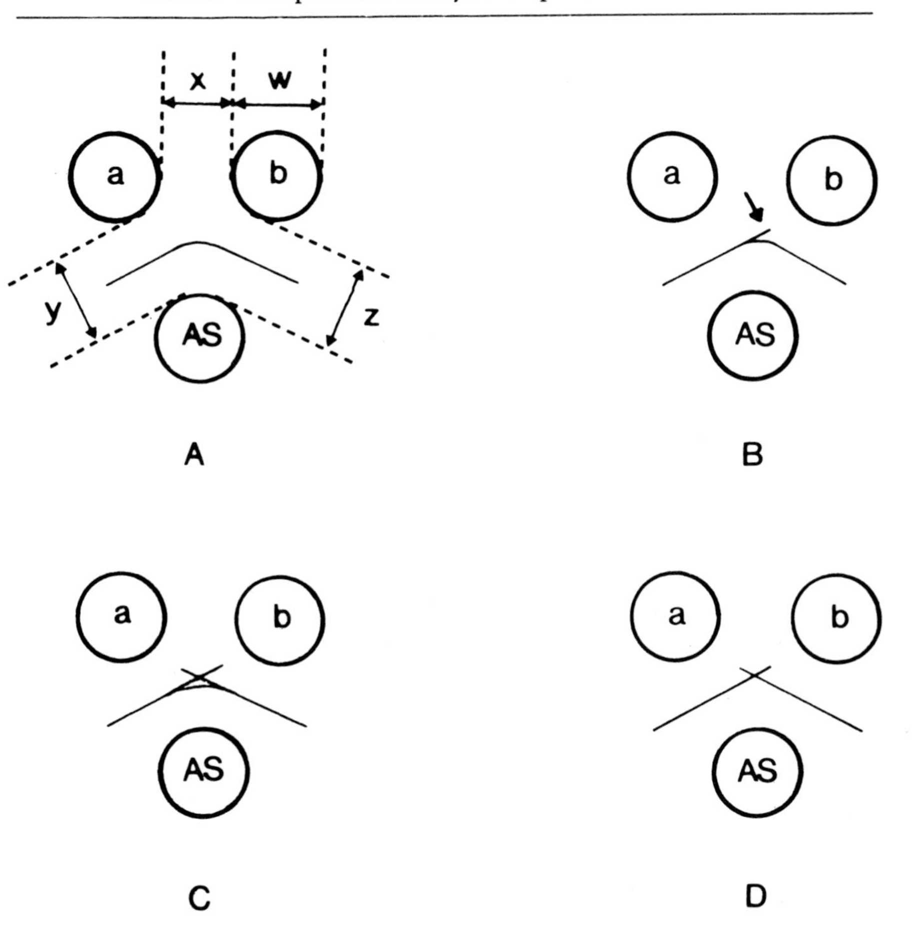

Figure 3. Diagram illustrating various patterns of precipitation lines between paired samples of homologous protein antigens, reacting with an antiserum (AS).

A – identity of determinant patterns.
B – sample *a* contains more determinants than sample *b*.
C – common determinants for *a* and *b* and one determinant, lacking on the other homologue, for both *a* and *b*.
D – *a* and *b* carry no common determinants.
In B, the arrow indicates a 'spur'.

(*b*) *Test*: The troughs were filled with the samples and the antiserum as shown in Fig. 3. Then the slides were kept at 37°C in a moist chamber. Evaluation followed after 1, 2 and 3 days of diffusion, by inspection of the precipitation lines for the occurrence of spurring phenomena (Fig. 3).

In order to apply comparative Ouchterlony analysis to the relationship between different primate species, we have systematically investigated – by use of antisera to the human homologues of some 25 different plasma proteins – pairs of plasmas from the following species:

Homo sapiens	*Erythrocebus patas*
Pan troglodytes	*Cercocebus galeritus*
Gorilla gorilla	*Cebus albifrons*
Pongo pygmaeus	*Galago crassicaudatus*
Macaca mulatta	*Nycticebus coucang*
Cercopithecus aethiops	*Perodicticus potto*

For comparative purposes, the plasmas of *Bos taurus, Ovis aries, Equus caballus, Cavia, Cricetus cricetus, Felis domestica, Lupus canis* were also examined. It must be emphasised that not all plasma proteins included did cross-react with all homologues mentioned. These differences between the various proteins as to their ranges of cross-reaction are observable without exact analyses of determinant structures. They allow a distinction of immunological evolution groups (IEG) of human proteins, which show parallels between ontogenetic and phylogenetic development.[14,15] (Fig. 4) A survey of the IEG system is given in Table 2.[16]

As a typical example of the results obtained by paired CDA, we show our data on the distribution of human fibrinogen determinants among subhuman primates (Table 3).[17] The determinants themselves are elucidated by combining all relations between pairs of homologues a and b of the types a ≡ b, a > b, a < b, or a (X) b (symbol adopted for reaction pattern C of Fig. 3).

The question of controls has assumed great importance since

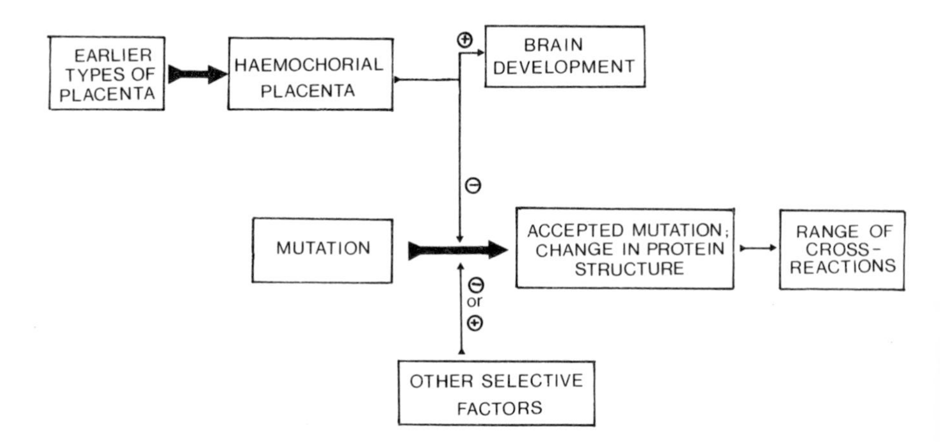

Figure 4. Diagram illustrating the theoretical concept developed by Goodman,[15] on the assumption that the emergence of the haemochorial placenta was essential for brain evolution. As a side effect, the acceptance of immunologically relevant mutations is affected in all proteins which are synthesised during early foetal life. This parallelism between ontogenetic and phylogenetic development has been reaffirmed in our laboratory.[14]

discrepancies were observed between determinant patterns of human α_2-macroglobulin, reported by James,[18] and by Hässig's group in Switzerland.[19] Because of the great interest in the immunogenetic treatment of primatological and other palaeontological problems we have re-examined this protein in our laboratory,[20] and obtained results reaffirming James' data. Both sets of findings are at variance with those of Roulet et al.[19] Apart from the fact that the same determinant formulas were estimated in two different laboratories, we have conducted extensive control experiments. The discrepancy demonstrates in general the importance of running appropriate controls with each and every experiment involving comparative

Table 2 The system of immunological evolution groups (IEG) of human plasma proteins

A stepwise increase in the range of cross-reactions observed is typical for the IEG system.[16]

Groups of species cross-reacting with human plasma protein homologue	*IEG*					
	I	*IIa*	*IIb*	*IIc*	*III*	*IV*
Pongids	+	+	+	+	+	+
Catarrhines	—	+	+	+	+	+
Simians	—	—	+	+	+	+
Primates	—	—	—	+	+	+
Eutheria	—	—	—	—	+	+
Mammalia	—	—	—	—	—	+

Table 3 The results of comparative Ouchterlony tests[17] with antiserum to human fibrinogen

The determinants derived from the reaction patterns are indicated for each pecies. Determinants which are present in non-primates as well as in primates are termed A, B . . .; those restricted to primates are designated P_1, P_2 . . . The guinea pig was selected because it possesses the largest number of human determinants among all the non-primates examined.

Species	*Determinants*	*Reaction of Plasma from*						
		Ho	*Go*	*Po*	*Ma*	*Ce*	*Ga*	*GP*
H. sapiens (Ho)	AB $P_1 P_2 P_3$	≡	≡	≡	≡	>	>	>
G. gorilla (Go)	AB $P_1 P_2 P_3$		≡	≡	≡	>	>	>
P. pygmaeus (Po)	AB $P_1 P_2 P_3$			≡	≡	>	>	>
M. nemestrina (Ma)	AB $P_1 P_2 P_3$				≡	>	>	>
C. albifrons (Ce)	AB $P_1 P_2$					≡	>	>
G. crassicaudatus (Ga)	AB P_1						≡	>
Guinea pig (GP)	AB							≡

KEY: ≡ identical determinant structure (Fig. 3*a*); > Spurring indicating a higher number of determinants on the homologue from the species to the left, compared with the one at the top (Fig. 3*b*).

Ouchterlony analysis, in order to exclude erroneous results. These can be caused by technical errors, evaluation faults and the phenomenon of 'false spurs' (spur formation which occurs in spite of the presence of equal determinant structures if the examination is conducted under conditions of extreme concentration).

To avoid such errors we have used three controls: firstly, only data obtained in three independent experiments were accepted; secondly, absorption studies were run after the Ouchterlony tests; and finally, immunoelectrophoresis was used as a further criterion that the cross-reactions of the subhuman primates were actually due to homologue activity and not to defective antisera. The latter test introduces a non-immunological parameter, electrophoretic mobility, to characterise the homologues.

Only the combination of Ouchterlony analysis and absorption control (and where possible immunoelectrophoresis) is accepted as a CDA.

2. *Absorption control*

We have mentioned that protein antisera are mixtures of determinant-specific antibodies. If one adds to the antiserum a homologue of the immunogen (human protein) carrying less determinants than this, all antibodies which find their respective determinants on the homologue will be bound and precipitated; the supernatant, 'absorbed' antiserum contains the rest of the antibodies and will react only with the corresponding determinants. Absorption with the immunogen (in this study: human plasma) totally destroys the

Table 4 Absorption test of anti-human fibrinogen serum from rabbit used for the Ouchterlony analysis shown in Table 3

The same stepwise decrease of cross-reactions with increasing taxonomic distance from man is apparent. The absorption is conducted exhaustively, i.e. for as long as any residual reaction with plasma from the species used for the absorption is observed. The first and the last line taken together indicate the specificity of the cross-reaction: Antibody populations the same as those effecting the cross-reaction are responsible for the reaction of human plasma.[28]

Antiserum to human homologue of	*Absorbed exhaustively with plasma of*	*Reaction with plasma from*						
		Ho	*Go*	*Po*	*Ma*	*Ce*	*Ga*	*GP*
fibrinogen	*H. sapiens* (Ho)	—	—	—	—	—	—	—
fibrinogen	*G. gorilla* (Go)	—	—	—	—	—	—	—
fibrinogen	*P. pygmaeus* (Po)	—	—	—	—	—	—	—
fibrinogen	*N. nemestrina* (Ma)	—	—	—	—	—	—	—
fibrinogen	*C. albifrons* (Ce)	+	+	+	+	—	—	—
fibrinogen	*G. crassicaudatus* (Ga)	+	+	+	+	+	—	—
fibrinogen	Guinea pig (GP)	+	+	+	+	+	+	—
fibrinogen	saline control	+	+	+	+	+	+	+

ability of the antiserum to react with any homologue, including the human, all residual reactions (if any) being unspecific in nature. Table 4 provides an example for such an absorption test.

3. *Immunoelectrophoresis*

Because of a limited supply of several subhuman primate plasmas and as a result of the somewhat lower sensitivity of this method, this test was not always conducted. Using the technical procedure described elsewhere,[13] in all cases examined electrophoretic mobilities were identified in the same range for the human and subhuman primate samples.

Evaluation

Determinant analysis of single proteins as described in the preceding paragraphs has several implications. Firstly, one can estimate the minimal number of determinants with different specificity on the human protein. At the same time, a certain characterisation and definition of these determinants is achieved (e.g. a certain determinant is always restricted to certain homologues and is missing in all other species). Secondly, the distribution of human determinants among proteins of other primates (and non-primates) is examined.[21] In the present connection, the third implication is the most important; namely, that it is possible to estimate evolutionary rates and subsequently to calculate the age of Latest Common Ancestors (LCAs).

Every spur in CDA (Fig. 3*B*) corresponds to at least one accepted mutation separating the two species compared. This is a direct consequence of the colinearity of antigenic structure and the respective structural gene. Those substitutions observable by immunological means are evidently a random sample of all substitutions occurring. It is thus possible to conclude from the resulting data, which have been compiled in Table 5, that differences in determinant structures tend to increase with taxonomic distance between the species involved. However, it is a matter for discussion whether, and to what extent, the number of mutations fixed is a function of time elapsed. Yet it is a precondition for the use of CDA for the estimation of numerical relationships among various species that such a function exists. The main argument against such a dependence upon time elapsed is that the acceptance of mutations is influenced by natural selection. The selective forces must have varied almost unceasingly during evolution, so that no continuous acceptance or acceptance rate would have resulted. This problem was faced by biochemical geneticists long before immunogeniticists, and the latter can therefore happily make use of biochemical arguments for

Table 5 Compilation of the determinants present on primate molecules of plasma proteins, detected by antisera to human homologues. (In addition to our own results, data from the literature have been employed. An earlier paper[11] provides references, while our experimental findings on the chimpanzee will be published elsewhere.)[29] Only P-determinants are considered.

	Gm, γ_3-chain	Plasminogen	Gm, γ_1-chain	β-lipoproteins	μ-chain (IgM)	β_2-glycoprotein I	Fibrinogen	Ceruloplasmin	α_1-lipoproteins	C4	C5	Inv-determ. (κ-chain)	Xh-factor	Myoglobin	Prealbumin	α-chain (IgA)	Transferrin	α_1-trypsin inhibitor
Homo sapiens	9	6	5	4	4	3	3	3	3	3	3	3	3	3	2	2	2	2
Gorilla gorilla	4	5	2	4	4	2	3	2	3	3		0	3	3	2	2		
Pongo pygmaeus	4	4	1	4	3	3	3	3	3	3	2	0		3	2			
Macaca mulatta	3	3	0	3	2	2	3	3		2	1	0	1	2	1	1	1	1
Cercopithecus aethiops	0	3	0	3	2	2	3	3		2	1	0		2	1		1	1
Erythrocebus patas	3	3	0	3	2	2	3	3	3	2	1	0			1		1	1
Cercocebus galeritus		3		3	2	2	3	3		2	1	0			1		1	1
Cebus albifrons	0	2	0	2	2	2	2		2	1		0				0		
Galago crassicaudatus	0	1	0	1	1	1	1		1	1		0				0		0
Nycticebus coucang	0		0			1						0						0
Perodicticus potto	0		0		1					1		0				0		0
Non-primate eutherians	0	0	0	0	0	0	0	0	0	0	0	0	0	0	0	0	0	0

	γ-chain (IgG)	C_1-esterase inhibitor	Haptoglobin	β_2-glycoprotein II	β_2-glycoprotein III	α_2-macroglobulin	Albumin	Lp (a) (β-lipoproteins)	Inter-α-Trypsin inhibitor	Cholinesterase	δ-chain (IgD)	Acid α_1-glycoprotein	α_{2HS}-glycoprotein	Haemopexin	Catalase (E.C.1.11.1.6)	GOT (E.C. 2.6.1.1)	Gc-globulin	Isf (1) (γ-chain)	Lt (a) (β-lipoproteins)
Homo sapiens	2	2	2	2	2	2	2	2	1	1	1	1	1	1	1	1	1	1	1
Gorilla gorilla				2	2				1	1	1						1	0	1
Pongo pygmaeus								2	1		1						1	0	
Macaca mulatta	1	1	1	1	1	1	1	1	1	1	1	1	1	1	1	1		0	
Cercopithecus aethiops	1		1	1	1	1						1	1	1	1	1		0	
Erythrocebus patas	1		1	1	1	1						1	1	1				0	
Cercocebus galeritus	1		1	1	1	1						1	1	1					
Cebus albifrons				0	0				0	0	0								
Galago crassicaudatus			0	0	0				0	0	0	0	0	0					
Nycticebus coucang			0										0	0					
Perodicticus potto													0						
Non-primate eutherians	0	0	0	0	0	0	0	0	0	0	0	0	0	0	0	0	0	0	0

an evolutionary rate which is constant over long periods of time.

1. It is possible to estimate evolutionary rates for two of the proteins (haemoglobins and cytochromes c) most thoroughly examined by sequence analyses, as well as for other proteins. It has been observed that the number of substitutions between homologues from pairs of taxonomic groups (e.g. mammals and birds) are more or less of the same magnitude. From findings of this kind, it is possible to calculate either 'unit evolutionary periods' – times for one substitution to occur – or changes per 100 residues and per 100 million years by simple comparison with palaeontologically estimated times of divergence among major species groups.[3,22,23]
2. New observations appear to indicate that neutral mutations may have played a more important role during evolution than would be expected as a result of purely selective forces,[24] and that the amino acid composition of an average vertebrate protein more or less reflects the genetic code.[25]

The calculation of the ages of Latest Common Ancestors for man and several other primates from CDA data. Our evaluation is based on the same procedure as that described for biochemical methods. We have used those determinants restricted to primate homologues and their distribution in the primate order, assuming a direct linear relation between time elapsed and numbers of immunologically detected mutations. The formula used was

$$T_i(RTU) = \frac{100}{n} \cdot \sum_{j=1}^{j=n} \frac{f_{ji}}{f_j\, HS} \qquad \text{– I}$$

where T_i (RTU) is the time since the eutherian common ancestor (T_i = O) to the LCA of man and species i in relative time units (RTU), defined as T_i = 100 (RTU) for man. n is the number of proteins included, and f_{ji} and $f_j\, HS$ are the numbers of (human) determinant present on the i and the human homologue of protein j. The relative immunological time scale (*R*ITS) of primate evolution, measuring time in RTU, can be transformed into an absolute one (*A*ITS) by taking

$$100\ RTU = x \text{ million years} \qquad \text{– II}$$

Since x (which is equivalent to the age of the eutherian common ancestor) cannot be determined exactly by classical means, uncertainties not inherent in the immunogenetical method of CDA must enter into an absolute scale as compared with the relative scale. Therefore we prefer the latter in cases where only the molecular relationships are concerned. If absolute values are needed, equation I can be transformed accordingly. With x = 70 million years:

$$T_i(MY) = \frac{70}{100} T_i\,(RTU) = \frac{70}{n} \sum_{j=1}^{j=n} \frac{f_{ji}}{f_j\,HS} \qquad \text{– Ia}$$

It should be mentioned that Zuckerkandl and Pauling propose an exponential function to be used for the same purpose (using biochemical data) in order to take account of multiple substitutions at the same site.[3] We believe however, that a linear relationship is appropriate for periods of time as 'short' (by phylogenetic standards) as that involved in primate evolution.

Results

The results obtained with equations I and Ia from CDA results are compiled in Table 6. All scale values prove to follow the classically established traditional pattern as to the successive divergence of prosimians, New World monkeys, Old World monkeys and the pongids from the line leading to man. In the same Table, classically determined values for times of divergence are given, and these are in the same range as are our CDA data. This concordance of results derived by two methods as different as the morphological and the molecular immunogenetic approaches further reaffirms the theoretical concepts discussed above.

Table 6 The T_i values for the primate species included in our time scale of primate evolution[11,29] compared with data derived from morphological criteria

Species	*Immunogenetically determined* RITS in RTU	AITS (in 10^6	age years)	*Classically derived data after* Pilbeam[26]	Simons[9]
H. sapiens	100.0	70.0	0.0	–	–
P. troglodytes	86.2	60.3	9.7 }		
G. gorilla	82.6	57.9	12.1 }	15	14-20
P. pygmaeus	77.5	54.3	15.7 }	(10-20)	
Old World monkeys				40	30
(mean of 4 species)	59.6	40.4	28.3	(30-50)	(27-33)
New World monkeys				55	50-55
(*Cebus albifrons*)	22.9	16.1	54.9	(45-60)	
Prosimians					
(*Galago crassicaudatus*)	9.5	6.7	63.3	60-80	70
LCA of eutherians	0.0	0.0	70.0	–	–

1. *Prosimians*

Three species were included in our study: *G. crassicaudatus, N. coucang* and *P. potto*, all three belonging to the Lorisiformes. Only the first-mentioned species was investigated according to our criterion of 15 different proteins per species, because of the limited supply of plasma available in the other two cases. However, in all cases where *N. coucang* or *P. potto* homologues were actually investigated, they proved to be identical to *Galago* proteins in their antigenic structure. Therefore, it may be justifiable to estimate the same value of 6.7 million years after the origin of the scale as the time of divergence of all three prosimians (*Nycticebus, Potto* and *Galago*) from the line of descent leading to higher primates.

The *Galago* time of divergence is the oldest observed by us in the primate order. On the other hand, all proteins examined in this species shared more determinants with man than did any non-primate. In this connection, investigation of insectivores and of other groups of prosimians regarded as diverging very close to the origin might provide interesting conclusions concerning the positions of prosimians relative both to other primates and to other eutherians.

2. *New World monkeys*

Only one species, *Cebus albifrons*, was investigated. In all cases, determinants present on *Cebus* homologues were also observed in all higher primate protein molecules, while the *Cebus* molecules carried at least the same number of determinants as the *Galago* proteins, and usually exhibited more.

3. *Old World monkeys*

Four species were included in the scale. They were in all cases found to share identical antigenic structures in their molecules.* Several more Old World monkeys (OWM) investigated showed the same behaviour, including *M. nemestrina* and *P. hamadryas.* OWM represent the second group of primates which is remarkably homogeneous as to the 'human' determinants their members possess. This finding is quite contrary to the pongid determinant patterns (see below) and this at first recalls Landsteiner's idea of *serological perspective.*[27] This 'perspective' analogy was employed to explain the observation that antisera from closely related species detected small antigenic differences which were otherwise not observable between homologues of other closely related species. However, on the basis of modern molecular immunogenetics,[28] we have been able to offer an alternative concept to this 'perspective effect', introduced by

* One of the values indicated in the literature (Gm factors of γ_3-chains) provides an exception to this rule, however, as has been confirmed in our own experiments.

classical serologists into early theory. With respect to the appearance of both prosimians and OWM as homogenous groups in terms of their patterns of 'human' determinants, we propose the following explanation: all antisera employed were raised in the rabbit, which recognises those surface areas of introduced protein molecules which are sufficiently substituted as determinants† in relation to its own homologue. Evidently, these determinants are not recognised as 'human' but as 'non-self'. On the other hand, the existence of nodal sequences for proteins of a common evolutionary group should have a parallel in antigenic structures. Therefore we suggest that determinants which are present in all primates, including prosimians, are actually *common primate determinants*, those found in all primates except prosimians are *common simian determinants*, those observed in all higher primates are *common catarrhine determinants.*

4. Pongids

With few exceptions, all pongids included in this study carried either more or an equal number of 'human' determinants on their plasma proteins than do OWM homologues. A general regularity observed, in addition, was that the orang-utan possesses less or an equal number of determinants of this group than does the gorilla, which in turn bears the same relationship to the chimpanzee. This pattern allows us to distinguish three more subgroups of 'human' determinants and – if any – species-specific human determinants proper. The number of the latter group was astonishingly small in comparison with structures common with pongids, especially the chimpanzee. The consequent implications for the evolution of pongids and man will be discussed in greater detail elsewhere.[29]

Conclusions and future aspects

The foregoing description of concepts and results of CDA in the study of primate evolution can be summarised as follows: it is possible by immunological techniques to measure mutations established during phylogeny and to derive for the LCAs of man and major subhuman primate groups time relations which are consistent with classical palaeontological evidence.[11] Of the three groups in which several species were investigated, prosimians and OWM are remarkably homogenous, while pongids show stepwise approximation to human antigenic structures from orang to gorilla to chimpanzee. This observation is explained by a distinction of common determinant groups (CDG), common to primates (present as the only determinants on prosimian molecules), common to

† The antiserum to C_1-esterase inhibitors was raised in the goat.

simians (on New World monkey proteins) and common to catarrhines (present in addition to other, primate and simian, determinants on OWM homologues). Similar determinant groups can be distinguished in pongids. The relationships thus established for the primate order by the immunogenetical approach are indicated in Fig. 5. It is apparent that much additional research is needed before results of comparable volume can be used to establish phylogenetic relations among primates on the basis of sequence analyses. It would therefore seem promising to investigate further primate species by CDA. Key species which were omitted from the present study because of lack of plasma samples are the gibbon (a more primitive pongid) and prosimians such as *Tarsius* or a representative of the Tupaiiformes. Study of these forms would yield additional information on the relationships of these primates, with their special branching positions at the root of the whole order.

A major advantage of CDA is that the same technique and the same antiserum can be applied to problems like this. Difficulties arise, however, whenever antisera to proteins from other species than man are employed. If our above-mentioned criteria are applied, 15

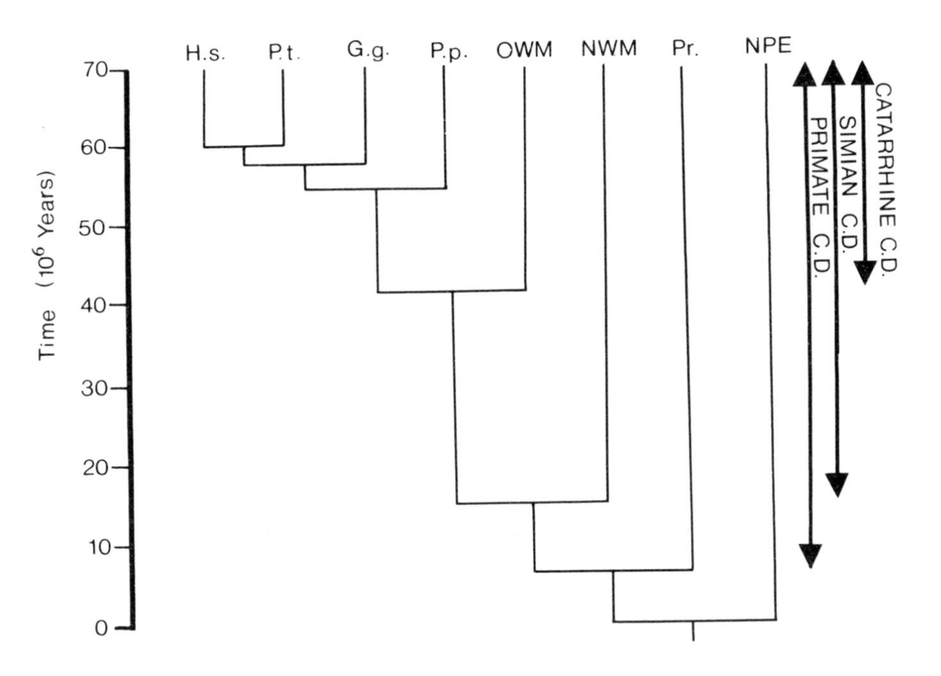

Figure 5. An evolutionary pedigree (AITS) derived from the CDA data given in Table 6. The arrows at the right indicate the ranges over which common primate, simian and catarrhine determinants are found. Key: H.s. = *Homo sapiens*; G.g. = *Gorilla gorilla*; P.p. = *Pongo pygmaeus*; OWM = Old World Monkeys; NWM = New World Monkeys (*Cebus albifrons*); Pr. = Prosimians; NPE = non-primate eutherian mammals. Also included are our results on the chimpanzee (P.t. = *Pan troglodytes*).[29]

different proteins of, say, *Galago* must be isolated and purified, antisera must be prepared and absorbed properly (since 'pure' proteins are rarely a reality), *before* the actual analysis can start. However, such antisera might eventually be employed to provide control over our present results. Moreover, they may be expected to distinguish more readily between the closer relatives of the species producing the immunogen. In the case of anti-human sera, differences between proteins from man and the various pongids become particularly apparent; accordingly a panel of anti-*Galago* sera might be used to investigate relations between various prosimian species.

In this connection, one further aspect should be briefly mentioned, though it is outside the subject of the present symposium, and that is the use of CDA outside the primate order. Human plasma proteins do not cross-react to a sufficient degree to permit phylogenetic conclusions to be drawn outside the primate order. But here again other antigen-producers might be used. Another possibility would be to employ as antigens proteins which show a wider range of cross-reactions than do plasma proteins. Possible candidates for such proteins are the eye-lens proteins investigated by Manski.[30] Interesting results for the new field of immunogenetics which has been created by CDA of homologous proteins may be expected as soon as the various components of this protein group are better characterized; at present it is already established that the range of cross-reactions for eye-lens proteins comprises all vertebrates.

The medical use of antisera to a wide variety of human plasma proteins has, however, led to a preponderance of investigations exploiting the practical availability of anti-human sera. One more application which deserves mention, as it may prove valuable in primate research, is the use of antisera to human proteins for the quantitative determination of a subhuman primate homologue. If, for example, human albumin cross-reacts strongly with its *Galago* homologue, it is also possible to determine the latter quantitatively in the blood of the animal, by making use of the anti-human serum or even of commercially available sets of Mancini plates which have been developed and are used in clinical diagnostics.[31]

In conclusion, it would seem that the construction of two time-scales of primate evolution consistent with fossil evidence, and the selection of likely further applications of cross-reactions (particularly through CDA) among proteins, adequately demonstrate that the approach described here should prove to be a powerful tool in primatological research.

Acknowledgments

The experiments described in this paper have been sponsored as part of a study of the determinant structures of human plasma proteins supported by the Deutsche Forschungsgemeinschaft, Bonn-Bad Godesberg, Germany. The skilful technical assistance of U. Reinle and R. Rauschnabel is gratefully acknowledged.

NOTES

1 Sanger, F. (1965), in Green, D.E. (ed.), *Currents in Biochemical Research*, New York.

2 Bauer, K. (1974), 'Cross-reactions between human and animal plasma proteins. V: The role of the antiserum and immunological specificity in relation to the phenomenon of cross-reaction', *Humangenetik*, 21, 179.

3 Zuckerkandl, E. and Pauling, L. (1965), 'Evolutionary divergence and convergence in proteins', in Bryson, V. and Vogel, H.J. (eds.), *Evolving Genes and Proteins*, New York.

4 Florkin, M. (1966), *A molecular approach to Phylogeny*, Amsterdam.

5 Bauer, K. and Rehberger, D. (1970), unpublished results.

6 Buettner-Janusch, J. (1965), comment in the discussion of part III of Bryson, V. and Vogel, H.J. (eds.), *Evolving Genes and Proteins*, New York.

7 Pauling, L., Pressman, D. and Campbell, D.H. (1944), 'The serological precipitation of simple substances. VI: The precipitation of a mixture of two specific antisera by a dihaptenic substance containing the two corresponding haptenic groups. Evidence for the framework theory of serological precipitation', *J. Am. Chem. Soc.* 66, 330.

8 Sarich, V.M. and Wilson, A.C. (1967), 'Immunological time scale of hominid evolution', *Science* 158, 1200.

9 Simons, E.L. (1969), 'The origins and radiation of the primates', *Ann. N.Y. Acad. Sci.* 167, 319.

10 Read, D.W. and Lestril, P.E. (1970), 'Hominid phylogeny and immunology: a critical appraisal', *Science* 168, 578.

11 Bauer, K. (1970), 'An immunological time scale for primate evolution consistent with fossil evidence', *Humangenetik* 10, 344.

12 Ouchterlony, O. (1962), 'Diffusion-in-gel methods for immunological analysis – II', *Progr. Allergy* 6, 30.

13 Clausen, J. (1969), *Immunochemical Techniques for the Identification of Estimation of Macromolecules*, Amsterdam and London.

14 Bauer, K. (1970), 'Cross-reactions between human and animal plasma proteins. II: The influence of structural and functional factors', *Humangenetik* 10, 1.

15 Goodman, M. (1964), 'The specificity of proteins and the process of primate evolution', *Protides of the Biol. Fluids* 1964, 70.

16 Bauer, K. (1971), 'Cross-reactions between human and animal plasma proteins, IV: Non-eutherian Mammalia', *Humangenetik* 13, 49.

17 Bauer, K. (1970), 'Investigations on the antigenic heterogeneity of human fibrinogen and plasminogen', *Klin. Wschr.* 48, 443.

18 James, K. (1965), 'A study of the alpha$_2$-macroglobulin homologues of various species', *Immunology* 8, 55.

19 Roulet, D.L.A., Gugler, E., Rosin, S., Renaud, N.M. and Hässig, A. (1960), 'Untersuchungen über die Determinantenstruktur menschlicher Serumproteine', *Vox Sang.* 5, 479.

20 Bauer, K. (1970), 'The antigenic determinants of several human plasma proteins: their determination from the investigation of the cross-reactions between human and other mammalian plasmas', *Intern. J. Prot. Res.* 2, 137.

21 Bauer, K. (1969), 'The occurrence of antigenic determinants of human blood proteins in mammalian plasmas', *Humangenetik* 8, 27.

22 Margoliash, E., Fitch, W.M. and Dickerson, R.E. (1971), 'Molecular expression of evolutionary phenomena in the primary and tertiary structures of cytochrome c', in Schoffeniels, E. (ed.), *Biochemical Evolution and the Origin of Life*, Amsterdam and London.

23 Dayhoff, M.O. (1969), *Atlas of Protein Sequence and Structure*, Silver Spring: Nat. Biomed. Res. Found.

24 Kimura, M. and Ohta, T. (1971), 'Protein polymorphism as a phase of molecular evolution', *Nature* 229, 467.

25 King, J.L. and Jukes, T.H. (1969), 'Non-Darwinian evolution', *Science* 164, 788.

26 Pilbeam, D. (1970), personal communication.

27 Landsteiner, K. (1947), 'The specificity of serological reactions', Cambridge, Mass.

28 Bauer, K. (1973), *Einführung in die Immungenetik*, Stuttgart.

29 Bauer, K. (1973), 'Age determination by immunological techniques of the last common ancestor of man and chimpanzee', *Humangenetik* 17, 253.

30 Manski, W. (1959), 'On the biological basis of organ-specific cross-reactions of tissue antigens', *Intern. Convoc. Immunol.*, Buffalo, N.Y., Basel.

31 Bauer, K. (1974), 'Cross-reactions between human animal plasma proteins VI: An assay method for ape and monkey plasma proteins using antihuman antisera', *Humangenetik* 21, 273.

C. N. COOK and D. HEWETT-EMMETT

The uses of protein sequence data in systematics

Introduction

Amino acid sequences, besides defining the structure and hence function of proteins, could provide an extremely large pool of taxonomic characters of equal weight. As such, they could be used to refine our current inferred phylogenetic trees to a greater extent than any other single type of character, except the base sequence of the DNA itself. The uniformity of these characters, derived from the fact that each amino acid is coded by three nucleotide units of chromosomal DNA, should eliminate the need for subjective weighting (or equally subjective lack of weighting) associated with classical character sources. We need: (1) to discuss the validity of the above postulates; (2) to ensure that the maximum amount of phylogenetic information is extracted from the data; (3) to examine the relevance of the methods to systematics as a whole. The rationale behind the use of characters at the molecular rather than the organismal level has already been adequately argued by Zuckerkandl and Pauling.[1]

Before considering the methods currently in use, it is desirable to discuss the relationship between phylogenetic trees and topologies. We follow Eck and Dayhoff and define a topology as a branched structure which can represent the relationships between the N taxa under consideration.[2,3] Proposed evolutionary events can be marked along the branches; but a topology has no time direction. A topology can acquire a time direction and be converted into a phylogenetic tree by folding it about one of its (2N-3) branches, such that all the branches progress forward in time from the point of folding, which will represent the common ancestor of all the taxa involved. It is arguable whether a phylogenetic tree, so produced, can have a time scale, although certain points on the tree might be fixed by use of dated fossils. To apply a time scale to a phylogenetic tree based on protein sequence data would require the anchoring in time of some of the ancestor nodes by use of fossils of a reliable nature, and knowledge either of the constancy of the rate of mutation of the protein or of those sites in the protein sequence which are free to change randomly because they are not subject to selection.

The problem of deciding where the topology should be folded is a common one in classical taxonomy and usually manifests itself as argument about the primary sub-division of an order, for example whether the Carnivora should be divided into Pinnipedia and Fissipedia or whether a division into Feloidea and Canoidea-Pinnipedia is more appropriate. Closer to home, there is still discussion about whether the Primates are best split into Prosimii and Simii or Strepsirhini and Haplorhini, i.e. whether the tarsiers should be included with the lemur-loris group or with the monkey group (Fig. 1). These are folding problems and not topological questions.

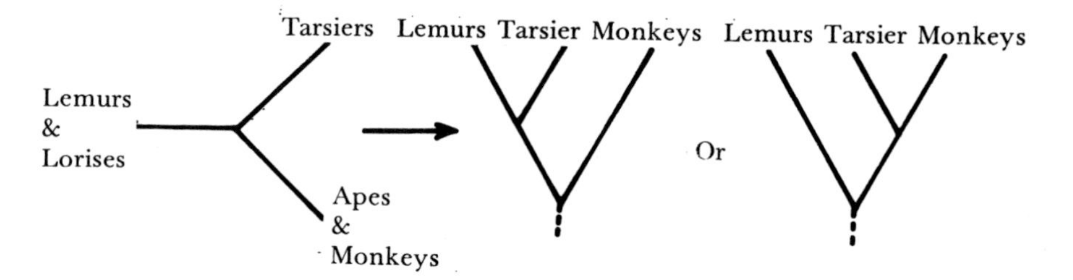

Figure 1. Derivation of alternative phylogenetic trees from a single topology.

The ancestral node of all the taxa being compared (i.e. the folding point of the topology) is not strictly determinable, but two approaches can give an approximation. The first invokes the somewhat dubious principle of non-reversibility of evolution, which might hold if a large number of complex characters were involved. This method cannot be used for protein sequences without bringing in functional considerations and, to date, we have very little data to associate function with the variable parts of sequences (apart from the physiological studies of Sullivan, and the crystallographic comparisons of human variant haemoglobins undertaken by Greer and Perutz).[5,6] In fact, some would say that these variable sites have no functional significance and are therefore not subject to natural selection (e.g. Kimura; King and Jukes).[7,8] The second method attempts to find the mid-point of the topology, such that the average branch lengths from this point are equal. This method is well suited to uniform characters such as sequence data; but it involves the assumption that the rate of evolution is more or less constant. If there is a radiation of branches near this point, its localisation could be suspect; but this problem may be side-stepped by the addition of an extra distantly related taxon, for instance to compare a group of primates with a clear non-primate such as a bird. In this way, a distinct ancestral primate node could be computed and the topology could be safely folded along the bird-ancestral primate branch using both of the above criteria.

The number of dichotomous topologies (T) which can be fitted to N taxa (N⩾ 3) conforms to a product series, viz:

$$T = \prod_{x=0}^{N-3} 2(N-x) - 5$$

T therefore rises very rapidly as the number of taxa is increased. As examples, 10 taxa can be arranged in 2,027,025 topologies, while the 51 recognised 'prosimian' species[9] can be arranged in nearly 10^{76} topologies: to get the number of trees, these figures are multiplied by (2N-3). It is essential, therefore, that some selection of promising structures be imposed and that the folding of the topology be delayed to a late stage in the work. The methods discussed all build a best fit topology and then refine the structure until no further improvement in the fit can be attained.

Matrix methods

A matrix method for amino acid sequences was first proposed by Fitch and Margoliash,[10] and used with slight modification by Barnabas, Goodman and Moore.[11] It derives from the application of cluster analysis to numerical taxonomy (Rao; Sokal and Sneath; Edwards and Cavalli-Sforza).[4,12,13]

The set of protein sequences are aligned so that supposedly homologous sites are compared, and a table of differences between the taxa is constructed. As a refinement, each difference is scored according to whether one, two or three nucleotide base mutations (it is necessary to distinguish base mutations at the DNA level from phenetic mutations, e.g. amino acid changes) would be required to interconvert the two amino acids using the genetic code (see Fig. 2).[14] This is called a minimum mutation distance. Since deletions have occurred over time, these distances are corrected according to the number of amino acids in each sequence to give a triangular or half-matrix of 'relative distances'.

The shortest distance, d_{AB}, in the matrix between any two taxa (A and B) is selected, and these taxa are assumed to form a terminal pair of branches. All other taxa are assigned to set C and the average of the distances from A and B, respectively, to the members of C are calculated to give d_{AC} and d_{BC}. It is now possible to solve the three simultaneous equations to obtain the lengths of branches a, b and c, as shown in Fig. 3. Whereas lengths a and b will correspond to actual branch lengths, c will be the average length of all the branches from the node to the taxa of set C.

The next shortest distance is selected and the branch lengths

First Nucleotide	Second nucleotide: U	C	A	G	Third Nucleotide
U	Phe	Ser	Tyr	Cys	U
	Phe	Ser	Tyr	Cys	C
	Leu	Ser	terminate	terminate	A
	Leu	Ser	terminate	Trp	G
C	Leu	Pro	His	Arg	U
	Leu	Pro	His	Arg	C
	Leu	Pro	Gln	Arg	A
	Leu	Pro	Gln	Arg	G
A	Ile	Thr	Asn	Ser	U
	Ile	Thr	Asn	Ser	C
	Ile	Thr	Lys	Arg	A
	Met	Thr	Lys	Arg	G
G	Val	Ala	Asp	Gly	U
	Val	Ala	Asp	Gly	C
	Val	Ala	Glu	Gly	A
	Val	Ala	Glu	Gly	G

Figure 2. The Genetic Code.

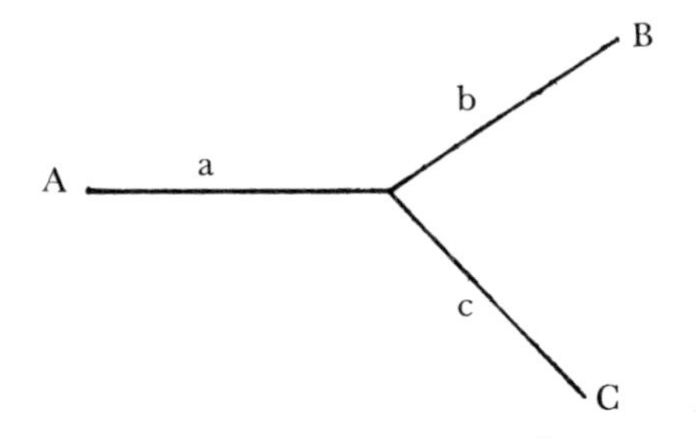

$$a = \tfrac{1}{2}(d_{AB} + d_{AC} - d_{BC})$$
$$b = \tfrac{1}{2}(d_{AB} + d_{BC} - d_{AC})$$
$$c = \tfrac{1}{2}(d_{AC} + d_{AC} - d_{AB})$$

Figure 3. Calculation of average branch lengths.

calculated. The branch lengths about an internal node of the topology may all be averages (see Appendix 1 for details) and the average distance AB, for example, will be the average of all the combinations of comparisons between all the members of sets A and B. This is an unweighted pair group method of clustering. Finally only two clusters of taxa will remain, and the overall common ancestor will be determined by finding the average mid-point such that $a = b = \tfrac{1}{2}d_{AB}$ (see Appendix 1). The lengths of the internal branches are found as follows: the length of any branch between two adjacent internal nodes x and y equals the average length from x to all its terminal elements, via y, minus the average length from y to these same elements. Examples of these calculations are given in Appendix 1.

Having assigned all the branch lengths, the original matrix may be reconstructed and the 'average per cent standard deviation' calculated:

$$\text{`average \%S.D.'} = 100 \times \sqrt{\frac{1}{\frac{1}{2}N(N-1)-1} \sum_{i>j} \left(\frac{d_{ij}\ \text{original} - d_{ij}\ \text{reconstructed}}{d_{ij}\ \text{original}} \right)^2}$$

where N = number of taxa
dij = mutation distance (original and reconstructed) between taxa i and j

The fraction before the sum is the reciprocal of the number of elements in the half matrix minus one. The 'average per cent standard deviation' is not in fact a standard deviation as used in statistics. It can be used as an arbitrary figure indicating the fit of the reconstructed matrix to the original matrix. By slightly relaxing the criteria for choosing the next pair of branches to be joined, the data may be fitted to alternative topologies, until no improvement of fit can be obtained by further alterations.

If there is to be an exact fit of data and reconstructed matrices, one must formulate the axiom that there exists a topology such that, for all pairs of taxa, the sum of the branch lengths between any two taxa i and j is always equal to the value of the matrix element of i and j.

In the event that this axiom does not hold (in the real case this could be due to the presence of convergence or multiple mutations at a single site), there will not be complete agreement and a positive 'standard deviation' will be noted.

The trees constructed using matrix methods appear to be in fairly good general agreement with the course of evolution as at present conceived on independent grounds. The major anomaly that may be observed is the occasional presence of branches with negative length. If the axiom formulated above holds good, then the presence of a negative branch would indicate that the clustering was at fault, and that an alternative tree should be tried.[15] If it is impossible to get an exact fit, then negative branches may have to be accepted; but it is difficult to see what they could represent in evolutionary terms. It is probably best to consider them as manifestations of the fact that the data do not conform to the initial axiom, i.e. that a matrix distance should be equal to the sum of the corresponding branch lengths. It might be more appropriate to load against negative branches by making them either zero or positive for the purposes of reconstructing the matrix. In Appendix 2, we show the results of constructing a few of the many possible trees from some α-chain haemoglobin sequences, and applying the criteria discussed above. It may be noted that the 'better-fitting' trees lacked or had insignificant negative branches.

The method described above uses an unweighted cluster analysis, since the length from a node along a branch to a group of taxa is the mean of all the lengths from that node to each of the taxa, irrespective of their arrangement on more terminal branches. Many biologists feel intuitively that one should give equal weight to the two branches which are being joined, irrespective of the number of taxa giving rise to these branches.

Hence in Fig. 4 the average length from node X is:

(i) *weighted* = $\frac{1}{2}$ d + n + $\frac{1}{2}$ [c + m + $\frac{1}{2}$(a + b)]
(ii) *unweighted* = $\frac{1}{4}$ d + (n + c) + (n + m + b) + (n + m + a)

A weighted clustering is performed initially as in an unweighted method; but when the two branches A and B are joined, the two rows of the matrix elements of A and B are replaced by a single row of (AB) elements, which are the average of A and B. This reduced matrix is taken as the starting-point for the next clustering cycle. The disadvantage of this weighted method is that for the same tree different values may be obtained for the branch lengths depending upon the order in which available pairs of branches are joined. The unweighted method generates a unique set of branch lengths for any given tree, so that, on balance, we prefer this method. Dayhoff (personal communication) has developed a weighted matrix method which does not suffer from this disadvantage. Both Fitch and Margoliash and Barnabas, Goodman and Moore use an unweighted method,[10,11] but in a later paper Fitch, Margoliash and Gould use a weighted method.[16]

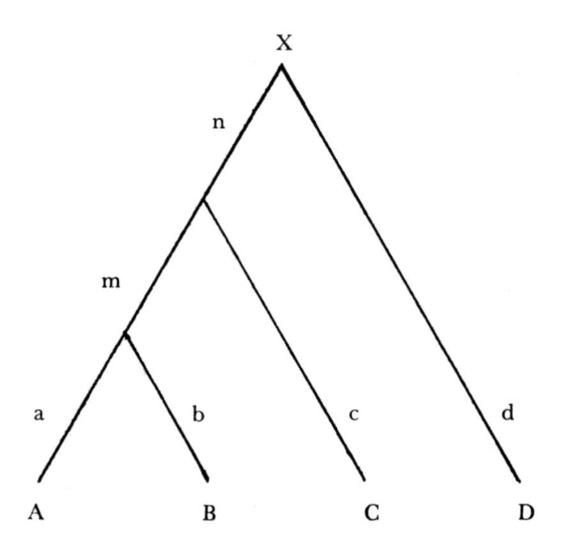

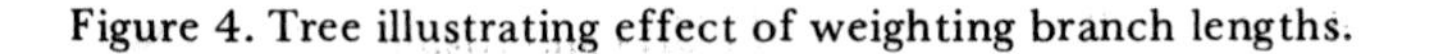
Figure 4. Tree illustrating effect of weighting branch lengths.

When, as often happens, two homologous proteins differ not by an amino acid change, but by a site having been deleted, it is difficult to compare them directly. To make allowance for this both Fitch and Margoliash and Barnabas, Goodman and Moore corrected their minimum mutation distances.[10,11] Fitch and Margoliash multiply their distances by the ratio of amino acids in each protein (with the larger number as numerator of the fraction) to give a relative distance. Barnabas, Goodman and Moore prefer to divide their distances by the number of shared amino acids and multiply by 100 to give a 'percentage mutational divergence'. Both these corrected values may give valid measures of affinity (that of Barnabas, Goodman and Moore being probably the better of the two), but it is clear that neither leaves the data with comparable units. We cannot therefore expect to obtain branch lengths in distinct units (i.e. mutations) either. It may also be noted that the amino acid sites in a set of sequences may be divided between the variable and so-called invariant groups. The latter provide us with no taxonomic information and in this context are superfluous and therefore should not be incorporated in the above correction factors. The only solution to the problem of deletions is to compare only those sites which are common to all the taxa under study.

The question of testing the fit of the tree to the data must remain a central one. We have already stated that the 'average per cent standard deviation' is not in this case a statistically valid one. It is clear that the data, if they can be used to produce a tree of affinities, cannot be considered as at all independent; for example, 20 'independent' sequences would generate 190 matrix comparisons, from which would be calculated the lengths of only 35 independent branches. Hence this type of statistical analysis cannot be invoked. Other workers have fallen into the trap of making non-independent comparisons when trying to compute rates of protein evolution, e.g. Air et al. and Kimura.[17,7] Having established the non-statistical nature of the test of fit, we may ask whether it may nevertheless be a useful arbitrary figure of merit. If the intention of the method is to obtain a tree which produces a reconstructed matrix as similar as possible to the original data matrix, then the 'average per cent standard deviation' is probably as good an indication as any. If, however, the intention is to build a minimum evolution tree from the data, then it may not provide the best answer. Consider the case (see Fig. 9 in Appendix 1) of an average deviation from the data by one unit in the terminal branch between S and node 1. This branch length would be used in computing six distances of the reconstructed matrix. A similar unit deviation on the internal branch between node 2 and node 3 would appear in twelve distances of the reconstructed matrix. In a 20-taxon tree, a single branch value could be used in reconstructing between 19 and 100 of the 190 elements of the matrix. It is clear that there will be a bias towards selecting a tree in which deviations from the data are found in the more terminal branches, even though that tree may not be the minimum evolution tree. A comparison between the trees produced by Eck and Dayhoff, and Fitch and Margoliash, for cytochrome-c showed that roughly equivalent middle branches were lower, and terminal branches greater in length in the tree of Fitch and Margoliash.[2,10]

A further point which should be noted is that the distances in the data matrix are *minimum* mutation distances. It is therefore axiomatic that the elements of the reconstructed matrix should all exceed or be equal to the corresponding data elements; yet an examination of the two matrices of Fitch and Margoliash shows the presence of a large number of negative deviations from the data, confirming that the best fit contains distortions.[7] The best test of fit for the minimal evolution model would be to find that tree which gave the lowest sum of the absolute branch lengths. In this test, each branch is used only once and, by taking absolute values, negative branches would be discouraged and occur only rarely in better fit trees.

Minimum path methods

This method attempts to find a minimum evolution fit at each homologous site, among all the taxa under consideration simultaneously,[2,18,19] rather than to try to fit all the data simultaneously to pairs of taxa as the matrix methods do. To achieve this, the method makes use of three computer routines:

Routine 1. For a given topology, the routine generates for each homologous site the nodal amino acids (as far as this is possible) and counts the mutations (i.e. phenetic changes) required to interconvert all members. This is repeated for the whole set of data and the total number of mutations required is calculated. This total is the 'minimum mutation value' for the topology.

Routine 2. This is a topology-building routine.

Routine 3. This is a topology-refining routine.

This method is best illustrated by working a simple example. Consider the following sequences of 5 amino acids from 5 species (in fact, the data are positions 8, 35, 68, 71 and 111 from haemoglobin α-chains selected purely to illustrate the method). Four of these species are chosen at random and Routine 2 sets out the three possible topologies (Fig. 5*a*):

Routine 1 fits the data to each topology in turn and, by reconstructing the most likely ancestral sequence at each node, counts the substitutions required to interconvert each node. It does this by listing all the possible amino acids that could occur on each

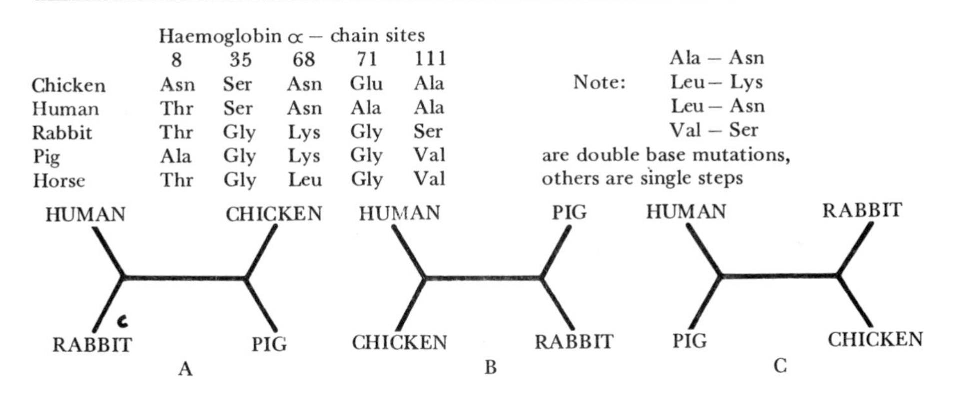

	Haemoglobin α – chain sites				
	8	35	68	71	111
Chicken	Asn	Ser	Asn	Glu	Ala
Human	Thr	Ser	Asn	Ala	Ala
Rabbit	Thr	Gly	Lys	Gly	Ser
Pig	Ala	Gly	Lys	Gly	Val
Horse	Thr	Gly	Leu	Gly	Val

Figure 5*a*. Example used to illustrate Eck and Dayhoff method.

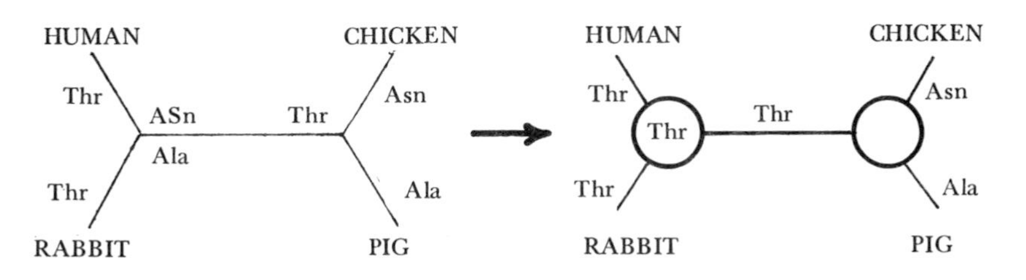

Figure 5*b*. Site 1 (α8) amino acids in topology A.

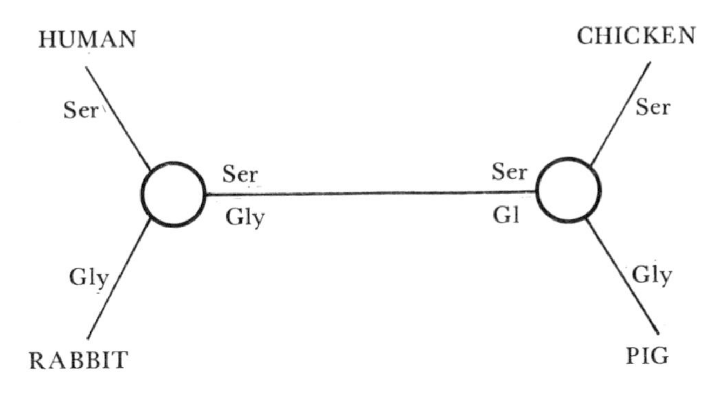

Figure 5*c*. Fit of site 2 (α35) amino acids to topology A.

of the three branches about a node and choosing the most abundant one as the nodal amino acid. For example, fitting the data from the first site (α8) to the branches about the 'left hand' node of topology A shows that two of the branches are Thr, which is therefore the nodal amino acid (see Fig. 5*b*). The 'right hand' node has three different amino acids, and therefore this site (α8) requires, for topology A, two of the three changes Thr-Ala, Thr-Asn, and Ala-Asn.

Site 2 (α35) presents the same situation at each node; each nodal amino acid being indeterminate, since there is an equal choice between glycine and serine (see Fig. 5*c*). The mutation value will be a minimum of two, there being two Gly-Ser changes (the exact branches on which they occur being indeterminate). The third site (α68) is an identical situation to site 2 (α35), requiring a minimum of two mutations, Ans-Lys, but with indeterminate nodal amino acids. The fourth (α71) and fifth (α111) sites both require the theoretical minimum of two mutations to interconvert the three amino acids. Thus, for topology A, the minimum number of mutations required to fit the sequences is 10. Topology C also requires 10 mutations. With topology B we find a different situation when the sequences are fitted. The species are now arranged such that only one Gly-Ser change is required for site 2 (α35), the nodes now being firmly allocated, and only one Asn-Lys change for site 3 (α68). The score for this topology is thus 8, which is also the theoretical minimum.

Routine 2 then takes this best fit topology and adds a fifth taxon, the new branch being added to each of the 5 existing branches, generating 5 topologies. The sequences are then fitted to each topology in turn by routine 1 as before. With larger numbers of taxa, the method described may leave several blank nodes, so the routine has been constructed so that the 3 neighbouring nodes to a blank node are examined. If two of these are the same, then the blank node can be filled in, and this cycling is repeated until no further nodes can be assigned. Any remaining blank nodes are left indeterminate. Routines 2 and 1 thus allocate branches and assess the resultant topologies until all the taxa have been added.

It is possible, however, that a chance convergence of the data has caused us to choose a topology which is not the best fit for the data when all the taxa have been added in. Routine 3 systematically cuts each branch and applies the 'cut end' to all the branches in the other part of the topology (see Fig. 6) and Routine 1 tests the fit until no topology with lower mutation value can be found. The nodal sequences are then determined (as described above) and the mutations interconverting them assigned to the branches.

Finally, the topology may be folded to convert it into a phylogenetic

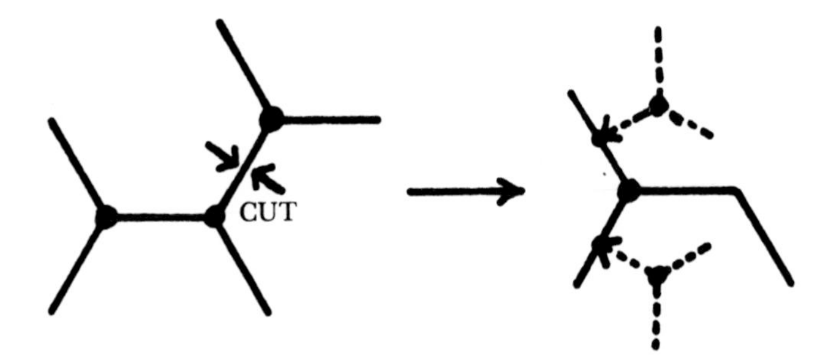

Figure 6. Showing how a topology may be cut (by routine 3) and alternative topologies produced.

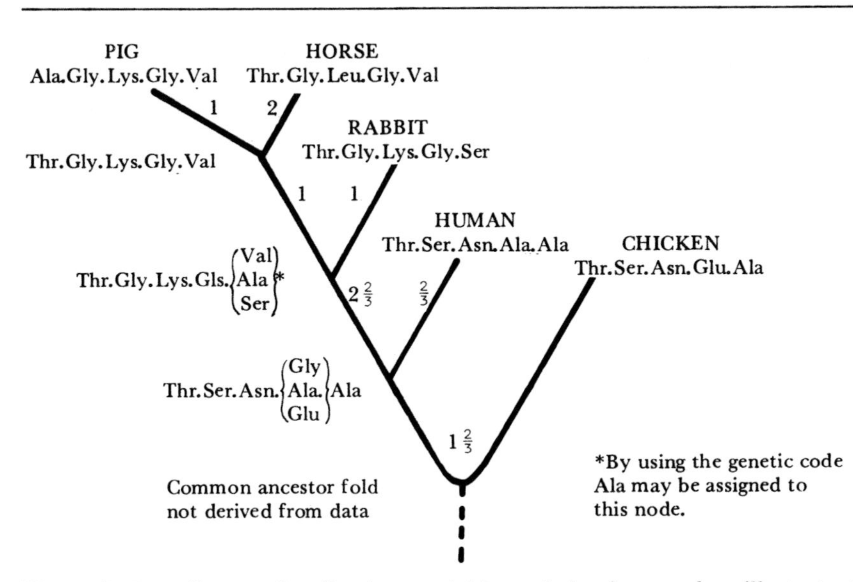

Figure 7. Best fit tree for five haemoglobin α-chain sites used to illustrate the method of Eck and Dayhoff.[2]

tree using either the criterion of constant rate of mutation or other criteria not derived from the data. Fig. 7 represents the tree produced from the 5 amino acid sequences of the 5 species used to explain the method.

Although the computer programme has not yet been published, the 3 routines would seem to provide a fully efficient method for arriving at a minimum evolution topology, using amino acids as unit characters.

In particular, Routine 3 represents an important advance in refining taxonomic structures, but a slight doubt must always remain that the final 'minimum' topology may not be the true minimum, but a local 'hollow' (to use a geographical analogy). This is unlikely, but cannot be proved not to occur unless all possible topologies are tested.

Examination of the genetic code (Fig. 2) reveals that a single amino acid mutation may require up to three nucleotide *base* mutations. This suggests that a greater precision might be obtained by inferring nucleotide base sequences from amino acid sequences, and processing these character states by the above method. In Fig. 7, the fifth site (α111) ancestral node for rabbit-pig-horse is ambiguous when amino acid unit characters are employed. However, the genetic code shows that a valine-serine mutation requires a double base mutation and that alanine is a possible intermediate. Hence alanine may be assigned to this site.

Fitch has produced an algorithm to fit inferred nucleotide sequences to a selected tree.[20] Although he has solved the problem posed by the alternative codons (triplets) for serine, leucine and arginine, his technique suffers in that it operates on a tree and not on a topology, and hence the potential work-load is increased by a factor of (2N-3).

Discussion

In comparing the matrix methods with the minimum path method of Eck and Dayhoff,[2] one is attracted to the latter in that it portrays events which could have taken place in the course of evolution of the proteins. The Eck and Dayhoff method handles variable amino acid sites as single characters which could equally well be included as additional characters in a numerical analysis of morphological character states. It is because of this approach that the method can be fully efficient in a minimum evolution model. This cannot be said of the matrix methods. In particular, the presence of convergence and/or multiple substitution sites will, with matrix methods, be realised as errors from the best fit. Such characters are acceptable to the minimum path method as long as their effects on the topology are not too localised or too extensive, as is also the case with classical studies.

Matrix methods are, however, the only ones that may be used for serological data. If the experimental data for a protein can be transformed into distance data, a matrix can be tabulated which could be considered as an approximation to a distance matrix of sequence data. We would recommend that in these circumstances an unweighted analysis (or a weighted one giving a unique solution) be used with a minimum total path length test of fit.

It might be worthwhile to point out that all the methods described have a tree-building programme which produces only an approximation to the best fit. This is because cluster analysis relies on an assumption of an even rate of evolution, and the computer cannot be programmed with hindsight. Accepting this as the case, any reasonable approximation such as one based on classical methods could be used to initiate the refining programme in the Eck and Dayhoff method and, in addition, this could be adopted in the matrix methods, provided that alternative trees were examined systematically in much the same manner.

Having devised adequate methods of finding a minimum evolution fit of the sequences, it would be nice to say that all we needed now were a lot of variable proteins, accurately sequenced, which could be lined up to generate a phylogeny, with high accuracy for all those species in which we were interested.

Although this is *the* major requirement, there remains one more problem which is still somewhat intractable – that of sequence homology. When proteins from two species are examined, although we may be sure that they are homologous (in that they were derived from a common ancestral *protein*), we may not be able to say whether that common ancestral protein was present in the most recent common ancestral species. For example, if we compared a human haemoglobin α-chain with a baboon β-chain, we would find ourselves faced with the ludicrous conclusion that the two species separated in the late pre-Cambrian! This is an extreme case; but others are more subtle. Boyer et al. have found haemoglobin chain variants in both the chimpanzee and gorilla.[21] Both of these variant chains differ from the chimpanzee (= human) α^A chain by at least 9 mutations, but from each other by only 2 mutations, suggesting the presence of an α-chain gene duplication which predated the divergence of the two species but which has remained 'silent' except in these two individuals. Boyer et al. have also examined the haemoglobins of ceboid monkeys.[22] In this group, all species examined contain a minor haemoglobin which shows resemblances to hominoid haemoglobin A_2 and consists of two α^A-chains and two δ-like chains. This type of minor haemoglobin component is not found in the Cercopithecoid monkeys.[23] Comparison of the δ-chain sequences from the two groups (Ceboidea and Hominoidea) makes it equally likely that they are either the result of a β-chain gene duplication before the separation of those two lineages, or the product of independent duplications after the superfamilies had separated. The significance of this situation is that if the β-duplication occurred separately (Fig. 8*a*), β-β, δ-δ and intergroup β-δ comparisons (including the cercopithecoid β-chains) may be made for the purpose of studying the superfamilial relationships. However, if the β-duplication occurred before the simian radiation (Fig. 8*b*), a β-δ

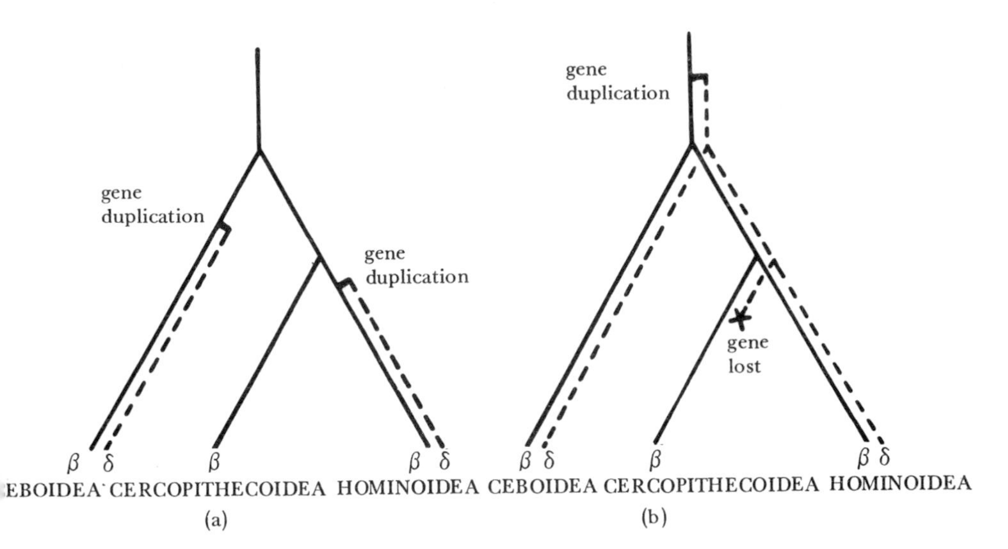

Figure 8. Alternative explanations of the origin of δ-like chains in the Anthropoidea.

comparison would lead to the erroneous assessment that there was an early separation of the superfamilies. The only slight hope that this problem might be resolved would be to examine the δ (or δ-like) chains of more species, particularly those with the more ancient ancestral relationships within each superfamily.

Buettner-Janusch and Twichell, Buettner-Janusch and Hill, and Goodman et al. have all suggested that the non-α chains of lemuroids are likely to be homologues of the human γ-chain (foetal chain), since lemuroid haemoglobins show several properties analogous to simian foetal haemoglobins: no separate foetal haemoglobin has been found in the lemuroids, and the sequences of the non-α globin chains of *Lemur fulvus* and *Propithecus verreauxi*, though not yet adequately published, are claimed by Goodman et al. to cluster with human γ and kangaroo non-α in their best globin phylogenetic tree.[24–28]

Duplications are also known in the chains of the buffalo, goat, mouse, *Macaca fascicularis* and horse.[29–34] The studies of the two latter species well illustrate the complexities that arise when both gene duplications and allelic variations are present. Lack of serial homology in the sequence set of a particular protein may normally be detected by considering sequence sets from other types of proteins, i.e. it is essentially a sampling problem, as in classical evolutionary studies.

Goodman et al. have also suggested that, in view of its marked difference from other mammalian haemoglobin α-chains, *Tupaia glis* α-chain should be considered as the survivor of an ancient duplication, but it could equally well be used as further evidence that the tupaioids are not primates.[26,27,35]

In Appendix 2 are some examples of trees which we have computed using the methods described in this paper. The data are sequences of haemoglobin α-chains from four prosimians, a macaque, man and representatives of some other orders. These trees illustrate several features of interest, some of which are relevant to tree building with more complex characters.

It will be seen that there is not a complete correspondence between the two criteria of fitness used. Conversely, there are several different trees giving very similar low values, so that a final choice of tree is not easy to make. The tree shrew, although initially placed as the most isolated mammal by the cluster analysis, gave a much better fit when placed with the mouse. (However, bearing in mind the range of species involved, this may not be taken as a suggestion that the tupaioids are members of the Rodentia!) This latter move also eliminated or gave insignificant negative branches.

Most of the internal branches are short, having very few mutations assigned to them. Although part of the reason for this is undoubtedly the short time duration of the branches, the longer branches computed by the Eck and Dayhoff method suggest that procedural factors may also be involved. Whatever the reason, they suggest that only a small proportion of the data are relevant to the determination of the branching structure of the trees. We feel that these short branches justify the use of the genetic code to gain some additional precision and also indicate that the weights applied to classical characters may be more critical than generally thought. Many of the changes are species-specific so that, although they show the distinctness of the taxa, they contribute nothing to elucidation of their relationships; within the context of the species chosen they are superfluous.

The trees in Appendix 2 are drawn to scale and not, as most authors have done, brought to a common 'now' plane. This format shows very clearly that there has been a very wide variation in the apparent evolutionary rates of these proteins. We therefore reject the use of protein sequences as 'evolutionary clocks'. A similar conclusion may be drawn from the fibrinopeptide data of Doolittle.[36,37] While the inconstancy of evolutionary rates in general is accepted by most biologists, it is often forgotten that constancy of rates is enshrined in the principle that phenetic similarity is an accurate measure of degree of kinship, which is the basis of most of our numerical methods.

The method of Eck and Dayhoff allows us to distribute the changes of character states upon the branches of a given tree. This implies a partition of characters into ancestral and group-specific, an approach proposed for morphological characters by Remane,[38] which allows a better assessment of the status of the higher taxa that we have erected. It may be suggested, for instance, that the Lemuridae are at present defined by mainly ancestral prosimian characters, and that the failure to discriminate between ancestral and group-specific characters has been responsible in the past for at least part of the controversies over the relationships of the higher apes with man. This approach makes it evident that further characters will be required to refine our present trees. It is likely that protein sequencing will be of great assistance in this task.

We have described how sequences may be fitted to a minimum evolution model. Although we feel intuitively that the minimum evolution model is the best approximation to the actual evolutionary tree, we must not regard the final phylogenetic tree that we compute as necessarily reflecting the precise course of the evolution of our proteins. For example, from an analysis of apparent single, double and triple mutations, Jukes has concluded that there is more multiple substitution than the minimum model shows.[39] The principle of minimum evolution has not yet been formally demonstrated to be the best estimator of the actual course of evolution.

We hope that this paper has shown that in practice the use of protein sequences in taxonomy is not so straightforward as the simple statement in our introductory remarks had suggested.[40] It appears, in fact, that the use of sequences is prone to many of the problems that beset those who use more complex polygenic characters, with the exception that the character states are of equal weight. The advantage of this uniformity should not be underrated. It allows us to quantify our analytical methods, and in so doing we are forced to re-consider the mechanisms involved in the act of reconstructing phylogeny.

We dance round in a ring and suppose
But the secret sits in the middle and knows

Robert Frost

Acknowledgments

We thank N.A. Barnicot, A.W.F. Edwards, R.D. Martin, and D.A. Coleman for stimulating discussion, and we are very grateful to M.O. Dayhoff and N.A. Barnicot for reading the manuscript and making valuable suggestions.

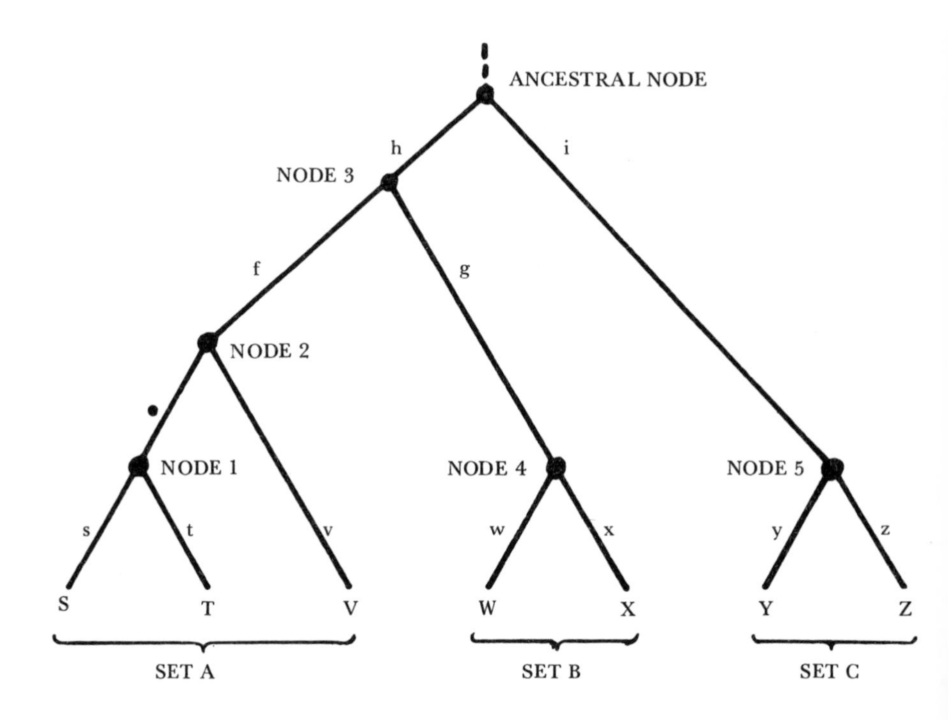

Figure 9. Hypothetical tree, labelled for use with the equations in Appendix 1.

Appendix 1

Derivation of branch lengths by the unweighted matrix methods used by Fitch and Margoliash (1967)[10], and Barnabas, Goodman and Moore (1971).[26] See Fig. 9.

1. Average distance d_{AB} = average distance between SET A and SET B

$$= \frac{1}{6}\,[d_{SW} + d_{SX} + d_{TW} + d_{TX} + d_{VW} + d_{VX}]$$

where d_{IJ} = a matrix element between taxa I and J.

2. Average branch length l_{A3} = average of NODE 3 to SET A.

$$= \frac{1}{2}\,[d_{AB} + d_{AC} + d_{BC}]$$

3. Internal branch length $f = l_{A3} - \frac{1}{3}\,[v + (s+e) + (t+e)]$

f will also equal average distance from NODE 2 to SETS B and C minus the average from NODE 3 to SETS B and C.

Thus $f = \frac{1}{2}\,(l_{B2} + l_{C2}) - \frac{1}{2}\,(l_{B3} + l_{C3})$

4. ANCESTRAL NODE is at $\frac{d_{(AB)C}}{2} = \frac{1}{2}\,[\frac{1}{10}(d_{SY} + d_{SZ} + d_{TY} +$

$d_{TZ} + d_{VY} + d_{VZ} + d_{WY} + d_{WZ} + d_{XY} + d_{XZ})]$

Hence $h = \frac{d_{(AB)C}}{2} - \frac{1}{5}\,[(e + f + s) + (e + f + t) + (f + v) + (g + w) + (g + x)]$

and $i = \frac{d_{(AB)C}}{2} - \frac{1}{2}\,(y + z)$

Appendix 2

This appendix contains the results of our tree-building using the methods described in the text. We used the unweighted matrix methods to test 12 trees using the data matrix of Goodman et al. for the haemoglobin α-chains of nine species.[26] The details are shown below. The trees include that generated by the unweighted pair group method (Fig. 10*a*) and the 'best-fit tree' of Goodman et al.[26] (Fig. 10*b*).sWe tested all the trees by the 'average per cent standard deviation' (APSD) and the 'total path length' (TPL) criteria, as shown on the diagrams (Figs. 10 and 11). If negative branches occurred, the tests were performed by summing distances by taking note of the sign (APSD-, TPL-) and by taking absolute values (APSD+, TPL+). The trees are drawn to a minimum mutation distance scale.

Man *Homo sapiens*	(Hs)	–							
Rhesus Monkey *Macaca mulatta*	(Mm)	5	–						
Greater Bush Baby *Galago crassicaudatus*	(Gc)	14	14	–					
Sifaka *Propithecus verreauxi*	(Pv)	23	24	25	–				
Brown Lemur *Lemur fulvus*	(Lf)	18	20	19	18	–			
Malayan Tree Shrew *Tupaia glis*	(Tg)	31	29	30	33	31	–		
House Mouse (Strain C57) *Mus musculus*	(Mus)	19	20	22	27	25	26	–	–
Horse (Hb fast 24F) *Equus caballus*	(Ec)	24	21	29	39	31	37	29	–
CHICKEN (Hb type AII) *Gallus domesticus*	(Gd)	45	46	49	57	54	60	51	53
		Hs	Mm	Gc	Pv	Lf	Tg	Mus	Ec

N.B. Some of these values do not agree with published sequences. The commonest cause of error is misassignment of acid-amides.

NOTES

1 Zuckerkandl, E. and Pauling, L. (1965), 'Evolutionary divergence and convergence in proteins', in Bryson, V. and Vogel, H.J. (eds.), *Evolving Genes and Proteins*, New York and London.

2 Eck, R.V. and Dayhoff, M.O. (1966), *Atlas of Protein Sequence and Structure*, Silver Spring: Nat. Biomed. Research Found.

3 *Taxon* is here used as a general term to cover those species, individuals, populations or other groupings under study. It is synonymous with Operational Taxonomic Unit as defined by Sokal and Sneath.[4]

4 Sokal, R. and Sneath, P.H.A. (1963), *Principles of Numerical Taxonomy*, San Francisco.

5 Sullivan, B. (1971), 'Structure, function and evolution of primate haemoglobins. II: A survey of the oxygen binding properties', *Comp. Biochem. Physiol.* 40, 359-80.

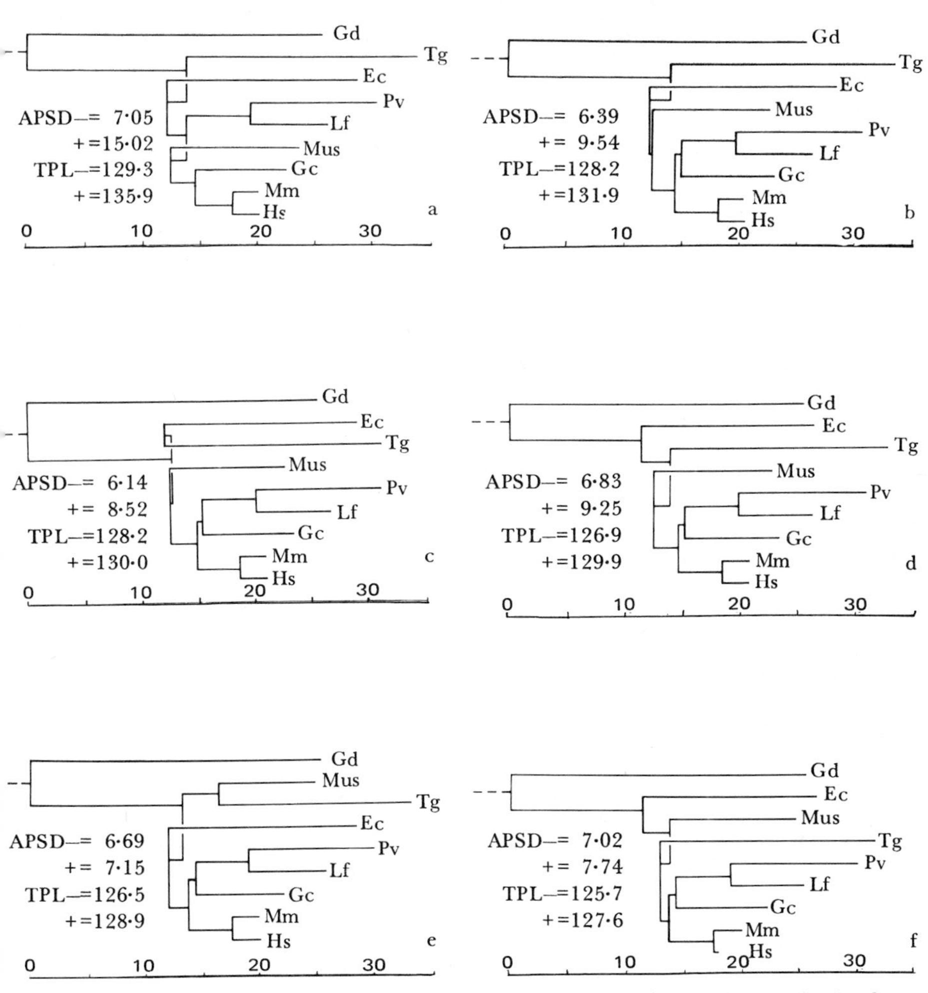

Figure 10. Examples of trees derived by unweighted matrix methods for haemoglobin α-chains.

6 Greer, J. and Perutz, M.F. (1971), 'Three dimensional structure of Haemoglobin Rainier', *Nat. New Biol.* 230, 261-4.

7 Kimura, M. (1968), 'Evolutionary rate at the molecular level', *Nature* 217, 624-6.

8 King, J.L. and Jukes, T.H. (1969), 'Non-Darwinian evolution', *Science*, 164, 788-98.

9 Napier, J.R. and Napier, P.H. (1967), *Handbook of Living Primates*, New York and London.

10 Fitch, W.M. and Margoliash, E. (1967), 'Construction of phylogenetic trees', *Science*, 155, 279-84.

11 Barnabas, J., Goodman, M. and Moore, G.W. (1971), 'Evolution of haemoglobin in primates and other therian mammals', *Comp. Biochem. Physiol.* 39, 455-82.

12 Rao, C.R. (1952), *Advanced Statistical Methods in Biometric Research*, New York.

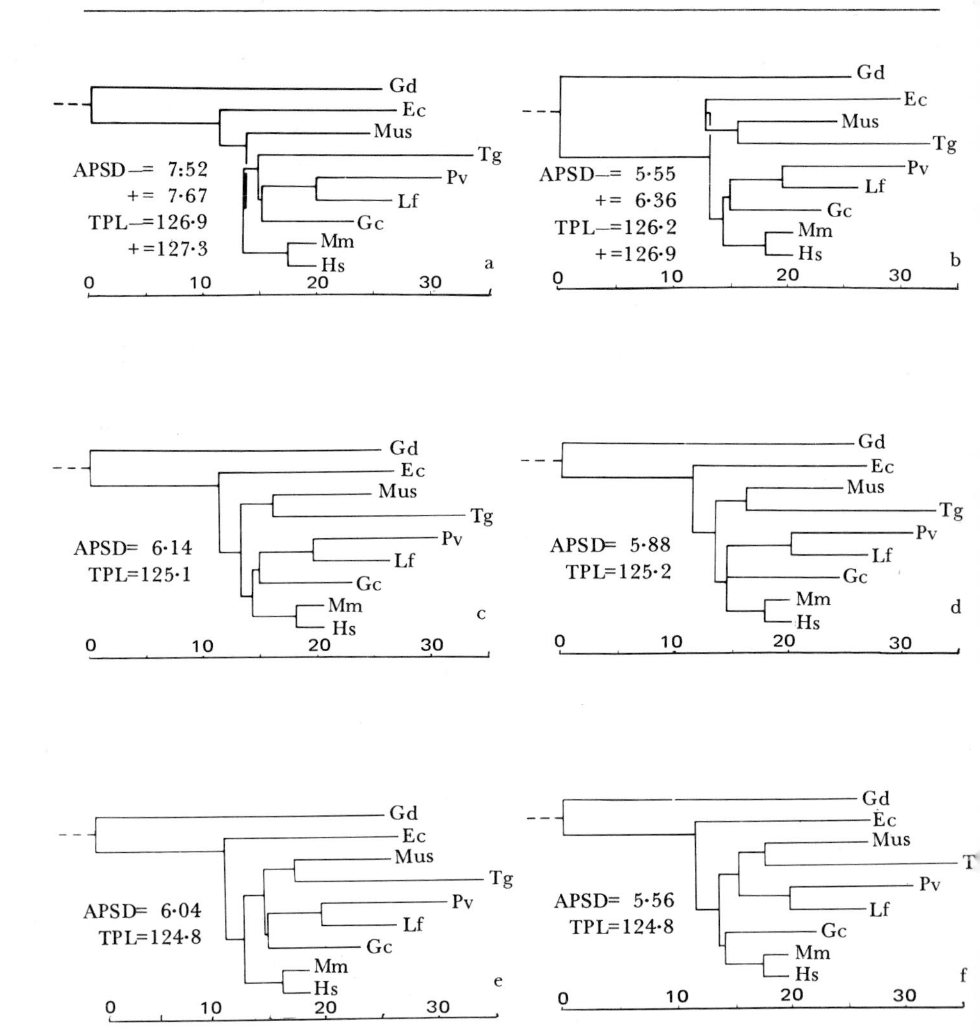

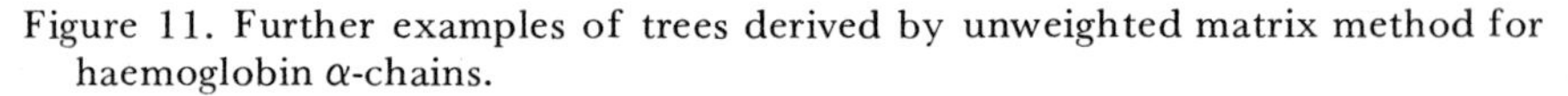
Figure 11. Further examples of trees derived by unweighted matrix method for haemoglobin α-chains.

13 Edwards, A.W.F. and Cavalli-Sforza, L.L. (1965), 'A method for cluster analysis', *Biometrics* 21, 362-75.

14 Caskey, C.T. (1970), 'The universal RNA genetic code', *Quart. Rev. Biophys.* 3, 295.

15 Cavalli-Sforza, L.L. and Edwards, A.W.F. (1967), 'Phylogenetic analysis, models and estimation procedures', *Am. J. Human Genet.* 19, 233-57.

16 Fitch, W.M., Margolish, E. and Gould, K.S. (1969), 'The construction of phylogenetic trees. II: How well do they reflect past history?', *Brookhaven Symp. Biol.* 21, 217-42.

17 Air, G.M., Thompson, E.O.P., Richardson, B.J. and Sharman, G.B. (1971), 'Amino acid sequences of kangaroo myoglobin and haemoglobin and the date of the Marsupial-Eutherian divergence', *Nature* 229, 391-4.

18 Dayhoff, M.O. (1969), *Atlas of Protein Sequence and Structure*, Silver Spring: Nat. Biomed. Research Found.

19 Dayhoff, M.O. (1969), 'Computer analysis of protein evolution', *Sci. Amer.* 221, 87.

20 Fitch, W.M. (1971), 'Towards defining the course of evolution: minimum change for a specific tree topology', *Syst. Zool.* 20, 406-16.

21 Boyer, S.H., Noyes, A.N., Vrablik, G.R., Donaldson, L.J., Shaefer, E.W., Gray, C.W. and Thurman, T.F. (1971), 'Silent haemoglobin alpha genes in apes: potential source of Thalassemia', *Science* 171, 182-5.

22 Boyer, S.H., Crosby, E.F., Noyes, A.N., Fuller, G.F., Leslie, S.E., Donaldson, L.J., Vrablik, G.R., Schaefer, E.W. and Thurman, T.F. (1971), 'Primate haemoglobins: some sequences and some proposals concerning evolution and mutation', *Biochem. Genet.* 5, 405-48.

23 Barnicot, N.A. and Wade, P.T. (1970), 'Protein structure and the systematics of Old World Monkeys', in Napier, J.R. and Napier, P.H. (eds.), *Old World Monkeys*, New York and London.

24 Buettner-Janusch, J. and Twichell, J.B. (1961), 'Alkali-resistant haemoglobins in prosimian primates', *Nature* 192, 669.

25 Buettner-Janusch, J. and Hill, R.L. (1965), 'Evolution of haemoglobin in primates', in Bryson, V. and Vogel, H.J. (eds.), *Evolving Genes and Proteins*, New York and London.

26 Goodman, M., Barnabas, J., Matsuda, G. and Moore, G.W. (1971), 'Molecular evolution in the descent of man', *Nature* 233, 604-13.

27 Buettner-Janusch, J. and Hill, R.L. (1965), 'Molecules and monkeys', *Science* 147, 836-42.

28 Hill, R.L. (1968), 'Unpublished sequences', in Sober, H. (ed.), *Handbook of Biochemistry*, Chemical Rubber Company. (N.B. There are errors in some sequences shown.)

29 Balani, A.S. and Barnabas, J. (1965), 'Polypeptide chains of buffalo haemoglobins', *Nature* 205, 1019-21.

30 Huisman, T.H.J., Adams, H.R., Dimock, M.O., Edwards, W.E. and Wilson, J.B. (1967), 'The structure of goat haemoglobins'. *J. Biol. Chem.* 242, 2534-41.

31 Hilse, K. and Popp, R.A. (1968), 'The structure of mouse haemoglobin', *Proc. Nat. Acad. Sci. U.S.* 61, 930-5.

32 Barnicot, N.A., Huehns, E.R. and Jolly, C.J. (1966), 'Biochemical studies on haemoglobin variants of the irus macaque', *Proc. Roy. Soc. Lond.* B. 165, 224-44.

33 Barnicot, N.A., Wade, P.T. and Cohen, P. (1970), 'Evidence for a second haemoglobin α-locus duplication in *Macaca irus*', *Nature* 228, 379-81.

34 Clegg, J.B. (1970), 'Horse haemoglobin polymorphism: evidence for two linked non-allelic α-chain genes', *Proc. Roy. Soc. Lond. B*, 176, 235-46.

35 Martin, R.D. (1968), 'Towards a new definition of primates', *Man* 3, 377-401.

36 Doolittle, R.F., this volume.

37 Doolittle, R.F., Wooding, G.L., Lin, Y. and Riley, M. (1971), 'Hominoid evolution as judged by fibrinopeptide structures', *J. Molec. Evolution* 1, 74-83.

38 Remane, A. (1956), 'Paläontologie und Evolution der Primaten', in Hofer, H., Schultz, A.H. and Stark, D. (eds.), *Primatologia*, Basel.

39 Jukes, T.H. (1971), 'Comparisons of the polypeptide chains of globins', *J. Molec. Evolution* 1, 46-62.

40 Since the seminar, a publication by Barnabas et al. has appeared (Barnabas, J., Goodman, M. and Moore, G.W. (1972), 'Descent of mammalian alpha globin chain sequences investigated by the maximum parsimony method', *J. Mol. Biol.* 69, 249-78). In this paper, many of the ideas suggested in our review are discussed. Topologies instead of trees are generated. A method for exchanging terminal branches has been devised such that, for a topology of N taxa 2(N-2) nearest neighbour interchange topologies exist (excluding the starting topology generated from the mutation distance matrix methods

described by Barnabas et al.).[11] A new 'statistical' test is used, named the Moore residual coefficient (MRC). This is essentially a 'sum of squared deviations' worked out for each of the interior branch points of the topology. Thus, the method generates a topology and estimates its MRC, after which the new ITERA programme generates the nearest neighbour topologies and their MRCs. The topology with lowest MRC is selected and the ITERA programme begun again, the cyclic procedure continuing until a 'final' topology is produced. When using *real* data, this is regarded as a 'valley' and the whole procedure begun again with a radically different initial topology, presumably selected by subjective criteria. They have found that several 'valleys' often exist. These 'best' topologies are subjected to an amplification of Fitch's 'maximum parsimony ancestral nucleotide reconstruction' method,[20] which itself owes much to Eck and Dayhoff's approach.[2,18,19] Ancestral reconstruction is probably near the limit of its operational efficiency. From the nucleotide ancestors produced at the nodes, they calculate the total mutation length of the topology. In doing so, they confirm some of our doubts by finding hidden mutations (e.g. Alanine (GAU or GAC) → Alanine (GAA or GAC) at α 53 (not α 54 as published) in mammalian haemoglobins).

Interestingly though, when constructing topologies from a wide range of haemoglobin α-chain sequences, three topologies, considered plausible on biological and paleontological grounds, are *not* produced by the iterative procedure (presumably having non-optimal MRC values or being undiscovered 'valleys'!). However they have total mutation lengths *lower* than the two best MRC topologies. This confirms our suspicions with regard to the choice of test of fit. We doubt whether such an oversimplified and incomplete data entry as the mutation distance matrix is worthy of so sophisticated a technique as the iterative procedure. The essential problem remains – how to select the 'best' topology, using only the sequences, and to be aware that you have it!

R. DOOLITTLE

Prosimian biology and protein evolution

Introduction

The biology of prosimians, like the biology of any group of organisms, is intrinsically interesting, if only because the creatures exist. Beyond that, however, prosimians are naturally of special interest because of their evolutionary relationship to our own species. In this article I am going to attempt to demonstrate that prosimians are also of special interest to students of protein evolution, and that the structure of their molecules, quite apart from shedding light on the molecular antiquity of man, may answer some fundamental questions about protein evolution in general.

Protein evolution

The elucidation of the structure of insulin in the early 1950s opened an entirely new aspect for comparative biology.[1] For the first time, it became possible to compare the amino acid sequences of proteins from different species. It was not surprising that, for any given protein, the more distant the relationship between the creatures compared, the more differences were found in their proteins. The unravelling of the genetic code firmly supported the general suppositions that single base substitutions in the genetic material (usually DNA) were directly reflected by single amino acid replacements in the proteins they prescribed. These were considered a major part of the grist upon which the mill of evolution turned.

It is generally assumed that the spontaneous occurrence of these single base substitutions is a random process which takes place at a more or less constant rate (although estimates of this rate vary from 10^{-6} to 10^{-9} in terms of the probability that a substitution will occur at any particular nucleotide location during a given replication).

Once a base substitution and amino acid replacement occur (the nature of the code is such that approximately 20% of base substitutions do not give rise to amino acid changes), the likelihood of persistence of that change must be considered. There are two schools of thought on this matter, which may be cast in their extreme forms as follows:

(*a*) most of the amino acid replacements which survive during the evolution of species are selected for in a Darwinian sense, the replacement being reproductively advantageous for that particular organism,

(*b*) most of those amino acid replacements which survive during the evolution of species are neutral changes, which are randomly fixed and have no bearing on reproductive success of the creatures in which they occur.[2,3]

It is important to know which of these arguments is nearer the truth. One question which can be posed to help settle this dispute is: Do primitive creatures have primitive proteins? This question can be better understood by reference to Fig. 1, in which the proteins of four related creatures (designated I, II, III and IV) are depicted upon a traditional phylogenetic tree. On the branch segments leading to these creatures are a series of ratios, m/M, n/N, o/O, etc., in which the denominators (capitals; M, N, etc.) represent the numbers of changes which *occur* in a particular protein, and the numerators (small letters; m, n, o, etc.) represent the fraction of these which *survive* during the evolution of the species (we would expect these to be very small fractions, regardless of the predominant process leading to survival). If the widely held assumption about the constancy of

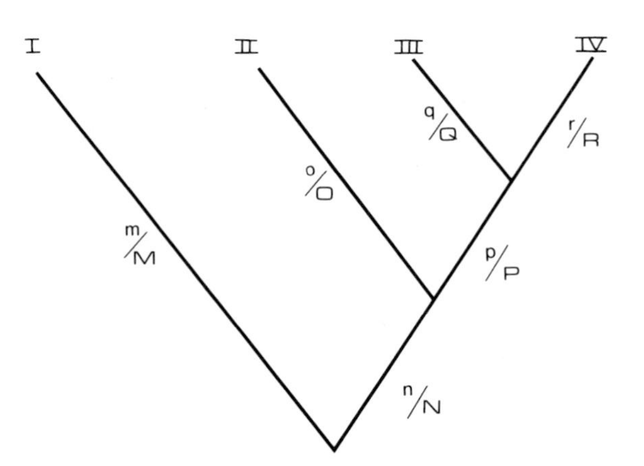

Figure 1. Cladogram showing lines of descent leading to a hypothetical protein in four different species (I, II, III and IV). Denominators (M, N, etc. represent changes which occur, whereas numerators (m, m, etc.) denote changes which actually survive evolution.

rates of occurrence is correct, then Σ M + N + O = Σ M + N + P + Q = Σ M + N + P + R, etc. The question we are trying to answer, however, has to do with the numerators of these ratios. We attempt to find if Σ m + n + o = Σ m + n + p + q = Σ m + n + p + r, etc., by comparing the amino acid sequences of proteins I, II, III and IV. If I is no more different from IV than it is from II and III, then it would appear that the number of surviving changes since II, III and IV had a common ancestor has been constant along their (summed) segments. On the other hand, let us assume that II is derived from a relatively primitive creature whose proteins have not significantly changed since it had a common ancestor with III and IV. In this case, the structures of the proteins I and II would be much more similar than I compared with III or IV. If 'old' organisms do indeed have old' proteins, then it is more difficult to accept the notion that the majority of amino acid replacements in proteins are randomly fixed, since the number of opportunities for fixation during historical time would necessarily have been equivalent to those for 'new' species.

Primitive primates

The argument can be cast in terms of primates and their traditional classification. It is generally agreed that primates can be divided into two major groups, the prosimians and the anthropoids, and that the latter can be subdivided into three subgroups, the New World monkeys, the Old World monkeys and the hominoids (Fig. 2). The question of 'primitiveness' is a delicate one, but, in simple terms, we would like to answer the same questions about the segmental lengths

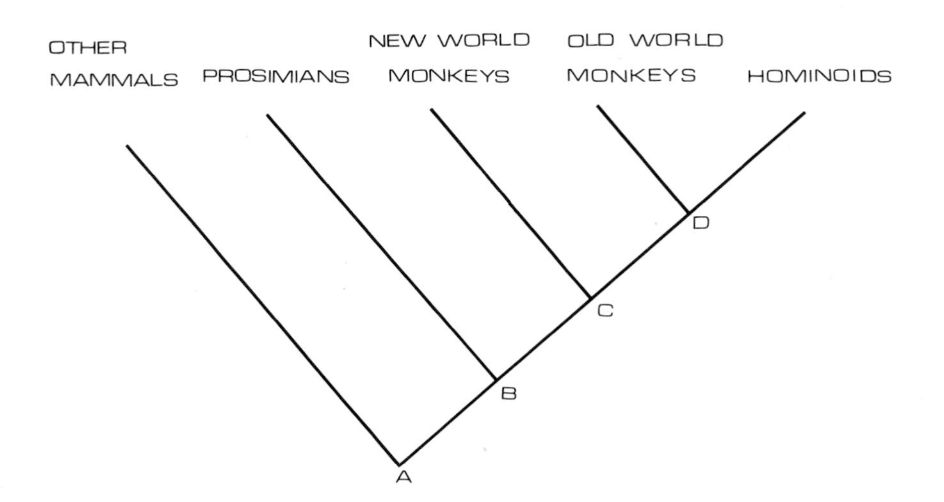

Figure 2. Simplified classification of major primate groups, A, B, C and D denote important common ancestors.

of the limbs of this classical phylogenetic tree as we posed for our protein comparisons. Do existing prosimians, for example, more closely resemble the common ancestor which existed at junction B than do the 'higher' primates? (Fig. 2). If the answer to this question is ever 'yes', then we would certainly like to examine that creature's proteins to find how divergent they are from the proteins of other primates on the one hand, and from those of non-primate mammals on the other.

Which proteins to examine?

Ideally, we would like to compare a number of different proteins for a large number of species from the groups we are studying. The practical limitations of the labour involved in amino acid sequence determinations usually demand that a choice be made restricting either the number of proteins examined or the number of species, or both. Different proteins evolve at different rates, and it is important to choose a protein for study consistent with the expected relationship of the groups being compared. For example, cytochrome c is a very conservative protein which changes very slowly during evolution (Fig. 3). On the other hand, it is a virtually ubiquitous material, and as such it is ideal for comparing distantly related creatures, and maps covering much of the animal and plant kingdoms have been prepared.[4] It is of little use for studying the relationships within a single order like the Primates, however, since there will probably be too few changes to arrive at any statistically significant conclusions. A faster-changing protein like haemoglobin (Fig. 3) would be much more suitable for the latter purpose.

At first glance, the fact that different proteins evolve at different rates (Fig. 3) might seem to argue against random fixation of amino acid replacements even before we begin our analysis. The point to be made, however, is that Darwinian *rejection* is certainly allowed by the random fixation hypothesis. In a slowly changing protein like cytochrome c, fewer of the spontaneously occurring amino acid replacements are neutral, and so fewer replacements are randomly fixed. This is quite different from thinking that a greater number of the replacements which occur in haemoglobins are selectively advantageous.

1. *Fibrinopeptides*

The fibrinopeptides play a unique role in the field of molecular evolution because of their extraordinary propensity for rapid change.[5] They are a pair of peptides which split off from vertebrate fibrinogen molecules during blood-clotting. Their relatively undemending rôle of preventing the spontaneous gelation of fibrinogen

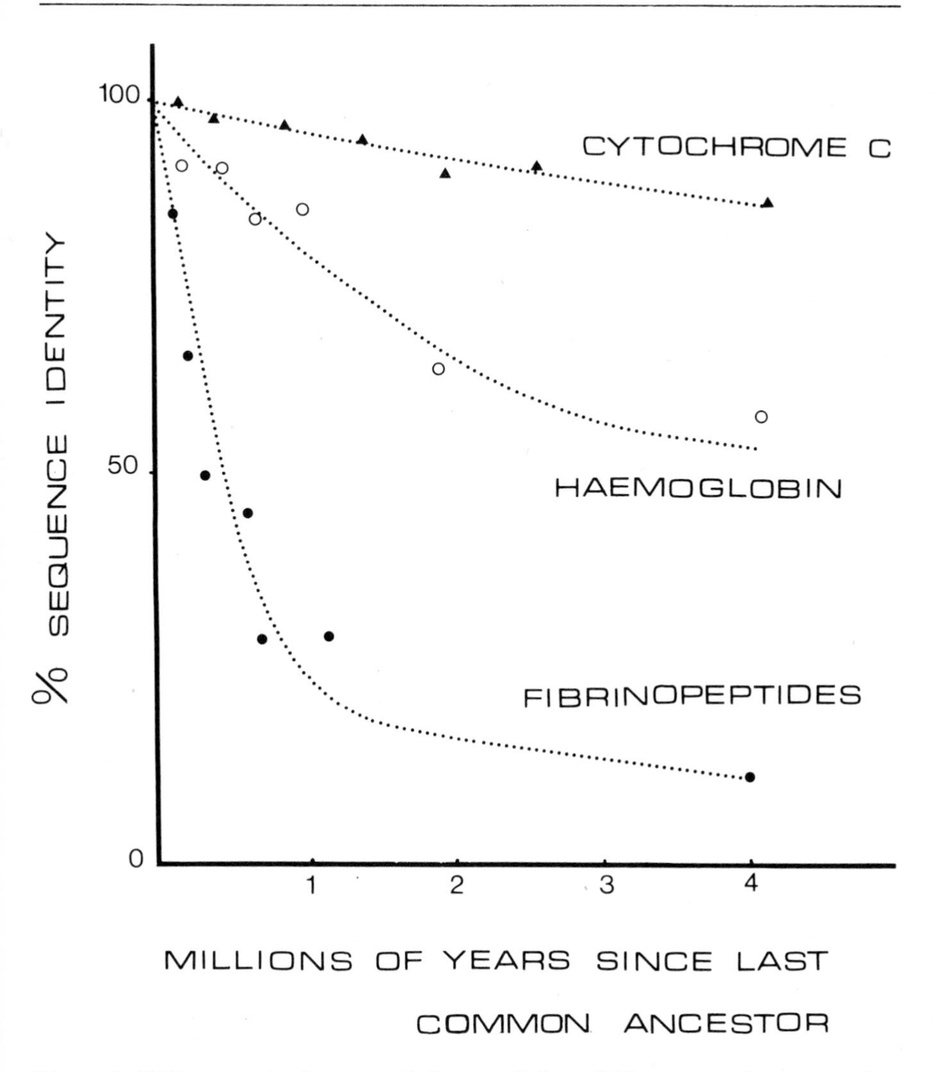

Figure 3. Differences in the rates of change of three different vertebrate proteins (only part of a protein in the case of fibrinopeptides).

molecules can evidently be accomplished by a wide variety of amino acid combinations, and amino acid replacements occur at a rate considerably faster than in the case of most other proteins or peptides (Fig. 3). To date, the fibrinopeptides of more than 45 mammalian species, representing 8 different orders, have been studied, either by Blombäck's group in Stockholm or in our laboratory in La Jolla,[6,7] where the emphasis has been on two particular orders: the artiodactyls (of which about 20 have now been studied) and the primates.[8,9] The advantage of studying fibrinopeptides for establishing the relationships between mammals stems from the fact that the changes occur frequently enough to allow step-by-step comparisons, whilst certain parts of the molecule are conservative enough to ensure proper alignment of the amino acid segments even in the face of considerable flux elsewhere.

The major drawback in the study of fibrinopeptides stems from the substantial amounts of blood necessary for their preparation. For example, in one of our early studies we required 30 litres of sheep's

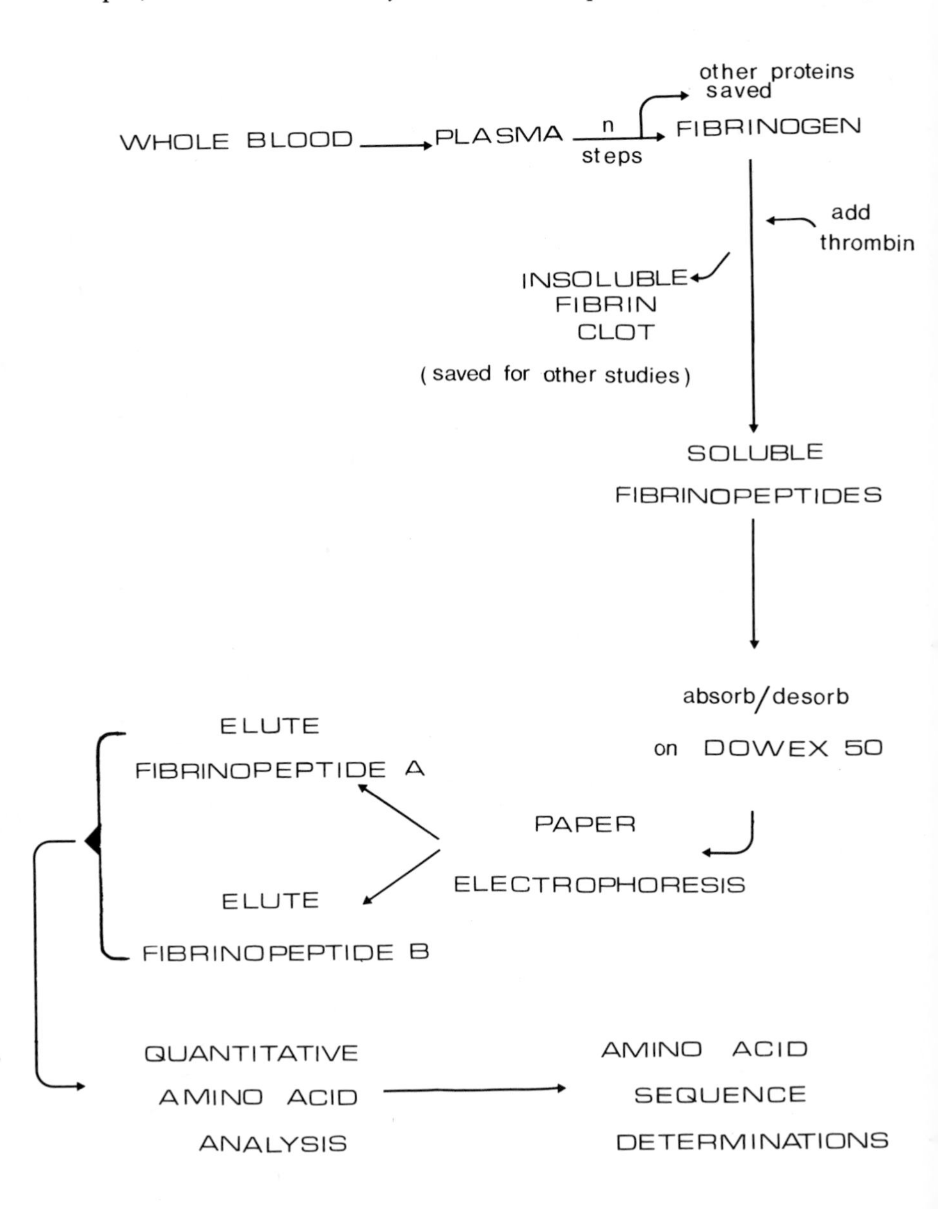

Figure 4. Diagram of steps involved in preparation of fibrinopeptides.

blood to prepare enough material to determine the sequences of the sheep fibrinopeptides A and B.[10] Techniques have improved over the years, however, and we are now able to establish the sequences of a new pair of fibrinopeptides with less than 100 ml. of blood plasma. However, this is still a considerable quantity when dealing with small, rare and expensive primates. Nevertheless, the blood plasma is quite stable when frozen, and pools from small animals can be accumulated over long periods of time. The general procedure used in our laboratory for the preparation of the fibrinopeptides is outlined in Fig. 4.

2. *Primate fibrinopeptides*

The structures of the twelve pairs of primate fibrinopeptides which have been studied so far are depicted cladistically in Figs. 5 and 6. The step-by-step amino acid replacements are readily localised, although the directions of the changes at the bases of the cladograms are necessarily equivocal. The relationships between the various species is completely consistent with other biochemical data, including immunologic criteria.[11,12,13] The advantage of these data lies in the dramatic visualisation of each of the genetic events leading to the present-day structures.

To date, only one prosimian has been studied in this survey.[9] The fibrinopeptides of the slow loris were prepared in our laboratory from pooled blood plasma stocks accumulated by Morris Goodman in Detroit. There are several very interesting features about the slow loris fibrinopeptides which have a bearing on our discussion of general protein evolution. In the first place, the two peptides, A and B, both differ considerably from the corresponding structures of the other 11 primates. Secondly, they contain certain structural features found in non-primate fibrinopeptides, including the rare amino acid tyrosine-O-sulfate – a residue found in most non-primate fibrinopeptide B, but never before found in a primate. The questions before us should now be clear: first, is the slow loris a primitive primate in a classical sense (i.e. does it resemble the prosimian-anthropoid common ancestor more than do existing anthropoid species), and, second, are its fibrinopeptides more primitive than those of the anthropoid fibrinopeptides which have been studied?

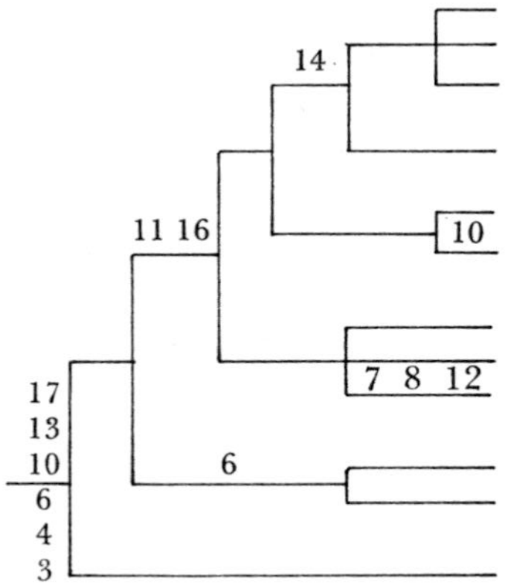

	18	17	16	15	14	13	12	11	10	9	8	7	6	5	4	3	2	1
Human			ALA	ASP	SER	GLY	GLU	GLY	ASP	PHE	LEU	ALA	GLU	GLY	GLY	GLY	VAL	ARG
Chimpanzee			ALA	ASP	SER	GLY	GLU	(GLY,	ASP)	PHE	(LEU,	ALA,	GLU,	GLY,	GLY,	GLY,	VAL)	ARG
Gorilla			ALA	ASP	SER	GLY	GLU	GLY	ASP	PHE	LEU	(ALA,	GLU,	GLY,	GLY,	GLY)	VAL	ARG
Orangutan			ALA	ASP	SER	GLY	GLU	GLY	ASP	PHE	LEU	(ALA	GLU	GLY	GLY	GLY	VAL	ARG
Siamang			ALA	ASP	THR	GLY	GLU	GLY	ASP	PHE	LEU	ALA	GLU	GLY	GLY	GLY	VAL	ARG
Gibbon			ALA	ASP	THR	GLY	GLU	(GLY,	GLU,	PHE	LEU	(ALA,	GLU,	GLY,	GLY,	GLY,	VAL)	ARG
Rhesus macaque			ALA	ASP	THR	GLY	GLU	GLY	ASP	PHE	LEU	ALA	GLU	GLY	GLY	GLY	VAL	ARG
Green monkey			ALA	ASP	THR	GLY	GLU	GLY	ASP	PHE	LEU	ALA	GLU	GLY	GLY	GLY	VAL	ARG
Drill			(ALA,	ASP,	THR,	GLY,	ASP,	GLY,	ASP,	PHE)	ILE	(THR,	GLU,	GLY,	GLY,	GLY)	VAL	ARG
Spider monkey			THR	ASP	THR	GLY	GLU	ASP	ASP	PHE	LEU	ALA	ALA	GLY	GLY	GLY	VAL	ARG
Cebus monkey			THR	(ASP,	THR,	GLY,	GLU,	ASP,	ASP,	PHE,	LEU,	ALA,	ALA,	GLY,	GLY,	GLY)	VAL	ARG
Slow loris	THR	ASP	THR	ASP	THR	ASP	GLU	ASP	GLY	PHE	LEU	ALA	LYS	GLY	ALA	ASP	VAL	ARG

Figure 5. Cladogram showing amino acid changes involved in the evolution of fibrinopeptides A of twelve primates. Numbers on left indicate position involved in change.

3 5
4
8d
9d 10
2 11
5 6 3
3 6 8 14 15
12 8
5 8
2

	15	14	13	12	11	10	9	8	7	6	5	4	3	2	1
Human		PCA	GLY	VAL	ASN	ASP	ASN	GLU	GLU	GLY	PHE	PHE	SER	ALA	ARG
Chimpanzee		PCA	(GLY,	VAL,	ASN,	ASP,	ASN,	GLU,	GLU,	GLY,	PHE)	PHE	SER	ALA	ARG
Gorilla		PCA	GLY	VAL	ASN	ASP	ASN	GLU	GLU	GLY	PHE	PHE	SER	ALA	ARG
Orangutan		PCA	GLY	VAL	ASN	ASP	ASN	GLU	GLU	GLY	LEU	PHE	GLY	ALA	ARG
Siamang		PCA	GLY	VAL	ASN	ASP	ASN	GLU	GLU	GLY	LEU	PHE	GLY	ALA	ARG
Gibbon		PCA	(GLY,	VAL,	ASX,	ASX,	ASX	–	GLX,	GLY,	LEU)	PHE	GLY*	ALA	ARG
Rhesus macaque		(———	———	?	———	———)	ASN	GLU	GLU	SER	PRO	PHE	SER	GLY	ARG
Green monkey		PCA	(GLY,	VAL,	ASX,	GLY)	ASN	GLU	GLU	GLY	LEU	PHE	GLY	GLY	ARG
Drill		PCA	(GLY,	VAL,	ASX,	GLY,	ASX,	GLX,	GLX,	GLY,	LEU)	PHE	GLY	GLY	ARG
Spider monkey		PCA	GLY	GLY	ASX	ASX	ASX	THR	GLX	GLY	ILE	LEU	GLY	ALA	ARG
Cebus monkey		PCA	(GLY,	GLY,	ASX,	ASX,	ASX)	LYS	GLU	GLY	LEU	LEU	GLY	VAL	ARG
Slow loris	ASP	GLX	VAL	ASX	ASX	*TYR	ASX	GLY	GLX	ARG	LEU	LEU	ASP	ALA	ARG

Figure 6. Cladogram showing amino acid changes involved in the evolution of fibrinopeptides B of twelve primates.

	18	17	16	15	14	13	12	11	10	9	8	7	6	5	4	3	2	1
HUMAN A			ALA	ASP	SER	GLY	GLU	GLY	ASP	PHE	LEU	ALA	GLU	GLY	GLY	GLY	VAL	ARG
SLOW LORIS A	THR	ASP	THR	ASP	THR	ASP	GLU	ASP	GLY	PHE	LEU	ALA	LYS	GLY	ALA	ASP	VAL	ARG
RABBIT A			VAL	ASP	PRO	GLY	GLU	SER	THR	PHE	ILE	ASP	GLU	GLY	ALA	THR	GLY	ARG
HUMAN B				PCA	GLY	VAL	ASN	ASP	ASN	–	GLU	GLU	GLY	PHE	PHE	SER	ALA	ARG
SLOW LORIS B				ASP	GLX	VAL	ASX	ASX	TYR*	ASX	GLY	GLX	ARG	LEU	LEU	ASP	ALA	ARG
RABBIT B						ALA	ASP	ASP	TYR*	ASP	ASP	GLU	VAL	LEU	PRO	ASP	ALA	ARG

Figure 7. Comparision of rabbit, slow loris and human fibrinopeptides.

Analysing the data

The analysis of comparative amino acid sequence data can take a variety of forms of greater or lesser sophistication. Certainly, the best of these utilise computers; but simple tabulations by hand are not particularly difficult. All of the approaches ultimately depend on the *Principle of Parsimony* for basic data-reduction in terms of positioning the amino acid replacements. Usually, a simple matrix is constructed of all the sequences being compared. In the case of the primate fibrinopeptides, a 12 x 12 matrix of amino acid replacements and the corresponding base substitutions can be readily prepared and the numbers of differences between all pairs of peptides tabulated. These data are easily transformed into phylogenetic trees of the sort discussed above, the limb-lengths of which are made proportional to the number of differences (Fig. 8). These 'mutational distances' necessarily involve certain arbitrary assumptions, especially with regard to the base of the tree, where the limbs are arbitrarily set to satisfy the sums expressed with regard to Fig. 1. This is just the issue we are trying to resolve, however, and it is obvious that we need to add another grouping (cf. Fig. 2) to answer our question vis-à-vis the slow loris. A non-judicious choice of non-primate group could be misleading, however, and ideally we would like a 'closely related' non-primate group for this comparison. (In theory, any non-primate should do; in practice, the numbers of neutral changes are limited. For example, human fibrinopeptides differ from ox fibrinopeptides by a minimum of 21 base changes; but

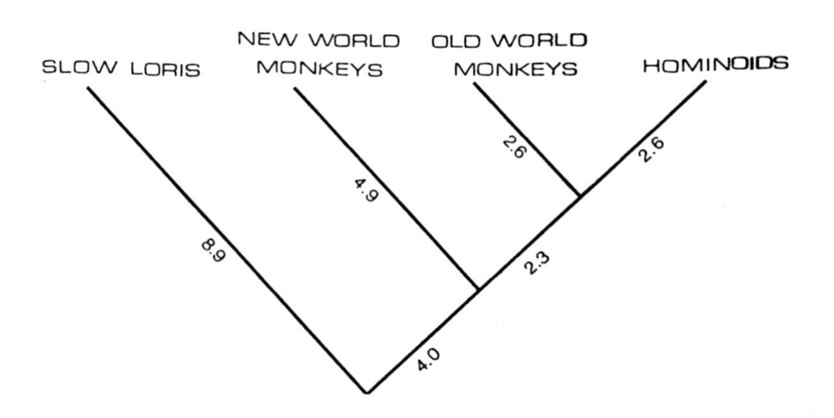

Figure 8. "Mutational distances" separating major primate groups based on a reconstruction of the amino acid sequences of 12 existing species.

pig fibrinopeptides differ from the ox peptides by 20 base changes, even though pigs and cattle are both in the same order.[7] It would be absurd to think that humans are only slightly more distantly related to the ox than is the pig, and a comparison of groups intermediate between primates and artiodactyls should confirm this.)

The slow loris fibrinopeptides are remarkably similar to rabbit fibrinopeptides (Fig. 7). This is especially true in the case of the B peptide, the slow loris have 9 of 13 amino acid residues in common with the rabbit, including the unusual tyrosine-O-sulfate residue. On the other hand, the loris B peptide has only 6 of 14 residues in common with the human B. The rabbit peptide B and human peptide B have only 4 of 13 residues in common. If the rabbit fibrinopeptides are used for the non-primate comparison on the phylogenetic tree, then, there is a suggestion that the slow loris does indeed have primitive fibrinopeptides. This would appear to be true, even though the comparison of the rabbit and slow loris fibrinopeptides is not entirely straightforward (Fig. 8).

Conclusion

There is an obvious need for a much greater body of data before one could conclude that prosimians are primitive primates with primitive proteins. The slow loris fibrinopeptide sequences are tantalising, however, because they fly in the face of trends of thought which are *au courant* among protein evolutionists. What one would like is to see if data from haemoglobin, for example, follow the same pattern. We would also like to examine the fibrinopeptides of more prosimians, including tarsiers, aye-ayes, other lemurs and tree shrews, as well as the classically interesting hedgehog. The most difficult aspect for us is obtaining the necessary blood samples, however. There is a certain amount of skill involved in the sequence determinations, of course, but this can be readily acquired.

In this article I hope to have made a convincing case for the usefulness of amino acid sequence studies on more prosimian fibrinopeptides. These conveniently small, fast-changing peptides offer a graphic demonstration of the step-by-step changes leading to the divergence of the various primate groups. They may also help to solve some of the basic problems plaguing the field of protein evolution.

Acknowledgments

This article was written while the author was a visiting scientist in the Department of Biochemistry, Oxford University, and was supported by a Career Development Award from the US Public Health Service.

NOTES

1 Brown, H., Sanger, F. and Kitai, R. (1955), 'Pig and sheep insulins', *Biochem. J.* 60, 556-65.

2 King, J.L. and Jukes, T.H. (1969), 'Non-Darwinian evolution', *Science* 164, 788-98.

3 Kimura, M. and Ohta, T. (1971), 'On the rate of molecular evolution', *J. Mol. Evol.* 1, 1-17.

4 Dayhoff, M.O. (1969), *Atlas of Protein Sequence and Structure*, vol. 4, Maryland.

5 Doolittle, R.F. and Blombäck, B. (1964), 'Amino acid sequence investigations of fibrinopeptides from various mammals: evolutionary implications', *Nature* 202, 147-52.

6 Blombäck, B. and Blombäck, M. (1968), 'Primary structure of animal proteins as a guide in taxonomic studies', in Hawkes, J.G. (ed.), *Chemotaxonomy and Serotaxonomy*, London.

7 Doolittle, R.F. (1970), 'Evolution of fibrinogen molecules', *Thromb. Diath. Haem.* Supp. 39, 25-42.

8 Doolittle, R.F., Wooding, G.L., Lin, Y. and Riley, M. (1971), 'Hominoid evolution as judged by fibrinopeptide structures', *J. Mol. Evol.* 1, 74-83.

9 Wooding, G.L. and Doolittle, R.F. (1972), 'Primate fibrinopeptides: evolutionary significance', *J. Hum. Evol.* 1, 553-63.

10 Blombäck, B. and Doolittle, R.F. (1963), 'Amino acid sequence studies on fibrinopeptides from several species', *Acta. Chem. Scand.* 17, 1819-22.

11 Goodman, M. (1963), 'Serological analysis of the systematics of recent hominoids', *Hum. Biol.* 35, 377-406.

12 Hafleigh, A.S. and Williams, C.A. Jr. (1966), 'Antigenic correspondence of serum albumins among the primates', *Science* 151, 1530-5.

13 Sarich, V. and Wilson, A.C. (1966), 'Quantitative immunochemistry and the evolution of primate albumins: micro-complement fixation', *Science* 154, 1563-6.

14 Mross, G.A. and Doolittle, R.F. (1967), 'Amino acid sequence studies on artiodactyl fibrinopeptides. II: Vicuna, elk, muntjak, pronghorn antelope, and water buffalo', *Arch. Biochem. Biophys.* 122, 674-84.

Index of authors

General index